Graduate Texts in Mathematics 250

Graduate Texts in Mathematics

Graduate Texts in Mathematics bridge the gap between passive study and creative understanding, offering graduate-level introductions to advanced topics in mathematics. The volumes are carefully written as teaching aids and highlight characteristic features of the theory. Although these books are frequently used as textbooks in graduate courses, they are also suitable for individual study.

For further volumes:
http://www.springer.com/series/136

Loukas Grafakos

Modern Fourier Analysis

Third Edition

Springer

Loukas Grafakos
Department of Mathematics
University of Missouri
Columbia, MO, USA

ISSN 0072-5285
ISSN 2197-5612 (electronic)
ISBN 978-1-4939-5078-2
ISBN 978-1-4939-1230-8 (eBook)
DOI 10.1007/978-1-4939-1230-8
Springer New York Heidelberg Dordrecht London

Mathematics Subject Classification (2010): 42Axx, 42Bxx

Printed on acid-free paper

Springer is part of Springer Science+Business Media (www.springer.com)

Για την Ιωάννα, την Κωνσταντίνα,
και την Θεοδώρα

Preface

I am truly elated to have had the opportunity to write the present third edition of this book, which is a sequel to GTM 249 *Classical Fourier Analysis, 3rd Edition.* This edition was born from my desire to improve the exposition, to fix a few inaccuracies, and to add a new chapter on multilinear operators. I am very fortunate that diligent readers across the globe have shared with me numerous corrections and suggestions for improvements.

Based on my experience as a graduate student, I decided to include great detail in the proofs presented. I hope that this will not make the reading unwieldy. First time readers may prefer to skim through the technical aspects of the presentation and concentrate on the flow of ideas.

This second volume *Modern Fourier Analysis* is addressed to graduate students who wish to delve deeper into Fourier analysis. I believe that after completing a study of this text, a student will be prepared to begin research in the topics covered by the book. While there is more material than can be covered in a semester course, the list of sections that could be taught in a semester without affecting the logical coherence of the book is: 1.1, 1.2, 1.3, 2.1, 2.2, 3.1, 3.2, 3.3, 4.1, 4.2, 4.3, and 5.1.

In such a large piece of work, it is impossible to have no mistakes or omissions. I encourage you to send your corrections to me directly (grafakosl@missouri.edu). The website

`http://math.missouri.edu/~loukas/FourierAnalysis.html`

will be updated with any significant corrections. Solutions to all of the exercises for the present edition will be available to instructors who teach a course out of this book.

Athens, Greece,
March 2014

Loukas Grafakos

Acknowledgments

I am extremely fortunate that several people have pointed out errors, misprints, and omissions in the previous three editions of the books in this series. All these individuals have provided me with invaluable help that resulted in the improved exposition of the text. For these reasons, I would like to express my deep appreciation and sincere gratitude to all the following people.

First edition acknowledgments: Georgios Alexopoulos, Nakhlé Asmar, Bruno Calado, Carmen Chicone, David Cramer, Geoffrey Diestel, Jakub Duda, Brenda Frazier, Derrick Hart, Mark Hoffmann, Steven Hofmann, Helge Holden, Brian Hollenbeck, Petr Honzík, Alexander Iosevich, Tunde Jakab, Svante Janson, Ana Jiménez del Toro, Gregory Jones, Nigel Kalton, Emmanouil Katsoprinakis, Dennis Kletzing, Steven Krantz, Douglas Kurtz, George Lobell, Xiaochun Li, José María Martell, Antonios Melas, Keith Mersman, Stephen Montgomery-Smith, Andrea Nahmod, Nguyen Cong Phuc, Krzysztof Oleszkiewicz, Cristina Pereyra, Carlos Pérez, Daniel Redmond, Jorge Rivera-Noriega, Dmitriy Ryabogin, Christopher Sansing, Lynn Savino Wendel, Shih-Chi Shen, Roman Shvidkoy, Elias M. Stein, Atanas Stefanov, Terence Tao, Erin Terwilleger, Christoph Thiele, Rodolfo Torres, Deanie Tourville, Nikolaos Tzirakis, Don Vaught, Igor Verbitsky, Brett Wick, James Wright, and Linqiao Zhao.

Second edition acknowledgments: Marco Annoni, Pascal Auscher, Andrew Bailey, Dmitriy Bilyk, Marcin Bownik, Juan Cavero de Carondelet Fiscowich, Leonardo Colzani, Simon Cowell, Mita Das, Geoffrey Diestel, Yong Ding, Jacek Dziubanski, Frank Ganz, Frank McGuckin, Wei He, Petr Honzík, Heidi Hulsizer, Philippe Jaming, Svante Janson, Ana Jiménez del Toro, John Kahl, Cornelia Kaiser, Nigel Kalton, Kim Jin Myong, Doowon Koh, Elena Koutcherik, David Kramer, Enrico Laeng, Sungyun Lee, Qifan Li, Chin-Cheng Lin, Liguang Liu, Stig-Olof Londen, Diego Maldonado, José María Martell, Mieczysław Mastyło, Parasar Mohanty, Carlo Morpurgo, Andrew Morris, Mihail Mourgoglou, Virginia Naibo, Tadahiro Oh, Marco Peloso, Maria Cristina Pereyra, Carlos Pérez, Humberto Rafeiro, Maria Carmen Reguera Rodríguez, Alexander Samborskiy, Andreas Seeger, Steven Senger, Sumi Seo, Christopher Shane, Shu Shen, Yoshihiro Sawano, Mark Spencer,

Vladimir Stepanov, Erin Terwilleger, Rodolfo H. Torres, Suzanne Tourville, Ignacio Uriarte-Tuero, Kunyang Wang, Huoxiong Wu, Kôzô Yabuta, Takashi Yamamoto, and Dachun Yang.

Third edition acknowledgments: Marco Annoni, Vinita Arokianathan, Mark Ashbaugh, Daniel Azagra, Andrew Bailey, Árpad Bényi, Dmitriy Bilyk, Nicholas Boros, Almut Burchard, María Carro, Jameson Cahill, Juan Cavero de Carondelet Fiscowich, Andrea Fraser, Shai Dekel, Fausto Di Biase, Zeev Ditzian, Jianfeng Dong, Oliver Dragičević, Sivaji Ganesh, Friedrich Gesztesy, Piotr Hajłasz, Danqing He, Andreas Heinecke, Steven Hofmann, Takahisa Inui, Junxiong Jia, Hans Koelsch, Richard Laugesen, Kaitlin Leach, Andrei Lerner, Yiyu Liang, Calvin Lin, Liguang Liu, Chao Lu, Richard Lynch, Diego Maldonado, Lech Maligranda, Mieczysław Mastyło, Mariusz Mirek, Carlo Morpurgo, Virginia Naibo, Hanh Van Nguyen, Seungly Oh, Tadahiro Oh, Yusuke Oi, Lucas da Silva Oliveira, Hesam Oveys, Manos Papadakis, Marco Peloso, Carlos Pérez, Jesse Peterson, Dmitry Prokhorov, Amina Ravi, Maria Carmen Reguera Rodríguez, Yoshihiro Sawano, Mirye Shin, Javier Soria, Patrick Spencer, Marc Strauss, Krystal Taylor, Naohito Tomita, Suzanne Tourville, Rodolfo H. Torres, Fujioka Tsubasa, Ignacio Uriarte-Tuero, Brian Tuomanen, Shibi Vasudevan, Michael Wilson, Dachun Yang, Kai Yang, Yandan Zhang, Fayou Zhao, and Lifeng Zhao.

Among all these people, I would like to give special thanks to an individual who has studied extensively the two books in the series and has helped me more than anyone else in the preparation of the third edition: Danqing He. I am indebted to him for all the valuable corrections, suggestions, and constructive help he has provided me with in this work. Without him, these books would have been a lot poorer.

Finally, I would also like to thank the University of Missouri for granting me a research leave during the academic year 2013-2014. This time off enabled me to finish the third edition of this book on time. I spent my leave in Greece.

Contents

Chapter 1
Smoothness and Function Spaces

We embark on the study of smoothness with a quick examination of differentiability properties of functions. There are several ways to measure differentiability and numerous ways to quantify smoothness. In this chapter we measure smoothness using the Laplacian, which is easily related to the Fourier transform. This relation becomes the foundation of a very crucial and deep connection between smoothness and Littlewood–Paley theory.

Certain spaces of functions are introduced to serve the purpose of measuring and fine-tuning smoothness. The main function spaces we study in this chapter are Sobolev and Lipschitz spaces. Before undertaking their study, we introduce relevant notation and we review basic facts about smooth functions and tempered distributions.

1.1 Smooth Functions and Tempered Distributions

We denote by $\mathbf{R}^n$ the Euclidean space of n tuples of real numbers. The magnitude of a point $x=(x_1,\dots,x_n)\in\mathbf{R}^n$ is $|x|=(x_1^2+\cdots+x_n^2)^{1/2}$. An open ball centered at $x_0\in\mathbf{R}^n$ of radius $R>0$ is denoted by $B(x_0,R)$. The partial derivative of a function f on $\mathbf{R}^n$ with respect to the jth variable x_j is denoted by $\partial_j f$. The mth partial derivative with respect to the jth variable is denoted by $\partial_j^m f$. The *gradient* of a function f is the vector $\nabla f=(\partial_1 f,\dots,\partial_n f)$. A *multi-index* α is an ordered n-tuple of nonnegative integers. Given α,β multi-indices, we write $\alpha\le\beta$ if $\alpha_j\le\beta_j$ for all $j=1,\dots,n$. For a multi-index $\alpha=(\alpha_1,\dots,\alpha_n)$, $\partial^\alpha f$ denotes the derivative $\partial_1^{\alpha_1}\cdots\partial_n^{\alpha_n}f$. If $\alpha=(\alpha_1,\dots,\alpha_n)$ is a multi-index, then the number $|\alpha|=\alpha_1+\cdots+\alpha_n$ is called the *size* of α and indicates the *total order of differentiation* of $\partial^\alpha f$. The space of functions in $\mathbf{R}^n$ all of whose derivatives of order at most $N\in\mathbf{Z}^+$ are continuous is denoted by $\mathscr{C}^N(\mathbf{R}^n)$ and the space of all *infinitely differentiable functions* on $\mathbf{R}^n$ by $\mathscr{C}^\infty(\mathbf{R}^n)$. The space of smooth functions with compact support on $\mathbf{R}^n$ is denoted by $\mathscr{C}_0^\infty(\mathbf{R}^n)$. The class of Schwartz functions $\mathscr{S}(\mathbf{R}^n)$ is the space of all $\mathscr{C}^\infty(\mathbf{R}^n)$ functions all of whose

L. Grafakos, *Modern Fourier Analysis*, Graduate Texts in Mathematics 250,
DOI 10.1007/978-1-4939-1230-8_1, © Springer Science+Business Media New York 2014

derivatives are bounded by $C_N(1+|\xi|^2)^{-N}$ for every $N\in\mathbf{Z}^+$. The space $\mathscr{S}(\mathbf{R}^n)$ can be canonically equipped with a topology induced by the family of seminorms

$$\rho_{\alpha,\beta}(\varphi)=\sup_{\xi\in\mathbf{R}^n}|\xi^\alpha\partial^\beta\varphi(\xi)|,$$

indexed by all multi-indices α,β, or the alternative family of seminorms

$$\rho'_{\alpha,\beta}(\varphi)=\sup_{\xi\in\mathbf{R}^n}\left|\partial^\beta\left(\xi^\alpha\varphi(\xi)\right)\right|,$$

also indexed by all multi-indices α,β (Exercise 1.1.1). According to this topology a sequence of Schwartz functions φ_j converges to another Schwartz function φ in $\mathscr{S}(\mathbf{R}^n)$ if and only if $\rho_{\alpha,\beta}(\varphi_j-\varphi)\to 0$ as $j\to\infty$ for all multi-indices α,β. This statement is equivalent to the statement that $\rho'_{\alpha,\beta}(\widehat{\varphi}_j-\widehat{\varphi})\to 0$ as $j\to\infty$.

The dual of $\mathscr{S}(\mathbf{R}^n)$ with respect to this topology is the space $\mathscr{S}'(\mathbf{R}^n)$ of tempered distributions on $\mathbf{R}^n$. The Fourier transform of a Schwartz function φ is defined by

$$\widehat{\varphi}(\xi)=\int_{\mathbf{R}^n}\varphi(x)e^{-2\pi i x\cdot\xi}\,dx$$

where $x\cdot\xi=x_1\xi_1+\cdots+x_n\xi_n$ if $x=(x_1,\dots,x_n)$ and $\xi=(\xi_1,\dots,\xi_n)$. The inverse Fourier transform of φ in $\mathscr{S}(\mathbf{R}^n)$ is defined by $\varphi^\vee(\xi)=\widehat{\varphi}(-\xi)$. The operations of differentiation and the Fourier (and inverse Fourier) transform, as well as many other operations on Schwartz functions can be passed on to tempered distributions via duality.

1.1.1 Space of Tempered Distributions Modulo Polynomials

We begin by introducing the space of polynomials. We denote by $\mathscr{P}(\mathbf{R}^n)$ the set of all polynomials of n real variables, i.e., functions of the form

$$\sum_{|\beta|\le m}c_\beta x^\beta=\sum_{\substack{\beta_j\in\mathbf{Z}^+\cup\{0\}\\ \beta_1+\cdots+\beta_n\le m}}c_{\beta_1,\dots,\beta_n}x_1^{\beta_1}\cdots x_n^{\beta_n},$$

where m is an arbitrary integer and c_β are complex coefficients. We then define an equivalence relation $\equiv$ on $\mathscr{S}'(\mathbf{R}^n)$ by setting

$$u\equiv v\iff u-v\in\mathscr{P}(\mathbf{R}^n).$$

The space of all resulting equivalence classes is denoted by $\mathscr{S}'(\mathbf{R}^n)/\mathscr{P}(\mathbf{R}^n)$ and is called the space of *tempered distributions modulo polynomials.* To avoid cumbersome notation, two elements u,v of the same equivalence class in $\mathscr{S}'/\mathscr{P}$ are identified, and in this case we write $u=v$ in $\mathscr{S}'/\mathscr{P}$.

Definition 1.1.1. We define $\mathscr{S}_0(\mathbf{R}^n)$ to be the space of all Schwartz functions φ with the property

$$\int_{\mathbf{R}^n} x^\gamma \varphi(x)\,dx = 0$$

for all multi-indices γ. This condition is equivalent to the statement that $\partial^\gamma(\widehat{\varphi})(0) = 0$ for all multi-indices γ. Then $\mathscr{S}_0(\mathbf{R}^n)$ is a subspace of $\mathscr{S}(\mathbf{R}^n)$ that inherits the same topology as $\mathscr{S}(\mathbf{R}^n)$.

Example 1.1.2. Let $\eta(\xi)$ be a compactly supported smooth function. The inverse Fourier transform of the function $e^{-1/|\xi|^2}\eta(\xi)$ lies in $\mathscr{S}_0(\mathbf{R}^n)$. Indeed, every derivative of $e^{-1/|\xi|^2}\eta(\xi)$ at the origin is equal to a finite linear combination of expressions of the form

$$\lim_{\xi\to 0}\partial_\xi^\beta(|\xi|^{-2})e^{-1/|\xi|^2} = 0,$$

and thus it is zero.

Proposition 1.1.3. *The dual space of $\mathscr{S}_0(\mathbf{R}^n)$ under the topology inherited from $\mathscr{S}(\mathbf{R}^n)$ is*

$$\mathscr{S}_0'(\mathbf{R}^n) = \mathscr{S}'(\mathbf{R}^n)/\mathscr{P}(\mathbf{R}^n).$$

Proof. To identify the dual of $\mathscr{S}_0(\mathbf{R}^n)$ we argue as follows. For each u in $\mathscr{S}'(\mathbf{R}^n)$, let $J(u)$ be the restriction of u on the subspace $\mathscr{S}_0(\mathbf{R}^n)$ of $\mathscr{S}(\mathbf{R}^n)$. Then J is a linear map from $\mathscr{S}'(\mathbf{R}^n)$ to $\mathscr{S}_0'(\mathbf{R}^n)$, and we claim that the kernel of J is exactly $\mathscr{P}(\mathbf{R}^n)$. Indeed, if $\langle u,\varphi\rangle = 0$ for all $\varphi\in\mathscr{S}_0(\mathbf{R}^n)$, then $\langle\widehat{u},\varphi^\vee\rangle = 0$ for all $\varphi\in\mathscr{S}_0(\mathbf{R}^n)$, that is, $\langle\widehat{u},\psi\rangle = 0$ for all ψ in $\mathscr{S}(\mathbf{R}^n)$ supported in $\mathbf{R}^n\setminus\{0\}$. It follows that $\widehat{u}$ is supported at the origin and thus u must be a polynomial; see Proposition 2.4.1 in [156]. This proves that the kernel of the map J is $\mathscr{P}(\mathbf{R}^n)$. We also claim that the range of J is the entire $\mathscr{S}_0'(\mathbf{R}^n)$. Indeed, given $v\in\mathscr{S}_0'(\mathbf{R}^n)$, v is a linear functional on $\mathscr{S}_0$, which is a subspace of the vector space $\mathscr{S}$, and $|\langle v,\phi\rangle|\le p(\phi)$ for all $\phi\in\mathscr{S}_0$, where $p(\phi)$ is equal to a constant times a finite sum of Schwartz seminorms of ϕ. By the Hanh–Banach theorem (Appendix G in [156]), v has an extension V on $\mathscr{S}$ such that $|\langle V,\Phi\rangle|\le p(\Phi)$ for all $\Phi\in\mathscr{S}$. Then $J(V) = v$, and this shows that J is surjective. Combining these two facts we conclude that there is an identification

$$\mathscr{S}'(\mathbf{R}^n)/\mathscr{P}(\mathbf{R}^n) = \mathscr{S}_0'(\mathbf{R}^n),$$

as claimed. □

In view of the identification in Proposition 1.1.3, we have that $u_j\to u$ in $\mathscr{S}'/\mathscr{P}$ if and only if u_j, u are elements of $\mathscr{S}'(\mathbf{R}^n)/\mathscr{P}(\mathbf{R}^n)$ and

$$\langle u_j,\varphi\rangle\to\langle u,\varphi\rangle$$

as $j\to\infty$ for all φ in $\mathscr{S}_0(\mathbf{R}^n)$. Note that convergence in $\mathscr{S}$ implies convergence in $\mathscr{S}_0$, and consequently, convergence in $\mathscr{S}'$ implies convergence in $\mathscr{S}'/\mathscr{P}$.

The Fourier transform of $\mathscr{S}_0(\mathbf{R}^n)$ functions can be multiplied by $|\xi|^z$, $z\in\mathbf{C}$, and still be smooth and vanish to infinite order at zero. Indeed, let φ be in $\mathscr{S}_0(\mathbf{R}^n)$. Then we show that $\partial_j(|\xi|^z\widehat{\varphi}(\xi))(0)$ exists. Since every Taylor polynomial of $\widehat{\varphi}$ at zero is identically equal to zero, it follows from Taylor's theorem that $|\widehat{\varphi}(\xi)| \le C_M|\xi|^M$ for every $M \in \mathbf{Z}^+$, whenever ξ lies in a compact set. Consequently, if $M > 1 - \operatorname{Re} z$,

$$\frac{|te_j|^z\widehat{\varphi}(te_j)}{t}$$

tends to zero as $t \to 0$ when e_j is the vector with 1 in the jth entry and zero elsewhere. This shows that all partials of $|\xi|^z\widehat{\varphi}(\xi)$ at zero exist and are equal to zero. By induction we assume that $\partial^\gamma(|\xi|^z\widehat{\varphi}(\xi))(0) = 0$, and we need to prove that $\partial_j\partial^\gamma(|\xi|^z\widehat{\varphi}(\xi))(0)$ also exists and equals zero. Applying Leibniz's rule, we express $\partial^\gamma(|\xi|^z\widehat{\varphi}(\xi))$ as a finite sum of derivatives of $|\xi|^z$ times derivatives of $\widehat{\varphi}(\xi)$. But for each $|\beta| \le |\gamma|$ we have $|\partial^\beta(\widehat{\varphi})(\xi)| \le C_{M,\beta}|\xi|^M$ for all $M \in \mathbf{Z}^+$ whenever $|\xi| \le 1$. Picking $M > |\gamma| + 1 - \operatorname{Re} z$ and using the fact that $|\partial^{\gamma-\beta}(|\xi|^z)| \le C_\alpha|\xi|^{\operatorname{Re} z - |\gamma| + |\beta|}$, we deduce that $\partial_j\partial^\gamma(|\xi|^z\widehat{\varphi}(\xi))(0)$ also exists and equals zero.

We have now proved that if φ belongs to $\mathscr{S}_0(\mathbf{R}^n)$, then so does $\big(|\xi|^z\widehat{\varphi}(\xi)\big)^\vee$ for all $z \in \mathbf{C}$. This allows us to introduce the operation of multiplication by $|\xi|^z$ on the Fourier transforms of distributions modulo polynomials. This is described in the following definition.

Definition 1.1.4. Let $s \in \mathbf{C}$ and $u \in \mathscr{S}(\mathbf{R}^n)/\mathscr{P}(\mathbf{R}^n)$. We define another distribution $(|\xi|^s\widehat{u})^\vee$ in $u \in \mathscr{S}(\mathbf{R}^n)/\mathscr{P}(\mathbf{R}^n)$ by setting for all φ in $\mathscr{S}_0(\mathbf{R}^n)$

$$\langle (|\cdot|^s\widehat{u})^\vee, \varphi \rangle = \langle u, (|\cdot|^s\varphi^\vee)^\wedge \rangle .$$

This definition is consistent with the corresponding operations on functions and makes sense since, as observed, φ in $\mathscr{S}_0(\mathbf{R}^n)$ implies that $(|\cdot|^s\widehat{\varphi})^\vee$ also lies in $\mathscr{S}_0(\mathbf{R}^n)$, and thus the action of u on this function is defined.

The next proposition allows us to deduce that an infinite sum of $\mathscr{C}^s$ functions is also in $\mathscr{C}^s$ under certain circumstances.

Proposition 1.1.5. *Let $N \in \mathbf{Z}^+$. Suppose that $\{g_i\}_{i\in\mathbf{Z}}$ are functions in $\mathscr{C}^{|\alpha|}(\mathbf{R}^n)$ for all multi-indices α with $|\alpha| \le N$ and that $\sum_{i\in\mathbf{Z}} \|\partial^\alpha g_i\|_{L^\infty} < \infty$ for all $|\alpha| \le N$. Then the function $g = \sum_{i\in\mathbf{Z}} g_i$ is in $\mathscr{C}^{|\alpha|}(\mathbf{R}^n)$ and*

$$\partial^\alpha g = \sum_{i\in\mathbf{Z}} \partial^\alpha g_i$$

for all $|\alpha| \le N$.

Proof. Let e_j be the vector in $\mathbf{R}^n$ with 1 in the jth coordinate and zero in the remaining ones. For $h \in \mathbf{R} \setminus \{0\}$ we have

$$\frac{g(x+he_j) - g(x)}{h} = \sum_{i\in\mathbf{Z}} \frac{g_i(x+he_j) - g_i(x)}{h} .$$

The functions

$$\frac{g_i(x+he_j)-g_i(x)}{h}$$

converge pointwise to $\partial_j g_i(x)$ as $h\to 0$ and are (uniformly in h) bounded above by $\|\partial_j g_i\|_{L^\infty}$, which satisfy $\sum_{i\in\mathbf{Z}}\|\partial_j g_i\|_{L^\infty}<\infty$. The Lebesgue dominated convergence theorem implies that

$$\sum_{i\in\mathbf{Z}}\frac{g_i(x+he_j)-g_i(x)}{h}\to\sum_{i\in\mathbf{Z}}\partial_j g_i(x)$$

as $h\to 0$. This shows that g has partial derivatives and these are continuous in view of the uniform convergence of the series. We can continue this process by induction for all multi-indices α with $|\alpha|\le N$, since for these multi-indices we have by assumption that $\sum_{i\in\mathbf{Z}}\|\partial^\alpha g_i\|_{L^\infty}<\infty$. □

1.1.2 Calderón Reproducing Formula

Given $t>0$ and a function g on $\mathbf{R}^n$, we denote by $g_t(x)=t^{-n}g(t^{-1}x)$ the L^1 dilation of g. Given an integrable function Ψ on $\mathbf{R}^n$ whose Fourier transform vanishes at the origin and $j\in\mathbf{Z}$, we denote by Δ_j^Ψ the *Littlewood–Paley operator* defined by

$$\Delta_j^\Psi(f)=f*\Psi_{2^{-j}}=\big(\widehat{f}(\xi)\widehat{\Psi}(2^{-j}\xi)\big)^\vee$$

for a distribution $f\in\mathscr{S}'(\mathbf{R}^n)$. Given an integrable function Φ whose Fourier transform does not vanish at the origin, for $j\in\mathbf{Z}$ we define the averaging operator

$$S_j^\Phi(f)=f*\Phi_{2^{-j}}=\big(\widehat{f}(\xi)\widehat{\Phi}(2^{-j}\xi)\big)^\vee$$

whenever $f\in\mathscr{S}'(\mathbf{R}^n)$.

Proposition 1.1.6. *(a) Let $\widehat{\Phi}$ be a $\mathscr{C}_0^\infty$ function that is equal to 1 on $\overline{B(0,1)}$. Then for all $\varphi\in\mathscr{S}(\mathbf{R}^n)$ we have*

$$S_N^\Phi(\varphi)\to\varphi \tag{1.1.1}$$

in $\mathscr{S}(\mathbf{R}^n)$ as $N\to\infty$. Also, for all $f\in\mathscr{S}'(\mathbf{R}^n)$,

$$S_N^\Phi(f)\to f \tag{1.1.2}$$

as $N\to\infty$ in the topology of $\mathscr{S}'(\mathbf{R}^n)$.
(b) Let Φ be a Schwartz function whose Fourier transform is supported in a compact set that contains an open ball centered at zero, and let Ψ be a Schwartz function whose Fourier transform is supported in an annulus that does not contain the origin and satisfies

$$\widehat{\Phi}(\xi)+\sum_{j=1}^{\infty}\widehat{\Psi}(2^{-j}\xi)=1$$

for all $\xi \in \mathbf{R}^n$. Then for all $\varphi \in \mathscr{S}(\mathbf{R}^n)$ we have

$$S_0^{\Phi}(\varphi) + \sum_{|j|<N} \Delta_j^{\Psi}(\varphi) \to \varphi \tag{1.1.3}$$

in $\mathscr{S}(\mathbf{R}^n)$ as $N \to \infty$. Also, for all $f \in \mathscr{S}'(\mathbf{R}^n)$,

$$S_0^{\Phi}(f) + \sum_{|j|<N} \Delta_j^{\Psi}(f) \to f \tag{1.1.4}$$

as $N \to \infty$ in the topology of $\mathscr{S}'(\mathbf{R}^n)$.
(c) Let Ψ be a Schwartz function whose Fourier transform is supported in an annulus that does not contain the origin and satisfies

$$\sum_{j\in\mathbf{Z}} \widehat{\Psi}(2^{-j}\xi) = 1$$

for all $\xi \neq 0$. Then for all φ in $\mathscr{S}_0(\mathbf{R}^n)$ we have

$$\sum_{|j|<N} \Delta_j^{\Psi}(\varphi) \to \varphi \tag{1.1.5}$$

in $\mathscr{S}_0(\mathbf{R}^n)$ as $N \to \infty$. Also for all f in $\mathscr{S}'(\mathbf{R}^n)/\mathscr{P}(\mathbf{R}^n)$ we have that

$$\sum_{|j|<N} \Delta_j^{\Psi}(f) \to f \tag{1.1.6}$$

in $\mathscr{S}'(\mathbf{R}^n)/\mathscr{P}(\mathbf{R}^n)$ as $N \to \infty$.

Proof. (a) Let $\widetilde{\Phi}(x) = \Phi(-x)$. We observe that for any $f \in \mathscr{S}'(\mathbf{R}^n)$ and $\varphi \in \mathscr{S}(\mathbf{R}^n)$ we have

$$\left\langle S_N^{\Phi}(f), \varphi \right\rangle = \left\langle f, S_N^{\widetilde{\Phi}}(\varphi) \right\rangle.$$

In view of this, (1.1.2) follows from (1.1.1) via duality, since $\widetilde{\Phi}$ and Φ have the same properties. To prove (1.1.1), we fix a function φ in $\mathscr{S}$. It is equivalent to show that $(S_N^{\Phi}(\varphi))\widehat{\ } \to \widehat{\varphi}$ in $\mathscr{S}(\mathbf{R}^n)$. Fix multi-indices α, β. It will suffice to show that

$$\rho'_{\alpha,\beta}((S_N^{\Phi}(\varphi))\widehat{\ } - \widehat{\varphi}) = \sup_{\xi\in\mathbf{R}^n} \left|\partial_\xi^\beta \left[(1-\widehat{\Phi}(2^{-N}\xi))\widehat{\varphi}(\xi)\xi^\alpha\right]\right| \to 0 \tag{1.1.7}$$

as $N \to \infty$. Since $\widehat{\Phi}$ is equal to 1 on the unit ball, it follows that the supremum in (1.1.7) is over the set $|\xi| \geq 2^N$. By Leibniz's rule, the ∂^β derivative in the preceding expression is equal to a sum of ∂^γ derivatives falling on $(1-\widehat{\Phi}(2^{-N}\xi))$ times $\partial^{\beta-\gamma}$ derivatives falling on $\widehat{\varphi}(\xi)\xi^\alpha$, where $\gamma \leq \beta$. If $\gamma \neq 0$, then then a factor of 2^{-N} appears from the differentiation in γ. If $\gamma = 0$, then then the conclusion follows in view of the rapid decay of $\partial^\beta(\widehat{\varphi}(\xi)\xi^\alpha)$ on the set $|\xi| \geq 2^N$.

The proof of (b) follows in the same way as the proof of (a) with the function $\Phi(\xi) + \sum_{j=1}^N \widehat{\Psi}(2^{-j}\xi)$ in place of $\widehat{\Phi}(2^{-N}\xi)$, which has similar support properties.

(c) Assertion (1.1.6) follows from (1.1.5) by duality. To prove (1.1.5), we use the Fourier transform. We have that if φ_N, φ lie in $\mathscr{S}_0$, then $\varphi_N \to \varphi$ in $\mathscr{S}_0$ if and only if $\varphi_N \to \varphi$ in $\mathscr{S}$ which happens if and only if $\widehat{\varphi_N} \to \widehat{\varphi}$ in $\mathscr{S}$. Thus to show that $\sum_{|j|<N} \Delta_j^{\Psi}(\varphi) \to \varphi$, it suffices to show that $\sum_{|j|\geq N} \widehat{\varphi}(\xi)\widehat{\Psi}(2^{-j}\xi) \to 0$ in $\mathscr{S}(\mathbf{R}^n)$. But $\partial_\xi^\beta(\xi^\alpha \widehat{\varphi}(\xi)\sum_{j\geq N}\widehat{\Psi}(2^{-j}\xi))$ is supported in $|\xi| \geq c2^N$, for some constant $c>0$, and decays rapidly at infinity, so

$$\sup_{\xi\in\mathbf{R}^n} |\partial_\xi^\beta(\xi^\alpha \widehat{\varphi}(\xi)\sum_{j\geq N}\widehat{\Psi}(2^{-j}\xi))| \to 0$$

as $N\to\infty$. Also, $\partial_\xi^\beta(\xi^\alpha \widehat{\varphi}(\xi)\sum_{j\leq -N}\widehat{\Psi}(2^{-j}\xi))$ is supported in $|\xi| \leq c'2^{-N}$ and vanishes at zero to infinite order; thus, it satisfies

$$\left|\partial_\xi^\beta(\xi^\alpha \widehat{\varphi}(\xi)\sum_{j\leq -N}\widehat{\Psi}(2^{-j}\xi))\right| \leq c_{\alpha,\beta,\varphi,\Psi} \sup_{|\xi|\leq 2^{-N+1}} |\xi|,$$

which tends to zero as $N\to\infty$. □

Corollary 1.1.7. ***(Calderón reproducing formula)*** *Let Ψ, Ω be Schwartz functions whose Fourier transforms are supported in annuli that do not contain the origin and satisfy*

$$\sum_{j\in\mathbf{Z}} \widehat{\Psi}(2^{-j}\xi)\widehat{\Omega}(2^{-j}\xi) = 1$$

for all $\xi \neq 0$. Then for all $f \in \mathscr{S}'(\mathbf{R}^n)/\mathscr{P}(\mathbf{R}^n)$ we have

$$\sum_{j\in\mathbf{Z}} \Psi_{2^{-j}} * \Omega_{2^{-j}} * f = \sum_{j\in\mathbf{Z}} \Delta_j^{\Psi}\Delta_j^{\Omega}(f) = f, \tag{1.1.8}$$

where the convergence is in $\mathscr{S}'(\mathbf{R}^n)/\mathscr{P}(\mathbf{R}^n)$.

Proof. The assertion is contained in the conclusion of Proposition 1.1.6(c) with $\Psi * \Omega$ in place of Ψ. □

Exercises

1.1.1. Given multi-indices α, β, show that there are constants C, C' such that

$$\rho_{\alpha,\beta}(\varphi) \leq C \sum_{|\gamma|\leq|\alpha|}\sum_{|\delta|\leq|\beta|} \rho'_{\gamma,\delta}(\varphi),$$

$$\rho'_{\alpha,\beta}(\varphi) \leq C' \sum_{|\gamma|\leq|\alpha|}\sum_{|\delta|\leq|\beta|} \rho_{\gamma,\delta}(\varphi).$$

for all Schwartz functions φ.

[*Hint:* The first inequality follows by Leibniz's rule. Conversely, to express $\xi^\alpha \partial^\beta \varphi$ in terms of linear combinations of $\partial^\beta(\xi^\gamma \varphi(\xi))$, proceed by induction on $|\alpha|$, using that $\xi_j \partial^\beta \varphi = \partial^\beta(\xi_j \varphi) - \partial^\beta \varphi - (\beta_j - 1)\partial^{\beta - e_j}\varphi$ if $\beta_j \geq 1$ and $\xi_j \partial^\beta \varphi = \partial^\beta(\xi_j \varphi)$ if $\beta_j = 0$. Here $\beta = (\beta_1, \dots, \beta_n)$ and $e_j = (0, \dots, 1, \dots, 0)$ with 1 in the jth entry.]

1.1.2. Suppose that a function φ lies in $\mathscr{C}^\infty(\mathbf{R}^n \setminus \{0\})$ and that for all multi-indices α there exist constants L_α such that φ satisfies

$$\lim_{t \to 0} \partial^\alpha \varphi(t) = L_\alpha .$$

Then φ lies in $\mathscr{C}^\infty(\mathbf{R}^n)$ and $\partial^\alpha \varphi(0) = L_\alpha$ for all multi-indices α.

1.1.3. Let $u_N \in \mathscr{S}'(\mathbf{R}^n)$. Suppose that $u_N \to u$ in $\mathscr{S}'/\mathscr{P}$ and $u_N \to v$ in $\mathscr{S}'$. Then prove that $u - v$ is a polynomial.
[*Hint:* Use Proposition 1.1.3 or directly Proposition 2.4.1 in [156].]

1.1.4. Suppose that Ψ is a Schwartz function whose Fourier transform is supported in an annulus that does not contain the origin and satisfies $\sum_{j \in \mathbf{Z}} \widehat{\Psi}(2^{-j}\xi) = 1$ for all $\xi \neq 0$. Show that for functions $g \in L^1(\mathbf{R}^n)$ with $\widehat{g} \in L^1(\mathbf{R}^n)$ we have $\sum_{j \in \mathbf{Z}} \Delta_j^\Psi(g) = g$ pointwise everywhere.

1.1.5. Let Θ and Φ be Schwartz functions whose Fourier transforms are compactly supported and let Ψ, Ω be Schwartz functions whose Fourier transforms are supported in annuli that do not contain the origin and satisfy

$$\widehat{\Phi}(\xi)\widehat{\Theta}(\xi) + \sum_{j=1}^{\infty} \widehat{\Psi}(2^{-j}\xi)\widehat{\Omega}(2^{-j}\xi) = 1$$

for all $\xi \in \mathbf{R}^n$. Then for all $f \in \mathscr{S}'(\mathbf{R}^n)$ we have

$$\Phi * \Theta * f + \sum_{j=1}^{\infty} \Delta_j^\Psi \Delta_j^\Omega(f) = f$$

where the series converges in $\mathscr{S}'(\mathbf{R}^n)$.

1.1.6. (a) Show that for any multi-index α on $\mathbf{R}^n$ there is a polynomial Q_α of n variables of degree $|\alpha|$ such that for all $\xi \in \mathbf{R}^n$ we have

$$\partial^\alpha(e^{-|\xi|^2}) = Q_\alpha(\xi)e^{-|\xi|^2} .$$

(b) Show that for all multi-indices $|\alpha| \geq 1$ and for each k in $\{0, 1, \dots, |\alpha| - 1\}$ there is a polynomial $P_{\alpha,k}$ of n variables of degree at most $|\alpha|$ such that

$$\partial^\alpha(e^{-|\xi|}) = \sum_{k=0}^{|\alpha|-1} \frac{1}{|\xi|^k} P_{\alpha,k}\left(\frac{\xi_1}{|\xi|}, \dots, \frac{\xi_n}{|\xi|}\right) e^{-|\xi|}$$

for every $\xi \in \mathbf{R}^n \setminus \{0\}$. Conclude that for $|\alpha| \geq 1$ we have

$$|\partial^\alpha(e^{-|\xi|})| \leq C_\alpha \left(1 + \frac{1}{|\xi|} + \cdots + \frac{1}{|\xi|^{|\alpha|-1}}\right) e^{-|\xi|}$$

for some constant C_α and all $\xi \neq 0$.
[*Hint:* For the two identities use induction on $|\alpha|$. Part (b): Use that the ∂_j derivative of a homogeneous polynomial of degree at most $|\alpha|$ is another homogeneous polynomial of degree at most $|\alpha|+1$ times $|\xi|^{-1}$.]

1.2 Laplacian, Riesz Potentials, and Bessel Potentials

The Laplacian is the operator

$$\Delta = \partial_1^2 + \cdots + \partial_n^2 ,$$

which may act on functions or tempered distributions. The Laplacian satisfies the following identity for all $f \in \mathscr{S}(\mathbf{R}^n)$:

$$-\widehat{\Delta f}(\xi) = 4\pi^2 |\xi|^2 \widehat{f}(\xi) .$$

Motivated by this identity, we replace the exponent 2 by a complex exponent z and we define $(-\Delta)^{z/2}$ as the operator given by the multiplication with the function $(2\pi|\xi|)^z$ on the Fourier transform. More precisely, for $z \in \mathbf{C}$ and Schwartz functions f we define

$$(-\Delta)^{\frac{z}{2}} f(x) = ((2\pi|\xi|)^z \widehat{f}(\xi))^\vee(x) . \tag{1.2.1}$$

Roughly speaking, the operator $(-\Delta)^{z/2}$ acts as a derivative of order z if z is an even integer. If z is a complex number with real part less than $-n$, then the function $|\xi|^z$ is not locally integrable on $\mathbf{R}^n$ and so (1.2.1) may not be well defined. For this reason, whenever we write (1.2.1), we assume that either $\operatorname{Re} z > -n$ or $\operatorname{Re} z \leq -n$ and that $\widehat{f}$ vanishes to sufficiently high order at the origin so that the expression $|\xi|^z \widehat{f}(\xi)$ is integrable. Note that the family of operators $(-\Delta)^z$ satisfies the semigroup property

$$(-\Delta)^z (-\Delta)^w = (-\Delta)^{z+w}$$

for all $z, w \in \mathbf{C}$ when acting on Schwartz functions whose Fourier transform vanishes in a neighborhood of the origin.

The operator $(-\Delta)^{z/2}$ is given by convolution with the inverse Fourier transform of $(2\pi)^z |\xi|^z$. Theorem 2.4.6 in [156] gives that this inverse Fourier transform is equal to

$$(2\pi)^z (|\xi|^z)^\vee(x) = (2\pi)^z \frac{\pi^{-\frac{z}{2}}}{\pi^{\frac{z+n}{2}}} \frac{\Gamma(\frac{n+z}{2})}{\Gamma(\frac{-z}{2})} |x|^{-z-n} , \tag{1.2.2}$$

provided $-n < \mathrm{Re}\, z < 0$, in which case both $|\xi|^z$ and $|x|^{-z-n}$ are locally integrable functions. Dividing both sides of (1.2.2) by $\Gamma(\frac{n+z}{2})$ allows one to extend (1.2.2) to all complex numbers z as an identity between distributions; see Theorem 2.4.6 in [156] for details.

1.2.1 Riesz Potentials

When s is a positive real number, the operation $f \mapsto (-\Delta)^{-s/2} f$ is not really *differentiating* f; rather, it is *integrating*. For this reason, we introduce a slightly different notation that better reflects the nature of this operator.

Definition 1.2.1. Let s be a complex number with $0 < \mathrm{Re}\, s < \infty$. The *Riesz potential operator* of order s is

$$\mathcal{I}_s = (-\Delta)^{-s/2}.$$

Clearly $\mathcal{I}_s$ is well defined on Schwartz functions whose Fourier transform vanishes in a neighborhood of the origin; if $\mathrm{Re}\, s < n$, the function $\xi \mapsto |\xi|^{-s}$ is locally integrable, and thus $\mathcal{I}_s$ is well defined on the entire Schwartz class. Using identity (1.2.2), we express

$$\mathcal{I}_s(f)(x) = 2^{-s}\pi^{-\frac{n}{2}} \frac{\Gamma(\frac{n-s}{2})}{\Gamma(\frac{s}{2})} \int_{\mathbf{R}^n} f(x-y)|y|^{-n+s}\, dy,$$

and since this integral is convergent for all functions f in the Schwartz class, $\mathcal{I}_s$ is well defined on this space for all s with $\mathrm{Re}\, s > 0$.

We begin with a simple remark concerning the homogeneity of the operator $\mathcal{I}_s$.

Remark 1.2.2. Suppose that for some $s \in \mathbf{C}$, with $\mathrm{Re}\, s > 0$, we had an estimate

$$\big\|\mathcal{I}_s(f)\big\|_{L^q(\mathbf{R}^n)} \le C(p,q,n,s)\big\|f\big\|_{L^p(\mathbf{R}^n)} \tag{1.2.3}$$

for some positive indices p, q and all $f \in \mathscr{S}(\mathbf{R}^n)$. Then p and q must be related by

$$\frac{1}{p} - \frac{1}{q} = \frac{\mathrm{Re}\, s}{n}. \tag{1.2.4}$$

This follows by applying (1.2.3) to the dilation $\delta^\lambda(f)(x) = f(\lambda x)$, $\lambda > 0$, in lieu of f. Indeed, replacing f by $\delta^\lambda(f)$ in (1.2.3) and using the identity

$$\mathcal{I}_s(\delta^\lambda(f)) = \lambda^{-\mathrm{Re}\, s}\delta^\lambda(\mathcal{I}_s(f))$$

which follows by a changes of variables, we obtain

$$\lambda^{-\frac{n}{q}-\mathrm{Re}\, s}\big\|\mathcal{I}_s(f)\big\|_{L^q(\mathbf{R}^n)} \le C(p,q,n,s)\lambda^{-\frac{n}{p}}\big\|f\big\|_{L^p(\mathbf{R}^n)} \tag{1.2.5}$$

or, equivalently,

$$\|\mathcal{I}_s(f)\|_{L^q(\mathbf{R}^n)} \le C(p,q,n,s)\lambda^{\frac{n}{q}-\frac{n}{p}+\mathrm{Re}\,s}\|f\|_{L^p(\mathbf{R}^n)}. \tag{1.2.6}$$

If $\frac{1}{p} > \frac{1}{q} + \frac{\mathrm{Re}\,s}{n}$, then we let $\lambda \to \infty$ in (1.2.6), whereas if $\frac{1}{p} < \frac{1}{q} + \frac{\mathrm{Re}\,s}{n}$, then we let $\lambda \to 0$ in (1.2.6). In both cases we obtain that $\mathcal{I}_s(f) = 0$ for all Schwartz functions f, but this is obviously not the case for the function $f(x) = e^{-\pi|x|^2}$. It follows that (1.2.4) must necessarily hold.

This example provides an excellent paradigm of situations where the homogeneity (or the dilation structure) of an operator dictates a relationship on the indices p and q for which it (may) map L^p to L^q.

As we saw in Remark 1.2.2, if the Riesz potentials map L^p to L^q for some p,q, then we must have $q > p$. Such operators that improve the integrability of a function are called *smoothing*. The importance of the Riesz potentials lies in the fact that they are indeed smoothing operators. This is the essence of the *Hardy–Littlewood–Sobolev theorem on fractional integration*, which we now formulate and prove. Since

$$|\mathcal{I}_s(f)| \le \mathcal{I}_{\mathrm{Re}\,s}(|f|),$$

one may restrict the study of $\mathcal{I}_s(f)$ to nonnegative functions f and $s > 0$.

Theorem 1.2.3. *Let s be a real number, with $0 < s < n$, and let $1 < p < q < \infty$ satisfy*

$$\frac{1}{p} - \frac{1}{q} = \frac{s}{n}.$$

Then there exist constants $C(n,s,p), C(s,n) < \infty$ such that for all f in $\mathscr{S}(\mathbf{R}^n)$ we have

$$\|\mathcal{I}_s(f)\|_{L^q} \le C(n,s,p)\|f\|_{L^p}$$

and

$$\|\mathcal{I}_s(f)\|_{L^{\frac{n}{n-s},\infty}} \le C(n,s)\|f\|_{L^1}.$$

Consequently $\mathcal{I}_s$ has a unique extension on $L^p(\mathbf{R}^n)$ for all p with $1 \le p < \infty$ such that the preceding estimates are valid.

Proof. For a given nonnegative (and nonzero) function f in the Schwartz class we write

$$\int_{\mathbf{R}^n} f(x-y)|y|^{s-n}\,dy = I_1(f)(x) + I_2(f)(x),$$

where I_1 and I_2 are defined by

$$\begin{aligned} I_1(f)(x) &= \int_{|y|<R(x)} f(x-y)|y|^{s-n}\,dy, \\ I_2(f)(x) &= \int_{|y|\ge R(x)} f(x-y)|y|^{s-n}\,dy, \end{aligned}$$

for some $R(x) > 0$ to be determined later. Note that the function $K(y) = |y|^{-n+s}\chi_{|y|<1}$ is radial, integrable, and symmetrically decreasing about the origin and that

$$I_1(f)(x) = R(x)^s(f * K_{R(x)})(x)$$

where $K_\varepsilon(x) = \varepsilon^{-n}K(x/\varepsilon)$. It follows from Theorem 2.1.10 in [156] that

$$I_1(f)(x) \le R(x)^s M(f)(x) \int_{|y|<1} |y|^{-n+s}\,dy = \tfrac{\omega_{n-1}}{s} R(x)^s M(f)(x), \tag{1.2.7}$$

where M is the Hardy–Littlewood maximal function, defined by

$$M(g)(x) = \sup_{\substack{B \ni x \\ B \text{ open ball in } \mathbf{R}^n}} \frac{1}{|B|}\int_B |g(y)|\,dy.$$

Let $p' = \frac{p}{p-1}$ if $1 < p < \infty$ and $1' = \infty$. Observe that $(n-s)p' = n + \frac{p'n}{q} > n$. Hölder's inequality gives that

$$\begin{aligned} |I_2(f)(x)| &\le \left(\int_{|y|\ge R(x)} |y|^{-(n-s)p'}\,dy\right)^{\frac{1}{p'}} \|f\|_{L^p(\mathbf{R}^n)} \\ &= \left(\frac{q\omega_{n-1}}{p'n}\right)^{\frac{1}{p'}} R(x)^{-\frac{n}{q}} \|f\|_{L^p(\mathbf{R}^n)}, \end{aligned} \tag{1.2.8}$$

and note that this estimate is also valid when $p = 1$ (in which case $q = \frac{n}{n-s}$), provided the $L^{p'}$ norm is interpreted as the L^∞ norm and the constant $\left(\frac{q\omega_{n-1}}{p'n}\right)^{\frac{1}{p'}}$ is replaced by 1. Combining (1.2.7) and (1.2.8), we obtain that

$$\mathcal{I}_s(f)(x) \le C'_{n,s,p}\left(R(x)^s M(f)(x) + R(x)^{-\frac{n}{q}} \|f\|_{L^p}\right). \tag{1.2.9}$$

We choose

$$R(x) = \|f\|_{L^p}^{\frac{p}{n}} \left(M(f)(x)\right)^{-\frac{p}{n}}$$

to minimize the expression on the right-hand side in (1.2.9). We observe that if f is nonzero, then $M(f)(x) > 0$ for all $x \in \mathbf{R}^n$ and thus $R(x)$ is well defined. This choice of $R(x)$ yields the estimate

$$\mathcal{I}_s(f)(x) \le C_{n,s,p} M(f)(x)^{\frac{p}{q}} \|f\|_{L^p}^{1-\frac{p}{q}}. \tag{1.2.10}$$

The required inequality for $p > 1$ follows by raising (1.2.10) to the power q, integrating over $\mathbf{R}^n$, and using the boundedness of the Hardy–Littlewood maximal operator M on $L^p(\mathbf{R}^n)$ (Theorem 2.1.6 in [156]). The case $p = 1$, $q = \frac{n}{n-s}$ also follows from (1.2.10) by the weak type $(1,1)$ property of M (see also Theorem 2.1.6 in [156]). Indeed for all $\lambda > 0$ we have

$$\begin{aligned}\left|\{C_{n,s,1}M(f)^{\frac{n-s}{n}}\|f\|_{L^1}^{\frac{s}{n}}>\lambda\}\right| &= \left|\left\{M(f)>\left(\frac{\lambda}{C_{n,s,1}\|f\|_{L^1}^{\frac{s}{n}}}\right)^{\frac{n}{n-s}}\right\}\right| \\ &\le 3^n\left(\frac{C_{n,s,1}\|f\|_{L^1}^{\frac{s}{n}}}{\lambda}\right)^{\frac{n}{n-s}}\|f\|_{L^1} \\ &= C(n,s)\left(\frac{\|f\|_{L^1}}{\lambda}\right)^{\frac{n}{n-s}}.\end{aligned}$$

This estimate says that T maps $L^1(\mathbf{R}^n)$ to weak $L^{\frac{n}{n-s}}(\mathbf{R}^n)$. □

1.2.2 Bessel Potentials

While the behavior of the kernels $|x|^{-n+s}$ as $|x|\to 0$ is well suited to their smoothing properties, their decay as $|x|\to\infty$ gets worse as s increases. We can slightly adjust the Riesz potentials so that we maintain their essential behavior near zero but achieve exponential decay at infinity. The simplest way to achieve this is by replacing the *nonnegative* operator $-\Delta$ by the *strictly positive* operator $I-\Delta$. Here the terms *nonnegative* and *strictly positive*, as one may have surmised, refer to the Fourier multipliers of these operators.

Definition 1.2.4. Let z be a complex number satisfying $0<\operatorname{Re} z<\infty$. The *Bessel potential operator* of order z is

$$\mathcal{J}_z=(I-\Delta)^{-z/2}.$$

This operator acts on functions f as follows:

$$\mathcal{J}_z(f)=\big(\widehat{f}\,\widehat{G_z}\big)^\vee=f*G_z,$$

where

$$G_z(x)=\big((1+4\pi^2|\xi|^2)^{-z/2}\big)^\vee(x).$$

The Bessel potential is obtained by replacing $4\pi^2|\xi|^2$ in the Riesz potential by the smooth term $1+4\pi^2|\xi|^2$. This adjustment creates smoothness, which yields rapid decay for G_z at infinity. The next result quantifies the behavior of G_z near zero and near infinity.

Proposition 1.2.5. *Let z be a complex number with $\operatorname{Re} z>0$. Then the function G_z is smooth on $\mathbf{R}^n\setminus\{0\}$. Moreover, if s is real, then G_s is strictly positive, $\|G_s\|_{L^1}=1$, and there exist positive finite constants $C(s,n)$, $c(s,n)$ such that*

$$G_s(x)\le C(s,n)e^{-\frac{|x|}{2}}\qquad \textit{when } |x|\ge 2 \tag{1.2.11}$$

and such that

$$\frac{1}{c(s,n)} \le \frac{G_s(x)}{H_s(x)} \le c(s,n) \qquad \text{when } |x| \le 2, \tag{1.2.12}$$

where H_s is equal to

$$H_s(x) = \begin{cases} |x|^{s-n} + 1 + O(|x|^{s-n+2}) & \text{for } 0 < s < n, \\ \log \frac{2}{|x|} + 1 + O(|x|^2) & \text{for } s = n, \\ 1 + O(|x|^{s-n}) & \text{for } s > n, \end{cases}$$

and $O(t)$ is a function with the property $|O(t)| \le |t|$ for $t \ge 0$.

Now let z be a complex number with $\operatorname{Re} z > 0$. *Then there exist finite positive constants $C'(\operatorname{Re} z, n)$ and $c'(\operatorname{Re} z, n)$ such that when $|x| \ge 2$, we have*

$$|G_z(x)| \le \frac{C'(\operatorname{Re} z, n)}{|\Gamma(\frac{z}{2})|} e^{-\frac{|x|}{2}} \tag{1.2.13}$$

and when $|x| \le 2$, we have

$$|G_z(x)| \le \frac{c'(\operatorname{Re} z, n)}{|\Gamma(\frac{z}{2})|} \begin{cases} |x|^{\operatorname{Re} z - n} & \text{for } \operatorname{Re} z < n, \\ \log \frac{2}{|x|} & \text{for } \operatorname{Re} z = n, \\ 1 & \text{for } \operatorname{Re} z > n. \end{cases}$$

Proof. For $A > 0$ and z with $\operatorname{Re} z > 0$ we have the gamma function identity

$$A^{-\frac{z}{2}} = \frac{1}{\Gamma(\frac{z}{2})} \int_0^\infty e^{-tA} t^{\frac{z}{2}} \frac{dt}{t},$$

which we use to obtain

$$(1 + 4\pi^2|\xi|^2)^{-\frac{z}{2}} = \frac{1}{\Gamma(\frac{z}{2})} \int_0^\infty e^{-t} e^{-\pi|2\sqrt{\pi t}\,\xi|^2} t^{\frac{z}{2}} \frac{dt}{t}.$$

Note that the preceding integral converges at both ends. Now take the inverse Fourier transform in ξ and use the fact that the function $e^{-\pi|\xi|^2}$ is equal to its Fourier transform (Example 2.2.9 in [156]) to obtain

$$G_z(x) = \frac{(2\sqrt{\pi})^{-n}}{\Gamma(\frac{z}{2})} \int_0^\infty e^{-t} e^{-\frac{|x|^2}{4t}} t^{\frac{z-n}{2}} \frac{dt}{t}.$$

This identity shows that G_z is smooth on $\mathbf{R}^n \setminus \{0\}$. Moreover, taking $z = s > 0$ proves that $G_s(x) > 0$ for all $x \in \mathbf{R}^n$. Consequently, $\|G_s\|_{L^1} = \int_{\mathbf{R}^n} G_s(x)\,dx = \widehat{G_s}(0) = 1$.

Now suppose $|x| \geq 2$. Then $t + \frac{|x|^2}{4t} \geq t + \frac{1}{t}$ and $t + \frac{|x|^2}{4t} \geq |x|$. This implies that

$$-t - \frac{|x|^2}{4t} \leq -\frac{t}{2} - \frac{1}{2t} - \frac{|x|}{2},$$

from which it follows that when $|x| \geq 2$,

$$|G_z(x)| \leq \frac{(2\sqrt{\pi})^{-n}}{|\Gamma(\frac{z}{2})|}\left(\int_0^\infty e^{-\frac{t}{2}} e^{-\frac{1}{2t}} t^{\frac{\mathrm{Re}\, z-n}{2}} \frac{dt}{t}\right) e^{-\frac{|x|}{2}} = \frac{C'(\mathrm{Re}\, z, n)}{|\Gamma(\frac{z}{2})|} e^{-\frac{|x|}{2}}.$$

This proves (1.2.13) and (1.2.11) with $C(s,n) = \Gamma(\frac{s}{2})^{-1} C'(s,n)$ when $s > 0$.

Suppose now that $|x| \leq 2$ and $s > 0$. Write $G_s(x) = G_s^1(x) + G_s^2(x) + G_s^3(x)$, where

$$\begin{aligned}
G_s^1(x) &= \frac{(2\sqrt{\pi})^{-n}}{\Gamma(\frac{s}{2})} \int_0^{|x|^2} e^{-t'} e^{-\frac{|x|^2}{4t'}} (t')^{\frac{s-n}{2}} \frac{dt'}{t'} \\
&= |x|^{s-n} \frac{(2\sqrt{\pi})^{-n}}{\Gamma(\frac{s}{2})} \int_0^1 e^{-t|x|^2} e^{-\frac{1}{4t}} t^{\frac{s-n}{2}} \frac{dt}{t}, \\
G_s^2(x) &= \frac{(2\sqrt{\pi})^{-n}}{\Gamma(\frac{s}{2})} \int_{|x|^2}^4 e^{-t} e^{-\frac{|x|^2}{4t}} t^{\frac{s-n}{2}} \frac{dt}{t}, \\
G_s^3(x) &= \frac{(2\sqrt{\pi})^{-n}}{\Gamma(\frac{s}{2})} \int_4^\infty e^{-t} e^{-\frac{|x|^2}{4t}} t^{\frac{s-n}{2}} \frac{dt}{t}.
\end{aligned}$$

Since $t|x|^2 \leq 4$, in G_s^1 we have $e^{-t|x|^2} = 1 + O(t|x|^2)$ by the mean value theorem, where $O(t)$ is a function with the property $|O(t)| \leq |t|$. Thus, we can write

$$G_s^1(x) = |x|^{s-n} \frac{(2\sqrt{\pi})^{-n}}{\Gamma(\frac{s}{2})} \int_0^1 e^{-\frac{1}{4t}} t^{\frac{s-n}{2}} \frac{dt}{t} + O(|x|^{s-n+2}) \frac{(2\sqrt{\pi})^{-n}}{\Gamma(\frac{s}{2})} \int_0^1 e^{-\frac{1}{4t}} t^{\frac{s-n}{2}} dt.$$

Since $0 \leq \frac{|x|^2}{4t} \leq \frac{1}{4}$ and $0 \leq t \leq 4$ in G_s^2, we have $e^{-\frac{17}{4}} \leq e^{-t-\frac{|x|^2}{4t}} \leq 1$; thus, we deduce

$$G_s^2(x) \approx \int_{|x|^2}^4 t^{\frac{s-n}{2}} \frac{dt}{t} = \begin{cases} \frac{2}{n-s}|x|^{s-n} - \frac{2^{s-n+1}}{n-s} & \text{for } s < n, \\ 2\log\frac{2}{|x|} & \text{for } s = n, \\ \frac{1}{s-n} 2^{s-n+1} - \frac{2}{s-n}|x|^{s-n} & \text{for } s > n. \end{cases}$$

Finally, we have $e^{-\frac{1}{4}} \leq e^{-\frac{|x|^2}{4t}} \leq 1$ in G_s^3, which yields that $G_s^3(x)$ is bounded above and below by fixed positive constants. Combining the estimates for $G_s^1(x)$, $G_s^2(x)$, and $G_s^3(x)$, we obtain (1.2.12).

When z is complex, with $\mathrm{Re}\, z > 0$, we write as before $G_z = G_z^1 + G_z^2 + G_z^3$. When $|x| \leq 2$, we have that $|G_z^1(x)| \leq c_1(\mathrm{Re}\, z, n) |\Gamma(\frac{z}{2})|^{-1} |x|^{\mathrm{Re}\, z - n}$. For G_z^2 we have

$$|G_z^2(x)| \leq \frac{(2\sqrt{\pi})^{-n}}{|\Gamma(\frac{z}{2})|}\begin{cases} \frac{2}{n-\mathrm{Re}\,z}|x|^{\mathrm{Re}\,z-n} - \frac{2^{\mathrm{Re}\,z-n+1}}{n-\mathrm{Re}\,z} & \text{for } \mathrm{Re}\,z < n,\\ 2\log\frac{2}{|x|} & \text{for } \mathrm{Re}\,z = n,\\ \frac{1}{\mathrm{Re}\,z-n}2^{\mathrm{Re}\,z-n+1} - \frac{2}{\mathrm{Re}\,z-n}|x|^{\mathrm{Re}\,z-n} & \text{for } \mathrm{Re}\,z > n \end{cases}$$

$$\leq \frac{c_2(\mathrm{Re}\,z,n)}{|\Gamma(\frac{z}{2})|}\begin{cases} |x|^{\mathrm{Re}\,z-n} & \text{for } \mathrm{Re}\,z < n,\\ \log\frac{2}{|x|} & \text{for } \mathrm{Re}\,z = n,\\ 1 & \text{for } \mathrm{Re}\,z > n, \end{cases}$$

when $|x| \leq 2$. Finally, $|G_z^3(x)| \leq c_3(\mathrm{Re}\,z,n)|\Gamma(\frac{z}{2})|^{-1}$ when $|x| \leq 2$. Combining these estimates we obtain the claimed conclusion for $G_z(x)$ when $|x| \leq 2$. □

We end this section with a result analogous to that of Theorem 1.2.3 for the operator $\mathcal{J}_s$.

Corollary 1.2.6. *(a) For all $0 < s < \infty$, the operator $\mathcal{J}_s$ maps $L^r(\mathbf{R}^n)$ to itself with norm 1 for all $1 \leq r \leq \infty$.*
(b) Let $0 < s < n$ and $1 \leq p < q < \infty$ satisfy (1.2.4). Then there exist constants $C_{p,q,n,s} < \infty$ such that for all f in $L^p(\mathbf{R}^n)$, with $p > 1$, we have

$$\big\|\mathcal{J}_s(f)\big\|_{L^q} \leq C_{p,q,n,s}\big\|f\big\|_{L^p}$$

and $\big\|\mathcal{J}_s(f)\big\|_{L^{q,\infty}} \leq C_{1,q,n,s}\big\|f\big\|_{L^1}$ when $p = 1$.

Proof. (a) Since $\widehat{G_s}(0) = 1$ and $G_s > 0$, it follows that G_s has L^1 norm 1. The operator $\mathcal{J}_s$ is given by convolution with the positive function G_s, which has L^1 norm 1; thus, it maps $L^r(\mathbf{R}^n)$ to itself with at most norm 1 for all $1 \leq r \leq \infty$; in fact this norm is exactly 1 in view of Exercise 1.2.9 in [156].
(b) In the special case $0 < s < n$, we have that the kernel G_s of $\mathcal{J}_s$ satisfies

$$G_s(x) \approx \begin{cases} |x|^{-n+s} & \text{when } |x| \leq 2,\\ e^{-\frac{|x|}{2}} & \text{when } |x| \geq 2. \end{cases}$$

Then we can write

$$\begin{aligned}\mathcal{J}_s(f)(x) &\leq C_{n,s}\left[\int_{|y|\leq 2}|f(x-y)|\,|y|^{-n+s}\,dy + \int_{|y|\geq 2}|f(x-y)|\,e^{-\frac{|y|}{2}}\,dy\right]\\ &\leq C_{n,s}\left[\mathcal{I}_s(|f|)(x) + \int_{\mathbf{R}^n}|f(x-y)|\,e^{-\frac{|y|}{2}}\,dy\right].\end{aligned}$$

Using that the function $y \mapsto e^{-|y|/2}$ is in L^r for all $r < \infty$, Young's inequality (Theorem 1.2.12 in [156]), and Theorem 1.2.3, we complete the proof. □

Exercises

1.2.1. (a) Let $0 < s,t < \infty$ be such that $s+t < n$. Show that $\mathcal{I}_s\mathcal{I}_t = \mathcal{I}_{s+t}$ as operators acting on $\mathscr{S}(\mathbf{R}^n)$.
(b) Let $\operatorname{Re} s > 2\operatorname{Re} z$. Prove the operator identities

$$\mathcal{I}_s(-\Delta)^z = (-\Delta)^z\mathcal{I}_s = \mathcal{I}_{s-2z} = (-\Delta)^{z-\frac{s}{2}}$$

on functions whose Fourier transforms vanish in a neighborhood of zero.
(c) Prove that for all $z \in \mathbf{C}$ we have

$$\langle (-\Delta)^z f \,|\, (-\Delta)^{-z} g\rangle = \langle f \,|\, g\rangle$$

whenever the Fourier transforms of f and g vanish to sufficiently high order at the origin.
(d) Given s with $\operatorname{Re} s > 0$, find an $\alpha \in \mathbf{C}$ such that the identity

$$\langle \mathcal{I}_s(f) \,|\, f\rangle = \big\|(-\Delta)^\alpha f\big\|_{L^2}^2$$

is valid for all functions f as in part (c).

1.2.2. Prove that for $-\infty < \alpha < n/2 < \beta < \infty$ and that for all $f \in \mathscr{S}(\mathbf{R}^n)$ we have

$$\big\|f\big\|_{L^\infty(\mathbf{R}^n)} \le C\big\|\Delta^{\alpha/2} f\big\|_{L^2(\mathbf{R}^n)}^{\frac{\beta-n/2}{\beta-\alpha}}\big\|\Delta^{\beta/2} f\big\|_{L^2(\mathbf{R}^n)}^{\frac{n/2-\alpha}{\beta-\alpha}},$$

where C depends only on α, n, β.
[*Hint:* You may want to use Exercise 2.2.14 in [156].]

1.2.3. Show that when $0 < s < n$ we have

$$\sup_{\|f\|_{L^1(\mathbf{R}^n)}=1} \big\|\mathcal{I}_s(f)\big\|_{L^{\frac{n}{n-s}}(\mathbf{R}^n)} = \sup_{\|f\|_{L^1(\mathbf{R}^n)}=1} \big\|\mathcal{J}_s(f)\big\|_{L^{\frac{n}{n-s}}(\mathbf{R}^n)} = \infty .$$

Thus, neither $\mathcal{I}_s$ nor $\mathcal{J}_s$ is of strong type $(1, \frac{n}{n-s})$.

1.2.4. Let $0 < s < n$. Consider the function $h(x) = |x|^{-s}(\log\frac{1}{|x|})^{-\frac{s}{n}(1+\delta)}$ for $|x| \le 1/e$ and zero otherwise. Prove that when $0 < \delta < \frac{n-s}{s}$ we have $h \in L^{\frac{n}{s}}(\mathbf{R}^n)$ but that $\lim_{x\to 0}\mathcal{I}_s(h)(x) = \infty$. Conclude that $\mathcal{I}_s$ does not map $L^{\frac{n}{s}}(\mathbf{R}^n)$ to $L^\infty(\mathbf{R}^n)$.

1.2.5. For $1 < p < \infty$ and $0 < s < \infty$ define the *Bessel potential space* $\mathscr{L}^p_s(\mathbf{R}^n)$ as the space of all functions $f \in L^p(\mathbf{R}^n)$ for which there exists another function f_0 in $L^p(\mathbf{R}^n)$ such that $\mathcal{J}_s(f_0) = f$. Define a norm on these spaces by setting $\|f\|_{\mathscr{L}^p_s} = \|f_0\|_{L^p}$. Prove the following properties of these spaces:

(a) $\|f\|_{L^p} \le \|f\|_{\mathscr{L}^p_s}$; hence, $\mathscr{L}^p_s(\mathbf{R}^n)$ is a subspace of $L^p(\mathbf{R}^n)$.
(b) For all $0 < t, s < \infty$ we have $G_s * G_t = G_{s+t}$ and thus

$$\mathscr{L}^p_s(\mathbf{R}^n) * \mathscr{L}^q_t(\mathbf{R}^n) \subseteq \mathscr{L}^r_{s+t}(\mathbf{R}^n),$$

where $1 < p, q, r < \infty$ and $\frac{1}{p} + \frac{1}{q} = \frac{1}{r} + 1$.
(c) The sequence of norms $\|f\|_{\mathscr{L}^p_s}$ increases, and therefore the spaces $\mathscr{L}^p_s(\mathbf{R}^n)$ decrease as s increases.
(d) The map $\mathcal{J}_t$ is a one-to-one and onto isometry from $\mathscr{L}^p_s(\mathbf{R}^n)$ to $\mathscr{L}^p_{s+t}(\mathbf{R}^n)$.
[The Bessel potential space $\mathscr{L}^p_s(\mathbf{R}^n)$ coincides with the Sobolev space $L^p_s(\mathbf{R}^n)$, introduced in Section 1.3.]

1.2.6. For $0 \le s < n$ define the *fractional maximal function*

$$M^s(f)(x) = \sup_{t>0} \frac{1}{(v_n t^n)^{\frac{n-s}{n}}} \int_{|y| \le t} |f(x-y)|\, dy,$$

where v_n is the volume of the unit ball in $\mathbf{R}^n$.
(a) Show that for some constant C we have

$$M^s(f) \le C \mathcal{I}_s(f)$$

for all $f \ge 0$ and conclude that M^s maps L^p to L^q whenever $\mathcal{I}_s$ does.
(b) ([1]) Let $0 < s < n$, $1 < p < \frac{n}{s}$, $1 \le q \le \infty$ be such that $\frac{1}{r} = \frac{1}{p} - \frac{s}{n} + \frac{sp}{nq}$. Show that there is a constant $C > 0$ (depending on the previous parameters) such that for all positive functions f we have

$$\big\|\mathcal{I}_s(f)\big\|_{L^r} \le C \big\|M^{\frac{n}{p}}(f)\big\|_{L^q}^{\frac{sp}{n}} \|f\|_{L^p}^{1-\frac{sp}{n}}.$$

[*Hint:* For $f \ne 0$, write $\mathcal{I}_s(f) = I_1 + I_2$, where

$$I_1 = \int_{|x-y| \le \delta} f(y)\, |x-y|^{s-n}\, dy, \qquad I_2 = \int_{|x-y| > \delta} f(y)\, |x-y|^{s-n}\, dy.$$

Show that $I_1 \le C\delta^s M^0(f)$ and that $I_2(f) \le C\delta^{s-\frac{n}{p}} M^{\frac{n}{p}}(f)$. Optimize over $\delta > 0$ to obtain

$$\mathcal{I}_s(f) \le C\big(M^{\frac{n}{p}}(f)\big)^{\frac{sp}{n}} \big(M^0(f)\big)^{1-\frac{sp}{n}},$$

from which the required conclusion follows easily.]

1.2.7. Suppose that a function K defined on $\mathbf{R}^n$ satisfies $|K(y)| \le C(1+|y|)^{s-n-\varepsilon}$, where $0 < s < n$ and $0 < C, \varepsilon < \infty$. Prove that the maximal operator

$$T_K(f)(x) = \sup_{t>0} t^{-n+s} \left| \int_{\mathbf{R}^n} f(x-y) K(y/t)\, dy \right|$$

maps $L^p(\mathbf{R}^n)$ to $L^q(\mathbf{R}^n)$ whenever $\mathcal{I}_s$ maps $L^p(\mathbf{R}^n)$ to $L^q(\mathbf{R}^n)$.
[*Hint:* Control T_K by the maximal function M^s of Exercise 1.2.6.]

1.2.8. Let $0 < s < n$. Use the following steps to obtain another proof of Theorem 1.2.3 via a more delicate interpolation.
(a) Prove that there is a constant $c_{s,n} < \infty$ such that for any measurable set E we have $\big\|\mathcal{I}_s(\chi_E)\big\|_{L^\infty} \le c_{s,n}|E|^{\frac{s}{n}}$.
(b) Prove that for some $c'_{s,n} < \infty$ and any two measurable sets E and F we have

$$\int_F |\mathcal{I}_s(\chi_E)(x)|\,dx \le c'_{s,n}|E|\,|F|^{\frac{s}{n}}.$$

(c) Use Exercise 1.1.12 in [156] to find a constant $C_{n,s}$ such that for all measurable sets E we have

$$\big\|\mathcal{I}_s(\chi_E)\big\|_{L^{\frac{n}{n-s},\infty}} \le C_{n,s}|E|.$$

(d) Use parts (a) and (c) and Theorem 1.4.19 in [156] to obtain another proof of Theorem 1.2.3.
[*Hint:* Parts (a) and (b): Use that when $\lambda > 0$, the integral $\int_E |y|^{-\lambda}\,dy$ becomes largest when E is a ball centered at the origin equimeasurable to E.]

1.2.9. ([366]) Let $0 < \alpha < n$, and suppose $0 < \varepsilon < \min(\alpha, n-\alpha)$. Show that there exists a constant depending only on α, ε, and n such that for all compactly supported bounded functions f we have

$$|\mathcal{I}_\alpha(f)| \le C\sqrt{M^{\alpha-\varepsilon}(f)M^{\alpha+\varepsilon}(f)},$$

where $M^\beta(f)$ is the fractional maximal function of Exercise 1.2.6.
[*Hint:* Write

$$|\mathcal{I}_\alpha(f)(x)| \le c\int_{|x-y|<s}\frac{|f(y)|\,dy}{|x-y|^{n-\alpha}} + c\int_{|x-y|\ge s}\frac{|f(y)|\,dy}{|x-y|^{n-\alpha}}$$

and split each integral into a sum of integrals over annuli centered at x to obtain the estimate

$$|\mathcal{I}_\alpha(f)| \le C\big(s^\varepsilon M^{\alpha-\varepsilon}(f) + s^{-\varepsilon}M^{\alpha+\varepsilon}(f)\big).$$

Then optimize over s.]

1.2.10. The *discrete maximal operator* of a sequence $a = \{a_j\}_{j\in\mathbf{Z}^n}$ is the sequence $\mathcal{M}_d(a) = \{\mathcal{M}_d(a)_m\}_{m\in\mathbf{Z}^n}$ whose terms are

$$\mathcal{M}_d(a)_m = \sup_{Q\ni m}\frac{1}{\#(Q\cap\mathbf{Z}^n)}\sum_{k\in Q\cap\mathbf{Z}^n}|a_k|,$$

where $m \in \mathbf{Z}^n$, and the supremum is taken over all cubes Q in $\mathbf{R}^n$.
(a) Show that $\mathcal{M}_d$ is bounded from $\ell^p(\mathbf{Z}^n)$ to itself.
(b) Prove that for all $m \in \mathbf{Z}^n$

$$(b*a)_m \le \|b\|_{\ell^1(\mathbf{Z}^n)}\mathcal{M}_d(a)_m$$

whenever $b = \{b_m\}_m$ is an ℓ^1 sequence of positive numbers that satisfies $b_m = b_{m'}$ when $|m| = |m'|$ and $b_m \ge b_{m'}$ when $|m| \le |m'|$.

(c) Use these results to prove that the *discrete fractional integral operator*

$$\{a_j\}_{j\in\mathbf{Z}^n} \mapsto \Big\{ \sum_{k\in\mathbf{Z}^n} \frac{a_k}{(|j-k|+1)^{n-\alpha}} \Big\}_{j\in\mathbf{Z}^n}$$

maps $\ell^s(\mathbf{Z}^n)$ to $\ell^t(\mathbf{Z}^n)$ when $0<\alpha<n$, $1<s<t$, and $\frac{1}{s}-\frac{1}{t}=\frac{\alpha}{n}$.

1.2.11. Show that the operator,

$$\mathcal{I}_{\alpha,\alpha}(f)(x_1,x_2) = \int_{\mathbf{R}^n}\int_{\mathbf{R}^n} f(x_1-y_1,x_2-y_2)|y_1|^{-n+\alpha}|y_2|^{-n+\alpha}dy_1dy_2\,,$$

acting on Schwartz functions f on $\mathbf{R}^{2n}$, maps $L^p(\mathbf{R}^{2n})$ to $L^q(\mathbf{R}^{2n})$ whenever $0<\alpha<n$, $\frac{\alpha}{n}+\frac{1}{q}=\frac{1}{p}$, and $1<p<q<\infty$.
[*Hint:* Write $\mathcal{I}_{\alpha,\alpha}(f)(x_1,x_2)=c_0\mathcal{I}_\alpha^{(1)}\big(\mathcal{I}_\alpha^{(2)}(f)(x_2)\big)(x_1)$, where $\mathcal{I}_\alpha^{(2)}$ is a fractional integral operator acting in the x_2 variable of the function $f(x_1,x_2)$ with x_1 frozen, and $\mathcal{I}_\alpha^{(1)}$ is defined analogously.]

1.2.12. Fill in the following steps to provide an alternative proof of Theorem 2.1.3 when $p=1$. Without loss of generality assume that f is nonnegative and smooth, has compact support, and satisfies $\|f\|_{L^1}=1$. Let $E_\lambda=\{x\in\mathbf{R}^n : \mathcal{I}_s(f)(x)>\lambda\}$ for $\lambda>0$. Prove that

$$\mathcal{I}_s(f)(x)\le \sum_{j\in\mathbf{Z}} 2^{(j-1)(s-n)}\int_{|y|\le 2^j} f(x-y)\,dy,$$

and that

$$\int_{E_\lambda}\int_{|y|\le 2^j} f(x-y)dy\,dx \le \min(|E_\lambda|, v_n 2^{jn})\,.$$

Using these facts and $|E_\lambda|\le\frac{1}{\lambda}\int_{E_\lambda}\mathcal{I}_s(f)(x)\,dx$ conclude that $|E_\lambda|\le C(n,s)\frac{1}{\lambda}|E_\lambda|^{\frac{s}{n}}$.

1.3 Sobolev Spaces

In this section we study a quantitative way of measuring the smoothness of functions. Sobolev spaces serve exactly this purpose. They measure the smoothness of functions in terms of the integrability of their derivatives. We begin with the classical definition of Sobolev spaces.

Definition 1.3.1. Let k be a nonnegative integer, and let $1<p<\infty$. The *Sobolev space* $L^p_k(\mathbf{R}^n)$ is defined as the space of functions f in $L^p(\mathbf{R}^n)$ all of whose distributional derivatives $\partial^\alpha f$ are also in $L^p(\mathbf{R}^n)$ for all multi-indices α that satisfy $|\alpha|\le k$. This space is normed by the expression

$$\|f\|_{L^p_k}=\sum_{|\alpha|\le k}\|\partial^\alpha f\|_{L^p}, \tag{1.3.1}$$

where $\partial^{(0,\dots,0)}f=f$.

Sobolev norms quantify smoothness. The index k indicates the *degree* of smoothness of a given function in L^p_k. As k increases the functions become smoother. Equivalently, these spaces form a decreasing sequence

$$L^p \supset L^p_1 \supset L^p_2 \supset L^p_3 \supset \cdots,$$

meaning that each $L^p_{k+1}(\mathbf{R}^n)$ is a proper subspace of $L^p_k(\mathbf{R}^n)$. This property, which coincides with our intuition of smoothness, is a consequence of the definition of Sobolev norms.

We next observe that the space $L^p_k(\mathbf{R}^n)$ is complete. Indeed, if f_j is a Cauchy sequence in the norm given by (1.3.1), then $\{\partial^\alpha f_j\}_j$ are Cauchy sequences for all $|\alpha| \le k$. By the completeness of L^p, there exist functions f_α in $L^p(\mathbf{R}^n)$ such that $\partial^\alpha f_j \to f_\alpha$ in L^p for all $|\alpha| \le k$, in particular $f_j \to f_0$ in L^p as $j \to \infty$. This implies that for all φ in the Schwartz class we have

$$(-1)^{|\alpha|} \int_{\mathbf{R}^n} f_j \,(\partial^\alpha \varphi)\, dx = \int_{\mathbf{R}^n} (\partial^\alpha f_j)\, \varphi\, dx \to \int_{\mathbf{R}^n} f_\alpha\, \varphi\, dx.$$

Since the first expression converges to

$$(-1)^{|\alpha|} \int_{\mathbf{R}^n} f_0 \,(\partial^\alpha \varphi)\, dx,$$

it follows that the distributional derivative $\partial^\alpha f_0$ is f_α. This implies that $f_j \to f_0$ in $L^p_k(\mathbf{R}^n)$ and proves the completeness of this space.

Our goal in this section is to investigate relations between these spaces and the Riesz and Bessel potentials discussed in the previous section and to obtain a Littlewood–Paley characterization of them. Before we embark on this study, we note that we can extend the definition of Sobolev spaces to the case in which the index k is not necessarily an integer. In fact, we extend the definition of the spaces $L^p_k(\mathbf{R}^n)$ to the case in which the number k is real.

1.3.1 Definition and Basic Properties of General Sobolev Spaces

Definition 1.3.2. Let s be a real number and let $1 < p < \infty$. The *inhomogeneous Sobolev space* $L^p_s(\mathbf{R}^n)$ is defined as the space of all tempered distributions u in $\mathscr{S}'(\mathbf{R}^n)$ with the property that

$$((1+|\xi|^2)^{\frac{s}{2}} \widehat{u})^\vee \tag{1.3.2}$$

is an element of $L^p(\mathbf{R}^n)$. For such distributions u we define

$$\|u\|_{L^p_s} = \|((1+|\cdot|^2)^{\frac{s}{2}} \widehat{u})^\vee\|_{L^p(\mathbf{R}^n)}.$$

Note that the function $(1+|\xi|^2)^{\frac{s}{2}}$ is $\mathscr{C}^\infty$ and has at most polynomial growth at infinity. Since $\widehat{u} \in \mathscr{S}'(\mathbf{R}^n)$, the product in (1.3.2) is well defined.

Several observations are in order. First, we note that when $s=0$, $L^p_s = L^p$. It is natural to ask whether elements of L^p_s are always L^p functions. We show that this is the case when $s \geq 0$ but not when $s<0$. We also show that the space L^p_s coincides with the space L^p_k given in Definition 1.3.1 when $s=k$ and k is an integer.

To prove that elements of L^p_s are indeed L^p functions when $s \geq 0$, we simply note that if $f_s = ((1+|\xi|^2)^{s/2}\widehat{f}\,)^\vee$, then

$$f = \big(\widehat{f_s}(\xi)\widehat{G_s}(\xi/2\pi)\big)^\vee = f_s * (2\pi)^n G_s(2\pi(\cdot)),$$

where G_s is given in Definition 1.2.4. Thus Corollary 1.2.6 yields that

$$c^{-1}\big\|f\big\|_{L^p} \leq \big\|f_s\big\|_{L^p} = \big\|f\big\|_{L^p_s} < \infty,$$

for some constant c.

We now prove that if $s=k$ is a nonnegative integer and $1<p<\infty$, then the norm of the space L^p_k as given in Definition 1.3.1 is comparable to that in Definition 1.3.2. Suppose that $f \in L^p_k$ according to Definition 1.3.2. Then for all $|\alpha| \leq k$ we have that the distributional derivatives $\partial^\alpha f$ are equal to

$$\partial^\alpha f = c_\alpha(\widehat{f}(\xi)\xi^\alpha)^\vee = c_\alpha\bigg(\widehat{f}(\xi)(1+|\xi|^2)^{\frac{k}{2}}\frac{\xi^\alpha}{(1+|\xi|^2)^{\frac{k}{2}}}\bigg)^\vee. \tag{1.3.3}$$

Theorem 6.2.7 in [156] gives that the function

$$\frac{\xi^\alpha}{(1+|\xi|^2)^{k/2}}$$

is an L^p multiplier. Since by assumption $\big(\widehat{f}(\xi)(1+|\xi|^2)^{\frac{k}{2}}\big)^\vee$ is in $L^p(\mathbf{R}^n)$, it follows from (1.3.3) that the distributional derivatives $\partial^\alpha f$ lie in $L^p(\mathbf{R}^n)$ and that

$$\sum_{|\alpha|\leq k}\big\|\partial^\alpha f\big\|_{L^p} \leq C_{p,n,k}\big\|((1+|\cdot|^2)^{\frac{k}{2}}\widehat{f}\,)^\vee\big\|_{L^p} < \infty.$$

Conversely, suppose that $f \in L^p_k$ according to Definition 1.3.1; then

$$(1+\xi_1^2+\cdots+\xi_n^2)^{\frac{k}{2}} = \sum_{|\alpha|\leq k}\frac{k!}{\alpha_1!\cdots\alpha_n!(k-|\alpha|)!}\xi^\alpha\frac{\xi^\alpha}{(1+|\xi|^2)^{\frac{k}{2}}}.$$

As we have observed, the functions $m_\alpha(\xi) = \xi^\alpha(1+|\xi|^2)^{-\frac{k}{2}}$ are L^p multipliers whenever $|\alpha| \leq k$. Since

$$((1+|\xi|^2)^{\frac{k}{2}}\widehat{f}\,)^\vee = \sum_{|\alpha|\leq k} c_{\alpha,k}\big(m_\alpha(\xi)\xi^\alpha\widehat{f}\,\big)^\vee = \sum_{|\alpha|\leq k} c'_{\alpha,k}\big(m_\alpha(\xi)\widehat{\partial^\alpha f}\,\big)^\vee,$$

it follows that

$$\|(\widehat{f}(\xi)(1+|\xi|^2)^{\frac{k}{2}})^\vee\|_{L^p} \leq C_{p,n,k} \sum_{|\gamma|\leq k} \|(\widehat{f}(\xi)\xi^\gamma)^\vee\|_{L^p} < \infty.$$

Example 1.3.3. Every Schwartz function lies in $L^p_s(\mathbf{R}^n)$ for s real. Sobolev spaces with negative indices can indeed contain tempered distributions that are not locally integrable functions. For example, consider the Dirac mass at the origin δ_0. Then $\|\delta_0\|_{L^p_{-s}(\mathbf{R}^n)} = \|G_s^\vee\|_{L^p}$ when $s>0$. For $s \geq n$ this quantity is always finite in view of Proposition 1.2.5. For $0<s<n$ the function $G_s(x) = \big((1+|\xi|^2)^{-\frac{s}{2}}\big)^\vee(x)$ is integrable to the power p as long as $(s-n)p > -n$, that is, when $1<p<\frac{n}{n-s}$. We conclude that δ_0 lies in $L^p_{-s}(\mathbf{R}^n)$ for $1<p<\frac{n}{n-s}$ when $0<s<n$ and in $L^p_{-s}(\mathbf{R}^n)$ for all $1<p<\infty$ when $s \geq n$.

Example 1.3.4. We consider the function $h(t) = 1-t$ for $0 \leq t \leq 1$, $h(t) = 1+t$ for $-1 \leq t < 0$, and $h(t) = 0$ for $|t| > 1$. Obviously, the distributional derivative of h is the function $h'(t) = \chi_{(-1,0)} - \chi_{(0,1)}$. The distributional second derivative h'' is equal to $\delta_1 + \delta_{-1} - 2\delta_0$; see Exercise 2.3.4(a) in [156]. Clearly, h'' does not belong to any L^p space; hence h is not in $L^p_2(\mathbf{R})$. But for $1<p<\infty$, h lies in $L^p_1(\mathbf{R})$, and we thus have an example of a function in $L^p_1(\mathbf{R})$ but not in $L^p_2(\mathbf{R})$.

Definition 1.3.2 allows us to fine-tune the smoothness of h by finding all s for which h lies in $L^p_s(\mathbf{R})$. An easy calculation gives

$$\widehat{h}(\xi) = \frac{e^{2\pi i x\xi} + e^{-2\pi i x\xi} - 2}{4\pi^2|\xi|^2}.$$

Fix a smooth function φ that is equal to one in a neighborhood of infinity and vanishes in the interval $[-2,2]$. Then $\big(\widehat{h}(\xi)(1+|\xi|^2)^{s/2}(1-\varphi(\xi))\big)^\vee$ is the inverse Fourier transform of a smooth function with compact support; hence it is a Schwartz function and belongs to all L^p spaces. It suffices to examine for which p the function

$$u = \left((1+|\xi|^2)^{s/2}\frac{e^{2\pi i x\xi} + e^{-2\pi i x\xi} - 2}{4\pi^2(1+|\xi|^2)}\varphi(\xi)\frac{1+|\xi|^2}{|\xi|^2}\right)^\vee \tag{1.3.4}$$

lies in $L^p(\mathbf{R})$. We first observe that the function u in (1.3.4) lies in $L^p(\mathbf{R})$ if and only if the function

$$v = \left((1+|\xi|^2)^{s/2}\frac{e^{2\pi i x\xi} + e^{-2\pi i x\xi} - 2}{4\pi^2(1+|\xi|^2)}\right)^\vee \tag{1.3.5}$$

lies in $L^p(\mathbf{R})$. Indeed, if v lies in L^p, then u lies in L^p for $1<p<\infty$ in view of Theorem 6.2.7 in [156], since the bounded function $m(\xi) = \varphi(\xi)\frac{1+|\xi|^2}{|\xi|^2}$ satisfies the Mihlin condition $|m'(\xi)| \leq C|\xi|^{-1}$. Conversely, if u lies in L^p, then

$$v = \big(\widehat{v}(\xi)\,(1-\varphi)(\xi)\big)^\vee + \left(\widehat{u}(\xi)\,\frac{|\xi|^2}{1+|\xi|^2}\right)^\vee,$$

which also lies in $L^p(\mathbf{R})$ since the function $m_0(\xi) = \frac{|\xi|^2}{1+|\xi|^2}$ also satisfies the Mihlin condition $|m_0'(\xi)| \le C|\xi|^{-1}$ and $\big(\widehat{v}(1-\varphi)\big)^\vee$ is in the Schwartz class. But the function v can be explicitly calculated. In fact, for $0 < s < 2$ one has

$$\begin{aligned} v(x) &= 2\pi G_{2-s}(2\pi(\cdot)) * (\delta_1 + \delta_{-1} - 2\delta_0)(x) \\ &= 2\pi G_{2-s}(2\pi(x-1)) + 2\pi G_{2-s}(2\pi(x+1)) + 2\pi G_{2-s}(2\pi x), \end{aligned}$$

where G_{2-s} is the Bessel potential of order $2-s$. If $2-s<1$, i.e., when $1<s<2$, then $G_{2-s}(x)$ has a spike at zero of order $|x|^{1-s}$; this is integrable to the power p only when $1 < s < 1+\frac{1}{p}$. The function $v(x)$ has similar spikes at the points $-1, 1, 0$ and it lies in L^p only when $1 < s < 1 + \frac{1}{p}$ as well. Thus, h lies in $L_s^p(\mathbf{R})$ if and only if $s < 1+\frac{1}{p}$.

Next we have a result concerning the embedding of Sobolev spaces.

Theorem 1.3.5. ***(Sobolev embedding theorem)*** *(a) Let $0 < s < \frac{n}{p}$ and $1 < p < \infty$. Then the Sobolev space $L_s^p(\mathbf{R}^n)$ continuously embeds in $L^q(\mathbf{R}^n)$ when*

$$\frac{1}{p} - \frac{1}{q} = \frac{s}{n}.$$

(b) Let $0 < s = \frac{n}{p}$ and $1 < p < \infty$. Then $L_s^p(\mathbf{R}^n)$ continuously embeds in $L^q(\mathbf{R}^n)$ for any $\frac{n}{s} < q < \infty$.
(c) Let $\frac{n}{p} < s < \infty$ and $1 < p < \infty$. Then every element of $L_s^p(\mathbf{R}^n)$ can be modified on a set of measure zero so that the resulting function is bounded and uniformly continuous.

Proof. (a) If $f \in L_s^p$, then $f_s(x) = ((1+|\xi|^2)^{\frac{s}{2}}\widehat{f}\,)^\vee(x)$ is in $L^p(\mathbf{R}^n)$. Thus,

$$f(x) = ((1+|\xi|^2)^{-\frac{s}{2}}\widehat{f_s}\,)^\vee(x);$$

hence, $f = G_s * f_s$. Since $s < n$, Proposition 1.2.5 gives that

$$|G_s(x)| \le C_{s,n}|x|^{s-n}$$

for all $x \in \mathbf{R}^n$. This implies that $|f| = |G_s * f_s| \le C_{s,n}\mathcal{I}_s(|f_s|)$. Theorem 1.2.3 now yields the required conclusion:

$$\|f\|_{L^q} \le C'_{s,n}\|\mathcal{I}_s(|f_s|)\|_{L^q} \le C''_{s,n}\|f\|_{L_s^p}.$$

(b) Given any $\frac{n}{s} < q < \infty$, we can find $t > 1$ such that

$$1 + \frac{1}{q} = \frac{s}{n} + \frac{1}{t} = \frac{1}{p} + \frac{1}{t}.$$

Then $1 < \frac{s}{n} + \frac{1}{t}$, which implies that $(-n+s)t > -n$. Thus, the function $|x|^{-n+s}\chi_{|x|\le 2}$ is integrable to the tth power, which implies that G_s is in $L^t(\mathbf{R}^n)$. Since $f = G_s * f_s$, Young's inequality (Theorem 1.2.12 in [156]) gives that

$$\big\|f\big\|_{L^q(\mathbf{R}^n)} \le \big\|f_s\big\|_{L^p(\mathbf{R}^n)}\big\|G_s\big\|_{L^t(\mathbf{R}^n)} = C_{n,s}\big\|f\big\|_{L^p_{n/p}}.$$

(c) As before, we have $f = G_s * f_s$. If $s \ge n$, then Proposition 1.2.5 gives that the function G_s is in $L^{p'}(\mathbf{R}^n)$. If $0 < s < n$, then $G_s(x)$ is bounded by a multiple of $|x|^{-n+s}$ near zero and has exponential decay at infinity. This function is integrable to the power p' near the origin if and only if $(-n+s)p' > -n$, i.e., $s > n/p$, which is what we are assuming; thus, $G_s \in L^{p'}(\mathbf{R}^n)$ when $0 < s < n$. Hence, f is given as the convolution of an L^p function and an $L^{p'}$ function, and thus it is bounded and can be identified with a uniformly continuous function (cf. Exercise 1.2.3 in [156]). □

1.3.2 Littlewood–Paley Characterization of Inhomogeneous Sobolev Spaces

We now present the first main result of this section, the characterization of the inhomogeneous Sobolev spaces using Littlewood–Paley theory.

For the purposes of the next theorem we need the following setup. We fix a Schwartz function Ψ on $\mathbf{R}^n$ whose Fourier transform is nonnegative, supported in the annulus $1 - \frac{1}{7} \le |\xi| \le 2$, equal to 1 on the smaller annulus $1 \le |\xi| \le 2 - \frac{2}{7}$, and satisfies $\widehat{\Psi}(\xi) + \widehat{\Psi}(\xi/2) = 1$ on the annulus $1 \le |\xi| \le 4 - \frac{4}{7}$. This function has the property

$$\sum_{j\in\mathbf{Z}} \widehat{\Psi}(2^{-j}\xi) = 1 \tag{1.3.6}$$

for all $\xi \ne 0$. We define the associated Littlewood–Paley operators Δ_j^Ψ given by multiplication on the Fourier transform side by the function $\widehat{\Psi}(2^{-j}\xi)$, that is,

$$\Delta_j^\Psi(f) = \Psi_{2^{-j}} * f. \tag{1.3.7}$$

Notice that the support properties of the operators Δ_j^Ψ yield the simple identity

$$\Delta_j^\Psi = (\Delta_{j-1}^\Psi + \Delta_j^\Psi + \Delta_{j+1}^\Psi)\Delta_j^\Psi$$

for all $j \in \mathbf{Z}$. We also define a Schwartz function Φ such that

$$\widehat{\Phi}(\xi) = \begin{cases} \sum_{j\le 0}\widehat{\Psi}(2^{-j}\xi) & \text{when } \xi \ne 0, \\ 1 & \text{when } \xi = 0. \end{cases} \tag{1.3.8}$$

Note that $\widehat{\Phi}(\xi)$ is equal to 1 for $|\xi| \le 2 - \frac{2}{7}$, vanishes when $|\xi| \ge 2$, and satisfies

$$\widehat{\Phi}(\xi) + \sum_{j=1}^{\infty} \widehat{\Psi}(2^{-j}\xi) = 1 \tag{1.3.9}$$

for all ξ in $\mathbf{R}^n$. We now introduce an operator S_0^{Φ} by setting

$$S_0(f) = \Phi * f, \tag{1.3.10}$$

for $f \in \mathscr{S}'(\mathbf{R}^n)$. Identity (1.3.9) yields the operator identity

$$S_0^{\Phi} + \sum_{j=1}^{\infty} \Delta_j^{\Psi} = I,$$

in which the series converges in $\mathscr{S}'(\mathbf{R}^n)$, in view of Proposition 1.1.6(b).

Having introduced the relevant background, we are now ready to state and prove the following result.

Theorem 1.3.6. *Let Ψ satisfy (1.3.6), Φ be as in (1.3.8), and let Δ_j^{Ψ}, S_0^{Φ} be as in (1.3.7) and (1.3.10), respectively. Fix $s \in \mathbf{R}$ and $1 < p < \infty$. Then there exists a constant C_1 that depends only on n, s, p, Φ, and Ψ such that for all $f \in L_s^p$ we have*

$$\big\|S_0^{\Phi}(f)\big\|_{L^p} + \Big\|\Big(\sum_{j=1}^{\infty}(2^{js}|\Delta_j^{\Psi}(f)|)^2\Big)^{\frac{1}{2}}\Big\|_{L^p} \le C_1 \big\|f\big\|_{L_s^p}. \tag{1.3.11}$$

Conversely, there exists a constant C_2 that depends on the parameters n, s, p, Φ, and Ψ such that every tempered distribution f that satisfies

$$\big\|S_0^{\Phi}(f)\big\|_{L^p} + \Big\|\Big(\sum_{j=1}^{\infty}(2^{js}|\Delta_j^{\Psi}(f)|)^2\Big)^{\frac{1}{2}}\Big\|_{L^p} < \infty$$

is an element of the Sobolev space L_s^p with norm

$$\big\|f\big\|_{L_s^p} \le C_2\Big(\big\|S_0^{\Phi}(f)\big\|_{L^p} + \Big\|\Big(\sum_{j=1}^{\infty}(2^{js}|\Delta_j^{\Psi}(f)|)^2\Big)^{\frac{1}{2}}\Big\|_{L^p}\Big). \tag{1.3.12}$$

Proof. We denote by C a generic constant that depends on the parameters n, s, p, Φ, and Ψ and that may vary in different occurrences. For a given tempered distribution f we define another tempered distribution f_s by setting

$$f_s = \big((1+|\cdot|^2)^{\frac{s}{2}}\widehat{f}\big)^{\vee},$$

so that we have $\|f\|_{L_s^p} = \|f_s\|_{L^p}$ if $f \in L_s^p$.

We first assume that the expression on the right-hand side in (1.3.12) is finite, and we show that the tempered distribution f lies in the space L^p_s by controlling the L^p norm of f_s by a multiple of this expression. We begin by writing

$$f_s = \big(\widehat{\Phi}\widehat{f_s}\big)^\vee + \big((1-\widehat{\Phi})\widehat{f_s}\big)^\vee,$$

and we plan to show that both quantities on the right are in L^p. Pick a smooth function with compact support η_0 that is equal to 1 on the support of $\widehat{\Phi}$. It is a simple fact that for all $s \in \mathbf{R}$ the function $(1+|\xi|^2)^{\frac{s}{2}}\eta_0(\xi)$ lies in $\mathcal{M}_p(\mathbf{R}^n)$ (i.e., it is an L^p Fourier multiplier). Since

$$\big(\widehat{\Phi}\widehat{f_s}\big)^\vee(x) = \{((1+|\xi|^2)^{\frac{s}{2}}\eta_0(\xi))\,\widehat{S_0^\Phi(f)}(\xi)\}^\vee(x), \tag{1.3.13}$$

we have the estimate

$$\big\|\big(\widehat{\Phi}\widehat{f_s}\big)^\vee\big\|_{L^p} \le C\|S_0^\Phi(f)\|_{L^p}. \tag{1.3.14}$$

We now introduce a smooth function η_∞ that vanishes in a neighborhood of the origin and is equal to 1 on the support of $1-\widehat{\Phi}$. Using Theorem 6.2.7 in [156], we easily see that the function

$$\frac{(1+|\xi|^2)^{\frac{s}{2}}}{|\xi|^s}\eta_\infty(\xi)$$

is in $\mathcal{M}_p(\mathbf{R}^n)$ (with norm depending on n, p, η_∞, and s). Since

$$\big((1+|\xi|^2)^{\frac{s}{2}}(1-\widehat{\Phi}(\xi))\widehat{f}\big)^\vee(x) = \Big(\frac{(1+|\xi|^2)^{\frac{s}{2}}\eta_\infty(\xi)}{|\xi|^s}\,|\xi|^s(1-\widehat{\Phi}(\xi))\widehat{f}\Big)^\vee(x),$$

we obtain the estimate

$$\big\|\big((1-\widehat{\Phi})\widehat{f_s}\big)^\vee\big\|_{L^p} \le C\big\|f_\infty\big\|_{L^p}, \tag{1.3.15}$$

where f_∞ is another tempered distribution defined via

$$f_\infty = \big(|\xi|^s(1-\widehat{\Phi}(\xi))\widehat{f}\big)^\vee.$$

We will show that the quantity $\|f_\infty\|_{L^p}$ is finite using Littlewood–Paley theory. To achieve this, we introduce a smooth bump $\widehat{\zeta}$ supported in the annulus $\frac{1}{2} \le |\xi| \le 4$ and equal to 1 on the support of $\widehat{\Psi}$. Then we define $\widehat{\theta}(\xi) = |\xi|^s\widehat{\zeta}(\xi)$ and introduce the Littlewood–Paley operators

$$\Delta_j^\theta(g) = g * \theta_{2^{-j}},$$

where $\theta_{2^{-j}}(t) = 2^{jn}\theta(2^j t)$. Recalling that

$$1-\widehat{\Phi}(\xi) = \sum_{k\ge 1}\widehat{\Psi}(2^{-k}\xi),$$

we obtain that

$$\widehat{f_\infty} = \sum_{j=1}^{\infty} |\xi|^s \widehat{\Psi}(2^{-j}\xi)\widehat{\zeta}(2^{-j}\xi)\widehat{f} = \sum_{j=1}^{\infty} 2^{js}\widehat{\Psi}(2^{-j}\xi)\widehat{\theta}(2^{-j}\xi)\widehat{f}$$

and hence

$$f_\infty = \sum_{j=1}^{\infty} \Delta_j^\theta(2^{js}\Delta_j^\Psi(f)),$$

where all series converge in $\mathscr{S}'(\mathbf{R}^n)$. We now invoke the estimate

$$\Big\| \sum_{j\in\mathbf{Z}} \Delta_j^\theta(f_j) \Big\|_{L^p(\mathbf{R}^n)} \le C(n,p,\theta) \Big\| \Big(\sum_{j\in\mathbf{Z}} |f_j|^2 \Big)^{\frac{1}{2}} \Big\|_{L^p(\mathbf{R}^n)}$$

proved in Remark 6.1.3 in [156]. Setting $f_j = 2^{js}\Delta_j^\Psi(f)$, we obtain

$$\big\|f_\infty\big\|_{L^p} \le C \big\| \big(\sum_{j=1}^{\infty} |2^{js}\Delta_j^\Psi(f)|^2 \big)^{\frac{1}{2}} \big\|_{L^p} < \infty. \tag{1.3.16}$$

Combining (1.3.14), (1.3.15), and (1.3.16), we deduce the estimate in (1.3.12). This argument also shows that f_∞ is a function.

To obtain the converse inequality (1.3.11), we must essentially reverse our steps. Here we assume that $f \in L_s^p$, and we show the validity of (1.3.11). First, we have the estimate

$$\big\|S_0^\Phi(f)\big\|_{L^p} \le C\big\|f_s\big\|_{L^p} = C\big\|f\big\|_{L_s^p}, \tag{1.3.17}$$

since we can obtain the Fourier transform of $S_0^\Phi(f) = \Phi * f$ by multiplying $\widehat{f_s}$ by the L^p Fourier multiplier $(1+|\xi|^2)^{-\frac{s}{2}}\widehat{\Phi}(\xi)$. Second, setting $\widehat{\sigma}(\xi) = |\xi|^{-s}\widehat{\Psi}(\xi)$ and letting Δ_j^σ be the Littlewood–Paley operator associated with the bump $\widehat{\sigma}(2^{-j}\xi)$, we have

$$2^{js}\widehat{\Psi}(2^{-j}\xi)\widehat{f} = \widehat{\sigma}(2^{-j}\xi)|\xi|^s\widehat{f} = \widehat{\sigma}(2^{-j}\xi)|\xi|^s(1-\widehat{\Phi}(\xi))\widehat{f}$$

when $j \ge 2$, since $\widehat{\Phi}$ vanishes on the support of $\widehat{\sigma}(2^{-j}\xi)$ when $j \ge 2$. This yields the operator identity

$$2^{js}\Delta_j^\Psi(f) = \Delta_j^\sigma(f_\infty). \tag{1.3.18}$$

Using identity (1.3.18) we obtain

$$\Big\| \Big(\sum_{j=2}^{\infty} |2^{js}\Delta_j^\Psi(f)|^2 \Big)^{\frac{1}{2}} \Big\|_{L^p} = \Big\| \Big(\sum_{j=2}^{\infty} |\Delta_j^\sigma(f_\infty)|^2 \Big)^{\frac{1}{2}} \Big\|_{L^p} \le C\big\|f_\infty\big\|_{L^p}, \tag{1.3.19}$$

where the last inequality follows by Theorem 6.1.2 in [156]. Notice that

$$f_\infty = \big(|\xi|^s(1-\widehat{\Phi}(\xi))\widehat{f}\big)^\vee = \Big(\frac{|\xi|^s(1-\widehat{\Phi}(\xi))}{(1+|\xi|^2)^{\frac{s}{2}}} \widehat{f_s} \Big)^\vee$$

and that the function $|\xi|^s(1-\widehat{\Phi}(\xi))(1+|\xi|^2)^{-\frac{s}{2}}$ is in $\mathcal{M}_p(\mathbf{R}^n)$ by Theorem 6.2.7 in [156]. It follows that

$$\big\|f_\infty\big\|_{L^p} \le C\big\|f_s\big\|_{L^p} = C\big\|f\big\|_{L^p_s},$$

which, combined with (1.3.19), yields

$$\Big\|\Big(\sum_{j=2}^{\infty}|2^{js}\Delta_j^{\Psi}(f)|^2\Big)^{\frac{1}{2}}\Big\|_{L^p} \le C\big\|f\big\|_{L^p_s}. \tag{1.3.20}$$

Finally, we have

$$2^s\Delta_1(f) = 2^s\big(\widehat{\Psi}(\tfrac{1}{2}\xi)(1+|\xi|^2)^{-\frac{s}{2}}(1+|\xi|^2)^{\frac{s}{2}}\widehat{f}\,\big)^\vee = 2^s\big(\widehat{\Psi}(\tfrac{1}{2}\xi)(1+|\xi|^2)^{-\frac{s}{2}}\widehat{f_s}\,\big)^\vee,$$

and since the function $\widehat{\Psi}(\frac{1}{2}\xi)(1+|\xi|^2)^{-\frac{s}{2}}$ being smooth with compact support lies in $\mathcal{M}_p(\mathbf{R}^n)$, it follows that

$$\big\|2^s\Delta_1^{\Psi}(f)\big\|_{L^p} \le C\big\|f_s\big\|_{L^p} = C\big\|f\big\|_{L^p_s}. \tag{1.3.21}$$

Combining estimates (1.3.17), (1.3.20), and (1.3.21), we conclude the proof of (1.3.11). □

1.3.3 Littlewood–Paley Characterization of Homogeneous Sobolev Spaces

We now introduce the homogeneous Sobolev spaces $\dot{L}^p_s$. The main difference with the inhomogeneous spaces L^p_s is that elements of $\dot{L}^p_s$ may not themselves be elements of L^p. Another point of differentiation is that elements of homogeneous Sobolev spaces whose differences are polynomials are identified.

For the purposes of the following definition, for $1<p<\infty$ we define $\dot{L}^p(\mathbf{R}^n)$ as the space of all elements in $\mathscr{S}'(\mathbf{R}^n)/\mathscr{P}(\mathbf{R}^n)$ such that every equivalence class $\big[$formed from the relationship $u\equiv v$ if $u-v\in\mathscr{P}(\mathbf{R}^n)\big]$ contains a unique representative that belongs to $L^p(\mathbf{R}^n)$. One defines the $\dot{L}^p(\mathbf{R}^n)$ norm of every element of the equivalence class to be the L^p norm of the unique L^p representative. Under this definition we have

$$\big\|f+P\big\|_{\dot{L}^p} = \big\|f\big\|_{\dot{L}^p} = \big\|f\big\|_{L^p}$$

whenever $f\in L^p$ and P is a polynomial.

Definition 1.3.7. Let s be a real number, and let $1<p<\infty$. The *homogeneous Sobolev space* $\dot{L}^p_s(\mathbf{R}^n)$ is defined as the space of all u in $\mathscr{S}'(\mathbf{R}^n)/\mathscr{P}(\mathbf{R}^n)$ for which the well-defined distribution

$$(|\xi|^s\widehat{u})^\vee$$

coincides with a function in $\dot{L}^p(\mathbf{R}^n)$. For distributions u in $\dot{L}^p_s(\mathbf{R}^n)$ we define

$$\|u\|_{\dot{L}^p_s} = \big\|(|\cdot|^s\widehat{u})^\vee\big\|_{\dot{L}^p(\mathbf{R}^n)}. \tag{1.3.22}$$

As noted earlier, to avoid working with equivalence classes of functions, we identify two distributions in $\dot{L}^p_s(\mathbf{R}^n)$ whose difference is a polynomial. Under this identification, the quantity in (1.3.22) is a norm.

Theorem 1.3.6 also has a homogeneous version.

Theorem 1.3.8. *Let Ψ satisfy (1.3.6), and let Δ_j^Ψ be the Littlewood–Paley operator associated with Ψ. Let $s \in \mathbf{R}$ and $1 < p < \infty$. Then there exists a constant C_1 that depends only on n, s, p, and Ψ such that for all $f \in \dot{L}^p_s(\mathbf{R}^n)$ we have*

$$\Big\|\Big(\sum_{j\in\mathbf{Z}}(2^{js}|\Delta_j^\Psi(f)|)^2\Big)^{\frac{1}{2}}\Big\|_{L^p} \le C_1\|f\|_{\dot{L}^p_s}. \tag{1.3.23}$$

Conversely, there exists a constant C_2 that depends on the parameters n, s, p, and Ψ such that every element f of $\mathscr{S}'(\mathbf{R}^n)/\mathscr{P}(\mathbf{R}^n)$ that satisfies

$$\Big\|\Big(\sum_{j\in\mathbf{Z}}(2^{js}|\Delta_j^\Psi(f)|)^2\Big)^{\frac{1}{2}}\Big\|_{L^p} < \infty$$

lies in the homogeneous Sobolev space $\dot{L}^p_s$ and we have

$$\|f\|_{\dot{L}^p_s} \le C_2\Big\|\Big(\sum_{j\in\mathbf{Z}}(2^{js}|\Delta_j^\Psi(f)|)^2\Big)^{\frac{1}{2}}\Big\|_{L^p}. \tag{1.3.24}$$

Proof. The proof of the theorem is similar to but a bit simpler than that of Theorem 1.3.6. To obtain (1.3.23), we start with $f \in \dot{L}^p_s$ and note that

$$2^{js}\Delta_j(f) = 2^{js}\big(|\xi|^s|\xi|^{-s}\widehat{\Psi}(2^{-j}\xi)\widehat{f}\,\big)^\vee = \big(\widehat{\sigma}(2^{-j}\xi)\widehat{f_s}\,\big)^\vee = \Delta_j^\sigma(f_s),$$

where $\widehat{\sigma}(\xi) = \widehat{\Psi}(\xi)|\xi|^{-s}$ and Δ_j^σ is the Littlewood–Paley operator given on the Fourier transform side by multiplication with the function $\widehat{\sigma}(2^{-j}\xi)$. We have

$$\Big\|\Big(\sum_{j\in\mathbf{Z}}|2^{js}\Delta_j^\Psi(f)|^2\Big)^{\frac{1}{2}}\Big\|_{L^p} = \Big\|\Big(\sum_{j\in\mathbf{Z}}|\Delta_j^\sigma(f_s)|^2\Big)^{\frac{1}{2}}\Big\|_{L^p} \le C\|f_s\|_{\dot{L}^p} = C\|f\|_{\dot{L}^p_s},$$

where the last inequality follows from Theorem 6.1.2 in [156]. This proves (1.3.23).

Next we show that if the expression on the right-hand side in (1.3.24) is finite, then the distribution f in $\mathscr{S}'(\mathbf{R}^n)/\mathscr{P}(\mathbf{R}^n)$ must lie in the homogeneous Sobolev space $\dot{L}^p_s$ with norm controlled by a multiple of this expression.

Define Littlewood–Paley operators Δ_j^{η} given by convolution with $\eta_{2^{-j}}$, where $\widehat{\eta}$ is a smooth bump supported in the annulus $\frac{4}{5} \le |\xi| \le 2$ that satisfies

$$\sum_{k\in\mathbf{Z}} \widehat{\eta}(2^{-k}\xi) = 1, \qquad \xi \neq 0 \tag{1.3.25}$$

or, in operator form,

$$\sum_{k\in\mathbf{Z}} \Delta_k^{\eta} = I,$$

where the convergence is in the sense of $\mathscr{S}'/\mathscr{P}$ in view of Proposition 1.1.6(c). We introduce another family of Littlewood–Paley operators Δ_j^{θ} given by convolution with $\theta_{2^{-j}}$, where $\widehat{\theta}(\xi) = \widehat{\eta}(\xi)|\xi|^s$. Given $f \in \mathscr{S}'(\mathbf{R}^n)/\mathscr{P}$, we set $f_s = \big(|\xi|^s \widehat{f}\,\big)^{\vee}$, which is also an element of $\mathscr{S}'(\mathbf{R}^n)/\mathscr{P}$. In view of (1.3.25) we use Theorem 6.1.2 in [156] to obtain the existence of a polynomial Q such that

$$\|f\|_{\dot{L}^p_s} = \|f_s - Q\|_{L^p} \le C\Big\|\Big(\sum_{j\in\mathbf{Z}} |\Delta_j^{\eta}(f_s)|^2\Big)^{\frac{1}{2}}\Big\|_{L^p} = C\Big\|\Big(\sum_{j\in\mathbf{Z}} |2^{js}\Delta_j^{\theta}(f)|^2\Big)^{\frac{1}{2}}\Big\|_{L^p}.$$

Recalling the definition of Δ_j (see the discussion before the statement of Theorem 1.3.6), we notice that the function

$$\widehat{\Psi}(\tfrac{1}{2}\xi) + \widehat{\Psi}(\xi) + \widehat{\Psi}(2\xi)$$

is equal to 1 on the support of $\widehat{\theta}$ (which is the same as the support of η). It follows that

$$\Delta_j^{\theta} = (\Delta_{j-1}^{\Psi} + \Delta_j^{\Psi} + \Delta_{j+1}^{\Psi})\Delta_j^{\theta}.$$

We therefore have the estimate

$$\Big\|\Big(\sum_{j\in\mathbf{Z}} |2^{js}\Delta_j^{\theta}(f)|^2\Big)^{\frac{1}{2}}\Big\|_{L^p} \le \sum_{r=-1}^{1} \Big\|\Big(\sum_{j\in\mathbf{Z}} |\Delta_j^{\theta}\Delta_{j+r}^{\Psi}(2^{js}f)|^2\Big)^{\frac{1}{2}}\Big\|_{L^p},$$

and, applying Proposition 6.1.4 in [156], we control the right-hand side of the preceding expression $\big($and thus $\|f\|_{\dot{L}^p_s}\big)$ by a constant multiple of

$$\Big\|\Big(\sum_{j\in\mathbf{Z}} |\Delta_j^{\Psi}(2^{js}f)|^2\Big)^{\frac{1}{2}}\Big\|_{L^p}.$$

This proves that the homogeneous Sobolev norm of f is controlled by a multiple of the expression in (1.3.24). In particular, the distribution f lies in the homogeneous Sobolev space $\dot{L}^p_s$. This ends the proof of the converse direction and completes the proof of the theorem. □

Exercises

1.3.1. Let $1 < p < \infty$ and $s \in \mathbf{R}$. Show that the spaces $\dot{L}^p_s$ and L^p_s are complete and that the latter decrease as s increases.

1.3.2. (a) Let $1 < p < \infty$ and $s \in \mathbf{Z}^+$. Suppose that $f \in L^p_s(\mathbf{R}^n)$ and that φ is in $\mathscr{S}(\mathbf{R}^n)$. Prove that φf is also an element of $L^p_s(\mathbf{R}^n)$.
(b) Let v be a function whose Fourier transform is a bounded compactly supported function. Prove that if f is in $L^2_s(\mathbf{R}^n)$, then so is vf.

1.3.3. Fix $s > 0$ and let α be a multi-index. Let δ_0 be the Dirac mass at the origin on $\mathbf{R}^n$.
(a) If $0 < s - |\alpha| < n$, show that $\partial^\alpha \delta_0 \in L^p_{-s}$ whenever $1 < p < \frac{n}{n+|\alpha|-s}$.
(b) If $n \le s - |\alpha|$, prove that $\partial^\alpha \delta_0 \in L^p_{-s}$ for all $p \in (1,\infty)$.
[*Hint:* Use Proposition 1.2.5.]

1.3.4. Let I be the identity operator, $\mathcal{I}_1$ the Riesz potential of order 1, and R_j the usual Riesz transform. Prove that

$$I = \sum_{j=1}^{n} \mathcal{I}_1 R_j \partial_j \,,$$

and use this identity to obtain Theorem 1.3.5(a) when $s = 1$.
[*Hint:* Take the Fourier transform.]

1.3.5. Let f be in L^p_s for some $1 < p < \infty$. Prove that $\partial^\alpha f$ is in $L^p_{s-|\alpha|}$.

1.3.6. Prove that for all $\mathscr{C}^1$ functions f that are supported in a ball B we have

$$|f(x)| \le \frac{1}{\omega_{n-1}} \int_B |\nabla f(y)| |x-y|^{-n+1} \, dy \,,$$

where $\omega_{n-1} = |\mathbf{S}^{n-1}|$. For such functions obtain the local Sobolev inequality

$$\big\| f \big\|_{L^q(B)} \le C_{q,r,n} \big\| \nabla f \big\|_{L^p(B)} \,,$$

where $1 < p < q < \infty$ and $1/p = 1/q + 1/n$.
[*Hint:* Start from $f(x) = \int_0^\infty \nabla f(x - t\theta) \cdot \theta \, dt$ and integrate over $\theta \in \mathbf{S}^{n-1}$.]

1.3.7. Show that there is a constant C such that for all $\mathscr{C}^1$ functions f that are supported in a ball B we have

$$\frac{1}{|B'|} \int_{B'} |f(x) - f(z)| \, dz \le C \int_B |\nabla f(y)| |x-y|^{-n+1} \, dy$$

for all B' balls contained in B and all $x \in B'$.
[*Hint:* Start with $f(z) - f(x) = \int_0^1 \nabla f(x + t(z-x)) \cdot (z-x) \, dt$.]

1.3.8. Let $1 < p < \infty$ and $s > 0$. Prove that $\dot{L}^p_s \cap L^p = L^p_s$ by showing that

$$\|f\|_{L^p_s} \approx \|f\|_{L^p} + \|f\|_{\dot{L}^p_s}.$$

[*Hint:* Write $(1+|\xi|^2)^{\frac{s}{2}} = (1+|\xi|^2)^{\frac{s}{2}}\phi(\xi) + \{\frac{(1+|\xi|^2)^{\frac{s}{2}}}{|\xi|^s}(1-\phi(\xi))\}|\xi|^s$ for some smooth compactly supported function ϕ equal to one on the ball $B(0,1)$. Then show that the term in the curly brackets is a Mihlin multiplier (i.e., it satisfies the hypotheses of Theorem 6.2.7 in [156]) to deduce that $\|f\|_{L^p_s}$ is controlled by $\|f\|_{L^p} + \|f\|_{\dot{L}^p_s}$. Conversely, use that $\frac{|\xi|^s}{(1+|\xi|^2)^{\frac{s}{2}}}$ is also a Mihlin multiplier.]

1.3.9. ([148], [285]) Prove that all Schwartz functions on $\mathbf{R}^n$ satisfy the estimate

$$\|f\|_{L^q} \le \prod_{j=1}^{n} \|\partial_j f\|_{L^1}^{1/n},$$

where $1/q + 1/n = 1$.
[*Hint:* Use induction beginning with the case $n = 1$. Assuming that the inequality is valid for $n-1$, set $I_j(x_1) = \int_{\mathbf{R}^{n-1}} |\partial_j f(x_1,x')|\,dx'$ for $j = 2,\dots,n$, where $x = (x_1,x') \in \mathbf{R}\times\mathbf{R}^{n-1}$ and $I_1(x') = \int_{\mathbf{R}^1} |\partial_1 f(x_1,x')|\,dx_1$. Apply the induction hypothesis to obtain

$$\|f(x_1,\cdot)\|_{L^{q'}} \le \prod_{j=2}^{n} I_j(x_1)^{1/(n-1)}$$

and use that $|f|^q \le I_1(x')^{1/(n-1)}|f|$ and Hölder's inequality to calculate $\|f\|_{L^q}$.]

1.3.10. Prove that there is a constant $c_n > 0$ such that for all $f \in L^2_1(\mathbf{R}^n)$ we have

$$\int_{\mathbf{R}^n}\int_{\mathbf{R}^n} \frac{|f(x+t)+f(x-t)-2f(x)|^2}{|t|^{n+2}}\,dx\,dt = c_n \int_{\mathbf{R}^n} \sum_{j=1}^{n} |\partial_j f(x)|^2\,dx.$$

1.3.11. ([75]) Let $0 \le \beta < \infty$, let $g \in L^2(\mathbf{R}^n)$, and suppose

$$C_0 = \int_{\mathbf{R}^n} |\widehat{g}(\xi)|^2 (1+|\xi|)^n \big(\log(2+|\xi|)\big)^{-\beta}\,d\xi < \infty.$$

(a) Prove that there is a constant $C(n,\beta,C_0)$ such that for every $q > 2$ we have

$$\|g\|_{L^q(\mathbf{R}^n)} \le C(n,\beta,C_0)\, q^{\frac{\beta+1}{2}}.$$

(b) Conclude that for any compact subset K of $\mathbf{R}^n$ we have

$$\int_K e^{|g(x)|^\gamma}\,dx < \infty$$

whenever $\gamma < \frac{2}{\beta+1}$.

[*Hint:* Part (a): For $q > 2$ control $\|g\|_{L^q(\mathbf{R}^n)}$ by $\|\widehat{g}\|_{L^{q'}(\mathbf{R}^n)}$ and apply Hölder's inequality with exponents $\frac{2}{q'}$ and $\frac{2(q-1)}{q-2}$. Part (b): Expand the exponential in Taylor series.]

1.3.12. Suppose that $m \in L^2_s(\mathbf{R}^n)$ for some $s > \frac{n}{2}$, and let $\lambda > 0$. Define the operator T_λ by setting $\widehat{T_\lambda(f)}(\xi) = m(\lambda\xi)\widehat{f}(\xi)$. Show that there exists a constant $C = C(n,s)$ such that for all f and $u \geq 0$ and $\lambda > 0$ we have

$$\int_{\mathbf{R}^n} |T_\lambda(f)(x)|^2 u(x)\,dx \leq C\|m\|^2_{L^2_s} \int_{\mathbf{R}^n} |f(y)|^2 M(u)(y)\,dy.$$

[*Hint:* Prove that $|T_\lambda(f)(x)|^2 \leq C\|m\|^2_{L^2_s} \int_{\mathbf{R}^n} \lambda^{-n}(1+4\pi^2|\lambda^{-1}(x-y)|^2)^{-s}|f(y)|^2\,dy$ using the Cauchy-Schwarz inequality.]

1.4 Lipschitz Spaces

Lipschitz spaces measure the degree of fractional smoothness of functions. According to the classical definition, a function f on $\mathbf{R}^n$ is Lipschitz (or Hölder) continuous of order $\gamma \in (0,1)$ if there is a constant $C < \infty$ such that for all $x, y \in \mathbf{R}^n$ we have

$$|f(x+y) - f(x)| \leq C|y|^\gamma. \tag{1.4.1}$$

Lipschitz norms are introduced to quantify the smoothness *measured* by the quantity γ in (1.4.1), and Lipschitz spaces are defined in terms of these norms. In this section we discuss analogs of condition (1.4.1) for $\gamma > 1$ and explore connections with the Fourier transform and orthogonality. The main achievement of this section is a characterization of Lipschitz spaces using Littlewood–Paley theory.

1.4.1 Introduction to Lipschitz Spaces

Definition 1.4.1. Let $0 \leq \gamma < 1$. A function f on $\mathbf{R}^n$ is said to be *Lipschitz of order* γ if it is continuous and bounded, and satisfies (1.4.1). In this case, we set

$$\|f\|_{\Lambda_\gamma(\mathbf{R}^n)} = \|f\|_{L^\infty} + \sup_{x\in\mathbf{R}^n}\sup_{h\in\mathbf{R}^n\setminus\{0\}} \frac{|f(x+h)-f(x)|}{|h|^\gamma}$$

and we define the space

$$\Lambda_\gamma(\mathbf{R}^n) = \{f : \mathbf{R}^n \to \mathbf{C} : \text{ continuous and } \|f\|_{\Lambda_\gamma(\mathbf{R}^n)} < \infty\}.$$

We call $\Lambda_\gamma(\mathbf{R}^n)$ an *inhomogeneous Lipschitz space* of order γ. We note that $\Lambda_0(\mathbf{R}^n) = L^\infty(\mathbf{R}^n)\cap\mathscr{C}(\mathbf{R}^n)$, where $\mathscr{C}(\mathbf{R}^n)$ is the space of all continuous functions on $\mathbf{R}^n$, and the $\|\cdot\|_{\Lambda_0}$ norm is comparable with the L^∞ norm; see Exercise 1.4.2.

Obviously, only constants satisfy

$$\sup_{x\in\mathbf{R}^n}\sup_{h\in\mathbf{R}^n\setminus\{0\}} |h|^{-\gamma}|f(x+h)-f(x)| < \infty$$

for $\gamma > 1$, and the preceding definition would not be applicable in this case. To extend Definition 1.4.1 for indices $\gamma \ge 1$, for $h \in \mathbf{R}^n$ we define the *difference operator* D_h by setting

$$D_h(f)(x) = f(x+h) - f(x)$$

for a continuous function $f:\ \mathbf{R}^n \to \mathbf{C}$. One easily verifies that

$$\begin{aligned} D_h^2(f)(x) &= D_h(D_h f)(x) = f(x+2h) - 2f(x+h) + f(x), \\ D_h^3(f)(x) &= D_h(D_h^2 f)(x) = f(x+3h) - 3f(x+2h) + 3f(x+h) - f(x), \end{aligned}$$

and in general, that $D_h^{k+1}(f) = D_h^k(D_h(f))$ is given by

$$D_h^{k+1}(f)(x) = \sum_{s=0}^{k+1} (-1)^{k+1-s}\binom{k+1}{s} f(x+sh) \tag{1.4.2}$$

for a nonnegative integer k. See Exercise 1.4.3.

Definition 1.4.2. For $\gamma > 0$ define

$$\big\|f\big\|_{\Lambda_\gamma} = \big\|f\big\|_{L^\infty} + \sup_{x\in\mathbf{R}^n}\sup_{h\in\mathbf{R}^n\setminus\{0\}} \frac{|D_h^{[\gamma]+1}(f)(x)|}{|h|^\gamma},$$

where $[\gamma]$ denotes the integer part of γ, and set

$$\Lambda_\gamma = \big\{f : \mathbf{R}^n \to \mathbf{C} \text{ continuous} :\ \|f\|_{\Lambda_\gamma} < \infty\big\}.$$

We call $\Lambda_\gamma(\mathbf{R}^n)$ an inhomogeneous *Lipschitz space* of order $\gamma \in \mathbf{R}^+$.

We note that functions in Λ_γ and $\dot{\Lambda}_\gamma$ are required to be continuous, since this does not necessarily follow from the definition when $\gamma \ge 1$. This is because of the axiom of choice, which implies the existence of a basis $\{v_i\}_{i\in I}$ of the vector space $\mathbf{R}$ over $\mathbf{Q}$. Without loss of generality we may assume that 1 is an element of the basis. Define a function f by setting $f(1) = 1$ and $f(v_i) = -1$ if $v_i \ne 1$, and extend f to $\mathbf{R}$ by linearity. Then f is everywhere discontinuous[1] but $D_h(f)(x) = f(h)$ for all $x,h \in \mathbf{R}$, and thus $D_h^k(f) = 0$ for all $k \ge 2$.

[1] If $v_i \ne 1$, then v_i is irrational. Let $q_k \in \mathbf{Q}$ such that $q_k \to v_i$ as $k \to \infty$. Then $f(q_k) = q_k \to v_i$ as $k \to \infty$ but $f(v_i) = -1 \ne v_i$; thus, f is discontinuous at v_i and by linearity everywhere else.

We now define the homogeneous Lipschitz spaces.

Definition 1.4.3. For $\gamma > 0$ we define

$$\|f\|_{\dot{\Lambda}_\gamma} = \sup_{x\in\mathbf{R}^n}\sup_{h\in\mathbf{R}^n\setminus\{0\}} \frac{|D_h^{[\gamma]+1}(f)(x)|}{|h|^\gamma}$$

and we let $\dot{\Lambda}_\gamma$ be the space of all continuous functions f on $\mathbf{R}^n$ that satisfy $\|f\|_{\dot{\Lambda}_\gamma} < \infty$. We call $\dot{\Lambda}_\gamma$ a *homogeneous Lipschitz space* of order γ.

We verify that elements of $\dot{\Lambda}_\gamma$ have at most polynomial growth at infinity. Indeed, identity (1.4.2) implies for all $h \in \mathbf{R}^n$

$$D_h^{k+1}(f - f(0))(0) = \sum_{s=1}^{k+1}(-1)^{k+1-s}\binom{k+1}{s}(f(sh) - f(0))$$

and thus

$$\begin{aligned}|f((k+1)h) - f(0)| &\le \sum_{s=1}^{k}\binom{k+1}{s}|f(sh) - f(0)| + \|f\|_{\dot{\Lambda}_\gamma}|h|^{k+1}\\ &\le 2^{k+1}\Big[\sup_{s\in\{1,\dots,k\}}|f(sh) - f(0)| + \|f\|_{\dot{\Lambda}_\gamma}|h|^{k+1}\Big].\end{aligned}$$

Iterating, we obtain for all $h \in \mathbf{R}^n$

$$\begin{aligned}|f((k+1)^2h) - f(0)| &\le 2^{k+1}\Big[\sup_{s\in\{1,\dots,k\}}|f(s(k+1)h) - f(0)| + \|f\|_{\dot{\Lambda}_\gamma}|h|^{k+1}\Big]\\ &\le 2^{k+1}\Big[2^{k+1}\sup_{s,s'\in\{1,\dots,k\}}|f(ss'h) - f(0)| + 2\|f\|_{\dot{\Lambda}_\gamma}|h|^{k+1}\Big]\\ &\le (2^{k+1})^2\Big[\sup_{s\in\{1,\dots,k^2\}}|f(sh) - f(0)| + \|f\|_{\dot{\Lambda}_\gamma}|h|^{k+1}\Big],\end{aligned}$$

and thus by induction for all $M \in \mathbf{Z}^+$ and $h \in \mathbf{R}^n$ we deduce

$$|f((k+1)^M h) - f(0)| \le (2^{k+1})^M\Big[\sup_{s\in\{1,\dots,k^M\}}|f(sh) - f(0)| + \|f\|_{\dot{\Lambda}_\gamma}|h|^{k+1}\Big].$$

It follows from this that

$$|f(h) - f(0)| \le (2^{k+1})^M\Big[\sup_{s\in\{1,\dots,k^M\}}|f(s(k+1)^{-M}h) - f(0)| + \|f\|_{\dot{\Lambda}_\gamma}\Big].$$

Given $|h| > 1$, there is an $M \in \mathbf{Z}^+$ such that $(\frac{k+1}{k})^{M-1} < |h| \le (\frac{k+1}{k})^M$. Then, if $c(k) = (k+1)/\log_2(\frac{k+1}{k})$, we have

$$(2^{k+1})^M = (\tfrac{k+1}{k})^{Mc(k)} \le (\tfrac{k+1}{k})^{c(k)}|h|^{c(k)}.$$

But f is continuous, so $\|f\|_{L^\infty(\overline{B(0,1)})} < \infty$, and consequently for all $|h| > 1$ we obtain

$$|f(h) - f(0)| \le (\tfrac{k+1}{k})^{c(k)} \big[2\|f\|_{L^\infty(\overline{B(0,1)})} + \|f\|_{\dot\Lambda_\gamma}\big] \, |h|^{c(k)} .$$

We conclude that functions in $\dot\Lambda_\gamma$ have at most polynomial growth at infinity and they can be thought of as elements of $\mathscr{S}'(\mathbf{R}^n)$.

Since elements of $\dot\Lambda_\gamma$ can be viewed as tempered distributions, we extend the definition of $D_h^k(u)$ to tempered distributions. For $u \in \mathscr{S}'(\mathbf{R}^n)$ we define another tempered distribution $D_h^k(u)$ via the identity

$$\langle D_h^k(u), \varphi\rangle = \langle u, D_{-h}^k(\varphi)\rangle$$

for all φ in the Schwartz class.

Constant functions f satisfy $D_h(f)(x) = 0$ for all $h, x \in \mathbf{R}^n$, and therefore the quantity $\|\cdot\|_{\dot\Lambda_\gamma}$ is insensitive to constants. Similarly, the expressions $D_h^{[\gamma]+1}(f)$ and $\|f\|_{\dot\Lambda_\gamma}$ do not recognize polynomials of degree up to $[\gamma]$. Moreover, polynomials are the only continuous functions with this property; see Exercise 1.4.1. This means that the quantity $\|\cdot\|_{\dot\Lambda_\gamma}$ is not a norm but only a seminorm. It can be made a norm if we consider equivalent classes of functions modulo polynomials. For this reason we often view $\dot\Lambda_\gamma$ as a subspace of $\mathscr{S}'(\mathbf{R}^n)/\mathscr{P}_{[\gamma]}(\mathbf{R}^n)$, where $\mathscr{P}_d$ is the space of polynomials of degree at most d for $d \ge 0$.

Examples 1.4.4. Let $a \in \mathbf{R}^n$, and let $0 < \gamma < 1$. Then the function $h(x) = \cos(x \cdot a)$ lies in $\Lambda_\gamma(\mathbf{R}^n)$ since $|h(x) - h(y)| \le \min(2, |a|\,|x-y|)$, and thus

$$|h(x) - h(y)| \le 2^{1-\gamma} |a|^\gamma |x-y|^\gamma .$$

Also, the function $x \mapsto |x|^\gamma$ lies in $\dot\Lambda_\gamma(\mathbf{R}^n)$ since $\big||x+h|^\gamma - |x|^\gamma\big| \le |h|^\gamma$ for $0 < \gamma < 1$.

Interesting examples of functions in Lipschitz spaces of higher order arise by the powers of the absolute value. Consider, for instance, the function $|x|^2$ on $\mathbf{R}^n$: we have $D_h(|x|^2) = 2|h|^2$, and thus $|x|^2 \in \dot\Lambda_\gamma(\mathbf{R}^n)$ if and only if $\gamma \ge 2$.

Another example is given by the function $|x|^{3/2}$ on $\mathbf{R}^n$ which has continuous partial derivatives at any point: $\partial_j |x|^{3/2} = \frac{3}{2} x_j |x|^{-1/2}$, $j = 1, \dots, n$, on $\mathbf{R}^n$ (with a value of 0 at the origin), while $(|x|^{3/2})' = \frac{3}{2}|x|^{1/2}$ when $n = 1$. We claim that the function $|x|^{3/2}$ lies in $\dot\Lambda_{3/2}(\mathbf{R}^n)$ and that the functions $x_j|x|^{-1/2}$ lie in $\dot\Lambda_{1/2}(\mathbf{R}^n)$. To verify these assertions, we first prove the inequality

$$\left| \frac{x_j + h_j}{|x+h|^{\frac12}} - \frac{x_j}{|x|^{\frac12}} \right| \le C |h|^{\frac12} \tag{1.4.3}$$

by considering the following three cases: (a) $x = 0$ and $h \neq 0$, which is trivial; (b) $x \neq 0$ and $2|h| < |x|$, in which case both functions are smooth and the mean value theorem yields a bound of the form $c|h|\,|x+\xi|^{-1/2}$ for some $|\xi| \le |h|$, proving

(1.4.3), since $|x+\xi| \geq |x| - |\xi| \geq |x| - |h| \geq |h|$; and (c) $2|h| \geq |x|$ and $h \neq -x \neq 0$, in which case the left-hand side of (1.4.3) is bounded by

$$|x+h|^{1/2} + |x|^{1/2} \leq C|h|^{1/2}.$$

Now for some $\xi, \xi' \in \mathbf{R}^n$, with $|\xi|, |\xi'| \leq |h|$, we have[2]

$$D_h^2(|x|^{3/2}) = \nabla(|x|^{3/2})(x+h+\xi) \cdot h - \nabla(|x|^{3/2})(x+\xi') \cdot h$$

and applying (1.4.3) we deduce that

$$|D_h^2(|x|^{3/2})| \leq C|h|^{3/2}.$$

We will make use of the following properties of the difference operators D_h^k.

Proposition 1.4.5. *Let f be a $\mathscr{C}^m$ function on $\mathbf{R}^n$ for some $m \in \mathbf{Z}^+$. Then for all $h = (h_1, \dots, h_n)$ and $x \in \mathbf{R}^n$ the following identity holds:*

$$D_h(f)(x) = \int_0^1 \sum_{j=1}^n h_j (\partial_j f)(x+sh)\, ds. \tag{1.4.4}$$

More generally, we have that

$$D_h^m(f)(x) = \int_0^1 \cdots \int_0^1 \sum_{j_1=1}^n \cdots \sum_{j_m=1}^n h_{j_1} \cdots h_{j_m} (\partial_{j_1} \cdots \partial_{j_m} f)(x+(s_1+\dots+s_m)h)\, ds_1 \cdots ds_m. \tag{1.4.5}$$

Consequently, if, for some $\gamma \in (0,1)$, $\partial^\alpha f$ lies in $\dot\Lambda_\gamma$ for all multi-indices $|\alpha| = m$, then f lies in $\dot\Lambda_{m+\gamma}$.

Proof. Identity (1.4.4) is a consequence of the fundamental theorem of calculus applied to the function $t \mapsto f((1-t)x + t(x+h))$ on $[0,1]$, whereas identity (1.4.5) follows from (1.4.4) by induction.

Now suppose that $\partial^\alpha f$ lie in $\dot\Lambda_\gamma$ for all multi-indices $|\alpha| = m$. Apply D_h on both sides of (1.4.5); using that

$$|D_h(\partial_{j_1} \cdots \partial_{j_m} f)(x+(s_1+\dots+s_m)h)| \leq \big\|\partial_{j_1} \cdots \partial_{j_m} f\big\|_{\dot\Lambda_\gamma} |h|^\gamma$$

we obtain

$$|D_h^{m+1}(f)(x)| \leq |h|^{m+\gamma} \sum_{j_1=1}^n \cdots \sum_{j_m=1}^n \big\|\partial_{j_1} \cdots \partial_{j_m} f\big\|_{\dot\Lambda_\gamma},$$

which proves that f lies in $\dot\Lambda_{m+\gamma}$. □

[2] We used that $g(b) - g(a) = \int_0^1 \nabla g((1-t)a + tb) \cdot (b-a)\, dt = \nabla g((1-t^*)a + t^*b) \cdot (b-a)$ for all $a, b \in \mathbf{R}^n$, for a $\mathscr{C}^1$ function g on $\mathbf{R}^n$ and some $t^* \in (0,1)$, depending on g, a, b.

1.4.2 Littlewood–Paley Characterization of Homogeneous Lipschitz Spaces

We now characterize the homogeneous Lipschitz spaces using the Littlewood–Paley operators Δ_j. As in the previous section, we fix a radial Schwartz function Ψ whose Fourier transform is nonnegative, is supported in the annulus $1-\frac{1}{7} \le |\xi| \le 2$, is equal to one on the annulus $1 \le |\xi| \le 2-\frac{2}{7}$, and satisfies

$$\sum_{j\in\mathbf{Z}} \widehat{\Psi}(2^{-j}\xi) = 1 \tag{1.4.6}$$

for all $\xi \neq 0$. The Littlewood–Paley operators Δ_j^{Ψ} associated with Ψ are given by multiplication on the Fourier transform side by the smooth bump $\widehat{\Psi}(2^{-j}\xi)$. Since a given f in $\dot{\Lambda}_\gamma$ has polynomial growth at infinity, it is a tempered distribution, and thus the convolution $\Psi_{2^{-j}} * f = \Delta_j^{\Psi}(f)$ is a well-defined smooth function of at most polynomial growth at infinity (cf. Theorem 2.3.20 in [156]).

Theorem 1.4.6. *Let Ψ, Δ_j^{Ψ} be as above and $\gamma > 0$. Then there is a constant $C = C(n,\gamma,\Psi)$ such that for all f in $\dot{\Lambda}_\gamma$ we have the estimate*

$$\sup_{j\in\mathbf{Z}} 2^{j\gamma} \big\|\Delta_j^{\Psi}(f)\big\|_{L^\infty} \le C \big\|f\big\|_{\dot{\Lambda}_\gamma}. \tag{1.4.7}$$

Conversely, given f in $\mathscr{S}'(\mathbf{R}^n)$ satisfying

$$\sup_{j\in\mathbf{Z}} 2^{j\gamma} \big\|\Delta_j^{\Psi}(f)\big\|_{L^\infty} = C_0 < \infty, \tag{1.4.8}$$

there is a polynomial Q such that $|f(x)-Q(x)| \le C_{n,\gamma} C_0 (1+|x|)^{[\gamma]+1}$ for all $x \in \mathbf{R}^n$ and some constant $C_{n,\gamma}$. Moreover, $f-Q$ lies in $\mathscr{C}^{[\gamma]}(\mathbf{R}^n)$ and in $\dot{\Lambda}_\gamma(\mathbf{R}^n)$ and satisfies

$$\big\|f-Q\big\|_{\dot{\Lambda}_\gamma} \le C'(n,\gamma,\Psi)\, C_0 \tag{1.4.9}$$

for some constant $C'(n,\gamma,\Psi)$.

In particular, functions in $\dot{\Lambda}_\gamma(\mathbf{R}^n)$ are in $\mathscr{C}^{[\gamma]}(\mathbf{R}^n)$.

Proof. We begin with the proof of (1.4.7). We first consider the case $0<\gamma<1$, which is very simple. Since each Δ_j^{Ψ} is given by convolution with a function with mean value zero, for a function $f \in \dot{\Lambda}_\gamma$ and every $x \in \mathbf{R}^n$ we write

$$\begin{aligned}
\Delta_j^{\Psi}(f)(x) &= \int_{\mathbf{R}^n} f(x-y)\Psi_{2^{-j}}(y)\,dy \\
&= \int_{\mathbf{R}^n} (f(x-y)-f(x))\Psi_{2^{-j}}(y)\,dy \\
&= 2^{-j\gamma} \int_{\mathbf{R}^n} \frac{D_{-y}(f)(x)}{|y|^\gamma} |2^j y|^\gamma 2^{jn}\Psi(2^j y)\,dy,
\end{aligned}$$

and the previous expression is easily seen to be controlled by a constant multiple of $2^{-j\gamma}\|f\|_{\dot{\Lambda}_\gamma}$. This proves (1.4.7) when $0<\gamma<1$. In the case $\gamma \geq 1$ we work a bit harder.

Notice that for any $u \in \mathscr{S}'(\mathbf{R}^n)$ we have the identity

$$D_h^{[\gamma]+1}(u) = \big((e^{2\pi i \xi\cdot h}-1)^{[\gamma]+1}\widehat{u}(\xi)\big)^\vee,$$

where inside the inverse Fourier transform we have the well-defined operation of multiplication of a tempered distribution by a bounded smooth function.

Let us now fix $f \in \dot{\Lambda}_\gamma$ for some $\gamma \geq 1$. To express $\Delta_j^\Psi(f)$ in terms of $D_h^{[\gamma]+1}(f)$, we need to introduce the function

$$\xi \mapsto \widehat{\Psi}(2^{-j}\xi)\,(e^{2\pi i\xi\cdot h}-1)^{-[\gamma]-1}.$$

But as the support of $\widehat{\Psi}(2^{-j}\xi)$ may intersect the set of all ξ for which $\xi\cdot h$ is an integer, the previous function is not well defined. To deal with this problem, we pick a finite family of unit vectors $\{u_r\}_r$ so that the annulus $\frac{1}{2}\leq|\xi|\leq 2$ is covered by the union of sets

$$U_r = \big\{\xi\in\mathbf{R}^n : \tfrac{1}{2}\leq|\xi|\leq 2,\quad \tfrac{1}{4}\leq|\xi\cdot u_r|\leq 2\big\}.$$

Then we write $\widehat{\Psi}$ as a finite sum of smooth functions $\widehat{\Psi^{(r)}}$, where each $\widehat{\Psi^{(r)}}$ is supported in U_r. Setting

$$h_{r,j} = \frac{1}{8}2^{-j}u_r,$$

we note that for the given $f\in\dot{\Lambda}_\gamma$ we have $\widehat{f}\in\mathscr{S}'(\mathbf{R}^n)$ and

$$\begin{aligned}(\Psi^{(r)})_{2^{-j}}*f &= \Big(\widehat{\Psi^{(r)}}(2^{-j}\xi)\,(e^{2\pi i\xi\cdot h_{r,j}}-1)^{-[\gamma]-1}(e^{2\pi i\xi\cdot h_{r,j}}-1)^{[\gamma]+1}\widehat{f}(\xi)\Big)^\vee\\ &= \Big(\widehat{\Psi^{(r)}}(2^{-j}\xi)\,(e^{2\pi i 2^{-j}\xi\cdot\frac{1}{8}u_r}-1)^{-[\gamma]-1}\widehat{D_{h_{r,j}}^{[\gamma]+1}(f)}(\xi)\Big)^\vee, \qquad (1.4.10)\end{aligned}$$

and we observe that the exponential is never equal to 1 since

$$2^{-j}\xi\in U_r \implies \tfrac{1}{32}\leq|2^{-j}\xi\cdot\tfrac{1}{8}u_r|\leq\tfrac{1}{4}.$$

Since the function $\widehat{\zeta^{(r)}} = \widehat{\Psi^{(r)}}(\xi)\,(e^{2\pi i\xi\cdot\frac{1}{8}u_r}-1)^{-[\gamma]-1}$ is well defined and smooth with compact support, it follows that

$$(\Psi^{(r)})_{2^{-j}}*f = (\zeta^{(r)})_{2^{-j}}*D_{h_{r,j}}^{[\gamma]+1}(f),$$

which implies that

$$\|(\Psi^{(r)})_{2^{-j}}*f\|_{L^\infty} \leq \|(\zeta^{(r)})_{2^{-j}}\|_{L^1}\|D_{2^{-j}\frac{1}{8}u_r}^{[\gamma]+1}(f)\|_{L^\infty} \leq \|\zeta^{(r)}\|_{L^1}\|f\|_{\dot{\Lambda}_\gamma}2^{-j\gamma}.$$

Summing over the finite number of r, we obtain the estimate

$$\big\|\Delta_j^{\Psi}(f)\big\|_{L^\infty} \le C\big\|f\big\|_{\dot\Lambda_\gamma} 2^{-j\gamma}, \tag{1.4.11}$$

where C depends on n,γ, and Ψ but is independent of j.

We now prove the converse statement (1.4.9) assuming (1.4.8). We pick a Schwartz function η on $\mathbf{R}^n$ whose Fourier transform is nonnegative, is supported in the annulus $\frac{4}{5} \le |\xi| \le 2$, and satisfies

$$\sum_{j\in\mathbf{Z}} \widehat{\eta}(2^{-j}\xi)^2 = 1 \tag{1.4.12}$$

for all $\xi \neq 0$. Associated with η, we define the Littlewood–Paley operators Δ_j^{η} given by multiplication on the Fourier transform side by the smooth bump $\widehat{\eta}(2^{-j}\xi)$. With Ψ as in (1.4.6) we set

$$\widehat{\Theta}(\xi) = \widehat{\Psi}(\tfrac{1}{2}\xi) + \widehat{\Psi}(\xi) + \widehat{\Psi}(2\xi),$$

and we denote by $\Delta_j^{\Theta} = \Delta_{j-1}^{\Psi} + \Delta_j^{\Psi} + \Delta_{j+1}^{\Psi}$ the Littlewood–Paley operator given by multiplication on the Fourier transform side by the smooth bump $\widehat{\Theta}(2^{-j}\xi)$. It follows from (1.4.8) that for all $j \in \mathbf{Z}$

$$\big\|\Delta_j^{\Theta}(f)\big\|_{L^\infty} \le C_0(2^\gamma + 1 + 2^{-\gamma})2^{-j\gamma}. \tag{1.4.13}$$

The fact that $\widehat{\Theta}$ is equal to 1 on the support of $\widehat{\eta}$, together with identity (1.4.12), yields the operator identity

$$I = \sum_{j\in\mathbf{Z}} (\Delta_j^{\eta})^2 = \sum_{j\in\mathbf{Z}} \Delta_j^{\Theta}\Delta_j^{\eta}\Delta_j^{\eta},$$

with convergence in the sense of the space $\mathscr{S}'(\mathbf{R}^n)/\mathscr{P}(\mathbf{R}^n)$; see (1.1.8).

Throughout the rest of the proof we fix $f \in \mathscr{S}'(\mathbf{R}^n)$ such that (1.4.8) holds. For $L = 1,2,3,\dots$ we define

$$f_L = \sum_{|j|\le L} \Delta_j^{\Theta}\Delta_j^{\eta}\Delta_j^{\eta}(f) = \sum_{|j|\le L} \Delta_j^{\Theta}(f) * \eta_{2^{-j}} * \eta_{2^{-j}}.$$

Obviously, f_L is a $\mathscr{C}^\infty$ function for all L and $f_L \to f$ in $\mathscr{S}'/\mathscr{P}$ as $L \to \infty$.

Since convolution is a linear operation, we have $D_h^{[\gamma]+1}(F*G) = F*D_h^{[\gamma]+1}(G)$, from which we deduce

$$\begin{aligned} D_h^{[\gamma]+1}(f_L) &= \sum_{|j|\le L} \Delta_j^{\Theta}(f) * D_h^{[\gamma]+1}(\eta_{2^{-j}}) * \eta_{2^{-j}} \\ &= \sum_{|j|\le L} D_h^{[\gamma]+1}(\Delta_j^{\Theta}(f)) * (\eta*\eta)_{2^{-j}}. \end{aligned} \tag{1.4.14}$$

Using (1.4.2), we easily obtain the estimate

$$\big\|D_h^{[\gamma]+1}(\Delta_j^{\Theta}(f))*(\eta*\eta)_{2^{-j}}\big\|_{L^\infty}\le 2^{[\gamma]+1}\big\|\eta*\eta\big\|_{L^1}\big\|\Delta_j^{\Theta}(f)\big\|_{L^\infty}. \tag{1.4.15}$$

Let $k=[\gamma]$. We first integrate over $(s_1,\dots,s_{k+1})\in[0,1]^{k+1}$ the identity

$$\sum_{r_1=1}^{n}\cdots\sum_{r_{k+1}=1}^{n}h_{r_1}\cdots h_{r_{k+1}}(\partial_{r_1}\cdots\partial_{r_{k+1}}\eta_{2^{-j}})(x+(s_1+\cdots+s_{k+1})h)$$
$$=2^{j(k+1)}\sum_{r_1=1}^{n}\cdots\sum_{r_{k+1}=1}^{n}h_{r_1}\cdots h_{r_{k+1}}(\partial_{r_1}\dots\partial_{r_{k+1}}\eta)_{2^{-j}}(x+(s_1+\cdots+s_{k+1})h).$$

We then use (1.4.5), with $m=k+1$, and we integrate over $x\in\mathbf{R}^n$ to obtain

$$\big\|D_h^{k+1}(\eta_{2^{-j}})\big\|_{L^1}\le 2^{j(k+1)}|h|^{k+1}\sum_{r_1=1}^{n}\cdots\sum_{r_{k+1}=1}^{n}\big\|\partial_{r_1}\cdots\partial_{r_{k+1}}\eta\big\|_{L^1}.$$

We deduce the validity of the estimate

$$\begin{aligned}&\big\|\Delta_j^{\Theta}(f)*D_h^{[\gamma]+1}(\eta_{2^{-j}})*\eta_{2^{-j}}\big\|_{L^\infty}\\&\quad\le\big\|\Delta_j^{\Theta}(f)\big\|_{L^\infty}\big\|D_h^{[\gamma]+1}(\eta_{2^{-j}})*\eta_{2^{-j}}\big\|_{L^1}\\&\quad\le\big\|\Delta_j^{\Theta}(f)\big\|_{L^\infty}|2^jh|^{[\gamma]+1}c_\gamma\sum_{|\alpha|\le[\gamma]+1}\big\|\partial^\alpha\eta\big\|_{L^1}\big\|\eta\big\|_{L^1}.\end{aligned}\tag{1.4.16}$$

Combining (1.4.15) and (1.4.16), we obtain

$$\big\|\Delta_j^{\Theta}(f)*D_h^{[\gamma]+1}(\eta_{2^{-j}})*\eta_{2^{-j}}\big\|_{L^\infty}\le C_{\eta,n,\gamma}\big\|\Delta_j^{\Theta}(f)\big\|_{L^\infty}\min\big(1,|2^jh|^{[\gamma]+1}\big). \tag{1.4.17}$$

We insert estimate (1.4.17) into (1.4.14) to deduce

$$\frac{\big\|D_h^{[\gamma]+1}(f_L)\big\|_{L^\infty}}{|h|^\gamma}\le\frac{C_{n,\gamma}}{|h|^\gamma}\sum_{|j|\le L}2^{j\gamma}\big\|\Delta_j^{\Theta}(f)\big\|_{L^\infty}\min\big(2^{-j\gamma},2^{j([\gamma]+1-\gamma)}|h|^{[\gamma]+1}\big),$$

from which it follows that

$$\begin{aligned}\big\|f_L\big\|_{\dot\Lambda_\gamma}&\le\sup_{h\in\mathbf{R}^n\setminus\{0\}}\frac{C_{n,\gamma}}{|h|^\gamma}\sum_{|j|\le L}2^{j\gamma}\big\|\Delta_j^{\Theta}(f)\big\|_{L^\infty}\min\big(2^{-j\gamma},2^{j([\gamma]+1-\gamma)}|h|^{[\gamma]+1}\big)\\&\le C'_{n,\gamma}\sup_{j\in\mathbf{Z}}2^{j\gamma}\big\|\Delta_j^{\Theta}(f)\big\|_{L^\infty}\sup_{h\neq0}\sum_{j\in\mathbf{Z}}\min\big(|h|^{-\gamma}2^{-j\gamma},2^{j([\gamma]+1-\gamma)}|h|^{[\gamma]+1-\gamma}\big)\\&\le C''_{n,\gamma}\sup_{j\in\mathbf{Z}}2^{j\gamma}\big\|\Delta_j^{\Theta}(f)\big\|_{L^\infty}\\&\le C''_{n,\gamma}(2^\gamma+1+2^{-\gamma})C_0,\end{aligned}\tag{1.4.18}$$

since the numerical series converges. In the last inequality we used (1.4.13).

We now write $f_L = f_L^1 + f_L^2$, where

$$f_L^1 = \sum_{j=-L}^{-1} \Delta_j^{\Theta}(f) * \eta_{2^{-j}} * \eta_{2^{-j}}, \qquad f_L^2 = \sum_{j=0}^{L} \Delta_j^{\Theta}(f) * \eta_{2^{-j}} * \eta_{2^{-j}}.$$

It follows from (1.4.13) that with $C_0' = (2^\gamma + 1 + 2^{-\gamma})C_0$ we have

$$\|\Delta_j^{\Theta}(f) * \eta_{2^{-j}} * \eta_{2^{-j}}\|_{L^\infty} \le \|\Delta_j^{\Theta}(f)\|_{L^\infty} \|\eta_{2^{-j}} * \eta_{2^{-j}}\|_{L^1} \le C_0' \|\eta * \eta\|_{L^1} 2^{-j\gamma};$$

thus, f_L^2 converges uniformly to a continuous and bounded function g_2 as $L \to \infty$. Also, $\partial^\beta f_L^2$ converges uniformly for all $|\beta| < \gamma$ as $L \to \infty$. Using Lemma 1.4.7 we conclude that g_2 is in $\mathscr{C}^{[\gamma]}$ and that $\partial^\beta f_L^2$ converges uniformly to $\partial^\beta g_2$ as $L \to \infty$ for all $|\beta| < \gamma$.

We now turn our attention to f_L^1. Obviously, f_L^1 is in $\mathscr{C}^\infty$ and

$$\partial^\alpha f_L^1 = \sum_{j=-L}^{-1} \Delta_j^{\Theta}(f) * 2^{j|\alpha|} (\partial^\alpha \eta)_{2^{-j}} * \eta_{2^{-j}}.$$

Thus for all multi-indices α with $|\alpha| \ge [\gamma] + 1$ we have

$$\sup_{L \in \mathbf{Z}^+} \|\partial^\alpha f_L^1\|_{L^\infty} \le \sum_{j=-\infty}^{-1} C_0' 2^{-j\gamma} 2^{j([\gamma]+i)} \|\partial^\alpha \eta * \eta\|_{L^1} = c_{\alpha,\gamma} C_0 < \infty. \tag{1.4.19}$$

Let P_L^d be the Taylor polynomial of f_L^1 of degree d. By Taylor's theorem we have

$$f_L^1(x) - P_L^{[\gamma]}(x) = ([\gamma] + 1) \sum_{|\alpha| = [\gamma]+1} \frac{x^\alpha}{\alpha!} \int_0^1 (1-t)^{[\gamma]} (\partial^\alpha f_L^1)(tx)\, dt. \tag{1.4.20}$$

Using (1.4.19), with $|\alpha| \in \{[\gamma]+1, \dots, [\gamma] + |\beta| + 2\}$, we obtain that the sequence $\{\nabla(\partial^\beta(f_L^1 - P_L^{[\gamma]}))\}_{L=1}^\infty$ is uniformly bounded on every ball $\overline{B(0,K)}$; thus, the sequence $\{\partial^\beta(f_L^1 - P_L^{[\gamma]})\}_{L=1}^\infty$ is equicontinuous on every such ball. By the Arzelà–Ascoli theorem, for every $K = 1, 2, \dots$ and for every $|\beta| < \gamma$ there is a subsequence of $\{\partial^\beta(f_L^1 - P_L^{[\gamma]})\}_{L=1}^\infty$ that converges uniformly on $\overline{B(0,K)}$. The diagonal subsequence of these subsequences converges uniformly on every compact subset of $\mathbf{R}^n$ for all $|\beta| < \gamma$. Hence, there is a continuous function g_1 on $\mathbf{R}^n$ and a subsequence L_m of $\mathbf{Z}^+$ such that $f_{L_m}^1 - P_{L_m}^{[\gamma]} \to g_1$ uniformly on compact sets as $m \to \infty$ and $\partial^\beta(f_{L_m}^1 - P_{L_m}^{[\gamma]})$ converges uniformly on compact sets for all $|\beta| \ge [\gamma]$. Using Lemma 1.4.7, stated at the end of this proof, we deduce that g_1 is $\mathscr{C}^{[\gamma]}$ and that $\partial^\beta(f_{L_m}^1 - P_{L_m}^{[\gamma]}) \to \partial^\beta g_1$ as $m \to \infty$ for all $|\beta| \le [\gamma]$.

Set $g = g_1 + g_2$. It follows from (1.4.20) and from $\sup_{L \in \mathbf{Z}^+} \|f_L^2\|_{L^\infty} < \infty$ that $|g(x)| \le C_{n,\gamma} C_0 (1 + |x|)^{[\gamma]+1}$ for all $x \in \mathbf{R}^n$. Thus, g can be viewed as an element of $\mathscr{S}'$, and one has $f_{L_m} - P_{L_m}^{[\gamma]} \to g$ in $\mathscr{S}'(\mathbf{R}^n)$. Since both g_1 and g_2 are in $\mathscr{C}^{[\gamma]}$, it follows that so is g.

But we know that $f_L \to f$ in $\mathscr{S}'/\mathscr{P}$ as $L \to \infty$; see (1.1.8). Thus, given φ in $\mathscr{S}(\mathbf{R}^n)$ whose Fourier transform is supported away from the origin, we have

$$\langle f_{L_m} - P_{L_m}^{[\gamma]}, \varphi \rangle = \langle f_{L_m}, \varphi \rangle \to \langle f, \varphi \rangle .$$

We also have $\langle f_{L_m} - P_{L_m}^{[\gamma]}, \varphi \rangle \to \langle g, \varphi \rangle$. Hence, $\langle f - g, \varphi \rangle = 0$ for all such φ, and thus $f - g$ is a distribution whose Fourier transform is supported at the origin. By Proposition 2.4.1 in [156] we conclude that $f - g$ is equal to a polynomial.

Using (1.4.18) we obtain that

$$|D_h^{[\gamma]+1}(f_L)(x)| \le C_{n,\gamma}''' C_0 |h|^\gamma$$

for all L and in particular for L_m. Since $D_h^{[\gamma]+1}(P_{L_m}^{[\gamma]}) = 0$, letting $m \to \infty$, we deduce that

$$|D_h^{[\gamma]+1}(g)(x)| \le C_{n,\gamma}''' C_0 |h|^\gamma$$

for all $h \neq 0$. This proves (1.4.9). □

Lemma 1.4.7. *Let h_k, $k = 1, 2, \dots$ be $\mathscr{C}^N$ functions on $\mathbf{R}^n$ such that $h_k \to h$ uniformly on compact subsets of $\mathbf{R}^n$ as $k \to \infty$. Suppose that there exist finite constants C_α, M_α such that $\sup_{k \in \mathbf{Z}^+} |\partial^\alpha h_k(x)| \le C_\alpha (1 + |x|)^{M_\alpha}$ for all $x \in \mathbf{R}^n$. Suppose also that for all multi-indices $|\alpha| \le N$, $\partial^\alpha h_k \to u_\alpha$ uniformly on compact subsets of $\mathbf{R}^n$ to some continuous function u_α. Then h lies in $\mathscr{C}^N$ and $\partial^\alpha h = u_\alpha$ for all $|\alpha| \le N$.*

Proof. It follows from the hypothesis that h has at most polynomial growth at infinity, and thus it can be thought of as an element on $\mathscr{S}'(\mathbf{R}^n)$. Then $\partial^\alpha h$ exist as elements of $\mathscr{S}'(\mathbf{R}^n)$ for all multi-indices α. We show that $\partial^\alpha h = u_\alpha$ for all $|\alpha| \le N$. Fix a function $\varphi \in \mathscr{S}(\mathbf{R}^n)$. Given $|\alpha| \le N$ and $\varepsilon > 0$, there exists an $R_\alpha > 0$ such that

$$\int_{|x| \ge R_\alpha} |\varphi(x)| 2C_\alpha (1 + |x|)^{M_\alpha} \, dx < \varepsilon/4$$

and

$$\int_{|x| \ge R_\alpha} |\partial^\alpha \varphi(x)| 2C_\alpha (1 + |x|)^{M_\alpha} \, dx < \varepsilon/4 .$$

On the compact ball $\overline{B(0, R_\alpha)}$ we have that $h_k \to h$ and $\partial^\alpha h_k \to u_\alpha$ uniformly. Thus there is a $k_0 \in \mathbf{Z}^+$ such that for all $k \ge k_0$ we have

$$\int_{|x| \le R_\alpha} |\varphi(x)| \, dx \, \big\| \partial^\alpha h_k - u_\alpha \big\|_{L^\infty(\overline{B(0,R_\alpha)})} < \varepsilon/4$$

and

$$\int_{|x| \le R_\alpha} |\partial^\alpha \varphi(x)| \, dx \, \big\| h_k - h \big\|_{L^\infty(\overline{B(0,R_\alpha)})} < \varepsilon/4 .$$

Combining these elements we obtain that for $k \ge k_0$

$$\int_{\mathbf{R}^n} |h_k(x) - h(x)| \, |\partial^\alpha \varphi(x)| \, dx + \int_{\mathbf{R}^n} |\partial^\alpha h_k(x) - u_\alpha(x)| \, |\varphi(x)| \, dx < \varepsilon$$

and thus

$$\begin{aligned}|\langle \partial^\alpha h - u_\alpha, \varphi\rangle| &\le |\langle \partial^\alpha h - \partial^\alpha h_k, \varphi\rangle| + |\langle \partial^\alpha h_k - u_\alpha, \varphi\rangle| \\ &= |\langle h - h_k, \partial^\alpha \varphi\rangle| + |\langle \partial^\alpha h_k - u_\alpha, \varphi\rangle| < \varepsilon.\end{aligned}$$

Since ε was arbitrary, we deduce that $\partial^\alpha h = u_\alpha$, in particular $h \in \mathscr{C}^N$. □

Corollary 1.4.8. *Any function f in $\dot\Lambda_\gamma$ lies in $\mathscr{C}^{|\beta|}$ for any $|\beta| < \gamma$, and its derivatives $\partial^\beta f$ lie in $\dot\Lambda_{\gamma-|\beta|}$ and satisfy*

$$\|\partial^\beta f\|_{\dot\Lambda_{\gamma-|\beta|}} \le C_{n,\gamma,\beta} \|f\|_{\dot\Lambda_\gamma}. \tag{1.4.21}$$

Proof. We proved in Theorem 1.4.6 that if f lies in $\dot\Lambda_\gamma$, then (1.4.7) holds, and that (1.4.7) implies that there exists a polynomial Q such that $f - Q$ lies in $\mathscr{C}^{[\gamma]}$ and in $\dot\Lambda_\gamma$. It follows that f lies in $\mathscr{C}^{[\gamma]}$. It also follows that Q lies in $\dot\Lambda_\gamma$, and this imposes a restriction on the degree of Q; in view of the result of Exercise 1.4.1, we have that Q must have degree at most $[\gamma]$; thus, $f \equiv f - Q$ in the space $\mathscr{S}'/\mathscr{P}_{[\gamma]}$, i.e., they belong to the same equivalence class.

Let Ψ be a Schwartz function on $\mathbf{R}^n$ whose Fourier transform is supported in $1 - \frac{1}{7} \le |\xi| \le 2$ and is equal to one on $1 \le |\xi| \le 2 - \frac{2}{7}$. Given a multi-index β with $|\beta| < \gamma$, we denote by $\Delta_j^{\partial^\beta \Psi}$ the Littlewood–Paley operator associated with $(\partial^\beta \Psi)_{2^{-j}}$. Then one has

$$\Delta_j^\Psi(\partial^\beta f) = 2^{j|\beta|} \Delta_j^{\partial^\beta \Psi}(f)$$

for all $f \in \Lambda_\gamma$. One can easily check that

$$2^{j(\gamma-|\beta|)} \Delta_j^\Psi(\partial^\beta f) = 2^{j\gamma} \Delta_j^{\partial^\beta \Psi}(\Delta_{j-1}^\Psi + \Delta_j^\Psi + \Delta_{j+1}^\Psi)(f),$$

and from this it easily follows that

$$\sup_{j\in\mathbf{Z}} 2^{j(\gamma-|\beta|)} \|\Delta_j^\Psi(\partial^\beta f)\|_{L^\infty} \le (2^\gamma + 1 + 2^{-\gamma}) \|\partial^\beta \Psi\|_{L^1} \sup_{j\in\mathbf{Z}} 2^{j\gamma} \|\Delta_j^\Psi(f)\|_{L^\infty},$$

which implies that $\partial^\beta f$ lies in $\dot\Lambda_{\gamma-|\beta|}$ when $|\beta| < \gamma$. □

1.4.3 Littlewood–Paley Characterization of Inhomogeneous Lipschitz Spaces

We have seen that quantities involving the Littlewood–Paley operators Δ_j characterize homogeneous Lipschitz spaces. We now address the same question for inhomogeneous Lipschitz spaces.

We fix a radial Schwartz function Ψ whose Fourier transform $\widehat{\Psi}$ is nonnegative, is supported in the annulus $1-\frac{1}{7} \le |\xi| \le 2$, is equal to one on the annulus $1 \le |\xi| \le 2-\frac{2}{7}$, and satisfies

$$\sum_{j\in\mathbf{Z}} \widehat{\Psi}(2^{-j}\xi) = 1$$

for all $\xi \neq 0$. We define a Schwartz function Φ introduced by setting

$$\widehat{\Phi}(\xi) = \begin{cases} \sum_{j\le 0} \widehat{\Psi}(2^{-j}\xi) & \text{when } \xi \neq 0, \\ 1 & \text{when } \xi = 0. \end{cases} \tag{1.4.22}$$

Note that $\widehat{\Phi}(\xi)$ is equal to 1 for $|\xi| \le 2-\frac{2}{7}$ and vanishes when $|\xi| \ge 2$. Finally, we define $\Delta_j^\Psi(f) = \Psi_{2^{-j}} * f$ and $S_0^\Phi(f) = \Phi * f$ for any $f \in \mathscr{S}'(\mathbf{R}^n)$.

Theorem 1.4.9. *Let Ψ, Φ, Δ_j^Ψ, and S_0^Φ be as above, and let $\gamma > 0$. Then there is a constant $C = C(n,\gamma)$ such that for every function f in Λ_γ the following estimate holds:*

$$\big\|S_0^\Phi(f)\big\|_{L^\infty} + \sup_{j\ge 1} 2^{j\gamma}\big\|\Delta_j^\Psi(f)\big\|_{L^\infty} \le C\big\|f\big\|_{\Lambda_\gamma}. \tag{1.4.23}$$

Conversely, suppose that a tempered distribution f satisfies

$$\big\|S_0^\Phi(f)\big\|_{L^\infty} + \sup_{j\ge 1} 2^{j\gamma}\big\|\Delta_j^\Psi(f)\big\|_{L^\infty} < \infty. \tag{1.4.24}$$

Then f is in $\mathscr{C}^{[\gamma]}$, and the derivatives $\partial^\alpha f$ are bounded for all $|\alpha| \le [\gamma]$. Moreover, f lies in Λ_γ, and there is a constant $C' = C'(n,\gamma)$ such that

$$\big\|f\big\|_{\Lambda_\gamma} \le C'\Big(\big\|S_0^\Phi(f)\big\|_{L^\infty} + \sup_{j\ge 1} 2^{j\gamma}\big\|\Delta_j^\Psi(f)\big\|_{L^\infty}\Big). \tag{1.4.25}$$

In particular, functions in Λ_γ are in $\mathscr{C}^{[\gamma]}$ and have bounded derivatives up to order $[\gamma]$. Also,

$$\big\|f\big\|_{\Lambda_\gamma} \approx \sum_{|\alpha|<[\gamma]} \big\|\partial^\alpha f\big\|_{L^\infty} + \sum_{|\alpha|=[\gamma]} \big\|\partial^\alpha f\big\|_{\Lambda_{\gamma-[\gamma]}}.$$

Proof. The proof of (1.4.23) is immediate since we trivially have

$$\big\|S_0^\Phi(f)\big\|_{L^\infty} = \big\|f*\Phi\big\|_{L^\infty} \le \big\|\Phi\big\|_{L^1}\big\|f\big\|_{L^\infty} \le C\big\|f\big\|_{\Lambda_\gamma},$$

and, in view of estimate (1.4.11), we have

$$\sup_{j\ge 1} 2^{j\gamma}\big\|\Delta_j^\Psi(f)\big\|_{L^\infty} \le C\big\|f\big\|_{\dot{\Lambda}_\gamma} \le C\big\|f\big\|_{\Lambda_\gamma}.$$

We may therefore focus on the proof of the converse estimate (1.4.25). We fix $f \in \mathscr{S}'(\mathbf{R}^n)$ which satisfies (1.4.24). We introduce Schwartz functions ζ, η such that

$$\widehat{\zeta}(\xi)^2 + \sum_{j=1}^{\infty} \widehat{\eta}(2^{-j}\xi)^2 = 1$$

and such that $\widehat{\eta}$ is supported in the annulus $\frac{2}{5} \le |\xi| \le 1$ and $\widehat{\zeta}$ is supported in the ball $|\xi| \le 1$. We associate Littlewood–Paley operators Δ_j^η given by convolution with the functions $\eta_{2^{-j}}$ and we let $\Delta_j^\Theta = \Delta_{j-1}^\Psi + \Delta_j^\Psi + \Delta_{j+1}^\Psi$. Using this identity and (1.4.24) we obtain for some $C_0 < \infty$

$$\|\Delta_j^\Theta(f)\|_{L^\infty} \le C_0 2^{-j\gamma}. \tag{1.4.26}$$

Note that $\widehat{\Phi}$ is equal to one on the support of $\widehat{\zeta}$. Moreover, $\Delta_j^\Theta \Delta_j^\eta = \Delta_j^\eta$; hence, for our given tempered distribution f we have the identity

$$f = \zeta * \zeta * \Phi * f + \sum_{j=1}^{\infty} \eta_{2^{-j}} * \eta_{2^{-j}} * \Delta_j^\Theta(f), \tag{1.4.27}$$

where the series converges in $\mathscr{S}'(\mathbf{R}^n)$, in view of the result of Exercise 1.1.5.

But this series also converges in L^∞ since, in view of (1.4.26),

$$\big\|\eta_{2^{-j}} * \eta_{2^{-j}} * \Delta_j^\Theta(f)\|_{L^\infty} \le \|\eta * \eta\|_{L^1} \|\Delta_j^\Theta(f)\|_{L^\infty} \le C_0 2^{-j\gamma},$$

and thus f is a continuous and bounded function. Also, for all $|\alpha| < \gamma$ we have

$$\big\|\partial^\alpha(\eta_{2^{-j}} * \eta_{2^{-j}} * \Delta_j^\Theta(f))\|_{L^\infty} \le 2^{j|\alpha|} \|\partial^\alpha(\eta * \eta)\|_{L^1} \|\Delta_j^\Theta(f)\|_{L^\infty} \le C_0 2^{-j(\gamma-|\alpha|)},$$

and thus Proposition 1.1.5 yields that our given tempered distribution f is a $\mathscr{C}^\alpha$ function whose derivatives are bounded for all $|\alpha| < \gamma$.

It remains to show that the function f is in Λ_γ. With $k = [\gamma]$ we write

$$\frac{D_h^{k+1}(f)}{|h|^\gamma} = \zeta * \frac{D_h^{k+1}(\zeta)}{|h|^\gamma} * \Phi * f + \sum_{j=1}^{\infty} \eta_{2^{-j}} * \frac{D_h^{k+1}(\eta_{2^{-j}})}{|h|^\gamma} * \Delta_j^\Theta(f). \tag{1.4.28}$$

We use Proposition 1.4.5 to estimate the L^∞ norm of the term $\zeta * \frac{D_h^{k+1}(\zeta)}{|h|^\gamma} * \Phi * f$ in the previous sum as follows:

$$\begin{aligned}\Big\|\zeta * \tfrac{D_h^{k+1}(\zeta)}{|h|^\gamma} * \Phi * f\Big\|_{L^\infty} &\le \Big\|\tfrac{D_h^{k+1}(\zeta)}{|h|^\gamma}\Big\|_{L^\infty} \Big\|\zeta * \Phi * f\Big\|_{L^1} \\ &\le C' \min\big(\tfrac{1}{|h|^\gamma}, \tfrac{|h|^{k+1}}{|h|^\gamma}\big) \big\|\Phi * f\big\|_{L^\infty} \\ &\le C' \big\|\Phi * f\big\|_{L^\infty}.\end{aligned} \tag{1.4.29}$$

The corresponding L^∞ estimates for $\Delta_j^\Theta(f) * \eta_{2^{-j}} * D_h^{k+1}(\eta_{2^{-j}})$ were already obtained in (1.4.17). Indeed, we obtained

$$\big\|D_h^{k+1}(\eta_{2^{-j}}) * \eta_{2^{-j}} * \Delta_j^\Theta(f)\big\|_{L^\infty} \le C_{\eta,n,k}\big\|\Delta_j^\Theta(f)\big\|_{L^\infty} \min\left(1, |2^j h|^{k+1}\right),$$

from which it follows that

$$\begin{aligned}
&\Big\|\sum_{j=1}^\infty \eta_{2^{-j}} * \frac{D_h^{k+1}(\eta_{2^{-j}})}{|h|^\gamma} * \Delta_j^\Theta(f)\Big\|_{L^\infty} \\
&\le C'\Big(\sup_{j\ge1} 2^{j\gamma}\big\|\Delta_j^\Theta(f)\big\|_{L^\infty}\Big)\sum_{j=1}^\infty 2^{-j\gamma}|h|^{-\gamma}\min\left(1, |2^j h|^{k+1}\right) \\
&\le C''\Big(\sup_{j\ge1} 2^{j\gamma}\big\|\Delta_j^\Psi(f)\big\|_{L^\infty}\Big)\sum_{j=1}^\infty \min\left(|2^j h|^{-\gamma}, |2^j h|^{k+1-\gamma}\right) \\
&\le C'''\sup_{j\ge1} 2^{j\gamma}\big\|\Delta_j^\Psi(f)\big\|_{L^\infty},
\end{aligned} \tag{1.4.30}$$

where the last series is easily seen to converge uniformly in $h \in \mathbf{R}^n$ since $k+1 = [\gamma]+1 > \gamma$. We now combine (1.4.28) with estimates (1.4.29) and (1.4.30) to deduce that our given distribution f is indeed an element of Λ_γ that satisfies (1.4.25). □

Next, we obtain consequences of the Littlewood–Paley characterization of Lipschitz spaces.

Corollary 1.4.10. *For $0 < \gamma < \delta < \infty$ there is a constant $C_{n,\gamma,\delta} < \infty$ such that for all $f \in \Lambda_\delta(\mathbf{R}^n)$ we have*

$$\big\|f\big\|_{\Lambda_\gamma} \le C_{n,\gamma,\delta}\big\|f\big\|_{\Lambda_\delta}.$$

In other words, the space $\Lambda_\delta(\mathbf{R}^n)$ can be identified with a subspace of $\Lambda_\gamma(\mathbf{R}^n)$.

Proof. If $0 < \gamma < \delta$ and $j \ge 1$, then we must have $2^{j\gamma} < 2^{j\delta}$, and thus

$$\sup_{j\ge1} 2^{j\gamma}\big\|\Delta_j^\Psi(f)\big\|_{L^\infty} \le \sup_{j\ge1} 2^{j\delta}\big\|\Delta_j^\Psi(f)\big\|_{L^\infty}.$$

Adding $\|S_0^\Phi(f)\|_{L^\infty}$ and using Theorem 1.4.9, we obtain the required conclusion. □

Corollary 1.4.11. *Let $\gamma > 0$, and let α be a multi-index with $|\alpha| < \gamma$.*
Then any function f in Λ_γ lies in $\mathscr{C}^\alpha$, $\partial^\alpha f$ lies in $\Lambda_{\gamma-|\alpha|}$, and the estimate

$$\big\|\partial^\alpha f\big\|_{\Lambda_{\gamma-|\alpha|}} \le C_{n,\gamma,\alpha}\big\|f\big\|_{\Lambda_\gamma} \tag{1.4.31}$$

holds for some constant $C_{n,\gamma,\alpha}$.

Proof. Let α be a multi-index with $|\alpha| < \gamma$. We denote by $\Delta_j^{\partial^\alpha\Psi}$ the Littlewood–Paley operator associated with the bump $(\partial^\alpha\Psi)_{2^{-j}}$. Let $f \in \Lambda_\gamma$. Theorem 1.4.9

implies that $f \in \mathscr{C}^\alpha$ for all $|\alpha| < \gamma$ and $\partial^\alpha f$ are bounded functions. It is straightforward to check the identity

$$\Delta_j^{\Psi}(\partial^\alpha f) = 2^{j|\alpha|}\Delta_j^{\partial^\alpha \Psi}(f).$$

Using the support properties of Ψ, we obtain

$$2^{j(\gamma-|\alpha|)}\Delta_j^{\Psi}(\partial^\alpha f) = 2^{j\gamma}\Delta_j^{\partial^\alpha \Psi}(\Delta_{j-1}^{\Psi} + \Delta_j^{\Psi} + \Delta_{j+1}^{\Psi})(f), \tag{1.4.32}$$

and from this it easily follows that

$$\sup_{j\ge 1} 2^{j(\gamma-|\alpha|)} \big\|\Delta_j^{\Psi}(\partial^\alpha f)\big\|_{L^\infty} \le (2^\gamma+2)\,\big\|\partial^\alpha \Psi\big\|_{L^1} \sup_{j\ge 1} 2^{j\gamma}\big\|\Delta_j^{\Psi}(f)\big\|_{L^\infty} < \infty. \tag{1.4.33}$$

Additionally, we note that

$$S_0^{\Phi}(\partial^\alpha f) = \Phi * (\partial^\alpha f) = \partial^\alpha \Phi * f = \partial^\alpha \Phi * (\Phi + \Psi_{2^{-1}}) * f,$$

since the function $\widehat{\Phi} + \widehat{\Psi_{2^{-1}}}$ is equal to 1 on the support of $\widehat{\partial^\alpha \Phi}$. Taking L^∞ norms, we obtain

$$\begin{aligned}\big\|S_0^{\Phi}(\partial^\alpha f)\big\|_{L^\infty} &\le \big\|\partial^\alpha \Phi\big\|_{L^1}\big(\big\|\Phi * f\big\|_{L^\infty} + \big\|\Psi_{2^{-1}} * f\big\|_{L^\infty}\big)\\ &\le \big\|\partial^\alpha \Phi\big\|_{L^1}\Big(\big\|S_0^{\Phi}(f)\big\|_{L^\infty} + \sup_{j\ge 1}\big\|\Delta_j^{\Psi}(f)\big\|_{L^\infty}\Big),\end{aligned}$$

which, combined with (1.4.33), yields (1.4.31) for all $|\alpha| < \gamma$. □

We end this section by noting that the specific choice of the functions Ψ and Φ is unimportant in the Littlewood–Paley characterization of the spaces Λ_γ. In particular, if we know (1.4.25) and (1.4.8) for some choice of Littlewood–Paley operators $\Delta_j^{\widetilde{\Psi}}$ and some Schwartz function $\widetilde{\Phi}$ whose Fourier transform is supported in a neighborhood of the origin, then (1.4.25) and (1.4.8) would also hold for our fixed choice of Δ_j^{Ψ} and Φ.

Exercises

1.4.1. Fix $k \in \mathbf{Z}^+$, $\alpha_1, \dots, \alpha_n \in \mathbf{Z}^+ \cup \{0\}$, and $\gamma > 0$. Set $|\alpha| = \alpha_1 + \cdots + \alpha_n$.
(a) Let $Q(x) = x_1^{\alpha_1} \cdots x_n^{\alpha_n}$ be a monomial on $\mathbf{R}^n$ of degree $|\alpha|$. Define $C(k,m) = \sum_{j=0}^{k} \binom{k}{j}(-1)^{k-j} j^m$ for $m \in \mathbf{Z}^+$. Show that when $|\alpha| \ge k$ for all $x, h \in \mathbf{R}^n$ we have

$$D_h^k(Q)(x) = \sum_{\substack{\beta \le \alpha\\ |\beta| \ge k}} \binom{\alpha_1}{\beta_1} \cdots \binom{\alpha_n}{\beta_n} C(k,|\beta|)\, h^\beta x^{\alpha-\beta}$$

and note that $C(k,|\beta|)=0$ if $|\beta|<k$ and $C(k,|\beta|)=k!$ if $|\beta|=k$. Conclude that $D_h^{|\alpha|}(Q)(x)=|\alpha|!\,Q(h)$. Also, observe that $D_h^m(Q)(x)=0$ if $m>|\alpha|$ for all $h,x\in\mathbf{R}^n$.
(b) Show that a continuous function f satisfies $D_h^k(f)(x)=0$ for all x,h in $\mathbf{R}^n$ if and only if f is a polynomial of degree at most $k-1$.
(c) Prove that a polynomial lies in $\dot{\Lambda}_\gamma$ if and only if it has degree at most $[\gamma]$.
[*Hint:* (a) Use the Fourier transform. (b) One direction follows by part (a) while for the converse direction, use a partition of unity as in the proof of Theorem 1.4.6 to show that $\widehat{f}$ is supported at the origin and use Proposition 2.4.1 in [156].]

1.4.2. (a) Show that for all continuous and bounded functions f we have

$$\big\|f\big\|_{L^\infty}\le\big\|f\big\|_{\Lambda_0}\le 3\big\|f\big\|_{L^\infty};$$

hence, the space $(\Lambda_0(\mathbf{R}^n),\|\cdot\|_{\Lambda_0})$ can be identified with $(L^\infty(\mathbf{R}^n)\cap\mathscr{C}(\mathbf{R}^n),\|\cdot\|_{L^\infty})$.
(b) Given a continuous function f on $\mathbf{R}^n$, we define

$$\big\|f\big\|_{\dot{L}^\infty}=\inf\big\{\|f+c\|_{L^\infty}:\,c\in\mathbf{C}\big\}.$$

Let $\dot{L}^\infty(\mathbf{R}^n)$ be the space of equivalence classes of continuous functions whose difference is a constant equipped with this norm. Show that for all continuous functions f on $\mathbf{R}^n$ we have

$$\big\|f\big\|_{\dot{L}^\infty}\le\sup_{x,h\in\mathbf{R}^n}|f(x+h)-f(x)|\le 2\big\|f\big\|_{\dot{L}^\infty}.$$

In other words, $(\dot{\Lambda}_0(\mathbf{R}^n),\|\cdot\|_{\dot{\Lambda}_0})$ can be identified with $(\dot{L}^\infty(\mathbf{R}^n)\cap\mathscr{C}(\mathbf{R}^n),\|\cdot\|_{\dot{L}^\infty})$.

1.4.3. Let f be a continuous function on $\mathbf{R}^n$.
(a) Prove the identity

$$D_h^{k+1}(f)(x)=\sum_{s=0}^{k+1}(-1)^{k+1-s}\binom{k+1}{s}f(x+sh)$$

for all $x,h\in\mathbf{R}^n$ and $k\in\mathbf{Z}^+\cup\{0\}$.
(b) Prove that $D_h^kD_h^l=D_h^{k+l}$ for all $k,l\in\mathbf{Z}^+\cup\{0\}$ and all $h\in\mathbf{R}^n$.
(c) Prove that $D_{h_1}D_{h_2}=D_{h_1+h_2}-D_{h_1}-D_{h_2}$ for all $h_1,h_2\in\mathbf{R}^n$.
(d) Suppose that $|D_t^2(f)(x)|\le C|t|^M$ for all $t,x\in\mathbf{R}^n$ and some constants $C,M>0$. Prove that

$$|D_tD_s(f)(x)|\le C'(|t|+|s|)^M$$

for all $t,s,x\in\mathbf{R}^n$ and some other constant C' that depends on C and M.

1.4.4. For $x\in\mathbf{R}$ let

$$f(x)=\sum_{k=1}^\infty 2^{-k}e^{2\pi i2^kx}.$$

(a) Prove that $f \in \Lambda_\gamma(\mathbf{R})$ for all $0<\gamma<1$.
(b) Prove that there is an $A<\infty$ such that

$$\sup_{x,t\neq 0} |f(x+t)+f(x-t)-2f(x)|\,|t|^{-1} \le A;$$

thus, $f \in \Lambda_1(\mathbf{R})$.
(c) Show, however, that

$$\sup_{0<t<1} |f(t)-f(0)|\,t^{-1} = \infty;$$

thus, f is not differentiable at zero.
[*Hint:* Part (c). Given $t>0$, cut the series at $k_0 = [\log_2(4t)^{-1}]$.]

1.4.5. For $0<a,b<\infty$ and $x \in \mathbf{R}$ let

$$g_{ab}(x) = \sum_{k=1}^{\infty} 2^{-ak} e^{2\pi i 2^{bk} x}.$$

Show that g_{ab} lies in $\Lambda_{\frac{a}{b}}(\mathbf{R})$.
[*Hint:* Use the estimate $|D_h^L(e^{2\pi i 2^{bk}x})| \le C\min\big(1,(2^{bk}|h|)^L\big)$, with $L=[a/b]+1$, and split the sum into two parts.]

1.4.6. Let $\gamma>0$, and let $k=[\gamma]$.
(a) Use Exercise 1.4.3(a) and (b) to prove that if $|D_h^k(f)(x)| \le C|h|^\gamma$ for all $x,h\in\mathbf{R}^n$, then $|D_h^{k+l}(f)(x)| \le C2^l|h|^\gamma$ for all $l\ge 1$.
(b) Conversely, assuming that for some $l\ge 1$ we have

$$\sup_{x,h\in\mathbf{R}^n} \frac{|D_h^{k+l}(f)(x)|}{|h|^\gamma} < \infty,$$

show that $f\in\dot{\Lambda}_\gamma$.
[*Hint:* Part (b): Use (1.4.10), but replace $[\gamma]+1$ by $k+l$.]

1.4.7. Let Ψ and Δ_j^Ψ be as in Theorem 1.4.6. Define a continuous operator Q_t by setting

$$Q_t(f) = f*\Psi_t, \qquad \Psi_t(x) = t^{-n}\Psi(t^{-1}x).$$

Show that all tempered distributions f satisfy

$$\sup_{t>0} t^{-\gamma}\big\|Q_t(f)\big\|_{L^\infty} \approx \sup_{j\in\mathbf{Z}} 2^{j\gamma}\big\|\Delta_j^\Psi(f)\big\|_{L^\infty}$$

with the interpretation that if either term is finite, then it controls the other term by a constant multiple of itself.
[*Hint:* Observe that $Q_t = Q_t(\Delta_{j-2}^\Psi+\Delta_{j-1}^\Psi+\Delta_j^\Psi+\Delta_{j+1}^\Psi)$ when $2^{-j}\le t\le 2^{1-j}$.]

1.4.8. (a) Let $0 \le \gamma < 1$, and suppose that $\partial_j f \in \dot{\Lambda}_\gamma$ for all $1 \le j \le n$. Show that

$$\|f\|_{\dot{\Lambda}_{\gamma+1}} \le \sum_{j=1}^{n} \|\partial_j f\|_{\dot{\Lambda}_\gamma},$$

and conclude that $f \in \dot{\Lambda}_{\gamma+1}$.
(b) Let $\gamma \ge 0$. If we have $\partial^\alpha f \in \dot{\Lambda}_\gamma$ for all multi-indices α with $|\alpha| = r$, then there is an estimate

$$\|f\|_{\dot{\Lambda}_{\gamma+r}} \le \sum_{|\alpha|=r} \|\partial^\alpha f\|_{\dot{\Lambda}_\gamma},$$

and thus $f \in \dot{\Lambda}_{\gamma+r}$.
(c) Use Corollary 1.4.8 to obtain that the estimates in both (a) and (b) can be reversed with the insertion of a multiplicative constant.
[*Hint:* Part (a): Write

$$D_h^2(f)(x) = \int_0^1 \sum_{j=1}^{n} \big[\partial_j f(x+th+2h) - \partial_j f(x+th+h)\big]\, h_j\, dt\,.$$

Part (b): Use induction.]

1.4.9. Introduce the difference operator

$$\mathscr{D}^\beta(f)(x) = \left[\int_{\mathbf{R}^n} \frac{|D_y^{[\beta]+1}(f)(x)|^2}{|y|^{n+2\beta}}\, dy\right]^{\frac{1}{2}},$$

where $\beta > 0$. Show that for some constant $c_0(n,\beta)$ we have

$$\|\mathscr{D}^\beta(f)\|_{L^2(\mathbf{R}^n)}^2 = c_0(n,\beta) \int_{\mathbf{R}^n} |\widehat{f}(\xi)|^2 |\xi|^{2\beta}\, d\xi$$

for all functions f in the Sobolev space $\dot{L}_\beta^2(\mathbf{R}^n)$.

1.4.10. Suppose that a continuous function $f(x)$ on the real line satisfies:

$$|D_h^2(f)(x)| \le C\,|h|^{1+\gamma}$$

for some $\gamma \in (0,1)$ and all $x, h \in \mathbf{R}$. Follow the steps below to show, without appealing to Theorem 1.4.6, that f is differentiable.
(a) The hypothesis implies that $|D_{2h}(f)(x) - 2D_h(f)(x+ih)| \le C\,|h|^{1+\gamma}$ for $i = 0, 1$. Iterate to obtain $|D_{2^j h}(f)(x) - 2^j D_h(f)(x+ih)| \le C_\gamma |2^j h|^{1+\gamma}$ for all $0 \le i < 2^j$.
(b) Given a positive integer m find a j such that $2^{j-1} \le m < 2^j$ and use the estimate in part (a) to conclude that $|D_h(f)(x+mh) - D_h(f)(x)| \le 2^{\gamma+1} C_\gamma |mh|^\gamma |h|$ for all integers m.

(c) Use the result of Exercise 1.4.3(d) and a telescoping argument to conclude that $f(x+nh)-f(x)=n(f(x+h)-f(x))+O(|nh|^{1+\gamma})$. Conclude that for all $h\neq 0$ and $n\in\mathbf{Z}\setminus\{0\}$ we have

$$\left|\frac{f(x+h)-f(x)}{h}-\frac{f(x+h/n)-f(x)}{h/n}\right|\leq C''|h|^{\gamma}.$$

(d) Deduce that for all nonzero rationals h,h' and all $x\in\mathbf{R}$ we have

$$\frac{f(x+h)-f(x)}{h}-\frac{f(x+h')-f(x)}{h'}=O(|h|^{\gamma}+|h'|^{\gamma}).$$

Use the continuity of f to extend this identity to all reals and obtain that $\frac{f(x+h)-f(x)}{h}$ satisfies the Cauchy criterion.

HISTORICAL NOTES

The strong type $L^p\to L^q$ estimates in Theorem 1.2.3 were obtained by Hardy and Littlewood [182] (see also [183]) when $n=1$ and by Sobolev [319] for general n. The weak type estimate $L^1\to L^{\frac{n}{n-s},\infty}$ first appeared in Zygmund [376]. The proof of Theorem 1.2.3 using estimate (1.2.10) is taken from Hedberg [188]. The best constants in this theorem when $p=\frac{2n}{n+s}$, $q=\frac{2n}{n-s}$, and $0<s<n$ were precisely evaluated by Lieb [246]. A generalization of Theorem 1.2.3 for nonconvolution operators was obtained by Folland and Stein [141].

The Riesz potentials were systematically studied by Riesz [304] on $\mathbf{R}^n$, although their one-dimensional version appeared in earlier work of Weyl [368]. The Bessel potentials were introduced by Aronszajn and Smith [7] and by Calderón [53], who was the first to observe that the potential space $\mathscr{L}_s^p$ (i.e., the Sobolev space L_s^p) coincides with the space L_k^p given in the classical Definition 1.3.1 when $s=k$ is an integer. Theorem 1.3.5 is due to Sobolev [319] when s is a positive integer. The case $p=1$ of Sobolev's theorem (Exercise 1.3.9) was later obtained independently by Gagliardo [148] and Nirenberg [285]. We refer to the books of Adams [2], Lieb and Loss [247], and Maz'ya [260] for a more systematic study of Sobolev spaces and their use in analysis.

An early characterization of Lipschitz spaces using Littlewood–Paley type operators (built from the Poisson kernel) appears in the work of Hardy and Littlewood [184]. These and other characterizations were obtained and extensively studied in higher dimensions by Taibleson [335], [336], [337] in his extensive study. Lipschitz spaces can also be characterized via mean oscillation over cubes. This idea originated in the simultaneous but independent work of Campanato [62], [63] and Meyers [266] and led to duality theorems for these spaces. Incidentally, the predual of the space $\dot{\Lambda}_\alpha$ is the Hardy space H^p, with $p=\frac{n}{n+\alpha}$, as shown by Duren, Romberg, and Shields [128] for the unit circle and by Walsh [363] for higher-dimensional spaces; see also Fefferman and Stein [139]. We refer to the book of García-Cuerva and Rubio de Francia [150] for a nice exposition of these results. An excellent expository reference on Lipschitz spaces is the article of Krantz [229]. The solution to Exercise 1.4.10 was suggested by T. Tao.

Chapter 2
Hardy Spaces, Besov Spaces, and Triebel–Lizorkin Spaces

The main function spaces we study in this chapter are Hardy spaces which measure smoothness within the realm of rough distributions. Hardy spaces also serve as a substitute for L^p when $p < 1$. We also take a quick look at Besov–Lipschitz and Triebel–Lizorkin spaces, which provide an appropriate framework that unifies the subject of function spaces.

One of the main achievements of this chapter is the characterization of these spaces using Littlewood–Paley theory. Another major accomplishment of this chapter is the atomic characterization of these function spaces. This is obtained from the Littlewood–Paley characterization of these spaces in a single way for all of them.

2.1 Hardy Spaces

The Hardy spaces $H^p(\mathbf{R}^n)$, $0 < p < \infty$, are spaces of distributions which become more singular as p decreases. These function spaces have remarkable similarities to L^p and, in many ways, serve as a substitute for L^p when $p < 1$. In previous sections, we have been able to characterize L^p spaces, Sobolev spaces, and Lipschitz spaces using Littlewood–Paley theory, and it should not come as a surprise that a similar characterization is available for the Hardy spaces as well.

There exists an abundance of equivalent characterizations for Hardy spaces, of which only a few representative ones are discussed in this section. A reader interested in going through the material quickly may define the Hardy space H^p as the space of all tempered distributions f modulo polynomials for which

$$\|f\|_{H^p} = \Big\| \Big(\sum_{j\in\mathbf{Z}} |\Delta_j(f)|^2 \Big)^{\frac{1}{2}} \Big\|_{L^p} < \infty \tag{2.1.1}$$

whenever $0 < p \le 1$. An atomic decomposition for Hardy spaces can be obtained from this definition (see Section 2.3), and once this is available, the analysis of these spaces is significantly simplified. For historical reasons, however, we choose

L. Grafakos, *Modern Fourier Analysis*, Graduate Texts in Mathematics 250,
DOI 10.1007/978-1-4939-1230-8_2,

to define Hardy spaces using a more classical approach, and, as a result, we have to go through a considerable amount of work to obtain the characterization in (2.1.1).

2.1.1 Definition of Hardy Spaces

To give the definition of Hardy spaces on $\mathbf{R}^n$, we need some background. We say that a tempered distribution v is *bounded* if $\varphi * v \in L^\infty(\mathbf{R}^n)$ whenever φ is in $\mathscr{S}(\mathbf{R}^n)$. We observe that if v is a bounded tempered distribution and $h \in L^1(\mathbf{R}^n)$, then the convolution $h*v$ can be defined as a distribution via the convergent integral

$$\langle h*v,\varphi\rangle = \langle \widetilde{\varphi}*v,\widetilde{h}\rangle = \int_{\mathbf{R}^n} (\widetilde{\varphi}*v)(x)\widetilde{h}(x)\,dx,$$

where φ is a Schwartz function and $\widetilde{\varphi}(x)=\varphi(-x)$, $\widetilde{h}(x)=h(-x)$.

The Poisson kernel P is the function

$$P(x) = \frac{\Gamma(\frac{n+1}{2})}{\pi^{\frac{n+1}{2}}}\frac{1}{(1+|x|^2)^{\frac{n+1}{2}}}\,. \tag{2.1.2}$$

For $t>0$, let $P_t(x)=t^{-n}P(t^{-1}x)$. If v is a bounded tempered distribution, then P_t*v is a well-defined distribution, since P_t is in L^1. We claim that P_t*v can be identified with a well-defined bounded function. To see this, write $1=\widehat{\varphi}(\xi)+1-\widehat{\varphi}(\xi)$, where $\widehat{\varphi}\in\mathscr{S}(\mathbf{R}^n)$ is equal to 1 in a neighborhood of the origin. Then $\delta_0=\varphi+(\delta_0-\varphi)$ and

$$P_t*v = P_t*(\varphi*v)+P_t*(\delta_0-\varphi)*v\,.$$

Since P_t lies in L^1 and $\varphi*v$ in L^∞, it follows that $P_t*(\varphi*v)$ is a bounded function. Also the Fourier transform of $P_t*(\delta_0-\varphi)$ is $e^{-2\pi t|\xi|}(1-\widehat{\varphi}(\xi))$ which is a Schwartz function. Thus, $P_t*(\delta_0-\varphi)$ is also a Schwartz function, and since v is a bounded distribution, it follows that $P_t*(\delta_0-\varphi)*v$ is a bounded function. These observations prove that P_t*v is a bounded function, whenever v is a bounded distribution.

An important property of bounded tempered distributions f is that

$$P_t*f\to f \qquad \text{in } \mathscr{S}'(\mathbf{R}^n) \text{ as } t\to 0. \tag{2.1.3}$$

For this, see Exercise 2.1.4.

Definition 2.1.1. Let f be a bounded tempered distribution on $\mathbf{R}^n$ and let $0<p<\infty$. We say that f lies in the *Hardy space* $H^p(\mathbf{R}^n)$ if the *Poisson maximal function*

$$M(f;P)(x) = \sup_{t>0}|(P_t*f)(x)| \tag{2.1.4}$$

lies in $L^p(\mathbf{R}^n)$. If this is the case, we set

$$\big\|f\big\|_{H^p} = \big\|M(f;P)\big\|_{L^p}.$$

It is quite easy to see that the Dirac mass δ_0 does not belong in any Hardy space; indeed, $\delta_0 * P_t = P_t$ and $\sup_{t>0} P_t(x)$ is comparable to $|x|^{-n}$ which does not lie in $L^p(\mathbf{R}^n)$ for any p. However, the difference of Dirac masses $\delta_1 - \delta_{-1}$ lies in $H^p(\mathbf{R})$ for $1/2 < p < 1$. To see this, notice that

$$\sup_{t>0} \Big| (\delta_1 * P_t)(x) - (\delta_{-1} * P_t)(x) \Big| = \sup_{t>0} \frac{4|x|}{\pi} \frac{t}{(t^2+|x-1|^2)(t^2+|x+1|^2)} . \tag{2.1.5}$$

Suppose that $|x+1| < |x-1|$, i.e., $x < 0$. Then we have

$$\sup_{t\le|x+1|} \frac{t\,|x|}{(t^2+|x-1|^2)(t^2+|x+1|^2)} \approx \sup_{t\le|x+1|} \frac{t\,|x|}{|x-1|^2|x+1|^2} = \frac{|x|}{|x-1|^2|x+1|} .$$

Also,

$$\sup_{|x+1|\le t\le|x-1|} \frac{t\,|x|}{(t^2+|x-1|^2)(t^2+|x+1|^2)} \approx \sup_{|x+1|\le t\le|x-1|} \frac{t\,|x|}{|x-1|^2 t^2} = \frac{|x|}{|x-1|^2|x+1|} ,$$

while

$$\sup_{t\ge|x-1|} \frac{t\,|x|}{(t^2+|x-1|^2)(t^2+|x+1|^2)} \approx \sup_{t\ge|x-1|} \frac{t\,|x|}{t^4} = \frac{|x|}{|x-1|^3} .$$

Thus (2.1.5) is comparable to $\frac{|x|}{|x-1|^2|x+1|}$ for $x<0$ and analogously to $\frac{|x|}{|x+1|^2|x-1|}$ for $x>0$. Consequently, (2.1.5) lies in $L^p(\mathbf{R})$ if and only if $1/2 < p < 1$.

At this point we don't know whether the H^p spaces coincide with any other known spaces for some values of p. In the next theorem we show that this is the case when $1 < p < \infty$.

Theorem 2.1.2. *(a) Let $1 < p < \infty$. Then every bounded tempered distribution f in H^p is an element of L^p. Moreover, there is a constant $C_{n,p}$ such that for all such f we have*

$$\big\|f\big\|_{L^p} \le \big\|f\big\|_{H^p} \le C_{n,p}\big\|f\big\|_{L^p} ,$$

and therefore $H^p(\mathbf{R}^n)$ coincides with $L^p(\mathbf{R}^n)$.
(b) When $p=1$, every element of H^1 is an integrable function. In other words, $H^1(\mathbf{R}^n) \subseteq L^1(\mathbf{R}^n)$ and for all $f \in H^1$ we have

$$\big\|f\big\|_{L^1} \le \big\|f\big\|_{H^1} . \tag{2.1.6}$$

Proof. (a) Let $f \in H^p(\mathbf{R}^n)$ for some $1 < p < \infty$. The set $\{P_t * f : t > 0\}$ lies in a multiple of the unit ball of $L^p(\mathbf{R}^n)$, which is the dual space of the separable Banach space $L^{p'}(\mathbf{R}^n)$, and hence it is sequentially compact by the Banach–Alaoglu theorem. Therefore, there exists a sequence $t_j \to 0$ such that $P_{t_j} * f$ converges to some L^p function f_0 in the weak* topology of L^p. On the other hand, in view of (2.1.3), $P_{t_j} * f$ in $\mathscr{S}'(\mathbf{R}^n)$ as $t_j \to 0$, and thus the bounded tempered distribution f coincides with the L^p function f_0. Since the family $\{P_t\}_{t>0}$ is an approximate identity, Theorem 1.2.19 in [156] gives that

$$\big\|P_t * f - f\big\|_{L^p} \to 0 \qquad \text{as } t \to 0,$$

from which it follows that

$$\|f\|_{L^p} \le \big\|\sup_{t>0} |P_t * f|\big\|_{L^p} = \|f\|_{H^p}. \tag{2.1.7}$$

The converse inequality is a consequence of the fact that

$$\sup_{t>0} |P_t * f| \le M(f),$$

where M is the Hardy–Littlewood maximal operator. (See Corollary 2.1.12 in [156].)

(b) The case $p = 1$ requires only a small modification of the case $p > 1$. We embed L^1 in the space of finite Borel measures $\mathscr{M}$ which is the dual of the separable space $C_{00}^{\infty}(\mathbf{R}^n)$ of all continuous functions on $\mathbf{R}^n$ that vanish at infinity. By the Banach-Alaoglu theorem, the unit ball of $\mathscr{M}$ is weak* sequentially compact, and we can extract a sequence $t_j \to 0$ such that $P_{t_j} * f$ converges to some measure μ in the topology of measures. In view of (2.1.3), it follows that the distribution f can be identified with the measure μ.

It remains to show that μ is absolutely continuous with respect to Lebesgue measure, which would imply that it coincides with some L^1 function. We show that μ is absolutely continuous with respect to Lebesgue measure by showing that for all subsets E of $\mathbf{R}^n$ we have $|E| = 0 \implies |\mu(E)| = 0$. Since $\sup_{t>0} |P_t * f|$ lies in $L^1(\mathbf{R}^n)$, given $\varepsilon > 0$, there exists a $\delta > 0$ such that for any measurable subset F of $\mathbf{R}^n$ we have

$$|F| < \delta \implies \int_F \sup_{t>0} |P_t * f|\, dx < \varepsilon.$$

Given E with $|E| = 0$, we can find an open set U such that $E \subseteq U$ and $|U| < \delta$. Let us denote by $\mathscr{C}_{00}(U)$ the space of continuous functions $g(x)$ that are supported in U and tend to zero as $|x| \to \infty$. Then for any g in $\mathscr{C}_{00}(U)$ we have

$$\begin{aligned}
\left|\int_{\mathbf{R}^n} g\, d\mu\right| &= \lim_{j\to\infty} \left|\int_{\mathbf{R}^n} g(x)\,(P_{t_j} * f)(x)\, dx\right| \\
&\le \|g\|_{L^\infty} \int_U \sup_{t>0} |(P_t * f)(x)|\, dx \\
&< \varepsilon \|g\|_{L^\infty}.
\end{aligned}$$

Let $|\mu|$ be the total variation of μ. Then we have (see [190] (20.49))

$$|\mu|(U) = \int_U 1\, d|\mu| = \sup\left\{ \left|\int_{\mathbf{R}^n} g\, d\mu\right| : g \in \mathscr{C}_{00}(U), \quad \|g\|_{L^\infty} \le 1 \right\},$$

which implies $|\mu|(U) < \varepsilon$. Since ε was arbitrary, it follows that $|\mu|(E) = 0$ and thus $\mu(E) = 0$; hence μ is absolutely continuous with respect to Lebesgue measure. Finally, (2.1.6) is a consequence of (2.1.7), which is also valid for $p = 1$. □

We may wonder whether H^1 coincides with L^1. We show in Corollary 2.4.8 that elements of H^1 have integral zero; thus H^1 is a proper subspace of L^1.

2.1.2 Quasi-norm Equivalence of Several Maximal Functions

We now obtain some characterizations of these spaces.

Definition 2.1.3. Let $a,b>0$. Let Φ be a Schwartz function and let f be a tempered distribution on $\mathbf{R}^n$. We define the *smooth maximal function of f with respect to Φ* as

$$M(f;\Phi)(x)=\sup_{t>0}|(\Phi_t*f)(x)|.$$

We define the *nontangential maximal function (with aperture a) of f with respect to Φ* as

$$M_a^*(f;\Phi)(x)=\sup_{t>0}\sup_{\substack{y\in\mathbf{R}^n\\|y-x|\le at}}|(\Phi_t*f)(y)|.$$

We also define the *auxiliary maximal function*

$$M_b^{**}(f;\Phi)(x)=\sup_{t>0}\sup_{y\in\mathbf{R}^n}\frac{|(\Phi_t*f)(x-y)|}{(1+t^{-1}|y|)^b}, \tag{2.1.8}$$

and we observe that

$$M(f;\Phi)\le M_a^*(f;\Phi)\le(1+a)^bM_b^{**}(f;\Phi) \tag{2.1.9}$$

for all $a,b>0$. We note that if Φ is merely integrable, for example, if Φ is the Poisson kernel, the maximal functions $M(f;\Phi)$, $M_a^*(f;\Phi)$, and $M_b^{**}(f;\Phi)$ are well defined only for bounded tempered distributions f on $\mathbf{R}^n$.

For a fixed positive integer N and a Schwartz function φ we define the quantity

$$\mathfrak{N}_N(\varphi)=\int_{\mathbf{R}^n}(1+|x|)^N\sum_{|\alpha|\le N+1}|\partial^\alpha\varphi(x)|\,dx. \tag{2.1.10}$$

We now define

$$\mathscr{F}_N=\Big\{\varphi\in\mathscr{S}(\mathbf{R}^n):\ \mathfrak{N}_N(\varphi)\le 1\Big\}, \tag{2.1.11}$$

and we also define the *grand maximal function of f (with respect to N)* as

$$\mathscr{M}_N(f)(x)=\sup_{\varphi\in\mathscr{F}_N}M_1^*(f;\varphi)(x).$$

It is a fact that all the maximal functions of the preceding subsection have comparable L^p quasi-norms for all $0<p<\infty$. This is the essence of the following theorem.

Theorem 2.1.4. *Let $0 < p < \infty$. Then the following statements are valid:*
(a) There exists a Schwartz function Φ^o with $\int_{\mathbf{R}^n} \Phi^o(x)\,dx = 1$ such that

$$\big\|M(f;\Phi^o)\big\|_{L^p} \le 500\,\big\|f\big\|_{H^p} \tag{2.1.12}$$

for all bounded distributions $f \in \mathscr{S}'(\mathbf{R}^n)$.
(b) For every $a > 0$ and every Φ in $\mathscr{S}(\mathbf{R}^n)$ there exists $C_2(n,p,a,\Phi) < \infty$ such that

$$\big\|M_a^*(f;\Phi)\big\|_{L^p} \le C_2(n,p,a,\Phi)\big\|M(f;\Phi)\big\|_{L^p} \tag{2.1.13}$$

for all distributions $f \in \mathscr{S}'(\mathbf{R}^n)$.
(c) For every $a > 0$, $b > n/p$, and every Φ in $\mathscr{S}(\mathbf{R}^n)$ there exists a constant $C_3(n,p,a,b,\Phi) < \infty$ such that

$$\big\|M_b^{**}(f;\Phi)\big\|_{L^p} \le C_3(n,p,a,b,\Phi)\big\|M_a^*(f;\Phi)\big\|_{L^p} \tag{2.1.14}$$

for all distributions $f \in \mathscr{S}'(\mathbf{R}^n)$.
(d) For every $b > 0$ and every Φ in $\mathscr{S}(\mathbf{R}^n)$ with $\int_{\mathbf{R}^n} \Phi(x)\,dx = 1$ there exists a constant $C_4(b,\Phi) < \infty$ such that if $N = [b]+1$ we have

$$\big\|\mathscr{M}_N(f)\big\|_{L^p} \le C_4(b,\Phi)\big\|M_b^{**}(f;\Phi)\big\|_{L^p} \tag{2.1.15}$$

for all distributions $f \in \mathscr{S}'(\mathbf{R}^n)$.
(e) For every positive integer N there exists a constant $C_5(n,N)$ such that every tempered distribution f with $\big\|\mathscr{M}_N(f)\big\|_{L^p} < \infty$ is a bounded distribution and satisfies

$$\big\|f\big\|_{H^p} \le C_5(n,N)\big\|\mathscr{M}_N(f)\big\|_{L^p}, \tag{2.1.16}$$

that is, it lies in the Hardy space H^p.

Choosing $\Phi = \Phi^o$ in parts (b), (c), and (d), $\frac{n}{p} < b < [\frac{n}{p}]+1$, and $N = [\frac{n}{p}]+1$, we conclude that for bounded distributions f we have

$$\big\|f\big\|_{H^p} \approx \big\|\mathscr{M}_N(f)\big\|_{L^p}.$$

Moreover, for any Schwartz function Φ with $\int_{\mathbf{R}^n} \Phi(x)\,dx = 1$ and any bounded distribution f in $\mathscr{S}'(\mathbf{R}^n)$, the following quasi-norms are equivalent

$$\big\|f\big\|_{H^p} \approx \big\|M(f;\Phi)\big\|_{L^p},$$

with constants that depend only on Φ, n, p.

Before we begin the proof of Theorem 2.1.4, we state and prove a useful lemma.

Lemma 2.1.5. *Let $m \in \mathbf{Z}^+$ and let Φ in $\mathscr{S}(\mathbf{R}^n)$ satisfy $\int_{\mathbf{R}^n} \Phi(x)\,dx = 1$. Then there exists a constant $C_0(\Phi,m)$ such that for any Ψ in $\mathscr{S}(\mathbf{R}^n)$, there are Schwartz functions $\Theta^{(s)}$, $0 \le s \le 1$, with the properties*

$$\Psi(x) = \int_0^1 (\Theta^{(s)} * \Phi_s)(x)\,ds \tag{2.1.17}$$

and

$$\int_{\mathbf{R}^n}(1+|x|)^m|\Theta^{(s)}(x)|\,dx \le C_0(\Phi,m)\,s^m\,\mathfrak{N}_m(\Psi). \tag{2.1.18}$$

Proof. We start with a smooth function ζ supported in $[0,1]$ that satisfies

$$\begin{aligned} 0\le \zeta(s) \le \frac{2s^m}{m!} \qquad & \text{for all } 0\le s\le 1\,,\\ \zeta(s) = \frac{s^m}{m!} \qquad & \text{for all } 0\le s\le \frac{1}{2}\,,\\ \frac{d^r\zeta}{dt^r}(1) = 0 \qquad & \text{for all } 0\le r\le m+1\,. \end{aligned}$$

We define

$$\Theta^{(s)} = \Xi^{(s)} - \frac{d^{m+1}\zeta}{ds^{m+1}}(s)\,\overbrace{\Phi_s*\cdots*\Phi_s}^{m+1\text{ terms}}*\Psi\,, \tag{2.1.19}$$

where

$$\Xi^{(s)} = (-1)^{m+1}\zeta(s)\frac{d^{m+1}}{ds^{m+1}}\Big(\overbrace{\Phi_s*\cdots*\Phi_s}^{m+2\text{ terms}}\Big)*\Psi\,,$$

and we claim that (2.1.17) holds for this choice of $\Theta^{(s)}$. To verify this assertion, we apply $m+1$ integration by parts to write

$$\begin{aligned} \int_0^1 \Theta^{(s)}*\Phi_s\,ds = \int_0^1 \Xi^{(s)}*\Phi_s\,ds + \frac{d^m\zeta}{ds^m}(0)\lim_{s\to 0+}\Big(\overbrace{\Phi*\cdots*\Phi}^{m+2\text{ terms}}\Big)_s*\Psi \\ -(-1)^{m+1}\int_0^1\zeta(s)\frac{d^{m+1}}{ds^{m+1}}\Big(\overbrace{\Phi_s*\cdots*\Phi_s}^{m+2\text{ terms}}\Big)*\Psi\,ds\,, \end{aligned}$$

noting that all the boundary terms vanish except for the term at $s=0$ in the first integration by parts. The first and the third terms in the previous expression on the right add up to zero, while the second term is equal to Ψ, since Φ has integral one. This implies that the family $\{(\Phi*\cdots*\Phi)_s\}_{s>0}$ is an approximate identity as $s\to 0$. Therefore, (2.1.17) holds.

We now prove estimate (2.1.18). Let Ω be the $(m+1)$-fold convolution of Φ. For the second term on the right in (2.1.19), we note that the $(m+1)$st derivative of $\zeta(s)$ vanishes on $[0,\frac{1}{2}]$, so that we may write

$$\begin{aligned} &\int_{\mathbf{R}^n}(1+|x|)^m\Big|\frac{d^{m+1}\zeta(s)}{ds^{m+1}}\Big|\,|\Omega_s*\Psi(x)|\,dx \\ &\le C_m\chi_{[\frac{1}{2},1]}(s)\int_{\mathbf{R}^n}(1+|x|)^m\Big[\int_{\mathbf{R}^n}\tfrac{1}{s^n}|\Omega(\tfrac{x-y}{s})|\,|\Psi(y)|\,dy\Big]\,dx \\ &\le C_m\chi_{[\frac{1}{2},1]}(s)\int_{\mathbf{R}^n}\int_{\mathbf{R}^n}(1+|y+sx|)^m|\Omega(x)|\,|\Psi(y)|\,dy\,dx \\ &\le C_m\chi_{[\frac{1}{2},1]}(s)\int_{\mathbf{R}^n}\int_{\mathbf{R}^n}(1+|sx|)^m|\Omega(x)|\,(1+|y|)^m|\Psi(y)|\,dy\,dx \end{aligned}$$

$$\le C_m\,\chi_{[\frac{1}{2},1]}(s)\Big(\int_{\mathbf{R}^n}(1+|x|)^m|\Omega(x)|\,dx\Big)\Big(\int_{\mathbf{R}^n}(1+|y|)^m|\Psi(y)|\,dy\Big)$$
$$\le C_0'(\Phi,m)\,s^m\,\mathfrak{N}_m(\Psi),$$

since $\chi_{[\frac{1}{2},1]}(s)\le 2^m s^m$. To obtain a similar estimate for the first term on the right in (2.1.19), we argue as follows:

$$\int_{\mathbf{R}^n}(1+|x|)^m|\zeta(s)|\left|\frac{d^{m+1}(\Omega_s*\Psi)}{ds^{m+1}}(x)\right|dx$$
$$=\int_{\mathbf{R}^n}(1+|x|)^m\,|\zeta(s)|\left|\frac{d^{m+1}}{ds^{m+1}}\int_{\mathbf{R}^n}\frac{1}{s^n}\Omega\Big(\frac{x-y}{s}\Big)\Psi(y)\,dy\right|dx$$
$$=\int_{\mathbf{R}^n}(1+|x|)^m\,|\zeta(s)|\left|\int_{\mathbf{R}^n}\Omega(y)\frac{d^{m+1}\Psi(x-sy)}{ds^{m+1}}\,dy\right|dx$$
$$\le C_m'\int_{\mathbf{R}^n}(1+|x|)^m\,|\zeta(s)|\int_{\mathbf{R}^n}|\Omega(y)|\Big[\sum_{|\alpha|\le m+1}|\partial^\alpha\Psi(x-sy)|\,|y|^{|\alpha|}\Big]\,dy\,dx$$
$$\le C_m'|\zeta(s)|\int_{\mathbf{R}^n}\int_{\mathbf{R}^n}(1+|x+sy|)^m|\Omega(y)|\sum_{|\alpha|\le m+1}|\partial^\alpha\Psi(x)|\,(1+|y|)^{m+1}\,dy\,dx$$
$$\le C_m'|\zeta(s)|\int_{\mathbf{R}^n}(1+|y|)^{m+1}\,|\Omega(y)|\,(1+|y|)^m\,dy\int_{\mathbf{R}^n}(1+|x|)^m\sum_{|\alpha|\le m+1}|\partial^\alpha\Psi(x)|\,dx$$
$$\le C_0''(\Phi,m)\,s^m\,\mathfrak{N}_m(\Psi).$$

We now set $C_0(\Phi,m)=C_0'(\Phi,m)+C_0''(\Phi,m)$ to conclude the proof of (2.1.18). □

Next, we discuss the proof of Theorem 2.1.4.

Proof. (a) We pick a continuous and integrable function $\psi(s)$ on the interval $[1,\infty)$ that decays faster than any negative power of s (i.e., $|\psi(s)|\le C_N s^{-N}$ for all $N>0$) and such that

$$\int_1^\infty s^k\,\psi(s)\,ds=\begin{cases}1 & \text{if } k=0,\\ 0 & \text{if } k=1,2,3,\dots.\end{cases}\tag{2.1.20}$$

Such a function exists; see Exercise 2.1.3. In fact, we may take

$$\psi(s)=\frac{e}{\pi}\frac{1}{s}\,e^{-\frac{\sqrt{2}}{2}(s-1)^{\frac14}}\sin\Big(\frac{\sqrt{2}}{2}(s-1)^{\frac14}\Big).\tag{2.1.21}$$

We now define the function

$$\Phi^o(x)=\int_1^\infty\psi(s)P_s(x)\,ds,\tag{2.1.22}$$

where P_s is the Poisson kernel. Note that the double integral

$$\int_{\mathbf{R}^n}\int_1^\infty\frac{s}{(s^2+|x|^2)^{\frac{n+1}{2}}}\,s^{-N}\,ds\,dx$$

converges and so it follows from (2.1.20) and (2.1.22) via Fubini's theorem that

$$\int_{\mathbf{R}^n} \Phi^o(x)\,dx = 1\,.$$

Moreover, another application of Fubini's theorem yields that

$$\widehat{\Phi^o}(\xi) = \int_1^\infty \psi(s)\widehat{P_s}(\xi)\,ds = \int_1^\infty \psi(s)e^{-2\pi s|\xi|}\,ds$$

using that $\widehat{P_s}(\xi) = e^{-2\pi s|\xi|}$ (cf. Exercise 2.2.11 in [156]). This function is rapidly decreasing as $|\xi| \to \infty$ and the same is true for all the derivatives

$$\partial^\alpha \widehat{\Phi^o}(\xi) = \int_1^\infty \psi(s)\partial_\xi^\alpha \big(e^{-2\pi s|\xi|}\big)\,ds\,. \tag{2.1.23}$$

Moreover, the function $\widehat{\Phi^o}$ is clearly smooth on $\mathbf{R}^n \setminus \{0\}$ and we will show that it is also smooth at the origin. Notice that for all multi-indices α we have

$$\partial_\xi^\alpha \big(e^{-2\pi s|\xi|}\big) = s^{|\alpha|} p_\alpha(\xi)|\xi|^{-m_\alpha} e^{-2\pi s|\xi|}$$

for some $m_\alpha \in \mathbf{Z}^+$ and some polynomial $p_\alpha(\xi)$. By Taylor's theorem, for some function $v(s,|\xi|)$ with $0 \le v(s,|\xi|) \le 2\pi s|\xi|$, we have

$$e^{-2\pi s|\xi|} = \sum_{k=0}^{L} (-2\pi)^k \frac{|\xi|^k}{k!} s^k + \frac{(-2\pi s|\xi|)^{L+1}}{(L+1)!} e^{-v(s,|\xi|)}\,.$$

Choosing $L > m_\alpha$ gives

$$\partial_\xi^\alpha \big(e^{-2\pi s|\xi|}\big) = \sum_{k=0}^{L} (-2\pi)^k \frac{|\xi|^k}{k!} s^{k+|\alpha|} \frac{p_\alpha(\xi)}{|\xi|^{m_\alpha}} + s^{|\alpha|} \frac{p_\alpha(\xi)}{|\xi|^{m_\alpha}} \frac{(-2\pi s|\xi|)^{L+1}}{(L+1)!} e^{-v(s,|\xi|)}\,,$$

which, inserted in (2.1.23) and in view of (2.1.20), yields that when $|\alpha| > 0$, the derivative $\partial^\alpha \widehat{\Phi^o}(\xi)$ tends to zero as $\xi \to 0$ and when $\alpha = 0$, $\widehat{\Phi^o}(\xi) \to 1$ as $\xi \to 0$. We conclude that $\widehat{\Phi^o}$ is continuously differentiable and hence smooth at the origin (cf. Exercise 1.1.1); hence it lies in the Schwartz class, and thus so does Φ^o.

Finally, we have the estimate

$$\begin{aligned} M(f;\Phi^o)(x) &= \sup_{t>0} |(\Phi_t^o * f)(x)| \\ &= \sup_{t>0} \left| \int_1^\infty \psi(s)(f * P_{ts})(x)\,ds \right| \\ &\le \int_1^\infty |\psi(s)|\,ds\, M(f;P)(x)\,, \end{aligned}$$

and the required conclusion follows since $\int_1^\infty |\psi(s)|\,ds \le 500$. Note that we actually obtained the stronger pointwise estimate

$$M(f;\Phi^o) \le 500 M(f;P)$$

rather than (2.1.12).

(b) The control of the nontagential maximal function $M_a^*(\cdot\,;\Phi)$ in terms of the vertical maximal function $M(\cdot\,;\Phi)$ is the hardest and most technical part of the proof. For matters of exposition, we present the proof only in the case that $a=1$ and we note that the case of general $a>0$ presents only notational differences. We derive (2.1.13) as a consequence of the estimate

$$\big\|M_1^*(f;\Phi)\big\|_{L^p}^p \le C_2''(n,p,\Phi)^p\big\|M(f;\Phi)\big\|_{L^p}^p + \frac{1}{2}\big\|M_1^*(f;\Phi)\big\|_{L^p}^p, \tag{2.1.24}$$

which is useful only if we know that $\|M_1^*(f;\Phi)\|_{L^p}<\infty$. This presents a significant hurdle that needs to be overcome by an approximation. For this reason we introduce a family of maximal functions $M_1^*(f;\Phi)^{\varepsilon,N}$ for $0\le \varepsilon,N<\infty$ such that $\|M_1^*(f;\Phi)^{\varepsilon,N}\|_{L^p}<\infty$ and such that $M_1^*(f;\Phi)^{\varepsilon,N}\uparrow M_1^*(f;\Phi)$ as $\varepsilon\downarrow 0$ and we prove (2.1.24) with $M_1^*(f;\Phi)^{\varepsilon,N}$ in place of $M_1^*(f;\Phi)^{\varepsilon,N}$. In other words we prove

$$\big\|\,M_1^*(f;\Phi)^{\varepsilon,N}\big\|_{L^p}^p \le C_2'(n,p,\Phi,N)^p\big\|M(f;\Phi)\big\|_{L^p}^p + \frac{1}{2}\big\|M_1^*(f;\Phi)^{\varepsilon,N}\big\|_{L^p}^p, \tag{2.1.25}$$

where there is an additional dependence on N in the constant $C_2'(n,p,\Phi,N)$, but there is no dependence on ε. The $M_1^*(f;\Phi)^{\varepsilon,N}$ are defined as follows: for a bounded distribution f in $\mathscr{S}'(\mathbf{R}^n)$ such that $M(f;\Phi)\in L^p$ we define

$$M_1^*(f;\Phi)^{\varepsilon,N}(x) = \sup_{0<t<\frac{1}{\varepsilon}}\ \sup_{|y-x|\le t} \big|(\Phi_t*f)(y)\big|\Big(\frac{t}{t+\varepsilon}\Big)^N\frac{1}{(1+\varepsilon|y|)^N}\,.$$

We first show that $M_1^*(f;\Phi)^{\varepsilon,N}$ lies in $L^p(\mathbf{R}^n)\cap L^\infty(\mathbf{R}^n)$ if N is large enough depending on f. Indeed, using that $(\Phi_t*f)(x)=\langle f,\Phi_t(x-\cdot)\rangle$ and the fact that f is in $\mathscr{S}'(\mathbf{R}^n)$, we obtain constants C_f and $m=m_f$ such that:

$$\begin{aligned}
|(\Phi_t*f)(y)| &\le C_f \sum_{|\gamma|\le m,\,|\beta|\le m}\ \sup_{w\in\mathbf{R}^n}|w^\gamma(\partial^\beta\Phi_t)(y-w)|\\
&\le C_f \sum_{|\beta|\le m}\sup_{z\in\mathbf{R}^n}(1+|y|^m+|z|^m)|(\partial^\beta\Phi_t)(z)|\\
&\le C_f(1+|y|^m)\sum_{|\beta|\le m}\sup_{z\in\mathbf{R}^n}(1+|z|^m)|(\partial^\beta\Phi_t)(z)|\\
&\le C_f\frac{(1+|y|^m)}{\min(t^n,t^{n+m})}\sum_{|\beta|\le m}\sup_{z\in\mathbf{R}^n}(1+|z|^m)|(\partial^\beta\Phi)(z/t)|
\end{aligned}$$

$$\leq C_f \frac{(1+|y|)^m}{\min(t^n, t^{n+m})}(1+t^m) \sum_{|\beta| \leq m} \sup_{z \in \mathbf{R}^n} (1+|z/t|^m)|(\partial^\beta \Phi)(z/t)|$$
$$\leq C_{f,\Phi}(1+\varepsilon|y|)^m \varepsilon^{-m}(1+t^m)(t^{-n}+t^{-n-m}).$$

Multiplying by $(\frac{t}{t+\varepsilon})^N(1+\varepsilon|y|)^{-N}$ for some $0<t<\frac{1}{\varepsilon}$ and $|y-x|<t$ yields

$$\big|(\Phi_t * f)(y)\big|\Big(\frac{t}{t+\varepsilon}\Big)^N \frac{1}{(1+\varepsilon|y|)^N} \leq C_{f,\Phi} \frac{\varepsilon^{-m}(1+\varepsilon^{-m})(\varepsilon^{n-N}+\varepsilon^{n+m-N})}{(1+\varepsilon|y|)^{N-m}},$$

and using that $1+\varepsilon|y| \geq \frac{1}{2}(1+\varepsilon|x|)$, we obtain for some $C''(f,\Phi,\varepsilon,n,m,N)<\infty$,

$$M_1^*(f;\Phi)^{\varepsilon,N}(x) \leq \frac{C''(f,\Phi,\varepsilon,n,m,N)}{(1+\varepsilon|x|)^{N-m}}.$$

Taking $N>m+n/p$, we have that $M_1^*(f;\Phi)^{\varepsilon,N}$ lies in $L^p(\mathbf{R}^n)$. This choice of N depends on m and hence on the distribution f.

We now introduce functions

$$U(f;\Phi)^{\varepsilon,N}(x) = \sup_{0<t<\frac{1}{\varepsilon}} \sup_{|y-x|<t} t\big|\nabla(\Phi_t * f)(y)\big|\Big(\frac{t}{t+\varepsilon}\Big)^N \frac{1}{(1+\varepsilon|y|)^N}$$

and

$$V(f;\Phi)^{\varepsilon,N}(x) = \sup_{0<t<\frac{1}{\varepsilon}} \sup_{y\in\mathbf{R}^n} \big|(\Phi_t * f)(y)\big|\Big(\frac{t}{t+\varepsilon}\Big)^N \frac{1}{(1+\varepsilon|y|)^N}\Big(\frac{t}{t+|x-y|}\Big)^{[\frac{2n}{p}]+1}.$$

Let $C(n)=\|M\|_{L^2(\mathbf{R}^n)\to L^2(\mathbf{R}^n)}$, where M is the Hardy–Littlewood maximal operator. We need the norm estimate

$$\big\|V(f;\Phi)^{\varepsilon,N}\big\|_{L^p} \leq C(n)^{\frac{2}{p}} \big\|M_1^*(f;\Phi)^{\varepsilon,N}\big\|_{L^p} \tag{2.1.26}$$

and the pointwise estimate

$$U(f;\Phi)^{\varepsilon,N} \leq A(n,p,\Phi,N)\, V(f;\Phi)^{\varepsilon,N}, \tag{2.1.27}$$

where

$$A(\Phi,N,n,p) = 2^{[\frac{2n}{p}]+1} C_0(\partial_j\Phi, N+[\tfrac{2n}{p}]+1)\, \mathfrak{N}_{N+[\frac{2n}{p}]+1}(\partial_j\Phi).$$

To prove (2.1.26) we observe that when $z\in B(y,t)\subseteq B(x,|x-y|+t)$ we have

$$\big|(\Phi_t * f)(y)\big|\Big(\frac{t}{t+\varepsilon}\Big)^N \frac{1}{(1+\varepsilon|y|)^N} \leq M_1^*(f;\Phi)^{\varepsilon,N}(z),$$

from which it follows that for any $y \in \mathbf{R}^n$

$$
\begin{aligned}
&\big|(\Phi_t * f)(y)\big|\Big(\frac{t}{t+\varepsilon}\Big)^N \frac{1}{(1+\varepsilon|y|)^N}\\
&\le \bigg(\frac{1}{|B(y,t)|}\int_{B(y,t)}\big[M_1^*(f;\Phi)^{\varepsilon,N}(z)\big]^{\frac{p}{2}}\,dz\bigg)^{\frac{2}{p}}\\
&\le \Big(\frac{|x-y|+t}{t}\Big)^{\frac{2n}{p}}\bigg(\frac{1}{|B(x,|x-y|+t)|}\int_{B(x,|x-y|+t)}\big[M_1^*(f;\Phi)^{\varepsilon,N}(z)\big]^{\frac{p}{2}}\,dz\bigg)^{\frac{2}{p}}\\
&\le \Big(\frac{|x-y|+t}{t}\Big)^{[\frac{2n}{p}]+1} M\Big(\big[M_1^*(f;\Phi)^{\varepsilon,N}\big]^{\frac{p}{2}}\Big)^{\frac{2}{p}}(x)\,.
\end{aligned}
$$

We now use the boundedness of the Hardy–Littlewood maximal operator M on L^2 to obtain (2.1.26).

In proving (2.1.27), we may assume that Φ has integral 1; otherwise we can multiply Φ by a suitable constant to arrange for this to happen. We note that

$$
t\big|\nabla(\Phi_t * f)\big| = \big|(\nabla\Phi)_t * f\big| \le \sqrt{n}\sum_{j=1}^{n}\big|(\partial_j\Phi)_t * f\big|,
$$

and it suffices to work with each partial derivative $\partial_j\Phi$ of Φ. Using Lemma 2.1.5 we write

$$
\partial_j\Phi = \int_0^1 \Theta^{(s)} * \Phi_s\,ds
$$

for suitable Schwartz functions $\Theta^{(s)}$. Fix $x \in \mathbf{R}^n$, $t>0$, and y with $|y-x|<t<1/\varepsilon$. Then we have

$$
\begin{aligned}
&\big|((\partial_j\Phi)_t * f)(y)\big|\Big(\frac{t}{t+\varepsilon}\Big)^N \frac{1}{(1+\varepsilon|y|)^N}\\
&= \Big(\frac{t}{t+\varepsilon}\Big)^N \frac{1}{(1+\varepsilon|y|)^N}\bigg|\int_0^1 \big((\Theta^{(s)})_t * \Phi_{st} * f\big)(y)\,ds\bigg| \qquad (2.1.28)\\
&\le \Big(\frac{t}{t+\varepsilon}\Big)^N \int_0^1\int_{\mathbf{R}^n} t^{-n}\big|\Theta^{(s)}(t^{-1}z)\big|\frac{\big|(\Phi_{st} * f)(y-z)\big|}{(1+\varepsilon|y|)^N}\,dz\,ds\,.
\end{aligned}
$$

Inserting the factor 1 written as

$$
\Big(\frac{ts}{ts+|x-(y-z)|}\Big)^{[\frac{2n}{p}]+1}\Big(\frac{ts}{ts+\varepsilon}\Big)^N\Big(\frac{ts+|x-(y-z)|}{ts}\Big)^{[\frac{2n}{p}]+1}\Big(\frac{ts+\varepsilon}{ts}\Big)^N
$$

in the preceding z-integral and using that

$$
\frac{1}{(1+\varepsilon|y|)^N} \le \frac{(1+\varepsilon|z|)^N}{(1+\varepsilon|y-z|)^N}
$$

and the fact that $|x-y|<t<1/\varepsilon$, we obtain the estimate

$$\begin{aligned}
&\Big(\frac{t}{t+\varepsilon}\Big)^N \int_0^1\int_{\mathbf{R}^n} t^{-n}\big|\Theta^{(s)}(t^{-1}z)\big|\frac{\big|(\Phi_{st}*f)(y-z)\big|}{(1+\varepsilon|y|)^N}\,dz\,ds\\
&\le V(f;\Phi)^{\varepsilon,N}(x)\int_0^1\int_{\mathbf{R}^n}(1+\varepsilon|z|)^N\Big(\frac{ts+|x-(y-z)|}{ts}\Big)^{[\frac{2n}{p}]+1} t^{-n}\big|\Theta^{(s)}(t^{-1}z)\big|\,dz\frac{ds}{s^N}\\
&\le V(f;\Phi)^{\varepsilon,N}(x)\int_0^1\int_{\mathbf{R}^n} s^{-[\frac{2n}{p}]-1-N}(1+\varepsilon t|z|)^N(s+1+|z|)^{[\frac{2n}{p}]+1}\big|\Theta^{(s)}(z)\big|\,dz\,ds\\
&\le 2^{[\frac{2n}{p}]+1}C_0(\partial_j\Phi,N+[\tfrac{2n}{p}]+1)\mathfrak{N}_{N+[\frac{2n}{p}]+1}(\partial_j\Phi)V(f;\Phi)^{\varepsilon,N}(x)
\end{aligned}$$

in view of conclusion (2.1.18) of Lemma 2.1.5. Combining this estimate with (2.1.28), we deduce (2.1.27). Estimates (2.1.26) and (2.1.27) together yield

$$\big\|U(f;\Phi)^{\varepsilon,N}\big\|_{L^p}\le C(n)A(n,p,\Phi,N)\big\|M_1^*(f;\Phi)^{\varepsilon,N}\big\|_{L^p}. \tag{2.1.29}$$

We now set

$$E_\varepsilon=\big\{x\in\mathbf{R}^n:\ U(f;\Phi)^{\varepsilon,N}(x)\le K\,M_1^*(f;\Phi)^{\varepsilon,N}(x)\big\}$$

for some constant K to be determined shortly. With $A=A(n,p,\Phi,N)$, we have

$$\begin{aligned}
\int_{(E_\varepsilon)^c}\big[M_1^*(f;\Phi)^{\varepsilon,N}(x)\big]^p dx &\le \frac{1}{K^p}\int_{(E_\varepsilon)^c}\big[U(f;\Phi)^{\varepsilon,N}(x)\big]^p dx\\
&\le \frac{1}{K^p}\int_{\mathbf{R}^n}\big[U(f;\Phi)^{\varepsilon,N}(x)\big]^p dx\\
&\le \frac{C(n)^pA^p}{K^p}\int_{\mathbf{R}^n}\big[M_1^*(f;\Phi)^{\varepsilon,N}(x)\big]^p dx\\
&\le \frac{1}{2}\int_{\mathbf{R}^n}\big[M_1^*(f;\Phi)^{\varepsilon,N}(x)\big]^p dx,
\end{aligned} \tag{2.1.30}$$

provided we choose K such that $K^p=2C(n)^pA(n,p,\Phi,N)^p$. Obviously K is a function of n,p,Φ,N and in particular depends on N.

It remains to estimate the contribution of the integral of $\big[M_1^*(f;\Phi)^{\varepsilon,N}(x)\big]^p$ over the set E_ε. We claim that the following pointwise estimate is valid:

$$M_1^*(f;\Phi)^{\varepsilon,N}(x)\le 4C'(n,N,K)^{\frac{1}{q}}\Big[M\big(M(f;\Phi)^q\big)(x)\Big]^{\frac{1}{q}} \tag{2.1.31}$$

for any $x\in E_\varepsilon$ and $0<q<\infty$ and some constant $C'(n,N,K)$, where M is the Hardy–Littlewood maximal operator. To prove (2.1.31) we fix $x\in E_\varepsilon$ and we also fix y such that $|y-x|<t$.

By the definition of $M_1^*(f;\Phi)^{\varepsilon,N}(x)$ there exists a point $(y_0,t)\in\mathbf{R}^{n+1}_+$ such that $|x-y_0|<t<\frac{1}{\varepsilon}$ and

$$\big|(\Phi_t*f)(y_0)\big|\Big(\frac{t}{t+\varepsilon}\Big)^N\frac{1}{(1+\varepsilon|y_0|)^N}\ge\frac{1}{2}M_1^*(f;\Phi)^{\varepsilon,N}(x). \tag{2.1.32}$$

Also by the definitions of E_ε and $U(f;\Phi)^{\varepsilon,N}$, for any $x \in E_\varepsilon$ we have

$$t\big|\nabla(\Phi_t * f)(\xi)\big|\Big(\frac{t}{t+\varepsilon}\Big)^N \frac{1}{(1+\varepsilon|\xi|)^N} \le K M_1^*(f;\Phi)^{\varepsilon,N}(x) \tag{2.1.33}$$

for all ξ satisfying $|\xi - x| < t < \frac{1}{\varepsilon}$. It follows from (2.1.32) and (2.1.33) that

$$t\big|\nabla(\Phi_t * f)(\xi)\big| \le 2K\big|(\Phi_t * f)(y_0)\big|\left(\frac{1+\varepsilon|\xi|}{1+\varepsilon|y_0|}\right)^N \tag{2.1.34}$$

for all ξ satisfying $|\xi - x| < t < \frac{1}{\varepsilon}$. We let z be such that $|z-x| < t$. Applying the mean value theorem and using (2.1.34), we obtain, for some ξ between y_0 and z,

$$\begin{aligned}
\big|(\Phi_t * f)(z) - (\Phi_t * f)(y_0)\big| &= \big|\nabla(\Phi_t * f)(\xi)\big|\,|z-y_0| \\
&\le \frac{2K}{t}\big|(\Phi_t * f)(\xi)\big|\left(\frac{1+\varepsilon|\xi|}{1+\varepsilon|y_0|}\right)^N |z-y_0| \\
&\le \frac{2^{N+1}K}{t}\big|(\Phi_t * f)(y_0)\big|\,|z-y_0| \\
&\le \frac{1}{2}\big|(\Phi_t * f)(y_0)\big|,
\end{aligned}$$

provided z also satisfies $|z-y_0| < 2^{-N-2}K^{-1}t$ in addition to $|z-x| < t$. Therefore, for z satisfying $|z-y_0| < 2^{-N-2}K^{-1}t$ and $|z-x| < t$ we have

$$\big|(\Phi_t * f)(z)\big| \ge \frac{1}{2}\big|(\Phi_t * f)(y_0)\big| \ge \frac{1}{4}M_1^*(f;\Phi)^{\varepsilon,N}(x),$$

where the last inequality uses (2.1.32). Thus we have

$$\begin{aligned}
M\big(M(f;\Phi)^q\big)(x) &\ge \frac{1}{|B(x,t)|}\int_{B(x,t)} \big[M(f;\Phi)(w)\big]^q\,dw \\
&\ge \frac{1}{|B(x,t)|}\int_{B(x,t)\cap B(y_0,2^{-N-2}K^{-1}t)} \big[M(f;\Phi)(w)\big]^q\,dw \\
&\ge \frac{1}{|B(x,t)|}\int_{B(x,t)\cap B(y_0,2^{-N-2}K^{-1}t)} \frac{1}{4^q}\big[M_1^*(f;\Phi)^{\varepsilon,N}(x)\big]^q\,dw \\
&\ge \frac{|B(x,t)\cap B(y_0,2^{-N-2}K^{-1}t)|}{|B(x,t)|}\frac{1}{4^q}\big[M_1^*(f;\Phi)^{\varepsilon,N}(x)\big]^q \\
&\ge C'(n,N,K)^{-1}4^{-q}\big[M_1^*(f;\Phi)^{\varepsilon,N}(x)\big]^q,
\end{aligned}$$

where we used the simple geometric fact that if $|x-y_0| \le t$ and $\delta > 0$, then

$$\frac{|B(x,t)\cap B(y_0,\delta t)|}{|B(x,t)|} \ge c_{n,\delta} > 0,$$

the minimum of this constant being obtained when $|x-y_0| = t$. See Figure 2.1.

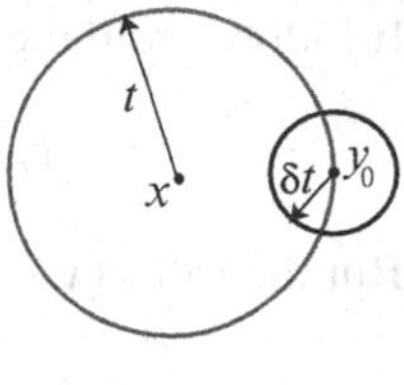

Fig. 2.1 The ball $B(y_0, \delta t)$ captures at least a fixed proportion of the ball $B(x,t)$.

This proves (2.1.31). Taking $q = p/2$ and applying the boundedness of the Hardy–Littlewood maximal operator on L^2 yields

$$\int_{E_\varepsilon} \left[M_1^*(f;\Phi)^{\varepsilon,N}(x)\right]^p dx \le C_2'(n,p,\Phi,N) \int_{\mathbf{R}^n} M(f;\Phi)(x)^p\,dx. \tag{2.1.35}$$

Combining this estimate with (2.1.30), we finally prove (2.1.25).

Recalling the fact (obtained earlier) that $\|M_1^*(f;\Phi)^{\varepsilon,N}\|_{L^p} < \infty$, we deduce from (2.1.25) that

$$\left\|M_1^*(f;\Phi)^{\varepsilon,N}\right\|_{L^p} \le 2^{\frac{1}{p}} C_2'(n,p,\Phi,N) \left\|M(f;\Phi)\right\|_{L^p}. \tag{2.1.36}$$

The previous constant depends on f but is independent of ε. Notice that

$$M_1^*(f;\Phi)^{\varepsilon,N}(x) \ge \frac{2^{-N}}{(1+\varepsilon|x|)^N} \sup_{0<t<1/\varepsilon} \left(\frac{t}{t+\varepsilon}\right)^N \sup_{|y-x|<t} \left|(\Phi_t * f)(y)\right|$$

and that the preceding expression on the right increases to

$$2^{-N} M_1^*(f;\Phi)(x)$$

as $\varepsilon \downarrow 0$. Since the constant in (2.1.36) does not depend on ε, an application of the Lebesgue monotone convergence theorem yields

$$\|M_1^*(f;\Phi)\|_{L^p} \le 2^{N+\frac{1}{p}} C_2'(n,p,\Phi,N) \|M(f;\Phi)\|_{L^p}. \tag{2.1.37}$$

The problem with this estimate is that the finite constant $2^N C_2'(n,p,\Phi,N)$ depends on N and thus on f. However, we have managed to show that under the assumption $\|M(f;\Phi)\|_{L^p} < \infty$, one must necessarily have $\|M_1^*(f;\Phi)\|_{L^p} < \infty$.

Keeping this significant observation in mind, we repeat the preceding argument from the point where the functions $U(f;\phi)^{\varepsilon,N}$ and $V(f;\phi)^{\varepsilon,N}$ are introduced, setting $\varepsilon = N = 0$. Then we arrive at (2.1.24) with a constant $C_2''(n,p,\Phi) = C_2'(n,p,\Phi,0)$ which is independent of N and thus of f. We conclude the validity of (2.1.13) with $C_2(n,p,1,\Phi) = 2^{1/p} C_2''(n,p,\Phi)$ when $a = 1$. A similar constant (depending on a) is obtained for different values of $a > 0$.

(c) As usual, $B(x,R)$ denotes a ball centered at x with radius R. Recall that

$$M_b^{**}(f;\Phi)(x) = \sup_{t>0} \sup_{y\in\mathbf{R}^n} \frac{|(\Phi_t * f)(x-y)|}{\left(\frac{|y|}{t}+1\right)^b}.$$

It follows from the definition of $M_a^*(f;\Phi)(z) = \sup_{t>0}\sup_{|w-z|<at}|(\Phi_t * f)(w)|$ that

$$|(\Phi_t * f)(x-y)| \le M_a^*(f;\Phi)(z) \qquad \text{if } z \in B(x-y,at).$$

But the ball $B(x-y,at)$ is contained in the ball $B(x,|y|+at)$; hence it follows that

$$\begin{aligned} |(\Phi_t * f)(x-y)|^{\frac{n}{b}} &\le \frac{1}{|B(y,at)|}\int_{B(y,at)} M_a^*(f;\Phi)(z)^{\frac{n}{b}}\,dz \\ &\le \frac{1}{|B(y,at)|}\int_{B(x,|y|+at)} M_a^*(f;\Phi)(z)^{\frac{n}{b}}\,dz \\ &\le \left(\frac{|y|+at}{at}\right)^n M\big(M_a^*(f;\Phi)^{\frac{n}{b}}\big)(x) \\ &\le \max(1,a^{-n})\left(\frac{|y|}{t}+1\right)^n M\big(M_a^*(f;\Phi)^{\frac{n}{b}}\big)(x), \end{aligned}$$

from which we conclude that for all $x \in \mathbf{R}^n$ we have

$$M_b^{**}(f;\Phi)(x) \le \max(1,a^{-b})\left\{M\big(M_a^*(f;\Phi)^{\frac{n}{b}}\big)(x)\right\}^{\frac{b}{n}}.$$

Raising to the power p and using the fact that $p > n/b$ and the boundedness of the Hardy–Littlewood maximal operator M on $L^{pb/n}$, we obtain the required conclusion (2.1.14).

(d) In proving (d) we may replace b by the integer $b_0 = [b]+1$. Let Φ be a Schwartz function with integral equal to 1. Applying Lemma 2.1.5 with $m = b_0$, we write any function φ in $\mathscr{F}_N$ as

$$\varphi(y) = \int_0^1 (\Theta^{(s)} * \Phi_s)(y)\,ds$$

for some choice of Schwartz functions $\Theta^{(s)}$. Then we have

$$\varphi_t(y) = \int_0^1 ((\Theta^{(s)})_t * \Phi_{ts})(y)\,ds$$

for all $t > 0$. Fix $x \in \mathbf{R}^n$. Then for y in $B(x,t)$ we have

$$\begin{aligned} |(\varphi_t * f)(y)| &\le \int_0^1\int_{\mathbf{R}^n} |(\Theta^{(s)})_t(z)|\,|(\Phi_{ts} * f)(y-z)|\,dz\,ds \\ &\le \int_0^1\int_{\mathbf{R}^n} |(\Theta^{(s)})_t(z)|\,M_{b_0}^{**}(f;\Phi)(x)\left(\frac{|x-(y-z)|}{st}+1\right)^{b_0} dz\,ds \\ &\le \int_0^1 s^{-b_0}\int_{\mathbf{R}^n} |(\Theta^{(s)})_t(z)|\,M_{b_0}^{**}(f;\Phi)(x)\left(\frac{|x-y|}{t}+\frac{|z|}{t}+1\right)^{b_0} dz\,ds \\ &\le 2^{b_0} M_{b_0}^{**}(f;\Phi)(x)\int_0^1 s^{-b_0}\int_{\mathbf{R}^n} |\Theta^{(s)}(w)|\,\big(|w|+1\big)^{b_0}\,dw\,ds \\ &\le 2^{b_0} M_{b_0}^{**}(f;\Phi)(x)\int_0^1 s^{-b_0} C_0(\Phi,b_0)\,s^{b_0}\,\mathfrak{N}_{b_0}(\varphi)\,ds, \end{aligned}$$

where we applied conclusion (2.1.18) of Lemma 2.1.5. Setting $N = b_0 = [b]+1$, we obtain for y in $B(x,t)$ and $\varphi \in \mathscr{F}_N$,

$$|(\varphi_t * f)(y)| \le 2^{b_0} C_0(\Phi, b_0) M_{b_0}^{**}(f;\Phi)(x).$$

Taking the supremum over all y in $B(x,t)$, over all $t>0$, and over all φ in $\mathscr{F}_N$, we obtain the pointwise estimate

$$\mathscr{M}_N(f)(x) \le 2^{b_0} C_0(\Phi, b_0) M_{b_0}^{**}(f;\Phi)(x), \qquad x \in \mathbf{R}^n,$$

where $N = b_0 + 1$. This clearly yields (2.1.15) if we set $C_4 = 2^{b_0} C_0(\Phi, b_0)$.

(e) We fix an $f \in \mathscr{S}'(\mathbf{R}^n)$ that satisfies $\|\mathscr{M}_N(f)\|_{L^p} < \infty$ for some fixed positive integer N. To show that f is a bounded distribution, we fix a Schwartz function φ and we observe that for some positive constant $c = c_\varphi$, we have that $c\,\varphi$ is an element of $\mathscr{F}_N$ and thus $M_1^*(f;c\,\varphi) \le \mathscr{M}_N(f)$. Then

$$\begin{aligned}
c^p\,|(\varphi * f)(x)|^p &\le \inf_{|y-x|\le 1} \sup_{|z-y|\le 1} |(c\,\varphi * f)(z)|^p \\
&\le \inf_{|y-x|\le 1} M_1^*(f;c\,\varphi)(y)^p \\
&\le \frac{1}{v_n}\int_{|y-x|\le 1} M_1^*(f;c\,\varphi)(y)^p\,dy \\
&\le \frac{1}{v_n}\int_{\mathbf{R}^n} M_1^*(f;c\,\varphi)(y)^p\,dy \\
&\le \frac{1}{v_n}\int_{\mathbf{R}^n} \mathscr{M}_N(f)(y)^p\,dy < \infty,
\end{aligned}$$

which implies that $\varphi * f$ is a bounded function. We conclude that f is a bounded distribution. We now proceed to show that f is an element of H^p. We fix a smooth radial nonnegative compactly supported function θ such that

$$\theta(x) = \begin{cases} 1 & \text{if} \quad |x| < 1, \\ 0 & \text{if} \quad |x| > 2. \end{cases}$$

We observe that the identity

$$\begin{aligned}
P(x) &= P(x)\theta(x) + \sum_{k=1}^{\infty}\big(\theta(2^{-k}x)P(x) - \theta(2^{-(k-1)}x)P(x)\big) \\
&= P(x)\theta(x) + \frac{\Gamma(\frac{n+1}{2})}{\pi^{\frac{n+1}{2}}}\sum_{k=1}^{\infty} 2^{-k}\left(\frac{\theta(\cdot)-\theta(2(\cdot))}{(2^{-2k}+|\cdot|^2)^{\frac{n+1}{2}}}\right)_{2^k}(x)
\end{aligned}$$

is valid for all $x \in \mathbf{R}^n$. We set

$$\Phi^{(k)}(x) = \big(\theta(x) - \theta(2x)\big)\frac{1}{(2^{-2k}+|x|^2)^{\frac{n+1}{2}}},$$

and we claim that for all bounded tempered distributions f and for all $t > 0$ we have

$$P_t * f = (\theta P)_t * f + \frac{\Gamma(\frac{n+1}{2})}{\pi^{\frac{n+1}{2}}} \sum_{k=1}^{\infty} 2^{-k} (\Phi^{(k)})_{2^k t} * f, \tag{2.1.38}$$

where the series converges in $\mathscr{S}'(\mathbf{R}^n)$; see Exercise 2.1.5.

Assuming (2.1.38), we claim that for some fixed constant $c_0 = c_0(n,N)$, the functions $c_0\,\theta P$ and $c_0\Phi^{(k)}$ lie in $\mathscr{F}_N$ uniformly in $k = 1,2,3,\dots$.

To verify this assertion for $|\alpha| \le N+1$, we apply Leibniz's rule to write

$$\begin{aligned} \left| \partial^\alpha \left[\frac{\theta(x) - \theta(2x)}{(2^{-2k} + |x|^2)^{\frac{n+1}{2}}} \right] \right| &= \left| \sum_{\beta \le \alpha} c_{\alpha,\beta} \partial_x^{\alpha-\beta} (\theta(x) - \theta(2x)) \partial_x^\beta \left(\frac{1}{(2^{-2k} + |x|^2)^{\frac{n+1}{2}}} \right) \right| \\ &\le \sum_{\beta \le \alpha} |c'_{\alpha,\beta}| \chi_{\frac12 \le |x| \le 2} \frac{K_\beta}{(2^{-2k} + |x|^2)^{\frac{n+1}{2} - |\beta|}}, \end{aligned}$$

where

$$K_\beta = \sup_{\substack{m,\gamma \\ m+|\gamma| = |\beta|}} \sup_{\substack{t,x \\ t^2+|x|^2}} \left| \frac{\partial^m}{\partial t^m} \frac{\partial^\gamma}{\partial x^\gamma} \frac{1}{(t^2 + |x|^2)^{\frac{n+1}{2}}} \right|,$$

and this estimate follows from the fact that the function $(t^2 + |x|^2)^{-\frac{n+1}{2}}$ is homogeneous of degree $-n-1$ on $\mathbf{R}^{n+1}$ and smooth on the sphere $\mathbf{S}^n$. These estimates are uniform in $k = 0,1,2,\dots$ and thus $\mathfrak{N}_N(\theta P) + \mathfrak{N}_N(\Phi^{(k)}) \le 1/c_0(n,N)$ for all some constant $c_0 = c_0(n,N)$ for all $k = 0,1,2,\dots$.

Then we obtain

$$\begin{aligned} \sup_{t>0} |P_t * f| &\le \sup_{t>0} |(\theta P)_t * f| + \frac{1}{c_0} \frac{\Gamma(\frac{n+1}{2})}{\pi^{\frac{n+1}{2}}} \sum_{k=1}^{\infty} 2^{-k} \sup_{t>0} \left| (c_0 \Phi^{(k)})_{2^k t} * f \right| \\ &\le C_5(n,N) \mathscr{M}_N(f), \end{aligned}$$

which proves the required conclusion (2.1.16).

We observe that the last estimate also yields the stronger estimate

$$M_1^*(f;P)(x) = \sup_{t>0} \sup_{\substack{y \in \mathbf{R}^n \\ |y-x| \le t}} |(P_t * f)(y)| \le C_5(n,N) \mathscr{M}_N(f)(x). \tag{2.1.39}$$

It follows that the quasi-norm $\|M_1^*(f;P)\|_{L^p(\mathbf{R}^n)}$ is also equivalent to $\|f\|_{H^p}$. □

Remark 2.1.6. To simplify the understanding of the equivalences just proved, a first-time reader may wish to define the H^p quasi-norm of a distribution f as

$$\|f\|_{H^p} = \left\| M_1^*(f;P) \right\|_{L^p}$$

and then study only the implications (a) $\Longrightarrow$ (c), (c) $\Longrightarrow$ (d), (d) $\Longrightarrow$ (e), and (e) $\Longrightarrow$ (a) in the proof of Theorem 2.1.4. In this way one avoids passing through the statement in part (b). For many applications, the identification of $\|f\|_{H^p}$ with $\|M_1^*(f;\Phi)\|_{L^p}$ for some Schwartz function Φ (with nonvanishing integral) suffices.

We also remark that the proof of Theorem 2.1.4 yields

$$\|f\|_{H^p(\mathbf{R}^n)} \approx \|\mathscr{M}_N(f)\|_{L^p(\mathbf{R}^n)},$$

where $N = [\frac{n}{p}] + 1$.

2.1.3 Consequences of the Characterizations of Hardy Spaces

In this subsection we look at a few consequences of Theorem 2.1.4. In many applications we need to be working with dense subspaces of H^p. It turns out that both $H^p \cap L^2$ and $H^p \cap L^1$ are dense in H^p.

Proposition 2.1.7. *Let $0 < p \le 1$ and let r satisfy $p \le r \le \infty$. Then $L^r \cap H^p$ is dense in H^p. Hence, $H^p \cap L^2$ and $H^p \cap L^1$ are dense in H^p.*

Proof. Let f be a distribution in $H^p(\mathbf{R}^n)$. Recall the Poisson kernel $P(x)$ and set $N = [\frac{n}{p}] + 1$. For any fixed $x \in \mathbf{R}^n$ and $t > 0$ we have

$$|(P_t * f)(x)| \le M_1^*(f;P)(y) \le C\mathscr{M}_N(f)(y) \tag{2.1.40}$$

for any $|y - x| \le t$. Indeed, the first estimate in (2.1.40) follows from the definition of $M_1^*(f;P)$, and the second estimate by (2.1.39). Raising (2.1.40) to the power p and averaging over the ball $B(x,t)$, we obtain

$$|(P_t * f)(x)|^p \le \frac{C^p}{v_n t^n} \int_{B(x,t)} \mathscr{M}_N(f)(y)^p \, dy \le \frac{C_1^p}{t^n} \|f\|_{H^p}^p. \tag{2.1.41}$$

It follows that the function $P_t * f$ is in $L^\infty(\mathbf{R}^n)$ with norm at most a constant multiple of $t^{-n/p}\|f\|_{H^p}$. Moreover, this function is also in $L^p(\mathbf{R}^n)$, since it is controlled by $M(f;P)$. Therefore, the functions $P_t * f$ lie in $L^r(\mathbf{R}^n)$ for all r with $p \le r \le \infty$. It remains to show that $P_t * f$ also lie in H^p and that $P_t * f \to f$ in H^p as $t \to 0$.

To see that $P_t * f$ lies in H^p, we use the semigroup formula $P_t * P_s = P_{t+s}$ for the Poisson kernel, which is a consequence of the fact that $\widehat{P_t}(\xi) = e^{-2\pi t|\xi|}$ by applying the Fourier transform. Therefore, for any $t > 0$ we have

$$\sup_{s>0} |P_s * P_t * f| = \sup_{s>0} |P_{s+t} * f| \le \sup_{s>0} |P_s * f|,$$

which implies that

$$\|P_t * f\|_{H^p} \le \|f\|_{H^p}$$

for all $t > 0$. We now need to show that $P_t * f \to f$ in H^p as $t \to 0$. This will be a consequence of the Lebesgue dominated convergence theorem once we know that

$$\sup_{s>0} |P_s * P_t * f - P_s * f| \le 2 \sup_{s>0} |P_s * f| \in L^p(\mathbf{R}^n) \tag{2.1.42}$$

and also that

$$\sup_{s>0} |(P_s * P_t * f - P_s * f)(x)| \to 0 \qquad \text{as} \quad t \to 0 \tag{2.1.43}$$

pointwise for all $x \in \mathbf{R}^n$. Statement (2.1.42) is a trivial consequence of the semigroup formula for the Poisson kernel.

The proof of (2.1.43) requires considerable more work. In proving (2.1.43), by a translation, we may assume that $x = 0$. Let us fix $\varepsilon > 0$. In view of (2.1.41), we have

$$\sup_{s \geq M} |(P_s * P_t * f - P_s * f)(0)| \leq C' M^{-n/p}$$

and we pick M such that $C' M^{-n/p} < \varepsilon$. It will suffice to show that

$$\sup_{0<s<M} |P_t * P_s * f(0) - P_s * f(0)| < 3\varepsilon \tag{2.1.44}$$

for t sufficiently close to zero. Let η_0 be a Schwartz function whose Fourier transform $\widehat{\eta_0}$ is equal to 1 on the ball $B(0,1)$ and vanishes outside $B(0,2)$. We write $1 = \widehat{\eta_0} + \widehat{\eta_\infty}$. Then $\eta_\infty = \delta_0 - \eta_0$,

$$P_s * f = P_s * \eta_0 * f + P_s * \eta_\infty * f,$$

and we will show that

$$\sup_{0<s<M} |P_t * P_s * \eta_0 * f(0) - P_s * \eta_0 * f(0)| < 2\varepsilon \tag{2.1.45}$$

and

$$\sup_{0<s<M} |P_t * P_s * \eta_\infty * f(0) - P_s * \eta_\infty * f(0)| < \varepsilon \tag{2.1.46}$$

for t sufficiently small. In order to prove (2.1.45), we write

$$\begin{aligned}
&P_t * P_s * \eta_0 * f(0) - P_s * \eta_0 * f(0) \\
&= \int_{\mathbf{R}^n} P_s(y)(P_t * \eta_0 * f(y) - \eta_0 * f(y))dy \\
&= \int_{\mathbf{R}^n} P_s(y)\left(\int_{\mathbf{R}^n} P_t(z)\big(\eta_0 * f(y-z) - \eta_0 * f(y)\big)\,dz\right)dy.
\end{aligned}$$

Note that $\eta_0 * f$ and $P_t * \eta_0 * f$ are in $L^\infty \cap \mathscr{C}^\infty$, since f is a bounded distribution. There is an $A > 0$ such that $\int_{|y| \geq A/M} P(y)\,dy < \varepsilon$ and so

$$\left| \int_{|y| \geq A} P_s(y)\big(P_t * \eta_0 * f(y) - \eta_0 * f(y)\big)dy \right| \leq \|\eta_0 * f\|_{L^\infty} \int_{|y| \geq A} P_s(y)dy < \varepsilon$$

for all $s \le M$. For $|y| \le A$, $\eta_0 * f$ is uniformly continuous in this region, so

$$\sup_{0<s<M} \int_{|y|\le A} P_s(y) \left\| P_t * \eta_0 * f - \eta_0 * f \right\|_{L^\infty} dy < \varepsilon \|P_s\|_{L^1} = \varepsilon$$

for t sufficiently small, since $\{P_t\}_{t>0}$ is an approximate identity; see Theorem 1.2.19 (2) in [156]. Therefore (2.1.45) holds.

Next, we write (2.1.46) as

$$\sup_{0<s<M} |\langle f, P_s * \eta_\infty * P_t - P_s * \eta_\infty \rangle|,$$

and since $f \in \mathscr{S}'(\mathbf{R}^n)$, this is controlled by a finite sum of expressions of the form:

$$\begin{aligned} &\sup_{0<s<M} \sup_{x\in\mathbf{R}^n} |x^\alpha \partial_x^\beta (P_s * \eta_\infty * P_t - P_s * \eta_\infty)(x)| \\ = &\sup_{0<s<M} \sup_{x\in\mathbf{R}^n} \left| x^\alpha \int_{\mathbf{R}^n} (\partial_x^\beta (P_s * \eta_\infty * P_t - P_s * \eta_\infty))^\frown(\xi) e^{2\pi i \xi\cdot x} d\xi \right| \\ =&(2\pi)^{|\beta|} \sup_{0<s<M} \sup_{x\in\mathbf{R}^n} \left| x^\alpha \int_{\mathbf{R}^n} \xi^\beta \widehat{\eta_\infty}(\xi)(e^{-2\pi t|\xi|} - 1) e^{-2\pi s|\xi|} e^{2\pi i \xi\cdot x} d\xi \right|. \qquad (2.1.47) \end{aligned}$$

For a fixed x, find a j such that $|x_j| = \sup_{1\le k\le n} |x_k|$. Set $N = |\alpha| + |\beta| + n + 2$. Integrate (2.1.47) by parts to rewrite it as

$$(2\pi)^{|\beta|} \sup_{0<s<M} \sup_{x\in\mathbf{R}^n} \left| \frac{x^\alpha}{(2\pi i x_j)^N} \int_{\mathbf{R}^n} \partial_j^N \left(\xi^\beta \widehat{\eta_\infty}(\xi)(e^{-2\pi t|\xi|} - 1) e^{-2\pi s|\xi|} \right) e^{2\pi i \xi\cdot x} d\xi \right|.$$

Note that the choice of N yields $\sup_{x\in\mathbf{R}^n} \frac{|x|^{|\alpha|}}{|x_j|^N} < \infty$. To compute the ∂_j^N derivative, we need the estimate for $0 \le m \le N$:

$$|\partial_j^m \xi^\beta| \le C|\xi|^{|\beta|-m}$$

and the following estimates for $1 \le m \le N$ (c.f. Exercise 1.1.6(b)):

$$\begin{aligned} |\partial_j^m \widehat{\eta_\infty}(\xi)| &\le C\chi_{[1,2]}(\xi) \\ |\partial_j^m e^{-s|\xi|}| &\le \frac{C}{|\xi|^m} \frac{s|\xi| + \cdots + (s|\xi|)^m}{e^{s|\xi|}} \\ |\partial_j^m (e^{-t|\xi|} - 1)| &\le \frac{C}{|\xi|^m} \frac{t|\xi| + \cdots + (t|\xi|)^m}{e^{t|\xi|}}. \end{aligned}$$

Let $N = a_1 + a_2 + a_3 + a_4$, where $a_1, a_2, a_3, a_4 \in \{0, 1, \ldots, N\}$. Then

$$\begin{aligned} &\partial_j^N \left(\xi^\beta \widehat{\eta_\infty}(\xi)(e^{-2\pi t|\xi|} - 1) e^{-2\pi s|\xi|} \right) \\ = &\sum_{a_1,a_2,a_3,a_4} c(a_1,a_2,a_3,a_4)(\partial_j^{a_1} \xi^\beta)(\partial_j^{a_2} \widehat{\eta_\infty})(\xi)(\partial_j^{a_3}(e^{-2\pi t|\xi|} - 1))(\partial_j^{a_4} e^{-2\pi s|\xi|}), \end{aligned}$$

for some suitable constants $c(a_1,a_2,a_3,a_4)$, in view of Leibniz's rule. We claim that for all $a_1,a_2,a_3,a_4 \in \{0,1,\dots,N\}$ we have

$$\int_{\mathbf{R}^n} \big|(\partial_j^{a_1}\xi^\beta)(\partial_j^{a_2}\widehat{\eta_\infty}(\xi))(\partial_j^{a_3}(e^{-2\pi t|\xi|}-1))(\partial_j^{a_4}e^{-2\pi s|\xi|})\big|\,d\xi \le C't \tag{2.1.48}$$

for all $0<s<M$. Obviously $C't$ multiplied by $(2\pi)^{|\beta|}\sup_{x\in\mathbf{R}^n}\frac{|x|^{|\alpha|}}{|x_j|^N}<\infty$ can be made smaller than the given ε if t is sufficiently close to zero.

Let us now prove (2.1.48). If $a_2>0$, then the integral is over the annulus $1\le|\xi|\le 2$ and we can easily derive (2.1.48), since for ξ in this range we have $|\partial_j^{a_3}(e^{-t|\xi|}-1)|\le C''t$. If $a_2=0$, $a_3>0$, $a_4>0$ then the integral is over the region $|\xi|\ge 1$, and using the preceding estimates we write

$$\begin{aligned}
&\int_{|\xi|\ge1} |(\partial_j^{a_1}\xi^\beta)\widehat{\eta_\infty}(\xi)(\partial_j^{a_3}(e^{-2\pi t|\xi|}-1))(\partial_j^{a_4}e^{-2\pi s|\xi|})|\,d\xi\\
=&\,t\int_{|\xi|\ge1} C|\xi|^{|\beta|-a_1-a_3-a_4+1}\frac{1+(t|\xi|)^2+\cdots+(t|\xi|)^{a_3-1}}{e^{t|\xi|}}\frac{s|\xi|+\cdots+(s|\xi|)^{a_4}}{e^{s|\xi|}}\,d\xi\\
\le&\, Ct\int_{|\xi|\ge1}|\xi|^{|\beta|-N+1}C_1C_2\,d\xi\\
\le&\, C't,
\end{aligned}$$

by the choice of N. In the case $a_2=a_3=0$, $a_4>0$ we use the inequality $|e^t-1|\le Ct$ and argue in a similar fashion to prove (2.1.48). The same argument is valid in the last case $a_2=a_3=a_4=0$. □

Next we observe the following consequence of Theorem 2.1.4.

Corollary 2.1.8. *For any two Schwartz functions Φ and Θ with nonvanishing integral we have*

$$\big\|\sup_{t>0}|\Theta_t*f|\big\|_{L^p}\approx\big\|\sup_{t>0}|\Phi_t*f|\big\|_{L^p}\approx\big\|f\big\|_{H^p}$$

for all $f\in\mathscr{S}'(\mathbf{R}^n)$, with constants depending only on n, p, Φ, and Θ.

Proof. See the discussion after Theorem 2.1.4. □

Next we define a *norm* on Schwartz functions relevant in the theory of Hardy spaces:

$$\mathfrak{N}_N(\varphi;x_0,R)=\int_{\mathbf{R}^n}\Big(1+\Big|\frac{x-x_0}{R}\Big|\Big)^N\sum_{|\alpha|\le N+1}R^{|\alpha|}|\partial^\alpha\varphi(x)|\,dx.$$

Note that $\mathfrak{N}_N(\varphi;0,1)=\mathfrak{N}_N(\varphi)$.

Corollary 2.1.9. *(a) For any $0<p\le1$, every $f\in H^p(\mathbf{R}^n)$, and any $\varphi\in\mathscr{S}(\mathbf{R}^n)$, we have*

$$|\langle f,\varphi\rangle|\le\mathfrak{N}_N(\varphi)\inf_{|z|\le1}\mathscr{M}_N(f)(z), \tag{2.1.49}$$

where $N = [\frac{n}{p}] + 1$, *and consequently there is a constant* $C_{n,p}$ *such that*

$$\big|\langle f,\varphi\rangle\big| \le \mathfrak{N}_N(\varphi)\, C_{n,p}\, \big\|f\big\|_{H^p}. \tag{2.1.50}$$

(b) Let $0 < p \le 1$, $N = [n/p] + 1$, *and* $p \le r \le \infty$. *Then there is a constant* $C(p,n,r)$ *such that for any* $f \in H^p$ *and* $\varphi \in \mathscr{S}(\mathbf{R}^n)$ *we have*

$$\big\|\varphi * f\big\|_{L^r} \le C(p,n,r)\mathfrak{N}_N(\varphi)\big\|f\big\|_{H^p}. \tag{2.1.51}$$

(c) For any $x_0 \in \mathbf{R}^n$, *for all* $R > 0$, *and any* $\psi \in \mathscr{S}(\mathbf{R}^n)$, *we have*

$$\big|\langle f,\psi\rangle\big| \le \mathfrak{N}_N(\psi; x_0, R) \inf_{|z-x_0|\le R} \mathscr{M}_N(f)(z). \tag{2.1.52}$$

Proof. (a) We use that $\langle f,\varphi\rangle = (\widetilde{\varphi} * f)(0)$, where $\widetilde{\varphi}(x) = \varphi(-x)$ and we observe that $\mathfrak{N}_N(\varphi) = \mathfrak{N}_N(\widetilde{\varphi})$. Then (2.1.49) follows from the inequality

$$|(\widetilde{\varphi} * f)(0)| \le \mathfrak{N}_N(\varphi) M_1^*\Big(f; \frac{\widetilde{\varphi}}{\mathfrak{N}_N(\varphi)}\Big)(z) \le \mathfrak{N}_N(\varphi)\mathscr{M}_N(f)(z)$$

for all $|z| < 1$, which is valid, since $\widetilde{\varphi}/\mathfrak{N}_N(\varphi)$ lies in $\mathscr{F}_N$. We deduce (2.1.50) as follows:

$$\begin{aligned}\big|\langle f,\varphi\rangle\big|^p &\le \mathfrak{N}_N(\varphi)^p \inf_{|z|\le 1} \mathscr{M}_N(f)(z)^p\\ &\le \mathfrak{N}_N(\varphi)^p \frac{1}{|B(0,1)|}\int_{|z|\le 1} \mathscr{M}_N(f)^p\,dz\\ &\le \mathfrak{N}_N(\varphi)^p C_{n,p}^p \big\|f\big\|_{H^p}^p.\end{aligned}$$

(b) For any fixed $x \in \mathbf{R}^n$ and $t > 0$ we have

$$|(\varphi_t * f)(x)| \le \mathfrak{N}_N(\varphi) M_1^*\Big(f; \frac{\varphi}{\mathfrak{N}_N(\varphi)}\Big)(y) \le \mathfrak{N}_N(\varphi)\mathscr{M}_N(f)(y) \tag{2.1.53}$$

for all y satisfying $|y - x| \le t$. Restricting to $t = 1$ yields

$$|(\varphi * f)(x)|^p \le \frac{\mathfrak{N}_N(\varphi)^p}{|B(x,1)|}\int_{B(x,1)} \mathscr{M}_N(f)^p(y)\,dy \le \mathfrak{N}_N(\varphi)^p C_{p,n}^p \big\|f\big\|_{H^p}^p.$$

This implies that $\|\varphi * f\|_{L^\infty} \le C_{p,n}\mathfrak{N}_N(\varphi)\|f\|_{H^p}$. Choosing $y = x$ and $t = 1$ in (2.1.53) and then taking L^p quasi-norms yields a similar estimate for $\|\varphi * f\|_{L^p}$. By interpolation we deduce $\|\varphi * f\|_{L^r} \le C(p,n,r)\mathfrak{N}_N(\varphi)\|f\|_{H^p}$, when $r \le p \le \infty$.

(c) To prove (2.1.52), given a Schwartz function ψ and $R > 0$, define $\varphi(y) = \psi(-Ry + x_0)$ so that $\psi(x) = \varphi(\frac{x_0 - x}{R}) = R^n\varphi_R(x_0 - x)$. In view of (2.1.53) we have

$$\big|\langle f,\psi\rangle\big| = R^n\big|(\varphi_R * f)(x_0)\big| \le R^n\,\mathfrak{N}_N(\varphi) \inf_{|z-x_0|\le R} \mathscr{M}_N(f)(z).$$

But a simple change of variables shows that $R^n \mathfrak{N}(\varphi) = \mathfrak{N}(\psi; x_0, R)$ and this combined with the preceding inequality yields (2.1.52).

□

Proposition 2.1.10. *Let $0 < p \le 1$. Then the following statements are valid:*
(a) Convergence in H^p implies convergence in $\mathscr{S}'$.
(b) If $f_k \in H^p$ satisfy $\sup_{k\in\mathbf{Z}^+} \|f_k\|_{H^p} \le C < \infty$ and $f_k \to f$ in $\mathscr{S}'(\mathbf{R}^n)$ as $k \to \infty$, then $f \in H^p$.
(c) H^p is a complete quasi-normed metrizable space.

Proof. (a) Let f_j, f in $H^p(\mathbf{R}^n)$ and suppose that $f_j \to f$ in $H^p(\mathbf{R}^n)$. Applying (2.1.50) we obtain that for any $\varphi \in \mathscr{S}(\mathbf{R}^n)$ we have $\langle f_j - f, \varphi\rangle \to 0$; hence $f_j \to f$ in $\mathscr{S}'(\mathbf{R}^n)$.

(b) For any $\Phi \in \mathscr{S}(\mathbf{R}^n)$ with integral one and $t > 0$ we have $\Phi_t * f_k \to \Phi_t * f$ as $k \to \infty$, since $f_k \to f$ in $\mathscr{S}'(\mathbf{R}^n)$. Thus

$$|\Phi_t * f| = \liminf_{k\to\infty} |\Phi_t * f_k| \le \liminf_{k\to\infty} \sup_{t>0} |\Phi_t * f_k| .$$

Taking the supremum over t, we obtain $\sup_{t>0} |\Phi_t * f| \le \liminf_{k\to\infty} \sup_{t>0} |\Phi_t * f_k|$. Then we apply L^p quasi-norms and Fatou's lemma to deduce that $\|M(f;\Phi)\|_{L^p}$ is bounded by a multiple of C; thus, $f \in H^p$.

(c) Suppose $\{f_j\}_{j=1}^\infty$ is a Cauchy sequence in $H^p(\mathbf{R}^n)$. Then there is a constant C_0 such that $\sup_{j\ge 1} \|f_j\|_{H^p} \le C_0$. Using (2.1.50) (with $f_j - f_k$ in place of f) we obtain that for every φ in $\mathscr{S}'(\mathbf{R}^n)$ the sequence $\{\langle f_j, \varphi\rangle\}_{j=1}^\infty$ is Cauchy in $\mathbf{C}$ and thus it converges to a complex number $f(\varphi)$. We claim that the mapping $\varphi \mapsto f(\varphi)$ is a tempered distribution. We clearly have

$$|f(\varphi)| = \lim_{k\to\infty} \big|\langle f_k, \varphi\rangle\big| \le C_{n,p} \mathfrak{N}_N(\varphi) C_0 .$$

But an easy calculation shows that $\mathfrak{N}_N(\varphi)$ is controlled by the finite sum of seminorms $\rho_{\alpha,\beta}(\varphi)$ with $|\alpha|, |\beta| \le N + n + 1$. This yields that f lies in $\mathscr{S}'(\mathbf{R}^n)$, in particular f is a bounded distribution, and obviously $f_j \to f$ in $\mathscr{S}'(\mathbf{R}^n)$. Part (b) implies that f is an element of $H^p(\mathbf{R}^n)$.

Next we show that $f_k \to f$ in H^p. Given $\Phi \in \mathscr{S}(\mathbf{R}^n)$ with integral 1, we have for any $t > 0$ and any $k \ge 1$

$$|(f_k - f) * \Phi_t| = \liminf_{\ell\to\infty} |(f_k - f_\ell) * \Phi_t| \le \liminf_{\ell\to\infty} \sup_{t>0} |(f_k - f_\ell) * \Phi_t| .$$

Taking the supremum over $t > 0$ on the left and then the L^p quasi-norm and applying Fatou's lemma we deduce that

$$\big\|M(f_k - f;\Phi)\big\|_{L^p} \le \liminf_{\ell\to\infty} \big\|M(f_k - f_\ell;\Phi)\big\|_{L^p} .$$

Letting $k \to \infty$ we obtain that

$$\limsup_{k\to\infty} \big\|M(f_k - f;\Phi)\big\|_{L^p} \le \limsup_{k,\ell\to\infty} \big\|M(f_k - f_\ell;\Phi)\big\|_{L^p} = 0;$$

thus $\|f_k - f\|_{H^p} \to 0$ as $k \to \infty$. Therefore H^p is complete. Finally we observe that the map $(f,g) \mapsto \|f-g\|_{H^p}^p$ is a metric on H^p that generates the same topology as the quasi-norm $f \mapsto \|f\|_{H^p}$; hence H^p is metrizable. $\square$

2.1.4 Vector-Valued H^p and Its Characterizations

We now obtain a vector-valued analogue of Theorem 2.1.4 crucial in the characterization of Hardy spaces using Littlewood–Paley theory. To state this analogue we need to extend the definitions of the maximal operators to finite sequences of distributions. Let $a,b>0$ and let Φ be a Schwartz function on $\mathbf{R}^n$. In accordance with Definition 2.1.3, we give the following definitions.

Definition 2.1.11. Let $L \in \mathbf{Z}^+$. We denote by ℓ_L^2 the space of all complex-valued sequences $\vec{a} = (a_1,\dots,a_L)$ of length L with norm $\|\vec{a}\|_{\ell_L^2} = (|a_1|^2+\cdots+|a_L|^2)^{1/2}$. For a sequence $\vec{f} = \{f_j\}_{j=1}^L$ of tempered distributions on $\mathbf{R}^n$ we define the *smooth maximal function of $\vec{f}$ with respect to Φ* as

$$M(\vec{f};\Phi)(x) = \sup_{t>0} \big\|\{(\Phi_t * f_j)(x)\}_j\big\|_{\ell_L^2}.$$

We define the *nontangential maximal function (with aperture a) of f with respect to Φ* as

$$M_a^*(\vec{f};\Phi)(x) = \sup_{t>0} \sup_{\substack{y\in\mathbf{R}^n \\ |y-x|\le at}} \big\|\{(\Phi_t * f_j)(y)\}_j\big\|_{\ell_L^2}.$$

We also define the *auxiliary maximal function*

$$M_b^{**}(\vec{f};\Phi)(x) = \sup_{t>0} \sup_{y\in\mathbf{R}^n} \frac{\big\|\{(\Phi_t * f_j)(x-y)\}_j\big\|_{\ell_L^2}}{(1+t^{-1}|y|)^b}.$$

We note that if the function Φ is not assumed to be Schwartz but merely integrable, for example, if Φ is the Poisson kernel, the maximal functions $M(\vec{f};\Phi)$, $M_a^*(\vec{f};\Phi)$, and $M_b^{**}(\vec{f};\Phi)$ are well defined for sequences $\vec{f} = \{f_j\}_{j=1}^L$ whose terms are bounded tempered distributions on $\mathbf{R}^n$.

For a fixed positive integer N we define the *grand maximal function of $\vec{f}$ (with respect to N)* as

$$\mathscr{M}_N(\vec{f}) = \sup_{\varphi\in\mathscr{F}_N} M_1^*(\vec{f};\varphi), \tag{2.1.54}$$

where

$$\mathscr{F}_N = \Big\{\varphi \in \mathscr{S}(\mathbf{R}^n) : \mathfrak{N}_N(\varphi) \le 1\Big\}$$

is as defined in (2.1.11) and

$$\mathfrak{N}_N(\varphi) = \int_{\mathbf{R}^n} (1+|x|)^N \sum_{|\alpha|\le N+1} |\partial^\alpha \varphi(x)|\,dx\,.$$

We note that as in the scalar case, we have the sequence of simple inequalities

$$M(\vec{f};\Phi) \le M_a^*(\vec{f};\Phi) \le (1+a)^b M_b^{**}(\vec{f};\Phi)\,. \tag{2.1.55}$$

We now define the vector-valued Hardy space $H^p(\mathbf{R}^n,\ell_L^2)$.

Definition 2.1.12. Let $\vec{f} = \{f_j\}_{j=1}^L$ be a finite sequence of bounded tempered distributions on $\mathbf{R}^n$ and let $0<p<\infty$. We say that $\vec{f}$ lies in the vector-valued Hardy space $H^p(\mathbf{R}^n,\ell_L^2)$ if the *Poisson maximal function*

$$M(\vec{f};P)(x) = \sup_{t>0} \big\|\{(P_t * f_j)(x)\}_j\big\|_{\ell_L^2}$$

lies in $L^p(\mathbf{R}^n)$. If this is the case, we set

$$\big\|\vec{f}\,\big\|_{H^p(\mathbf{R}^n,\ell_L^2)} = \big\|M(\vec{f};P)\big\|_{L^p(\mathbf{R}^n)} = \bigg\|\sup_{\varepsilon>0}\Big(\sum_{j=1}^L |f_j * P_\varepsilon|^2\Big)^{\frac12}\bigg\|_{L^p(\mathbf{R}^n)}.$$

The next theorem provides a vector-valued analogue of Theorem 2.1.4.

Theorem 2.1.13. *Let $0<p<\infty$, $L\in\mathbf{Z}^+$. Then the following statements are valid:*
(a) There exists a Schwartz function Φ^o with $\int_{\mathbf{R}^n}\Phi^o(x)\,dx = 1$ and a constant C_1 ($C_1 = 500$ works) such that

$$\big\|M(\vec{f};\Phi^o)\big\|_{L^p(\mathbf{R}^n)} \le C_1 \big\|\vec{f}\,\big\|_{H^p(\mathbf{R}^n,\ell_L^2)} \tag{2.1.56}$$

for every sequence $\vec{f} = \{f_j\}_{j=1}^L$ of bounded tempered distributions.
(b) For every $a>0$ and Φ in $\mathscr{S}(\mathbf{R}^n)$ there exists a constant $C_2(n,p,a,\Phi)$ such that

$$\big\|M_a^*(\vec{f};\Phi)\big\|_{L^p(\mathbf{R}^n)} \le C_2(n,p,a,\Phi)\big\|M(\vec{f};\Phi)\big\|_{L^p(\mathbf{R}^n,\ell_L^2)} \tag{2.1.57}$$

for every sequence $\vec{f} = \{f_j\}_{j=1}^L$ of tempered distributions.
(c) For every $a>0$, $b>n/p$, and Φ in $\mathscr{S}(\mathbf{R}^n)$ there exists a constant $C_3(n,p,a,b,\Phi)$ such that

$$\big\|M_b^{**}(\vec{f};\Phi)\big\|_{L^p(\mathbf{R}^n)} \le C_3(n,p,a,b,\Phi)\big\|M_a^*(\vec{f};\Phi)\big\|_{L^p(\mathbf{R}^n,\ell_L^2)} \tag{2.1.58}$$

for every sequence $\vec{f} = \{f_j\}_{j=1}^L$ of tempered distributions.
(d) For every $b>0$ and Φ in $\mathscr{S}(\mathbf{R}^n)$ with $\int_{\mathbf{R}^n}\Phi(x)\,dx \ne 0$ there exists a constant $C_4(b,\Phi)$ such that if $N = [\frac{n}{p}]+1$ we have

$$\big\|\mathscr{M}_N(\vec{f})\big\|_{L^p(\mathbf{R}^n)} \le C_4(b,\Phi)\big\|M_b^{**}(\vec{f};\Phi)\big\|_{L^p(\mathbf{R}^n,\ell_L^2)} \tag{2.1.59}$$

for every sequence $\vec{f} = \{f_j\}_{j=1}^L$ of tempered distributions.

(e) For every positive integer N there exists a constant $C_5(n,N)$ such that for all $f_j \in \mathscr{S}'(\mathbf{R}^n)$, $j = 1,\dots,L$ with $\|\mathscr{M}_N(\vec{f})\|_{L^p(\mathbf{R}^n,\ell^2_L)} < \infty$ (where $\vec{f} = \{f_j\}_{j=1}^L$) we must have that f_j are bounded distributions and satisfy

$$\|\vec{f}\|_{H^p(\mathbf{R}^n)} \le C_5(n,N)\big\|\mathscr{M}_N(\vec{f})\big\|_{L^p(\mathbf{R}^n,\ell^2_L)}, \tag{2.1.60}$$

that is, $\vec{f}$ lies in the Hardy space $H^p(\mathbf{R}^n,\ell^2_L)$.

Proof. The proof of this theorem is obtained via a step-by-step repetition of the proof of Theorem 2.1.4 in which the scalar absolute values of complex numbers are replaced by ℓ^2_L norms. The verification of the details of this extension is omitted. The crucial observation in the adaptation of the proof of Theorem 2.1.4 is that the constants that appear in all inequalities do not depend on L. □

We end this subsection by observing the validity of the following vector-valued analogue of (2.1.52):

$$\Big(\sum_{j=1}^L |\langle f_j,\varphi\rangle|^2\Big)^{\frac12} \le \mathfrak{N}_N(\varphi;x_0,R) \inf_{|z-x_0|\le R} \mathscr{M}_N(\vec{f})(z). \tag{2.1.61}$$

The proof of (2.1.61) is identical to the corresponding estimate for scalar-valued functions. Set $\psi(x) = \varphi(-Rx+x_0)$. It follows directly from Definition 2.1.11 that for any fixed z with $|z-x_0| \le R$ we have

$$\begin{aligned}\Big(\sum_{j=1}^L |\langle f_j,\varphi\rangle|^2\Big)^{\frac12} &= R^n\big\|\{(f_j*\psi_R)(x_0)\}_j\big\|_{\ell^2_L}\\ &\le \sup_{y:\,|y-z|\le R} R^n\big\|\{(f_j*\psi_R)(y)\}_j\big\|_{\ell^2_L}\\ &\le R^n\,\mathfrak{N}_N(\psi)\,\mathscr{M}_N(\vec{f})(z),\end{aligned}$$

which, combined with the observation

$$R^n\,\mathfrak{N}_N(\psi) = \mathfrak{N}_N(\varphi;x_0,R),$$

yields (2.1.61) when we take the infimum over all z with $|z-x_0| \le R$.

2.1.5 Singular Integrals on vector-valued Hardy Spaces

To obtain the Littlewood–Paley characterization of Hardy spaces, we need a multiplier theorem for vector-valued Hardy spaces.

Fix $L \in \mathbf{Z}^+$. Suppose that $\{K_j(x)\}_{j=1}^L$ is a family of functions defined on $\mathbf{R}^n \setminus \{0\}$ with the following properties: There exist constants $A, B < \infty$ and an integer N such that for all multi-indices α with $|\alpha| \le N$ and $x \ne 0$ we have

$$\sum_{j=1}^{L} \big|\partial^\alpha K_j(x)\big| \le A\,|x|^{-n-|\alpha|} < \infty \tag{2.1.62}$$

and also

$$\sup_{\xi \in \mathbf{R}^n} \sum_{j=1}^{L} \big|\widehat{K_j}(\xi)\big| \le B < \infty. \tag{2.1.63}$$

Note that for $h \in L^1(\mathbf{R}^n)$, $K_j * h$ is a well-defined function in $L^{1,\infty}(\mathbf{R}^n)$.

An example of such a sequence of kernels is given by $K_j(x) = \Psi_{2^{-j}}(x)$, where Ψ is a fixed Schwartz function on $\mathbf{R}^n$ whose Fourier transform is supported in a compact annulus that does not contain the origin.

Theorem 2.1.14. *Suppose that a finite sequence of kernels $\{K_j\}_{j=1}^L$ satisfies (2.1.62) and (2.1.63) with $N = [\frac{n}{p}] + 1$, for some $0 < p \le 1$. Then there exists a constant $C_{n,p}$ that depends only on the dimension n and on p such that for all sequences of integrable functions $\{f_j\}_{j=1}^L$ we have the estimate*

$$\Big\| \sum_{j=1}^{L} K_j * f_j \Big\|_{H^p(\mathbf{R}^n)} \le C_{n,p}(A+B) \big\| \{f_j\}_j \big\|_{H^p(\mathbf{R}^n, \ell^2_L)} .$$

Moreover, the space $L^1(\mathbf{R}^n, \ell^2_L)$ is dense in $H^p(\mathbf{R}^n, \ell^2_L)$ and thus there is a unique bounded extension of the operator

$$\{f_j\}_{j=1}^L \mapsto \sum_{j=1}^{L} K_j * f_j \tag{2.1.64}$$

from $H^p(\mathbf{R}^n, \ell^2_L)$ to $H^p(\mathbf{R}^n)$.

Proof. We fix a smooth positive function Φ supported in the unit ball $B(0,1)$ with $\int_{\mathbf{R}^n} \Phi(x)\,dx = 1$ and we consider the maximal function

$$M\Big(\sum_{j=1}^{L} K_j * f_j ; \Phi\Big) = \sup_{\varepsilon > 0} \Big|\Phi_\varepsilon * \sum_{j=1}^{L} K_j * f_j\Big|,$$

defined for $f_j \in L^1(\mathbf{R}^n)$.We will show that this maximal function lies in $L^p(\mathbf{R}^n)$.

We now fix a $\lambda > 0$ and we set $N = [\frac{n}{p}] + 1$. We also fix $\gamma > 0$ to be chosen later and we define the set

$$\Omega_\lambda = \{x \in \mathbf{R}^n : \mathscr{M}_N(\vec{f})(x) > \gamma\lambda\}.$$

The set Ω_λ is open, and we may use the Whitney decomposition (Appendix J in [156]) to write it as a union of cubes Q_k such that

(a) $\bigcup_k Q_k = \Omega_\lambda$ and the Q_k's have disjoint interiors;

(b) $\sqrt{n}\,\ell(Q_k) \le \operatorname{dist}(Q_k, (\Omega_\lambda)^c) \le 4\sqrt{n}\,\ell(Q_k)$.

We denote by $c(Q_k)$ the center of the cube Q_k. For each k we set

$$d_k = \operatorname{dist}(Q_k, (\Omega_\lambda)^c) + 2\sqrt{n}\,\ell(Q_k) \approx \ell(Q_k),$$

so that

$$B(c(Q_k), d_k) \cap (\Omega_\lambda)^c \neq \emptyset .$$

We now introduce a partition of unity $\{\varphi_k\}_k$ adapted to the sequence of cubes $\{Q_k\}_k$ such that

(c) $\chi_{\Omega_\lambda} = \sum_k \varphi_k$ and each φ_k satisfies $0 \le \varphi_k \le 1$;

(d) each φ_k is supported in $\frac{6}{5}Q_k$ and satisfies $I_k = \int_{\mathbf{R}^n} \varphi_k \, dx \approx d_k^n$;

(e) $\|\partial^\alpha \varphi_k\|_{L^\infty} \le C_\alpha d_k^{-|\alpha|}$ for all multi-indices α and some constants C_α independent of k.

We fix a sequence of integrable functions f_j and we decompose each function as

$$f_j = g_j + \sum_k b_{j,k},$$

where g_j is the *good function* of the decomposition given by

$$g_j = f_j \chi_{\mathbf{R}^n \setminus \Omega_\lambda} + \sum_k \frac{\int_{\mathbf{R}^n} f_j \varphi_k \, dx}{I_k} \varphi_k$$

and $b_j = \sum_k b_{j,k}$ is the *bad function* of the decomposition given by

$$b_{j,k} = \left(f_j - \frac{\int_{\mathbf{R}^n} f_j \varphi_k \, dx}{I_k} \right) \varphi_k .$$

We note that each $b_{j,k}$ has integral zero. We define $\vec{g} = \{g_j\}_{j=1}^L$ and $\vec{b} = \{b_j\}_{j=1}^L$. At this point we appeal to (2.1.61) and to properties (d) and (e) to obtain

$$\left(\sum_{j=1}^L \left| \frac{\int_{\mathbf{R}^n} f_j \varphi_k \, dx}{I_k} \right|^2 \right)^{\frac{1}{2}} \le \frac{\mathfrak{N}_N(\varphi_k; c(Q_k), d_k)}{I_k} \inf_{|z - c(Q_k)| \le d_k} \mathscr{M}_N(\vec{f})(z) . \tag{2.1.65}$$

But since

$$\frac{\mathfrak{N}_N(\varphi_k; c(Q_k), d_k)}{I_k} \le \left[\int_{\frac{6}{5}Q_k} \left(1 + \frac{|x - c(Q_k)|}{d_k} \right)^N \sum_{|\alpha| \le N+1} \frac{d_k^{|\alpha|} C_\alpha d_k^{-|\alpha|}}{I_k} \, dx \right] \le C_{N,n},$$

it follows that (2.1.65) is at most a constant multiple of λ, since the ball $B(c(Q_k),d_k)$ meets the complement of Ω_λ. We conclude that

$$\big\|\vec{g}\big\|_{L^\infty(\Omega_\lambda,\ell^2_L)} \le C_{N,n}\gamma\lambda\,. \tag{2.1.66}$$

We now estimate $M(\sum_{j=1}^L K_j * b_{j,k};\Phi)$. For fixed k and $\varepsilon>0$ we have

$$\begin{aligned}
&\Big(\Phi_\varepsilon * \sum_{j=1}^L K_j * b_{j,k}\Big)(x)\\
&= \int_{\mathbf{R}^n}\Big(\Phi_\varepsilon * \sum_{j=1}^L K_j\Big)(x-y)\Big[f_j(y)\varphi_k(y) - \frac{\int_{\mathbf{R}^n} f_j\varphi_k\,dx}{I_k}\varphi_k(y)\Big]dy\\
&= \int_{\mathbf{R}^n}\sum_{j=1}^L\Big\{(\Phi_\varepsilon * K_j)(x-z) - \int_{\mathbf{R}^n}(\Phi_\varepsilon * K_j)(x-y)\frac{\varphi_k(y)}{I_k}dy\Big\}\varphi_k(z)f_j(z)\,dz\\
&= \int_{\mathbf{R}^n}\sum_{j=1}^L R_{j,k}(x,z)\varphi_k(z)f_j(z)\,dz,
\end{aligned}$$

where we set $R^\varepsilon_{j,k}(x,z)$ for the expression inside the curly brackets. Using (2.1.52), we obtain

$$\begin{aligned}
&\Big|\int_{\mathbf{R}^n}\sum_{j=1}^L R^\varepsilon_{j,k}(x,z)\varphi_k(z)f_j(z)\,dz\Big|\\
&\qquad\le \sum_{j=1}^L \mathfrak{N}_N(R^\varepsilon_{j,k}(x,\cdot\,)\varphi_k;c(Q_k),d_k)\inf_{|z-c(Q_k)|\le d_k}\mathscr{M}_N(f_j)(z)\\
&\qquad\le \sum_{j=1}^L \mathfrak{N}_N(R^\varepsilon_{j,k}(x,\cdot\,)\varphi_k;c(Q_k),d_k)\inf_{|z-c(Q_k)|\le d_k}\mathscr{M}_N(\vec{f})(z)\,.
\end{aligned} \tag{2.1.67}$$

Since $\varphi_k(z)$ is supported in $\frac{6}{5}Q_k$, the term $(1+\frac{|z-c(Q_k)|}{d_k})^N$ contributes only a constant factor in the integral defining $\mathfrak{N}_N(R^\varepsilon_{j,k}(x,\cdot\,)\varphi_k;c(Q_k),d_k)$, and we obtain

$$\begin{aligned}
&\mathfrak{N}_N(R^\varepsilon_{j,k}(x,\cdot\,)\varphi_k;c(Q_k),d_k)\\
&\qquad\le C_{N,n}\int_{\frac{6}{5}Q_k}\sum_{|\alpha|\le N+1} d_k^{|\alpha|}\Big|\frac{\partial^\alpha}{\partial z^\alpha}\big(R^\varepsilon_{j,k}(x,z)\varphi_k(z)\big)\Big|\,dz\,.
\end{aligned} \tag{2.1.68}$$

For notational convenience we set $K^\varepsilon_j = \Phi_\varepsilon * K_j$. We observe that the family $\{K^\varepsilon_j\}_j$ satisfies (2.1.62) and (2.1.63) with constants A' and B' that are only multiples of A and B, respectively, uniformly in ε; see Exercise 2.1.13. We now obtain a pointwise estimate for $\mathfrak{N}_N(R^\varepsilon_{j,k}(x,\cdot\,)\varphi_k;c(Q_k),d_k)$ when $x\in\mathbf{R}^n\setminus\Omega_\lambda$. For fixed $x\in\mathbf{R}^n\setminus\Omega_\lambda$ we have

$$R^\varepsilon_{j,k}(x,z)\varphi_k(z) = \int_{\mathbf{R}^n}\varphi_k(z)\Big\{K^\varepsilon_j(x-z) - K^\varepsilon_j(x-y)\Big\}\frac{\varphi_k(y)\,dy}{I_k},$$

from which it follows that

$$\left|\frac{\partial^\alpha}{\partial z^\alpha}R^\varepsilon_{j,k}(x,z)\varphi_k(z)\right| \le \int_{\mathbf{R}^n}\left|\frac{\partial^\alpha}{\partial z^\alpha}\left\{\varphi_k(z)\Big[K^\varepsilon_j(x-z)-K^\varepsilon_j(x-y)\Big]\right\}\right|\frac{\varphi_k(y)\,dy}{I_k}.$$

Using hypothesis (2.1.62), we can obtain the estimate

$$\sum_{j=1}^{L}\left|\frac{\partial^\alpha}{\partial z^\alpha}\left\{\varphi_k(z)\left\{K^\varepsilon_j(x-z)-K^\varepsilon_j(x-y)\right\}\right\}\right| \le C_{N,n}A\frac{d_k d_k^{-|\alpha|}}{|x-c(Q_k)|^{n+1}} \tag{2.1.69}$$

for all $|\alpha|\le N$, for all $y,z\in Q_k$ and all $x\in\mathbf{R}^n\setminus\Omega_\lambda$. Indeed, by Leibniz's rule, the left-hand side of (2.1.69) is controlled by

$$\begin{aligned}
&C'_\alpha\sum_{|\beta|\le|\alpha|} d_k^{-|\alpha|+|\beta|}\sum_{j=1}^{L}\left|\frac{\partial^\beta}{\partial z^\beta}\left\{K^\varepsilon_j(x-z)-K^\varepsilon_j(x-y)\right\}\right|\\
&\le C''_\alpha\left[\sum_{\substack{|\beta|\le|\alpha|\\ \beta\neq 0}}\frac{Ad_k^{-|\alpha|+|\beta|}}{|x-z|^{n+|\beta|}}+\frac{d_k^{-|\alpha|}A\,d_k}{|x-z|^{n+1}}\right]\\
&= C''_\alpha A\left[\sum_{\substack{|\beta|\le|\alpha|\\ \beta\neq 0}}\frac{d_k^{-|\alpha|+1}}{|x-z|^{n+1}}\left(\frac{d_k}{|x-z|}\right)^{|\beta|-1}+\frac{d_k^{-|\alpha|}\,d_k}{|x-z|^{n+1}}\right]\\
&\le C_{N,n}\frac{d_k^{-|\alpha|}Ad_k}{|x-c(Q_k)|^{n+1}}
\end{aligned}$$

since $|x-z|\ge cd_k$ and $|x-z|\approx|x-c(Q_k)|$. This proves (2.1.69).

It follows from (2.1.69) that

$$d_k^{|\alpha|}\sum_{j=1}^{L}\left|\frac{\partial^\alpha}{\partial z^\alpha}\{R^\varepsilon_{j,k}(x,z)\varphi_k(z)\}\right| \le C_{N,n}A\frac{d_k}{|x-c(Q_k)|^{n+1}}.$$

Inserting this estimate in (2.1.68) and summing over all j yields for $x\in\mathbf{R}^n\setminus\Omega_\lambda$

$$\sum_{j=1}^{L}\mathfrak{N}_N(R^\varepsilon_{j,k}(x,\cdot)\varphi_k;c(Q_k),d_k) \le C_{N,n}A\frac{d_k^{n+1}}{|x-c(Q_k)|^{n+1}}. \tag{2.1.70}$$

Combining (2.1.70) with (2.1.67) gives for $x\in\mathbf{R}^n\setminus\Omega_\lambda$,

$$\sum_{j=1}^{L}\left|\int_{\mathbf{R}^n}R^\varepsilon_{j,k}(x,z)\varphi_k(z)f_j(z)\,dz\right| \le \frac{C_{N,n}A\,d_k^{n+1}}{|x-c(Q_k)|^{n+1}}\inf_{|z-c(Q_k)|\le d_k}\mathscr{M}_N(\vec{f})(z).$$

This provides the estimate

$$\sup_{\varepsilon>0}\Big|\sum_{j=1}^{L}(K_j^{\varepsilon}*b_{j,k})(x)\Big| \le \frac{C_{N,n}A\,d_k^{n+1}}{|x-c(Q_k)|^{n+1}}\gamma\lambda$$

for all $x\in\mathbf{R}^n\setminus\Omega_\lambda$, since the ball $B(c(Q_k),d_k)$ intersects $(\Omega_\lambda)^c$. Summing over k and using the sublinearity of $M(\cdot;\Phi)$ results in

$$M\Big(\sum_{j=1}^{L}K_j*b_j;\Phi\Big)(x)\le\sum_k\frac{C_{N,n}A\,\gamma\lambda\,d_k^{n+1}}{|x-c(Q_k)|^{n+1}}\le\sum_k\frac{C'_{N,n}A\,\gamma\lambda\,d_k^{n+1}}{(d_k+|x-c(Q_k)|)^{n+1}}$$

for all $x\in(\Omega_\lambda)^c$. It is a simple fact that the *Marcinkiewicz function* below satisfies

$$\int_{\mathbf{R}^n}\sum_k\frac{d_k^{n+1}}{(d_k+|x-c(Q_k)|)^{n+1}}\,dx\le C_n\sum_k|Q_k|=C_n\,|\Omega_\lambda|;$$

see Exercise 5.6.6 in [156]. We have therefore shown that

$$\begin{aligned}\tfrac{\lambda}{2}\big|(\Omega_\lambda)^c\cap\{M(\vec K*\vec b;\Phi)>\tfrac{\lambda}{2}\}\big| &\le\int_{(\Omega_\lambda)^c}M(\vec K*\vec b;\Phi)(x)\,dx\\ &\le C_{N,n}A\gamma\lambda\,|\Omega_\lambda|,\end{aligned}\tag{2.1.71}$$

where we used the notation $\vec K*\vec b=\sum_{j=1}^{L}K_j*b_j$. Also define $\vec K*\vec g=\sum_{j=1}^{L}K_j*g_j$.

We now combine the information we have acquired so far. First we have

$$\big|\{M(\vec K*\vec f;\Phi)>\lambda\}\big|\le\big|\{M(\vec K*\vec g;\Phi)>\tfrac{\lambda}{2}\}\big|+\big|\{M(\vec K*\vec b;\Phi)>\tfrac{\lambda}{2}\}\big|.$$

For the good function $\vec g$ we have the estimate

$$\begin{aligned}\big|\{M(\vec K*\vec g;\Phi)>\tfrac{\lambda}{2}\}\big| &\le\frac{4}{\lambda^2}\int_{\mathbf{R}^n}M(\vec K*\vec g;\Phi)(x)^2\,dx\\ &\le\frac{4}{\lambda^2}\int_{\mathbf{R}^n}M(\vec K*\vec g)(x)^2\,dx\\ &\le\frac{C_n}{\lambda^2}\int_{\mathbf{R}^n}|(\vec K*\vec g)(x)|^2\,dx\\ &=\frac{C_n}{\lambda^2}\int_{\mathbf{R}^n}\Big|\sum_{j=1}^{L}\widehat{K_j}(\xi)\widehat{g_j}(\xi)\Big|^2\,d\xi\\ &\le\frac{C_n}{\lambda^2}\int_{\mathbf{R}^n}\Big(\sum_{j=1}^{L}|\widehat{K_j}(\xi)|\Big)^2\Big(\sum_{j=1}^{L}|\widehat{g_j}(\xi)|^2\Big)\,d\xi\\ &\le\frac{C_nB^2}{\lambda^2}\int_{\mathbf{R}^n}\sum_{j=1}^{L}|g_j(x)|^2\,dx\end{aligned}$$

$$\leq \frac{C_n B^2}{\lambda^2}\int_{\Omega_\lambda}\sum_{j=1}^{L}|g_j(x)|^2\,dx + \frac{C_n B^2}{\lambda^2}\int_{(\Omega_\lambda)^c}\sum_{j=1}^{L}|f_j(x)|^2\,dx$$

$$\leq B^2 C_{N,n}\gamma^2|\Omega_\lambda| + \frac{C_n B^2}{\lambda^2}\int_{(\Omega_\lambda)^c}\mathscr{M}_N(\vec{f})(x)^2\,dx,$$

where we used Corollary 2.1.12 in [156], the L^2 boundedness of the Hardy–Littlewood maximal operator, hypothesis (2.1.63), the fact that $f_j = g_j$ on $(\Omega_\lambda)^c$, estimate (2.1.66), and the fact that $\|\vec{f}\|_{\ell^2_L} \leq \mathscr{M}_N(\vec{f})$ in the preceding sequence of estimates.

On the other hand, estimate (2.1.71) gives

$$\big|\{M(\vec{K}*\vec{b};\Phi) > \tfrac{\lambda}{2}\}\big| \leq |\Omega_\lambda| + 2C_{N,n}A\gamma|\Omega_\lambda|,$$

which, combined with the previously obtained estimate for $\vec{g}$, gives

$$\big|\{M(\vec{K}*\vec{f};\Phi) > \lambda\}\big| \leq 2C_{N,n}(1+A\gamma+B^2\gamma^2)|\Omega_\lambda| + \frac{C_n B^2}{\lambda^2}\int_{(\Omega_\lambda)^c}\mathscr{M}_N(\vec{f})(x)^2\,dx.$$

Multiplying this estimate by $p\lambda^{p-1}$, recalling that $\Omega_\lambda = \{\mathscr{M}_N(\vec{f}) > \gamma\lambda\}$, and integrating in λ from 0 to ∞, we can easily obtain

$$\big\|M(\vec{K}*\vec{f};\Phi)\big\|^p_{L^p(\mathbf{R}^n)} \leq 2C_{N,n}(1+A\gamma+B^2\gamma^2)\gamma^{-p}\big\|\mathscr{M}_N(\vec{f})\big\|^p_{L^p(\mathbf{R}^n,\ell^2_L)}. \tag{2.1.72}$$

Choosing $\gamma = (A+B)^{-1}$ and recalling that $N = [\frac{n}{p}]+1$ gives the required conclusion for some constant $C_{n,p}$ that depends only on n and p.

Finally, we discuss the extension of the operator (2.1.64) to the entire $H^p(\mathbf{R}^n,\ell^2_L)$. In view of Proposition 2.1.7, $L^1(\mathbf{R}^n)\cap H^p(\mathbf{R}^n)$ is dense in $H^p(\mathbf{R}^n)$. It follows that $L^1(\mathbf{R}^n,\ell^2_L)\cap H^p(\mathbf{R}^n,\ell^2_L)$ is dense in $H^p(\mathbf{R}^n,\ell^2_L)$. Indeed, given $\vec{f} = (f_1,\dots,f_L)$ in $H^p(\mathbf{R}^n,\ell^2_L)$, find sequences $h_j^{(k)}$ in $L^1(\mathbf{R}^n)$ such that $h_j^{(k)} \to f_j$ in $H^p(\mathbf{R}^n)$ as $k\to\infty$. Set $\vec{h}^{(k)} = (h_1^{(k)},\dots,h_L^{(k)})$. Then for any $\Phi\in\mathscr{S}(\mathbf{R}^n)$ with integral one we have

$$M(\vec{f}-\vec{h}^{(k)};\Phi) \leq M(f_1-\vec{h}_1^{(k)};\Phi)+\cdots+M(f_L-\vec{h}_L^{(k)};\Phi).$$

Apply the L^p quasi-norm on both sides of the preceding expression and then let $k\to\infty$ to obtain the density of $L^1(\mathbf{R}^n,\ell^2_L)\cap H^p(\mathbf{R}^n,\ell^2_L)$ in $H^p(\mathbf{R}^n,\ell^2_L)$. In view of this, the operator in (2.1.64) admits a unique bounded extension from $H^p(\mathbf{R}^n,\ell^2_L)$ to $H^p(\mathbf{R}^n)$. □

Exercises

2.1.1. Prove that if v is a bounded tempered distribution and h_1, h_2 are in $\mathscr{S}(\mathbf{R}^n)$, then

$$(h_1 * h_2) * v = h_1 * (h_2 * v).$$

2.1.2. (a) Show that the H^1 norm remains invariant under the L^1 dilation $f_t(x) = t^{-n} f(t^{-1}x)$.
(b) Show that the H^p norm remains invariant under the L^p dilation $t^{n-n/p} f_t(x)$ interpreted in the sense of distributions.

2.1.3. Show that the continuous function $\psi(s) = \frac{e}{\pi}\frac{1}{s} e^{-\frac{\sqrt{2}}{2}(s-1)^{\frac{1}{4}}} \sin\big(\frac{\sqrt{2}}{2}(s-1)^{\frac{1}{4}}\big)$ defined on $[1,\infty)$ satisfies

$$\int_1^\infty s^k \psi(s)\,ds = \begin{cases} 1 & \text{if } k=0, \\ 0 & \text{if } k=1,2,3,\dots. \end{cases}$$

[*Hint:* Consider the analytic function $F(z) = \frac{4e}{\pi} z^3 (z^4+1)^{k-1} e^{-\frac{\sqrt{2}}{2}z} e^{i\frac{\sqrt{2}}{2}z}$ integrated over the boundary of the domain formed by the disc of radius R intersected with the quadrant $\operatorname{Re} z \geq 0, \operatorname{Im} z \geq 0$. Apply Cauchy's residue theorem to F over this contour, noting that a pole appears only when $k=0$. In this case a simple pole at the point $\frac{\sqrt{2}}{2} + i\frac{\sqrt{2}}{2}$ produces a nonzero residue.]

2.1.4. Let P_t be the Poisson kernel. Show that for any bounded tempered distribution f we have

$$P_t * f \to f \qquad \text{in } \mathscr{S}'(\mathbf{R}^n) \text{ as } t \to 0.$$

[*Hint:* Fix a smooth function ϕ whose Fourier transform is equal to 1 in a neighborhood of zero. Show that $P_t * (\phi * f) \to \phi * f$ in $\mathscr{S}'(\mathbf{R}^n)$ and that $\widehat{P_t}(1-\widehat{\phi})\widehat{f} \to (1-\widehat{\phi})\widehat{f}$ in $\mathscr{S}'(\mathbf{R}^n)$ as $t \to 0$.]

2.1.5. Fix a smooth radial nonnegative compactly supported function θ on $\mathbf{R}^n$ such that $\theta = 1$ on the unit ball and vanishing outside the ball of radius 2. Set $\Phi^{(k)}(x) = \big(\theta(x) - \theta(2x)\big)(2^{-2k} + |x|^2)^{-\frac{n+1}{2}}$ for $k \geq 1$. Prove that for all bounded tempered distributions f and for all $t > 0$ we have

$$P_t * f = (\theta P)_t * f + \frac{\Gamma(\frac{n+1}{2})}{\pi^{\frac{n+1}{2}}} \sum_{k=1}^\infty 2^{-k} (\Phi^{(k)})_{2^k t} * f,$$

where the series converges in $\mathscr{S}'(\mathbf{R}^n)$. Here $P(x) = \Gamma(\frac{n+1}{2})/\pi^{\frac{n+1}{2}}(1+|x|^2)^{\frac{n+1}{2}}$ is the Poisson kernel.
[*Hint:* Fix a function $\phi \in \mathscr{S}(\mathbf{R}^n)$ whose Fourier transform is equal to 1 in a neighborhood of zero and prove the required conclusion for $\phi * f$ and for $(\delta_0 - \phi) * f$. In the first case use the Lebesgue dominated convergence theorem and in the second case the Fourier transform.]

2.1.6. Let $0 < p < \infty$ be fixed. Show that a bounded tempered distribution f lies in H^p if and only if the nontangential Poisson maximal function

$$M_1^*(f;P)(x) = \sup_{t>0} \sup_{\substack{y \in \mathbf{R}^n \\ |y-x| \leq t}} |(P_t * f)(y)|$$

lies in L^p, and in this case we have $\|f\|_{H^p} \approx \|M_1^*(f;P)\|_{L^p}$.

[*Hint:* Observe that $M(f;P)$ can be replaced with $M_1^*(f;P)$ in the proof of part (e) of Theorem 2.1.4.]

2.1.7. (a) Let $1 < q \le \infty$ and let g in $L^q(\mathbf{R}^n)$ be a compactly supported function with integral zero. Show that g lies in the Hardy space $H^1(\mathbf{R}^n)$.
(b) Prove the same conclusion when L^q is replaced by $L\log^+ L$.
[*Hint:* Part (a): Pick a $\mathscr{C}_0^\infty$ function Φ supported in the unit ball with nonvanishing integral and suppose that the support of g is contained in the ball $B(0,R)$. For $|x| \le 2R$ we have that $M(f;\Phi)(x) \le C_\Phi M(g)(x)$, and since $M(g)$ lies in L^q, it also lies in $L^1(B(0,2R))$. For $|x| > 2R$, write $(\Phi_t * g)(x) = \int_{\mathbf{R}^n} \big(\Phi_t(x-y) - \Phi_t(x)\big) g(y)\, dy$ and use the mean value theorem to estimate this expression by $t^{-n-1}\|\nabla\Phi\|_{L^\infty}\|g\|_{L^1} \le |x|^{-n-1} C_\Phi \|g\|_{L^q}$, since $t \ge |x-y| \ge |x| - |y| \ge |x|/2$ whenever $|x| \ge 2R$ and $|y| \le R$. Thus $M(f;\Phi)$ lies in $L^1(\mathbf{R}^n)$. Part (b): You may use Exercise 2.1.4(b) in [156] to deduce that $M(g)$ is integrable over $B(0,2R)$.]

2.1.8. Show that for every integrable function g with mean value zero and support inside a ball B, we have $M(g;\Phi) \in L^p((3B)^c)$ for $p > n/(n+1)$. Here Φ is in $\mathscr{S}$.

2.1.9. Show that the space of all Schwartz functions whose Fourier transform is supported away from a neighborhood of the origin is dense in H^p.
[*Hint:* Use the square function characterization of H^p.]

2.1.10. (a) Suppose that $f \in H^p(\mathbf{R}^n)$ for some $0 < p \le 1$ and Φ in $\mathscr{S}(\mathbf{R}^n)$. Then show that for all $t > 0$ the function $\Phi_t * f$ belongs to $L^r(\mathbf{R}^n)$ for all $p \le r \le \infty$. Find an estimate for the L^r norm of $\Phi_t * f$ in terms of $\|f\|_{H^p}$ and $t > 0$.
(b) Let $0 < p \le 1$. Show that for all f in $H^p(\mathbf{R}^n)$, $\widehat{f}$ is a continuous function and prove that there exists a constant $C_{n,p}$ such that for all $\xi \ne 0$

$$|\widehat{f}(\xi)| \le C_{n,p} |\xi|^{\frac{n}{p}-n} \|f\|_{H^p}.$$

[*Hint:* Part (a): Use Proposition 2.1.7. Part (b): Use part (a) with $r = 1$.]

2.1.11. Show that $H^p(\mathbf{R}^n, \ell^2) = L^p(\mathbf{R}^n, \ell^2)$ whenever $1 < p < \infty$ and that $H^1(\mathbf{R}^n, \ell^2)$ is contained in $L^1(\mathbf{R}^n, \ell^2)$.
[*Hint:* Prove these assertions for ℓ^2_L first for some $L \in \mathbf{Z}^+$.]

2.1.12. For a sequence of tempered distributions $\vec{f} = \{f_j\}_j$, define the following variant of the grand maximal function:

$$\widetilde{\mathscr{M}}_N(\vec{f})(x) = \sup_{\{\varphi_j\}_j \in \widetilde{\mathscr{F}}_N} \sup_{\varepsilon>0} \sup_{\substack{y \in \mathbf{R}^n \\ |y-x|<\varepsilon}} \Big(\sum_j \big|((\varphi_j)_\varepsilon * f_j)(y)\big|^2\Big)^{\frac{1}{2}},$$

where $N \ge [\frac{n}{p}] + 1$ and

$$\widetilde{\mathscr{F}}_N = \left\{ \{\varphi_j\}_j \in \mathscr{S}(\mathbf{R}^n) : \Big(\sum_j \mathfrak{N}_N(\varphi_j)^2\Big)^{1/2} \le 1 \right\}.$$

Show that for all sequences of tempered distributions $\vec{f} = \{f_j\}_j$ we have

$$\big\|\widetilde{\mathcal{M}_N}(\vec{f})\big\|_{L^p(\mathbf{R}^n,\ell^2)} \approx \big\|\mathcal{M}_N(\vec{f})\big\|_{L^p(\mathbf{R}^n,\ell^2)}$$

with constants depending only on n and p.
[*Hint:* Fix Φ in $\mathscr{S}(\mathbf{R}^n)$ with integral 1. Using Lemma 2.1.5, write

$$(\varphi_j)_t(y) = \int_0^1 ((\Theta_j^{(s)})_t * \Phi_{ts})(y)\,ds$$

and adapt the proof of part (d) of Theorem 2.1.4 to obtain the pointwise estimate

$$\widetilde{\mathcal{M}_N}(\vec{f}) \le C_{n,p} M_m^{**}(\vec{f};\Phi),$$

where $m > n/p$.]

2.1.13. Suppose that the family $\{K_j\}_{j=1}^L$ satisfies (2.1.62) and (2.1.63) and let Φ be a smooth function supported in the unit ball $B(0,1)$. If $\Phi_\varepsilon(x) = \varepsilon^{-n}\Phi(x/\varepsilon)$, then the family $\{\Phi_\varepsilon * K_j\}_{j=1}^L$ also satisfies (2.1.62) and (2.1.63) with constants A' and B' proportional to $A+B$ and B, respectively.
[*Hint:* In the proof of (2.1.62) consider the cases $|x| \ge 2\varepsilon$ and $|x| \le 2\varepsilon$. In the second case write

$$\text{p.v.}\int_{\mathbf{R}^n} \Phi_\varepsilon(x-y)K_j(y)\,dy = \int_{\mathbf{R}^n} \big(\Phi_\varepsilon(x-y) - \Phi_\varepsilon(x)\big)K_j(y)\Phi_0(y/\varepsilon)\,dy$$

$$+\left(\text{p.v.}\int_{\mathbf{R}^n} K_j(y)\Phi_0(y/\varepsilon)\,dy\right)\Phi_\varepsilon(x),$$

where $\Phi_0(y)$ is a smooth function which is equal to 1 on the ball $|y| \le 3$ and vanishes outside the ball $|y| \le 4$.]

2.2 Function Spaces and the Square Function Characterization of Hardy Spaces

In Sections 1.2 and 1.3 we obtained a remarkable characterization of Sobolev and Lipschitz using the Littlewood–Paley operators Δ_j. In this section we achieve a similar characterization for the Hardy spaces. These characterizations motivate the introduction of classes of spaces defined in terms of mixed (discrete and continuous) quasi-norms of the sequences $\Delta_j^\Psi(f)$, for a suitable $\Psi \in \mathscr{S}(\mathbf{R}^n)$. Within the general framework of these classes, one can launch a study of function spaces from a unified perspective.

We have encountered two expressions involving the operators Δ_j^{Ψ} in the characterizations of Sobolev and Lipschitz spaces. Sobolev spaces were characterized by an L^p norm of the Littlewood–Paley square function

$$\Big(\sum_j |2^{j\alpha}\Delta_j^{\Psi}(f)|^2\Big)^{\frac{1}{2}},$$

but Lipschitz spaces were characterized by an ℓ^q norm of the sequence of quantities $\big\|2^{j\alpha}\Delta_j^{\Psi}(f)\big\|_{L^p}$. These examples motivate the introduction of two fundamental scales of function spaces, called the Triebel–Lizorkin and Besov–Lipschitz spaces, respectively.

2.2.1 Introduction to Function Spaces

Before we give the pertinent definitions, we recall the setup that we developed in Section 1.3 and used in Section 1.4. Throughout this section we fix a radial Schwartz function Ψ on $\mathbf{R}^n$ whose Fourier transform is nonnegative, is supported in the annulus $1-\frac{1}{7}\le|\xi|\le 2$, is equal to one on the smaller annulus $1\le|\xi|\le 2-\frac{2}{7}$, and satisfies

$$\sum_{j\in\mathbf{Z}}\widehat{\Psi}(2^{-j}\xi)=1, \qquad \xi\neq 0. \tag{2.2.1}$$

Associated with this bump, we define the Littlewood–Paley operators Δ_j^{Ψ} given by multiplication on the Fourier transform side by the function $\widehat{\Psi}(2^{-j}\xi)$. We also define a Schwartz function Φ such that

$$\widehat{\Phi}(\xi)=\begin{cases}\sum_{j\le 0}\widehat{\Psi}(2^{-j}\xi) & \text{when } \xi\neq 0,\\ 1 & \text{when } \xi=0.\end{cases} \tag{2.2.2}$$

Note that $\widehat{\Phi}(\xi)$ is equal to 1 for $|\xi|\le 2-\frac{2}{7}$ and vanishes when $|\xi|\ge 2$. It follows from these definitions that

$$S_0+\sum_{j=1}^{\infty}\Delta_j^{\Psi}=I, \tag{2.2.3}$$

where S_0 is the operator given by convolution with the bump Φ and the convergence of the series in (2.2.3) is in $\mathscr{S}'(\mathbf{R}^n)$. Moreover, we also have the identity

$$\sum_{j\in\mathbf{Z}}\Delta_j^{\Psi}=I, \tag{2.2.4}$$

where the convergence of the series in (2.2.4) is in the sense of $\mathscr{S}'(\mathbf{R}^n)/\mathscr{P}$.

Definition 2.2.1. Let $\alpha \in \mathbf{R}$ and $0 < p, q \le \infty$. For $f \in \mathscr{S}'(\mathbf{R}^n)$ we set

$$\|f\|_{B_p^{\alpha,q}} = \|S_0(f)\|_{L^p} + \Big(\sum_{j=1}^{\infty} (2^{j\alpha} \|\Delta_j^{\Psi}(f)\|_{L^p})^q \Big)^{\frac{1}{q}}$$

with the obvious modification when $p, q = \infty$. When $p, q < \infty$ we also define

$$\|f\|_{F_p^{\alpha,q}} = \|S_0(f)\|_{L^p} + \Big\| \Big(\sum_{j=1}^{\infty} (2^{j\alpha} |\Delta_j^{\Psi}(f)|)^q \Big)^{\frac{1}{q}} \Big\|_{L^p}.$$

The space of all tempered distributions f for which the quantity $\|f\|_{B_p^{\alpha,q}}$ is finite is called the (inhomogeneous) *Besov–Lipschitz* space with indices α, p, q and is denoted by $B_p^{\alpha,q}$. The space of all tempered distributions f for which the quantity $\|f\|_{F_p^{\alpha,q}}$ is finite is called the (inhomogeneous) *Triebel–Lizorkin* space with indices α, p, q and is denoted by $F_p^{\alpha,q}$.

We now define the corresponding homogeneous versions of these spaces. For an element f of $\mathscr{S}'(\mathbf{R}^n)/\mathscr{P}$ we let

$$\|f\|_{\dot{B}_p^{\alpha,q}} = \Big(\sum_{j \in \mathbf{Z}} (2^{j\alpha} \|\Delta_j^{\Psi}(f)\|_{L^p})^q \Big)^{\frac{1}{q}}$$

and

$$\|f\|_{\dot{F}_p^{\alpha,q}} = \Big\| \Big(\sum_{j \in \mathbf{Z}} (2^{j\alpha} |\Delta_j^{\Psi}(f)|)^q \Big)^{\frac{1}{q}} \Big\|_{L^p}.$$

The space of all f in $\mathscr{S}'(\mathbf{R}^n)/\mathscr{P}$ for which the quantity $\|f\|_{\dot{B}_p^{\alpha,q}}$ is finite is called the (homogeneous) *Besov–Lipschitz* space with indices α, p, q and is denoted by $\dot{B}_p^{\alpha,q}$. The space of f in $\mathscr{S}'(\mathbf{R}^n)/\mathscr{P}$ such that $\|f\|_{\dot{F}_p^{\alpha,q}} < \infty$ is called the (homogeneous) *Triebel–Lizorkin* space with indices α, p, q and is denoted by $\dot{F}_p^{\alpha,q}$.

We now make several observations related to these definitions. First we note that the expressions $\|\cdot\|_{\dot{F}_p^{\alpha,q}}$, $\|\cdot\|_{F_p^{\alpha,q}}$, $\|\cdot\|_{\dot{B}_p^{\alpha,q}}$, and $\|\cdot\|_{B_p^{\alpha,q}}$ are built in terms of L^p quasi-norms of ℓ^q quasi-norms of $2^{j\alpha}\Delta_j$ or ℓ^q quasi-norms of L^p quasi-norms of the same expressions. As a result, we can see that these quantities satisfy the triangle inequality with a constant (which may be taken to be 1 when $1 \le p, q < \infty$). To determine whether these quantities are indeed quasi-norms, we need to check whether the following property holds:

$$\|f\|_X = 0 \Longrightarrow f = 0, \tag{2.2.5}$$

where X is one of the $\dot{F}_p^{\alpha,q}$, $F_p^{\alpha,q}$, $\dot{B}_p^{\alpha,q}$, and $B_p^{\alpha,q}$. Since these are spaces of distributions, the identity $f = 0$ in (2.2.5) should be interpreted in the sense of distributions. If $\|f\|_X = 0$ for some inhomogeneous space X, then $S_0(f) = 0$ and $\Delta_j^{\Psi}(f) = 0$ for all $j \ge 1$. Using (2.2.3), we conclude that $f = 0$; thus the quantities $\|\cdot\|_{F_p^{\alpha,q}}$ and $\|\cdot\|_{B_p^{\alpha,q}}$ are indeed quasi-norms. Let us investigate what happens when $\|f\|_X = 0$ for some homogeneous space X. In this case we must have $\Delta_j(f) = 0$, and using (2.2.4) we conclude that $\widehat{f}$ must be supported at the origin. Proposition 2.4.1 in [156] yields that f must be a polynomial and thus f must be zero (since distributions whose difference is a polynomial are identified in homogeneous spaces).

Remark 2.2.2. We interpret the previous definition in certain cases. According to what we have seen so far, we have

$$\begin{aligned}
\dot{F}_p^{0,2} &\approx F_p^{0,2} \approx L^p, && 1<p<\infty,\\
F_p^{s,2} &\approx L_s^p, && 1<p<\infty,\\
\dot{F}_p^{s,2} &\approx \dot{L}_s^p, && 1<p<\infty,\\
B_\infty^{\gamma,\infty} &\approx \Lambda_\gamma, && \gamma>0,\\
\dot{B}_\infty^{\gamma,\infty} &\approx \dot{\Lambda}_\gamma, && \gamma>0,
\end{aligned}$$

where $\approx$ indicates that the corresponding norms are equivalent. Moreover, later in this section we will see that

$$\dot{F}_p^{0,2} \approx H^p \qquad 0<p\le 1.$$

Although in this text we restrict attention to the case $p<\infty$, it is worth noting that when $p=\infty$, $\dot{F}_\infty^{\alpha,q}$ can be defined as the space of all $f\in\mathscr{S}'/\mathscr{P}$ that satisfy

$$\|f\|_{\dot{F}_\infty^{\alpha,q}} = \sup_{Q \text{ dyadic cube}} \int_Q \frac{1}{|Q|}\bigg(\sum_{j=-\log_2 \ell(Q)}^{\infty} (2^{j\alpha}|\Delta_j^\Psi(f)|)^q\bigg)^{\frac{1}{q}} < \infty.$$

In the particular case $q=2$ and $\alpha=0$, the space obtained in this way is called *BMO* and coincides with the space introduced and studied in Chapter 3; this space serves as a substitute for L^∞ and plays a fundamental role in analysis. It should now be clear that several important spaces in analysis can be thought of as elements of the scale of Triebel–Lizorkin spaces.

It would have been more natural to denote Besov–Lipschitz and Triebel–Lizorkin spaces by $B^p_{\alpha,q}$ and $F^p_{\alpha,q}$ to maintain the upper and lower placements of the corresponding indices analogous to those in the previously defined Lebesgue, Sobolev, Lipschitz, and Hardy spaces. However, the notation in Definition 2.2.1 is more or less prevalent in the field of function spaces, and we adhere to it.

2.2.2 Properties of Functions with Compactly Supported Fourier Transforms

The definitions of the quasi-norms of the spaces $B_p^{\alpha,q}$, $F_p^{\alpha,q}$, $\dot{B}_p^{\alpha,q}$, and $\dot{F}_p^{\alpha,q}$ depend on the function Ψ (and Φ which is defined in terms of Ψ). It is not clear from Definition 2.2.1 whether a different choice of bump Ψ produces equivalent quasi-norms for these spaces. In this subsection we show that if Ω is another function that satisfies (2.2.1) and Θ is defined in terms of Ω in the same way that Φ is defined in terms of Ψ, [i.e., via (2.2.2)], then the norms defined in Definition 2.2.1 with respect to the pairs (Φ,Ψ) and (Θ,Ω) are comparable. To prove this assertion we need the following lemma.

Lemma 2.2.3. *Let $0<r<\infty$. Then there exist constants C_1 and C_2 such that for all $t>0$ and for all $\mathscr{C}^1$ functions u on $\mathbf{R}^n$ whose distributional Fourier transform is supported in the ball $|\xi|\le t$ we have*

$$\sup_{z\in\mathbf{R}^n}\frac{1}{t}\frac{|\nabla u(x-z)|}{(1+t|z|)^{\frac{n}{r}}}\le C_1\sup_{z\in\mathbf{R}^n}\frac{|u(x-z)|}{(1+t|z|)^{\frac{n}{r}}}, \tag{2.2.6}$$

$$\sup_{z\in\mathbf{R}^n}\frac{|u(x-z)|}{(1+t|z|)^{\frac{n}{r}}}\le C_2 M(|u|^r)(x)^{\frac{1}{r}}, \tag{2.2.7}$$

where M denotes the Hardy–Littlewood maximal operator. The constants C_1 and C_2 depend only on the dimension n and r; in particular they are independent of t.

Proof. Select a Schwartz function Φ whose Fourier transform is supported in the ball $|\xi|\le 2$ and is equal to 1 on the unit ball $|\xi|\le 1$. Then $\widehat{\Phi}(\frac{\xi}{t})$ is equal to 1 on the support of $\widehat{u}$ and we can write

$$u(x-z)=(\Phi*u)(x-z)=\int_{\mathbf{R}^n}t^n\Phi(t(x-z-y))u(y)\,dy.$$

Taking partial derivatives and using that Φ is a Schwartz function, we obtain

$$|\nabla u(x-z)|\le C_N\int_{\mathbf{R}^n}t^{n+1}(1+t|x-z-y|)^{-N}|u(y)|\,dy,$$

where N is arbitrarily large. Using that for all $x,y,z\in\mathbf{R}^n$ we have

$$1\le(1+t|x-z-y|)^{\frac{n}{r}}\frac{(1+t|z|)^{\frac{n}{r}}}{(1+t|x-y|)^{\frac{n}{r}}},$$

we obtain

$$\frac{1}{t}\frac{|\nabla u(x-z)|}{(1+t|z|)^{\frac{n}{r}}}\le C_N\int_{\mathbf{R}^n}t^n(1+t|x-z-y|)^{\frac{n}{r}-N}\frac{|u(y)|}{(1+t|x-y|)^{\frac{n}{r}}}\,dy,$$

from which (2.2.6) follows easily by choosing $N=n+1+n/r$.

We now turn to the proof of (2.2.7). We first prove this estimate under the additional assumption that u is a bounded function. Let $|y|\le\delta$ for some $\delta>0$ to be chosen later. We now apply the mean value theorem to write

$$u(x-z)=(\nabla u)(x-z-\xi_y)\cdot y+u(x-z-y)$$

for some ξ_y satisfying $|\xi_y| \le |y| \le \delta$. This implies that

$$|u(x-z)| \le \sup_{|w|\le |z|+\delta} |(\nabla u)(x-w)|\,\delta + |u(x-z-y)|.$$

Raising the preceding inequality to the power r, averaging over the ball $|y| \le \delta$, and then raising to the power $\frac{1}{r}$ yields

$$|u(x-z)| \le c_r \left[\sup_{|w|\le |z|+\delta} |(\nabla u)(x-w)|\,\delta + \left(\frac{1}{v_n \delta^n} \int_{|y|\le\delta} |u(x-z-y)|^r\, dy \right)^{\frac{1}{r}} \right]$$

with $c_r = \max(2^{1/r}, 2^r)$. Here v_n is the volume of the unit ball in $\mathbf{R}^n$. Then

$$\frac{|u(x-z)|}{(1+t|z|)^{\frac{n}{r}}} \le c_r \left[\sup_{|w|\le |z|+\delta} \frac{|(\nabla u)(x-w)|}{(1+t|z|)^{\frac{n}{r}}}\,\delta + \frac{\left(\frac{1}{v_n \delta^n} \int_{|y|\le\delta+|z|} |u(x-y)|^r\, dy \right)^{\frac{1}{r}}}{(1+t|z|)^{\frac{n}{r}}} \right].$$

We now set $\delta = \varepsilon/t$ for some $\varepsilon \le 1$. Then we have

$$|w| \le |z| + \frac{\varepsilon}{t} \implies \frac{1}{1+t|z|} \le \frac{2}{1+t|w|},$$

and we can use this to obtain the estimate

$$\frac{|u(x-z)|}{(1+t|z|)^{\frac{n}{r}}} \le c_{r,n} \left[\sup_{w\in\mathbf{R}^n} \frac{1}{t} \frac{|(\nabla u)(x-w)|}{(1+t|w|)^{\frac{n}{r}}}\,\varepsilon + \frac{\left(\frac{t^n}{v_n \varepsilon^n} \int_{|y|\le\frac{1}{t}+|z|} |u(x-y)|^r\, dy \right)^{\frac{1}{r}}}{(1+t|z|)^{\frac{n}{r}}} \right]$$

with $c_{r,n} = \max(2^{1/r}, 2^r)2^{n/r}$. It follows that

$$\sup_{z\in\mathbf{R}^n} \frac{|u(x-z)|}{(1+t|z|)^{\frac{n}{r}}} \le c_{r,n} \left[\sup_{w\in\mathbf{R}^n} \frac{1}{t} \frac{|(\nabla u)(x-w)|}{(1+t|w|)^{\frac{n}{r}}}\,\varepsilon + \varepsilon^{-\frac{n}{r}} M(|u|^r)(x)^{\frac{1}{r}} \right].$$

We apply inequality (2.2.6) and we select $\varepsilon = \frac{1}{2}(c_{r,n} C_1)^{-1}$, where C_1 is the constant in the inequality in (2.2.6). We obtain

$$\sup_{z\in\mathbf{R}^n} \frac{|u(x-z)|}{(1+t|z|)^{\frac{n}{r}}} \le \frac{1}{2} \sup_{z\in\mathbf{R}^n} \frac{|u(x-z)|}{(1+t|z|)^{\frac{n}{r}}} + c_{r,n} \varepsilon^{-\frac{n}{r}} M(|u|^r)(x)^{\frac{1}{r}}.$$

Using that

$$\sup_{z\in\mathbf{R}^n} \frac{|u(x-z)|}{(1+t|z|)^{\frac{n}{r}}} \le \|u\|_{L^\infty} < \infty,$$

we deduce (2.2.7) with constant $C_2 = 2c_{r,n}\varepsilon^{-n/r}$, where $\varepsilon = \frac{1}{2}(c_{r,n} C_1)^{-1}$.

We now discuss inequality (2.2.7) when u is not a bounded function. Since

$$u = (\widehat{u})^{\vee} = \big(\widehat{u}\widehat{\Phi}(\tfrac{\cdot}{t})\big)^{\vee} = u * t^n \Phi(t(\cdot))$$

and $t^n \Phi(tx)$ is a Schwartz function, we have that $|u(x)| \le C'(t,u)(1+|x|)^{Q_u}$ for some constant $C'(t,u)$ and some $Q_u \in \mathbf{Z}^+$; see Theorem 2.3.20 in [156]. We pick a function ϕ in $\mathscr{S}(\mathbf{R}^n)$ whose Fourier transform is nonnegative, is supported in the unit ball, and has integral one. For $\delta \le \min(1,t)$ consider the $\mathscr{C}^1$ function $x \mapsto \phi(\delta x)u(x)$ whose Fourier transform is supported in $B(0,\delta)+B(0,t)$, which is contained in the ball $B(0,2t)$. Certainly ϕ is a Schwartz function, and so for every $N > 0$ there is a constant $C_0(N)$ such that $|\phi(y)| \le C_0(N)(1+|y|)^{-N}$ for all $y \in \mathbf{R}^n$. For $N = Q_u$ and $y = \delta x$, $\delta \le 1$, we have

$$|\phi(\delta x)u(x)| \le C_0(Q_u)\frac{C'(t,u)(1+|x|)^{Q_u}}{(1+|\delta x|)^{Q_u}} \le C_0(Q_u)C'(t,u)\frac{1}{\delta^{Q_u}}\frac{(1+|x|)^{Q_u}}{(1+|x|)^{Q_u}}$$

and this is a bounded function with L^∞ norm $C_0(Q_u)C'(t,u)\delta^{-Q_u}$. By the preceding case, we have

$$\frac{\phi(\delta(x-z))|u(x-z)|}{(1+2t|z|)^{\frac{n}{r}}} \le C_2 M(|u|^r)(x)^{\frac{1}{r}} \|\phi\|_{L^\infty}$$

for every $x,z \in \mathbf{R}^n$. Letting $\delta \to 0$ and using that $\phi(0) = 1$ we deduce (2.2.7) with the constant $2^{n/r}C_2\|\phi\|_{L^\infty}$ in place of C_2. □

Corollary 2.2.4. *Let $0 < p \le \infty$ and α a multi-index. Then there are constants $C = C(\alpha,n,p)$ and $C' = C(\alpha,n,p)$ such that for all Schwartz functions u on $\mathbf{R}^n$ whose Fourier transform is supported in the ball $B(0,t)$, for some $t > 0$, we have*

$$\big\|\partial^\alpha u\big\|_{L^p(\mathbf{R}^n)} \le C t^{|\alpha|} \|u\|_{L^p(\mathbf{R}^n)} \tag{2.2.8}$$

and

$$\big\|\partial^\alpha u\big\|_{L^\infty(\mathbf{R}^n)} \le C' t^{|\alpha|+\frac{n}{p}} \|u\|_{L^p(\mathbf{R}^n)}\,. \tag{2.2.9}$$

Proof. Given $0 < p \le \infty$, pick $0 < r < p$. Then (2.2.6) and (2.2.7) imply that

$$\frac{1}{t}|\nabla u(x)| \le \sup_{z\in\mathbf{R}^n} \frac{1}{t}\frac{|\nabla u(x-z)|}{(1+t|z|)^{\frac{n}{r}}} \le C_1C_2 M(|u|^r)(x)^{\frac{1}{r}}\,, \tag{2.2.10}$$

where M is the Hardy–Littlewood maximal operator and C_1 and C_2 depend only on n and r. Taking L^p quasi-norms and using the boundedness of M on $L^{p/r}$ we obtain (2.2.8) when $|\alpha| = 1$. Since every derivative of u also has Fourier transform supported in $B(0,t)$, we obtain (2.2.8) for $|\alpha| \ge 2$ by iteration.

Select a Schwartz function Φ whose Fourier transform is supported in the ball $|\xi| \le 2$ and is equal to 1 on the unit ball $|\xi| \le 1$. Then $\widehat{\Phi}(\frac{\xi}{t})$ is equal to 1 on the support of $\widehat{u}$ and we can write $u = u * t^n \Phi(t(\cdot))$, hence

$$\partial^\alpha u(x) = \int_{\mathbf{R}^n} t^{n+|\alpha|}(\partial^\alpha \Phi)(t(x-y))u(y)\,dy\,. \tag{2.2.11}$$

If $1 \le p \le \infty$, Hölder's inequality gives that

$$\|\partial^\alpha u\|_{L^\infty} \le t^{|\alpha|+\frac{n}{p}} \|u\|_{L^p} \|\partial^\alpha \Phi\|_{L^{p'}}.$$

When $0 < p < 1$ we obtain from (2.2.11) that

$$|\partial^\alpha u(x)| \le t^{n+|\alpha|} \|\partial^\alpha \Phi\|_{L^\infty} \|u\|_{L^\infty}^{1-p} \int_{\mathbf{R}^n} |u(y)|^p \, dy, \tag{2.2.12}$$

which certainly implies (2.2.9) when $t = 1$, by taking the supremum over all x in $\mathbf{R}^n$. If $\widehat{u}$ is supported in $B(0,t)$ for some $t \neq 1$, we apply (2.2.9) when $t = 1$ to the Schwartz function $u_t(x) = t^{-n}u(t^{-1}x)$ whose Fourier transform is supported in $B(0,1)$. The inequality

$$\big\|\partial^\alpha u_t\big\|_{L^\infty(\mathbf{R}^n)} \le C' \, \|u_t\|_{L^p(\mathbf{R}^n)}.$$

transforms into (2.2.9) by changing variables. We note that if $p < 1$ and $t < 1$, then (2.2.12) implies the estimate $\|\partial^\alpha u\|_{L^\infty(\mathbf{R}^n)} \le C' t^{\frac{|\alpha|}{p}+\frac{n}{p}} \|u\|_{L^p(\mathbf{R}^n)}$, which is stronger than (2.2.9). □

2.2.3 Equivalence of Function Space Norms

We now derive other consequences of Lemma 2.2.3 that will allow us to prove that different norms in Triebel–Lizorkin spaces are equivalent.

Corollary 2.2.5. *Let $\Phi, \Omega, \Psi \in \mathscr{S}(\mathbf{R}^n)$. Suppose that the Fourier transforms of Ω, Ψ are supported in the annulus $1 - \frac{1}{7} \le |\xi| \le 2$. Let $0 < r < \infty$. Then for all f in $\mathscr{S}'(\mathbf{R}^n)/\mathscr{P}(\mathbf{R}^n)$ and for all $x \in \mathbf{R}^n$ and $t > 0$ we have*

$$|\Phi_t * \Delta_j^\Psi(f)(x)| \le C_{\Phi,n,r}(M(|\Delta_j^\Psi(f)|^r)(x))^{\frac{1}{r}}. \tag{2.2.13}$$

In particular, for any $k, j \in \mathbf{Z}$ and $x \in \mathbf{R}^n$ we have

$$\big|\Delta_k^\Omega \Delta_j^\Psi(f)(x)\big| \le C_{\Omega,n,r}(M(|\Delta_j^\Psi(f)|^r)(x))^{\frac{1}{r}}. \tag{2.2.14}$$

Proof. Given r pick $N = \frac{n}{r} + n + 1$. Then we have

$$\begin{aligned}
\big|(\Phi_t * \Delta_j^\Psi(f))(x)\big| &\le C_{\Phi,N} \int_{\mathbf{R}^n} \frac{\big|\Delta_j^\Psi(f)(x-z)\big|}{(1+t^{-1}|z|)^{\frac{n}{r}}} \frac{t^{-n}dz}{(1+t^{-1}|z|)^{N-\frac{n}{r}}} \\
&\le C'_{\Phi,n,r} \sup_{z\in\mathbf{R}^n} \frac{\big|\Delta_j^\Psi(f)(x-z)\big|}{(1+t^{-1}|z|)^{\frac{n}{r}}} \int_{\mathbf{R}^n} \frac{t^{-n}dz}{(1+t^{-1}|z|)^{N-\frac{n}{r}}} \qquad (2.2.15) \\
&\le C_{\Phi,n,r}(M(|\Delta_j^\Psi(f)|^r)(x))^{\frac{1}{r}},
\end{aligned}$$

in view of Lemma 2.2.3, since $\Delta_j^{\Psi}(f)$ is a $\mathscr{C}^1$ function whose Fourier transform is supported in the ball $B(0,2^{j+1})$. This proves (2.2.13), which implies (2.2.14). □

We now return to a point alluded to earlier, that replacing Ψ by another function Ω with similar properties yields equivalent quasi-norms for the function spaces in Definition 2.2.1.

Corollary 2.2.6. *Let Ψ, Ω be Schwartz functions whose Fourier transforms are supported in the annulus $1-\frac{1}{7}\le|\xi|\le 2$ and satisfy (2.2.1). Let Φ be as in (2.2.2) and let*

$$\widehat{\Theta}(\xi)=\begin{cases}\sum_{j\le 0}\widehat{\Omega}(2^{-j}\xi) & \text{when } \xi\neq 0,\\ 1 & \text{when } \xi=0.\end{cases}$$

Then the homogeneous Triebel–Lizorkin and Besov–Lipschitz quasi-norms defined with respect to Ψ and Ω are equivalent. Likewise, the inhomogeneous Triebel–Lizorkin and Besov–Lipschitz quasi-norms defined with respect to the pairs (Ψ,Φ) and (Ω,Θ) are also equivalent.

Proof. The support properties of Ψ and Ω imply the identity

$$\Delta_j^{\Omega}=\Delta_j^{\Omega}(\Delta_{j-1}^{\Psi}+\Delta_j^{\Psi}+\Delta_{j+1}^{\Psi}). \tag{2.2.16}$$

Thus for any $f\in\mathscr{S}'(\mathbf{R}^n)/\mathscr{P}(\mathbf{R}^n)$, the L^p quasi-norm of $\Delta_j^{\Omega}(f)$ is controlled by the finite sum of the L^p quasi-norms of $\Delta_j^{\Omega}\Delta_{j+i}^{\Psi}(f)$ over $i\in\{-1,0,1\}$. Using (2.2.14) with $r<p$ and applying the boundedness of the Hardy–Littlewood maximal operator on $L^{p/r}(\mathbf{R}^n)$, we deduce that any homogeneous Besov–Lipschitz quasi-norm defined in terms of Ω is controlled by the corresponding norm defined in terms of Ψ.

The corresponding result for Triebel–Lizorkin quasi-norms is as follows:

$$\begin{aligned}\Big\|\Big(\sum_{j\in\mathbf{Z}}|2^{j\alpha}\Delta_j^{\Omega}(f)|^q\Big)^{\frac{1}{q}}\Big\|_{L^p} &\le C_{p,q}\sum_{i\in\{-1,0,1\}}\Big\|\Big(\sum_{j\in\mathbf{Z}}|2^{j\alpha}\Delta_j^{\Omega}\Delta_{j+i}^{\Psi}(f)|^q\Big)^{\frac{1}{q}}\Big\|_{L^p}\\ &\le C_{p,q,n,r,\Omega}\Big\|\Big(\sum_{j\in\mathbf{Z}}|M(|2^{j\alpha}\Delta_j^{\Psi}(f)|^r)|^{\frac{q}{r}}\Big)^{\frac{1}{q}}\Big\|_{L^p}\\ &= C_{p,q,n,r,\Omega}\Big\|\Big(\sum_{j\in\mathbf{Z}}|M(|2^{j\alpha}\Delta_j^{\Psi}(f)|^r)|^{\frac{q}{r}}\Big)^{\frac{r}{q}}\Big\|_{L^{p/r}}^{\frac{1}{r}}\end{aligned}$$

for all $f\in\mathscr{S}'(\mathbf{R}^n)$. Picking $r<p,q$, we use the $L^{p/r}(\mathbf{R}^n,\ell^{q/r})$ to $L^{p/r}(\mathbf{R}^n,\ell^{q/r})$ boundedness of the Hardy–Littlewood maximal operator (Theorem 5.6.6 in [156]) to complete the proof of the equivalence of the Triebel–Lizorkin quasi-norms in the homogeneous case.

In the case of the inhomogeneous spaces, we let S_0^{Φ} and S_0^{Θ} be the operators given by convolution with the bumps Φ and Θ, respectively. Then for $f\in\mathscr{S}'(\mathbf{R}^n)$ we have

$$\Theta*f=\Theta*(\Phi*f)+\Theta*(\Psi_{2^{-1}}*f), \tag{2.2.17}$$

since the Fourier transform of the function $\Phi+\Psi_{2^{-1}}$ is equal to 1 on the support of $\widehat{\Theta}$. Applying Corollary 2.2.5 (with $t=1$), we obtain that

$$|\Theta*(\Phi*f)|\le C_r M(|\Phi*f|^r)^{\frac{1}{r}}$$

and also

$$|\Theta*(\Psi_{2^{-1}}*f)|\le C_r M(|\Psi_{2^{-1}}*f|^r)^{\frac{1}{r}}$$

for any $0<r<\infty$. Picking $r<p$, we obtain that

$$\big\|\Theta*(\Phi*f)\big\|_{L^p}\le C\big\|S_0^{\Phi}(f)\big\|_{L^p}$$

and also

$$\big\|\Theta*(\Psi_{2^{-1}}*f)\big\|_{L^p}\le C\big\|\Delta_1^{\Psi}(f)\big\|_{L^p}.$$

Inserting the last two estimates in (2.2.17), we obtain that $\|S_0^{\Theta}(f)\|_{L^p}$ is controlled by a multiple of

$$\|S_0^{\Phi}(f)\|_{L^p}+\|\Delta_1^{\Psi}(f)\|_{L^p}$$

which is in turn bounded by a multiple of the $F_p^{\alpha,q}$ quasi-norm of f defined in terms of the pair (Ψ,Φ). This gives the equivalence of quasi-norms in the inhomogeneous case. □

The idea behind the proof of the equivalence of function space quasi-norms defined in terms of different bumps is quite useful. In the rest of this subsection, we take this idea a bit further.

Definition 2.2.7. Let $\Psi\in\mathscr{S}(\mathbf{R}^n)$. For $b>0$, $j\in\mathbf{R}$, and $f\in\mathscr{S}'(\mathbf{R}^n)$, we introduce the notation

$$M_{b,j}^{**}(f;\Psi)(x)=\sup_{y\in\mathbf{R}^n}\frac{|(\Psi_{2^{-j}}*f)(x-y)|}{(1+2^j|y|)^b}.$$

Note that

$$\sup_{j>0}M_{b,j}^{**}(f;\Psi)\le M_b^{**}(f;\Psi),$$

where M_b^{**} was introduced in (2.1.8). The operator $M_{b,j}^{**}(f;\Psi)$ is called the *Peetre maximal function of f (with respect to Ψ).*

We clearly have

$$|\Delta_j^{\Psi}(f)|\le M_{b,j}^{**}(f;\Psi),$$

but the next result shows that a certain converse of this inequality is also valid.

Theorem 2.2.8. *Let $\alpha\in\mathbf{R}$, $b>n(\min(p,q))^{-1}$, and $0<p,q<\infty$. Let Ψ be a Schwartz function whose Fourier transform is supported in the annulus $1-\frac{1}{7}\le|\xi|\le 2$, is equal to 1 on the annulus $1\le|\xi|\le 2-\frac{2}{7}$, and satisfies (2.2.1). Let Ω be another Schwartz function which has vanishing moments of all order, i.e., $\int\Omega(y)y^{\gamma}dy=0$ for all multi-indices γ. Then there is a constant $C=C_{\alpha,p,q,n,b,\Psi,\Omega}$, such that*

$$\Big\|\Big(\sum_{j\in\mathbf{Z}}|2^{j\alpha}M^{**}_{b,j}(f;\Omega)|^q\Big)^{\frac{1}{q}}\Big\|_{L^p}\le C\Big\|\Big(\sum_{j\in\mathbf{Z}}|2^{j\alpha}\Delta^{\Psi}_j(f)|^q\Big)^{\frac{1}{q}}\Big\|_{L^p} \tag{2.2.18}$$

for all $f\in\mathscr{S}'(\mathbf{R}^n)/\mathscr{P}(\mathbf{R}^n)$.

Proof. We start with a Schwartz function Θ whose Fourier transform is nonnegative, supported in the annulus $1-\frac{2}{7}\le|\xi|\le 2$, and satisfies

$$\sum_{j\in\mathbf{Z}}\widehat{\Theta}(2^{-j}\xi)^2=1,\qquad \xi\in\mathbf{R}^n\setminus\{0\}. \tag{2.2.19}$$

Using (2.2.19), we have

$$\Omega_{2^{-k}}*f=\sum_{j\in\mathbf{Z}}(\Omega_{2^{-k}}*\Theta_{2^{-j}})*(\Theta_{2^{-j}}*f).$$

It follows that

$$\begin{aligned}
&2^{k\alpha}\frac{|(\Omega_{2^{-k}}*f)(x-z)|}{(1+2^k|z|)^b}\\
&\le\sum_{j\in\mathbf{Z}}2^{k\alpha}\int_{\mathbf{R}^n}|(\Omega_{2^{-k}}*\Theta_{2^{-j}})(y)|\frac{|(\Theta_{2^{-j}}*f)(x-z-y)|}{(1+2^k|z|)^b}\,dy\\
&=\sum_{j\in\mathbf{Z}}2^{k\alpha}\int_{\mathbf{R}^n}2^{kn}|(\Omega*\Theta_{2^{-(j-k)}})(2^ky)|\frac{(1+2^j|y+z|)^b}{(1+2^k|z|)^b}\frac{|(\Theta_{2^{-j}}*f)(x-z-y)|}{(1+2^j|y+z|)^b}\,dy\\
&\le\sum_{j\in\mathbf{Z}}2^{k\alpha}\int_{\mathbf{R}^n}|(\Omega*\Theta_{2^{-(j-k)}})(y)|\frac{(1+2^j|2^{-k}y+z|)^b}{(1+2^k|z|)^b}\frac{|(\Theta_{2^{-j}}*f)(x-z-2^{-k}y)|}{(1+2^j|2^{-k}y+z|)^b}\,dy\\
&\le\sum_{j\in\mathbf{Z}}2^{(k-j)\alpha}\int_{\mathbf{R}^n}|(\Omega*\Theta_{2^{-(j-k)}})(y)|\frac{(1+2^{j-k}|y|+2^j|z|)^b}{(1+2^k|z|)^b}\,dy\,2^{j\alpha}M^{**}_{b,j}(f;\Theta)(x)\\
&\le\sum_{j\in\mathbf{Z}}2^{(k-j)\alpha}\int_{\mathbf{R}^n}|(\Omega*\Theta_{2^{-(j-k)}})(y)|\,(1+2^{j-k})^b(1+2^{j-k}|y|)^b\,dy\,2^{j\alpha}M^{**}_{b,j}(f;\Theta)(x).
\end{aligned}$$

We conclude that

$$2^{k\alpha}M^{**}_{b,k}(f;\Omega)(x)\le\sum_{j\in\mathbf{Z}}V_{j-k}\,2^{j\alpha}M^{**}_{b,j}(f;\Theta)(x), \tag{2.2.20}$$

where

$$V_j=2^{-j\alpha}(1+2^j)^b\int_{\mathbf{R}^n}|(\Omega*\Theta_{2^{-j}})(y)|\,(1+2^j|y|)^b\,dy.$$

We now use the facts that both Ω and Θ have vanishing moments of all orders and the result in Appendix B.4 to obtain

$$|(\Omega*\Theta_{2^{-j}})(y)|\le C_{L,N,n,\Theta,\Omega}\frac{2^{jn}2^{-|j|L}}{(1+2^{\min(0,j)}|y|)^N}$$

for all $L,N>0$. We deduce the estimate

$$|V_j| \le C_{L,M,n,\Theta,\Omega} 2^{-|j|M}$$

for all M sufficiently large, which, in turn, yields the estimate

$$\sum_{j\in\mathbf{Z}} |V_j|^{\min(1,q)} < \infty.$$

We deduce from (2.2.20) that for all $x \in \mathbf{R}^n$ we have

$$\big\|\{2^{k\alpha} M^{**}_{b,k}(f;\Omega)(x)\}_k\big\|_{\ell^q} \le C_{\alpha,p,q,n,\Theta,\Omega} \big\|\{2^{k\alpha} M^{**}_{b,k}(f;\Theta)(x)\}_k\big\|_{\ell^q}. \tag{2.2.21}$$

Lemma 2.2.3 gives

$$2^{k\alpha} M^{**}_{b,k}(f;\Theta) \le C_2 2^{k\alpha} M(|\Delta_k^\Theta(f)|^r)^{\frac{1}{r}} = C_2 M(|2^{k\alpha}\Delta_k^\Theta(f)|^r)^{\frac{1}{r}}. \tag{2.2.22}$$

In view of (2.2.1) we have the identity

$$\Delta_k^\Theta = \Delta_k^\Theta\left(\Delta_{k-1}^\Psi + \Delta_k^\Psi + \Delta_{k+1}^\Psi\right),$$

and applying (2.2.14) to each term of the preceding sum yields

$$M(|2^{k\alpha}\Delta_k^\Theta(f)|^r)^{\frac{1}{r}} \le C'\left(MM(|2^{k\alpha}\Delta_k^\Psi(f)|^r)\right)^{\frac{1}{r}}. \tag{2.2.23}$$

We now choose $r<\min(p,q)$, we combine (2.2.21), (2.2.22), (2.2.23), and we use twice the $L^{p/r}(\mathbf{R}^n,\ell^{q/r})$ to $L^{p/r}(\mathbf{R}^n,\ell^{q/r})$ boundedness of the Hardy–Littlewood maximal operator (Theorem 5.6.6 in [156]) to complete the proof. □

2.2.4 The Littlewood–Paley Characterization of Hardy Spaces

We discuss an important characterization of Hardy spaces in terms of Littlewood–Paley square functions. The vector-valued Hardy spaces and the action of singular integrals on them are crucial tools in obtaining this characterization.

We have the following.

Theorem 2.2.9. *Let Ψ be a Schwartz function on $\mathbf{R}^n$ whose Fourier transform is nonnegative, supported in $\frac{6}{7}\le|\xi|\le 2$, equal to 1 on $1\le|\xi|\le\frac{12}{7}$, and satisfies for all $\xi\neq 0$*

$$\sum_{j\in\mathbf{Z}} \widehat{\Psi}(2^{-j}\xi) = 1. \tag{2.2.24}$$

Let Δ_j^Ψ be the Littlewood–Paley operators associated with Ψ and let $0<p\le 1$. Then there exists a constant $C=C_{n,p,\Psi}$ such that for all $f\in H^p(\mathbf{R}^n)$ we have

$$\Big\|\Big(\sum_{j\in\mathbf{Z}} |\Delta_j^\Psi(f)|^2\Big)^{\frac{1}{2}}\Big\|_{L^p} \le C\|f\|_{H^p}. \tag{2.2.25}$$

Conversely, suppose that a tempered distribution f satisfies

$$\Big\|\Big(\sum_{j\in\mathbf{Z}}|\Delta_j^\Psi(f)|^2\Big)^{\frac{1}{2}}\Big\|_{L^p}<\infty. \tag{2.2.26}$$

Then there exists a unique polynomial $Q(x)$ such that $f-Q$ lies in the Hardy space H^p and satisfies for some constant $C=C_{n,p,\Psi}$

$$\frac{1}{C}\big\|f-Q\big\|_{H^p}\leq\Big\|\Big(\sum_{j\in\mathbf{Z}}|\Delta_j^\Psi(f)|^2\Big)^{\frac{1}{2}}\Big\|_{L^p}. \tag{2.2.27}$$

Proof. We fix $\Phi\in\mathscr{S}(\mathbf{R}^n)$ with integral equal to 1 and we take $f\in H^p\cap L^1$ and M in $\mathbf{Z}^+$. Let r_j be the Rademacher functions, defined in Appendix C.1 in [156], reindexed so that their index set is the set of all integers (not the set of nonnegative integers). We begin with the estimate

$$\Big|\sum_{j=-M}^{M}r_j(\omega)\Delta_j^\Psi(f)\Big|\leq\sup_{\varepsilon>0}\Big|\Phi_\varepsilon*\sum_{j=-M}^{M}r_j(\omega)\Delta_j^\Psi(f)\Big|,$$

which holds since $\{\Phi_\varepsilon\}_{\varepsilon>0}$ is an approximate identity. We raise this inequality to the power p, we integrate over $x\in\mathbf{R}^n$ and $\omega\in[0,1]$, and we use the maximal function characterization of H^p [Theorem 2.1.4(a)] to obtain

$$\int_0^1\int_{\mathbf{R}^n}\Big|\sum_{j=-M}^{M}r_j(\omega)\Delta_j^\Psi(f)(x)\Big|^p dx\,d\omega\leq C_{p,n}^p\int_0^1\Big\|\sum_{j=-M}^{M}r_j(\omega)\Delta_j^\Psi(f)\Big\|_{H^p}^p d\omega.$$

Applying Fubini's theorem and the lower inequality for the Rademacher functions in Appendix C.2 in [156], yields

$$\int_{\mathbf{R}^n}\Big(\sum_{j=-M}^{M}|\Delta_j^\Psi(f)(x)|^2\Big)^{\frac{p}{2}}dx\leq C_p^pC_{p,n}^p\int_0^1\Big\|\sum_{j=-M}^{M}r_j(\omega)\Delta_j^\Psi(f)\Big\|_{H^p}^p d\omega. \tag{2.2.28}$$

Next, for any fixed $M\in\mathbf{Z}^+$ and $\omega\in[0,1]$, we consider the mapping

$$f\mapsto\sum_{j=-M}^{M}r_j(\omega)\Delta_j^\Psi(f)$$

whose kernel

$$\sum_{k=-M}^{M}r_k(\omega)\Psi_{2^{-k}}(x)$$

satisfies (2.1.62) and (2.1.63) with constants A and B depending only on n and Ψ (thus, independent of ω and M). Applying Theorem 2.1.14 (the scalar version, i.e., the case where $L=1$) we obtain

$$\Big\|\sum_{j=-M}^{M}r_j(\omega)\Delta_j^\Psi(f)\Big\|_{H^p}^p\leq C(n,p,\Psi)\big\|f\big\|_{H^p}^p.$$

Using this fact and (2.2.28), we conclude that

$$\Big\|\Big(\sum_{j=-M}^{M}|\Delta_j^{\Psi}(f)|^2\Big)^{\frac{1}{2}}\Big\|_{L^p}\le C_{n,p,\Psi}\big\|f\big\|_{H^p},$$

from which (2.2.25) follows directly by letting $M\to\infty$. We have now established (2.2.25) for $f\in H^p\cap L^1$. Using density, we can extend this estimate to all $f\in H^p$.

We now turn to the converse statement of the theorem. Assume that (2.2.26) holds for some tempered distribution f.

Set $\widehat{\eta}(\xi)=\widehat{\Psi}(\frac{1}{2}\xi)+\widehat{\Psi}(\xi)+\widehat{\Psi}(2\xi)$. Then $\widehat{\eta}$ is supported in an annulus and is equal to 1 on the support of $\widehat{\Psi}$. Using Theorem 2.1.14 we obtain that for any $L\in\mathbf{Z}^+$ and $L'\subset\mathbf{Z}^+\cup\{0\}$ with $L'<L$ the mapping

$$\{f_j\}_{L'\le|j|<L}\mapsto\sum_{L'\le|j|<L}\Delta_j^{\eta}(f_j)$$

maps $H^p(\mathbf{R}^n,\ell^2_{2L-2L'})$ to $H^p(\mathbf{R}^n)$; note that if $L'=0$, then $\ell^2_{2L-2L'}$ should be ℓ^2_{2L-1}. Indeed, Theorem 2.1.14 can be applied, since the family of kernels $\{\eta_{2^{-j}}\}_{L'\le|j|<L}$ satisfies $\sum_{L'\le|j|<L}|\partial_x^{\alpha}(\eta_{2^{-j}})(x)|\le C_\alpha|x|^{-n-|\alpha|}$, $x\neq0$, for all multilindices α and $\sum_{L'\le|j|<L}|\widehat{\eta_{2^{-j}}}|\le c'$ with constants independent of L,L'. Thus we have

$$\Big\|\sum_{L'\le|j|<L}\Delta_j^{\eta}(f_j)\Big\|_{H^p}\le C_{p,n,\Phi}\Big\|\sup_{t>0}\Big(\sum_{L'\le|j|<L}|\Phi_t*f_j|^2\Big)^{\frac{1}{2}}\Big\|_{L^p}$$

for any Φ Schwartz function with nonvanishing integral and any $f_j\in H^p$. Taking[1] $f_j=\Delta_j^{\Psi}(f)$ and using that $\Delta_j^{\eta}\Delta_j^{\Psi}=\Delta_j^{\Psi}$, we deduce that for all $L\in\mathbf{Z}^+$ we have

$$\Big\|\sum_{L'\le|j|<L}\Delta_j^{\Psi}(f)\Big\|_{H^p}\le C_{p,n,\Phi}\Big\|\sup_{t>0}\Big(\sum_{L'\le|j|<L}|\Phi_t*\Delta_j^{\Psi}(f)|^2\Big)^{\frac{1}{2}}\Big\|_{L^p}.$$

Applying Corollary 2.2.5 for some $r<p$ we arrive at the estimate

$$\begin{aligned}\Big\|\sum_{L'\le|j|<L}\Delta_j^{\Psi}(f)\Big\|_{H^p}&\le C_{p,n}\Big\|\Big(\sum_{L'\le|j|<L}|M(|\Delta_j^{\Psi}(f)|^r)|^{\frac{2}{r}}\Big)^{\frac{1}{2}}\Big\|_{L^p}\\&=C_{p,n}\Big\|\Big(\sum_{L'\le|j|<L}|M(|\Delta_j^{\Psi}(f)|^r)|^{\frac{2}{r}}\Big)^{\frac{r}{2}}\Big\|_{L^{\frac{p}{r}}}^{\frac{1}{r}}.\end{aligned}$$

Since $r<\min(2,p)$, we use the $L^{p/r}(\mathbf{R}^n,\ell^{2/r}_{2L-2L'})$ to $L^{p/r}(\mathbf{R}^n,\ell^{2/r}_{2L-2L'})$ boundedness of the Hardy–Littlewood maximal operator (Theorem 5.6.6 in [156]) to obtain the inequality

[1] $f_j\in H^p$ since $\sup_{t>0}|\Phi_t*\Delta_j^{\Psi}(f)|\le C'M(|\Delta_j^{\Psi}(f)|^r)^{1/r}\in L^p$ for $r<p$ in view of (2.2.26).

$$\sup_{L\in\mathbf{Z}^+}\sup_{0\le L'<L}\Big\|\sum_{L'\le|j|<L}\Delta_j^{\Psi}(f)\Big\|_{H^p}\le C'_{p,n}\Big\|\Big(\sum_{L'\le|j|<L}|\Delta_j^{\Psi}(f)|^2\Big)^{\frac{1}{2}}\Big\|_{L^p}. \tag{2.2.29}$$

Thus the sequence $S^L(f)=\sum_{|j|<L}\Delta_j^{\Psi}(f)$, $L=1,2,\dots$ is Cauchy in H^p and by the completeness of H^p [Proposition 2.1.10(c)] it converges to an element $u_f\in H^p$. Obviously (2.2.29) has as a consequence that

$$\|u_f\|_{H^p}\le C'_{p,n}\Big\|\Big(\sum_{j\in\mathbf{Z}}|\Delta_j^{\Psi}(f)|^2\Big)^{\frac{1}{2}}\Big\|_{L^p}. \tag{2.2.30}$$

It remains to relate u_f and f. In view of (1.1.6) we know that $S^L(f)\to f$ in $\mathscr{S}'(\mathbf{R}^n)/\mathscr{P}(\mathbf{R}^n)$; thus for any Schwartz function ψ whose support is disjoint from $\{0\}$ we have $\langle S^L(f)\widehat{\ },\psi\rangle\to\langle\widehat{f},\psi\rangle$. Thus $\langle\widehat{u_f},\psi\rangle=\langle\widehat{f},\psi\rangle$ and this implies that the support of $\widehat{u_f}-\widehat{f}$ is $\{0\}$. Proposition 2.4.1 in [156] gives the existence of a unique polynomial Q such that $u_f=f-Q$. Then clearly (2.2.30) implies (2.2.27). □

The preceding proof can be modified to provide the following extension.

Corollary 2.2.10. *Fix Ψ in $\mathscr{S}(\mathbf{R}^n)$ with Fourier transform supported in $\frac{6}{7}\le|\xi|\le 2$, equal 1 on the annulus $1\le|\xi|\le\frac{12}{7}$, and satisfying $\sum_{j\in\mathbf{Z}}\widehat{\Psi}(2^{-j}\xi)=1$ for $\xi\neq 0$. Fix b_1,b_2 with $b_1<b_2$ and define a Schwartz function Ω via*

$$\widehat{\Omega}(\xi)=\sum_{j=b_1}^{b_2}\widehat{\Psi}(2^{-j}\xi).$$

Define $\Delta_k^{\Omega}(g)\widehat{\ }(\xi)=\widehat{g}(\xi)\widehat{\Omega}(2^{-k}\xi)$, $k\in\mathbf{Z}$. Let $q=b_2-b_1+1$, $0<p\le 1$, and fix $r\in\{0,1,\dots,q-1\}$. Then there exists a constant $C=C_{n,p,b_1,b_2,\Psi}$ such that for all $f\in H^p(\mathbf{R}^n)$ we have

$$\Big\|\Big(\sum_{j=r \bmod q}|\Delta_j^{\Omega}(f)|^2\Big)^{\frac{1}{2}}\Big\|_{L^p}\le C\|f\|_{H^p}. \tag{2.2.31}$$

Conversely, suppose that a tempered distribution f satisfies

$$\Big\|\Big(\sum_{j=r \bmod q}|\Delta_j^{\Omega}(f)|^2\Big)^{\frac{1}{2}}\Big\|_{L^p}<\infty. \tag{2.2.32}$$

Then there exists a unique polynomial $Q(x)$ such that $f-Q$ lies in the Hardy space H^p and satisfies for some constant $C=C_{n,p,b_1,b_2,\Psi}$

$$\frac{1}{C}\|f-Q\|_{H^p}\le\Big\|\Big(\sum_{j=r \bmod q}|\Delta_j^{\Omega}(f)|^2\Big)^{\frac{1}{2}}\Big\|_{L^p}. \tag{2.2.33}$$

Proof. Inequality (2.2.31) is a direct consequence of (2.2.25) since Δ_k^{Ω} can be written as a finite sum of Δ_j^{Ω}'s. Conversely, we introduce a Schwartz function η whose Fourier transform $\widehat{\eta}$ is supported in an annulus of the form $0 < c_1 \le |\xi| \le c_2 < \infty$ and is equal to 1 on the support of Ω. Then (2.2.30) with Ω in place of Ψ follows as in the preceding proof. Since $\sum_{j=r \bmod q} \Delta_j^{\Omega}(f) = f$ in $\mathscr{S}'(\mathbf{R}^n)/\mathscr{P}(\mathbf{R}^n)$, which is a consequence of the fact that $\sum_{j=r \bmod q} \widehat{\Omega}(2^{-j}\xi) = 1$ for all $\xi \neq 0$, we conclude that there is a unique polynomial such that $f - Q$ lies in H^p and satisfies (2.2.33).

Exercises

2.2.1. Let $0 < q_0 \le q_1 < \infty$, $0 < p < \infty$, $\varepsilon > 0$, and $\alpha \in \mathbf{R}$. Prove the embeddings

$$B_p^{\alpha,q_0} \subseteqq B_p^{\alpha,q_1},$$
$$F_p^{\alpha,q_0} \subseteqq F_p^{\alpha,q_1},$$
$$B_p^{\alpha+\varepsilon,q_0} \subseteqq B_p^{\alpha,q_1},$$
$$F_p^{\alpha+\varepsilon,q_0} \subseteqq F_p^{\alpha,q_1},$$

where p and q_1 are allowed to be infinite in the case of Besov spaces.

2.2.2. Let $0 < q < \infty$, $0 < p < \infty$, and $\alpha \in \mathbf{R}$. Prove that the embeddings

$$B_p^{\alpha,\min(p,q)} \subseteqq F_p^{\alpha,q} \subseteqq B_p^{\alpha,\max(p,q)}$$

hold with norm one, if the norms in the spaces are defined with respect to the same Schwartz function Ψ.
[*Hint:* When $p \ge q$ use Minkowski's inequality for $L^{p/q}$ for one embedding and the embedding $\ell^q \subseteqq \ell^p$ for the other. When $p < q$ use the reverse Minkowski inequality for $L^{p/q}$ for one embedding and the fact $(\sum_k |a_k|)^{p/q} \le \sum_k |a_k|^{p/q}$ for the other.]

2.2.3. Let $-\infty < \alpha < \infty$ and $0 < p, \beta < \infty$. Let $1' = \infty$ and $p' = p/(p-1)$ for $p \neq 1$.
(a) Suppose that the Fourier transform of function g is $\mathscr{C}^\infty$ and is equal to $|\xi|^{-\alpha}$ for $|\xi| \ge 10$. Show that g lies in $B_p^{\gamma,q}(\mathbf{R}^n)$ if and only if $0 < q < \infty$ and $\gamma < \alpha - \frac{n}{p'}$ or $q = \infty$ and $\gamma \le \alpha - \frac{n}{p'}$.
(b) If the Fourier transform of function g is $\mathscr{C}^\infty$ and is equal to $|\xi|^{-\alpha}(\log|\xi|)^{-\beta}$ for $|\xi| \ge 10$, then show that g lies in $B_p^{\alpha-\frac{n}{p'},q}(\mathbf{R}^n)$ if and only if $q > 1/\beta$.

2.2.4. Let $0 < p, q < \infty$ and $\alpha \in \mathbf{R}$. Show that the space of Schwartz functions is dense in all the spaces $B_p^{\alpha,q}(\mathbf{R}^n)$ and $F_p^{\alpha,q}(\mathbf{R}^n)$.

[*Hint:* Fix a function $\varphi \in \mathscr{S}(\mathbf{R}^n)$ whose Fourier transform has integral one and is supported in a ball of radius 1 centered at zero. Given $f \in F_p^{\alpha,q}(\mathbf{R}^n)$ consider the family of Schwartz functions

$$f_{N,\delta}(x) = S_0^{\Phi}(f)(x)\varphi(\delta x) + \sum_{j=1}^{N} \Delta_j^{\Psi}(f)(x)\varphi(\delta x)$$

for $0 < \delta < 1/10$.]

2.2.5. Let $\alpha \in \mathbf{R}$, let $0 < p,q < \infty$, and let $N = [\frac{n}{2} + \frac{n}{\min(p,q)}] + 1$. Assume that m is a $\mathscr{C}^N$ function on $\mathbf{R}^n \setminus \{0\}$ that satisfies

$$|\partial^\gamma m(\xi)| \le C_\gamma |\xi|^{-|\gamma|}$$

for all $|\gamma| \le N$. Show that there exists a constant C such that for all $f \in \mathscr{S}'/\mathscr{P}'$ we have

$$\big\|(m\widehat{f})^\vee\big\|_{\dot{B}_p^{\alpha,q}} \le C\big\|f\big\|_{\dot{B}_p^{\alpha,q}}.$$

[*Hint:* Pick $r < \min(p,q)$ such that $N > \frac{n}{2} + \frac{n}{r}$. Write $m_j(\xi) = m(\xi)(\widehat{\Psi}(2^{-j+1}\xi) + \widehat{\Psi}(2^{-j}\xi) + \widehat{\Psi}(2^{-j-1}\xi))$. Then $\Delta_j^{\Psi}((m\widehat{f})^\vee) = m_j^\vee * \Delta_j^{\Psi}(f)$. Obtain the estimate

$$\begin{aligned}\big|(m_j^\vee * \Delta_j^{\Psi}(f))(x)\big| &\le C \sup_{y\in\mathbf{R}^n} \frac{|\Delta_j^{\Psi}(f)(x-y)|}{(1+2^j|y|)^{\frac{n}{r}}} \int_{\mathbf{R}^n} |m_j^\vee(y)|(1+2^j|y|)^{\frac{n}{r}}\,dy \\ &\le C' M(|\Delta_j^{\Psi}(f)|^r)^{\frac{1}{r}}(x)\bigg(\int_{\mathbf{R}^n} |m_j(2^j(\cdot))^\vee(y)|^2(1+|y|)^{2N}\,dy\bigg)^{\frac{1}{2}}.\end{aligned}$$

The hypothesis on m implies that the preceding integral is bounded by a constant.]

2.2.6. ([293]) Let m be as in Exercise 2.2.5. Show that there exists a constant C such that for all $f \in \mathscr{S}'(\mathbf{R}^n)/\mathscr{P}'(\mathbf{R}^n)$ we have

$$\big\|(m\widehat{f})^\vee\big\|_{\dot{F}_p^{\alpha,q}} \le C\big\|f\big\|_{\dot{F}_p^{\alpha,q}}.$$

[*Hint:* Use the hint of Exercise 2.2.5 and Theorem 5.6.6 in [156].]

2.2.7. (a) Suppose that $B_{p_0}^{\alpha_0,q_0} = B_{p_1}^{\alpha_1,q_1}$ with equivalent norms. Prove that $\alpha_0 = \alpha_1$ and $p_0 = p_1$. Prove the same result for the scale of Triebel–Lizorkin spaces.
(b) Suppose that $B_{p_0}^{\alpha_0,q_0} = B_{p_1}^{\alpha_1,q_1}$ with equivalent norms. Prove that $q_0 = q_1$. Argue similarly with the scale of Triebel–Lizorkin spaces.
[*Hint:* Part (a): Test the corresponding norms on the function $\eta(2^j x)$, where η is chosen so that its Fourier transform is supported in $1 \le |\xi| \le \frac{12}{7}$. Part (b): Try a function f of the form $\widehat{f}(\xi) = \sum_{j=1}^{N} a_j \widehat{\varphi}(\xi_1 - 2^j, \xi_2, \dots, \xi_n)$, where φ is a Schwartz function whose Fourier transform is supported in a small neighborhood of the origin.]

2.3 Atomic Decomposition of Homogeneous Triebel–Lizorkin Spaces

In this section we focus attention on the homogeneous Triebel–Lizorkin spaces $\dot{F}_p^{\alpha,q}$, which include the Hardy spaces discussed in Section 2.1. Most results discussed in this section are also valid for the inhomogeneous Triebel–Lizorkin spaces and for the Besov–Lipschitz spaces via a similar or simpler analysis.

2.3.1 Embeddings and Completeness of Triebel–Lizorkin Spaces

Proposition 2.3.1. *Let $0 < p,q < \infty$, and $\alpha \in \mathbf{R}$. The homogeneous Triebel–Lizorkin space $\dot{F}_p^{\alpha,q}(\mathbf{R}^n)$ is continuously embedded in the Besov space $\dot{B}_p^{\alpha,\infty}(\mathbf{R}^n)$ which is in turn continuously embedded in $\mathscr{S}'(\mathbf{R}^n)/\mathscr{P}(\mathbf{R}^n)$. Moreover, the space $\dot{F}_p^{\alpha,q}(\mathbf{R}^n)$ is complete.*

Proof. Given $f \in \mathscr{S}'(\mathbf{R}^n)/\mathscr{P}(\mathbf{R}^n)$ we have the sequence of inequalities

$$\sup_{j\in\mathbf{Z}} 2^{j\alpha}\|\Delta_j^\Psi(f)\|_{L^p} \le \Big\|\sup_{j\in\mathbf{Z}} |2^{j\alpha}\Delta_j^\Psi(f)|\Big\|_{L^p} \le \Big\|\Big(\sum_{j\in\mathbf{Z}} |2^{j\alpha}\Delta_j^\Psi(f)|^q\Big)^{\frac{1}{q}}\Big\|_{L^p}, \tag{2.3.1}$$

which shows that $\|f\|_{\dot{B}_p^{\alpha,\infty}} \le \|f\|_{\dot{F}_p^{\alpha,q}}$. Thus we proved the embedding $\dot{F}_p^{\alpha,q} \subseteq \dot{B}_p^{\alpha,\infty}$.

Next we prove that $\dot{B}_p^{\alpha,\infty}(\mathbf{R}^n)$ continuously embeds in $\mathscr{S}'(\mathbf{R}^n)/\mathscr{P}(\mathbf{R}^n)$. Let ψ be in $\mathscr{S}_0(\mathbf{R}^n)$. Then given Ψ as in (2.2.1), let $\widehat{\Omega}(\xi) = \widehat{\Psi}(\frac{1}{2}\xi) + \widehat{\Psi}(\xi) + \widehat{\Psi}(2\xi)$. Given $f \in \mathscr{S}'(\mathbf{R}^n)/\mathscr{P}(\mathbf{R}^n)$ we have

$$\langle f,\psi\rangle = \sum_{j\in\mathbf{Z}} \langle \Delta_j^\Psi(f),\psi\rangle = \sum_{j\in\mathbf{Z}} \langle \Delta_j^\Psi(f),\Delta_j^\Omega(\psi)\rangle,$$

where the first identity is due to the fact that the series $\sum_{j\in\mathbf{Z}}\Delta_j^\Psi$ converges in $\mathscr{S}'(\mathbf{R}^n)/\mathscr{P}(\mathbf{R}^n)$ and the second identity to the fact that $\widehat{\Omega}$ is equal to one on the support of $\widehat{\Psi}$. It follows that

$$\begin{aligned}
\big|\langle f,\psi\rangle\big| &\le \sum_{j\in\mathbf{Z}} \big\|\Delta_j^\Psi(f)\big\|_{L^\infty}\big\|\Delta_j^\Omega(\psi)\big\|_{L^1}\\
&\le C\sum_{j\in\mathbf{Z}} 2^{\frac{jn}{p}-j\alpha}\big\|2^{j\alpha}\Delta_j^\Psi(f)\big\|_{L^p}\big\|\Delta_j^\Omega(\psi)\big\|_{L^1}\\
&\le \|f\|_{\dot{F}_p^{\alpha,q}} C\sum_{j\in\mathbf{Z}} 2^{\frac{jn}{p}-j\alpha}\big\|\Delta_j^\Omega(\psi)\big\|_{L^1}\\
&= C\|f\|_{\dot{F}_p^{\alpha,q}}\|\psi\|_{\dot{B}_1^{\frac{n}{p}-\alpha,1}},
\end{aligned}$$

where we used Corollary 2.2.4 in the second inequality and (2.3.1) in the last inequality.

Next, we show that $\|\psi\|_{\dot{B}_1^{\frac{n}{p}-\alpha,1}}$ is controlled by a finite sum of Schwartz seminorms of $\psi \in \mathscr{S}_0(\mathbf{R}^n)$. Using the result in Appendix B.4, we obtain the following estimate for all $L \in \mathbf{Z}^+$ and $N > 0$ satisfying $N < N' - (L+1+n)$

$$\big|\Delta_j^{\Omega}(\psi)(x)\big| \le C''_{N,N',L,n}\Big[\sup_{|\gamma|\le L}\sup_{x\in\mathbf{R}^n}|\partial^\gamma\psi(x)|(1+|x|)^{N'}\Big]\frac{2^{\min(j,0)n-|j|(L+1)}}{(1+2^{\min(j,0)}|x|)^N},$$

where the constant $C''_{N,N',L,n}$ also depends on Ω. Consequently we obtain that

$$\big\|\Delta_j^{\Omega}(\psi)\big\|_{L^p} \le C'''_{N,N',L,n}\Big[\sup_{|\gamma|\le L}\sup_{x\in\mathbf{R}^n}|\partial^\gamma\psi(x)|(1+|x|)^{N'}\Big]2^{\min(j,0)\frac{n}{p'}-|j|(L+1)}$$

if $N > n/p$. Choosing $L > n+|\alpha|$, it follows that

$$\big\|\psi\big\|_{\dot{B}_1^{\frac{n}{p}-\alpha,1}} = \sum_{j\in\mathbf{Z}} 2^{j(\frac{n}{p}-\alpha)}\big\|\Delta_j^{\Omega}(\psi)\big\|_{L^p}$$

is bounded by a constant multiple of the expression

$$\sup_{|\gamma|\le L}\sup_{x\in\mathbf{R}^n}|\partial^\gamma\psi(x)|(1+|x|)^{N'}$$

which is controlled by a finite sum of seminorms $\rho_{\alpha,\beta}(\psi)$. This proves that $\dot{B}_p^{\alpha,\infty}(\mathbf{R}^n)$ is continuously embedded in $\mathscr{S}'(\mathbf{R}^n)/\mathscr{P}(\mathbf{R}^n)$.

Finally, we turn to the last assertion that the space $\dot{F}_p^{\alpha,q}(\mathbf{R}^n)$ is complete. Since $\dot{F}_p^{\alpha,q}(\mathbf{R}^n)$ is continuously embedded in $\mathscr{S}'/\mathscr{P}$, every Cauchy sequence $\{u_M\}_{M=0}^\infty$ in $\dot{F}_p^{\alpha,q}(\mathbf{R}^n)$ is Cauchy in $\mathscr{S}'/\mathscr{P}$ and thus it converges to an element $u \in \mathscr{S}'/\mathscr{P}$, defined by $\langle u,\psi\rangle = \lim_{M\to\infty}\langle u_M,\psi\rangle$ for all $\psi\in\mathscr{S}_0(\mathbf{R}^n)$.

Since $u_M \to u$ in $\mathscr{S}'/\mathscr{P}$, it follows that for every $j \in \mathbf{Z}$

$$\Delta_j^{\Psi}(u_M - u_{M'}) \to \Delta_j^{\Psi}(u - u_{M'})$$

as $M \to \infty$. Thus for any $J \in \mathbf{Z}^+$ we have

$$\begin{aligned}\Big(\sum_{|j|\le J}(2^{j\alpha}|\Delta_j^{\Psi}(u-u_{M'})|)^q\Big)^{\frac{1}{q}} &= \liminf_{M\to\infty}\Big(\sum_{|j|\le J}(2^{j\alpha}|\Delta_j^{\Psi}(u_M-u_{M'})|)^q\Big)^{\frac{1}{q}}\\ &\le \liminf_{M\to\infty}\Big(\sum_{j\in\mathbf{Z}}(2^{j\alpha}|\Delta_j^{\Psi}(u_M-u_{M'})|)^q\Big)^{\frac{1}{q}}.\end{aligned}$$

First we let $J \to \infty$, then we take L^p quasi-norms and we apply Fatou's lemma. We obtain

$$\|u-u_{M'}\|_{\dot{F}_p^{\alpha,q}} \le \liminf_{M\to\infty}\|u_M-u_{M'}\|_{\dot{F}_p^{\alpha,q}}.$$

If we replace $u_{M'}$ by 0, this implies that u lies in $\dot{F}_p^{\alpha,q}$ since $\sup_{M\geq 0}\|u_M\|_{\dot{F}_p^{\alpha,q}} < \infty$. Then we have

$$\limsup_{M'\to\infty}\|u-u_{M'}\|_{\dot{F}_p^{\alpha,q}} \leq \limsup_{M'\to\infty}\limsup_{M\to\infty}\|u_M-u_{M'}\|_{\dot{F}_p^{\alpha,q}},$$

but the expression on the right is zero since the sequence $\{u_M\}_{M=0}^{\infty}$ is Cauchy in $\dot{F}_p^{\alpha,q}$. It follows that $u_M \to u$ in $\dot{F}_p^{\alpha,q}$ as $M\to\infty$; thus $\dot{F}_p^{\alpha,q}$ is complete. □

2.3.2 The Space of Triebel–Lizorkin Sequences

To provide further intuition into the understanding of the homogeneous Triebel–Lizorkin spaces we introduce a related space consisting of sequences of scalars. This space is denoted by $\dot{f}_p^{\alpha,q}$ and is related to $\dot{F}_p^{\alpha,q}$ in a way similar to that in which $\ell^2(\mathbf{Z})$ is related to $L^2([0,1])$.

Definition 2.3.2. Let $0<q\leq\infty$ and $\alpha\in\mathbf{R}$. Let $\mathscr{D}$ be the set of all dyadic cubes in $\mathbf{R}^n$. We consider the set of all sequences $\{s_Q\}_{Q\in\mathscr{D}}$ such that the function

$$g^{\alpha,q}(\{s_Q\}_Q)=\Big(\sum_{Q\in\mathscr{D}}(|Q|^{-\frac{\alpha}{n}-\frac{1}{2}}|s_Q|\chi_Q)^q\Big)^{\frac{1}{q}} \tag{2.3.2}$$

is in $L^p(\mathbf{R}^n)$. For such sequences $s=\{s_Q\}_Q$ we set

$$\|s\|_{\dot{f}_p^{\alpha,q}}=\big\|g^{\alpha,q}(s)\big\|_{L^p(\mathbf{R}^n)}.$$

2.3.3 The Smooth Atomic Decomposition of Homogeneous Triebel–Lizorkin Spaces

We discuss the smooth atomic decomposition of homogeneous Triebel–Lizorkin spaces. We denote by $\mathscr{D}$ the space of all dyadic cubes on $\mathbf{R}^n$. For any fixed $j\in\mathbf{Z}$ we let $\mathscr{D}_j=\{Q\in\mathscr{D}:\ \ell(Q)=2^{-j}\}$. We begin with the definition of smooth atoms on $\mathbf{R}^n$.

Definition 2.3.3. Let Q be a dyadic cube and let L be a nonnegative integer. A $\mathscr{C}^\infty$ function a_Q on $\mathbf{R}^n$ is called a *smooth L-atom for Q* if it satisfies the following properties:

(a) a_Q is supported in $3Q$ (the cube concentric with Q having three times its side length);

(b) $\displaystyle\int_{\mathbf{R}^n} x^\gamma a_Q(x)\,dx=0$ for all multi-indices γ with $|\gamma|\leq L$;

(c) $|\partial^\gamma a_Q|\leq|Q|^{-\frac{|\gamma|}{n}-\frac{1}{2}}$ for all multi-indices γ satisfying $|\gamma|\leq L+1$.

In view of properties (a) and (c) of Definition 1.3.2, for every $M > 0$ there is a constant $C(n,M,L)$ such that every smooth L-atom a_Q supported in Q with center c_Q and side length $\ell(Q)$ satisfies

$$|\partial^\gamma a_Q(x)| \le C(n,M,L)\ell(Q)^{\frac{n}{2}} \frac{\ell(Q)^{-n-|\gamma|}}{\left(1+\frac{|x-c_Q|}{\ell(Q)}\right)^M} \tag{2.3.3}$$

for all $x \in \mathbf{R}^n$ and for all multi-indices γ with $|\gamma| \le L+1$.

We now prove a theorem stating that elements of $\dot F_p^{\alpha,q}$ can be decomposed as sums of smooth atoms.

Theorem 2.3.4. *Let $0 < p,q < \infty$, $\alpha \in \mathbf{R}$, and let*

$$L = \left[\max\left(n\max(1,\tfrac{1}{p},\tfrac{1}{q}) - n - \alpha, \alpha\right)\right].$$

Then there is a constant $C_{n,p,q,\alpha}$ such that for every sequence of smooth L-atoms $\{a_Q\}_{Q\in\mathscr{D}}$ and every sequence of complex scalars $\{s_Q\}_{Q\in\mathscr{D}}$ in $\dot f_p^{\alpha,q}$ we have that the series $\sum_{\mu\in\mathbf{Z}}\left(\sum_{Q\in\mathscr{D}_\mu} s_Q a_Q\right)$ converges in $\dot F_p^{\alpha,q}(\mathbf{R}^n)$ to an element f of $\dot F_p^{\alpha,q}(\mathbf{R}^n)$ with quasi-norm

$$\big\|f\big\|_{\dot F_p^{\alpha,q}} \le C_{n,p,q,\alpha}\big\|\{s_Q\}_Q\big\|_{\dot f_p^{\alpha,q}}. \tag{2.3.4}$$

Conversely, there is a constant $C'_{n,p,q,\alpha}$ such that given any distribution f in $\dot F_p^{\alpha,q}$ and any $L \in \mathbf{Z}^+$, there exist a sequence of smooth L-atoms $\{a_Q\}_{Q\in\mathscr{D}}$ and a sequence of complex scalars $\{s_Q\}_{Q\in\mathscr{D}}$ such that the series $\sum_{\mu\in\mathbf{Z}}\left(\sum_{Q\in\mathscr{D}_\mu} s_Q a_Q\right)$ converges to f in $\dot F_p^{\alpha,q}(\mathbf{R}^n)$ and

$$\big\|\{s_Q\}_Q\big\|_{\dot f_p^{\alpha,q}} \le C'_{n,p,q,\alpha}\big\|f\big\|_{\dot F_p^{\alpha,q}}. \tag{2.3.5}$$

We observe that for any given x the expression $\sum_{Q\in\mathscr{D}_\mu} s_Q a_Q(x)$ is a finite sum with at most 3^n summands, so the convergence concerns the series in μ.

Proof. We prove the first assertion of the theorem. We let Δ_j^Ψ be the Littlewood–Paley operator associated with a Schwartz function Ψ whose Fourier transform is compactly supported away from the origin in $\mathbf{R}^n$. Let a_Q be a smooth L-atom supported in a cube $3Q$ with center c_Q and let the side length of Q be $\ell(Q) = 2^{-\mu}$. It follows from (2.3.3) that a_Q satisfies

$$|\partial_y^\gamma a_Q(y)| \le C_{N',n} 2^{-\frac{\mu n}{2}} \frac{2^{\mu|\gamma|+\mu n}}{(1+2^\mu|y-c_Q|)^{N'}} \tag{2.3.6}$$

for all $N' > 0$ and for all multi-indices γ satisfying $|\gamma| \le L+1$. Moreover, the function $y \mapsto \Psi_{2^{-j}}(y-x)$ satisfies

$$|\partial_y^\beta \Psi_{2^{-j}}(y-x)| \le C_{N',n,\beta} \frac{2^{j|\beta|+jn}}{(1+2^j|y-x|)^{N'}} \tag{2.3.7}$$

for all $N' > 0$ and for all multi-indices β.

The function a_Q has vanishing moments of all orders up to and including $L = (L+1)-1$ and satisfies (2.3.6) for all multi-indices γ with $|\gamma| \le L+1$. The function $y \mapsto \Psi_{2^{-j}}(y-x)$ has vanishing moments of all orders and satisfies (2.3.7) for all multi-indices β. Using the result in Appendix B.4, we deduce the following estimate for all $N>0$ satisfying $N < N' - (L+1+n)$

$$\big|\Delta_j^{\Psi}(a_Q)(x)\big| \le C_{N,n,L}\, 2^{-\frac{\mu n}{2}} \frac{2^{\min(j,\mu)n - |\mu - j|(L+1)}}{(1+2^{\min(j,\mu)}|x-c_Q|)^N}\,. \tag{2.3.8}$$

Now fix $0 < b < \min(1,p,q)$ so that

$$L+1 > \frac{n}{b} - n - \alpha\,. \tag{2.3.9}$$

This can be achieved by taking b close enough to $\min(1,p,q)$, since our assumption $L = \big[\max\big(n\max\big(1,\frac{1}{p},\frac{1}{q}\big) - n - \alpha, \alpha\big)\big]$ yields that $L+1 > n\max\big(1,\frac{1}{p},\frac{1}{q}\big) - n - \alpha$ and also that $L+1 > \alpha$. These two conditions imply that the function $d(k)$ defined for $k \in \mathbf{Z}$ by

$$d(k) = 2^{\min(k,0)(n-\frac{n}{b}) + k\alpha - |k|(L+1)}$$

satisfies for some $\delta > 0$

$$d(k) \le C 2^{-|k|\delta} \tag{2.3.10}$$

for all $k \in \mathbf{Z}$. Using Exercise 2.3.6, we obtain

$$\sum_{Q \in \mathscr{D}_\mu} \frac{|s_Q|}{(1+2^{\min(j,\mu)}|x-c_Q|)^N} \le c\, 2^{\max(\mu - j, 0)\frac{n}{b}} \Big\{ M\Big(\sum_{Q\in\mathscr{D}_\mu} |s_Q|^b \chi_Q\Big)(x)\Big\}^{\frac{1}{b}}$$

whenever $N > n/b$, where M is the Hardy–Littlewood maximal operator. It follows from the preceding estimate and (2.3.8) that

$$2^{j\alpha} \sum_{\mu \in \mathbf{Z}} \sum_{Q\in\mathscr{D}_\mu} |s_Q| |\Delta_j^{\Psi}(a_Q)| \le C_0 \sum_{\mu\in\mathbf{Z}} d(j-\mu) \Big\{ M\Big(\sum_{Q\in\mathscr{D}_\mu} \big(|s_Q|\,|Q|^{-\frac{1}{2}-\frac{\alpha}{n}}\big)^b \chi_Q\Big)\Big\}^{\frac{1}{b}},$$

where $C_0 = cC_{N,n,L}$. In particular this estimate is valid for any finite subset $\mathbf{Z}'$ of $\mathbf{Z}$. For such a subset we have

$$2^{j\alpha} \Delta_j^{\Psi}\Big(\sum_{\mu\in\mathbf{Z}'} \sum_{Q\in\mathscr{D}_\mu} s_Q a_Q\Big) = 2^{j\alpha} \sum_{\mu\in\mathbf{Z}'} \sum_{Q\in\mathscr{D}_\mu} s_Q \Delta_j^{\Psi}(a_Q)\,. \tag{2.3.11}$$

Raise the last displayed inequality to the power q and sum over $j \in \mathbf{Z}$; then raise to the power $1/q$ and take $\|\cdot\|_{L^p}$ quasi-norms. We obtain

$$\Big\| \sum_{\mu\in\mathbf{Z}'} \sum_{Q\in\mathscr{D}_\mu} s_Q a_Q \Big\|_{\dot F_p^{\alpha,q}}$$

$$\le C_0 \bigg\| \bigg\{ \sum_{j\in\mathbf{Z}} \bigg[\sum_{\mu\in\mathbf{Z}'} d(j-\mu) \Big\{ M\Big(\sum_{Q\in\mathscr{D}_\mu} \big(|s_Q|\,|Q|^{-\frac{1}{2}-\frac{\alpha}{n}}\big)^b \chi_Q\Big)\Big\}^{\frac{1}{b}} \bigg]^q \bigg\}^{\frac{1}{q}} \bigg\|_{L^p}.$$

We now estimate the expression inside the preceding L^p norm by

$$\Big\{\sum_{j\in\mathbf{Z}} d(j)^{\min(1,q)}\Big\}^{\frac{1}{\min(1,q)}}\Big\{\sum_{\mu\in\mathbf{Z}'}\Big\{M\Big(\sum_{Q\in\mathscr{D}_\mu}(|s_Q|\,|Q|^{-\frac{1}{2}-\frac{\alpha}{n}})^b\chi_Q\Big)\Big\}^{\frac{q}{b}}\Big\}^{\frac{1}{q}},$$

and we note that the first term is a constant in view of (2.3.10). We conclude that

$$\begin{aligned}\Big\|\sum_{\mu\in\mathbf{Z}'}\sum_{Q\in\mathscr{D}_\mu}s_Q\Big\|_{\dot F_p^{\alpha,q}} &\le C_0C\Big\|\Big\{\sum_{\mu\in\mathbf{Z}'}\Big\{M\Big(\sum_{Q\in\mathscr{D}_\mu}(|s_Q|\,|Q|^{-\frac{1}{2}-\frac{\alpha}{n}})^b\chi_Q\Big)\Big\}^{\frac{q}{b}}\Big\}^{\frac{1}{q}}\Big\|_{L^p}\\ &= C_0C\Big\|\Big\{\sum_{\mu\in\mathbf{Z}'}\Big\{M\Big(\sum_{Q\in\mathscr{D}_\mu}(|s_Q|\,|Q|^{-\frac{1}{2}-\frac{\alpha}{n}})^b\chi_Q\Big)\Big\}^{\frac{q}{b}}\Big\}^{\frac{b}{q}}\Big\|_{L^{\frac{p}{b}}}^{\frac{1}{b}}\\ &\le C_0C'\Big\|\Big\{\sum_{\mu\in\mathbf{Z}'}\Big\{\sum_{Q\in\mathscr{D}_\mu}(|s_Q|\,|Q|^{-\frac{1}{2}-\frac{\alpha}{n}})^b\chi_Q\Big\}^{\frac{q}{b}}\Big\}^{\frac{b}{q}}\Big\|_{L^{\frac{p}{b}}}^{\frac{1}{b}}\\ &= C_0C'\Big\|\Big\{\sum_{\mu\in\mathbf{Z}'}\sum_{Q\in\mathscr{D}_\mu}(|s_Q|\,|Q|^{-\frac{1}{2}-\frac{\alpha}{n}})^q\chi_Q\Big\}^{\frac{1}{q}}\Big\|_{L^p},\end{aligned} \tag{2.3.12}$$

where in the last inequality we used Theorem 5.6.6 in [156], which is valid under the assumption $1<\frac{p}{b},\frac{q}{b}<\infty$. We now take $\mathbf{Z}'=\{\mu\in\mathbf{Z}:\ M'<|\mu|\le M\}$, for some integers $M'<M$, and we use the following consequence of the Lebesgue dominated convergence theorem

$$\lim_{M',M\to\infty}\Big\|\Big\{\sum_{M'<|\mu|\le M}\sum_{Q\in\mathscr{D}_\mu}(|s_Q|\,|Q|^{-\frac{1}{2}-\frac{\alpha}{n}})^q\chi_Q\Big\}^{\frac{1}{q}}\Big\|_{L^p}=0,$$

since $s=\{s_Q\}_{Q\in\mathscr{D}}\in\dot f_p^{\alpha,q}$. We obtain that the sequence

$$F_M=\sum_{|\mu|\le M}\sum_{Q\in\mathscr{D}_\mu}s_Qa_Q$$

is Cauchy in $\dot F_p^{\alpha,q}$. Proposition 2.3.1 yields that it converges to an element f in $\dot F_p^{\alpha,q}$.

We now repeat the preceding argument replacing $\mathbf{Z}'$ by $\mathbf{Z}$ and $\sum_{\mu\in\mathbf{Z}'}\sum_{Q\in\mathscr{D}_\mu}s_Q$ by f noting that (2.3.11) holds for $\mathbf{Z}$ in place of $\mathbf{Z}'$ since we can interchange Δ_j^Ψ with the infinite sum over μ (and certainly with the finite sum in $Q\in\mathscr{D}_\mu$) in view of the convergence of the sequence $\sum_{|\mu|\le M}\big(\sum_{Q\in\mathscr{D}_\mu}s_Qa_Q\big)$ to f in $\dot F_p^{\alpha,q}$ (and thus in $\mathscr{S}'/\mathscr{P}$). This proves (2.3.4) since (2.3.12) is controlled by $\|s\|_{\dot f_p^{\alpha,q}}$.

We now turn to the converse statement of the theorem. It is not difficult to see that given $L\in\mathbf{Z}^+$ there exist Schwartz functions Θ and Ψ (unrelated to the previous one) such that $\widehat{\Psi}$ is supported in the annulus $\frac{1}{2}\le|\xi|\le 2$ and Θ is supported in the ball $|x|\le 1$ and satisfies $\int_{\mathbf{R}^n}x^\gamma\Theta(x)\,dx=0$ for all $|\gamma|\le L$, such that the identity

$$\sum_{j\in\mathbf{Z}}\widehat{\Psi}(2^{-j}\xi)\widehat{\Theta}(2^{-j}\xi)=1 \tag{2.3.13}$$

holds for all $\xi\in\mathbf{R}^n\setminus\{0\}$; see Exercise 2.3.1.

Given a distribution $f \in \dot{F}_p^{\alpha,q}$, using identity (2.3.13), we write

$$f = \sum_{j\in\mathbf{Z}} \Psi_{2^{-j}} * \Theta_{2^{-j}} * f,$$

where the convergence is in $\mathscr{S}'(\mathbf{R}^n)/\mathscr{P}(\mathbf{R}^n)$ in view of Corollary 1.1.7.

For each Q in $\mathscr{D}_j$ define a constant

$$s_Q = |Q|^{\frac{1}{2}} \sup_{y\in Q} |(\Psi_{\ell(Q)} * f)(y)| \sup_{|\gamma|\le L+1} \big\|\partial^\gamma \Theta\big\|_{L^1}$$

and a function

$$a_Q(x) = \frac{1}{s_Q}\int_Q \Theta_{\ell(Q)}(x-y)(\Psi_{\ell(Q)} * f)(y)\,dy. \tag{2.3.14}$$

It is straightforward to verify that a_Q is supported in $3Q$ and that it has vanishing moments up to and including order L, since θ does so. Moreover, using (2.3.14) we obtain for all $|\gamma| \le L+1$

$$|\partial^\gamma a_Q| \le \frac{1}{s_Q}\big\|\partial^\gamma \Theta\big\|_{L^1} \ell(Q)^{-|\gamma|} \sup_Q |\Psi_{\ell(Q)} * f| \le |Q|^{-\frac{1}{2}-\frac{|\gamma|}{n}},$$

which makes the function a_Q a smooth L-atom.

Using this notation, we write

$$f = \sum_{j\in\mathbf{Z}} \sum_{Q\in\mathscr{D}_j} \int_Q \Theta_{2^{-j}}(x-y)(\Psi_{2^{-j}} * f)(y)\,dy = \sum_{j\in\mathbf{Z}} \Big(\sum_{Q\in\mathscr{D}_j} s_Q a_Q\Big),$$

where the series in j converges in $\mathscr{S}'(\mathbf{R}^n)/\mathscr{P}(\mathbf{R}^n)$.

Let b be as in (2.3.9). Now note that

$$\begin{aligned}
&\sum_{\ell(Q)=2^{-j}} \big(|Q|^{-\frac{\alpha}{n}-\frac{1}{2}} s_Q \chi_Q(x)\big)^q \\
&\quad = C \sum_{\ell(Q)=2^{-j}} \big(2^{j\alpha} \sup_{y\in Q} |(\Psi_{2^{-j}} * f)(y)| \chi_Q(x)\big)^q \\
&\quad \le C \sup_{|z|\le\sqrt{n}2^{-j}} \big(2^{j\alpha}(1+2^j|z|)^{-b}|(\Psi_{2^{-j}} * f)(x-z)|\big)^q (1+2^j|z|)^{bq} \\
&\quad \le C\big(2^{j\alpha} M_{b,j}^{**}(f,\Psi)(x)\big)^q,
\end{aligned}$$

where we used the fact that in the first inequality there is only one nonzero term in the sum because of the appearance of the characteristic function. Summing over all $j \in \mathbf{Z}^n$, raising to the power $1/q$, and taking L^p norms yields the estimate

$$\big\|\{s_Q\}_Q\big\|_{\dot{f}_p^{\alpha,q}} \le C \Big\| \Big(\sum_{j\in\mathbf{Z}} |2^{j\alpha} M_{b,j}^{**}(f;\Psi)|^q\Big)^{\frac{1}{q}} \Big\|_{L^p} \le C \big\|f\big\|_{\dot{F}_p^{\alpha,q}},$$

where the last inequality follows from Theorem 2.2.8. This proves (2.3.5). It follows from (2.3.5) that $\{s_Q\}_{Q\in\mathscr{D}}$ lies in $\dot{f}_p^{\alpha,q}$ and thus by the first assertion of the theorem we have that the series

$$\sum_\mu \Big(\sum_{Q\in\mathscr{D}_\mu} s_Q a_Q \Big)$$

converges to some element in $\dot{F}_p^{\alpha,q}$. Since it converges to f in $\mathscr{S}'/\mathscr{P}$, it follows that $\sum_\mu (\sum_{Q\in\mathscr{D}_\mu} s_Q a_Q)$ converges to f in $\dot{F}_p^{\alpha,q}$, and this completes the proof. □

2.3.4 The Nonsmooth Atomic Decomposition of Homogeneous Triebel–Lizorkin Spaces

We now discuss the main theorem of this section, the nonsmooth atomic decomposition of the homogeneous Triebel–Lizorkin spaces $\dot{F}_p^{\alpha,q}$, which in particular includes that of the Hardy spaces H^p. We begin with a definition.

Definition 2.3.5. Let $0 < p,q < \infty$. A sequence of complex numbers $r = \{r_Q\}_{Q\in\mathscr{D}}$ is called an ∞-atom for $\dot{f}_p^{\alpha,q}$ if there exists a dyadic cube Q_0 such that

(a) $r_Q = 0$ if $Q \not\subseteq Q_0$;

(b) $\big\| g^{\alpha,q}(r) \big\|_{L^\infty} \le |Q_0|^{-\frac{1}{p}}$,

where, recalling from (2.3.2),

$$g^{\alpha,q}(\{r_Q\}_Q) = \Big(\sum_{Q\in\mathscr{D}} (|Q|^{-\frac{\alpha}{n}-\frac{1}{2}} |r_Q| \chi_Q)^q \Big)^{\frac{1}{q}}.$$

We observe that every ∞-atom $r = \{r_Q\}$ for $\dot{f}_p^{\alpha,q}$ satisfies $\|r\|_{\dot{f}_p^{\alpha,q}} \le 1$. Indeed,

$$\|r\|_{\dot{f}_p^{\alpha,q}}^p = \int_{Q_0} |g^{\alpha,q}(r)|^p \, dx \le |Q_0|^{-1} |Q_0| = 1.$$

The following theorem concerns the atomic decomposition of the spaces $\dot{f}_p^{\alpha,q}$.

Theorem 2.3.6. *Let $\alpha \in \mathbf{R}$, $0 < p,q < \infty$, and $s = \{s_Q\}_{Q\in\mathscr{D}}$ be in $\dot{f}_p^{\alpha,q}$. Then there exist $C_{n,p,q} > 0$, a sequence of scalars λ_j, and a sequence of ∞-atoms $r_j = \{r_{j,Q}\}_{Q\in\mathscr{D}}$ for $\dot{f}_p^{\alpha,q}$ such that for each $Q \in \mathscr{D}$ the series $\sum_{j=1}^\infty \lambda_j r_{j,Q}$ is absolutely convergent and equal to s_Q, i.e.,*

$$s = \{s_Q\}_{Q\in\mathscr{D}} = \sum_{j=1}^\infty \lambda_j \{r_{j,Q}\}_{Q\in\mathscr{D}} = \sum_{j=1}^\infty \lambda_j r_j,$$

and such that

$$\Big(\sum_{j=1}^\infty |\lambda_j|^p \Big)^{\frac{1}{p}} \le C_{n,p,q} \|s\|_{\dot{f}_p^{\alpha,q}}. \tag{2.3.15}$$

Proof. We fix α, p, q, and a sequence $s = \{s_Q\}_{Q\in\mathscr{D}}$ in $\dot{f}_p^{\alpha,q}$. For a dyadic cube R in $\mathscr{D}$ we define the function

$$g_R^{\alpha,q}(s)(x) = \Big(\sum_{\substack{Q\in\mathscr{D}\\ R\subseteq Q}} \big(|Q|^{\frac{\alpha}{n}-\frac{1}{2}}|s_Q|\chi_Q(x)\big)^q\Big)^{\frac{1}{q}}$$

and we observe that this function is constant on R. We also note that for dyadic cubes R_1 and R_2 with $R_1 \subseteq R_2$ we have

$$g_{R_2}^{\alpha,q}(s) \le g_{R_1}^{\alpha,q}(s).$$

Finally, we observe that

$$\lim_{\substack{\ell(R)\to\infty\\ x\in R}} g_R^{\alpha,q}(s)(x) = 0$$

and

$$\lim_{\substack{\ell(R)\to 0\\ x\in R}} g_R^{\alpha,q}(s)(x) = g^{\alpha,q}(s)(x),$$

where $g^{\alpha,q}(s)$ is the function defined in (2.3.2).

For $k \in \mathbf{Z}$ we set

$$\mathscr{A}_k = \big\{R\in\mathscr{D}:\ g_R^{\alpha,q}(s)(x) > 2^k \quad \text{for all } x\in R\big\}.$$

We note that $\mathscr{A}_{k+1} \subseteq \mathscr{A}_k$ for all k in $\mathbf{Z}$ and that

$$\{x\in\mathbf{R}^n:\ g^{\alpha,q}(s)(x) > 2^k\} = \bigcup_{R\in\mathscr{A}_k} R. \tag{2.3.16}$$

Moreover, we have for all $k \in \mathbf{Z}$,

$$\Big(\sum_{Q\in\mathscr{D}\setminus\mathscr{A}_k} \big(|Q|^{-\frac{\alpha}{n}-\frac{1}{2}}|s_Q|\chi_Q(x)\big)^q\Big)^{\frac{1}{q}} \le 2^k, \tag{2.3.17}$$

for all $x \in \mathbf{R}^n$.

To prove (2.3.17) we assume that $g^{\alpha,q}(s)(x) > 2^k$; otherwise, the conclusion is trivial. Then there exists a maximal dyadic cube $R_{\max}$ in $\mathscr{A}_k$ such that $x \in R_{\max}$. Letting R_0 be the unique dyadic cube that contains $R_{\max}$ and has twice its side length, we have that the left-hand side of (2.3.17) is equal to $g_{R_0}^{\alpha,q}(s)(x)$, which is at most 2^k, since R_0 is not contained in $\mathscr{A}_k$.

Since $g^{\alpha,q}(s) \in L^p(\mathbf{R}^n)$, by our assumption, and $g^{\alpha,q}(s) > 2^k$ for all $x \in Q$ if $Q \in \mathscr{A}_k$, the cubes in $\mathscr{A}_k$ must have size bounded above by some constant. We set

$$\mathscr{B}_k = \big\{J\in\mathscr{D}:\quad J \text{ is a maximal dyadic cube in } \mathscr{A}_k\setminus\mathscr{A}_{k+1}\big\}.$$

For J in $\mathscr{B}_k$ we define a sequence $t(k,J) = \{t(k,J)_Q\}_{Q\in\mathscr{D}}$ by setting

$$t(k,J)_Q = \begin{cases} s_Q & \text{if } Q \subseteq J \text{ and } Q \in \mathscr{A}_k \setminus \mathscr{A}_{k+1}, \\ 0 & \text{otherwise.} \end{cases}$$

Notice that

$$\text{if} \quad Q \notin \bigcup_{k\in\mathbf{Z}} \mathscr{A}_k, \qquad \text{then} \qquad s_Q = 0.$$

Moreover, the identity

$$s = \sum_{k\in\mathbf{Z}} \sum_{J\in\mathscr{B}_k} t(k,J) \tag{2.3.18}$$

is valid and it is worth noticing that for each $Q \in \mathscr{D}$, there is at most one $k \in \mathbf{Z}$ and at most one $J \in \mathscr{B}_k$ such that $t(k,J)_Q$ is nonzero, i.e., the sum in (2.3.18) evaluated at Q has at most one nonzero term.

For all $x \in \mathbf{R}^n$ we have

$$\begin{aligned} g^{\alpha,q}(t(k,J))(x) &= \Big(\sum_{\substack{Q\subseteq J \\ Q\in\mathscr{A}_k\setminus\mathscr{A}_{k+1}}} \big(|Q|^{-\frac{\alpha}{n}-\frac{1}{2}} |s_Q|\, \chi_Q(x)\big)^q \Big)^{\frac{1}{q}} \\ &\le \Big(\sum_{\substack{Q\subseteq J \\ Q\in\mathscr{D}\setminus\mathscr{A}_{k+1}}} \big(|Q|^{-\frac{\alpha}{n}-\frac{1}{2}} |s_Q|\, \chi_Q(x)\big)^q \Big)^{\frac{1}{q}} \\ &\le 2^{k+1}, \end{aligned} \tag{2.3.19}$$

where we used (2.3.17) in the last estimate. We define atoms $r(k,J) = \{r(k,J)_Q\}_{Q\in\mathscr{D}}$ by setting

$$r(k,J)_Q = 2^{-k-1} |J|^{-\frac{1}{p}} t(k,J)_Q, \tag{2.3.20}$$

and we also define scalars

$$\lambda_{k,J} = 2^{k+1} |J|^{\frac{1}{p}}.$$

To see that each $r(k,J)$ is an ∞-atom for $\dot{f}_p^{\alpha,q}$, we observe that $r(k,J)_Q = 0$ if $Q \not\subseteq J$ and that

$$g^{\alpha,q}(r(k,J))(x) \le |J|^{-\frac{1}{p}}, \qquad \text{for all } x \in \mathbf{R}^n,$$

in view of (2.3.19) and (2.3.20). Also using (2.3.18) and (2.3.20), we obtain that

$$s = \sum_{k\in\mathbf{Z}} \sum_{J\in\mathscr{B}_k} \lambda_{k,J}\, r(k,J), \tag{2.3.21}$$

which says that s can be written as a countably infinite sum of atoms. We now reindex the countable set $\mathscr{U} = \{(k,J) : k \in \mathbf{Z},\ J \in \mathscr{B}_k\}$ by $\mathbf{Z}^+$ and write

$$s = \sum_{j=1}^{\infty} \lambda_j r_j, \tag{2.3.22}$$

where $\{\lambda_1, \lambda_2, \dots\} = \{\lambda_{k,J} : (k,J) \in \mathscr{U}\}$ and $\{r_1, r_2, \dots\} = \{r(k,J) : (k,J) \in \mathscr{U}\}$. As observed the sum in (2.3.21) has the property that for each $Q \in \mathscr{D}$, there is at most one $k \in \mathbf{Z}$ and at most one $J \in \mathscr{B}_k$ such that $\lambda_{k,J}\, r(k,J)_Q = t(k,J)_Q$ is nonzero. Thus for each $Q \in \mathscr{D}$, at most one term in the sum $\sum_{j=1}^{\infty} \lambda_j r_{j,Q}$ is nonzero; in particular, this series is absolutely convergent.

Finally, we estimate the sum of the pth power of the coefficients $\lambda_{k,J}$. We have

$$
\begin{aligned}
\sum_{j=1}^{\infty} |\lambda_j|^p &= \sum_{k\in\mathbf{Z}} \sum_{J\in\mathscr{B}_k} \lambda_{k,J}^p \\
&= \sum_{k\in\mathbf{Z}} 2^{(k+1)p} \sum_{J\in\mathscr{B}_k} |J| \\
&\le 2^p \sum_{k\in\mathbf{Z}} 2^{kp} \Big| \bigcup_{Q\in\mathscr{A}_k} Q \Big| \\
&= 2^p \sum_{k\in\mathbf{Z}} 2^{k(p-1)} 2^k |\{x \in \mathbf{R}^n : g^{\alpha,q}(s)(x) > 2^k\}| \\
&\le 2^p \sum_{k\in\mathbf{Z}} \int_{2^k}^{2^{k+1}} 2^{k(p-1)} |\{x \in \mathbf{R}^n : g^{\alpha,q}(s)(x) > \tfrac{\lambda}{2}\}|\, d\lambda \\
&\le 2^p \sum_{k\in\mathbf{Z}} \int_{2^k}^{2^{k+1}} \lambda^{p-1} |\{x \in \mathbf{R}^n : g^{\alpha,q}(s)(x) > \tfrac{\lambda}{2}\}|\, d\lambda \\
&= \frac{2^{2p}}{p} \big\| g^{\alpha,q}(s) \big\|_{L^p}^p \\
&= \frac{2^{2p}}{p} \|s\|_{\dot f_p^{\alpha,q}}^p .
\end{aligned}
$$

Taking the pth root yields (2.3.15). The proof of the theorem is now complete. □

We now deduce a corollary concerning a new characterization of the space $\dot f_p^{\alpha,q}$.

Corollary 2.3.7. *Suppose $\alpha \in \mathbf{R}$, $0 < p \le 1$, and $p \le q < \infty$. Then for a given sequence $s \in \dot f_p^{\alpha,q}$ we have the following equivalence:*

$$
\|s\|_{\dot f_p^{\alpha,q}} \approx \inf\bigg\{ \Big(\sum_{j=1}^{\infty} |\lambda_j|^p\Big)^{\frac{1}{p}} : \lim_{N\to\infty} \Big\| s - \sum_{j=1}^{N} \lambda_j r_j \Big\|_{\dot f_p^{\alpha,q}} = 0,\ r_j \textit{ are } \infty\textit{-atoms for } \dot f_p^{\alpha,q} \bigg\}.
$$

Remark 2.3.8. Notice that $\dot f_p^{\alpha,q}$ is complete (Exercise 2.3.6(b)), so if r_j are ∞-atoms for $\dot f_p^{\alpha,q}$, if $(\sum_{j=1}^{\infty} |\lambda_j|^p)^{\frac{1}{p}} < \infty$ and if

$$
\Big\| s - \sum_{j=1}^{N} \lambda_j r_j \Big\|_{\dot f_p^{\alpha,q}} \to 0
$$

as $N \to \infty$, then s must be an element of $\dot f_p^{\alpha,q}$.

Proof. Given $s \in \dot{f}_p^{\alpha,q}$, let r_j be ∞-atoms for $\dot{f}_p^{\alpha,q}$ and λ_j be as in Theorem 2.3.6. Then $\sum_{j=1}^{\infty} |\lambda_j|^p \le C_{n,p,q}^p \|s\|_{\dot{f}_p^{\alpha,q}}^p < \infty$ and $s_Q = \sum_{j=1}^{\infty} \lambda_j r_{j,Q}$, where the series converges absolutely. Then as $N \to \infty$ we have

$$\Big\|s - \sum_{j=1}^{N} \lambda_j r_j\Big\|_{\dot{f}_p^{\alpha,q}}^p = \Big\|\sum_{j=1}^{\infty} \lambda_j r_j - \sum_{j=1}^{N} \lambda_j r_j\Big\|_{\dot{f}_p^{\alpha,q}}^p \le \Big\|\sum_{j=N+1}^{\infty} |\lambda_j r_j|\Big\|_{\dot{f}_p^{\alpha,q}}^p \le \sum_{j=N+1}^{\infty} |\lambda_j|^p \to 0,$$

where we used Exercise 2.3.6(a) in the last inequality together with the observation made after Definition 2.3.5 that every ∞-atom r for $\dot{f}_p^{\alpha,q}$ satisfies $\|r\|_{\dot{f}_p^{\alpha,q}} \le 1$. (Here $|r_j| = \{|r_{j,Q}|\}_{Q\in\mathscr{D}}$.) Then λ_j and s are as in the statement of the corollary and (2.3.15) implies the $\ge$ inequality in the claimed equivalence.

The converse inequality ($\le$) is easier since for any r_j ∞-atoms for $\dot{f}_p^{\alpha,q}$ we have

$$\|s\|_{\dot{f}_p^{\alpha,q}}^p \le \Big\|s - \sum_{j=1}^{N} \lambda_j r_j\Big\|_{\dot{f}_p^{\alpha,q}}^p + \Big\|\sum_{j=1}^{N} \lambda_j r_j\Big\|_{\dot{f}_p^{\alpha,q}}^p \le \Big\|s - \sum_{j=1}^{N} \lambda_j r_j\Big\|_{\dot{f}_p^{\alpha,q}}^p + \sum_{j=1}^{\infty} |\lambda_j|^p;$$

thus letting $N \to \infty$ and taking the infimum over all $(\sum_{j=1}^{\infty} |\lambda_j|^p)^{1/p}$ yields the desired inequality. □

Theorem 2.3.6 allows us to obtain an atomic decomposition for the space $\dot{F}_p^{\alpha,q}$ as well. Indeed, we have the following result.

Corollary 2.3.9. *Let $\alpha \in \mathbf{R}$, $0 < p \le 1$, $L \ge [\max(\frac{n}{p} - n - \alpha, \alpha)]$, and let q satisfy $p \le q < \infty$. Then for a given $f \in \dot{F}_p^{\alpha,q}$ we have the following equivalence:*

$$\|f\|_{\dot{F}_p^{\alpha,q}} \approx \inf\Big\{\Big(\sum_{j=1}^{\infty} |\lambda_j|^p\Big)^{\frac{1}{p}} : \lim_{N\to\infty} \|f - \sum_{j=1}^{N} \lambda_j A_j\|_{\dot{F}_p^{\alpha,q}} = 0,$$
$$\text{where } A_j = \sum_{\mu\in\mathbf{Z}} \Big(\sum_{Q\in\mathscr{D}_\mu} r_{j,Q} a_Q\Big) \text{ converging in } \dot{F}_p^{\alpha,q},$$
$$a_Q \text{ are smooth } L\text{-atoms, and } r_j = \{r_{j,Q}\}_{Q\in\mathscr{D}} \text{ are } \infty\text{-atoms for } \dot{f}_p^{\alpha,q}\Big\}.$$

Proof. Let λ_j, A_j be as above such that

$$\lim_{N\to\infty} \|f - \sum_{j=1}^{N} \lambda_j A_j\|_{\dot{F}_p^{\alpha,q}} = 0.$$

In view of the subadditivity of the expression $h \mapsto \|f\|_{\dot{F}_p^{\alpha,q}}^p$ (Exercise 2.3.2) we have that

$$\|f\|_{\dot{F}_p^{\alpha,q}}^p \le \sum_{j=1}^{N} \|\lambda_j A_j\|_{\dot{F}_p^{\alpha,q}}^p + \|f - \sum_{j=1}^{N} \lambda_j A_j\|_{\dot{F}_p^{\alpha,q}}^p.$$

It follows from this that

$$\|f\|_{\dot F_p^{\alpha,q}}^p \le \limsup_{N\to\infty}\sum_{j=1}^N |\lambda_j|^p \|A_j\|_{\dot F_p^{\alpha,q}}^p \le \sum_{j=1}^\infty |\lambda_j|^p \|A_j\|_{\dot F_p^{\alpha,q}}^p \le c_{n,p}\sum_{j=1}^\infty |\lambda_j|^p \|r_j\|_{\dot f_p^{\alpha,q}}^p ,$$

where the last inequality is due to Theorem 2.3.4. Using the fact that every ∞-atom $r=\{r_Q\}$ for $\dot f_p^{\alpha,q}$ satisfies $\|r\|_{\dot f_p^{\alpha,q}}\le 1$, we take the infimum over all representations of f as above to deduce the $\le$ part of the claimed equivalence.

Conversely, given f in $\dot F_p^{\alpha,q}$, we use Theorem 2.3.4 to write

$$f=\sum_{\mu\in\mathbf Z}\Big(\sum_{Q\in\mathscr D_\mu} s_Q a_Q\Big)$$

where $s=\{s_Q\}_{Q\in\mathscr D}$ lies in $\dot f_p^{\alpha,q}$, each a_Q is a smooth L-atom for the cube Q and the series converges in $\dot F_p^{\alpha,q}$. Now Theorem 2.3.6 gives that $s=\{s_Q\}_Q$ can be written as a sum of r_j, ∞-atoms for $\dot f_p^{\alpha,q}$, that is,

$$s=\sum_{j=1}^\infty \lambda_j r_j ,$$

where

$$\Big(\sum_{j=1}^\infty |\lambda_j|^p\Big)^{\frac1p}\le c\|s\|_{\dot f_p^{\alpha,q}} .$$

Since $\|s\|_{\dot f_p^{\alpha,q}}\le C_{p,q,n,\alpha}\|f\|_{\dot F_p^{\alpha,q}}$, Theorem 2.3.4 implies that

$$\Big(\sum_{j=1}^\infty |\lambda_j|^p\Big)^{\frac1p}\le c'\|f\|_{\dot F_p^{\alpha,q}} . \tag{2.3.23}$$

For $j=1,2,\dots$ set

$$A_j=\sum_{\mu\in\mathbf Z}\Big(\sum_{Q\in\mathscr D_\mu} r_{j,Q} a_Q\Big) \tag{2.3.24}$$

and note that Theorem 2.3.4 gives that the series in μ in (2.3.24) converges in $\dot F_p^{\alpha,q}$ and the $\dot F_p^{\alpha,q}$ quasi-norm of A_j is bounded by a constant in view of (2.3.4), since $\|r_j\|_{\dot f_p^{\alpha,q}}\le 1$. Appealing again to (2.3.4) in Theorem 2.3.4 we obtain

$$\Big\|\sum_{j=1}^N \lambda_j A_j - f\Big\|_{\dot F_p^{\alpha,q}}^p \le C_{n,p,q,\alpha}^p \Big\|\sum_{j=1}^N \lambda_j r_j - s\Big\|_{\dot f_p^{\alpha,q}}^p \le C_{n,p,q,\alpha}^p \sum_{j=N+1}^\infty |\lambda_j|^p \to 0$$

as $N\to\infty$. The $\ge$ part of the claimed equivalence is a consequence of (2.3.23). □

2.3.5 Atomic Decomposition of Hardy Spaces

We now pass to one of the main theorems of this chapter, the atomic decomposition of $H^p(\mathbf{R}^n)$ for $0 < p \le 1$. We begin by defining atoms for H^p.

Definition 2.3.10. Let $1 < q \le \infty$. A function A is called *an* L^q*-atom for* $H^p(\mathbf{R}^n)$ if there exists a cube Q such that

(a) A is supported in Q;

(b) $\big\|A\big\|_{L^q} \le |Q|^{\frac{1}{q}-\frac{1}{p}}$;

(c) $\displaystyle\int x^\gamma A(x)\,dx = 0$ for all multi-indices γ with $|\gamma| \le [\frac{n}{p} - n]$.

Notice that any L^r-atom for H^p is also an L^q-atom for H^p whenever $0 < p \le 1$ and $1 < q < r \le \infty$. It is also simple to verify that an L^q-atom A for H^p is in fact in H^p. We prove this result in the next theorem for $q = 2$, and we refer to Exercise 2.3.4 for a general q.

Theorem 2.3.11. *Let* $0 < p \le 1$. *There is a constant* $C_{n,p} < \infty$ *such that every* L^2*-atom* A *for* $H^p(\mathbf{R}^n)$ *satisfies*

$$\|A\|_{H^p} \le C_{n,p}\,.$$

Proof. We could prove this theorem either by showing that the smooth maximal function $M(A;\Phi)$ is in L^p or by showing that the square function $\big(\sum_{j\in\mathbf{Z}} |\Delta_j^\Psi(A)|^2\big)^{1/2}$ is in L^p. Both proofs are similar and we choose to present the second.

Let $A(x)$ be an atom supported in a cube Q centered at the origin; otherwise apply the argument to the atom $A(x - c_Q)$, where c_Q is the center of Q. We control the L^p quasi-norm of $\big(\sum_{j\in\mathbf{Z}} |\Delta_j^\Psi(A)|^2\big)^{1/2}$ by estimating it over the cube Q^* and over $(Q^*)^c$, where $Q^* = 2\sqrt{n}\,Q$. We have

$$\bigg(\int_{Q^*}\big(\sum_{j\in\mathbf{Z}} |\Delta_j^\Psi(A)|^2\big)^{\frac{p}{2}}\,dx\bigg)^{\frac{1}{p}} \le \bigg(\int_{Q^*}\sum_{j\in\mathbf{Z}} |\Delta_j^\Psi(A)|^2\,dx\bigg)^{\frac{1}{2}} |Q^*|^{\frac{1}{p(2/p)'}}\,.$$

Using that the square function $f \mapsto \big(\sum_{j\in\mathbf{Z}} |\Delta_j^\Psi(f)|^2\big)^{\frac{1}{2}}$ is L^2 bounded, we obtain

$$\begin{aligned}\bigg(\int_{Q^*}\big(\sum_{j\in\mathbf{Z}} |\Delta_j^\Psi(A)|^2\big)^{\frac{p}{2}}\,dx\bigg)^{\frac{1}{p}} &\le C_n \|A\|_{L^2} |Q^*|^{\frac{1}{p(2/p)'}} \\ &\le C_n (2\sqrt{n})^{\frac{n}{p}-\frac{n}{2}} |Q|^{\frac{1}{2}-\frac{1}{p}} |Q|^{\frac{1}{p}-\frac{1}{2}} \\ &= C_n'\,.\end{aligned} \tag{2.3.25}$$

To estimate the contribution of the square function outside Q^*, we use the cancellation of the atoms. Let $k = [\frac{n}{p} - n] + 1$. We have

$$
\begin{aligned}
\Delta_j^\Psi(A)(x) &= \int_Q A(y)\Psi_{2^{-j}}(x-y)\,dy \\
&= 2^{jn}\int_Q A(y)\left[\Psi(2^jx-2^jy) - \sum_{|\beta|\le k-1}(\partial^\beta\Psi)(2^jx)\frac{(-2^jy)^\beta}{\beta!}\right]dy \\
&= 2^{jn}\int_Q A(y)\left[\sum_{|\beta|=k}(\partial^\beta\Psi)(2^jx-2^j\theta y)\frac{(-2^jy)^\beta}{\beta!}\right]dy,
\end{aligned}
$$

where $0 \le \theta \le 1$. Taking absolute values, using the fact that $\partial^\beta\Psi$ are Schwartz functions, and that $|x-\theta y| \ge |x|-|y| \ge \frac{1}{2}|x|$ whenever $y \in Q$ and $x \notin Q^*$, we obtain the estimate

$$
\begin{aligned}
|\Delta_j^\Psi(A)(x)| &\le 2^{jn}\int_Q |A(y)|\sum_{|\beta|=k}\frac{C_N}{(1+2^j\frac{1}{2}|x|)^N}\frac{|2^jy|^k}{\beta!}dy \\
&\le \frac{C_{N,p,n}2^{j(k+n)}}{(1+2^j|x|)^N}\left(\int_Q|A(y)|^2\,dy\right)^{\frac{1}{2}}\left(\int_Q|y|^{2k}\,dy\right)^{\frac{1}{2}} \\
&\le \frac{C'_{N,p,n}2^{j(k+n)}}{(1+2^j|x|)^N}|Q|^{\frac{1}{2}-\frac{1}{p}}|Q|^{\frac{k}{n}+\frac{1}{2}} \\
&= \frac{C_{N,p,n}2^{j(k+n)}}{(1+2^j|x|)^N}|Q|^{1+\frac{k}{n}-\frac{1}{p}}
\end{aligned}
$$

for $j \subset \mathbf{Z}$ and $x \in (Q^*)^c$. For such x we now have

$$
\left(\sum_{j\in\mathbf{Z}}|\Delta_j^\Psi(A)(x)|^2\right)^{\frac{1}{2}} \le C_{N,p,n}|Q|^{1+\frac{k}{n}-\frac{1}{p}}\left(\sum_{j\in\mathbf{Z}}\frac{2^{2j(k+n)}}{(1+2^j|x|)^{2N}}\right)^{\frac{1}{2}}. \tag{2.3.26}
$$

It is a simple fact that the series in (2.3.26) converges. Indeed, considering the cases $2^j \le 1/|x|$ and $2^j > 1/|x|$ we see that the series on the right in (2.3.26) contributes at most a fixed multiple of $|x|^{-2k-2n}$. It remains to estimate the L^p quasi-norm of the expression on the right in (2.3.26) over $(Q^*)^c$. This is bounded by a constant multiple of

$$
\left(\int_{(Q^*)^c}\frac{(|Q|^{1+\frac{k}{n}-\frac{1}{p}})^p}{|x|^{p(k+n)}}\,dx\right)^{\frac{1}{p}} \le C_{n,p}|Q|^{1+\frac{k}{n}-\frac{1}{p}}\left(\int_{c|Q|^{\frac{1}{n}}}^\infty r^{-p(k+n)+n-1}\,dr\right)^{\frac{1}{p}},
$$

for some constant c, and the latter is easily seen to be bounded above by an absolute constant. Here we used the fact that $p(k+n) > n$ or, equivalently, $k > \frac{n}{p} - n$, which is certainly true, since k was chosen to be $[\frac{n}{p} - n] + 1$. Combining this estimate with that in (2.3.25), we conclude the proof of the theorem. □

We have now proved that L^q-atoms for H^p are indeed elements of H^p. We now obtain the converse statement, i.e., every element of H^p can be decomposed as a sum of L^2-atoms for H^p. Applying the same idea as in Corollary 2.3.9 to H^p, we obtain the following result.

Theorem 2.3.12. *Let $0 < p \le 1$. Given a distribution $f \in H^p(\mathbf{R}^n)$, there exists a sequence of L^2-atoms for H^p, $\{A_j\}_{j=1}^\infty$, and a sequence of scalars $\{\lambda_j\}_{j=1}^\infty$ such that*

$$\sum_{j=1}^{N} \lambda_j A_j \to f \qquad \text{in } H^p. \tag{2.3.27}$$

Thus the space of all finite linear combinations of L^2-atoms for H^p is dense in H^p. Moreover, we have

$$\begin{aligned}\big\|f\big\|_{H^p} \approx \inf\Big\{\Big(\sum_{j=1}^{\infty} |\lambda_j|^p\Big)^{\frac{1}{p}} : \lim_{N\to\infty}\Big\|\sum_{j=1}^{N} \lambda_j A_j - f\Big\|_{H^p} = 0 \\ \text{where } A_j \text{ are } L^2\text{-atoms for } H^p.\Big\}.\end{aligned} \tag{2.3.28}$$

Proof. Fix $f \in H^p(\mathbf{R}^n)$. Let A_j be L^2-atoms for H^p and $\sum_{j=1}^\infty |\lambda_j|^p < \infty$ such that (2.3.27) holds. It follows from Theorem 2.3.11 and the sublinearity of the expression $g \mapsto \|g\|_{H^p}^p$ that

$$\Big\|\sum_{j=1}^{N} \lambda_j A_j\Big\|_{H^p}^p \le C_{n,p}^p \sum_{j=1}^{N} |\lambda_j|^p .$$

Thus if the sequence $\sum_{j=1}^N \lambda_j A_j$ converges to f in H^p, then

$$\big\|f\big\|_{H^p}^p \le \Big\|f - \sum_{j=1}^{N} \lambda_j A_j\Big\|_{H^p}^p + C_{n,p}^p \sum_{j=1}^{N} |\lambda_j|^p ,$$

and letting $N \to \infty$ proves the direction $\le$ in (2.3.28).

We now focus on the converse statement, which is similar to the analogous statement in Corollary 2.3.9. Let $L = [\frac{n}{p} - n]$. Given f in $\dot F_p^{0,2} = H^p$, via Theorem 2.3.4 we write

$$f = \sum_{\mu\in\mathbf{Z}}\Big(\sum_{Q\in\mathscr{D}_\mu} s_Q a_Q\Big)$$

where $s = \{s_Q\}_{Q\in\mathscr{D}}$ lies in $\dot f_p^{0,2}$, each a_Q is a smooth L-atom for the cube Q and the series converges in H^p. Theorem 2.3.6 gives that $s = \{s_Q\}_Q$ can be written as a sum of r_j, ∞-atoms for $\dot f_p^{0,2}$, that is, $s = \sum_{j=1}^\infty \lambda_j r_j$, where

$$\Big(\sum_{j=1}^{\infty} |\lambda_j|^p\Big)^{\frac{1}{p}} \le c\|s\|_{\dot f_p^{0,2}} .$$

Since $\|s\|_{\dot{f}_p^{0,2}} \le C_{n,p}\|f\|_{H^p}$, Theorem 2.3.4 implies that

$$\Big(\sum_{j=1}^{\infty}|\lambda_j|^p\Big)^{\frac{1}{p}} \le c'\|f\|_{H^p}. \tag{2.3.29}$$

For $j = 1,2,\dots$ set

$$A_j = \sum_{\mu\in\mathbf{Z}}\Big(\sum_{Q\in\mathscr{D}_\mu} r_{j,Q}a_Q\Big)$$

where the series converges in H^p (see Theorem 2.3.4) and the H^p quasi-norm of A_j is bounded by a constant in view of (2.3.4), since $\|r_j\|_{\dot{f}_p^{0,2}} \le 1$. Using again (2.3.4) in Theorem 2.3.4 we obtain

$$\begin{aligned}
\Big\|\sum_{j=1}^{N}\lambda_jA_j - f\Big\|_{H^p}^p &\le C_{n,p}^p\Big\|\sum_{j=1}^{N}\lambda_jr_j - s\Big\|_{\dot{f}_p^{0,2}}^p\\
&\le C_{n,p}^p\Big\|\sum_{j=N+1}^{\infty}|\lambda_j|\,|r_j|\Big\|_{\dot{f}_p^{0,2}}^p\\
&\le C_{n,p}^p\sum_{j=N+1}^{\infty}|\lambda_j|^p \to 0,
\end{aligned}$$

as $N\to\infty$, where the last inequality follows from Exercise 2.3.6(a).

Next we show that each A_j is a fixed multiple of an L^2-atom for H^p. Let us fix an index j. By the definition of the ∞-atom for $\dot{f}_p^{0,2}$, there exists a dyadic cube Q_0^j such that $r_{j,Q} = 0$ for all dyadic cubes Q not contained in Q_0^j. Then the support of each a_Q is contained in $3Q$, hence in $3Q_0^j$. This implies that the function A_j is supported in $3Q_0^j$. The same is true for the function $g^{0,2}(r_j)$ defined in (2.3.2). Using this fact, we have

$$\begin{aligned}
\|A_j\|_{L^2} &\approx \|A_j\|_{\dot{F}_2^{0,2}}\\
&\le c\,\|r_j\|_{\dot{f}_2^{0,2}}\\
&= c\,\|g^{0,2}(r_j)\|_{L^2}\\
&\le c\,\|g^{0,2}(r_j)\|_{L^\infty}|3Q_0^j|^{\frac{1}{2}}\\
&\le c\,|3Q_0^j|^{-\frac{1}{p}+\frac{1}{2}}.
\end{aligned}$$

Since

$$g^{0,2}(r_j) = \Big(\sum_{Q\in\mathscr{D}}|Q|^{-1}|r_{j,Q}|^2\chi_Q\Big)^{\frac{1}{2}}$$

the estimate $\|g^{0,2}(r_j)\|_{L^2} \le |3Q_0^j|^{-\frac{1}{p}+\frac{1}{2}}$ we proved implies that

$$\sum_{Q\in\mathscr{D}} |r_{j,Q}|^2 < \infty.$$

Let $M' < M$ be positive integers. Then

$$\begin{aligned}\Big\|\sum_{M'<|\mu|\le M}\sum_{Q\in\mathscr{D}_\mu} r_{j,Q}a_Q\Big\|_{L^1} &\le |3Q_0^j|^{\frac{1}{2}}\Big\|\sum_{M'<|\mu|\le M}\sum_{Q\in\mathscr{D}_\mu} r_{j,Q}a_Q\Big\|_{L^2}\\ &= |3Q_0^j|^{\frac{1}{2}}\Big(\sum_{M'<|\mu|\le M}\sum_{Q\in\mathscr{D}_\mu}|r_{j,Q}|^2\Big)^{\frac{1}{2}} \to 0\end{aligned}$$

as $M',M\to\infty$. Therefore the sequence $\sum_{|\mu|\le M}\sum_{Q\in\mathscr{D}_\mu} r_{j,Q}a_Q$ is Cauchy in L^1 and hence it converges in L^1. But this sequence converges in H^p to A_j by Theorem 2.3.4, so finally it converges to A_j in L^1.

The fact that $A_j = \sum_{\mu\in\mathbf{Z}}\sum_{Q\in\mathscr{D}_\mu} r_{j,Q}a_Q$ with convergence in L^1 allows us to deduce that vanishing moments of a_Q pass on to A_j. We conclude that each A_j is a fixed multiple of an L^2-atom for H^p. The $\ge$ direction in (2.3.28) now follows from (2.3.29), given that we have now established all the remaining properties. □

Remark 2.3.13. Property (c) in Definition 2.3.10 can be replaced by

$$\int x^\gamma A(x)\,dx = 0 \qquad \text{for all multi-indices } \gamma \text{ with } |\gamma|\le L,$$

for any $L \ge [\frac{n}{p}-n]$, and the atomic decomposition of H^p holds unchanged. In fact, in the proof of Theorem 2.3.12 we may take $L \ge [\frac{n}{p}-n]$ instead of $L=[\frac{n}{p}-n]$ and then apply Theorem 2.3.4 for this L. Note that Theorem 2.3.4 was valid for all $L\ge[\frac{n}{p}-n]$. This observation turns out to be quite useful in certain applications.

Exercises

2.3.1. (a) Given $N\in\mathbf{Z}^+$, prove that there exists a Schwartz function Θ supported in the unit ball $|x|\le 1$ such that $\int_{\mathbf{R}^n} x^\gamma\Theta(x)\,dx=0$ for all multi-indices γ with $|\gamma|\le N$ and such that $|\widehat{\Theta}(\xi)|\ge\frac{1}{2}$ for all ξ in the annulus $\frac{1}{2}\le|\xi|\le 2$.
(b) Prove there exists a Schwartz function Ψ whose Fourier transform is supported in the annulus $\frac{1}{2}\le|\xi|\le 2$ and is at least $c>0$ in the smaller annulus $\frac{3}{5}\le|\xi|\le\frac{5}{3}$ and which satisfies for all $\xi\in\mathbf{R}^n\setminus\{0\}$

$$\sum_{j\in\mathbf{Z}}\widehat{\Psi}(2^{-j}\xi)\widehat{\Theta}(2^{-j}\xi)=1.$$

[*Hint:* Part (a): Let θ be an even real-valued Schwartz function supported in the ball $|x|\le 1$ and such that $\widehat{\theta}(0)=1$. Then for some $\varepsilon\in(0,\frac{1}{2})$ we have $\widehat{\theta}(\xi)\ge\frac{1}{2}$ for all

ξ satisfying $|\xi| < 2\varepsilon$. Set $\Theta = (-\Delta)^N(\theta_\varepsilon)$. Part (b): Define the function $\widehat{\Psi}(\xi) = \widehat{\eta}(\xi)\big(\sum_{j\in\mathbf{Z}}\widehat{\eta}(2^{-j}\xi)\widehat{\Theta}(2^{-j}\xi)\big)^{-1}$, where $\widehat{\eta}(\xi)$ is a Schwartz function supported in the annulus $\frac{1}{2} \le |\xi| \le 2$ and equal to 1 on the smaller annulus $\frac{3}{5} \le |\xi| \le \frac{5}{3}$.]

2.3.2. Let $\alpha \in \mathbf{R}$, $0 < p \le 1$, $p \le q < +\infty$.
(a) For all f,g in $\dot{F}_p^{\alpha,q}$ show that

$$\|f+g\|_{\dot{F}_p^{\alpha,q}}^p \le \|f\|_{\dot{F}_p^{\alpha,q}}^p + \|g\|_{\dot{F}_p^{\alpha,q}}^p .$$

(b) For all sequences $\{s_Q\}_{Q\in\mathscr{D}}$ and $\{t_Q\}_{Q\in\mathscr{D}}$ show that

$$\|\{s_Q\}_Q + \{t_Q\}_Q\|_{\dot{f}_p^{\alpha,q}}^p \le \|\{s_Q\}_Q\|_{\dot{f}_p^{\alpha,q}}^p + \|\{t_Q\}_Q\|_{\dot{f}_p^{\alpha,q}}^p .$$

[*Hint:* Use $|a+b|^p \le |a|^p + |b|^p$ and apply Minkowski's inequality on $L^{q/p}$ (or on $\ell^{q/p}$).]

2.3.3. Let Φ be a smooth function supported in the unit ball of $\mathbf{R}^n$. Use the same idea as in Theorem 2.3.11 to show directly (without appealing to any other theorem) that the smooth maximal function $M(\,\cdot\,;\Phi)$ of an L^2-atom for H^p lies in L^p when $p<1$. Recall that $M(f;\Phi) = \sup_{t>0}|\Phi_t * f|$.

2.3.4. Extend Theorem 2.3.11 to the case $1 < q \le \infty$. Precisely, prove that there is a constant $C_{n,p,q}$ such that every L^q-atom A for H^p satisfies

$$\|A\|_{H^p} \le C_{n,p,q} .$$

[*Hint:* If $1<q<2$ use the boundedness of the square function on L^q while for $2 \le q \le \infty$ use its boundedness on L^2.]

2.3.5. (a) Suppose that $s_Q^k \ge 0$ for all $Q \in \mathscr{D}$ and $k = 1,2,\dots$. Prove that

$$\|\{\sum_{k=1}^\infty s_Q^k\}_Q\|_{\dot{f}_p^{\alpha,q}}^p \le \sum_{k=1}^\infty \|\{s_Q^k\}_Q\|_{\dot{f}_p^{\alpha,q}}^p .$$

(b) Prove the completeness of the spaces $\dot{f}_p^{\alpha,q}$ when $\alpha\in\mathbf{R}$, $0<p\le 1$, $p \le q < \infty$.
[*Hint:* Part (b): You may want to use part (a) together with the fact that if a quasi-normed space $(X,\|\cdot\|)$ has the property $\|x+y\|^p \le \|x\|^p + \|y\|^p$ for all $x,y\in X$, then $(X,\|\cdot\|)$ is complete if and only if for every sequence $x_k\in X$ with the property $\sum_{k=1}^\infty \|x_k\|^p < \infty$ there is an x_* such that $\|\sum_{k=1}^N x_k - x_*\| \to 0$ as $N\to\infty$.]

2.3.6. Show that for all $\mu, j \in \mathbf{Z}$, all $N,b>0$ satisfying $N > n/b$ and $b<1$, all scalars s_Q (indexed by dyadic cubes Q with centers c_Q), and all $x\in\mathbf{R}^n$ we have

$$\sum_{Q\in\mathscr{D}_\mu} \frac{|s_Q|}{(1+2^{\min(j,\mu)}|x-c_Q|)^N}$$
$$\leq c(n,N,b)\, 2^{\max(\mu-j,0)\frac{n}{b}} \left\{ M\Big(\sum_{Q\in\mathscr{D}_\mu} |s_Q|^b \chi_Q\Big)(x)\right\}^{\frac{1}{b}},$$

where M is the Hardy–Littlewood maximal operator and $c(n,N,b)$ is a constant.
[*Hint:* Fix $x\in\mathbf{R}^n$ and define $\mathscr{F}_0=\{Q\in\mathscr{D}_\mu:\ |c_Q-x|\,2^{\min(j,\mu)}\le 1\}$ and for $k\ge 1$ $\mathscr{F}_k=\{Q\in\mathscr{D}_\mu:\ 2^{k-1}<|c_Q-x|\,2^{\min(j,\mu)}\le 2^k\}$. Break up the sum on the left as a sum over the families $\mathscr{F}_k$ and use that $\sum_{Q\in\mathscr{F}_k}|s_Q|\le\big(\sum_{Q\in\mathscr{F}_k}|s_Q|^b\big)^{1/b}$ and the fact that $\big|\bigcup_{Q\in\mathscr{F}_k}Q\big|\le c_n 2^{-\min(j,\mu)n+kn}$.]

2.3.7. Let A be an L^2-atom for $H^p(\mathbf{R}^n)$ for some $0<p<1$. Show that there is a constant C such that for all multi-indices α with $|\alpha|\le k=[\frac{n}{p}-n]$ we have

$$\sup_{\xi\in\mathbf{R}^n} |\xi|^{|\alpha|-k-1}\big|(\partial^\alpha\widehat{A})(\xi)\big|\le C\big\|A\big\|_{L^2(\mathbf{R}^n)}^{-\frac{2p}{2-p}(\frac{k+1}{n}+\frac{1}{2})+1}.$$

[*Hint:* Subtract the Taylor polynomial of degree $k-|\alpha|$ at 0 of the function $x\mapsto e^{-2\pi i x\cdot\xi}$.]

2.3.8. Let A be an L^2-atom for $H^p(\mathbf{R}^n)$ for some $0<p<1$. Show that for all multi-indices α and all $1\le r\le\infty$ there is a constant C such that

$$\big\|\,|\partial^\alpha\widehat{A}|^2\big\|_{L^{r'}(\mathbf{R}^n)}\le C\big\|A\big\|_{L^2(\mathbf{R}^n)}^{-\frac{2p}{2-p}(\frac{2|\alpha|}{n}+\frac{1}{r})+2}.$$

[*Hint:* In the case $r=1$ use the $L^1\to L^\infty$ boundedness of the Fourier transform and in the case $r=\infty$ use Plancherel's theorem. For general r use interpolation.]

2.3.9. Let f be in $H^p(\mathbf{R}^n)$ for some $0<p\le 1$. Then the Fourier transform of f, originally defined as a tempered distribution, is a continuous function that satisfies

$$|\widehat{f}(\xi)|\le C_{n,p}\big\|f\big\|_{H^p(\mathbf{R}^n)}|\xi|^{\frac{n}{p}-n}$$

for some constant $C_{n,p}$ independent of f.
[*Hint:* If f is an L^2-atom for H^p, combine the estimates of Exercises 2.3.7 and 2.3.8 with $\alpha=0$ (and $r=1$). In general, apply Theorem 2.3.12.]

2.3.10. Let A be an L^∞-atom for $H^p(\mathbf{R}^n)$ for some $0<p<1$ and let $\alpha=\frac{n}{p}-n$. Show that there is a constant $C_{n,p}$ such that for all g in $\dot{\Lambda}_\alpha(\mathbf{R}^n)$ we have

$$\left|\int_{\mathbf{R}^n} A(x)g(x)\,dx\right|\le C_{n,p}\,\big\|g\big\|_{\dot{\Lambda}_\alpha(\mathbf{R}^n)}.$$

[*Hint:* Suppose that A is supported in a cube Q of side length $2^{-\nu}$ and center c_Q. Write the previous integrand as $\sum_{j\in\mathbf{Z}}\Delta_j^\Omega(A)\Delta_j^\Psi(g)$ for a Littlewood–Paley operator

Δ_j^Ψ associated with a function $\Psi \in \mathscr{S}$ whose Fourier transform is nonnegative, supported in $\frac{6}{7} < |\xi| < 2$, and satisfies $\sum_{j\in\mathbf{Z}} \widehat{\Psi}(2^{-j}\xi) = 1$ for all $\xi \neq 0$, while $\widehat{\Omega}(\xi)$ is $\mathscr{C}^\infty$, supported in $\frac{6}{14} < |\xi| < 4$, and is equal to one on the support of $\widehat{\Psi}(\xi)$. Then apply the result of Appendix B.4 to obtain the estimate

$$\left|\Delta_j^\Omega(\overline{A})(x)\right| \le C_N |Q|^{-\frac{1}{p}+1} \frac{2^{\min(j,v)n} 2^{-|j-v|D}}{\left(1+2^{\min(j,v)}|x-c_Q|\right)^N},$$

where $D = [\alpha]+1$ when $v \ge j$ and $D = 0$ when $v < j$. Use Theorem 1.4.6.]

2.3.11. Let $\varepsilon > 0$. Show that the function

$$h(x) = \frac{\chi_{|x|<1/2}}{x\left(\log\frac{1}{|x|}\right)^{1+\varepsilon}}$$

lies in the Hardy space $H^1(\mathbf{R})$ although

$$\int_{-1/2}^{1/2} |h(t)| \log|h(t)|\, dt = \infty.$$

[*Hint:* For $j = 1, 2, \dots$ define atoms $a_j(x) = c\, j^{1+\varepsilon}\left(h\chi_{I_j} - \operatorname{Avg}_{I_j}(h\chi_{I_j})\right)$ supported in $I_j = (2^{-j}, 2^{-j+1})$ and $b_j(x) = c\, j^{1+\varepsilon}\left(h\chi_{L_j} - \operatorname{Avg}_{I_j}(h\chi_{L_j})\right)$ supported in $L_j = (-2^{-j+1}, -2^{-j})$ for a suitable $c > 0$. Then write $h = \sum_{j=1}^\infty \frac{1}{c\, j^{1+\varepsilon}}(a_j + b_j)$.]

2.4 Singular Integrals on Function Spaces

Our final task in this chapter is to investigate the action of singular integrals on function spaces. The emphasis of our study focuses on Hardy spaces, although with no additional effort the action of singular integrals on other function spaces can also be obtained.

2.4.1 Singular Integrals on the Hardy Space H^1

Before we discuss the main results in this topic, we review some background on singular integrals.

Let $K(x)$ be a function defined away from the origin on $\mathbf{R}^n$ that satisfies the size estimate

$$\sup_{0<R<\infty} \frac{1}{R} \int_{|x|\le R} |K(x)|\, |x|\, dx \le A_1, \tag{2.4.1}$$

the smoothness condition,

$$\sup_{y\in\mathbf{R}^n\setminus\{0\}} \int_{|x|\geq 2|y|} |K(x-y)-K(x)|\,dx \leq A_2\,, \tag{2.4.2}$$

and the cancellation condition

$$\sup_{0<R_1<R_2<\infty} \left| \int_{R_1<|x|<R_2} K(x)\,dx \right| \leq A_3\,, \tag{2.4.3}$$

for some $A_1, A_2, A_3 < \infty$. Condition (2.4.3) implies that there exists a sequence $\varepsilon_j \downarrow 0$ as $j \to \infty$ such that the following limit exists:

$$\lim_{j\to\infty} \int_{\varepsilon_j \leq |x| \leq 1} K(x)\,dx = L_0.$$

This gives that for a smooth and compactly supported function f on $\mathbf{R}^n$, the limit

$$\lim_{j\to\infty} \int_{|x-y|>\varepsilon_j} K(x-y)f(y)\,dy = T(f)(x) \tag{2.4.4}$$

exists and defines a linear operator T. This operator T is given by convolution with a tempered distribution W that coincides with the function K on $\mathbf{R}^n \setminus \{0\}$.

We know that such a T, initially defined on $\mathscr{C}_0^\infty(\mathbf{R}^n)$, admits an extension that is L^p bounded for all $1 < p < \infty$ and is also of weak type $(1,1)$. All these norms are bounded above by dimensional constant multiples of the quantity $A_1 + A_2 + A_3$ (cf. Theorem 5.4.1 in [156]). Therefore, such a T is well defined on $L^1(\mathbf{R}^n)$ and in particular on $H^1(\mathbf{R}^n)$, which is contained in $L^1(\mathbf{R}^n)$. The following result concerns the H^1 to L^1 boundedness of T.

Theorem 2.4.1. *Let K satisfy (2.4.1), (2.4.2), and (2.4.3), and let T be defined as in (2.4.4). Then there is a constant C_n such that for all f in $H^1(\mathbf{R}^n)$ we have*

$$\big\|T(f)\big\|_{L^1} \leq C_n(A_1+A_2+A_3)\big\|f\big\|_{H^1}. \tag{2.4.5}$$

Proof. To prove this theorem we have a powerful tool at our disposal, the atomic decomposition of $H^1(\mathbf{R}^n)$. It is therefore natural to start by checking the validity of (2.4.5) whenever f is an L^2-atom for H^1.

Since T is a convolution operator (i.e., it commutes with translations), it suffices to take the atom f supported in a cube Q centered at the origin. Let $f = a$ be such an atom, supported in Q, and let $Q^* = 2\sqrt{n}\,Q$. We write

$$\int_{\mathbf{R}^n} |T(a)(x)|\,dx = \int_{Q^*} |T(a)(x)|\,dx + \int_{(Q^*)^c} |T(a)(x)|\,dx \tag{2.4.6}$$

and we estimate each term separately. We have

$$\begin{aligned}\int_{Q^*}|T(a)(x)|\,dx &\le |Q^*|^{\frac{1}{2}}\Big(\int_{Q^*}|T(a)(x)|^2\,dx\Big)^{\frac{1}{2}}\\ &\le C_n(A_1+A_2+A_3)|Q^*|^{\frac{1}{2}}\Big(\int_Q|a(x)|^2\,dx\Big)^{\frac{1}{2}}\\ &\le C_n(A_1+A_2+A_3)|Q^*|^{\frac{1}{2}}|Q|^{\frac{1}{2}-\frac{1}{1}}\\ &= C_n'(A_1+A_2+A_3),\end{aligned}$$

where we used the L^2 boundedness of T and property (b) of atoms in Definition 2.3.10. Now note that if $x\notin Q^*$ and $y\in Q$, then $|x|\ge 2|y|$ and $x-y$ stays away from zero; thus $K(x-y)$ is well defined. Moreover, in this case $T(a)(x)$ can be expressed as a convergent integral of $a(y)$ against $K(x-y)$. We have

$$\begin{aligned}\int_{(Q^*)^c}|T(a)(x)|\,dx &= \int_{(Q^*)^c}\Big|\int_Q K(x-y)a(y)\,dy\Big|\,dx\\ &= \int_{(Q^*)^c}\Big|\int_Q \big(K(x-y)-K(x)\big)a(y)\,dy\Big|\,dx\\ &\le \int_Q\int_{(Q^*)^c}\big|K(x-y)-K(x)\big|\,dx\,|a(y)|\,dy\\ &\le \int_Q\int_{|x|\ge 2|y|}\big|K(x-y)-K(x)\big|\,dx\,|a(y)|\,dy\\ &\le A_2\int_Q|a(x)|\,dx\\ &\le A_2|Q|^{\frac{1}{2}}\Big(\int_Q|a(x)|^2\,dx\Big)^{\frac{1}{2}}\\ &\le A_2|Q|^{\frac{1}{2}}|Q|^{\frac{1}{2}-\frac{1}{1}} = A_2.\end{aligned}$$

Combining this calculation with the previous one and inserting the final conclusions in (2.4.6) we deduce that L^2-atoms a for H^1 satisfy

$$\big\|T(a)\big\|_{L^1}\le (C_n'+1)(A_1+A_2+A_3). \tag{2.4.7}$$

We now pass to general functions in H^1. In view of Theorem 2.3.12 we can write an $f\in H^1$ as

$$f=\sum_{j=1}^{\infty}\lambda_j a_j,$$

where the series converges in H^1, the a_j are L^2-atoms for H^1, and

$$\big\|f\big\|_{H^1}\approx\sum_{j=1}^{\infty}|\lambda_j|<\infty. \tag{2.4.8}$$

Since T maps L^1 to $L^{1,\infty}$ (Theorem 5.3.3 in [156]), $T(f)$ is already a well-defined $L^{1,\infty}$ function.

We claim that

$$T(f) = \sum_{j=1}^{\infty} \lambda_j T(a_j) \qquad \text{a.e.} \tag{2.4.9}$$

noting that the series in (2.4.9) converges in L^1 and produces a well-defined integrable function. Once (2.4.9) is established, the required conclusion (2.4.5) follows easily by taking L^1 norms in (2.4.9) and using (2.4.7) and (2.4.8).

To prove (2.4.9), we make use of the fact that T is of weak type $(1,1)$. For a given $\delta > 0$ we have

$$\begin{aligned}
&\Big|\Big\{\Big|T(f) - \sum_{j=1}^{\infty} \lambda_j T(a_j)\Big| > \delta\Big\}\Big| \\
&\le \Big|\Big\{\Big|T(f) - \sum_{j=1}^{N} \lambda_j T(a_j)\Big| > \delta/2\Big\}\Big| + \Big|\Big\{\Big|\sum_{j=N+1}^{\infty} \lambda_j T(a_j)\Big| > \delta/2\Big\}\Big| \\
&\le \frac{2}{\delta}\big\|T\big\|_{L^1 \to L^{1,\infty}} \Big\|f - \sum_{j=1}^{N} \lambda_j a_j\Big\|_{L^1} + \frac{2}{\delta}\Big\|\sum_{j=N+1}^{\infty} \lambda_j T(a_j)\Big\|_{L^1} \\
&\le \frac{2}{\delta}\big\|T\big\|_{L^1 \to L^{1,\infty}} \Big\|f - \sum_{j=1}^{N} \lambda_j a_j\Big\|_{H^1} + \frac{2}{\delta}(C_n' + 1)(A_1 + A_2 + A_3) \sum_{j=N+1}^{\infty} |\lambda_j| .
\end{aligned}$$

Since $\sum_{j=1}^{N} \lambda_j a_j$ converges to f in H^1 and $\sum_{j=1}^{\infty} |\lambda_j| < \infty$, both terms in the sum converge to zero as $N \to \infty$. We conclude that

$$\Big|\Big\{x \in \mathbf{R}^n : \Big|T(f)(x) - \sum_{j=1}^{\infty} \lambda_j T(a_j)(x)\Big| > \delta\Big\}\Big| = 0$$

for all $\delta > 0$, which implies (2.4.9). □

2.4.2 Singular Integrals on Besov–Lipschitz Spaces

We continue with a corollary concerning Besov–Lipschitz spaces.

Corollary 2.4.2. *Let K satisfy (2.4.1), (2.4.2), and (2.4.3), and let T be defined as in (2.4.4). Let $1 \le p \le \infty$, $0 < q \le \infty$, and $\alpha \in \mathbf{R}$. Then there is a constant $C_{n,p,q,\alpha}$ such that for all f in $\mathscr{S}(\mathbf{R}^n)$ we have*

$$\big\|T(f)\big\|_{\dot{B}_p^{\alpha,q}} \le C_n(A_1 + A_2 + A_3)\big\|f\big\|_{\dot{B}_p^{\alpha,q}} . \tag{2.4.10}$$

Therefore, T admits a bounded extension on all homogeneous Besov–Lipschitz spaces $\dot{B}_p^{\alpha,q}$ with $p \ge 1$, in particular, on all homogeneous Lipschitz spaces.

Proof. Let Ψ be a Schwartz function whose Fourier transform is supported in the annulus $1-\frac{1}{7}\le|\xi|\le 2$ and that satisfies

$$\sum_{j\in\mathbf{Z}}\widehat{\Psi}(2^{-j}\xi)=1, \qquad \xi\neq 0.$$

Pick a Schwartz function ζ whose Fourier transform $\widehat{\zeta}$ is supported in the annulus $\frac{1}{4}<|\xi|<8$ and that is equal to one on the support of $\widehat{\Psi}$. Let W be the tempered distribution that coincides with K on $\mathbf{R}^n\setminus\{0\}$ so that $T(f)=f*W$. Then we have $\zeta_{2^{-j}}*\Psi_{2^{-j}}=\Psi_{2^{-j}}$ for all j and hence

$$\begin{aligned}\big\|\Delta_j(T(f))\big\|_{L^p}&=\big\|\zeta_{2^{-j}}*\Psi_{2^{-j}}*W*f\big\|_{L^p}\\&\le\big\|\zeta_{2^{-j}}*W\big\|_{L^1}\big\|\Delta_j(f)\big\|_{L^p},\end{aligned}\tag{2.4.11}$$

since $1\le p\le\infty$. It is not hard to check that the function $\zeta_{2^{-j}}$ is in H^1 with norm independent of j. Therefore, $\zeta_{2^{-j}}$ is in H^1. Using Theorem 2.4.1, we conclude that

$$\big\|T(\zeta_{2^{-j}})\big\|_{L^1}=\big\|\zeta_{2^{-j}}*W\big\|_{L^1}\le C\big\|\zeta_{2^{-j}}\big\|_{H^1}=C'.$$

Inserting this in (2.4.11), multiplying by $2^{j\alpha}$, and taking ℓ^q quasi-norms, we obtain the required conclusion. □

2.4.3 Singular Integrals on $H^p(\mathbf{R}^n)$

It is possible to extend Theorem 2.4.1 to $H^p(\mathbf{R}^n)$ for $p<1$, provided the kernel K has additional smoothness.

For the purposes of this subsection, we fix a $\mathscr{C}^\infty$ function $K(x)$ on $\mathbf{R}^n\setminus\{0\}$. We suppose that there is a positive integer N (to be specified later) such that

$$|\partial^\beta K(x)|\le A\,|x|^{-n-|\beta|}\qquad\text{for all }|\beta|\le N\tag{2.4.12}$$

and that

$$\sup_{0<R_1<R_2<\infty}\left|\int_{R_1<|x|<R_2}K(x)\,dx\right|\le A,\tag{2.4.13}$$

for some $A<\infty$.

We fix a nonnegative smooth function η on $\mathbf{R}^n$ that vanishes in the unit ball of $\mathbf{R}^n$ and is equal to 1 outside the ball $B(0,2)$ and for $0<\varepsilon<1/2$ we define the smoothly truncated family of kernels

$$K^{(\varepsilon)}(x)=K(x)\eta(x/\varepsilon)$$

and the doubly smoothly truncated family of kernels

$$K_{(\varepsilon)}(x) = K^{(\varepsilon)}(x) - K^{(1/\varepsilon)}(x).$$

Condition (2.4.12) with $\beta = 0$ and (2.4.13) imply that

$$\left| \int_{|x|\le 1} K(x)\eta(x/\varepsilon)\,dx \right| \le (1+\omega_{n-1}\log 2)A$$

for all $\varepsilon < 1/2$; hence there exists a sequence $\varepsilon_j < 1/2$ with $\varepsilon_j \downarrow 0$ as $j \to \infty$ such that the following limit exists:

$$\lim_{j\to\infty} \int_{|x|\le 1} K(x)\eta(x/\varepsilon_j)\,dx = L_0.$$

We now define W in $\mathscr{S}'(\mathbf{R}^n)$ by setting

$$\begin{aligned} \langle W, \varphi\rangle &= \lim_{j\to\infty} \int_{\mathbf{R}^n} K_{(\varepsilon_j)}(x)\varphi(x)\,dx \qquad (2.4.14)\\ &= L_0\varphi(0) + \int_{|x|\le 1} K(x)(\varphi(x)-\varphi(0))\,dx + \int_{|x|\ge 1} K(x)\varphi(x)\,dx \end{aligned}$$

for φ in $\mathscr{S}$. It is quite easy to verify that the preceding expression is bounded by a constant multiple of a finite sum of Schwartz seminorms of φ. Note that this distribution[2] depends on the number L_0 and hence on the bump η.

We define the associated doubly smoothly truncated singular integral by setting

$$T_{(\varepsilon)}(\varphi)(x) = \int_{\mathbf{R}^n} K_{(\varepsilon)}(y)\varphi(x-y)\,dy \qquad (2.4.15)$$

for Schwartz functions φ on $\mathbf{R}^n$.

We also define an operator T given by convolution with W by setting

$$T(\varphi) = \lim_{j\to\infty} T_{(\varepsilon_j)}(\varphi) = \varphi * W \qquad (2.4.16)$$

for $\varphi \in \mathscr{S}(\mathbf{R}^n)$. Observe that W coincides with K on $\mathbf{R}^n \setminus \{0\}$, since if φ is supported in $|x| \ge t_0 > 0$, (2.4.14) implies that the action of W on $\varphi \in \mathscr{S}$ coincides with that of $K^{(\varepsilon_j)}$ on φ when $\varepsilon_j < t_0/2$. Condition (2.4.12) with $|\beta| = 1$ implies

$$\sup_{y\neq 0} \int_{|x|\ge 2|y|} |K(x-y) - K(x)|\,dx \le cA; \qquad (2.4.17)$$

hence Theorem 5.4.1 in [156] yields the L^2 boundedness of T. Note that (2.4.17) also holds for $K_{(\varepsilon)}$ in place of K uniformly in ε; thus again by Theorem 5.4.1 in [156] the operators $T_{(\varepsilon)}$ are uniformly bounded on $L^2(\mathbf{R}^n)$.

[2] Alternatively, we could have defined W as an element of $\mathscr{S}'(\mathbf{R}^n)/\mathscr{P}(\mathbf{R}^n)$ acting on functions $\varphi \in \mathscr{S}_0$; in this case W would have been independent of L_0 and η.

We summarize these and other observations about $K_{(\varepsilon)}$, $T_{(\varepsilon)}$, and T.

(i) The kernels $K_{(\varepsilon)}$ satisfy the same estimates as K uniformly in ε with constant A' in place of A, where A' is comparable to A.
(ii) $T_{(\varepsilon)}$ are uniformly bounded on L^2.
(iii) $T_{(\varepsilon_j)}(g)$ tends to $T(g)$ in L^2 for any $g \in L^2(\mathbf{R}^n)$.
(iv) T is L^2 bounded with norm $\|\widehat{W}\|_{L^\infty} \le CA$.
(v) For any $f \in H^p$, $T(f)$ is a well-defined element of $\mathscr{S}'$.

We have already explained assertions (i) and (ii) and (iv).
We explain (iii). Theorem 5.3.4 in [156] gives that for all $g \in L^2$ we have

$$\sup_{\varepsilon>0} |T_{(\varepsilon)}(g)| \le M(T(f)) + C_n A M(g);$$

hence the maximal operator $T^{(**)}(g) = \sup_{\varepsilon>0} |T_{(\varepsilon)}(g)|$ is L^2 bounded. Moreover, as an easy consequence of (2.4.14), for each $\varphi \in \mathscr{S}$ we have $T_{(\varepsilon_j)}(\varphi) \to T(\varphi)$ pointwise everywhere. In view of Theorem 2.1.14 in [156], for every $g \in L^2(\mathbf{R}^n)$ we have $T_{(\varepsilon_j)}(g) - T(g) \to 0$ a.e. as $j \to \infty$. Since

$$|T_{(\varepsilon_j)}(g) - T(g)| \le 2T^{(**)}(g) \in L^2,$$

the Lebesgue dominated convergence theorem yields that $T_{(\varepsilon_j)}(g) - T(g) \to 0$ in L^2.

To verify the validity of (v) we write $W = W_0 + K_\infty$, where $W_0 = \Phi W$ and $K_\infty = (1-\Phi)K$, where Φ is a smooth function equal to one on the ball $B(0,1)$ and vanishing outside the ball $B(0,2)$. Then for f in $H^p(\mathbf{R}^n)$, $0 < p \le 1$, we define a tempered distribution $T(f) = W * f$ by setting

$$\langle T(f), \phi \rangle = \langle f, \phi * \widetilde{W_0} \rangle + \langle \widetilde{\phi} * f, \widetilde{K_\infty} \rangle \tag{2.4.18}$$

for ϕ in $\mathscr{S}(\mathbf{R}^n)$. (Here $\widetilde{\varphi}(x) = \varphi(-x)$ for functions and analogously for distributions.) The function $\phi * \widetilde{W_0}$ is in $\mathscr{S}$, so the action of f on it is well defined. Also $\widetilde{\phi} * f$ is in L^1 (see Proposition 2.1.9), while $\widetilde{K_\infty}$ is in L^∞; hence the second term on the right in (2.4.18) represents an absolutely convergent integral. Moreover, in view of Theorem 2.3.20 in [156] and Corollary 2.1.9, both terms on the right in (2.4.18) are controlled by a finite sum of seminorms $\rho_{\alpha,\beta}(\phi)$ (cf. Definition 2.2.1 in [156]). This defines $T(f)$ as a tempered distribution for every $f \in H^p$.

The following is an extension of Theorem 2.4.1 for $p < 1$.

Theorem 2.4.3. *Let $0 < p < 1$ and $N = [\frac{n}{p} - n] + 1$. Let K be a $\mathscr{C}^\infty$ function on $\mathbf{R}^n \setminus \{0\}$ that satisfies* (2.4.13) *and* (2.4.12) *for some $A < \infty$ for all multi-indices $|\beta| \le N$ and all $x \neq 0$. Let W be a tempered distribution that coincides with K on $\mathbf{R}^n \setminus \{0\}$, as defined in* (2.4.14). *Then there is a constant $C_{n,p}$ such that for all $f \in H^p$ the distribution $T(f)$ defined in* (2.4.18) *coincides with an L^p function that satisfies*

$$\big\|T(f)\big\|_{L^p} \le C_{n,p} A \big\|f\big\|_{H^p}.$$

Proof. The proof of this theorem is based on the atomic decomposition of H^p.

We first take $f = a$ to be an L^2-atom for H^p, and without loss of generality we may assume that a is supported in a cube Q centered at the origin. We let Q^* be the cube with side length $2\sqrt{n}\ell(Q)$, where $\ell(Q)$ is the side length of Q. We have

$$\begin{aligned}\left(\int_{Q^*}|T(a)(x)|^p\,dx\right)^{\frac{1}{p}} &\le C|Q^*|^{\frac{1}{p}-\frac{1}{2}}\left(\int_{Q^*}|T(a)(x)|^2\,dx\right)^{\frac{1}{2}}\\ &\le C''A|Q|^{\frac{1}{p}-\frac{1}{2}}\left(\int_{Q}|a(x)|^2\,dx\right)^{\frac{1}{2}}\\ &\le C_nA\,|Q|^{\frac{1}{p}-\frac{1}{2}}|Q|^{\frac{1}{2}-\frac{1}{p}}\\ &= C_nA\,,\end{aligned}$$

where we used that T is L^2 bounded with norm at most a constant multiple of A.

For $x \notin Q^*$ and $y \in Q$, we have $|x| \ge 2|y|$, and thus $x-y$ stays away from zero and $K(x-y)$ is well defined. We have

$$T(a)(x) = \int_Q K(x-y)\,a(y)\,dy.$$

Recall that $N = [\frac{n}{p} - n] + 1$. Using the cancellation of atoms in H^p, we deduce

$$\begin{aligned}T(a)(x) &= \int_Q a(y)K(x-y)\,dy\\ &= \int_Q a(y)\left[K(x-y) - \sum_{|\beta|\le N-1}(\partial^\beta K)(x)\frac{y^\beta}{\beta!}\right]dy\\ &= \int_Q a(y)\left[\sum_{|\beta|=N}(\partial^\beta K)(x-\theta_y y)\frac{y^\beta}{\beta!}\right]dy,\end{aligned}$$

for some $0 \le \theta_y \le 1$. The fact that $|x| \ge 2|y|$ implies that $|x-\theta_y y| \ge \frac{1}{2}|x|$ and using (2.4.12) we obtain the estimate

$$|T(a)(x)| \le c_{n,N}\frac{A}{|x|^{N+n}}\int_Q |a(y)|\,|y|^N\,dy,$$

from which it follows that for $x \notin Q^*$ we have

$$|T(a)(x)| \le c_{n,p}\frac{A}{|x|^{N+n}}|Q|^{1+\frac{N}{n}-\frac{1}{p}}$$

via a calculation using Hölder's inequality and the fact that $\|a\|_{L^q} \le |Q|^{\frac{1}{q}-\frac{1}{p}}$. Integrating over $(Q^*)^c$, we obtain that

$$\left(\int_{(Q^*)^c}|T(a)(x)|^p dx\right)^{\frac{1}{p}} \le c_{n,p}A\,|Q|^{1+\frac{N}{n}-\frac{1}{p}}\left(\int_{(Q^*)^c}\frac{1}{|x|^{p(N+n)}}\,dx\right)^{\frac{1}{p}} \le c'_{n,p}A\,.$$

We have now shown that there exists a constant $C_{n,p}$ such that

$$\big\|T(a)\big\|_{L^p} \le C_{n,p} A \tag{2.4.19}$$

whenever a is an L^2-atom for H^p.

To replace a by general $f \in H^p$ we need Lemma 2.4.4 that follows immediately, which we apply to the family Schwartz functions $\zeta_{\varepsilon_j}(x) = K(x)\big(\eta(x/\varepsilon_j) - \eta(\varepsilon_j x)\big)$ and the associated operators $T_{(\varepsilon_j)}$. Fix $f \in H^p \cap L^2$, with atomic decomposition $f = \sum_{j=1}^{\infty} \lambda_j a_j$ where a_j are L^2-atoms for H^p and $\sum_{j=1}^{\infty} |\lambda_j|^p \le 2^p \|f\|_{H^p}^p$. Then Lemma 2.4.4 yields

$$T(f) = \sum_{j=1}^{\infty} \lambda_j T(a_j) \qquad \text{a.e.,} \tag{2.4.20}$$

where the series converges in L^p. We apply L^p quasi-norms on both sides of (2.4.20), we use the sublinearity of $h \mapsto \|h\|_{L^p}^p$, and (2.4.19) to deduce that

$$\big\|T(f)\big\|_{L^p}^p \le C_{n,p}^p A^p \sum_{j=1}^{\infty} |\lambda_j|^p \le 2^p C_{n,p}^p A^p \|f\|_{H^p}^p \tag{2.4.21}$$

for all $f \in L^2 \cap H^p$. Recall that $T(f)$ is well defined for all $f \in H^p$, as observed in item (v) in the introductory comments of this subsection. Then by the density of $L^2 \cap H^p$ in H^p, the estimate

$$\big\|T(f)\big\|_{L^p}^p \le 2^p C_{n,p}^p A^p \|f\|_{H^p}^p$$

obtained in (2.4.21) for $f \in L^2 \cap H^p$ extends to any $f \in H^p$. □

Lemma 2.4.4. *Let $\{\zeta_\varepsilon\}_{\varepsilon>0}$ be a family of Schwartz functions and for each $\varepsilon > 0$ let T_ε be the operator given by convolution with ζ_ε. Suppose that the T_ε's are uniformly (in $\varepsilon > 0$) bounded on $L^2(\mathbf{R}^n)$ and that there is an $L^2(\mathbf{R}^n)$-bounded operator T such that for each $g \in L^2(\mathbf{R}^n)$, we have*

$$\big\|T_\varepsilon(g) - T(g)\big\|_{L^2} \to 0 \qquad \text{as } \varepsilon \to 0. \tag{2.4.22}$$

Suppose moreover that for a given $0 < p \le 1$ there is a constant C_0 such that for all a that are L^2-atoms for H^p we have

$$\sup_{\varepsilon>0} \|T_\varepsilon(a)\|_{L^p} \le C_0\,. \tag{2.4.23}$$

Then for every $f \in L^2 \cap H^p$ with atomic decomposition $f = \sum_{j=1}^{\infty} \lambda_j a_j$, where a_j are L^2-atoms for H^p and $\sum_{j=1}^{\infty} |\lambda_j|^p \le 2^p \|f\|_{H^p}^p$, the sequence $\sum_{j=1}^{N} \lambda_j T(a_j)$ is Cauchy in L^p and converges in L^p to a well-defined L^p function $\sum_{j=1}^{\infty} \lambda_j T(a_j)$ which is equal almost everywhere to $T(f)$, i.e., we have

$$T(f) = \sum_{j=1}^{\infty} \lambda_j T(a_j) \qquad \text{a.e.} \tag{2.4.24}$$

Proof. We begin the proof by observing that as a consequence of (2.4.23) we have

$$\|T(a)\|_{L^p} \le C_0 \tag{2.4.25}$$

for all a that are L^2-atoms for H^p. Indeed, (2.4.22) implies that for a given L^2 atom a for H^p, there is sequence $\varepsilon_k \downarrow 0$ such that

$$T(a) = \lim_{k\to\infty} T_{\varepsilon_k}(a) = \liminf_{k\to\infty} T_{\varepsilon_k}(a) \qquad \text{a.e.}$$

Then Fatou's lemma on L^p together with (2.4.23) imply (2.4.25).

Given $f \in H^p \cap L^2$, we write $f = \sum_{j=1}^{\infty} \lambda_j a_j$ in an atomic decomposition, where a_j are L^2-atoms for H^p, the series converges to f in H^p, and $\sum_{j=1}^{\infty} |\lambda_j|^p \le 2^p \|f\|_{H^p}^p$.

We observe that the sequence $\big\{\sum_{j=1}^{N} \lambda_j T(a_j)\big\}_{N=1}^{\infty}$ is Cauchy in L^p since

$$\Big\| \sum_{j=N'}^{N} \lambda_j T(a_j) \Big\|_{L^p}^p \le \sum_{j=N'}^{N} |\lambda_j|^p C_0^p,$$

which tends to zero as $N', N \to \infty$. Thus the sequence $\sum_{j=1}^{N} \lambda_j T(a_j)(x)$ converges in L^p to a well-defined L^p function. We set

$$\sum_{j=1}^{\infty} \lambda_j T(a_j) = L^p \text{ limit of } \sum_{j=1}^{N} \lambda_j T(a_j).$$

To prove (2.4.24), we first prove an analogous result about T_ε, namely,

$$T_\varepsilon(f) = \sum_{j=1}^{\infty} \lambda_j T_\varepsilon(a_j) \qquad \text{a.e.} \tag{2.4.26}$$

where $\sum_{j=1}^{\infty} \lambda_j T_\varepsilon(a_j)$ denotes the L^p limit of the Cauchy sequence $\sum_{j=1}^{N} \lambda_j T_\varepsilon(a_j)$. We fix $\varepsilon, \delta > 0$. Then by the linearity of T_ε for each $L \in \mathbf{Z}^+$ we have

$$\begin{aligned}
&\Big|\Big\{x \in \mathbf{R}^n : |T_\varepsilon(f)(x) - \sum_{j=1}^{\infty} \lambda_j T_\varepsilon(a_j)(x)| > \delta\Big\}\Big| \\
&\le \Big|\Big\{x \in \mathbf{R}^n : |T_\varepsilon\Big(\sum_{j=L+1}^{\infty} \lambda_j a_j\Big)(x) - \sum_{j=L+1}^{\infty} \lambda_j T_\varepsilon(a_j)(x)| > \delta\Big\}\Big| \\
&\le \Big|\Big\{x \in \mathbf{R}^n : |T_\varepsilon\Big(\sum_{j=L}^{\infty} \lambda_j a_j\Big)(x)| > \frac{\delta}{2}\Big\}\Big| + \Big|\Big\{x \in \mathbf{R}^n : |\sum_{j=L+1}^{\infty} \lambda_j T_\varepsilon(a_j)(x)| > \frac{\delta}{2}\Big\}\Big| \\
&\le \frac{2^p}{\delta^p} \Big\| T_\varepsilon\Big(\sum_{j=L+1}^{\infty} \lambda_j a_j\Big)\Big\|_{L^p}^p + \frac{2^p}{\delta^p} \sum_{j=L+1}^{\infty} |\lambda_j|^p \|T_\varepsilon(a_j)\|_{L^p}^p.
\end{aligned} \tag{2.4.27}$$

By assumption (2.4.23) the second term in the sum in (2.4.27) is controlled by $C_0^p(\frac{2}{\delta})^p\sum_{j=L+1}^{\infty}|\lambda_j|^p$ which tends to zero as $L\to\infty$.

To show the same conclusion for the first sum in (2.4.27) we recall the grand maximal function

$$\mathscr{M}_N(f)(x) = \sup_{\varphi\in\mathscr{F}_N}\sup_{t>0}\sup_{\substack{y\in\mathbf{R}^n\\|y-x|\le t}}|(\varphi_t*f)(y)|$$

where

$$\mathscr{F}_N = \Big\{\varphi\in\mathscr{S}(\mathbf{R}^n):\ \int_{\mathbf{R}^n}(1+|x|)^N\sum_{|\alpha|\le N+1}|\partial^\alpha\varphi(x)|\,dx\le 1\Big\}.$$

The function ζ_ε lies in $\mathscr{S}(\mathbf{R}^n)$; thus there is a constant $c_{\varepsilon,N}$ such that $c_{\varepsilon,N}\zeta_\varepsilon$ lies in $\mathscr{F}_N$. Then we have

$$|T_\varepsilon(\sum_{j=L+1}^{\infty}\lambda_j a_j)|\le\frac{1}{c_{\varepsilon,N}}\mathscr{M}_N(f-\sum_{j=1}^{L}\lambda_j a_j).$$

Taking L^p quasi-norms we obtain

$$\Big\|T_\varepsilon(\sum_{j=L+1}^{\infty}\lambda_j a_j)\Big\|_{L^p}^p\le\frac{1}{c_{\varepsilon,N}^p}\Big\|\mathscr{M}_N(f-\sum_{j=1}^{L}\lambda_j a_j)\Big\|_{L^p}^p\le\frac{C_{n,p}^p}{c_{\varepsilon,N}^p}\Big\|f-\sum_{j=1}^{L}\lambda_j a_j\Big\|_{L^p}^p,$$

and since $\sum_{j=1}^{L}\lambda_j a_j\to f$ in H^p as $L\to\infty$, we deduce that the first sum in (2.4.27) tends to zero as $L\to\infty$. This proves that for any $\varepsilon,\delta>0$ we have

$$\Big|\Big\{x\in\mathbf{R}^n:\ |T_\varepsilon(f)(x)-\sum_{j=1}^{\infty}\lambda_j T_\varepsilon(a_j)(x)|>\delta\Big\}\Big|=0;$$

hence (2.4.26) holds.

Next, we claim that $\sum_{j=1}^{\infty}\lambda_j T_\varepsilon(a_j)\to\sum_{j=1}^{\infty}\lambda_j T(a_j)$ in measure as $\varepsilon\to 0$. Indeed, given $\delta>0$, write

$$\begin{aligned}
&\Big|\Big\{\sum_{j=1}^{\infty}\lambda_j T_\varepsilon(a_j)-\sum_{j=1}^{\infty}\lambda_j T(a_j)|>\delta\Big\}\Big|\\
&\le\Big|\Big\{\Big|\sum_{j=1}^{L}\lambda_j(T_\varepsilon(a_j)-T(a_j))\Big|>\frac{\delta}{2}\Big\}\Big|+\Big|\Big\{\Big|\sum_{j=L+1}^{\infty}\lambda_j T_\varepsilon(a_j)-\sum_{j=L+1}^{\infty}\lambda_j T(a_j)\Big|>\frac{\delta}{2}\Big\}\Big|\\
&\le\frac{2^2}{\delta^2}\Big\|\sum_{j=1}^{L}\lambda_j(T_\varepsilon(a_j)-T(a_j))\Big\|_{L^2}^2+\frac{2^p}{\delta^p}\sum_{j=L+1}^{\infty}|\lambda_j|^p\Big[\|T_\varepsilon(a_j)\|_{L^p}^p+\|T(a_j)\|_{L^p}^p\Big]\\
&\le\frac{2^2}{\delta^2}\Big\|T_\varepsilon(\sum_{j=1}^{L}\lambda_j a_j)-T(\sum_{j=1}^{L}\lambda_j a_j)\Big\|_{L^2}^2+\frac{2^{p+1}C_0^p}{\delta^p}\sum_{j=L+1}^{\infty}|\lambda_j|^p,
\end{aligned}\tag{2.4.28}$$

where we made use of (2.4.23) and (2.4.25) in the last estimate above. The second term in (2.4.28) can be made less than any given number $\tau > 0$ if L is chosen to be large enough. Once we fix L, then there is a $\varepsilon_0 > 0$ such that for $\varepsilon_0 < \varepsilon$ the first term in (2.4.28) is controlled by τ too, since $T_\varepsilon(\sum_{j=1}^L \lambda_j a_j)$ converges to $T(\sum_{j=1}^L \lambda_j a_j)$ in L^2 in view of (2.4.22). Therefore (2.4.28) can be made arbitrarily small for ε sufficiently small, and the claimed convergence in measure is valid. By Theorem 1.1.11 in [156] there is sequence ε_i (subsequence of $\varepsilon > 0$) such that

$$\sum_{k=1}^{\infty} \lambda_k T_{\varepsilon_i}(a_k)(x) \to \sum_{k=1}^{\infty} \lambda_k T(a_k)(x) \qquad \text{a.e. as } i \to \infty. \tag{2.4.29}$$

Since $T_{\varepsilon_i}(f)$ tends to $T(f)$ in L^2, we can find a subsequence $\{\varepsilon_{i_\ell}\}$ of the subsequence $\{\varepsilon_i\}$ such that

$$T_{\varepsilon_{i_\ell}}(f)(x) \to T(f)(x) \qquad \text{a.e. as } \ell \to \infty. \tag{2.4.30}$$

Using identity (2.4.26) with ε_{i_ℓ} in place of ε, together with (2.4.29) with i_ℓ in place of i, along with (2.4.30), letting $\ell \to \infty$, we deduce (2.4.24). □

We discuss a version of the Theorem 2.4.3 in which the target space is H^p.

Theorem 2.4.5. *Under the hypotheses of Theorem 2.4.3, there is a constant $C_{n,p}$ such that for all $f \in H^p$,*

$$\big\|T(f)\big\|_{H^p} \le C_{n,p} A \,\big\|f\big\|_{H^p} . \tag{2.4.31}$$

Proof. We fix a smooth function Φ supported in the unit ball $B(0,1)$ in $\mathbf{R}^n$ whose mean value is not equal to zero. For $t > 0$ we define the smooth functions

$$K^{(t)} = \Phi_t * W$$

and for $f \in H^p$, we define an operator

$$T^{(t)}(f) = \Phi_t * T(f)$$

noting that the convolution is well defined since $T(f)$ lies in $\mathscr{S}'$ and Φ_t is in $\mathscr{C}_0^\infty$.

We observe that the family of kernels $K^{(t)}$ satisfies

$$\sup_{t>0} \big|\widehat{K^{(t)}}(\xi)\big| \le \big\|\widehat{\Phi}\big\|_{L^\infty} \big\|\widehat{W}\big\|_{L^\infty} \le CA \big\|\widehat{\Phi}\big\|_{L^\infty} \tag{2.4.32}$$

and that

$$\sup_{t>0} |\partial^\beta K^{(t)}(x)| \le C_\Phi A\, |x|^{-n-|\beta|} \tag{2.4.33}$$

for all $|\beta| \le N$, where

$$C_\Phi = \sup_{|\gamma| \le N} \int_{\mathbf{R}^n} |\xi|^{|\gamma|} |\widehat{\Phi}(\xi)|\, d\xi .$$

Indeed, assertion (2.4.32) is easily verified. When $|x| \le 2t$ assertion (2.4.33) follows from the identity

$$K^{(t)}(x) = ((\Phi_t * W)^{\widehat{}})^{\vee}(x) = \int_{\mathbf{R}^n} e^{2\pi i x\cdot\xi}\, \widehat{W}(\xi)\, \widehat{\Phi}(t\xi)\, d\xi\,,$$

while whenever $|x| \ge 2t$, (2.4.33) follows from (2.4.12) and from the integral representation

$$\partial^\beta K^{(t)}(x) = \int_{|y|\le t} \partial^\beta K(x-y)\, \Phi_t(y)\, dy\,.$$

We now take $f = a$ to be an L^2-atom for H^p, and without loss of generality we may assume that a is supported in a cube Q centered at the origin. We let Q^* be the cube with side length $2\sqrt{n}\ell(Q)$, where $\ell(Q)$ is the side length of Q. Recall the smooth maximal function $M(f;\Phi)$ from Section 2.1. Then $M(T(a);\Phi)$ is pointwise controlled by the Hardy–Littlewood maximal function of $T(a)$. Using an argument similar to that in Theorem 2.4.1, we have

$$\begin{aligned}
\left(\int_{Q^*} |M(T(a);\Phi)(x)|^p\, dx\right)^{\frac{1}{p}} &\le \|\Phi\|_{L^1}\left(\int_{Q^*} |M(T(a))(x)|^p\, dx\right)^{\frac{1}{p}}\\
&\le C|Q^*|^{\frac{1}{p}-\frac{1}{2}}\left(\int_{Q^*} |M(T(a))(x)|^2\, dx\right)^{\frac{1}{2}}\\
&\le C'|Q|^{\frac{1}{p}-\frac{1}{2}}\left(\int_{\mathbf{R}^n} |T(a)(x)|^2\, dx\right)^{\frac{1}{2}}\\
&\le C''A|Q|^{\frac{1}{p}-\frac{1}{2}}\left(\int_{Q} |a(x)|^2\, dx\right)^{\frac{1}{2}}\\
&\le C_n A\, |Q|^{\frac{1}{p}-\frac{1}{2}}\, |Q|^{\frac{1}{2}-\frac{1}{p}}\\
&= C_n A\,.
\end{aligned}$$

It therefore remains to estimate the contribution of $M(T(a);\Phi)$ on the complement of Q^*.

For $x \notin Q^*$ we write

$$T^{(t)}(a)(x) = (a * K^{(t)})(x) = \int_Q K^{(t)}(x-y)\, a(y)\, dy\,.$$

Recall that $N = [\frac{n}{p} - n] + 1$. Using the cancellation of L^2 atoms for H^p we deduce

$$\begin{aligned}
T^{(t)}(a)(x) &= \int_Q a(y)\left[K^{(t)}(x-y) - \sum_{|\beta|\le N-1} (\partial^\beta K^{(t)})(x)\frac{y^\beta}{\beta!}\right] dy\\
&= \int_Q a(y)\left[\sum_{|\beta|=N} (\partial^\beta K^{(t)})(x-\theta_y y)\frac{y^\beta}{\beta!}\right] dy
\end{aligned}$$

for some $0 \le \theta_y \le 1$. Since $x \notin Q^*$ and $y \in Q$ we have $|x - \theta_y y| \ge |x| - |y| \ge \frac{1}{2}|x|$; thus using (2.4.33) we obtain the estimate

$$|T^{(t)}(a)(x)| \le c_{n,N} \frac{A}{|x|^{N+n}} \int_Q |a(y)|\, |y|^N \, dy,$$

from which it follows that for $x \notin Q^*$ we have

$$|T^{(t)}(a)(x)| \le c_{n,p} \frac{A}{|x|^{N+n}} |Q|^{1+\frac{N}{n}-\frac{1}{p}}$$

via a calculation using properties of atoms (see the proof of Theorem 2.3.11). Taking the supremum over all $t > 0$ and integrating over $(Q^*)^c$, we obtain that

$$\left(\int_{(Q^*)^c} \sup_{t>0} |(T(a) * \Phi_t)(x)|^p dx \right)^{\frac{1}{p}} \le c_{n,p} A\, |Q|^{1+\frac{N}{n}-\frac{1}{p}} \left(\int_{(Q^*)^c} \frac{1}{|x|^{p(N+n)}} dx \right)^{\frac{1}{p}},$$

and the latter is easily seen to be finite and controlled by a constant multiple of A. Combining this estimate with the previously obtained estimate for the integral of $M(T(a); \Phi) = \sup_{t>0} |T^{(t)}(a)|$ over Q^* yields the conclusion of the theorem when $f = a$ is an atom.

We have now shown that there exists a constant $C_{n,p}$ such that

$$\big\| T^{(t)}(a) \big\|_{L^p} \le \big\| T(a) \big\|_{H^p} \le C_{n,p} A \tag{2.4.34}$$

whenever a is an L^2-atom for H^p. We now extend this estimate to arbitrary f in $L^2 \cap H^p$. To achieve this, we verify that the assumptions of Lemma 2.4.4 are valid for the family of Schwartz functions $\zeta_\varepsilon = \Phi_t * K_{(\varepsilon)}$ and the family of operators $T_\varepsilon^{(t)}(g) = \Phi_t * K_{(\varepsilon)} * g$, which are uniformly bounded on L^2 and converge in L^2 to $T^{(t)}(g) = \Phi_t * g$ for any $g \in L^2(\mathbf{R}^n)$.

Fix $f \in L^2 \cap H^p$, with atomic decomposition $f = \sum_{j=1}^\infty \lambda_j a_j$, where $\sum_{j=1}^\infty |\lambda_j|^p \le 2^p \|f\|_{H^p}^p$. Observe that the sequences $\sum_{j=1}^N |\lambda_j T^{(t)}(a_j)|$ and $\sum_{j=1}^N \lambda_j T^{(t)}(a_j)$ are Cauchy in L^p and thus they converge in L^p. We set

$$\sum_{j=1}^\infty \lambda_j T^{(t)}(a_j) = L^p \text{ limit of } \sum_{j=1}^N \lambda_j T^{(t)}(a_j) \text{ as } N \to \infty$$

$$\sum_{j=1}^\infty |\lambda_j T^{(t)}(a_j)| = L^p \text{ limit of } \sum_{j=1}^N |\lambda_j T^{(t)}(a_j)| \text{ as } N \to \infty,$$

and extracting subsequences that converge almost everywhere, we notice that

$$\Big| \sum_{j=1}^\infty \lambda_j T^{(t)}(a_j) \Big| \le \sum_{j=1}^\infty |\lambda_j T^{(t)}(a_j)| \qquad \text{a.e.} \tag{2.4.35}$$

To apply Lemma 2.4.4, we consider the family of Schwartz functions $\zeta_{\varepsilon_j}(x) = \Phi_t * K_{(\varepsilon_j)}$ and the associated operators $T^{(t)}_{(\varepsilon_j)} = \Phi_t * T_{(\varepsilon_j)}$, which are L^2-bounded uniformly in ε_j and $t > 0$. It is easy to see that for $g \in L^2$, $\Phi_t * T_{(\varepsilon_j)}(g) \to T^{(t)}(g)$ in L^2 as $j \to \infty$, and furthermore (2.4.34) also holds if $T^{(t)}$ is replaced by $T^{(t)}_{(\varepsilon_j)}$ uniformly in ε_j via a similar argument; hence the hypotheses of Lemma 2.4.4 are valid. Using the conclusion of Lemma 2.4.4 we write

$$T^{(t)}(f) = \sum_{j=1}^{\infty} \lambda_j T^{(t)}(a_j) \qquad \text{a.e.} \tag{2.4.36}$$

It follows from this fact and (2.4.35) that

$$|T^{(t)}(f)| \le \sum_{j=1}^{\infty} |\lambda_j T^{(t)}(a_j)| \le \sum_{j=1}^{\infty} |\lambda_j| M(T(a_j);\Phi) \qquad \text{a.e.} \tag{2.4.37}$$

Taking the supremum over $t > 0$ in (2.4.37) we deduce

$$M(T(f);\Phi) \le \sum_{j=1}^{\infty} |\lambda_j| M(T(a_j);\Phi) \qquad \text{a.e.} \tag{2.4.38}$$

and applying L^p quasi-norms on both sides and using (2.4.34) yields the desired conclusion (2.4.31) for $f \in L^2 \cap H^p$. The extension to general $f \in H^p$ follows by density and the fact that $T(f)$ is well defined for all $f \in H^p$, as observed in item (v) in the introduction of this subsection. □

2.4.4 A Singular Integral Characterization of $H^1(\mathbf{R}^n)$

We showed in Section 2.4.1 that singular integrals map H^1 to L^1. In particular, the Riesz transforms have this property. In this subsection we obtain a converse to this statement. We show that if $R_j(f)$ are integrable functions for some $f \in L^1$ and all $j = 1, \dots, n$, then f must be an element of the Hardy space H^1. This provides a characterization of $H^1(\mathbf{R}^n)$ in terms of the Riesz transforms.

Theorem 2.4.6. *For $n \ge 2$, there exists a constant C_n such that for f in $L^1(\mathbf{R}^n)$, we have*

$$C_n \big\|f\big\|_{H^1} \le \big\|f\big\|_{L^1} + \sum_{k=1}^{n} \big\|R_k(f)\big\|_{L^1}. \tag{2.4.39}$$

When $n = 1$ the corresponding statement is

$$C_1 \big\|f\big\|_{H^1} \le \big\|f\big\|_{L^1} + \big\|H(f)\big\|_{L^1} \tag{2.4.40}$$

for all $f \in L^1(\mathbf{R})$. Naturally, these statements are interesting when the expressions on the right in (2.4.39) and (2.4.40) are finite.

Before we prove this theorem we discuss a couple of corollaries.

Corollary 2.4.7. *An integrable function on the line lies in the Hardy space $H^1(\mathbf{R})$ if and only if its Hilbert transform is integrable. For $n \geq 2$, an integrable function on $\mathbf{R}^n$ lies in the Hardy space $H^1(\mathbf{R}^n)$ if and only its Riesz transforms are also in $L^1(\mathbf{R}^n)$.*

Proof. The corollary follows by combining Theorems 2.4.1 and 2.4.6. □

Corollary 2.4.8. *Functions in $H^1(\mathbf{R}^n)$, $n \geq 1$, have integral zero.*

Proof. Indeed, if $f \in H^1(\mathbf{R}^n)$, we must have $R_1(f) \in L^1(\mathbf{R}^n)$ by Theorem 2.4.1; thus $\widehat{R_1(f)}$ is uniformly continuous. But since

$$\widehat{R_1(f)}(\xi) = -i\widehat{f}(\xi)\frac{\xi_1}{|\xi|},$$

it follows that $\widehat{R_1(f)}$ is continuous at zero if and only if $\widehat{f}(0) = 0$. But this happens exactly when f has integral zero. □

We now discuss the proof of Theorem 2.4.6.

Proof. We consider the case $n \geq 2$, although the argument below also works in the case $n = 1$ with a suitable change of notation. Let P_t be the Poisson kernel. In the proof we may assume that f is real-valued, since it can be written as $f = f_1 + i f_2$, where f_k are real-valued and $R_j(f_k)$ are also integrable. Given a real-valued function $f \in L^1(\mathbf{R}^n)$ such that $R_j(f)$ are integrable over $\mathbf{R}^n$ for every $j = 1, \dots, n$, we associate with it the $n+1$ functions

$$\begin{aligned} u_1(x,t) &= (P_t * R_1(f))(x), \\ \dots &= \dots \\ u_n(x,t) &= (P_t * R_n(f))(x), \\ u_{n+1}(x,t) &= (P_t * f)(x), \end{aligned}$$

which are harmonic on the space $\mathbf{R}^{n+1}_+$ (see Example 2.1.13 in [156]). It is convenient to denote the last variable t by x_{n+1}. One may check using the Fourier transform that these harmonic functions satisfy the following system:

$$\begin{aligned} \sum_{j=1}^{n+1} \frac{\partial u_j}{\partial x_j} &= 0, \\ \frac{\partial u_j}{\partial x_k} - \frac{\partial u_k}{\partial x_j} &= 0, \qquad k, j \in \{1, \dots, n+1\}, \quad k \neq j. \end{aligned} \tag{2.4.41}$$

This system of equations may also be expressed as $\operatorname{div} F = 0$ and $\operatorname{curl} F = \vec{0}$, where $F = (u_1, \dots, u_{n+1})$ is a vector field in $\mathbf{R}^{n+1}_+$. Note that when $n = 1$, the equations in (2.4.41) are the usual Cauchy–Riemann equations, which assert that the function

$F = (u_1, u_2) = u_1 + iu_2$ is holomorphic in the upper half-space. For this reason, when $n \geq 2$, the equations in (2.4.41) are often referred to as the *system of generalized Cauchy–Riemann equations.*

The function $|F|$ enjoys a crucial property in the study of this problem.

Lemma 2.4.9. *Let u_j be real-valued harmonic functions on $\mathbf{R}^{n+1}$ satisfying the system of equations* (2.4.41) *and let $F = (u_1, \dots, u_{n+1})$. Then the function*

$$|F|^q = \Big(\sum_{j=1}^{n+1} |u_j|^2\Big)^{q/2}$$

is subharmonic when $q \geq (n-1)/n$, i.e., it satisfies $\Delta(|F|^q) \geq 0$, on $\mathbf{R}^{n+1}_+$.

Lemma 2.4.10. *Let $0 < q < p < \infty$. Suppose that the function $|F(x,t)|^q$ defined on $\mathbf{R}^{n+1}_+$ is subharmonic and satisfies*

$$\sup_{t>0} \left(\int_{\mathbf{R}^n} |F(x,t)|^p \, dx \right)^{1/p} \leq A < \infty. \tag{2.4.42}$$

Then there is a constant $C_{n,p,q} < \infty$ such that the nontangential maximal function $|F|^*(x) = \sup_{t>0} \sup_{|y-x|<t} |F(y,t)|$, $x \in \mathbf{R}^n$, (cf. Definition 3.3.1) satisfies

$$\big\| |F|^* \big\|_{L^p(\mathbf{R}^n)} \leq C_{n,p,q} A .$$

Assuming these lemmas, whose proofs are postponed until the end of this section, we return to the proof of the theorem.

Without loss of generality, we may assume that the given integrable function f is real-valued, so that $R_j(f)$ are also real-valued and we are able to apply Lemma 2.4.9. Since the Poisson kernel is an approximate identity, the function $x \mapsto u_{n+1}(x,t)$ converges to $f(x)$ in L^1 as $t \to 0$. To show that $f \in H^1(\mathbf{R}^n)$, it suffices to show that the Poisson maximal function

$$M(f;P)(x) = \sup_{t>0} |(P_t * f)(x)| = \sup_{t>0} |u_{n+1}(x,t)|$$

is integrable. But this maximal function is pointwise controlled by

$$\sup_{t>0} |F(x,t)| \leq \sup_{t>0} \Big[|(P_t * f)(x)| + \sum_{j=1}^{n} |(P_t * R_j(f))(x)| \Big],$$

and certainly it satisfies

$$\sup_{t>0} \int_{\mathbf{R}^n} |F(x,t)| \, dx \leq A_f , \tag{2.4.43}$$

where

$$A_f = \big\| f \big\|_{L^1} + \sum_{k=1}^{n} \big\| R_k(f) \big\|_{L^1} .$$

We now have

$$M(f;P)(x) \le \sup_{t>0} |u_{n+1}(x,t)| \le \sup_{t>0} |F(x,t)| \le |F|^*(x), \tag{2.4.44}$$

and using Lemma 2.4.9 with $q = \frac{n-1}{n}$ and Lemma 2.4.10 with $p = 1$ we obtain that

$$\big\| |F|^* \big\|_{L^1(\mathbf{R}^n)} \le C_n A_f. \tag{2.4.45}$$

Combining (2.4.43), (2.4.44), and (2.4.45), we deduce that

$$\big\| M(f;P)(x) \big\|_{L^1(\mathbf{R}^n)} \le C_n \bigg(\big\| f \big\|_{L^1} + \sum_{k=1}^{n} \big\| R_k(f) \big\|_{L^1} \bigg),$$

from which (2.4.39) follows. This proof is also valid when $n = 1$, provided one replaces the Riesz transforms with the Hilbert transform; hence the proof of (2.4.40) is subsumed in that of (2.4.39). □

We now give a proof of Lemma 2.4.9

Proof. Denoting the variable t by x_{n+1}, we have

$$\frac{\partial}{\partial x_j} |F|^q = q|F|^{q-2} \Big(F \cdot \frac{\partial F}{\partial x_j} \Big)$$

and also

$$\frac{\partial^2}{\partial x_j^2} |F|^q = q|F|^{q-2} \bigg[F \cdot \frac{\partial^2 F}{\partial x_j^2} + \frac{\partial F}{\partial x_j} \cdot \frac{\partial F}{\partial x_j} \bigg] + q(q-2)|F|^{q-4} \Big(F \cdot \frac{\partial F}{\partial x_j} \Big)^2$$

for all $j = 1, 2, \dots, n+1$. Summing over all these j's, we obtain

$$\Delta(|F|^q) = q|F|^{q-4} \bigg[|F|^2 \sum_{j=1}^{n+1} \Big| \frac{\partial F}{\partial x_j} \Big|^2 + (q-2) \sum_{j=1}^{n+1} \Big| F \cdot \frac{\partial F}{\partial x_j} \Big|^2 \bigg], \tag{2.4.46}$$

since the term containing $F \cdot \Delta(F) = \sum_{j=1}^{n+1} u_j \Delta(u_j)$ vanishes because each u_j is harmonic. The only term that could be negative in (2.4.46) is that containing the factor $q-2$ and naturally, if $q \ge 2$, the conclusion is obvious. Let us assume that $\frac{n-1}{n} \le q < 2$. Since $q \ge \frac{n-1}{n}$, we must have that $2 - q \le \frac{n+1}{n}$. Thus (2.4.46) is nonnegative if

$$\sum_{j=1}^{n+1} \Big| F \cdot \frac{\partial F}{\partial x_j} \Big|^2 \le \frac{n}{n+1} |F|^2 \sum_{j=1}^{n+1} \Big| \frac{\partial F}{\partial x_j} \Big|^2. \tag{2.4.47}$$

This is certainly valid for points (x,t) such that $F(x,t) = 0$. To prove (2.4.47) for points (x,t) with $F(x,t) \ne 0$, it suffices to show that for every vector $v \in \mathbf{R}^{n+1}$ with Euclidean norm $|v| = 1$, we have

$$\sum_{j=1}^{n+1}\left|v\cdot\frac{\partial F}{\partial x_j}\right|^2\le\frac{n}{n+1}\sum_{j=1}^{n+1}\left|\frac{\partial F}{\partial x_j}\right|^2. \tag{2.4.48}$$

Denoting by A the $(n+1)\times(n+1)$ matrix whose entries are $a_{j,k}=\partial u_k/\partial x_j$, we rewrite (2.4.48) as

$$|Av|^2\le\frac{n}{n+1}\|A\|^2, \tag{2.4.49}$$

where

$$\|A\|^2=\sum_{j=1}^{n+1}\sum_{k=1}^{n+1}|a_{j,k}|^2.$$

By assumption, the functions u_j are real-valued and thus the numbers $a_{j,k}$ are real. In view of identities (2.4.41), the matrix A is real symmetric and has zero trace (i.e., $\sum_{j=1}^{n+1}a_{j,j}=0$). A real symmetric matrix A can be written as $A=PDP^t$, where P is an orthogonal matrix and D is a real diagonal matrix. Since orthogonal matrices preserve the Euclidean distance, estimate (2.4.49) follows from the corresponding one for a diagonal matrix D. If $A=PDP^t$, then the traces of A and D are equal; hence $\sum_{j=1}^{n+1}\lambda_j=0$, where λ_j are entries on the diagonal of D. Notice that estimate (2.4.49) with the matrix D in the place of A is equivalent to

$$\sum_{j=1}^{n+1}|\lambda_j|^2|v_j|^2\le\frac{n}{n+1}\Big(\sum_{j=1}^{n+1}|\lambda_j|^2\Big), \tag{2.4.50}$$

where we set $v=(v_1,\dots,v_{n+1})$ and we are assuming that $|v|^2=\sum_{j=1}^{n+1}|v_j|^2=1$. Estimate (2.4.50) is certainly a consequence of

$$\sup_{1\le j\le n+1}|\lambda_j|^2\le\frac{n}{n+1}\Big(\sum_{j=1}^{n+1}|\lambda_j|^2\Big). \tag{2.4.51}$$

But this is easy to prove. Let $|\lambda_{j_0}|=\max_{1\le j\le n+1}|\lambda_j|$. Then

$$|\lambda_{j_0}|^2=\Big|-\sum_{j\neq j_0}\lambda_j\Big|^2\le\Big(\sum_{j\neq j_0}|\lambda_j|\Big)^2\le n\sum_{j\neq j_0}|\lambda_j|^2. \tag{2.4.52}$$

Adding $n|\lambda_{j_0}|^2$ to both sides of (2.4.52), we deduce (2.4.51) and thus (2.4.47). □

We now give the proof of Lemma 2.4.10.

Proof. A consequence of the subharmonicity of $|F|^q$ is that

$$|F(x,t+\varepsilon)|^q\le(|F(\cdot,\varepsilon)|^q*P_t)(x) \tag{2.4.53}$$

for all $x\in\mathbf{R}^n$ and $t,\varepsilon>0$. To prove (2.4.53), fix $\varepsilon>0$ and consider the functions

$$U(x,t)=|F(x,t+\varepsilon)|^q,\qquad V(x,t)=(|F(\cdot,\varepsilon)|^q*P_t)(x).$$

Given $\eta > 0$, we find a half-ball

$$B_{R_0} = \{(x,t) \in \mathbf{R}^{n+1}_+ : \, |x|^2 + t^2 < R_0^2\}$$

such that for $(x,t) \in \mathbf{R}^{n+1}_+ \setminus B_{R_0}$ we have

$$U(x,t) - V(x,t) \leq \eta \,. \tag{2.4.54}$$

Suppose that this is possible. Since $U(x,0) = V(x,0)$, then (2.4.54) actually holds on the entire boundary of B_{R_0}. The function V is harmonic and U is subharmonic; thus $U - V$ is subharmonic. The maximum principle for subharmonic functions implies that (2.4.54) holds in the interior of B_{R_0}, and since it also holds on the exterior, it must be valid for all (x,t) with $x \in \mathbf{R}^n$ and $t \geq 0$. Since η was arbitrary, letting $\eta \to 0+$ implies (2.4.53).

We now prove that R_0 exists such that (2.4.54) is possible for $(x,t) \in \mathbf{R}^{n+1}_+ \setminus B_{R_0}$. Let $B((x,t),t/2)$ be the $(n+1)$-dimensional ball of radius $t/2$ centered at (x,t). The subharmonicity of $|F|^q$ is reflected in the inequality

$$|F(x,t)|^q \leq \frac{1}{|B((x,t),\, t/2)|} \int_{B((x,t),t/2)} |F(y,s)|^q \, dyds \,,$$

which by Hölder's inequality and the fact $p > q$ gives

$$|F(x,t)|^q \leq \left(\frac{1}{|B((x,t),\, t/2)|} \int_{B((x,t),t/2)} |F(y,s)|^p \, dyds \right)^{\frac{q}{p}} .$$

From this we deduce that

$$|F(x,t+\varepsilon)|^q \leq \left[\frac{2^{n+1}/v_{n+1}}{(t+\varepsilon)^{n+1}} \int_{\frac{1}{2}(t+\varepsilon)}^{\frac{3}{2}(t+\varepsilon)} \int_{|y| \geq |x| - \frac{1}{2}(t+\varepsilon)} |F(y,s)|^p \, dy ds \right]^{\frac{q}{p}} . \tag{2.4.55}$$

If $t + \varepsilon \geq |x|$, using (2.4.42), we see that the expression on the right in (2.4.55) is bounded by $c' A^q (t+\varepsilon)^{-(n+1)q/p}$, and thus it can be made smaller than $\eta/2$ by taking $t \geq R_1 = \max\big(\varepsilon, (\eta/2c'A^q)^{-p/q(n+1)}\big)$. Since $R_1 \geq \varepsilon$, we must have $2t \geq t + \varepsilon \geq |x|$, which implies that $t \geq |x|/2$, and thus with $R_0' = \sqrt{5} R_1$, if $|(x,t)| > R_0'$ then $t \geq R_1$. Hence, the expression in (2.4.55) can be made smaller than $\eta/2$ for $|(x,t)| > R_0'$.

If $t + \varepsilon < |x|$ we estimate the expression on the right in (2.4.55) by

$$\left(\frac{2^{n+1}}{v_{n+1}} \frac{1}{(t+\varepsilon)^{n+1}} \int_{\frac{1}{2}(t+\varepsilon)}^{\frac{3}{2}(t+\varepsilon)} \left[\int_{|y| \geq \frac{1}{2}|x|} |F(y,s)|^p \, dy \right] ds \right)^{\frac{q}{p}} ,$$

and we notice that the preceding expression is bounded by

$$\left(\frac{3^{n+1}}{v_{n+1}} \int_{\frac{1}{2}\varepsilon}^{\infty} \left[\int_{|y| \geq \frac{1}{2}|x|} |F(y,s)|^p \, dy \right] \frac{ds}{s^{n+1}} \right)^{\frac{q}{p}} . \tag{2.4.56}$$

Let $G_{|x|}(s)$ be the function inside the square brackets in (2.4.56). Then $G_{|x|}(s) \to 0$ as $|x| \to \infty$ for all s. The hypothesis (2.4.42) implies that $G_{|x|}$ is bounded by a constant and it is therefore integrable over the interval $\left[\frac{1}{2}\varepsilon, \infty\right)$ with respect to the measure $s^{-n-1}ds$. By the Lebesgue dominated convergence theorem we deduce that the expression in (2.4.56) converges to zero as $|x| \to \infty$ and thus it can be made smaller that $\eta/2$ for $|x| \geq R_2$, for some constant R_2. Then with $R_0'' = \sqrt{2}R_2$ we have that if $|(x,t)| \geq R_0''$ then (2.4.56) is at most $\eta/2$. Since $U - V \leq U$, we deduce the validity of (2.4.54) for $|(x,t)| > R_0 = \max(R_0', R_0'')$.

Let $r = p/q > 1$. Assumption (2.4.42) implies that the functions $x \mapsto |F(x,t)|^q$ are in L^r uniformly in t. Since any closed ball of L^r is weak* compact, there is a sequence $\varepsilon_k \to 0$ such that $|F(x,\varepsilon_k)|^q \to h$ weakly in L^r as $k \to \infty$ to some function $h \in L^r$. Since $P_t \in L^{r'}$, this implies that

$$(|F(\cdot,\varepsilon_k)|^q * P_t)(x) \to (h * P_t)(x)$$

for all $x \in \mathbf{R}^n$. Using (2.4.53) we obtain

$$|F(x,t)|^q = \limsup_{k\to\infty} |F(x,t+\varepsilon_k)|^p \leq \limsup_{k\to\infty} \big(|F(x,\varepsilon_k)|^q * P_t\big)(x) = (h * P_t)(x),$$

which gives for all $x \in \mathbf{R}^n$,

$$|F|^*(x) \leq \Big[\sup_{t>0} \sup_{|y-x|<t} (|h| * P_t)(x)\Big]^{1/q} \leq C_n' M(h)(x)^{1/q}. \tag{2.4.57}$$

Let $g \in L^{r'}(\mathbf{R}^n)$ with $L^{r'}$ norm at most one. The weak convergence yields

$$\int_{\mathbf{R}^n} |F(x,\varepsilon_k)|^q g(x)\,dx \to \int_{\mathbf{R}^n} h(x)\,g(x)\,dx$$

as $k \to \infty$, and consequently we have

$$\left|\int_{\mathbf{R}^n} h(x)\,g(x)\,dx\right| \leq \sup_k \int_{\mathbf{R}^n} |F(x,\varepsilon_k)|^q |g(x)|\,dx \leq \|g\|_{L^{r'}} \sup_{t>0}\left(\int_{\mathbf{R}^n} |F(x,t)|^p\,dx\right)^{\frac{1}{r}}.$$

Since g is arbitrary with $L^{r'}$ norm at most one, this implies that

$$\|h\|_{L^r} \leq \sup_{t>0}\left(\int_{\mathbf{R}^n} |F(x,t)|^p\,dx\right)^{\frac{1}{r}}. \tag{2.4.58}$$

Putting things together, we have

$$\begin{aligned}
\big\||F|^*\big\|_{L^p} &\leq C_n' \big\|M(h)^{1/q}\big\|_{L^p} \\
&= C_n' \big\|M(h)\big\|_{L^r}^{1/q} \\
&\leq C_{n,p,q} \big\|h\big\|_{L^r}^{1/q}
\end{aligned}$$

$$
\begin{aligned}
&= C_{n,p,q} \sup_{t>0} \left(\int_{\mathbf{R}^n} |F(x,t)|^p \, dx \right)^{1/qr} \\
&\le C_{n,p,q} A \, ,
\end{aligned}
$$

where we have used (2.4.57) and (2.4.58) in the last two displayed inequalities. □

Exercises

2.4.1. Let f be an integrable function on the line whose Fourier transform vanishes on $(-\infty,0)$. Show that f lies in $H^1(\mathbf{R})$.

2.4.2. (a) Let h be a function on $\mathbf{R}$ such that $h(x)$ and $xh(x)$ are in $L^2(\mathbf{R})$. Show that h is integrable over $\mathbf{R}$ and satisfies

$$
\|h\|_{L^1}^2 \le 8 \, \|h\|_{L^2} \big\| xh(x) \big\|_{L^2} \, .
$$

(b) Suppose that g is an integrable function on $\mathbf{R}$ with vanishing integral and $g(x)$ and $xg(x)$ are in $L^2(\mathbf{R})$. Show that g lies in $H^1(\mathbf{R})$ and that for some constant C we have

$$
\|g\|_{H^1}^2 \le C \, \|g\|_{L^2} \big\| xg(x) \big\|_{L^2} \, .
$$

$\big[$*Hint:* Part (a): Split the integral of $|h(x)|$ over the regions $|x| \le R$ and $|x| > R$ and pick a suitable R. Part (b): Show that both $H(g)$ and $H(yg(y))$ lie in L^2. But since g has vanishing integral, we have $xH(g)(x) = H(yg(y))(x)$.$\big]$

2.4.3. (a) Let H be the Hilbert transform on the real line. Prove the identity

$$
H(fg - H(f)H(g)) = fH(g) + gH(f)
$$

for all f,g real-valued Schwartz functions. (b) Show that the bilinear operators

$$
\begin{aligned}
(f,g) &\mapsto f H(g) + H(f) \, g \, , \\
(f,g) &\mapsto f g - H(f) \, H(g) \, ,
\end{aligned}
$$

map $L^p(\mathbf{R}) \times L^{p'}(\mathbf{R}) \to H^1(\mathbf{R})$ whenever $1 < p < \infty$.
$\big[$*Hint:* Part (a): Consider product $U_f(z)U_g(z)$, where $U_f(z) = \frac{i}{\pi} \int_{\mathbf{R}} \frac{f(t)}{z-t} \, dt$ is holomorphic on the upper half space and has boundary values $f + iH(f)$. Part (b): Use part (a) and Theorem 2.4.6.$\big]$

2.4.4. Follow the steps given to prove the following interpolation result. Let $1 < p_1 \le \infty$ and let T be a subadditive operator that maps $H^1(\mathbf{R}^n) + L^{p_1}(\mathbf{R}^n)$ into measurable functions on $\mathbf{R}^n$. Suppose that there is $A_0 < \infty$ such that for all $f \in H^1(\mathbf{R}^n)$ we have

$$
\sup_{\lambda>0} \lambda \, \big| \{ x \in \mathbf{R}^n : \, |T(f)(x)| > \lambda \} \big| \le A_0 \big\| f \big\|_{H^1}
$$

and that it also maps $L^{p_1}(\mathbf{R}^n)$ to $L^{p_1,\infty}(\mathbf{R}^n)$ with norm at most A_1. Show that for any $1 < p < p_1$, T maps $L^p(\mathbf{R}^n)$ to itself with norm at most

$$C A_0^{\frac{\frac{1}{p}-\frac{1}{p_1}}{1-\frac{1}{p_1}}} A_1^{\frac{1-\frac{1}{p}}{1-\frac{1}{p_1}}},$$

where $C = C(n,p,p_1)$.
(a) Fix $1 < q < p < p_1 < \infty$ and f and let Q_j be the family of all maximal dyadic cubes such that $\lambda^q < |Q_j|^{-1}\int_{Q_j}|f|^q\,dx$. Write $E_\lambda = \bigcup Q_j$ and note that $E_\lambda \subseteq \{M(|f|^q)^{\frac{1}{q}} > \lambda\}$ and that $|f| \le \lambda$ a.e. on $(E_\lambda)^c$. Write f as the sum of the *good function*

$$g_\lambda = f\chi_{(E_\lambda)^c} + \sum_j (\operatorname*{Avg}_{Q_j} f)\,\chi_{Q_j}$$

and the *bad function*

$$b_\lambda = \sum_j b_\lambda^j, \qquad \text{where} \qquad b_\lambda^j = (f - \operatorname*{Avg}_{Q_j} f)\,\chi_{Q_j}.$$

(b) Show that g_λ lies in $L^{p_1}(\mathbf{R}^n)\cap L^\infty(\mathbf{R}^n)$, $\|g_\lambda\|_{L^\infty} \le 2^{\frac{n}{q}}\lambda$, and that

$$\|g_\lambda\|_{L^{p_1}}^{p_1} \le \int_{|f|\le\lambda} |f(x)|^{p_1}\,dx + 2^{\frac{np_1}{q}}\lambda^{p_1}|E_\lambda| < \infty.$$

(c) Show that for $c = 2^{\frac{n}{q}+1}$, each $c^{-1}\lambda^{-1}|Q_j|^{-1}b_\lambda^j$ is an L^q-atom for H^1. Conclude that b_λ lies in $H^1(\mathbf{R}^n)$ and satisfies

$$\|b_\lambda\|_{H^1} \le c\lambda\sum_j |Q_j| \le c\lambda\,|E_\lambda| < \infty.$$

(d) Start with

$$\begin{aligned}\|T(f)\|_{L^p}^p \le{}& p\gamma^p\int_0^\infty \lambda^{p-1}\big|\{T(g_\lambda)| > \tfrac{1}{2}\gamma\lambda\}\big|\,d\lambda\\ &+ p\gamma^p\int_0^\infty \lambda^{p-1}\big|\{T(b_\lambda)| > \tfrac{1}{2}\gamma\lambda\}\big|\,d\lambda\end{aligned}$$

and use the results in parts (b) and (c) to obtain that the preceding expression is at most $C(n,p,q,p_1)\max(A_1\gamma^{p-p_1}, \gamma^{p-1}A_0)$. Select $\gamma = A_1^{\frac{p_1}{p_1-1}} A_0^{-\frac{1}{p_1-1}}$ to obtain the required conclusion.
(e) In the case $p_1 = \infty$ we have $|T(g_\lambda)| \le A_1 2^{\frac{n}{q}}\lambda$ and pick $\gamma > 2A_1 2^{\frac{n}{q}}$ to make the integral involving g_λ vanishing.

2.4.5. Let P_t be the Poisson kernel and K_j be the kernel of the Riesz transform R_j. Let $\widehat{\varphi} \in \mathscr{S}$ be equal to 1 in a neighborhood of the origin. Then $\delta_0 = \varphi + (\delta_0 - \varphi)$ and for a bounded distribution f (cf. Section 2.1.1) and $t > 0$ write

$$(P_t * K_j) * f = (P_t * K_j) * (\varphi * f) + (P_t * K_j) * (\delta_0 - \varphi) * f.$$

Since P_t lies in L^1 and $\varphi * f$ in L^∞, $(K_j * P_t) * (\varphi * f) = R_j(P_t * \varphi * f)$ is a BMO function. The Fourier transform of $(P_t * K_j) * (\delta_0 - \varphi)$ is $-i\frac{\xi_j}{|\xi|}e^{-2\pi t|\xi|}(1 - \widehat{\varphi}(\xi))$, which is a Schwartz function. Thus $(P_t * K_j) * (\delta_0 - \varphi)$ is also a Schwartz function and $(P_t * K_j) * (\delta_0 - \varphi) * f$ is a smooth function. Hence $(P_t * K_j) * f = P_t * R_j(f)$ is a well-defined function for all $t > 0$ and $j = 1, \dots, n$. Let $\frac{n-1}{n} < p < 1$.
(a) Show that there are constants C_n, C_1 such that for any $f \in H^p(\mathbf{R}^n)$ we have

$$\sup_{\delta>0}\Big[\big\|P_\delta * f\big\|_{L^p} + \sum_{k=1}^n \big\|P_\delta * R_k(f)\big\|_{L^p}\Big] \le C_n\big\|f\big\|_{H^p}$$

when $n \ge 2$ and

$$\sup_{\delta>0}\Big[\big\|P_\delta * f\big\|_{L^p} + \big\|P_\delta * H(f)\big\|_{L^p}\Big] \le C_1\big\|f\big\|_{H^p}$$

when $n = 1$.
(b) Prove that there are constants C_1, C_n such that for any bounded tempered distribution f on $\mathbf{R}^n$ we have

$$c_n\big\|f\big\|_{H^p} \le \sup_{\delta>0}\Big[\big\|P_\delta * f\big\|_{L^p} + \sum_{k=1}^n \big\|P_\delta * R_k(f)\big\|_{L^p}\Big]$$

when $n \ge 2$ and

$$c_1\big\|f\big\|_{H^p} \le \sup_{\delta>0}\Big[\big\|P_\delta * f\big\|_{L^p} + \big\|P_\delta * H(f)\big\|_{L^p}\Big]$$

when $n = 1$.
[*Hint:* Part (a): This is a consequence of Theorem 2.4.5. Part (b): Define $F_\delta = (P_\delta * u_1, \dots, P_\delta * u_n, P_\delta * u_{n+1})$, where $u_j(x,t) = (P_t * R_j(f))(x)$, $j = 1, \dots, n$, and $u_{n+1}(x,t) = (P_t * f)(x)$. Each $P_\delta * u_j$ is a harmonic function on $\mathbf{R}^{n+1}_+$ and continuous up to the boundary. The subharmonicity of $|F_\delta(x,t)|^p$ has as a consequence that $|F_\delta(x,t+\varepsilon)|^p \le (|F_\delta(\cdot,\varepsilon)|^p * P_t)(x)$ in view of (2.4.53). Letting $\varepsilon \to 0$ implies that $|F_\delta(x,t)|^p \le (|F_\delta(\cdot,0)|^p * P_t)(x)$, by the continuity of F_δ up to the boundary. Since $F_\delta(x,0) = (P_\delta * R_1(f), \dots, P_\delta * R_n(f), P_\delta * f)$, the hypothesis that $P_\delta * f, P_\delta * R_j(f)$ are in L^p uniformly in $\delta > 0$ yields $\sup_{t,\delta>0}\int_{\mathbf{R}^n}|F_\delta(x,t)|^p\,dx < \infty$. Fatou's lemma implies (2.4.42) for $F(x,t) = (u_1, \dots, u_{n+1})$ and then Lemma 2.4.10 yields the claim.]

HISTORICAL NOTES

The theory of Hardy spaces is vast and complicated. In classical complex analysis, the Hardy spaces H^p were spaces of analytic functions and were introduced to characterize boundary values of analytic functions on the unit disk.

Hardy [180] proved that the mean value of the pth power of the modulus of an analytic function on the unit disc is an increasing function of the radius and its logarithm is a convex function of the logarithm of the radius. The first systematic study of the class $H_p(\mathbb{D})$ of all analytic functions F on the unit disk $\mathbb{D}$ with the property that $\sup_{0<r<1}\int_0^1 |F(re^{2\pi i\theta})|^p\,d\theta<\infty$, $0<p<\infty$, can be traced to F. Riesz's article [303]. In this article Riesz proved the factorization theorem, the existence of boundary values, and other basic properties of such functions and adopted the symbol H_p, honoring Hardy for the fact that the aforementioned mean values are increasing as a function of the radius r. When $1<p<\infty$, the space $H_p(\mathbb{D})$ coincides with the space of analytic functions whose real parts are Poisson integrals of functions in $L^p(\mathbf{T}^1)$. But for $0<p\le 1$ this characterization fails and for several years a satisfactory characterization was missing. For a systematic treatment of these spaces we refer to the books of Duren [127] and Koosis [226].

With the illuminating work of Stein and Weiss [327] on systems of conjugate harmonic functions the road opened to higher-dimensional extensions of Hardy spaces. Burkholder, Gundy, and Silverstein [52] proved the fundamental theorem that an analytic function F lies in $H^p(\mathbf{R}^2_+)$ [i.e., $\sup_{y>0}\int_{\mathbf{R}}|F(x+iy)|^p\,dx<\infty$] if and only if the nontangential maximal function of its real part lies in $L^p(\mathbf{R})$. This result was proved using Brownian motion, but later Koosis [225] obtained another proof using complex analysis. This theorem spurred the development of the modern theory of Hardy spaces by providing the first characterization without the notion of conjugacy and indicating that Hardy spaces are intrinsically defined. The pioneering article of Fefferman and Stein [139] furnished three new characterizations of Hardy spaces: using a maximal function associated with a general approximate identity, using the grand maximal function, and using the area function of Luzin. From this point on, the role of the Poisson kernel faded into the background, when it turned out that it was not essential in the study of Hardy spaces. A previous characterization of Hardy spaces using the g-function, a radial analogue of the Luzin area function, was obtained by Calderón [54]. Two alternative characterizations of Hardy spaces were obtained by Uchiyama in terms of the generalized Littlewood–Paley g-function [356] and in terms of Fourier multipliers [357]. A characterization of $H^1(\mathbf{R})$ in terms of the variation of the function $m_f(y)=\int_{\mathbf{R}}f(x)\ln|x-y|^{-1}\,dx$ was obtained by Stefanov [321]. An extension of this result in higher dimensions was provided by Wang [364]. Necessary and sufficient conditions for systems of singular integral operators to characterize $H^1(\mathbf{R}^n)$ were also obtained by Uchiyama [355]. The characterization of H^p using Littlewood–Paley theory was observed by Peetre [292]. The case $p=1$ was later independently obtained by Rubio de Francia, Ruiz, and Torrea [308].

The one-dimensional atomic decomposition of Hardy spaces is due to Coifman [86] and its higher-dimensional extension to Latter [239]. A simplification of some of the technical details in Latter's proof was subsequently obtained by Latter and Uchiyama [240]. Using the atomic decomposition Coifman and Weiss [97] extended the definition of Hardy spaces to more general structures. The idea of obtaining the atomic decomposition from the reproducing formula (2.3.13) goes back to Calderón [56]. Another simple proof of the L^2-atomic decomposition for H^p (starting from the nontangential Poisson maximal function) was obtained by Wilson [370]. With only a little work, one can show that L^q-atoms for H^p can be written as sums of L^∞-atoms for H^p. We refer to the book of García-Cuerva and Rubio de Francia [150] for a proof of this fact. Although finite sums of atoms are dense in H^1, an example due to Y. Meyer (contained in [265]) shows that the H^1 norm of a function may not be comparable to $\inf\sum_{j=1}^N|\lambda_j|$, where the infimum is taken over all representations of the function as finite linear combinations $\sum_{j=1}^N\lambda_j a_j$ with the a_j being L^∞-atoms for H^1. Based on this idea, Bownik [48] constructed an example of a linear functional on a dense subspace of H^1 that is uniformly bounded on L^∞-atoms for H^1 but does not extend to a bounded linear functional on the whole H^1. However, if a Banach-valued linear operator is bounded uniformly on all L^q-atoms for H^p with $1<q<\infty$ and $0<p\le 1$, then it is bounded on the entire H^p as shown by Meda, Sjögren, and Vallarino [261]. This fact is also valid for quasi-Banach-valued linear operators, and when $q=2$ it was obtained independently by Yang and Zhou [374]. A related general result says that a sublinear operator maps the Triebel–Lizorkin space $\dot{F}^s_{p,q}(\mathbf{R}^n)$ to a quasi-Banach space if and only if it is uniformly bounded on certain infinitely differentiable atoms of the space; see Liu and Yang [250]. Atomic decompositions of general function spaces were obtained in the fundamental work of Frazier and Jawerth [143], [144]. The exposition in Section 2.3 is based on the article of Frazier and Jawerth [145]. The work of these authors provides a solid

manifestation that atomic decompositions are intrinsically related to Littlewood–Paley theory and not wedded to a particular space. Littlewood–Paley theory therefore provides a comprehensive and unifying perspective on function spaces.

Main references on H^p spaces and their properties are the books of Baernstein and Sawyer [14], Folland and Stein [142] in the context of homogeneous groups, Lu [252] (on which the proofs of Lemma 2.1.5 and Theorem 2.1.4 are based), Strömberg and Torchinsky [333] (on weighted Hardy spaces), and Uchiyama [358]. The articles of Calderón and Torchinsky [58], [59] develop and extend the theory of Hardy spaces to the nonisotropic setting. Hardy spaces can also be defined in terms of nonstandard convolutions, such as the "twisted convolution" on $\mathbf{R}^{2n}$. Characterizations of the space H^1 in this context have been obtained by Mauceri, Picardello, and Ricci [259].

The localized Hardy spaces h_p, $0 < p \le 1$, were introduced by Goldberg [155] as spaces of distributions for which the maximal operator $\sup_{0<t<1} |\Phi_t * f|$ lies in $L^p(\mathbf{R}^n)$ (here Φ is a Schwartz function with nonvanishing integral). These spaces can be characterized in ways analogous to those of the homogeneous Hardy spaces H^p; in particular, they admit an atomic decomposition. It was shown by Bui [50] that the space h^p coincides with the Triebel–Lizorkin space $F_p^{0,2}(\mathbf{R}^n)$; see also Meyer [263]. For the local theory of Hardy spaces one may consult the articles of Dafni [108] and Chang, Krantz, and Stein [73].

Interpolation of operators between Hardy spaces was originally based on complex function theory; see the articles of Calderón and Zygmund [57] and Weiss [365]. The real-interpolation approach discussed in Exercise 2.4.4 can be traced to the article of Igari [201]. Interpolation between Hardy spaces was further studied and extended by Riviere and Sagher [305]; Fefferman, Riviere, and Sagher [137]; and He [187].

The action of singular integrals on periodic spaces was studied by Calderón and Zygmund [61]. The preservation of Lipschitz spaces under singular integral operators is due to Taibleson [334]. The case $0 < \alpha < 1$ was earlier considered by Privalov [301] for the conjugate function on the circle. Fefferman and Stein [139] were the first to show that singular integrals map Hardy spaces to themselves. The boundedness of fractional integrals on H^p was obtained by Krantz [228]. The case $p = 1$ was earlier considered by Stein and Weiss [327]. An exposition on the subject of function spaces and the action of singular integrals on them was written by Frazier, Jawerth, and Weiss [146]. For a careful study of the action of singular integrals on function spaces, we refer to the book of Torres [352]. The study of anisotropic function spaces and the action of singular integrals on them has been studied by Bownik [47]. Weighted anisotropic Hardy spaces have been studied by Bownik, Li, Yang, and Zhou [49].

Besov spaces are named after Besov, who obtained a trace theorem and embeddings for them [34], [35]. The spaces $B_p^{\alpha,q}$, as defined in Section 2.2, were introduced by Peetre [290], although the case $p = q = 2$ was earlier considered by Hörmander [194]. The connection of Besov spaces with modern Littlewood–Paley theory was brought to the surface by Peetre [290]. The extension of the definition of Besov spaces to the case $p < 1$ is also due to Peetre [291], but there was a forerunner by Flett [140]. Peetre's monograph [294] contains an excellent exposition on the topic of Besov spaces. The spaces $F_p^{\alpha,q}$ with $1 < p,q < \infty$ were introduced by Triebel [353] and independently by Lizorkin [251]. The extension of the spaces $F_p^{\alpha,q}$ to the case $0 < p < \infty$ and $0 < q \le \infty$ first appeared in Peetre [293], who also obtained a maximal characterization for all of these spaces. Lemma 2.2.3 originated in Peetre [293]; the version given in the text is based on a refinement of Triebel [354]. The article of Lions, Lizorkin, and Nikol'skij [249] presents an account of the treatment of the spaces $F_p^{\alpha,q}$ introduced by Triebel and Lizorkin as well as the equivalent characterizations obtained by Lions, using interpolation between Banach spaces, and by Nikol'skij, using best approximation.

Chapter 3
BMO and Carleson Measures

If the deviation of a function from its averages over all cubes is bounded, then the function is called of bounded mean oscillation (BMO). Bounded functions are of bounded mean oscillation, but there exist unbounded *BMO* functions. Such functions are slowly growing, and they typically have at most logarithmic blowup. The space *BMO* shares similar properties with the space L^∞, and often serves as a substitute for it. For instance, classical singular integrals do not map L^∞ to L^∞ but L^∞ to *BMO*. And in many instances interpolation between L^p and *BMO* works just as well between L^p and L^∞. But the role of the space *BMO* is deeper and more far-reaching than that. This space crucially arises in many situations in analysis, such as in the characterization of the L^2 boundedness of nonconvolution singular integral operators with standard kernels.

Carleson measures are among the most important tools in harmonic analysis. These measures capture essential orthogonality properties and exploit properties of extensions of functions on the upper half-space. There exists a natural and deep connection between Carleson measures and *BMO* functions; indeed, certain types of measures defined in terms of functions are Carleson if and only if the underlying functions are in *BMO*. Carleson measures are especially crucial in the study of L^2 problems, where the Fourier transform cannot be used in conjunction with Plancherel's theorem. The power of the Carleson measure techniques becomes apparent in certain important topics studied in Chapter 4.

3.1 Functions of Bounded Mean Oscillation

What exactly is bounded mean oscillation and what kind of functions have this property? The mean of a (locally integrable) function over a set is another word for its average over that set. The oscillation of a function over a set is the absolute value of the difference of the function from its mean over this set. Mean oscillation is therefore the average of this oscillation over a set. A function is said to be of bounded

L. Grafakos, *Modern Fourier Analysis*, Graduate Texts in Mathematics 250,

DOI 10.1007/978-1-4939-1230-8_3,

mean oscillation if its mean oscillation over all cubes is bounded. Precisely, given a locally integrable function f on $\mathbf{R}^n$ and a measurable set Q in $\mathbf{R}^n$, denote by

$$\operatorname*{Avg}_Q f = \frac{1}{|Q|}\int_Q f(x)\,dx$$

the mean (or average) of f over Q. Then the *oscillation* of f over Q is the function $|f - \mathrm{Avg}_Q f|$, and the *mean oscillation* of f over Q is

$$\frac{1}{|Q|}\int_Q \big|f(x) - \operatorname*{Avg}_Q f\big|\,dx.$$

3.1.1 Definition and Basic Properties of *BMO*

Definition 3.1.1. For f a complex-valued locally integrable function on $\mathbf{R}^n$, set

$$\|f\|_{BMO} = \sup_Q \frac{1}{|Q|}\int_Q \big|f(x) - \operatorname*{Avg}_Q f\big|\,dx,$$

where the supremum is taken over all cubes Q in $\mathbf{R}^n$. The function f is of bounded mean oscillation if $\|f\|_{BMO} < \infty$ and $BMO(\mathbf{R}^n)$ is the set of all locally integrable functions f on $\mathbf{R}^n$ with $\|f\|_{BMO} < \infty$.

Several remarks are in order. First it is a simple fact that $BMO(\mathbf{R}^n)$ is a linear space, that is, if $f, g \in BMO(\mathbf{R}^n)$ and $\lambda \in \mathbf{C}$, then $f+g$ and λf are also in $BMO(\mathbf{R}^n)$ and

$$\begin{aligned}\|f+g\|_{BMO} &\le \|f\|_{BMO} + \|g\|_{BMO},\\ \|\lambda f\|_{BMO} &= |\lambda|\,\|f\|_{BMO}.\end{aligned}$$

But $\|\ \|_{BMO}$ is not a norm. The problem is that if $\|f\|_{BMO} = 0$, this does not imply that $f = 0$ but that f is a constant. See Proposition 3.1.2. Moreover, every constant function c satisfies $\|c\|_{BMO} = 0$. Consequently, functions f and $f + c$ have the same *BMO* norms whenever c is a constant. In the sequel, we keep in mind that elements of *BMO* whose difference is a constant are identified. Although $\|\ \|_{BMO}$ is only a seminorm, we occasionally refer to it as a norm when there is no possibility of confusion.

We begin with a list of basic properties of *BMO*.

Proposition 3.1.2. *The following properties of the space $BMO(\mathbf{R}^n)$ are valid:*

(1) If $\|f\|_{BMO} = 0$, then f is a.e. equal to a constant.

(2) $L^\infty(\mathbf{R}^n)$ is contained in $BMO(\mathbf{R}^n)$ and $\|f\|_{BMO} \le 2\|f\|_{L^\infty}$.

(3) Suppose that there exists an $A>0$ such that for all cubes Q in $\mathbf{R}^n$ there exists a constant c_Q such that

$$\sup_Q \frac{1}{|Q|}\int_Q |f(x)-c_Q|\,dx \le A. \tag{3.1.1}$$

Then $f\in BMO(\mathbf{R}^n)$ and $\|f\|_{BMO}\le 2A$.

(4) For all f locally integrable we have

$$\frac{1}{2}\big\|f\big\|_{BMO} \le \sup_Q \frac{1}{|Q|}\inf_{c_Q}\int_Q |f(x)-c_Q|\,dx \le \big\|f\big\|_{BMO}.$$

(5) If $f\in BMO(\mathbf{R}^n)$, $h\in\mathbf{R}^n$, and $\tau^h(f)$ is given by $\tau^h(f)(x)=f(x-h)$, then $\tau^h(f)$ is also in $BMO(\mathbf{R}^n)$ and

$$\big\|\tau^h(f)\big\|_{BMO} = \big\|f\big\|_{BMO}.$$

(6) If $f\in BMO(\mathbf{R}^n)$ and $\lambda>0$, then the function $\delta^\lambda(f)$ defined by $\delta^\lambda(f)(x)=f(\lambda x)$ is also in $BMO(\mathbf{R}^n)$ and

$$\big\|\delta^\lambda(f)\big\|_{BMO} = \big\|f\big\|_{BMO}.$$

(7) If $f\in BMO$, then so is $|f|$. Similarly, if f,g are real-valued BMO functions, then so are $\max(f,g)$ and $\min(f,g)$. In other words, BMO is a lattice. Moreover,

$$\begin{aligned}
\big\||f|\big\|_{BMO} &\le 2\big\|f\big\|_{BMO},\\
\big\|\max(f,g)\big\|_{BMO} &\le \frac{3}{2}\Big(\big\|f\big\|_{BMO}+\big\|g\big\|_{BMO}\Big),\\
\big\|\min(f,g)\big\|_{BMO} &\le \frac{3}{2}\Big(\big\|f\big\|_{BMO}+\big\|g\big\|_{BMO}\Big).
\end{aligned}$$

(8) For locally integrable functions f define

$$\big\|f\big\|_{BMO_{\text{balls}}} = \sup_B \frac{1}{|B|}\int_B \Big|f(x)-\operatorname*{Avg}_B f\Big|\,dx, \tag{3.1.2}$$

where the supremum is taken over all balls B in $\mathbf{R}^n$. Then there are positive constants c_n, C_n such that

$$c_n\big\|f\big\|_{BMO} \le \big\|f\big\|_{BMO_{\text{balls}}} \le C_n\big\|f\big\|_{BMO}.$$

Proof. To prove (1) note that f has to be a.e. equal to its average c_N over every cube $[-N,N]^n$. Since $[-N,N]^n$ is contained in $[-N-1,N+1]^n$, it follows that $c_N=c_{N+1}$ for all N. This implies the required conclusion. To prove (2) observe that

$$\operatorname*{Avg}_Q\Big|f-\operatorname*{Avg}_Q f\Big| \le 2\operatorname*{Avg}_Q|f| \le 2\big\|f\big\|_{L^\infty}.$$

For part (3) note that

$$\big|f-\operatorname*{Avg}_Q f\big|\le |f-c_Q|+\big|\operatorname*{Avg}_Q f-c_Q\big|\le |f-c_Q|+\frac{1}{|Q|}\int_Q |f(t)-c_Q|\,dt.$$

Averaging over Q and using (3.1.1), we obtain that $\|f\|_{BMO}\le 2A$. The lower inequality in (4) follows from (3) while the upper one is trivial. Property (5) is immediate. For (6) note that $\operatorname{Avg}_Q \delta^\lambda(f)=\operatorname{Avg}_{\lambda Q} f$ and thus

$$\frac{1}{|Q|}\int_Q \big|f(\lambda x)-\operatorname*{Avg}_Q \delta^\lambda(f)\big|\,dx=\frac{1}{|\lambda Q|}\int_{\lambda Q}\big|f(x)-\operatorname*{Avg}_{\lambda Q} f\big|\,dx.$$

The first inequality in (7) is a consequence of the fact that

$$\big||f|-\operatorname*{Avg}_Q |f|\big|\le \big|f-\operatorname*{Avg}_Q f\big|+\operatorname*{Avg}_Q\big|f-\operatorname*{Avg}_Q f\big|.$$

The second and the third inequalities in (7) follow from the first inequality in (7) and the facts that

$$\max(f,g)=\frac{f+g+|f-g|}{2},\qquad \min(f,g)=\frac{f+g-|f-g|}{2}.$$

We now turn to (8). Given any cube Q in $\mathbf{R}^n$, we let B be the smallest ball that contains it. Then $|B|/|Q|=2^{-n}v_n\sqrt{n^n}$, where v_n is the volume of the unit ball, and

$$\frac{1}{|Q|}\int_Q\big|f(x)-\operatorname*{Avg}_B f\big|\,dx\le\frac{|B|}{|Q|}\frac{1}{|B|}\int_B\big|f(x)-\operatorname*{Avg}_B f\big|\,dx\le\frac{v_n\sqrt{n^n}}{2^n}\,\|f\|_{BMO_{\text{balls}}}.$$

It follows from (3) that

$$\|f\|_{BMO}\le 2^{1-n}v_n\sqrt{n^n}\,\|f\|_{BMO_{\text{balls}}}.$$

To obtain the reverse conclusion, given any ball B find the smallest cube Q that contains it and argue similarly using a version of (3) for the space BMO_{balls}. □

Example 3.1.3. We indicate why $L^\infty(\mathbf{R}^n)$ is a proper subspace of $BMO(\mathbf{R}^n)$. We claim that the unbounded function $\log|x|$ is in $BMO(\mathbf{R}^n)$. To prove this, for every $x_0\in\mathbf{R}^n$ and $R>0$, we find a constant $C_{x_0,R}$ such that the average of $|\log|x|-C_{x_0,R}|$ over the ball $\overline{B(0,R)}=\{x\in\mathbf{R}^n:\,|x-x_0|\le R\}$ is uniformly bounded. The constant $C_{x_0,R}=\log|x_0|$ if $|x_0|>2R$ and $C_{x_0,R}=\log R$ if $|x_0|\le 2R$ has this property. Indeed, if $|x_0|>2R$, then

$$\begin{aligned}\frac{1}{v_nR^n}\int_{|x-x_0|\le R}\big|\log|x|-C_{x_0,R}\big|\,dx&=\frac{1}{v_nR^n}\int_{|z-x_0|\le R}\Big|\log\frac{|z|}{|x_0|}\Big|\,dz\\&\le\max\Big(\log\frac{3}{2},\Big|\log\frac{1}{2}\Big|\Big)\\&=\log 2,\end{aligned}$$

since $\frac{1}{2}|x_0| \le |z| \le \frac{3}{2}|x_0|$ when $|z-x_0| \le R$ and $|x_0| > 2R$. Also, if $|x_0| \le 2R$, then

$$\begin{aligned}\frac{1}{v_n R^n} \int\limits_{|x-x_0|\le R} \big|\log|x| - C_{x_0,R}\big|\,dx &= \frac{1}{v_n R^n} \int\limits_{|z-x_0|\le R} \Big|\log\frac{|z|}{R}\Big|\,dz \\ &\le \frac{1}{v_n R^n} \int\limits_{|z|\le 3R} \Big|\log\frac{|z|}{R}\Big|\,dz \\ &= \frac{1}{v_n} \int\limits_{|z|\le 3} \big|\log|z|\big|\,dz.\end{aligned}$$

Thus $\log|x|$ is in BMO.

The function $\log|x|$ turns out to be a typical element of BMO, but we make this statement a bit more precise later. It is interesting to observe that an abrupt cutoff of a BMO function may not give a function in the same space.

Example 3.1.4. The function $h(x) = \chi_{x>0}\log\frac{1}{x}$ is not in $BMO(\mathbf{R})$. Indeed, the problem is at the origin. Consider the intervals $(-\varepsilon,\varepsilon)$, where $0<\varepsilon<\frac{1}{2}$. We have that

$$\operatorname*{Avg}_{(-\varepsilon,\varepsilon)} h = \frac{1}{2\varepsilon}\int_{-\varepsilon}^{+\varepsilon} h(x)\,dx = \frac{1}{2\varepsilon}\int_0^{\varepsilon} \log\frac{1}{x}\,dx = \frac{1+\log\frac{1}{\varepsilon}}{2}.$$

But then

$$\frac{1}{2\varepsilon}\int_{-\varepsilon}^{+\varepsilon} \Big|h(x) - \operatorname*{Avg}_{(-\varepsilon,\varepsilon)} h\Big|\,dx \ge \frac{1}{2\varepsilon}\int_{-\varepsilon}^{0} \Big|\operatorname*{Avg}_{(-\varepsilon,\varepsilon)} h\Big|\,dx = \frac{1+\log\frac{1}{\varepsilon}}{4},$$

and the latter is clearly unbounded as $\varepsilon \to 0$.

Let us now look at some basic properties of BMO functions. Observe that if a cube Q_1 is contained in a cube Q_2, then

$$\begin{aligned}\Big|\operatorname*{Avg}_{Q_1} f - \operatorname*{Avg}_{Q_2} f\Big| &\le \frac{1}{|Q_1|}\int_{Q_1} \Big|f - \operatorname*{Avg}_{Q_2} f\Big|\,dx \\ &\le \frac{1}{|Q_1|}\int_{Q_2} \Big|f - \operatorname*{Avg}_{Q_2} f\Big|\,dx \\ &\le \frac{|Q_2|}{|Q_1|}\,\|f\|_{BMO}.\end{aligned} \tag{3.1.3}$$

The same estimate holds if the sets Q_1 and Q_2 are balls.

A version of this inequality is the first statement in the following proposition. For simplicity, we denote by $\|f\|_{BMO}$ the expression given by $\|f\|_{BMO_{\text{balls}}}$ in (3.1.2), since these quantities are comparable. For a ball B and $a>0$, aB denotes the ball that is concentric with B and whose radius is a times the radius of B.

Proposition 3.1.5. *(i) Let f be in $BMO(\mathbf{R}^n)$. Given a ball B and a positive integer m, we have*

$$\Big|\operatorname*{Avg}_{B} f - \operatorname*{Avg}_{2^m B} f\Big| \le 2^n m \big\|f\big\|_{BMO}. \tag{3.1.4}$$

(ii) For any $\delta > 0$ there is a constant $C_{n,\delta}$ such that for any ball $B(x_0,R)$ we have

$$R^\delta \int_{\mathbf{R}^n} \frac{\big|f(x) - \operatorname{Avg}_{B(x_0,R)} f\big|}{(R+|x-x_0|)^{n+\delta}}\,dx \le C_{n,\delta}\big\|f\big\|_{BMO}. \tag{3.1.5}$$

An analogous estimate holds for cubes with center x_0 and side length R.
(iii) There exists a constant C_n such that for all $f \in BMO(\mathbf{R}^n)$ we have

$$\sup_{y\in\mathbf{R}^n}\sup_{t>0}\int_{\mathbf{R}^n} |f(x) - (P_t * f)(y)|P_t(x-y)\,dx \le C_n\big\|f\big\|_{BMO}. \tag{3.1.6}$$

Here

$$P_t(x) = \Gamma\Big(\frac{n+1}{2}\Big)\pi^{-\frac{n+1}{2}}t(t^2+|x|^2)^{-\frac{n+1}{2}}$$

denotes the Poisson kernel.
(iv) Conversely, there is a constant C_n' such that for all $f \in L^1_{\mathrm{loc}}(\mathbf{R}^n)$ for which

$$\int_{\mathbf{R}^n} \frac{|f(x)|}{(1+|x|)^{n+1}}\,dx < \infty$$

we have $f * P_t$ is well defined and

$$C_n'\big\|f\big\|_{BMO} \le \sup_{y\in\mathbf{R}^n}\sup_{t>0}\int_{\mathbf{R}^n} |f(x) - (P_t * f)(y)|P_t(x-y)\,dx. \tag{3.1.7}$$

Proof. (i) We have

$$\begin{aligned}
\Big|\operatorname*{Avg}_{B} f - \operatorname*{Avg}_{2B} f\Big| &= \frac{1}{|B|}\Big|\int_B \big(f(t) - \operatorname*{Avg}_{2B} f\big)\,dt\Big| \\
&\le \frac{2^n}{|2B|}\int_{2B} \Big|f(t) - \operatorname*{Avg}_{2B} f\Big|\,dt \\
&\le 2^n\big\|f\big\|_{BMO}.
\end{aligned}$$

Using this inequality, we derive (3.1.4) by adding and subtracting the terms

$$\operatorname*{Avg}_{2B} f,\quad \operatorname*{Avg}_{2^2B} f,\quad \dots,\quad \operatorname*{Avg}_{2^{m-1}B} f.$$

(ii) In the proof below we take $B(x_0,R)$ to be the ball $B = B(0,1)$ with radius 1 centered at the origin. Once this case is known, given a ball $B(x_0,R)$, we replace the function f by the function $f(Rx+x_0)$. When $B = B(0,1)$ we have

$$
\begin{aligned}
&\int_{\mathbf{R}^n} \frac{\big|f(x) - \operatorname*{Avg}_B f\big|}{(1+|x|)^{n+\delta}}\,dx \\
&\le \int_B \frac{\big|f(x) - \operatorname*{Avg}_B f\big|}{(1+|x|)^{n+\delta}}\,dx + \sum_{k=0}^{\infty} \int\limits_{2^{k+1}B\setminus 2^k B} \frac{\big|f(x) - \operatorname*{Avg}_{2^{k+1}B} f\big| + \big|\operatorname*{Avg}_{2^{k+1}B} f - \operatorname*{Avg}_B f\big|}{(1+|x|)^{n+\delta}}\,dx \\
&\le \int_B \big|f(x) - \operatorname*{Avg}_B f\big|\,dx \\
&\qquad + \sum_{k=0}^{\infty} 2^{-k(n+\delta)} \int\limits_{2^{k+1}B} \Big(\big|f(x) - \operatorname*{Avg}_{2^{k+1}B} f\big| + \big|\operatorname*{Avg}_{2^{k+1}B} f - \operatorname*{Avg}_B f\big|\Big)\,dx \\
&\le v_n\|f\|_{BMO} + \sum_{k=0}^{\infty} 2^{-k(n+\delta)}\big(1+2^n(k+1)\big)(2^{k+1})^n v_n \|f\|_{BMO} \\
&= C'_{n,\delta}\|f\|_{BMO}.
\end{aligned}
$$

(iii) The proof of (3.1.6) is a reprise of the argument given in (ii). Set $B_t = B(y,t)$. We first prove a version of (3.1.6) in which the expression $(P_t * f)(y)$ is replaced by $\operatorname{Avg}_{B_t} f$. For fixed y,t we have

$$
\begin{aligned}
&\int_{\mathbf{R}^n} \frac{t\,\big|f(x) - \operatorname*{Avg}_{B_t} f\big|}{(t^2+|x-y|^2)^{\frac{n+1}{2}}}\,dx \\
&\le \int_{B_t} \frac{t\,\big|f(x) - \operatorname*{Avg}_{B_t} f\big|}{(t^2+|x-y|^2)^{\frac{n+1}{2}}}\,dx \\
&\qquad + \sum_{k=0}^{\infty} \int_{2^{k+1}B_t\setminus 2^k B_t} \frac{t\Big(\big|f(x) - \operatorname*{Avg}_{2^{k+1}B_t} f\big| + \big|\operatorname*{Avg}_{2^{k+1}B_t} f - \operatorname*{Avg}_{B_t} f\big|\Big)}{(t^2+|x-y|^2)^{\frac{n+1}{2}}}\,dx \\
&\le \int_{B_t} \frac{\big|f(x) - \operatorname*{Avg}_{B_t} f\big|}{t^n}\,dx \\
&\qquad + \sum_{k=0}^{\infty} \frac{2^{-k(n+1)}}{t^n} \int_{2^{k+1}B_t} \Big(\big|f(x) - \operatorname*{Avg}_{2^{k+1}B_t} f\big| + \big|\operatorname*{Avg}_{2^{k+1}B_t} f - \operatorname*{Avg}_{B_t} f\big|\Big)\,dx \\
&\le v_n\|f\|_{BMO} + \sum_{k=0}^{\infty} 2^{-k(n+1)}\big(1+2^n(k+1)\big)(2^{k+1})^n v_n \|f\|_{BMO} \\
&= C'_n\|f\|_{BMO}.
\end{aligned}
$$

Thus we proved

$$
\frac{\Gamma(\frac{n+1}{2})}{\pi^{\frac{n+1}{2}}} \int_{\mathbf{R}^n} \frac{t\,\big|f(x) - \operatorname{Avg}_{B_t} f\big|}{(t^2+|x-y|^2)^{\frac{n+1}{2}}}\,dx \le C''_n\|f\|_{BMO}. \tag{3.1.8}
$$

Moving the absolute value outside, this inequality implies

$$\begin{aligned}\int_{\mathbf{R}^n} \big|(P_t * f)(y) - \operatorname*{Avg}_{B_t} f\big| P_t(x-y)\,dx &= \big|(P_t * f)(y) - \operatorname*{Avg}_{B_t} f\big| \\ &\le \int_{\mathbf{R}^n} P_t(x-y)\big|f(x) - \operatorname*{Avg}_{B_t} f\big|\,dx \\ &\le C_n'' \big\|f\big\|_{BMO}.\end{aligned}$$

Combining this last inequality with (3.1.8) yields (3.1.6) with constant $C_n = 2C_n''$.
(iv) Conversely, let A be the expression on the right in (3.1.7). For $|x-y| \le t$ we have $P_t(x-y) \ge c_n t(2t^2)^{-\frac{n+1}{2}} = c_n' t^{-n}$, which gives

$$A \ge \int_{\mathbf{R}^n} |f(x) - (P_t * f)(y)| P_t(x-y)\,dx \ge \frac{c_n'}{t^n} \int_{|x-y|\le t} |f(x) - (P_t * f)(y)|\,dx.$$

Proposition 3.1.2 (3) now implies that

$$\big\|f\big\|_{BMO} \le 2A/(v_n c_n').$$

This concludes the proof of the proposition. □

3.1.2 The John–Nirenberg Theorem

Having set down some basic facts about *BMO*, we now turn to a deeper property of *BMO* functions: their exponential integrability. We begin with a preliminary remark. As we saw in Example 3.1.3, the function $g(x) = \log(|x|^{-1})$ is in $BMO(\mathbf{R}^n)$. This function is exponentially integrable over any compact subset K of $\mathbf{R}^n$ in the sense that

$$\int_K e^{c|g(x)|}\,dx < \infty$$

for any $c < n$. It turns out that this is a general property of *BMO* functions, and this is the content of the next theorem.

Theorem 3.1.6. *For all $f \in BMO(\mathbf{R}^n)$, for all cubes Q, and all $\alpha > 0$ we have*

$$\Big|\Big\{x \in Q:\ \big|f(x) - \operatorname*{Avg}_{Q} f\big| > \alpha\Big\}\Big| \le e|Q|e^{-A\alpha/\|f\|_{BMO}} \tag{3.1.9}$$

with $A = (2^n e)^{-1}$.

Proof. Since inequality (3.1.9) is not altered when we multiply both f and α by the same constant, it suffices to assume that $\big\|f\big\|_{BMO} = 1$. Let us now fix a closed cube Q and a constant $b > 1$ to be chosen later.

We apply the Calderón–Zygmund decomposition to the function $f - \operatorname{Avg}_Q f$ inside the cube Q. We introduce the following selection criterion for a cube R:

$$\frac{1}{|R|}\int_R \big|f(x) - \operatorname*{Avg}_Q f\big|\,dx > b. \tag{3.1.10}$$

Since

$$\frac{1}{|Q|}\int_Q \big|f(x) - \operatorname*{Avg}_Q f\big|\,dx \le \|f\|_{BMO} = 1 < b,$$

the cube Q does not satisfy the selection criterion (3.1.10). Set $Q^{(0)} = Q$ and subdivide $Q^{(0)}$ into 2^n equal closed subcubes of side length equal to half of the side length of Q. Select such a subcube R if it satisfies the selection criterion (3.1.10). Now subdivide all nonselected cubes into 2^n equal subcubes of half their side length by bisecting the sides, and select among these subcubes those that satisfy (3.1.10). Continue this process indefinitely. We obtain a countable collection of cubes $\{Q_j^{(1)}\}_j$ satisfying the following properties:

(A-1) The interior of every $Q_j^{(1)}$ is contained in $Q^{(0)}$.

(B-1) $b < |Q_j^{(1)}|^{-1}\displaystyle\int_{Q_j^{(1)}} \big|f(x) - \operatorname*{Avg}_{Q^{(0)}} f\big|\,dx \le 2^n b.$

(C-1) $\big|\operatorname*{Avg}_{Q_j^{(1)}} f - \operatorname*{Avg}_{Q^{(0)}} f\big| \le 2^n b.$

(D-1) $\displaystyle\sum_j |Q_j^{(1)}| \le \frac{1}{b}\sum_j \int_{Q_j^{(1)}} \big|f(x) - \operatorname*{Avg}_{Q^{(0)}} f\big|\,dx \le \frac{1}{b}|Q^{(0)}|.$

(E-1) $\big|f - \operatorname*{Avg}_{Q^{(0)}} f\big| \le b$ a.e. on the set $Q^{(0)} \setminus \bigcup_j Q_j^{(1)}$.

We call the cubes $Q_j^{(1)}$ of first generation. Note that the second inequality in (D-1) requires (B-1) and the fact that $Q^{(0)}$ does not satisfy (3.1.10).

We now fix a selected first-generation cube $Q_j^{(1)}$ and we introduce the following selection criterion for a cube R:

$$\frac{1}{|R|}\int_R \big|f(x) - \operatorname*{Avg}_{Q_j^{(1)}} f\big|\,dx > b. \tag{3.1.11}$$

Observe that $Q_j^{(1)}$ does not satisfy the selection criterion (3.1.11). We apply a similar Calderón–Zygmund decomposition to the function

$$f - \operatorname*{Avg}_{Q_j^{(1)}} f$$

inside the cube $Q_j^{(1)}$. Subdivide $Q_j^{(1)}$ into 2^n equal closed subcubes of side length equal to half of the side length of $Q_j^{(1)}$ by bisecting the sides, and select such a subcube R if it satisfies the selection criterion (3.1.11). Continue this process indefinitely. Also repeat this process for any other cube $Q_j^{(1)}$ of the first generation. We obtain a collection of cubes $\{Q_l^{(2)}\}_l$ of second generation each contained in some $Q_j^{(1)}$ such that versions of (A-1)–(E-1) are satisfied, with the superscript (2) replacing (1) and the superscript (1) replacing (0). We use the superscript (k) to denote the generation of the selected cubes.

For a fixed selected cube $Q_l^{(2)}$ of second generation, introduce the selection criterion

$$\frac{1}{|R|}\int_R \Big|f(x) - \operatorname*{Avg}_{Q_l^{(2)}} f\Big|\,dx > b$$

and repeat the previous process to obtain a collection of cubes of third generation inside $Q_l^{(2)}$. Repeat this procedure for any other cube $Q_j^{(2)}$ of the second generation. Denote by $\{Q_s^{(3)}\}_s$ the thus obtained collection of all cubes of the third generation.

We iterate this procedure indefinitely to obtain a doubly indexed family of cubes $Q_j^{(k)}$ satisfying the following properties:

(A-k) The interior of every $Q_j^{(k)}$ is contained in a unique $Q_{j'}^{(k-1)}$.

(B-k) $b < |Q_j^{(k)}|^{-1}\displaystyle\int_{Q_j^{(k)}} \Big|f(x) - \operatorname*{Avg}_{Q_{j'}^{(k-1)}} f\Big|\,dx \le 2^n b.$

(C-k) $\Big|\operatorname*{Avg}_{Q_j^{(k)}} f - \operatorname*{Avg}_{Q_{j'}^{(k-1)}} f\Big| \le 2^n b.$

(D-k) $\displaystyle\sum_j |Q_j^{(k)}| \le \frac{1}{b}\sum_{j'} |Q_{j'}^{(k-1)}|.$

(E-k) $\Big|f - \operatorname*{Avg}_{Q_{j'}^{(k-1)}} f\Big| \le b$ a.e. on the set $Q_{j'}^{(k-1)} \setminus \bigcup_j Q_j^{(k)}$.

We prove (A-k)–(E-k). Note that (A-k) and the lower inequality in (B-k) are satisfied by construction. The upper inequality in (B-k) is a consequence of the fact that the unique cube $Q_{j_0}^{(k)}$ with double the side length of $Q_j^{(k)}$ that contains it was not selected in the process. Now (C-k) follows from the upper inequality in (B-k). (E-k) is a consequence of the Lebesgue differentiation theorem, since for every point in $Q_{j'}^{(k-1)} \setminus \bigcup_j Q_j^{(k)}$ there is a sequence of cubes shrinking to it and the averages of

$$\Big|f - \operatorname*{Avg}_{Q_{j'}^{(k-1)}} f\Big|$$

over all these cubes is at most b. It remains to prove (D-k). We have

$$\begin{aligned}
\sum_j |Q_j^{(k)}| &< \frac{1}{b}\sum_j \int_{Q_j^{(k)}} \Big|f(x) - \operatorname*{Avg}_{Q_{j'}^{(k-1)}} f\Big|\,dx \\
&= \frac{1}{b}\sum_{j'} \sum_{j \text{ corresp. to } j'} \int_{Q_j^{(k)}} \Big|f(x) - \operatorname*{Avg}_{Q_{j'}^{(k-1)}} f\Big|\,dx \\
&\le \frac{1}{b}\sum_{j'} \int_{Q_{j'}^{(k-1)}} \Big|f(x) - \operatorname*{Avg}_{Q_{j'}^{(k-1)}} f\Big|\,dx \\
&\le \frac{1}{b}\sum_{j'} |Q_{j'}^{(k-1)}|\,\|f\|_{BMO} \\
&= \frac{1}{b}\sum_{j'} |Q_{j'}^{(k-1)}|.
\end{aligned}$$

Having established (A-k)–(E-k) we turn to some consequences. Applying (D-k) successively $k-1$ times, we obtain

$$\sum_j |Q_j^{(k)}| \le b^{-k}|Q^{(0)}|. \tag{3.1.12}$$

For any fixed j we have that $\big|\operatorname{Avg}_{Q_j^{(1)}} f - \operatorname{Avg}_{Q^{(0)}} f\big| \le 2^n b$ and $\big|f - \operatorname{Avg}_{Q_j^{(1)}} f\big| \le b$ a.e. on $Q_j^{(1)} \setminus \bigcup_l Q_l^{(2)}$. This gives

$$\Big|f - \operatorname*{Avg}_{Q^{(0)}} f\Big| \le 2^n b + b \qquad \text{a.e.} \quad \text{on} \quad Q_j^{(1)} \setminus \bigcup_l Q_l^{(2)},$$

which, combined with (E-1), yields

$$\Big|f - \operatorname*{Avg}_{Q^{(0)}} f\Big| \le 2^n 2b \qquad \text{a.e.} \quad \text{on} \quad Q^{(0)} \setminus \bigcup_l Q_l^{(2)}. \tag{3.1.13}$$

For every fixed l we also have that $\big|f - \operatorname{Avg}_{Q_l^{(2)}} f\big| \le b$ a.e. on $Q_l^{(2)} \setminus \bigcup_s Q_s^{(3)}$, which combined with $\big|\operatorname{Avg}_{Q_l^{(2)}} f - \operatorname{Avg}_{Q_{l'}^{(1)}} f\big| \le 2^n b$ and $\big|\operatorname{Avg}_{Q_{l'}^{(1)}} f - \operatorname{Avg}_{Q^{(0)}} f\big| \le 2^n b$ yields

$$\Big|f - \operatorname*{Avg}_{Q^{(0)}} f\Big| \le 2^n 3b \qquad \text{a.e.} \quad \text{on} \quad Q_l^{(2)} \setminus \bigcup_s Q_s^{(3)}.$$

In view of (3.1.13), the same estimate is valid on $Q^{(0)} \setminus \bigcup_s Q_s^{(3)}$. Continuing this reasoning, we obtain by induction that for all $k \ge 1$ we have

$$\Big|f - \operatorname*{Avg}_{Q^{(0)}} f\Big| \le 2^n k b \qquad \text{a.e.} \quad \text{on} \quad Q^{(0)} \setminus \bigcup_s Q_s^{(k)}. \tag{3.1.14}$$

This proves the almost everywhere inclusion

$$\Big\{x \in Q : \big|f(x) - \operatorname*{Avg}_Q f\big| > 2^n k b\Big\} \subseteq \bigcup_j Q_j^{(k)}$$

for all $k = 1, 2, 3, \dots$. (This also holds when $k = 0$.) We now use (3.1.12) and (3.1.14) to prove (3.1.9). We fix an $\alpha > 0$. If

$$2^n k b < \alpha \le 2^n (k+1) b$$

for some $k \ge 0$, then

$$\begin{aligned}
\Big|\Big\{x \in Q : |f - \operatorname*{Avg}_Q f| > \alpha\Big\}\Big| &\le \Big|\Big\{x \in Q : |f - \operatorname*{Avg}_Q f| > 2^n k b\Big\}\Big| \\
&\le \sum_j |Q_j^{(k)}| \le \frac{1}{b^k} |Q^{(0)}| \\
&= |Q| \, e^{-k \log b} \\
&\le |Q| \, b \, e^{-\alpha \log b / (2^n b)},
\end{aligned}$$

since $-k \le 1 - \frac{\alpha}{2^n b}$. Choosing $b = e > 1$ yields (3.1.9). □

3.1.3 Consequences of Theorem 3.1.6

Having proved the important distribution inequality (3.1.9), we are now in a position to deduce from it a few corollaries.

Corollary 3.1.7. *Every BMO function is exponentially integrable over any cube. Precisely, for any $\gamma < 1/(2^n e)$, for all $f \in BMO(\mathbf{R}^n)$, and all cubes Q we have*

$$\frac{1}{|Q|} \int_Q e^{\gamma |f(x) - \operatorname{Avg}_Q f| / \|f\|_{BMO}} \, dx \le 1 + \frac{2^n e^2 \gamma}{1 - 2^n e \gamma}.$$

Proof. Using identity

$$\int_X e^{|f|} - 1 \, d\mu = \int_0^\infty e^\alpha \mu(\{x \in X : |f(x)| > \alpha\}) \, d\alpha,$$

proved in Proposition 1.1.4 in [156] for a σ-finite measure space (X, μ), we write

$$\frac{1}{|Q|} \int_Q e^h \, dx = 1 + \frac{1}{|Q|} \int_Q (e^h - 1) \, dx = 1 + \frac{1}{|Q|} \int_0^\infty e^\alpha |\{x \in Q : |h(x)| > \alpha\}| \, d\alpha$$

for a measurable function h on $\mathbf{R}^n$. Then we take $h = \gamma|f - \mathrm{Avg}_Q f|/\|f\|_{BMO}$ and we use inequality (3.1.9) with $\gamma < A = (2^n e)^{-1}$ to obtain

$$\frac{1}{|Q|}\int_Q e^{\gamma|f(x)-\mathrm{Avg}_Q f|/\|f\|_{BMO}}\,dx \le \int_0^\infty e^\alpha\, e\, e^{-A(\frac{\alpha}{\gamma}\|f\|_{BMO})/\|f\|_{BMO}}\,d\alpha = C_{n,\gamma},$$

where $C_{n,\gamma}$ is a unit less than the constant in the statement of the inequality. □

As a consequence of Corollary 3.1.7 we deduce that for any compact set K in $\mathbf{R}^n$,

$$\int_K e^{c|f(x)|}\,dx < \infty,$$

whenever $f \in BMO$ and $c < (2^n e\|f\|_{BMO})^{-1}$. Another important corollary of Theorem 3.1.6 is the following.

Corollary 3.1.8. *For all $0 < p < \infty$, there exists a finite constant $B_{p,n}$ such that for all $f \in BMO$ we have*

$$\sup_Q \left(\frac{1}{|Q|}\int_Q |f(x) - \mathrm{Avg}_Q f|^p\,dx\right)^{\frac{1}{p}} \le B_{p,n}\|f\|_{BMO(\mathbf{R}^n)}. \tag{3.1.15}$$

Proof. This result can be obtained from the one in the preceding corollary or directly in the following way:

$$\begin{aligned}\frac{1}{|Q|}\int_Q |f(x) - \mathrm{Avg}_Q f|^p\,dx &= \frac{p}{|Q|}\int_0^\infty \alpha^{p-1}|\{x \in Q:\ |f(x) - \mathrm{Avg}_Q f| > \alpha\}|\,d\alpha \\ &\le \frac{p}{|Q|}\, e|Q| \int_0^\infty \alpha^{p-1} e^{-A\alpha/\|f\|_{BMO}}\,d\alpha \\ &= p\Gamma(p)\frac{e}{A^p}\|f\|_{BMO}^p,\end{aligned}$$

where $A = (2^n e)^{-1}$. Setting $B_{p,n} = (p\Gamma(p)\,eA^{-p})^{\frac{1}{p}} = (p\Gamma(p))^{\frac{1}{p}} e^{\frac{1}{p}+1} 2^n$, we conclude the proof of (3.1.15). □

Since the inequality in Corollary 3.1.8 can be reversed when $p > 1$ via Hölder's inequality, we obtain the following important L^p characterization of BMO norms.

Corollary 3.1.9. *For all $1 < p < \infty$ and f in $L^1_{\mathrm{loc}}(\mathbf{R}^n)$ we have*

$$\sup_Q \left(\frac{1}{|Q|}\int_Q |f(x) - \mathrm{Avg}_Q f|^p\,dx\right)^{\frac{1}{p}} \approx \|f\|_{BMO}. \tag{3.1.16}$$

Proof. One direction follows from Corollary 3.1.8. Conversely, the supremum in (3.1.16) is bigger than or equal to the corresponding supremum with $p = 1$, which is equal to the BMO norm of f, by definition. □

Exercises

3.1.1. Prove that *BMO* is a complete space, that is, every *BMO*-Cauchy sequence converges in *BMO*.
[*Hint:* Use Proposition 3.1.5 (ii) to show first that such a sequence is Cauchy in L^1 of every compact set.]

3.1.2. Find an example showing that the product of two *BMO* functions may not be in *BMO*.

3.1.3. Prove that

$$\big\| |f|^{\alpha} \big\|_{BMO} \leq 2 \big\| f \big\|_{BMO}^{\alpha}$$

whenever $0 < \alpha \leq 1$.

3.1.4. Let f be a real-valued *BMO* function on $\mathbf{R}^n$. Prove that the functions

$$f_{KL}(x) = \begin{cases} K & \text{if } f(x) < K, \\ f(x) & \text{if } K \leq f(x) \leq L, \\ L & \text{if } f(x) > L, \end{cases}$$

satisfy $\|f_{KL}\|_{BMO} \leq \frac{9}{4}\|f\|_{BMO}$.

3.1.5. Let $a > 1$, let B be a ball (or a cube) in $\mathbf{R}^n$, and let aB be a concentric ball whose radius is a times the radius of B. Show that there is a dimensional constant C_n such that for all f in *BMO* we have

$$\big| \operatorname*{Avg}_{aB} f - \operatorname*{Avg}_{B} f \big| \leq C_n \log(a+1) \big\| f \big\|_{BMO}.$$

3.1.6. Let $a > 1$ and let f be a *BMO* function on $\mathbf{R}^n$. Show that there exist dimensional constants C_n, C_n' such that
(a) for all balls B_1 and B_2 in $\mathbf{R}^n$ with radius R whose centers are at distance aR we have

$$\big| \operatorname*{Avg}_{B_1} f - \operatorname*{Avg}_{B_2} f \big| \leq C_n' \log(a+1) \big\| f \big\|_{BMO}.$$

(b) Conclude that

$$\big| \operatorname*{Avg}_{(a+1)B_1} f - \operatorname*{Avg}_{B_2} f \big| \leq C_n \log(a+1) \big\| f \big\|_{BMO}.$$

[*Hint:* Part (a): Replace $\operatorname{Avg}_{B_1} f$ by $\operatorname{Avg}_{2aB_1} f$ and $\operatorname{Avg}_{B_2} f$ by $\operatorname{Avg}_{aB_2} f$ and use the fact that aB_2 is contained in $2aB_1$ and use Exercise 3.1.5.]

3.1.7. Let f be locally integrable on $\mathbf{R}^n$. Suppose that there exist positive constants m and b such that for all cubes Q in $\mathbf{R}^n$ and for all $0 < p < \infty$ we have

$$\alpha \Big| \Big\{ x \in Q : \, \big| f(x) - \operatorname*{Avg}_{Q} f \big| > \alpha \Big\} \Big|^{\frac{1}{p}} \leq b\, p^m \, |Q|^{\frac{1}{p}}.$$

Show that f satisfies the estimate

$$\left|\left\{x \in Q : \left|f(x) - \operatorname*{Avg}_{Q} f\right| > \alpha\right\}\right| \leq |Q|\, e^{-c\alpha^{1/m}}$$

with $c = (2b)^{-1/m} \log 2$.
[*Hint:* Try $p = (\alpha/2b)^{1/m}$.]

3.1.8. Prove that $\big|\log|x|\big|^p$ is not in $BMO(\mathbf{R})$ when $1 < p < \infty$.
[*Hint:* Show that if $\big|\log|x|\big|^p$ were in BMO, then estimate (3.1.9) would be violated for large α.]

3.1.9. Given $1 < p < \infty$ and f locally integrable on $\mathbf{R}^n$ prove that

$$\sup_Q \frac{1}{|Q|}\left(\inf_{c_Q} \int_Q |f(x) - c_Q|^p\, dx\right)^{\frac{1}{p}} \approx \|f\|_{BMO}.$$

[*Hint:* Use Proposition 3.1.2 (4) and Corollary 3.1.9.]

3.1.10. Let $f \in BMO(\mathbf{R})$ have mean value equal to zero on a fixed closed interval I. Find a BMO function g on $\mathbf{R}$ such that

(1) $g = f$ on I;

(2) $g = 0$ on $\mathbf{R} \setminus \frac{5}{3}I$;

(3) $\|g\|_{BMO} \leq 12\, \|f\|_{BMO}$.

[*Hint:* Let I_0 be the closed middle third of I. Write the interior of I as $\bigcup_{k\in\mathbf{Z}} I_k$, where for $|k| > 0$, I_k are closed subintervals of I such that the right endpoint of I_k coincides with the left endpoint of I_{k+1} and $\operatorname{dist}(I_k, \partial I) = |I_k| = \frac{1}{3}2^{-|k|}$. For $|k| \geq 1$, let J_k be the reflection of I_k with respect to the closest endpoint of I and set $g = \operatorname{Avg}_{I_k} f$ on J_k for $|k| > 1$, $g = f$ on I, and zero otherwise. To prove property (3), given an arbitrary interval Q on the real line, consider the cases where $|Q| \geq \frac{1}{3}|I|$ and $|Q| < \frac{1}{3}|I|$.]

3.2 Duality between H^1 and BMO

The next result we discuss is a remarkable duality relationship between the Hardy space H^1 and BMO. Precisely, we show that BMO is the dual space of H^1. This means that every continuous linear functional on the Hardy space H^1 can be realized as integration against a fixed BMO function, where *integration* in this context is an abstract operation, not necessarily given by an absolutely convergent integral. Restricting our attention, however, to a dense subspace of H^1 such as the space of all finite sums of atoms, the use of the word *integration* is well justified. Indeed, first we note that an important consequence of (3.1.15) is that any BMO function b lies in $L^p(Q)$ for any Q in $\mathbf{R}^n$ and any p satisfying $1 < p < \infty$; in particular it is square integrable over any cube. Thus given a BMO function b on $\mathbf{R}^n$ and an L^2 function g with integral zero on $\mathbf{R}^n$, the integral $\int_{\mathbf{R}^n} g(x)\,b(x)\,dx$ converges absolutely by the Cauchy–Schwarz inequality.

Definition 3.2.1. Denote by $H^1_0(\mathbf{R}^n)$ the space of all finite linear combinations of L^2 atoms for $H^1(\mathbf{R}^n)$ and fix $b \in BMO(\mathbf{R}^n)$. Given $g \in H^1_0$ we define a linear functional

$$L_b(g) = \int_{\mathbf{R}^n} g(x)b(x)\,dx \tag{3.2.1}$$

as an absolutely convergent integral. Observe that the integral in (3.2.1) and thus the definition of L_b on H^1_0 remain the same if b is replaced by $b+c$, where c is an additive constant. Additionally, we observe that (3.2.1) is also an absolutely convergent integral when g is a general element of $H^1(\mathbf{R}^n)$ and the BMO function b is bounded.

To extend the definition of L_b on the entire H^1 for all functions b in BMO we need to know that

$$\big\|L_b\big\|_{H^1\to\mathbf{C}} \le C_n \|b\|_{BMO}, \qquad \text{whenever } b \text{ is bounded,} \tag{3.2.2}$$

a fact that will be proved momentarily. Assuming (3.2.2), take $b \in BMO$ and let $b_M(x) = b\chi_{|b|\le M}$ for $M = 1,2,3,\dots$. Since $\|b_M\|_{BMO} \le \frac{9}{4}\|b\|_{BMO}$ (Exercise 3.1.4), the sequence of linear functionals $\{L_{b_M}\}_M$ lies in a multiple of the unit ball of $(H^1)^*$ and by the Banach–Alaoglou theorem there is a subsequence $M_j \to \infty$ as $j \to \infty$ such that $L_{b_{M_j}}$ converges weakly to a bounded linear functional $\widetilde{L}_b$ on H^1. This means that for all f in $H^1(\mathbf{R}^n)$ we have

$$L_{b_{M_j}}(f) \to \widetilde{L}_b(f)$$

as $j \to \infty$.

If a_Q is a fixed L^2 atom for H^1, the difference $|L_{b_{M_j}}(a_Q) - L_b(a_Q)|$ is bounded by $\|a_Q\|_{L^2}\,(\|b_{M_j} - \operatorname{Avg}_Q b_{M_j} - b + \operatorname{Avg}_Q b\|_{L^2(Q)})$ which is in turn bounded by $\|a_Q\|_{L^2}(\|b_{M_j} - b\|_{L^2(Q)} + |Q|^{1/2}|\operatorname{Avg}_Q(b_{M_j} - b)|)$, and this expression tends to zero as $j \to \infty$ by the Lebesgue dominated convergence theorem. The same conclusion holds for any finite linear combination of a_Q's. Thus for all $g \in H^1_0$ we have

$$L_{b_{M_j}}(g) \to L_b(g),$$

and consequently, $L_b(g) = \widetilde{L}_b(g)$ for all $g \in H^1_0$. Since H^1_0 is dense in H^1 and L_b and $\widetilde{L}_b$ coincide on H^1_0, it follows that $\widetilde{L}_b$ is the *unique* bounded extension of L_b on H^1. We have therefore defined L_b on the entire H^1 as a weak limit of bounded linear functionals.

Having set the definition of L_b, we proceed by showing the validity of (3.2.2). Let b be a bounded BMO function. Given f in H^1, find a sequence a_k of L^2 atoms for H^1 supported in cubes Q_k such that

$$f = \sum_{k=1}^{\infty} \lambda_k a_k \tag{3.2.3}$$

and

$$\sum_{k=1}^{\infty} |\lambda_k| \le 2\big\|f\big\|_{H^1}.$$

Since the series in (3.2.3) converges in H^1, it must converge in L^1, and then we have

$$\begin{aligned} |L_b(f)| &= \left| \int_{\mathbf{R}^n} f(x)b(x)\,dx \right| \\ &= \left| \sum_{k=1}^{\infty} \lambda_k \int_{Q_k} a_k(x)\big(b(x) - \operatorname*{Avg}_{Q_k} b\big)\,dx \right| \\ &\le \sum_{k=1}^{\infty} |\lambda_k| \, \big\|a_k\big\|_{L^2} |Q_k|^{\frac{1}{2}} \left(\frac{1}{|Q_k|} \int_{Q_k} \big|b(x) - \operatorname*{Avg}_{Q_k} b\big|^2 \,dx \right)^{\frac{1}{2}} \\ &\le 2\big\|f\big\|_{H^1} B_{2,n} \big\|b\big\|_{BMO}, \end{aligned}$$

where in the last step we used Corollary 3.1.8 and the fact that L^2 atoms for H^1 satisfy $\|a_k\|_{L^2} \le |Q_k|^{-\frac{1}{2}}$. This proves (3.2.2) for bounded functions b in BMO.

We have proved that every BMO function b gives rise to a bounded linear functional $\widetilde{L_b}$ on $H^1(\mathbf{R}^n)$ (from now on denoted by L_b) that satisfies

$$\big\|L_b\big\|_{H^1 \to \mathbf{C}} \le C_n \, \|b\|_{BMO}.$$

The fact that every bounded linear functional on H^1 arises in this way is the gist of the equivalence of the next theorem.

Theorem 3.2.2. *There exist finite constants C_n and C_n' such that the following statements are valid:*
(a) Given $b \in BMO(\mathbf{R}^n)$, the linear functional L_b lies in $(H^1(\mathbf{R}^n))^*$ and has norm at most $C_n\|b\|_{BMO}$. Moreover, the mapping $b \mapsto L_b$ from BMO to $(H^1)^*$ is injective.
(b) For every bounded linear functional L on H^1 there exists a BMO function b such that for all $f \in H_0^1$ we have $L(f) = L_b(f)$ and also

$$\big\|b\big\|_{BMO} \le C_n' \big\|L_b\big\|_{H^1 \to \mathbf{C}}.$$

Proof. We have already proved that for all $b \in BMO(\mathbf{R}^n)$, L_b lies in $(H^1(\mathbf{R}^n))^*$ and has norm at most $C_n\|b\|_{BMO}$. The embedding $b \mapsto L_b$ is injective as a consequence of Exercise 3.2.2. It remains to prove (b). Fix a bounded linear functional L on $H^1(\mathbf{R}^n)$ and also fix a cube Q. Consider the space $L^2(Q)$ of all square integrable functions supported in Q with norm

$$\big\|g\big\|_{L^2(Q)} = \left(\int_Q |g(x)|^2 \,dx \right)^{\frac{1}{2}}.$$

We denote by $L_0^2(Q)$ the closed subspace of $L^2(Q)$ consisting of all functions in $L^2(Q)$ with mean value zero. We show that every element in $L_0^2(Q)$ is in $H^1(\mathbf{R}^n)$ and we have the inequality

$$\big\|g\big\|_{H^1} \le c_n |Q|^{\frac{1}{2}} \big\|g\big\|_{L^2}. \tag{3.2.4}$$

To prove (3.2.4) we use the square function characterization of H^1. We fix a Schwartz function Ψ on $\mathbf{R}^n$ whose Fourier transform is supported in the annulus $\frac{1}{2} \le |\xi| \le 2$ and that satisfies (1.3.6) for all $\xi \neq 0$ and we let $\Delta_j(g) = \Psi_{2^{-j}} * g$. To estimate the L^1 norm of $\big(\sum_j |\Delta_j(g)|^2\big)^{1/2}$ over $\mathbf{R}^n$, consider the part of the integral over $3\sqrt{n}\,Q$ and the integral over $(3\sqrt{n}\,Q)^c$. First we use Hölder's inequality and an L^2 estimate to prove that

$$\int_{3\sqrt{n}Q} \Big(\sum_j |\Delta_j(g)(x)|^2\Big)^{\frac{1}{2}} dx \le c_n |Q|^{\frac{1}{2}} \big\|g\big\|_{L^2}.$$

Now for $x \notin 3\sqrt{n}\,Q$ we use the mean value property of g to obtain

$$|\Delta_j(g)(x)| \le \frac{c_n \big\|g\big\|_{L^2} 2^{nj+j} |Q|^{\frac{1}{n}+\frac{1}{2}}}{(1+2^j|x-c_Q|)^{n+2}}, \tag{3.2.5}$$

where c_Q is the center of Q. Estimate (3.2.5) is obtained in a way similar to that we obtained the corresponding estimate for one atom; see Theorem 2.3.11 for details. Now (3.2.5) implies that

$$\int_{(3\sqrt{n}Q)^c} \Big(\sum_j |\Delta_j(g)(x)|^2\Big)^{\frac{1}{2}} dx \le c_n |Q|^{\frac{1}{2}} \big\|g\big\|_{L^2},$$

which proves (3.2.4).

Since $L^2_0(Q)$ is a subspace of H^1, it follows from (3.2.4) that the linear functional $L : H^1 \to \mathbf{C}$ is also a bounded linear functional on $L^2_0(Q)$ with norm

$$\big\|L\big\|_{L^2_0(Q)\to\mathbf{C}} \le c_n |Q|^{1/2} \big\|L\big\|_{H^1\to\mathbf{C}}. \tag{3.2.6}$$

By the Riesz representation theorem for the Hilbert space $L^2_0(Q)$, there is an element F^Q in $(L^2_0(Q))^* = L^2(Q)/\{\text{constants}\}$ such that

$$L(g) = \int_Q F^Q(x) g(x)\, dx, \tag{3.2.7}$$

for all $g \in L^2_0(Q)$, and this F^Q satisfies

$$\big\|F^Q\big\|_{L^2(Q)} \le \big\|L\big\|_{L^2_0(Q)\to\mathbf{C}}. \tag{3.2.8}$$

Thus for any cube Q in $\mathbf{R}^n$, there is square integrable function F^Q supported in Q such that (3.2.7) is satisfied. We observe that if a cube Q is contained in another cube Q', then F^Q differs from $F^{Q'}$ by a constant on Q. Indeed, for all $g \in L^2_0(Q)$ we have

$$\int_Q F^{Q'}(x) g(x)\, dx = L(g) = \int_Q F^Q(x) g(x)\, dx$$

and thus

$$\int_Q (F^{Q'}(x) - F^Q(x))g(x)\,dx = 0.$$

Consequently,

$$g \to \int_Q (F^{Q'}(x) - F^Q(x))g(x)\,dx$$

is the zero functional on $L_0^2(Q)$; hence $F^{Q'} - F^Q$ must be the zero function in the space $(L_0^2(Q))^*$, i.e., $F^{Q'} - F^Q$ is a constant on Q.

Let

$$Q_m = [-m/2, m/2]^n$$

for $m = 1, 2, \dots$. Then $|Q_1| = 1$. We define a locally integrable function $b(x)$ on $\mathbf{R}^n$ by setting

$$b(x) = F^{Q_m}(x) - \frac{1}{|Q_1|}\int_{Q_1} F^{Q_m}(t)\,dt \tag{3.2.9}$$

whenever $x \in Q_m$. We check that this definition is unambiguous. Let $1 \le \ell < m$. Then for $x \in Q_\ell$, $b(x)$ is also defined as in (3.2.9) with ℓ in the place of m. The difference of these two functions is

$$F^{Q_m} - F^{Q_\ell} - \operatorname*{Avg}_{Q_1}(F^{Q_m} - F^{Q_\ell}) = 0,$$

since the function $F^{Q_m} - F^{Q_\ell}$ is constant in the cube Q_ℓ (which is contained in Q_m), as indicated earlier.

Next we claim that for any cube Q there is a constant C_Q such that

$$F^Q = b - C_Q \qquad \text{on } Q. \tag{3.2.10}$$

Indeed, given a cube Q pick the smallest m such that Q is contained in Q^m and observe that

$$F^Q = \underbrace{F^Q - F^{Q_m}}_{\text{constant on } Q} + \underbrace{F^{Q_m} - \operatorname*{Avg}_{Q_1} F^{Q_m}}_{b(x)} + \underbrace{\operatorname*{Avg}_{Q_1} F^{Q_m}}_{\text{constant on } Q}$$

and let $-C_Q$ be the sum of the two preceding constant expressions on Q.

We have now found a locally integrable function b such that for all cubes Q and all $g \in L_0^2(Q)$ we have

$$\int_Q b(x)g(x)\,dx = \int_Q (F^Q(x) + C_Q)g(x)\,dx = \int_Q F^Q(x)g(x)\,dx = L(g), \tag{3.2.11}$$

as follows from (3.2.7) and (3.2.10). We conclude the proof by showing that b lies in $BMO(\mathbf{R}^n)$. By (3.2.10), (3.2.8), and (3.2.6) we have

$$\begin{aligned}\sup_Q \frac{1}{|Q|}\int_Q |b(x)-C_Q|\,dx &= \sup_Q \frac{1}{|Q|}\int_Q |F^Q(x)|\,dx \\ &\le \sup_Q |Q|^{-1}|Q|^{\frac{1}{2}}\big\|F^Q\big\|_{L^2(Q)} \\ &\le \sup_Q |Q|^{-\frac{1}{2}}\big\|L\big\|_{L_0^2(Q)\to\mathbf{C}} \\ &\le c_n\big\|L\big\|_{H^1\to\mathbf{C}} < \infty.\end{aligned}$$

Using Proposition 3.1.2 (3), we deduce that $b \in BMO$ and $\|b\|_{BMO} \le 2c_n\|L\|_{H^1\to\mathbf{C}}$. Finally, (3.2.11) implies that

$$L(g) = \int_{\mathbf{R}^n} b(x)g(x)\,dx = L_b(g)$$

for all $g \in H_0^1(\mathbf{R}^n)$, proving that the linear functional L coincides with L_b on a dense subspace of H^1. Consequently, $L = L_b$, and this conclude the proof of part (b). □

Exercises

3.2.1. Given b in BMO, let L_b be as in Definition 3.2.1. Prove that for b in BMO we have

$$\big\|b\big\|_{BMO} \approx \sup_{\|f\|_{H^1}\le 1} \big|L_b(f)\big|,$$

and for a given f in H^1 we have

$$\big\|f\big\|_{H^1} \approx \sup_{\|b\|_{BMO}\le 1} \big|L_b(f)\big|.$$

[*Hint:* Use $\|T\|_{X^*} = \sup_{\substack{x\in X\\ \|x\|_X\le 1}} |T(x)|$ for all T in the dual of a Banach space X.]

3.2.2. Suppose that a locally integrable function u is supported in a cube Q in $\mathbf{R}^n$ and satisfies

$$\int_Q u(x)g(x)\,dx = 0$$

for all square integrable functions g on Q with mean value zero. Show that u is almost everywhere equal to a constant.

3.3 Nontangential Maximal Functions and Carleson Measures

Many properties of functions defined on $\mathbf{R}^n$ are related to corresponding properties of associated functions defined on $\mathbf{R}_+^{n+1}$ in a natural way. A typical example of this situation is the relation between an $L^p(\mathbf{R}^n)$ function f and its Poisson integral

$f * P_t$ or more generally $f * \Phi_t$, where $\{\Phi_t\}_{t>0}$ is an approximate identity. Here Φ is a Schwartz function on $\mathbf{R}^n$ with integral 1. A maximal operator associated to the approximate identity $\{f * \Phi_t\}_{t>0}$ is

$$f \to \sup_{t>0} |f * \Phi_t|,$$

which we know is pointwise controlled by a multiple of the Hardy–Littlewood maximal function $M(f)$. Another example of a maximal operator associated to the previous approximate identity is the nontangential maximal function

$$f \to M^*(f;\Phi)(x) = \sup_{t>0} \sup_{|y-x|<t} |(f * \Phi_t)(y)|.$$

To study nontangential behavior we consider general functions F defined on $\mathbf{R}^{n+1}_+$ that are not necessarily given as an average of functions defined on $\mathbf{R}^n$. Throughout this section we use capital letters to denote functions defined on $\mathbf{R}^{n+1}_+$. When we write $F(x,t)$ we mean that $x \in \mathbf{R}^n$ and $t > 0$.

3.3.1 Definition and Basic Properties of Carleson Measures

Definition 3.3.1. Let F be a measurable function on $\mathbf{R}^{n+1}_+$. For x in $\mathbf{R}^n$ let $\Gamma(x)$ be the cone with vertex x defined by

$$\Gamma(x) = \{(y,t) \in \mathbf{R}^n \times \mathbf{R}^+ : \; |y-x| < t\}.$$

A picture of this cone is shown in Figure 3.1. The *nontangential maximal function* of F is the function

$$F^*(x) = \sup_{(y,t)\in\Gamma(x)} |F(y,t)|$$

defined on $\mathbf{R}^n$. This function is obtained by taking the supremum of the values of F inside the cone $\Gamma(x)$.

We observe that if $F^*(x) = 0$ for almost all $x \in \mathbf{R}^n$, then F is identically equal to zero on $\mathbf{R}^{n+1}_+$. To establish this claim, suppose that $F^*(x_0) > 0$. Then there exists $(y_0,t_0) \in \Gamma(x_0) = \{(y,t) : \; |y-x_0| < t\}$ such that $|F(y_0,t_0)| > \frac{1}{2}F^*(x_0)$. Then for all z with $|z-y_0| < t_0 - |y_0-x_0| = \delta_0$ we have $|y_0-z_0| < t_0$, hence $F^*(z) \geq |F(y_0,t_0)| > \frac{1}{2}F^*(x_0) > 0$. Thus $F^* > \frac{1}{2}F^*(x_0)$ on the ball $B(y_0,\delta_0)$, which is a contradiction.

Definition 3.3.2. Given a ball $B = B(x_0,r)$ in $\mathbf{R}^n$ we define the *cylindrical tent* over B to be the "cylindrical set"

$$T(B) = \{(x,t) \in \mathbf{R}^{n+1}_+ : \; x \in B, \quad 0 < t \leq r\}.$$

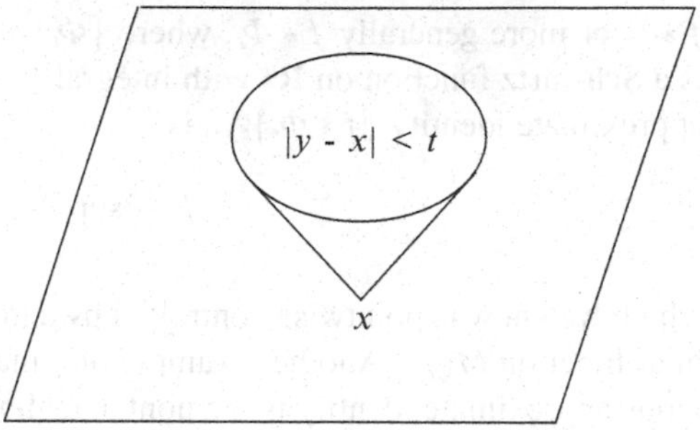

Fig. 3.1 The cone $\Gamma(x)$ truncated at height t.

For a cube Q in $\mathbf{R}^n$ we define the *tent* over Q to be the cube

$$T(Q) = Q \times (0, \ell(Q)].$$

A tent over a ball and over a cube are shown in Figure 3.2. A positive measure μ on $\mathbf{R}^{n+1}_+$ is called a *Carleson measure* if

$$\|\mu\|_{\mathscr{C}} = \sup_Q \frac{1}{|Q|}\mu(T(Q)) < \infty, \tag{3.3.1}$$

where the supremum in (3.3.1) is taken over all cubes Q in $\mathbf{R}^n$. The *Carleson function* of the measure μ is defined as

$$\mathscr{C}(\mu)(x) = \sup_{Q \ni x} \frac{1}{|Q|}\mu(T(Q)), \tag{3.3.2}$$

where the supremum in (3.3.2) is taken over all cubes in $\mathbf{R}^n$ containing the point x. Observe that $\|\mathscr{C}(\mu)\|_{L^\infty} = \|\mu\|_{\mathscr{C}}$.

We also define

$$\|\mu\|_{\mathscr{C}}^{\text{cylinder}} = \sup_B \frac{1}{|B|}\mu(T(B)), \tag{3.3.3}$$

where the supremum is taken over all balls B in $\mathbf{R}^n$. One can easily verify that there exist dimensional constants c_n and C_n such that

$$c_n\|\mu\|_{\mathscr{C}} \le \|\mu\|_{\mathscr{C}}^{\text{cylinder}} \le C_n\|\mu\|_{\mathscr{C}}$$

for all measures μ on $\mathbf{R}^{n+1}_+$, that is, a measure satisfies the Carleson condition (3.3.1) with respect to cubes if and only if it satisfies the analogous condition (3.3.3) with respect to balls. Likewise, the Carleson function $\mathscr{C}(\mu)$ defined with respect to tents over cubes is comparable to

$$\mathscr{C}^{\text{cylinder}}(\mu)(x) = \sup_{B \ni x} \frac{1}{|B|}\mu(T(B)),$$

defined with respect to cylindrical tents over balls B in $\mathbf{R}^n$.

Examples 3.3.3. The Lebesgue measure on $\mathbf{R}^{n+1}_+$ is not a Carleson measure. Indeed, it is not difficult to see that condition (3.3.1) cannot hold for large balls.

Let L be a line in $\mathbf{R}^2$. For A measurable subsets of $\mathbf{R}^2_+$ define $\mu(A)$ to be the linear Lebesgue measure of the set $L \cap A$. Then μ is a Carleson measure on $\mathbf{R}^2_+$. Indeed, the linear measure of the part of a line inside the box $[x_0 - r, x_0 + r] \times (0, r]$ is at most equal to the diagonal of the box, that is, $\sqrt{5}r$.

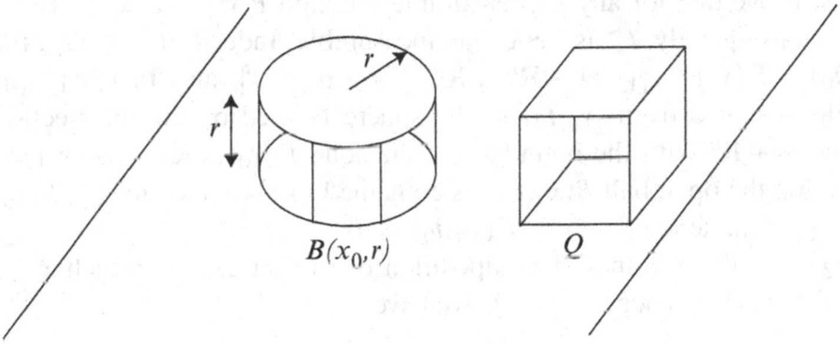

Fig. 3.2 The tents over the ball $B(x_0, r)$ and over a cube Q in $\mathbf{R}^2$.

Likewise, let P be an affine plane in $\mathbf{R}^{n+1}$ and define a measure ν by setting $\nu(A)$ to be the n-dimensional Lebesgue measure of the set $A \cap P$ for any $A \subseteq \mathbf{R}^{n+1}_+$. A similar idea shows that ν is a Carleson measure on $\mathbf{R}^{n+1}_+$.

We now turn to the study of some interesting boundedness properties of functions on $\mathbf{R}^{n+1}_+$ with respect to Carleson measures.

A useful tool in this study is the *Whitney decomposition* of an open set in $\mathbf{R}^n$. This is a decomposition of a general open set Ω in $\mathbf{R}^n$ as a union of disjoint cubes whose lengths are proportional to their distance from the boundary of the open set. For a given cube Q in $\mathbf{R}^n$, we denote by $\ell(Q)$ its length.

Proposition 3.3.4. ***(Whitney decomposition)*** *Let Ω be an open nonempty proper subset of $\mathbf{R}^n$. Then there exists a family of closed cubes $\{Q_j\}_j$ such that*

(a) $\bigcup_j Q_j = \Omega$ and the Q_j's have disjoint interiors;

(b) $\sqrt{n}\,\ell(Q_j) \le \text{dist}\,(Q_j, \Omega^c) \le 4\sqrt{n}\,\ell(Q_j)$;

(c) if the boundaries of two cubes Q_j and Q_k touch, then

$$\frac{1}{4} \le \frac{\ell(Q_j)}{\ell(Q_k)} \le 4;$$

(d) for a given Q_j there exist at most 12^n Q_k's that touch it.

The proof of Proposition 3.3.4 is given in Appendix J in [156].

Theorem 3.3.5. *There exists a dimensional constant C_n such that for all $\alpha > 0$, all measures $\mu \ge 0$ on $\mathbf{R}^{n+1}_+$, and all μ-measurable functions F on $\mathbf{R}^{n+1}_+$, the set $\Omega_\alpha = \{F^* > \alpha\}$ is open (thus Lebesgue measurable) and we have*

$$\mu(\{(x,t) \in \mathbf{R}^{n+1}_+ : |F(x,t)| > \alpha\}) \le C_n \int_{\{F^* > \alpha\}} \mathscr{C}(\mu)(x)\,dx. \tag{3.3.4}$$

In particular, if μ is a Carleson measure, then

$$\mu(\{|F| > \alpha\}) \le C_n \|\mu\|_{\mathscr{C}} |\{F^* > \alpha\}|. \tag{3.3.5}$$

Proof. We prove this theorem by working with the equivalent definition of Carleson measures and Carleson functions using balls and cylinders over balls. As observed earlier, these quantities are comparable to the corresponding quantities using cubes.

We first prove that for any μ-measurable function F the set $\Omega_\alpha = \{F^* > \alpha\}$ is open, and consequently, F^* is Lebesgue measurable. Indeed, if $x_0 \in \Omega_\alpha$, then there is a $(y_0,t_0) \in \Gamma(x_0) = \{(y,t) \in \mathbf{R}^n \times \mathbf{R}^+ : |y - x_0| < t\}$ such that $|F(y_0,t_0)| > \alpha$. If d_0 is the distance from (y_0,t_0) to the sphere formed by the intersection of the hyperplane $t_0 + \mathbf{R}^n$ with the boundary of the cone $\Gamma(x_0)$, then $|x_0 - y_0| = t_0 - d_0$. It follows that the open ball $B(x_0,d_0)$ is contained in Ω_α, since for $z \in B(x_0,d_0)$ we have $|z - y_0| < t_0$; hence $F^*(z) \ge |F(y_0,t_0)| > \alpha$.

Let $\{Q_k\}$ be the Whitney decomposition of the set Ω_α. For each $x \in \Omega_\alpha$, set $\delta_\alpha(x) = \operatorname{dist}(x, \Omega_\alpha^c)$. Then for $z \in Q_k$ we have

$$\delta_\alpha(z) \le \sqrt{n}\,\ell(Q_k) + \operatorname{dist}(Q_k, \Omega_\alpha^c) \le 5\sqrt{n}\,\ell(Q_k) \tag{3.3.6}$$

in view of Proposition 3.3.4(b). For each Q_k, let B_k be the smallest ball that contains Q_k. Then the radius of B_k is $\sqrt{n}\,\ell(Q_k)/2$. Combine this observation with (3.3.6) to obtain that

$$z \in Q_k \quad \Longrightarrow \quad B(z, \delta_\alpha(z)) \subseteq 12B_k.$$

This implies that

$$\bigcup_{z\in\Omega_\alpha} T(B(z,\delta_\alpha(z))) \subseteq \bigcup_k T(12B_k). \tag{3.3.7}$$

Next we claim that

$$\{|F| > \alpha\} \subseteq \bigcup_{z\in\Omega_\alpha} T(B(z,\delta_\alpha(z))). \tag{3.3.8}$$

Indeed, let $(x,t) \in \mathbf{R}_+^{n+1}$ such that $|F(x,t)| > \alpha$. Then by the definition of F^* we have that $F^*(y) > \alpha$ for all $y \in \mathbf{R}^n$ satisfying $|x - y| < t$. Thus $B(x,t) \subseteq \Omega_\alpha$ and so $\delta_\alpha(x) \ge t$. This gives that $(x,t) \in T\big(B(x,\delta_\alpha(x))\big)$, which proves (3.3.8).

Combining (3.3.7) and (3.3.8) we obtain

$$\{|F| > \alpha\} \subseteq \bigcup_k T(12B_k).$$

Applying the measure μ and using the definition of the Carleson function, we obtain

$$\begin{aligned}
\mu(\{|F| > \alpha\}) &\le \sum_k \mu(T(12B_k)) \\
&\le \sum_k |12B_k| \inf_{x\in 12B_k} \mathscr{C}^{\text{cylinder}}(\mu)(x) \\
&\le \sum_k |12B_k| \inf_{x\in Q_k} \mathscr{C}^{\text{cylinder}}(\mu)(x)
\end{aligned}$$

$$\leq 12^n \sum_k \frac{|B_k|}{|Q_k|} \int_{Q_k} \mathscr{C}^{\text{cylinder}}(\mu)(x)\, dx$$

$$\leq (6\sqrt{n})^n v_n \int_{\Omega_\alpha} \mathscr{C}^{\text{cylinder}}(\mu)(x)\, dx$$

$$\leq C_n \int_{\Omega_\alpha} \mathscr{C}(\mu)(x)\, dx,$$

since $\mathscr{C}^{\text{cylinder}}(\mu)$ is pointwise comparable to $\mathscr{C}(\mu)$. This proves (3.3.4). □

Corollary 3.3.6. *For any Carleson measure μ and every μ-measurable function F on $\mathbf{R}^{n+1}_+$ we have*

$$\int_{\mathbf{R}^{n+1}_+} |F(x,t)|^p \, d\mu(x,t) \leq C_n \|\mu\|_{\mathscr{C}} \int_{\mathbf{R}^n} (F^*(x))^p \, dx \tag{3.3.9}$$

for all $0 < p < \infty$.

Proof. Start with (3.3.5), multiply by $p\alpha^{p-1}$ and integrate in α from zero to infinity. We obtain

$$\int_0^\infty p\alpha^{p-1} \mu(\{|F| > \alpha\})\, d\alpha \leq C_n \|\mu\|_{\mathscr{C}} \int_0^\infty p\alpha^{p-1} |\{F^* > \alpha\}|\, d\alpha,$$

which is a restatement of (3.3.9). □

A particular example of this situation arises when $F(x,t) = f * \Phi_t(x)$ for some nice integrable function Φ. Here and in the sequel, $\Phi_t(x) = t^{-n}\Phi(t^{-1}x)$. For instance one may take Φ_t to be the Poisson kernel P_t.

Theorem 3.3.7. *Let Φ be a function on $\mathbf{R}^n$ that satisfies for some $0 < C, \delta < \infty$,*

$$|\Phi(x)| \leq \frac{C}{(1+|x|)^{n+\delta}}. \tag{3.3.10}$$

Let μ be a Carleson measure on $\mathbf{R}^{n+1}_+$. Then for every $1 < p < \infty$ there is a constant $C_{p,n}(\mu)$ such that for all $f \in L^p(\mathbf{R}^n)$ we have

$$\int_{\mathbf{R}^{n+1}_+} |(\Phi_t * f)(x)|^p \, d\mu(x,t) \leq C_{p,n}(\mu) \int_{\mathbf{R}^n} |f(x)|^p \, dx, \tag{3.3.11}$$

where $C_{p,n}(\mu) \leq C(p,n)\|\mu\|_{\mathscr{C}}$.

Conversely, suppose that Φ is a nonnegative function that satisfies (3.3.10) *and $\int_{|x|\leq 1} \Phi(x)\, dx > 0$. If μ is a measure on $\mathbf{R}^{n+1}_+$ such that for some $1 < p < \infty$ there is a constant $C_{p,n}(\mu)$ such that* (3.3.11) *holds for all $f \in L^p(\mathbf{R}^n)$, then μ is a Carleson measure with norm at most a multiple of $C_{p,n}(\mu)$.*

Proof. If μ is a Carleson measure, we may obtain (3.3.11) as a consequence of Corollary 3.3.6. Indeed, for $F(x,t) = (\Phi_t * f)(x)$, we have

$$F^*(x) = \sup_{t>0} \sup_{\substack{y\in\mathbf{R}^n \\ |y-x|<t}} |(\Phi_t * f)(y)|.$$

Using (3.3.10) and Corollary 2.1.12 in [156], this is easily seen to be pointwise controlled by the Hardy–Littlewood maximal operator, which is L^p bounded. See also Exercise 3.3.4.

Conversely, if (3.3.11) holds, then we fix a ball $B = B(x_0, r)$ in $\mathbf{R}^n$ with center x_0 and radius $r > 0$. Then for (x,t) in $T(B)$ we have

$$(\Phi_t * \chi_{2B})(x) = \int_{x-2B} \Phi_t(y)\,dy \ge \int_{B(0,t)} \Phi_t(y)\,dy = \int_{B(0,1)} \Phi(y)\,dy = c_n > 0,$$

since $B(0,t) \subseteq x - 2B(x_0, r)$ whenever $t \le r$. Therefore, we have

$$\begin{aligned} \mu(T(B)) &\le \frac{1}{c_n^p} \int_{\mathbf{R}^{n+1}_+} |(\Phi_t * \chi_{2B})(x)|^p \, d\mu(x,t) \\ &\le \frac{C_{p,n}(\mu)}{c_n^p} \int_{\mathbf{R}^n} |\chi_{2B}(x)|^p \, dx \\ &= \frac{2^n C_{p,n}(\mu)}{c_n^p} |B|. \end{aligned}$$

This proves that μ is a Carleson measure with $\|\mu\|_{\mathscr{C}} \le 2^n c_n^{-p} C_{p,n}(\mu)$. □

3.3.2 *BMO* Functions and Carleson Measures

We now turn to an interesting connection between *BMO* functions and Carleson measures. We have the following.

Theorem 3.3.8. *Let b be a BMO function on* $\mathbf{R}^n$ *and let* Ψ *be an integrable function with mean value zero on* $\mathbf{R}^n$ *that satisfies*

$$|\Psi(x)| \le A(1+|x|)^{-n-\delta} \tag{3.3.12}$$

for some $0 < A, \delta < \infty$. *Consider the dilations* $\Psi_t = t^{-n}\Psi(t^{-1}x)$ *and define the Littlewood–Paley operators* $\Delta_j(f) = f * \Psi_{2^{-j}}$.
(a) Suppose that

$$\sup_{\xi\in\mathbf{R}^n} \sum_{j\in\mathbf{Z}} |\widehat{\Psi}(2^{-j}\xi)|^2 \le B^2 < \infty \tag{3.3.13}$$

and let $\delta_{2^{-j}}(t)$ be Dirac mass at the point $t = 2^{-j}$. Then there is a constant $C_{n,\delta}$ such that

$$d\mu(x,t) = \sum_{j\in\mathbf{Z}} |(\Psi_{2^{-j}} * b)(x)|^2 \, dx \, \delta_{2^{-j}}(t)$$

is a Carleson measure on $\mathbf{R}^{n+1}_+$ with norm at most $C_{n,\delta}(A+B)^2\|b\|^2_{BMO}$.
(b) Suppose that

$$\sup_{\xi\in\mathbf{R}^n} \int_0^\infty |\widehat{\Psi}(t\xi)|^2 \frac{dt}{t} \le B^2 < \infty. \tag{3.3.14}$$

Then the continuous version $d\nu(x,t)$ of $d\mu(x,t)$ defined by

$$d\nu(x,t) = |(\Psi_t * b)(x)|^2 \, dx \frac{dt}{t}$$

is a Carleson measure on $\mathbf{R}^{n+1}_+$ with norm at most $C_{n,\delta}(A+B)^2\|b\|^2_{BMO}$ for some constant $C_{n,\delta}$.
(c) Let $\delta, A > 0$. Suppose that $\{K_t\}_{t>0}$ are functions on $\mathbf{R}^n \times \mathbf{R}^n$ that satisfy

$$|K_t(x,y)| \le \frac{A t^\delta}{(t+|x-y|)^{n+\delta}} \tag{3.3.15}$$

for all $t > 0$ and all $x, y \in \mathbf{R}^n$. Let R_t be the linear operator

$$R_t(f)(x) = \int_{\mathbf{R}^n} K_t(x,y) f(y) \, dy,$$

which is well defined for all $f \in \bigcup_{1\le p\le\infty} L^p(\mathbf{R}^n)$. Suppose that $R_t(1) = 0$ for all $t > 0$ and that there is a constant $B > 0$ such that

$$\int_0^\infty \int_{\mathbf{R}^n} |R_t(f)(x)|^2 \frac{dx\,dt}{t} \le B^2 \|f\|^2_{L^2(\mathbf{R}^n)} \tag{3.3.16}$$

for all $f \in L^2(\mathbf{R}^n)$. Then for all b in BMO the measure

$$|R_t(b)(x)|^2 \frac{dx\,dt}{t}$$

is Carleson with norm at most a constant multiple of $(A+B)^2\|b\|^2_{BMO}$.

We note that if, in addition to (3.3.12), the function Ψ has mean value zero and satisfies $|\nabla\Psi(x)| \le A(1+|x|)^{-n-\delta}$, then (3.3.13) and (3.3.14) hold and therefore conclusions (a) and (b) of Theorem 3.3.8 follow.

Proof. We prove (a). The measure μ is defined so that for every μ-integrable function F on $\mathbf{R}^{n+1}_+$ we have

$$\int_{\mathbf{R}^{n+1}_+} F(x,t) \, d\mu(x,t) = \sum_{j\in\mathbf{Z}} \int_{\mathbf{R}^n} |(\Psi_{2^{-j}} * b)(x)|^2 F(x, 2^{-j}) \, dx. \tag{3.3.17}$$

For a cube Q in $\mathbf{R}^n$ we let Q^* be the cube with the same center and orientation whose side length is $3\sqrt{n}\,\ell(Q)$, where $\ell(Q)$ is the side length of Q. Fix a cube Q in $\mathbf{R}^n$, take F to be the characteristic function of the tent of Q, and split b as

$$b = (b - \operatorname*{Avg}_Q b)\chi_{Q^*} + (b - \operatorname*{Avg}_Q b)\chi_{(Q^*)^c} + \operatorname*{Avg}_Q b.$$

Since Ψ has mean value zero, $\Psi_{2^{-j}} * \operatorname{Avg}_Q b = 0$. Then (3.3.17) gives

$$\mu(T(Q)) = \sum_{2^{-j}\le \ell(Q)} \int_Q |\Delta_j(b)(x)|^2\,dx \le 2\Sigma_1 + 2\Sigma_2,$$

where

$$\begin{aligned}
\Sigma_1 &= \sum_{j\in\mathbf{Z}} \int_{\mathbf{R}^n} \big|\Delta_j\big((b - \operatorname*{Avg}_Q b)\chi_{Q^*}\big)(x)\big|^2\,dx,\\
\Sigma_2 &= \sum_{2^{-j}\le \ell(Q)} \int_Q \big|\Delta_j\big((b - \operatorname*{Avg}_Q b)\chi_{(Q^*)^c}\big)(x)\big|^2\,dx.
\end{aligned}$$

Using Plancherel's theorem and (3.3.13), we obtain

$$\begin{aligned}
\Sigma_1 &\le \sup_\xi \sum_{j\in\mathbf{Z}} |\widehat{\Psi}(2^{-j}\xi)|^2 \int_{\mathbf{R}^n} \big|\big((b - \operatorname*{Avg}_Q b)\chi_{Q^*}\big)\widehat{\ }(\xi)\big|^2\,d\xi\\
&\le B^2 \int_{Q^*} \big|b(x) - \operatorname*{Avg}_Q b\big|^2\,dx\\
&\le 2B^2 \int_{Q^*} \big|b(x) - \operatorname*{Avg}_{Q^*} b\big|^2\,dx + 2B^2|Q^*|\,\big|\operatorname*{Avg}_{Q^*} b - \operatorname*{Avg}_Q b\big|^2\\
&\le B^2 \int_{Q^*} \big|b(x) - \operatorname*{Avg}_{Q^*} b\big|^2\,dx + c_n 2B^2 \|b\|_{BMO}^2\,|Q|\\
&\le C_n B^2 \|b\|_{BMO}^2\,|Q|,
\end{aligned}$$

in view of Proposition 3.1.5 (i) and Corollary 3.1.8. To estimate Σ_2, we use the size estimate of the function Ψ. We obtain

$$\big|\big(\Psi_{2^{-j}} * (b - \operatorname*{Avg}_Q b)\chi_{(Q^*)^c}\big)(x)\big| \le \int_{(Q^*)^c} \frac{A 2^{-j\delta}\,|b(y) - \operatorname{Avg}_Q b|}{(2^{-j} + |x-y|)^{n+\delta}}\,dy. \tag{3.3.18}$$

But note that if c_Q is the center of Q, then

$$\begin{aligned}
2^{-j} + |x-y| &\ge |y-x|\\
&\ge |y - c_Q| - |c_Q - x|\\
&\ge \frac{1}{2}|c_Q - y| + \frac{3\sqrt{n}}{4}\ell(Q) - |c_Q - x|
\end{aligned}$$

$$\geq \frac{1}{2}|c_Q - y| + \frac{3\sqrt{n}}{4}\ell(Q) - \frac{\sqrt{n}}{2}l(Q)$$
$$= \frac{1}{2}\Big(|c_Q - y| + \frac{\sqrt{n}}{2}\ell(Q)\Big)$$

when $y \in (Q^*)^c$ and $x \in Q$. Inserting this estimate in (3.3.18), integrating over Q, and summing over j with $2^{-j} \leq \ell(Q)$, we obtain

$$\begin{aligned}\Sigma_2 &\leq C_n \sum_{j:\, 2^{-j}\leq \ell(Q)} 2^{-2j\delta} \int_Q \bigg(A \int_{\mathbf{R}^n} \frac{|b(y) - \operatorname{Avg}_Q b|}{(\ell(Q) + |c_Q - y|)^{n+\delta}}\,dy \bigg)^2 dx \\ &\leq C_n A^2 |Q| \bigg(\int_{\mathbf{R}^n} \frac{\ell(Q)^\delta |b(y) - \operatorname{Avg}_Q b|}{(\ell(Q) + |y - c_Q|)^{n+\delta}}\,dy \bigg)^2 \\ &\leq C'_{n,\delta} A^2 |Q|\, \|b\|_{BMO}^2\end{aligned}$$

in view of (3.1.5). This proves that

$$\Sigma_1 + \Sigma_2 \leq C_{n,\delta}(A^2 + B^2)|Q|\, \|b\|_{BMO}^2,$$

which implies that $\mu(T(Q)) \leq C_{n,\delta}(A+B)^2 \|b\|_{BMO}^2 |Q|$.

The proof of part (b) of the theorem is obtained in a similar fashion. Finally, part (c) is a generalization of part (b) and is proved likewise. We sketch its proof. Write

$$b = \big(b - \operatorname*{Avg}_Q b\big)\chi_{Q^*} + \big(b - \operatorname*{Avg}_Q b\big)\chi_{(Q^*)^c} + \operatorname*{Avg}_Q b$$

and note that $R_t(\operatorname{Avg}_Q b) = 0$. We handle the term containing $R_t\big((b - \operatorname{Avg}_Q b)\chi_{Q^*}\big)$ using an L^2 estimate over Q^* and condition (3.3.16), while for the term containing $R_t\big((b - \operatorname{Avg}_Q b)\chi_{(Q^*)^c}\big)$ we use an L^1 estimate and condition (3.3.15). In both cases we obtain the required conclusion in a way analogous to that in part (a). □

Exercises

3.3.1. Let $\{a_j\}_{j=-\infty}^{\infty}$ be an increasing sequence of positive real numbers and let $\{b_j\}_{j=-\infty}^{\infty}$ be another sequence of positive real numbers such that $\sum_{j\in\mathbf{Z}} b_j < \infty$. Define a measure μ on $\mathbf{R}^{n+1}_+$ by setting

$$\mu(E) = \sum_{j\in\mathbf{Z}} b_j |E \cap \{(x, a_j) :\, x \in \mathbf{R}^n\}|,$$

where E is a subset of $\mathbf{R}^{n+1}_+$ and $|\ \ |$ denotes n-dimensional Lebesgue measure on the affine planes $t = a_j$ of $\mathbf{R}^n \times \mathbf{R}^+ = \{(x,t) :\, x \in \mathbf{R}^n, t > 0\}$. Show that μ is a Carleson measure with norm

$$\|\mu\|_{\mathscr{C}}^{\text{cylinder}} = \|\mu\|_{\mathscr{C}} = \sum_{j\in\mathbf{Z}} b_j.$$

3.3.2. Let $x_0 \in \mathbf{R}^n$ and $\mu = \delta_{(x_0,1)}$ be the Dirac mass at the point $(x_0, 1)$. Show that μ is Carleson measure and compute $\|\mu\|_{\mathscr{C}}^{\text{cylinder}}$ and $\|\mu\|_{\mathscr{C}}$. Which of these norms is larger?

3.3.3. Define *conical* and *hemispherical* tents over balls in $\mathbf{R}^n$ as well as *pyramidal* tents over cubes in $\mathbf{R}^n$ and define the expressions $\|\mu\|_{\mathscr{C}}^{\text{cone}}$, $\|\mu\|_{\mathscr{C}}^{\text{hemisphere}}$, and $\|\mu\|_{\mathscr{C}}^{\text{pyramid}}$. Show that

$$\|\mu\|_{\mathscr{C}}^{\text{cone}} \approx \|\mu\|_{\mathscr{C}}^{\text{hemisphere}} \approx \|\mu\|_{\mathscr{C}}^{\text{pyramid}} \approx \|\mu\|_{\mathscr{C}},$$

where all the implicit constants in the previous estimates depend only on the dimension.

3.3.4. Suppose that Φ has a radial, bounded, symmetrically decreasing integrable majorant. Set $F(x,t) = (f * \Phi_t)(x)$, where f is a locally integrable function on $\mathbf{R}^n$. Prove that

$$F^*(x) \le C_n M(f)(x),$$

where M is the Hardy–Littlewood maximal operator and C_n is a constant that depends only on the dimension.
[*Hint:* If $\varphi(|x|)$ is the claimed majorant of $\Phi(x)$, then the function $\psi(|x|) = \varphi(0)$ for $|x| \le 1$ and $\psi(|x|) = \varphi(|x|-1)$ for $|x| \ge 1$ is a majorant for the function $\Psi(x) = \sup_{|u|\le 1} |\Phi(x-u)|$.]

3.3.5. Let F be a function on $\mathbf{R}^{n+1}_+$, let F^* be the nontangential maximal function derived from F, and let $\mu \ge 0$ be a measure on $\mathbf{R}^{n+1}_+$. Prove that

$$\big\|F\big\|_{L^r(\mathbf{R}^{n+1}_+,\mu)} \le C_n^{1/r}\bigg(\int_{\mathbf{R}^n} \mathscr{C}(\mu)(x) F^*(x)^r\,dx\bigg)^{1/r},$$

where C_n is the constant of Theorem 3.3.5 and $0 < r < \infty$.

3.3.6. (a) Given A a closed subset of $\mathbf{R}^n$ and $0 < \gamma < 1$, define

$$A_\gamma^* = \bigg\{x \in \mathbf{R}^n : \inf_{r>0} \frac{|A \cap B(x,r)|}{|B(x,r)|} \ge \gamma\bigg\}.$$

Show that A^* is a closed subset of A and that it satisfies

$$|(A_\gamma^*)^c| \le \frac{3^n}{1-\gamma}|A^c|.$$

[*Hint:* Consider the Hardy–Littlewood maximal function of χ_{A^c}.]

(b) For a function F on $\mathbf{R}^{n+1}_+$ and $0 < a < \infty$, set

$$F_a^*(x) = \sup_{t>0} \sup_{|y-x|<at} |F(y,t)|.$$

Let $0 < a < b < \infty$ be given. Prove that for all $\lambda > 0$ we have

$$|\{F_a^* > \lambda\}| \le |\{F_b^* > \lambda\}| \le 3^n a^{-n} (a+b)^n |\{F_a^* > \lambda\}|.$$

3.3.7. Let μ be a Carleson measure on $\mathbf{R}^{n+1}_+$. Show that for any $z_0 \in \mathbf{R}^n$ and $t > 0$ we have

$$\iint_{\mathbf{R}^n \times (0,t)} \frac{t}{(|z-z_0|^2 + t^2 + s^2)^{\frac{n+1}{2}}} \, d\mu(z,s) \le \|\mu\|_{\mathscr{C}}^{\text{cylinder}} \frac{\pi^{\frac{n+1}{2}}}{\Gamma(\frac{n+1}{2})}.$$

[*Hint:* Begin by writing

$$\frac{t}{(|z-z_0|^2 + t^2 + s^2)^{\frac{n+1}{2}}} = (n+1)t \int_Q^\infty \frac{dr}{r^{n+2}},$$

where $Q = \sqrt{|z-z_0|^2 + t^2 + s^2}$. Apply Fubini's theorem to estimate the required expression by

$$t(n+1) \int_t^\infty \int_{T\left(B(z_0, \sqrt{r^2 - t^2})\right)} d\mu(z,s) \frac{dr}{r^{n+2}} \le t(n+1) v_n \|\mu\|_{\mathscr{C}}^{\text{cylinder}} \int_t^\infty (r^2 - t^2)^{\frac{n}{2}} \frac{dr}{r^{n+2}},$$

where v_n is the volume of the unit ball in $\mathbf{R}^n$. Reduce the last integral to a beta function.]

3.3.8. ([361]) Let μ be a Carleson measure on $\mathbf{R}^{n+1}_+$. Show that for all $p > 2$ there exists a dimensionless constant C_p such that

$$\int_{\mathbf{R}^{n+1}_+} |(P_t * f)(x)|^p \, d\mu(x,t) \le C_p \|\mu\|_{\mathscr{C}}^{\text{cylinder}} \int_{\mathbf{R}^n} |f(x)|^p \, dx.$$

[*Hint:* It suffices to prove that the operator $f \mapsto P_t * f$ maps $L^2(\mathbf{R}^n)$ to $L^{2,\infty}(\mathbf{R}^{n+1}_+, d\mu)$ with a dimensionless constant C, since then the conclusion follows by interpolation with the corresponding L^∞ estimate, which holds with constant 1. By duality and Exercise 1.4.7 in [156] this is equivalent to showing that

$$\int_{\mathbf{R}^n} \left[\iint_E P_t(x-y) \, d\mu(y,t) \iint_E P_s(x-z) \, d\mu(z,s) \right] dx \le C \mu(E)$$

for any set E in $\mathbf{R}^{n+1}_+$ with $\mu(E) < \infty$. Apply Fubini's theorem, use the identity

$$\int_{\mathbf{R}^n} P_t(x-y)P_s(x-z)\,dx = P_{t+s}(y-z),$$

and consider the cases $t \le s$ and $s \le t$.]

3.4 The Sharp Maximal Function

In Section 3.1 we defined BMO as the space of all locally integrable functions on $\mathbf{R}^n$ whose mean oscillation is at most a finite constant. In this section we introduce a quantitative way to measure the mean oscillation of a function near any point.

3.4.1 Definition and Basic Properties of the Sharp Maximal Function

The local behavior of the mean oscillation of a function is captured to a certain extent by the sharp maximal function. This is a device that enables us to relate integrability properties of a function to those of its mean oscillations.

Definition 3.4.1. Given a locally integrable function f on $\mathbf{R}^n$, we define its *sharp maximal function* $M^{\#}(f)$ as

$$M^{\#}(f)(x) = \sup_{Q \ni x} \frac{1}{|Q|} \int_Q \big| f(t) - \operatorname*{Avg}_Q f \big|\, dt,$$

where the supremum is taken over all cubes Q in $\mathbf{R}^n$ that contain the given point x.

The sharp maximal function is an analogue of the Hardy–Littlewood maximal function, but it has some advantages over it, especially in dealing with the endpoint space L^∞. The very definition of $M^{\#}(f)$ brings up a connection with BMO that is crucial in interpolation. Precisely, we have

$$BMO(\mathbf{R}^n) = \{ f \in L^1_{\text{loc}}(\mathbf{R}^n) : M^{\#}(f) \in L^\infty(\mathbf{R}^n) \},$$

and in this case

$$\big\| f \big\|_{BMO} = \big\| M^{\#}(f) \big\|_{L^\infty}.$$

We summarize some properties of the sharp maximal function.

Proposition 3.4.2. *Let f, g be a locally integrable functions on $\mathbf{R}^n$. Then*

(1) $M^{\#}(f) \le 2M_c(f)$, where M_c is the Hardy–Littlewood maximal operator with respect to cubes in $\mathbf{R}^n$.

(2) For all cubes Q in $\mathbf{R}^n$ we have

$$\frac{1}{2}M^{\#}(f)(x) \le \sup_{Q \ni x} \inf_{a \in \mathbf{C}} \frac{1}{|Q|}\int_Q |f(y)-a|\,dy \le M^{\#}(f)(x).$$

(3) $M^{\#}(|f|) \le 2M^{\#}(f)$.

(4) We have $M^{\#}(f+g) \le M^{\#}(f)+M^{\#}(g)$.

Proof. The proof of (1) is trivial. To prove (2) we fix $\varepsilon > 0$ and for any cube Q we pick a constant a_Q such that

$$\frac{1}{|Q|}\int_Q |f(y)-a_Q|\,dy \le \inf_{a \in Q} \frac{1}{|Q|}\int_Q |f(y)-a|\,dy + \varepsilon\,.$$

Then

$$\begin{aligned}
\frac{1}{|Q|}\int_Q \Big|f(y)-\operatorname*{Avg}_Q f\Big|\,dy &\le \frac{1}{|Q|}\int_Q |f(y)-a_Q|\,dy + \frac{1}{|Q|}\int_Q \Big|\operatorname*{Avg}_Q f - a_Q\Big|\,dy \\
&\le \frac{1}{|Q|}\int_Q |f(y)-a_Q|\,dy + \frac{1}{|Q|}\int_Q |f(y)-a_Q|\,dy \\
&\le 2\inf_{a \in Q} \frac{1}{|Q|}\int_Q |f(y)-a|\,dy + 2\varepsilon\,.
\end{aligned}$$

Taking the supremum over all cubes Q in $\mathbf{R}^n$, we obtain the first inequality in (2), since $\varepsilon > 0$ was arbitrary. The other inequality in (2) is simple. The proofs of (3) and (4) are immediate. □

We saw that $M^{\#}(f) \le 2M_c(f)$, which implies that

$$\big\|M^{\#}(f)\big\|_{L^p} \le C_n p(p-1)^{-1}\big\|f\big\|_{L^p} \tag{3.4.1}$$

for $1 < p < \infty$. Thus the sharp function of an L^p function is also in L^p whenever $1 < p < \infty$. The fact that the converse inequality is also valid is one of the main results in this section. We obtain this estimate via a distributional inequality for the sharp function called a *good lambda* inequality.

3.4.2 A Good Lambda Estimate for the Sharp Function

A useful tool in obtaining the converse inequality to (3.4.1) is the dyadic maximal function.

Definition 3.4.3. A dyadic cube is a set of the form $\prod_{j=1}^{m}[m_j 2^{-k},(m_1+1)2^{-k})$, where $m_1,\dots,m_n,k \in \mathbf{Z}$. Given a locally integrable function f on $\mathbf{R}^n$, we define its *dyadic maximal function* $M_d(f)$ by

$$M_d(f)(x) = \sup_{\substack{Q \ni x \\ Q \text{ dyadic cube}}} \frac{1}{|Q|} \int_Q |f(t)|\,dt\,.$$

The supremum is taken over all dyadic cubes Q in $\mathbf{R}^n$ that contain a given point x.

Obviously, one has the pointwise estimate

$$M_d(f) \le M_c(f) \tag{3.4.2}$$

for all locally integrable functions. This yields the boundedness of M_d on L^p for $1 < p \le \infty$ and the weak type $(1,1)$ property of M_d. More precise estimates on the norm of M_d can be derived. In fact, M_d is of weak type $(1,1)$ with norm at most 1; see Exercise 2.1.12 in [156]. By interpolation (precisely Exercise 1.3.3(a) in [156]), it follows that M_d maps $L^p(\mathbf{R}^n)$ to itself with norm at most

$$\big\|M_d\big\|_{L^p(\mathbf{R}^n)\to L^p(\mathbf{R}^n)} \le \frac{p}{p-1}$$

when $1 < p < \infty$.

One may wonder whether an estimate converse to (3.4.2) holds. But a quick observation shows that for a nonzero locally integrable function f that vanishes on certain open sets, $M_d(f)$ could have zeros, but $M_c(f)$ never vanishes. For instance if a function f on the line vanishes on $\mathbf{R}^-$, then $M_d(f)$ also vanishes on $\mathbf{R}^-$, since no dyadic interval that contains a point in the support of f can also contain a negative number. Therefore, there is no hope for $M_d(f)$ and $M_c(f)$ to be pointwise comparable. However, we show below that the functions $M_d(f)$ and $M(f)$ are comparable in norm.

The next result provides an example of a *good lambda distributional inequality*.

Theorem 3.4.4. *For all $\gamma > 0$, all $\lambda > 0$, and all locally integrable functions f on $\mathbf{R}^n$, we have the estimate*

$$\big|\{x \in \mathbf{R}^n : M_d(f)(x) > 2\lambda,\ M^\#(f)(x) \le \gamma\lambda\}\big| \le 2^n\gamma\big|\{x \in \mathbf{R}^n : M_d(f)(x) > \lambda\}\big|.$$

Moreover, for any q with $1 \le q \le \infty$ and $w \in A_q$, there are constants C and ε_0 depending on n, q, and $[w]_{A_q}$ such that for all $\gamma, \lambda > 0$ we have

$$\begin{aligned} w\big(\{x \in \mathbf{R}^n : M_d(f)(x) > 2\lambda, M^\#(f)(x) \le \gamma\lambda\}\big) \\ \le C\gamma^{\varepsilon_0}\, w\big(\{x \in \mathbf{R}^n : M_d(f)(x) > \lambda\}\big). \end{aligned} \tag{3.4.3}$$

Proof. We may suppose that the set $\Omega_\lambda = \{x \in \mathbf{R}^n : M_d(f)(x) > \lambda\}$ has finite measure; otherwise, there is nothing to prove. Then for each $x \in \Omega_\lambda$ there is a maximal dyadic cube Q^x that contains x such that

$$\frac{1}{|Q^x|}\int_{Q^x} |f(y)|\,dy > \lambda\,; \tag{3.4.4}$$

otherwise, Ω_λ would have infinite measure. Let Q_j be the collection of all such maximal dyadic cubes containing all x in Ω_λ, i.e., $\{Q_j\}_j = \{Q^x : x \in \Omega_\lambda\}$. Maximal dyadic cubes are disjoint; hence any two different Q_j are disjoint. Moreover, we note that if $x, y \in Q_j$, then $Q_j = Q^x = Q^y$. It follows that $\Omega_\lambda = \bigcup_j Q_j$. To prove the required estimate, it suffices to show that for all Q_j we have

$$\big|\{x \in Q_j : M_d(f)(x) > 2\lambda,\, M^\#(f)(x) \le \gamma\lambda\}\big| \le 2^n \gamma |Q_j|, \tag{3.4.5}$$

for once (3.4.5) is established, the conclusion follows by summing on j.

We fix j and $x \in Q_j$ such that $M_d(f)(x) > 2\lambda$. Then the supremum

$$M_d(f)(x) = \sup_{R \ni x} \frac{1}{|R|} \int_R |f(y)|\, dy \tag{3.4.6}$$

is taken over all dyadic cubes R that either contain Q_j or are contained in Q_j (since $Q_j \cap R \neq \emptyset$). If $R \supsetneqq Q_j$, the maximality of Q_j implies that (3.4.4) does not hold for R; thus the average of $|f|$ over R is at most λ. Thus, if $M_d(f)(x) > 2\lambda$, then the average in (3.4.6) is bigger than 2λ for some dyadic cube R contained (not properly) in Q_j. Therefore, if $x \in Q_j$ and $M_d(f)(x) > 2\lambda$, then we can replace f by $f\chi_{Q_j}$ in (3.4.6) and we must have $M_d(f\chi_{Q_j})(x) > 2\lambda$. We let Q'_j be the unique dyadic cube of twice the side length of Q_j that contains Q_j. Therefore, for $x \in Q_j$ we have

$$M_d\Big(\big(f - \operatorname*{Avg}_{Q'_j} f\big)\chi_{Q_j}\Big)(x) \ge M_d(f\chi_{Q_j})(x) - \big|\operatorname*{Avg}_{Q'_j} f\big| > 2\lambda - \lambda = \lambda,$$

since $|\operatorname{Avg}_{Q'_j} f| \le \operatorname{Avg}_{Q'_j} |f| \le \lambda$ because of the maximality of Q_j. We conclude that

$$\big|\{x \in Q_j : M_d(f)(x) > 2\lambda\}\big| \le \Big|\Big\{x \in Q_j : M_d\big((f - \operatorname*{Avg}_{Q'_j} f)\chi_{Q_j}\big)(x) > \lambda\Big\}\Big|, \tag{3.4.7}$$

and using the fact that M_d is of weak type $(1,1)$ with constant 1, we control the last expression in (3.4.7) by

$$\begin{aligned} \frac{1}{\lambda}\int_{Q_j} \big|f(y) - \operatorname*{Avg}_{Q'_j} f\big|\, dy &\le \frac{2^n|Q_j|}{\lambda} \frac{1}{|Q'_j|} \int_{Q'_j} \big|f(y) - \operatorname*{Avg}_{Q'_j} f\big|\, dy \\ &\le \frac{2^n|Q_j|}{\lambda} M^\#(f)(\xi_j) \end{aligned} \tag{3.4.8}$$

for all $\xi_j \in Q_j$. In proving (3.4.5) we may assume that for some $\xi_j \in Q_j$ we have $M^\#(f)(\xi_j) \le \gamma\lambda$; otherwise, there is nothing to prove. For this ξ_j, using (3.4.7) and (3.4.8) we obtain (3.4.5).

Given $w \in A_\infty$, it follows from (3.4.5) and Theorem 7.3.3(d) in [156] when $q = \infty$ or Proposition 7.2.8 in [156] when $q < \infty$ that there exist constants C_2, ε_0 depending on n and $[w]_{A_q}$ such that

$$\frac{w(\{x \in Q_j : M_d(f)(x) > 2\lambda,\, M^\#(f)(x) \le \gamma\lambda\})}{w(Q_j)} \le C_2\,(2^n\gamma)^{\varepsilon_0}.$$

Summing over j we obtain (3.4.3) with $C = C_2\,2^{n\gamma}$. □

Good lambda inequalities can be used to obtain L^p bounds for quantities they contain. For example, we use Theorem 3.4.4 to obtain the equivalence of the L^p norms of $M_d(f)$ and $M^\#(f)$. Since $M^\#(f)$ is pointwise controlled by $2M_c(f)$ and

$$\big\|M_c(f)\big\|_{L^p} \le C(p,n)\big\|f\big\|_{L^p} \le C(p,n)\big\|M_d(f)\big\|_{L^p},$$

we have the estimate

$$\big\|M^\#(f)\big\|_{L^p(\mathbf{R}^n)} \le 2C(p,n)\big\|M_d(f)\big\|_{L^p(\mathbf{R}^n)}$$

for all f in $L^p(\mathbf{R}^n)$. The next theorem says that the converse estimate is valid.

Theorem 3.4.5. *Let $0 < p_0 \le p < \infty$. Then there is a constant $C_n(p)$ such that for all functions f in $L^1_{\mathrm{loc}}(\mathbf{R}^n)$ with $M_d(f) \in L^{p_0}(\mathbf{R}^n)$ we have*

$$\big\|M_d(f)\big\|_{L^p(\mathbf{R}^n)} \le C_n(p)\big\|M^\#(f)\big\|_{L^p(\mathbf{R}^n)}. \tag{3.4.9}$$

If $w \in A_q$ for some q satisfying $1 \le q \le \infty$, there is a constant $C_n(p,q,[w]_{A_q})$, such that if $M_d(f) \in L^{p_0}(\mathbf{R}^n, w)$, then

$$\big\|M_d(f)\big\|_{L^p(\mathbf{R}^n,w)} \le C_n(p,q,[w]_{A_q})\big\|M^\#(f)\big\|_{L^p(\mathbf{R}^n,w)}. \tag{3.4.10}$$

Proof. Fix $p \ge p_0$ with $p < \infty$. For a positive real number N we set

$$I_N = \int_0^N p\lambda^{p-1}\big|\{x \in \mathbf{R}^n : M_d(f)(x) > \lambda\}\big|\,d\lambda.$$

We note that I_N is finite, since $p \ge p_0$ and it is bounded by

$$\frac{pN^{p-p_0}}{p_0}\int_0^N p_0\lambda^{p_0-1}\big|\{x \in \mathbf{R}^n : M_d(f)(x) > \lambda\}\big|\,d\lambda \le \frac{pN^{p-p_0}}{p_0}\big\|M_d(f)\big\|_{L^{p_0}}^{p_0} < \infty.$$

We now write

$$I_N = 2^p \int_0^{\frac{N}{2}} p\lambda^{p-1}\big|\{x \in \mathbf{R}^n : M_d(f)(x) > 2\lambda\}\big|\,d\lambda$$

and we use Theorem 3.4.4 to obtain the following sequence of inequalities:

$$\begin{aligned} I_N &\le 2^p \int_0^{\frac{N}{2}} p\lambda^{p-1}\big|\{x \in \mathbf{R}^n : M_d(f)(x) > 2\lambda,\, M^\#(f)(x) \le \gamma\lambda\}\big|\,d\lambda \\ &\quad + 2^p \int_0^{\frac{N}{2}} p\lambda^{p-1}\big|\{x \in \mathbf{R}^n : M^\#(f)(x) > \gamma\lambda\}\big|\,d\lambda \end{aligned}$$

$$\leq 2^p 2^n \gamma \int_0^{\frac{N}{2}} p\lambda^{p-1} \big|\{x \in \mathbf{R}^n : M_d(f)(x) > \lambda\}\big| \, d\lambda$$

$$+ 2^p \int_0^{\frac{N}{2}} p\lambda^{p-1} \big|\{x \in \mathbf{R}^n : M^{\#}(f)(x) > \gamma\lambda\}\big| \, d\lambda$$

$$\leq 2^p 2^n \gamma I_N + \frac{2^p}{\gamma^p} \int_0^{\frac{N\gamma}{2}} p\lambda^{p-1} \big|\{x \in \mathbf{R}^n : M^{\#}(f)(x) > \lambda\}\big| \, d\lambda .$$

At this point we pick a γ such that $2^p 2^n \gamma = 1/2$. Since I_N is finite, we can subtract from both sides of the inequality the quantity $\frac{1}{2} I_N$ to obtain

$$I_N \leq 2^{p+1} 2^{p(n+p+1)} \int_0^{\frac{N\gamma}{2}} p\lambda^{p-1} \big|\{x \in \mathbf{R}^n : M^{\#}(f)(x) > \lambda\}\big| \, d\lambda ,$$

from which we obtain (3.4.9) with $C_n(p) = 2^{n+p+2+\frac{1}{p}}$ letting $N \to \infty$.

To prove (3.4.10), for a fixed $w \in A_\infty$, we introduce

$$I_N(w) = \int_0^N p\lambda^{p-1} w\big(\{x \in \mathbf{R}^n : M_d(f)(x) > \lambda\}\big) \, d\lambda .$$

Since $M_d(f) \in L^{p_0}(w)$, as before we deduce $I_N(w) < \infty$. Using (3.4.3) we obtain

$$I_N(w) \leq 2^p C \gamma^{\varepsilon_0} I_N(w) + \frac{2^p}{\gamma^p} \int_0^{\frac{N\gamma}{2}} p\lambda^{p-1} w\big(\{x \in \mathbf{R}^n : M^{\#}(f)(x) > \lambda\}\big) \, d\lambda .$$

Selecting γ such that $2^p C \gamma^{\varepsilon_0} = 1/2$, subtracting the finite expression $\frac{1}{2} I_N(w)$ from both sides of the preceding inequality and letting $N \to \infty$ we deduce (3.4.10). □

Corollary 3.4.6. *Let $0 < p_0 < \infty$. Then for any p with $p_0 \leq p < \infty$ and for all locally integrable functions f on $\mathbf{R}^n$ with $M_d(f) \in L^{p_0}(\mathbf{R}^n)$ we have*

$$\big\|f\big\|_{L^p(\mathbf{R}^n)} \leq C_n(p) \big\|M^{\#}(f)\big\|_{L^p(\mathbf{R}^n)} , \tag{3.4.11}$$

where $C_n(p)$ is the constant in Theorem 3.4.5.

Proof. Since for every point in $\mathbf{R}^n$ there is a sequence of dyadic cubes shrinking to it, the Lebesgue differentiation theorem yields that for almost every point x in $\mathbf{R}^n$ the averages of the locally integrable function f over the dyadic cubes containing x converge to $f(x)$. Consequently,

$$|f| \leq M_d(f) \qquad \text{a.e.}$$

Using this fact, the proof of (3.4.11) is immediate, since

$$\big\|f\big\|_{L^p(\mathbf{R}^n)} \leq \big\|M_d(f)\big\|_{L^p(\mathbf{R}^n)} ,$$

and by Theorem 3.4.5 the latter is controlled by $C_n(p)\|M^{\#}(f)\|_{L^p(\mathbf{R}^n)}$. □

Estimate (3.4.11) provides the sought converse to (3.4.1).

3.4.3 Interpolation Using BMO

We continue this section by proving an interpolation result in which the space L^∞ is replaced by BMO. The sharp function plays a key role in the following theorem.

Theorem 3.4.7. *Let $1 \le p_0 < \infty$. Let T be a linear operator that maps $L^{p_0}(\mathbf{R}^n)$ to $L^{p_0}(\mathbf{R}^n)$ with bound A_0, and $L^\infty(\mathbf{R}^n)$ to $BMO(\mathbf{R}^n)$ with bound A_1. Then for all p with $p_0 < p < \infty$ there is a constant $C_{n,p}$ such that for all $f \in L^p$ we have*

$$\big\|T(f)\big\|_{L^p(\mathbf{R}^n)} \le C_{n,p,p_0} A_0^{\frac{p_0}{p}} A_1^{1-\frac{p_0}{p}} \big\|f\big\|_{L^p(\mathbf{R}^n)}. \tag{3.4.12}$$

Remark 3.4.8. In certain applications, the operator T may not be a priori defined on all of $L^{p_0} + L^\infty$ but only on some subspace of it. In this case one may state that the hypotheses and the conclusion of the preceding theorem hold for a subspace of these spaces.

Proof. We consider the operator

$$S(f) = M^\#(T(f))$$

defined for $f \in L^{p_0} + L^\infty$. It is easy to see that S is a sublinear operator. We prove that S maps L^∞ to itself and L^{p_0} to itself if $p_0 > 1$ or L^1 to $L^{1,\infty}$ if $p_0 = 1$. For $f \in L^{p_0}$ we have

$$\begin{aligned} \big\|S(f)\big\|_{L^{p_0}} &= \big\|M^\#(T(f))\big\|_{L^{p_0}} \le 2\big\|M_c(T(f))\big\|_{L^{p_0}} \\ &\le C_{n,p_0}\big\|T(f)\big\|_{L^{p_0}} \le C_{n,p_0}A_0\big\|f\big\|_{L^{p_0}}, \end{aligned}$$

where the three L^{p_0} norms on the top line should be replaced by $L^{1,\infty}$ if $p_0 = 1$. For $f \in L^\infty$ one has

$$\big\|S(f)\big\|_{L^\infty} = \big\|M^\#(T(f))\big\|_{L^\infty} = \big\|T(f)\big\|_{BMO} \le A_1\big\|f\big\|_{L^\infty}.$$

Interpolating between these estimates using Theorem 1.3.2 in [156], we deduce

$$\big\|M^\#(T(f))\big\|_{L^p} = \big\|S(f)\big\|_{L^p} \le C_{p,p_0} A_0^{\frac{p_0}{p}} A_1^{1-\frac{p_0}{p}} \big\|f\big\|_{L^p}$$

for all $f \in L^p$, where $p_0 < p < \infty$.

Consider now a function $h \in L^p \cap L^{p_0}$. In the case $p_0 > 1$, $M_d(T(h)) \in L^{p_0}$; hence Corollary 3.4.6 is applicable and gives

$$\big\|T(h)\big\|_{L^p} \le C_n(p)\, C_{p,p_0} A_0^{\frac{p_0}{p}} A_1^{1-\frac{p_0}{p}} \big\|h\big\|_{L^p}.$$

Density yields the same estimate for all $f \in L^p(\mathbf{R}^n)$. If $p_0 = 1$, one applies the same idea but needs the endpoint estimate of Exercise 3.4.6, since $M_d(T(h)) \in L^{1,\infty}$. □

3.4.4 Estimates for Singular Integrals Involving the Sharp Function

We use the sharp function to obtain pointwise estimates for singular integrals. These enable us to recover previously obtained estimates for singular integrals, but also to deduce a new endpoint boundedness result from L^∞ to BMO.

We recall some facts about singular integral operators. Suppose that K is a function defined on $\mathbf{R}^n \setminus \{0\}$ that satisfies

$$|K(x)| \le A_1 |x|^{-n}, \tag{3.4.13}$$

$$|K(x-y) - K(x)| \le A_2 |y|^\delta |x|^{-n-\delta} \qquad \text{when } |x| \ge 2|y| > 0, \tag{3.4.14}$$

$$\sup_{r<R<\infty} \left| \int_{r \le |x| \le R} K(x)\,dx \right| \le A_3 . \tag{3.4.15}$$

Let W be a tempered distribution that coincides with K on $\mathbf{R}^n \setminus \{0\}$ and let T be the linear operator given by convolution with W.

Under these assumptions we have that T is L^2 bounded with norm at most a constant multiple of $A_1 + A_2 + A_3$ (Theorem 5.4.1 in [156]), and hence it is also L^p bounded with a similar norm on L^p for $1 < p < \infty$ (Theorem 5.3.3 in [156]).

Theorem 3.4.9. *Let T be given by convolution with a distribution W that coincides with a function K on $\mathbf{R}^n \setminus \{0\}$ satisfying (3.4.14). Assume that T has an extension that is L^2 bounded with a norm B. Then there is a constant C_n such that for any $s > 1$ the estimate*

$$M^\#(T(f))(x) \le C_n (A_2 + B) \max(s, (s-1)^{-1}) M(|f|^s)^{\frac{1}{s}}(x) \tag{3.4.16}$$

is valid for all f in $\bigcup_{s \le p < \infty} L^p$ and all $x \in \mathbf{R}^n$.

Proof. In view of Proposition 3.4.2 (2), given any cube Q, it suffices to find a constant a_Q such that

$$\frac{1}{|Q|} \int_Q |T(f)(y) - a_Q|\,dy \le C_n \max(s, (s-1)^{-1})(A_2 + B) M(|f|^s)^{\frac{1}{s}}(x) \tag{3.4.17}$$

for almost all $x \in Q$. To prove this estimate we employ a well-known theme. We write $f = f_Q^0 + f_Q^\infty$, where $f_Q^0 = f\chi_{6\sqrt{n}Q}$ and $f_Q^\infty = f\chi_{(6\sqrt{n}Q)^c}$. Here $6\sqrt{n}\,Q$ denotes the cube that is concentric with Q, has sides parallel to those of Q, and has side length $6\sqrt{n}\,\ell(Q)$, where $\ell(Q)$ is the side length of Q.

We now fix an f in $\bigcup_{s \le p < \infty} L^p$ and we select $a_Q = T(f_Q^\infty)(x)$. Then a_Q is finite (and thus well defined) for all $x \in Q$. Indeed, for all $x \in Q$, (3.4.13) yields

$$|T(f_Q^\infty)(x)| = \left| \int_{(Q^*)^c} f(y) K(x-y)\,dy \right| \le \|f\|_{L^p} \left(\int_{|x-y| \ge c_n \ell(Q)} \frac{A_1^{p'}\,dy}{|x-y|^{np'}} \right)^{\frac{1}{p'}} < \infty,$$

where c_n is a positive constant. It follows that

$$\frac{1}{|Q|}\int_Q |T(f)(y)-a_Q|\,dy$$
$$\leq \frac{1}{|Q|}\int_Q |T(f_Q^0)(y)|\,dy + \frac{1}{|Q|}\int_Q |T(f_Q^\infty)(y)-T(f_Q^\infty)(x)|\,dy. \tag{3.4.18}$$

In view of Theorem 5.3.3 in [156], T maps L^s to L^s with norm at most a dimensional constant multiple of $\max(s,(s-1)^{-1})(B+A_2)$. The first term in (3.4.18) is controlled by

$$\left(\frac{1}{|Q|}\int_Q |T(f_Q^0)(y)|^s\,dy\right)^{\frac{1}{s}} \leq C_n \max(s,(s-1)^{-1})(B+A_2)\left(\frac{1}{|Q|}\int_{\mathbf{R}^n} |f_Q^0(y)|^s\,dy\right)^{\frac{1}{s}}$$
$$\leq C_n' \max(s,(s-1)^{-1})(B+A_2)M(|f|^s)^{\frac{1}{s}}(x).$$

To estimate the second term in (3.4.18), we first note that

$$\int_Q |T(f_Q^\infty)(y)-T(f_Q^\infty)(x)|\,dy \leq \int_Q \left| \int_{(6\sqrt{n}Q)^c} (K(y-z)-K(x-z))f(z)\,dz\right| dy.$$

We make a few geometric observations. Since both x and y are in Q, we have $|x-y|\leq \sqrt{n}\,\ell(Q)$. Also (see Figure 3.3), since $z\notin 6\sqrt{n}\,Q$ and $x\in Q$, we must have

$$|x-z| \geq \operatorname{dist}\left(Q,(6\sqrt{n}Q)^c\right) \geq (3\sqrt{n}-\tfrac{1}{2})\,\ell(Q) \geq 2\sqrt{n}\,\ell(Q) \geq 2|x-y|.$$

Therefore, we have $|x-z|\geq 2|x-y|$, and this allows us to conclude that

$$\big|K(y-z)-K(x-z)\big| = \big|K((x-z)-(x-y))-K(x-z)\big| \leq A_2\frac{|x-y|^\delta}{|x-z|^{n+\delta}}$$

using condition (3.4.14). Using these observations, we bound the second term in (3.4.18) by

$$\frac{1}{|Q|}\int_Q\int_{(6\sqrt{n}Q)^c} \frac{A_2|x-y|^\delta}{|x-z|^{n+\delta}}|f(z)|\,dz\,dy \leq C_n\frac{A_2}{|Q|}\int_{(6\sqrt{n}Q)^c}\frac{\ell(Q)^{n+\delta}}{|x-z|^{n+\delta}}|f(z)|\,dz$$
$$\leq C_nA_2\int_{\mathbf{R}^n}\frac{\ell(Q)^\delta}{(\ell(Q)+|x-z|)^{n+\delta}}|f(z)|\,dz$$
$$\leq C_nA_2M(f)(x)$$
$$\leq C_nA_2\,(M(|f|^s)(x))^{\frac{1}{s}},$$

where we used the fact that $|x-z|$ is at least $\ell(Q)$ and Theorem 2.1.10 in [156]. This proves (3.4.17) and hence (3.4.16). □

The inequality (3.4.16) in Theorem 3.4.9 is noteworthy, since it provides a pointwise estimate for $T(f)$ in terms of a maximal function. This clearly strengthens the L^p boundedness of T. As a consequence of this estimate, we deduce the following result.

Corollary 3.4.10. *Let T be given by convolution with a distribution W that coincides with a function K on $\mathbf{R}^n\setminus\{0\}$ that satisfies (3.4.14). Assume that T has an extension that is L^2 bounded with norm B. Then there is a constant C_n such that the estimate*

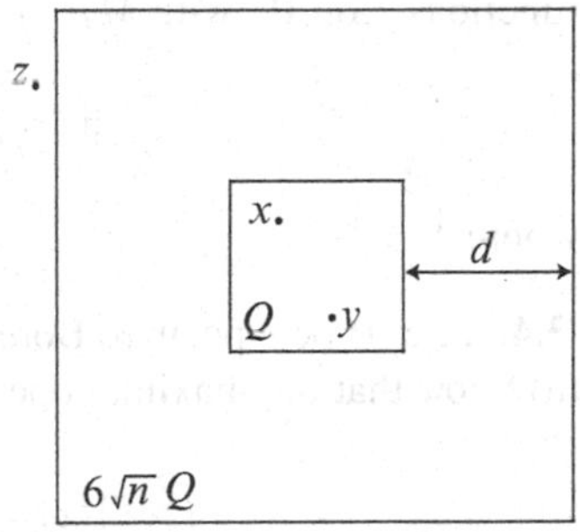

Fig. 3.3 The cubes Q and $6\sqrt{n}Q$. The distance d is equal to $(3\sqrt{n}-\frac{1}{2})\ell(Q)$.

$$\big\|T(f)\big\|_{BMO}\le C_n(A_2+B)\big\|f\big\|_{L^\infty} \tag{3.4.19}$$

is valid for all $f\in L^\infty\bigcap\big(\bigcup_{1\le p<\infty}L^p\big)$.

Proof. We take $s=2$ in Theorem 3.4.9 and we observe that

$$\big\|T(f)\big\|_{BMO}=\big\|M^\#(T(f))\big\|_{L^\infty}\le C_n(A_2+B)\big\|M(|f|^2)^{\frac{1}{2}}\big\|_{L^\infty},$$

and the last expression is easily controlled by $C_n(A_2+B)\|f\|_{L^\infty}$. □

Remark 3.4.11. At this point we have not defined the action of $T(f)$ when f lies merely in L^∞; and for this reason we restricted the functions f in Corollary 3.4.10 to be also in L^p for some $p\in[1,\infty)$. There is, however, a way to define T on L^∞ abstractly via duality. Theorem 2.4.1 gives that T and its transpose T^t map H^1 to L^1. Then for $f\in L^\infty(\mathbf{R}^n)$ we define a linear functional $T(f)$ on $H^1(\mathbf{R}^n)$ by setting

$$\big\langle T(f),\varphi\big\rangle=\big\langle f,T^t(\varphi)\big\rangle=\int_{\mathbf{R}^n}f(x)T^t(\varphi)(x)\,dx,\qquad \varphi\in H^1(\mathbf{R}^n),$$

noting that the expression on the right is a convergent integral, since $f\in L^\infty$ and $T^t(\varphi)$ lies in $L^1(\mathbf{R}^n)$. The preceding integral is bounded by $\|f\|_{L^\infty}C_n(A_2+B)\|\varphi\|_{H^1}$, since T^t maps H^1 to L^1 and consequently, the linear functional $T(f)$ is continuous. It can therefore be identified with a BMO function (Theorem 3.2.2) that satisfies

$$\big\|T(f)\big\|_{BMO}\le C_n'(A_2+B)\|f\|_{L^\infty}.$$

In this way, one defines $T(f)$ for $f \in L^\infty$ as a *BMO* function, but $T(f)$ is not explicitly defined. An explicit definition is given in the next chapter in a slightly more general setting. Using this definition, inequality (3.4.19) extends to hold for all $f \in L^\infty$.

Exercises

3.4.1. Let $1 < q < p < \infty$. Prove that there is a constant $C_{n,p,q}$ such that for all functions f on $\mathbf{R}^n$ with $M_d(f) \in L^q(\mathbf{R}^n)$ we have

$$\|f\|_{L^p} \le C_{n,p,q} \|f\|_{L^q}^{1-\theta} \|f\|_{BMO}^{\theta},$$

where $\frac{1}{p} = \frac{1-\theta}{q}$.

3.4.2. Let μ be a positive Borel measure on $\mathbf{R}^n$.
(a) Show that the maximal operator

$$M_\mu^d(f)(x) = \sup_{\substack{Q \ni x \\ Q \text{ dyadic cube}}} \frac{1}{\mu(Q)} \int_Q |f(t)| \, d\mu(t)$$

maps $L^1(\mathbf{R}^n, d\mu)$ to $L^{1,\infty}(\mathbf{R}^n, d\mu)$ with constant 1.
(b) For a μ-locally integrable function f, define the *sharp maximal function with respect to* μ,

$$M_\mu^\#(f)(x) = \sup_{Q \ni x} \frac{1}{\mu(Q)} \int_Q \Big| f(t) - \operatorname*{Avg}_{Q,\mu} f \Big| \, d\mu(t),$$

where $\operatorname{Avg}_{Q,\mu} f$ denotes the average of f over Q with respect to μ. Assume that μ is a doubling measure with doubling constant $C(\mu)$ [this means that $\mu(3Q) \le C(\mu)\mu(Q)$ for all cubes Q]. Prove that for all $\gamma > 0$, all $\lambda > 0$, and all μ-locally integrable functions f on $\mathbf{R}^n$ we have the estimate

$$\mu\big(\{x : M_\mu^d(f)(x) > 2\lambda, \, M_\mu^\#(f)(x) \le \gamma\lambda\}\big) \le C(\mu)\,\gamma\,\mu\big(\{x : M_\mu^d(f)(x) > \lambda\}\big).$$

[*Hint:* Part (a): For any x in the set $\{x \in \mathbf{R}^n : M_\mu^d(f)(x) > \lambda\}$, choose a maximal dyadic cube $Q = Q(x)$ such that $\int_Q |f(t)| \, d\mu(t) > \lambda \mu(Q)$. Part (b): Mimic the proof of Theorem 3.4.4.]

3.4.3. Let $0 < p_0 < \infty$ and let M_μ^d and $M_\mu^\#$ be as in Exercise 3.4.2. Prove that for any p with $p_0 \le p < \infty$ there is a constant $C_n(p,\mu)$ such that for all locally integrable functions f with $M_\mu^d(f) \in L^{p_0}(\mathbf{R}^n, d\mu)$ we have

$$\|M_\mu^d(f)\|_{L^p(\mathbf{R}^n, d\mu)} \le C_n(p,\mu) \|M_\mu^\#(f)\|_{L^p(\mathbf{R}^n, d\mu)}.$$

3.4.4. We say that a function f on $\mathbf{R}^n$ is in $BMO_d(\mathbf{R}^n)$ (or dyadic BMO) if

$$\|f\|_{BMO_d} = \sup_{Q \text{ dyadic cube}} \frac{1}{|Q|}\int_Q \Big|f(x) - \operatorname*{Avg}_Q f\Big|\,dx < \infty.$$

(a) Show that BMO is a proper subset of BMO_d.
(b) Two dyadic cubes of the same length are called adjacent if they are different and their closures have nonempty intersection. Suppose that A is a finite constant and that a function f in $BMO_d(\mathbf{R}^n)$ satisfies

$$\Big|\operatorname*{Avg}_{Q_1} f - \operatorname*{Avg}_{Q_2} f\Big| \le A$$

for all adjacent dyadic cubes of the same length. Show that f is in $BMO(\mathbf{R}^n)$.
[*Hint:* Part (b): Consider first the case $n=1$. Given an interval I, find adjacent dyadic intervals of the same length I_1 and I_2 such that $I \subsetneqq I_1 \bigcup I_2$ and $|I_1| \le |I| < 2|I_1|$.]

3.4.5. Suppose that K is a function on $\mathbf{R}^n \setminus \{0\}$ that satisfies (3.4.13), (3.4.14), and (3.4.15). Let η be a smooth function that vanishes in a neighborhood of the origin and is equal to 1 in a neighborhood of infinity. For $\varepsilon > 0$ let $K_\eta^{(\varepsilon)}(x) = K(x)\eta(x/\varepsilon)$ and let $T_\eta^{(\varepsilon)}$ be the operator given by convolution with $K_\eta^{(\varepsilon)}$. Prove that for any $1 < s < \infty$ there is a constant $C_{n,s}$ such that for all p with $s < p < \infty$ and f in L^p we have

$$\Big\|\sup_{\varepsilon>0} M^\#(T_\eta^{(\varepsilon)}(f))\Big\|_{L^p(\mathbf{R}^n)} \le C_{n,s}(A_1 + A_2 + A_3)\big\|f\big\|_{L^p(\mathbf{R}^n)}.$$

[*Hint:* Observe that the kernels $K_\eta^{(\varepsilon)}$ satisfy (3.4.13), (3.4.14), and (3.4.15) uniformly in $\varepsilon > 0$ and use Theorem 3.4.9 and Theorem 5.4.1 in [156].]

3.4.6. Let $0 < p_0 < \infty$ and suppose that for some locally integrable function f we have that $M_d(f)$ lies in $L^{p_0,\infty}(\mathbf{R}^n)$. Show that for any p in (p_0,∞) there exists a constant $C_n(p)$ such that

$$\big\|f\big\|_{L^p(\mathbf{R}^n)} \le \big\|M_d(f)\big\|_{L^p(\mathbf{R}^n)} \le C_n(p)\big\|M^\#(f)\big\|_{L^p(\mathbf{R}^n)},$$

where $C_n(p)$ depends only on n and p.
[*Hint:* With the same notation as in the proof of Theorem 3.4.5, use the hypothesis $\|M_d(f)\|_{L^{p_0,\infty}} < \infty$ to prove that $I_N < \infty$ whenever $p > p_0$. Then the arguments in the proofs of Theorem 3.4.5 and Corollary 3.4.6 remain unchanged.]

3.4.7. (a) Let b_N, $N = 1,2,\dots$ be BMO functions on $\mathbf{R}^n$ such that

$$\sup_{N\ge 1}\|b_N\|_{BMO} = C < \infty.$$

Suppose that $b_N \to b$ a.e. Show that b lies in $BMO(\mathbf{R}^n)$ and that $\|b\|_{BMO} \le 2C$.

(b) Prove that the functions

$$\Sigma_N(x) = \sum_{k=1}^{N} \frac{e^{2\pi i k x}}{k}$$

have uniformly bounded imaginary parts in N and x. Also show that $\Sigma_N(x)$ converges as $N \to \infty$ to a function $b(x)$ for all $x \in \mathbf{R} \setminus \mathbf{Z}$. Use Corollary 3.4.10 and Remark 3.4.11 to prove that the real parts of Σ_N are in BMO uniformly in N and consequently,

$$\sup_{N \geq 1} \big\| \Sigma_N \big\|_{BMO(\mathbf{R})} < \infty .$$

Deduce that b lies in $BMO(\mathbf{R})$.

$\big[$*Hint:* Part (a): For each cube Q prove that $\frac{1}{|Q|}\int_Q |b(x) - \liminf_N \mathrm{Avg}_Q\, b_N|\, dx \leq C$. Part (b) Use summation by parts and the fact that the Hilbert transform of $\sin(2\pi k x)$ is $-\cos(2\pi k x)$.$\big]$

3.5 Commutators of Singular Integrals with *BMO* Functions

The mean value zero property of $H^1(\mathbf{R}^n)$ is often manifested by the pairing with BMO. It is therefore natural to expect that BMO can be utilized to express the cancellation of H^1. We give an example to indicate this assertion. If H is the Hilbert transform, then the bilinear operator

$$(f,g) \mapsto fH(g) + H(f)g$$

maps $L^2(\mathbf{R}^n) \times L^2(\mathbf{R}^n)$ to $H^1(\mathbf{R}^n)$; see Exercise 2.4.3. Pairing with a BMO function b and using that $H^t = -H$, we obtain that

$$\big\langle fH(g) + H(f)g\, , b \big\rangle = \big\langle f\, , H(g)b - H(gb) \big\rangle ,$$

and hence the operator $g \mapsto H(g)b - H(gb)$ should be L^2 bounded. This expression $H(g)b - H(gb)$ is called the *commutator* of H with the BMO function b. More generally, we give the following definition.

Definition 3.5.1. The *commutator* of a singular integral operator T with a function b is defined as

$$[b,T](f) = bT(f) - T(bf) .$$

If the function b is locally integrable and has at most polynomial growth at infinity, then the operation $[b,T]$ is well defined when acting on Schwartz functions f.

In view of the preceding remarks, the L^p boundedness of the commutator $[b,T]$ for b in BMO exactly captures the cancellation property of the bilinear expression

$$(f,g) \mapsto T(f)g - fT^t(g) .$$

As in the case with the Hilbert transform, it is natural to expect that the commutator $[b,T]$ of a general singular integral T is L^p bounded for all $1 < p < \infty$. This fact is proved in this section. Since *BMO* functions are unbounded in general, one may surmise that the presence of the negative sign in the definition of the commutator plays a crucial cancellation role.

We introduce some material needed in the study of the boundedness of the commutator.

3.5.1 An Orlicz-Type Maximal Function

We can express the L^p norm ($1 \le p < \infty$) of a function f on a measure space X by

$$\big\|f\big\|_{L^p(X)} = \left(\int_X |f|^p \, d\mu\right)^{\frac{1}{p}} = \inf\Big\{\lambda > 0 : \int_X \Big|\frac{|f|}{\lambda}\Big|^p d\mu \le 1\Big\}.$$

Motivated by the second expression, we may replace the function t^p by a general increasing convex function $\Phi(t)$. We give the following definition.

Definition 3.5.2. A *Young's function* is a continuous increasing convex function Φ on $[0,\infty)$ that satisfies $\Phi(0) = 0$ and $\lim_{t\to\infty}\Phi(t) = \infty$. The *Orlicz norm* of a measurable function f on a measure space (X,μ) with respect to a Young's function Φ is defined as

$$\big\|f\big\|_{\Phi(L)(X,\mu)} = \inf\Big\{\lambda > 0 : \int_X \Phi(|f|/\lambda)\, d\mu \le 1\Big\}.$$

The *Orlicz space* $\Phi(L)(X,\mu)$ is then defined as the space of all measurable functions f on X such that $\|f\|_{\Phi(L)(X,\mu)} < \infty$.

We are mostly concerned with the case in which the measure space X is a cube in $\mathbf{R}^n$ with normalized Lebesgue measure $|Q|^{-1}dx$. For a measurable function f on a cube Q in $\mathbf{R}^n$, the Orlicz norm of f is therefore

$$\big\|f\big\|_{\Phi(L)(Q,\frac{dx}{|Q|})} = \inf\Big\{\lambda > 0 : \frac{1}{|Q|}\int_Q \Phi(|f|/\lambda)\, dx \le 1\Big\},$$

which is simply denoted by $\|f\|_{\Phi(L)(Q)}$, since the measure is understood to be normalized Lebesgue whenever the ambient space is a cube.

Since for $C > 1$ convexity gives $\Phi(t/C) \le \Phi(t)/C$ for all $t \ge 0$, it follows that

$$\big\|f\big\|_{C\Phi(Q)} \le C\big\|f\big\|_{\Phi(Q)}, \tag{3.5.1}$$

which implies that the norms with respect to Φ and $C\Phi$ are comparable.

A case of particular interest arises when $\Phi(t) = t\log(e+t)$. This function is pointwise comparable to $t(1+\log^+ t)$ for $t \ge 0$. We make use in the sequel of a certain maximal operator defined in terms of the corresponding Orlicz norm.

Definition 3.5.3. We define the *Orlicz maximal operator*

$$M_{L\log(e+L)}(f)(x) = \sup_{Q\ni x} \big\|f\big\|_{L\log(e+L)(Q)},$$

where the supremum is taken over all cubes Q with sides parallel to the axes that contain the given point x.

The boundedness properties of this maximal operator are a consequence of the following lemma.

Lemma 3.5.4. *There is a positive constant $c(n)$ such that for any cube Q in $\mathbf{R}^n$ and any nonnegative locally integrable function w, we have*

$$\big\|w\big\|_{L\log(e+L)(Q)} \le \frac{c(n)}{|Q|}\int_Q M_c(w)\,dx, \tag{3.5.2}$$

where M_c is the Hardy–Littlewood maximal operator with respect to cubes. Hence, for some other dimensional constant $c'(n)$ and all nonnegative w in $L^1_{\mathrm{loc}}(\mathbf{R}^n)$ the inequality

$$M_{L\log(e+L)}(w)(x) \le c'(n)M^2(w)(x) \tag{3.5.3}$$

is valid, where $M^2 = M\circ M$ and M is the Hardy–Littlewood maximal operator.

Proof. Fix a cube Q in $\mathbf{R}^n$ with sides parallel to the axes. We introduce a *maximal operator associated with Q* as follows:

$$M_c^Q(f)(x) = \sup_{\substack{R\ni x\\ R\subseteq Q}} \frac{1}{|R|}\int_R |f(y)|\,dy,$$

where the supremum is taken over cubes R in $\mathbf{R}^n$ with sides parallel to the axes. The key estimate follows from the following local version of the reverse weak-type $(1,1)$ estimate of the Hardy–Littlewood maximal function (see Exercise 2.1.4(c) in [156]). For each nonnegative function f on $\mathbf{R}^n$ and $\alpha \ge \mathrm{Avg}_Q f$, we have

$$\frac{1}{\alpha}\int_{Q\cap\{f>\alpha\}} f\,dx \le 2^n\big|\{x\in Q:\ M_c^Q(f)(x)>\alpha\}\big|. \tag{3.5.4}$$

Indeed, to prove (3.5.4), we apply Proposition 2.1.20 in [156] to the function f and the number $\alpha>0$. Then there exists a collection of disjoint (possibly empty) open cubes Q_j such that for almost all $x\in\big(\bigcup_j Q_j\big)^c$ we have $f(x)\le\alpha$ and

$$\alpha < \frac{1}{|Q_j|}\int_{Q_j} f(t)\,dt \le 2^n\alpha. \tag{3.5.5}$$

According to Corollary 2.1.21 in [156] we have $Q\setminus(\bigcup_j Q_j)\subseteq\{f\le\alpha\}$. This implies that $Q\cap\{f>\alpha\}\subseteq\bigcup_j Q_j$, which is contained in $\{x\in Q:\ M_c^Q(f)(x)>\alpha\}$. Multiplying both sides of (3.5.5) by $|Q_j|/\alpha$ and summing over j we obtain (3.5.4).

Using the definition of $M_{L\log(e+L)}$, (3.5.2) follows from the fact that for some constant $c > 1$ independent of w we have

$$\frac{1}{|Q|}\int_Q \frac{w}{\lambda_Q}\log\Big(e+\frac{w}{\lambda_Q}\Big)\,d\mu \le 1, \tag{3.5.6}$$

where

$$\lambda_Q = \frac{c}{|Q|}\int_Q M_c(w)\,dx = c\,\operatorname*{Avg}_Q M_c(w).$$

We let $f = w/\lambda_Q$; by the Lebesgue differentiation theorem we have that $0 \le \operatorname{Avg}_Q f \le 1/c$. It is true that

$$\int_X \phi(f)\,d\nu = \int_0^\infty \phi'(t)\,\nu(\{x\in X : f(x)>t\})\,dt\,,$$

where $\nu \ge 0$, (X,ν) is a σ-finite measure space, ϕ is an increasing continuously differentiable function with $\phi(0) = 0$, and $f \in L^p(X)$; see Proposition 1.1.4 in [156]. We take $X = Q$, $d\nu = |Q|^{-1} f\chi_Q\,dx$, and $\phi(t) = \log(e+t)-1$ to deduce

$$\begin{aligned}\frac{1}{|Q|}\int_Q f\log(e+f)\,dx &= \frac{1}{|Q|}\int_Q f\,dx + \frac{1}{|Q|}\int_0^\infty \frac{1}{e+t}\Big(\int_{Q\cap\{f>t\}} f\,dx\Big)\,dt\\ &= I_0 + I_1 + I_2\,,\end{aligned}$$

where

$$\begin{aligned}I_0 &= \frac{1}{|Q|}\int_Q f\,dx,\\ I_1 &= \frac{1}{|Q|}\int_0^{\operatorname{Avg}_Q f} \frac{1}{e+t}\Big(\int_{Q\cap\{f>t\}} f\,dx\Big)\,dt,\\ I_2 &= \frac{1}{|Q|}\int_{\operatorname{Avg}_Q f}^\infty \frac{1}{e+t}\Big(\int_{Q\cap\{f>t\}} f\,dx\Big)\,dt.\end{aligned}$$

We now clearly have that $I_0 = \operatorname{Avg}_Q f \le 1/c$, while $I_1 \le (\operatorname{Avg}_Q f)^2 \le 1/c^2$. For I_2 we use estimate (3.5.4). Indeed, one has

$$\begin{aligned}I_2 &= \frac{1}{|Q|}\int_{\operatorname{Avg}_Q f}^\infty \frac{1}{e+t}\Big(\int_{Q\cap\{f>t\}} f\,dx\Big)\,dt\\ &\le \frac{2^n}{|Q|}\int_{\operatorname{Avg}_Q f}^\infty \frac{t}{e+t}|\{x\in Q : M_c^Q(f)(x)>t\}|\,dt\\ &\le \frac{2^n}{|Q|}\int_0^\infty |\{x\in Q : M_c^Q(f)(x)>\lambda\}|\,d\lambda\\ &= \frac{2^n}{|Q|}\int_Q M_c^Q(f)\,dx\\ &= \frac{2^n}{|Q|}\int_Q M_c(w)\,dx\,\frac{1}{\lambda_Q} = \frac{2^n}{c}\end{aligned}$$

using the definition of λ_Q. Combining all the estimates obtained, we deduce that

$$I_0 + I_1 + I_2 \le \frac{1}{c} + \frac{1}{c^2} + \frac{2^n}{c} \le 1,$$

provided c is large enough. □

3.5.2 A Pointwise Estimate for the Commutator

We introduce certain modifications of the sharp maximal operator $M^\#$ defined in Section 3.4. We have the centered version

$$\mathcal{M}^\#(f)(x) = \sup_{\substack{Q \text{ cube in } \mathbf{R}^n \\ \text{center of } Q \text{ is } x}} \frac{1}{|Q|} \int_Q |f(y) - \operatorname{Avg}_Q f| \, dy,$$

which is pointwise equivalent to $M^\#(f)(x)$ by a simple argument based on the fact that the smallest cube that contains a fixed cube and is centered at point in its interior has comparable size with the fixed cube.

Then we introduce the "smaller" sharp maximal function

$$\mathcal{M}^{\#\#}(f)(x) = \sup_{\substack{Q \text{ cube in } \mathbf{R}^n \\ \text{center of } Q \text{ is } x}} \inf_c \frac{1}{|Q|} \int_Q |f(y) - c| \, dy, \tag{3.5.7}$$

which is pointwise equivalent to $\mathcal{M}^\#(f)(x)$ [and thus to $M^\#(f)(x)$] by an argument similar with that given in Proposition 3.4.2 (2). For $\delta > 0$ we also introduce the maximal operators

$$\begin{aligned} M_\delta(f) &= M(|f|^\delta)^{1/\delta} \\ M_\delta^\#(f) &= M^\#(|f|^\delta)^{1/\delta} \\ \mathcal{M}_\delta^\#(f) &= \mathcal{M}^\#(|f|^\delta)^{1/\delta} \\ \mathcal{M}_\delta^{\#\#}(f) &= \mathcal{M}^{\#\#}(|f|^\delta)^{1/\delta}, \end{aligned}$$

where M is the Hardy–Littlewood maximal operator. Of these four maximal functions, the last three are pointwise comparable to each other.

The next lemma states a pointwise estimate for commutators of singular integral operators with *BMO* functions in terms of the maximal functions and maximal functions of singular integrals.

Lemma 3.5.5. *Let T be a linear operator given by convolution with a tempered distribution on $\mathbf{R}^n$ that coincides with a function $K(x)$ on $\mathbf{R}^n \setminus \{0\}$ satisfying (3.4.13), (3.4.14), and (3.4.15). Let b be in $BMO(\mathbf{R}^n)$, and let $0 < \delta < \varepsilon$. Then there exists a positive constant $C = C_{\delta,\varepsilon,n}$ such that for every smooth function f with compact support we have*

$$M_\delta^\#([b,T](f)) \le C\|b\|_{BMO}\{M_\varepsilon(T(f)) + M^2(f)\}. \tag{3.5.8}$$

Proof. We will prove (3.5.8) for the equivalent operator $\mathcal{M}_\delta^{\#\#}([b,T](f))$. Fix a cube Q in $\mathbf{R}^n$ with sides parallel to the axes centered at the point x. Since for $0<\delta<1$ we have $\big||\alpha|^\delta - |\beta|^\delta\big| \le |\alpha-\beta|^\delta$ for $\alpha,\beta \in \mathbf{R}$, it is enough to show for some complex constant $c = c_Q$ that there exists $C = C_\delta > 0$ such that

$$\left(\frac{1}{|Q|}\int_Q \big|[b,T](f)(y) - c\big|^\delta dy\right)^{\frac{1}{\delta}} \le C\|b\|_{BMO}\{M_\varepsilon(T(f))(x) + M^2(f)(x)\}. \tag{3.5.9}$$

Denote by Q^* the cube $5\sqrt{n}\,Q$ that has side length $5\sqrt{n}$ times the side length of Q and the same center x as Q. Let $f = f_1 + f_2$, where $f_1 = f\chi_{Q^*}$. For an arbitrary constant a we write

$$[b,T](f) = (b-a)T(f) - T((b-a)f_1) - T((b-a)f_2).$$

Selecting

$$c = \operatorname*{Avg}_{Q} T((b-a)f_2) \qquad \text{and} \qquad a = \operatorname*{Avg}_{Q^*} b,$$

we can estimate the left-hand side of (3.5.9) by a multiple of $L_1 + L_2 + L_3$, where

$$\begin{aligned}
L_1 &= \left(\frac{1}{|Q|}\int_Q \Big|(b(y) - \operatorname*{Avg}_{Q^*} b)\, T(f)(y)\Big|^\delta dy\right)^{\frac{1}{\delta}}, \\
L_2 &= \left(\frac{1}{|Q|}\int_Q \Big|T\big((b - \operatorname*{Avg}_{Q^*} b) f_1\big)(y)\Big|^\delta dy\right)^{\frac{1}{\delta}}, \\
L_3 &= \left(\frac{1}{|Q|}\int_Q \Big|T\big((b - \operatorname*{Avg}_{Q^*} b) f_2\big) - \operatorname*{Avg}_{Q} T\big((b - \operatorname*{Avg}_{Q^*} b) f_2\big)\Big|^\delta dy\right)^{\frac{1}{\delta}}.
\end{aligned}$$

To estimate L_1, we use Hölder's inequality with exponents r and r' for some $1 < r < \varepsilon/\delta$:

$$\begin{aligned}
L_1 &\le \left(\frac{1}{|Q|}\int_Q \Big|b(y) - \operatorname*{Avg}_{Q^*} b\Big|^{\delta r'} dy\right)^{\frac{1}{\delta r'}} \left(\frac{1}{|Q|}\int_Q |T(f)(y)|^{\delta r} dy\right)^{\frac{1}{\delta r}} \\
&\le C\,\|b\|_{BMO} M_{\delta r}(T(f))(x) \\
&\le C\,\|b\|_{BMO} M_\varepsilon(T(f))(x),
\end{aligned}$$

recalling that x is the center of Q. Since $T : L^1(\mathbf{R}^n) \to L^{1,\infty}(\mathbf{R}^n)$ and $0 < \delta < 1$, Kolmogorov's inequality (Exercise 2.1.5 in [156]) yields

$$
\begin{aligned}
L_2 &\le \frac{C}{|Q|}\int_{\mathbf{R}^n}\big|(b(y)-\operatorname*{Avg}_{Q^*}b)f_1(y)\big|\,dy\\
&= \frac{C'}{|Q^*|}\int_{Q^*}\big|(b(y)-\operatorname*{Avg}_{Q^*}b)f(y)\big|\,dy\\
&\le 2C'\big\|b-\operatorname*{Avg}_{Q^*}b\big\|_{(e^L-1)(Q^*)}\big\|f\big\|_{L\log(1+L)(Q^*)},
\end{aligned}
$$

using Exercise 3.5.2(c).

For some $0<\gamma<(2^ne)^{-1}$, let $C_{n,\gamma}>2$ be a constant larger than that appearing on the right-hand side of the inequality in Corollary 3.1.7. We set $c_0=C_{n,\gamma}-1>1$. We use (3.5.1) and we claim that

$$
\big\|b-\operatorname*{Avg}_{Q^*}b\big\|_{(e^L-1)(Q^*)}\le c_0\big\|b-\operatorname*{Avg}_{Q^*}b\big\|_{c_0^{-1}(e^L-1)(Q^*)}\le\frac{c_0}{\gamma}\big\|b\big\|_{BMO}. \tag{3.5.10}
$$

Indeed, the last inequality is equivalent to

$$
\frac{1}{|Q^*|}\int_{Q^*}c_0^{-1}\Big[e^{\gamma|b(y)-\operatorname{Avg}_{Q^*}b|/\|b\|_{BMO}}-1\Big]\,dy\le 1,
$$

which is a restatement of Corollary 3.1.7. We therefore conclude that

$$
L_2\le C\|b\|_{BMO}\,M_{L\log(1+L)}(f)(x).
$$

Finally, we turn our attention to the term L_3. Note that if $z,y\in Q$ and $w\notin Q^*$, then $|z-w|\ge 2|z-y|$. Using Fubini's theorem and property (3.4.14) successively, we control L_3 pointwise by

$$
\begin{aligned}
&\frac{1}{|Q|}\int_Q\Big|T\big((b-\operatorname*{Avg}_{Q^*}b)f_2\big)(y)-\operatorname*{Avg}_{Q}T\big((b-\operatorname*{Avg}_{Q^*}b)f_2\big)\Big|\,dy\\
&\le\frac{1}{|Q|^2}\int_Q\int_Q\int_{\mathbf{R}^n\setminus Q^*}|K(y-w)-K(z-w)|\big|(b(w)-\operatorname*{Avg}_{Q^*}b)f(w)\big|\,dw\,dz\,dy\\
&\le\frac{1}{|Q|^2}\int_Q\int_Q\sum_{j=0}^\infty\int_{2^{j+1}Q^*\setminus 2^jQ^*}\frac{A_2|y-z|^\delta}{|z-w|^{n+\delta}}\big|b(w)-\operatorname*{Avg}_{Q^*}b\big|\,|f(w)|\,dw\,dz\,dy\\
&\le CA_2\sum_{j=0}^\infty\frac{\ell(Q)^\delta}{(2^j\ell(Q))^{n+\delta}}\int_{2^{j+1}Q^*}\big|b(w)-\operatorname*{Avg}_{Q^*}b\big|\,|f(w)|\,dw\\
&\le CA_2\Bigg(\sum_{j=0}^\infty\frac{2^{-j\delta}}{(2^j\ell(Q))^n}\int_{2^{j+1}Q^*}\big|b(w)-\operatorname*{Avg}_{2^{j+1}Q^*}b\big|\,|f(w)|\,dw\\
&\qquad+\sum_{j=0}^\infty 2^{-j\delta}\big|\operatorname*{Avg}_{2^{j+1}Q^*}b-\operatorname*{Avg}_{Q^*}b\big|\frac{1}{(2^j\ell(Q))^n}\int_{2^{j+1}Q^*}|f(w)|\,dw\Bigg)
\end{aligned}
$$

$$\leq C'A_2 \sum_{j=0}^{\infty} 2^{-j\delta} \big\|b - \operatorname*{Avg}_{2^{j+1}Q^*} b\big\|_{(e^L-1)(2^{j+1}Q^*)} \big\|f\big\|_{L\log(1+L)(2^{j+1}Q^*)}$$
$$+ C'A_2 \|b\|_{BMO} \sum_{j=1}^{\infty} \frac{j}{2^{j\delta}} M(f)(x)$$
$$\leq C''A_2 \|b\|_{BMO} M_{L\log(1+L)}(f)(x) + C''A_2 \|b\|_{BMO} M(f)(x)$$
$$\leq C'''A_2 \|b\|_{BMO} M^2(f)(x),$$

where we have used inequality (3.5.10), Lemma 3.5.4, and the simple estimate

$$\Big|\operatorname*{Avg}_{2^{j+1}Q^*} b - \operatorname*{Avg}_{Q^*} b\Big| \leq C_n j \|b\|_{BMO}$$

of Exercise 3.1.5. □

3.5.3 L^p Boundedness of the Commutator

We note that if f has compact support and b is in BMO, then bf lies in $L^q(\mathbf{R}^n)$ for all $q < \infty$ and therefore $T(bf)$ is well defined whenever T is a singular integral operator. Likewise, $[b,T]$ is a well-defined operator on $\mathscr{C}_0^\infty$ for all b in BMO.

Having obtained the crucial Lemma 3.5.5, we now pass to an important result concerning its L^p boundedness.

Theorem 3.5.6. *Let T be as in Lemma 3.5.5. Then for any $1 < p < \infty$ there exists a constant $C = C_{p,n}$ such that for all smooth functions with compact support f and all BMO functions b, the following estimate is valid:*

$$\big\|[b,T](f)\big\|_{L^p(\mathbf{R}^n)} \leq C \big\|b\big\|_{BMO} \big\|f\big\|_{L^p(\mathbf{R}^n)}. \tag{3.5.11}$$

Consequently, the linear operator

$$f \mapsto [b,T](f)$$

admits a bounded extension from $L^p(\mathbf{R}^n)$ to $L^p(\mathbf{R}^n)$ for all $1 < p < \infty$ with norm at most a multiple of $\|b\|_{BMO}$.

Proof. Using the inequality of Theorem 3.4.4, we obtain for functions g, with $|g|^\delta$ locally integrable,

$$\big|\{M_d(|g|^\delta)^{\frac{1}{\delta}} > 2^{\frac{1}{\delta}}\lambda\} \cap \{M_\delta^\#(g) \leq \gamma\lambda\}\big| \leq 2^n \gamma^\delta \big|\{M_d(|g|^\delta)^{\frac{1}{\delta}} > \lambda\}\big| \tag{3.5.12}$$

for all $\lambda,\gamma,\delta>0$. Then a repetition of the proof of Theorem 3.4.5 yields the second inequality:

$$\big\|M(|g|^\delta)^{\frac{1}{\delta}}\big\|_{L^p}\le C_n\big\|M_d(|g|^\delta)^{\frac{1}{\delta}}\big\|_{L^p}\le C_n(p)\big\|M_\delta^\#(g)\big\|_{L^p} \tag{3.5.13}$$

for all $p\in(p_0,\infty)$, provided $M_d(|g|^\delta)^{\frac{1}{\delta}}\in L^{p_0}(\mathbf{R}^n)$ for some $p_0>0$.

For the following argument, it is convenient to replace b by the bounded function

$$b_k(x)=\begin{cases} k & \text{if } b(x)>k,\\ b(x) & \text{if } -k\le b(x)\le k,\\ -k & \text{if } b(x)<-k,\end{cases}$$

which satisfies $\|b_k\|_{BMO}\le\frac{9}{4}\|b\|_{BMO}$ for any $k>0$; see Exercise 3.1.4.

For given $1<p<\infty$, select p_0 such that $1<p_0<p$. Given a smooth function with compact support f, we note that the function b_kf lies in L^{p_0}; thus $T(b_kf)$ also lies in L^{p_0}. Likewise, $b_kT(f)$ also lies in L^{p_0}. Since M_δ is bounded on L^{p_0} for $0<\delta<1$, we conclude that

$$\big\|M_\delta([b_k,T](f))\big\|_{L^{p_0}}\le C_\delta\big(\big\|M_\delta(b_kT(f))\big\|_{L^{p_0}}+\big\|M_\delta(T(b_kf))\big\|_{L^{p_0}}\big)<\infty.$$

This allows us to obtain (3.5.13) with $g=[b_k,T](f)$. We now turn to Lemma 3.5.5, in which we pick $0<\delta<\varepsilon<1$. Taking L^p norms on both sides of (3.5.8) and using (3.5.13) with $g=[b_k,T](f)$ and the boundedness of M_ε, T, and M^2 on $L^p(\mathbf{R}^n)$, we deduce the a priori estimate (3.5.11) for smooth functions with compact support f and the truncated *BMO* functions b_k.

The Lebesgue dominated convergence theorem gives that $b_k\to b$ in L^2 of every compact set and, in particular, in $L^2(\mathrm{supp} f)$. It follows that $b_kf\to bf$ in L^2 and therefore $T(b_kf)\to T(bf)$ in L^2 by the boundedness of T on L^2. We deduce that for some subsequence of integers k_j, $T(b_{k_j}f)\to T(bf)$ a.e. For this subsequence we have $[b_{k_j},T](f)\to[b,T](f)$ a.e. Letting $j\to\infty$ and using Fatou's lemma, we deduce that (3.5.11) holds for all *BMO* functions b and smooth functions f with compact support.

Since smooth functions with compact support are dense in L^p, it follows that the commutator admits a bounded extension on L^p that satisfies (3.5.11). □

We refer to Exercise 3.5.4 for an analogue of Theorem 3.5.6 when $p=1$.

Exercises

3.5.1. Use Jensen's inequality to show that the Hardy–Littlewood maximal operator M is pointwise controlled by a constant multiple of $M_{L\log(1+L)}$.

3.5.2. (a) (*Young's inequality for Orlicz spaces*) Let φ be a continuous, real-valued, strictly increasing function defined on $[0,\infty)$ which satisfies $\varphi(0)=0$ and $\lim_{t\to\infty}\varphi(t)=\infty$. Let $\psi=\varphi^{-1}$ and for $x\in[0,\infty)$ define

$$\Phi(x)=\int_0^x\varphi(t)\,dt\,,\qquad \Psi(x)=\int_0^x\psi(t)\,dt\,.$$

Show that for $s,t\in[0,\infty)$ we have $st\le\Phi(s)+\Psi(t)$.
(b) Choose a suitable function φ in part (a) to deduce for s,t in $[0,\infty)$ the inequality

$$st\le(t+1)\log(t+1)-t+e^s-s-1\le t\log(t+1)+e^s-1\,.$$

(c) (*Hölder's inequality for Orlicz spaces*) Deduce the inequality

$$\big|\langle f,g\rangle\big|\le 2\big\|f\big\|_{\Phi(L)}\big\|g\big\|_{\Psi(L)}\,.$$

[*Hint:* Give a geometric proof distinguishing the cases $t>\varphi(s)$ and $t\le\varphi(s)$. Use that for $u\ge0$ we have $\int_0^u\varphi(t)\,dt+\int_0^{\varphi(u)}\psi(s)\,ds=u\varphi(u)$.]

3.5.3. Let T be as in Lemma 3.5.5. Show that there is a constant $C_n<\infty$ such that for all $f\in L^p(\mathbf{R}^n)$ and $g\in L^{p'}(\mathbf{R}^n)$ we have

$$\big\|T(f)g-fT^t(g)\big\|_{H^1(\mathbf{R}^n)}\le C\big\|f\big\|_{L^p(\mathbf{R}^n)}\big\|g\big\|_{L^{p'}(\mathbf{R}^n)}\,.$$

In other words, show that the bilinear operator $(f,g)\mapsto T(f)g-fT^t(g)$ maps $L^p(\mathbf{R}^n)\times L^{p'}(\mathbf{R}^n)$ to $H^1(\mathbf{R}^n)$.

3.5.4. ([295]) Let $\Phi(t)=t\log(1+t)$. Then there exists a positive constant C, depending on the *BMO* constant of b, such that for any smooth function f with compact support the following is valid:

$$\sup_{\alpha>0}\frac{1}{\Phi(\frac{1}{\alpha})}\big|\{|[b,T](f)|>\alpha\}\big|\le C\sup_{\alpha>0}\frac{1}{\Phi(\frac{1}{\alpha})}\big|\{M^2(f)>\alpha\}\big|\,.$$

3.5.5. Let R_1,R_2 be the Riesz transforms in $\mathbf{R}^2$. Show that there is a constant $C<\infty$ such that for all square integrable functions g_1,g_2 on $\mathbf{R}^2$ the following is valid:

$$\big\|R_1(g_1)R_2(g_2)-R_1(g_2)R_2(g_1)\big\|_{H^1}\le C_p\|g_1\|_{L^2}\|g_2\|_{L^2}\,.$$

[*Hint:* Consider the pairing $\big\langle g_1,R_2([b,R_1](g_2))-R_1([b,R_2](g_2))\big\rangle$ with $b\in BMO$.]

3.5.6. ([89]) Use Exercise 3.5.5 to prove that the Jacobian

$$J_f=\det\begin{pmatrix}\partial_1 f_1 & \partial_2 f_1\\ \partial_1 f_2 & \partial_2 f_2\end{pmatrix},$$

of a map $f=(f_1,f_2):\mathbf{R}^2\to\mathbf{R}^2$, lies in $H^1(\mathbf{R}^2)$ whenever $f_1,f_2\in\dot{L}^2_1(\mathbf{R}^2)$.
[*Hint:* Set $g_j=\Delta^{1/2}(f_j)$.]

3.5.7. Let $\Phi(t) = t(1+\log^+ t)^\alpha$, where $0 \le \alpha < \infty$. Let T be a linear (or sublinear) operator that maps $L^{p_0}(\mathbf{R}^n)$ to $L^{p_0,\infty}(\mathbf{R}^n)$ with norm B for some $1 < p_0 \le \infty$ and also satisfies the following *weak type Orlicz estimate*: for all functions f in $\Phi(L)$,

$$|\{x \in \mathbf{R}^n : |T(f)(x)| > \lambda\}| \le A \int_{\mathbf{R}^n} \Phi\Big(\frac{|f(x)|}{\lambda}\Big)\, dx,$$

for some $A < \infty$ and all $\lambda > 0$. Prove that T is bounded from $L^p(\mathbf{R}^n)$ to itself, whenever $1 < p < p_0$.
[*Hint:* Set $f^\lambda = f\chi_{|f|>\lambda}$ and $f_\lambda = f - f^\lambda$. When $p_0 < \infty$, estimate $|\{|T(f)| > 2\lambda\}|$ by $|\{|T(f^\lambda)| > \lambda\}| + |\{|T(f_\lambda)| > \lambda\}| \le A\int_{|f|>\lambda} \Phi(\frac{|f(x)|}{\lambda})dx + B^{p_0}\int_{|f|\le\lambda} \frac{|f(x)|^{p_0}}{\lambda^{p_0}} dx$. Multiply by p, integrate with respect to the measure $\lambda^{p-1}d\lambda$ from 0 to infinity, apply Fubini's theorem, and use that $\int_0^1 \Phi(1/\lambda)\lambda^{p-1}\,d\lambda < \infty$ to deduce that T maps L^p to $L^{p,\infty}$. When $p_0 = \infty$, use that $|\{|T(f)| > 2B\lambda\}| \le |\{|T(f^\lambda)| > B\lambda\}|$ and argue as in the case $p_0 < \infty$. Boundedness from L^p to L^p follows by interpolation.]

HISTORICAL NOTES

The space of functions of bounded mean oscillation first appeared in the work of John and Nirenberg [205] in the context of nonlinear partial differential equations that arise in the study of minimal surfaces. Theorem 3.1.6 was obtained by John and Nirenberg [205]. The relationship of *BMO* functions and Carleson measures is due to Fefferman and Stein [139]. For a variety of issues relating *BMO* to complex function theory one may consult the book of Garnett [151]. The duality of H^1 and *BMO* (Theorem 3.2.2) was announced by Fefferman in [133], but its first proof appeared in the article of Fefferman and Stein [139]. This article actually contains two proofs of this result. The proof of Theorem 3.2.2 is based on the atomic decomposition of H^1, which was obtained subsequently. An alternative proof of the duality between H^1 and *BMO* was given by Carleson [71]. Dyadic *BMO* (Exercise 3.4.4) in relation to *BMO* is studied in Garnett and Jones [153]. The same authors studied the distance in *BMO* to L^∞ in [152].

Carleson measures first appeared in the work of Carleson [67] and [68]. Corollary 3.3.6 was first proved by Carleson, but the proof given here is due to Stein. The characterization of Carleson measures in Theorem 3.3.8 was obtained by Carleson [67]. A theory of balayage for studying *BMO* was developed by Varopoulos [360]. The space *BMO* can also be characterized in terms Carleson measures via Theorem 3.3.8. The converse of Theorem 3.3.8 (see Fefferman and Stein [139]) states that if the function Ψ satisfies a nondegeneracy condition and $|f * \Psi_t|^2 \frac{dx\,dt}{t}$ is a Carleson measure, then f must be a *BMO* function. We refer to Stein [326] (page 159) for a proof of this fact, which uses a duality idea related to tent spaces. The latter were introduced by Coifman, Meyer, and Stein [95] to systematically study the connection between square functions and Carleson measures.

The sharp maximal function was introduced by Fefferman and Stein [139], who first used it to prove Theorem 3.4.5 and derive interpolation for analytic families of operators when one endpoint space is *BMO*. Theorem 3.4.7 provides the main idea why L^∞ can be replaced by *BMO* in this context. The fact that L^2-bounded singular integrals also map L^∞ to *BMO* was independently obtained by Peetre [289], Spanne [320], and Stein [324]. Peetre [289] also observed that translation-invariant singular integrals (such as the ones in Corollary 3.4.10) actually map *BMO* to itself. Another interesting property of *BMO* is that it is preserved under the action of the Hardy–Littlewood maximal operator. This was proved by Bennett, DeVore, and Sharpley [21]; see also the almost simultaneous proof of Chiarenza and Frasca [74]. The decomposition of open sets given in Proposition 3.3.4 is due to Whitney [369].

An alternative characterization of *BMO* can be obtained in terms of commutators of singular integrals. Precisely, we have that the commutator $[b,T](f)$ is L^p bounded for $1 < p < \infty$ if and only if the function b is in *BMO*. The sufficiency of this result (Theorem 3.5.6) is due to Coifman, Rochberg, and Weiss [96], who used it to extend the classical theory of H^p spaces to higher dimensions. The necessity was obtained by Janson [203], who also obtained a simpler proof of the sufficiency. The exposition in Section 3.5 is based on the article of Pérez [295]. This approach is not the shortest available, but the information derived in Lemma 3.5.5 is often useful; for instance, it is used in the substitute of the weak type $(1,1)$ estimate of Exercise 3.5.4. The inequality (3.5.3) in Lemma 3.5.4 can be reversed as shown by Pérez and Wheeden [297]. Weighted L^p estimates for the commutator in terms of the double iteration of the Hardy–Littlewood maximal operator can be deduced as a consequence of Lemma 3.5.5; see the article of Pérez [296].

Orlicz spaces were introduced by Birbaum and Orlicz [39] and further elaborated by Orlicz [287], [288]. For a modern treatment one may consult the book of Rao and Ren [302]. Bounded mean oscillation with Orlicz norms was considered by Strömberg [332].

The space of functions of vanishing mean oscillation (*VMO*) was introduced by Sarason [309] as the set of integrable functions f on $\mathbf{T}^1$ satisfying $\lim_{\delta\to 0}\sup_{I:\,|I|\le\delta}|I|^{-1}\int_I|f-\mathrm{Avg}_I f|\,dx=0$. This space is the closure in the *BMO* norm of the subspace of $BMO(\mathbf{T}^1)$ consisting of all uniformly continuous functions on $\mathbf{T}^1$. One may define $VMO(\mathbf{R}^n)$ as the space of functions on $\mathbf{R}^n$ that satisfy $\lim_{\delta\to 0}\sup_{Q:\,|Q|\le\delta}|Q|^{-1}\int_Q|f-\mathrm{Avg}_Q f|\,dx=0$, $\lim_{N\to\infty}\sup_{Q:\,\ell(Q)\ge N}|Q|^{-1}\int_Q|f-\mathrm{Avg}_Q f|\,dx=0$, and $\lim_{R\to\infty}\sup_{Q:\,Q\cap B(0,R)=\emptyset}|Q|^{-1}\int_Q|f-\mathrm{Avg}_Q f|\,dx=0$; here I denotes intervals in $\mathbf{T}^1$ and Q cubes in $\mathbf{R}^n$. Then $VMO(\mathbf{R}^n)$ is the closure of the the space of continuous functions that vanish at infinity in the $BMO(\mathbf{R}^n)$ norm. One of the important features of $VMO(\mathbf{R}^n)$ is that it is the predual of $H^1(\mathbf{R}^n)$, as was shown by Coifman and Weiss [97]. As a companion to Corollary 3.4.10, singular integral operators can be shown to map the space of continuous functions that vanish at infinity into *VMO*. We refer to the article of Dafni [109] for a short and elegant exposition of these results as well as for a local version of the *VMO*-H^1 duality.

Chapter 4
Singular Integrals of Nonconvolution Type

We study singular integrals whose kernels do not necessarily commute with translations. Such operators appear in many places in harmonic analysis and partial differential equations. For instance, a large class of pseudodifferential operators falls under the scope of this theory.

This broader point of view does not necessarily bring additional complications in the development of the subject except at the study of L^2 boundedness, where Fourier transform techniques are lacking. The L^2 boundedness of convolution operators is easily understood via a careful examination of the Fourier transform of the kernel, but for nonconvolution operators different tools are required in this study. The main result of this chapter is the derivation of a set of necessary and sufficient conditions for nonconvolution singular integrals to be L^2 bounded. This result is referred to as the $T(1)$ theorem and owes its name to a condition expressed in terms of the action of the operator T on the function 1.

An extension of the $T(1)$ theorem, called the $T(b)$ theorem, is obtained in Section 4.6 and is used to deduce the L^2 boundedness of the Cauchy integral along Lipschitz curves. A variant of the $T(b)$ theorem is also used in the boundedness of the square root of a divergence form elliptic operator discussed in the last section of the chapter.

4.1 General Background and the Role of *BMO*

We begin by recalling the notion of the adjoint and transpose operator. One may choose to work with either a real or a complex inner product on pairs of functions. For f,g complex-valued functions with integrable product, we denote the real inner product by

$$\langle f,g\rangle = \int_{\mathbf{R}^n} f(x)g(x)\,dx.$$

L. Grafakos, *Modern Fourier Analysis*, Graduate Texts in Mathematics 250,
DOI 10.1007/978-1-4939-1230-8_4, © Springer Science+Business Media New York 2014

This notation is suitable when we think of f as a distribution acting on a test function g. We also have the complex inner product

$$\langle f \,|\, g\rangle = \int_{\mathbf{R}^n} f(x)\overline{g(x)}\,dx,$$

which is an appropriate notation when we think of f and g as elements of a Hilbert space over the complex numbers. Now suppose that T is a linear operator bounded on L^p. Then the *adjoint* operator T^* of T is uniquely defined via the identity

$$\langle T(f) \,|\, g\rangle = \langle f \,|\, T^*(g)\rangle$$

for all f in L^p and g in $L^{p'}$. The *transpose* operator T^t of T is uniquely defined via the identity

$$\langle T(f), g\rangle = \langle f, T^t(g)\rangle = \langle T^t(g), f\rangle$$

for all functions f in L^p and g in $L^{p'}$. The name *transpose* comes from matrix theory, where if A^t denotes the transpose of a complex $n \times n$ matrix A, then we have the identity

$$\langle Ax, y\rangle = \sum_{j=1}^{n} (Ax)_j y_j = Ax \cdot y = x \cdot A^t y = \sum_{j=1}^{n} x_j (A^t y)_j = \langle x, A^t y\rangle$$

for all column vectors $x = (x_1, \dots, x_n)$, $y = (y_1, \dots, y_n)$ in $\mathbf{C}^n$. We may easily check the following intimate relationship between the transpose and the adjoint of a linear operator T:

$$T^*(f) = \overline{T^t(\overline{f}\,)},$$

indicating that they have almost interchangeable use. Because of this, in many cases, it is convenient to avoid complex conjugates and work with the transpose operator for simplicity. Observe that if a linear operator T has kernel $K(x,y)$, that is,

$$T(f)(x) = \int K(x,y) f(y)\,dy,$$

then the kernel of T^t is $K^t(x,y) = K(y,x)$ and that of T^* is $K^*(x,y) = \overline{K(y,x)}$.

An operator is called *self-adjoint* if $T = T^*$ and *self-transpose* if $T = T^t$. For example, the operator iH, where H is the Hilbert transform, is self-adjoint but not self-transpose, and the operator with kernel $i(x+y)^{-1}$ is self-transpose but not self-adjoint.

4.1.1 Standard Kernels

The singular integrals we study in this chapter have kernels that satisfy size and regularity properties similar to those of the classical convolution-type Calderón–

Zygmund operators. We introduce the relevant background. We consider functions $K(x,y)$ defined on $\mathbf{R}^n \times \mathbf{R}^n \setminus \{(x,x) : x \in \mathbf{R}^n\}$ that satisfy for some $A > 0$ the size condition

$$|K(x,y)| \leq \frac{A}{|x-y|^n} \tag{4.1.1}$$

and for some $\delta > 0$ the regularity conditions

$$|K(x,y) - K(x',y)| \leq \frac{A\,|x-x'|^\delta}{(|x-y|+|x'-y|)^{n+\delta}}, \tag{4.1.2}$$

whenever $|x-x'| \leq \frac{1}{2} \max\big(|x-y|, |x'-y|\big)$ and

$$|K(x,y) - K(x,y')| \leq \frac{A\,|y-y'|^\delta}{(|x-y|+|x-y'|)^{n+\delta}}, \tag{4.1.3}$$

whenever $|y-y'| \leq \frac{1}{2} \max\big(|x-y|, |x-y'|\big)$.

Remark 4.1.1. Observe that if

$$|x-x'| \leq \frac{1}{2} \max\big(|x-y|, |x'-y|\big),$$

then

$$\max\big(|x-y|, |x'-y|\big) \leq 2 \min\big(|x-y|, |x'-y|\big),$$

which implies that the numbers $|x-y|$ and $|x'-y|$ are comparable. Likewise if the roles of x and y are interchanged. These facts are useful in specific calculations.

Another important observation is that if (4.1.1) holds and we have

$$|\nabla_x K(x,y)| + |\nabla_y K(x,y)| \leq \frac{A}{|x-y|^{n+1}}$$

for all $x \neq y$, then K is in $SK(1, 4^{n+1}A)$.

Definition 4.1.2. Functions on $\mathbf{R}^n \times \mathbf{R}^n \setminus \{(x,x) : x \in \mathbf{R}^n\}$ that satisfy (4.1.1), (4.1.2), and (4.1.3) are called *standard kernels* with constants δ, A. The class of all standard kernels with constants δ, A is denoted by $SK(\delta, A)$. Given a kernel $K(x,y)$ in $SK(\delta, A)$, we observe that the functions $K(y,x)$ and $\overline{K(y,x)}$ are also in $SK(\delta, A)$. These kernels have special names. The function

$$K^t(x,y) = K(y,x)$$

is called the *transpose kernel* of K, while the function

$$K^*(x,y) = \overline{K(y,x)}$$

is called the *adjoint kernel* of K.

Example 4.1.3. The function $K(x,y) = |x-y|^{-n}$ defined away from the diagonal of $\mathbf{R}^n \times \mathbf{R}^n$ is in $SK(1, n4^{n+1})$. Indeed, for

$$|x-x'| \le \frac{1}{2}\max\big(|x-y|, |x'-y|\big)$$

the mean value theorem gives

$$\big|\,|x-y|^{-n} - |x'-y|^{-n}\big| \le \frac{n|x-x'|}{|\theta-y|^{n+1}}$$

for some θ that lies on the line segment joining x and x'. But then we have $|\theta-y| \ge \frac{1}{2}\max\big(|x-y|, |x'-y|\big)$, which gives (4.1.2) with $A = n4^{n+1}$.

Remark 4.1.4. Notice that if $K(x,y)$ satisfies

$$|\nabla_x K(x,y)| \le A'|x-y|^{-n-1}$$

for all $x \ne y$ in $\mathbf{R}^n$, then $K(x,y)$ also satisfies (4.1.2) with $\delta = 1$ and $A = cA'$, for some constant c. Likewise, if

$$|\nabla_y K(x,y)| \le A'|x-y|^{-n-1}$$

for all $x \ne y$ in $\mathbf{R}^n$, then $K(x,y)$ satisfies (4.1.3) with with $\delta = 1$ and $A = cA'$.

We are interested in standard kernels K in $SK(\delta, A)$ for which there are tempered distributions W on $\mathbf{R}^n \times \mathbf{R}^n$ that coincide with K on $\mathbf{R}^n \times \mathbf{R}^n \setminus \{(x,x) : x \in \mathbf{R}^n\}$. This means that the convergent integral representation

$$\langle W, F\rangle = \int_{\mathbf{R}^n}\int_{\mathbf{R}^n} K(x,y)F(x,y)\,dx\,dy \tag{4.1.4}$$

is valid whenever the Schwartz function F on $\mathbf{R}^n \times \mathbf{R}^n$ is supported away from the diagonal $\{(x,x) : x \in \mathbf{R}^n\}$. Note that the integral in (4.1.4) is well defined and absolutely convergent whenever F is a Schwartz function whose support does not intersect the set $\{(x,x) : x \in \mathbf{R}^n\}$. Also observe that there may be several distributions W coinciding with a fixed function $K(x,y)$. In fact, if W is such a distribution, then so is $W + \delta_{x=y}$, where $\delta_{x=y}$ denotes Lebesgue measure on the diagonal of $\mathbf{R}^{2n}$. (This is some kind of a Dirac distribution.)

We now consider continuous linear operators

$$T : \mathscr{S}(\mathbf{R}^n) \to \mathscr{S}'(\mathbf{R}^n)$$

from the space of Schwartz functions $\mathscr{S}(\mathbf{R}^n)$ to the space of all tempered distributions $\mathscr{S}'(\mathbf{R}^n)$. By the *Schwartz kernel theorem* ([196, p. 129]), for such an operator T there is a distribution W in $\mathscr{S}'(\mathbf{R}^{2n})$ that satisfies

$$\langle T(f), \varphi\rangle = \langle W, \varphi \otimes f\rangle \tag{4.1.5}$$

when $f,\varphi \in \mathscr{S}(\mathbf{R}^n)$, where $(\varphi \otimes f)(x,y) = \varphi(x)f(y)$ for all $x,y \in \mathbf{R}^n$, and there exist constants C,N,M such that for all $f,g \in \mathscr{S}(\mathbf{R}^n)$ we have

$$|\langle T(f),g\rangle| = |\langle W, g\otimes f\rangle| \le C\Big[\sum_{|\alpha|,|\beta|\le N} \rho_{\alpha,\beta}(g)\Big]\Big[\sum_{|\alpha|,|\beta|\le M} \rho_{\alpha,\beta}(f)\Big]. \tag{4.1.6}$$

Here $\rho_{\alpha,\beta}(\varphi) = \sup_{x\in\mathbf{R}^n} |\partial_x^\alpha (x^\beta \varphi)(x)|$ are the seminorms for the topology in $\mathscr{S}$. A distribution W that satisfies (4.1.5) and (4.1.6) is called a *Schwartz kernel* or the *distributional kernel* of T.

Here we study continuous linear operators $T : \mathscr{S}(\mathbf{R}^n) \to \mathscr{S}'(\mathbf{R}^n)$ whose distributional kernels coincide with standard kernels $K(x,y)$ on $\mathbf{R}^n \times \mathbf{R}^n \setminus \{(x,x) : x \in \mathbf{R}^n\}$. This means that (4.1.5) admits the absolutely convergent integral representation

$$\langle T(f),\varphi\rangle = \int_{\mathbf{R}^n}\int_{\mathbf{R}^n} K(x,y)f(y)\varphi(x)\,dx\,dy \tag{4.1.7}$$

whenever f and φ are Schwartz functions whose supports do not intersect.

We make some remarks concerning duality in this context. Given a continuous linear operator $T : \mathscr{S}(\mathbf{R}^n) \to \mathscr{S}'(\mathbf{R}^n)$ with distributional kernel W, we can define another distribution W^t as follows:

$$\langle W^t, F\rangle = \langle W, F^t\rangle,$$

where $F^t(x,y) = F(y,x)$. This implies that for all $f,\varphi \in \mathscr{S}(\mathbf{R}^n)$ we have

$$\langle W, \varphi\otimes f\rangle = \langle W^t, f\otimes\varphi\rangle.$$

It is a simple fact that the transpose operator T^t of T, which satisfies

$$\langle T(\varphi), f\rangle = \langle T^t(f),\varphi\rangle \tag{4.1.8}$$

for all f,φ in $\mathscr{S}(\mathbf{R}^n)$, is the unique continuous linear operator from $\mathscr{S}(\mathbf{R}^n)$ to $\mathscr{S}'(\mathbf{R}^n)$ whose Schwartz kernel is the distribution W^t, that is, we have

$$\langle T^t(f),\varphi\rangle = \langle T(\varphi),f\rangle = \langle W, f\otimes\varphi\rangle = \langle W^t, \varphi\otimes f\rangle. \tag{4.1.9}$$

We now observe that a large class of standard kernels admits extensions to tempered distributions W on $\mathbf{R}^{2n}$.

Example 4.1.5. Suppose that $K(x,y)$ satisfies (4.1.1) and (4.1.2) and is *antisymmetric*, in the sense that

$$K(x,y) = -K(y,x)$$

for all $x \neq y$ in $\mathbf{R}^n$. Then K also satisfies (4.1.3), and moreover, there is a distribution W on $\mathbf{R}^{2n}$ that extends K on $\mathbf{R}^n \times \mathbf{R}^n$.

Indeed, define

$$\langle W,F\rangle = \lim_{\varepsilon\to 0} \iint_{|x-y|>\varepsilon} K(x,y)F(x,y)\,dy\,dx \tag{4.1.10}$$

for all F in the Schwartz class of $\mathbf{R}^{2n}$. In view of antisymmetry, we may write

$$\iint_{|x-y|>\varepsilon} K(x,y)F(x,y)\,dy\,dx = \frac{1}{2}\iint_{|x-y|>\varepsilon} K(x,y)\big(F(x,y)-F(y,x)\big)\,dy\,dx.$$

In view of (4.1.1), the observation that

$$|F(x,y)-F(y,x)| \le \frac{2|x-y|}{(1+|x|^2+|y|^2)^{n+1}} \sup_{(x,y)\in\mathbf{R}^{2n}} \left|\nabla_{x,y}\Big((1+|x|^2+|y|^2)^{n+1}F(x,y)\Big)\right|,$$

and the fact that the preceding supremum is controlled by a finite sum of Schwartz seminorms of F, the limit in (4.1.10) exists and gives a tempered distribution on $\mathbf{R}^{2n}$. We can therefore define an operator $T:\mathscr{S}(\mathbf{R}^n)\to\mathscr{S}'(\mathbf{R}^n)$ with kernel W via

$$\begin{aligned}\langle T(f),\varphi\rangle &= \lim_{\varepsilon\to 0}\iint_{|x-y|>\varepsilon} K(x,y)f(y)\varphi(x)\,dy\,dx \\ &= \frac{1}{2}\int_{\mathbf{R}^n}\int_{\mathbf{R}^n} K(x,y)[f(y)\varphi(x)-f(x)\varphi(y)]\,dy\,dx,\end{aligned} \tag{4.1.11}$$

for all $f,\varphi\in\mathscr{S}(\mathbf{R}^n)$.

Example 4.1.6. Let A be a real-valued Lipschitz function on $\mathbf{R}$. This means that it satisfies the estimate $|A(x)-A(y)|\le L|x-y|$ for some $L<\infty$ and all $x,y\in\mathbf{R}$. For $x,y\in\mathbf{R}$, $x\neq y$, we let

$$K_A(x,y) = \frac{1}{x-y+i(A(x)-A(y))}. \tag{4.1.12}$$

A simple calculation gives that when $|y-y'|\le\frac{1}{2}\max\big(|x-y|,|x-y'|\big)$ then

$$|K_A(x,y)-K_A(x,y')| \le \frac{|y-y'|+|A(y)-A(y')|}{|x-y||x-y'|} \le \frac{(1+L)|y-y'|}{\frac{1}{8}(|x-y|+|x-y'|)}$$

where the last inequality uses the observation in Remark 4.1.1. Since K_A is antisymmetric, it follows that it is a standard kernel in $SK(1,8(1+L))$.

Example 4.1.7. Let the function A be as in the previous example. For each integer $m\ge 1$ and $x,y\in\mathbf{R}$ we set

$$K_m(x,y) = \left(\frac{A(x)-A(y)}{x-y}\right)^m \frac{1}{x-y}. \tag{4.1.13}$$

Clearly, K_m is an antisymmetric function. To see that each K_m is a standard kernel, notice that when $|y-y'| \le \frac{1}{2}\max\big(|x-y|,|x-y'|\big)$ we have

$$\begin{aligned}\left|\frac{A(x)-A(y)}{x-y}-\frac{A(x)-A(y')}{x-y'}\right| &= \left|\frac{(x-y)(A(y')-A(y))+(y-y')(A(x)-A(y))}{(x-y)(x-y')}\right| \\ &\le 2L\frac{|y-y'|}{|x-y'|}.\end{aligned}$$

Combining this fact with $|a^m-b^m| \le |a-b|(|a|^{m-1}+|a|^{m-2}|b|+\cdots+|b|^{m-1})$ we obtain

$$\begin{aligned}&|K_m(x,y)-K_m(x,y')| \\ &\le \left|\left(\frac{A(x)-A(y)}{x-y}\right)^m-\left(\frac{A(x)-A(y')}{x-y'}\right)^m\right|\frac{1}{|x-y|}+\left|\frac{A(x)-A(y')}{x-y'}\right|^m\left|\frac{1}{x-y}-\frac{1}{x-y'}\right| \\ &\le \frac{2L|y-y'|}{|x-y'|}mL^{m-1}\frac{1}{|x-y|}+L^m\frac{|y-y'|}{|x-y|\,|x-y'|} \\ &= \frac{(2m+1)L^m|y-y'|}{|x-y|\,|x-y'|} \\ &\le \frac{8(2m+1)L^m|y-y'|}{|x-y|+|x-y'|}.\end{aligned}$$

It follows that K_m lies in $SK(\delta,C)$ with $\delta=1$ and $C=8(2m+1)L^m$. The linear operator with kernel $(\pi i)^{-1}K_m$ is called the mth *Calderón commutator*.

4.1.2 Operators Associated with Standard Kernels

Having introduced standard kernels, we are in a position to define linear operators associated with them.

Definition 4.1.8. Let $0<\delta,A<\infty$, and K in $SK(\delta,A)$. A continuous linear operator T from $\mathscr{S}(\mathbf{R}^n)$ to $\mathscr{S}'(\mathbf{R}^n)$ is said to be *associated with K* if it satisfies

$$T(f)(x)=\int_{\mathbf{R}^n}K(x,y)f(y)\,dy \tag{4.1.14}$$

for all $f\in\mathscr{C}_0^\infty$ and x not in the support of f. If T is associated with K, then the Schwartz kernel W of T coincides with K on $\mathbf{R}^n\times\mathbf{R}^n\setminus\{(x,x):\,x\in\mathbf{R}^n\}$.

If T is associated with K and satisfies

$$\big\|T(\varphi)\big\|_{L^2}\le B\,\|\varphi\|_{L^2} \tag{4.1.15}$$

for all $\varphi\in\mathscr{S}(\mathbf{R}^n)$, then T is called a *Calderón–Zygmund operator* associated with the standard kernel K. Such operators T admit a bounded extension on $L^2(\mathbf{R}^n)$, i.e., given any f in $L^2(\mathbf{R}^n)$ one can define $T(f)$ as the unique L^2 limit of the Cauchy sequence $\{T(\varphi_k)\}_k$, where $\varphi_k \in \mathscr{S}(\mathbf{R}^n)$ and φ_k converges to f in L^2. In this case we keep the same notation for the L^2 extension of T.

In the sequel we denote by $CZO(\delta,A,B)$ the class of all Calderón–Zygmund operators associated with standard kernels in $SK(\delta,A)$ that admit L^2–bounded extensions with norm at most B.

We make the point that there may be several Calderón–Zygmund operators associated with a given standard kernel K. For instance, we may check that the zero operator and the identity operator have the same kernel $K(x,y)=0$. We investigate connections between any two such operators in Proposition 4.1.11. Next we discuss the important fact that once an operator T admits an extension that is L^2 bounded, then (4.1.14) holds for all f that are bounded and compactly supported whenever the point x does not lie in its support.

Proposition 4.1.9. *Let T be an element of $CZO(\delta,A,B)$ associated with a standard kernel K. Then for every f and φ bounded and compactly supported functions that satisfy*

$$\operatorname{dist}(\operatorname{supp}\varphi,\operatorname{supp}f)>0, \tag{4.1.16}$$

then we have the (absolutely convergent) integral representation

$$\int_{\mathbf{R}^n} T(f)(x)\,\varphi(x)\,dx=\int_{\mathbf{R}^n}\int_{\mathbf{R}^n} K(x,y)f(y)\varphi(x)\,dy\,dx. \tag{4.1.17}$$

Moreover, given any bounded function with compact support f, there is a set of measure zero $E(f)$ such that $x_0 \notin E(f)\cap\operatorname{supp}f$ we have the (absolutely convergent) integral representation

$$T(f)(x_0)=\int_{\mathbf{R}^n} K(x_0,y)f(y)\,dy. \tag{4.1.18}$$

Proof. We first prove (4.1.17). Given f and φ bounded functions with compact support select $f_j,\varphi_j\in\mathscr{C}_0^\infty$ such that φ_j are uniformly bounded and supported in a small neighborhood of the support of φ, $\varphi_j\to\varphi$ in L^2 and almost everywhere, $f_j\to f$ in L^2 and almost everywhere, and

$$\operatorname{dist}(\operatorname{supp}\varphi_j,\operatorname{supp}f_j)\geq\frac{1}{2}\operatorname{dist}(\operatorname{supp}\varphi,\operatorname{supp}f)=c>0$$

for all $j\in\mathbf{Z}^+$. In view of (4.1.7), identity (4.1.17) is valid for the functions f_j and φ_j in place of f and φ, i.e.,

$$\int_{\mathbf{R}^n}\int_{\mathbf{R}^n} K(x,y)f_j(y)\varphi_j(x)\,dy\,dx=\int_{\mathbf{R}^n} T(f_j)(x)\varphi_j(x)\,dx. \tag{4.1.19}$$

By the boundedness of T, it follows that $T(f_j)$ converges to $T(f)$ in L^2 and thus as $j \to \infty$ we have

$$\int_{\mathbf{R}^n} T(f_j)(x)\varphi_j(x)\,dx \to \int_{\mathbf{R}^n} T(f)(x)\varphi(x)\,dx. \tag{4.1.20}$$

Now write $f_j(y)\varphi_j(x) - f(y)\varphi(x) = (f_j(y) - f(y))\varphi_j(x) + f(y)(\varphi_j(x) - \varphi(x))$ and observe that

$$\left| \int_{\mathbf{R}^n}\int_{\mathbf{R}^n} K(x,y)f(y)(\varphi_j(x) - \varphi(x))\,dy\,dx \right| \le Ac^{-n}\|f\|_{L^1}\|\varphi_j - \varphi\|_{L^1} \to 0,$$

since $\|\varphi_j - \varphi\|_{L^1} \le C\|\varphi_j - \varphi\|_{L^2} \to 0$ as $j \to \infty$, and

$$\left| \int_{\mathbf{R}^n}\int_{\mathbf{R}^n} K(x,y)(f_j(y) - f(y))\varphi_j(x)\,dy\,dx \right| \le Ac^{-n}\|f_j - f\|_{L^1}\|\varphi\|_{L^1} \to 0,$$

as $j \to \infty$. Combining these facts with (4.1.19) and (4.1.20) we obtain

$$\int_{\mathbf{R}^n}\int_{\mathbf{R}^n} K(x,y)f_j(y)\varphi_j(x)\,dy\,dx \to \int_{\mathbf{R}^n}\int_{\mathbf{R}^n} K(x,y)f(y)\varphi(x)\,dy\,dx$$

as $j \to \infty$ and proves the validity of (4.1.17). Note that the double integral on the right is absolutely convergent and bounded by $A\,(2c)^{-n}\|f\|_{L^1}\|\varphi\|_{L^1}$.

To prove (4.1.18) we fix a compactly supported and bounded function f and we pick f_j as before. Then $T(f_j)$ converges to $T(f)$ in L^2 and thus a subsequence $T(f_{j_l})$ converges pointwise on $\mathbf{R}^n \setminus E(f)$, for some measurable set $E(f)$ with $|E(f)| = 0$. Given $x_0 \notin E(f) \cap \operatorname{supp} f$ we have

$$T(f_{j_l})(x_0) = \int_{\mathbf{R}^n} K(x_0,y)f_{j_l}(y)\,dy$$

and letting $l \to \infty$ we obtain (4.1.18) since $T(f_{j_l})(x_0) \to T(f)(x_0)$ and

$$\left| \int_{\mathbf{R}^n} K(x_0,y)f_{j_l}(y)\,dy - \int_{\mathbf{R}^n} K(x_0,y)f(y)\,dy \right| \le Ac^{-n}\|f_{j_l} - f\|_{L^1} \to 0.$$

as $l \to \infty$. Thus (4.1.18) holds. □

We now define truncated kernels and operators.

Definition 4.1.10. Given a kernel K in $SK(\delta,A)$ and $\varepsilon > 0$, we define the *truncated kernel*

$$K^{(\varepsilon)}(x,y) = K(x,y)\chi_{|x-y|>\varepsilon}\,.$$

Given a continuous linear operator T from $\mathscr{S}(\mathbf{R}^n)$ to $\mathscr{S}'(\mathbf{R}^n)$ and $\varepsilon > 0$, we define the *truncated operator* $T^{(\varepsilon)}$ by

$$T^{(\varepsilon)}(f)(x) = \int_{\mathbf{R}^n} K^{(\varepsilon)}(x,y)\,f(y)\,dy$$

and the *maximal singular operator* associated with T as follows:

$$T^{(*)}(f)(x) = \sup_{\varepsilon>0} \left| T^{(\varepsilon)}(f)(x) \right|.$$

Note that both $T^{(\varepsilon)}(f)$ and $T^{(*)}(f)$ are well defined for f in $\bigcup_{1\le p<\infty} L^p(\mathbf{R}^n)$, by an application of Hölder's inequality.

We investigate a certain connection between the boundedness of T and the boundedness of the family $\{T^{(\varepsilon)}\}_{\varepsilon>0}$ uniformly in $\varepsilon > 0$.

Proposition 4.1.11. *Let K be a kernel in $SK(\delta,A)$ and let T in $CZO(\delta,A,B)$ be associated with K. For $\varepsilon > 0$, let $T^{(\varepsilon)}$ be the truncated operators obtained from T. Assume that there exists a constant $B' < \infty$ such that*

$$\sup_{\varepsilon>0} \left\| T^{(\varepsilon)} \right\|_{L^2\to L^2} \le B'. \tag{4.1.21}$$

Then there exists a linear operator T_0 defined on $L^2(\mathbf{R}^n)$ such that

(1) The distributional kernel of T_0 coincides with K on

$$\mathbf{R}^n \times \mathbf{R}^n \setminus \{(x,x) : x \in \mathbf{R}^n\}.$$

(2) For some subsequence $\varepsilon_j \downarrow 0$, we have

$$\int_{\mathbf{R}^n} T^{(\varepsilon_j)}(f)(x)g(x)\,dx \to \int_{\mathbf{R}^n} T_0(f)(x)\,g(x)\,dx \tag{4.1.22}$$

as $j \to \infty$ for all f,g in $L^2(\mathbf{R}^n)$.
(3) T_0 is bounded on $L^2(\mathbf{R}^n)$ with norm

$$\left\| T_0 \right\|_{L^2\to L^2} \le B'.$$

(4) There exists a measurable function b on $\mathbf{R}^n$ with $\|b\|_{L^\infty} \le B + B'$ such that

$$T(f) - T_0(f) = bf,$$

for all $f \in L^2(\mathbf{R}^n)$.

Proof. Since $L^2(\mathbf{R}^n)$ is separable, by the Banach-Alaoglu theorem the unit ball of its dual is weak* compact and metrizable for the weak* topology. Let $\{f_k\}_{k=1}^\infty$ be a dense countable subset of $L^2(\mathbf{R}^n)$. By (4.1.21), the functions $T^{(\varepsilon)}(f_k)$ lie in multiple of the unit ball of $(L^2)^*$, which is weak* compact, and hence for each f_k we find a sequence $\{\varepsilon_j^k\}_{j=1}^\infty$ such that for each $g \in L^2(\mathbf{R}^n)$ we have

$$\lim_{j\to\infty} \int_{\mathbf{R}^n} T^{(\varepsilon_j^k)}(f_k)(x)g(x)\,dx = \int_{\mathbf{R}^n} T_0^{f_k}(x)g(x)\,dx, \tag{4.1.23}$$

for some function $T_0^{f_k}$ in $L^2(\mathbf{R}^n)$. Moreover, each $\{\varepsilon_j^k\}_{j=1}^\infty$ can be chosen to be a subsequence of $\{\varepsilon_j^{k-1}\}_{j=1}^\infty$, $k \ge 2$. Then the diagonal sequence $\{\varepsilon_j^j\}_{j=1}^\infty = \{\varepsilon_j\}_{j=1}^\infty$ satisfies

$$\lim_{j\to\infty} \int_{\mathbf{R}^n} T^{(\varepsilon_j)}(f_k)(x)g(x)\,dx = \int_{\mathbf{R}^n} T_0^{f_k}(x)g(x)\,dx \tag{4.1.24}$$

for each k and $g \in L^2$. Since $\{f_k\}_{k=1}^\infty$ is dense in $L^2(\mathbf{R}^n)$, a standard $\varepsilon/3$ argument gives that the sequence of complex numbers

$$\int_{\mathbf{R}^n} T^{(\varepsilon_j)}(f_k)(x)g(x)\,dx$$

is Cauchy and thus it converges. Now L^2 is complete[1] in the weak* topology; therefore for each $f \in L^2(\mathbf{R}^n)$ there is a function $T_0(f)$ such that (4.1.22) holds for all f,g in $L^2(\mathbf{R}^n)$ as $j \to \infty$. It is easy to see that T_0 is a linear operator with the property $T_0(f_k) = T_0^{f_k}$ for each $k = 1,2,\dots$. This proves (2).

The L^2 boundedness of T_0 is a consequence of (4.1.22), (4.1.21), and duality, since

$$\big\|T_0(f)\big\|_{L^2} \le \sup_{\|g\|_{L^2}\le 1} \limsup_{j\to\infty} \left| \int_{\mathbf{R}^n} T^{(\varepsilon_j)}(f)(x)g(x)\,dx \right| \le B'\big\|f\big\|_{L^2}.$$

This proves (3). Finally, (1) is a consequence of the integral representation

$$\int_{\mathbf{R}^n} T^{(\varepsilon_j)}(f)(x)g(x)\,dx = \int_{\mathbf{R}^n}\int_{\mathbf{R}^n} K^{(\varepsilon_j)}(x,y)f(y)\,dy\,g(x)\,dx,$$

whenever f, g are Schwartz functions with disjoint supports, by letting $j \to \infty$.

We finally prove (4). We first observe that if g is a bounded function with compact support and Q is an open cube in $\mathbf{R}^n$, we have

$$(T^{(\varepsilon_j)} - T)(g\chi_Q)(x) = \chi_Q(x)\,(T^{(\varepsilon_j)} - T)(g)(x), \tag{4.1.25}$$

for almost all $x \notin \partial Q$ whenever ε_j is small enough (depending on x). Indeed, since $g\chi_Q$ is bounded and has compact support, by the integral representation formula (4.1.18) in Proposition 4.1.9 there is a null set $E(g\chi_Q)$ such that for $x \notin \overline{Q} \cap E(g\chi_Q)$ and for $\varepsilon_j < \operatorname{dist}(x, \operatorname{supp} g\chi_Q)$, the left-hand side in (4.1.25) is zero, since in this case x is not in the support of $g\chi_Q$. Moreover, since $g\chi_{Q^c}$ is also bounded and compactly supported, there is a null set $E(g\chi_{Q^c})$ such that for $x \in Q \cap E(g\chi_{Q^c})$ and $\varepsilon_j < \operatorname{dist}(x, \operatorname{supp} g\chi_{Q^c})$ we have that x does not lie in the support of $g\chi_{Q^c}$, and thus $(T^{(\varepsilon_j)} - T)(g\chi_{Q^c})(x) = 0$; hence (4.1.25) holds in this case as well. This proves (4.1.25) for almost x not in the boundary ∂Q. Taking weak limits in (4.1.25) as $\varepsilon_j \to 0$, we obtain that

$$(T_0 - T)(g\chi_Q) = \chi_Q\,(T_0 - T)(g) \qquad \text{a.e.} \tag{4.1.26}$$

[1] the unit ball of L^2 in the weak* topology is compact and metrizable, hence complete.

for all open cubes Q in $\mathbf{R}^n$. This means that for any g bounded function with compact support and cube Q in $\mathbf{R}^n$ there is a set of measure zero $E_{Q,g}$ such that (4.1.26) holds on $\mathbf{R}^n \setminus E_{Q,g}$. Consider the countable family $\mathscr{F}$ of all cubes in $\mathbf{R}^n$ with corners in $\mathbf{Q}^n$ and set $E_g = \cup_{Q\in\mathscr{F}} E_{Q,g}$. Then $|E_g| = 0$ and by linearity we obtain

$$(T_0 - T)(gh) = h(T_0 - T)(g) \qquad \text{on } \mathbf{R}^n \setminus E_g$$

whenever h is a finite linear combination of characteristic functions of cubes in $\mathscr{F}$, which is a dense subspace of L^2. Via a simple density argument, using the fact that $T_0 - T$ is L^2 bounded, we obtain that for all f in L^2 and g bounded with compact support there is a null set $E_{f,g}$ such that

$$(T_0 - T)(gf) = f(T_0 - T)(g) \qquad \text{on } \mathbf{R}^n \setminus E_{f,g}. \tag{4.1.27}$$

Now if $B(0,j)$ is the open ball with center 0 and radius j, when $j \le j'$ we have

$$(T_0 - T)(\chi_{B(0,j)}) = (T_0 - T)(\chi_{B(0,j)}\chi_{B(0,j')}) = \chi_{B(0,j)}\,(T_0 - T)(\chi_{B(0,j')}) \ \text{ a.e.}$$

Therefore, the functions $(T_0 - T)(\chi_{B(0,j)})$ satisfy the "consistency" property

$$(T_0 - T)(\chi_{B(0,j)}) = (T_0 - T)(\chi_{B(0,j')}) \quad \text{a.e. on } B(0,j)$$

when $j \le j'$. It follows that there exists a well-defined measurable function b such that

$$b\chi_{B(0,j)} = (T_0 - T)(\chi_{B(0,j)}) \quad \text{a.e.}$$

Applying (4.1.27) with $f \in L^2$ and $g = \chi_{B(0,j)}$, we obtain

$$(T_0 - T)(f\chi_{B(0,j)}) = f(T_0 - T)(\chi_{B(0,j)}) = fb \quad \text{a.e. on } B(0,j). \tag{4.1.28}$$

Since the norm of $T - T_0$ on L^2 is at most $B + B'$, we obtain from (4.1.28) that

$$B + B' \ge \sup_{j\ge 1} \sup_{\substack{0\neq f\in L^2 \\ \text{supp } f\subseteq B(0,j)}} \frac{\|(T_0 - T)(f\chi_{B(0,j)})\|_{L^2}}{\|f\|_{L^2}} = \sup_{\substack{0\neq f\in L^2 \\ \text{supp } f \text{ compact}}} \frac{\|fb\|_{L^2}}{\|f\|_{L^2}} = \|b\|_{L^\infty}.$$

The fact that $b \in L^\infty$ together with (4.1.28) easily yields

$$(T_0 - T)(f) = bf \quad \text{a.e.}$$

for all $f \in L^2$. This identifies $T_0 - T$ and concludes the proof of (4). □

Remark 4.1.12. We show in the next section (cf. Corollary 4.2.5) that if a Calderón–Zygmund operator maps L^2 to L^2, then so do all of its truncations $T^{(\varepsilon)}$ uniformly in $\varepsilon > 0$. By Proposition 4.1.11, there exists a linear operator T_0 that has the form

$$T_0(f)(x) = \lim_{j\to\infty} \int_{|x-y|>\varepsilon_j} K(x,y)f(y)\,dy,$$

where the limit is taken in the weak topology of L^2, so that T is equal to T_0 plus a bounded function times the identity operator.

We give a special name to operators of this form.

Definition 4.1.13. Suppose that for a given T in $CZO(\delta,A,B)$ there is a sequence ε_j of positive numbers that tends to zero as $j\to\infty$ such that for all $f\in L^2(\mathbf{R}^n)$,

$$T^{(\varepsilon_j)}(f)\to T(f)$$

weakly in L^2. Then T is called a *Calderón–Zygmund singular integral operator.* Thus Calderón–Zygmund singular integral operators are special kinds of Calderón–Zygmund operators. The subclass of $CZO(\delta,A,B)$ consisting of all Calderón–Zygmund singular integral operators is denoted by $CZSIO(\delta,A,B)$.

In view of Proposition 4.1.11 and Remark 4.1.12, a Calderón–Zygmund operator is equal to a Calderón–Zygmund singular integral operator plus a bounded function times the identity operator. For this reason, the study of Calderón–Zygmund operators is equivalent to the study of Calderón–Zygmund singular integral operators, and in almost all situations it suffices to restrict attention to the latter.

4.1.3 Calderón–Zygmund Operators Acting on Bounded Functions

We are now interested in defining the action of a Calderón–Zygmund operator T on bounded and smooth functions. To achieve this we first need to define the space of special test functions $\mathscr{D}_0$.

Definition 4.1.14. We denote by $\mathscr{D}(\mathbf{R}^n)=\mathscr{C}_0^\infty(\mathbf{R}^n)$ the space of all smooth functions with compact support on $\mathbf{R}^n$. We define $\mathscr{D}_0(\mathbf{R}^n)$ to be the space of all smooth functions with compact support and integral zero. We equip $\mathscr{D}_0(\mathbf{R}^n)$ with the same topology as the space $\mathscr{D}(\mathbf{R}^n)$. This means that a linear functional $u\in\mathscr{D}_0(\mathbf{R}^n)$ is continuous if for any compact set K in $\mathbf{R}^n$ there is a constant C_K and an integer M such that

$$\big|\langle u,\varphi\rangle\big|\le C_K\sum_{|\alpha|\le M}\|\partial^\alpha\varphi\|_{L^\infty}$$

for all φ smooth functions supported in K. The dual space of $\mathscr{D}_0(\mathbf{R}^n)$ under this topology is denoted by $\mathscr{D}_0'(\mathbf{R}^n)$. This is a space of distributions larger than $\mathscr{D}'(\mathbf{R}^n)$.

Example 4.1.15. *BMO* functions are examples of elements of $\mathscr{D}_0'(\mathbf{R}^n)$. Indeed, given $b\in BMO(\mathbf{R}^n)$, for any compact set K there is a constant $C_K=\|b\|_{L^1(K)}$ such that

$$\left|\int_{\mathbf{R}^n}b(x)\varphi(x)\,dx\right|\le C_K\big\|\varphi\big\|_{L^\infty}$$

for any $\varphi\in\mathscr{D}_0(\mathbf{R}^n)$. Moreover, observe that the preceding integral remains unchanged if the *BMO* function b is replaced by $b+c$, where c is a constant.

Definition 4.1.16. Let T be a continuous linear operator from $\mathscr{S}(\mathbf{R}^n)$ to $\mathscr{S}'(\mathbf{R}^n)$ that satisfies (4.1.5) for some distribution W that coincides with a standard kernel $K(x,y)$ satisfying (4.1.1), (4.1.2), and (4.1.3). Given f bounded and smooth, we define an element $T(f)$ of $\mathscr{D}_0'(\mathbf{R}^n)$ as follows: For a given φ in $\mathscr{D}_0(\mathbf{R}^n)$, select η in $\mathscr{C}_0^\infty$ with $0 \le \eta \le 1$ and equal to 1 in a neighborhood of the support of φ. Since T maps $\mathscr{S}$ to $\mathscr{S}'$, the expression $T(f\eta)$ is a tempered distribution, and its action on φ is well defined. We define the action of $T(f)$ on φ via the identity

$$\langle T(f),\varphi\rangle = \langle T(f\eta),\varphi\rangle + \int_{\mathbf{R}^n}\left[\int_{\mathbf{R}^n} K(x,y)\varphi(x)\,dx\right] f(y)(1-\eta(y))\,dy, \quad (4.1.29)$$

provided we make sense of the double integral as an absolutely convergent integral. To do this, we pick x_0 in the support of φ and we split the y-integral in (4.1.29) into the sum of integrals over the regions $I_0 = \{y \in \mathbf{R}^n : |x-x_0| > \frac{1}{2}|x_0-y|\}$ and $I_\infty = \{y \in \mathbf{R}^n : |x-x_0| \le \frac{1}{2}|x_0-y|\}$. By the choice of η we must necessarily have $\operatorname{dist}(\operatorname{supp}(1-\eta), \operatorname{supp}\varphi) > 0$, and hence the part of the double integral in (4.1.29) when y is restricted to I_0 is absolutely convergent in view of (4.1.1). For $y \in I_\infty$ we use the mean value property of φ to write the expression inside the square brackets in (4.1.29) as

$$\int_{\mathbf{R}^n} \big(K(x,y) - K(x_0,y)\big)\varphi(x)\,dx.$$

With the aid of (4.1.2) we deduce the absolute convergence of the double integral in (4.1.29) as follows:

$$\begin{aligned}
&\iint_{|y-x_0|\ge 2|x-x_0|} |K(x,y)-K(x_0,y)|\,|\varphi(x)|\,(1-\eta(y))\,|f(y)|\,dx\,dy \\
&\qquad \le \int_{\mathbf{R}^n} A|x-x_0|^\delta \int_{|y-x_0|\ge 2|x-x_0|} |x_0-y|^{-n-\delta}\,|f(y)|\,dy\,|\varphi(x)|\,dx \\
&\qquad \le A\,\frac{\omega_{n-1}}{\delta 2^\delta}\,\|\varphi\|_{L^1}\,\|f\|_{L^\infty} < \infty .
\end{aligned}$$

This completes the definition of $T(f)$ as an element of $\mathscr{D}_0'$ when $f \in \mathscr{C}^\infty \cap L^\infty$, and certainly (4.1.29) is independent of x_0, but leaves two points open. First, we need to show that this definition is independent of η and secondly that whenever f is a Schwartz function, the distribution $T(f)$ defined in (4.1.29) coincides with the original element of $\mathscr{S}'(\mathbf{R}^n)$ given in Definition 4.1.8.

Remark 4.1.17. We show that the definition of $T(f)$ is independent of the choice of the function η. Indeed, if ζ is another function satisfying $0 \le \zeta \le 1$ that is also equal to 1 in a neighborhood of the support of φ, then $f(\eta-\zeta)$ and φ have disjoint supports, and by (4.1.7) we have the absolutely convergent integral realization

$$\langle T(f(\eta-\zeta)),\varphi\rangle = \int_{\mathbf{R}^n}\int_{\mathbf{R}^n} K(x,y)f(y)(\eta-\zeta)(y)\,dy\,\varphi(x)\,dx.$$

It follows that the expression in (4.1.29) coincides with the corresponding expression obtained when η is replaced by ζ.

Next, if f is a Schwartz function, then both ηf and $(1-\eta)f$ are Schwartz functions; by the linearity of T one has $\langle T(f),\varphi\rangle = \langle T(\eta f),\varphi\rangle + \langle T((1-\eta)f),\varphi\rangle$, and by (4.1.7) the second expression can be written as the double absolutely convergent integral in (4.1.29), since φ and $(1-\eta)f$ have disjoint supports. Thus the distribution $T(f)$ defined in (4.1.29) coincides with the original element of $\mathscr{S}'(\mathbf{R}^n)$ given in Definition 4.1.8.

Remark 4.1.18. When T has a bounded extension that maps L^2 to itself, we may define $T(f)$ for all $f \in L^\infty(\mathbf{R}^n)$, not necessarily smooth. Simply observe that under this assumption, the expression $T(f\eta)$ is a well-defined L^2 function and thus

$$\langle T(f\eta),\varphi\rangle = \int_{\mathbf{R}^n} T(f\eta)(x)\varphi(x)\,dx$$

is given by an absolutely convergent integral for all $\varphi \in \mathscr{D}_0$.

Finally, observe that although $\langle T(f),\varphi\rangle$ is defined for f in L^∞ and φ in $\mathscr{D}_0$, this definition is valid for all square integrable functions φ with compact support and integral zero; indeed, the smoothness of φ was never an issue in the definition of $\langle T(f),\varphi\rangle$.

In summary, if T is a Calderón–Zygmund operator and f lies in $L^\infty(\mathbf{R}^n)$, then $T(f)$ has a well-defined *action* $\langle T(f),\varphi\rangle$ on square integrable functions φ with compact support and integral zero. This action satisfies

$$\big|\langle T(f),\varphi\rangle\big| \le \big\|T(f\eta)\big\|_{L^2}\big\|\varphi\big\|_{L^2} + C_{n,\delta}A\,\big\|\varphi\big\|_{L^1}\big\|f\big\|_{L^\infty} < \infty. \tag{4.1.30}$$

In the next section we show that in this case, $T(f)$ is in fact an element of *BMO*.

Exercises

4.1.1. Suppose that K is a function defined away from the diagonal on $\mathbf{R}^n \times \mathbf{R}^n$ that satisfies for some $\delta > 0$ the condition

$$|K(x,y) - K(x',y)| \le A' \frac{|x-x'|^\delta}{|x-y|^{n+\delta}}$$

whenever $|x-x'| \le \frac{1}{2}|x-y|$. Prove that K satisfies (4.1.2) with constant $A = 2^{n+\delta}A'$. Obtain an analogous statement for condition (4.1.3).

4.1.2. (a) Show that the tempered distribution W on $\mathbf{R}^2$ defined for $F \in \mathscr{S}(\mathbf{R}^2)$ by

$$\langle W,F\rangle = -\int_{\mathbf{R}}\int_{\mathbf{R}} \left(\frac{\partial^2}{\partial t^2}F(y+t,y)\right)\log|t|\,dt\,dy$$

coincides with the function $|x-y|^{-2}$ on $\mathbf{R}^2 \setminus \{(x,x):\, x \in \mathbf{R}\}$.

(b) Show that the tempered distribution W_β defined for $\varphi \in \mathscr{S}(\mathbf{R})$ by

$$\langle W_\beta, \varphi \rangle = -\frac{1}{\beta!}\int_{\mathbf{R}} \varphi^{(\beta+1)}(x)\log|x|(\operatorname{sgn} x)^{\beta+1}\,dx$$

when $\beta \in \mathbf{Z}^+ \cup \{0\}$ coincides with the function $|x|^{-1-\beta}$ on $\mathbf{R}\setminus\{0\}$.
(c) Show that the tempered distribution W_β defined for $\varphi \in \mathscr{S}(\mathbf{R})$ by

$$\langle W_\beta, \varphi \rangle = \frac{1}{\beta(\beta-1)\cdots(\beta-[\beta])}\int_{\mathbf{R}} \varphi^{([\beta]+1)}(x)|x|^{-\beta+[\beta]}(\operatorname{sgn} x)^{[\beta]+1}\,dx$$

when $\beta \in \mathbf{R}^+ \setminus \mathbf{Z}^+$ coincides with the function $|x|^{-1-\beta}$ on $\mathbf{R}\setminus\{0\}$.

4.1.3. Let $\varphi(x)$ be a smooth radial function that is equal to 1 when $|x| \ge 1$ and vanishes when $|x| \le \frac{1}{2}$. Let $0 < \delta \le 1$. Show that there is a constant $c > 0$ that depends only on n, φ, and δ such that if K lies in $SK(\delta, A)$, then all the smooth truncations $K_\varphi^{(\varepsilon)}(x,y) = K(x,y)\varphi(\frac{x-y}{\varepsilon})$ lie in $SK(\delta, cA)$ uniformly in $\varepsilon > 0$.

4.1.4. Suppose that A is a Lipschitz map from $\mathbf{R}^n$ to $\mathbf{R}^m$. This means that there exists a constant L such that $|A(x) - A(y)| \le L|x-y|$ for all $x, y \in \mathbf{R}^n$. Suppose that F is a $\mathscr{C}^1$ odd function defined on $\mathbf{R}^m$. Show that the kernel

$$K(x,y) = \frac{1}{|x-y|^n}F\left(\frac{A(x)-A(y)}{|x-y|}\right)$$

is in $SK(1, C)$ for some $C > 0$.

4.1.5. Let T be an operator as in Definition 4.1.16, and assume that T admits a unique bounded extension from $L^2(\mathbf{R}^n)$ to itself.
(a) Prove that for every smooth compactly supported function φ with integral zero, we have that $T(\varphi)$ and $T^t(\varphi)$ are integrable.
(b) Show that the condition $T(1) = 0$ is equivalent to the statement that for all φ smooth with compact support and integral zero we have $\int_{\mathbf{R}^n} T^t(\varphi)(x)\,dx = 0$. Formulate an analogous statement for T^t.

4.1.6. Let V be a compact subset of $\mathbf{R}^n$. Suppose that $K(x,y)$ is continuous, bounded, and strictly positive function on $V \times V$ that satisfies $\int_V K(x,y)\,dy = 1$ for all $x \in V$. Define a linear operator by setting $T(f)(x) = \int_V K(x,y)f(y)\,dy$ for $f \in L^1(V)$.
(a) Show that T preserves the set of integrable functions that are bounded below by a fixed constant.
(b) Suppose that h is a continuous function on V that satisfies $T(h)(x) = h(x)$ for all $x \in V$. Prove that h is a constant function.
(c) Suppose that $T(T(f)) = f$ for some continuous strictly positive function f on V. Show that $T(f) = f$.

[*Hint:* Part (b) Consider the minimum $h(x_0)$ of h on V. Part (c): Let $L(x,y)$ be the kernel of $T \circ T$. Show that

$$\int_V L(x,y)\frac{f(y)}{f(x)}\frac{T(f)(y)}{f(y)}\,dy = \frac{T(f)(x)}{f(x)}$$

and conclude by part (a) that $\frac{T(f)(y)}{f(y)}$ is a constant.]

4.2 Consequences of L^2 Boundedness

Calderón–Zygmund singular integral operators admit L^2-bounded extensions. As in the case of convolution operators, L^2 boundedness has several consequences. In this section we are concerned with consequences of the L^2 boundedness of Calderón–Zygmund singular integral operators. Throughout the entire discussion, we assume that $K(x,y)$ is a kernel defined away from the diagonal in $\mathbf{R}^{2n}$ that satisfies the standard size and regularity conditions (4.1.1), (4.1.2), and (4.1.3). These conditions may be relaxed; see the exercises at the end of this section.

4.2.1 Weak Type $(1,1)$ and L^p Boundedness of Singular Integrals

We recall the Calderón–Zygmund decomposition of a function.

Proposition 4.2.1. *Let $f \in L^1(\mathbf{R}^n)$ and $\alpha > 0$. Then there exist functions g and b on $\mathbf{R}^n$ such that*

(1) $f = g + b$.

(2) $\|g\|_{L^1} \le \|f\|_{L^1}$ *and* $\|g\|_{L^\infty} \le 2^n\alpha$.

(3) $b = \sum_j b_j$, *where each b_j is supported in a dyadic cube Q_j. The cubes Q_k and Q_j are disjoint when $j \neq k$.*

(4) $\displaystyle\int_{Q_j} b_j(x)\,dx = 0$.

(5) $\|b_j\|_{L^1} \le 2^{n+1}\alpha|Q_j|$.

(6) $\sum_j |Q_j| \le \alpha^{-1}\|f\|_{L^1}$.

A proof of Proposition 4.2.1 can be given by considering the level set of the uncentered maximal function with respect to cubes at height α; see Exercise 4.2.6.

Another proof is given in Proposition 5.3.1 in [156]. We note that the construction of the functions g and b yields that, if f is a finite linear combination of characteristic functions of dyadic cubes, then the collection of cubes $\{Q_j\}_j$ in Proposition 4.2.1 is finite. We now prove that operators in $CZO(\delta,A,B)$ have bounded extensions from L^1 to $L^{1,\infty}$.

Theorem 4.2.2. *Assume that $K(x,y)$ is in $SK(\delta,A)$ and let T be an element of $CZO(\delta,A,B)$ associated with the kernel K. Then T has a bounded extension that maps $L^1(\mathbf{R}^n)$ to $L^{1,\infty}(\mathbf{R}^n)$ with norm*

$$\|T\|_{L^1\to L^{1,\infty}} \le C_n(A+B),$$

and also maps $L^p(\mathbf{R}^n)$ to itself for $1<p<\infty$ with norm

$$\|T\|_{L^p\to L^p} \le C_n \max(p,(p-1)^{-1})(A+B),$$

where C_n is a dimensional constant.

Proof. Fix $\alpha>0$ and let f be in $L^1(\mathbf{R}^n)$. Since T may not be defined on general integrable functions, we work with the class $\mathscr{F}_0$ of finite linear combination of characteristic functions of dyadic cubes. The class $\mathscr{F}_0$ is dense in L^1 and also contained in L^2, on which the operator is already defined. Once we obtain a weak type $(1,1)$ estimate for $\mathscr{F}_0$, by density this extends to the entire L^1.

We apply the Calderón–Zygmund decomposition to f in $\mathscr{F}_0$ at height $\gamma\alpha$, where γ is a positive constant to be chosen later. Write $f=g+b$, where $b=\sum_j b_j$ and conditions (1)–(6) of Proposition 4.2.1 are satisfied with the constant α replaced by $\gamma\alpha$. Since f lies in $\mathscr{F}_0$ the sum $b=\sum_j b_j$ extends over a finite set of indices. Moreover, since f is bounded, each bad function b_j is bounded and by construction is also compactly supported. Thus $T(b_j)$ is an L^2 function, and for almost all x not in the support of b_j we have the integral representation

$$T(b_j)(x) = \int_{Q_j} b_j(y)K(x,y)\,dy$$

in view of Proposition 4.1.9.

We denote by $\ell(Q)$ the side length of a cube Q. Let Q_j^* be the unique cube with sides parallel to the axes having the same center as Q_j and having side length

$$\ell(Q_j^*) = 2\sqrt{n}\,\ell(Q_j).$$

We have

$$\begin{aligned}
&|\{x\in\mathbf{R}^n : |T(f)(x)|>\alpha\}| \\
&\quad \le \left|\left\{x\in\mathbf{R}^n : |T(g)(x)|>\frac{\alpha}{2}\right\}\right| + \left|\left\{x\in\mathbf{R}^n : |T(b)(x)|>\frac{\alpha}{2}\right\}\right| \\
&\quad \le \frac{2^2}{\alpha^2}\|T(g)\|_{L^2}^2 + \left|\bigcup_j Q_j^*\right| + \left|\left\{x\notin\bigcup_j Q_j^* : |T(b)(x)|>\frac{\alpha}{2}\right\}\right|
\end{aligned}$$

$$
\begin{aligned}
&\le \frac{2^2}{\alpha^2}B^2\|g\|_{L^2}^2+\sum_j |Q_j^*|+\frac{2}{\alpha}\int_{(\bigcup_j Q_j^*)^c}|T(b)(x)|\,dx\\
&\le \frac{2^2}{\alpha^2}2^nB^2(\gamma\alpha)\|f\|_{L^1}+(2\sqrt{n})^n\frac{\|f\|_{L^1}}{\gamma\alpha}+\frac{2}{\alpha}\sum_j\int_{(Q_j^*)^c}|T(b_j)(x)|\,dx\\
&\le \left(\frac{(2^{n+1}B\gamma)^2}{2^n\gamma}+\frac{(2\sqrt{n})^n}{\gamma}\right)\frac{\|f\|_{L^1}}{\alpha}+\frac{2}{\alpha}\sum_j\int_{(Q_j^*)^c}|T(b_j)(x)|\,dx.
\end{aligned}
$$

It suffices to show that the last sum is bounded by some constant multiple of $\|f\|_{L^1}$. Let y_j be the center of the cube Q_j. For $x \in (Q_j^*)^c$, we have $|x-y_j| \ge \frac{1}{2}\ell(Q_j^*) = \sqrt{n}\,\ell(Q_j)$. But if $y \in Q_j$ we have $|y-y_j| \le \sqrt{n}\,\ell(Q_j)/2$; thus $|y-y_j| \le \frac{1}{2}|x-y_j|$, since the diameter of a cube is equal to $\sqrt{n}$ times its side length. We now estimate the last displayed sum as follows:

$$
\begin{aligned}
\sum_j\int_{(Q_j^*)^c}|T(b_j)(x)|\,dx &= \sum_j\int_{(Q_j^*)^c}\left|\int_{Q_j} b_j(y)K(x,y)\,dy\right|dx\\
&= \sum_j\int_{(Q_j^*)^c}\left|\int_{Q_j} b_j(y)\big(K(x,y)-K(x,y_j)\big)\,dy\right|dx\\
&\le \sum_j\int_{Q_j}|b_j(y)|\int_{(Q_j^*)^c}|K(x,y)-K(x,y_j)|\,dx\,dy\\
&\le \sum_j\int_{Q_j}|b_j(y)|\int_{|x-y_j|\ge 2|y-y_j|}|K(x,y)-K(x,y_j)|\,dx\,dy\\
&\le A_2\sum_j\int_{Q_j}|b_j(y)|\,dy\\
&= A_2\sum_j\|b_j\|_{L^1}\\
&\le A_2 2^{n+1}\|f\|_{L^1}.
\end{aligned}
$$

Combining these facts and choosing $\gamma = B^{-1}$, we deduce the claimed inequality for f in $\mathscr{F}_0$. By density, we obtain that T has a bounded extension from L^1 to $L^{1,\infty}$ with bound at most $C_n(A+B)$. The L^p result for $1<p<2$ follows by interpolation, while the fact that the constant blows up like $(p-1)^{-1}$ as $p \to 1$ can be deduced from the result of Exercise 1.3.2 in [156]. The result for $2<p<\infty$ follows by duality; one uses here that the dual operator T^t has a kernel $K^t(x,y)=K(y,x)$ that satisfies the same estimates as K, and by the result just proved, it is also bounded on L^p for $1<p<2$ with norm at most $C_n(A+B)$. Thus T must be bounded on L^p for $2<p<\infty$ with norm at most a constant multiple of $A+B$. □

Consequently, for operators T in $CZO(\delta,A,B)$ and L^p functions f, $1 \le p < \infty$, the expressions $T(f)$ make sense as L^p (or $L^{1,\infty}$ when $p=1$) functions. The following result addresses the question whether these functions can be expressed as integrals.

Proposition 4.2.3. *Let T be an operator in $CZO(\delta,A,B)$ associated with a kernel K. Then for $g \in L^p(\mathbf{R}^n)$, $1 \le p < \infty$, the following absolutely convergent integral representation is valid:*

$$T(g)(x) = \int_{\mathbf{R}^n} K(x,y)\,g(y)\,dy \tag{4.2.1}$$

for almost all $x \in \mathbf{R}^n \setminus$ supp g, provided that supp $g \subsetneq \mathbf{R}^n$.

Proof. Set $g_k(x) = g(x)\chi_{|g(x)|\le k}\chi_{|x|\le k}$. These are L^p functions with compact support contained in the support of g. Also, the g_k converge to g in L^p as $k \to \infty$. In view of Proposition 4.1.9, for every k we have

$$T(g_k)(x) = \int_{\mathbf{R}^n} K(x,y)\,g_k(y)\,dy$$

for almost all $x \in \mathbf{R}^n \setminus$ supp g. Since T maps L^p to L^p (or to weak L^1 when $p = 1$), it follows that $T(g_k)$ converges to $T(g)$ in weak L^p and hence in measure. By Proposition 1.1.9 in [156], a subsequence of $T(g_k)$ converges to $T(g)$ almost everywhere. On the other hand, for $x \in \mathbf{R}^n \setminus$ supp g we have

$$\int_{\mathbf{R}^n} K(x,y)\,g_k(y)\,dy \to \int_{\mathbf{R}^n} K(x,y)\,g(y)\,dy$$

when $k \to \infty$, since the absolute value of the difference is bounded by $B\|g_k - g\|_{L^p}$, which tends to zero. The constant B is the $L^{p'}$ norm of the function $|x-y|^{-n-\delta}$ on the support of g; one has $|x-y| \ge c > 0$ for all y in the support of g and thus $B < \infty$. Therefore $T(g_k)(x)$ converges a.e. to both sides of the identity (4.2.1) for x not in the support of g. This concludes the proof of this identity. □

4.2.2 Boundedness of Maximal Singular Integrals

We pose the question whether there is a result concerning the maximal singular integral operator $T^{(*)}$ analogous to Theorem 4.2.2. We note that given f in $L^p(\mathbf{R}^n)$ for some $1 \le p < \infty$, the expression $T^{(*)}(f)(x)$ is well defined for all $x \in \mathbf{R}^n$. This is a simple consequence of estimate (4.1.1) and Hölder's inequality.

Theorem 4.2.4. *Let K be in $SK(\delta,A)$ and T in $CZO(\delta,A,B)$ be associated with K. Let $r \in (0,1)$. Then there is a constant $C(n,r)$ such that* Cotlar's inequality

$$|T^{(*)}(f)(x)| \le C(n,r)\Big[M(|T(f)|^r)(x)^{\frac{1}{r}} + (A+B)\,M(f)(x)\Big] \tag{4.2.2}$$

is valid for all functions in $\bigcup_{1\le p<\infty} L^p(\mathbf{R}^n)$. *Also, there exist dimensional constants* C_n, C_n' *such that*

$$\big\|T^{(*)}(f)\big\|_{L^{1,\infty}(\mathbf{R}^n)} \le C_n'(A+B)\big\|f\big\|_{L^1(\mathbf{R}^n)}, \tag{4.2.3}$$

$$\big\|T^{(*)}(f)\big\|_{L^p(\mathbf{R}^n)} \le C_n(A+B)\max(p,(p-1)^{-1})\big\|f\big\|_{L^p(\mathbf{R}^n)}, \tag{4.2.4}$$

for all $1\le p<\infty$ *and all* f *in* $L^p(\mathbf{R}^n)$.

Proof. We fix r so that $0<r<1$ and $f\in L^p(\mathbf{R}^n)$ for some p satisfying $1\le p<\infty$. To prove (4.2.2), we also fix $\varepsilon>0$ and we set $f_0^{\varepsilon,x}=f\chi_{B(x,\varepsilon)}$ and $f_\infty^{\varepsilon,x}=f\chi_{B(x,\varepsilon)^c}$. Since $x\notin \operatorname{supp} f_\infty^{\varepsilon,x}$ whenever $|x-y|\ge\varepsilon$, using Proposition 4.2.3 we can write

$$T(f_\infty^{\varepsilon,x})(x)=\int_{\mathbf{R}^n}K(x,y)\,f_\infty^{\varepsilon,x}(y)\,dy=\int_{|x-y|\ge\varepsilon}K(x,y)\,f(y)\,dy=T^{(\varepsilon)}(f)(x).$$

In view of (4.1.2), for $z\in B(x,\frac{\varepsilon}{2})$ we have $|z-x|\le\frac12|x-y|$ whenever $|x-y|\ge\varepsilon$ and thus

$$\begin{aligned}
|T(f_\infty^{\varepsilon,x})(x)-T(f_\infty^{\varepsilon,x})(z)| &= \left|\int_{|x-y|\ge\varepsilon}\big(K(z,y)-K(x,y)\big)f(y)\,dy\right|\\
&\le |z-x|^\delta\int_{|x-y|\ge\varepsilon}\frac{A\,|f(y)|}{(|x-y|+|y-z|)^{n+\delta}}\,dy\\
&\le \left(\frac{\varepsilon}{2}\right)^\delta\int_{|x-y|\ge\varepsilon}\frac{A\,|f(y)|}{(|x-y|+\varepsilon/2)^{n+\delta}}\,dy\\
&\le C_{n,\delta}\,A\,M(f)(x),
\end{aligned}$$

where the last estimate is a consequence of Theorem 2.1.10 in [156]. We conclude that for all $z\in B(x,\frac{\varepsilon}{2})$, we have

$$\begin{aligned}
|T^{(\varepsilon)}(f)(x)| &= |T(f_\infty^{\varepsilon,x})(x)|\\
&\le |T(f_\infty^{\varepsilon,x})(x)-T(f_\infty^{\varepsilon,x})(z)|+|T(f_\infty^{\varepsilon,x})(z)|\\
&\le C_{n,\delta}\,A\,M(f)(x)+|T(f_0^{\varepsilon,x})(z)|+|T(f)(z)|.
\end{aligned}\tag{4.2.5}$$

For $0<r<1$ it follows from (4.2.5) that for $z\in B(x,\frac{\varepsilon}{2})$ we have

$$|T^{(\varepsilon)}(f)(x)|^r\le C_{n,\delta}^r A^r M(f)(x)^r+|T(f_0^{\varepsilon,x})(z)|^r+|T(f)(z)|^r. \tag{4.2.6}$$

Integrating over $z\in B(x,\frac{\varepsilon}{2})$, dividing by $|B(x,\frac{\varepsilon}{2})|$, and raising to the power $\frac1r$, we obtain

$$\begin{aligned}
|T^{(\varepsilon)}(f)(x)| \le 3^{\frac1r}\Bigg[C_{n,\delta}\,A\,M(f)(x)&+\left(\frac{1}{|B(x,\frac{\varepsilon}{2})|}\int_{B(x,\frac{\varepsilon}{2})}|T(f_0^{\varepsilon,x})(z)|^r dz\right)^{\frac1r}\\
&+M(|T(f)|^r)(x)^{\frac1r}\Bigg].
\end{aligned}$$

The middle term on the right-hand side of the preceding equation can be estimated, via Exercise 2.1.5 in [156] (Kolmogorov's inequality), by

$$\left(\frac{1}{|B(x,\frac{\varepsilon}{2})|}\frac{\|T\|_{L^1\to L^{1,\infty}}^r}{1-r}|B(x,\tfrac{\varepsilon}{2})|^{1-r}\|f_0^{\varepsilon,x}\|_{L^1}^r\right)^{\frac{1}{r}} \le C_{n,r}(A+B)M(f)(x)\,.$$

This proves (4.2.2).

We now use estimate (4.2.2) to show that T is L^p bounded and of weak type $(1,1)$. To obtain the weak type $(1,1)$ estimate for $T^{(*)}$ we need to use that the Hardy–Littlewood maximal operator maps $L^{p,\infty}$ to $L^{p,\infty}$ for all $1<p<\infty$; see Exercise 2.1.13 in [156]. We also use the trivial fact that for all $0<p,q<\infty$ we have

$$\big\||f|^q\big\|_{L^{p,\infty}} = \|f\|_{L^{pq,\infty}}^q\,.$$

Take any $r<1$ in (4.2.2). Then we have

$$\begin{aligned}
\big\|M(|T(f)|^r)^{\frac{1}{r}}\big\|_{L^{1,\infty}} &= \big\|M(|T(f)|^r)\big\|_{L^{\frac{1}{r},\infty}}^{\frac{1}{r}}\\
&\le C_{n,r}\big\||T(f)|^r\big\|_{L^{\frac{1}{r},\infty}}^{\frac{1}{r}}\\
&= C_{n,r}\big\|T(f)\big\|_{L^{1,\infty}}\\
&\le \widetilde{C}_{n,r}(A+B)\big\|f\big\|_{L^1}\,,
\end{aligned}$$

where we used the weak type $(1,1)$ bound for T in the last estimate.

To obtain the L^p boundedness of $T^{(*)}$ for $1<p<\infty$, we use the same argument as before. We fix $r=\frac{1}{2}$. Recall that the maximal function is bounded on L^{2p} with norm at most $3^{\frac{n}{2p}}\frac{2p}{2p-1}\le 2\cdot 3^{\frac{n}{2}}$ (Theorem 2.1.6 in [156]). We have

$$\begin{aligned}
\big\|M(|T(f)|^{\frac{1}{2}})^2\big\|_{L^p} &= \big\|M(|T(f)|^{\frac{1}{2}})\big\|_{L^{2p}}^2\\
&\le \left(3^{\frac{n}{2p}}\tfrac{2p}{2p-1}\right)^2\big\||T(f)|^{\frac{1}{2}}\big\|_{L^{2p}}^2\\
&\le 4\cdot 3^n\big\|T(f)\big\|_{L^p}\\
&\le C_n\max(\tfrac{1}{p-1},p)(A+B)\big\|f\big\|_{L^p}\,,
\end{aligned}$$

where we used the L^p boundedness of T in the last estimate. □

We end this section with a corollary which confirms a fact mentioned in Remark 4.1.12.

Corollary 4.2.5. *Let K be in $SK(\delta,A)$ and T in $CZO(\delta,A,B)$ be associated with K. Then there exists a dimensional constant C_n such that*

$$\sup_{\varepsilon>0}\big\|T^{(\varepsilon)}\big\|_{L^2\to L^2}\le C_n\Big(A+\big\|T\big\|_{L^2\to L^2}\Big)\,.$$

4.2.3 $H^1 \to L^1$ and $L^\infty \to BMO$ Boundedness of Singular Integrals

We discuss a couple of endpoint results concerning operators in $CZO(\delta,A,B)$.

Theorem 4.2.6. *Let T be an operator in $CZO(\delta,A,B)$. Then T has an extension that maps $H^1(\mathbf{R}^n)$ to $L^1(\mathbf{R}^n)$. Precisely, there is a constant $C_{n,\delta}$ such that*

$$\|T\|_{H^1\to L^1} \le C_{n,\delta}\big(A+\|T\|_{L^2\to L^2}\big).$$

Proof. The proof is analogous to that of Theorem 2.4.1. Let $B=\|T\|_{L^2\to L^2}$. We start by examining the action of T on L^2 atoms for H^1. Let $f=a$ be such an atom, supported in a cube Q. Let c_Q be the center of Q and let $Q^*=2\sqrt{n}\,Q$. We write

$$\int_{\mathbf{R}^n}|T(a)(x)|\,dx=\int_{Q^*}|T(a)(x)|\,dx+\int_{(Q^*)^c}|T(a)(x)|\,dx \tag{4.2.7}$$

and we estimate each term separately. We have

$$\begin{aligned}\int_{Q^*}|T(a)(x)|\,dx &\le |Q^*|^{\frac12}\Big(\int_{Q^*}|T(a)(x)|^2\,dx\Big)^{\frac12}\\ &\le B|Q^*|^{\frac12}\Big(\int_{Q}|a(x)|^2\,dx\Big)^{\frac12}\\ &\le B|Q^*|^{\frac12}|Q|^{-\frac12}\\ &= C_nB,\end{aligned}$$

where we used property (b) of atoms in Definition 2.3.10. Now observe that if $x\notin Q^*$ and $y\in Q$, then

$$|y-c_Q|\le\frac{1}{2}|x-c_Q|;$$

hence $x-y$ stays away from zero and $T(a)(x)$ can be expressed as a convergent integral by Proposition 4.2.3. We have

$$\begin{aligned}\int_{(Q^*)^c}|T(a)(x)|\,dx &= \int_{(Q^*)^c}\Big|\int_Q K(x,y)a(y)\,dy\Big|\,dx\\ &= \int_{(Q^*)^c}\Big|\int_Q (K(x,y)-K(x,c_Q))a(y)\,dy\Big|\,dx\\ &\le \int_Q\int_{(Q^*)^c}|K(x,y)-K(x,c_Q)|\,dx\,|a(y)|\,dy\\ &\le \int_Q\int_{(Q^*)^c}\frac{A|y-c_Q|^\delta}{|x-c_Q|^{n+\delta}}\,dx\,|a(y)|\,dy\\ &\le C'_{n,\delta}A\int_Q|a(y)|\,dy\end{aligned}$$

$$
\begin{aligned}
&\le C'_{n,\delta} A |Q|^{\frac{1}{2}} \|a\|_{L^2} \\
&\le C'_{n,\delta} A |Q|^{\frac{1}{2}} |Q|^{-\frac{1}{2}} \\
&= C'_{n,\delta} A\,.
\end{aligned}
$$

Combining this calculation with the previous one and inserting the final conclusions in (4.2.7), we deduce that L^2 atoms for H^1 satisfy

$$\big\|T(a)\big\|_{L^1} \le C_{n,\delta}\,(A+B)\,. \tag{4.2.8}$$

To pass to general functions in H^1, we use Theorem 2.3.12 to write an $f \in H^1$ as

$$f = \sum_{j=1}^{\infty} \lambda_j a_j\,,$$

where the series converges in H^1, the a_j are L^2 atoms for H^1, and

$$\big\|f\big\|_{H^1} \approx \sum_{j=1}^{\infty} |\lambda_j|\,. \tag{4.2.9}$$

Since T maps L^1 to weak L^1 by Theorem 4.2.2, $T(f)$ is already a well-defined $L^{1,\infty}$ function. We plan to prove that

$$T(f) = \sum_{j=1}^{\infty} \lambda_j T(a_j) \qquad \text{a.e.} \tag{4.2.10}$$

Note that the series in (4.2.10) converges in L^1 and defines an integrable function almost everywhere. Once (4.2.10) is established, the required conclusion (2.4.5) follows easily by taking L^1 norms in (4.2.10) and using (4.2.8) and (4.2.9).

To prove (4.2.10), we use that T is of weak type $(1,1)$. For a given $\mu > 0$ we have

$$
\begin{aligned}
&\Big|\Big\{\Big|T(f) - \sum_{j=1}^{\infty} \lambda_j T(a_j)\Big| > \mu\Big\}\Big| \\
&\le \Big|\Big\{\Big|T(f) - \sum_{j=1}^{N} \lambda_j T(a_j)\Big| > \mu/2\Big\}\Big| + \Big|\Big\{\Big|\sum_{j=N+1}^{\infty} \lambda_j T(a_j)\Big| > \mu/2\Big\}\Big| \\
&\le \frac{2}{\mu} \big\|T\big\|_{L^1\to L^{1,\infty}} \Big\|f - \sum_{j=1}^{N} \lambda_j a_j\Big\|_{L^1} + \frac{2}{\mu} \Big\|\sum_{j=N+1}^{\infty} \lambda_j T(a_j)\Big\|_{L^1} \\
&\le \frac{2}{\mu} \big\|T\big\|_{L^1\to L^{1,\infty}} \Big\|f - \sum_{j=1}^{N} \lambda_j a_j\Big\|_{H^1} + \frac{2}{\mu} C_{n,\delta}\,(A+B) \sum_{j=N+1}^{\infty} |\lambda_j|\,.
\end{aligned}
$$

Since $\sum_{j=1}^{N} \lambda_j a_j$ converges to f in H^1 and $\sum_{j=1}^{\infty} |\lambda_j| < \infty$, both terms in the sum converge to zero as $N \to \infty$. We conclude that

$$\left|\left\{\left|T(f)-\sum_{j=1}^{\infty}\lambda_j T(a_j)\right|>\mu\right\}\right|=0$$

for all $\mu > 0$, which implies (4.2.10). □

Theorem 4.2.7. *Let T be in $CZO(\delta,A,B)$. Then for any bounded function f, the distribution $T(f)$ can be identified with a BMO function that satisfies*

$$\left\|T(f)\right\|_{BMO}\leq C'_{n,\delta}(A+B)\left\|f\right\|_{L^\infty}, \tag{4.2.11}$$

where $C_{n,\delta}$ is a constant.

Proof. Let $L^2_{0,c}$ be the space of all square integrable functions with compact support and integral zero on $\mathbf{R}^n$. This space is contained in $H^1(\mathbf{R}^n)$ (cf. Exercise 2.1.7) and contains the set of finite sums of L^2 atoms for H^1, which is dense in H^1; thus $L^2_{0,c}$ is dense in H^1. Recall that for $f\in L^\infty$, $T(f)$ has a well-defined action $\langle T(f),\varphi\rangle$ on functions φ in $L^2_{0,c}$ and (4.1.30) holds.

Suppose we have proved the identity

$$\langle T(f),\varphi\rangle=\int_{\mathbf{R}^n}T^t(\varphi)(x)f(x)\,dx, \tag{4.2.12}$$

for all bounded functions f and all φ in $L^2_{0,c}$. Since such a φ is in H^1, Theorem 4.2.6 yields that $T^t(\varphi)$ is in L^1, and consequently, thc integral in (4.2.12) converges absolutely. Assuming (4.2.12) and using Theorem 4.2.6 we obtain that

$$\left|\langle T(f),\varphi\rangle\right|\leq\left\|T^t(\varphi)\right\|_{L^1}\left\|f\right\|_{L^\infty}\leq C_{n,\delta}(A+B)\left\|\varphi\right\|_{H^1}\left\|f\right\|_{L^\infty}.$$

We conclude that $L(\varphi)=\langle T(f),\varphi\rangle$ is a bounded linear functional on $L^2_{0,c}$ with norm at most $C_{n,\delta}(A+B)\left\|f\right\|_{L^\infty}$. Obviously, L has a bounded extension on H^1 with the same norm. By Theorem 3.2.2 there exists a BMO function b_f that satisfies $\|b_f\|_{BMO}\leq C'_n\|L\|_{H^1\to\mathbf{C}}$ such that the linear functional L has the form L_{b_f}, using the notation of Theorem 3.2.2. In other words, the distribution $T(f)$ can be identified with a BMO function that satisfies (4.2.11) with $C_{n,\delta}=C'_nC_{n,\delta}$, i.e.,

$$\left\|T(f)\right\|_{BMO}\leq C'_nC_{n,\delta}(A+B)\left\|f\right\|_{L^\infty}.$$

We return to the proof of identity (4.2.12). Pick a smooth function with compact support η that satisfies $0\leq\eta\leq 1$ and is equal to 1 in a neighborhood of the support of φ. We write the right-hand side of (4.2.12) as

$$\int_{\mathbf{R}^n}T^t(\varphi)\eta f\,dx+\int_{\mathbf{R}^n}T^t(\varphi)(1-\eta)f\,dx=\langle T(\eta f),\varphi\rangle+\int_{\mathbf{R}^n}T^t(\varphi)(1-\eta)f\,dx.$$

In view of Definition 4.1.16, to prove (4.2.12) it will suffice to show that

$$\int_{\mathbf{R}^n}T^t(\varphi)(1-\eta)f\,dx=\int_{\mathbf{R}^n}\int_{\mathbf{R}^n}(K(x,y)-K(x_0,y))\varphi(x)\,dx(1-\eta(y))f(y)\,dy,$$

where x_0 lies in the support of φ. In the outer integral above we have $y \notin \text{supp } \varphi$ and the inner integral above is absolutely convergent and equal to

$$\int_{\mathbf{R}^n} (K(x,y) - K(x_0,y))\,\varphi(x)\,dx = \int_{\mathbf{R}^n} K^t(y,x)\varphi(x)\,dx = T^t(\varphi)(y),$$

by Proposition 4.1.9, since $y \notin \text{supp } \varphi$. Thus (4.2.12) is valid. □

Exercises

4.2.1. Let $T : \mathscr{S}(\mathbf{R}^n) \to \mathscr{S}'(\mathbf{R}^n)$ be a continuous linear operator whose Schwartz kernel coincides with a function $K(x,y)$ on $\mathbf{R}^n \times \mathbf{R}^n$ minus its diagonal. Suppose that the function $K(x,y)$ satisfies

$$\sup_{R>0} \int_{R \le |x-y| \le 2R} |K(x,y)|\,dy = A < \infty.$$

(a) Show that the previous condition is equivalent to

$$\sup_{R>0} \frac{1}{R} \int_{|x-y| \le R} |x-y|\,|K(x,y)|\,dy = A' < \infty$$

by proving that $A' \le A \le 2A'$.
(b) For $\varepsilon > 0$, let $T^{(\varepsilon)}$ be the truncated linear operators with kernels $K^{(\varepsilon)}(x,y) = K(x,y)\chi_{|x-y|>\varepsilon}$. Show that the integral defining $T^{(\varepsilon)}(f)$ converges absolutely for Schwartz functions f.
[*Hint:* Part (b): Consider the annuli $\varepsilon 2^j \le |x| \le \varepsilon 2^{j+1}$ for $j \ge 0$.]

4.2.2. Let T be as in Exercise 4.2.1. Prove that the limit $T^{(\varepsilon)}(f)(x)$ exists for all f in the Schwartz class for almost all $x \in \mathbf{R}^n$ as $\varepsilon \to 0$ if and only if the limit

$$\lim_{\varepsilon \to 0} \int_{\varepsilon < |x-y| < 1} K(x,y)\,dy$$

exists for almost all $x \in \mathbf{R}^n$.

4.2.3. Let $K(x,y)$ be a function defined away from the diagonal in $\mathbf{R}^{2n}$ that satisfies

$$\sup_{R>0} \int_{R \le |x-y| \le 2R} |K(x,y)|\,dx \le A < \infty$$

and also *Hörmander's condition*

$$\sup_{\substack{y,y' \in \mathbf{R}^n \\ y \ne y'}} \int_{|x-y| \ge 2|y-y'|} |K(x,y) - K(x,y')|\,dx \le A'' < \infty. \tag{4.2.13}$$

Show that the truncations $K^{(\varepsilon)}(x,y)$ also satisfy Hörmander's condition uniformly in $\varepsilon > 0$ with a constant $A+A''$. The same conclusion is valid for the truncations $K^{(\varepsilon)}(x,y) - K^{(M)}(x,y)$ for $0 < \varepsilon < M < \infty$.

4.2.4. Let T be as in Exercise 4.2.1 and assume that T maps $L^r(\mathbf{R}^n)$ to itself for some $1 < r < \infty$ with norm B.
(a) Assume that $K(x,y)$ satisfies condition (4.2.13). Show that T has an extension that maps $L^1(\mathbf{R}^n)$ to $L^{1,\infty}(\mathbf{R}^n)$ with norm

$$\|T\|_{L^1 \to L^{1,\infty}} \le C_n(A + A'' + B).$$

Conclude that T has a bounded extension from $L^p(\mathbf{R}^n)$ to itself for $1 < p < r$ with norm

$$\|T\|_{L^p \to L^p} \le C_n(p-1)^{-\frac{1}{p}}(A + A'' + B),$$

where C_n is a dimensional constant.
(b) Assuming that $K^t(x,y) = K(y,x)$ also satisfies (4.2.13), prove that T has a bounded extension from $L^p(\mathbf{R}^n)$ to itself for $r < p < \infty$ with norm

$$\|T\|_{L^p \to L^p} \le C_n\,(p-1)^{1-\frac{1}{p}}\,(A + A'' + B),$$

where C_n is independent of p.
[*Hint:* Use Exercise 1.3.2 in [156].]

4.2.5. Let K and T be as in Theorem 4.2.4. Show that estimate (4.2.2) also holds when $r = 1$.
[*Hint:* Estimate (4.2.6) holds when $r = 1$. For fixed $\varepsilon > 0$, take $0 < b < |T^{(\varepsilon)}(f)(x)|$ and define $B_1^\varepsilon(x) = B(x, \frac{\varepsilon}{2}) \cap \{|T(f)| > \frac{b}{3}\}$, $B_2^\varepsilon(x) = B(x, \frac{\varepsilon}{2}) \cap \{|T(f_0^{\varepsilon,x})| > \frac{b}{3}\}$, and $B_3^\varepsilon(x) = B(x, \frac{\varepsilon}{2})$ if $C_{n,\delta} A M(f)(x) > \frac{b}{3}$ and empty otherwise. Then $|B(x, \frac{\varepsilon}{2})| \le |B_1^\varepsilon(x)| + |B_2^\varepsilon(x)| + |B_3^\varepsilon(x)|$. Use the weak type $(1,1)$ property of T to show that $b \le C(n)\big(M(|T(f)|)(x) + (A+B)M(f)(x)\big)$, and take the supremum over all $b < |T^{(\varepsilon)}(f)(x)|$.]

4.2.6. (Calderón–Zygmund decomposition with bounded overlap) Let $f \in L^1(\mathbf{R}^n)$ and $\alpha > 0$. Prove that there exist functions g and b on $\mathbf{R}^n$ such that

(1) $f = g + b$.

(2) $\|g\|_{L^1} \le \|f\|_{L^1}$, $\|g\|_{L^\infty} \le (10\sqrt{n})^n \alpha$.

(3) $b = \sum_j b_j$, where each b_j is supported in a dyadic cube Q_j. Furthermore, the interiors of Q_k and Q_j are disjoint when $j \ne k$.

(4) $\displaystyle\int_{Q_j} b_j(x)\,dx = 0$.

(5) $\|b_j\|_{L^1} \le 2(10\sqrt{n})^n \alpha |Q_j|$.

(6) $\sum_j |Q_j| \le \alpha^{-1} \|f\|_{L^1}$.

(7) $\sum_j \chi_{Q_j^*} \le 12^n$, where Q_j^* has the same center as Q_j and $\ell(Q_j^*) = (1+\varepsilon)\ell(Q_j)$, for any ε with $0 < \varepsilon < 1/4$.

[*Hint:* Let M_c be the uncentered Hardy–Littlewood maximal operator with respect to cubes on $\mathbf{R}^n$. Let Q_j be the Whitney cubes of $\Omega = \{M_c(f) > \alpha\}$ (see Appendix J in [156]). Define $b_j = \big(f - \frac{1}{|Q_j|}\int_{Q_j} f\,dx\big)\chi_{Q_j}$ and $b = \sum_j b_j$. Use that $10\sqrt{n}Q_j$ intersects Ω^c and that on Ω^c, $g = f \le M_c(f) \le \alpha$ a.e.]

4.3 The $T(1)$ Theorem

We now turn to one of the main results of this chapter, the so-called $T(1)$ theorem. This theorem gives necessary and sufficient conditions for linear operators T with standard kernels to be bounded on $L^2(\mathbf{R}^n)$. In this section we obtain several such equivalent conditions. The name of theorem $T(1)$ is due to the fact that one of the equivalent ways to characterize boundedness is expressed in terms of properties of the distribution $T(1)$, which was introduced in Definition 4.1.16.

4.3.1 Preliminaries and Statement of the Theorem

We begin with some preliminary facts and definitions.

Definition 4.3.1. A *normalized bump* is a smooth function φ supported in the ball $B(0,10)$ that satisfies

$$|(\partial_x^\alpha \varphi)(x)| \le 1$$

for all multi-indices $|\alpha| \le 2\left[\frac{n}{2}\right]+2$, where $[x]$ denotes here the integer part of x.

Observe that every smooth function supported inside the ball $B(0,10)$ is a constant multiple of a normalized bump. Also note that if a normalized bump is supported in a compact subset of $B(0,10)$, then small translations of it are also normalized bumps.

Given a function f on $\mathbf{R}^n$, $R>0$, and $x_0 \in \mathbf{R}^n$, we use the notation f_R to denote the function $f_R(x) = R^{-n} f(R^{-1}x)$ and $\tau^{x_0} f$ to denote the function $\tau^{x_0} f(x) = f(x - x_0)$. Thus

$$\tau^{x_0} f_R(y) = f_R(y - x_0) = R^{-n} f\big(R^{-1}(y - x_0)\big).$$

Set $N = \left[\frac{n}{2}\right]+1$. Using that all derivatives up to order $2N$ of normalized bumps are bounded by 1, we easily deduce that for all $x_0 \in \mathbf{R}^n$, all $R>0$, and all normalized bumps φ we have the estimate

$$
\begin{aligned}
R^n \int_{\mathbf{R}^n} \big|\widehat{\tau^{x_0}\varphi_R}(\xi)\big|\,d\xi
&= \int_{\mathbf{R}^n} \big|\widehat{\varphi}(\xi)\big|\,d\xi \\
&= \int_{\mathbf{R}^n} \bigg|\int_{\mathbf{R}^n} \varphi(y)e^{-2\pi i y\cdot\xi}\,dy\bigg|\,d\xi \\
&= \int_{\mathbf{R}^n} \bigg|\int_{\mathbf{R}^n} (I-\Delta)^N(\varphi)(y)e^{-2\pi i y\cdot\xi}\,dy\bigg| \frac{d\xi}{(1+4\pi^2|\xi|^2)^N} \\
&\le C_n,
\end{aligned}
\tag{4.3.1}
$$

since $|(\partial_x^\alpha \varphi)(x)| \le 1$ for all multi-indices α with $|\alpha| \le [\frac{n}{2}]+1$, and C_n is independent of the bump φ. Here $I-\Delta$ denotes the operator

$$
(I-\Delta)(\varphi) = \varphi + \sum_{j=1}^{n} \frac{\partial^2 \varphi}{\partial x_j^2}.
$$

Definition 4.3.2. We say that a continuous linear operator

$$
T : \mathscr{S}(\mathbf{R}^n) \to \mathscr{S}'(\mathbf{R}^n)
$$

satisfies the *weak boundedness property* (WBP) if there is a constant C such that for all f and g normalized bumps and for all $x_0 \in \mathbf{R}^n$ and $R > 0$ we have

$$
|\langle T(\tau^{x_0} f_R), \tau^{x_0} g_R\rangle| \le C R^{-n}. \tag{4.3.2}
$$

The smallest constant C in (4.3.2) is denoted by $\|T\|_{WB}$.

Note that $\|\tau^{x_0} f_R\|_{L^2} = \|f_R\|_{L^2} = \|f\|_{L^2} R^{-n/2}$ and thus if T has a bounded extension from $L^2(\mathbf{R}^n)$ to itself, then T satisfies the weak boundedness property with bound

$$
\|T\|_{WB} \le 10^n v_n \|T\|_{L^2\to L^2},
$$

where v_n is the volume of the unit ball in $\mathbf{R}^n$.

We now state one of the main theorems in this chapter.

Theorem 4.3.3. *Let T be a continuous linear operator from $\mathscr{S}(\mathbf{R}^n)$ to $\mathscr{S}'(\mathbf{R}^n)$ whose Schwartz kernel coincides with a function K on $\mathbf{R}^n \times \mathbf{R}^n \setminus \{(x,x) : x \in \mathbf{R}^n\}$ that satisfies (4.1.1), (4.1.2), and (4.1.3) for some $0 < A < \infty$ and $0 < \delta \le 1$. Let $K^{(\varepsilon)}$ and $T^{(\varepsilon)}$ be the usual truncated kernel and operator for $\varepsilon > 0$. Assume that there exists a sequence $\varepsilon_j \downarrow 0$ such that for all $\varphi, \psi \in \mathscr{S}(\mathbf{R}^n)$ we have*

$$
\langle T^{(\varepsilon_j)}(\varphi), \psi\rangle \to \langle T(\varphi), \psi\rangle. \tag{4.3.3}
$$

Consider the assertions:

(i) The following statement is valid:

$$B_1 = \sup_{B}\sup_{\varepsilon>0}\left[\frac{\big\|T^{(\varepsilon)}(\chi_B)\big\|_{L^2}}{|B|^{\frac{1}{2}}} + \frac{\big\|(T^{(\varepsilon)})^t(\chi_B)\big\|_{L^2}}{|B|^{\frac{1}{2}}}\right] < \infty,$$

where the first supremum is taken over all balls B in $\mathbf{R}^n$.

(ii) We have that

$$B_2 = \sup_{\varepsilon,N,x_0}\left[\frac{1}{N^n}\int_{B(x_0,N)}\left|\int_{|x-y|<N} K^{(\varepsilon)}(x,y)\,dy\right|^2 dx\right.$$
$$\left. + \frac{1}{N^n}\int_{B(x_0,N)}\left|\int_{|x-y|<N} K^{(\varepsilon)}(y,x)\,dy\right|^2 dx\right]^{\frac{1}{2}} < \infty,$$

where the supremum is taken over all $0<\varepsilon,N<\infty$ with $\varepsilon<N$, and all $x_0\in\mathbf{R}^n$.

(iii) The following statement is valid:

$$B_3 = \sup_{\varphi}\sup_{x_0\in\mathbf{R}^n}\sup_{R>0} R^{\frac{n}{2}}\left[\big\|T(\tau^{x_0}\varphi_R)\big\|_{L^2} + \big\|T^t(\tau^{x_0}\varphi_R)\big\|_{L^2}\right] < \infty,$$

where the first supremum is taken over all normalized bumps φ.

(iv) The operator T satisfies the weak boundedness property and the distributions $T(1)$ and $T^t(1)$ coincide with BMO functions, that is,

$$B_4 = \big\|T(1)\big\|_{BMO} + \big\|T^t(1)\big\|_{BMO} + \big\|T\big\|_{WB} < \infty.$$

(v) For every $\xi\in\mathbf{R}^n$ the distributions $T(e^{2\pi i(\cdot)\cdot\xi})$ and $T^t(e^{2\pi i(\cdot)\cdot\xi})$ coincide with BMO functions such that

$$B_5 = \sup_{\xi\in\mathbf{R}^n}\big\|T(e^{2\pi i(\cdot)\cdot\xi})\big\|_{BMO} + \sup_{\xi\in\mathbf{R}^n}\big\|T^t(e^{2\pi i(\cdot)\cdot\xi})\big\|_{BMO} < \infty.$$

(vi) The following statement is valid:

$$B_6 = \sup_{\varphi}\sup_{x_0\in\mathbf{R}^n}\sup_{R>0} R^n\left[\big\|T(\tau^{x_0}\varphi_R)\big\|_{BMO} + \big\|T^t(\tau^{x_0}\varphi_R)\big\|_{BMO}\right] < \infty,$$

where the first supremum is taken over all normalized bumps φ.

Then assertions (i)–(vi) are all equivalent to each other and to the L^2 boundedness of T, and we have the following equivalence of the previous quantities:

$$c_{n,\delta}(A+B_j) \le \big\|T\big\|_{L^2\to L^2} \le C_{n,\delta}(A+B_j),$$

for all $j \in \{1,2,3,4,5,6\}$, for some constants $c_{n,\delta}, C_{n,\delta}$ that depend only on the dimension n and on the parameter $\delta > 0$.

Remark 4.3.4. Condition (4.3.3) says that the operator T is the weak limit of a sequence of its truncations. We already know from Proposition 4.1.11 that if T is bounded on L^2, then it must be equal to an operator that satisfies (4.3.3) plus a bounded function times the identity operator. Therefore, it is not a serious restriction to assume condition (4.3.3). In Remark 4.3.6 we discuss versions of Theorem 4.3.3 in which this assumption is not imposed.

One should always keep in mind the following pathological situation: consider the distribution $W_0 \in \mathscr{S}(\mathbf{R}^n \times \mathbf{R}^n)$ defined for F in $\mathscr{S}(\mathbf{R}^{2n})$ by

$$\langle W_0, F \rangle = \int_{\mathbf{R}^n} F(t,t) h(t)\, dt\,,$$

where $h(t) = |t|^2$. In this case, $T^{(\varepsilon)} = 0$ for all $\varepsilon > 0$; hence $T^{(\varepsilon)}$ are uniformly bounded on L^2, but $\langle T(f), \varphi \rangle = \int_{\mathbf{R}^n} \varphi(t) f(t) h(t)\, dt$; thus $T(f)$ can be identified with fh for all $f \in \mathscr{S}$, which is certainly an unbounded operator on $L^2(\mathbf{R}^n)$. Notice that (4.3.3) fails in this case.

Before we begin the lengthy proof of this theorem, we state a lemma that we need.

Lemma 4.3.5. *Under assumptions (4.1.1), (4.1.2), and (4.1.3), there is a constant C_n such that for all normalized bumps φ we have*

$$\sup_{x_0 \in \mathbf{R}^n} \int_{|x-x_0| \geq 20R} \left| \int_{\mathbf{R}^n} K(x,y) \tau^{x_0} \varphi_R(y)\, dy \right|^2 dx \leq \frac{C_n A^2}{R^n}. \tag{4.3.4}$$

Proof. Note that the interior integral in (4.3.4) is absolutely convergent, since $\tau^{x_0}\varphi_R$ is supported in the ball $B(x_0, 10R)$ and x lies in the complement of the double of this ball. To prove (4.3.4), simply observe that since $|K(x,y)| \leq A|x-y|^{-n}$, we have that

$$|T(\tau^{x_0}\varphi_R)(x)| \leq \frac{C_n A}{|x-x_0|^n}$$

whenever $|x - x_0| \geq 20R$. The estimate follows easily. □

4.3.2 The Proof of Theorem 4.3.3

This subsection is dedicated to the proof of Theorem 4.3.3.

Proof. The proof is based on a series of steps. We begin by showing that condition (iii) implies condition (iv).

(iii) $\Longrightarrow$ (iv)

Fix a $\mathscr{C}_0^\infty$ function ϕ with $0 \le \phi \le 1$, supported in the ball $B(0,4)$, and equal to 1 on the ball $B(0,2)$. We consider the functions $\phi(\cdot/R)$ that tend to 1 as $R \to \infty$ and we show that $T(1)$ is the weak limit of the functions $T(\phi(\cdot/R))$. This means that for all $g \in \mathscr{D}_0$ (smooth functions with compact support and integral zero) one has

$$\langle T(\phi(\cdot/R)),g\rangle \to \langle T(1),g\rangle \tag{4.3.5}$$

as $R \to \infty$. To prove (4.3.5) we fix a $\mathscr{C}_0^\infty$ function η that is equal to one on the support of g. Then we write

$$\begin{aligned}\langle T(\phi(\cdot/R)),g\rangle &= \langle T(\eta\phi(\cdot/R)),g\rangle + \langle T((1-\eta)\phi(\cdot/R)),g\rangle \\ &= \langle T(\eta\phi(\cdot/R)),g\rangle \\ &\quad + \int_{\mathbf{R}^n}\int_{\mathbf{R}^n} \big(K(x,y)-K(x_0,y)\big)g(x)(1-\eta(y))\phi(y/R)\,dy\,dx,\end{aligned}$$

where x_0 is a point in the support of g. There exists an $R_0 > 0$ such that for $R \ge R_0$, $\phi(\cdot/R)$ is equal to 1 on the support of η, and moreover the expressions

$$\int_{\mathbf{R}^n}\int_{\mathbf{R}^n} \big(K(x,y)-K(x_0,y)\big)g(x)(1-\eta(y))\phi(y/R)\,dy\,dx$$

converge to

$$\int_{\mathbf{R}^n}\int_{\mathbf{R}^n} \big(K(x,y)-K(x_0,y)\big)g(x)(1-\eta(y))\,dy\,dx$$

as $R \to \infty$ by the Lebesgue dominated convergence theorem. Using Definition 4.1.16, we obtain the validity of (4.3.5).

Next we observe that the functions $\phi(\cdot/R)$ are in L^2. We show that

$$\big\|T(\phi(\cdot/R))\big\|_{BMO} \le C_{n,\delta}(A+B_3) \tag{4.3.6}$$

uniformly in $R > 0$. Once (4.3.6) is established, then the sequence $\{T(\phi(\cdot/j))\}_{j=1}^\infty$ lies in a multiple of the unit ball of $BMO = (H^1)^*$, and by the Banach–Alaoglou theorem, there is a subsequence of the positive integers R_j such that $T(\phi(\cdot/R_j))$ converges weakly to an element b in BMO. This means that

$$\langle T(\phi(\cdot/R_j)),g\rangle \to \langle b,g\rangle \tag{4.3.7}$$

as $j \to \infty$ for all $g \in \mathscr{D}_0$. Using (4.3.5), we conclude that $T(1)$ can be identified with the BMO function b, and as a consequence of (4.3.6) it satisfies

$$\big\|T(1)\big\|_{BMO} \le C_{n,\delta}(A+B_3).$$

In a similar fashion, we identify $T^t(1)$ with a BMO function with norm satisfying

$$\big\|T^t(1)\big\|_{BMO} \le C_{n,\delta}(A+B_3).$$

We return to the proof of (4.3.6). We fix a ball $B = B(x_0, r)$ with radius $r > 0$ centered at $x_0 \in \mathbf{R}^n$. If we had a constant c_B such that

$$\frac{1}{|B|}\int_B |T(\phi(\cdot/R))(x) - c_B|\,dx \le c_{n,\delta}\,(A + B_3) \tag{4.3.8}$$

for all $R > 0$, then property (3) in Proposition 3.1.2 (adapted to balls) would yield (4.3.6). Obviously, (4.3.8) is a consequence of the two estimates

$$\frac{1}{|B|}\int_B \left|T\left[\phi(\tfrac{\cdot - x_0}{r})\phi(\tfrac{\cdot}{R})\right](x)\right|dx \le c_n B_3\,, \tag{4.3.9}$$

$$\frac{1}{|B|}\int_B \left|T\left[(1-\phi(\tfrac{\cdot - x_0}{r}))\phi(\tfrac{\cdot}{R})\right](x) - T\left[(1-\phi(\tfrac{\cdot - x_0}{r}))\phi(\tfrac{\cdot}{R})\right](x_0)\right|dx \le \frac{c_n}{\delta}A\,. \tag{4.3.10}$$

We bound the double integral in (4.3.10) by

$$\frac{1}{|B|}\int_B \int_{|y-x_0|\ge 2r} |K(x,y) - K(x_0,y)|\,\phi(y/R)\,dy\,dx\,, \tag{4.3.11}$$

since $1 - \phi((y-x_0)/r) = 0$ when $|y - x_0| \le 2r$. Since $|x - x_0| \le r \le \frac{1}{2}|y - x_0|$, condition (4.1.2) gives that (4.3.11) holds with $c_n = \omega_{n-1} = |\mathbf{S}^{n-1}|$.

It remains to prove (4.3.9). It is easy to verify that there is a constant $C_0 = C_0(n,\phi)$ such that for $0 < \varepsilon \le 1$ and for all $a \in \mathbf{R}^n$ the functions

$$C_0^{-1}\phi(\varepsilon(x+a))\phi(x), \qquad C_0^{-1}\phi(x)\phi(-a+\varepsilon x) \tag{4.3.12}$$

are normalized bumps. The important observation is that with $a = x_0/r$ we have

$$\phi(\tfrac{x}{R})\phi(\tfrac{x-x_0}{r}) = r^n \tau^{x_0}\left[\left(\phi\left(\tfrac{r}{R}(\cdot + a)\right)\phi(\cdot)\right)_r\right](x) \tag{4.3.13}$$

$$= R^n\left(\phi(\cdot)\phi\left(-a + \tfrac{R}{r}(\cdot)\right)\right)_R(x), \tag{4.3.14}$$

and thus in either case $r \le R$ or $R \le r$, one may express the product $\phi(\frac{x}{R})\phi(\frac{x-x_0}{r})$ as a multiple of a translation of an L^1 dilation of a normalized bump.

Let us suppose that $r \le R$. In view of (4.3.13) we write

$$T\left[\phi(\tfrac{\cdot - x_0}{r})\phi(\tfrac{\cdot}{R})\right](x) = C_0 r^n T\left[\tau^{x_0}\varphi_r\right](x)$$

for some normalized bump φ. Using this fact and the Cauchy–Schwarz inequality, we estimate the expression on the left in (4.3.9) by

$$\frac{C_0\, r^{n/2}}{|B|^{\frac{1}{2}}} r^{n/2}\left(\int_B \left|T\left[\tau^{x_0}\varphi_r\right](x)\right|^2 dx\right)^{\frac{1}{2}} \le \frac{C_0\, r^{n/2}}{|B|^{\frac{1}{2}}} B_3 = c_n B_3\,,$$

where the first inequality follows by applying hypothesis (iii).

We now consider the case $R \le r$. In view of (4.3.14) we write

$$T\left[\phi\left(\tfrac{\cdot - x_0}{r}\right)\phi\left(\tfrac{\cdot}{R}\right)\right](x) = C_0 R^n T\left(\varphi_R\right)(x)$$

for some other normalized bump φ. Using this fact and the Cauchy–Schwarz inequality, we estimate the expression on the left in (4.3.9) by

$$\frac{C_0 R^{n/2}}{|B|^{\frac{1}{2}}} R^{n/2} \left(\int_B |T(\zeta_R)(x)|^2 \, dx \right)^{\frac{1}{2}} \le \frac{C_0 R^{n/2}}{|B|^{\frac{1}{2}}} B_3 \le c_n B_3$$

applying hypothesis (iii) and recalling that $R \le r$. This proves (4.3.9).

To finish the proof of (iv), we need to prove that T satisfies the weak boundedness property. But this is elementary, since for all normalized bumps φ and ψ and all $x \in \mathbf{R}^n$ and $R > 0$ we have

$$\begin{aligned} \left|\left\langle T(\tau^x \psi_R), \tau^x \varphi_R \right\rangle\right| &\le \left\|T(\tau^x \psi_R)\right\|_{L^2} \left\|\tau^x \varphi_R\right\|_{L^2} \\ &\le B_3 R^{-\frac{n}{2}} \left\|\tau^x \varphi_R\right\|_{L^2} \\ &\le C_n B_3 R^{-n}. \end{aligned}$$

This gives $\|T\|_{WB} \le C_n B_3$, which implies the estimate $B_4 \le C_{n,\delta}(A + B_3)$ and concludes the proof of the fact that condition (iii) implies (iv).

(iv) $\Longrightarrow$ (L^2 boundedness of T)

We now assume condition (iv) and we present the most important step of the proof, establishing the fact that T has an extension that maps $L^2(\mathbf{R}^n)$ to itself. The assumption that the distributions $T(1)$ and $T^t(1)$ coincide with BMO functions leads to the construction of Carleson measures that provide the key tool in the boundedness of T.

We pick a smooth radial function Φ with compact support that is supported in the ball $B(0, \frac{1}{2})$ and that satisfies $\int_{\mathbf{R}^n} \Phi(x)\, dx = 1$. For $t > 0$ we define $\Phi_t(x) = t^{-n}\Phi(\frac{x}{t})$. Since Φ is a radial function, the operator

$$P_t(f) = f * \Phi_t \tag{4.3.15}$$

is self-transpose. The operator P_t is a continuous analogue of $S_j = \sum_{k \le j} \Delta_k$, where the Δ_j's are the Littlewood–Paley operators.

We now fix a Schwartz function f whose Fourier transform is supported away from a neighborhood of the origin. We discuss an integral representation for $T(f)$. We begin with the facts, which can be found in Exercises 4.3.1 and 4.3.2, that

$$\begin{aligned} T(f) &= \lim_{s \to 0} P_s^2 T P_s^2(f), \\ 0 &= \lim_{s \to \infty} P_s^2 T P_s^2(f), \end{aligned}$$

where the first limit is in the topology of $\mathscr{S}'(\mathbf{R}^n)$ and the second one is in the topology of $\mathscr{S}'(\mathbf{R}^n)/\mathscr{P}(\mathbf{R}^n)$. Thus, with the use of the fundamental theorem of calculus and the product rule, we are able to write

$$\begin{aligned} T(f) &= \lim_{s\to 0} P_s^2 T P_s^2(f) - \lim_{s\to\infty} P_s^2 T P_s^2(f) \\ &= -\lim_{\varepsilon\to 0}\int_\varepsilon^{\frac{1}{\varepsilon}} s\frac{d}{ds}(P_s^2 T P_s^2)(f)\frac{ds}{s} \\ &= -\lim_{\varepsilon\to 0}\int_\varepsilon^{\frac{1}{\varepsilon}} \left[s\left(\frac{d}{ds}P_s^2\right)TP_s^2(f) + P_s^2\left(Ts\frac{d}{ds}P_s^2\right)(f)\right]\frac{ds}{s}, \end{aligned} \tag{4.3.16}$$

where the limit is in the sense of $\mathscr{S}'(\mathbf{R}^n)/\mathscr{P}(\mathbf{R}^n)$. For a Schwartz function g we have

$$\begin{aligned} \left(s\frac{d}{ds}P_s^2(g)\right)^{\widehat{}}(\xi) &= \widehat{g}(\xi)\, s\frac{d}{ds}\widehat{\Phi}(s\xi)^2 \\ &= \widehat{g}(\xi)\,\widehat{\Phi}(s\xi)\,(2s\xi\cdot\nabla\widehat{\Phi}(s\xi)) \\ &= \widehat{g}(\xi)\sum_{k=1}^n \widehat{\Psi_k}(s\xi)\widehat{\Theta_k}(s\xi) \\ &= \sum_{k=1}^n \left(\widetilde{Q}_{k,s}Q_{k,s}(g)\right)^{\widehat{}}(\xi) = \sum_{k=1}^n \left(Q_{k,s}\widetilde{Q}_{k,s}(g)\right)^{\widehat{}}(\xi), \end{aligned}$$

where for $1 \le k \le n$, $\widehat{\Psi_k}(\xi) = 2\xi_k\widehat{\Phi}(\xi)$, $\widehat{\Theta_k}(\xi) = \partial_k\widehat{\Phi}(\xi)$, and $Q_{k,s}$, $\widetilde{Q}_{k,s}$ are operators defined by

$$Q_{k,s}(g) = g*(\Psi_k)_s\,, \qquad \widetilde{Q}_{k,s}(g) = g*(\Theta_k)_s\,;$$

here $(\Theta_k)_s(x) = s^{-n}\Theta_k(s^{-1}x)$ and $(\Psi_k)_s$ are defined similarly. Observe that Ψ_k and Θ_k are smooth odd bumps supported in $B(0,\frac{1}{2})$ and have integral zero. Since Ψ_k and Θ_k are odd, they are anti-self-transpose, meaning that $(Q_{k,s})^t = -Q_{k,s}$ and $(\widetilde{Q}_{k,s})^t = -\widetilde{Q}_{k,s}$. We now write the expression in (4.3.16) as

$$-\lim_{\varepsilon\to 0}\sum_{k=1}^n\left[\int_\varepsilon^{\frac{1}{\varepsilon}}\widetilde{Q}_{k,s}Q_{k,s}TP_sP_s(f)\frac{ds}{s} + \int_\varepsilon^{\frac{1}{\varepsilon}}P_sP_sTQ_{k,s}\widetilde{Q}_{k,s}(f)\frac{ds}{s}\right], \tag{4.3.17}$$

where the limit is in the sense of $\mathscr{S}'(\mathbf{R}^n)/\mathscr{P}(\mathbf{R}^n)$. We set

$$T_{k,s} = Q_{k,s}TP_s\,,$$

and we observe that the operator $P_sTQ_{k,s}$ is equal to $-((T^t)_{k,s})^t$.

Recall the notation $\tau^x h(z) = h(z-x)$. For a given $\varphi \in \mathscr{S}(\mathbf{R}^n)$ we have

$$\begin{aligned} Q_{k,s}TP_s(\varphi)(x) &= \left\langle TP_s(\varphi), \tau^x\Psi_{k,s}\right\rangle \\ &= \left\langle T(\Phi_s*\varphi), \tau^x\Psi_{k,s}\right\rangle \end{aligned}$$

$$
\begin{aligned}
&= \Big\langle T\Big(\int_{\mathbf{R}^n} \varphi(y)\,(\tau^y\Phi_s)\,dy\Big)\,,\,\tau^x\Psi_{k,s}\Big\rangle \\
&= \int_{\mathbf{R}^n} \langle T(\tau^y\Phi_s)\,,\,\tau^x\Psi_{k,s}\rangle \varphi(y)\,dy. \qquad (4.3.18)
\end{aligned}
$$

The last equality is justified by the convergence of the Riemann sums R_N of the integral $I = \int_{\mathbf{R}^n} \varphi(y)\,(\tau^y\Phi_s)(\cdot)\,dy$ to itself in the topology of $\mathscr{S}$ (this is contained in the proof of Theorem 2.3.20 in [156]); by the continuity of T, $T(R_N)$ converges to $T(I)$ in $\mathscr{S}'$ and thus $\langle T(R_N), \tau^x\Psi_{k,s}\rangle$ converges to $\langle T(I), \tau^x\Psi_{k,s}\rangle$. But $\langle T(R_N), \tau^x\Psi_{k,s}\rangle$ is also a Riemann sum for the rapidly convergent integral in (4.3.18); hence it converges to it as well.

We deduce that the operator $T_{k,s} = Q_{k,s}TP_s$ has kernel

$$
K_{k,s}(x,y) = \big\langle T(\tau^y\Phi_s), \tau^x(\Psi_k)_s\big\rangle = \big\langle T^t(\tau^x(\Psi_k)_s), \tau^y\Phi_s\big\rangle. \qquad (4.3.19)
$$

Hence, the operator $P_sTQ_{k,s} = -((T^t)_{k,s})^t$ has kernel

$$
-\big\langle T^t(\tau^x\Phi_s), \tau^y(\Psi_k)_s\big\rangle = -\big\langle T(\tau^y(\Psi_k)_s), \tau^x\Phi_s\big\rangle.
$$

For $1 \le k \le n$ we need the following facts regarding the kernels of these operators:

$$
\begin{aligned}
\big|\big\langle T(\tau^y(\Psi_k)_s), \tau^x\Phi_s\big\rangle\big| &\le C_{n,\delta}\big(\|T\|_{WB}+A\big)\,p_s(x-y), \qquad (4.3.20)\\
\big|\big\langle T^t(\tau^x(\Psi_k)_s), \tau^y\Phi_s\big\rangle\big| &\le C_{n,\delta}\big(\|T\|_{WB}+A\big)\,p_s(x-y), \qquad (4.3.21)
\end{aligned}
$$

where

$$
p_t(u) = \frac{1}{t^n}\frac{1}{(1+|\frac{u}{t}|)^{n+\delta}}
$$

is the L^1 dilation of the function $p(u) = (1+|u|)^{-n-\delta}$.

To prove (4.3.21), we consider the following two cases: If $|x-y| \le 5s$, then the weak boundedness property gives

$$
\big|\big\langle T(\tau^y\Phi_s), \tau^x(\Psi_k)_s\big\rangle\big| = \big|\big\langle T(\tau^x((\tau^{\frac{y-x}{s}}\Phi)_s)), \tau^x(\Psi_k)_s\big\rangle\big| \le \frac{C_n\|T\|_{WB}}{s^n},
$$

since both Ψ_k and $\tau^{\frac{y-x}{s}}\Phi$ are multiples of normalized bumps. Notice here that both of these functions are supported in $B(0,10)$, since $\frac{1}{s}|x-y| \le 5$. This estimate proves (4.3.21) when $|x-y| \le 5s$.

We now turn to the case $|x-y| \ge 5s$. Then the functions $\tau^y\Phi_s$ and $\tau^x(\Psi_k)_s$ have disjoint supports and so we have the integral representation

$$
\big\langle T^t(\tau^x(\Psi_k)_s), \tau^y\Phi_s\big\rangle = \int_{\mathbf{R}^n}\int_{\mathbf{R}^n} \Phi_s(v-y)K(u,v)(\Psi_k)_s(u-x)\,du\,dv.
$$

Using that Ψ_k has mean value zero, we can write the previous expression as

$$
\int_{\mathbf{R}^n}\int_{\mathbf{R}^n} \Phi_s(v-y)\big(K(u,v)-K(x,v)\big)(\Psi_k)_s(u-x)\,du\,dv.
$$

We observe that $|u-x| \le s$ and $|v-y| \le s$ in the preceding double integral. Since $|x-y| \ge 5s$, this makes $|u-v| \ge |x-y| - 2s \ge 3s$, which implies that $|u-x| \le \frac{1}{2}|u-v|$. Using (4.1.2), we obtain

$$|K(u,v) - K(x,v)| \le \frac{A|x-u|^{\delta}}{(|u-v|+|x-v|)^{n+\delta}} \le C_{n,\delta} A \frac{s^{\delta}}{|x-y|^{n+\delta}},$$

where we used the fact that $|u-v| \approx |x-y|$. Inserting this estimate in the double integral, we obtain (4.3.21). Estimate (4.3.20) is proved similarly.

At this point we drop the dependence of $Q_{k,s}$ and $\widetilde{Q}_{k,s}$ on the index k, since we can concentrate on one term of the sum in (4.3.17). We have managed to express $T(f)$ as a finite sum of operators of the form

$$\int_0^{\infty} \widetilde{Q}_s T_s P_s(f) \frac{ds}{s} \tag{4.3.22}$$

and of the form

$$\int_0^{\infty} P_s T_s \widetilde{Q}_s(f) \frac{ds}{s}, \tag{4.3.23}$$

where the preceding integrals converge in $\mathscr{S}'(\mathbf{R}^n)/\mathscr{P}(\mathbf{R}^n)$ and the T_s's have kernels $K_s(x,y)$, which are pointwise dominated by a constant multiple of $(A+B_4)p_s(x-y)$.

It suffices to obtain L^2 bounds for an operator of the form (4.3.22) with constant at most a multiple of $A+B_4$. Then by duality the same estimate also holds for the operators of the form (4.3.23). We make one more observation. Using (4.3.19) (recall that we have dropped the indices k), we obtain

$$T_s(1)(x) = \int_{\mathbf{R}^n} K_s(x,y)\,dy = \langle T^t(\tau^x \Psi_s), 1\rangle = (\Psi_s * T(1))(x), \tag{4.3.24}$$

where all integrals converge absolutely.

We can therefore concentrate on the L^2 boundedness of the operator in (4.3.22). We pair this operator with a Schwartz function g in $\mathscr{S}_0(\mathbf{R}^n)$ and we use the convergence of the integral in $\mathscr{S}'/\mathscr{P}(\mathbf{R}^n)$ and the property $(\widetilde{Q}_s)^t = -\widetilde{Q}_s$ to obtain

$$\left\langle \int_0^{\infty} \widetilde{Q}_s T_s P_s(f) \frac{ds}{s}, g \right\rangle = \int_0^{\infty} \langle \widetilde{Q}_s T_s P_s(f), g\rangle \frac{ds}{s} = -\int_0^{\infty} \langle T_s P_s(f), \widetilde{Q}_s(g)\rangle \frac{ds}{s}.$$

The intuition here is as follows: T_s is an averaging operator at scale s and $P_s(f)$ is essentially constant on that scale. Therefore, the expression $T_s P_s(f)$ must look like $T_s(1)P_s(f)$. To be precise, we introduce this term and try to estimate the error that occurs. We have

$$T_s P_s(f) = T_s(1)P_s(f) + \big[T_s P_s(f) - T_s(1)P_s(f)\big]. \tag{4.3.25}$$

We estimate the terms that arise from this splitting. Recalling (4.3.24), we write

$$\left| \int_0^\infty \left\langle (\Psi_s * T(1)) P_s(f), \widetilde{Q}_s(g) \right\rangle \frac{ds}{s} \right| \tag{4.3.26}$$

$$\leq \left(\int_0^\infty \|P_s(f)\,(\Psi_s * T(1))\|_{L^2}^2 \frac{ds}{s} \right)^{\frac{1}{2}} \left(\int_0^\infty \|\widetilde{Q}_s(g)\|_{L^2}^2 \frac{ds}{s} \right)^{\frac{1}{2}}$$

$$= \left\| \left(\int_0^\infty |P_s(f)\,(\Psi_s * T(1))|^2 \frac{ds}{s} \right)^{\frac{1}{2}} \right\|_{L^2} \left\| \left(\int_0^\infty |\widetilde{Q}_s(g)|^2 \frac{ds}{s} \right)^{\frac{1}{2}} \right\|_{L^2}. \tag{4.3.27}$$

Since $T(1)$ is a BMO function, $|(\Psi_s * T(1))(x)|^2 dx \frac{ds}{s}$ is a Carleson measure on $\mathbf{R}^{n+1}_+$. Using Theorem 3.3.8 and the continuous version of the Littlewood–Paley theorem (Exercise 6.1.4 in [156]), we obtain that (4.3.27) is controlled by

$$C_n \|T(1)\|_{BMO} \|f\|_{L^2} \|g\|_{L^2} \leq C_n B_4 \|f\|_{L^2} \|g\|_{L^2} .$$

This gives the sought estimate for the first term in (4.3.25). For the second term in (4.3.25) we have

$$\left| \int_0^\infty \int_{\mathbf{R}^n} \widetilde{Q}_s(g)(x) \left[T_s P_s(f) - T_s(1) P_s(f) \right](x)\, dx \frac{ds}{s} \right|$$

$$\leq \left(\int_0^\infty \int_{\mathbf{R}^n} |\widetilde{Q}_s(g)(x)|^2 dx \frac{ds}{s} \right)^{\frac{1}{2}} \left(\int_0^\infty \int_{\mathbf{R}^n} |(T_s P_s(f) - T_s(1) P_s(f))(x)|^2 dx \frac{ds}{s} \right)^{\frac{1}{2}}$$

$$\leq C_n \|g\|_{L^2} \left(\int_0^\infty \int_{\mathbf{R}^n} \left| \int_{\mathbf{R}^n} K_s(x,y) [P_s(f)(y) - P_s(f)(x)]\, dy \right|^2 dx \frac{ds}{s} \right)^{\frac{1}{2}}$$

$$\leq C_n (A + B_4) \|g\|_{L^2} \left(\int_0^\infty \int_{\mathbf{R}^n} \int_{\mathbf{R}^n} p_s(x-y) |P_s(f)(y) - P_s(f)(x)|^2 dy\, dx \frac{ds}{s} \right)^{\frac{1}{2}},$$

where in the last estimate we used the fact that the measure $p_s(x-y)\,dy$ is a multiple of a probability measure. It suffices to estimate the last displayed square root. Changing variables $u = x - y$ and applying Plancherel's theorem, we express this square root as

$$\left(\int_0^\infty \int_{\mathbf{R}^n} \int_{\mathbf{R}^n} p_s(u) |P_s(f)(y) - P_s(f)(y+u)|^2 du\, dy \frac{ds}{s} \right)^{\frac{1}{2}}$$

$$= \left(\int_0^\infty \int_{\mathbf{R}^n} \int_{\mathbf{R}^n} p_s(u) |\widehat{\Phi}(s\xi) - \widehat{\Phi}(s\xi) e^{2\pi i u \cdot \xi}|^2 |\widehat{f}(\xi)|^2 du\, d\xi \frac{ds}{s} \right)^{\frac{1}{2}}$$

$$\leq \left(\int_0^\infty \int_{\mathbf{R}^n} \int_{\mathbf{R}^n} p_s(u) |\widehat{\Phi}(s\xi)|^2 4\pi^{\frac{\delta}{2}} |u|^{\frac{\delta}{2}} |\xi|^{\frac{\delta}{2}} |\widehat{f}(\xi)|^2 du\, d\xi \frac{ds}{s} \right)^{\frac{1}{2}}$$

$$= 2\pi^{\frac{\delta}{4}} \left(\int_{\mathbf{R}^n} \int_0^\infty \left(\int_{\mathbf{R}^n} p_s(u) |\tfrac{u}{s}|^{\frac{\delta}{2}} du \right) |\widehat{\Phi}(s\xi)|^2 |s\xi|^{\frac{\delta}{2}} \frac{ds}{s} |\widehat{f}(\xi)|^2 d\xi \right)^{\frac{1}{2}},$$

and we claim that this last expression is bounded by $C_{n,\delta}\|f\|_{L^2}$. Indeed, we first bound the quantity $\int_{\mathbf{R}^n} p_s(u)\left|\frac{u}{s}\right|^{\delta/2} du$ by a constant, and then we use the estimate

$$\int_0^\infty |\widehat{\Phi}(s\xi)|^2 |s\xi|^{\frac{\delta}{2}} \frac{ds}{s} = \int_0^\infty |\widehat{\Phi}(se_1)|^2 s^{\frac{\delta}{2}} \frac{ds}{s} \le C'_{n,\delta} < \infty$$

and Plancherel's theorem to obtain the claim. [Here $e_1 = (1,0,\dots,0)$.] Taking g to be an arbitrary function in $\mathscr{S}_0(\mathbf{R}^n)$ with L^2 norm at most 1 and using duality and the fact that $\mathscr{S}_0(\mathbf{R}^n)$ is dense in $L^2(\mathbf{R}^n)$, we deduce the estimate $\|T(f)\|_{L^2} \le C_{n,\delta}(A+B_4)\|f\|_{L^2}$ for all Schwartz functions f whose Fourier transform does not contain a neighborhood of the origin. Such functions are dense in $L^2(\mathbf{R}^n)$ (cf. Exercise 6.2.9 in [156]) and thus T admits an extension on L^2 that satisfies $\|T\|_{L^2\to L^2} \le C_{n,\delta}(A+B_4)$.

(L^2 boundedness of T) $\implies$ (v)

If T has an extension that maps L^2 to itself, then by Theorem 4.2.7 we have

$$B_5 \le C_{n,\delta}\big(A + \|T\|_{L^2\to L^2}\big) < \infty.$$

Thus the boundedness of T on L^2 implies condition (v).

(v) $\implies$ (vi)

At a formal level the proof of this fact is clear, since we can write a normalized bump as the inverse Fourier transform of its Fourier transform and interchange the integrations with the action of T to obtain

$$T(\tau^{x_0}\varphi_R) = \int_{\mathbf{R}^n} \widehat{\tau^{x_0}\varphi_R}(\xi)\, T(e^{2\pi i\xi\cdot(\cdot)})\, d\xi. \tag{4.3.28}$$

The conclusion follows by taking BMO norms. To make identity (4.3.28) precise we provide the following argument.

Let us fix a normalized bump φ and a smooth and compactly supported function g with mean value zero. We pick a smooth function η with compact support that is equal to 1 on the unit ball and vanishes outside the double of that ball. Define $\eta_k(\xi) = \eta(\xi/k)$ and note that η_k tends pointwise to 1 as $k\to\infty$. Observe that $\eta_k\tau^{x_0}\varphi_R$ converges to $\tau^{x_0}\varphi_R$ in $\mathscr{S}(\mathbf{R}^n)$ as $k\to\infty$, and by the continuity of T we obtain

$$\lim_{k\to\infty} \langle T(\eta_k\tau^{x_0}\varphi_R), g\rangle = \langle T(\tau^{x_0}\varphi_R), g\rangle.$$

We have

$$\begin{aligned} T\big(\eta_k e^{2\pi i\xi\cdot(\cdot)}\big) &= T\bigg(\int_{\mathbf{R}^n} \widehat{\tau^{x_0}\varphi_R}(\xi)\eta_k(\cdot)e^{2\pi i\xi\cdot(\cdot)}d\xi\bigg) \\ &= \int_{\mathbf{R}^n} \widehat{\tau^{x_0}\varphi_R}(\xi)\, T\big(\eta_k(\cdot)e^{2\pi i\xi\cdot(\cdot)}\big)\, d\xi, \end{aligned} \tag{4.3.29}$$

where the second equality is justified by the continuity and linearity of T along with the fact that the Riemann sums of the integral in (4.3.29) converge to that integral in $\mathscr{S}$ (a proof of this fact is essentially contained in the proof of Theorem 2.3.21 in [156]). Consequently,

$$\left\langle T(\tau^{x_0}\varphi_R),g\right\rangle = \lim_{k\to\infty}\int_{\mathbf{R}^n}\widehat{\tau^{x_0}\varphi_R}(\xi)\left\langle T\left(\eta_k e^{2\pi i\xi\cdot(\cdot)}\right),g\right\rangle d\xi. \tag{4.3.30}$$

Let W be the distributional kernel of T. By (4.1.5) we have

$$\left\langle T\left(\eta_k e^{2\pi i\xi\cdot(\cdot)}\right),g\right\rangle = \left\langle W,g\otimes\eta_k e^{2\pi i\xi\cdot(\cdot)}\right\rangle. \tag{4.3.31}$$

Using (4.1.6), we obtain that the expression in (4.3.31) is controlled by a finite sum of L^∞ norms of derivatives of the function

$$g(x)\,\eta_k(y)e^{2\pi i\xi\cdot y}$$

on some compact set (that depends on g). Then for some $M>0$ and some constant $C(g)$ depending on g, we have that this sum of L^∞ norms of derivatives is controlled by

$$C(g)\,(1+|\xi|)^M$$

uniformly in $k\geq 1$. Since $\widehat{\tau^{x_0}\varphi_R}$ is integrable, the Lebesgue dominated convergence theorem allows us to pass the limit inside the integrals in (4.3.30) to obtain

$$\left\langle T(\tau^{x_0}\varphi_R),g\right\rangle = \int_{\mathbf{R}^n}\widehat{\tau^{x_0}\varphi_R}(\xi)\left\langle T\left(e^{2\pi i\xi\cdot(\cdot)}\right),g\right\rangle d\xi.$$

We now use assumption (v). The distributions $T\left(e^{2\pi i\xi\cdot(\cdot)}\right)$ coincide with BMO functions whose norm is at most B_5. It follows that

$$\begin{aligned}\left|\left\langle T(\tau^{x_0}\varphi_R),g\right\rangle\right| &\leq \left\|\widehat{\tau^{x_0}\varphi_R}\right\|_{L^1}\sup_{\xi\in\mathbf{R}^n}\left\|T\left(e^{2\pi i\xi\cdot(\cdot)}\right)\right\|_{BMO}\left\|g\right\|_{H^1}\\ &\leq C_nB_5R^{-n}\left\|g\right\|_{H^1},\end{aligned} \tag{4.3.32}$$

where the constant C_n is independent of the normalized bump φ in view of (4.3.1). It follows from (4.3.32) that

$$g\mapsto\left\langle T(\tau^{x_0}\varphi_R),g\right\rangle$$

is a bounded linear functional on BMO with norm at most a multiple of B_5R^{-n}. It follows from Theorem 3.2.2 that $T(\tau^{x_0}\varphi_R)$ coincides with a BMO function that satisfies

$$R^n\left\|T(\tau^{x_0}\varphi_R)\right\|_{BMO}\leq C_nB_5.$$

The same argument is valid for T^t, and this shows that

$$B_6\leq C_{n,\delta}(A+B_5)$$

and concludes the proof that (v) implies (vi).

(vi) $\Longrightarrow$ (iii)

We fix $x_0 \in \mathbf{R}^n$ and $R > 0$. Pick z_0 in $\mathbf{R}^n$ such that $|x_0 - z_0| = 40R$. Then if $|y-x_0| \le 10R$ and $|x-z_0| \le 20R$ we have

$$\begin{aligned} 10R &\le |z_0-x_0| - |x-z_0| - |y-x_0| \\ &\le |x-y| \\ &\le |x-z_0| + |z_0-x_0| + |x_0-y| \le 70R. \end{aligned}$$

From this it follows that when $|x-z_0| \le 20R$ we have

$$\left|\int_{|y-x_0|\le 10R} K(x,y)\tau^{x_0}\varphi_R(y)\,dy\right| \le \int_{10R\le|x-y|\le 70R} |K(x,y)|\frac{dy}{R^n} \le \frac{C_{n,\delta}A}{R^n}$$

and thus

$$\left|\operatorname*{Avg}_{B(z_0,20R)} T(\tau^{x_0}\varphi_R)\right| \le \frac{C_{n,\delta}A}{R^n}, \tag{4.3.33}$$

where $\mathrm{Avg}_B\, g$ denotes the average of g over B. Because of assumption (vi), the BMO norm of the function $T(\tau^{x_0}\varphi_R)$ is bounded by a multiple of B_6R^{-n}, a fact used in the following sequence of implications. We have

$$\begin{aligned} &\big\|T(\tau^{x_0}\varphi_R)\big\|_{L^2(B(x_0,20R))} \\ &\le \left\|T(\tau^{x_0}\varphi_R) - \operatorname*{Avg}_{B(x_0,20R)} T(\tau^{x_0}\varphi_R)\right\|_{L^2(B(x_0,20R))} \\ &\qquad + v_n^{\frac{1}{2}}(20R)^{\frac{n}{2}}\left|\operatorname*{Avg}_{B(x_0,20R)} T(\tau^{x_0}(\varphi_R) - \operatorname*{Avg}_{B(z_0,20R)} T(\tau^{x_0}\varphi_R)\right| \\ &\qquad + v_n^{\frac{1}{2}}(20R)^{\frac{n}{2}}\left|\operatorname*{Avg}_{B(z_0,20R)} T(\tau^{x_0}\varphi_R)\right| \\ &\le C_{n,\delta}\Big(R^{\frac{n}{2}}\big\|T(\tau^{x_0}\varphi_R)\big\|_{BMO} + R^{\frac{n}{2}}\big\|T(\tau^{x_0}\varphi_R)\big\|_{BMO} + R^{-\frac{n}{2}}A\Big) \\ &\le C_{n,\delta}R^{-\frac{n}{2}}(B_6+A), \end{aligned}$$

where we used (4.3.33) and Exercise 3.1.6. Now we have that

$$\big\|T(\tau^{x_0}\varphi_R)\big\|_{L^2(B(x_0,20R)^c)} \le C_{n,\delta}AR^{-\frac{n}{2}}$$

in view of Lemma 4.3.5. Since the same computations apply to T^t, it follows that

$$R^{\frac{n}{2}}\big(\|T(\tau^{x_0}\varphi_R)\|_{L^2} + \|T^t(\tau^{x_0}\varphi_R)\|_{L^2}\big) \le C_{n,\delta}(A+B_6), \tag{4.3.34}$$

which proves that $B_3 \le C_{n,\delta}(A+B_6)$ and hence (iii). This concludes the proof of the fact that (vi) implies (iii)

We have now completed the proof of the following equivalence of statements:

$$\big(L^2 \text{ boundedness of } T\big) \iff \text{(iii)} \iff \text{(iv)} \iff \text{(v)} \iff \text{(vi)} \qquad (4.3.35)$$

and we have established that

$$\big\|T\big\|_{L^2\to L^2} \approx A+B_3 \approx A+B_4 \approx A+B_5 \approx A+B_6\,.$$

(i) $\implies$ (ii)

We show that the quantity B_2 is bounded by a multiple of $A+B_1$; if so then so would do the quantity $A+B_2$. We set

$$I_{\varepsilon,N}(x) = \int_{\varepsilon<|x-y|<N} K(x,y)\,dy \qquad \text{and} \qquad I^t_{\varepsilon,N}(x) = \int_{\varepsilon<|x-y|<N} K^t(x,y)\,dy\,.$$

It suffices to show that there is a constant C_n such that for any x_1 in $\mathbf{R}^n$ we have

$$\sup_{\varepsilon,N}\left[\frac{1}{N^n}\int_{|x-x_1|<\frac{N}{2}} |I_{\varepsilon,N}(x)|^2\,dx\right]^{\frac12} \le C_n\,(A+B_1)\,. \qquad (4.3.36)$$

If (4.3.36) holds, then we can cover the ball $B(x_0,N)$ by finitely many balls $B(x_1,N/2)$ and thus deduce

$$\sup_{x_0\in\mathbf{R}^n}\sup_{\varepsilon,N}\left[\frac{1}{N^n}\int_{|x-x_0|<N} |I_{\varepsilon,N}(x)|^2\,dx\right]^{\frac12} \le C'_n\,(A+B_1) \qquad (4.3.37)$$

with a larger constant C'_n in place of C_n.

We estimate the expression on the left in (4.3.36) by $I+II$, where

$$I = \sup_{\varepsilon,N}\left[\frac{1}{N^n}\int_{|x-x_1|<\frac{N}{2}} |I_{\varepsilon,N}(x) - T^{(\varepsilon)}(\chi_{B(x_1,N)})(x)|^2\,dx\right]^{\frac12}$$

$$II = \sup_{\varepsilon,N}\left[\frac{1}{N^n}\int_{|x-x_1|<N} |T^{(\varepsilon)}(\chi_{B(x_1,N)})(x)|^2\,dx\right]^{\frac12}.$$

By hypothesis, we have that II is bounded by B_1. Also for $|x-x_1|<\frac{N}{2}$ we have

$$\Big|I_{\varepsilon,N}(x) - T^{(\varepsilon)}(\chi_{B(x_1,N)})(x)\Big| = \Bigg|\int_{\varepsilon<|x-y|<N} K(x,y)\,dy - \int_{\substack{\varepsilon<|x-y|\\ |x_0-y|<N}} K(x,y)\,dy\Bigg|$$

$$\leq \int_{\frac{N}{2}\leq|x-y|\leq\frac{3N}{2}} |K(x,y)|\,dy$$
$$\leq A\omega_{n-1}\log 3,$$

where the first inequality is due to the fact that the symmetric difference of the sets $\{y\in\mathbf{R}^n : \varepsilon<|x-y|<N\}$ and $\{y\in\mathbf{R}^n : \varepsilon<|x_1-y|<N\}$ is contained in the annulus $\frac{N}{2}\leq|x-y|\leq\frac{3N}{2}$ and the second inequality is a consequence of (4.1.1).

Thus I is bounded by $\omega_{n-1}(\log 3)A2^{-n/2}$. Combining the estimates for I and II yields the proof of (4.3.36) and hence of (4.3.37). Similarly, we can prove that

$$\sup_{x_0\in\mathbf{R}^n}\sup_{\varepsilon,N}\left[\frac{1}{N^n}\int_{|x-x_0|<N}|I^t_{\varepsilon,N}(x)|^2\,dx\right]^{\frac12}\leq C'_n\,(A+B_1),$$

which together with (4.3.37) implies that $B_2\leq 2C'_n\,(A+B_1)$.

We now consider the following condition analogous to (iii):

$$\text{(iii)}'\quad B'_3=\sup_{\varphi}\sup_{x_0\in\mathbf{R}^n}\sup_{\substack{\varepsilon>0\\R>0}}R^{\frac{n}{2}}\left[\big\|T^{(\varepsilon)}(\tau^{x_0}\varphi_R)\big\|_{L^2}+\big\|(T^{(\varepsilon)})^t(\tau^{x_0}\varphi_R)\big\|_{L^2}\right]<\infty,$$

where the first supremum is taken over all normalized bumps φ. We continue the proof by showing that this condition is a consequence of (ii).

<u>(ii) $\Longrightarrow$ (iii)$'$</u>

More precisely, we prove that $B'_3\leq C_{n,\delta}(A+B_2)$. To prove (iii)$'$, fix a normalized bump φ, a point $x_0\in\mathbf{R}^n$, and $R>0$. Also fix $x\in\mathbf{R}^n$ with $|x-x_0|\leq 20R$. Then we have

$$T^{(\varepsilon)}(\tau^{x_0}\varphi_R)(x)=\int_{\varepsilon<|x-y|\leq 30R}K^{(\varepsilon)}(x,y)\tau^{x_0}\varphi_R(y)\,dy=U_1(x)+U_2(x),$$

where

$$U_1(x)\;=\;\int_{\varepsilon<|x-y|\leq 30R}K(x,y)\big(\tau^{x_0}\varphi_R(y)-\tau^{x_0}\varphi_R(x)\big)\,dy,$$
$$U_2(x)\;=\;\tau^{x_0}\varphi_R(x)\int_{\varepsilon<|x-y|\leq 30R}K(x,y)\,dy.$$

But we have that $|\tau^{x_0}\varphi_R(y)-\tau^{x_0}\varphi_R(x)|\leq C_nR^{-1-n}|x-y|$; thus we obtain

$$|U_1(x)|\leq C_nAR^{-n}$$

on $B(x_0,20R)$ hence $\big\|U_1\big\|_{L^2(B(x_0,20R))}\leq C_nAR^{-\frac{n}{2}}$. Condition (ii) gives that

$$\big\|U_2\big\|_{L^2(B(x_0,20R))}\leq R^{-n}\big\|I_{\varepsilon,30R}\big\|_{L^2(B(x_0,30R))}\leq B_2(30R)^{\frac{n}{2}}R^{-n}.$$

Combining these two, we obtain

$$\left\|T^{(\varepsilon)}(\tau^{x_0}\varphi_R)\right\|_{L^2(B(x_0,20R))} \le C_n(A+B_2)R^{-\frac{n}{2}} \tag{4.3.38}$$

and likewise for $(T^{(\varepsilon)})^t$. It follows from Lemma 4.3.5 that

$$\left\|T^{(\varepsilon)}(\tau^{x_0}\varphi_R)\right\|_{L^2(B(x_0,20R)^c)} \le C_{n,\delta}AR^{-\frac{n}{2}},$$

which combined with (4.3.38) gives condition (iii)$'$ with constant $B_3' \le C_{n,\delta}(A+B_2)$. This concludes the proof that condition (ii) implies (iii)$'$.

(iii)$'$ $\implies$ $[T^{(\varepsilon)} : L^2 \to L^2$ uniformly in $\varepsilon > 0]$

For $\varepsilon > 0$ we introduce the smooth truncations $T_\zeta^{(\varepsilon)}$ of T by setting

$$T_\zeta^{(\varepsilon)}(f)(x) = \int_{\mathbf{R}^n} K(x,y)\zeta(\tfrac{x-y}{\varepsilon})\, f(y)\,dy,$$

where $\zeta(x)$ is a smooth function that is equal to 1 for $|x| \ge 1$ and vanishes for $|x| \le \frac{1}{2}$. We observe that

$$\left|T_\zeta^{(\varepsilon)}(f) - T^{(\varepsilon)}(f)\right| \le C_n A\, M(f); \tag{4.3.39}$$

thus the uniform boundedness of $T^{(\varepsilon)}$ on L^2 is equivalent to the uniform boundedness of $T_\zeta^{(\varepsilon)}$. In view of Exercise 4.1.3, the kernels of the operators $T_\zeta^{(\varepsilon)}$ lie in $SK(\delta, cA)$ uniformly in $\varepsilon > 0$ (for some constant c), since $\delta \le 1$. Moreover, because of (4.3.39), we see that the operators $T_\zeta^{(\varepsilon)}$ satisfy (iii)$'$ with constant $C_nA + B_3'$. The point to be noted here is that condition (iii) for T (with constant B_3) is identical to condition (iii)$'$ for the operators $T_\zeta^{(\varepsilon)}$ uniformly in $\varepsilon > 0$ (with constant $C_nA + B_3'$).

A careful examination of the proof of the implications

$$\text{(iii)} \implies \text{(iv)} \implies (L^2 \text{ boundedness of } T)$$

reveals that all the estimates obtained depend only on the constants B_3, B_4, and A, but not on the specific operator T. Therefore, these estimates are valid for the operators $T_\zeta^{(\varepsilon)}$ that satisfy condition (iii)$'$. This gives the uniform boundedness of the $T_\zeta^{(\varepsilon)}$ on $L^2(\mathbf{R}^n)$ with bounds at most a constant multiple of $A + B_3'$. The same conclusion also holds for the operators $T^{(\varepsilon)}$.

$[T^{(\varepsilon)} : L^2 \to L^2$ uniformly in $\varepsilon > 0]$ $\implies$ (i)

This implication holds trivially.

We have now established the equivalence of the following statements

$$\text{(i)} \iff \text{(ii)} \iff \text{(iii)}' \iff [T^{(\varepsilon)} : L^2 \to L^2 \text{ uniformly in } \varepsilon > 0] \tag{4.3.40}$$

so that

$$A+B_1 \approx A+B_2 \approx A+B_3' \approx \sup_{\varepsilon>0} \big\|T^{(\varepsilon)}\big\|_{L^2\to L^2}.$$

Finally, it remains to link the sets of equivalent conditions (4.3.35) and (4.3.40). We do this by proving the equivalence of (iii) and (iii)$'$

(iii) $\iff$ (iii)$'$

We first observe that (iii)$'$ implies (iii). Indeed, using duality and (4.3.3), we obtain

$$\begin{aligned}\big\|T(\tau^{x_0}\varphi_R)\big\|_{L^2} &= \sup_{\substack{h\in\mathscr{S}\\ \|h\|_{L^2}\le 1}} \left|\int_{\mathbf{R}^n} T(\tau^{x_0}\varphi_R)(x)\,h(x)\,dx\right| \\ &\le \sup_{\substack{h\in\mathscr{S}\\ \|h\|_{L^2}\le 1}} \limsup_{j\to\infty} \left|\int_{\mathbf{R}^n} T^{(\varepsilon_j)}(\tau^{x_0}\varphi_R)(x)\,h(x)\,dx\right| \\ &\le B_3' R^{-\frac{n}{2}},\end{aligned}$$

which gives $B_3 \le B_3'$.

We have shown that (iii) implies the L^2 boundedness of T. But in view of Corollary 4.2.5, the boundedness of T on L^2 implies the boundedness of $T^{(\varepsilon)}$ on L^2 uniformly in $\varepsilon>0$, which implies (iii)$'$. Moreover B_3' is bounded by a constant multiple of $A+B_3$.

This completes the proof of the equivalence of the six statements (i)–(vi) in such a way that

$$\big\|T\big\|_{L^2\to L^2} \approx (A+B_j)$$

for all $j\in\{1,2,3,4,5,6\}$. The proof of the theorem is now complete. □

Remark 4.3.6. Suppose that condition (4.3.3) is removed from the hypothesis of Theorem 4.3.3. Then the given proof of Theorem 4.3.3 actually shows that (i) and (ii) are equivalent to each other and to the statement that the $T^{(\varepsilon)}$'s have bounded extensions on $L^2(\mathbf{R}^n)$ that satisfy

$$\sup_{\varepsilon>0} \big\|T^{(\varepsilon)}\big\|_{L^2\to L^2} < \infty.$$

Additionally, without hypothesis (4.3.3), the proof of Theorem 4.3.3 also shows that conditions (iii), (iv), (v), and (vi) are equivalent to each other and to the statement that T has an extension that maps $L^2(\mathbf{R}^n)$ to $L^2(\mathbf{R}^n)$.

4.3.3 An Application

We end this section with one application of the $T(1)$ theorem. We begin with the following observation.

Corollary 4.3.7. *Let K be a standard kernel that is antisymmetric, i.e., it satisfies $K(x,y) = -K(y,x)$ for all $x \neq y$. Then a linear continuous operator T associated with K is L^2 bounded if and only if $T(1)$ is in BMO.*

Proof. In view of Exercise 4.3.3, T automatically satisfies the weak boundedness property. Moreover, $T^t = -T$. Therefore, the three conditions of Theorem 4.3.3 (iv) reduce to the single condition $T(1) \in BMO$. □

Example 4.3.8. Let us recall the kernels K_m of Example 4.1.7. These arise in the expansion of the kernel in Example 4.1.6 in geometric series

$$\frac{1}{x-y+i(A(x)-A(y))} = \frac{1}{x-y}\sum_{m=0}^{\infty}\left(i\frac{A(x)-A(y)}{x-y}\right)^m \tag{4.3.41}$$

when $L = \sup_{x\neq y}\frac{|A(x)-A(y)|}{|x-y|} < 1$. The operator with kernel $(i\pi)^{-1}K_m(x,y)$ is called the mth *Calderón commutator* and is denoted by $\mathscr{C}_m$. This operator satisfies

$$\langle \mathscr{C}_m(\varphi),\psi\rangle = \lim_{\varepsilon\to 0}\langle \mathscr{C}_m^{(\varepsilon)}(\varphi),\psi\rangle = \frac{1}{2}\int_{\mathbf{R}}\int_{\mathbf{R}} K_m(x,y)[\varphi(y)\psi(x)-\varphi(x)\psi(y)]\,dxdy$$

for all $\varphi,\psi\in\mathscr{S}(\mathbf{R})$, in view of of (4.1.11), where

$$\mathscr{C}_m^{(\varepsilon)}(\varphi)(x) = \frac{1}{\pi i}\int_{|x-y|>\varepsilon}\left(\frac{A(x)-A(y)}{x-y}\right)^m\frac{1}{x-y}\varphi(y)\,dy. \tag{4.3.42}$$

Thus condition (4.3.3) is satisfied.

We use the $T(1)$ theorem to show that the operators $\mathscr{C}_m$ are L^2 bounded.

We show that there exists a constant $R>0$ such that for all $m\geq 0$ we have

$$\big\|\mathscr{C}_m\big\|_{L^2\to L^2} \leq R^m L^m. \tag{4.3.43}$$

We prove (4.3.43) by induction. We note that (4.3.43) is trivially true when $m=0$, since $\mathscr{C}_0 = -iH$, where H is the Hilbert transform.

Assume that (4.3.43) holds for a certain m. We show its validity for $m+1$. Recall that K_m is a kernel in $SK(1,8(2m+1)L^m)$ by the discussion in Example 4.1.7. We need the following estimate proved in Theorem 4.2.7:

$$\big\|\mathscr{C}_m\big\|_{L^\infty\to BMO} \leq C_2\Big[8(2m+1)L^m + \big\|\mathscr{C}_m\big\|_{L^2\to L^2}\Big], \tag{4.3.44}$$

which holds for some absolute constant C_2.

We start with the following consequence of Theorem 4.3.3:

$$\big\|\mathscr{C}_{m+1}\big\|_{L^2\to L^2} \le C_1\Big[\big\|\mathscr{C}_{m+1}(1)\big\|_{BMO} + \big\|(\mathscr{C}_{m+1})^t(1)\big\|_{BMO} + \big\|\mathscr{C}_{m+1}\big\|_{WB}\Big], \tag{4.3.45}$$

valid for some absolute constant C_1. The key observation is that

$$\mathscr{C}_{m+1}(1) = \mathscr{C}_m(A'), \tag{4.3.46}$$

for which we refer to Exercise 4.3.4. Here A' denotes the derivative of A, which exists almost everywhere, since Lipschitz functions are differentiable almost everywhere. Note that the kernel of $\mathscr{C}_m$ is antisymmetric; consequently, $(\mathscr{C}_m)^t = -\mathscr{C}_m$ and Exercise 4.3.3 gives that $\|\mathscr{C}_m\|_{WB} \le C_3\, 8\,(2m+1)\,L^m$ for some other absolute constant C_3. Using all these facts we deduce from (4.3.45) that

$$\big\|\mathscr{C}_{m+1}\big\|_{L^2\to L^2} \le C_1\big[2\big\|\mathscr{C}_m(A')\big\|_{BMO} + C_3\, 8\,(2m+3)\,L^{m+1}\big].$$

Using (4.3.44) and the fact that $\|A'\|_{L^\infty} \le L$ we obtain that

$$\big\|\mathscr{C}_{m+1}\big\|_{L^2\to L^2} \le C_1\Big[2C_2L\Big\{8\,(2m+1)L^m + \big\|\mathscr{C}_m\big\|_{L^2\to L^2}\Big\} + C_3\, 8\,(2m+3)\,L^{m+1}\Big].$$

Combining this estimate with the induction hypothesis (4.3.43), we obtain

$$\big\|\mathscr{C}_{m+1}(1)\big\|_{BMO} \le R^{m+1}L^{m+1},$$

provided that R is chosen so that

$$\begin{aligned}
16C_1C_2(2m+1) &\le \frac{1}{3}R^{m+1},\\
2C_1C_2 &\le \frac{1}{3}R,\\
8C_1C_3(2m+3) &\le \frac{1}{3}R^{m+1}
\end{aligned}$$

for all $m \ge 0$. Such an R exists independent of m. This completes the proof of (4.3.43) by induction.

Exercises

4.3.1. Let T be a continuous linear operator from $\mathscr{S}(\mathbf{R}^n)$ to $\mathscr{S}'(\mathbf{R}^n)$ and let f be in $\mathscr{S}(\mathbf{R}^n)$. Let P_t be as in (4.3.15).
(a) Show that $P_t(f)$ converges to f in $\mathscr{S}(\mathbf{R}^n)$ as $t \to 0$.
(b) Conclude that $TP_t(f) \to T(f)$ in $\mathscr{S}'(\mathbf{R}^n)$ as $t \to 0$.
(c) Conclude that $P_tTP_t(f) \to T(f)$ in $\mathscr{S}'(\mathbf{R}^n)$ as $t \to 0$.

(d) Observe that (a)–(c) are also valid if P_t is replaced by P_t^2.
[*Hint:* Part (a): Use that $g_k \to g$ in $\mathscr{S}$ if and only if $\widehat{g_k} \to \widehat{g}$ in $\mathscr{S}$.]

4.3.2. Let T and P_t be as in Exercise 4.3.1 and let f be a Schwartz function whose Fourier transform vanishes in a neighborhood of the origin.
(a) Show that $P_t(f)$ converges to 0 in $\mathscr{S}(\mathbf{R}^n)$ as $t \to \infty$.
(b) Conclude that $TP_t(f) \to 0$ in $\mathscr{S}'(\mathbf{R}^n)$ as $t \to \infty$.
(c) Conclude that $P_tTP_t(f) \to 0$ in $\mathscr{S}'(\mathbf{R}^n)/\mathscr{P}(\mathbf{R}^n)$ as $t \to \infty$.
(d) Observe that (a)–(c) are also valid if P_t is replaced by P_t^2.
[*Hint:* Part (a): Use the hint in Exercise 4.3.1 and the observation that $|\widehat{\Phi}(t\xi)\widehat{f}(\xi)| \le C(1+tc_0)^{-1}|\widehat{f}(\xi)|$ if $\widehat{f}$ is supported outside the ball $B(0,c_0)$. Part (c): Pair with a function g in $\mathscr{S}_0(\mathbf{R}^n)$ and use part (a) and the fact that all Schwartz seminorms of $P_t(g)$ are bounded uniformly in $t > 0$. To prove the latter you may need that all Schwartz seminorms of $P_t(g)$ are bounded uniformly in $t > 0$ if and only if all Schwartz seminorms of $P_t(g)\widehat{\ }$ are bounded uniformly in $t > 0$.]

4.3.3. (a) Prove that every linear operator T from $\mathscr{S}(\mathbf{R}^n)$ to $\mathscr{S}'(\mathbf{R}^n)$ associated with an antisymmetric kernel in $SK(\delta,A)$ satisfies the weak boundedness property. Precisely, for some dimensional constant C_n, we have

$$\|T\|_{WB} \le C_n A\,.$$

(b) Conclude that for some constant $c < \infty$, the Calderón commutators satisfy

$$\big\|\mathscr{C}_m\big\|_{WB} \le c\,(2m+1)\,L^m\,.$$

[*Hint:* For given f,g normalized bumps, write $\langle T(\tau^{x_0} f_R), \tau^{x_0} g_R\rangle$ as

$$\frac{1}{2}\int_{\mathbf{R}^n}\int_{\mathbf{R}^n} K(x,y)\big(\tau^{x_0} f_R(y)\,\tau^{x_0} g_R(x) - \tau^{x_0} f_R(x)\,\tau^{x_0} g_R(y)\big)\,dy\,dx$$

and use the mean value theorem.]

4.3.4. (a) Let η be a smooth function on $\mathbf{R}^n$ which is equal to 1 on $B(0,1)$ and vanishes outside $B(0,2)$. Let T be as in Definition 4.1.16. For $f \in L^\infty(\mathbf{R}^n)$ show that $T(f\eta_k) \to T(f)$ in $\mathscr{D}_0'$ as $k \to \infty$, where $\eta_k(x) = \eta(x/k)$ for any $k \ge 1$ and $x \in \mathbf{R}^n$.
(b) Prove identity (4.3.46), i.e., show that $\mathscr{C}_{m+1}(1) = \mathscr{C}_m(A')$ for all $m \in \mathbf{Z}^+ \cup \{0\}$.
[*Hint:* Part (b). Start with

$$0 = \int_{\mathbf{R}} \frac{d}{dy}\left\{\frac{\eta_k(y)}{m+1}\left(\frac{A(x)-A(y)}{x-y}\right)^{m+1}\right\}dy$$

and use the antisymmetry of the kernels of $\mathscr{C}_{m+1}, \mathscr{C}_m$ to split the integral and the assertion in part (a).]

4.3.5. Suppose that a standard kernel $K(x,y)$ has the form $k(x-y)$ for some function k on $\mathbf{R}^n \setminus \{0\}$. Suppose that k extends to a tempered distribution on $\mathbf{R}^n$ whose Fourier transform is a bounded function defined on $\mathbf{R}^n$. Let T be a continuous linear operator from $\mathscr{S}(\mathbf{R}^n)$ to $\mathscr{S}'(\mathbf{R}^n)$ given by $T(\psi)(x) = \lim_{j\to\infty} \int_{|x-y|\geq \varepsilon_j} \psi(y)k(x-y)\,dy$ for some $\varepsilon_j \downarrow 0$ as $j \to \infty$ for all $\psi \in \mathscr{S}(\mathbf{R}^n)$.
(a) Identify the functions $T(e^{2\pi i \xi \cdot ()})$ and $T^t(e^{2\pi i \xi \cdot ()})$ and restrict to $\xi = 0$ to obtain $T(1)$ and $T^t(1)$.
(b) Use Theorem 4.3.3 to obtain the L^2 boundedness of T.
(c) What are $H(1)$ and $H^t(1)$ equal to when H is the Hilbert transform?

4.3.6. (*A. Calderón*) Let A be a Lipschitz function on $\mathbf{R}$. Use expansion (4.3.41) and estimate (4.3.43) to show that the operator

$$\mathscr{C}_A(f)(x) = \frac{1}{\pi i} \lim_{\varepsilon \to 0} \int_{|x-y|>\varepsilon} \frac{f(y)\,dy}{x-y+i(A(x)-A(y))}$$

is bounded on $L^2(\mathbf{R})$ when $\|A'\|_{L^\infty} < R^{-1}$, where R satisfies (4.3.44).

4.3.7. Prove that condition (i) of Theorem 4.3.3 is equivalent to the statement that

$$\sup_Q \sup_{\varepsilon>0} \left(\frac{\|T^{(\varepsilon)}(\chi_Q)\|_{L^2}}{|Q|^{\frac{1}{2}}} + \frac{\|(T^{(\varepsilon)})^t(\chi_Q)\|_{L^2}}{|Q|^{\frac{1}{2}}} \right) = B_1' < \infty,$$

where the first supremum is taken over all cubes Q in $\mathbf{R}^n$.
[*Hint:* You may repeat the argument that (i) $\Longrightarrow$ (ii) replacing the ball $B(x_0,N)$ by a cube centered at x_0 with side length $N/2$. The other direction was proved in Theorem 4.3.3.]

4.4 Paraproducts

In this section we study a useful class of operators called *paraproducts*. Their name suggests that these operators are related with products but in fact, their properties go *beyond* the desirable properties of products, hence the choice of the Greek preposition para. For instance, differentiating a paraproduct yields another paraproduct in which the derivative falls in only one of two functions, while differentiating a classical product of two functions yields two terms, each consisting of the derivative of one function times the other.

Paraproducts provide interesting examples of nonconvolution operators with standard kernels whose L^2 boundedness was discussed in Section 4.3. They have found use in many situations, including a proof of the main implication in Theorem 4.3.3. This proof is discussed in the present section.

4.4.1 Introduction to Paraproducts

Throughout this section we fix a radial Schwartz function Ψ whose Fourier transform is supported in the annulus $\frac{1}{2} \le |\xi| \le 2$ and that satisfies

$$\sum_{j\in\mathbf{Z}} \widehat{\Psi}(2^{-j}\xi) = 1, \qquad \text{when} \quad \xi \in \mathbf{R}^n \setminus \{0\}. \tag{4.4.1}$$

Associated with this Ψ we define the *Littlewood–Paley operator* $\Delta_j(f) = f * \Psi_{2^{-j}}$, where $\Psi_t(x) = t^{-n}\Psi(t^{-1}x)$. Using (4.4.1), we easily obtain

$$\sum_{j\in\mathbf{Z}} \Delta_j = I, \tag{4.4.2}$$

where (4.4.2) is interpreted as an identity on Schwartz functions with mean value zero. See Exercise 4.4.1. Note that by construction, the function Ψ is radial and thus even. This makes the operator Δ_j equal to its transpose.

We now observe that in view of the properties of Ψ, the function

$$\xi \mapsto \sum_{j\le 0} \widehat{\Psi}(2^{-j}\xi) \tag{4.4.3}$$

is supported in $|\xi| \le 2$, and is equal to 1 when $0 < |\xi| \le \frac{1}{2}$. But $\widehat{\Psi}(0) = 0$, which implies that the function in (4.4.3) also vanishes at the origin. We can easily fix this discontinuity by introducing the Schwartz function whose Fourier transform is equal to

$$\widehat{\Phi}(\xi) = \begin{cases} \sum_{j\le 0} \widehat{\Psi}(2^{-j}\xi) & \text{when } \xi \neq 0, \\ 1 & \text{when } \xi = 0. \end{cases}$$

Definition 4.4.1. We define the *partial sum operator* S_j as

$$S_j = \sum_{k\le j} \Delta_k. \tag{4.4.4}$$

In view of the preceding discussion, S_j is given by convolution with $\Phi_{2^{-j}}$, that is,

$$S_j(f)(x) = (f * \Phi_{2^{-j}})(x), \tag{4.4.5}$$

and the expression in (4.4.5) is well defined for all f in $\bigcup_{1\le p\le\infty} L^p(\mathbf{R}^n)$. Since Φ is a radial function by construction, the operator S_j is self-transpose.

Similarly, $\Delta_j(g)$ is also well defined for all g in $\bigcup_{1\le p\le\infty} L^p(\mathbf{R}^n)$. Moreover, since Δ_j is given by convolution with a function with mean value zero, it also follows that $\Delta_j(b)$ is well defined when $b \in BMO(\mathbf{R}^n)$. See Exercise 4.4.2 for details.

Definition 4.4.2. Given a function g on $\mathbf{R}^n$, we define the *paraproduct operator* P_g as follows:

$$P_g(f) = \sum_{j\in\mathbf{Z}} \Delta_j(g)\, S_{j-3}(f) = \sum_{j\in\mathbf{Z}} \sum_{k\le j-3} \Delta_j(g)\, \Delta_k(f), \tag{4.4.6}$$

for a suitable function f on $\mathbf{R}^n$. It is not clear for which functions g and in what sense the series in (4.4.6) converges even when f is a Schwartz function. One may verify that the series in (4.4.6) converges absolutely almost everywhere when g is a Schwartz function with mean value zero; in this case, by Exercise 4.4.1 the series $\sum_j |\Delta_j(g)|$ converges (everywhere) and $S_j(f)$ is uniformly bounded by the Hardy–Littlewood maximal function $M(f)$, which is finite almost everywhere.

One of the main goals of this section is to show that the series in (4.4.6) converges in L^2 when f is in $L^2(\mathbf{R}^n)$ and g is a BMO function.

The paraproduct $P_g(f)$ contains essentially "half" the product of fg. Indeed, in view of the identity in (4.4.2), the product fg can be written as

$$fg = \sum_j \sum_k \Delta_j(f)\,\Delta_k(g)\,.$$

Restricting the summation of the indices to $k < j$ defines an operator that corresponds to "half" the product of fg. It is only for minor technical reasons that we take $k \le j-3$ in (4.4.6).

The main feature of the paraproduct operator P_g is that it is essentially a sum of orthogonal L^2 functions. Indeed, the Fourier transform of the function $\widehat{\Delta_j(g)}$ is supported in the annulus

$$\{\xi \in \mathbf{R}^n :\, 2^{j-1} \le |\xi| \le 2^{j+1}\},$$

while the Fourier transform of the function $\widehat{S_{j-3}(f)}$ is supported in the ball

$$\bigcup_{k\le j-3} \{\xi \in \mathbf{R}^n :\, 2^{k-1} \le |\xi| \le 2^{k+1}\} \cup \{0\}.$$

This implies that the Fourier transform of the function $\Delta_j(g)\,S_{j-3}(f)$ is supported in the algebraic sum

$$\{\xi \in \mathbf{R}^n :\, 2^{j-1} \le |\xi| \le 2^{j+1}\} + \{\xi \in \mathbf{R}^n :\, |\xi| \le 2^{j-2}\}.$$

But this sum is contained in the set

$$\{\xi \in \mathbf{R}^n :\, 2^{j-2} \le |\xi| \le 2^{j+2}\}, \tag{4.4.7}$$

and the family of sets in (4.4.7) is "almost disjoint" as j varies. This means that every point in $\mathbf{R}^n$ belongs to at most four annuli of the form (4.4.7). Therefore, the paraproduct $P_g(f)$ can be written as a sum of functions h_j such that the families $\{h_j :\, j \in 4\mathbf{Z}+r\}$ are mutually orthogonal in L^2, for all $r \in \{0,1,2,3\}$. This orthogonal decomposition of the paraproduct has as an immediate consequence its L^2 boundedness when g is an element of BMO.

4.4.2 L^2 Boundedness of Paraproducts

The following theorem is the main result of this subsection.

Theorem 4.4.3. *For fixed $b \in BMO(\mathbf{R}^n)$ and $f \in L^2(\mathbf{R}^n)$ the series*

$$\sum_{|j|\le M} \Delta_j(b)\, S_{j-3}(f)$$

converges in L^2 as $M \to \infty$ to a function denoted by $P_b(f)$. The operator P_b thus defined is bounded on $L^2(\mathbf{R}^n)$, and there is a dimensional constant C_n such that for all $b \in BMO(\mathbf{R}^n)$ we have

$$\big\|P_b\big\|_{L^2\to L^2} \le C_n \big\|b\big\|_{BMO}.$$

Proof. The proof of this result follows by putting together some of the powerful ideas developed in Chapter 3. First we define a measure on $\mathbf{R}^{n+1}_+$ by setting

$$d\mu(x,t) = \sum_{j\in\mathbf{Z}} |\Delta_j(b)(x)|^2\, dx\, \delta_{2^{-(j-3)}}(t)\,.$$

By Theorem 3.3.8 we have that μ is a Carleson measure on $\mathbf{R}^{n+1}_+$ whose norm is controlled by a constant multiple of $\|b\|_{BMO}^2$. Now fix $f \in L^2(\mathbf{R}^n)$ and recall that $\Phi(x) = \sum_{r\le 0} \Psi_{2^{-r}}(x)$. We define a function $F(x,t)$ on $\mathbf{R}^{n+1}_+$ by setting

$$F(x,t) = (\Phi_t * f)(x)\,.$$

Observe that $F(x,2^{-k}) = S_k(f)(x)$ for all $k \in \mathbf{Z}$. We estimate the L^2 norm of a finite sum of terms of the form $\Delta_j(b)\, S_{j-3}(f)$. For $M,N \in \mathbf{Z}^+$ with $M \ge N$ we have

$$\begin{aligned}\int_{\mathbf{R}^n}\Big|\sum_{N\le|j|\le M} \Delta_j(b)(x)\, S_{j-3}(f)(x)\Big|^2 dx \\ = \int_{\mathbf{R}^n}\Big|\sum_{N\le|j|\le M} (\Delta_j(b)\, S_{j-3}(f))\widehat{\ }(\xi)\Big|^2 d\xi\,.\end{aligned} \tag{4.4.8}$$

It is a simple fact that every $\xi \in \mathbf{R}^n$ belongs to at most four annuli of the form (4.4.7). It follows that at most four terms in the last sum in (4.4.8) are nonzero. Thus

$$\begin{aligned}\int_{\mathbf{R}^n}\Big|\sum_{N\le|j|\le M} (\Delta_j(b) S_{j-3}(f))\widehat{\ }(\xi)\Big|^2 d\xi \\ \le 4 \sum_{N\le|j|\le M} \int_{\mathbf{R}^n} \big|(\Delta_j(b)\, S_{j-3}(f))\widehat{\ }(\xi)\big|^2 d\xi \\ \le 4 \sum_{j\in\mathbf{Z}} \int_{\mathbf{R}^n} |\Delta_j(b)(x) S_{j-3}(f)(x)|^2 dx\end{aligned} \tag{4.4.9}$$

$$= 4\int_{\mathbf{R}^n} |F(x,t)|^2\, d\mu(x,t)$$
$$= 4\int_{\mathbf{R}^n} |(\Phi_t * f)(x)|^2\, d\mu(x,t)$$
$$\leq C_n \|b\|_{BMO}^2 \int_{\mathbf{R}^n} |f(x)|^2\, dx, \tag{4.4.10}$$

where we used Theorem 3.3.7 in the last inequality.

Since the expression in (4.4.10) is finite, given $\varepsilon > 0$, we can find an $N_0 > 0$ such that

$$M \geq N \geq N_0 \implies \sum_{N\leq |j|\leq M} \int_{\mathbf{R}^n} \big|(\Delta_j(b)\,S_{j-3}(f))\widehat{\ }(\xi)\big|^2\, d\xi < \varepsilon.$$

Recalling that

$$\int_{\mathbf{R}^n} \Big| \sum_{N\leq |j|\leq M} \Delta_j(b)(x)\,S_{j-3}(f)(x)\Big|^2 dx \leq 4 \sum_{N\leq |j|\leq M} \int_{\mathbf{R}^n} \big|(\Delta_j(b)S_{j-3}(f))\widehat{\ }(\xi)\big|^2\, d\xi\,,$$

we conclude that the sequence

$$\Big\{ \sum_{|j|\leq M} \Delta_j(b)S_{j-3}(f) \Big\}_M$$

is Cauchy in $L^2(\mathbf{R}^n)$, and therefore it converges in L^2 to a function $P_b(f)$. The boundedness of P_b on L^2 follows by setting $N = 0$ and letting $M \to \infty$ in (4.4.9). □

4.4.3 Fundamental Properties of Paraproducts

Having established the L^2 boundedness of paraproducts, we turn to their properties. We begin by studying their kernels. The paraproducts P_b are examples of integral operators of the form discussed in Section 4.1. Since P_b is L^2 bounded, it has a distributional kernel W_b. We show that for each b in BMO the distribution W_b coincides with a standard kernel L_b defined on $\mathbf{R}^n \times \mathbf{R}^n \setminus \{(x,x) : x \in \mathbf{R}^n\}$.

First we study the kernel of the operator $f \mapsto \Delta_j(b)S_{j-3}(f)$ for any $j \in \mathbf{Z}$. We have that

$$\Delta_j(b)(x)S_{j-3}(f)(x) = \int_{\mathbf{R}^n} L_j(x,y) f(y)\, dy,$$

where L_j is the integrable function

$$L_j(x,y) = (b * \Psi_{2^{-j}})(x)2^{(j-3)n}\Phi(2^{j-3}(x-y)).$$

Next we can easily verify the following size and regularity estimates for L_j:

$$|L_j(x,y)| \le C_n \|b\|_{BMO} \frac{2^{nj}}{(1+2^j|x-y|)^{n+1}}, \tag{4.4.11}$$

$$|\partial_x^\alpha \partial_y^\beta L_j(x,y)| \le C_{n,\alpha,\beta,N} \|b\|_{BMO} \frac{2^{j(n+|\alpha|+|\beta|)}}{(1+2^j|x-y|)^{n+1+N}}, \tag{4.4.12}$$

for all multi-indices α and β and all $N \ge |\alpha|+|\beta|$.

It follows from (4.4.11) that when $x \neq y$ the series

$$\sum_{j\in\mathbf{Z}} L_j(x,y) \tag{4.4.13}$$

converges absolutely and is controlled in absolute value by

$$C_n \|b\|_{BMO} \sum_{j\in\mathbf{Z}} \frac{2^{nj}}{(1+2^j|x-y|)^{n+1}} \le \frac{C_n' \|b\|_{BMO}}{|x-y|^n}.$$

Similarly, by taking $N \ge |\alpha|+|\beta|$, we show that the series

$$\sum_{j\in\mathbf{Z}} \partial_x^\alpha \partial_y^\beta L_j(x,y) \tag{4.4.14}$$

converges absolutely when $x \neq y$ and the absolute value of (4.4.14) is bounded by

$$C_{n,\alpha,\beta,N} \|b\|_{BMO} \sum_{j\in\mathbf{Z}} \frac{2^{j(n+|\alpha|+|\beta)}}{(1+2^j|x-y|)^{n+1+N}} \le \frac{C_{n,\alpha,\beta}' \|b\|_{BMO}}{|x-y|^{n+|\alpha|+|\beta|}}$$

for all multi-indices α and β. It follows that the function

$$L_b(x,y) = \sum_{j\in\mathbf{Z}} L_j(x,y)$$

defined $\mathbf{R}^n \times \mathbf{R}^n \setminus \{(x,x) : x \in \mathbf{R}^n\}$ is $\mathscr{C}^\infty$ and satisfies the estimates

$$|\partial_x^\alpha \partial_y^\beta L_b(x,y)| \le \frac{C_{n,\alpha,\beta}' \|b\|_{BMO}}{|x-y|^{n+|\alpha|+|\beta|}} \tag{4.4.15}$$

away from the diagonal $x=y$ for all multi-indices α,β. (This fact is a consequence of the following: if F_m satisfy $\sum_m \|\partial^\gamma F_m\|_{L^\infty(\Omega)} < \infty$ for some open set Ω and all multi-indices γ, then $F=\sum_m F_m$ is a $\mathscr{C}^\infty$ function and $\partial^\gamma F = \sum_m \partial^\gamma F_m$.)

To show that the distributional kernel W_b of P_b coincides with the function L_b on $\mathbf{R}^n \times \mathbf{R}^n \setminus \{(x,x) : x \in \mathbf{R}^n\}$, we note that if f and ϕ are Schwartz functions with disjoint compact supports we have

$$\int_{\mathbf{R}^n}\int_{\mathbf{R}^n} L_b(x,y) f(y)\phi(x)\,dy\,dx = \langle P_b(f),\phi\rangle = \langle W_b, \phi\otimes f\rangle. \tag{4.4.16}$$

Every $\mathscr{C}_0^\infty(\mathbf{R}^{2n})$ function $\Phi(x,y)$ whose support does not contain the diagonal can be written as a rapidly convergent sum of functions of the form $\phi_j \otimes f_j$ and thus (4.4.16) holds with Φ in place of $\phi \otimes f$. Thus L_b and W_b coincide outside the diagonal.

We note that the transpose of the operator P_b is formally given by the identity

$$P_b^t(f) = \sum_{j\in\mathbf{Z}} S_{j-3}(f\Delta_j(b)).$$

As remarked in the previous section, the kernel of the operator P_b^t is a distribution W_b^t that coincides with the function

$$L_b^t(x,y) = L_b(y,x)$$

away from the diagonal of $\mathbf{R}^{2n}$. It is trivial to observe that L_b^t satisfies the same size and regularity estimates (4.4.15) as L_b. Moreover, it follows from Theorem 4.4.3 that the operator P_b^t is bounded on $L^2(\mathbf{R}^n)$ with norm at most a multiple of the BMO norm of b.

We now turn to two important properties of paraproducts. In view of Definition 4.1.16, we have a meaning for $P_b(1)$ and $P_b^t(1)$, where P_b is the paraproduct operator. The first property we prove is that $P_b(1) = b$. Observe that this statement is trivially valid at a formal level, since $S_j(1) = 1$ for all j and $\sum_j \Delta_j(b) = b$. The second property is that $P_b^t(1) = 0$. This is also trivially checked at a formal level, since $S_{j-3}(\Delta_j(b)) = 0$ for all j, as a Fourier transform calculation shows. We make both of these statements precise in the following proposition.

Proposition 4.4.4. *Given $b \in BMO(\mathbf{R}^n)$, let P_b be the paraproduct operator defined as in (4.4.6). Then the distributions $P_b(1)$ and $P_b^t(1)$ coincide with elements of BMO. Precisely, we have*

$$P_b(1) = b \qquad \textit{and} \qquad P_b^t(1) = 0. \tag{4.4.17}$$

Proof. Let φ be an element of $\mathscr{D}_0(\mathbf{R}^n)$. Suppose that φ is supported in the ball $B(0,R)$. Fix a smooth function with compact support η equal to 1 on the ball $B(0,3R)$ and vanishing outside the double of this ball. Set $\eta_N(x) = \eta(x/N)$. As we observed in Remark 4.1.17, the definition of $P_b(1)$ is independent of the choice of sequence η_N, so we have the following identity for all $N \geq 1$:

$$\begin{aligned}\big\langle P_b(1), \varphi\big\rangle = &\int_{\mathbf{R}^n} \sum_{j\in\mathbf{Z}} \Delta_j(b)(x)\, S_{j-3}(\eta_N)(x)\, \varphi(x)\, dx \\ &+ \int_{\mathbf{R}^n} \left[\int_{\mathbf{R}^n} L_b(x,y)\, \varphi(x)\, dx\right] (1-\eta_N(y))\, dy.\end{aligned} \tag{4.4.18}$$

Since φ has mean value zero, we can subtract the constant $L_b(y_0,y)$ from $L_b(x,y)$ in the integral inside the square brackets in (4.4.18), for some y_0 in the support of φ. Then we estimate the absolute value of the double integral in (4.4.18) by

$$\int_{|y-y_0|\geq 3R} \int_{|x-y_0|\leq R} A\frac{|y_0-x|}{|y_0-y|^{n+1}} |1-\eta_N(y)|\, |\varphi(x)|\, dx\, dy,$$

which tends to zero as $N \to \infty$ by the Lebesgue dominated convergence theorem.

It suffices to prove that the first integral in (4.4.18) tends to $\int_{\mathbf{R}^n} b(x)\varphi(x)\,dx$ as $N \to \infty$. Let us make some preliminary observations. Since the Fourier transform of the product $\Delta_j(b)S_{j-3}(\eta_N)$ is supported in the annulus

$$\{\xi \in \mathbf{R}^n : 2^{j-2} \le |\xi| \le 2^{j+2}\}, \tag{4.4.19}$$

we may introduce a smooth and compactly supported function $\widehat{Z}(\xi)$ such that for all $j \in \mathbf{Z}$ the function $\widehat{Z}(2^{-j}\xi)$ is equal to 1 on the annulus (4.4.19) and vanishes outside the annulus $\{\xi \in \mathbf{R}^n : 2^{j-3} \le |\xi| \le 2^{j+3}\}$. Let us denote by Q_j the operator given by multiplication on the Fourier transform by the function $\widehat{Z}(2^{-j}\xi)$.

Note that $S_j(1)$ is well defined and equal to 1 for all j. This is because Φ has integral equal to 1. Also, since $\widehat{\Phi}$ is radial, the duality identity

$$\int_{\mathbf{R}^n} f(x)\,S_j(g)(x)\,dx = \int_{\mathbf{R}^n} g(x)\,S_j(f)(x)\,dx \tag{4.4.20}$$

holds for all $f \in L^1$ and $g \in L^\infty$. For φ in $\mathscr{D}_0(\mathbf{R}^n)$ we have

$$\begin{aligned}
&\int_{\mathbf{R}^n} \sum_{j\in\mathbf{Z}} \Delta_j(b)\,S_{j-3}(\eta_N)\,\varphi\,dx \\
&= \sum_{j\in\mathbf{Z}} \int_{\mathbf{R}^n} \Delta_j(b)\,S_{j-3}(\eta_N)\,\varphi\,dx && \text{(series converges in } L^2 \text{ and } \varphi \in L^2) \\
&= \sum_{j\in\mathbf{Z}} \int_{\mathbf{R}^n} \Delta_j(b)\,S_{j-3}(\eta_N)\,Q_j(\varphi)\,dx && \big[\widehat{Q_j(\varphi)} = \widehat{\varphi} \text{ on the} \\
& && \text{support of } \big((\Delta_j(b)\,S_{j-3}(\eta_N))\widehat{\ }\big] \\
&= \sum_{j\in\mathbf{Z}} \int_{\mathbf{R}^n} \eta_N\, S_{j-3}\big(\Delta_j(b)Q_j(\varphi)\big)\,dx && \text{(duality)} \\
&= \int_{\mathbf{R}^n} \eta_N \sum_{j\in\mathbf{Z}} S_{j-3}\big(\Delta_j(b)Q_j(\varphi)\big)\,dx && \text{(series converges in } L^1 \text{ and } \eta_N \in L^\infty).
\end{aligned}$$

We now explain why the last series of the foregoing expression converges in L^1. Since φ is in $\mathscr{D}_0(\mathbf{R}^n)$, Exercise 4.4.1 gives that the series $\sum_{j\in\mathbf{Z}} Q_j(\varphi)$ converges in L^1. Since S_j preserves L^1 and Exercise 4.4.2 gives

$$\sup_j \big\|\Delta_j(b)\big\|_{L^\infty} \le C_n \|b\|_{BMO},$$

it follows that the series $\sum_{j\in\mathbf{Z}} S_{j-3}\big(\Delta_j(b)Q_j(\varphi)\big)$ also converges in L^1.

We now use the Lebesgue dominated convergence theorem to obtain that the expression

$$\int_{\mathbf{R}^n} \eta_N \sum_{j\in\mathbf{Z}} S_{j-3}\big(\Delta_j(b)Q_j(\varphi)\big)\,dx$$

converges as $N \to \infty$ to

$$\begin{aligned}
&\int_{\mathbf{R}^n} \sum_{j\in\mathbf{Z}} S_{j-3}\big(\Delta_j(b)Q_j(\varphi)\big)\,dx \\
&\quad = \sum_{j\in\mathbf{Z}} \int_{\mathbf{R}^n} S_{j-3}\big(\Delta_j(b)Q_j(\varphi)\big)\,dx && \text{(series converges in } L^1) \\
&\quad = \sum_{j\in\mathbf{Z}} \int_{\mathbf{R}^n} S_{j-3}(1)\,\Delta_j(b)\,Q_j(\varphi)\,dx && \text{(in view of (4.4.20))} \\
&\quad = \sum_{j\in\mathbf{Z}} \int_{\mathbf{R}^n} \Delta_j(b)\,Q_j(\varphi)\,dx && \text{(since } S_{j-3}(1)=1) \\
&\quad = \sum_{j\in\mathbf{Z}} \int_{\mathbf{R}^n} \Delta_j(b)\,\varphi\,dx && \Big[\widehat{Q_j(\varphi)} = \widehat{\varphi} \text{ on support } \widehat{\Delta_j(b)}\Big] \\
&\quad = \sum_{j\in\mathbf{Z}} \big\langle b, \Delta_j(\varphi)\big\rangle && \text{(duality)} \\
&\quad = \Big\langle b, \sum_{j\in\mathbf{Z}} \Delta_j(\varphi)\Big\rangle && \text{(series converges in } H^1,\ b\in BMO) \\
&\quad = \big\langle b, \varphi\big\rangle && \text{(Exercise 4.4.1(a)).}
\end{aligned}$$

Regarding the fact that the series $\sum_j \Delta_j(\varphi)$ converges in H^1, we refer to Exercise 4.4.1. We now obtain that the first integral in (4.4.18) tends to $\langle b, \varphi\rangle$ as $N \to \infty$. We have therefore proved that

$$\big\langle P_b(1), \varphi\big\rangle = \big\langle b, \varphi\big\rangle$$

for all φ in $\mathscr{D}_0(\mathbf{R}^n)$. In other words, we have now identified $P_b(1)$ as an element of $\mathscr{D}_0'$ with the BMO function b.

For the transpose operator P_b^t we observe that we have the identity

$$\begin{aligned}
\big\langle P_b^t(1), \varphi\big\rangle = &\int_{\mathbf{R}^n} \sum_{j\in\mathbf{Z}} S_{j-3}^t\big(\Delta_j(b)\,\eta_N\big)(x)\,\varphi(x)\,dx \\
&+ \int_{\mathbf{R}^n}\int_{\mathbf{R}^n} L_b^t(x,y)(1-\eta_N(y))\,\varphi(x)\,dy\,dx.
\end{aligned} \tag{4.4.21}$$

As before, we can use the Lebesgue dominated convergence theorem to show that the double integral in (4.4.21) tends to zero. As for the first integral in (4.4.21), we have the identity

$$\int_{\mathbf{R}^n} P_b^t(\eta_N)\,\varphi\,dx = \int_{\mathbf{R}^n} \eta_N\,P_b(\varphi)\,dx.$$

Since φ is a multiple of an L^2-atom for H^1, Theorem 4.2.6 gives that $P_b(\varphi)$ is an L^1 function. The Lebesgue dominated convergence theorem now implies that

$$\int_{\mathbf{R}^n} \eta_N\,P_b(\varphi)\,dx \to \int_{\mathbf{R}^n} P_b(\varphi)\,dx = \int_{\mathbf{R}^n} \sum_{j\in\mathbf{Z}} \Delta_j(b)\,S_{j-3}(\varphi)\,dx$$

as $N \to \infty$. The required conclusion would follow if we could prove that the function $P_b(\varphi)$ has integral zero. Since $\Delta_j(b)$ and $S_{j-3}(\varphi)$ have disjoint Fourier transforms, it follows that

$$\int_{\mathbf{R}^n} \Delta_j(b)\, S_{j-3}(\varphi)\, dx = 0$$

for all j in $\mathbf{Z}$. But the series

$$\sum_{j\in\mathbf{Z}} \Delta_j(b)\, S_{j-3}(\varphi) \tag{4.4.22}$$

defining $P_b(\varphi)$ converges in L^2 and not necessarily in L^1, and for this reason we need to justify the interchange of the following integrals:

$$\int_{\mathbf{R}^n} \sum_{j\in\mathbf{Z}} \Delta_j(b)\, S_{j-3}(\varphi)\, dx = \sum_{j\in\mathbf{Z}} \int_{\mathbf{R}^n} \Delta_j(b)\, S_{j-3}(\varphi)\, dx. \tag{4.4.23}$$

To complete the proof, it suffices to show that when φ is in $\mathscr{D}_0(\mathbf{R}^n)$, the series in (4.4.22) converges in L^1. To prove this, consider a ball $B(0,R)$ which contains the support of φ. The series in (4.4.22) converges in $L^2(3B(0,R))$ and hence converges in $L^1(3B(0,R))$. It remains to prove that it converges in $L^1((3B(0,R))^c)$. For a fixed $x \in (3B(0,R))^c$, $y_0 \in B(0,R)$, and a finite subset F of $\mathbf{Z}$, we have

$$\sum_{j\in F} \int_{\mathbf{R}^n} L_j(x,y)\varphi(y)\, dy = \sum_{j\in F} \int_{B(0,R)} \big(L_j(x,y) - L_j(x,y_0)\big)\, \varphi(y)\, dy. \tag{4.4.24}$$

Using estimates (4.4.12), we obtain that the expression in (4.4.24) is controlled by a constant multiple of

$$\int_{B(0,R)} \sum_{j\in F} \frac{|y-y_0| 2^{nj} 2^j}{(1+2^j|x-y_0|)^{n+2}} |\varphi(y)|\, dy \le c \frac{1}{|x-y_0|^{n+1}} \int_{\mathbf{R}^n} |y-y_0|\, |\varphi(y)|\, dy.$$

Integrating this estimate with respect to $x \in (3B(0,R))^c$, we obtain that

$$\sum_{j\in F} \big\|\Delta_j(b) S_{j-3}(\varphi)\big\|_{L^1((3B(0,R))^c)} \le C_n \big\|\varphi\big\|_{L^1} < \infty$$

for all finite subsets F of $\mathbf{Z}$. This proves that the series in (4.4.22) converges in L^1.

We have now proved that $\langle P_b^t(1), \varphi\rangle = 0$ for all $\varphi \in \mathscr{D}_0(\mathbf{R}^n)$. This shows that the distribution $P_b^t(1)$ is a constant function, which is of course identified with zero if considered as an element of BMO. □

Remark 4.4.5. The boundedness of P_b on L^2 is a consequence of Theorem 4.3.3, since hypothesis (iv) is satisfied. Indeed, $P_b(1) = b$, $P_b^t(1) = 0$ are both BMO functions; see Exercise 4.4.4 for a sketch of a proof of the estimate $\|P_b\|_{WB} \le C_n\|b\|_{BMO}$. This provides another proof of the fact that $\|P_b\|_{L^2\to L^2} \le C_n\|b\|_{BMO}$, without using Theorem 4.3.3. We use this result in the next section to obtain a different proof of the main direction in Theorem 4.3.3.

Exercises

4.4.1. Let Ψ be a radial Schwartz function that satisfies (4.4.1) and define $\Delta_j(f) = f * \Psi_{2^{-j}}$ for $j \in \mathbf{Z}$. Fix a function f in $\mathscr{S}(\mathbf{R}^n)$ with mean value zero.
(a) Show that $\lim_{N\to\infty} \sum_{|j|\le N} \Delta_j(f) = f$ pointwise everywhere.
(b) Prove that the convergence in part (a) also holds L^1.
(c) Show that the convergence in part (a) is also valid in H^1.
(d) Show that $\sum_{j\in\mathbf{Z}} \|\Delta_j(f)\|_{L^\infty} < \infty$.
[*Hint:* To obtain convergence in L^1 for $j \ge 0$ use the estimate

$$\|\Delta_j(f)\|_{L^1} \le 2^{-j} \int_{\mathbf{R}^n} \int_{\mathbf{R}^n} 2^{jn} |\Psi(2^j y)| \, |2^j y| \, |(\nabla f)(x - \theta y)| \, dy \, dx$$

for some θ in $[0,1]$ and consider the cases $|x| \ge 2|y|$ and $|x| < 2|y|$. When $j \le 0$ use the simple identity $f * \Psi_{2^{-j}} = (f_{2^j} * \Psi)_{2^{-j}}$ and reverse the roles of f and Ψ. To show convergence in H^1, use the square function characterization of H^1 in terms of the Δ_j's and the fact that $\Delta_k \Delta_j$ is zero when $|j-k| \ge 2$.]

4.4.2. Without appealing to the H^1-BMO duality theorem, prove that there is a dimensional constant C_n such that for all $b \in BMO(\mathbf{R}^n)$ we have

$$\sup_{j\in\mathbf{Z}} \big\|\Delta_j(b)\big\|_{L^\infty} \le C_n \|b\|_{BMO}.$$

4.4.3. (a) Show that for all $1 < p,q,r < \infty$ with $\frac{1}{p} + \frac{1}{q} = \frac{1}{r}$ there is a constant C_{pqr} such that for all Schwartz functions f,g on $\mathbf{R}^n$ we have

$$\big\|P_g(f)\big\|_{L^r} \le C_{pqr} \big\|f\big\|_{L^p} \big\|g\big\|_{L^q}.$$

(b) Obtain the same conclusion for the bilinear operator

$$\widetilde{P}_g(f) = \sum_j \sum_{k\le j} \Delta_j(g) \Delta_k(f).$$

[*Hint:* Part (a): Estimate the L^r norm using duality. Part (b): Use part (a).]

4.4.4. Let f,g be normalized bumps on $\mathbf{R}^n$ (see Definition 4.3.1) and let $R > 0$.
(a) Let Ψ a Schwartz function with integral zero. If $\Delta_j(f_R) = f_R * \Psi_{2^{-j}}$, prove that

$$\big\|\Delta_j(f_R)\big\|_{L^\infty} \le C(n,\Psi) \min\big(2^{-j} R^{-(n+1)}, 2^{nj}\big)$$

for all $R > 0$ and $j \in \mathbf{Z}$. Then interpolate between L^1 and L^∞ to obtain

$$\big\|\Delta_j(f_R)\big\|_{L^2} \le C(n,\Psi) \min\Big(2^{-\frac{j}{2}} R^{-\frac{n+1}{2}}, 2^{\frac{nj}{2}}\Big).$$

(b) Observe that the same result is valid for the operators Q_j given by convolution with the function $2^{jn}Z(2^j\xi)$; here $\widehat{Z}(\xi)$ is a smooth function whose Fourier transform is equal to 1 on the annulus $\{\xi \in \mathbf{R}^n : 2^{-2} \le |\xi| \le 2^2\}$ and vanishes outside $\{\xi \in \mathbf{R}^n : 2^{-3} \le |\xi| \le 2^3\}$. Conclude that for some constant C_n we have

$$\sum_{j\in\mathbf{Z}} \big\|Q_j(g_R)\big\|_{L^2} \le C_n R^{-\frac{n}{2}}.$$

(c) Let P_b be defined as in (4.4.6). Show that there is a constant C_n' such that for all normalized bumps f and g and all $R > 0$ we have

$$\big|\langle P_b(\tau^{x_0} f_R), \tau^{x_0} g_R\rangle\big| \le C_n' R^{-n} \|b\|_{BMO}.$$

[*Hint:* Part (a): Use the cancellation of the functions f and Ψ. Part (c): Write

$$\langle P_b(\tau^{x_0} f_R), \tau^{x_0} g_R\rangle = \sum_{j\in\mathbf{Z}} \int_{\mathbf{R}^n} \Delta_j(\tau^{-x_0} b) Q_j(g_R) S_{j-3}(f_R)\, dx.$$

Apply the Cauchy–Schwarz inequality, and use the boundedness of S_{j-3} on L^2, Exercise 4.4.2, and part (b).]

4.4.5. (*Continuous paraproducts*) Let Φ and Ψ be radial Schwartz functions on $\mathbf{R}^n$ with $\int_{\mathbf{R}^n} \Phi(x)\,dx = 1$ and $\int_{\mathbf{R}^n} \Psi(x)\,dx = 0$. For $t > 0$ define operators $P_t(f) = \Phi_t * f$ and $Q_t(f) = \Psi_t * f$ for $f \in L^2(\mathbf{R}^n)$. Let $b \in BMO(\mathbf{R}^n)$ and $f \in L^2(\mathbf{R}^n)$. Show that the limit

$$\lim_{\substack{\varepsilon\to 0\\ N\to\infty}} \int_\varepsilon^N Q_t\big(Q_t(b) P_t(f)\big)\,\frac{dt}{t}$$

converges in $L^2(\mathbf{R}^n)$ and defines an operator $\Pi_b(f)$ that satisfies

$$\big\|\Pi_b\big\|_{L^2\to L^2} \le C_n \|b\|_{BMO}$$

for some dimensional constant C_n.
[*Hint:* Suitably adapt the proof of Theorem 4.4.3.]

4.5 An Almost Orthogonality Lemma and Applications

In this section we discuss an important lemma concerning orthogonality of operators and some of its applications.

It is often the case that a linear operator T is given as an infinite sum of other linear operators T_j such that the T_j's are uniformly bounded on L^2. This sole condition is not enough to imply that the sum of the T_j's is also L^2 bounded, although this is often the case. Let us consider, for instance, the linear operators $\{T_j\}_{j\in\mathbf{Z}}$ given by convolution with the smooth functions $e^{2\pi ijt}$ on the circle $\mathbf{T}^1$. Each T_j can be written as

$$T_j(f) = (\widehat{f} \otimes \delta_j)^\vee,$$

where $\widehat{f}$ is the sequence of Fourier coefficients of f; here δ_j is the infinite sequence with zeros in every entry except the jth entry, in which it has 1, and $\otimes$ denotes term-by-term multiplication of infinite sequences. It follows that each operator T_j is bounded on $L^2(\mathbf{T}^1)$ with norm 1. Moreover, the sum of the T_j's is the identity operator, which is also L^2 bounded with norm 1.

It is apparent from the preceding discussion that the crucial property of the T_j's that makes their sum bounded is their orthogonality. In the preceding example we have $T_jT_k = 0$ unless $j = k$. It turns out that this orthogonality condition is a bit too strong, and it can be weakened significantly.

4.5.1 The Cotlar–Knapp–Stein Almost Orthogonality Lemma

The next result provides a sufficient orthogonality criterion for boundedness of sums of linear operators on a Hilbert space.

Lemma 4.5.1. *Let $\{T_j\}_{j\in\mathbf{Z}}$ be a family of operators mapping a Hilbert space H to itself. Assume that there is a a function $\gamma : \mathbf{Z} \to \mathbf{R}^+$ such that*

$$\big\|T_j^* T_k\big\|_{H\to H} + \big\|T_j T_k^*\big\|_{H\to H} \le \gamma(j-k) \tag{4.5.1}$$

for all j,k in $\mathbf{Z}$. Suppose that

$$A = \sum_{j\in\mathbf{Z}} \sqrt{\gamma(j)} < \infty.$$

Then the following three conclusions are valid:

(i) For all finite subsets Λ of $\mathbf{Z}$ we have

$$\Big\|\sum_{j\in\Lambda} T_j\Big\|_{H\to H} \le A.$$

(ii) For all $x \in H$ we have

$$\sum_{j\in\mathbf{Z}} \big\|T_j(x)\big\|_H^2 \le A^2 \big\|x\big\|_H^2.$$

(iii) For all $x \in H$ the sequence $\sum_{|j|\le N} T_j(x)$ converges to some $T(x)$ as $N \to \infty$ in the norm topology of H. The linear operator T defined in this way is bounded from H to H with norm

$$\big\|T\big\|_{H\to H} \le A.$$

Proof. As usual we denote by S^* the adjoint of a linear operator S. It is a simple fact that any bounded linear operator $S:\ H\to H$ satisfies

$$\big\|S\big\|_{H\to H}^2=\big\|SS^*\big\|_{H\to H}. \tag{4.5.2}$$

See Exercise 4.5.1. By taking $j=k$ in (4.5.1) and using (4.5.2), we obtain

$$\big\|T_j\big\|_{H\to H}\le\sqrt{\gamma(0)} \tag{4.5.3}$$

for all $j\in\mathbf{Z}$. It also follows from (4.5.2) that if an operator S is self-adjoint, then $\big\|S\big\|_{H\to H}^2=\big\|S^2\big\|_{H\to H}$, and more generally,

$$\big\|S\big\|_{H\to H}^m=\big\|S^m\big\|_{H\to H} \tag{4.5.4}$$

for m that are powers of 2. Now observe that the linear operator

$$\Big(\sum_{j\in\Lambda}T_j\Big)\Big(\sum_{j\in\Lambda}T_j^*\Big)$$

is self-adjoint. Applying (4.5.2) and (4.5.4) to this operator, we obtain

$$\Big\|\sum_{j\in\Lambda}T_j\Big\|_{H\to H}^2=\Big\|\Big[\Big(\sum_{j\in\Lambda}T_j\Big)\Big(\sum_{j\in\Lambda}T_j^*\Big)\Big]^m\Big\|_{H\to H}^{\frac{1}{m}}, \tag{4.5.5}$$

where m is a power of 2. We now expand the mth power of the expression in (4.5.5). So we write the right side of the identity in (4.5.5) as

$$\Big\|\sum_{j_1,\cdots,j_{2m}\in\Lambda}T_{j_1}T_{j_2}^*\cdots T_{j_{2m-1}}T_{j_{2m}}^*\Big\|_{H\to H}^{\frac{1}{m}}, \tag{4.5.6}$$

which is controlled by

$$\Big(\sum_{j_1,\cdots,j_{2m}\in\Lambda}\big\|T_{j_1}T_{j_2}^*\cdots T_{j_{2m-1}}T_{j_{2m}}^*\big\|_{H\to H}\Big)^{\frac{1}{m}}. \tag{4.5.7}$$

We estimate the expression inside the sum in (4.5.7) in two different ways. First we group j_1 with j_2, j_3 with j_4, $\dots$, j_{2m-1} with j_{2m} and we apply (4.5.3) and (4.5.1) to control this expression by

$$\gamma(j_1-j_2)\gamma(j_3-j_4)\cdots\gamma(j_{2m-1}-j_{2m}).$$

Grouping j_2 with j_3, j_4 with j_5, $\dots$, j_{2m-2} with j_{2m-1} and leaving j_1 and j_{2m} alone, we also control the expression inside the sum in (4.5.7) by

$$\sqrt{\gamma(0)}\gamma(j_2-j_3)\gamma(j_4-j_5)\cdots\gamma(j_{2m-2}-j_{2m-1})\sqrt{\gamma(0)}.$$

Taking the geometric mean of these two estimates, we obtain the following bound for (4.5.7):

$$\Big(\sum_{j_1,\dots,j_{2m}\in\Lambda}\sqrt{\gamma(0)}\sqrt{\gamma(j_1-j_2)}\sqrt{\gamma(j_2-j_3)}\cdots\sqrt{\gamma(j_{2m-1}-j_{2m})}\Big)^{\frac{1}{m}}.$$

Summing first over j_1, then over j_2, etc, and finally over j_{2m-1}, we obtain the estimate

$$\gamma(0)^{\frac{1}{2m}}A^{\frac{2m-1}{m}}\Big(\sum_{j_{2m}\in\Lambda}1\Big)^{\frac{1}{m}}$$

for (4.5.7). Using (4.5.5), we conclude that

$$\Big\|\sum_{j\in\Lambda}T_j\Big\|^2_{H\to H}\le\gamma(0)^{\frac{1}{2m}}A^{\frac{2m-1}{m}}|\Lambda|^{\frac{1}{m}},$$

and letting $m\to\infty$, we obtain conclusion (i) of the proposition.

To prove (ii) we use the Rademacher functions r_j of Appendix C.1 in [156]. These functions are defined for nonnegative integers j, but we can reindex them so that the subscript j runs through the integers. The fundamental property of these functions is their orthogonality, that is,

$$\int_0^1 r_j(\omega)r_k(\omega)\,d\omega=0$$

when $j\neq k$. Using the fact that the norm $\|\cdot\|_H$ comes from an inner product, for every finite subset Λ of $\mathbf{Z}$ and x in H we obtain

$$\begin{aligned}\int_0^1\Big\|\sum_{j\in\Lambda}r_j(\omega)T_j(x)\Big\|_H^2\,d\omega\\ &=\sum_{j\in\Lambda}\|T_j(x)\|_H^2+\int_0^1\sum_{\substack{j,k\in\Lambda\\ j\neq k}}r_j(\omega)r_k(\omega)\left\langle T_j(x),T_k(x)\right\rangle_H d\omega \qquad (4.5.8)\\ &=\sum_{j\in\Lambda}\|T_j(x)\|_H^2.\end{aligned}$$

For any fixed $\omega\in[0,1]$ we now use conclusion (i) of the proposition for the operators $r_j(\omega)T_j$, which also satisfy assumption (4.5.1), since $r_j(\omega)=\pm1$. We obtain that

$$\Big\|\sum_{j\in\Lambda}r_j(\omega)T_j(x)\Big\|_H^2\le A^2\|x\|_H^2,$$

which, combined with (4.5.8), gives conclusion (ii).

We now prove (iii). First we show that given $x \in H$ the sequence

$$\Big\{ \sum_{j=-N}^{N} T_j(x) \Big\}_N$$

is Cauchy in H. Suppose that this is not the case. This means that there is some $\varepsilon > 0$ and a subsequence of integers $1 \le N_1 < N_2 < N_3 < \cdots$ such that

$$\big\|\widetilde{T}_k(x)\big\|_H \ge \varepsilon \,, \tag{4.5.9}$$

where we set

$$\widetilde{T}_k(x) = \sum_{N_k \le |j| < N_{k+1}} T_j(x).$$

For any fixed $\omega \in [0,1]$, apply conclusion (i) to the operators $S_j = r_k(\omega) T_j$ whenever $N_k \le |j| < N_{k+1}$, since these operators clearly satisfy hypothesis (4.5.1). Taking $N_1 \le |j| \le N_{K+1}$, we obtain

$$\Big\| \sum_{k=1}^{K} r_k(\omega) \sum_{N_k \le |j| < N_{k+1}} T_j(x) \Big\|_H = \Big\| \sum_{k=1}^{K} r_k(\omega) \widetilde{T}_k(x) \Big\|_H \le A \, \|x\|_H .$$

Squaring and integrating this inequality with respect to ω in $[0,1]$, and using (4.5.8) with $\widetilde{T}_k$ in the place of T_k and $\{1,2,\dots,K\}$ in the place of Λ, we obtain

$$\sum_{k=1}^{K} \big\|\widetilde{T}_k(x)\big\|_H^2 \le A^2 \|x\|_H^2 .$$

But this clearly contradicts (4.5.9) as $K \to \infty$.

We conclude that every sequence

$$\Big\{ \sum_{j=-N}^{N} T_j(x) \Big\}_N$$

is Cauchy in H and thus it converges to Tx for some linear operator T. In view of conclusion (i), it follows that T is a bounded operator on H with norm at most A. □

Remark 4.5.2. At first sight, it appears strange that the norm of the operator T is independent of the norm of every piece T_j and depends only on the quantity A in (4.5.1). But as observed in the proof, if we take $j = k$ in (4.5.1), we obtain

$$\|T_j\|_{H \to H}^2 = \|T_j T_j^*\|_{H \to H} \le \gamma(0) \le A^2 \,;$$

thus the norm of each individual T_j is also controlled by the constant A.

We also note that there wasn't anything special about the role of the index set $\mathbf{Z}$ in Lemma 4.5.1. Indeed, the set $\mathbf{Z}$ can be replaced by any countable group, such as $\mathbf{Z}^k$ for some k. For instance, see Theorem 4.5.7, in which the index set is $\mathbf{Z}^{2n}$.

See also Exercises 4.5.7 and 4.5.8, in which versions of Lemma 4.5.1 are given with no group structure on the set of indices.

4.5.2 An Application

We now discuss an application of the almost orthogonality lemma just proved concerning sums of nonconvolution operators on $L^2(\mathbf{R}^n)$. We begin with the following version of Theorem 4.3.3, in which it is assumed that $T(1) = T^t(1) = 0$.

Proposition 4.5.3. *Suppose that $K_j(x,y)$ are functions on $\mathbf{R}^n \times \mathbf{R}^n$ indexed by $j \in \mathbf{Z}$ that satisfy*

$$|K_j(x,y)| \le \frac{A2^{nj}}{(1+2^j|x-y|)^{n+\delta}}, \tag{4.5.10}$$

$$|K_j(x,y) - K_j(x,y')| \le A2^{\gamma j}2^{nj}|y-y'|^{\gamma}, \tag{4.5.11}$$

$$|K_j(x,y) - K_j(x',y)| \le A2^{\gamma j}2^{nj}|x-x'|^{\gamma}, \tag{4.5.12}$$

for some $0 < A, \gamma, \delta < \infty$ and all $x,y,x',y' \in \mathbf{R}^n$. Suppose also that

$$\int_{\mathbf{R}^n} K_j(z,y)\,dz = 0 = \int_{\mathbf{R}^n} K_j(x,z)\,dz, \tag{4.5.13}$$

for all $x,y \in \mathbf{R}^n$ and all $j \in \mathbf{Z}$. For $j \in \mathbf{Z}$ define integral operators

$$T_j(f)(x) = \int_{\mathbf{R}^n} K_j(x,y)\, f(y)\,dy$$

for $f \in L^2(\mathbf{R}^n)$. Then the series

$$\sum_{j\in\mathbf{Z}} T_j(f)$$

converges in L^2 to some $T(f)$ for all $f \in L^2(\mathbf{R}^n)$, and the linear operator T defined in this way is L^2 bounded.

Proof. It is a consequence of (4.5.10) that the kernels K_j are in $L^1(dy)$ uniformly in $x \in \mathbf{R}^n$ and $j \in \mathbf{Z}$ and hence the operators T_j map $L^2(\mathbf{R}^n)$ to $L^2(\mathbf{R}^n)$ uniformly in j. Our goal is to show that the sum of the T_j's is also bounded on $L^2(\mathbf{R}^n)$. We achieve this using the orthogonality considerations of Lemma 4.5.1. To be able to use Lemma 4.5.1, we need to prove (4.5.1). Indeed, we show that for all $k, j \in \mathbf{Z}$ we have

$$\big\|T_jT_k^*\big\|_{L^2\to L^2} \le CA^2 2^{-\frac{1}{4}\frac{\delta}{n+\delta}\min(\gamma,\delta)|j-k|}, \tag{4.5.14}$$

$$\big\|T_j^*T_k\big\|_{L^2\to L^2} \le CA^2 2^{-\frac{1}{4}\frac{\delta}{n+\delta}\min(\gamma,\delta)|j-k|}, \tag{4.5.15}$$

for some $0 < C = C_{n,\gamma,\delta} < \infty$. We prove only (4.5.15), since the proof of (4.5.14) is similar. In fact, since the kernels of T_j and T_j^* satisfy similar size, regularity, and cancellation estimates, (4.5.15) is directly obtained from (4.5.14) when T_j are replaced by T_j^*.

It suffices to prove (4.5.15) under the extra assumption that $k \le j$. Once (4.5.15) is established under this assumption, taking $j \le k$ yields

$$\big\|T_j^* T_k\big\|_{L^2\to L^2} = \big\|(T_k^* T_j)^*\big\|_{L^2\to L^2} = \big\|T_k^* T_j\big\|_{L^2\to L^2} \le CA^2 2^{-\frac{1}{2}\min(\gamma,\delta)|j-k|},$$

thus proving (4.5.15) also under the assumption $j \le k$.

We therefore take $k \le j$ in the proof of (4.5.15). Note that the kernel of $T_j^* T_k$ is

$$L_{jk}(x,y) = \int_{\mathbf{R}^n} \overline{K_j(z,x)} K_k(z,y)\, dz.$$

We prove that

$$\sup_{x\in\mathbf{R}^n} \int_{\mathbf{R}^n} |L_{kj}(x,y)|\, dy \le CA^2 2^{-\frac{1}{4}\frac{\delta}{n+\delta}\min(\gamma,\delta)|k-j|}, \tag{4.5.16}$$

$$\sup_{y\in\mathbf{R}^n} \int_{\mathbf{R}^n} |L_{kj}(x,y)|\, dx \le CA^2 2^{-\frac{1}{4}\frac{\delta}{n+\delta}\min(\gamma,\delta)|k-j|}. \tag{4.5.17}$$

Once (4.5.16) and (4.5.17) are established, (4.5.15) follows directly from the classical Schur lemma in Appendix A.1.

We need to use the following estimate, valid for $k \le j$:

$$\int_{\mathbf{R}^n} \frac{2^{nj}\min(1,(2^k|u|)^\gamma)}{(1+2^j|u|)^{n+\delta}}\, du \le C_{n,\delta} 2^{-\frac{1}{2}\min(\gamma,\delta)(j-k)}. \tag{4.5.18}$$

Indeed, to prove (4.5.18), we observe that by changing variables we may assume that $j=0$ and $k \le 0$. Taking $r = k-j \le 0$, we establish (4.5.18) as follows:

$$\begin{aligned}
\int_{\mathbf{R}^n} \frac{\min(1,(2^r|u|)^\gamma)}{(1+|u|)^{n+\delta}}\, du &\le \int_{\mathbf{R}^n} \frac{\min\big(1,(2^r|u|)^{\frac{1}{2}\min(\gamma,\delta)}\big)}{(1+|u|)^{n+\delta}}\, du \\
&\le \int_{|u|\le 2^{-r}} \frac{(2^r|u|)^{\frac{1}{2}\min(\gamma,\delta)}}{(1+|u|)^{n+\delta}}\, du + \int_{|u|\ge 2^{-r}} \frac{1}{(1+|u|)^{n+\delta}}\, du \\
&\le 2^{\frac{1}{2}\min(\gamma,\delta)r} \int_{\mathbf{R}^n} \frac{1}{(1+|u|)^{n+\frac{\delta}{2}}}\, du + \int_{|u|\ge 2^{-r}} \frac{1}{|u|^{n+\delta}}\, du \\
&\le C'_{n,\delta}\big[2^{\frac{1}{2}\min(\gamma,\delta)r} + 2^{\delta r}\big] \\
&\le C_{n,\delta}\, 2^{-\frac{1}{2}\min(\gamma,\delta)|r|},
\end{aligned}$$

We now obtain estimates for L_{jk} in the case $k \le j$. Using (4.5.13), we write

$$\begin{aligned}
|L_{jk}(x,y)| &= \left| \int_{\mathbf{R}^n} K_k(z,y)\overline{K_j(z,x)}\,dz \right| \\
&= \left| \int_{\mathbf{R}^n} \big[K_k(z,y) - K_k(x,y)\big]\overline{K_j(z,x)}\,dz \right| \\
&\le A^2 \int_{\mathbf{R}^n} 2^{nk}\min(1,(2^k|x-z|)^\gamma)\frac{2^{nj}}{(1+2^j|z-x|)^{n+\delta}}\,dz \\
&\le CA^2\,2^{kn}\,2^{-\frac{1}{2}\min(\gamma,\delta)(j-k)}
\end{aligned}$$

using estimate (4.5.18). Combining this estimate with

$$|L_{jk}(x,y)| \le \int_{\mathbf{R}^n} |K_j(z,x)|\,|K_k(z,y)|\,dz \le \frac{CA^2 2^{kn}}{(1+2^k|x-y|)^{n+\delta}},$$

which follows from (4.5.10) and the result in Appendix B.1 (since $k \le j$), yields

$$|L_{jk}(x,y)| \le C_{n,\gamma,\delta}A^2\,2^{-\frac{1}{2}\frac{\delta/2}{n+\delta}\min(\gamma,\delta)(j-k)}\frac{2^{kn}}{(1+2^k|x-y|)^{n+\frac{\delta}{2}}},$$

which easily implies (4.5.16) and (4.5.17). This concludes the proof of the proposition. □

4.5.3 Almost Orthogonality and the $T(1)$ Theorem

We now give an important application of the proposition just proved. We re-prove the difficult direction of the $T(1)$ theorem proved in Section 4.3. We have the following:

Theorem 4.5.4. *Let K be in $SK(\delta,A)$ and let T be a continuous linear operator from $\mathscr{S}(\mathbf{R}^n)$ to $\mathscr{S}'(\mathbf{R}^n)$ associated with K. Assume that*

$$\big\|T(1)\big\|_{BMO} + \big\|T^t(1)\big\|_{BMO} + \big\|T\big\|_{WB} = B_4 < \infty.$$

Then T extends to bounded linear operator on $L^2(\mathbf{R}^n)$ with norm at most a constant multiple of $A+B_4$.

Proof. Consider the paraproduct operators $P_{T(1)}$ and $P_{T^t(1)}$ introduced in the previous section. Then, as we showed in Proposition 4.4.4, we have

$$\begin{aligned}
P_{T(1)}(1) &= T(1), & (P_{T(1)})^t(1) &= 0, \\
P_{T^t(1)}(1) &= T^t(1), & (P_{T^t(1)})^t(1) &= 0.
\end{aligned}$$

Let us define an operator

$$L = T - P_{T(1)} - (P_{T^t(1)})^t .$$

Using Proposition 4.4.4, we obtain that

$$L(1) = L^t(1) = 0.$$

In view of (4.4.15), we have that L is an operator whose kernel satisfies the estimates (4.1.1), (4.1.2), and (4.1.3) with constants controlled by a dimensional constant multiple of

$$A + \|T(1)\|_{BMO} + \|T^t(1)\|_{BMO} .$$

Both of these numbers are controlled by $A + B_4$. We also have

$$\begin{aligned}
\|L\|_{WB} &\le C_n\Big(\|T\|_{WB} + \|P_{T(1)}\|_{L^2\to L^2} + \|(P_{T^t(1)})^t\|_{L^2\to L^2}\Big)\\
&\le C_n\Big(\|T\|_{WB} + \|T(1)\|_{BMO} + \|T^t(1)\|_{BMO}\Big)\\
&\le C_n(A + B_4),
\end{aligned}$$

which is a very useful fact.

Next we introduce operators Δ_j and S_j; one should be cautious as these are not the operators Δ_j and S_j introduced in Section 4.4 but rather discrete analogues of those introduced in the proof of Theorem 4.3.3. We pick a smooth radial real-valued function Φ with compact support contained in the unit ball $B(0, \frac{1}{2})$ that satisfies $\int_{\mathbf{R}^n} \Phi(x)\,dx = 1$ and we define

$$\Psi(x) = \Phi(x) - 2^{-n}\Phi(\tfrac{x}{2}). \tag{4.5.19}$$

Notice that Ψ has mean value zero. We define

$$\Phi_{2^{-j}}(x) = 2^{nj}\Phi(2^j x) \qquad \text{and} \qquad \Psi_{2^{-j}}(x) = 2^{nj}\Psi(2^j x)$$

and we observe that both Φ and Ψ are supported in $B(0,1)$ and are multiples of normalized bumps. We then define Δ_j to be the operator given by convolution with the function $\Psi_{2^{-j}}$ and S_j the operator given by convolution with the function $\Phi_{2^{-j}}$. In view of identity (4.5.19) we have that $\Delta_j = S_j - S_{j-1}$. Notice that

$$S_j L S_j = S_{j-1} L S_{j-1} + \Delta_j L S_j + S_{j-1} L \Delta_j ,$$

which implies that for all integers $N < M$ we have

$$\begin{aligned}
S_M L S_M - S_{N-1} L S_{N-1} &= \sum_{j=N}^{M} \big(S_j L S_j - S_{j-1} L S_{j-1}\big)\\
&= \sum_{j=N}^{M} \Delta_j L S_j + \sum_{j=N}^{M} S_{j-1} L \Delta_j .
\end{aligned} \tag{4.5.20}$$

Until the end of the proof we fix a Schwartz function f whose Fourier transform vanishes in a neighborhood of the origin. Such functions are dense in L^2; see Exercise 6.2.9 in [156]. We would like to use Proposition 4.5.3 to conclude that

$$\sup_{M\in\mathbf{Z}}\sup_{N<M}\big\|S_MLS_M(f)-S_{N-1}LS_{N-1}(f)\big\|_{L^2}\le C_n(A_2+B_4)\big\|f\big\|_{L^2} \tag{4.5.21}$$

and that $S_MLS_M(f)-S_{N-1}LS_{N-1}(f)\to\widetilde{L}(f)$ in L^2 as $M\to\infty$ and $N\to-\infty$. Once these statements are proved, we deduce that $\widetilde{L}(f)=L(f)$. To see this, it suffices to prove that $S_MLS_M(f)-S_{N-1}LS_{N-1}(f)$ converges to $L(f)$ weakly in L^2. Indeed, let g be another Schwartz function. Then

$$\begin{aligned}\big\langle S_MLS_M(f)&-S_{N-1}LS_{N-1}(f),g\big\rangle-\big\langle L(f),g\big\rangle\\&=\big\langle S_MLS_M(f)-L(f),g\big\rangle-\big\langle S_{N-1}LS_{N-1}(f),g\big\rangle.\end{aligned} \tag{4.5.22}$$

We first prove that the first term in (4.5.22) tends to zero as $M\to\infty$. Indeed,

$$\begin{aligned}\big\langle S_MLS_M(f)-L(f),g\big\rangle&=\big\langle LS_M(f),S_Mg\big\rangle-\big\langle L(f),g\big\rangle\\&=\big\langle L(S_M(f)-f),S_M(g)\big\rangle+\big\langle L(f),S_M(g)-g\big\rangle,\end{aligned}$$

and both terms converge to zero, since $S_M(f)-f\to 0$ and $S_M(g)-g$ tend to zero in $\mathscr{S}$, L is continuous from $\mathscr{S}$ to $\mathscr{S}'$, and all Schwartz seminorms of $S_M(g)$ are bounded uniformly in M; see also Exercise 4.3.1.

The second term in (4.5.22) is $\big\langle S_{N-1}LS_{N-1}(f),g\big\rangle=\big\langle LS_{N-1}(f),S_{N-1}(g)\big\rangle$. Since $\widehat{f}$ is supported away from the origin, $S_N(f)\to 0$ in $\mathscr{S}$ as $N\to-\infty$; see Exercise 4.3.2(a). By the continuity of L, $LS_{N-1}(f)\to 0$ in $\mathscr{S}'$, and since all Schwartz seminorms of $S_{N-1}(g)$ are bounded uniformly in N, we conclude that the term $\big\langle LS_{N-1}(f),S_{N-1}(g)\big\rangle$ tends to zero as $N\to-\infty$. We thus deduce that $\widetilde{L}(f)=L(f)$.

It remains to prove (4.5.21). We now define

$$L_j=\Delta_jLS_j \qquad\text{and}\qquad L_j'=S_{j-1}L\Delta_j$$

for $j\in\mathbf{Z}$. In view of the convergence of the Riemann sums to the integral defining $f*\Phi_{2^{-j}}$ in the topology of $\mathscr{S}$ (this is contained in the proof of Theorem 2.3.20 in [156]), we have

$$\big(L(f*\Phi_{2^{-j}})*\Psi_{2^{-j}}\big)(x)=\int_{\mathbf{R}^n}\big\langle L(\tau^y\Phi_{2^{-j}}),\tau^x\Psi_{2^{-j}}\big\rangle f(y)\,dy,$$

where $\tau^yg(u)=g(u-y)$. Thus the kernel K_j of L_j is

$$K_j(x,y)=\big\langle L(\tau^y\Phi_{2^{-j}}),\tau^x\Psi_{2^{-j}}\big\rangle$$

and the kernel K_j' of L_j' is

$$K_j'(x,y)=\big\langle L(\tau^y\Psi_{2^{-j}}),\tau^x\Phi_{2^{-(j-1)}}\big\rangle.$$

We plan to prove that

$$|K_j(x,y)|+2^{-j}|\nabla K_j(x,y)| \le C_n(A+B_4)2^{nj}(1+2^j|x-y|)^{-n-\delta}\,, \tag{4.5.23}$$

noting that an analogous estimate holds for $K_j'(x,y)$. Once (4.5.23) is established, Exercise 4.5.2 and the conditions

$$L_j(1)=\Delta_j LS_j(1)=\Delta_j L(1)=0\,, \qquad L_j'(1)=S_{j-1}L\Delta_j(1)=0\,,$$

yield the hypotheses of Proposition 4.5.3. Recalling (4.5.20), the conclusion of this proposition yields (4.5.21).

To prove (4.5.23) we quickly repeat the corresponding argument from the proof of Theorem 4.3.3. We consider the following two cases: If $|x-y|\le 5\cdot 2^{-j}$, then the weak boundedness property gives

$$\begin{aligned}\left|\langle L(\tau^y\Phi_{2^{-j}}),\tau^x\Psi_{2^{-j}}\rangle\right| &= \left|\langle L(\tau^x(\tau^{2^j(y-x)}(\Phi)_{2^{-j}})),\tau^x\Psi_{2^{-j}}\rangle\right| \\ &\le C_n\|L\|_{WB}2^{jn}\,,\end{aligned}$$

since Ψ and $\tau^{2^j(y-x)}\Phi$, whose support is contained in $B(0,\frac{1}{2})+B(0,5)\subseteq B(0,10)$, are multiples of normalized bumps. This proves the first of the two estimates in (4.5.23) when $|x-y|\le 5\cdot 2^{-j}$.

We now turn to the case $|x-y|\ge 5\cdot 2^{-j}$. Then the functions $\tau^y\Phi_{2^{-j}}$ and $\tau^x\Psi_{2^{-j}}$ have disjoint supports, and so we have the integral representation

$$K_j(x,y)=\int_{\mathbf{R}^n}\int_{\mathbf{R}^n}\Phi_{2^{-j}}(v-y)K(u,v)\Psi_{2^{-j}}(u-x)\,du\,dv\,.$$

Using that Ψ has mean value zero, we can write the previous expression as

$$\int_{\mathbf{R}^n}\int_{\mathbf{R}^n}\Phi_{2^{-j}}(v-y)\big(K(u,v)-K(x,v)\big)\Psi_{2^{-j}}(u-x)\,du\,dv\,.$$

We observe that $|u-x|\le 2^{-j}$ and $|v-y|\le 2^{-j}$ in the preceding integral. Since $|x-y|\ge 5\cdot 2^{-j}$, this makes $|u-v|\ge |x-y|-2\cdot 2^{-j}\ge 2\cdot 2^{-j}$, which implies that $|u-x|\le \frac{1}{2}|u-v|$. Using the regularity condition (4.1.2), we deduce

$$|K(u,v)-K(x,v)|\le A\frac{|x-u|^{\delta}}{|u-v|^{n+\delta}}\le C_{n,\delta}A\frac{2^{-j\delta}}{|x-y|^{n+\delta}}\,.$$

Inserting this estimate in the preceding double integral, we obtain the first estimate in (4.5.23). The second estimate in (4.5.23) is proved similarly. □

4.5.4 Pseudodifferential Operators

We now turn to another elegant application of Lemma 4.5.1 regarding pseudodifferential operators. We first introduce pseudodifferential operators.

Definition 4.5.5. Let $m \in \mathbf{R}$ and $0 \le \rho, \delta \le 1$. A $\mathscr{C}^\infty$ function $\sigma(x,\xi)$ on $\mathbf{R}^n \times \mathbf{R}^n$ is called a *symbol of class* $S^m_{\rho,\delta}$ if for all multi-indices α and β there is a constant $B_{\alpha,\beta}$ such that

$$|\partial_x^\alpha \partial_\xi^\beta \sigma(x,\xi)| \le B_{\alpha,\beta}(1+|\xi|)^{m-\rho|\beta|+\delta|\alpha|}. \tag{4.5.24}$$

For $\sigma \in S^m_{\rho,\delta}$, the linear operator

$$T_\sigma(f)(x) = \int_{\mathbf{R}^n} \sigma(x,\xi)\widehat{f}(\xi)e^{2\pi i x\cdot\xi}\,d\xi$$

initially defined for f in $\mathscr{S}(\mathbf{R}^n)$ is called a *pseudodifferential operator* with symbol $\sigma(x,\xi)$.

Example 4.5.6. Let b be a bounded function on $\mathbf{R}^n$. Consider the symbol

$$\sigma_b(x,\xi) = \sum_{j\in\mathbf{Z}} (b * \Psi_{2^{-j}})(x)\widehat{\Psi}(2^{-j}\xi), \tag{4.5.25}$$

where $\widehat{\Psi}$ is a smooth function supported in the annulus $1/ \le |\xi| \le 2$. It is not hard to see that the symbol σ_b satisfies

$$|\partial_x^\alpha \partial_\xi^\beta \sigma_b(x,\xi)| \le C_{\alpha,\beta}\|b\|_{L^\infty}|\xi|^{-|\beta|+|\alpha|} \tag{4.5.26}$$

for all multi-indices α and β. Indeed, every differentiation in x produces a factor of 2^j, while every differentiation in ξ produces a factor of 2^{-j}. But since $\widehat{\Psi}$ is supported in $\frac{1}{2}\cdot 2^j \le |\xi| \le 2\cdot 2^j$, it follows that $|\xi| \approx 2^j$, which yields (4.5.26). It follows that σ_b is not in any of the classes $S^m_{\rho,\delta}$ introduced in Definition 4.5.5, since σ_b is not necessarily smooth at the origin. However, if we restrict the indices of summation in (4.5.25) to $j \ge 0$, then $|\xi| \approx 1+|\xi|$ and we obtain a symbol of class $S^0_{1,1}$. Note that not all symbols in $S^0_{1,1}$ give rise to bounded operators on L^2. See Exercise 4.5.6.

An example of a symbol in $S^m_{1,0}$ is $(1+|\xi|^2)^{\frac{1}{2}(m+it)}$ when $m,t \in \mathbf{R}$.

We do not plan to embark on a systematic study of pseudodifferential operators here, but we would like to study the L^2 boundedness of symbols of class $S^0_{0,0}$.

Theorem 4.5.7. *Suppose that a symbol σ belongs to the class $S^0_{0,0}$. Then the pseudodifferential operator T_σ with symbol σ, initially defined on $\mathscr{S}(\mathbf{R}^n)$, has a bounded extension on $L^2(\mathbf{R}^n)$.*

Proof. In view of Plancherel's theorem, it suffices to obtain the L^2 boundedness of the linear operator

$$\widetilde{T}_\sigma(f)(x) = \int_{\mathbf{R}^n} \sigma(x,\xi) f(\xi) e^{2\pi i x\cdot\xi}\, d\xi\,, \tag{4.5.27}$$

defined for f in $\mathscr{S}(\mathbf{R}^n)$. We fix a nonnegative smooth function $\varphi(\xi)$ supported in a small multiple of the unit cube $Q_0 = [0,1]^n$ (say in $[-\frac{1}{9}, \frac{10}{9}]^n$) that satisfies

$$\sum_{j\in\mathbf{Z}^n} \varphi(x-j) = 1\,, \qquad x \in \mathbf{R}^n\,. \tag{4.5.28}$$

For $j,k \in \mathbf{Z}^n$ we define symbols

$$\sigma_{j,k}(x,\xi) = \varphi(x-j)\sigma(x,\xi)\varphi(\xi-k)$$

and corresponding operators $T_{j,k}$ given by (4.5.27) in which $\sigma(x,\xi)$ is replaced by $\sigma_{j,k}(x,\xi)$. Using (4.5.28), for any $f \in \mathscr{S}(\mathbf{R}^n)$, we obtain that

$$\widetilde{T}_\sigma(f) = \sum_{j,k\in\mathbf{Z}^n} T_{j,k}(f)\,,$$

where the double sum is easily shown to converge pointwise. Our goal is to show that for all $N \in \mathbf{Z}^+$ we have

$$\big\|T_{j,k}^* T_{j',k'}\big\|_{L^2\to L^2} \le C_N(1+|j-j'|+|k-k'|)^{-2N}\,, \tag{4.5.29}$$

$$\big\|T_{j,k} T_{j',k'}^*\big\|_{L^2\to L^2} \le C_N(1+|j-j'|+|k-k'|)^{-2N}\,, \tag{4.5.30}$$

where C_N depends on N and n but is independent of j, j', k, k'.

We note that

$$T_{j,k}^* T_{j',k'}(f)(x) = \int_{\mathbf{R}^n} K_{j,k,j',k'}(x,y) f(y)\, dy\,,$$

where

$$K_{j,k,j',k'}(x,y) = \int_{\mathbf{R}^n} \overline{\sigma_{j,k}(z,x)}\,\sigma_{j',k'}(z,y) e^{2\pi i(y-x)\cdot z}\, dz\,. \tag{4.5.31}$$

We integrate by parts in (4.5.31) using the identity

$$e^{2\pi i z\cdot(y-x)} = \frac{(I-\Delta_z)^N (e^{2\pi i z\cdot(y-x)})}{(1+4\pi^2|x-y|^2)^N}\,,$$

and we express $K_{j,k,j',k'}(x,y)$ as

$$\frac{\varphi(x-k)\varphi(y-k')}{(1+4\pi^2|x-y|^2)^N} \int_{\mathbf{R}^n} (I-\Delta_z)^N \Big(\varphi(z-j)\overline{\sigma(z,x)}\sigma(z,y)\varphi(z-j') \Big) e^{2\pi i(y-x)\cdot z}\, dz$$

The support property of φ forces $|j-j'| \le c_n$ for some dimensional constant c_n; indeed, $c_n = 2\sqrt{n}$ suffices. Moreover, all derivatives of σ and φ are controlled by constants, and φ is supported in a cube of finite measure. We also have $1+|x-y| \approx 1+|k-k'|$. It follows that

$$|K_{j,k,j',k'}(x,y)| \le \begin{cases} \dfrac{C_N \varphi(x-k)\varphi(y-k')}{(1+|k-k'|)^{2N}} & \text{when } |j-j'| \le c_n, \\ 0 & \text{otherwise.} \end{cases}$$

We can rewrite the preceding estimates in a more compact (and symmetric) form as

$$|K_{j,k,j',k'}(x,y)| \le \frac{C_{n,N}\varphi(x-k)\varphi(y-k')}{(1+|j-j'|+|k-k'|)^{2N}},$$

from which we easily obtain that

$$\sup_{x\in\mathbf{R}^n}\int_{\mathbf{R}^n}|K_{j,k,j',k'}(x,y)|\,dy \le \frac{C_{n,N}}{(1+|j-j'|+|k-k'|)^{2N}}, \tag{4.5.32}$$

$$\sup_{y\in\mathbf{R}^n}\int_{\mathbf{R}^n}|K_{j,k,j',k'}(x,y)|\,dx \le \frac{C_{n,N}}{(1+|j-j'|+|k-k'|)^{2N}}. \tag{4.5.33}$$

Using the classical Schur lemma in Appendix A.1, we obtain that

$$\big\|T_{j,k}^* T_{j',k'}\big\|_{L^2\to L^2} \le \frac{C_{n,N}}{(1+|j-j'|+|k-k'|)^{2N}},$$

which proves (4.5.29). Since $\rho = \delta = 0$, the roles of the variables x and ξ are symmetric, and (4.5.30) can be proved in exactly the same way as (4.5.29). The almost orthogonality Lemma 4.5.1 now applies, since

$$\sum_{j,k\in\mathbf{Z}^n}\sqrt{\frac{1}{(1+|j|+|k|)^{2N}}} \le \sum_{j\in\mathbf{Z}^n}\sum_{k\in\mathbf{Z}^n}\frac{1}{(1+|j|)^{\frac{N}{2}}}\frac{1}{(1+|k|)^{\frac{N}{2}}} < \infty$$

for $N \ge 2n+2$, and the boundedness of $\widetilde{T}_\sigma$ on L^2 follows. □

Remark 4.5.8. The reader may want to check that the argument in Theorem 4.5.7 is also valid for symbols of the class $S^0_{\rho,\rho}$ whenever $0<\rho<1$.

Exercises

4.5.1. Prove that any bounded linear operator S from a Hilbert space H to itself satisfies $\|S\|^2_{H\to H} = \|SS^*\|_{H\to H}$.

4.5.2. Show that if a family of kernels K_j satisfy (4.5.10) and

$$|\nabla_x K_j(x,y)| + |\nabla_y K_j(x,y)| \le \frac{A2^{(n+1)j}}{(1+2^j|x-y|)^{n+\delta}}$$

for all $x, y \in \mathbf{R}^n$, then conditions (4.5.11) and (4.5.12) hold with $\gamma = 1$.

4.5.3. Prove the boundedness of the Hilbert transform using Lemma 4.5.1 and without using the Fourier transform.
[*Hint:* Pick a smooth function η supported in $[1/2,2]$ such that $\sum_{j\in\mathbf{Z}} \eta(2^{-j}x) = 1$ for $x \neq 0$ and set $K_j(x) = x^{-1}\eta(2^{-j}|x|)$ and $H_j(f) = f * K_j$. Note that $H_j^* = -H_j$. Estimate $\|H_k H_j\|_{L^2\to L^2}$ by $\|K_k * K_j\|_{L^1} \le \|K_k * K_j\|_{L^\infty} |\text{supp}\,(K_k * K_j)|$. When $j<k$, use the mean value property of K_j and that $\|K_k'\|_{L^\infty} \le C2^{-2k}$ to obtain that $\|K_k * K_j\|_{L^\infty} \le C2^{-2k+j}$. Conclude that $\|H_k H_j\|_{L^2\to L^2} \le C2^{-|j-k|}$.]

4.5.4. For a symbol $\sigma(x,\xi)$ in $S^0_{1,0}$, let $k(x,z)$ denote the inverse Fourier transform (evaluated at z) of the function $\sigma(x,\cdot)$ with x fixed. Show that for all $x \in \mathbf{R}^n$, the distribution $k(x,\cdot)$ coincides with a smooth function away from the origin in $\mathbf{R}^n$ that satisfies the estimates

$$|\partial_x^\alpha \partial_z^\beta k(x,z)| \le C_{\alpha,\beta}|z|^{-n-|\beta|},$$

and conclude that the kernels $K(x,y) = k(x,x-y)$ are well defined and smooth functions away from the diagonal in $\mathbf{R}^{2n}$ that belong to $SK(1,A)$ for some $A>0$. Conclude that pseudodifferential operators with symbols in $S^0_{1,0}$ are associated with standard kernels.
[*Hint:* Consider the distribution $(\partial^\gamma \sigma(x,\cdot))^\vee = (-2\pi i z)^\gamma k(x,\cdot)$. Since $\partial_\xi^\gamma \sigma(x,\xi)$ is integrable in ξ when $|\gamma| \ge n+1$, it follows that $k(x,\cdot)$ coincides with a smooth function on $\mathbf{R}^n \setminus \{0\}$. Next, set $\sigma_j(x,\xi) = \sigma(x,\xi)\widehat{\Psi}(2^{-j}\xi)$, where Ψ is as in Section 4.4 and k_j the inverse Fourier transform of σ_j in z. For $|\gamma| = M$ use that

$$(-2\pi i z)^\gamma \partial_x^\alpha \partial_\xi^\beta k_j(x,z) = \int_{\mathbf{R}^n} \partial_\xi^\gamma\big((2\pi i\xi)^\beta \partial_x^\alpha \sigma_j(x,\xi)\big) e^{2\pi i \xi\cdot z}\, d\xi$$

to obtain $|\partial_x^\alpha \partial_z^\beta k_j(x,z)| \le B_{M,\alpha,\beta} 2^{jn} 2^{j|\alpha|} (2^j n|z|)^{-M}$ and sum over $j \in \mathbf{Z}$.]

4.5.5. Prove that pseudodifferential operators with symbols in $S^0_{1,0}$ that have compact support in x are elements of $CZO(1,A,B)$ for some $A,B>0$.
[*Hint:* Write

$$T_\sigma(f)(x) = \int_{\mathbf{R}^n} \left(\int_{\mathbf{R}^n} \widehat{\sigma}(a,\xi) \widehat{f}(\xi) e^{2\pi i x\cdot\xi}\, d\xi \right) e^{2\pi i x\cdot a}\, da,$$

where $\widehat{\sigma}(a,\xi)$ denotes the Fourier transform of $\sigma(x,\xi)$ in the variable x. Use integration by parts to obtain $\sup_\xi |\widehat{\sigma}(a,\xi)| \le C_N(1+|a|)^{-N}$ and pass the L^2 norm inside the integral in a to obtain the required conclusion using the translation-invariant case.]

4.5.6. Let $\widehat{\eta}(\xi)$ be a smooth bump on $\mathbf{R}$ that is supported in $2^{-\frac{1}{2}} \le |\xi| \le 2^{\frac{1}{2}}$ and is equal to 1 on $2^{-\frac{1}{4}} \le |\xi| \le 2^{\frac{1}{4}}$. Let

$$\sigma(x,\xi) = \sum_{k=1}^{\infty} e^{-2\pi i 2^k x} \widehat{\eta}(2^{-k}\xi).$$

Show that σ is an element of $S^0_{1,1}$ on the line but the corresponding pseudodifferential operator T_σ is not L^2 bounded.
[*Hint:* To see the latter statement, consider the sequence of functions $f_N(x) = \sum_{k=5}^{N} \frac{1}{k} e^{2\pi i 2^k x} h(x)$, where $h(x)$ is a Schwartz function whose Fourier transform is supported in the set $|\xi| \le \frac{1}{4}$. Show that $\|f_N\|_{L^2} \le C\|h\|_{L^2}$ but $\|T_\sigma(f_N)\|_{L^2} \ge c\,(\log N)\,\|h\|_{L^2}$ for some positive constants c, C.]

4.5.7. Prove conclusions (i) and (ii) of Lemma 4.5.1 if hypothesis (4.5.1) is replaced by

$$\big\|T_j^* T_k\big\|_{H\to H} + \big\|T_j T_k^*\big\|_{H\to H} \le \Gamma(j,k),$$

where Γ is a nonnegative function on $\mathbf{Z}\times\mathbf{Z}$ such that

$$\sup_j \sum_{k\in\mathbf{Z}} \sqrt{\Gamma(j,k)} = A < \infty.$$

4.5.8. Let $\{T_t\}_{t\in\mathbf{R}^+}$ be a family of operators mapping a Hilbert space H to itself. Assume that there is $c>0$ and a continuous function $\gamma : \mathbf{R}^+\times\mathbf{R}^+ \to \mathbf{R}^+\cup\{0\}$ satisfying

$$A_\gamma = \sup_{t>0} \int_0^\infty \sqrt{\gamma(t,s)}\,\frac{ds}{s} < \infty$$

and $\gamma(t,t)\le c$ for all $t>0$, such that

$$\big\|T_t^* T_s\big\|_{H\to H} + \big\|T_t T_s^*\big\|_{H\to H} \le \gamma(t,s)$$

for all t,s in $\mathbf{R}^+$. [An example of a function with $A_\gamma<\infty$ is $\gamma(t,s) = \min\left(\frac{s}{t},\frac{t}{s}\right)^\varepsilon$ for some $\varepsilon>0$.] Then prove that for all $0<\varepsilon<N$ we have

$$\left\| \int_\varepsilon^N T_t\,\frac{dt}{t} \right\|_{H\to H} \le A_\gamma .$$

4.6 The Cauchy Integral of Calderón and the $T(b)$ Theorem

The Cauchy integral is almost as old as complex analysis itself. In the classical theory of complex analysis, if Γ is a curve in $\mathbf{C}$ and f is a function on the curve, the Cauchy integral of f is given by

$$\frac{1}{2\pi i}\int_\Gamma \frac{f(\zeta)}{\zeta - z}\,d\zeta .$$

One situation in which this operator appears is the following: If Γ is a closed simple curve (i.e., a Jordan curve), Ω_+ is the interior-connected component of $\mathbf{C}\setminus\Gamma$, Ω_- is the exterior-connected component of $\mathbf{C}\setminus\Gamma$, and f is a smooth complex function on Γ, is it possible to find analytic functions F_+ on Ω_+ and F_- on Ω_-, respectively, that have continuous extensions on Γ such that their difference is equal to the given f on Γ? It turns out that a solution of this problem is given by the functions

$$F_+(w) = \frac{1}{2\pi i}\int_\Gamma \frac{f(\zeta)}{\zeta - w}\,d\zeta\,, \quad w\in\Omega_+\,,$$

and

$$F_-(w) = \frac{1}{2\pi i}\int_\Gamma \frac{f(\zeta)}{\zeta - w}\,d\zeta\,, \quad w\in\Omega_-\,.$$

We are would like to study the case in which the Jordan curve Γ passes through infinity, in particular, when it is the graph of a Lipschitz function on $\mathbf{R}$. In this case we compute the boundary limits of F_+ and F_- and we see that they give rise to a very interesting operator on the curve Γ. To fix notation we let

$$A:\ \mathbf{R}\to\mathbf{R}$$

be a Lipschitz function. This means that there is a constant $L>0$ such that for all $x,y\in\mathbf{R}$ we have $|A(x)-A(y)|\le L|x-y|$. We define a curve

$$\gamma:\ \mathbf{R}\to\mathbf{C}$$

by setting

$$\gamma(x) = x + iA(x)$$

and we denote by

$$\Gamma = \{\gamma(x):\ x\in\mathbf{R}\} \tag{4.6.1}$$

the graph of γ. Given a smooth function f on Γ we set

$$F(w) = \frac{1}{2\pi i}\int_\Gamma \frac{f(\zeta)}{\zeta - w}\,d\zeta\,, \quad w\in\mathbf{C}\setminus\Gamma\,. \tag{4.6.2}$$

We now show that for $z\in\Gamma$, both $F(z+i\delta)$ and $F(z-i\delta)$ have limits as $\delta\downarrow 0$, and these limits give rise to an operator on the curve Γ that we would like to study.

4.6.1 Introduction of the Cauchy Integral Operator along a Lipschitz Curve

Let $f(\zeta)$ be a $\mathscr{C}^1$ function on the curve Γ that decays faster than $C|\zeta|^{-1}$ as $|\zeta|\to\infty$. For $z\in\Gamma$ we define the *Cauchy integral of f at z* as

$$\mathfrak{C}_\Gamma(f)(z) = \lim_{\varepsilon\to 0+} \frac{1}{\pi i} \int\limits_{\substack{\zeta\in\Gamma \\ |\operatorname{Re}\zeta - \operatorname{Re} z|>\varepsilon}} \frac{f(\zeta)}{\zeta - z}\, d\zeta\,, \tag{4.6.3}$$

assuming that the limit exists. The decay assumption of f makes the integral in (4.6.3) converge when $|\operatorname{Re}\zeta - \operatorname{Re} z| \ge 1$. In the next proposition we show that the limit in (4.6.3) exists as $\varepsilon \to 0$ for almost all $z \in \Gamma$.

Proposition 4.6.1. *Let Γ be as in (4.6.1). Let $f(\zeta)$ be a $\mathscr{C}^1$ function on Γ that decays faster than $C|\zeta|^{-1}$ as $|\zeta| \to \infty$. Then the limit in (4.6.3) exists as $\varepsilon \to 0$ for all $z \in \Gamma$ such that A is differentiable at* $\operatorname{Re} z$. *Moreover, given f, we define a function F as in (4.6.2) related to f. Then for all such $z \in \Gamma$ such that A is differentiable at* $\operatorname{Re} z$ *(thus for almost all $z \in \Gamma$) we have that*

$$\lim_{\delta\downarrow 0} F(z+i\delta) = \frac{1}{2}\mathfrak{C}_\Gamma(f)(z) + \frac{1}{2} f(z)\,, \tag{4.6.4}$$

$$\lim_{\delta\downarrow 0} F(z-i\delta) = \frac{1}{2}\mathfrak{C}_\Gamma(f)(z) - \frac{1}{2} f(z)\,. \tag{4.6.5}$$

Proof. We show first that the limit in (4.6.3) exists as $\varepsilon \to 0$. For $z \in \Gamma$ and $0 < \varepsilon < 1$ we write

$$\begin{aligned} \frac{1}{\pi i} \int\limits_{\substack{\zeta\in\Gamma \\ |\operatorname{Re}\zeta - \operatorname{Re} z|>\varepsilon}} \frac{f(\zeta)\,d\zeta}{\zeta - z} = {} & \frac{1}{\pi i} \int\limits_{\substack{\zeta\in\Gamma \\ |\operatorname{Re}\zeta - \operatorname{Re} z|>1}} \frac{f(\zeta)\,d\zeta}{\zeta - z} \\ & + \frac{1}{\pi i} \int\limits_{\substack{\zeta\in\Gamma \\ \varepsilon\le|\operatorname{Re}\zeta - \operatorname{Re} z|\le 1}} \frac{(f(\zeta)-f(z))\,d\zeta}{\zeta - z} \\ & + \frac{f(z)}{\pi i} \int\limits_{\substack{\zeta\in\Gamma \\ \varepsilon\le|\operatorname{Re}\zeta - \operatorname{Re} z|\le 1}} \frac{d\zeta}{\zeta - z}\,. \end{aligned} \tag{4.6.6}$$

By the smoothness of f, the middle term of the sum in (4.6.6) has a limit as $\varepsilon \to 0$. We therefore study the third (last) term of this sum.

We denote by U_+ the (open) angle centered at the origin whose bisector is the positive imaginary axis and whose width is $2\arctan L^{-1}$. We let U_- be the (open) angle centered at the origin whose bisector is the negative imaginary axis and whose width is also $\arctan L^{-1}$. Obviously U_+ and U_- are symmetric about the origin. We let V_+ be the connected component of $\mathbf{C}\setminus(\overline{U_+\cup U_-})$ that contains the positive real axis and V_- be the connected component of $\mathbf{C}\setminus(\overline{U_+\cup U_-})$ that contains the negative real axis. The angles V_+ and V_- are open sets and notice that $\overline{V_+}\cup\overline{V_-}$ contains the range of the map $t \mapsto \gamma(t+\tau) - \gamma(\tau) = t + i(A(t+\tau) - A(\tau))$ for any real τ.

We consider two branches of the complex logarithm: $\log_{upper}(z)$ defined on $\mathbf{C}\setminus\{iy:\ y\le 0\}$ and $\log_{lower}(z)$ defined on $\mathbf{C}\setminus\{iy:\ y\ge 0\}$. These functions are defined to satisfy $\log_{upper}(1) = \log_{lower}(1) = 0$ and they coincide on $\overline{V_+}$, i.e.,

$\log_{upper} = \log_{lower}$ on $\overline{V_+}$. Note however that $\log_{upper} = 2\pi i + \log_{lower}$ on $\overline{V_-}$. For instance $\log_{lower}(-1) = -\pi i$ while $\log_{upper}(-1) = \pi i$. Also, $\log_{lower}(-i) = -\frac{\pi i}{2}$ and $\log_{upper}(i) = \frac{\pi i}{2}$.

Let $\tau = \operatorname{Re} z$ and $t = \operatorname{Re} \zeta$; then $z = \gamma(\tau) = \tau + iA(\tau)$ and $\zeta = \gamma(t)$. The function A is Lipschitz and thus differentiable almost everywhere; consequently, the function $\gamma(\tau) = \tau + iA(\tau)$ is differentiable for almost all $\tau \in \mathbf{R}$. Moreover, $\gamma'(\tau) = 1 + iA'(\tau) \neq 0$ and $\gamma'(\tau)$ lies in $\overline{V_+}$ whenever γ is differentiable at τ.

Fix a point $\tau = \operatorname{Re} z$ at which γ is differentiable. Using the change of variables $t = \tau + s$, we rewrite the last term in the sum in (4.6.6) as

$$\frac{f(z)}{\pi i} \int_{\varepsilon \le |s| \le 1} \frac{\gamma'(s+\tau)}{\gamma(s+\tau) - \gamma(\tau)} \, ds. \tag{4.6.7}$$

We evaluate (4.6.7) as follows:

$$\begin{aligned}
&\frac{f(z)}{\pi i} \Big[\log_{upper}\big(\gamma(1+\tau) - \gamma(\tau)\big) - \log_{upper}\big(\gamma(\varepsilon+\tau) - \gamma(\tau)\big) \\
&\qquad - \log_{upper}\big(\gamma(-1+\tau) - \gamma(\tau)\big) + \log_{upper}\big(\gamma(-\varepsilon+\tau) - \gamma(\tau)\big) \Big] \\
&= \frac{f(z)}{\pi i} \Big[\log_{upper}\big(\gamma(1+\tau) - \gamma(\tau)\big) - \log_{upper}\Big(\frac{\gamma(\varepsilon+\tau) - \gamma(\tau)}{\varepsilon}\Big) \\
&\qquad - \log_{upper}\big(\gamma(-1+\tau) - \gamma(\tau)\big) + \log_{upper}\Big(\frac{\gamma(-\varepsilon+\tau) - \gamma(\tau)}{\varepsilon}\Big) \Big].
\end{aligned}$$

This expression converges as $\varepsilon \to 0+$ to

$$\begin{aligned}
&\frac{f(z)}{\pi i} \Big[\log_{upper}\big(\gamma(1+\tau) - \gamma(\tau)\big) - \log_{upper}\big(\gamma(-1+\tau) - \gamma(\tau)\big) \\
&\qquad\qquad + \log_{upper}(\gamma'(\tau)) - \log_{upper}(-\gamma'(\tau)) \Big] \\
&= \frac{f(z)}{\pi i} \Big[\log_{upper}\big(\gamma(1+\tau) - \gamma(\tau)\big) - \log_{upper}\big(\gamma(-1+\tau) - \gamma(\tau)\big) - \pi i \Big],
\end{aligned}$$

since

$$\log_{upper}(\gamma'(\tau)) - \log_{upper}(-\gamma'(\tau)) = -\pi i. \tag{4.6.8}$$

Note that (4.6.7) also converges to

$$\frac{f(z)}{\pi i} \Big[\log_{lower}\big(\gamma(1+\tau) - \gamma(\tau)\big) - \log_{lower}\big(\gamma(-1+\tau) - \gamma(\tau)\big) + \pi i \Big],$$

as $\varepsilon \to 0+$, since one may equivalently evaluate the integral with $\log_{lower}$ instead of $\log_{upper}$ and use that

$$\log_{lower}(\gamma'(\tau)) - \log_{lower}(-\gamma'(\tau)) = \pi i. \tag{4.6.9}$$

Thus the limit in (4.6.6), and hence in (4.6.3), exists as $\varepsilon \to 0$ for almost all z on the curve Γ. Hence $\mathfrak{C}_\Gamma(f)$ is a well-defined operator whenever $f(\zeta)$ is a $\mathscr{C}^1$ function that decays faster than $C|\zeta|^{-1}$ at infinity.

We proceed with the proof of (4.6.4). For fixed $\delta > 0$ and $\varepsilon > 0$ we write

$$\begin{aligned} F(z+i\delta) &= \frac{1}{2\pi i} \int\limits_{\substack{\zeta\in\Gamma \\ |\mathrm{Re}\,\zeta - \mathrm{Re}\,z|>\varepsilon}} \frac{f(\zeta)}{\zeta - z - i\delta}\, d\zeta \\ &\quad + \frac{1}{2\pi i} \int\limits_{\substack{\zeta\in\Gamma \\ |\mathrm{Re}\,\zeta - \mathrm{Re}\,z|\le\varepsilon}} \frac{f(\zeta)-f(z)}{\zeta - z - i\delta}\, d\zeta \\ &\quad + \frac{f(z)}{2\pi i} \int\limits_{\substack{\zeta\in\Gamma \\ |\mathrm{Re}\,\zeta - \mathrm{Re}\,z|\le\varepsilon}} \frac{1}{\zeta - z - i\delta}\, d\zeta\,. \end{aligned} \tag{4.6.10}$$

The curve $s \mapsto \gamma(s+\tau) - \gamma(\tau) - i\delta$ lies strictly below the curve $s \mapsto \gamma(s+\tau) - \gamma(\tau)$; thus it avoids $\overline{U_+}$ and hence the last term in (4.6.10) can be evaluated via the use of $\log_{lower}$ as follows:

$$\frac{f(z)}{2\pi i}\big[\log_{lower}(\gamma(\tau+\varepsilon) - \gamma(\tau) - i\delta) - \log_{lower}(\gamma(\tau-\varepsilon) - \gamma(\tau) - i\delta)\big].$$

By the continuity of $\log_{lower}$ in this region, this converges to

$$\frac{f(z)}{2\pi i}\big[\log_{lower}(\gamma(\tau+\varepsilon) - \gamma(\tau)) - \log_{lower}(\gamma(\tau-\varepsilon) - \gamma(\tau))\big]$$

as $\delta \to 0+$. Thus we obtain from (4.6.10) that

$$\begin{aligned} \lim_{\delta\to 0+} F(z+i\delta) &= \frac{1}{2\pi i} \int\limits_{\substack{\zeta\in\Gamma \\ |\mathrm{Re}\,\zeta - \mathrm{Re}\,z|>\varepsilon}} \frac{f(\zeta)}{\zeta - z}\, d\zeta \\ &\quad + \frac{1}{2\pi i} \int\limits_{\substack{\zeta\in\Gamma \\ |\mathrm{Re}\,\zeta - \mathrm{Re}\,z|\le\varepsilon}} \frac{f(\zeta)-f(z)}{\zeta - z}\, d\zeta \\ &\quad + \frac{f(z)}{2\pi i}\Big[\log_{lower}\Big(\frac{\gamma(\tau+\varepsilon)-\gamma(\tau)}{\varepsilon}\Big) - \log_{lower}\Big(\frac{\gamma(\tau-\varepsilon)-\gamma(\tau)}{\varepsilon}\Big)\Big]. \end{aligned} \tag{4.6.11}$$

Letting $\varepsilon \to 0+$ in (4.6.11) we obtain

$$\begin{aligned} \lim_{\delta\to 0+} F(z+i\delta) &= \frac{1}{2}\mathfrak{C}_\Gamma(f)(z) + \frac{f(z)}{2\pi i}\Big[\log_{lower}(\gamma'(\tau)) - \log_{lower}(-\gamma'(\tau))\Big] \\ &= \frac{1}{2}\mathfrak{C}_\Gamma(f)(z) + \frac{f(z)}{2}, \end{aligned}$$

where we made use of (4.6.9). This proves (4.6.4).

In the proof of (4.6.5), the curve $s \mapsto \gamma(s+\tau) - \gamma(\tau) + i\delta$ lies strictly above the curve $s \mapsto \gamma(s+\tau) - \gamma(\tau)$; thus it avoids $\overline{U_-}$ and hence one needs to use $\log_{upper}$ to evaluate the integral. The proof proceeds along the same way with the only difference being that $\log_{lower}$ is replaced by $\log_{upper}$ and (4.6.9) is replaced by (4.6.8); this explains the presence of the minus sign in (4.6.5). □

Remark 4.6.2. Let F_+ be the restriction of F [as defined in (4.6.2)] on the region above the graph Γ and let F_- be the restriction of F on the region below the graph Γ. Then for all $z \in \Gamma$ the functions $t \mapsto F_+(z+it)$ and $t \mapsto F_-(z-it)$ are continuous on $(0,\infty)$ and for almost all $z \in \Gamma$ they have limits as $t \to 0+$ which satisfy

$$\lim_{t\to 0+} F_+(z+it) - \lim_{t\to 0+} F_-(z-it) = f(z),$$

where f is the given $\mathscr{C}^1$ function on the curve.

4.6.2 Resolution of the Cauchy Integral and Reduction of Its L^2 Boundedness to a Quadratic Estimate

Having introduced the Cauchy integral $\mathfrak{C}_\Gamma$ as an operator defined on smooth functions on the graph Γ of a Lipschitz function A, we turn to some of its properties. We are mostly interested in obtaining an a priori L^2 estimate for $\mathfrak{C}_\Gamma$. Before we achieve this goal, we make some observations. First we can write $\mathfrak{C}_\Gamma$ as

$$\mathfrak{C}_\Gamma(H)(x+iA(x)) = \lim_{\varepsilon\to 0} \frac{1}{\pi i} \int_{|x-y|>\varepsilon} \frac{H(y+iA(y))(1+iA'(y))}{y+iA(y)-x-iA(x)}\,dy, \tag{4.6.12}$$

where the integral is over the real line and H is a function on the curve Γ. (Recall that Lipschitz functions are differentiable almost everywhere.) To any function H on Γ we can associate a function h on the line $\mathbf{R}$ by setting

$$h(y) = H(y+iA(y)).$$

We have that

$$\int_\Gamma |H(y)|^2\,dy = \int_{\mathbf{R}} |h(y)|^2 (1+|A'(y)|^2)^{\frac{1}{2}}\,dy \approx \int_{\mathbf{R}} |h(y)|^2\,dy$$

for some constants that depend on the Lipschitz constant L of A. Therefore, the boundedness of the operator in (4.6.12) is equivalent to that of the operator

$$\mathcal{C}_\Gamma(h)(x) = \lim_{\varepsilon\to 0} \frac{1}{\pi i} \int_{|x-y|>\varepsilon} \frac{h(y)(1+iA'(y))}{y-x+i(A(y)-A(x))}\,dy \tag{4.6.13}$$

acting on Schwartz functions h on the line. It is this operator that we concentrate on in the remainder of this section. We recall that (see Example 4.1.6) the function

$$\frac{1}{y-x+i(A(y)-A(x))}$$

defined on $\mathbf{R}\times\mathbf{R}\setminus\{(x,x):\, x\in\mathbf{R}\}$ is a standard kernel in $SK(1,cL)$ for some $c>0$. We note that this is not the case with the kernel

$$\frac{1+iA'(y)}{y-x+i(A(y)-A(x))}, \tag{4.6.14}$$

for conditions (4.1.2) and (4.1.3) fail for this kernel, since the function $1+iA'$ does not possess any smoothness. [Condition (4.1.1) trivially holds for the function in (4.6.14).] We note, however, that the L^p boundedness of the operator in (4.6.13) is equivalent to that of

$$\widetilde{\mathcal{C}}_\Gamma(h)(x)=\lim_{\varepsilon\to 0}\frac{1}{\pi i}\int\limits_{|x-y|>\varepsilon}\frac{h(y)}{y-x+i(A(y)-A(x))}\,dy, \tag{4.6.15}$$

since the function $1+iA'$ is bounded above and below and can be absorbed in h. Therefore, the L^2 boundedness of $\mathcal{C}_\Gamma$ is equivalent to that of $\widetilde{\mathcal{C}}_\Gamma$, which has a kernel that satisfies standard estimates. This equivalence, however, is not as useful in the approach we take in the sequel. We choose to work with the operator $\mathcal{C}_\Gamma$, in which the appearance of the term $1+iA'(y)$ plays a crucial cancellation role.

In the proof of Theorem 4.3.3 we used a *resolution* of an operator T with standard kernel of the form

$$\int_0^\infty P_sT_sQ_s\frac{ds}{s},$$

where P_s and Q_s are nice averaging operators that approximate the identity and the zero operator, respectively. Our goal is to achieve a similar resolution for the operator $\mathcal{C}_\Gamma$ defined in (4.6.13). To achieve this, for every $s>0$, we introduce the auxiliary operator

$$\mathcal{C}_\Gamma(h)(x;s)=\frac{1}{\pi i}\int\limits_{\mathbf{R}}\frac{h(y)(1+iA'(y))}{y-x+i(A(y)-A(x))+is}\,dy \tag{4.6.16}$$

defined for Schwartz functions h on the line. We make two preliminary observations regarding this operator: For almost all $x\in\mathbf{R}$ we have

$$\lim_{s\to\infty}\mathcal{C}_\Gamma(h)(x;s) = 0, \tag{4.6.17}$$

$$\lim_{s\to 0}\mathcal{C}_\Gamma(h)(x;s) = \mathcal{C}_\Gamma(h)(x)+h(x). \tag{4.6.18}$$

Identity (4.6.17) is trivial. To obtain (4.6.18), for a fixed $\varepsilon > 0$ we write

$$\begin{aligned}\mathcal{C}_\Gamma(h)(x;s) &= \frac{1}{\pi i}\int_{|x-y|>\varepsilon} \frac{h(y)(1+iA'(y))}{y-x+i(A(y)-A(x))+is}\,dy \\ &\quad + \frac{1}{\pi i}\int_{|x-y|\le\varepsilon} \frac{(h(y)-h(x))(1+iA'(y))}{y-x+i(A(y)-A(x))+is}\,dy \\ &\quad + h(x)\,\frac{1}{\pi i}\log_{upper}\frac{\varepsilon+i(A(x+\varepsilon)-A(x))+is}{-\varepsilon+i(A(x-\varepsilon)-A(x))+is}\,,\end{aligned} \tag{4.6.19}$$

where $\log_{upper}$ denotes the analytic branch of the complex logarithm defined in the proof of Proposition 4.6.1. We used this branch of the logarithm, since for $s > 0$, the graph of the function $y \mapsto y+i(A(y+x)-A(x))+is$ lies outside a small angle centered at the origin that contains the negative imaginary axis (for instance the angle $\overline{U_-}$ as defined in Proposition 4.6.1).

We now take successive limits first as $s \to 0$ and then as $\varepsilon \to 0$ in (4.6.19). We obtain that

$$\begin{aligned}\lim_{s\to 0}\mathcal{C}_\Gamma(h)(x;s) &= \lim_{\varepsilon\to 0}\frac{1}{\pi i}\int_{|x-y|>\varepsilon} \frac{h(y)(1+iA'(y))}{y-x+i(A(y)-A(x))}\,dy \\ &\quad + h(x)\lim_{\varepsilon\to 0}\frac{1}{\pi i}\log_{upper}\frac{\varepsilon+i(A(x+\varepsilon)-A(x))}{-\varepsilon+i(A(x-\varepsilon)-A(x))}\,.\end{aligned}$$

Since this expression inside the logarithm tends to -1 as $\varepsilon \to 0$, this logarithm tends to πi, and this concludes the proof of (4.6.18).

We now consider the second derivative in s of the auxiliary operator $\mathcal{C}_\Gamma(h)(x;s)$.

$$\begin{aligned}&\int_0^\infty s^2\frac{d^2}{ds^2}\mathcal{C}_\Gamma(h)(x;s)\,\frac{ds}{s} \\ &\quad = \int_0^\infty s\frac{d^2}{ds^2}\mathcal{C}_\Gamma(h)(x;s)\,ds \\ &\quad = \lim_{s\to\infty} s\frac{d}{ds}\mathcal{C}_\Gamma(h)(x;s) - \lim_{s\to 0} s\frac{d}{ds}\mathcal{C}_\Gamma(h)(x;s) - \int_0^\infty \frac{d}{ds}\mathcal{C}_\Gamma(h)(x;s)\,ds \\ &\quad = 0-0+\lim_{s\to 0}\mathcal{C}_\Gamma(h)(x;s) - \lim_{s\to\infty}\mathcal{C}_\Gamma(h)(x;s) \\ &\quad = \mathcal{C}_\Gamma(h)(x)+h(x)\,,\end{aligned}$$

where we used integration by parts, the fact that for almost all $x \in \mathbf{R}$ we have

$$\lim_{s\to\infty} s\frac{d}{ds}\mathcal{C}_\Gamma(h)(x;s) = \lim_{s\to 0} s\frac{d}{ds}\mathcal{C}_\Gamma(h)(x;s) = 0\,, \tag{4.6.20}$$

and identities (4.6.17) and (4.6.18) whenever h is a Schwartz function. One may consult Exercise 4.6.2 for a proof of the identities in (4.6.20). So we have succeeded in writing the operator $\mathcal{C}_\Gamma(h)+h$ as an average of smoother operators. Precisely, we have shown that for $h\in\mathscr{S}(\mathbf{R})$ we have

$$\mathcal{C}_\Gamma(h)(x)+h(x)=\int_0^\infty s^2\frac{d^2}{ds^2}\mathcal{C}_\Gamma(h)(x;s)\,\frac{ds}{s}\,, \tag{4.6.21}$$

and it remains to understand what the operator

$$\frac{d^2}{ds^2}\mathcal{C}_\Gamma(h)(x;s)=\mathcal{C}_\Gamma(h)''(x;s)$$

really is. Differentiating (4.6.16) twice, we obtain

$$\begin{aligned}\mathcal{C}_\Gamma(h)(x)+h(x) &= \int_0^\infty s^2\mathcal{C}_\Gamma(h)''(x;s)\,\frac{ds}{s}\\ &= 4\int_0^\infty s^2\mathcal{C}_\Gamma(h)''(x;2s)\,\frac{ds}{s}\\ &= -\frac{8}{\pi i}\int_0^\infty\int_{\mathbf{R}}\frac{s^2h(y)(1+iA'(y))}{(y-x+i(A(y)-A(x))+2is)^3}\,dy\,\frac{ds}{s}\\ &= -\frac{8}{\pi i}\int_0^\infty\int_\Gamma\frac{s^2H(\zeta)}{(\zeta-z+2is)^3}\,d\zeta\,\frac{ds}{s}\,,\end{aligned}$$

where in the last step we set $z=x+iA(x)$, $H(z)=h(x)$, and we switched to complex integration over the curve Γ. We now use the following identity from complex analysis. For $\zeta,z\in\Gamma$ we have

$$\frac{1}{(\zeta-z+2is)^3}=-\frac{1}{4\pi i}\int_\Gamma\frac{1}{(\zeta-w+is)^2}\frac{1}{(w-z+is)^2}\,dw\,, \tag{4.6.22}$$

for which we refer to Exercise 4.6.3. Inserting this identity in the preceding expression for $\mathcal{C}_\Gamma(h)(x)+h(x)$, we obtain

$$\mathcal{C}_\Gamma(h)(x)+h(x)=-\frac{2}{\pi^2}\int_0^\infty\left[\int_\Gamma\frac{s}{(w-z+is)^2}\left(\int_\Gamma\frac{s\,H(\zeta)}{(\zeta-w+is)^2}\,d\zeta\right)dw\right]\frac{ds}{s}\,,$$

recalling that $z=x+iA(x)$. Introducing the linear operator

$$\Theta_s(h)(x)=\int_{\mathbf{R}}\theta_s(x,y)\,h(y)\,dy\,, \tag{4.6.23}$$

where

$$\theta_s(x,y)=\frac{s}{(y-x+i(A(y)-A(x))+is)^2}\,, \tag{4.6.24}$$

we may therefore write

$$\mathcal{C}_\Gamma(h)(x)+h(x)=-\frac{2}{\pi^2}\int_0^\infty\Theta_s\big((1+iA')\Theta_s\big((1+iA')h\big)\big)(x)\,\frac{ds}{s}\,. \tag{4.6.25}$$

We also introduce the multiplication operator

$$M_b(h) = bh,$$

which enables us to write (4.6.25) in a more compact form as

$$\mathcal{C}_\Gamma(h) = -h - \frac{2}{\pi^2}\int_0^\infty \Theta_s M_{1+iA'}\Theta_s M_{1+iA'}(h)\,\frac{ds}{s}. \tag{4.6.26}$$

This gives us the desired resolution of the operator $\mathcal{C}_\Gamma$. It suffices to obtain an L^2 estimate for the integral expression in (4.6.26). Using duality, we write

$$\Big\langle \int_0^\infty \Theta_s M_{1+iA'}\Theta_s M_{1+iA'}(h)\,\frac{ds}{s}, g\Big\rangle = \int_0^\infty \big\langle M_{1+iA'}\Theta_s M_{1+iA'}(h), \Theta_s^t(g)\big\rangle\,\frac{ds}{s},$$

which is easily bounded by

$$\begin{aligned}\sqrt{1+L^2}&\int_0^\infty \big\|\Theta_s M_{1+iA'}(h)\big\|_{L^2}\big\|\Theta_s^t(g)\big\|_{L^2}\frac{ds}{s}\\ &\le \sqrt{1+L^2}\Big(\int_0^\infty \big\|\Theta_s M_{1+iA'}(h)\big\|_{L^2}^2\frac{ds}{s}\Big)^{\frac12}\Big(\int_0^\infty \big\|\Theta_s(g)\big\|_{L^2}^2\frac{ds}{s}\Big)^{\frac12}.\end{aligned}$$

We have now reduced matters to the following estimate:

$$\Big(\int_0^\infty \big\|\Theta_s(h)\big\|_{L^2}^2\frac{ds}{s}\Big)^{\frac12} \le C\big\|h\big\|_{L^2}. \tag{4.6.27}$$

We derive (4.6.27) as a consequence of Theorem 4.6.6 discussed in Section 4.6.4.

4.6.3 A Quadratic $T(1)$ Type Theorem

We review what we have achieved so far and we introduce definitions that place matters into a new framework.

For the purposes of the subsequent exposition we can switch to $\mathbf{R}^n$, since there are no differences from the one-dimensional argument. Suppose that for all $s > 0$, there is a family of functions θ_s defined on $\mathbf{R}^n \times \mathbf{R}^n$ such that

$$|\theta_s(x,y)| \le \frac{1}{s^n}\,\frac{A}{\big(1+\frac{|x-y|}{s}\big)^{n+\delta}} \tag{4.6.28}$$

and

$$|\theta_s(x,y) - \theta_s(x,y')| \le \frac{A}{s^n}\,\frac{|y-y'|^\gamma}{s^\gamma} \tag{4.6.29}$$

for all $x,y,y' \in \mathbf{R}^n$ and some $0<\gamma,\delta,A<\infty$. Let Θ_s be the operator with kernel θ_s, that is,

$$\Theta_s(h)(x) = \int_{\mathbf{R}^n} \theta_s(x,y)\,h(y)\,dy\,, \tag{4.6.30}$$

which is well defined for all h in $\bigcup_{1\le p\le\infty} L^p(\mathbf{R}^n)$ in view of (4.6.28).

At this point we observe that both (4.6.28) and (4.6.29) hold for the θ_s defined in (4.6.24) with $\gamma=\delta=1$ and A a constant multiple of L. We leave the details of this calculation to the reader but we note that (4.6.29) can be obtained quickly using the mean value theorem. Our goal is to figure out under what additional conditions on Θ_s the quadratic estimate (4.6.27) holds. If we can find such a condition that is easily verifiable for the Θ_s associated with the Cauchy integral, this will conclude the proof of its L^2 boundedness.

We first consider a simple condition that implies the quadratic estimate (4.6.27).

Theorem 4.6.3. *For $s>0$, let θ_s be a family of kernels satisfying (4.6.28) and (4.6.29) and let Θ_s be the linear operator whose kernel is θ_s. Suppose that for all $s>0$ we have*

$$\Theta_s(1) = 0\,. \tag{4.6.31}$$

Then there is a constant $C_{n,\delta}$ such that for all $f\in L^2$ we have

$$\left(\int_0^\infty \big\|\Theta_s(f)\big\|_{L^2}^2 \frac{ds}{s}\right)^{\frac12} \le C_{n,\delta} A \big\|f\big\|_{L^2}\,. \tag{4.6.32}$$

We note that condition (4.6.31) is not satisfied for the operators Θ_s associated with the Cauchy integral as defined in (4.6.23). However, Theorem 4.6.3 gives us an idea of what we are looking for, something like the action of Θ_s on a specific function. We also observe that condition (4.6.31) is "basically" saying that $\Theta(1)=0$, where

$$\Theta = \int_0^\infty \Theta_s \frac{ds}{s}\,.$$

Proof. We introduce Littlewood–Paley operators Q_s given by convolution with a smooth function $\Psi_s = \frac{1}{s^n}\Psi(\frac{\cdot}{s})$ whose Fourier transform is supported in the annulus $s/2 \le |\xi| \le 2s$ that satisfies

$$\int_0^\infty Q_s^2 \frac{ds}{s} = \lim_{\substack{\varepsilon\to 0\\ N\to\infty}} \int_\varepsilon^N Q_s^2 \frac{ds}{s} = I\,, \tag{4.6.33}$$

where the limit is taken in the sense of distributions and the identity holds in $\mathscr{S}'(\mathbf{R}^n)/\mathscr{P}$. This identity and properties of Θ_t imply the operator identity

$$\Theta_t = \Theta_t \int_0^\infty Q_s^2 \frac{ds}{s} = \int_0^\infty \Theta_t Q_s^2 \frac{ds}{s}\,.$$

The key fact is the following estimate:

$$\big\|\Theta_t Q_s\big\|_{L^2\to L^2} \le A\,C_{n,\Psi} \min\left(\frac{s}{t},\frac{t}{s}\right)^{\varepsilon}, \tag{4.6.34}$$

which holds for some $\varepsilon = \varepsilon(\gamma,\delta,n) > 0$. [Recall that A, γ, and δ are as in (4.6.28) and (4.6.29).] Assuming momentarily estimate (4.6.34), we can quickly prove Theorem 4.6.3 using duality. Indeed, let us take a function $G(x,t)$ such that

$$\int_0^\infty \int_{\mathbf{R}^n} |G(x,t)|^2\,dx\,\frac{dt}{t} \le 1. \tag{4.6.35}$$

Then we have

$$\begin{aligned}
&\int_0^\infty \int_{\mathbf{R}^n} G(x,t)\,\Theta_t(f)(x)\,dx\,\frac{dt}{t} \\
&\quad= \int_0^\infty \int_{\mathbf{R}^n} G(x,t) \int_0^\infty \Theta_t Q_s^2(f)(x)\,\frac{ds}{s}\,dx\,\frac{dt}{t} \\
&\quad= \int_0^\infty \int_0^\infty \int_{\mathbf{R}^n} G(x,t)\,\Theta_t Q_s^2(f)(x)\,dx\,\frac{dt}{t}\,\frac{ds}{s} \\
&\quad\le \left(\int_0^\infty \int_0^\infty \int_{\mathbf{R}^n} |G(x,t)|^2 dx \min\left(\frac{s}{t},\frac{t}{s}\right)^{\varepsilon} \frac{dt}{t}\frac{ds}{s}\right)^{\frac12} \\
&\qquad\times \left(\int_0^\infty \int_0^\infty \int_{\mathbf{R}^n} |\Theta_t Q_s(Q_s(f))(x)|^2\,dx \min\left(\frac{s}{t},\frac{t}{s}\right)^{-\varepsilon} \frac{dt}{t}\frac{ds}{s}\right)^{\frac12}.
\end{aligned}$$

But we have the estimate

$$\sup_{t>0} \int_0^\infty \min\left(\frac{s}{t},\frac{t}{s}\right)^{\varepsilon} \frac{ds}{s} \le C_\varepsilon,$$

which, combined with (4.6.35), yields that the first term in the product of the two preceding square functions is controlled by $\sqrt{C_\varepsilon}$. Using this fact and (4.6.34), we write

$$\begin{aligned}
&\int_0^\infty \int_{\mathbf{R}^n} G(x,t)\,\Theta_t(f)(x)\,dx\,\frac{dt}{t} \\
&\quad\le \sqrt{C_\varepsilon}\left(\int_0^\infty \int_0^\infty \int_{\mathbf{R}^n} |\Theta_t Q_s(Q_s(f))(x)|^2\,dx \min\left(\frac{s}{t},\frac{t}{s}\right)^{-\varepsilon} \frac{dt}{t}\frac{ds}{s}\right)^{\frac12} \\
&\quad\le A\sqrt{C_\varepsilon}\left(\int_0^\infty \int_0^\infty \int_{\mathbf{R}^n} |Q_s(f)(x)|^2\,dx \min\left(\frac{s}{t},\frac{t}{s}\right)^{2\varepsilon} \min\left(\frac{s}{t},\frac{t}{s}\right)^{-\varepsilon} \frac{dt}{t}\frac{ds}{s}\right)^{\frac12} \\
&\quad\le A\sqrt{C_\varepsilon}\left(\int_0^\infty \int_0^\infty \int_{\mathbf{R}^n} |Q_s(f)(x)|^2\,dx \min\left(\frac{s}{t},\frac{t}{s}\right)^{\varepsilon} \frac{dt}{t}\frac{ds}{s}\right)^{\frac12}
\end{aligned}$$

$$\leq C_\varepsilon A\left(\int_0^\infty \int_{\mathbf{R}^n} |Q_s(f)(x)|^2\,dx\,\frac{ds}{s}\right)^{\frac{1}{2}}$$
$$\leq C_{n,\varepsilon} A\big\|f\big\|_{L^2},$$

where in the last step we used the continuous version of Theorem 6.1.2 in [156] (cf. Exercise 6.1.4 in [156]). Taking the supremum over all functions $G(x,t)$ that satisfy (4.6.35) yields estimate (4.6.32).

It remains to prove (4.6.34). What is crucial here is that both Θ_t and Q_s satisfy the cancellation conditions $\Theta_t(1) = 0$ and $Q_s(1) = 0$. The proof of estimate (4.6.34) is similar to that of estimates (4.5.14) and (4.5.15) in Proposition 4.5.3. Using ideas from the proof of Proposition 4.5.3, we quickly dispose of the proof of (4.6.34).

The kernel of $\Theta_t Q_s$ is seen easily to be

$$L_{t,s}(x,y) = \int_{\mathbf{R}^n} \theta_t(x,z)\Psi_s(z-y)\,dz\,.$$

Notice that the function $(y,z) \mapsto \Psi_s(z-y)$ satisfies (4.6.28) with $\delta = 1$ and $A = C_\Psi$ and satisfies

$$|\Psi_s(z-y) - \Psi_s(z'-y)| \leq \frac{C_\Psi}{s^n}\frac{|z-z'|}{s}$$

for all $z, z', y \in \mathbf{R}^n$ for some $C_\Psi < \infty$. We prove that

$$\sup_{x\in\mathbf{R}^n} \int_{\mathbf{R}^n} |L_{t,s}(x,y)|\,dy \leq C_\Psi A \min\left(\frac{t}{s},\frac{s}{t}\right)^{\frac{1}{4}\frac{\min(\delta,1)}{n+\min(\delta,1)}\min(\gamma,\delta,1)}, \tag{4.6.36}$$

$$\sup_{y\in\mathbf{R}^n} \int_{\mathbf{R}^n} |L_{t,s}(x,y)|\,dx \leq C_\Psi A \min\left(\frac{t}{s},\frac{s}{t}\right)^{\frac{1}{4}\frac{\min(\delta,1)}{n+\min(\delta,1)}\min(\gamma,\delta,1)}. \tag{4.6.37}$$

Once (4.6.36) and (4.6.37) are established, (4.6.34) follows directly from the lemma in Appendix A.1 with $\varepsilon = \frac{1}{4}\frac{\min(\delta,1)}{n+\min(\delta,1)}\min(\gamma,\delta,1)$.

We begin by observing that when $s \leq t$ we have the estimate

$$\int_{\mathbf{R}^n} \frac{s^{-n}\min(2,(t^{-1}|u|)^\gamma)}{(1+s^{-1}|u|)^{n+1}}\,du \leq C_n\left(\frac{s}{t}\right)^{\frac{1}{2}\min(\gamma,1)}. \tag{4.6.38}$$

Also when $t \leq s$ we have the analogous estimate

$$\int_{\mathbf{R}^n} \frac{t^{-n}\min(2,s^{-1}|u|)}{(1+t^{-1}|u|)^{n+\delta}}\,du \leq C_n\left(\frac{t}{s}\right)^{\frac{1}{2}\min(\delta,1)}. \tag{4.6.39}$$

Both (4.6.38) and (4.6.39) are trivial reformulations or consequences of (4.5.18).

We now take $s \le t$ and we use that $Q_s(1) = 0$ for all $s > 0$ to obtain

$$\begin{aligned}
|L_{t,s}(x,y)| &= \left| \int_{\mathbf{R}^n} \theta_t(x,z)\Psi_s(z-y)\,dz \right| \\
&= \left| \int_{\mathbf{R}^n} \big[\theta_t(x,z) - \theta_t(x,y)\big]\Psi_s(z-y)\,dz \right| \\
&\le CA \int_{\mathbf{R}^n} \frac{\min(2,(t^{-1}|z-y|)^{\gamma})}{t^n} \frac{s^{-n}}{(1+s^{-1}|z-y|)^{n+1}}\,dz \\
&\le C_n' A \frac{1}{t^n}\left(\frac{s}{t}\right)^{\frac{1}{2}\min(\gamma,1)} \\
&\le C_n' A \min\left(\frac{1}{t},\frac{1}{s}\right)^n \min\left(\frac{t}{s},\frac{s}{t}\right)^{\frac{1}{2}\min(\gamma,\delta,1)}
\end{aligned}$$

using estimate (4.6.38). Similarly, using (4.6.39) and the hypothesis that $\Theta_t(1) = 0$ for all $t > 0$, we obtain for $t \le s$,

$$\begin{aligned}
|L_{t,s}(x,y)| &= \left| \int_{\mathbf{R}^n} \theta_t(x,z)\Psi_s(z-y)\,dz \right| \\
&= \left| \int_{\mathbf{R}^n} \theta_t(x,z)\big[\Psi_s(z-y) - \Psi_s(x-y)\big]\,dz \right| \\
&\le xCA \int_{\mathbf{R}^n} \frac{t^{-n}}{(1+t^{-1}|x-z|)^{n+\delta}} \frac{\min(2,s^{-1}|x-z|)}{s^n}\,dz \\
&\le C_n' A \frac{1}{s^n}\left(\frac{t}{s}\right)^{\frac{1}{2}\min(\delta,1)} \\
&\le C_n' A \min\left(\frac{1}{t},\frac{1}{s}\right)^n \min\left(\frac{t}{s},\frac{s}{t}\right)^{\frac{1}{2}\min(\gamma,\delta,1)}.
\end{aligned}$$

Combining the estimates for $|L_{t,s}(x,y)|$ in the preceding cases $t \le s$ and $s \le t$ with the estimate

$$|L_{t,s}(x,y)| \le \int_{\mathbf{R}^n} |\theta_t(x,z)|\,|\Psi_s(z-y)|\,dz \le \frac{CA\min(\frac{1}{t},\frac{1}{s})^n}{\big(1+\min(\frac{1}{t},\frac{1}{s})|x-y|\big)^{n+\min(\delta,1)}},$$

which is a consequence of the result in Appendix B.1, gives

$$|L_{t,s}(x,y)| \le \frac{C\min(\frac{t}{s},\frac{s}{t})^{\frac{1}{2}\min(\gamma,\delta,1)(1-\beta)} A\min(\frac{1}{t},\frac{1}{s})^n}{\Big(\big(1+\min(\frac{1}{t},\frac{1}{s})|x-y|\big)^{n+\min(\delta,1)}\Big)^{\beta}}$$

for any $0 < \beta < 1$. Choosing $\beta = (n+\frac{1}{2}\min(\delta,1))(n+\min(\delta,1))^{-1}$ and integrating over x or y yields (4.6.36) and (4.6.37), respectively, and thus concludes the proof of estimate (4.6.34). □

We end this subsection with a small generalization of the previous theorem that follows by an examination of its proof. The simple details are left to the reader.

Corollary 4.6.4. *For $s>0$ let Θ_s be linear operators that are uniformly bounded on $L^2(\mathbf{R}^n)$ by a constant B. Suppose that each Θ_s has a kernel θ_s which satisfies (4.6.28) and (4.6.29). Let Ψ be a Schwartz function whose Fourier transform is supported in the annulus $1/2 \le |x| \le 2$ such that the Littlewood–Paley operator Q_s given by convolution with $\Psi_s(x) = s^{-n}\Psi(s^{-1}x)$ satisfies (4.6.33). Suppose that for some $C_{n,\Psi}, A, \varepsilon < \infty$,*

$$\big\|\Theta_t Q_s\big\|_{L^2\to L^2} \le A\, C_{n,\Psi} \min\left(\frac{s}{t},\frac{t}{s}\right)^{\varepsilon} \tag{4.6.40}$$

is satisfied for all $t,s>0$. Then there is a constant $C_{n,\Psi,\varepsilon}$ such that for all $f \in L^2(\mathbf{R}^n)$ we have

$$\left(\int_0^\infty \big\|\Theta_s(f)\big\|_{L^2}^2 \frac{ds}{s}\right)^{\frac{1}{2}} \le C_{n,\Psi,\varepsilon}(A+B)\big\|f\big\|_{L^2}.$$

4.6.4 A $T(b)$ Theorem and the L^2 Boundedness of the Cauchy Integral

The operators Θ_s defined in (4.6.23) and (4.6.24) that appear in the resolution of the Cauchy integral operator $\mathcal{C}_\Gamma$ do not satisfy the condition $\Theta_s(1)=0$ of Theorem 4.6.3. It turns out that a certain variant of this theorem is needed for the purposes of the application we have in mind, the L^2 boundedness of the Cauchy integral operator. This variant is a quadratic type $T(b)$ theorem discussed in this subsection. Before we state the main theorem, we need a definition.

Definition 4.6.5. A bounded complex-valued function b on $\mathbf{R}^n$ is said to be *accretive* if there is a constant $c_0>0$ such that $\operatorname{Re} b(x) \ge c_0$ for almost all $x \in \mathbf{R}^n$.

The following theorem is the main result of this section.

Theorem 4.6.6. *Let θ_s be a complex-valued function on $\mathbf{R}^n \times \mathbf{R}^n$ that satisfies (4.6.28) and (4.6.29), and let Θ_s be the linear operator in (4.6.30) whose kernel is θ_s. If there is an accretive function b such that*

$$\Theta_s(b) = 0 \tag{4.6.41}$$

for all $s>0$, then there is a constant $C_n(b)$ such that the estimate

$$\left(\int_0^\infty \big\|\Theta_s(f)\big\|_{L^2}^2 \frac{ds}{s}\right)^{\frac{1}{2}} \le C_n(b)\big\|f\big\|_{L^2} \tag{4.6.42}$$

holds for all $f \in L^2$.

Corollary 4.6.7. *The Cauchy integral operator $\mathcal{C}_\Gamma$ maps $L^2(\mathbf{R})$ to itself.*

The corollary is a consequence of Theorem 4.6.6. Indeed, the crucial and important cancellation property

$$\Theta_s(1+iA')=0 \tag{4.6.43}$$

is valid for the accretive function $1+iA'$, when Θ_s and θ_s are as in (4.6.23) and (4.6.24). To prove (4.6.43) we simply note that

$$\begin{aligned}\Theta_s(1+iA')(x) &= \int_{\mathbf{R}} \frac{s(1+iA'(y))\,dy}{(y-x+i(A(y)-A(x))+is)^2}\\ &= \left[\frac{-s}{y-x+i(A(y)-A(x))+is}\right]_{y=-\infty}^{y=+\infty}\\ &= 0-0=0.\end{aligned}$$

This condition plays exactly the role of (4.6.31), which may fail in general. The necessary "internal cancellation" of the family of operators Θ_s is exactly captured by the single condition (4.6.43).

It remains to prove Theorem 4.6.6.

Proof. We fix an approximation of the identity operator, such as

$$P_s(f)(x)=\int_{\mathbf{R}^n}\Phi_s(x-y)\,f(y)\,dy,$$

where $\Phi_s(x)=s^{-n}\Phi(s^{-1}x)$, and Φ is a nonnegative Schwartz function with integral 1. Then P_s is a nice positive averaging operator that satisfies $P_s(1)=1$ for all $s>0$. The key idea is to decompose the operator Θ_s as

$$\Theta_s=\big(\Theta_s-M_{\Theta_s(1)}P_s\big)+M_{\Theta_s(1)}P_s, \tag{4.6.44}$$

where $M_{\Theta_s(1)}$ is the operator given by multiplication by $\Theta_s(1)$. We begin with the first term in (4.6.44), which is essentially an error term. We simply observe that

$$\big(\Theta_s-M_{\Theta_s(1)}P_s\big)(1)=\Theta_s(1)-\Theta_s(1)\,P_s(1)=\Theta_s(1)-\Theta_s(1)=0.$$

Therefore, Theorem 4.6.3 is applicable once we check that the kernel of the operator $\Theta_s-M_{\Theta_s(1)}P_s$ satisfies (4.6.28) and (4.6.29). But these are verified easily, since the kernels of both Θ_s and P_s satisfy these estimates and $\Theta_s(1)$ is a bounded function uniformly in s. The latter statement is a consequence of condition (4.6.28).

We now need to obtain the required quadratic estimate for the term $M_{\Theta_s(1)}P_s$. With the use of Theorem 3.3.7, this follows once we prove that the measure

$$\big|\Theta_s(1)(x)\big|^2\frac{dx\,ds}{s}$$

is Carleson. It is here that we use condition (4.6.41). Since $\Theta_s(b) = 0$ we have

$$P_s(b)\,\Theta_s(1) = \big(P_s(b)\,\Theta_s(1) - \Theta_s P_s(b)\big) + \big(\Theta_s P_s(b) - \Theta_s(b)\big). \tag{4.6.45}$$

Suppose we could show that the measures

$$\big|\Theta_s(b)(x) - \Theta_s P_s(b)(x)\big|^2 \frac{dx\,ds}{s}, \tag{4.6.46}$$

$$\big|\Theta_s P_s(b)(x) - P_s(b)(x)\,\Theta_s(1)(x)\big|^2 \frac{dx\,ds}{s}, \tag{4.6.47}$$

are Carleson. Then it would follow from (4.6.45) that the measure

$$\big|P_s(b)(x)\,\Theta_s(1)(x)\big|^2 \frac{dx\,ds}{s}$$

is also Carleson. Using the accretivity condition on b and the positivity of P_s we obtain

$$\big|P_s(b)\big| \ge \operatorname{Re} P_s(b) = P_s(\operatorname{Re} b) \ge P_s(c_0) = c_0,$$

from which it follows that $|\Theta_s(1)(x)|^2 \le c_0^{-2}|P_s(b)(x)\,\Theta_s(1)(x)|^2$. Thus the measure $|\Theta_s(1)(x)|^2 dx\,ds/s$ must be Carleson.

Therefore, the proof will be complete if we can show that both measures (4.6.46) and (4.6.47) are Carleson. Theorem 3.3.8 plays a key role here.

We begin with the measure in (4.6.46). First we observe that the kernel

$$L_s(x,y) = \int_{\mathbf{R}^n} \theta_s(x,z)\,\Phi_s(z-y)\,dz$$

of $\Theta_s P_s$ satisfies (4.6.28) and (4.6.29). The verification of (4.6.28) is a straightforward consequence of the estimate in Appendix B.1, while (4.6.29) follows easily from the mean value theorem. It follows that the kernel of

$$R_s = \Theta_s - \Theta_s P_s$$

satisfies the same estimates. Moreover, it is easy to see that $R_s(1) = 0$ and thus the quadratic estimate (4.6.32) holds for R_s in view of Theorem 4.6.3. Therefore, the hypotheses of Theorem 3.3.8(c) are satisfied, and this gives that the measure in (4.6.46) is Carleson.

We now continue with the measure in (4.6.47). Here we set

$$T_s(f)(x) = \Theta_s P_s(f)(x) - P_s(f)(x)\Theta_s(1)(x).$$

The kernel of T_s is $L_s(x,y) - \Theta_s(1)(x)\Phi_s(x-y)$, which clearly satisfies (4.6.28) and (4.6.29), since $\Theta_s(1)(x)$ is a bounded function uniformly in $s > 0$. We also observe that $T_s(1) = 0$. Using Theorem 4.6.3, we conclude that the quadratic estimate (4.6.32) holds for T_s. Therefore, the hypotheses of Theorem 3.3.8(c) are satisfied; hence the measure in (4.6.46) is Carleson. □

We conclude by observing that if we attempt to replace Θ_s with $\widetilde{\Theta}_s = \Theta_s M_{1+iA'}$ in the resolution identity (4.6.26), then $\widetilde{\Theta}_s(1) = 0$ would hold, but the kernel of $\widetilde{\Theta}_s$ would not satisfy the regularity estimate (4.6.29). The whole purpose of Theorem 4.6.6 was to find a certain balance between regularity and cancellation.

Exercises

4.6.1. Given a function H on a Lipschitz graph Γ, we associate a function h on the line by setting $h(t) = H(t + iA(t))$. Prove that for all $0 < p < \infty$ we have

$$\|h\|_{L^p(\mathbf{R})}^p \leq \|H\|_{L^p(\Gamma)}^p \leq \sqrt{1+L^2}\, \|h\|_{L^p(\mathbf{R})}^p,$$

where L is the Lipschitz constant of the defining function A of the graph Γ.

4.6.2. Let $A : \mathbf{R} \to \mathbf{R}$ satisfy $|A(y) - A(y')| \leq L|y - y'|$ for all $y, y' \in \mathbf{R}$ for some $L > 0$. Let h be a Schwartz function on $\mathbf{R}$.
(a) Show that for all $s > 0$ and $x, y \in \mathbf{R}$ we have

$$\frac{s^2 + |x-y|^2}{|x-y|^2 + |A(x) - A(y) + s|^2} \leq 4L^2 + 2.$$

(b) Use the Lebesgue dominated convergence theorem to prove that for all $x \in \mathbf{R}$

$$\lim_{s \to 0} \int_{|x-y| > \sqrt{s}} \frac{s(1 + iA'(y))h(y)}{(y - x + i(A(y) - A(x)) + is)^2}\, dy = 0.$$

(c) Integrate directly to show that for all $x \in \mathbf{R}$ we have

$$\lim_{s \to 0} \int_{|x-y| \leq \sqrt{s}} \frac{s(1 + iA'(y))}{(y - x + i(A(y) - A(x)) + is)^2}\, dy = 0.$$

(d) Use part (a) to prove that for all $x \in \mathbf{R}$ we have

$$\lim_{s \to 0} \int_{|x-y| \leq \sqrt{s}} \frac{s(1 + iA'(y))(h(y) - h(x))}{(y - x + i(A(y) - A(x)) + is)^2}\, dy = 0.$$

(e) Show that for all $x \in \mathbf{R}$ we have

$$\lim_{s \to \infty} \int_{\mathbf{R}} \frac{s(1 + iA'(y))h(y)}{(y - x + i(A(y) - A(x)) + is)^2}\, dy = 0.$$

Conclude the validity of the statements in (4.6.20) for all $x \in \mathbf{R}$.

4.6.3. Prove identity (4.6.22).
[*Hint:* Write the identity in (4.6.22) as

$$\frac{-2}{((\zeta+is)-(z-is))^3}=\frac{1}{2\pi i}\int_\Gamma \frac{\frac{1}{(w-(z-is))^2}}{(w-(\zeta+is))^2}\,dw$$

and interpret it as Cauchy's integral formula for the derivative of the analytic function $w\mapsto (w-(z-is))^{-2}$ defined on the region above Γ. If Γ were a closed curve containing $\zeta+is$ but not $z-is$, then the previous assertion would be immediate. In general, consider a circle of radius R centered at the point $\zeta+is$ and the region U_R inside this circle and above Γ. See Figure 4.1. Integrate over the boundary of U_R and let $R\to\infty$.]

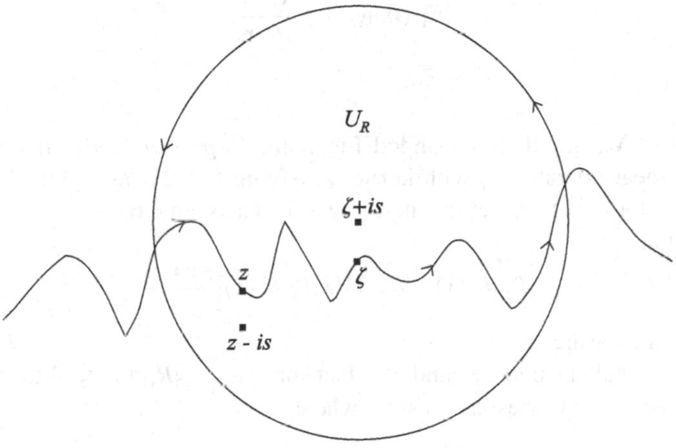

Fig. 4.1 The region U_R inside the circle and above the curve.

4.6.4. Given an accretive function b, define a pseudo-inner product

$$\langle f,g\rangle_b=\int_{\mathbf{R}^n} f(x)\,g(x)\,b(x)\,dx$$

on L^2. For an interval I, set $b_I=\int_I b(x)\,dx$. Let I_L denote the left half of a dyadic interval I and let I_R denote its right half. For a complex number z, let $z^{\frac12}=e^{\frac12\log_{right} z}$, where $\log_{right}$ is the branch of the logarithm defined on the complex plane minus the negative real axis normalized so that $\log_{right} 1=0$ [and $\log_{right}(\pm i)=\pm\frac{\pi}{2}i$]. Show that the family of functions

$$h_I=\frac{-1}{b(I)^{\frac12}}\left(\frac{b(I_R)^{\frac12}}{b(I_L)^{\frac12}}\chi_{I_L}-\frac{b(I_L)^{\frac12}}{b(I_R)^{\frac12}}\chi_{I_R}\right),$$

where I runs over all dyadic intervals, is an orthonormal family on $L^2(\mathbf{R})$ with respect to the preceding inner product. (This family of functions is called a *pseudo-Haar basis associated with b*.)

4.6.5. Let $I=(a,b)$ be a dyadic interval and let $3I$ be its triple. For a given $x\in\mathbf{R}$, let

$$d_I(x)=\min\left(|x-a|,|x-b|,|x-\tfrac{a+b}{2}|\right).$$

Show that there exists a constant C such that

$$\left|\mathfrak{C}_\Gamma(h_I)(x)\right|\le C\,|I|^{-\frac{1}{2}}\log\frac{10|I|}{d_I(x)}$$

whenever $x\in 3I\setminus\{a,b,\frac{a+b}{2}\}$ and also

$$\left|\mathfrak{C}_\Gamma(h_I)(x)\right|\le\frac{C\,|I|^{\frac{3}{2}}}{d_I(x)^2}$$

for $x\notin 3I$.

4.6.6. ([314]) We say that a bounded function b is *para-accretive* if for all $s>0$ there is a linear operator R_s with kernel satisfying (4.6.28) and (4.6.29) such that $|R_s(b)|\ge c_0$ for all $s>0$. Let Θ_s and P_s be as in Theorem 4.6.6.
(a) Prove that

$$\left|R_s(b)(x)-R_s(1)(x)P_s(b)(x)\right|^2\frac{dx\,ds}{s}$$

is a Carleson measure.
(b) Use the result in part (a) and the fact that $\sup_{s>0}|R_s(1)|\le C$ to obtain that $\chi_\Omega(x,s)\,dx\,ds/s$ is a Carleson measure, where

$$\Omega=\left\{(x,s):\;|P_s(b)(x)|\le\frac{c_0}{2}\left(\sup_{s>0}|R_s(1)|\right)^{-1}\right\}.$$

(c) Conclude that the measure $\left|\Theta_s(1)(x)\right|^2 dx\,ds/s$ is Carleson, thus obtaining a generalization of Theorem 4.6.6 for para-accretive functions.

4.6.7. Using the operator $\widetilde{\mathfrak{C}_\Gamma}$ defined in (4.6.15), obtain that $\mathfrak{C}_\Gamma$ is of weak type $(1,1)$ and bounded on $L^p(\mathbf{R})$ for all $1<p<\infty$.

4.7 Square Roots of Elliptic Operators

In this section we prove an L^2 estimate for the square root of a divergence form second-order elliptic operator on $\mathbf{R}^n$. This estimate is based on an approach in the spirit of the $T(b)$ theorem discussed in the previous section. However, matters here are significantly more complicated for two main reasons: the roughness of the variable coefficients of the aforementioned elliptic operator and the higher-dimensional nature of the problem.

4.7.1 Preliminaries and Statement of the Main Result

For $\xi = (\xi_1,\dots,\xi_n) \in \mathbf{C}^n$ we denote its complex conjugate $(\overline{\xi_1},\dots,\overline{\xi_n})$ by $\overline{\xi}$. Moreover, for $\xi,\zeta \in \mathbf{C}^n$ we use the inner product notation

$$\xi\cdot\zeta = \sum_{k=1}^{n} \xi_k\,\zeta_k\,.$$

Throughout this section, $A = A(x)$ is an $n\times n$ matrix of complex-valued L^∞ functions, defined on $\mathbf{R}^n$, that satisfies the *ellipticity* (or *accretivity*) conditions for some $0<\lambda\le\Lambda<\infty$, that

$$\begin{aligned} \lambda|\xi|^2 &\le \operatorname{Re}(A(x)\,\xi\cdot\overline{\xi}),\\ |A(x)\,\xi\cdot\overline{\zeta}| &\le \Lambda|\xi|\,|\zeta|, \end{aligned} \tag{4.7.1}$$

for all $x\in\mathbf{R}^n$ and $\xi,\zeta\in\mathbf{C}^n$. We interpret an element ξ of $\mathbf{C}^n$ as a column vector in $\mathbf{C}^n$ when the matrix A acts on it.

Associated with such a matrix A, we define a second-order *divergence form operator*

$$L(f) = -\operatorname{div}(A\nabla f) = -\sum_{j=1}^{n}\partial_j\big((A\nabla f)_j\big)\,, \tag{4.7.2}$$

which we interpret in the weak sense whenever f is a distribution.

The accretivity condition (4.7.1) enables us to define a square root operator $L^{1/2} = \sqrt{L}$ so that the operator identity $L=\sqrt{L}\sqrt{L}$ holds. The *square root operator* can be written in several ways, one of which is

$$\sqrt{L}(f) = \frac{16}{\pi}\int_0^{+\infty}(I+t^2L)^{-3}t^3L^2(f)\,\frac{dt}{t}\,. \tag{4.7.3}$$

We refer the reader to Exercise 4.7.2 for the existence of the square root operator and the validity of identity (4.7.3).

An important problem in the subject is to determine whether the estimate

$$\big\|\sqrt{L}(f)\big\|_{L^2} \le C_{n,\lambda,\Lambda}\big\|\nabla f\big\|_{L^2} \tag{4.7.4}$$

holds for functions f in a dense subspace of the homogeneous Sobolev space $\dot{L}^2_1(\mathbf{R}^n)$, where $C_{n,\lambda,\Lambda}$ is a constant depending only on n, λ, and Λ. Once (4.7.4) is known for a dense subspace of $\dot{L}^2_1(\mathbf{R}^n)$, then it can be extended to the entire space by density. The main purpose of this section is to discuss a detailed proof of the following result.

Theorem 4.7.1. *Let L be as in (4.7.2). Then there is a constant $C_{n,\lambda,\Lambda}$ such that for all smooth functions f with compact support, estimate (4.7.4) is valid.*

The proof of this theorem requires certain estimates concerning elliptic operators. These are presented in the next subsection, while the proof of the theorem follows in the remaining four subsections.

4.7.2 Estimates for Elliptic Operators on $\mathbf{R}^n$

The following lemma provides a quantitative expression for the mean decay of the resolvent kernel.

Lemma 4.7.2. *Let E and F be two closed sets of $\mathbf{R}^n$. Assume that the distance $d = \operatorname{dist}(E,F)$ between them is positive. Then for all complex-valued functions f supported in E and all vector-valued functions $\vec{f}$ supported in E, we have*

$$\int_F |(I+t^2L)^{-1}(f)(x)|^2\,dx \le Ce^{-c\frac{d}{t}}\int_E |f(x)|^2\,dx, \tag{4.7.5}$$

$$\int_F |t\nabla(I+t^2L)^{-1}(f)(x)|^2\,dx \le Ce^{-c\frac{d}{t}}\int_E |f(x)|^2\,dx, \tag{4.7.6}$$

$$\int_F |(I+t^2L)^{-1}(t\operatorname{div}\vec{f}\,)(x)|^2\,dx \le Ce^{-c\frac{d}{t}}\int_E |\vec{f}(x)|^2\,dx, \tag{4.7.7}$$

where $c = c(\lambda,\Lambda)$, $C = C(n,\lambda,\Lambda)$ are finite constants.

Proof. It suffices to obtain these inequalities whenever $d \ge t > 0$. Let us set $u_t = (I+t^2L)^{-1}(f)$. For all $v \in L^2_1(\mathbf{R}^n)$ we have

$$\int_{\mathbf{R}^n} u_t v\,dx + t^2\int_{\mathbf{R}^n} A\nabla u_t\cdot\nabla v\,dx = \int_{\mathbf{R}^n} fv\,dx.$$

Let η be a nonnegative smooth function with compact support that does not meet E and that satisfies $\|\eta\|_{L^\infty} = 1$. Taking $v = \overline{u_t}\,\eta^2$ and using that f is supported in E, we obtain

$$\int_{\mathbf{R}^n} |u_t|^2\eta^2\,dx + t^2\int_{\mathbf{R}^n} A\nabla u_t\cdot\overline{\nabla u_t}\;\eta^2\,dx = -2t^2\int_{\mathbf{R}^n} A(\eta\nabla u_t)\cdot\overline{u_t\nabla\eta}\,dx.$$

Using (4.7.1) and the inequality $2ab \le \varepsilon|a|^2+\varepsilon^{-1}|b|^2$, we obtain for all $\varepsilon > 0$,

$$\begin{aligned}\int_{\mathbf{R}^n} |u_t|^2\eta^2\,dx + \lambda t^2\int_{\mathbf{R}^n} |\nabla u_t|^2\,\eta^2\,dx \\ \le \Lambda\varepsilon t^2\int_{\mathbf{R}^n} |\nabla u_t|^2\,\eta^2\,dx + \Lambda\varepsilon^{-1}t^2\int_{\mathbf{R}^n} |u_t|^2|\nabla\eta|^2\,dx,\end{aligned}$$

and this reduces to

$$\int_{\mathbf{R}^n} |u_t|^2 |\eta|^2 \, dx \le \frac{\Lambda^2 t^2}{\lambda} \int_{\mathbf{R}^n} |u_t|^2 |\nabla \eta|^2 \, dx \tag{4.7.8}$$

by choosing $\varepsilon = \frac{\lambda}{\Lambda}$. Replacing η by $e^{k\eta} - 1$ in (4.7.8), where

$$k = \frac{\sqrt{\lambda}}{2\Lambda t \|\nabla \eta\|_{L^\infty}},$$

yields

$$\int_{\mathbf{R}^n} |u_t|^2 |e^{k\eta} - 1|^2 \, dx \le \frac{1}{4} \int_{\mathbf{R}^n} |u_t|^2 |e^{k\eta}|^2 \, dx. \tag{4.7.9}$$

Using that $|e^{k\eta} - 1|^2 \ge \frac{1}{2} |e^{k\eta}|^2 - 1$, we obtain

$$\int_{\mathbf{R}^n} |u_t|^2 |e^{k\eta}|^2 \, dx \le 4 \int_{\mathbf{R}^n} |u_t|^2 \, dx \le 4C \int_E |f|^2 \, dx,$$

where in the last estimate we use the uniform boundedness of $(I + t^2 L)^{-1}$ on $L^2(\mathbf{R}^n)$ (Exercise 4.7.1). If, in addition, we have $\eta = 1$ on F, then

$$|e^k|^2 \int_F |u_t|^2 \, dx \le \int_{\mathbf{R}^n} |u_t|^2 |e^{k\eta}|^2 \, dx,$$

and picking η so that $\|\nabla \eta\|_{L^\infty} \approx 1/d$, we conclude (4.7.5).

Next, choose $\varepsilon = \lambda / 2\Lambda$ and η as before to obtain

$$\begin{aligned} \int_F |t \nabla u_t|^2 \, dx &\le \int_{\mathbf{R}^n} |t \nabla u_t|^2 \eta^2 \, dx \\ &\le \frac{2\Lambda^2 t^2}{\lambda} \int_{\mathbf{R}^n} |u_t|^2 |\nabla \eta|^2 \, dx \\ &\le C t^2 d^{-2} e^{-c\frac{d}{t}} \int_E |f|^2 \, dx, \end{aligned}$$

which gives (4.7.6). Finally, (4.7.7) is obtained by duality from (4.7.6) applied to $L^* = -\mathrm{div}\,(A^* \nabla)$ when the roles of E and F are interchanged. □

Lemma 4.7.3. *Let M_f be the operator given by multiplication by a Lipschitz function f. Then there is a constant C that depends only on n, λ, and Λ such that*

$$\big\| \big[(I + t^2 L)^{-1}, M_f \big] \big\|_{L^2 \to L^2} \le C t \|\nabla f\|_{L^\infty} \tag{4.7.10}$$

and

$$\big\| \nabla \big[(I + t^2 L)^{-1}, M_f \big] \big\|_{L^2 \to L^2} \le C \|\nabla f\|_{L^\infty} \tag{4.7.11}$$

for all $t > 0$. Here $[T, S] = TS - ST$ is the commutator *of the operators T and S.*

Proof. Set $\vec{b} = A\nabla f$, $\vec{d} = A^t\nabla f$ and note that the operators given by pointwise multiplication by these vectors are L^2 bounded with norms at most a multiple of $C\|\nabla f\|_{L^\infty}$. Write

$$\begin{aligned}\big[(I+t^2L)^{-1}, M_f\big] &= -(I+t^2L)^{-1}\big[(I+t^2L), M_f\big](I+t^2L)^{-1}\\ &= -(I+t^2L)^{-1}t^2(\operatorname{div}\vec{b} + \vec{d}\cdot\nabla)(1+t^2L)^{-1}.\end{aligned}$$

The uniform L^2 boundedness of $(I+t^2L)^{-1}\, t\nabla(I+t^2L)^{-1}$ and $(I+t^2L)^{-1}t\,\mathrm{div}$ on L^2 (see Exercise 4.7.1) implies (4.7.10). Finally, using the L^2 boundedness of the operator $t^2\nabla(I+t^2L)^{-1}\mathrm{div}$ yields (4.7.11). □

Next we have a technical lemma concerning the mean square deviation of f from $(I+t^2L)^{-1}$.

Lemma 4.7.4. *There exists a constant C depending only on n, λ, and Λ such that for all Q cubes in $\mathbf{R}^n$ with sides parallel to the axes, for all $t \le \ell(Q)$, and all Lipschitz functions f on $\mathbf{R}^n$ we have*

$$\frac{1}{|Q|}\int_Q |(I+t^2L)^{-1}(f) - f|^2\,dx \le Ct^2\|\nabla f\|^2_{L^\infty}, \tag{4.7.12}$$

$$\frac{1}{|Q|}\int_Q |\nabla((I+t^2L)^{-1}(f) - f)|^2\,dx \le C\|\nabla f\|^2_{L^\infty}. \tag{4.7.13}$$

Proof. We begin by proving (4.7.12), while we omit the proof of (4.7.13), since it is similar. By a simple rescaling, we may assume that $\ell(Q) = 1$ and that $\|\nabla f\|_{L^\infty} = 1$. Set $Q_0 = 2Q$ (i.e., the cube with the same center as Q with twice its side length) and write $\mathbf{R}^n$ as a union of cubes Q_k of side length 2 with disjoint interiors and sides parallel to the axes. Lemma 4.7.2 implies that

$$(I+t^2L)^{-1}(1) = 1$$

in the sense that

$$\lim_{R\to\infty}(I+t^2L)^{-1}(\eta_R) = 1$$

in $L^2_{\mathrm{loc}}(\mathbf{R}^n)$, where $\eta_R(x) = \eta(x/R)$ and η is a smooth bump function with $\eta \equiv 1$ near 0. Hence, we may write

$$(I+t^2L)^{-1}(f)(x) - f(x) = \sum_{k\in\mathbf{Z}^n}(I+t^2L)^{-1}((f-f(x))\chi_{Q_k})(x) = \sum_{k\in\mathbf{Z}^n} g_k(x).$$

The term for $k = 0$ in the sum is $[(I+t^2L)^{-1}, M_f](\chi_{Q_0})(x)$. Hence, its $L^2(Q)$ norm is controlled by $Ct\|\chi_{Q_0}\|_{L^2}$ by (4.7.10). The terms for $k \ne 0$ are dealt with using the further decomposition

$$g_k(x) = (I+t^2L)^{-1}((f-f(x_k))\chi_{Q_k})(x) + (f(x_k)-f(x))(I+t^2L)^{-1}(\chi_{Q_k})(x),$$

where x_k is the center of Q_k. Applying Lemma 4.7.2 for $(I+t^2L)^{-1}$ on the sets $E = Q_k$ and $F = Q$ and using that f is a Lipschitz function, we obtain

$$\int_Q |g_k|^2\,dx \le Ct^2 e^{-c\frac{|x_k|}{t}} \big\|\chi_{Q_k}\big\|_{L^2}^2 = Ct^2 e^{-c\frac{|x_k|}{t}} 2^n |Q|\,.$$

The desired bound on the $L^2(Q)$ norm of $(I+t^2L)^{-1}(f) - f$ follows from these estimates, Minkowski's inequality, and the fact that $t \le 1 = \ell(Q)$. □

4.7.3 Reduction to a Quadratic Estimate

We are given a divergence form elliptic operator as in (4.7.2) with ellipticity constants λ and Λ in (4.7.1). Our goal is to obtain the a priori estimate (4.7.4) for functions f in some dense subspace of $\dot{L}_1^2(\mathbf{R}^n)$.

To obtain this estimate we need to resolve the operator $\sqrt{L}$ as an average of simpler operators that are uniformly bounded from $\dot{L}_1^2(\mathbf{R}^n)$ to $L^2(\mathbf{R}^n)$. In the sequel we use the following resolution of the square root:

$$\sqrt{L}(f) = \frac{16}{\pi}\int_0^\infty (I+t^2L)^{-3} t^3 L^2(f)\,\frac{dt}{t}\,,$$

in which the integral converges in $L^2(\mathbf{R}^n)$ for $f \in \mathscr{C}_0^\infty(\mathbf{R}^n)$. Take $g \in \mathscr{C}_0^\infty(\mathbf{R}^n)$ with $\|g\|_{L^2} = 1$. Using duality and the Cauchy–Schwarz inequality, we can control the quantity $\big|\big\langle \sqrt{L}(f)\,|\,g\big\rangle\big|^2$ by

$$\frac{256}{\pi^2}\bigg(\int_0^\infty \big\|(I+t^2L)^{-1}tL(f)\big\|_2^2\,\frac{dt}{t}\bigg)\bigg(\int_0^\infty \big\|V_t(g)\big\|_{L^2}^2\,\frac{dt}{t}\bigg), \tag{4.7.14}$$

where we set

$$V_t = t^2L^*(I+t^2L^*)^{-2}\,.$$

Here L^* is the adjoint operator to L and note that the matrix corresponding to L^* is the conjugate-transpose matrix A^* of A (i.e., the transpose of the matrix whose entries are the complex conjugates of the matrix A). We explain why the estimate

$$\int_0^\infty \big\|V_t(g)\big\|_{L^2}^2\,\frac{dt}{t} \le C\,\|g\|_{L^2}^2 \tag{4.7.15}$$

is valid. Fix a real-valued function $\Psi \in \mathscr{C}_0^\infty(\mathbf{R}^n)$ with mean value zero normalized so that

$$\int_0^\infty |\widehat{\Psi}(s\xi)|^2\,\frac{ds}{s} = 1$$

for all $\xi \in \mathbf{R}^n$ and define $\Psi_s(x) = \frac{1}{s^n}\Psi(\frac{x}{s})$. Throughout the proof, Q_s denotes the operator

$$Q_s(h) = h * \Psi_s\,. \tag{4.7.16}$$

Obviously we have

$$\int_0^\infty \|Q_s(g)\|_{L^2}^2 \frac{ds}{s} = \|g\|_{L^2}^2$$

for all L^2 functions g.

We obtain estimate (4.7.15) as a consequence of Corollary 4.6.4 applied to the operators V_t that have uniform (in t) bounded extensions on $L^2(\mathbf{R}^n)$. To apply Corollary 4.6.4, we need to check that condition (4.6.40) holds for $\Theta_t = V_t$. Since

$$V_t Q_s = -(I+t^2L^*)^{-2}t^2 \operatorname{div}(A^*\nabla Q_s),$$

we have

$$\big\|V_t Q_s\big\|_{L^2\to L^2} \le \big\|(I+t^2L^*)^{-2}t^2\operatorname{div}\big\|_{L^2\to L^2}\big\|A^*\nabla Q_s\big\|_{L^2\to L^2} \le c\frac{t}{s}, \tag{4.7.17}$$

with c depending only on n, λ, and Λ. This estimate is proved by observing two facts: first $(I+t^2L^*)^{-2}t\operatorname{div}$ is equal to the composition of L_1 and L_3 as defined in Exercise 4.7.1. Since L_1 and L_3 are uniformly bounded on L^2, the $L^2 \to L^2$ operator norm of $(I+t^2L^*)^{-2}t^2\operatorname{div}$ is bounded by a constant multiple of t. Also the kernel of ∇Q_s is $\nabla\Psi_s$ which has L^1 norm bounded by a constant multiple of $1/s$.

Choose $\Psi = \Delta\varphi$ with $\varphi \in \mathscr{C}_0^\infty(\mathbf{R}^n)$ radial so that in particular, $\Psi = \operatorname{div}(\nabla\varphi)$. This yields $\Psi_s(x) = s^{-n}\operatorname{div}(\nabla\varphi)(x/s) = s\operatorname{div}(s^{-n}\nabla\varphi(\cdot/s))(x)$, hence $Q_s = s\operatorname{div}\vec{R}_s$ with $\vec{R}_s$ uniformly bounded on L^2. Then we can write

$$\begin{aligned} V_t Q_s &= -(I+t^2L^*)^{-2}t^2 s\operatorname{div}(A^*\nabla\operatorname{div}\vec{R}_s) \\ &= -t^2 s\,(I+t^2L^*)^{-2}L^*\operatorname{div}(\vec{R}_s) \\ &= -\frac{s}{t}(I+t^2L^*)^{-1}(t^2L^*)(I+t^2L^*)^{-1}t\operatorname{div}(\vec{R}_s). \end{aligned}$$

But $(I+t^2L^*)^{-1}t\operatorname{div}$ is uniformly bounded on L^2 (operator L_3 in Exercise 4.7.1) and so is $\vec{R}_s$ and $(I+t^2L^*)^{-1}(t^2L^*) = -(I+t^2L^*)^{-1}$ (operator L_1 in Exercise 4.7.1). It follows that

$$\big\|V_t Q_s\big\|_{L^2\to L^2} \le c\frac{s}{t}, \tag{4.7.18}$$

with c depending only on n, λ, and Λ.

Combining (4.7.17) and (4.7.18) proves (4.6.40) with $\Theta_t = V_t$. Hence Corollary 4.6.4 is applicable and (4.7.15) follows.

Therefore, the second integral on the right-hand side of (4.7.14) is bounded, and estimate (4.7.4) is reduced to proving

$$\int_0^\infty \big\|(I+t^2L)^{-1}tL(f)\big\|_2^2\frac{dt}{t} \le C\int_{\mathbf{R}^n}|\nabla f|^2\,dx \tag{4.7.19}$$

for all $f \in \mathscr{C}_0^\infty(\mathbf{R}^n)$.

4.7.4 Reduction to a Carleson Measure Estimate

Our next goal is to reduce matters to a Carleson measure estimate. We first introduce some notation to be used throughout. For $\mathbf{C}^n$-valued functions $\vec{f} = (f_1, \dots, f_n)$ define

$$Z_t(\vec{f}) = -\sum_{k=1}^{n}\sum_{j=1}^{n}(I+t^2L)^{-1}t\partial_j(a_{j,k}f_k).$$

In short, we write $Z_t = -(I+t^2L)^{-1}t\,\mathrm{div}A$. With this notation, we reformulate (4.7.19) as

$$\int_0^\infty \|Z_t(\nabla f)\|_2^2\frac{dt}{t} \le C\int_{\mathbf{R}^n}|\nabla f|^2\,dx. \tag{4.7.20}$$

Also, define

$$\gamma_t(x) = Z_t(\mathbf{1})(x) = \Big(-\sum_{j=1}^{n}(I+t^2L)^{-1}t\,\partial_j(a_{j,k})(x)\Big)_{1\le k\le n},$$

where $\mathbf{1}$ is the $n\times n$ identity matrix and the action of Z_t on $\mathbf{1}$ is columnwise.

The reduction to a Carleson measure estimate and to a $T(b)$ argument requires the following inequality:

$$\int_{\mathbf{R}^n}\int_0^\infty |\gamma_t(x)\cdot P_t^2(\nabla g)(x) - Z_t(\nabla g)(x)|^2\frac{dx\,dt}{t} \le C\int_{\mathbf{R}^n}|\nabla g|^2\,dx, \tag{4.7.21}$$

where C depends only on n, λ, and Λ. Here, P_t denotes the operator

$$P_t(h) = h * p_t, \tag{4.7.22}$$

where $p_t(x) = t^{-n}p(t^{-1}x)$ and p denotes a nonnegative smooth function supported in the unit ball of $\mathbf{R}^n$ with integral equal to 1. To prove this, we need to handle Littlewood–Paley theory in a setting a bit more general than the one encountered in the previous section.

Lemma 4.7.5. *For $t>0$, let U_t be integral operators defined on $L^2(\mathbf{R}^n)$ with measurable kernels $L_t(x,y)$. Suppose that for some $m>n$ and for all $y\in\mathbf{R}^n$ and $t>0$ we have*

$$\int_{\mathbf{R}^n}\Big(1+\frac{|x-y|}{t}\Big)^{2m}|L_t(x,y)|^2\,dx \le t^{-n}. \tag{4.7.23}$$

Assume that for any ball $B(y,t)$, U_t has a bounded extension from $L^\infty(\mathbf{R}^n)$ to $L^2(B(y,t))$ such that for all f in $L^\infty(\mathbf{R}^n)$ and $y\in\mathbf{R}^n$ we have

$$\frac{1}{t^n}\int_{B(y,t)}|U_t(f)(x)|^2\,dx \le \|f\|_{L^\infty}^2. \tag{4.7.24}$$

Finally, assume that $U_t(1) = 0$ *in the sense that*

$$U_t(\chi_{B(0,R)}) \to 0 \qquad in \quad L^2(B(y,t)) \tag{4.7.25}$$

as $R \to \infty$ *for all* $y \in \mathbf{R}^n$ *and* $t > 0$.

Let Q_s *and* P_t *be as in (4.7.16) and (4.7.22), respectively. Then for some* $\alpha > 0$ *and* C *depending on* n *and* m *we have*

$$\big\|U_t P_t Q_s\big\|_{L^2\to L^2} \le C \min\left(\frac{t}{s}, \frac{s}{t}\right)^{\alpha} \tag{4.7.26}$$

and also

$$\big\|U_t Q_s\big\|_{L^2\to L^2} \le C\left(\frac{t}{s}\right)^{\alpha}, \qquad t \le s. \tag{4.7.27}$$

Proof. We begin by observing that $U_t^* U_t$ has a kernel $K_t(x,y)$ given by

$$K_t(x,y) = \int_{\mathbf{R}^n} \overline{L_t(z,x)} L_t(z,y)\, dz.$$

The simple inequality $(1+a+b) \le (1+a)(1+b)$ for $a,b>0$ combined with the Cauchy–Schwarz inequality and (4.7.23) yields that $\left(1+\frac{|x-y|}{t}\right)^m |K_t(x,y)|$ is bounded by

$$\int_{\mathbf{R}^n} \left(1+\frac{|x-z|}{t}\right)^m |L_t(z,x)|\, |L_t(z,y)| \left(1+\frac{|z-y|}{t}\right)^m dy \le t^{-n}.$$

We conclude that

$$|K_t(x,y)| \le \frac{1}{t^n}\left(1+\frac{|x-y|}{t}\right)^{-m}. \tag{4.7.28}$$

Hence $U_t^* U_t$ is bounded on all L^p, $1 \le p \le +\infty$ and, in particular, for $p=2$. Since L^2 is a Hilbert space, it follows that U_t is bounded on $L^2(\mathbf{R}^n)$ uniformly in $t>0$.

For $s \le t$ we use that $\|U_t\|_{L^2\to L^2} \le B < \infty$ and basic estimates to deduce that

$$\big\|U_t P_t Q_s\big\|_{L^2\to L^2} \le B\big\|P_t Q_s\big\|_{L^2\to L^2} \le CB\left(\frac{s}{t}\right)^{\alpha}.$$

Next, we consider the case $t \le s$. Since P_t has an integrable kernel, and the kernel of $U_t^* U_t$ satisfies (4.7.28), it follows that $W_t = U_t^* U_t P_t$ has a kernel that satisfies a similar estimate. If we prove that $W_t(1) = 0$, then we can deduce from standard arguments that when $t \le s$ we have

$$\big\|W_t Q_s\big\|_{L^2\to L^2} \le C\left(\frac{t}{s}\right)^{2\alpha} \tag{4.7.29}$$

for $0 < \alpha < m-n$. This would imply the required estimate (4.7.26), since

$$\big\|U_t P_t Q_s\big\|_{L^2\to L^2}^2 = \big\|Q_s^* P_t U_t^* U_t P_t Q_s\big\|_{L^2\to L^2} \le C\big\|U_t^* U_t P_t Q_s\big\|_{L^2\to L^2}.$$

We have that $W_t(1) = U_t^* U_t(1)$. Suppose that a function φ in $L^2(\mathbf{R}^n)$ is compactly supported. Then φ is integrable over $\mathbf{R}^n$ and we have

$$\langle U_t^* U_t(1) \,|\, \varphi\rangle = \lim_{R\to\infty} \langle U_t^* U_t(\chi_{B(0,R)}) \,|\, \varphi\rangle = \lim_{R\to\infty} \langle U_t(\chi_{B(0,R)}) \,|\, U_t(\varphi)\rangle .$$

We have

$$\langle U_t(\chi_{B(0,R)}) \,|\, U_t(\varphi)\rangle = \int_{\mathbf{R}^n}\int_{\mathbf{R}^n} U_t(\chi_{B(0,R)})(x)\overline{U_t(x,y)\varphi(y)}\,dy\,dx,$$

and this is in absolute value at most a constant multiple of

$$\left(t^{-n}\int_{\mathbf{R}^n}\int_{\mathbf{R}^n}\left(1+\frac{|x-y|}{t}\right)^{-2m}|U_t(\chi_{B(0,R)})(x)|^2|\varphi(y)|\,dy\,dx\right)^{\frac{1}{2}}\|\varphi\|_{L^1}^{\frac{1}{2}}$$

by (4.7.23) and the Cauchy–Schwarz inequality for the measure $|\varphi(y)|\,dy\,dx$. Using a covering in the x variable by a family of balls $B(y+ckt,t)$, $k\in\mathbf{Z}^n$, we deduce easily that the last displayed expression is at most

$$C_\varphi\left(\sum_{k\in\mathbf{Z}^n}\int_{\mathbf{R}^n}(1+|k|)^{-2m}c_R(y,k)|\varphi(y)|\,dy\right)^{\frac{1}{2}},$$

where C_φ is a constant that depends on φ and

$$c_R(y,k) = t^{-n}\int_{B(y+ckt,t)}|U_t(\chi_{B(0,R)})(x)|^2\,dx.$$

Applying thc dominated convergence theorem and invoking (4.7.24) and (4.7.25) as $R\to\infty$, we conclude that $\langle U_t^* U_t(1)\,|\,\varphi\rangle = 0$. The latter implies that $U_t^* U_t(1) = 0$. The same conclusion follows for W_t, since $P_t(1) = 1$.

To prove (4.7.27) when $t \le s$ we repeat the previous argument with $W_t = U_t^* U_t$. Since $W_t(1) = 0$ and W_t has a nice kernel, it follows that (4.7.29) holds. Thus

$$\|U_t Q_s\|_{L^2\to L^2}^2 = \|Q_s^* U_t^* U_t Q_s\|_{L^2\to L^2} \le C\|U_t^* U_t Q_s\|_{L^2\to L^2} \le C\left(\frac{t}{s}\right)^{2\alpha}.$$

This concludes the proof of the lemma. □

Lemma 4.7.6. *Let P_t be as in Lemma 4.7.5. Then the operator U_t defined by $U_t(\vec{f})(x) = \gamma_t(x)\cdot P_t(\vec{f})(x) - Z_t P_t(\vec{f})(x)$ satisfies*

$$\int_0^\infty \|U_t P_t(\vec{f})\|_{L^2}^2\frac{dt}{t} \le C\|\vec{f}\|_{L^2}^2,$$

where C depends only on n, λ, and Λ. Here the action of P_t on $\vec{f}$ is componentwise.

Proof. By the off-diagonal estimates of Lemma 4.7.2 for Z_t and the fact that p has support in the unit ball, it is simple to show that there is a constant C depending on n, λ, and Λ such that for all $y \in \mathbf{R}^n$,

$$\frac{1}{t^n}\int_{B(y,t)} |\gamma_t(x)|^2\,dx \le C \tag{4.7.30}$$

and that the kernel of $C^{-1}U_t$ satisfies the hypotheses in Lemma 4.7.5. The conclusion follows from Corollary 4.6.4 applied to U_tP_t. □

We now return to (4.7.21). We begin by writing

$$\gamma_t(x)\cdot P_t^2(\nabla g)(x) - Z_t(\nabla g)(x) = U_tP_t(\nabla g)(x) + Z_t(P_t^2 - I)(\nabla g)(x),$$

and we prove (4.7.21) for each term that appears on the right. For the first term we apply Lemma 4.7.6. Since P_t commutes with partial derivatives, we may use that

$$\|Z_t\nabla\|_{L^2\to L^2} = \|(I+t^2L)^{-1}tL\|_{L^2\to L^2} \le Ct^{-1},$$

and therefore we obtain for the second term

$$\begin{aligned}\int_{\mathbf{R}^n}\int_0^\infty |Z_t(P_t^2-I)(\nabla g)(x)|^2\,\frac{dx\,dt}{t} &\le C^2\int_{\mathbf{R}^n}\int_0^\infty |(P_t^2-I)(g)(x)|^2\,\frac{dt}{t^3}\,dx\\ &\le C^2c(p)\|\nabla g\|_2^2\end{aligned}$$

by Plancherel's theorem, where C depends only on n, λ, and Λ. This concludes the proof of (4.7.21).

Lemma 4.7.7. *The required estimate (4.7.4) follows from the Carleson measure estimate*

$$\sup_Q \frac{1}{|Q|}\int_Q\int_0^{\ell(Q)} |\gamma_t(x)|^2\,\frac{dx\,dt}{t} < \infty, \tag{4.7.31}$$

where the supremum is taken over all cubes in $\mathbf{R}^n$ with sides parallel to the axes.

Proof. Indeed, (4.7.31) and Theorem 3.3.7 imply

$$\int_{\mathbf{R}^n}\int_0^\infty |P_t^2(\nabla g)(x)\cdot\gamma_t(x)|^2\,\frac{dx\,dt}{t} \le C\int_{\mathbf{R}^n}|\nabla g|^2\,dx,$$

and together with (4.7.21) we deduce that (4.7.20) holds. □

Next we introduce an auxiliary averaging operator. We define a dyadic averaging operator S_t^Q as follows:

$$S_t^Q(\vec f)(x) = \left(\frac{1}{|Q_x'|}\int_{Q_x'} \vec f(y)\,dy\right)\chi_{Q_x'}(x),$$

where Q'_x is the unique dyadic cube contained in Q that contains x and satisfies $\frac{1}{2}\ell(Q'_x) < t \le \ell(Q'_x)$. Notice that S_t^Q is a projection, i.e., it satisfies $S_t^Q S_t^Q = S_t^Q$. We have the following technical lemma concerning S_t^Q.

Lemma 4.7.8. *For some C depending only on n, λ, and Λ, we have*

$$\int_Q \int_0^{\ell(Q)} |\gamma_t(x) \cdot (S_t^Q - P_t^2)(\vec{f})(x)|^2 \frac{dx\,dt}{t} \le C \int_{\mathbf{R}^n} |\vec{f}|^2\,dx. \tag{4.7.32}$$

Proof. We actually obtain a stronger version of (4.7.32) in which the t-integration on the left is taken over $(0,+\infty)$. Let Q_s be as in (4.7.16). Set $\Theta_t = \gamma_t \cdot (S_t^Q - P_t^2)$. The proof of (4.7.32) is based on Corollary 4.6.4 provided we show that for some $\alpha > 0$,

$$\big\|\Theta_t Q_s\big\|_{L^2 \to L^2} \le C \min\Big(\frac{t}{s}, \frac{s}{t}\Big)^{\alpha}.$$

Suppose first that $t \le s$. Notice that $\Theta_t(1) = 0$, and thus (4.7.25) holds. With the aid of (4.7.30), we observe that Θ_t satisfies the hypotheses (4.7.23) and (4.7.24) of Lemma 4.7.5. Conclusion (4.7.27) of this lemma yields that for some $\alpha > 0$ we have

$$\big\|\Theta_t Q_s\big\|_{L^2 \to L^2} \le C\Big(\frac{t}{s}\Big)^{\alpha}.$$

We now turn to the case $s \le t$. Since the kernel of P_t is bounded by $ct^{-n}\chi_{|x-y|\le t}$, condition (4.7.30) yields that $\gamma_t P_t$ is uniformly bounded on L^2 and thus

$$\big\|\gamma_t P_t^2 Q_s\big\|_{L^2 \to L^2} \le C\big\|P_t Q_s\big\|_{L^2 \to L^2} \le C'\frac{s}{t}.$$

It remains to consider the case $s \le t$ for the operator $U_t = \gamma_t \cdot S_t^Q$. We begin by observing that U_t is L^2 bounded uniformly in $t > 0$; this follows from a standard $U_t^* U_t$ argument using condition (4.7.23). Secondly, as already observed, S_t^Q is an orthogonal projection. Therefore, we have

$$\begin{aligned}
\big\|(\gamma_t \cdot S_t^Q) Q_s\big\|_{L^2 \to L^2} &\le \big\|(\gamma_t \cdot S_t^Q) S_t^Q Q_s\big\|_{L^2 \to L^2} \\
&\le \big\|S_t^Q Q_s\big\|_{L^2 \to L^2} \\
&\le \big\|S_t^Q\big\|_{L^2 \to \dot{L}^2_\alpha} \big\|Q_s\big\|_{\dot{L}^2_\alpha \to L^2} \\
&\le C s^{\alpha} t^{-\alpha}.
\end{aligned}$$

The last inequality follows from the facts that for any α in $(0, \frac{1}{2})$, Q_s maps the homogeneous Sobolev space $\dot{L}^2_\alpha$ to L^2 with norm at most a multiple of Cs^{α} and that the dyadic averaging operator S_t^Q maps $L^2(\mathbf{R}^n)$ to $\dot{L}^2_\alpha(\mathbf{R}^n)$ with norm $Ct^{-\alpha}$. The former of these statements is trivially verified by taking the Fourier transform, while the latter statement requires some explanation.

Fix an $\alpha \in (0, \frac{1}{2})$ and take $h, g \in L^2(\mathbf{R}^n)$. Also fix $j \in \mathbf{Z}$ such that $2^{-j-1} \le t < 2^{-j}$. We then have

$$\langle S_t^Q(-\Delta)^{\frac{\alpha}{2}}(h), g\rangle = \sum_{J_{j,k} \subseteqq Q} \Big\langle (-\Delta)^{\frac{\alpha}{2}}(h), \chi_{J_{j,k}}(x)\big(\operatorname*{Avg}_{J_{j,k}} g\big)\Big\rangle,$$

where $J_{j,k} = \prod_{r=1}^n [2^{-j}k_r, 2^{-j}(k_r+1))$ and $k = (k_1, \dots, k_n)$. It follows that

$$\begin{aligned}\langle S_t^Q(-\Delta)^{\frac{\alpha}{2}}(h), g\rangle &= \sum_{J_{j,k} \subseteqq Q} \Big\langle h, \big(\operatorname*{Avg}_{J_{j,k}} \overline{g}\big)\,(-\Delta)^{\frac{\alpha}{2}}(\chi_{J_{j,k}})(x)\Big\rangle \\ &= \Big\langle h, \sum_{J_{j,k} \subseteqq Q} 2^{\alpha j}\big(\operatorname*{Avg}_{J_{j,k}} \overline{g}\big)\,(-\Delta)^{\frac{\alpha}{2}}(\chi_{[0,1)^n})(2^j(\cdot)-k)\Big\rangle.\end{aligned}$$

Set $\chi_\alpha = (-\Delta)^{\frac{\alpha}{2}}(\chi_{[0,1)^n})$. We estimate the L^2 norm of the preceding sum. We have

$$\begin{aligned}&\int_{\mathbf{R}^n} \Big| \sum_{J_{j,k} \subseteqq Q} 2^{\alpha j}\big(\operatorname*{Avg}_{J_{j,k}} \overline{g}\big)\,\chi_\alpha(2^j x - k)\Big|^2 dx \\ &= 2^{2\alpha j - nj} \int_{\mathbf{R}^n} \Big| \sum_{J_{j,k} \subseteqq Q} \big(\operatorname*{Avg}_{J_{j,k}} \overline{g}\big)\,\chi_\alpha(x-k)\Big|^2 dx \\ &= 2^{2\alpha j - nj} \int_{\mathbf{R}^n} \Big| \sum_{J_{j,k} \subseteqq Q} e^{-2\pi i k\cdot\xi}\big(\operatorname*{Avg}_{J_{j,k}} \overline{g}\big)\Big|^2 |\widehat{\chi_\alpha}(\xi)|^2 d\xi \\ &= 2^{2\alpha j - nj} \int_{[0,1]^n} \Big| \sum_{J_{j,k} \subseteqq Q} e^{-2\pi i k\cdot\xi}\big(\operatorname*{Avg}_{J_{j,k}} \overline{g}\big)\Big|^2 \sum_{l \in \mathbf{Z}^n} |\widehat{\chi_\alpha}(\xi+l)|^2 d\xi \\ &\le 2^{2\alpha j - nj} \int_{[0,1]^n} \Big| \sum_{J_{j,k} \subseteqq Q} e^{-2\pi i k\cdot\xi}\big(\operatorname*{Avg}_{J_{j,k}} \overline{g}\big)\Big|^2 d\xi \sup_{\xi \in [0,1]^n} \sum_{l \in \mathbf{Z}^n} |\widehat{\chi_\alpha}(\xi+l)|^2 \\ &= 2^{2\alpha j - nj} \sum_{k \in \mathbf{Z}^n} \Big|\operatorname*{Avg}_{J_{j,k}} \overline{g}\Big|^2 C(n,\alpha)^2,\end{aligned}$$

where we used Plancherel's identity on the torus and we set

$$C(n,\alpha)^2 = \sup_{\xi \in [0,1]^n} \sum_{l \in \mathbf{Z}^n} |\widehat{\chi_\alpha}(\xi+l)|^2.$$

Since

$$\widehat{\chi_\alpha}(\xi) = |\xi|^\alpha \prod_{r=1}^n \frac{1-e^{-2\pi i \xi_r}}{2\pi i \xi_r},$$

it follows that $C(n,\alpha) < \infty$ when $0 < \alpha < \frac{1}{2}$. In this case we conclude that

$$\begin{aligned}\big|\langle S_t^Q(-\Delta)^{\frac{\alpha}{2}}(h), g\rangle\big| &\le C(n,\alpha)\big\|h\big\|_{L^2} 2^{j\alpha}\Big(2^{-nj}\sum_{k \in \mathbf{Z}^n} \Big|\operatorname*{Avg}_{J_{j,k}} \overline{g}\Big|^2\Big)^{\frac{1}{2}} \\ &\le C'\big\|h\big\|_{L^2} t^{-\alpha}\big\|g\big\|_{L^2},\end{aligned}$$

and this implies that $\|S_t^Q\|_{L^2 \to \dot{L}^2_\alpha} \le C t^{-\alpha}$ and hence the required conclusion. □

4.7.5 The $T(b)$ Argument

To obtain (4.7.31), we adapt the $T(b)$ theorem of the previous section for square roots of divergence form elliptic operators. We fix a cube Q with center c_Q, an $\varepsilon \in (0,1)$, and a unit vector w in $\mathbf{C}^n$. We define a scalar-valued function

$$f_{Q,w}^{\varepsilon} = (1+(\varepsilon\ell(Q))^2 L)^{-1}(\Phi_Q \cdot \overline{w}), \tag{4.7.33}$$

where

$$\Phi_Q(x) = x - c_Q.$$

We begin by observing that the following estimates are consequences of Lemma 4.7.4:

$$\int_{5Q} |f_{Q,w}^{\varepsilon} - \Phi_Q \cdot \overline{w}|^2 \, dx \le C_1 \varepsilon^2 \ell(Q)^2 |Q| \tag{4.7.34}$$

and

$$\int_{5Q} |\nabla(f_{Q,w}^{\varepsilon} - \Phi_Q \cdot \overline{w})|^2 \, dx \le C_2 |Q|, \tag{4.7.35}$$

where C_1, C_2 depend on n, λ, Λ and not on ε, Q, and w. It is important to observe that the constants C_1, C_2 are independent of ε.

The proof of (4.7.31) follows by combining the next two lemmas. The rest of this section is devoted to their proofs.

Lemma 4.7.9. *There exists an $\varepsilon > 0$ depending on n, λ, Λ, and a finite set $\mathscr{F}$ of unit vectors in $\mathbf{C}^n$ whose cardinality depends on ε and n, such that*

$$\sup_Q \frac{1}{|Q|} \int_Q \int_0^{\ell(Q)} |\gamma_t(x)|^2 \frac{dx\,dt}{t}$$
$$\le C \sum_{w \in \mathscr{F}} \sup_Q \frac{1}{|Q|} \int_Q \int_0^{\ell(Q)} |\gamma_t(x) \cdot (S_t^Q \nabla f_{Q,w}^{\varepsilon})(x)|^2 \frac{dx\,dt}{t},$$

where C depends only on ε, n, λ, and Λ. The suprema are taken over all cubes Q in $\mathbf{R}^n$ with sides parallel to the axes.

Lemma 4.7.10. *For C depending only on n, λ, Λ, and $\varepsilon > 0$, we have*

$$\int_Q \int_0^{\ell(Q)} |\gamma_t(x) \cdot (S_t^Q \nabla f_{Q,w}^{\varepsilon})(x)|^2 \frac{dx\,dt}{t} \le C|Q|. \tag{4.7.36}$$

We begin with the proof of Lemma 4.7.10, which is the easiest of the two.

Proof of Lemma 4.7.10. Pick a smooth bump function $\mathscr{X}_Q$ localized on $4Q$ and equal to 1 on $2Q$ with $\|\mathscr{X}_Q\|_{L^\infty} + \ell(Q)\|\nabla \mathscr{X}_Q\|_{L^\infty} \le c_n$. By Lemma 4.7.5 and estimate (4.7.21), the left-hand side of (4.7.36) is bounded by

$$C\int_{\mathbf{R}^n}\left|\nabla(\mathscr{X}_Q f^{\varepsilon}_{Q,w})\right|^2dx+2\int_Q\int_0^{\ell(Q)}\left|\gamma_t(x)\cdot(P_t^2\nabla(\mathscr{X}_Q f^{\varepsilon}_{Q,w}))(x)\right|^2\frac{dx\,dt}{t}$$
$$\leq C\int_{\mathbf{R}^n}\left|\nabla(\mathscr{X}_Q f^{\varepsilon}_{Q,w})\right|^2dx+4\int_Q\int_0^{\ell(Q)}\left|(Z_t\nabla(\mathscr{X}_Q f^{\varepsilon}_{Q,w}))(x)\right|^2\frac{dx\,dt}{t}.$$

It remains to control the last displayed expression by $C|Q|$.

First, it follows easily from (4.7.34) and (4.7.35) that

$$\int_{\mathbf{R}^n}|\nabla(\mathscr{X}_Q f^{\varepsilon}_{Q,w})|^2\,dx\leq C|Q|,$$

where C is independent of Q and w (but it may depend on ε). Next, we write

$$Z_t\nabla(\mathscr{X}_Q f^{\varepsilon}_{Q,w})=W_t^1+W_t^2+W_t^3,$$

where

$$\begin{aligned}W_t^1&=(I+t^2L)^{-1}t\left(\mathscr{X}_Q L(f^{\varepsilon}_{Q,w})\right),\\ W_t^2&=-(I+t^2L)^{-1}t\left(\operatorname{div}(Af^{\varepsilon}_{Q,w}\nabla\mathscr{X}_Q)\right),\\ W_t^3&=-(I+t^2L)^{-1}t\left(A\nabla f^{\varepsilon}_{Q,w}\cdot\nabla\mathscr{X}_Q\right),\end{aligned}$$

and we use different arguments to treat each term W_t^j.

To handle W_t^1, observe that

$$L(f^{\varepsilon}_{Q,w})=\frac{f^{\varepsilon}_{Q,w}-\Phi_Q\cdot\overline{w}}{\varepsilon^2\ell(Q)^2},$$

and therefore it follows from (4.7.34) that

$$\int_{\mathbf{R}^n}|\mathscr{X}_Q L(f^{\varepsilon}_{Q,w})|^2\leq C|Q|(\varepsilon\ell(Q))^{-2},$$

where C is independent of Q and w. Using the (uniform in t) boundedness of the operator $(I+t^2L)^{-1}$ on $L^2(\mathbf{R}^n)$, we obtain

$$\int_Q\int_0^{\ell(Q)}\left|W_t^1(x)\right|^2\frac{dx\,dt}{t}\leq\int_0^{\ell(Q)}\frac{C|Q|t^2}{(\varepsilon\ell(Q))^2}\frac{dt}{t}\leq\frac{C|Q|}{\varepsilon^2},$$

which establishes the required quadratic estimate for W_t^1.

To obtain a similar quadratic estimate for W_t^2, we apply Lemma 4.7.2 for the operator $(I+t^2L)^{-1}t\operatorname{div}$ with sets $F=Q$ and $E=\operatorname{supp}(f^{\varepsilon}_{Q,w}\nabla\mathscr{X}_Q)\subseteq 4Q\setminus 2Q$. We obtain that

$$\int_Q\int_0^{\ell(Q)}|W_t^2(x)|^2\frac{dx\,dt}{t}\leq C\int_0^{\ell(Q)}e^{-\frac{\ell(Q)}{ct}}\frac{dt}{t}\int_{4Q\setminus 2Q}|Af^{\varepsilon}_{Q,w}\nabla\mathscr{X}_Q|^2\,dx.$$

The first integral on the right provides at most a constant factor, while we handle the second integral by writing

$$f^{\varepsilon}_{Q,w} = (f^{\varepsilon}_{Q,w} - \Phi_Q \cdot \overline{w}) + \Phi_Q \cdot \overline{w}.$$

Using (4.7.34) and the facts that $\|\nabla \mathscr{X}_Q\|_{L^\infty} \le c_n \ell(Q)^{-1}$ and that $|\Phi_Q| \le c_n \ell(Q)$ on the support of $\mathscr{X}_Q$, we obtain that

$$\int_{4Q\setminus 2Q} |A f^{\varepsilon}_{Q,w} \nabla \mathscr{X}_Q|^2\, dx \le C\,|Q|\,,$$

where C depends only on n, λ, and Λ. This yields the required result for W_t^2.

To obtain a similar estimate for W_t^3, we use the (uniform in t) boundedness of $(I+t^2L)^{-1}$ on $L^2(\mathbf{R}^n)$ (Exercise 4.7.1) to obtain that

$$\int_Q \int_0^{\ell(Q)} |W_t^3(x)|^2 \frac{dx\,dt}{t} \le C \int_0^{\ell(Q)} t^2 \frac{dt}{t} \int_{4Q\setminus 2Q} |A\nabla f^{\varepsilon}_{Q,w} \cdot \nabla \mathscr{X}_Q|^2\, dx.$$

But the last integral is shown easily to be bounded by $C|Q|$ by writing $f^{\varepsilon}_{Q,w}$, as in the previous case, and using (4.7.35) and the properties of $\mathscr{X}_Q$ and Φ_Q. Note that C here depends only on n, λ, and Λ. This concludes the proof of Lemma 4.7.10. □

4.7.6 Proof of Lemma 4.7.9

It remains to prove Lemma 4.7.9. The main ingredient in the proof of Lemma 4.7.9 is the following proposition, which we state and prove first.

Proposition 4.7.11. *There exists an $\varepsilon > 0$ depending on n, λ, and Λ, and $\eta = \eta(\varepsilon) > 0$ such that for each unit vector w in $\mathbf{C}^n$ and each cube Q with sides parallel to the axes, there exists a collection $\mathscr{S}'_w = \{Q'\}$ of nonoverlapping dyadic subcubes of Q such that*

$$\Big|\bigcup_{Q'\in\mathscr{S}'_w} Q'\Big| \le (1-\eta)|Q|\,, \tag{4.7.37}$$

and moreover, if $\mathscr{S}''_w$ is the collection of all dyadic subcubes of Q not contained in any $Q' \in \mathscr{S}'_w$, then for any $Q'' \in \mathscr{S}''_w$ we have

$$\frac{1}{|Q''|}\int_{Q''} \operatorname{Re}\,(\nabla f^{\varepsilon}_{Q,w}(y)\cdot w)\, dy \ge \frac{3}{4} \tag{4.7.38}$$

and

$$\frac{1}{|Q''|}\int_{Q''} |\nabla f^{\varepsilon}_{Q,w}(y)|^2\, dy \le (4\varepsilon)^{-2}. \tag{4.7.39}$$

Proof. We begin by proving the following crucial estimate:

$$\left| \int_Q (1 - \nabla f^{\varepsilon}_{Q,w}(x) \cdot w)\, dx \right| \le C\varepsilon^{\frac{1}{2}} |Q|, \tag{4.7.40}$$

where C depends on n, λ, and Λ, but not on ε, Q, and w. Indeed, we observe that

$$\nabla(\Phi_Q \cdot \overline{w})(x) \cdot w = |w|^2 = 1\,,$$

so that

$$1 - \nabla f^{\varepsilon}_{Q,w}(x) \cdot w = \nabla g^{\varepsilon}_{Q,w}(x) \cdot w\,,$$

where we set

$$g^{\varepsilon}_{Q,w}(x) = \Phi_Q(x) \cdot \overline{w} - f^{\varepsilon}_{Q,w}(x)\,.$$

Next we state another useful lemma, whose proof is postponed until the end of this subsection.

Lemma 4.7.12. *There exists a constant $C = C_n$ such that for all $h \in \dot{L}^2_1$ we have*

$$\left| \int_Q \nabla h(x)\, dx \right| \le C \ell(Q)^{\frac{n-1}{2}} \left(\int_Q |h(x)|^2\, dx \right)^{\frac{1}{4}} \left(\int_Q |\nabla h(x)|^2\, dx \right)^{\frac{1}{4}}.$$

Applying Lemma 4.7.12 to the function $g^{\varepsilon}_{Q,w}$, we deduce (4.7.40) as a consequence of (4.7.34) and (4.7.35).

We now proceed with the proof of Proposition 4.7.11. First we deduce from (4.7.40) that

$$\frac{1}{|Q|} \int_Q \mathrm{Re}\,(\nabla f^{\varepsilon}_{Q,w}(x) \cdot w)\, dx \ge \frac{7}{8}\,,$$

provided that ε is small enough. We also observe that as a consequence of (4.7.35) we have

$$\frac{1}{|Q|} \int_Q |\nabla f^{\varepsilon}_{Q,w}(x)|^2\, dx \le C_3\,,$$

where C_3 is independent of ε. Now we perform a stopping-time decomposition to select a collection $\mathscr{S}'_w$ of dyadic subcubes of Q that are maximal with respect to either one of the following conditions:

$$\frac{1}{|Q'|} \int_{Q'} \mathrm{Re}\,(\nabla f^{\varepsilon}_{Q,w}(x) \cdot w)\, dx \;\le\; \frac{3}{4}\,, \tag{4.7.41}$$

$$\frac{1}{|Q'|} \int_{Q'} |\nabla f^{\varepsilon}_{Q,w}(x)|^2\, dx \;\ge\; (4\varepsilon)^{-2}\,. \tag{4.7.42}$$

This is achieved by subdividing Q dyadically and by selecting those cubes Q' for which either (4.7.41) or (4.7.42) holds, subdividing all the nonselected cubes, and repeating the procedure. The validity of (4.7.38) and (4.7.39) now follows from the construction and (4.7.41) and (4.7.42).

It remains to establish (4.7.37). Let B_1 be the union of the cubes in $\mathscr{S}'_w$ for which (4.7.41) holds. Also, let B_2 be the union of those cubes in $\mathscr{S}'_w$ for which (4.7.42) holds. We then have

$$\Big|\bigcup_{Q'\in\mathscr{S}'_w} Q'\Big| \le |B_1| + |B_2|.$$

The fact that the cubes in $\mathscr{S}'_w$ do not overlap yields

$$|B_2| \le (4\varepsilon)^2 \int_Q |\nabla f^{\varepsilon}_{Q,w}(x)|^2\,dx \le (4\varepsilon)^2 C_3|Q|.$$

Setting $b^{\varepsilon}_{Q,w}(x) = 1 - \mathrm{Re}\,(\nabla f^{\varepsilon}_{Q,w}(x)\cdot w)$, we also have

$$|B_1| \le 4\sum \int_{Q'} b^{\varepsilon}_{Q,w}\,dx = 4\int_Q b^{\varepsilon}_{Q,w}\,dx - 4\int_{Q\setminus B_1} b^{\varepsilon}_{Q,w}\,dx, \tag{4.7.43}$$

where the sum is taken over all cubes Q' that comprise B_1. The first term on the right in (4.7.43) is bounded above by $C\varepsilon^{\frac12}|Q|$ in view of (4.7.40). The second term on the right in (4.7.43) is controlled in absolute value by

$$4|Q\setminus B_1| + 4|Q\setminus B_1|^{\frac12}(C_3|Q|)^{\frac12} \le 4|Q\setminus B_1| + 4C_3\varepsilon^{\frac12}|Q| + \varepsilon^{-\frac12}|Q\setminus B_1|.$$

Since $|Q\setminus B_1| = |Q| - |B_1|$, we obtain

$$(5+\varepsilon^{-\frac12})|B_1| \le (4 + C\varepsilon^{\frac12} + \varepsilon^{-\frac12})|Q|,$$

which yields $|B_1| \le (1-\varepsilon^{\frac12} + o(\varepsilon^{\frac12}))|Q|$ if ε is small enough. Hence

$$|B| \le (1-\eta(\varepsilon))|Q|$$

with $\eta(\varepsilon)\approx\varepsilon^{\frac12}$ for small ε. This concludes the proof of Proposition 4.7.11. □

Next, we need the following simple geometric fact.

Lemma 4.7.13. *Let w,u,v be in $\mathbf{C}^n$ such that $|w|=1$ and let $0<\varepsilon\le 1$ be such that*

$$|u-(u\cdot\overline{w})w| \le \varepsilon\,|u\cdot\overline{w}|, \tag{4.7.44}$$

$$\mathrm{Re}\,(v\cdot w) \ge \frac{3}{4}, \tag{4.7.45}$$

$$|v| \le (4\varepsilon)^{-1}. \tag{4.7.46}$$

Then we have $|u|\le 4\,|u\cdot v|$.

Proof. It follows from (4.7.45) that

$$\tfrac34\,|u\cdot\overline{w}| \le |(u\cdot\overline{w})(v\cdot w)|. \tag{4.7.47}$$

Moreover, (4.7.44) and the triangle inequality imply that

$$|u| \le (1+\varepsilon)|u\cdot\overline{w}| \le 2\,|u\cdot\overline{w}|. \tag{4.7.48}$$

Also, as a consequence of (4.7.44) and (4.7.46), we obtain

$$|(u-(u\cdot\overline{w})w)\cdot v| \le \tfrac{1}{4}\,|u\cdot\overline{w}|. \tag{4.7.49}$$

Finally, using (4.7.47) and (4.7.49) together with the triangle inequality, we deduce that

$$|u\cdot v| \ge |(u\cdot\overline{w})(v\cdot w)| - |(u-(u\cdot\overline{w})w)\cdot v| \ge (\tfrac{3}{4}-\tfrac{1}{4})\,|u\cdot\overline{w}| \ge \tfrac{1}{4}\,|u|,$$

where in the last inequality we used (4.7.48). □

We now proceed with the proof of Lemma 4.7.9. We fix an $\varepsilon>0$ to be chosen later and we choose a finite number of cones $\mathscr{C}_w$ indexed by a finite set $\mathscr{F}$ of unit vectors w in $\mathbf{C}^n$ defined by

$$\mathscr{C}_w = \big\{u\in\mathbf{C}^n:\ |u-(u\cdot\overline{w})w| \le \varepsilon\,|u\cdot\overline{w}|\big\}, \tag{4.7.50}$$

so that

$$\mathbf{C}^n = \bigcup_{w\in\mathscr{F}} \mathscr{C}_w.$$

Note that the size of the set $\mathscr{F}$ can be chosen to depend only on ε and the dimension n.

It suffices to show that for each fixed $w\in\mathscr{F}$ we have a Carleson measure estimate for $\gamma_{t,w}(x)\equiv\chi_{\mathscr{C}_w}(\gamma_t(x))\gamma_t(x)$, where $\chi_{\mathscr{C}_w}$ denotes the characteristic function of $\mathscr{C}_w$. To achieve this we define

$$A_w \equiv \sup_Q \frac{1}{|Q|}\int_Q\int_0^{\ell(Q)} |\gamma_{t,w}(x)|^2\,\frac{dx\,dt}{t}, \tag{4.7.51}$$

where the supremum is taken over all cubes Q in $\mathbf{R}^n$ with sides parallel to the axes. By truncating $\gamma_{t,w}(x)$ for t small and t large, we may assume that this quantity is finite. Once an a priori bound independent of these truncations is obtained, we can pass to the limit by monotone convergence to deduce the same bound for $\gamma_{t,w}(x)$.

We now fix a cube Q and let $\mathscr{S}''_w$ be as in Proposition 4.7.11. We pick Q'' in $\mathscr{S}''_w$ and we set

$$v = \frac{1}{|Q''|}\int_{Q''} \nabla f^{\varepsilon}_{Q,w}(y)\,dy \in \mathbf{C}^n.$$

It is obvious that statements (4.7.38) and (4.7.39) in Proposition 4.7.11 yield conditions (4.7.45) and (4.7.46) of Lemma 4.7.13. Set $u=\gamma_{t,w}(x)$ and note that if $x\in Q''$ and $\frac{1}{2}\ell(Q'')<t\le\ell(Q'')$, then $v=S^Q_t(\nabla f^{\varepsilon}_{Q,w})(x)$; hence

$$\big|\gamma_{t,w}(x)\big| \le 4\,\big|\gamma_{t,w}(x)\cdot S^Q_t(\nabla f^{\varepsilon}_{Q,w})(x)\big| \le 4\,\big|\gamma_t(x)\cdot S^Q_t(\nabla f^{\varepsilon}_{Q,w})(x)\big| \tag{4.7.52}$$

from Lemma 4.7.13 and the definition of $\gamma_{t,w}(x)$.

We partition the Carleson region $Q\times(0,\ell(Q)]$ as a union of boxes $Q'\times(0,\ell(Q')]$ for Q' in $\mathscr{S}'_w$ and Whitney rectangles $Q''\times(\frac{1}{2}\ell(Q''),\ell(Q'')]$ for Q'' in $\mathscr{S}''_w$. This allows us to write

$$\int_Q \int_0^{\ell(Q)} |\gamma_{t,w}(x)|^2 \frac{dx\,dt}{t} = \sum_{Q' \in \mathscr{S}'_w} \int_{Q'} \int_0^{\ell(Q')} |\gamma_{t,w}(x)|^2 \frac{dx\,dt}{t} + \sum_{Q'' \in \mathscr{S}''_w} \int_{Q''} \int_{\frac{1}{2}\ell(Q'')}^{\ell(Q'')} |\gamma_{t,w}(x)|^2 \frac{dx\,dt}{t}.$$

First observe that

$$\sum_{Q' \in \mathscr{S}'_w} \int_{Q'} \int_0^{\ell(Q')} |\gamma_{t,w}(x)|^2 \frac{dx\,dt}{t} \le \sum_{Q' \in \mathscr{S}'_w} A_w |Q'| A_w (1-\eta)|Q|.$$

Second, using (4.7.52), we obtain

$$\begin{aligned}
&\sum_{Q'' \in \mathscr{S}''_w} \int_{Q''} \int_{\frac{1}{2}\ell(Q'')}^{\ell(Q'')} |\gamma_{t,w}(x)|^2 \frac{dx\,dt}{t} \\
&\qquad \le 16 \sum_{Q'' \in \mathscr{S}''_w} \int_{Q''} \int_{\frac{1}{2}\ell(Q'')}^{\ell(Q'')} |\gamma_t(x) \cdot S_t^Q(\nabla f^{\varepsilon}_{Q,w})(x)|^2 \frac{dx\,dt}{t} \\
&\qquad \le 16 \int_Q \int_0^{\ell(Q)} |\gamma_t(x) \cdot S_t^Q(\nabla f^{\varepsilon}_{Q,w})(x)|^2 \frac{dx\,dt}{t}.
\end{aligned}$$

Altogether, we obtain the bound

$$\begin{aligned}
&\int_Q \int_0^{\ell(Q)} |\gamma_{t,w}(x)|^2 \frac{dx\,dt}{t} \\
&\qquad \le A_w(1-\eta)|Q| + 16 \int_Q \int_0^{\ell(Q)} |\gamma_t(x) \cdot S_t^Q(\nabla f^{\varepsilon}_{Q,w})(x)|^2 \frac{dx\,dt}{t}.
\end{aligned}$$

We divide by $|Q|$, we take the supremum over all cubes Q with sides parallel to the axes, and we use the definition and the finiteness of A_w to obtain the required estimate

$$A_w \le 16\eta^{-1} \sup_Q \frac{1}{|Q|} \int_Q \int_0^{\ell(Q)} |\gamma_t(x) \cdot S_t^Q(\nabla f^{\varepsilon}_{Q,w})(x)|^2 \frac{dx\,dt}{t},$$

thus concluding the proof of the lemma. □

We end by verifying the validity of Lemma 4.7.12 used earlier.

Proof of Lemma 4.7.12. For simplicity we may take Q to be the cube $[-1,1]^n$. Once this case is established, the case of a general cube follows by translation and rescaling. Set

$$M = \left(\int_Q |h(x)|^2 \, dx \right)^{\frac{1}{2}}, \qquad M' = \left(\int_Q |\nabla h(x)|^2 \, dx \right)^{\frac{1}{2}}.$$

If $M \geq M'$, there is nothing to prove, so we may assume that $M < M'$. Take $t \in (0,1)$ and $\varphi \in \mathscr{C}_0^\infty(Q)$ with $\varphi(x) = 1$ when $\operatorname{dist}(x, \partial Q) \geq t$ and $0 \leq \varphi \leq 1$, $\|\nabla\varphi\|_{L^\infty} \leq C/t$, $C = C(n)$; here the distance is taken in the L^∞ norm of $\mathbf{R}^n$. Then

$$\int_Q \nabla h(x)\,dx = \int_Q (1-\varphi(x))\nabla h(x)\,dx - \int_Q h(x)\nabla\varphi(x)\,dx,$$

and the Cauchy–Schwarz inequality yields

$$\left|\int_Q \nabla h(x)\,dx\right| \leq C(M' t^{\frac{1}{2}} + M t^{-\frac{1}{2}}).$$

Choosing $t = M/M'$, we conclude the proof of the lemma. □

The proof of Theorem 4.7.1 is now complete. □

Exercises

4.7.1. Let L be as in (4.7.2). Given $t > 0$ and $f \in L^2(\mathbf{R}^n)$ define $(I + t^2L)^{-1}(f)$ to be the unique weak solution u of the inhomogeneous partial differential equation $u + t^2L(u) = f$. Show that the operators

$$\begin{aligned} L_1 &= (I + t^2L)^{-1}, \\ L_2 &= t\nabla(I + t^2L)^{-1}, \\ L_3 &= (I + t^2L)^{-1} t\,\mathrm{div} \end{aligned}$$

are bounded on $L^2(\mathbf{R}^n)$ uniformly in $t > 0$ with bounds depending only on n, λ, Λ.
[*Hint:* Given $f \in L^2(\mathbf{R}^n)$, by the Lax-Milgram theorem, there is a unique u_t in $L^2_1(\mathbf{R}^n)$ such that $\int_{\mathbf{R}^n} u\overline{v} + t^2\int_{\mathbf{R}^n} A\nabla u\,\nabla\overline{v}\,dx = \int_{\mathbf{R}^n} f\,\overline{v}\,dx$ for all $v \in L^2_1(\mathbf{R}^n)$. Then $u_t = L_1(f) = (I+t^2L)^{-1}(f)$. Taking $v = u_t$ yields $\int_{\mathbf{R}^n} |u_t|^2\,dx + t^2\int_{\mathbf{R}^n} A\nabla u_t\,\nabla\overline{u_t}\,dx = \int_{\mathbf{R}^n} \overline{u_t}\,f\,dx$ via the definition of L and integration by parts. Apply the ellipticity condition to bound the left side of this identity from below by $\int_{\mathbf{R}^n} |u_t|^2\,dx + \lambda\int_{\mathbf{R}^n} |t\nabla u_t|^2\,dx$. Also $\int_{\mathbf{R}^n} \overline{u_t}\,f\,dx \leq \varepsilon^{-1}\int_{\mathbf{R}^n} |f|^2\,dx + \varepsilon\int_{\mathbf{R}^n} |u_t|^2\,dx$ by the Cauchy–Schwarz inequality. Use that $\|u_t\|_{L^2} \leq \|u_t\|_{L^2_1} < \infty$. To bound L_2 estimate $\|t\nabla u_t\|^2_{L^2}$ in terms of $\|u_t\|_{L^2}$ and $\|f\|_{L^2}$. The L^2 boundedness of L_3 follows from that of L_2 (defined with L^* in place of L) via duality and integration by parts.]

4.7.2. Let L be as in the proof of Theorem 4.7.1.
(a) Show that for all $t > 0$ we have

$$(I + t^2L^2)^{-2} = \int_0^\infty e^{-u(I+t^2L)}\,u\,du$$

by checking the identities

$$\int_0^\infty (I+t^2L)^2 e^{-u(I+t^2L)}\,u\,du = \int_0^\infty e^{-u(I+t^2L)}(I+t^2L)^2\,u\,du = I\,.$$

(b) Prove that the operator

$$T = \frac{4}{\pi}\int_0^\infty L(I+t^2L)^{-2}\,dt$$

satisfies $TT = L$.
(c) Conclude that the operator

$$S = \frac{16}{\pi}\int_0^{+\infty} t^3L^2(I+t^2L)^{-3}\,\frac{dt}{t}$$

satisfies $SS = L$, that is, S is the square root of L. Moreover, all the integrals converge in $L^2(\mathbf{R}^n)$ when restricted to $f \in D(L\circ L) = \{h \in L^2(\mathbf{R}^n) : L\circ L(h) \in L^2(\mathbf{R}^n)\}$, which is a dense subset of $L^2(\mathbf{R}^n)$.
[*Hint:* You may use for free that (i) L and e^{-tL} commute, (ii) $e^{-tL}(g) \to g$ as $t \to 0$ in $L^2(\mathbf{R}^n)$, (iii) $\frac{d}{dt}(e^{-tL}) = -Le^{-tL} = -Le^{-tL} = -e^{-tL}L$, and (iv) $D(L\circ L)$ is dense in $L^2(\mathbf{R}^n)$. Part (a): Write $(I+t^2L)e^{-u(I+t^2L)} = -\frac{d}{du}(e^{-u(I+t^2L)})$ and apply integration by parts twice. Part (b): Write the integrand as in part (a) and use the identity

$$\int_0^\infty\int_0^\infty e^{-(ut^2+vs^2)L}L^2\,dt\,ds = \frac{\pi}{4}(uv)^{-\frac{1}{2}}\int_0^\infty e^{-r^2L}L^2\,2r\,dr.$$

Set $\rho = r^2$ and use $e^{-\rho L}L = \frac{d}{d\rho}(e^{-\rho L})$. Part (c): Show that $T = S$ using an integration by parts starting with the identity $L = \frac{d}{dt}(tL)$. Make use for free of the fact that $\frac{d}{dt}(I+t^2L)^{-2} = -4tL(I+t^2L)^{-3}$.]

4.7.3. Suppose that μ is a measure on $\mathbf{R}^{n+1}_+$. For a cube Q in $\mathbf{R}^n$ we define the tent $T(Q)$ of Q as the set $Q\times(0,\ell(Q))$. Suppose that there exist two positive constants $\alpha < 1$ and β such that for all cubes Q in $\mathbf{R}^n$ there exist subcubes Q_j of Q with disjoint interiors such that

1. $\left|Q\setminus\bigcup_j Q_j\right| > \alpha\,|Q|$,
2. $\mu\Big(T(Q)\setminus\bigcup_j T(Q_j)\Big) \le \beta\,|Q|$.

Then μ is a Carleson measure with constant

$$\|\mu\|_{\mathscr{C}} \le \frac{\beta}{\alpha}\,.$$

[*Hint:* We have

$$\begin{aligned}\mu(T(Q)) &\leq \mu\Big(T(Q)\setminus\bigcup_j T(Q_j)\Big)+\sum_j \mu(T(Q_j))\\ &\leq \beta|Q|+\|\mu\|_{\mathscr{C}}\sum_j|Q_j|,\end{aligned}$$

and the last expression is at most $(\beta+(1-\alpha)\|\mu\|_{\mathscr{C}})|Q|$. Assuming that $\|\mu\|_{\mathscr{C}}<\infty$, we obtain the required conclusion. In general, approximate the measure by a sequence of truncated measures.]

HISTORICAL NOTES

Most of the material in Sections 4.1 and 4.2 has been in the literature since the early development of the subject. Theorem 4.2.7 was independently obtained by Peetre [289], Spanne [320], and Stein [324].

The original proof of the $T(1)$ theorem obtained by David and Journé [111] stated that if $T(1)$, $T^t(1)$ are in BMO and T satisfies the weak boundedness property, then T is L^2 bounded. This proof is based on the boundedness of paraproducts and is given in Theorem 4.5.4. Paraproducts were first exploited by Bony [41] and Coifman and Meyer [93]. The proof of L^2 boundedness using condition (iv) given in the proof of Theorem 4.3.3 was later obtained by Coifman and Meyer [94]. The equivalent conditions (ii), (iii), and (vi) first appeared in Stein [326], while condition (iv) is also due to David and Journé [111]. Condition (i) appears in the article of Nazarov, Volberg, and Treil [283] in the context of nondoubling measures. The same authors [284] obtained a proof of Theorems 4.2.2 and 4.2.4 for Calderón–Zygmund operators on nonhomogeneous spaces. Multilinear versions of the $T(1)$ theorem were obtained by Christ and Journé [84], Grafakos and Torres [177], and Bényi, Demeter, Nahmod, Thiele, Torres, and Villaroya [23]. The article [84] also contains a proof of the quadratic $T(1)$ type Theorem 4.6.3. Smooth paraproducts viewed as bilinear operators have been studied by Bényi, Maldonado, Nahmod, and Torres [24] and Dini-continuous versions of them by Maldonado and Naibo [258].

The orthogonality Lemma 4.5.1 was first proved by Cotlar [105] for self-adjoint and mutually commuting operators T_j. The case of general noncommuting operators was obtained by Knapp and Stein [220]. Theorem 4.5.7 is due to Calderón and Vaillancourt [60] and is also valid for symbols of class $S^0_{\rho,\rho}$ when $0\leq\rho<1$. For additional topics on pseudodifferential operators we refer to the books of Coifman and Meyer [93], Journé [207], Stein [326], Taylor [344], Torres [352], and the references therein. The last reference presents a careful study of the action of linear operators with standard kernels on general function spaces. The continuous version of the orthogonality Lemma 4.5.1 given in Exercise 4.5.8 is due to Calderón and Vaillancourt [60]. Conclusion (iii) in the orthogonality Lemma 4.5.1 follows from a general principle saying that if $\sum x_j$ is a series in a Hilbert space such that $\|\sum_{j\in F}x_j\|\leq M$ for all finite sets F, then the series $\sum x_j$ converges in norm. This is a consequence of the Orlicz–Pettis theorem, which states that in any Banach space, if $\sum x_{n_j}$ converges weakly for every subsequence of integers n_j, then $\sum x_j$ converges in norm.

A nice exposition on the Cauchy integral that presents several historical aspects of its study is the book of Muskhelishvili [281]. Another book on this topic is that of Journé [207]. Proposition 4.6.1 is due to Plemelj [298] when Γ is a closed Jordan curve. The L^2 boundedness of the first commutator $\mathscr{C}_1$ in Example 4.3.8 is due to Calderón [54]. The L^2 boundedness of the remaining commutators $\mathscr{C}_m$, $m\geq 2$, is due to Coifman and Meyer [92], but with bounds of order $m!\,\|A'\|^m_{L^\infty}$. These bounds are not as good as those obtained in Example 4.3.8 and do not suffice in obtaining the boundedness of the Cauchy integral by summing the series of commutators. The L^2 boundedness of the Cauchy integral when $\|A'\|_{L^\infty}$ is small enough is due to Calderón [55]. The first proof of the boundedness of the Cauchy integral with arbitrary $\|A'\|_{L^\infty}$ was obtained by Coifman, McIntosh, and Meyer [90]. This proof is based on an improved operator norm for the commutators $\|\mathscr{C}_m\|_{L^2\to L^2} \le C_0 m^4 \|A'\|_{L^\infty}^m$. The quantity m^4 was improved by Christ and Journé [84] to $m^{1+\delta}$ for any $\delta > 0$; it is announced in Verdera [362] that Mateu and Verdera have improved this result by taking $\delta = 0$. Another proof of the L^2 boundedness of the Cauchy integral was given by David [110] by employing the following bootstrapping argument: If the Cauchy integral is L^2 bounded whenever $\|A'\|_{L^\infty} \le \varepsilon$, then it is also L^2 bounded whenever $\|A'\|_{L^\infty} \le \frac{10}{9}\varepsilon$. A refinement of this bootstrapping technique was independently obtained by Murai [274], who was also able to obtain the best possible bound for the operator norm $\|\widetilde{\mathcal{C}}_\Gamma\|_{L^2\to L^2} \le C(1+\|A'\|_{L^\infty})^{1/2}$ in terms of $\|A'\|_{L^\infty}$. Here $\widetilde{\mathcal{C}}_\Gamma$ is the operator defined in (4.6.15). Note that the corresponding estimate for $\mathcal{C}_\Gamma$ involves the power $3/2$ instead of $1/2$. See the book of Murai [275] for this result and a variety of topics related to the commutators and the Cauchy integral. Two elementary proofs of the L^2 boundedness of the Cauchy integral were given by Coifman, Jones, and Semmes [88]. The first of these proofs uses complex variables and the second a pseudo-Haar basis of L^2 adapted to the accretive function $1+iA'$. A geometric proof was given by Melnikov and Verdera [262]. Other proofs were obtained by Verdera [362] and Tchamitchian [345]. The proof of boundedness of the Cauchy integral given in Section 4.6 is taken from Semmes [314]. The book of Christ [81] contains an insightful exposition of many of the preceding results and discusses connections between the Cauchy integral and analytic capacity. The book of David and Semmes [113] presents several extensions of the results in this chapter to singular integrals along higher-dimensional surfaces.

The $T(1)$ theorem is applicable to many problems only after a considerable amount of work; see, for instance, Christ [81] for the case of the Cauchy integral. A more direct approach to many problems was given by McIntosh and Meyer [257], who replaced the function 1 by an accretive function b and showed that any operator T with standard kernel that satisfies $T(b) = T^t(b) = 0$ and $\|M_b T M_b\|_{WB} < \infty$ must be L^2 bounded. (M_b here is the operator given by multiplication by b.) This theorem easily implies the boundedness of the Cauchy integral. David, Journé, and Semmes [112] generalized this theorem even further as follows: If b_1 and b_2 are para-accretive functions such that T maps $b_1\mathscr{C}_0^\infty \to (b_2\mathscr{C}_0^\infty)'$ and is associated with a standard kernel, then T is L^2 bounded if and only if $T(b_1) \in BMO$, $T^t(b_2) \in BMO$, and $\|M_{b_1} T M_{b_2}\|_{WB} < \infty$. This is called the $T(b)$ theorem. The article of Semmes [314] contains a different proof of this theorem in the special case $T(b) = 0$ and $T^t(1) = 0$ (Exercise 4.6.6). Our proof of Theorem 4.6.6 is based on ideas from [314]. An alternative proof of the $T(b)$ theorem was given by Fabes, Mitrea, and Mitrea [131] based on a lemma due to Krein [230]. Another version of the $T(b)$ theorem that is applicable to spaces with no Euclidean structure was obtained by Christ [80].

Theorem 4.7.1 was posed as a problem by Kato [209] for maximal accretive operators and reformulated by McIntosh [254], [255] for square roots of elliptic operators. The reformulation was motivated by counterexamples found to Kato's original abstract formulation, first by Lions [248] for maximal accretive operators and later by McIntosh [253] for regularly accretive ones. The one-dimensional Kato problem and the boundeness of the Cauchy integral along Lipschitz curves are equivalent problems as shown by Kenig and Meyer [217]. See also Auscher, McIntosh, and Nahmod [11]. Coifman, Deng, and Meyer [87] and independently Fabes, Jerison, and Kenig [129], [130] solved the square root problem for small perturbations of the identity matrix. This method used multilinear expansions and can be extended to operators with smooth coefficients. McIntosh [256] considered coefficients in Sobolev spaces, Escauriaza in VMO (unpublished), and Alexopoulos [3] real Hölder coefficients using homogenization techniques. Peturbations of real symmetric matrices with L^∞ coefficients were treated in Auscher, Hofmann, Lewis, and Tchamitchian [9]. The solution of the two-dimensional Kato problem was obtained by Hofmann and McIntosh [192] using a previously derived $T(b)$ type reduction due to Auscher and Tchamitchian [12]. Hofmann, Lacey, and McIntosh [191] extended this theorem to the case in which the heat kernel of e^{-tL} satisfies Gaussian bounds. Theorem 4.7.1 was obtained by Auscher, Hofmann, Lacey, McIntosh, and Tchamitchian [8]; the exposition in the text is based on this reference. Combining Theorem 4.7.1 with a theorem of Lions [248], it follows that the domain of $\sqrt{L}$ is $\dot{L}_1^2(\mathbf{R}^n)$ and that for functions f in this space the equivalence of norms $\|\sqrt{L}(f)\|_{L^2} \approx \|\nabla f\|_{L^2}$ is valid.

Chapter 5
Boundedness and Convergence of Fourier Integrals

In this chapter we return to fundamental questions in Fourier analysis related to convergence of Fourier series and Fourier integrals. Our main goal is to understand in what sense the inversion property of the Fourier transform

$$f(x) = \int_{\mathbf{R}^n} \widehat{f}(\xi) e^{2\pi i x \cdot \xi} \, d\xi$$

holds when f is a function on $\mathbf{R}^n$. This question is equivalent to the corresponding question for the Fourier series

$$f(x) = \sum_{m \in \mathbf{Z}^n} \widehat{f}(m) e^{2\pi i x \cdot m}$$

when f is a function on $\mathbf{T}^n$. The main problem is that the function (or sequence) $\widehat{f}$ may not be integrable and the convergence of the preceding integral (or series) needs to be suitably interpreted. To address this issue, a summability method is employed. This is achieved via the introduction of a localizing factor $\Phi(\xi/R)$, leading to the study of the convergence of the expressions

$$\int_{\mathbf{R}^n} \Phi(\xi/R) \widehat{f}(\xi) e^{2\pi i x \cdot \xi} \, d\xi$$

as $R \to \infty$. Here Φ is a function on $\mathbf{R}^n$ that decays sufficiently rapidly at infinity and satisfies $\Phi(0) = 1$. For instance, we may take $\Phi = \chi_{B(0,1)}$, where $B(0,1)$ is the unit ball in $\mathbf{R}^n$. Analogous summability methods arise in the torus.

An interesting case arises when $\Phi(\xi) = (1 - |\xi|^2)_+^\lambda$, $\lambda \geq 0$, in which we obtain the Bochner–Riesz means introduced by Riesz when $n = 1$ and $\lambda = 0$ and Bochner for $n \geq 2$ and general $\lambda > 0$. The question is whether the Bochner–Riesz means

$$\sum_{m_1^2 + \cdots + m_n^2 \leq R^2} \left(1 - \frac{m_1^2 + \cdots + m_n^2}{R^2} \right)^\lambda \widehat{f}(m_1, \ldots, m_n) \, e^{2\pi i (m_1 x_1 + \cdots + m_n x_n)}$$

L. Grafakos, *Modern Fourier Analysis*, Graduate Texts in Mathematics 250,
DOI 10.1007/978-1-4939-1230-8_5, © Springer Science+Business Media New York 2014

converge in L^p. This question is equivalent to whether the function $(1-|\xi|^2)_+^\lambda$ is an L^p multiplier on $\mathbf{R}^n$ and is investigated in this chapter. Analogous questions concerning the almost everywhere convergence of these families are also studied.

5.1 The Multiplier Problem for the Ball

In this section we show that the characteristic function of the unit disk in $\mathbf{R}^2$ is not an L^p multiplier when $p \neq 2$. This implies the same conclusion in dimensions $n \geq 3$, since sections of higher-dimensional balls are disks and by Theorem 2.5.16 in [156] we have that if $\chi_{B(0,r)} \notin \mathscr{M}_p(\mathbf{R}^2)$ for all $r > 0$, then $\chi_{B(0,1)} \notin \mathscr{M}_p(\mathbf{R}^n)$ for any $n \geq 3$.

5.1.1 Sprouting of Triangles

We begin with a certain geometric construction that at first sight has no apparent relationship to the multiplier problem for the ball in $\mathbf{R}^n$. Given a triangle ABC with base $b = AB$ and height h_0 we let M be the midpoint of AB. We construct two other triangles AMF and BME from ABC as follows. We fix a height $h_1 > h_0$ and we extend the sides AC and BC in the direction away from its base until they reach a certain height h_1. We let E be the unique point on the line passing through the points B and C such that the triangle EMB has height h_1. Similarly, F is uniquely chosen on the line through A and C so that the triangle AMF has height h_1.

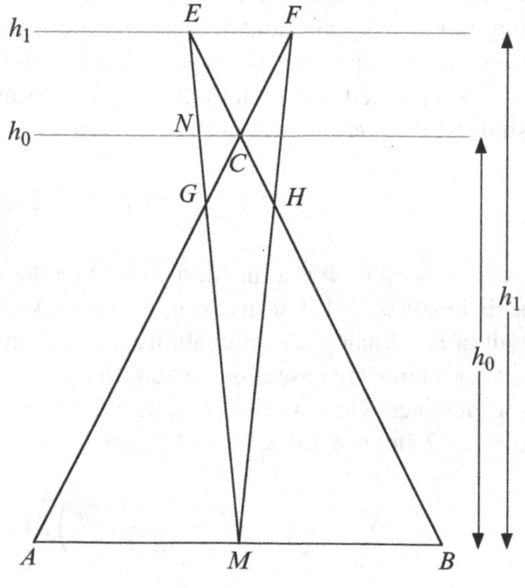

Fig. 5.1 The sprouting of the triangle ABC.

The triangle ABC now gives rise to two triangles AMF and BME called the *sprouts* of ABC. The union of the two sprouts AMF and BME is called the *sprouted figure* obtained from ABC and is denoted by $\mathrm{Spr}(ABC)$. Clearly $\mathrm{Spr}(ABC)$ contains ABC. We call the difference

$$\mathrm{Spr}(ABC)\setminus ABC$$

the *arms* of the sprouted figure. The sprouted figure $\mathrm{Spr}(ABC)$ has two arms of equal area, the triangles EGC and FCH as shown in Figure 5.1, and we can precisely compute the area of each arm. One may easily check (see Exercise 5.1.1) that

$$\text{Area (each arm of } \mathrm{Spr}(ABC)) = \frac{b}{2}\frac{(h_1-h_0)^2}{2h_1-h_0}, \tag{5.1.1}$$

where $b = AB$.

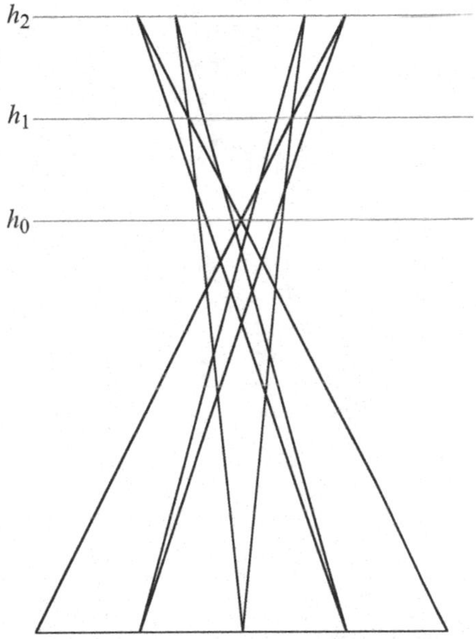

Fig. 5.2 The second step of the construction.

We start with an isosceles triangle $\Lambda = ABC$ in $\mathbf{R}^2$ with base AB of length $b_0 = \varepsilon$ and height $MC = h_0 = \varepsilon$, where M is the midpoint of AB. We define the heights

$$\begin{aligned}
h_1 &= \left(1+\frac{1}{2}\right)\varepsilon,\\
h_2 &= \left(1+\frac{1}{2}+\frac{1}{3}\right)\varepsilon,\\
&\dots\\
h_j &= \left(1+\frac{1}{2}+\dots+\frac{1}{j+1}\right)\varepsilon.
\end{aligned}$$

We apply the previously described sprouting procedure to Λ to obtain two sprouts $\Lambda_1 = AMF$ and $\Lambda_2 = EMB$, as in Figure 5.1, each with height h_1 and base length $b_0/2$. We now apply the same procedure to the triangles Λ_1 and Λ_2. We then obtain two sprouts Λ_{11} and Λ_{12} from Λ_1 and two sprouts Λ_{21} and Λ_{22} from Λ_2, a total of four sprouts with height h_2. See Figure 5.2. We continue this process, obtaining at the jth step 2^j sprouts $\Lambda_{r_1 \dots r_j}$, $r_1, \dots, r_j \in \{1,2\}$ each with base length $b_j = 2^{-j} b_0$ and height h_j. We stop this process when the kth step is completed. The third step is shown in Fig. 5.3.

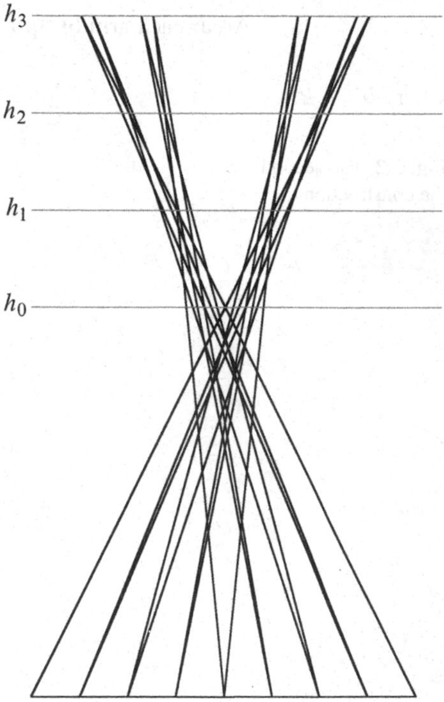

Fig. 5.3 The third step of the construction.

We let $E(\varepsilon, k)$ be the union of the triangles $\Lambda_{r_1 \dots r_k}$ over all sequences r_j of 1's and 2's. We obtain an estimate for the area of $E(\varepsilon, k)$ by adding to the area of Λ the areas of the arms of all the sprouted figures obtained during the construction. By (5.1.1) we have that each of the 2^j arms obtained at the jth step has area

$$\frac{b_{j-1}}{2} \frac{(h_j - h_{j-1})^2}{2h_j - h_{j-1}}.$$

Summing over all these areas and adding the area of the original triangle, we obtain the estimate

$$
\begin{aligned}
|E(\varepsilon,k)| &= \frac{1}{2}\varepsilon^2 + \sum_{j=1}^{k} 2^j \frac{b_{j-1}}{2} \frac{(h_j - h_{j-1})^2}{2h_j - h_{j-1}} \\
&\leq \frac{1}{2}\varepsilon^2 + \sum_{j=1}^{k} 2^j \frac{2^{-(j-1)} b_0}{2} \frac{\varepsilon^2}{(j+1)^2 \varepsilon} \\
&\leq \frac{1}{2}\varepsilon^2 + \sum_{j=2}^{\infty} \frac{\varepsilon^2}{j^2} \\
&= \left(\frac{1}{2} + \frac{\pi^2}{6} - 1\right)\varepsilon^2 \\
&\leq \frac{3}{2}\varepsilon^2,
\end{aligned}
$$

where we used the fact that $2h_j - h_{j-1} \geq \varepsilon$ for all $j \geq 1$.

Having completed the construction of the set $E(\varepsilon,k)$, we are now in a position to indicate some of the ideas that appear in the solution of the Kakeya problem. We first observe that no matter what k is, the measure of the set $E(\varepsilon,k)$ can be made as small as we wish if we take ε small enough. Our purpose is to make a needle of infinitesimal width and unit length move continuously from one side of this angle to the other utilizing each sprouted triangle in succession. To achieve this, we need to apply a similar construction to any of the 2^k triangles that make up the set $E(\varepsilon,k)$ and repeat the sprouting procedure a large enough number of times. We refer to [106] for details. An elaborate construction of this sort yields a set within which the needle can be turned only through a fixed angle. But adjoining a few such sets together allows us to rotate a needle through a half-turn within a set that still has arbitrarily small area. This is the idea used to solve the aforementioned needle problem.

5.1.2 The counterexample

We now return to the multiplier problem for the ball, which has an interesting connection with the Kakeya needle problem.

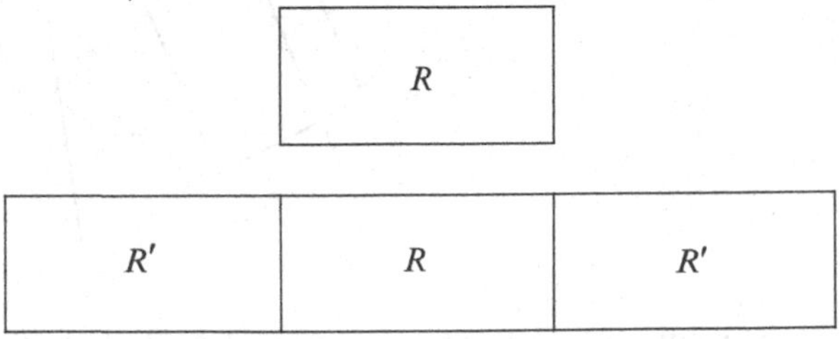

Fig. 5.4 A rectangle R and its adjacent rectangles R'.

In the discussion that follows we employ the following notation. Given a rectangle R in $\mathbf{R}^2$, we let R' be two copies of R adjacent to R along its shortest side so that $R \cup R'$ has the same width as R but three times its length. See Figure 5.4.

We need the following lemma.

Lemma 5.1.1. *Let $\delta > 0$ be a given number. Then there exist a measurable subset E of $\mathbf{R}^2$ and a finite collection of rectangles R_j in $\mathbf{R}^2$ such that*

(1) The R_j's are pairwise disjoint.
(2) We have $1/2 \le |E| \le 3/2$.
(3) We have $|E| \le \delta \sum_j |R_j|$.
(4) For all j we have $|R'_j \cap E| \ge \frac{1}{12}|R_j|$.

Proof. We start with an isosceles triangle ABC in the plane with height 1 and base AB, where $A = (0,0)$ and $B = (1,0)$. Given $\delta > 0$, we find a positive integer k such that $k+2 > e^{1/\delta}$. For this k we set $E = E(1,k)$, the set constructed earlier with $\varepsilon = 1$. We then have $1/2 \le |E| \le 3/2$; thus (2) is satisfied.

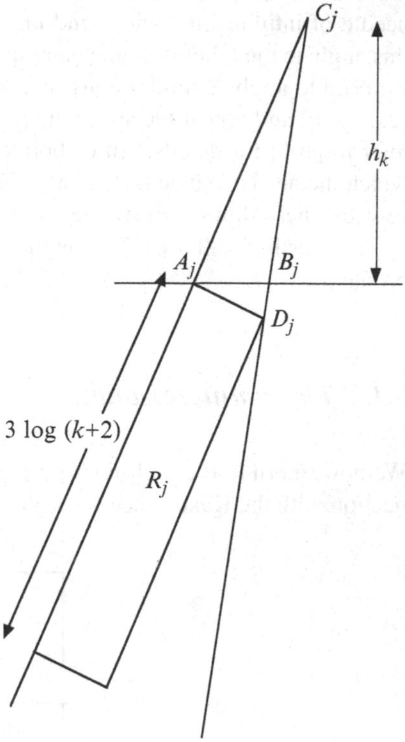

Fig. 5.5 A closer look at R_j.

Recall that each dyadic interval $[j2^{-k},(j+1)2^{-k}]$ in $[0,1]$ is the base of exactly one sprouted triangle $A_jB_jC_j$, where $j \in \{0,1,\dots,2^k-1\}$. Here we set $A_j = (j2^{-k},0)$, $B_j = ((j+1)2^{-k},0)$, and C_j the other vertex of the sprouted triangle.

We define a rectangle R_j inside the angle $\angle A_jC_jB_j$ as in Figure 5.5. The rectangle R_j is defined so that one of its vertices is either A_j or B_j and the length of its longest side is $3\log(k+2)$.

We now make some calculations. First we observe that the longest possible length that either A_jC_j or B_jC_j can achieve is $\sqrt{5}h_k/2$. By symmetry we may assume that the length of A_jC_j is larger than that of B_jC_j as in Figure 5.5. We now have that

$$\frac{\sqrt{5}}{2}h_k < \frac{3}{2}\Big(1+\frac{1}{2}+\cdots+\frac{1}{k+1}\Big) < \frac{3}{2}(1+\log(k+1)) < 3\log(k+2),$$

since $k \geq 1$ and $e < 3$. Hence R'_j contains the triangle $A_jB_jC_j$. We also have that

$$h_k = 1+\frac{1}{2}+\cdots+\frac{1}{k+1} > \log(k+2).$$

Using these two facts, we obtain

$$|R'_j \cap E| \geq \operatorname{Area}(A_jB_jC_j) = \frac{1}{2}2^{-k}h_k > 2^{-k-1}\log(k+2). \qquad (5.1.2)$$

Fig. 5.6 The rectangles R_j.

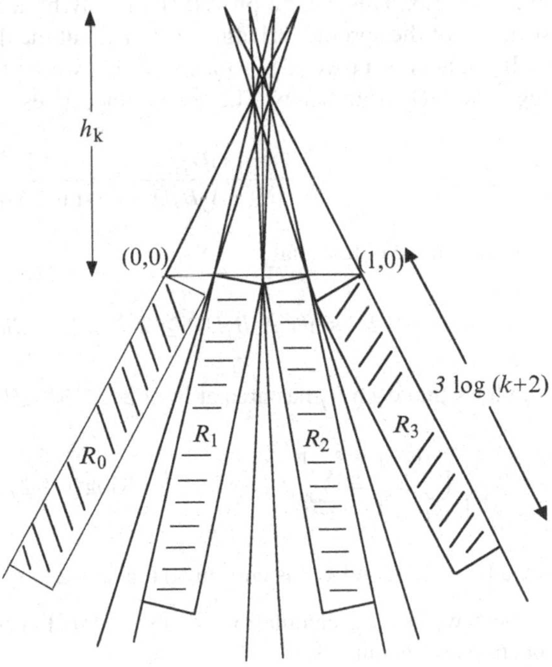

Denote by $|XY|$ the length of the line segment through the points X and Y. The law of sines applied to the triangle $A_jB_jD_j$ gives

$$|A_jD_j| = 2^{-k}\frac{\sin(\angle A_jB_jD_j)}{\sin(\angle A_jD_jB_j)} = 2^{-k}\frac{\sin(\angle A_jB_jD_j)}{\cos(\angle A_jC_jB_j)} \leq \frac{2^{-k}}{\cos(\angle A_jC_jB_j)}. \tag{5.1.3}$$

But the law of cosines applied to the triangle $A_jB_jC_j$, combined with the facts $h_k \leq |A_jC_j|, |B_jC_j| \leq \sqrt{5}h_k/2$, and $h_k > \log(k+2) > 2^{-k-1}$ for $k \geq 1$, yields that

$$\cos(\angle A_jC_jB_j) = \frac{|A_jC_j|^2 + |B_jC_j|^2 - 2|A_jB_j|}{2|A_jC_j|\,|B_jC_j|} \geq \frac{h_k^2 + h_k^2 - (2^{-k})^2}{2\frac{5}{4}h_k^2} \geq \frac{4}{5} - \frac{2}{5}\cdot\frac{1}{4} \geq \frac{1}{2}.$$

Combining this estimate with (5.1.3) we obtain

$$|A_jD_j| \leq 2^{-k+1} = 2\,|A_jB_j|.$$

Using this fact and (5.1.2), we deduce

$$|R_j' \cap E| \geq 2^{-k-1}\log(k+2) = \frac{1}{12}2^{-k+1}3\log(k+2) \geq \frac{1}{12}|R_j|,$$

which proves the required conclusion (4).

Conclusion (1) in Lemma 5.1.1 follows from the fact that the regions inside the angles $\angle A_jC_jB_j$ and under the triangles $A_jC_jB_j$ are pairwise disjoint. This is shown in Figure 5.6. This can be proved rigorously by a careful examination of the construction of the sprouted triangles $A_jC_jB_j$, but the details are omitted.

It remains to prove (3). To achieve this we first estimate the length of the line segment A_jD_j from below. The law of sines gives

$$\frac{|A_jD_j|}{\sin(\angle A_jB_jD_j)} = \frac{2^{-k}}{\sin(\angle A_jD_jB_j)},$$

from which we obtain that

$$|A_jD_j| \geq 2^{-k}\sin(\angle A_jB_jD_j) \geq 2^{-k}\sin(\angle A_0B_0D_0) = 2^{-k}\frac{\sqrt{5}}{2} > 2^{-k-1}.$$

It follows that each R_j has area at least $2^{-k-1}3\log(k+2)$. Therefore,

$$\left|\bigcup_{j=0}^{2^k-1} R_j\right| = \sum_{j=0}^{2^k-1}|R_j| \geq 2^k 2^{-k-1}3\log(k+2) \geq |E|\log(k+2) \geq \frac{|E|}{\delta},$$

since $|E| \leq 3/2$ and k was chosen so that $k+2 > e^{1/\delta}$. □

Next we have a calculation involving the Fourier transforms of characteristic functions of rectangles.

Proposition 5.1.2. *Let R be a rectangle whose center is the origin in $\mathbf{R}^2$ and let v be a unit vector parallel to its longest side. Consider the half-plane*

$$\mathscr{H} = \{x \in \mathbf{R}^2 : x \cdot v \geq 0\}$$

and the multiplier operator

$$S_{\mathscr{H}}(f) = (\widehat{f}\chi_{\mathscr{H}})^{\vee}.$$

Then we have $|S_{\mathscr{H}}(\chi_R)| \geq \frac{1}{10}\chi_{R'}$.

Remark 5.1.3. Applying a translation, we see that the same conclusion is valid for any rectangle in $\mathbf{R}^2$ whose longest side is parallel to v.

Proof. Applying a rotation, we reduce the problem to the case $R = [-a,a] \times [-b,b]$, where $0 < a \leq b < \infty$, and $v = e_2 = (0,1)$. Since the Fourier transform acts in each variable independently, we have the identity

$$\begin{aligned} S_{\mathscr{H}}(\chi_R)(x_1,x_2) &= \chi_{[-a,a]}(x_1)\big(\widehat{\chi_{[-b,b]}}\chi_{[0,\infty)}\big)^{\vee}(x_2) \\ &= \chi_{[-a,a]}(x_1)\frac{I+iH}{2}(\chi_{[-b,b]})(x_2). \end{aligned}$$

It follows that for $(x_1,x_2) \in R'$ we have

$$\begin{aligned} |S_{\mathscr{H}}(\chi_R)(x_1,x_2)| &\geq \frac{1}{2}\chi_{[-a,a]}(x_1)|H(\chi_{[-b,b]})(x_2)| \\ &= \frac{1}{2\pi}\chi_{[-a,a]}(x_1)\left|\log\left|\frac{x_2+b}{x_2-b}\right|\right|. \end{aligned}$$

But for $(x_1,x_2) \in R'$ we have $\chi_{[-a,a]}(x_1) = 1$ and $b < |x_2| < 3b$. So we have two cases, $b < x_2 < 3b$ and $-3b < x_2 < -b$. When $b < x_2 < 3b$ we see that

$$\left|\frac{x_2+b}{x_2-b}\right| = \frac{x_2+b}{x_2-b} > 2,$$

and similarly, when $-3b < x_2 < -b$ we have

$$\left|\frac{x_2-b}{x_2+b}\right| = \frac{b-x_2}{-b-x_2} > 2.$$

It follows that for $(x_1,x_2) \in R'$ the lower estimate is valid:

$$|S_{\mathscr{H}}(\chi_R)(x_1,x_2)| \geq \frac{\log 2}{2\pi} \geq \frac{1}{10}.$$

□

Next we have a lemma regarding vector-valued inequalities of half-plane multipliers.

Lemma 5.1.4. *Let* $v_1, v_2, \dots, v_j, \dots$ *be a sequence of unit vectors in* $\mathbf{R}^2$. *Define the half-planes*

$$\mathscr{H}_j = \{x \in \mathbf{R}^2 : x \cdot v_j \geq 0\} \tag{5.1.4}$$

and linear operators

$$S_{\mathscr{H}_j}(f) = (\widehat{f}\,\chi_{\mathscr{H}_j})^\vee .$$

Let $1 < p < \infty$. *Assume that the disk multiplier operator*

$$T(f) = (\widehat{f}\,\chi_{B(0,1)})^\vee$$

maps $L^p(\mathbf{R}^2)$ *to itself with norm* $B_p < \infty$. *Then we have the inequality*

$$\Big\| \Big(\sum_j |S_{\mathscr{H}_j}(f_j)|^2 \Big)^{\frac{1}{2}} \Big\|_{L^p} \le B_p \Big\| \Big(\sum_j |f_j|^2 \Big)^{\frac{1}{2}} \Big\|_{L^p} \tag{5.1.5}$$

for all functions f_j *in* L^p.

Proof. We prove the lemma for Schwartz functions f_j and we obtain the general case by a simple limiting argument. We define disks $D_{j,R} = \{x \in \mathbf{R}^2 : |x - Rv_j| \le R\}$ and we let

$$T_{j,R}(f) = (\widehat{f}\,\chi_{D_{j,R}})^\vee$$

be the associated multiplier operator. We observe that $\chi_{D_{j,R}} \to \chi_{\mathscr{H}_j}$ pointwise as $R \to \infty$, as shown in Figure 5.7.

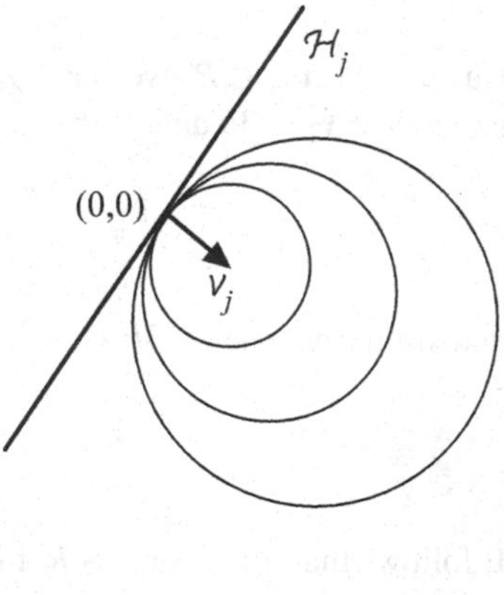

Fig. 5.7 A sequence of disks converging to a half-plane.

For $f \in \mathscr{S}(\mathbf{R}^2)$ and every $x \in \mathbf{R}^2$ we have

$$\lim_{R\to\infty} T_{j,R}(f)(x) = S_{\mathscr{H}_j}(f)(x)$$

by passing the limit inside the convergent integral. Fatou's lemma now yields

$$\Big\| \Big(\sum_j |S_{\mathscr{H}_j}(f_j)|^2 \Big)^{\frac{1}{2}} \Big\|_{L^p} \le \liminf_{R\to\infty} \Big\| \Big(\sum_j |T_{j,R}(f_j)|^2 \Big)^{\frac{1}{2}} \Big\|_{L^p} . \tag{5.1.6}$$

Next we observe that

$$\int_{|\xi - Rv_j| \le R} \widehat{f_j}(\xi) e^{2\pi i x\cdot\xi}\, d\xi = e^{2\pi i R v_j \cdot x} \int_{|\xi| \le R} \widehat{f_j}(\xi + Rv_j) e^{2\pi i x\cdot\xi}\, d\xi ,$$

hence the the multiplier operator $T_R(f) = (\widehat{f}\chi_{B(0,R)})^\vee$ satisfies

$$T_{j,R}(f_j)(x) = e^{2\pi iRv_j\cdot x}T_R(e^{-2\pi iRv_j\cdot(\cdot)}f_j)(x). \tag{5.1.7}$$

Setting $g_j = e^{-2\pi iRv_j\cdot(\cdot)}f_j$ and using (5.1.6) and (5.1.7), we deduce

$$\Big\|\Big(\sum_j |S_{\mathscr{H}_j}(f_j)|^2\Big)^{\frac{1}{2}}\Big\|_{L^p} \le \liminf_{R\to\infty}\Big\|\Big(\sum_j |T_R(g_j)|^2\Big)^{\frac{1}{2}}\Big\|_{L^p}. \tag{5.1.8}$$

Observe that $\|T_R\|_{L^p\to L^p} = \|T\|_{L^p\to L^p} = B_p < \infty$ for all $R > 0$, in view of fact that multipliers $m(\xi)$ and $m(t\xi)$ have the same $\mathscr{M}_p$ norm for any $t > 0$; see Proposition 2.5.14 in [156]. Applying Theorem 5.5.1 in [156], we obtain that the last term in (5.1.8) is bounded by

$$\liminf_{R\to\infty}\|T_R\|_{L^p\to L^p}\Big\|\Big(\sum_j |g_j|^2\Big)^{\frac{1}{2}}\Big\|_{L^p} = B_p\Big\|\Big(\sum_j |f_j|^2\Big)^{\frac{1}{2}}\Big\|_{L^p}.$$

Combining this inequality with (5.1.8), we obtain (5.1.5). □

We have now completed all the preliminary material we need to prove that the characteristic function of the unit disk in $\mathbf{R}^2$ is not an L^p multiplier if $p \neq 2$.

Theorem 5.1.5. *Let $n \ge 2$. The characteristic function of the unit ball in $\mathbf{R}^n$ is not an L^p multiplier when $1 < p \neq 2 < \infty$.*

Proof. In view of Theorem 2.5.16 in [156], it suffices to prove the result in dimension $n = 2$. By duality, matters reduce to the case $p > 2$. To reach a contradiction, suppose that $\chi_{B(0,1)} \in \mathscr{M}_p(\mathbf{R}^2)$ for some $p > 2$, say with norm $B_p < \infty$.

Suppose that $\delta > 0$ is given. Let E and R_j be as in Lemma 5.1.1. We let $f_j = \chi_{R_j}$. Let v_j be the unit vector parallel to the long side of R_j and let H_j be the half-plane defined as in (5.1.4). Using Proposition 5.1.2, we obtain

$$\begin{aligned}\int_E \sum_j |S_{\mathscr{H}_j}(f_j)(x)|^2\,dx &= \sum_j \int_E |S_{\mathscr{H}_j}(f_j)(x)|^2\,dx\\ &\ge \sum_j \int_E \frac{1}{10^2}\chi_{R'_j}(x)\,dx\\ &= \frac{1}{100}\sum_j |E\cap R'_j|\\ &\ge \frac{1}{1200}\sum_j |R_j|,\end{aligned} \tag{5.1.9}$$

where we used condition (4) of Lemma 5.1.1 in the last inequality. Hölder's inequality with exponents $p/2$ and $(p/2)' = p/(p-2)$ gives

$$
\begin{aligned}
\int_E \sum_j |S_{\mathscr{H}_j}(f_j)(x)|^2\,dx &\le |E|^{\frac{p-2}{p}} \Big\| \Big(\sum_j |S_{\mathscr{H}_j}(f_j)|^2\Big)^{\frac{1}{2}} \Big\|_{L^p}^2 \\
&\le B_p^2 |E|^{\frac{p-2}{p}} \Big\| \Big(\sum_j |f_j|^2\Big)^{\frac{1}{2}} \Big\|_{L^p}^2 \\
&= B_p^2 |E|^{\frac{p-2}{p}} \Big(\sum_j |R_j|\Big)^{\frac{2}{p}} \\
&\le B_p^2 \delta^{\frac{p-2}{p}} \Big(\sum_j |R_j|\Big)^{\frac{p-2}{p}} \Big(\sum_j |R_j|\Big)^{\frac{2}{p}},
\end{aligned}
\tag{5.1.10}
$$

where we used Lemma 5.1.4, the disjointness of the R_j's, and condition (3) of Lemma 5.1.1 successively. Combining (5.1.9) with (5.1.10), we obtain the inequality

$$
\sum_j |R_j| \le 1200 B_p \delta^{\frac{p-2}{p}} \sum_j |R_j|,
$$

which provides a contradiction when δ is very small. □

Exercises

5.1.1. Prove identity (5.1.1).
[*Hint:* With the notation of Figure 5.1, first prove

$$
\frac{h_1 - h_0}{h_1} = \frac{NC}{b/2}, \qquad \frac{\text{height}\,(NGC)}{h_0} = \frac{NC}{NC + b/2}
$$

using similar triangles.]

5.1.2. Prove that for any $1 \le p \le \infty$, the disk multiplier operator $T(f) = (\widehat{f}\,\chi_{B(0,1)})^\vee$ does not map $L^p(\mathbf{R}^n)$ to $L^{p,\infty}(\mathbf{R}^n)$, unless $p = 2$.

5.1.3. Is the characteristic function of the cylinder

$$
\{(\xi_1, \xi_2, \xi_3) \in \mathbf{R}^3 : \xi_1^2 + \xi_2^2 < 1\}
$$

a Fourier multiplier on $L^p(\mathbf{R}^3)$ for $1 < p < \infty$ and $p \ne 2$?

5.1.4. Modify the ideas of the proof of Lemma 5.1.4 to show that the characteristic function of the set

$$
\{(\xi_1, \xi_2) \in \mathbf{R}^2 : \xi_2 > \xi_1^2\}
$$

is not in $\mathscr{M}_p(\mathbf{R}^2)$ when $p \ne 2$.

[*Hint:* Let $\mathscr{H}_j = \{(\xi_1,\xi_2)\in\mathbf{R}^2:\ \xi_2 > s_j\xi_1\}$ for some $s_j>0$. The parabolic regions $\{(\xi_1,\xi_2)\in\mathbf{R}^2:\ \xi_2 + R\frac{s_j^2}{4} > \frac{1}{R}(\xi_1+R\frac{s_j}{2})^2\}$ are contained in $\mathscr{H}_j$, are translates of the region $\{(\xi_1,\xi_2)\in\mathbf{R}^2:\ \xi_2 > \frac{1}{R}\xi_1^2\}$, and tend to $\mathscr{H}_j$ as $R\to\infty$.]

5.1.5. Let $a_1,\dots,a_n>0$. Show that the characteristic function of the ellipsoid

$$\Big\{(\xi,\dots,\xi_n)\in\mathbf{R}^n:\ \frac{\xi_1^2}{a_1^2}+\cdots+\frac{\xi_n^2}{a_n^2}<1\Big\}$$

is not in $\mathscr{M}_p(\mathbf{R}^n)$ when $p\neq 2$.

5.2 Bochner–Riesz Means and the Carleson–Sjölin Theorem

We now address the problem of norm convergence for the Bochner–Riesz means. In this section we provide a satisfactory answer in dimension $n=2$, although a key ingredient required in the proof is left for the next section.

Definition 5.2.1. For a function f on $\mathbf{R}^n$ we define its *Bochner–Riesz means* of complex order λ with $\operatorname{Re}\lambda\geq 0$ to be the family of operators

$$B_R^\lambda(f)(x) = \int_{\mathbf{R}^n}(1-|\xi/R|^2)_+^\lambda \widehat{f}(\xi)e^{2\pi i x\cdot\xi}\,d\xi\,,\qquad R>0.$$

We are interested in the convergence of the family $B_R^\lambda(f)$ as $R\to\infty$. Observe that when $R\to\infty$ and f is a Schwartz function, the sequence $B_R^\lambda(f)$ converges pointwise to f. Does it converge in norm? This would be the case if the function $(1-|\xi|^2)_+^\lambda$ is an L^p multiplier, i.e., the linear operator

$$B^\lambda(f)(x) = \int_{\mathbf{R}^n}(1-|\xi|^2)_+^\lambda \widehat{f}(\xi)e^{2\pi i x\cdot\xi}\,d\xi$$

maps $L^p(\mathbf{R}^n)$ to itself; see Exercise 5.2.1. The question that arises is given λ with $\operatorname{Re}\lambda>0$ find the range of p's for which $(1-|\xi|^2)_+^\lambda$ is an $L^p(\mathbf{R}^n)$ Fourier multiplier; this question is investigated in this section when $n=2$.

An analogous question can be studied on the n-torus and this turns out to be equivalent with Euclidean problem; see Corollary 4.3.11 in [156]. Here we focus attention on the Euclidean case, and we start our investigation by studying the kernel of the operator B^λ.

5.2.1 The Bochner–Riesz Kernel and Simple Estimates

In view of the last identity in Appendix B.5 in [156], B^λ is a convolution operator with kernel

$$K_\lambda(x) = \frac{\Gamma(\lambda+1)}{\pi^\lambda}\frac{J_{\frac{n}{2}+\lambda}(2\pi|x|)}{|x|^{\frac{n}{2}+\lambda}}\,. \tag{5.2.1}$$

According to the result at the end of Appendix B.6 in [156], we have for $|x| \le 1$,

$$|K_\lambda(x)| = \frac{|\Gamma(\lambda+1)|}{|\pi^\lambda|}\frac{|J_{\frac{n}{2}+\lambda}(2\pi|x|)|}{|x|^{\frac{n}{2}+\mathrm{Re}\,\lambda}} \le \widetilde{C}_0(\tfrac{n}{2}+\mathrm{Re}\,\lambda)\,e^{|\frac{n}{2}+\mathrm{Im}\,\lambda|^2},$$

where $\widetilde{C}_0(t)$ is a constant that depends smoothly on t when $t \ge 0$.

For $|x| \ge 1$, following the result at the end of Appendix B.7 in [156], we have

$$|K_\lambda(x)| = \frac{|\Gamma(\lambda+1)|}{|\pi^\lambda|}\frac{|J_{\frac{n}{2}+\lambda}(2\pi|x|)|}{|x|^{\frac{n}{2}+\mathrm{Re}\,\lambda}} \le \widetilde{C}_1(\tfrac{n}{2}+\mathrm{Re}\,\lambda)\,\frac{e^{(1+\frac{\pi}{2})|\frac{n}{2}+\mathrm{Im}\,\lambda|^2}}{|x|^{\frac{1}{2}}}\frac{1}{|x|^{\frac{n}{2}+\mathrm{Re}\,\lambda}},$$

where $\widetilde{C}_1(t)$ depends smoothly on t when $t \ge 0$. In both cases we took the ν in the Appendices to be equal to $\frac{n}{2}+\lambda$, which satisfies $\mathrm{Re}\,\nu+\frac{1}{2} \ge 1$, since $\mathrm{Re}\,\lambda \ge 0$.

Combining these two observations, we obtain that for $\mathrm{Re}\,\lambda > \frac{n-1}{2}$, K_λ is a smooth integrable function on $\mathbf{R}^n$. Hence B^λ is a bounded operator on L^p for $1 \le p \le \infty$.

Proposition 5.2.2. *For all $1 \le p \le \infty$ and $\mathrm{Re}\,\lambda > \frac{n-1}{2}$, B^λ is a bounded operator on $L^p(\mathbf{R}^n)$ with norm at most $C_1\,e^{6|\mathrm{Im}\,\lambda|^2}$, where C_1 depends smoothly on $n, \mathrm{Re}\,\lambda \ge 0$.*

Proof. The ingredients of the proof have already been discussed. □

We refer to Exercise 5.2.8 for an analogous result for the maximal Bochner–Riesz operator.

According to the asymptotics for Bessel functions in Appendix B.8 in [156], K_λ is a smooth function equal to

$$\frac{\Gamma(\lambda+1)}{\pi^{\lambda+1}}\frac{\cos(2\pi|x|-\frac{\pi(n+1)}{4}-\frac{\pi\lambda}{2})}{|x|^{\frac{n+1}{2}+\lambda}} + O(|x|^{-\frac{n+3}{2}-\lambda}) \tag{5.2.2}$$

for $|x| \ge 1$. It is natural to examine whether the operators B^λ are bounded on certain L^p spaces by testing them on specific functions. This may provide some indication as to the range of p's for which these operators may be bounded on L^p.

Proposition 5.2.3. *When $\lambda > 0$ and $p \le \frac{2n}{n+1+2\lambda}$ or $p \ge \frac{2n}{n-1-2\lambda}$, the operators B^λ are not bounded on $L^p(\mathbf{R}^n)$.*

Proof. Let h be a Schwartz function whose Fourier transform is equal to 1 on the ball $B(0,2)$ and vanishes off the ball $B(0,3)$. Then

$$B^\lambda(h)(x) = \int_{|\xi|\le 1}(1-|\xi|^2)^\lambda e^{2\pi i\xi\cdot x}\,d\xi = K_\lambda(x),$$

and it suffices to show that K_λ is not in $L^p(\mathbf{R}^n)$ for the claimed range of p's. Notice that

$$\tfrac{\sqrt{2}}{2} \le \cos(2\pi|x|-\tfrac{\pi(n+1)}{4}-\tfrac{\pi\lambda}{2}) \le 1 \tag{5.2.3}$$

for all x lying in the annuli

$$A_k = \left\{x \in \mathbf{R}^n : k + \frac{n+2\lambda}{8} \le |x| \le k + \frac{n+2\lambda}{8} + \frac{1}{4}\right\}, \qquad k \in \mathbf{Z}^+,$$

since in this range, the argument of the cosine in (5.2.2) lies in $[2\pi k - \frac{\pi}{4}, 2\pi k + \frac{\pi}{4}]$.

Consider the range of p's that satisfy

$$\frac{2n}{n+1+2\lambda} \ge p > \frac{2n}{n+3+2\lambda}. \tag{5.2.4}$$

If we can show that B^λ is unbounded in this range, it will also have to be unbounded in the bigger range $\frac{2n}{n+1+2\lambda} \ge p$. This follows by interpolation between the values $p = \frac{2n}{n+3+2\lambda} - \delta$ and $p = \frac{2n}{n+1+2\lambda} + \delta$, $\delta > 0$, for λ fixed.

In view of (5.2.2) and (5.2.3), we have that

$$\|K_\lambda\|_{L^p}^p \ge C' \sum_{k=n}^{\infty} \int_{A_k} |x|^{-p\frac{n+1}{2} - p\lambda} dx - C'' - C''' \int_{|x|\ge 1} |x|^{-p\frac{n+3}{2} - p\lambda} dx, \tag{5.2.5}$$

where C'' is the integral of K_λ in the unit ball. It is easy to see that for p in the range (5.2.4), the integral outside the unit ball converges, while the series diverges in (5.2.5).

The unboundedness of B^λ on $L^p(\mathbf{R}^n)$ in the range of $p \ge \frac{2n}{n-1-2\lambda}$ follows by duality. $\square$

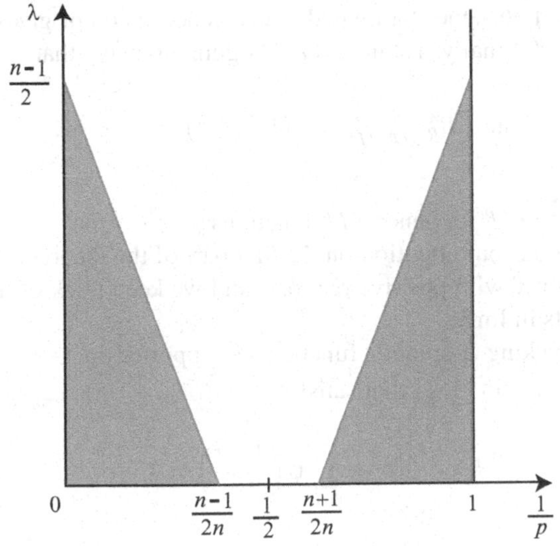

Fig. 5.8 The operator B^λ is unbounded on $L^p(\mathbf{R}^n)$ when $(1/p, \lambda)$ lies in the shaded region.

In Figure 5.8 the shaded region is the set of all pairs $(\frac{1}{p}, \lambda)$ for which the operators B^λ are known to be unbounded on $L^p(\mathbf{R}^n)$.

5.2.2 The Carleson–Sjölin Theorem

We now pass to the main result in this section. We prove the boundedness of the operators B^λ, $\lambda > 0$, in the range of p's not excluded by the previous proposition in dimension $n = 2$.

Theorem 5.2.4. *Suppose that* $0 < \operatorname{Re}\lambda \le 1/2$. *Then the Bochner–Riesz operator* B^λ *maps* $L^p(\mathbf{R}^2)$ *to itself when* $\frac{4}{3+2\operatorname{Re}\lambda} < p < \frac{4}{1-2\operatorname{Re}\lambda}$. *Moreover, for this range of p's and for all* $f \in L^p(\mathbf{R}^2)$ *we have that*

$$B_R^\lambda(f) \to f$$

in $L^p(\mathbf{R}^2)$ *as* $R \to \infty$.

Proof. Once the first assertion of the theorem is established, the second assertion will be a direct consequence of it and of the fact that the means $B_R^\lambda(h)$ converge to h in L^p for h in a dense subclass of L^p. Such a dense class is the space of all Schwartz functions h whose Fourier transforms are compactly supported (Exercise 6.2.9 in [156]). For a function h in this class, we see easily that $B_R^\lambda(h) \to h$ pointwise as $R \to \infty$. But if $\widehat{h}$ is supported in $|\xi| \le c$, then for $R \ge 2c$, integration by parts gives that the functions $B_R^\lambda(h)(x)$ are pointwise controlled by the function $(1+|x|)^{-N}$ with N large; then the Lebesgue dominated convergence theorem gives that the $B_R^\lambda(h)$ converge to h in L^p. Finally, a standard $\varepsilon/3$ argument, using that

$$\sup_{R>0} \|B_R^\lambda\|_{L^p \to L^p} = \|(1-|\xi|^2)_+^\lambda\|_{\mathscr{M}_p} < \infty,$$

yields $B_R^\lambda(f) \to f$ in L^p for general L^p functions f.

It suffices to focus our attention on the first part of the theorem. We therefore fix a complex number λ with positive real part and we keep track of the growth of all involved constants in $\operatorname{Im}\lambda$.

We start by picking a smooth function φ supported in $[-\frac{1}{2}, \frac{1}{2}]$ and a smooth function ψ supported in $[\frac{1}{8}, \frac{5}{8}]$ that satisfy

$$\varphi(t) + \sum_{k=0}^{\infty} \psi\Big(\frac{1-t}{2^{-k}}\Big) = 1$$

for all $t \in [0,1)$. We now decompose the multiplier $(1-|\xi|^2)_+^\lambda$ as

$$(1-|\xi|^2)_+^\lambda = m_{00}(\xi) + \sum_{k=0}^{\infty} 2^{-k\lambda} m_k(\xi), \tag{5.2.6}$$

where $m_{00}(\xi) = \varphi(|\xi|)(1-|\xi|^2)^\lambda$ and for $k \geq 0$, m_k is defined by

$$m_k(\xi) = \Big(\frac{1-|\xi|}{2^{-k}}\Big)^\lambda \psi\Big(\frac{1-|\xi|}{2^{-k}}\Big)(1+|\xi|)^\lambda .$$

Note that m_{00} is a smooth function with compact support; hence the multiplier m_{00} lies in $\mathscr{M}_p(\mathbf{R}^2)$ for all $1 \leq p \leq \infty$. Each function m_k is also smooth, radial, and supported in the small annulus

$$1 - \tfrac{5}{8} 2^{-k} \leq |\xi| \leq 1 - \tfrac{1}{8} 2^{-k}$$

and therefore also lies in $\mathscr{M}_p$; nevertheless the $\mathscr{M}_p$ norms of the m_k's grow as k increases, and it is crucial to determine how this growth depends on k so that we can sum the series in (5.2.6).

Next we show that the Fourier multiplier norm of each m_k on $L^4(\mathbf{R}^2)$ is at most $C(1+|k|)^{1/2}(1+|\lambda|)^3$. This implies that

$$\|m_k\|_{L^4(\mathbf{R}^2)\to L^4(\mathbf{R}^2)} \leq C(1+|\mathrm{Re}\,\lambda|)^3(1+|k|)^{1/2}(1+|\mathrm{Im}\,\lambda|)^3 .$$

Summing over $k \geq 0$ implies that B^λ maps $L^4(\mathbf{R}^2)$ to itself with norm at most a multiple of $(1+|\mathrm{Re}\,\lambda|)^3(1+|\mathrm{Im}\,\lambda|)^3$. Given this bound, we obtain that B^λ maps $L^p(\mathbf{R}^2)$ to itself when

$$\frac{4}{3+2\mathrm{Re}\,\lambda} < p < \frac{4}{1-2\mathrm{Re}\,\lambda}$$

via Theorem 1.3.7 in [156] (precisely Exercise 1.3.4 in [156]). Indeed, to apply this theorem, we note that the family $\{B^\lambda\}_{\mathrm{Re}\,\lambda\geq 0}$ is admissible, since for any pair E, F of measurable sets of finite measure we have

$$\left|\int_{\mathbf{R}^n} \chi_F B^\lambda(\chi_E)\,dx\right| \leq (|E|\,|F|)^{1/2},$$

which is independent of λ, when $\mathrm{Re}\,\lambda \geq 0$. Moreover, the function

$$z \mapsto \int_{\mathbf{R}^n} \widehat{\chi_F}(\xi)(1-|\xi|^2)_+^z\, \overline{\widehat{\chi_E}(\xi)}\,d\xi$$

is analytic in the region $\mathrm{Re}\,z > 0$, since

$$\sup_{|\xi|<1} |(1-|\xi|^2)_+^z|\,|\log(1-|\xi|^2)^{-1}| < \infty$$

for any z in a small neighborhood of a point z_0 with $\mathrm{Re}\, z_0 > 0$. The interpolation for the analytic family of operators $\lambda \mapsto B^\lambda$ is based on the estimates

$$\big\|B^\lambda\big\|_{L^4(\mathbf{R}^2)\to L^4(\mathbf{R}^2)} \le C_4\,(1+|\mathrm{Im}\,\lambda|)^3 \qquad \text{when } \mathrm{Re}\,\lambda = \varepsilon,$$

$$\big\|B^\lambda\big\|_{L^\infty(\mathbf{R}^2)\to L^\infty(\mathbf{R}^2)} \le C_\infty\, e^{c_1|\mathrm{Im}\,\lambda|^2} \qquad \text{when } \mathrm{Re}\,\lambda = \tfrac{1}{2}+\varepsilon,$$

where C_4, C_∞, c_1 depend only on $\varepsilon > 0$. The second estimate above is proved in Proposition 5.2.2 while the set of points $(1/p, \lambda)$ obtained by interpolation can be seen in Figure 5.8.

To estimate the norm of each m_k in $\mathscr{M}_4(\mathbf{R}^2)$, we need an additional decomposition of the operator m_k that takes into account the radial nature of m_k. For each $k \ge 0$ we define the sectorial arcs (parts of a sector between two arcs)

$$\Gamma_{k,\ell} = \big\{re^{2\pi i\theta} \in \mathbf{R}^2 :\ |\theta - \ell 2^{-\frac{k}{2}}| < 2^{-\frac{k}{2}},\quad 1-\tfrac{5}{8}2^{-k} \le r \le 1-\tfrac{1}{8}2^{-k}\big\}$$

for all $\ell \in \{0,1,2,\dots,[2^{k/2}]-1\}$. We now introduce a smooth function ω supported in $[-1,1]$ and equal to 1 on $[-1/4,1/4]$ such that for all $x \in \mathbf{R}$ we have

$$\sum_{\ell\in\mathbf{Z}} \omega(x-\ell) = 1\,.$$

Then we define

$$m_{k,\ell}(re^{2\pi i\theta}) = m_k(re^{2\pi i\theta})\,\omega(2^{k/2}\theta-\ell) = m_k(r,0)\,\omega(2^{k/2}\theta-\ell)$$

for integers ℓ in the set $\{0,1,2,\dots,[2^{k/2}]-1\}$. If k is an even integer, it follows from the construction that

$$m_k(\xi) = \sum_{\ell=0}^{[2^{k/2}]-1} m_{k,\ell}(\xi) \tag{5.2.7}$$

for all ξ in $\mathbf{R}^2$. If k is odd we replace the function $\theta \mapsto \omega(2^{k/2}\theta - ([2^{k/2}]-1))$ by a function $\omega_k(\theta)$ supported in the bigger interval $\big[([2^{k/2}]-2)2^{-k/2},1\big]$ that satisfies

$$\omega_k(\theta) + \omega(2^{k/2}\theta) = 1$$

on the interval $\big[([2^{k/2}]-1)2^{-k/2},1\big]$. This leads to a new definition of the function $m_{k,[2^{k/2}]-1}$ so that (5.2.7) is satisfied.

This provides the circular (angular) decomposition of m_k. Observe that for all positive integers α and β there exist constants $C_{\alpha,\beta}$ such that

$$|\partial_r^\alpha \partial_\theta^\beta m_{k,\ell}(re^{2\pi i\theta})| \le C_{\alpha,\beta}(1+|\lambda|)^\alpha 2^{k\alpha} 2^{\frac{k}{2}\beta} \tag{5.2.8}$$

and such that each $m_{k,\ell}$ is a smooth function supported in the sectorial arcs $\Gamma_{k,\ell}$.

We fix $k \ge 0$ and we group the set of all $\{m_{k,\ell}\}_\ell$ into five subsets: (a) those whose supports are contained in $Q = \{(x,y) \in \mathbf{R}^2 :\ x > 0,\ |y| < |x|\}$; (b) those $m_{k,\ell}$ whose supports are contained in the sector $Q' = \{(x,y) \in \mathbf{R}^2 : x < 0, \ |y| < |x|\}$; (c) those whose supports are contained in $Q'' = \{(x,y) \in \mathbf{R}^2 : y > 0, \ |y| > |x|\}$; (d) the $m_{k,\ell}$ with supports contained in $Q''' = \{(x,y) \in \mathbf{R}^2 : y < 0, \ |y| > |x|\}$; and finally (e) those $m_{k,\ell}$ whose supports intersect the lines $|y| = |x|$.

There are only at most eight $m_{k,\ell}$ in case (e), and their sum is easily shown to be an L^4 Fourier multiplier with a constant that grows like $(1+|\lambda|)^3$, as shown below. The remaining cases are symmetric, and we focus attention on case (a).

Let I be the set of all indices ℓ in the set $\{0,1,2,\dots,[2^{k/2}]-1\}$ corresponding to case (a), i.e., the sectorial arcs $\Gamma_{k,\ell}$ are contained in the quarter-plane Q. Let $T_{k,\ell}$ be the operator given on the Fourier transform by multiplication by the function $m_{k,\ell}$. We have

$$\begin{aligned}\Big\|\sum_{\ell\in I} T_{k,\ell}(f)\Big\|_{L^4}^4 &= \int_{\mathbf{R}^2}\Big|\sum_{\ell\in I} T_{k,\ell}(f)\Big|^4 dx \\ &= \int_{\mathbf{R}^2}\Big|\sum_{\ell\in I}\sum_{\ell'\in I} T_{k,\ell}(f)T_{k,\ell'}(f)\Big|^2 dx \\ &= \int_{\mathbf{R}^2}\Big|\sum_{\ell\in I}\sum_{\ell'\in I} \widehat{T_{k,\ell}(f)} * \widehat{T_{k,\ell'}(f)}\Big|^2 d\xi\,,\end{aligned} \tag{5.2.9}$$

where we used Plancherel's identity in the last equality. Each function $\widehat{T_{k,\ell}(f)}$ is supported in the sectorial arc $\Gamma_{k,\ell}$. Therefore, the function $\widehat{T_{k,\ell}(f)} * \widehat{T_{k,\ell'}(f)}$ is supported in $\Gamma_{k,\ell}+\Gamma_{k,\ell'}$ and we write the last integral as

$$\int_{\mathbf{R}^2}\Big|\sum_{\ell\in I}\sum_{\ell'\in I} \big(\widehat{T_{k,\ell}(f)} * \widehat{T_{k,\ell'}(f)}\big)\chi_{\Gamma_{k,\ell}+\Gamma_{k,\ell'}}\Big|^2 d\xi\,.$$

In view of the Cauchy–Schwarz inequality, the last expression is controlled by

$$\int_{\mathbf{R}^2}\Big(\sum_{\ell\in I}\sum_{\ell'\in I} \big|\widehat{T_{k,\ell}(f)} * \widehat{T_{k,\ell'}(f)}\big|^2\Big)\Big(\sum_{\ell\in I}\sum_{\ell'\in I} \big|\chi_{\Gamma_{k,\ell}+\Gamma_{k,\ell'}}\big|^2\Big)\, d\xi\,. \tag{5.2.10}$$

At this point we make use of the following lemma, in which the curvature of the circle is manifested.

Lemma 5.2.5. *There exists a constant C_0 such that for all $k \ge 0$ the following estimate holds:*

$$\sum_{\ell\in I}\sum_{\ell'\in I} \chi_{\Gamma_{k,\ell}+\Gamma_{k,\ell'}} \le C_0\,.$$

We postpone the proof of this lemma until the end of this section. Using Lemma 5.2.5, we control the expression in (5.2.10) by

$$C_0\int_{\mathbf{R}^2}\sum_{\ell\in I}\sum_{\ell'\in I} \big|\widehat{T_{k,\ell}(f)} * \widehat{T_{k,\ell'}(f)}\big|^2 d\xi = C_0\Big\|\Big(\sum_{\ell\in I}|T_{k,\ell}(f)|^2\Big)^{\frac12}\Big\|_{L^4}^4. \tag{5.2.11}$$

We examine each $T_{k,\ell}$ a bit more carefully. We have that $m_{k,0}$ is supported in a rectangle with sides parallel to the axes and dimensions 2^{-k} (along the ξ_1-axis) and $2^{-\frac{k}{2}+1}$ (along the ξ_2-axis). Moreover, in that rectangle, $\partial_{\xi_1} \approx \partial_r$ and $\partial_{\xi_2} \approx \partial_\theta$, and it follows that the smooth function $m_{k,0}$ satisfies

$$|\partial_{\xi_1}^{\alpha}\partial_{\xi_2}^{\beta} m_{k,0}(\xi_1,\xi_2)| \le C_{\alpha,\beta}(1+|\lambda|)^{\alpha}2^{k\alpha}2^{\frac{k}{2}\beta} \tag{5.2.12}$$

for all positive integers α and β. To give a precise proof of (5.2.12) we use the relation $r^2 = \xi_1^2+\xi_2^2$ and $\tan(2\pi\theta) = \xi_2/\xi_1$. We have

$$\frac{\partial m_{k,0}}{\partial \xi_1} = \frac{\partial m_{k,0}}{\partial r}\frac{\partial r}{\partial \xi_1} + \frac{\partial m_{k,0}}{\partial \theta}\frac{\partial \theta}{\partial \xi_1} = \frac{\partial m_{k,0}}{\partial r}\frac{\xi_1}{r} + \frac{\partial m_{k,0}}{\partial \theta}\frac{-\xi_2}{2\pi(\xi_2^2+\xi_1^2)}$$

which is pointwise bounded by

$$C(1+|\lambda|)2^k\frac{|\xi_1|}{(\xi_2^2+\xi_1^2)^{1/2}} + C2^{\frac{k}{2}}\frac{|\xi_2|}{\xi_2^2+\xi_1^2} \le C'(1+|\lambda|)2^k$$

in view of (5.2.8), since $|\xi_1|, |\xi_2| \le r$ and $\frac{1}{2} \le r \le 2$. Likewise, we have

$$\frac{\partial m_{k,0}}{\partial \xi_2} = \frac{\partial m_{k,0}}{\partial r}\frac{\partial r}{\partial \xi_2} + \frac{\partial m_{k,0}}{\partial \theta}\frac{\partial \theta}{\partial \xi_2} = \frac{\partial m_{k,0}}{\partial r}\frac{\xi_2}{r} + \frac{\partial m_{k,0}}{\partial \theta}\frac{\xi_1}{2\pi(\xi_2^2+\xi_1^2)},$$

which is controlled by

$$C(1+|\lambda|)2^k\frac{|\xi_2|}{(\xi_2^2+\xi_1^2)^{1/2}} + C2^{\frac{k}{2}}\frac{|\xi_1|}{\xi_2^2+\xi_1^2} \le C'(1+|\lambda|)2^{\frac{k}{2}}$$

since $|\xi_2| \lesssim 2^{-\frac{k}{2}}$ and $\frac{1}{2} \le r \le 2$. For arbitrary indices α, β, we use a similar procedure to prove (5.2.12).

Estimate (5.2.12) can also be written as

$$\left|\partial_{\xi_1}^{\alpha}\partial_{\xi_2}^{\beta}\left[m_{k,0}(2^{-k}\xi_1, 2^{-\frac{k}{2}}\xi_2)\right]\right| \le C_{\alpha,\beta}(1+|\lambda|)^{\alpha+\beta},$$

which easily implies that for some constant C we have

$$2^{\frac{3}{2}k}|m_{k,0}^{\vee}(2^k x_1, 2^{\frac{k}{2}}x_2)| \le C(1+|\lambda|)^3(1+|x_1|+|x_2|)^{-3}.$$

Let V_ℓ be the unit vector representing the point $e^{2\pi i\ell 2^{-k/2}}$ and $V_\ell^{\perp}$ the unit vector representing the point $ie^{2\pi i\ell 2^{-k/2}}$. Applying a rotation, we obtain that the functions $m_{k,\ell}^{\vee}$ satisfy

$$\left|m_{k,\ell}^{\vee}(x_1,x_2)\right| \le C(1+|\lambda|)^3 2^{-\frac{3k}{2}}(1+2^{-k}|x\cdot V_\ell|+2^{-\frac{k}{2}}|x\cdot V_\ell^{\perp}|)^{-3} \tag{5.2.13}$$

and hence

$$\sup_{k\geq 0}\sup_{\ell\in I}\big\|m_{k,\ell}^{\vee}\big\|_{L^1}\leq C(1+|\lambda|)^3. \tag{5.2.14}$$

The crucial fact is that the constant C in (5.2.14) is independent of ℓ and k.

At this point, for each fixed $k\geq 0$ and $\ell\in I$ we let $J_{k,\ell}$ be the ξ_2-projection of the support of $m_{k,\ell}$. Based on the earlier definition of $m_{k,\ell}$, we easily see that when $\ell>0$,

$$J_{k,\ell}=\big[(1-\tfrac{5}{8}2^{-k})\sin(2\pi 2^{-\frac{k}{2}}(\ell-1)),(1-\tfrac{1}{8}2^{-k})\sin(2\pi 2^{-\frac{k}{2}}(\ell+1))\big].$$

A similar formula holds for $\ell<0$ in I. The crucial observation is that for any fixed $k\geq 0$ the sets $J_{k,\ell}$ are "almost disjoint" for different $\ell\in I$. Indeed, the sets $J_{k,\ell}$ are contained in the intervals

$$\widetilde{J}_{k,\ell}=\big[(1-\tfrac{3}{8}2^{-k})\sin(2\pi 2^{-\frac{k}{2}}\ell)-10\cdot 2^{-\frac{k}{2}},(1-\tfrac{3}{8}2^{-k})\sin(2\pi 2^{-\frac{k}{2}}\ell)+10\cdot 2^{-\frac{k}{2}}\big],$$

which have length $20\cdot 2^{-\frac{k}{2}}$ and are centered at the points $(1-\frac{3}{8}2^{-k})\sin(2\pi 2^{-\frac{k}{2}}\ell)$. For $\sigma\in\mathbf{Z}$ and $\tau\in\{0,1,\dots,39\}$ we define the strips

$$S_{k,\sigma,\tau}=\big\{(\xi_1,\xi_2)\in\mathbf{R}^2:\ \xi_2\in[40\sigma 2^{-\frac{k}{2}}+\tau 2^{-\frac{k}{2}},40(\sigma+1)2^{-\frac{k}{2}}+\tau 2^{-\frac{k}{2}})\big\}.$$

These strips have length $40\cdot 2^{-\frac{k}{2}}$ and have the property that each $\widetilde{J}_{k,\ell}$ is contained in one of them; say $\widetilde{J}_{k,\ell}$ is contained in some $S_{k,\sigma_\ell,\tau_\ell}$, which we call $B_{k,\ell}$. Then we have

$$T_{k,\ell}(f)=T_{k,\ell}(f_{k,\ell}),$$

where we set

$$f_{k,\ell}=\big(\chi_{B_{k,\ell}}\widehat{f}\,\big)^{\vee}=\chi_{B_{k,\ell}}^{\vee}*f.$$

As a consequence of the Cauchy–Schwarz inequality (with respect to the measure $|m_{k,\ell}^{\vee}|\,dx$), we obtain

$$\begin{aligned}|T_{k,\ell}(f_{k,\ell})|^2 &\leq \big\|m_{k,\ell}^{\vee}\big\|_{L^1}\big(|m_{k,\ell}^{\vee}|*|f_{k,\ell}|^2\big)\\ &\leq C(1+|\lambda|)^3\big(|m_{k,\ell}^{\vee}|*|f_{k,\ell}|^2\big)\end{aligned}$$

in view of (5.2.14). We now return to (5.2.11), which controls (5.2.10) and hence (5.2.9). Using this estimate, we bound the term in (5.2.11) by

$$\begin{aligned}\Big\|\Big(\sum_{\ell\in I}|T_{k,\ell}(f)|^2\Big)^{\frac{1}{2}}\Big\|_{L^4}^4 &= \Big\|\sum_{\ell\in I}|T_{k,\ell}(f_{k,\ell})|^2\Big\|_{L^2}^2\\ &\leq C^2(1+|\lambda|)^6\Big\|\sum_{\ell\in I}|m_{k,\ell}^{\vee}|*|f_{k,\ell}|^2\Big\|_{L^2}^2\\ &= C^2(1+|\lambda|)^6\Big(\int_{\mathbf{R}^2}\sum_{\ell\in I}(|m_{k,\ell}^{\vee}|*|f_{k,\ell}|^2)\,g\,dx\Big)^2\end{aligned}$$

$$= C^2(1+|\lambda|)^6\Big(\sum_{\ell\in I}\int_{\mathbf{R}^2}(|\widehat{m_{k,\ell}}|*g)\,|f_{k,\ell}|^2\,dx\Big)^2$$

$$\le C^2(1+|\lambda|)^6\Big(\int_{\mathbf{R}^2}\sup_{\ell\in I}(|\widehat{m_{k,\ell}}|*g)\sum_{\ell\in I}|f_{k,\ell}|^2\,dx\Big)^2$$

$$\le C^2(1+|\lambda|)^6\Big\|\sup_{\ell\in I}(|\widehat{m_{k,\ell}}|*g)\Big\|_{L^2}^2\Big\|\Big(\sum_{\ell\in I}|f_{k,\ell}|^2\Big)^{\frac{1}{2}}\Big\|_{L^4}^4,$$

where g is an appropriate nonnegative function in $L^2(\mathbf{R}^2)$ of norm 1.

If we knew the validity of the estimates

$$\Big\|\sup_{\ell\in I}(|\widehat{m_{k,\ell}}|*g)\Big\|_{L^2}\le C(1+|\lambda|)^3(1+k)\big\|g\big\|_{L^2} \tag{5.2.15}$$

and

$$\Big\|\Big(\sum_{\ell\in I}|f_{k,\ell}|^2\Big)^{\frac{1}{2}}\Big\|_{L^4}\le C\big\|f\big\|_{L^4}, \tag{5.2.16}$$

then we would be able to conclude that

$$\big\|m_k\big\|_{\mathscr{M}_4}\le C(1+|\lambda|)^3(1+k)^{\frac{1}{2}} \tag{5.2.17}$$

and hence we could sum the series in (5.2.6).

Estimates (5.2.15) and (5.2.16) are discussed in the next two subsections. □

5.2.3 *The Kakeya Maximal Function*

We showed in the previous subsection that $m_{k,0}^\vee$ is integrable over $\mathbf{R}^2$ and satisfies the estimate

$$2^{\frac{3}{2}k}|m_{k,0}^\vee(2^kx_1,2^{\frac{k}{2}}x_2)|\le\frac{C(1+|\lambda|)^3}{(1+|x|)^3}.$$

Since

$$\frac{1}{(1+|x|)^3}\le C\sum_{s=0}^{\infty}\frac{2^{-s}}{2^{2s}}\chi_{[-2^s,2^s]\times[-2^s,2^s]}(x),$$

it follows that

$$|\widehat{m_{k,0}}(x)|\le C'(1+|\lambda|)^3\sum_{s=0}^{\infty}2^{-s}\frac{1}{|R_s|}\chi_{R_s}(x),$$

where $R_s=[-2^s2^k,2^s2^k]\times[-2^s2^{\frac{k}{2}},2^s2^{\frac{k}{2}}]$. Since a general $\widehat{m_{k,\ell}}$ is obtained from $\widehat{m_{k,0}}$ via a rotation, a similar estimate holds for it. Precisely, we have

$$|\widehat{m_{k,\ell}}(x)|\le C'(1+|\lambda|)^3\sum_{s=0}^{\infty}2^{-s}\frac{1}{|R_{s,\ell}|}\chi_{R_{s,\ell}}(x), \tag{5.2.18}$$

where $R_{s,\ell}$ is a rectangle with principal axes along the directions V_ℓ and $V_\ell^\perp$ and side lengths $2^s 2^k$ and $2^s 2^{\frac{k}{2}}$, respectively. Using (5.2.18), we obtain the following pointwise estimate for the maximal function in (5.2.15):

$$\sup_{\ell \in I} \big(|\widehat{m_{k,\ell}}| * g \big)(x) \le C' \sum_{s=0}^{\infty} 2^{-s} \sup_{\ell \in I} \frac{1}{|R_{s,\ell}|} \int_{R_{s,\ell}} g(x-y)\, dy, \tag{5.2.19}$$

where $R_{s,\ell}$ are rectangles with dimensions $2^s 2^k$ and $2^s 2^{\frac{k}{2}}$.

Motivated by (5.2.19), for fixed $N \ge 10$ and $a > 0$, we introduce the *Kakeya maximal operator without dilations*

$$\mathscr{K}_N^a(g)(x) = \sup_{R \ni x} \frac{1}{|R|} \int_R |g(y)|\, dy, \tag{5.2.20}$$

acting on functions $g \in L^1_{\mathrm{loc}}$, where the supremum is taken over all rectangles R in $\mathbf{R}^2$ of dimensions a and aN and arbitrary orientation. What makes this maximal operator interesting is that the rectangles R that appear in the supremum in (5.2.21) are allowed to have arbitrary orientation. We also define the *Kakeya maximal operator* $\mathscr{K}_N$ by setting

$$\mathscr{K}_N(w)(x) = \sup_{a>0} \mathscr{K}_N^a(w)(x), \tag{5.2.21}$$

for w locally integrable and $x \in \mathbf{R}^n$. The maximal function $\mathscr{K}_N(w)(x)$ is therefore obtained as the supremum of the averages of a function w over all rectangles in $\mathbf{R}^2$ that contain the point x and have arbitrary orientation but fixed eccentricity equal to N. (The eccentricity of a rectangle is the ratio of its longer side to its shorter side.)

We see that for some $c > 0$, $\mathscr{K}_N(f)$ is pointwise controlled by $cNM(f)$, where M is the Hardy–Littlewood maximal operator M. This implies that $\mathscr{K}_N$ is of weak type $(1,1)$ with bound at most a multiple of N. Since $\mathscr{K}_N$ is bounded on L^∞ with norm 1, it follows that $\mathscr{K}_N$ maps $L^p(\mathbf{R}^2)$ to itself with norm at most a multiple of $N^{1/p}$. However, we show in the next section that this estimate is very rough and can be improved significantly. In fact, we obtain an L^p estimate for $\mathscr{K}_N$ with norm that grows logarithmically in N (when $p \ge 2$), and this is very crucial, since $N = 2^{k/2}$ in the following application.

Using this new terminology, we write the estimate in (5.2.19) as

$$\sup_{\ell \in I} \big(|\widehat{m_{k,\ell}}| * g \big) \le C'(1+|\lambda|)^3 \sum_{s=0}^{\infty} 2^{-s} \mathscr{K}_{2^{k/2}}^{2^{s+k/2}}(g). \tag{5.2.22}$$

The required estimate (5.2.15) is a consequence of (5.2.22) and of the following theorem, whose proof is discussed in the next section.

Theorem 5.2.6. *There exists a constant C such that for all $N \ge 10$ and all f in $L^2(\mathbf{R}^2)$ the following norm inequality is valid:*

$$\sup_{a>0} \big\| \mathscr{K}_N^a(f) \big\|_{L^2(\mathbf{R}^2)} \le C (\log N) \big\| f \big\|_{L^2(\mathbf{R}^2)} .$$

Theorem 5.2.6 is a consequence of Theorem 5.3.5, in which the preceding estimate is proved for a more general maximal operator $\mathfrak{M}_{\Sigma_N}$, which in particular controls $\mathscr{K}_N$ and hence $\mathscr{K}_N^a$ for all $a > 0$. This maximal operator is introduced in the next section.

5.2.4 Boundedness of a Square Function

We now turn to the proof of estimate (5.2.16). This is a consequence of the following result, which is a version of the Littlewood–Paley theorem for intervals of equal length.

Theorem 5.2.7. *For $j \in \mathbf{Z}$, let I_j be intervals of equal length with disjoint interior whose union is $\mathbf{R}$. We define operators P_j with multipliers χ_{I_j}. Then for $2 \le p < \infty$, there is a constant C_p such that for all $f \in L^p(\mathbf{R})$ we have*

$$\Big\| \Big(\sum_j |P_j(f)|^2 \Big)^{\frac{1}{2}} \Big\|_{L^p(\mathbf{R})} \le C_p \big\| f \big\|_{L^p(\mathbf{R})} . \tag{5.2.23}$$

In particular, the same estimate holds if the intervals I_j have disjoint interiors and equal length but do not necessarily cover $\mathbf{R}$.

Proof. Multiplying the function f by a suitable exponential, we may assume that the intervals I_j have the form $\big((j-\frac{1}{2})a, (j+\frac{1}{2})a\big)$ for some $a > 0$. Applying a dilation to f reduces matters to the case $a = 1$. We conclude that the constant C_p is independent of the common size of the intervals I_j and it suffices to obtain estimate (5.2.23) in the case $a = 1$.

We assume therefore that $I_j = (j - \frac{1}{2}, j + \frac{1}{2})$ for all $j \in \mathbf{Z}$. Next, our goal is to replace the operators P_j by smoother analogues of them. To achieve this we introduce a smooth function ψ with compact support that is identically equal to 1 on the interval $[-\frac{1}{2}, \frac{1}{2}]$ and vanishes off the interval $[-\frac{3}{4}, \frac{3}{4}]$. We introduce operators S_j by setting

$$\widehat{S_j(f)}(\xi) = \widehat{f}(\xi)\psi(\xi - j)$$

and we note that the identity

$$P_j = P_j S_j \tag{5.2.24}$$

is valid for all $j \in \mathbf{Z}$. For $t \in \mathbf{R}$ we define multipliers m_t as

$$m_t(\xi) = \sum_{j \in \mathbf{Z}} e^{-2\pi i j t} \psi(\xi - j),$$

and we set $k_t = m_t^{\vee}$. With $I_0 = (-1/2, 1/2)$, we have

$$\begin{aligned} \int_{I_0} |(k_t * f)(x)|^2 \, dt &= \int_{I_0} \Big| \sum_{j \in \mathbf{Z}} e^{-2\pi i j t} S_j(f)(x) \Big|^2 dt \\ &= \sum_{j \in \mathbf{Z}} |S_j(f)(x)|^2, \end{aligned} \tag{5.2.25}$$

where the last equality is just Plancherel's identity on $I_0 = [-\frac{1}{2}, \frac{1}{2}]$. In view of the last identity, it suffices to analyze the operator given by convolution with the family of kernels k_t. By the Poisson summation formula (Theorem 3.2.8 in [156]) applied to the function $x \mapsto \psi(x)e^{2\pi ixt}$, we obtain

$$\begin{aligned} m_t(\xi) &= e^{-2\pi i\xi t} \sum_{j\in\mathbf{Z}} \psi(\xi - j)e^{2\pi i(\xi-j)t} \\ &= \sum_{j\in\mathbf{Z}} \big(\psi(\cdot)e^{2\pi i(\cdot)t}\big)\widehat{\ }(j)\, e^{2\pi ij\xi} e^{-2\pi i\xi t} \\ &= \sum_{j\in\mathbf{Z}} e^{2\pi i(j-t)\xi}\, \widehat{\psi}(j-t). \end{aligned}$$

Taking inverse Fourier transforms, we obtain

$$k_t = \sum_{j\in\mathbf{Z}} \widehat{\psi}(j-t)\delta_{-j+t},$$

where δ_b denotes Dirac mass at the point b. Therefore, k_t is a sum of Dirac masses with rapidly decaying coefficients. Since each Dirac mass has Borel norm at most 1, we conclude that for some constant C we have

$$\|k_t\|_{\mathcal{M}} \le \sum_{j\in\mathbf{Z}} |\widehat{\psi}(j-t)| \le C \sum_{j\in\mathbf{Z}} (1+|j-t|)^{-10} \le c_0, \tag{5.2.26}$$

where c_0 is independent of t. This says that the measures k_t have uniformly bounded norms. Take now $f \in L^p(\mathbf{R})$ and $p \ge 2$. Using identity (5.2.24), we obtain

$$\begin{aligned} \int_{\mathbf{R}} \Big(\sum_{j\in\mathbf{Z}} |P_j(f)(x)|^2\Big)^{\frac{p}{2}} dx &= \int_{\mathbf{R}} \Big(\sum_{j\in\mathbf{Z}} |P_jS_j(f)(x)|^2\Big)^{\frac{p}{2}} dx \\ &\le c_p \int_{\mathbf{R}} \Big(\sum_{j\in\mathbf{Z}} |S_j(f)(x)|^2\Big)^{\frac{p}{2}} dx, \end{aligned}$$

and the last inequality follows from Exercise 5.6.1(a) in [156]. The constant c_p depends only on p. Recalling identity (5.2.25), we write

$$\begin{aligned} c_p \int_{\mathbf{R}} \Big(\sum_{j\in\mathbf{Z}} |S_j(f)(x)|^2\Big)^{\frac{p}{2}} dx &\le c_p \int_{\mathbf{R}} \Big(\int_{I_0} |(k_t * f)(x)|^2\, dt\Big)^{\frac{p}{2}} dx \\ &\le c_p \int_{\mathbf{R}} \Big(\int_{I_0} |(k_t * f)(x)|^p dt\Big)^{\frac{p}{p}} dx \\ &= c_p \int_{I_0} \int_{\mathbf{R}} |(k_t * f)(x)|^p\, dx\, dt \\ &\le c_0 c_p \int_{I_0} \int_{\mathbf{R}} |f(x)|^p\, dx\, dt \\ &= c_0 c_p \|f\|_{L^p}^p, \end{aligned}$$

where we used Hölder's inequality on the interval I_0 together with the fact that $p \geq 2$ and (5.2.26). The proof of the theorem is complete with constant $C_p = (c_0 c_p)^{1/p}$. □

We now return to estimate (5.2.16). First recall the strips

$$S_{k,\sigma,\tau} = \{(\xi_1,\xi_2):\ \xi_2 \in [40\sigma 2^{-\frac{k}{2}} + 2^{-\frac{k}{2}}\tau, 40(\sigma+1)2^{-\frac{k}{2}} + 2^{-\frac{k}{2}}\tau)\}$$

defined for $\sigma \in \mathbf{Z}$ and $\tau \in \{0,1,\dots,39\}$. These strips have length $40 \cdot 2^{-\frac{k}{2}}$, and each $\widetilde{J}_{k,\ell}$ is contained in one of them, which we called $S_{k,\sigma_\ell,\tau_\ell} = B_{k,\ell}$.

The family $\{B_{k,\ell}\}_{\ell \in I}$ does not consist of disjoint sets, but we split it into 40 subfamilies by placing $B_{k,\ell}$ in different subfamilies if the indices τ_ℓ and $\tau_{\ell'}$ are different. We now write the set I as

$$I = I^1 \cup I^2 \cup \cdots \cup I^{40},$$

where for each $\ell, \ell' \in I^j$ the sets $B_{k,\ell}$ and $B_{k,\ell'}$ are disjoint.

We now use Theorem 5.2.7 to obtain the required quadratic estimate (5.2.16). Things now are relatively simple. We observe that the multiplier operators $f \mapsto (\chi_{B_{k,\ell}}\widehat{f})^\vee$ on $\mathbf{R}^2$ obey the estimates (5.2.23), in which $L^p(\mathbf{R})$ is replaced by $L^p(\mathbf{R}^2)$, since they are the identity operators in the ξ_1-variable.

We conclude that

$$\Big\| \Big(\sum_{\ell \in I^j} |T_{k,\ell}(f)|^2 \Big)^{\frac{1}{2}} \Big\|_{L^p(\mathbf{R}^2)} \leq C_p \big\| f \big\|_{L^p(\mathbf{R}^2)} \tag{5.2.27}$$

holds for all $p \geq 2$ and, in particular, for $p = 4$. This proves (5.2.16) for a single I^j, and the same conclusion follows for I with a constant 40 times as big.

5.2.5 The Proof of Lemma 5.2.5

We finally discuss the proof of Lemma 5.2.5.

Proof. If $k = 0, 1, \dots, k_0$ up to a fixed integer k_0, then there exist only finitely many pairs of sets $\Gamma_{k,\ell} + \Gamma_{k,\ell'}$ depending on k_0, and the lemma is trivially true. We may therefore assume that k is a large integer; in particular we may take $\delta = 2^{-k} \leq 2400^{-2}$. In the sequel, for simplicity we replace 2^{-k} by δ and we denote the set $\Gamma_{k,\ell}$ by Γ_ℓ. In the proof that follows we are working with a fixed $\delta \in [0, 2400^{-2}]$. Elements of the set $\Gamma_\ell + \Gamma_{\ell'}$ have the form

$$r e^{2\pi i(\ell+\alpha)\delta^{1/2}} + r' e^{2\pi i(\ell'+\alpha')\delta^{1/2}}, \tag{5.2.28}$$

where α, α' range in the interval $[-1,1]$ and r, r' range in $[1 - \frac{5}{8}\delta, 1 - \frac{1}{8}\delta]$. We set

$$w(\ell,\ell') = e^{2\pi i \ell \delta^{1/2}} + e^{2\pi i \ell' \delta^{1/2}} = 2\cos(\pi|\ell - \ell'|\delta^{\frac{1}{2}}) e^{\pi i(\ell+\ell')\delta^{1/2}}, \tag{5.2.29}$$

where the last equality is a consequence of a trigonometric identity that can be found in Appendix E in [156]. Using similar identities (also found in Appendix E in [156]) and performing algebraic manipulations, one may verify that the general element (5.2.28) of the set $\Gamma_\ell + \Gamma_{\ell'}$ can be written as

$$\begin{aligned} w(\ell,\ell') &+ \Big\{ r\frac{\cos(2\pi\alpha\delta^{\frac{1}{2}})+\cos(2\pi\alpha'\delta^{\frac{1}{2}})-2}{2}\Big\} w(\ell,\ell') \\ &+ \Big\{ r\frac{\sin(2\pi\alpha\delta^{\frac{1}{2}})+\sin(2\pi\alpha'\delta^{\frac{1}{2}})}{2}\Big\} iw(\ell,\ell') \\ &+ \mathrm{E}(r,r',\ell,\ell',\alpha,\alpha',\delta), \end{aligned}$$

where

$$\begin{aligned} \mathrm{E}(r,r',\ell,\ell',\alpha,\alpha',\delta) &= (r-1)\Big(e^{2\pi i\ell\delta^{1/2}}+e^{2\pi i\ell'\delta^{1/2}}\Big) \\ &+ (r'-r)e^{2\pi i(\ell'+\alpha')\delta^{1/2}} \\ &+ r\Big(e^{2\pi i\ell\delta^{1/2}}-e^{2\pi i\ell'\delta^{1/2}}\Big)\Big(\frac{\cos(2\pi\alpha\delta^{\frac{1}{2}})-\cos(2\pi\alpha'\delta^{\frac{1}{2}})}{2}\Big) \\ &+ ri\Big(e^{2\pi i\ell\delta^{1/2}}-e^{2\pi i\ell'\delta^{1/2}}\Big)\Big(\frac{\sin(2\pi\alpha\delta^{\frac{1}{2}})-\sin(2\pi\alpha'\delta^{\frac{1}{2}})}{2}\Big). \end{aligned}$$

The coefficients in the curly brackets are real, and $\mathrm{E}(r,r',\ell,\ell',\alpha,\alpha',\delta)$ is an error of magnitude at most $2\delta+8\pi^2|\ell-\ell'|\delta$. These observations and the facts $|\sin x|\le|x|$ and $|1-\cos x|\le|x|^2/2$ (see Appendix E in [156]) imply that the set $\Gamma_\ell+\Gamma_{\ell'}$ is contained in the rectangle $R(\ell,\ell')$ centered at the point $w(\ell,\ell')$ with half-width

$$4\pi^2\delta+(2\delta+8\pi^2|\ell-\ell'|\delta)\le 80\,(1+|\ell-\ell'|)\delta$$

in the direction along $w(\ell,\ell')$ and half-length

$$4\pi\delta^{\frac{1}{2}}+(2\delta+8\pi^2|\ell-\ell'|\delta)\le 40\,\delta^{\frac{1}{2}}$$

in the direction along $iw(\ell,\ell')$, which is perpendicular to that along $w(\ell,\ell')$, since $2\pi|\ell-\ell'|\delta^{\frac{1}{2}}<\frac{\pi}{2}$. Using this inequality we show that the rectangle $R(\ell,\ell')$ is contained in a disk of radius $105\,\delta^{\frac{1}{2}}$ centered at the point $w(\ell,\ell')$.

We immediately deduce that if $|w(\ell,\ell')-w(m,m')|$ is bigger than $210\,\delta^{\frac{1}{2}}$, then the sets $\Gamma_\ell+\Gamma_{\ell'}$ and $\Gamma_m+\Gamma_{m'}$ do not intersect. Therefore, if these sets intersect, we should have

$$|w(\ell,\ell')-w(m,m')|\le 210\,\delta^{\frac{1}{2}}.$$

In view of Exercise 5.2.2, the left-hand side of the last expression is at least

$$2\,\tfrac{2}{\pi}\cos(\tfrac{\pi}{4})|\pi(\ell+\ell')-\pi(m+m')|\delta^{\frac{1}{2}}$$

(here we use the hypothesis that $|2\pi\ell\delta^{\frac{1}{2}}| < \frac{\pi}{4}$ twice). We conclude that if the sets $\Gamma_\ell + \Gamma_{\ell'}$ and $\Gamma_m + \Gamma_{m'}$ intersect, then

$$|(\ell+\ell') - (m+m')| \le \frac{210}{4\cos(\frac{\pi}{4})} \le 150. \tag{5.2.30}$$

In this case the angle between the vectors $w(\ell,\ell')$ and $w(m,m')$ is

$$\varphi_{\ell,\ell',m,m'} = \pi|(\ell+\ell') - (m+m')|\delta^{\frac{1}{2}},$$

which is smaller than $\pi/16$, provided (5.2.30) holds and $\delta < 2400^{-2}$. This says that in this case, the rectangles $R(\ell,\ell')$ and $R(m,m')$ are essentially parallel to each other (the angle between them is smaller than $\pi/16$).

Let us fix a rectangle $R(\ell,\ell')$, and for another rectangle $R(m,m')$ we denote by $\widetilde{R}(m,m')$ the smallest rectangle containing $R(m,m')$ with sides parallel to the corresponding sides of $R(\ell,\ell')$. An easy trigonometric argument shows that $\widetilde{R}(m,m')$ has the same center as $R(m,m')$ and has half-sides at most

$$40\delta^{\frac{1}{2}}\cos(\varphi_{\ell,\ell',m,m'}) + 80(1+|\ell-\ell'|)\delta\sin(\varphi_{\ell,\ell',m,m'}),$$
$$80(1+|\ell-\ell'|)\delta\cos(\varphi_{\ell,\ell',m,m'}) + 40\delta^{\frac{1}{2}}\sin(\varphi_{\ell,\ell',m,m'}),$$

in view of Exercise 5.2.3. Then $\widetilde{R}(m,m')$ has half-sides at most $60000\delta^{\frac{1}{2}}$ and $16000(1+|\ell-\ell'|)\delta$. If $\Gamma_\ell+\Gamma_{\ell'}$ and $\Gamma_m+\Gamma_{m'}$ intersect, then so do $\widetilde{R}(m,m')$ and $R(\ell,\ell')$, and both of these rectangles have sides parallel to the vectors $w(\ell,\ell')$ and $iw(\ell,\ell')$. But in the direction of $w(\ell,\ell')$, these rectangles have sides with half-lengths at most $80(1+|\ell-\ell'|)\delta$ and $16000(1+|m-m'|)\delta$.

Assume that the sets $R(m,m')$ and $R(\ell,\ell')$ intersect and $(\ell,\ell') \neq (m,m')$. Without loss of generality we may suppose that $|w(\ell,\ell')| \ge |w(m,m')|$. In this case the distance of the lines parallel to the direction $i\,w(\ell,\ell')$ and passing through the centers of the rectangles $\widetilde{R}(m,m')$ and $R(\ell,\ell')$ is at least

$$2\big|\cos(\pi|\ell-\ell'|\delta^{\frac{1}{2}}) - \cos(\pi|m-m'|\delta^{\frac{1}{2}})\big|,$$

as we easily see using (5.2.29). Since these rectangles intersect, we must have

$$2\big|\cos(\pi|\ell-\ell'|\delta^{\frac{1}{2}}) - \cos(\pi|m-m'|\delta^{\frac{1}{2}})\big| \le 16080(2+|\ell-\ell'|+|m-m'|)\delta.$$

We conclude that if the sets $R(m,m')$ and $R(\ell,\ell')$ intersect and $(\ell,\ell') \neq (m,m')$, then

$$\big|\cos(\pi|\ell-\ell'|\delta^{\frac{1}{2}}) - \cos(\pi|m-m'|\delta^{\frac{1}{2}})\big| \le 50000(|\ell-\ell'|+|m-m'|)\delta.$$

But the expression on the left is equal to

$$2\big|\sin(\pi\tfrac{|\ell-\ell'|-|m-m'|}{2}\delta^{\frac{1}{2}})\sin(\pi\tfrac{|\ell-\ell'|+|m-m'|}{2}\delta^{\frac{1}{2}})\big|,$$

which is at least

$$2\big||\ell-\ell'|-|m-m'|\big|\left(|\ell-\ell'|+|m-m'|\right)\delta$$

in view of the simple estimate $|\sin t| \geq \frac{2}{\pi}|t|$ for $|t| < \frac{\pi}{2}$. We conclude that if the sets $R(m,m')$ and $R(\ell,\ell')$ intersect and $(\ell,\ell') \neq (m,m')$, then

$$\big||\ell-\ell'|-|m-m'|\big| \leq 25000. \tag{5.2.31}$$

Combining (5.2.30) with (5.2.31), we see that if $\Gamma_m+\Gamma_{m'}$ and $\Gamma_\ell+\Gamma_{\ell'}$ intersect, then

$$\max\Big(\big|\min(m,m')-\min(\ell,\ell')\big|,\big|\max(m,m')-\max(\ell,\ell')\big|\Big) \leq \frac{25150}{2}.$$

We conclude that the set $\Gamma_m+\Gamma_{m'}$ intersects the fixed set $\Gamma_\ell+\Gamma_{\ell'}$ for at most $(25151)^2$ pairs (m,m'). This finishes the proof of the lemma. □

Exercises

5.2.1. For $\lambda \geq 0$ show that if $(1-|\xi|^2)_+^\lambda$ lies in $\mathscr{M}_p(\mathbf{R}^n)$, for some $p \in (1,\infty)$, then for all $f \in L^p(\mathbf{R}^n)$ the Bochner–Riesz means of f, $B_R^\lambda(f)$ converge to f in $L^p(\mathbf{R}^n)$.

5.2.2. Let $|\theta_1|,|\theta_2| < \frac{\pi}{4}$ be two angles. Show geometrically that

$$|r_1e^{i\theta_1}-r_2e^{i\theta_2}| \geq \min(r_1,r_2)\sin|\theta_1-\theta_2|$$

and use the estimate $|\sin t| \geq \frac{2|t|}{\pi}$ for $|t| < \frac{\pi}{2}$ to obtain a lower bound for the second expression in terms of $|\theta_1-\theta_2|$.

5.2.3. Let R be a rectangle in $\mathbf{R}^2$ having length $b>0$ along a direction $\vec{v}=(\xi_1,\xi_2)$ and length $a>0$ along the perpendicular direction $\vec{v}^\perp=(-\xi_2,\xi_1)$. Let $\vec{w}$ be another vector that forms an angle $\varphi < \frac{\pi}{2}$ with $\vec{v}$. Show that the smallest rectangle R' that contains R and has sides parallel to $\vec{w}$ and $\vec{w}^\perp$ has side lengths $a\sin(\varphi)+b\cos(\varphi)$ along the direction $\vec{w}$ and $a\cos(\varphi)+b\sin(\varphi)$ along the direction $\vec{w}^\perp$.

5.2.4. Prove that Theorem 5.2.7 does not hold when $p<2$.
[*Hint:* Try the intervals $I_j=[j,j+1]$ and $\widehat{f}=\chi_{[0,N]}$ as $N\to\infty$.]

5.2.5. Let $\{I_k\}_k$ be a family of intervals in the real line with $|I_k|=|I_{k'}|$ and $I_k\cap I_{k'}=\emptyset$ for all $k\neq k'$. Define the sets

$$S_k=\big\{(\xi_1,\dots,\xi_n)\in\mathbf{R}^n:\ \xi_1\in I_k\big\}.$$

Prove that for all $p\geq 2$ and all $f\in L^p(\mathbf{R}^n)$, we have

$$\Big\|\Big(\sum_k|(\widehat{f}\chi_{S_k})^\vee|^2\Big)^{\frac{1}{2}}\Big\|_{L^p(\mathbf{R}^n)} \leq C_p\|f\|_{L^p(\mathbf{R}^n)},$$

where C_p is the constant of Theorem 5.2.7.

5.2.6. Let $\{I_k\}_k$, $\{J_\ell\}_\ell$ be two families of intervals in the real line with $|I_k| = |I_{k'}|$, $I_k \cap I_{k'} = \emptyset$ for all $k \neq k'$, and $|J_\ell| = |J_{\ell'}|$, $J_\ell \cap J_{\ell'} = \emptyset$ for all ℓ, ℓ'. Prove that for all $p \geq 2$ there is a constant C_p such that

$$\Big\| \Big(\sum_k \sum_\ell |(\widehat{f}\chi_{I_k \times J_\ell})^\vee|^2 \Big)^{\frac{1}{2}} \Big\|_{L^p(\mathbf{R}^2)} \leq C_p \big\| f \big\|_{L^p(\mathbf{R}^2)},$$

for all $f \in L^p(\mathbf{R}^2)$.
[*Hint:* Use the double Rademacher functions (Appendix C.5 in [156]) and apply Theorem 5.2.7 twice.]

5.2.7. ([307]) On $\mathbf{R}^n$ consider the points $x_\ell = \ell\sqrt{\delta}$, $\ell \in \mathbf{Z}^n$. Fix a Schwartz function h whose Fourier transform is supported in the unit ball in $\mathbf{R}^n$. Given a function f on $\mathbf{R}^n$, define

$$\widehat{f_\ell}(\xi) = \widehat{f}(\xi)\widehat{h}(\delta^{-\frac{1}{2}}(\xi - x_\ell)).$$

Prove that for some constant C (which depends only on h and n) the estimate

$$\Big(\sum_{\ell \in \mathbf{Z}^n} |f_\ell|^2 \Big)^{\frac{1}{2}} \leq CM(|f|^2)^{\frac{1}{2}}$$

holds for all functions f. Deduce the $L^p(\mathbf{R}^n)$ boundedness of the preceding square function for all $p > 2$.
[*Hint:* For a sequence λ_ℓ with $\sum_\ell |\lambda_\ell|^2 = 1$, set

$$G(f)(x) = \sum_{\ell \in \mathbf{Z}^n} \lambda_\ell f_\ell(x) = \int_{\mathbf{R}^n} \Big[\sum_{\ell \in \mathbf{Z}^n} \lambda_\ell e^{2\pi i \frac{x_\ell \cdot y}{\sqrt{\delta}}} \Big] f(x - \tfrac{y}{\sqrt{\delta}}) h(y)\, dy.$$

Split $\mathbf{R}^n$ as the union of $Q_0 = [-\frac{1}{2}, \frac{1}{2}]^n$ and $2^{j+1}Q_0 \setminus 2^j Q_0$ for $j \geq 0$ and control the integral on each such set using the decay of h and the $L^2(2^{j+1}Q_0)$ norms of the other two terms. Finally, exploit the orthogonality of the functions $e^{2\pi i \ell \cdot y}$ to estimate the $L^2(2^{j+1}Q_0)$ norm of the expression inside the square brackets by $C2^{nj/2}$. Sum over $j \geq 0$ to obtain the required conclusion.]

5.2.8. For $\lambda > 0$ define the *maximal Bochner–Riesz operator*

$$B_*^\lambda(f)(x) = \sup_{R>0} \Big| \int_{\mathbf{R}^n} (1 - |\xi/R|^2)_+^\lambda \widehat{f}(\xi) e^{2\pi i x \cdot \xi}\, d\xi \Big|.$$

Prove that B_*^λ maps $L^p(\mathbf{R}^n)$ to itself when $\lambda > \frac{n-1}{2}$ for $1 \leq p \leq \infty$.
[*Hint:* You may want to use Corollary 2.1.12 in [156].]

5.3 Kakeya Maximal Operators

We recall the Hardy–Littlewood maximal operator with respect to cubes on $\mathbf{R}^n$ defined as

$$M_c(f)(x) = \sup_{\substack{Q \in \mathscr{F} \\ Q \ni x}} \frac{1}{|Q|} \int_Q |f(y)|\, dy, \tag{5.3.1}$$

where $\mathscr{F}$ is the set of all closed cubes in $\mathbf{R}^n$ (with sides not necessarily parallel to the axes). The operator M_c is equivalent (bounded above and below by constants) to the corresponding maximal operator M_c' in which the family $\mathscr{F}$ is replaced by the more restrictive family $\mathscr{F}'$ of cubes in $\mathbf{R}^n$ with sides parallel to the coordinate axes.

It is interesting to observe that if the family of all cubes $\mathscr{F}$ in (5.3.1) is replaced by the family of all rectangles (or parallelepipeds) $\mathscr{R}$ in $\mathbf{R}^n$, then we obtain an operator M_0 that is unbounded on $L^p(\mathbf{R}^n)$; see Exercise 2.1.9 in [156]. If we substitute the family of all parallelepipeds $\mathscr{R}$, however, with the more restrictive family $\mathscr{R}'$ of all parallelepipeds with sides parallel to the coordinate axes, then we obtain the so-called strong maximal function

$$M_s(f)(x) = \sup_{\substack{R \in \mathscr{R}' \\ R \ni x}} \frac{1}{|R|} \int_R |f(y)|\, dy. \tag{5.3.2}$$

The operator M_s is bounded on $L^p(\mathbf{R}^n)$ for $1 < p < \infty$ but it is not of weak type $(1,1)$. See Exercise 5.3.1.

These examples indicate that averaging over long and skinny rectangles is quite different than averaging over squares. In general, the direction and the dimensions of the averaging rectangles play a significant role in the boundedness properties of the maximal functions. In this section we investigate aspects of this topic.

5.3.1 Maximal Functions Associated with a Set of Directions

Definition 5.3.1. Let Σ be a set of unit vectors in $\mathbf{R}^2$, i.e., a subset of the unit circle $\mathbf{S}^1$. Associated with Σ, we define $\mathscr{R}_\Sigma$ to be the set of all closed rectangles in $\mathbf{R}^2$ whose longest side is parallel to some vector in Σ. We also define a maximal operator $\mathfrak{M}_\Sigma$ associated with Σ as follows:

$$\mathfrak{M}_\Sigma(f)(x) = \sup_{\substack{R \in \mathscr{R}_\Sigma \\ R \ni x}} \frac{1}{|R|} \int_R |f(y)|\, dy,$$

where f is a locally integrable function on $\mathbf{R}^2$.

We also recall the definition given in (5.2.21) of the *Kakeya maximal operator*

$$\mathscr{K}_N(w)(x) = \sup_{R \ni x} \frac{1}{|R|} \int_R |w(y)|\, dy, \tag{5.3.3}$$

where the supremum is taken over all rectangles R in $\mathbf{R}^2$ of dimensions a and aN where $a > 0$ is arbitrary. Here N is a fixed real number that is at least 10.

Example 5.3.2. Let $\Sigma = \{v\}$ consist of only one vector $v = (a,b)$. Then

$$\mathfrak{M}_\Sigma(f)(x) = \sup_{0<r\le 1}\sup_{N>0} \frac{1}{rN^2}\int_{-N}^{N}\int_{-rN}^{rN} |f(x-t(a,b)-s(-b,a))|\,ds\,dt\,.$$

If $\Sigma = \{(1,0),(0,1)\}$ consists of the two unit vectors along the axes, then

$$\mathfrak{M}_\Sigma = M_s\,,$$

where M_s is the strong maximal function defined in (5.3.2).

It is obvious that for each $\Sigma \subseteq \mathbf{S}^1$, the maximal function $\mathfrak{M}_\Sigma$ maps $L^\infty(\mathbf{R}^2)$ to itself with constant 1. But $\mathfrak{M}_\Sigma$ may not always be of weak type $(1,1)$, as the example M_s indicates; see Exercise 5.3.1. The boundedness of $\mathfrak{M}_\Sigma$ on $L^p(\mathbf{R}^2)$ in general depends on the set Σ.

An interesting case arises in the following example as well.

Example 5.3.3. For $N \in \mathbf{Z}^+$, let

$$\Sigma = \Sigma_N = \big\{\big(\cos(\tfrac{2\pi j}{N}),\sin(\tfrac{2\pi j}{N})\big)\,:\, j = 0,1,2,\dots,N-1\big\}$$

be the set of N uniformly spread directions on the circle. Then we expect $\mathfrak{M}_{\Sigma_N}$ to be L^p bounded with constant depending on N. There is a connection between the operator $\mathfrak{M}_{\Sigma_N}$ previously defined and the Kakeya maximal operator $\mathscr{K}_N$ defined in (5.2.21). In fact, Exercise 5.3.3 says that

$$\mathscr{K}_N(f) \le 20\,\mathfrak{M}_{\Sigma_N}(f) \tag{5.3.4}$$

for all locally integrable functions f on $\mathbf{R}^2$.

We now indicate why the norms of $\mathscr{K}_N$ and $\mathfrak{M}_{\Sigma_N}$ on $L^2(\mathbf{R}^2)$ grow as $N \to \infty$. We refer to Exercises 5.3.4 and 5.3.7 for the corresponding result for $p \neq 2$.

Proposition 5.3.4. *There is a constant c such that for any $N \ge 10$ we have*

$$\big\|\mathscr{K}_N\big\|_{L^2(\mathbf{R}^2)\to L^2(\mathbf{R}^2)} \ge c\,\log N \tag{5.3.5}$$

and

$$\big\|\mathscr{K}_N\big\|_{L^2(\mathbf{R}^2)\to L^{2,\infty}(\mathbf{R}^2)} \ge c\,(\log N)^{\frac{1}{2}}\,. \tag{5.3.6}$$

Therefore, a similar conclusion follows for $\mathfrak{M}_{\Sigma_N}$.

Proof. We consider the family of functions $f_N(x) = \frac{1}{|x|}\chi_{3\le|x|\le N}$ defined on $\mathbf{R}^2$ for $N \ge 10$. Then we have

$$\big\|f_N\big\|_{L^2(\mathbf{R}^2)} \le c_1(\log N)^{\frac{1}{2}}\,. \tag{5.3.7}$$

On the other hand, for every x in the annulus $6 < |x| < N$, we consider the rectangle R_x of dimensions $|x|-3$ and $\frac{|x|-3}{N}$, one of whose shorter sides touches the circle $|y| = 3$ and the other has midpoint x. Then

$$\mathscr{K}_N(f_N)(x) \geq \frac{1}{|R_x|}\int_{R_x} |f_N(y)|\,dy \geq \frac{c_2 N}{(|x|-3)^2} \iint\limits_{\substack{3\leq y_1\leq |x| \\ |y_2|\leq \frac{|x|-3}{2N}}} \frac{dy_1 dy_2}{y_1} \geq c_3 \frac{\log|x|}{|x|}.$$

It follows that

$$\big\|\mathscr{K}_N(f_N)\big\|_{L^2(\mathbf{R}^2)} \geq c_3 \bigg(\int\limits_{6\leq|x|\leq N} \Big(\frac{\log|x|}{|x|}\Big)^2 dx\bigg)^{\frac{1}{2}} \geq c_4 (\log N)^{\frac{3}{2}}. \tag{5.3.8}$$

Combining (5.3.7) with (5.3.8) we obtain (5.3.5) with $c = c_4/c_1$.

We now turn to estimate (5.3.6). Since for all $6 < |x| < N$ we have

$$\mathscr{K}_N(f_N)(x) \geq c_3 \frac{\log|x|}{|x|} > c_3 \frac{\log N}{N},$$

it follows that $\big|\{\mathscr{K}_N(f_N) > c_3 \frac{\log N}{N}\}\big| \geq \pi(N^2-6^2) \geq c_5 N^2$ and hence

$$\begin{aligned}
\frac{\big\|\mathscr{K}_N(f_N)\big\|_{L^{2,\infty}}}{\big\|f_N\big\|_{L^2}} &\geq \frac{\sup\limits_{\lambda>0} \lambda \big|\{\mathscr{K}_N(f_N) > \lambda\}\big|^{\frac{1}{2}}}{c_1 (\log N)^{\frac{1}{2}}} \\
&\geq c_3 \frac{\log N}{N} \frac{\big|\{\mathscr{K}_N(f_N) > c_3 \frac{\log N}{N}\}\big|^{\frac{1}{2}}}{c_1 (\log N)^{\frac{1}{2}}} \\
&\geq \frac{c_3\sqrt{c_5}}{c_1} (\log N)^{\frac{1}{2}}.
\end{aligned}$$

This completes the proof. □

5.3.2 *The Boundedness of $\mathfrak{M}_{\Sigma_N}$ on $L^p(\mathbf{R}^2)$*

It is rather remarkable that both estimates of Proposition 5.3.4 are sharp in terms of their behavior as $N \to \infty$, as the following result indicates.

Theorem 5.3.5. *There exist constants $0 < B, C < \infty$ such that for every $N \geq 1000$ and all $f \in L^2(\mathbf{R}^2)$ we have*

$$\big\|\mathfrak{M}_{\Sigma_N}(f)\big\|_{L^{2,\infty}(\mathbf{R}^2)} \leq B (\log N)^{\frac{1}{2}} \big\|f\big\|_{L^2(\mathbf{R}^2)} \tag{5.3.9}$$

and

$$\left\|\mathfrak{M}_{\Sigma_N}(f)\right\|_{L^2(\mathbf{R}^2)} \le C(\log N)\left\|f\right\|_{L^2(\mathbf{R}^2)}. \tag{5.3.10}$$

In view of (5.3.4), similar estimates also hold for $\mathscr{K}_N$.

Proof. We deduce (5.3.10) from the weak type estimate (5.3.9), which we rewrite as

$$\left|\left\{x \in \mathbf{R}^2 : \mathfrak{M}_{\Sigma_N}(f)(x) > \lambda\right\}\right| \le B^2(\log N)\frac{\|f\|_{L^2}^2}{\lambda^2}. \tag{5.3.11}$$

We prove this estimate for some constant $B > 0$ independent of N. But prior to doing this we indicate why (5.3.11) implies (5.3.10).

Using Exercise 5.3.2, we have that $\mathfrak{M}_{\Sigma_N}$ maps $L^p(\mathbf{R}^2)$ to $L^p(\mathbf{R}^2)$ (and hence into $L^{p,\infty}$) with constant at most a multiple of $N^{1/p}$ for all $1 < p < \infty$. Using this with $p = 3/2$, we have

$$\left\|\mathfrak{M}_{\Sigma_N}\right\|_{L^{\frac{3}{2}} \to L^{\frac{3}{2},\infty}} \le \left\|\mathfrak{M}_{\Sigma_N}\right\|_{L^{\frac{3}{2}} \to L^{\frac{3}{2}}} \le A N^{\frac{2}{3}} \tag{5.3.12}$$

for some constant $A > 0$. Now split f as the sum $f = f_1 + f_2 + f_3$, where

$$\begin{aligned} f_1 &= f\chi_{|f| \le \frac{1}{4}\lambda}, \\ f_2 &= f\chi_{\frac{1}{4}\lambda < |f| \le N^2\lambda}, \\ f_3 &= f\chi_{N^2\lambda < |f|}. \end{aligned}$$

It follows that

$$\left|\left\{\mathfrak{M}_{\Sigma_N}(f) > \lambda\right\}\right| \le \left|\left\{\mathfrak{M}_{\Sigma_N}(f_2) > \tfrac{\lambda}{3}\right\}\right| + \left|\left\{\mathfrak{M}_{\Sigma_N}(f_3) > \tfrac{\lambda}{3}\right\}\right|, \tag{5.3.13}$$

since the set $\left\{\mathfrak{M}_{\Sigma_N}(f_1) > \frac{\lambda}{3}\right\}$ is empty. To obtain the required result we use the $L^{2,\infty}$ estimate (5.3.11) for f_2 and the $L^{\frac{3}{2},\infty}$ estimate (5.3.12) for f_3. We have

$$\begin{aligned} &\left\|\mathfrak{M}_{\Sigma_N}(f)\right\|_{L^2}^2 \\ &= 2\int_0^\infty \lambda\left|\left\{\mathfrak{M}_{\Sigma_N}(f) > \lambda\right\}\right| d\lambda \\ &\le \int_0^\infty 2\lambda\left|\left\{\mathfrak{M}_{\Sigma_N}(f_2) > \tfrac{\lambda}{3}\right\}\right| d\lambda + \int_0^\infty 2\lambda\left|\left\{\mathfrak{M}_{\Sigma_N}(f_3) > \tfrac{\lambda}{3}\right\}\right| d\lambda \\ &\le \int_0^\infty \frac{2\lambda B^2(\log N)}{\lambda^2}\int_{\frac{1}{4}\lambda < |f| \le N^2\lambda} |f|^2\,dx\,d\lambda + \int_0^\infty \frac{2\lambda A^{\frac{3}{2}} N}{\lambda^{\frac{3}{2}}}\int_{|f| > N^2\lambda} |f|^{\frac{3}{2}}\,dx\,d\lambda \\ &\le 2B^2(\log N)\int_{\mathbf{R}^2}|f(x)|^2\int_{\frac{|f(x)|}{N^2}}^{4|f(x)|}\frac{d\lambda}{\lambda}\,dx + 2A^{\frac{3}{2}}N\int_{\mathbf{R}^2}|f(x)|^{\frac{3}{2}}\int_0^{\frac{|f(x)|}{N^2}}\frac{d\lambda}{\lambda^{\frac{1}{2}}}\,dx \\ &= \left(4B^2(\log 2N)(\log N) + 4A^{\frac{3}{2}}\right)\|f\|_{L^2}^2 \\ &\le C(\log N)^2\|f\|_{L^2}^2 \end{aligned}$$

using Fubini's theorem for integrals. This proves (5.3.10).

To avoid problems with antipodal points, it is convenient to split Σ_N as the union of sixty four sets, in each of which the angle between any two vectors does not exceed $2\pi/64$. It suffices therefore to obtain (5.3.11) for each such subset of Σ_N. By rotational invariance, it will suffice to work with only one of these 64 subsets. We choose to work with Σ_N^1 the part of Σ_N contained in $\{e^{i\theta} : 0 \le \theta < \pi/32\}$. To prove (5.3.11), we fix a $\lambda > 0$ and we start with a compact subset K of the set $\{x \in \mathbf{R}^2 : \mathfrak{M}_{\Sigma_N^1}(f)(x) > \lambda\}$. Then, for every $x \in K$, there exists an open rectangle R_x that contains x and whose longest side is parallel to a vector in Σ_N^1. By compactness of K, there exists a finite subfamily $\{R_\alpha\}_{\alpha\in\mathscr{A}}$ of the family $\{R_x\}_{x\in K}$ such that

$$\int_{R_\alpha} |f(y)|\,dy > \lambda\,|R_\alpha|$$

for all $\alpha \in \mathscr{A}$ and such that the union of the R_α's covers K.

We claim that there is a constant C such that for any finite family $\{R_\alpha\}_{\alpha\in\mathscr{A}}$ of rectangles whose longest side is parallel to a vector in Σ_N^1 there is a subset $\mathscr{B}$ of $\mathscr{A}$ such that

$$\int_{\mathbf{R}^2} \Big(\sum_{\beta\in\mathscr{B}} \chi_{R_\beta}(x)\Big)^2 dx \le 3 \sum_{\beta\in\mathscr{B}} |R_\beta| \tag{5.3.14}$$

and that

$$\Big|\bigcup_{\alpha\in\mathscr{A}} R_\alpha\Big| \le C(\log N) \sum_{\beta\in\mathscr{B}} |R_\beta|\,. \tag{5.3.15}$$

Assuming (5.3.14) and (5.3.15), we easily deduce (5.3.11). Indeed,

$$\begin{aligned}
\sum_{\beta\in\mathscr{B}} |R_\beta| &< \frac{1}{\lambda} \sum_{\beta\in\mathscr{B}} \int_{R_\beta} |f(y)|\,dy \\
&= \frac{1}{\lambda} \int_{\mathbf{R}^2} \Big(\sum_{\beta\in\mathscr{B}} \chi_{R_\beta}(y)\Big) |f(y)|\,dy \\
&\le \frac{1}{\lambda} \Big(\int_{\mathbf{R}^2} \Big(\sum_{\beta\in\mathscr{B}} \chi_{R_\beta}(y)\Big)^2 dy\Big)^{\frac{1}{2}} \|f\|_{L^2} \\
&\le \frac{1}{\lambda} \Big[3 \sum_{\beta\in\mathscr{B}} |R_\beta|\Big]^{\frac{1}{2}} \|f\|_{L^2}\,,
\end{aligned}$$

from which it follows that

$$\sum_{\beta\in\mathscr{B}} |R_\beta| \le \frac{3}{\lambda^2} \|f\|_{L^2}^2\,.$$

Then, using (5.3.15), we obtain

$$|K| \le \Big|\bigcup_{\alpha\in\mathscr{A}} R_\alpha\Big| \le C(\log N) \sum_{\beta\in\mathscr{B}} |R_\beta| \le \frac{3C}{\lambda^2} (\log N) \|f\|_{L^2}^2\,,$$

and since K was an arbitrary compact subset of $\{x : \mathfrak{M}_{\Sigma_N^1}(f)(x) > \lambda\}$, the same estimate is valid for the latter set.

We now turn to the selection of the subfamily $\{R_\beta\}_{\beta\in\mathscr{B}}$ and the proof of (5.3.14) and (5.3.15).

Let R_{β_1} be the rectangle in $\{R_\alpha\}_{\alpha\in\mathscr{A}}$ with the longest side. Suppose we have chosen $R_{\beta_1}, R_{\beta_2}, \dots, R_{\beta_{j-1}}$ for some $j \geq 2$. Then among all rectangles R_α that satisfy

$$\sum_{k=1}^{j-1} |R_{\beta_k} \cap R_\alpha| \leq \frac{1}{2}|R_\alpha|, \tag{5.3.16}$$

we choose a rectangle R_{β_j} such that its longer side is as large as possible. Since the collection $\{R_\alpha\}_{\alpha\in\mathscr{A}}$ is finite, this selection stops after m steps. Define

$$\mathscr{B} = \{\beta_1, \beta_2, \dots, \beta_m\}.$$

Using (5.3.16), we obtain

$$\begin{aligned}
\int_{\mathbf{R}^2} \Big(\sum_{\beta\in\mathscr{B}} \chi_{R_\beta}\Big)^2 dx &\leq 2\sum_{j=1}^{m}\sum_{k=1}^{j} |R_{\beta_k} \cap R_{\beta_j}| \\
&= 2\sum_{j=1}^{m}\Big[\Big(\sum_{k=1}^{j-1} |R_{\beta_k} \cap R_{\beta_j}|\Big) + |R_{\beta_j}|\Big] \\
&\leq 2\sum_{j=1}^{m}\Big[\frac{1}{2}|R_{\beta_j}| + |R_{\beta_j}|\Big] \\
&= 3\sum_{j=1}^{m} |R_{\beta_j}|.
\end{aligned} \tag{5.3.17}$$

which implies inequality (5.3.14).

We now turn to the proof of (5.3.15). Let M_c be the usual Hardy–Littlewood maximal operator with squares in $\mathbf{R}^2$; recall $n = 2$. Since M_c is of weak type $(1,1)$, (5.3.15) is a consequence of the estimate

$$\bigcup_{\alpha\in\mathscr{A}\setminus\mathscr{B}} R_\alpha \subseteq \Big\{x \in \mathbf{R}^2 : M_c\Big(\sum_{\beta\in\mathscr{B}} \chi_{(R_\beta)^*}\Big)(x) > c(\log N)^{-1}\Big\} \tag{5.3.18}$$

for some absolute constant c, where $(R_\beta)^*$ is the rectangle R_β expanded 30 times in both directions. Indeed, if (5.3.18) holds, then

$$\begin{aligned}
\Big|\bigcup_{\alpha\in\mathscr{A}} R_\alpha\Big| &\leq \Big|\bigcup_{\beta\in\mathscr{B}} R_\beta\Big| + \Big|\bigcup_{\alpha\in\mathscr{A}\setminus\mathscr{B}} R_\alpha\Big| \\
&\leq \sum_{\beta\in\mathscr{B}} |R_\beta| + \frac{10}{c}(\log N) \sum_{\beta\in\mathscr{B}} |(R_\beta)^*|
\end{aligned}$$

$$
\begin{aligned}
&= \sum_{\beta\in\mathscr{B}} |R_\beta| + \frac{9000}{c}(\log N) \sum_{\beta\in\mathscr{B}} |R_\beta| \\
&\le C\,(\log N) \sum_{\beta\in\mathscr{B}} |R_\beta|,
\end{aligned}
$$

since $N \ge 1000$.

It remains to prove (5.3.18). At this point we need the following lemma. In the sequel we denote by θ_α the angle between the x axis and the vector pointing in the longer direction of R_α for any $\alpha \in \mathscr{A}$. We also denote by l_α the shorter side of R_α and by L_α the longer side of R_α for any $\alpha \in \mathscr{A}$.

Lemma 5.3.6. *Let $N \in \mathbf{Z}^+$ satisfy $N \ge 1000$. Let $\omega_0 = 0$ and define $\omega_k = 2\pi 2^k/N$ for $1 \le k < [\log_2(N/64)]$ and $\omega_{[\log_2(N/64)]} = \pi/32$. Let R_α be a rectangle in the family $\{R_\alpha\}_{\alpha\in\mathscr{A}}$. Suppose that $\beta \in \mathscr{B}$ is such that $L_\beta \ge L_\alpha$ and such that*

$$
\omega_k \le |\theta_\alpha - \theta_\beta| < \omega_{k+1}
$$

for some $0 \le k < [\log_2(N/64)]$. Let

$$
s_\alpha = 16\max(l_\alpha, \omega_k L_\alpha).
$$

For an arbitrary $x \in R_\alpha$, let Q be a square centered at x with sides of length s_α parallel to the sides of R_α. Then we have

$$
\frac{|R_\beta \cap R_\alpha|}{|R_\alpha|} \le 64\,\frac{|(R_\beta)^* \cap Q|}{|Q|}. \tag{5.3.19}
$$

Assuming Lemma 5.3.6, we conclude the proof of (5.3.18). Fix $\alpha \in \mathscr{A} \setminus \mathscr{B}$. Then the rectangle R_α was not selected in the selection procedure. This means that for all $l \in \{2, \dots, m+1\}$ we have exactly one of the following: either

$$
\sum_{j=1}^{l-1} |R_{\beta_j} \cap R_\alpha| > \frac{1}{2}|R_\alpha| \tag{5.3.20}
$$

or

$$
\sum_{j=1}^{l-1} |R_{\beta_j} \cap R_\alpha| \le \frac{1}{2}|R_\alpha| \qquad \text{and} \qquad L_\alpha \le L_{\beta_l}. \tag{5.3.21}
$$

If (5.3.21) holds for $l = 2$, we let $\mu \le m$ be the largest integer such that (5.3.21) holds for all $l \le \mu$. Then (5.3.21) fails for $l = \mu + 1$; hence (5.3.20) holds for $l = \mu + 1$; thus

$$
\frac{1}{2}|R_\alpha| < \sum_{j=1}^{\mu} |R_{\beta_j} \cap R_\alpha| \le \sum_{\substack{\beta\in\mathscr{B}\\ L_\beta \ge L_\alpha}} |R_\beta \cap R_\alpha|. \tag{5.3.22}
$$

If (5.3.21) fails for $l=2$, then (5.3.20) holds for $l=2$, and this implies that

$$\frac{1}{2}|R_\alpha| < |R_{\beta_1}\cap R_\alpha| \le \sum_{\substack{\beta\in\mathscr{B}\\ L_\beta\ge L_\alpha}} |R_\beta \cap R_\alpha|.$$

In either case we have

$$\frac{1}{2}|R_\alpha| < \sum_{\substack{\beta\in\mathscr{B}\\ L_\beta\ge L_\alpha}} |R_\beta \cap R_\alpha|,$$

and from this it follows that there exists a k with $0\le k<\left[\frac{\log(N/64)}{\log 2}\right]$ such that

$$\frac{\log 2}{2\log(N/64)}|R_\alpha| < \sum_{\substack{\beta\in\mathscr{B}\\ L_\beta\ge L_\alpha\\ \omega_k\le|\theta_\beta-\theta_\alpha|<\omega_{k+1}}} |R_\beta\cap R_\alpha|. \tag{5.3.23}$$

By Lemma 5.3.6, for any $x\in R_\alpha$ there is a square Q such that (5.3.19) holds for any R_β with $\beta\in\mathscr{B}$ satisfying $L_\beta\ge L_\alpha$ and $\omega_k\le|\theta_\beta-\theta_\alpha|<\omega_{k+1}$. It follows that

$$\frac{\log 2}{2\log(N/64)} < 64 \sum_{\substack{\beta\in\mathscr{B}\\ L_\beta\ge L_\alpha\\ \omega_k\le|\theta_\beta-\theta_\alpha|<\omega_{k+1}}} \frac{|(R_\beta)^*\cap Q|}{|Q|},$$

which implies for some $c>0$

$$\frac{c}{\log N} < \frac{\log 2}{128\log(N/64)} < \frac{1}{|Q|}\int_Q \sum_{\beta\in\mathscr{B}} \chi_{(R_\beta)^*}\,dx.$$

This proves (5.3.18), since for $\alpha\in\mathscr{A}\setminus\mathscr{B}$, any $x\in R_\alpha$ must be an element of the set $\{x\in\mathbf{R}^2:\ M_c\big(\sum_{\beta\in\mathscr{B}}\chi_{(R_\beta)^*}\big)(x)>c(\log N)^{-1}\}$. □

It remains to prove Lemma 5.3.6.

Proof. We fix R_α and R_β so that $L_\beta\ge L_\alpha$ and we assume that $\overline{R_\beta}$ intersects $\overline{R_\alpha}$; otherwise, (5.3.19) is obvious. Let τ be the angle between the directions of the rectangles R_α and R_β, that is,

$$\tau = |\theta_\alpha-\theta_\beta|.$$

By assumption we have $\tau<\omega_{k+1}\le\frac{\pi}{32}$ for all integers $k<\left[\frac{\log(N/64)}{\log 2}\right]$.

Let R_β^∞ denote the smallest closed infinite strip in the direction of the longer side of R_β that contains it. We make the following observation: if

$$\tan\tau \le \frac{\frac{1}{2}s_\alpha - l_\alpha}{\frac{1}{2}s_\alpha + L_\alpha}, \tag{5.3.24}$$

then the strip R_β^∞ intersects the upper side (according to Figure 5.9) of the square Q. Indeed, the worst possible case is drawn in Figure 5.9, in which equality holds in (5.3.24). For $\tau \le \pi/32$ we have $\tan\tau < 3\tau/2$, and since $\tau < 2\omega_k$, it follows that $\tan\tau < 3\omega_k$. Our choice of s_α implies

$$s_\alpha \ge 12\,\omega_k L_\alpha + 4l_\alpha \implies 3\omega_k \le \frac{\frac{1}{2}s_\alpha - l_\alpha}{\frac{1}{2}s_\alpha + L_\alpha};$$

hence (5.3.24) holds.

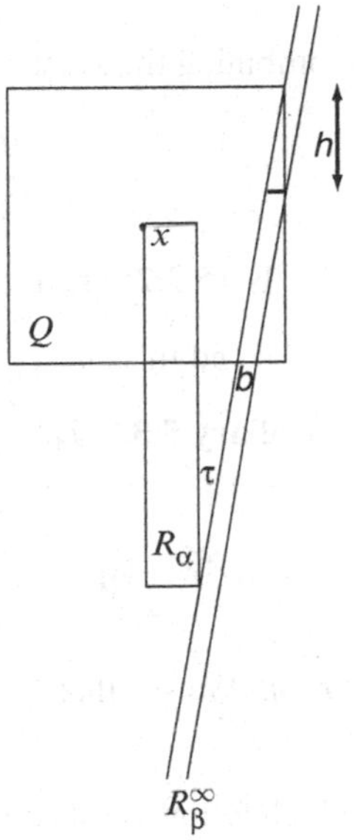

Fig. 5.9 For angles τ less than that displayed, the strip R_β^∞ meets the upper side of Q. The length of the intersection of R_β^∞ with the lower side of Q is denoted by b.

We have now proved that R_β^∞ meets the upper side of Q. We examine the size of the intersection $R_\beta^\infty \cap Q$. According to the picture in Figure 5.9, this intersection contains a parallelogram of base $b = l_\beta / \cos\tau$ and height $s_\alpha - h$ and a right triangle with base b and height h (with $0 \le h \le s_\alpha$). Then we have

$$\frac{|R_\beta^\infty \cap Q|}{|Q|} \ge \frac{1}{s_\alpha^2}\frac{l_\beta}{\cos\tau}\Big(s_\alpha - h + \frac{1}{2}h\Big) \ge \frac{1}{s_\alpha^2}\frac{l_\beta}{\cos\tau}\Big(\frac{1}{2}s_\alpha\Big) \ge \frac{1}{2}\frac{l_\beta}{s_\alpha}.$$

Since $(R_\beta)^*$ has length $30L_\beta$ and R_β meets R_α, we have that $R_\beta^\infty \cap Q \subseteq (R_\beta)^* \cap Q$ and therefore

$$\frac{|(R_\beta)^* \cap Q|}{|Q|} \ge \frac{1}{2}\frac{l_\beta}{s_\alpha}. \tag{5.3.25}$$

On the other hand, let $R_{\alpha,\beta}$ be the smallest parallelogram two of whose opposite sides are parallel to the shorter sides of R_α and whose remaining two sides are contained in the boundary lines of R_β^∞. Then

$$|R_\alpha \cap R_\beta| \le |R_{\alpha,\beta}| \le \frac{l_\beta}{\cos\tau} L_\alpha \le 2 l_\beta L_\alpha .$$

Another geometric argument shows that

$$|R_\alpha \cap R_\beta| \le l_\beta \frac{l_\alpha}{\sin(\tau)} \le l_\alpha l_\beta \frac{\pi}{2\tau} \le l_\alpha l_\beta \frac{\pi}{2\omega_k} \le 2\frac{l_\alpha l_\beta}{\omega_k} .$$

Combining these estimates, we deduce

$$\frac{|R_\alpha \cap R_\beta|}{|R_\alpha|} \le 2\min\left(\frac{l_\beta}{l_\alpha}, \frac{l_\beta}{\omega_k L_\alpha}\right) \le 32\frac{l_\beta}{s_\alpha} . \tag{5.3.26}$$

Finally, (5.3.25) and (5.3.26) yield (5.3.19). □

We end this subsection with an immediate corollary of the theorem just proved.

Corollary 5.3.7. *For every* $1 < p < \infty$ *there exists a constant* c_p *such that*

$$\big\|\mathscr{K}_N\big\|_{L^p(\mathbf{R}^2)\to L^p(\mathbf{R}^2)} \le c_p \begin{cases} N^{\frac{2}{p}-1}(\log N)^{\frac{1}{p'}} & \text{when } 1 < p < 2, \\ (\log N)^{\frac{1}{p}} & \text{when } 2 < p < \infty. \end{cases} \tag{5.3.27}$$

Proof. We see that

$$\big\|\mathscr{K}_N\big\|_{L^1(\mathbf{R}^2)\to L^{1,\infty}(\mathbf{R}^2)} \le CN , \tag{5.3.28}$$

which follows replacing a rectangle of dimensions $a \times aN$ by a square of side length aN that contains it. Interpolating between (5.3.9) and (5.3.28), we obtain the first statement in (5.3.27). The second statement in (5.3.27) follows by interpolation between (5.3.9) and the trivial $L^\infty \to L^\infty$ estimate. (In both cases we use Theorem 1.3.2 in [156].) □

5.3.3 The Higher-Dimensional Kakeya Maximal Operator

The Kakeya maximal operator without dilations $\mathscr{K}_N^a$ on $L^2(\mathbf{R}^2)$ was crucial in the study of the boundedness of the Bochner–Riesz operator B^λ on $L^4(\mathbf{R}^2)$. An analogous maximal operator could be introduced on $\mathbf{R}^n$.

Definition 5.3.8. Given fixed $a > 0$ and $N \ge 10$, we introduce the *Kakeya maximal operator without dilations* on $\mathbf{R}^n$ as

$$\mathscr{K}_N^a(f)(x) = \sup_R \frac{1}{|R|}\int_R |f(y)|\,dy ,$$

where the supremum is taken over all rectangular parallelepipeds (boxes) of arbitrary orientation in $\mathbf{R}^n$ that contain the point x and have dimensions

$$\underbrace{a\times a\times\cdots\times a}_{n-1\text{ times}}\times aN\,.$$

We also define the centered version $\mathfrak{K}_N^a$ of $\mathscr{K}_N^a$ as follows:

$$\mathfrak{K}_N^a(f)(x)=\sup_R\frac{1}{|R|}\int_R|f(y)|\,dy,$$

where the supremum is restricted to those rectangles among the previous ones that are centered at x. These two maximal operators are comparable, and we have

$$\mathfrak{K}_N^a\le\mathscr{K}_N^a\le 2^n\mathfrak{K}_N^a$$

by a simple geometric argument.

We also define the higher-dimensional analogue of the Kakeya maximal operator $\mathscr{K}_N$ introduced in (5.3.3).

Definition 5.3.9. Let $N\ge 10$. We denote by $\mathscr{R}(N)$ the set of all rectangular parallelepipeds (boxes) in $\mathbf{R}^n$ with arbitrary orientation and dimensions

$$\underbrace{a\times a\times\cdots\times a}_{n-1\text{ times}}\times aN$$

with arbitrary $a>0$. Given a locally integrable function f on $\mathbf{R}^n$, we define

$$\mathscr{K}_N(f)(x)=\sup_{\substack{R\in\mathscr{R}(N)\\ R\ni x}}\frac{1}{|R|}\int_R|f(y)|\,dy$$

and

$$\mathfrak{K}_N(f)(x)=\sup_{\substack{R\in\mathscr{R}(N)\\ R\text{ has center }x}}\frac{1}{|R|}\int_R|f(y)|\,dy;$$

$\mathfrak{K}_N$ and $\mathscr{K}_N$ are called the centered and uncentered *nth-dimensional Kakeya maximal operators*, respectively.

For convenience we call rectangular parallelepipeds, i.e., elements of $\mathscr{R}(N)$, higher-dimensional rectangles, or simply rectangles. We clearly have

$$\sup_{a>0}\mathscr{K}_N^a=\mathscr{K}_N\qquad\text{and}\qquad\sup_{a>0}\mathfrak{K}_N^a=\mathfrak{K}_N\,;$$

hence the boundedness of $\mathscr{K}_N^a$ can be deduced from that of $\mathscr{K}_N$; however, this deduction can essentially be reversed with only logarithmic loss in N (see the references at the end of this chapter). In the sequel we restrict attention to the operator

$\mathscr{K}_N^a$, whose study already presents all the essential difficulties and requires a novel set of ideas in its analysis. We consider a specific value of a, since a simple dilation argument yields that the norms of $\mathscr{K}_N^a$ and $\mathscr{K}_N^b$ on a fixed $L^p(\mathbf{R}^n)$ are equal for all $a,b>0$.

Concerning $\mathscr{K}_N^1$, we know that

$$\big\|\mathscr{K}_N^1\big\|_{L^1(\mathbf{R}^n)\to L^{1,\infty}(\mathbf{R}^n)}\le c_n N^{n-1}. \tag{5.3.29}$$

This estimate follows by replacing a rectangle of dimensions $\overbrace{1\times 1\times\cdots\times 1}^{n-1\text{ times}}\times N$ by the smallest cube of side length N that contains it. This estimate is sharp; see Exercise 5.3.7.

It would be desirable to know the following estimate for $\mathscr{K}_N^1$:

$$\big\|\mathscr{K}_N^1\big\|_{L^n(\mathbf{R}^n)\to L^{n,\infty}(\mathbf{R}^n)}\le c_n'(\log N)^{\frac{n-1}{n}} \tag{5.3.30}$$

for some dimensional constant c_n'. It would then follow that

$$\big\|\mathscr{K}_N^1\big\|_{L^n(\mathbf{R}^n)\to L^n(\mathbf{R}^n)}\le c_n''\log N \tag{5.3.31}$$

for some other dimensional constant c_n''; see Exercise 5.3.8(b). Moreover, if estimate (5.3.30) were true, then interpolating between (5.3.29) and (5.3.30) would yield the bound

$$\big\|\mathscr{K}_N^1\big\|_{L^p(\mathbf{R}^n)\to L^p(\mathbf{R}^n)}\le c_{n,p}N^{\frac{n}{p}-1}(\log N)^{\frac{1}{p'}}, \qquad 1<p<n. \tag{5.3.32}$$

It is estimate (5.3.32) that we would like to concentrate on. We have the following result for a certain range of p's in the interval $(1,n)$.

Theorem 5.3.10. *Let $p_n=\frac{n+1}{2}$ and $N\ge 10$. Then there exists a constant C_n such that*

$$\big\|\mathscr{K}_N^1\big\|_{L^{p_n,1}(\mathbf{R}^n)\to L^{p_n,\infty}(\mathbf{R}^n)}\le C_n N^{\frac{n}{p_n}-1}, \tag{5.3.33}$$

$$\big\|\mathscr{K}_N^1\big\|_{L^{p_n}(\mathbf{R}^n)\to L^{p_n,\infty}(\mathbf{R}^n)}\le C_n N^{\frac{n}{p_n}-1}(\log N)^{\frac{1}{p_n}}, \tag{5.3.34}$$

$$\big\|\mathscr{K}_N^1\big\|_{L^{p_n}(\mathbf{R}^n)\to L^{p_n}(\mathbf{R}^n)}\le C_n N^{\frac{n}{p_n}-1}(\log N). \tag{5.3.35}$$

Moreover, for every $1<p<p_n$, there exists a constant $C_{n,p}$ such that

$$\big\|\mathscr{K}_N^1\big\|_{L^p(\mathbf{R}^n)\to L^p(\mathbf{R}^n)}\le C_{n,p}N^{\frac{n}{p}-1}(\log N)^{\frac{1}{p'}}. \tag{5.3.36}$$

Proof. We begin by observing that (5.3.36) is a consequence of (5.3.29) and (5.3.34) using Theorem 1.3.2 in [156]. We also observe that (5.3.35) is a consequence of (5.3.34), while (5.3.34) is a consequence of (5.3.33) (see Exercise 5.3.8). We therefore concentrate on estimate (5.3.33).

We choose to work with the centered version $\mathfrak{K}_N^1$ of $\mathscr{K}_N^1$, which is comparable to it. To make the geometric idea of the proof a bit more transparent, we pick $\delta < 1/10$, we set $N = 1/\delta$, and we work with the equivalent operator $\mathfrak{K}_{1/\delta}^\delta$, whose norm is the same as that of $\mathscr{K}_N^1$. Since the operators in question are positive, we work with nonnegative functions.

The proof is based on a linearization of the operator $\mathscr{K}_{1/\delta}^\delta$. Let us call a rectangle of dimensions $\delta \times \delta \times \cdots \times \delta \times 1$ a *δ-tube*. We call the line segment parallel to the longest edges that joins the centers of its two smallest faces, a δ-tube's *axis of symmetry*.

For every x in $\mathbf{R}^n$ we select (in some measurable way) a δ-tube $\tau(x)$ that contains x such that

$$\frac{1}{2}\mathscr{K}_{1/\delta}^\delta(f)(x) \le \frac{1}{|\tau(x)|}\int_{\tau(x)} f(y)\,dy\,.$$

Suppose we have a grid of cubes in $\mathbf{R}^n$ each of side length $\delta' = \delta/(2\sqrt{n})$, and let Q_j be a cube in that grid with center c_{Q_j}. Then any δ-tube centered at a point $z \in Q_j$ must contain the entire Q_j, and it follows that

$$\mathfrak{K}_{1/\delta}^\delta(f)(z) \le \mathscr{K}_{1/\delta}^\delta(f)(c_{Q_j}) \le \frac{2}{|\tau(c_{Q_j})|}\int_{\tau(c_{Q_j})} f(y)\,dy\,. \tag{5.3.37}$$

This observation motivates the introduction of a grid of width $\delta' = \delta/(2\sqrt{n})$ in $\mathbf{R}^n$ so that for every cube Q_j in the grid there is an associated δ-tube τ_j satisfying

$$\tau_j \cap Q_j \neq \emptyset.$$

Then we define a linear operator

$$L^\delta(f) = \sum_j \left(\frac{1}{|\tau_j|}\int_{\tau_j} f(y)\,dy\right)\chi_{Q_j}\,,$$

which certainly satisfies

$$L^\delta(f) \le 2^n \mathscr{K}_{1/\delta}^{2\delta}(f) \le 4^n \mathfrak{K}_{1/\delta}^{2\delta}(f)\,,$$

and in view of (5.3.37), it also satisfies

$$\mathfrak{K}_{1/\delta}^\delta(f) \le 2L^\delta(f)\,.$$

It suffices to show that L^δ is bounded from $L^{p_n,1}$ to $L^{p_n,\infty}$ with constant $C_n(\delta^{-1})^{\frac{n}{p_n}-1}$, which is independent of the choice of δ-tubes τ_j.

Our next reduction is to take f to be the characteristic function of a set. The space $L^{p_n,\infty}$ is normable, i.e., it has an equivalent norm under which it is a Banach space (Exercise 1.1.12 in [156]). Then the boundedness of L^δ from $L^{p_n,1}$ to $L^{p_n,\infty}$ is a consequence of the restricted weak type estimate

$$\sup_{\lambda>0} \lambda \big|\{L^\delta(\chi_A) > \lambda\}\big|^{\frac{1}{p_n}} \le C_n'(\delta^{-1})^{\frac{n}{p_n}-1}|A|^{\frac{1}{p_n}}\,, \tag{5.3.38}$$

for some dimensional constant C_n and all sets A of finite measure (Exercise 1.4.7 in [156]). This estimate can be written as

$$\lambda^{\frac{n+1}{2}}\delta^{\frac{n-1}{2}}|E_\lambda| \le C_n|A|, \tag{5.3.39}$$

where

$$E_\lambda = \{x\in\mathbf{R}^n:\ L^\delta(\chi_A)(x)>\lambda\} = \{L^\delta(\chi_A)>\lambda\}.$$

Our final reduction stems from the observation that the operator L^δ is "local." This means that if f is supported in a cube Q, say of side length one, then $L^\delta(f)$ is supported in a fixed multiple of Q. Indeed, it is simple to verify that if $x\notin 10\sqrt{n}Q$ and f is supported in Q, then $L^\delta(f)(x)=0$, since no δ-tube containing x can reach Q. For "local" operators, it suffices to prove their boundedness for functions supported in cubes of side length one; see Exercise 5.4.4. We may therefore work with a measurable set A contained in a cube in $\mathbf{R}^n$ of side length one. This assumption has as a consequence that E_λ is contained in a fixed multiple of Q, such as $10\sqrt{n}\,Q$.

Having completed all the required reductions, we proceed by proving the restricted weak type estimate (5.3.39) for sets A supported in a cube of side length one. In proving (5.3.39) we may take $\lambda\le 1$; otherwise, the set E_λ is empty. We consider the cases $c_0(n)\,\delta\le\lambda$ and $c_0(n)\,\delta>\lambda$, for some large constant $c_0(n)$ to be determined later. If $c_0(n)\,\delta>\lambda$, then

$$|E_\lambda| \le C_n^1\,(1/\delta)^{n-1}\,\frac{|A|}{\lambda} \tag{5.3.40}$$

by the weak type $(1,1)$ boundedness of L^δ with constant $C_n^1\delta^{1-n}$. It follows from (5.3.40) that

$$C_n^1|A| \ge |E_\lambda|\delta^{n-1}\lambda > c_0(n)^{-\frac{n-1}{2}}\,|E_\lambda|\lambda^{\frac{n+1}{2}}\delta^{\frac{n-1}{2}},$$

which proves (5.3.39) in this case.

We now assume $c_0(n)\,\delta\le\lambda\le 1$. Since $L^\delta(\chi_A)$ is constant on each Q_j, we have that each Q_j is either entirely contained in the set E_λ or disjoint from it. Consequently, setting

$$\mathscr{E} = \{j:\ Q_j\subseteq E_\lambda\},$$

we have

$$E_\lambda = \bigcup_{j\in\mathscr{E}} Q_j.$$

Hence

$$|\mathscr{E}| = \#\{j:\ j\in\mathscr{E}\} = |E_\lambda|(\delta')^{-n},$$

and for all $j\in\mathscr{E}$ we have

$$|\tau_j\cap A| > \lambda|\tau_j| = \lambda\,\delta^{n-1}.$$

It follows that

$$\begin{aligned}|A| \sup_x \Big[\sum_{j\in\mathscr{E}} \chi_{\tau_j}(x) \Big] &\geq \int_A \sum_{j\in\mathscr{E}} \chi_{\tau_j}\, dx \\ &= \sum_{j\in\mathscr{E}} |\tau_j \cap A| \\ &> \lambda\, \delta^{n-1} |\mathscr{E}| \\ &= \lambda\, \delta^{n-1} \frac{|E_\lambda|}{(\delta')^n} \\ &= (2\sqrt{n})^n \frac{\lambda\, |E_\lambda|}{\delta}\,.\end{aligned}$$

Therefore, there exists an x_0 in A such that

$$\#\{ j \in \mathscr{E} : x_0 \in \tau_j \} > (2\sqrt{n})^n \frac{\lambda\, |E_\lambda|}{\delta |A|}\,.$$

Let $S(x_0, \frac{1}{2})$ be a sphere of radius $\frac{1}{2}$ centered at the point x_0. We find on this sphere a finite set of points $\Theta = \{\theta_k\}_k$ that is maximal with respect to the property that the balls $B(\theta_k, \delta)$ are at distance at least $10\sqrt{n}\,\delta$ from each other. Define spherical caps

$$S_k = S(x_0, \tfrac{1}{2}) \cap B(\theta_k, \delta)\,.$$

Since the S_k's are disjoint and have surface measure a constant multiple of δ^{n-1}, it follows that there are about δ^{1-n} such points θ_k.

We count the number of δ-tubes that contain x_0 and intersect a fixed cap S_k. All these δ-tubes are contained in a cylinder of length 3 and diameter $c_1(n)\delta$ whose axis of symmetry contains x_0 and the center of the cap S_k. This cylinder has volume $3v_{n-1}c_1(n)^{n-1}\delta^{n-1}$, and thus it intersects at most $c_2(n)\delta^{-1}$ cubes of the family Q_j, since the Q_j's are disjoint and all have volume equal to $(\delta')^n$. We deduce then that given such a cap S_k, there exist at most $c_3(n)\delta^{-1}$ δ-tubes (from the initial family) that contain the point x_0 and intersect S_k.

Let us call a set of δ-tubes *ε-separated* if for every τ and τ' in the set with $\tau \neq \tau'$ we have that the angle between the axis of symmetry of τ and τ' is at least $\varepsilon > 0$. Since we have at least $\frac{(2\sqrt{n})^n \lambda |E_\lambda|}{\delta |A|}$ δ-tubes that contain the given point x_0, and each cap S_k is intersected by at most $c_3(n)\delta^{-1}$ δ-tubes that contain x_0, it follows that at least $c_4(n)\frac{\lambda |E_\lambda|}{|A|}$ of these δ-tubes have to intersect different caps S_k. But δ-tubes that intersect different caps S_k and contain x_0 are δ-separated. We have therefore shown that there exist at least $c_4(n)\frac{\lambda |E_\lambda|}{|A|}$ δ-separated tubes from the original family that contain the point x_0. Call $\mathscr{T}$ the family of these δ-tubes.

We find a maximal subset Θ' of the θ_k's such that the balls $B(\theta_k, \delta)$, $\theta_k \in \Theta'$, have distance at least $\frac{30\sqrt{n}\,\delta}{\lambda}$ from each other. This is possible if $\lambda/\delta \geq c_0(n)$ for some large constant $c_0(n)$ [such as $c_0(n) = 1000\sqrt{n}$]. We "thin out" the family $\mathscr{T}$ by removing all the δ-tubes that intersect the caps S_k with $\theta_k \in \Theta \setminus \Theta'$. In other words, we essentially keep in $\mathscr{T}$ one out of every $1/\lambda^{n-1}$ δ-tubes. In this way we extract at least $c_5(n)\frac{\lambda^n |E_\lambda|}{|A|}$ δ-tubes from $\mathscr{T}$ that are $\frac{60\sqrt{n}\,\delta}{\lambda}$-separated and contain the point x_0. We denote these tubes by $\{\tau_j : j \in \mathscr{F}\}$.

We have therefore found a subset $\mathscr{F}$ of $\mathscr{E}$ such that

$$x_0 \in \tau_j \quad \text{for all} \quad j \in \mathscr{F}, \tag{5.3.41}$$

$$\tau_k, \tau_j \quad \text{are} \quad 60\sqrt{n}\frac{\delta}{\lambda}\text{ - separated} \quad \text{when} \quad j,k \in \mathscr{F},\ j \neq k, \tag{5.3.42}$$

$$|\mathscr{F}| \geq c_5(n)\frac{|E_\lambda|\lambda^n}{|A|}. \tag{5.3.43}$$

Notice that

$$\big|A \cap \tau_j \cap B(x_0, \tfrac{\lambda}{3})\big| \leq \big|\tau_j \cap B(x_0, \tfrac{\lambda}{3})\big| \leq \frac{2}{3}\lambda\,\delta^{n-1},$$

and since for any $j \in \mathscr{E}$ (and thus for $j \in \mathscr{F}$) we have $|A \cap \tau_j| > \lambda\,\delta^{n-1}$, it must be the case that

$$\big|A \cap \tau_j \cap B(x_0, \tfrac{\lambda}{3})^c\big| > \frac{1}{3}\lambda\,\delta^{n-1}. \tag{5.3.44}$$

Moreover, it is crucial to note that the sets

$$A \cap \tau_j \cap B(x_0, \tfrac{\lambda}{3})^c, \qquad j \in \mathscr{F}, \tag{5.3.45}$$

are pairwise disjoint. In fact, if x_j and x_k are points on the axes of symmetry of two $60\sqrt{n}\frac{\delta}{\lambda}$-separated δ-tubes τ_j and τ_k in $\mathscr{F}$ such that $|x_j - x_0| = |x_k - x_0| = \frac{\lambda}{3}$, then the distance from x_k to x_j must be at least $10\sqrt{n}\,\delta$. This implies that the distance between $\tau_j \cap B(x_0, \frac{\lambda}{3})^c$ and $\tau_k \cap B(x_0, \frac{\lambda}{3})^c$ is at least $6\sqrt{n}\,\delta > 0$. We now conclude the proof of the theorem as follows:

$$\begin{aligned}
|A| &\geq \Big|A \cap \bigcup_{j \in \mathscr{F}} \big(\tau_j \cap B(x_0, \tfrac{\lambda}{3})^c\big)\Big| \\
&= \sum_{j \in \mathscr{F}} \big|A \cap \tau_j \cap B(x_0, \tfrac{\lambda}{3})^c\big| \\
&\geq \sum_{j \in \mathscr{F}} \frac{\lambda\,\delta^{n-1}}{3} \\
&= |\mathscr{F}|\,\frac{\lambda\,\delta^{n-1}}{3} \\
&\geq c_5(n)\frac{|E_\lambda|\lambda^n}{|A|}\frac{\lambda\,\delta^{n-1}}{3},
\end{aligned}$$

using that the sets in (5.3.45) are disjoint, (5.3.44), and (5.3.43). We conclude that

$$|A|^2 \geq \frac{1}{3}c_5(n)\,\lambda^{n+1}\delta^{n-1}|E_\lambda| \geq c_6(n)\lambda^{n+1}\delta^{n-1}|E_\lambda|^2,$$

since, as observed earlier, the set E_λ is contained in a cube of side length 10. Taking square roots, we obtain (5.3.39). This proves (5.3.38) and hence (5.3.35). □

Exercises

5.3.1. Let h be the characteristic function of the square $[0,1]^2$ in $\mathbf{R}^2$. Prove that for any $0<\lambda<1$ we have

$$\big|\{x\in\mathbf{R}^2:\ M_s(h)(x)>\lambda\}\big|\ge\frac{1}{\lambda}\log\frac{1}{\lambda}.$$

Use this to show that M_s is not of weak type $(1,1)$.

5.3.2. (a) Given a unit vector v in $\mathbf{R}^2$ define the *directional maximal function along* $\vec{v}$ by

$$M_{\vec{v}}(f)(x)=\sup_{\varepsilon>0}\frac{1}{2\varepsilon}\int_{-\varepsilon}^{+\varepsilon}|f(x-t\vec{v})|\,dt$$

wherever f is locally integrable over $\mathbf{R}^2$. Prove that for such f, $M_{\vec{v}}(f)(x)$ is well defined for almost all x contained in any line not parallel to $\vec{v}$.
(b) For $1<p<\infty$, use the method of rotations to show that $M_{\vec{v}}$ maps $L^p(\mathbf{R}^2)$ to itself with norm the same as that of the centered Hardy–Littlewood maximal operator M on $L^p(\mathbf{R})$.
(c) Let Σ be a finite set of directions. Prove that for all $1<p\le\infty$, there is a constant $C_p>0$ such that

$$\big\|\mathfrak{M}_\Sigma(f)\big\|_{L^p(\mathbf{R}^2)}\le C_p\,|\Sigma|^{\frac{1}{p}}\,\big\|f\big\|_{L^p(\mathbf{R}^2)}$$

for all f in $L^p(\mathbf{R}^2)$.
[*Hint:* Use the inequality $\mathfrak{M}_\Sigma(f)^p\le\sum_{\vec{v}\in\Sigma}[M_{\vec{v}}M_{\vec{v}^\perp}(f)]^p$.]

5.3.3. Show that

$$\mathscr{K}_N\le 20\,\mathfrak{M}_{\Sigma_N},$$

where Σ_N is a set of N uniformly distributed vectors in $\mathbf{S}^1$.
[*Hint:* Use Exercise 5.2.3.]

5.3.4. This exercise indicates a connection between the Besicovitch construction in Section 5.1 and the Kakeya maximal function. Recall the set E of Lemma 5.1.1, which satisfies $\frac{1}{2}\le|E|\le\frac{3}{2}$.
(a) Show that there is a positive constant c such that for all $N\ge 10$ we have

$$\big|\{x\in\mathbf{R}^2:\ \mathscr{K}_N(\chi_E)(x)>\tfrac{1}{240}\}\big|\ge c\log\log N.$$

(b) Conclude that for all $2<p<\infty$ there is a constant c_p such that

$$\big\|\mathscr{K}_N\big\|_{L^p(\mathbf{R}^2)\to L^p(\mathbf{R}^2)}\ge c_p(\log\log N)^{\frac{1}{p}}.$$

[*Hint:* Using the notation of Lemma 5.1.1, if $6\log(k+2) < 2^{-k+1}N < 15\log(k+2)$, prove that

$$\left|\{x\in\mathbf{R}^2:\ \mathscr{K}_N(\chi_E)(x) > \tfrac{1}{240}\}\right| \geq \log(k+2),$$

by showing that the previous set contains all the disjoint rectangles R_j for $j = 1,2,\dots,2^k$; here k is a large positive integer. To show this, for x in $\bigcup_{j=1}^{2^k} R_j$ consider the unique rectangle R_{j_x} that contains x. Then $|R_{j_x}| = N(2^{-k+1})^2$, and we have

$$\frac{1}{|R_{j_x}|}\int_{R_{j_x}} |\chi_E(y)|\,dy \geq \frac{|E\cap R_{j_x}|}{|R_{j_x}|} \geq \frac{1}{240},$$

in view of conclusion (4) in Lemma 5.1.1. Part (b): Express the L^p norm of $\mathscr{K}_N(\chi_E)$ in terms of its distribution function.]

5.3.5. Show that $\mathfrak{M}_{\mathbf{S}^1}$ is unbounded on $L^p(\mathbf{R}^2)$ for any $p<\infty$.
[*Hint:* You may use Proposition 5.3.4 when $p\leq 2$. When $p>2$ one may need Exercise 5.3.4.]

5.3.6. Consider the n-dimensional Kakeya maximal operator $\mathscr{K}_N$. Show that there exist dimensional constants c_n and c_n' such that for N sufficiently large we have

$$\begin{aligned}\left\|\mathscr{K}_N\right\|_{L^n(\mathbf{R}^n)\to L^n(\mathbf{R}^n)} &\geq c_n\,(\log N),\\ \left\|\mathscr{K}_N\right\|_{L^n(\mathbf{R}^n)\to L^{n,\infty}(\mathbf{R}^n)} &\geq c_n'\,(\log N)^{\frac{n-1}{n}}.\end{aligned}$$

[*Hint:* Consider the functions $f_N(x) = \frac{1}{|x|}\chi_{3\leq|x|\leq N}$ and adapt the argument in Proposition 5.3.4 to an n-dimensional setting.]

5.3.7. For all $1\leq p<n$ show that there exist constants $c_{n,p}$ such that the n-dimensional Kakeya maximal operator $\mathscr{K}_N$ satisfies

$$\left\|\mathscr{K}_N\right\|_{L^p(\mathbf{R}^n)\to L^p(\mathbf{R}^n)} \geq \left\|\mathscr{K}_N\right\|_{L^p(\mathbf{R}^n)\to L^{p,\infty}(\mathbf{R}^n)} \geq c_{n,p}N^{\frac{n}{p}-1}.$$

[*Hint:* Consider the functions $h_N(x) = |x|^{-\frac{n+1}{p}}\chi_{3\leq|x|\leq N}$ and show that $\mathscr{K}_N(h_N)(x) > c/|x|$ for all x in the annulus $6<|x|<N$.]

5.3.8. ([65]) Let T be a sublinear operator defined on $L^1(\mathbf{R}^n)+L^\infty(\mathbf{R}^n)$ and taking values in a set of measurable functions. Let $10\leq N<\infty$, $1<p<\infty$, and $a,M>0$.
(a) Suppose that

$$\begin{aligned}\left\|T\right\|_{L^1\to L^{1,\infty}} &\leq C_1N^a,\\ \left\|T\right\|_{L^{p,1}\to L^{p,\infty}} &\leq M,\\ \left\|T\right\|_{L^\infty\to L^\infty} &\leq 1.\end{aligned}$$

Show that

$$\left\|T\right\|_{L^p\to L^{p,\infty}} \leq C(a,p,C_1)\,M\,(\log N)^{\frac{1}{p'}}.$$

(b) Suppose that

$$\begin{aligned}\big\|T\big\|_{L^1\to L^{1,\infty}} &\le C_1 N^a,\\ \big\|T\big\|_{L^p\to L^{p,\infty}} &\le M,\\ \big\|T\big\|_{L^\infty\to L^\infty} &\le 1.\end{aligned}$$

Show that

$$\big\|T\big\|_{L^p\to L^p} \le C'(a,p,C_1)\,M(\log N)^{\frac{1}{p}}.$$

[*Hint:* Part (a): Split $f = f_1 + f_2 + f_3$, where $f_3 = f\chi_{|f|\le\frac{\lambda}{4}}$, $f_2 = f\chi_{\frac{\lambda}{4}<|f|\le L\lambda}$, and $f_1 = f\chi_{|f|>L\lambda}$, where $L^{p-1} = N^a$. Use the weak type $(1,1)$ estimate for f_1 and the restricted weak type (p,p) estimate for f_2 and note that the measure of the set $\{|T(f_3)| > \lambda/3\}$ is zero. One needs the auxiliary result

$$\big\|f\chi_{a<|f|\le b}\big\|_{L^{p,1}} \le C(p)\,(1+\log\tfrac{b}{a})^{\frac{1}{p'}}\,\big\|f\big\|_{L^p},$$

which can be proved as follows. Use the identity of Proposition 1.4.9 in [156]. Note that $d_{f\chi_{a<|f|\le b}}(s) = d_f(a)$ when $s \le a$, $d_{f\chi_{a<|f|\le b}}(s) = d_f(s) - d_f(b)$ when $a < s \le b$, and $d_{f\chi_{a<|f|\le b}}(s) = 0$ when $s > b$. It follows that

$$\big\|f\chi_{a<|f|\le b}\big\|_{L^{p,1}} \le a\,d_f(a)^{\frac{1}{p}} + \int_a^b d_f(t)^{\frac{1}{p}}\,dt \le 2\int_{\frac{a}{2}}^a d_f(t)^{\frac{1}{p}}\,dt + \int_a^b d_f(t)^{\frac{1}{p}}\,dt,$$

from which the claimed estimate follows by Hölder's inequality. Part (b): Use the same splitting and the method employed in the proof of Theorem 5.3.5.]

5.4 Fourier Transform Restriction and Bochner–Riesz Means

If g is a continuous function on $\mathbf{R}^n$, its restriction to a hypersurface $S \subseteq \mathbf{R}^n$ is a well-defined function. By a hypersurface we mean a submanifold of $\mathbf{R}^n$ of dimension $n-1$. So, if f is an integrable function on $\mathbf{R}^n$, its Fourier transform $\widehat{f}$ is continuous and hence its restriction $\widehat{f}\big|_S$ on S is well defined.

Definition 5.4.1. Let $1 \le p,q \le \infty$. We say that a compact hypersurface S in $\mathbf{R}^n$ satisfies a (p,q) *restriction theorem* if the restriction operator

$$f \to \widehat{f}\big|_S,$$

which is initially defined on $L^1(\mathbf{R}^n) \cap L^p(\mathbf{R}^n)$, has an extension that maps $L^p(\mathbf{R}^n)$ boundedly into $L^q(S)$. The norm of this extension may depend on p,q,n, and S. If S satisfies a (p,q) restriction theorem, we write that property $R_{p\to q}(S)$ holds. We say that property $R_{p\to q}(S)$ holds with constant C if for all $f \in L^1(\mathbf{R}^n) \cap L^p(\mathbf{R}^n)$ we have

$$\big\|\widehat{f}\big\|_{L^q(S)} \le C\big\|f\big\|_{L^p(\mathbf{R}^n)}.$$

Example 5.4.2. Property $R_{1\to\infty}(S)$ holds for any compact hypersurface S.

We denote by $\mathscr{R}(f)=\widehat{f}\big|_S$ the restriction of the Fourier transform on a hypersurface S. Let $d\sigma$ be the canonically induced surface measure on S. Then for a function φ defined on S we have

$$\int_S \widehat{f}\,\varphi\,d\sigma=\int_{\mathbf{R}^n}\widehat{f}\,(\widehat{\varphi\,d\sigma})^\vee\,d\xi=\int_{\mathbf{R}^n} f\,\widehat{\varphi\,d\sigma}\,dx,$$

which says that the transpose of the linear operator $\mathscr{R}$ is the linear operator

$$\mathscr{R}^t(\varphi)=\widehat{\varphi\,d\sigma}. \tag{5.4.1}$$

By duality, we easily see that a (p,q) restriction theorem for a compact hypersurface S is equivalent to the following (q',p') *extension theorem* for S:

$$\mathscr{R}^t: L^{q'}(S)\to L^{p'}(\mathbf{R}^n).$$

Our objective is to determine all pairs of indices (p,q) for which the sphere $\mathbf{S}^{n-1}$ satisfies a (p,q) restriction theorem. It becomes apparent in this section that this problem is relevant in the understanding of the norm convergence of the Bochner–Riesz means.

5.4.1 Necessary Conditions for $R_{p\to q}(\mathbf{S}^{n-1})$ to Hold

We look at basic examples that impose restrictions on the indices p,q in order for $R_{p\to q}(\mathbf{S}^{n-1})$ to hold. We first make an observation. If $R_{p\to q}(\mathbf{S}^{n-1})$ holds, then $R_{p\to s}(\mathbf{S}^{n-1})$ for any $s\le q$.

Example 5.4.3. Let $d\sigma$ be surface measure on the unit sphere $\mathbf{S}^{n-1}$. Using the identity in Appendix B.4 in [156], we have

$$\widehat{d\sigma}(\xi)=\frac{2\pi}{|\xi|^{\frac{n-2}{2}}}J_{\frac{n-2}{2}}(2\pi|\xi|).$$

In view of the asymptotics in Appendix B.8 in [156], the last expression is equal to

$$\frac{2}{|\xi|^{\frac{n-1}{2}}}\cos(2\pi|\xi|-\tfrac{\pi(n-1)}{4})+O(|\xi|^{-\frac{n+1}{2}})$$

as $|\xi|\to\infty$. It follows that $\mathscr{R}^t(1)(\xi)=\widehat{d\sigma}(\xi)$ does not lie in $L^{p'}(\mathbf{R}^n)$ if $\frac{n-1}{2}p'\le n$ and $\frac{n+1}{2}p'>n$. Thus $R_{p\to q}(\mathbf{S}^{n-1})$ fails when $\frac{2n}{n+1}\le p<\frac{2n}{n-1}$. Since $R_{1\to q}(\mathbf{S}^{n-1})$ holds for all $q\in[1,\infty]$, by interpolation we deduce that $R_{p\to q}(\mathbf{S}^{n-1})$ fails when $p\ge\frac{2n}{n+1}$. We conclude that a necessary condition for $R_{p\to q}(\mathbf{S}^{n-1})$ to hold is that

$$1 \le p < \frac{2n}{n+1}. \tag{5.4.2}$$

In addition to this condition, there is another necessary condition for $R_{p\to q}(\mathbf{S}^{n-1})$ to hold. This is a consequence of the following revealing example.

Example 5.4.4. Let φ be a Schwartz function on $\mathbf{R}^n$ such that $\widehat{\varphi} \ge 0$ and $\widehat{\varphi}(\xi) \ge 1$ for all ξ in the closed ball $|\xi| \le 2$. For $N \ge 1$ define functions

$$f_N(x_1, x_2, \dots, x_{n-1}, x_n) = \varphi\Big(\frac{x_1}{N}, \frac{x_2}{N}, \dots, \frac{x_{n-1}}{N}, \frac{x_n}{N^2}\Big).$$

To test property $R_{p\to q}(\mathbf{S}^{n-1})$, instead of working with $\mathbf{S}^{n-1}$, we may work with the translated sphere $S = \mathbf{S}^{n-1} + (0,0,\dots,0,1)$ in $\mathbf{R}^n$ (cf. Exercise 5.4.2(a)). We have

$$\widehat{f_N}(\xi) = N^{n+1}\widehat{\varphi}(N\xi_1, N\xi_2, \dots, N\xi_{n-1}, N^2\xi_n).$$

We note that for all $\xi = (\xi_1, \dots, \xi_n)$ in the spherical cap

$$S' = S \cap \{\xi \in \mathbf{R}^n : \ \xi_1^2 + \dots + \xi_{n-1}^2 \le N^{-2} \quad \text{and} \quad \xi_n < 1\}, \tag{5.4.3}$$

we have $\xi_n \le 1 - (1 - \frac{1}{N^2})^{\frac{1}{2}} \le \frac{1}{N^2}$ and therefore

$$|(N\xi_1, N\xi_2, \dots, N\xi_{n-1}, N^2\xi_n)| \le 2.$$

This implies that for all ξ in S' we have $\widehat{f_N}(\xi) \ge N^{n+1}$. But the spherical cap S' in (5.4.3) has surface measure $c(N^{-1})^{n-1}$. We obtain

$$\big\|\widehat{f_N}\big\|_{L^q(S)} \ge \big\|\widehat{f_N}\big\|_{L^q(S')} \ge c^{\frac{1}{q}} N^{n+1} N^{\frac{1-n}{q}}.$$

On the other hand, $\|f_N\|_{L^p(\mathbf{R}^n)} = \|\varphi\|_{L^p(\mathbf{R}^n)} N^{\frac{n+1}{p}}$. Therefore, if $R_{p\to q}(\mathbf{S}^{n-1})$ holds, we must have

$$\|\varphi\|_{L^p(\mathbf{R}^n)} N^{\frac{n+1}{p}} \ge C c^{\frac{1}{q}} N^{n+1} N^{\frac{1-n}{q}},$$

and letting $N \to \infty$, we obtain the following necessary condition on p and q for $R_{p\to q}(\mathbf{S}^{n-1})$ to hold:

$$\frac{1}{q} \ge \frac{n+1}{n-1}\frac{1}{p'}. \tag{5.4.4}$$

We have seen that the restriction property $R_{p\to q}(\mathbf{S}^{n-1})$ fails in the shaded region of Figure 5.10 but obviously holds on the closed line segment CD. It remains to investigate the validity of property $R_{p\to q}(\mathbf{S}^{n-1})$ for $(\frac{1}{p}, \frac{1}{q})$ in the unshaded region of Figure 5.10.

It is a natural question to ask whether the restriction property $R_{p\to q}(\mathbf{S}^{n-1})$ holds on the line segment BD minus the point B in Figure 5.10, i.e., the set

$$\Big\{(p,q) : \ \frac{1}{q} = \frac{n+1}{n-1}\frac{1}{p'} \qquad 1 \le p < \frac{2n}{n+1}\Big\}. \tag{5.4.5}$$

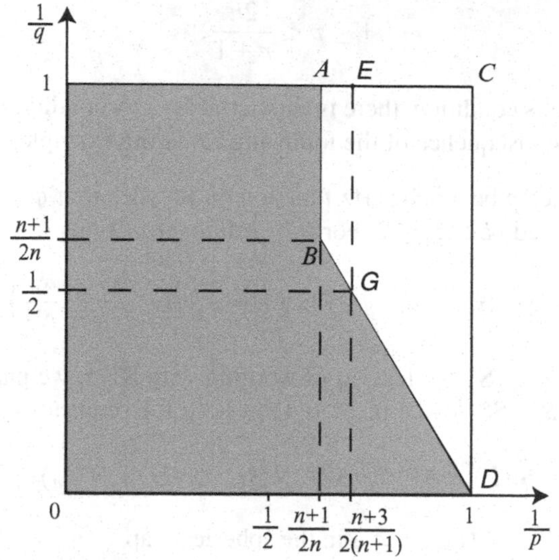

Fig. 5.10 The restriction property $R_{p\to q}(\mathbf{S}^{n-1})$ fails in the shaded region and on the closed line segment AB but holds on the closed line segment CD and could hold on the open line segment BD and inside the unshaded region.

If property $R_{p\to q}(\mathbf{S}^{n-1})$ holds for all points in this set, then it will also hold in the closure of the quadrilateral $ABDC$ minus the closed segment AB.

5.4.2 A Restriction Theorem for the Fourier Transform

In this subsection we establish the following restriction theorem for the Fourier transform.

Theorem 5.4.5. *Property $R_{p\to q}(\mathbf{S}^{n-1})$ holds for the set*

$$\left\{(p,q): \frac{1}{q} = \frac{n+1}{n-1}\frac{1}{p'}, \qquad 1 \le p \le \frac{2(n+1)}{n+3}\right\} \tag{5.4.6}$$

and therefore for the closure of the quadrilateral with vertices E, G, D, and C in Figure 5.10.

Proof. The case $p=1$ and $q=\infty$ is trivial. Therefore, we need to establish only the case $p=\frac{2(n+1)}{n+3}$ and $q=2$, since the remaining cases follow by interpolation and by the fact that the sphere has finite measure.

Using Plancherel's identity and Hölder's inequality, we obtain

$$\begin{aligned}\|\widehat{f}\|_{L^2(\mathbf{S}^{n-1})}^2 &= \int_{\mathbf{S}^{n-1}} \overline{\widehat{f}(\xi)}\,\widehat{f}(\xi)\,d\sigma(\xi)\\ &= \int_{\mathbf{R}^n} \overline{f(x)}\,(f*d\sigma^\vee)(x)\,dx\\ &\le \|f\|_{L^p(\mathbf{R}^n)}\|f*d\sigma^\vee\|_{L^{p'}(\mathbf{R}^n)}.\end{aligned}$$

To establish the required conclusion it is enough to show that

$$\|f*d\sigma^\vee\|_{L^{p'}(\mathbf{R}^n)} \le C_n\|f\|_{L^p(\mathbf{R}^n)} \qquad \text{when} \quad p=\frac{2(n+1)}{n+3}. \tag{5.4.7}$$

To obtain this estimate we need to split the sphere into pieces. Each hyperplane $\xi_k=0$ cuts the sphere $\mathbf{S}^{n-1}$ into two hemispheres, which we denote by H_k^1 and H_k^2. We introduce a finite smooth partition of unity $\{\varphi_j\}_j$ of $\mathbf{R}^n$ with the property that for any j there exist $k\in\{1,2,\dots,n\}$ and $l\in\{1,2\}$ such that

$$(\text{support}\,\varphi_j)\cap\mathbf{S}^{n-1}\subsetneqq H_k^l;$$

that is, the support of each φ_j intersected with the sphere $\mathbf{S}^{n-1}$ is properly contained in some hemisphere H_k^l. Then the family of all φ_j whose support meets $\mathbf{S}^{n-1}$ forms a finite partition of unity of the sphere when restricted to it. We therefore write

$$d\sigma=\sum_{j\in F}\varphi_j\,d\sigma,$$

where F is a finite set. If we obtain (5.4.7) for each measure $\varphi_j d\sigma$ instead of $d\sigma$, then (5.4.7) follows by summing on j. We fix such a measure $\varphi_j\,d\sigma$, which, without loss of generality, we assume is supported in $\{\xi\in\mathbf{S}^{n-1}:\ \xi_n>c\}\subsetneqq H_n^1$ for some $c\in(0,1)$. In the sequel we write elements $x\in\mathbf{R}^n$ as $x=(x',t)$, where $x'\in\mathbf{R}^{n-1}$ and $t\in\mathbf{R}$. Then for $x\in\mathbf{R}^n$ we have

$$(\varphi_j\,d\sigma)^\vee(x)=\int_{\mathbf{S}^{n-1}}\varphi_j(\xi)e^{2\pi i x\cdot\xi}\,d\sigma(\xi)=\int\limits_{\substack{\xi'\in\mathbf{R}^{n-1}\\ |\xi'|^2\le 1-c^2}} e^{2\pi i x\cdot\xi}\,\frac{\varphi_j(\xi',\sqrt{1-|\xi'|^2})\,d\xi'}{\sqrt{1-|\xi'|^2}},$$

where $\xi=(\xi',\xi_n)$; for the last identity we refer to Appendix D.5 in [156]. Writing $x=(x',t)\in\mathbf{R}^{n-1}\times\mathbf{R}$, we have

$$\begin{aligned}(\varphi_j\,d\sigma)^\vee(x',t) &= \int\limits_{\substack{\xi'\in\mathbf{R}^{n-1}\\ |\xi'|^2\le 1-c^2}} e^{2\pi i x'\cdot\xi'}\,e^{2\pi i t\sqrt{1-|\xi'|^2}}\,\frac{\varphi_j(\xi',\sqrt{1-|\xi'|^2})}{\sqrt{1-|\xi'|^2}}\,d\xi'\\ &=\left(e^{2\pi i t\sqrt{1-|\xi'|^2}}\,\frac{\varphi_j(\xi',\sqrt{1-|\xi'|^2})}{\sqrt{1-|\xi'|^2}}\right)^{\triangledown}(x'),\end{aligned} \tag{5.4.8}$$

where $^{\triangledown}$ indicates the inverse Fourier transform in the ξ' variable. For each $t \in \mathbf{R}$ we introduce a function on $\mathbf{R}^{n-1}$ by setting

$$K_t(x') = (\varphi_j d\sigma)^{\vee}(x',t).$$

We observe that identity (5.4.8) and the fact that $1 - |\xi'|^2 \ge c^2 > 0$ on the support of φ_j imply that

$$\sup_{t\in\mathbf{R}} \sup_{\xi'\in\mathbf{R}^{n-1}} |(K_t)^{\triangle}(\xi')| \le C_n < \infty, \tag{5.4.9}$$

where $^{\triangle}$ indicates the Fourier transform on $\mathbf{R}^{n-1}$. We also have that

$$K_t(x') = (\varphi_j d\sigma)^{\vee}(x',t) = \big(\varphi_j^{\vee} * d\sigma^{\vee}\big)(x',t).$$

Since $\varphi_j^{\vee}$ is a Schwartz function on $\mathbf{R}^n$ and

$$|d\sigma^{\vee}(x',t)| \le C(1+|(x',t)|)^{-\frac{n-1}{2}}$$

(see Appendices B.4, B.6, and B.7 in [156]), it follows that

$$|K_t(x')| \le C(1+|(x',t)|)^{-\frac{n-1}{2}} \le C(1+|t|)^{-\frac{n-1}{2}} \tag{5.4.10}$$

for all $x' \in \mathbf{R}^{n-1}$ (cf. Exercise 2.2.4 in [156]). Estimate (5.4.9) says that the operator given by convolution with K_t maps $L^2(\mathbf{R}^{n-1})$ to itself with norm at most a constant, while (5.4.10) says that the same operator maps $L^1(\mathbf{R}^{n-1})$ to $L^\infty(\mathbf{R}^{n-1})$ with norm at most a constant multiple of $(1+|t|)^{-\frac{n-1}{2}}$. Interpolating between these two estimates yields

$$\big\|K_t \star g\big\|_{L^{p'}(\mathbf{R}^{n-1})} \le C_{p,n}|t|^{-(n-1)(\frac{1}{p}-\frac{1}{2})}\big\|g\big\|_{L^p(\mathbf{R}^{n-1})}$$

for all $1 \le p \le 2$, where $\star$ denotes convolution on $\mathbf{R}^{n-1}$ (and $*$ convolution on $\mathbf{R}^n$).

We now return to the proof of the required estimate (5.4.7) in which $d\sigma^{\vee}$ is replaced by $(\varphi_j d\sigma)^{\vee}$. Let $f(x) = f(x',t)$ be a function on $\mathbf{R}^n$. We have

$$\begin{aligned}
\big\|f * (\varphi_j d\sigma)^{\vee}\big\|_{L^{p'}(\mathbf{R}^n)} &= \bigg\|\Big\|\int_{\mathbf{R}} f(\,\cdot\,,\tau) \star K_{t-\tau}\, d\tau\Big\|_{L^{p'}(\mathbf{R}^{n-1})}\bigg\|_{L^{p'}(\mathbf{R})} \\
&\le \bigg\|\int_{\mathbf{R}} \big\|f(\,\cdot\,,\tau) \star K_{t-\tau}\big\|_{L^{p'}(\mathbf{R}^{n-1})}\, d\tau\bigg\|_{L^{p'}(\mathbf{R})} \\
&\le C_{p,n}\bigg\|\int_{\mathbf{R}} \frac{\big\|f(\,\cdot\,,\tau)\big\|_{L^p(\mathbf{R}^{n-1})}}{|t-\tau|^{(n-1)(\frac{1}{p}-\frac{1}{2})}}\, d\tau\bigg\|_{L^{p'}(\mathbf{R})} \\
&= C_{p,n}\Big\|I_\beta\Big(\big\|f(\,\cdot\,,t)\big\|_{L^p(\mathbf{R}^{n-1})}\Big)\Big\|_{L^{p'}(\mathbf{R},dt)},
\end{aligned}$$

where

$$\beta = 1-(n-1)(\frac{1}{p}-\frac{1}{2})$$

and I_β is the Riesz potential (or fractional integral) given in Definition 1.2.1. Using Theorem 1.2.3 with $s=\beta$, $n=1$, and $q=p'$, we obtain that the last displayed equation is bounded by a constant multiple of

$$\Big\| \big\| f(\cdot,t)\big\|_{L^p(\mathbf{R}^{n-1})}\Big\|_{L^p(\mathbf{R},dt)} = \big\| f\big\|_{L^p(\mathbf{R}^n)}\,.$$

The condition $\frac{1}{p}-\frac{1}{q}=\frac{s}{n}$ on the indices p,q,s,n assumed in Theorem 1.2.3 translates exactly to

$$\frac{1}{p}-\frac{1}{p'}=\frac{\beta}{1}=1-\frac{n-1}{p}-\frac{n-1}{2}\,,$$

which is equivalent to $p=\frac{2(n+1)}{n+3}$. This concludes the proof of estimate (5.4.7) in which the measure $\sigma^\vee$ is replaced by $(\varphi_j\,d\sigma)^\vee$. Estimates for the remaining $(\varphi_j\,d\sigma)^\vee$ follow by a similar argument in which the role of the last coordinate is played by some other coordinate. The final estimate (5.4.7) follows by summing j over the finite set F. The proof of the theorem is now complete. □

5.4.3 Applications to Bochner–Riesz Multipliers

We now apply the restriction theorem obtained in the previous subsection to the Bochner–Riesz problem. In this subsection we prove the following result.

Theorem 5.4.6. *For* $\operatorname{Re}\lambda > \frac{n-1}{2(n+1)}$, *the Bochner–Riesz operator* B^λ *is bounded on* $L^p(\mathbf{R}^n)$ *for* p *in the optimal range*

$$\frac{2n}{n+1+2\operatorname{Re}\lambda} < p < \frac{2n}{n-1-2\operatorname{Re}\lambda}\,.$$

Proof. The proof is based on the following two estimates:

$$\big\|B^\lambda\big\|_{L^1(\mathbf{R}^n)\to L^1(\mathbf{R}^n)} \le C_1(\operatorname{Re}\lambda)\, e^{c_0|\operatorname{Im}\lambda|^2} \qquad \text{when } \operatorname{Re}\lambda > \tfrac{n-1}{2}\,, \tag{5.4.11}$$

$$\big\|B^\lambda\big\|_{L^p(\mathbf{R}^n)\to L^p(\mathbf{R}^n)} \le C_2(\operatorname{Re}\lambda)\, e^{c_0|\operatorname{Im}\lambda|^2} \qquad \text{when } \operatorname{Re}\lambda > \tfrac{n-1}{2(n+1)}\,, \tag{5.4.12}$$

where $p=\frac{2(n+1)}{n+3}$ and C_1, C_2 are constants that depend on n and $\operatorname{Re}\lambda$, while c_0 is an absolute constant. Once (5.4.11) and (5.4.12) are known, the required conclusion is a consequence of Theorem 1.3.7 in [156]. Recall that B^λ is given by convolution with the kernel K_λ defined in (5.2.1). This kernel satisfies

$$|K_\lambda(x)| \le C_3(\operatorname{Re}\lambda)\, e^{c_0|\operatorname{Im}\lambda|^2}(1+|x|)^{-\frac{n+1}{2}-\operatorname{Re}\lambda} \tag{5.4.13}$$

in view of the estimates in Appendices B.6 and B.7 in [156]. Then (5.4.11) follows easily from (5.4.13) and we focus our attention on (5.4.12).

The key ingredient in the proof of (5.4.12) is a decomposition of the kernel. But first we isolate the smooth part of the multiplier near the origin and we focus attention on the part of it near the boundary of the unit disk. Precisely, we start with a Schwartz function $0 \le \eta \le 1$ supported in the ball $B(0, \frac{3}{4})$ that is equal to 1 on the smaller ball $B(0, \frac{1}{2})$. Then we write

$$m_\lambda(\xi) = (1 - |\xi|^2)_+^\lambda = (1 - |\xi|^2)_+^\lambda \eta(\xi) + (1 - |\xi|^2)_+^\lambda (1 - \eta(\xi)).$$

Since the function $(1 - |\xi|^2)_+^\lambda \eta(\xi)$ is smooth and compactly supported, it is an L^p Fourier multiplier for all $1 < p < \infty$, with norm that is easily seen to grow polynomially in $|\lambda|$ (via Theorem 6.2.7 in [156]). We therefore need to concentrate on the nonsmooth piece of the multiplier $(1 - |\xi|^2)_+^\lambda (1 - \eta(\xi))$, which is supported in $B(0, \frac{1}{2})^c$. Let

$$K^\lambda(x) = \Big((1 - |\xi|^2)_+^\lambda (1 - \eta(\xi))\Big)^\vee(x)$$

be the kernel of the nonsmooth piece of the multiplier.

We pick a smooth *radial* function φ with support inside the ball $B(0,2)$ that is equal to 1 on the closed unit ball $\overline{B(0,1)}$. For $j = 1, 2, \dots$ we introduce functions

$$\psi_j(x) = \varphi(2^{-j}x) - \varphi(2^{-j+1}x)$$

supported in the annuli $2^{j-1} \le |x| \le 2^{j+1}$. Then we write

$$K^\lambda * f = T_0^\lambda(f) + \sum_{j=1}^\infty T_j^\lambda(f), \tag{5.4.14}$$

where T_0^λ is given by convolution with φK^λ and each T_j^λ is given by convolution with $\psi_j K^\lambda$.

We begin by examining the kernel φK^λ. Introducing a compactly supported function ζ that is equal to 1 on $B(0, \frac{3}{2})$, we write

$$\begin{aligned} K^\lambda &= \Big((1 - |\cdot|^2)_+^\lambda (1 - \eta)\zeta\Big)^\vee \\ &= ((1 - |\cdot|^2)_+^\lambda)^\vee * ((1 - \eta)\zeta)^\vee \\ &= K_\lambda * ((1 - \eta)\zeta)^\vee. \end{aligned}$$

Using this and (5.4.13) implies that K^λ is a bounded function, and thus φK^λ is bounded and compactly supported. Thus the operator T_0^λ is bounded on all the L^p spaces, $1 \le p \le \infty$, with a bound that grows at most exponentially in $|\mathrm{Im}\,\lambda|^2$.

Next we study the boundedness of the operators T_j^λ; here the dependence on the index j plays a role. Fix $p < 2$ as in the statement of the theorem. Our goal is to show that there exist positive constants C, δ (depending only on n and $\mathrm{Re}\,\lambda$) such that for all functions f in $L^p(\mathbf{R}^n)$ we have

$$\big\|T_j^\lambda(f)\big\|_{L^p(\mathbf{R}^n)} \le C\, e^{c_0|\mathrm{Im}\,\lambda|^2} 2^{-j\delta}\big\|f\big\|_{L^p(\mathbf{R}^n)}. \tag{5.4.15}$$

Once (5.4.15) is established, the L^p boundedness of the operator $f \mapsto K^\lambda * f$ follows by summing the series in (5.4.14).

As a consequence of (5.4.13) we obtain that

$$\begin{aligned} |K_j^\lambda(x)| &\le C_3(\mathrm{Re}\,\lambda)\, e^{c_0|\mathrm{Im}\,\lambda|^2}(1+|x|)^{-\frac{n+1}{2}-\mathrm{Re}\,\lambda}|\psi_j(x)| \\ &\le C' 2^{-(\frac{n+1}{2}+\mathrm{Re}\,\lambda)j}, \end{aligned} \tag{5.4.16}$$

since $\psi_j(x) = \psi(2^{-j}x)$ and ψ is supported in the annulus $\frac{1}{2} \le |x| \le 2$. From this point on, we tacitly assume that the constants containing a prime grow at most exponentially in $|\mathrm{Im}\,\lambda|^2$. Since K_j^λ is supported in a ball of radius 2^{j+1} and satisfies (5.4.16), we deduce the estimate

$$\big\|\widehat{K_j^\lambda}\big\|_{L^2}^2 = \big\|K_j^\lambda\big\|_{L^2}^2 \le C'' 2^{-(n+1+2\mathrm{Re}\,\lambda)j} 2^{nj} = C'' 2^{-(1+2\mathrm{Re}\,\lambda)j}. \tag{5.4.17}$$

We need another estimate for $\widehat{K_j^\lambda}$. We claim that for all $M \ge n+1$ there is a constant C_M such that

$$\int_{|\xi|\le\frac{1}{8}} |\widehat{K_j^\lambda}(\xi)|^2|\xi|^{-\beta}\,d\xi \le C_{M,n,\beta}\, 2^{-2j(M-n)}, \quad \beta < n. \tag{5.4.18}$$

Indeed, since $\widehat{K^\lambda}(\xi)$ is supported in $|\xi| \ge \frac{1}{2}$ [recall that the function η was chosen equal to 1 on $B(0,\frac{1}{2})$], we have

$$|\widehat{K_j^\lambda}(\xi)| = |(\widehat{K^\lambda} * \widehat{\psi_j})(\xi)| \le 2^{jn}\int_{\frac{1}{2}\le|\xi-\omega|\le 1}(1-|\xi-\omega|^2)_+^{\mathrm{Re}\,\lambda}|\widehat{\psi}(2^j\omega)|\,d\omega.$$

Suppose that $|\xi| \le \frac{1}{8}$. Since $|\xi-\omega| \ge \frac{1}{2}$, we must have $|\omega| \ge \frac{3}{8}$. Then

$$|\widehat{\psi}(2^j\omega)| \le C_M(2^j|\omega|)^{-M} \le (8/3)^M C_M 2^{-jM},$$

from which it follows easily that

$$\sup_{|\xi|\le\frac{1}{8}} |\widehat{K_j^\lambda}(\xi)| \le C_M' 2^{-j(M-n)}. \tag{5.4.19}$$

Then (5.4.18) is a consequence of (5.4.19) and of the fact that the function $|\xi|^{-\beta}$ is integrable near the origin.

We now return to estimate (5.4.15). A localization argument (Exercise 5.4.4) allows us to reduce estimate (5.4.15) to functions f that are supported in a cube of side length 2^j. Let us therefore assume that f is supported in some cube Q of side length 2^j. Then $T_j^\lambda(f)$ is supported in $5Q$ and we have for $1 \le p < 2$ by Hölder's inequality

$$\begin{aligned} \big\|T_j^\lambda(f)\big\|_{L^p(5Q)}^2 &\le |5Q|^{2(\frac{1}{p}-\frac{1}{2})}\big\|T_j^\lambda(f)\big\|_{L^2(5Q)}^2 \\ &\le C_n 2^{(\frac{1}{p}-\frac{1}{2})2nj}\big\|\widehat{K_j^\lambda}\widehat{f}\big\|_{L^2}^2 . \end{aligned} \tag{5.4.20}$$

Having returned to L^2, we are able to use the $L^p \to L^2$ restriction theorem obtained in the previous subsection. To this end we use polar coordinates and the fact that K_j^λ is a radial function to write

$$\big\|\widehat{K_j^\lambda}\widehat{f}\big\|_{L^2}^2 = \int_0^\infty |\widehat{K_j^\lambda}(re_1)|^2\bigg(\int_{\mathbf{S}^{n-1}}|\widehat{f}(r\theta)|^2\,d\theta\bigg)r^{n-1}dr, \tag{5.4.21}$$

where $e_1 = (1,0,\dots,0) \in \mathbf{S}^{n-1}$. Since the restriction of the function $x \mapsto r^{-n}f(x/r)$ on the sphere $\mathbf{S}^{n-1}$ is $\widehat{f}(r\theta)$, we have

$$\int_{\mathbf{S}^{n-1}}|\widehat{f}(r\theta)|^2\,d\theta \le C_{p,n}^2\bigg[\int_{\mathbf{R}^n} r^{-np}|f(x/r)|^p\,dx\bigg]^{\frac{2}{p}} = C_{p,n}^2 r^{-\frac{2n}{p'}}\big\|f\big\|_{L^p}^2, \tag{5.4.22}$$

where $C_{p,n}$ is the constant in Theorem 5.4.5 that holds whenever $p \le \frac{2(n+1)}{n+3}$. So assuming $p \le \frac{2(n+1)}{n+3}$ and inserting estimate (5.4.22) in (5.4.21) yields

$$\begin{aligned} \big\|\widehat{K_j^\lambda}\widehat{f}\big\|_{L^2}^2 &\le C_{p,n}^2\big\|f\big\|_{L^p}^2\int_0^\infty |\widehat{K_j^\lambda}(re_1)|^2 r^{n-1-\frac{2n}{p'}}\,dr \\ &\le \frac{C_{p,n}^2}{\omega_{n-1}}\big\|f\big\|_{L^p}^2\int_{\mathbf{R}^n}|\widehat{K_j^\lambda}(\xi)|^2|\xi|^{-\frac{2n}{p'}}\,d\xi , \end{aligned} \tag{5.4.23}$$

where $\omega_{n-1} = |\mathbf{S}^{n-1}|$. Appealing to estimate (5.4.18) for $|\xi| \le \frac{1}{8}$ with $\beta = \frac{2n}{p'} < n$ (since $p < 2$) and to estimate (5.4.17) for $|\xi| \ge \frac{1}{8}$, we obtain

$$\big\|\widehat{K_j^\lambda}\widehat{f}\big\|_{L^2}^2 \le C''' 2^{-(1+2\mathrm{Re}\,\lambda)j}\big\|f\big\|_{L^p}^2 .$$

Combining this inequality with the one previously obtained in (5.4.20) yields (5.4.15) with

$$\delta = \frac{n+1}{2} + \mathrm{Re}\,\lambda - \frac{n}{p} .$$

This number is positive exactly when $\frac{2n}{n+1+2\mathrm{Re}\,\lambda} < p$. This was the condition assumed by the theorem when $p < 2$. The other condition $\mathrm{Re}\,\lambda > \frac{n-1}{2(n+1)}$ is naturally imposed by the restriction $p \le \frac{2(n+1)}{n+3}$. Finally, the analogous result in the range $p > 2$ follows by duality. □

5.4.4 The Full Restriction Theorem on $\mathbf{R}^2$

In this section we prove the validity of the restriction condition $R_{p\to q}(\mathbf{S}^1)$ in dimension $n=2$, for the full range of exponents suggested by Figure 5.10.

To achieve this goal, we "fatten" the circle by a small amount 2δ. Then we obtain a restriction theorem for the "fattened circle" and then obtain the required estimate by taking the limit as $\delta \to 0$. Precisely, we use the fact

$$\int_{\mathbf{S}^1} |\widehat{f}(\omega)|^q \, d\omega = \lim_{\delta\to 0} \frac{1}{2\delta} \int_{1-\delta}^{1+\delta} \int_{\mathbf{S}^1} |\widehat{f}(r\theta)|^q \, d\theta \, r\, dr \tag{5.4.24}$$

to recover the restriction theorem for the circle from a restriction theorem for annuli of width 2δ.

Throughout this subsection, δ is a number satisfying $0<\delta<\frac{1}{1000}$, and for simplicity we use the notation

$$\chi^\delta(\xi) = \chi_{(1-\delta,1+\delta)}(|\xi|), \qquad \xi \in \mathbf{R}^2.$$

We note that in view of identity (5.4.24), the restriction property $R_{p\to q}(\mathbf{S}^1)$ is a trivial consequence of the estimate

$$\frac{1}{2\delta} \int_0^\infty \int_{\mathbf{S}^1} |\chi^\delta(r\theta)\widehat{f}(r\theta)|^q \, d\theta \, r\, dr \le C^q \|f\|_{L^p}^q, \tag{5.4.25}$$

or, equivalently, of

$$\|\chi^\delta \widehat{f}\|_{L^q(\mathbf{R}^2)} \le (2\delta)^{\frac{1}{q}} C \|f\|_{L^p(\mathbf{R}^2)}. \tag{5.4.26}$$

We have the following result.

Theorem 5.4.7. *(a) Given $1\le p<\frac{4}{3}$, set $q=\frac{p'}{3}$. Then there is a constant C_p such that for all L^p functions f on $\mathbf{R}^2$ and all small positive δ we have*

$$\|\chi^\delta \widehat{f}\|_{L^q(\mathbf{R}^2)} \le C_p \delta^{\frac{1}{q}} \|f\|_{L^p(\mathbf{R}^2)}. \tag{5.4.27}$$

(b) When $p=q=4/3$, there is a constant C such that for all $L^{4/3}$ functions f on $\mathbf{R}^2$ and all small $\delta>0$ we have

$$\|\chi^\delta \widehat{f}\|_{L^{\frac{4}{3}}(\mathbf{R}^2)} \le C\delta^{\frac{3}{4}} (\log \tfrac{1}{\delta})^{\frac{1}{4}} \|f\|_{L^{\frac{4}{3}}(\mathbf{R}^2)}. \tag{5.4.28}$$

Proof. To prove this theorem, we work with the *extension operator*

$$E^\delta(g) = \widehat{\chi^\delta g} = \widehat{\chi^\delta} * \widehat{g},$$

which is dual (i.e., transpose) to $f \mapsto \chi^\delta \widehat{f}$, and we need to show that

$$\|E^\delta(f)\|_{L^{p'}(\mathbf{R}^2)} \le C\delta^{\frac{1}{q}} (\log \tfrac{1}{\delta})^\beta \|f\|_{L^{q'}(\mathbf{R}^2)}, \tag{5.4.29}$$

where $\beta=\frac{1}{4}$ when $p=\frac{4}{3}$ and $\beta=0$ when $p<\frac{4}{3}$.

We employ a splitting similar to that used in Theorem 5.2.4, with the only difference being that the present partition of unity is nonsmooth and hence simpler. We define functions

$$\chi_\ell^\delta(\xi) = \chi^\delta(\xi)\chi_{2\pi\ell\delta^{1/2}\le \operatorname{Arg}\xi < 2\pi(\ell+1)\delta^{1/2}}$$

for $\ell \in \{0,1,\dots,[\delta^{-1/2}]\}$. We suitably adjust the support of the function $\chi^\delta_{[\delta^{-1/2}]}$ so that the sum of all these functions equals χ^δ. We now split the indices that appear in the set $\{0,1,\dots,[\delta^{-1/2}]\}$ into nine different subsets so that the supports of the functions indexed by them are properly contained in some sector centered at the origin of amplitude $\pi/4$. We therefore write E^δ as a sum of nine pieces, each properly supported in a sector of amplitude $\pi/4$. Let I be the set of indices that correspond to one of these nine sectors and let

$$E_I^\delta(f) = \sum_{\ell\in I} \widehat{\chi_\ell^\delta f}\,.$$

It suffices therefore to obtain (5.4.29) for each E_I^δ in lieu of E^δ. Let us fix such an index set I and without loss of generality we assume that

$$I = \{0,1,\dots,[\tfrac{1}{8}\delta^{-1/2}]\}\,.$$

Since the theorem is trivial when $p=1$, to prove part (a) we fix a number p with $1<p<\frac{4}{3}$. We set

$$r = (p'/2)'$$

and we observe that this r satisfies $\frac{1}{r} = \frac{1}{p'}+\frac{1}{q'}$. We note that $1<r<2$ and we apply the Hausdorff–Young inequality $\|h\|_{L^{r'}} \le \|h^\vee\|_{L^r}$. We have

$$\begin{aligned}
\|E_I^\delta(f)\|_{L^{p'}(\mathbf{R}^2)}^{p'} &= \int_{\mathbf{R}^2} |E_I^\delta(f)^2|^{r'}\,dx \\
&\le \left(\int_{\mathbf{R}^2} |(E_I^\delta(f)^2)^\vee|^r\,dx\right)^{\frac{r'}{r}} \\
&= \left(\int_{\mathbf{R}^2} \left|\sum_{\ell\in I}\sum_{\ell'\in I}(\chi_\ell^\delta f)*(\chi_{\ell'}^\delta f)\right|^r dx\right)^{\frac{r'}{r}}.
\end{aligned} \tag{5.4.30}$$

We obtain the estimate

$$\left(\int_{\mathbf{R}^2} \left|\sum_{\ell\in I}\sum_{\ell'\in I}(\chi_\ell^\delta f)*(\chi_{\ell'}^\delta f)\right|^r dx\right)^{\frac{r'}{r}} \le C\delta^{\frac{p'}{q}} \|f\|_{L^{q'}(\mathbf{R}^2)}^{p'}, \tag{5.4.31}$$

which suffices to prove the theorem.

Denote by $S_{\delta,\ell,\ell'}$ the support of $\chi_\ell^\delta * \chi_{\ell'}^\delta$. Then we write the left-hand side of (5.4.31) as

$$\left(\int_{\mathbf{R}^2}\Big|\sum_{\ell\in I}\sum_{\ell'\in I}((\chi_\ell^\delta f)*(\chi_{\ell'}^\delta f))\chi_{S_{\delta,\ell,\ell'}}\Big|^r dx\right)^{\frac{r'}{r}}, \tag{5.4.32}$$

which, via Hölder's inequality, is controlled by

$$\left(\int_{\mathbf{R}^2}\Big(\sum_{\ell\in I}\sum_{\ell'\in I}|(\chi_\ell^\delta f)*(\chi_{\ell'}^\delta f)|^r\Big)^{\frac{r}{r}}\Big(\sum_{\ell\in I}\sum_{\ell'\in I}|\chi_{S_{\delta,\ell,\ell'}}|^{r'}\Big)^{\frac{r}{r'}}dx\right)^{\frac{r'}{r}}. \tag{5.4.33}$$

We now recall Lemma 5.2.5, in which the curvature of the circle was crucial. In view of that lemma, the second factor of the integrand in (5.4.33) is bounded by a constant independent of δ. We have therefore obtained the estimate

$$\|E_I^\delta(f)\|_{L^{p'}}^{p'} \le C\left(\sum_{\ell\in I}\sum_{\ell'\in I}\int_{\mathbf{R}^2}|(\chi_\ell^\delta f)*(\chi_{\ell'}^\delta f)|^r dx\right)^{\frac{r'}{r}}. \tag{5.4.34}$$

We prove at the end of this section the following auxiliary result.

Lemma 5.4.8. *With the same notation as in the proof of Theorem 5.4.7, for any $1<r<\infty$, there is a constant C (independent of δ and f) such that*

$$\|(\chi_\ell^\delta f)*(\chi_{\ell'}^\delta f)\|_{L^r} \le C\left(\frac{\delta^{\frac{3}{2}}}{|\ell-\ell'|+1}\right)^{\frac{1}{r'}}\|\chi_\ell^\delta f\|_{L^r}\|\chi_{\ell'}^\delta f\|_{L^r} \tag{5.4.35}$$

for all $\ell,\ell'\in I=\{0,1,\dots,[\frac{1}{8}\delta^{-1/2}]\}$.

Assuming Lemma 5.4.8 and using (5.4.34), we write

$$\begin{aligned}\|E_I^\delta(f)\|_{L^{p'}}^{p'} &\le C\delta^{\frac{3}{2}}\left[\sum_{\ell\in I}\|\chi_\ell^\delta f\|_{L^r}^r\Big(\sum_{\ell'\in I}\frac{\|\chi_{\ell'}^\delta f\|_{L^r}^r}{(|\ell-\ell'|+1)^{\frac{r}{r'}}}\Big)\right]^{\frac{r'}{r}}\\ &\le C\delta^{\frac{3}{2}}\left[\sum_{\ell\in I}\|\chi_\ell^\delta f\|_{L^r}^{rs}\right]^{\frac{r'}{rs}}\left[\sum_{\ell\in I}\Big(\sum_{\ell'\in I}\frac{\|\chi_{\ell'}^\delta f\|_{L^r}^r}{(|\ell-\ell'|+1)^{\frac{r}{r'}}}\Big)^{s'}\right]^{\frac{r'}{rs'}},\end{aligned} \tag{5.4.36}$$

where we used Hölder's inequality for some $1<s<\infty$. We now recall the discrete fractional integral operator

$$\{a_j\}_j \mapsto \Big\{\sum_{j'}\frac{a_{j'}}{(|j-j'|+1)^{1-\alpha}}\Big\}_j,$$

which maps $\ell^s(\mathbf{Z})$ to $\ell^{s'}(\mathbf{Z})$ (see Exercise 1.2.10) when

$$\frac{1}{s}-\frac{1}{s'}=\alpha, \qquad 0<\alpha<1. \tag{5.4.37}$$

When $1 < p < \frac{4}{3}$, we have $1 < r < 2$, and choosing $\alpha = 2 - r = 1 - \frac{r}{r'}$, we obtain from (5.4.36) that

$$\begin{aligned}\|E_I^\delta(f)\|_{L^{p'}}^{p'} &\le C'\delta^{\frac{3}{2}}\left[\sum_{\ell\in I}\|\chi_\ell^\delta f\|_{L^r}^{rs}\right]^{\frac{r'}{rs}}\left[\sum_{\ell\in I}\|\chi_\ell^\delta f\|_{L^r}^{rs}\right]^{\frac{r'}{rs}} \\ &= C'\delta^{\frac{3}{2}}\left[\sum_{\ell\in I}\|\chi_\ell^\delta f\|_{L^r}^{rs}\right]^{\frac{2r'}{rs}}. \end{aligned} \tag{5.4.38}$$

The unique s that solves equation (5.4.37) is seen easily to be $s = q'/r$. Moreover, since $q = p'/3$, we have $1 < s < 2$. We use again Hölder's inequality to pass from $\|\chi_\ell^\delta f\|_{L^r}$ to $\|\chi_\ell^\delta f\|_{L^{q'}}$. Indeed, recalling that the support of χ_ℓ^δ has measure $\approx \delta^{\frac{3}{2}}$, we have

$$\|\chi_\ell^\delta f\|_{L^r} \le C(\delta^{\frac{3}{2}})^{\frac{1}{r}-\frac{1}{q'}}\|\chi_\ell^\delta f\|_{L^{q'}}.$$

Inserting this in (5.4.38) yields

$$\begin{aligned}\|E_I^\delta(f)\|_{L^{p'}}^{p'} &\le C\delta^{\frac{3}{2}}\left[\sum_{\ell\in I}\Big(C(\delta^{\frac{3}{2}})^{\frac{1}{r}-\frac{1}{q'}}\|\chi_\ell^\delta f\|_{L^{q'}}\Big)^{rs}\right]^{\frac{2r'}{rs}} \\ &= C'\delta^{\frac{3}{2}}(\delta^{\frac{3}{2}})^{2r'(\frac{1}{r}-\frac{1}{q'})}\left[\sum_{\ell\in I}\|\chi_\ell^\delta f\|_{L^{q'}}^{q'}\right]^{\frac{2r'}{q'}} \\ &\le C\delta^3\|f\|_{L^{q'}}^{p'} \\ &= C\delta^{\frac{p'}{q}}\|f\|_{L^{q'}}^{p'}, \end{aligned}$$

which is the required estimate since $\frac{1}{r} = \frac{1}{p'} + \frac{1}{q'}$ and $p' = 2r'$. In the last inequality we used the fact that the supports of the functions χ_ℓ^δ are disjoint and that these add up to a function that is at most 1.

To prove part (b) of the theorem, we need to adjust the previous argument to obtain the case $p = \frac{4}{3}$. Here we repeat part of the preceding argument taking $r = r' = s = s' = 2$.

Using (5.4.34) with $p = \frac{4}{3}$ (which forces r to be equal to 2) and Lemma 5.4.8 with $r = 2$ we write

$$\begin{aligned}\|E_I(f)\|_{L^4(\mathbf{R}^2)}^4 &\le C\delta^{\frac{3}{2}}\left[\sum_{\ell\in I}\|\chi_\ell^\delta f\|_{L^2}^2\Big(\sum_{\ell'\in I}\frac{\|\chi_{\ell'}^\delta f\|_{L^2}^2}{|\ell-\ell'|+1}\Big)\right] \\ &\le C\delta^{\frac{3}{2}}\left[\sum_{\ell\in I}\|\chi_\ell^\delta f\|_{L^2}^4\right]^{\frac{1}{2}}\left[\sum_{\ell\in I}\Big(\sum_{\ell'\in I}\frac{\|\chi_{\ell'}^\delta f\|_{L^2}^2}{|\ell-\ell'|+1}\Big)^2\right]^{\frac{1}{2}} \\ &\le C\delta^{\frac{3}{2}}\left[\sum_{\ell\in I}\|\chi_\ell^\delta f\|_{L^2}^4\right]^{\frac{1}{2}}\left[\sum_{\ell\in I}\|\chi_\ell^\delta f\|_{L^2}^4\right]^{\frac{1}{2}}\left[\sum_{\ell\in I}\frac{1}{|\ell|+1}\right] \end{aligned}$$

$$
\begin{aligned}
&\le C\delta^{\frac{3}{2}}\Big[\sum_{\ell\in I}\|\chi_\ell^\delta f\|_{L^2}^4\Big]\log(\delta^{-\frac{1}{2}})\\
&\le C\delta^{\frac{3}{2}}(\delta^{\frac{3}{2}})^{(\frac{1}{2}-\frac{1}{4})4}\Big[\sum_{\ell\in I}\|\chi_\ell^\delta f\|_{L^4}^4\Big]\log\tfrac{1}{\delta}\\
&\le C\delta^3\big(\log\tfrac{1}{\delta}\big)\|f\|_{L^4}^4 .
\end{aligned}
$$

□

We now prove Lemma 5.4.8, which we had left open.

Proof. The proof is based on interpolation. For fixed $\ell,\ell'\in I$ we define the bilinear operator

$$T_{\ell,\ell'}(g,h)=(g\chi_\ell^\delta)*(h\chi_{\ell'}^\delta).$$

As we have previously observed, it is a simple geometric fact that the support of χ_ℓ^δ is contained in a rectangle of side length $\approx\delta$ in the direction $e^{2\pi i\delta^{1/2}\ell}$ and of side length $\approx\delta^{\frac{1}{2}}$ in the direction $ie^{2\pi i\delta^{1/2}\ell}$. Any two rectangles with these dimensions in the aforementioned directions have an intersection that depends on the angle between them. Indeed, if $\ell\neq\ell'$, this intersection is contained in a parallelogram of sides δ and $\delta/\sin(2\pi\delta^{\frac{1}{2}}|\ell-\ell'|)$, and hence the measure of the intersection is seen easily to be at most a constant multiple of

$$\delta\cdot\frac{\delta}{\sin(2\pi\delta^{\frac{1}{2}}|\ell-\ell'|)}.$$

As for ℓ,ℓ' in the index set I we have $2\pi\delta^{\frac{1}{2}}|\ell-\ell'|<\pi/4$, the sine is comparable to its argument, and we conclude that the measure of the intersection is at most

$$C\delta^{\frac{3}{2}}(1+|\ell-\ell'|)^{-1}.$$

It follows that

$$\|\chi_\ell^\delta*\chi_{\ell'}^\delta\|_{L^\infty}=\sup_{z\in\mathbf{R}^2}|(z-\text{supp }(\chi_\ell^\delta))\cap\text{supp }(\chi_{\ell'}^\delta)|\le\frac{C\delta^{\frac{3}{2}}}{1+|\ell-\ell'|},$$

which implies the estimate

$$
\begin{aligned}
\|T_{\ell,\ell'}(g,h)\|_{L^\infty}&\le\|\chi_\ell^\delta*\chi_{\ell'}^\delta\|_{L^\infty}\|g\|_{L^\infty}\|h\|_{L^\infty}\\
&\le\frac{C\delta^{\frac{3}{2}}}{1+|\ell-\ell'|}\|g\|_{L^\infty}\|h\|_{L^\infty}.
\end{aligned}
\tag{5.4.39}
$$

Also, the estimate

$$\|T_{\ell,\ell'}(g,h)\|_{L^1}\le\|g\chi_\ell^\delta\|_{L^1}\|h\chi_{\ell'}^\delta\|_{L^1}\le\|g\|_{L^1}\|h\|_{L^1}\tag{5.4.40}$$

holds trivially. Interpolating between (5.4.39) and (5.4.40) yields the required estimate (5.4.35); see Corollary 7.2.11. □

Example 5.4.9. The presence of the logarithmic factor in estimate (5.4.28) is necessary. In fact, this estimate is sharp. We prove this by showing that the corresponding estimate for the "dual" extension operator E^δ is sharp. Let I be the set of indices we worked with in Theorem 5.4.7 (i.e., $I = \{0,1,\dots,[\frac{1}{8}\delta^{-1/2}]\}$.) Let

$$f^\delta = \sum_{\ell\in I} \chi_\ell^\delta .$$

Then

$$\big\|f^\delta\big\|_{L^4} \approx \delta^{\frac{1}{4}} .$$

However,

$$E^\delta(f^\delta) = \sum_{\ell\in I} \widehat{\chi_\ell^\delta},$$

and we have

$$\begin{aligned}
\|E^\delta(f^\delta)\|_{L^4} &= \Big(\int_{\mathbf{R}^2} \Big|\sum_{\ell\in I}\sum_{\ell'\in I} \widehat{\chi_\ell^\delta}\widehat{\chi_{\ell'}^\delta}\Big|^2 d\xi\Big)^{\frac{1}{4}} \\
&= \Big(\int_{\mathbf{R}^2} \Big|\sum_{\ell\in I}\sum_{\ell'\in I} \chi_\ell^\delta * \chi_{\ell'}^\delta\Big|^2 dx\Big)^{\frac{1}{4}} \\
&\ge \Big(\sum_{\ell\in I}\sum_{\ell'\in I} \int_{\mathbf{R}^2} \big|\chi_\ell^\delta * \chi_{\ell'}^\delta\big|^2 dx\Big)^{\frac{1}{4}} .
\end{aligned}$$

At this point observe that the function $\chi_\ell^\delta * \chi_{\ell'}^\delta$ is at least a constant multiple of $\delta^{\frac{3}{2}}(|\ell-\ell'|+1)^{-1}$ on a set of measure $c\delta^{\frac{3}{2}}(|\ell-\ell'|+1)$. (See Exercise 5.4.5.) Using this fact and the previous estimates, we deduce easily that

$$\|E^\delta(f^\delta)\|_{L^4} \ge c\Big(\sum_{\ell\in I}\sum_{\ell'\in I} \frac{\delta^3}{(|\ell-\ell'|+1)^2}\delta^{\frac{3}{2}}(|\ell-\ell'|+1)\Big)^{\frac{1}{4}} \approx \delta(\log\tfrac{1}{\delta})^{\frac{1}{4}},$$

since $|I| \approx \delta^{-\frac{1}{2}}$. It follows that

$$\frac{\|E^\delta(f^\delta)\|_{L^4}}{\|f^\delta\|_{L^4}} \ge c\delta^{\frac{3}{4}}(\log\tfrac{1}{\delta})^{\frac{1}{4}},$$

which justifies the sharpness of estimate (5.4.28).

Exercises

5.4.1. Let S be a compact hypersurface in $\mathbf{R}^n$ and let $d\sigma$ be surface measure on it. Suppose that for some $0 < b < n$ we have

$$|\widehat{d\sigma}(\xi)| \le C(1+|\xi|)^{-b}$$

for all $\xi \in \mathbf{R}^n$. Prove that $R_{p\to q}(S)$ does not hold for any $1 \le q \le \infty$ when $p \ge \frac{n}{n-b}$.

5.4.2. Let S be a compact hypersurface and let $1 \le p, q \le \infty$.
(a) Suppose that $R_{p\to q}(S)$ holds for S. Show that $R_{p\to q}(\tau+S)$ holds for the translated hypersurface $\tau+S$.
(b) For $r > 0$ let $r\mathbf{S}^{n-1} = \{r\xi : \xi \in \mathbf{S}^{n-1}\}$. Suppose that $R_{p\to q}(\mathbf{S}^{n-1})$ holds with constant C_{pqn}. Show that $R_{p\to q}(r\mathbf{S}^{n-1})$ holds with constant $C_{pqn} r^{\frac{n-1}{q}-\frac{n}{p'}}$.

5.4.3. Obtain a different proof of estimate (5.4.7) (and hence of Theorem 5.4.5) by following the sequence of steps outlined here:
(a) Consider the analytic family of functions

$$(K_z)^\vee(\xi) = 2\pi^{1-z}\frac{J_{\frac{n-2}{2}+z}(2\pi|\xi|)}{|\xi|^{\frac{n-2}{2}+z}}$$

and observe that in view of the identity in Appendix B.4 in [156], $(K_z)^\vee(\xi)$ reduces to $d\sigma^\vee(\xi)$ when $z = 0$, where $d\sigma$ is surface measure on $\mathbf{S}^{n-1}$.
(b) Use that the Bessel function $J_{-\frac{1}{2}+i\theta}$, $\theta \in \mathbf{R}$, satisfies

$$|J_{-\frac{1}{2}+i\theta}(x)| \le c(1+|\theta|)e^{|\theta|^2}|x|^{-\frac{1}{2}},$$

for some $c > 0$ (Appendix B.9 in [156]) to obtain that the family of operators given by convolution with $(K_z)^\vee$ maps $L^1(\mathbf{R}^n)$ to $L^\infty(\mathbf{R}^n)$ when $z = -\frac{n-1}{2} + i\theta$.
(c) Appeal to the result in Appendix B.5 in [156] to obtain that for $z \in \mathbf{C}$ we have

$$K_z(x) = \frac{2}{\Gamma(z)}(1-|x|^2)^{z-1}.$$

Use this identity to deduce that for $z = 1+i\theta$ the family of operators given by convolution with $(K_z)^\vee$ map $L^2(\mathbf{R}^n)$ to itself with constants that grow at most exponentially in $|\theta|$. (Appendix A.7 in [156] contains a useful lower estimate for $|\Gamma(1+i\theta)|$.)
(d) Use complex interpolation for analytic families (cf. Exercise 1.3.4 in [156]) to obtain that for $z = 0$ the operator given by convolution with $d\sigma^\vee$ maps $L^p(\mathbf{R}^n)$ to $L^{p'}(\mathbf{R}^n)$ when $p = \frac{2(n+1)}{n+3}$.

5.4.4. Suppose that T is a linear operator defined on a subspace of measurable functions on $\mathbf{R}^n$ with the property that whenever f is supported in a cube Q of side length s, then $T(f)$ is supported in aQ for some $a > 1$. Prove the following:
(a) If T is defined on $L^p(\mathbf{R}^n)$ for some $0 < p < \infty$ and

$$\big\|T(f)\big\|_{L^p} \le B\big\|f\big\|_{L^p}$$

for all f supported in a cube of side length s, then the same estimate holds (with a larger constant) for all functions in $L^p(\mathbf{R}^n)$.
(b) If T satisfies for some $0 < p < \infty$,

$$\left\|T(\chi_A)\right\|_{L^{p,\infty}} \leq B|A|^{\frac{1}{p}}$$

for all measurable sets A contained in a cube of side length s, then the same estimate holds (with a larger constant) for all measurable sets A in $\mathbf{R}^n$.

5.4.5. Using the notation of Theorem 5.4.7, show that there exist constants c, c' such that the function $\chi_\ell^\delta * \chi_{\ell'}^\delta$ is at least $c'\delta^{\frac{3}{2}}(|\ell-\ell'|+1)^{-1}$ on a set of measure $c\delta^{\frac{3}{2}}(|\ell-\ell'|+1)$.
[*Hint:* Prove the required conclusion for characteristic functions of rectangles with the same orientation and comparable dimensions. Then use that the support of each χ_ℓ^δ contains such a rectangle.]

5.5 Almost Everywhere Convergence of Bochner–Riesz Means

We recall the Bochner–Riesz means B_R^λ of complex order λ given in Definition 5.2.1. In this section we study the problem of almost everywhere convergence of $B_R^\lambda(f) \to f$ as $R \to \infty$. There is an intimate relationship between the almost everywhere convergence of a family of operators and boundedness properties of the associated maximal family (cf. Theorem 2.1.14 in [156]).[1]

For $f \in L^p(\mathbf{R}^n)$, the *maximal Bochner–Riesz operator* or order λ is defined by

$$B_*^\lambda(f) = \sup_{R>0} \left|B_R^\lambda(f)\right|.$$

5.5.1 A Counterexample for the Maximal Bochner–Riesz Operator

We have the following result.

Theorem 5.5.1. *Let $n \geq 2$, $\lambda > 0$, and let $1 < p < 2$ be such that*

$$\lambda < \frac{2n-1}{2p} - \frac{n}{2}.$$

Then B_*^λ does not map $L^p(\mathbf{R}^n)$ to weak $L^p(\mathbf{R}^n)$.

---

[1] In certain cases, this theorem can essentially be reversed. Given a $1 \leq p \leq 2$ and a family of distributions u_j with the mild continuity property that $u_j * f_k \to u_j * f$ in measure whenever $f_k \to f$ in $L^p(\mathbf{R}^n)$ such that the maximal operator $\mathcal{M}(f) = \sup_j |f * u_j| < \infty$ whenever $f \in L^p(\mathbf{R}^n)$, then $\mathcal{M}$ maps $L^p(\mathbf{R}^n)$ to $L^{p,\infty}(K)$ for any compact subset K of $\mathbf{R}^n$. See Stein [323], [326].

Proof. Figure 5.11 shows the region in which B_*^λ is known to be unbounded; this region contains the set of points $(1/p,\lambda)$ strictly below the line that joins the points $(1,(n-1)/2)$ and $(n/(2n-1),0)$.

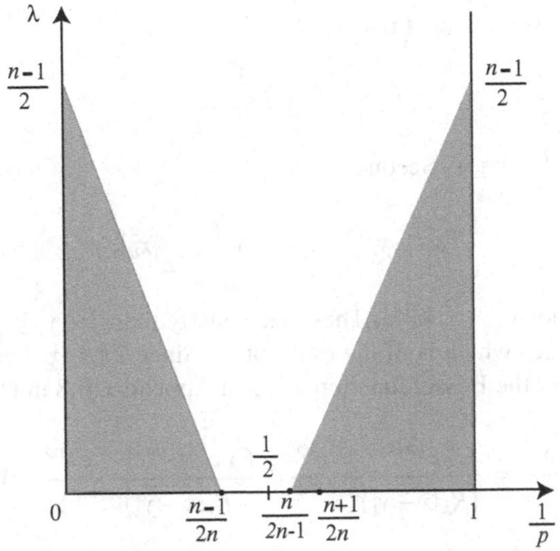

Fig. 5.11 The operators B_*^λ are unbounded on $L^p(\mathbf{R}^n)$ when $(1/p,\lambda)$ lies in the interior of the shaded region.

We denote points x in $\mathbf{R}^n$ by $x=(x',x_n)$, where $x'\in\mathbf{R}^{n-1}$, and we fix $M\geq 100$ and $\varepsilon<1/100$. We let $\psi(y)=\chi_{|y'|\leq 1}(y')\,\zeta(y_n)$, where ζ is a smooth bump supported in the interval $[-1,1]$ that is equal to 1 on $[-1/2,1/2]$ and satisfies $0\leq\zeta\leq 1$. We define

$$\psi_{\varepsilon,M}(y)=\psi(\varepsilon^{-1}y',\varepsilon^{-1}M^{-\frac{1}{2}}y_n)=\chi_{|y'|\leq\varepsilon}(y')\,\zeta(\varepsilon^{-1}M^{-\frac{1}{2}}y_n)$$

and we note that $\psi_{\varepsilon,M}(y)$ is supported in the set of y's that satisfy $|y'|\leq\varepsilon$ and $|y_n|\leq\varepsilon M^{\frac{1}{2}}$. We also define

$$f_M(y)=e^{2\pi i y_n}\,\psi_{\varepsilon,M}(y)$$

and

$$S_M=\{(x',x_n):\ M\leq|x'|\leq 2M,\ M\leq|x_n|\leq 2M\}.$$

Then

$$\big\|f_M\big\|_{L^p}\approx M^{\frac{1}{2p}}\,\varepsilon^{\frac{n}{p}}\qquad\text{and}\qquad |S_M|\approx M^n\,. \tag{5.5.1}$$

Every point $x\in S_M$ must satisfy $M\leq|x|\leq 3M$. We fix $x\in S_M$ and we estimate

$$B_*^\lambda(f_M)(x)=\sup_{R>0}\big|B_R^\lambda(f_M)(x)\big|$$

from below by picking $R = R_x = |x|/x_n$. Then $1 \le R_x \le 3$ and we have

$$B_*^\lambda(f_M)(x) \ge \frac{\Gamma(\lambda+1)}{\pi^\lambda}\left| \int_{\mathbf{R}^n} \frac{J_{\frac{n}{2}+\lambda}(2\pi R_x|x-y|)}{(R_x|x-y|)^{\frac{n}{2}+\lambda}} e^{2\pi i y_n} \psi_{\varepsilon,M}(y)\,dy \right|.$$

We make some observations. First

$$|x'-y'| \ge \frac{1}{2}|x'|,$$

since $|x'| \ge M$ and $|y'| \le \varepsilon$. Second,

$$|x_n - y_n| \ge |x_n| - |y_n| \ge \frac{1}{2}|x_n|,$$

since $|x_n| \ge M$ and $|y_n| \le \varepsilon M^{1/2}$. These facts imply that $|x-y| \ge \frac{1}{2}|x|$; thus $|x-y|$ is comparable to $|x|$, which is of the order of M. Since $2\pi R_x|x-y|$ is large, we use the asymptotics for the Bessel function $J_{\frac{n}{2}+\lambda}$ in Appendix B.8 in [156] to write

$$\frac{J_{\frac{n}{2}+\lambda}(2\pi R_x|x-y|)}{(R_x|x-y|)^{\frac{n}{2}+\lambda}} = \frac{C_\lambda\, e^{2\pi i R_x|x-y|} e^{i\varphi}}{(R_x|x-y|)^{\frac{n+1}{2}+\lambda}} + \frac{C_\lambda\, e^{-2\pi i R_x|x-y|} e^{-i\varphi}}{(R_x|x-y|)^{\frac{n+1}{2}+\lambda}} + V_{n,\lambda}(R_x|x-y|),$$

where $\varphi = -\frac{\pi}{2}(\frac{n}{2}+\lambda) - \frac{\pi}{4}$ and

$$|V_{n,\lambda}(R_x|x-y|)| \le \frac{C_{n,\lambda}}{(R_x|x-y|)^{\frac{n+3}{2}+\lambda}} \le \frac{C'_{n,\lambda}}{M^{\frac{n+3}{2}+\lambda}}, \tag{5.5.2}$$

since $R_x = \frac{|x|}{x_n} \approx 1$ and $|x-y| \ge \frac{1}{2}M$. Using the preceding expression for the Bessel function, we write

$$\begin{aligned} B_*^\lambda(f_M)(x) \ge C'_\lambda &\left| \int_{\mathbf{R}^n} \frac{e^{2\pi i R_x|x|} e^{i\varphi}}{(R_x|x-y|)^{\frac{n+1}{2}+\lambda}} \psi_{\varepsilon,M}(y)\,dy \right| \\ - C'_\lambda &\left| \int_{\mathbf{R}^n} \frac{(e^{2\pi i(R_x|x-y|+y_n)} - e^{2\pi i R_x|x|}) e^{i\varphi}}{(R_x|x-y|)^{\frac{n+1}{2}+\lambda}} \psi_{\varepsilon,M}(y)\,dy \right| \\ - C'_\lambda &\left| \int_{\mathbf{R}^n} \frac{e^{2\pi i(-R_x|x-y|+y_n)} e^{-i\varphi}}{(R_x|x-y|)^{\frac{n+1}{2}+\lambda}} \psi_{\varepsilon,M}(y)\,dy \right| \\ - &\left| \int_{\mathbf{R}^n} V_{n,\lambda}(R_x|x-y|) e^{2\pi i y_n} \psi_{\varepsilon,M}(y)\,dy \right|. \end{aligned}$$

The positive term is the main term and is bounded from below by

$$C'_\lambda\,(6M)^{-\frac{n+1}{2}-\lambda} \int_{\mathbf{R}^n} \psi_{\varepsilon,M}(y)\,dy = \frac{c_1\,\varepsilon^n M^{\frac{1}{2}}}{M^{\frac{n+1}{2}+\lambda}}. \tag{5.5.3}$$

The three terms with the minus signs are errors and are bounded in absolute value by smaller expressions. We notice that

$$\big|R_x|x-y|+y_n-R_x|x|\big|=\frac{|x|}{x_n}\Big||x-y|+\frac{x_ny_n}{|x|}-|x|\Big|=\frac{|x|}{x_n}\big|F_x(y)-F_x(0)\big|,$$

where $F_x(y)=|x-y|+|x|^{-1}x_ny_n$. Taylor's expansion yields

$$F_x(y)-F_x(0)=\nabla_yF_x(0)\cdot y+O\big(|y|^2\sup_{j,k}|\partial_j\partial_kF_x(y)|\big),$$

and a calculation gives

$$\nabla_yF_x(0)=(-|x|^{-1}x',0),$$

while

$$|\partial_j\partial_kF_x(y)|\le C|x-y|^{-1}.$$

It follows that

$$\frac{|x|}{x_n}\big|F_x(y)-F_x(0)\big|\le 3\left[\frac{|x'\cdot y'|}{|x|}+C'\frac{|y|^2}{|x-y|}\right]\le C''\left[\varepsilon+\frac{(\varepsilon M^{1/2})^2}{M}\right]\le 2C''\varepsilon.$$

Using this fact and the support properties of ψ, we obtain

$$C'_\lambda\left|\int_{\mathbf{R}^n}\frac{(e^{2\pi i(R_x|x-y|+y_n)}-e^{2\pi iR_x|x|})e^{i\varphi}}{(R_x|x-y|)^{\frac{n+1}{2}+\lambda}}\,\psi_{\varepsilon,M}(y)\,dy\right|\le\frac{c_2\,\varepsilon(\varepsilon^nM^{\frac{1}{2}})}{M^{\frac{n+1}{2}+\lambda}}.\tag{5.5.4}$$

Next we examine the phase $-R_x|x-y|+y_n$ as a function of y_n. Its derivative with respect to y_n is

$$\frac{\partial}{\partial y_n}\big(-R_x|x-y|+y_n\big)=R_x\frac{x_n-y_n}{|x-y|}+1\ge 1,$$

since $x_n\ge M$ and $|y_n|\le\varepsilon M^{1/2}$, which implies that $x_n-y_n>0$. Also note that

$$\left|\frac{\partial}{\partial y_n}\left(R_x\frac{x_n-y_n}{|x-y|}+1\right)^{-1}\right|\le\frac{C'''}{M}$$

and that

$$\left|\frac{\partial}{\partial y_n}\frac{1}{|x-y|^{\frac{n+1}{2}+\lambda}}\right|\le\frac{C'''}{M^{\frac{n+3}{2}+\lambda}},$$

while the derivative of $\zeta(\varepsilon^{-1}M^{-\frac{1}{2}}y_n)$ with respect to y_n gives only a factor of $\varepsilon^{-1}M^{-\frac{1}{2}}$. We integrate by parts one time with respect to y_n in the integral

$$\int_{\mathbf{R}^{n-1}}\int_{\mathbf{R}}\frac{e^{2\pi i(-R_x|x-y|+y_n)}e^{-i\varphi}}{(R_x|x-y|)^{\frac{n+1}{2}+\lambda}}\,\psi_{\varepsilon,M}(y)\,dy_n\,dy'$$

to obtain an additional factor of $\varepsilon^{-1}M^{-\frac{1}{2}}$. Thus

$$\left| \int_{\mathbf{R}^n} \frac{e^{2\pi i(-R_x|x-y|+y_n)}e^{-i\varphi}}{(R_x|x-y|)^{\frac{n+1}{2}+\lambda}} \psi_{\varepsilon,M}(y)\,dy \right| \le \frac{c_3\,\varepsilon^n M^{\frac{1}{2}}(\varepsilon^{-1}M^{-\frac{1}{2}})}{M^{\frac{n+1}{2}+\lambda}}. \tag{5.5.5}$$

Finally, using (5.5.2) we obtain that

$$\left| \int_{\mathbf{R}^n} V_{n,\lambda}(R_x|x-y|)e^{2\pi i y_n} \psi_{\varepsilon,M}(y)\,dy \right| \le \frac{c_4\,\varepsilon^n M^{\frac{1}{2}}}{M^{\frac{n+3}{2}+\lambda}}. \tag{5.5.6}$$

We combine (5.5.3), (5.5.4), (5.5.5), and (5.5.6) to deduce for $x \in S_M$,

$$B_*^\lambda(f_M)(x) \ge \frac{c_1\,\varepsilon^n}{M^{\frac{n}{2}+\lambda}} - \frac{c_2\,\varepsilon^{n+1}}{M^{\frac{n}{2}+\lambda}} - \frac{c_3\,\varepsilon^{n-1}}{M^{\frac{n+1}{2}+\lambda}} - \frac{c_4\,\varepsilon^n}{M^{\frac{n+2}{2}+\lambda}}.$$

We pick ε sufficiently small, say $\varepsilon \le c_1/(2c_2)$, and M_0 sufficiently large (depending on the constants c_1, c_2, c_3, c_4) that

$$x \in S_M \implies B_*^\lambda(f_M)(x) > c_0 \frac{1}{M^{\frac{n}{2}+\lambda}}$$

whenever $M \ge M_0$. This fact together with (5.5.1) gives

$$\frac{\big\|B_*^\lambda(f_M)\big\|_{L^{p,\infty}}}{\big\|f_M\big\|_{L^p}} \ge \frac{c_0\,M^{-\frac{n}{2}-\lambda}|S_M|^{\frac{1}{p}}}{c'\,M^{\frac{1}{2p}}} = c\,M^{\frac{2n-1}{2p}-\frac{n}{2}-\lambda},$$

and the required conclusion follows by letting $M \to \infty$. □

5.5.2 *Almost Everywhere Summability of the Bochner–Riesz Means*

We now investigate whether the Bochner–Riesz means B_R^λ converge almost everywhere for functions in $L^p(\mathbf{R}^n)$ for $p \ge 2$. The almost convergence holds when $\lambda > \frac{n-1}{2}$ in view of the result of Exercise 5.2.8. So, in the sequel we fix λ such that $0 < \lambda \le \frac{n-1}{2}$ and we focus our attention to this case. We begin with the following result.

Proposition 5.5.2. *Let $\lambda > 0$ and $0 \le \alpha < 1+2\lambda \le n$. Then there is a constant $C = C(\alpha,\lambda,n)$ such that*

$$\int_{\mathbf{R}^n} |B_*^\lambda(f)(x)|^2 |x|^{-\alpha}\,dx \le C^2 \int_{\mathbf{R}^n} |f(x)|^2 |x|^{-\alpha}\,dx \tag{5.5.7}$$

for all functions $f \in \mathscr{S}(\mathbf{R}^n)$. Moreover, B_*^λ has a unique bounded sublinear extension $\overline{B_*^\lambda}$ on $L^2(\mathbf{R}^n, |x|^{-\alpha}dx)$ which also satisfies (5.5.7). Also, for each $R > 0$, B_R^λ

has a unique bounded linear extension $\overline{B_R^\lambda}$ *on* $L^2(\mathbf{R}^n,|x|^{-\alpha}dx)$ *and the relationship* $\overline{B_*^\lambda}(g)=\sup_{R>0}|\overline{B_R^\lambda}(g)|$ *holds for all* $g\in L^2(\mathbf{R}^n,|x|^{-\alpha}dx)$.

Assuming the result of Proposition 5.5.2, given p such that

$$2\le p<p_\lambda=\frac{2n}{n-1-2\lambda},$$

choose α satisfying

$$0\le n\Big(1-\frac{2}{p}\Big)<\alpha<1+2\lambda=n\Big(1-\frac{2}{p_\lambda}\Big).$$

For simplicity, denote $\overline{B_R^\lambda}$ by B_R^λ and $\overline{B_*^\lambda}$ by B_*^λ. Then B_*^λ is well defined and bounded on L^2 and also on $L^2(|x|^{-\alpha}dx)$. Moreover, $B_R^\lambda(h)\to h$ everywhere as $R\to\infty$ when $h\in\mathscr{S}(\mathbf{R}^n)$ by an easy argument based on the Lebesgue dominated convergence theorem. Using these facts and Theorem 2.1.14 in [156], we deduce that $B_R^\lambda(f)\to f$ a.e. as $R\to\infty$ when f lies in L^2 and also in $L^2(|x|^{-\alpha}dx)$. Since $0\le\alpha<n$, we have

$$L^p\subseteq L^2+L^2(|x|^{-\alpha}),$$

in view of Exercise 5.5.1. Thus for a given function $f\in L^p(\mathbf{R}^n)$, we have that $B_R^\lambda(f)$ is well defined and converges almost everywhere to f as $R\to\infty$. These observations are stated below in the following theorem which is the main result of this section.

Theorem 5.5.3. *Let* $\lambda>0$ *and* $n\ge 2$. *Then for all* f *in* $L^p(\mathbf{R}^n)$ *with* $2\le p<\frac{2n}{n-1-2\lambda}$ *we have*

$$\lim_{R\to\infty}B_R^\lambda(f)(x)=f(x)$$

for almost all $x\in\mathbf{R}^n$.

For the rest of this section we focus attention on Proposition 5.5.2, which requires considerable work. We begin by explaining the last assertions of the theorem. In view of (5.5.7), each B_R^λ is bounded on $L^2(|x|^{-\alpha})$ and it has a unique bounded linear extension $\overline{B_R^\lambda}$ on $L^2(|x|^{-\alpha})$, since $\mathscr{S}(\mathbf{R}^n)$ is dense in $L^2(|x|^{-\alpha})$; for this see Exercise 7.4.1 in [156]. We notice that B_R^λ is given on $\mathscr{S}(\mathbf{R}^n)$ by convolution with the kernel $\Gamma(\lambda+1)\pi^{-\lambda}J_{\frac{n}{2}+\lambda}(2\pi|y|)|y|^{-\frac{n}{2}-\lambda}$ which lies in $L^2(|x|^{\alpha})$ when $2\lambda+1>\alpha$, in view of the asymptotics for Bessel functions in Appendix B.6 and B.7 in [156]. It follows that for a given $g\in L^2(|x|^{-\alpha})$, one may also define $\overline{B_R^\lambda}(g)$ as an absolutely convergent convolution of $\Gamma(\lambda+1)\pi^{-\lambda}J_{\frac{n}{2}+\lambda}(2\pi|y|)|y|^{-\frac{n}{2}-\lambda}$ with g. Moreover, if g_j is a sequence of functions $g_j\in\mathscr{S}(\mathbf{R}^n)$ such that $g_j\to g$ in $L^2(|x|^{-\alpha})$, then for any $x\in\mathbf{R}^n$ we have

$$\begin{aligned}|\overline{B_R^\lambda}(g)(x)-B_R^\lambda(g_j)(x)|&=\left|\int_{\mathbf{R}^n}(g(x-y)-g_j(x-y))\frac{\Gamma(\lambda+1)}{\pi^\lambda}\frac{R^nJ_{\frac{n}{2}+\lambda}(2\pi R|y|)}{|Ry|^{\frac{n}{2}+\lambda}}dy\right|\\&\le C\,\|g_j-g\|_{L^2(|x|^{-\alpha})}R^n\left(\int_{\mathbf{R}^n}\left|\frac{J_{\frac{n}{2}+\lambda}(2\pi R|y|)}{|Ry|^{\frac{n}{2}+\lambda}}\right|^2|y|^\alpha dy\right)^{\frac12}\end{aligned}$$

which tends to zero as $j \to \infty$, since the integral produces a constant, as observed. We conclude that the unique extension $\overline{B_R^\lambda}$ of B_R^λ on $L^2(|x|^{-\alpha})$ is indeed $\overline{B_R^\lambda}$ which was previously defined as an absolutely convergent convolution on $L^2(|x|^{-\alpha})$.

We continue with a similar discussion for B_*^λ. Clearly B_*^λ is a sublinear operator with nonnegative values, and hence it satisfies

$$|B_*^\lambda(f) - B_*^\lambda(h)| \le B_*^\lambda(f-h) \tag{5.5.8}$$

for all $f, h \in \mathscr{S}(\mathbf{R}^n)$. We define $\overline{B_*^\lambda}(f) = \lim_{j\to\infty} B_*^\lambda(f_j)$, where f_j is a sequence of Schwartz functions that converges to f in $L^2(|x|^{-\alpha})$. This limit exists in $L^2(|x|^{-\alpha})$, since the sequence $\{B_*^\lambda(f_j)\}_j$ is Cauchy in this space, in view of (5.5.8) and (5.5.7). By Fatou's lemma $\overline{B_*^\lambda}$ is bounded from $L^2(|x|^{-\alpha})$ to $L^2(|x|^{-\alpha})$ and is the unique extension of B_*^λ: indeed, if $\overline{\overline{B_*^\lambda}}$ is another sublinear bounded extension of B_*^λ on $L^2(|x|^{-\alpha})$ that coincides with B_*^λ on $\mathscr{S}$, then given f in $L^2(|x|^{-\alpha})$ and f_j as before, $\overline{\overline{B_*^\lambda}}(f)$ is the limit of $\overline{\overline{B_*^\lambda}}(f_j) = B_*^\lambda(f_j)$ in $L^2(|x|^{-\alpha})$ and likewise $\overline{B_*^\lambda}(f)$ is the limit of $\overline{B_*^\lambda}(f_j) = B_*^\lambda(f_j)$ in $L^2(|x|^{-\alpha})$. Thus $\overline{\overline{B_*^\lambda}}(f)$ and $\overline{B_*^\lambda}(f)$ are the limits of the same sequence and must coincide. To verify that

$$\overline{B_*^\lambda}(g) = \sup_{R>0} \overline{B_R^\lambda}(g)$$

for all $g \in L^2(|x|^{-\alpha})$, it will suffice to show that $g \mapsto \sup_{R>0} \overline{B_R^\lambda}(g)$ is a bounded extension of B_*^λ on $L^2(|x|^{-\alpha})$; then by the preceding observation, it must coincide with $\overline{B_*^\lambda}$ on $L^2(|x|^{-\alpha})$. Indeed, if $g_j \in \mathscr{S}$ converge to g in $L^2(|x|^{-\alpha})$, then

$$\begin{aligned}
\Big\| \sup_{R>0} |\overline{B_R^\lambda}(g)| \Big\|_{L^2(|x|^{-\alpha})} &= \Big\| \sup_{R>0} |\lim_{j\to\infty} B_R^\lambda(g_j)| \Big\|_{L^2(|x|^{-\alpha})} \\
&= \Big\| \sup_{R>0} |\liminf_{j\to\infty} B_R^\lambda(g_j)| \Big\|_{L^2(|x|^{-\alpha})} \\
&\le \Big\| \liminf_{j\to\infty} \sup_{R>0} |B_R^\lambda(g_j)| \Big\|_{L^2(|x|^{-\alpha})} \\
&\le \liminf_{j\to\infty} \Big\| \sup_{R>0} |B_R^\lambda(g_j)| \Big\|_{L^2(|x|^{-\alpha})} \\
&\le \liminf_{j\to\infty} C \|g_j\|_{L^2(|x|^{-\alpha})} \\
&= C \|g\|_{L^2(|x|^{-\alpha})} .
\end{aligned}$$

This concludes the proof of the last last assertion of Proposition 5.5.2.

We therefore fix a Schwartz function f on $\mathbf{R}^n$ and we focus on proving (5.5.7). We decompose the multiplier $(1-|\xi|^2)_+^\lambda$ as an infinite sum of smooth bumps supported in small concentric annuli in the interior of the sphere $|\xi| = 1$ as we did in the proof of Theorem 5.2.4.

We pick a smooth function φ supported in $[-\frac{1}{2}, \frac{1}{2}]$ and a smooth function ψ supported in $[\frac{1}{8}, \frac{5}{8}]$ and with values in $[0,1]$ that satisfy

$$\varphi(t) + \sum_{k=0}^{\infty} \psi\Big(\frac{1-t}{2^{-k}}\Big) = 1$$

for all $t \in [0,1)$. We decompose the multiplier $(1-|\xi|^2)_+^\lambda$ as

$$(1-|\xi|^2)_+^\lambda = m_{00}(\xi) + \sum_{k=0}^{\infty} 2^{-k\lambda} m_k(\xi), \tag{5.5.9}$$

where $m_{00}(\xi) = \varphi(|\xi|)(1-|\xi|^2)^\lambda$, and for $k \geq 0$, m_k is defined by

$$m_k(\xi) = \Big(\frac{1-|\xi|}{2^{-k}}\Big)^\lambda \psi\Big(\frac{1-|\xi|}{2^{-k}}\Big)(1+|\xi|)^\lambda .$$

Then we define maximal operators associated with the multipliers m_{00} and m_k,

$$S_*^{m_k}(f)(x) = \sup_{R>0} \big|\big(\widehat{f}(\xi) m_k(\xi/R)\big)^\vee(x)\big| ,$$

for $k \geq 0$, and analogously we define $S_*^{m_{00}}$. Using (5.5.9) we have

$$B_*^\lambda(f) \leq S_*^{m_{00}}(f) + \sum_{k=0}^{\infty} 2^{-k\lambda} S_*^{m_k}(f) . \tag{5.5.10}$$

Since $S_*^{m_{00}}$, $S_*^{m_0}$, $S_*^{m_1}$ and any finite number of them are pointwise controlled by the Hardy–Littlewood maximal operator, which is bounded on $L^2(|x|^b)$ whenever $-n < b < n$ (cf. Theorem 7.1.9 and Example 7.1.7 in [156]), we focus attention on the remaining terms.

We make a small change of notation. Thinking of 2^{-k} as roughly being δ (precisely $\delta = 2^{-k-3}$), for $\delta < 1/10$ we let $m^\delta(t)$ be a smooth function supported in the interval $[1-5\delta, 1-\delta]$ and taking values in the interval $[0,1]$ that satisfies

$$\sup_{1-5\delta \leq t \leq 1-\delta} \Big|\frac{d^\ell}{dt^\ell} m^\delta(t)\Big| \leq C_\ell \delta^{-\ell} \tag{5.5.11}$$

for all $\ell \in \mathbf{Z}^+ \cup \{0\}$. We define a related function

$$\widetilde{m}^\delta(t) = \delta t \frac{d}{dt} m^\delta(t) ,$$

which obviously satisfies estimates (5.5.11) with another constant $\widetilde{C}_\ell$ in place of C_ℓ.

Next we introduce the multiplier operators

$$S_t^\delta(f)(x) = \big(\widehat{f}(\xi) m^\delta(t|\xi|)\big)^\vee(x) , \qquad \widetilde{S}_t^\delta(f)(x) = \big(\widehat{f}(\xi) \widetilde{m}^\delta(t|\xi|)\big)^\vee(x) ,$$

and the $L^2(|x|^{-\alpha})$-bounded maximal multiplier operator

$$S_*^\delta(f) = \sup_{t>0} |S_t^\delta(f)|,$$

as well as the continuous square functions

$$G^\delta(f)(x) = \left(\int_0^\infty |S_t^\delta(f)(x)|^2 \frac{dt}{t} \right)^{\frac{1}{2}}, \quad \widetilde{G}^\delta(f)(x) = \left(\int_0^\infty |\widetilde{S}_t^\delta(f)(x)|^2 \frac{dt}{t} \right)^{\frac{1}{2}}.$$

The operators S_t^δ and $\widetilde{S}_t^\delta$ are related as follows

$$\frac{d}{dt} S_t^\delta(f) = \frac{1}{\delta t} \widetilde{S}_t^\delta(f),$$

an identity easily obtained by passing the differentiation inside the integral.

An application of the fundamental theorem of calculus yields

$$\begin{aligned} |S_t^\delta(f)(x)|^2 &= 2\operatorname{Re} \int_0^t \overline{S_u^\delta(f)(x)} \frac{d}{du} S_u^\delta(f)(x)\, du \\ &= \frac{2}{\delta} \operatorname{Re} \int_0^t \overline{S_u^\delta(f)(x)}\, \widetilde{S}_u^\delta(f)(x) \frac{du}{u}. \end{aligned}$$

Consequently,

$$|S_t^\delta(f)(x)|^2 \le \frac{2}{\delta} \int_0^t |S_u^\delta(f)(x)|\,|\widetilde{S}_u^\delta(f)(x)| \frac{du}{u} \le \frac{2}{\delta} |G^\delta(f)(x)|\,|\widetilde{G}^\delta(f)(x)|$$

for all $t > 0$ and all $x \in \mathbf{R}^n$, since $f \in \mathscr{S}(\mathbf{R}^n)$. It follows that

$$\big\| S_*^\delta(f) \big\|_{L^2(|x|^{-\alpha})}^2 \le \frac{2}{\delta} \big\| G^\delta(f) \big\|_{L^2(|x|^{-\alpha})} \big\| \widetilde{G}^\delta(f) \big\|_{L^2(|x|^{-\alpha})}, \tag{5.5.12}$$

and the asserted boundedness of S_*^δ reduces to that of the continuous square functions G^δ and $\widetilde{G}^\delta$ on weighted L^2 spaces with suitable constants depending on δ.

The boundedness of G^δ on $L^2(|x|^{-\alpha})$ is a consequence of the following lemma.

Lemma 5.5.4. *For $0 < \delta < 1/10$ and $0 \le \alpha < n$ we have*

$$\int_{\mathbf{R}^n} \int_1^2 |S_t^\delta(f)(x)|^2 \frac{dt}{t} \frac{dx}{|x|^\alpha} \le C_{n,\alpha} A_\alpha(\delta) \int_{\mathbf{R}^n} |f(x)|^2 \frac{dx}{|x|^\alpha} \tag{5.5.13}$$

for f in $\mathscr{S}(\mathbf{R}^n)$, where for $\varepsilon > 0$, $A_\alpha(\varepsilon)$ is defined by

$$A_\alpha(\varepsilon) = \begin{cases} \varepsilon^{2-\alpha} & \text{when } 1 < \alpha < n, \\ \varepsilon(|\log \varepsilon| + 1) & \text{when } \alpha = 1, \\ \varepsilon & \text{when } 0 \le \alpha < 1. \end{cases} \tag{5.5.14}$$

Assuming the statement of the lemma, we conclude the proof of Proposition 5.5.2 as follows. We take a Schwartz function ζ such that $\widehat{\zeta}$ vanishes in a neighborhood of the origin with $\widehat{\zeta}(\xi) = 1$ whenever $1/2 \le |\xi| \le 2$ and we let $\zeta_{2^k}(x) = 2^{-kn}\zeta(2^{-k}x)$. We make the observation that if $1 - 5\delta \le t|\xi| \le 1 - \delta$ and $2^{k-1} \le t \le 2^k$, then $1/2 \le 2^k|\xi| \le 2$, since $\delta < 1/10$. This implies that $\widehat{\zeta}(2^k\xi) = 1$ on the support of the function $\xi \mapsto m^\delta(t|\xi|)$. Hence

$$S_t^\delta(f) = S_t^\delta(\zeta_{2^k} * f)$$

whenever $2^{k-1} \le t \le 2^k$, and Lemma 5.5.4 (in conjunction with Exercise 5.5.2) yields

$$\int_{\mathbf{R}^n} \int_{2^{k-1}}^{2^k} |S_t^\delta(f)(x)|^2 \frac{dt}{t} \frac{dx}{|x|^\alpha} \le C_{n,\alpha} A_\alpha(\delta) \int_{\mathbf{R}^n} |\zeta_{2^k} * f(x)|^2 \frac{dx}{|x|^\alpha}.$$

Summing over $k \in \mathbf{Z}$ we obtain

$$\big\|G^\delta(f)\big\|^2_{L^2(|x|^{-\alpha})} \le C_{n,\alpha} A_\alpha(\delta) \Big\| \Big(\sum_{k\in\mathbf{Z}} |\zeta_{2^k} * f|^2 \Big)^{\frac{1}{2}} \Big\|^2_{L^2(|x|^{-\alpha})}.$$

A randomization argument allows us to linearize the problem as follows

$$\Big\| \Big(\sum_{k\in\mathbf{Z}} |\zeta_{2^k} * f|^2 \Big)^{\frac{1}{2}} \Big\|^2_{L^2(|x|^{-\alpha})} = \int_0^1 \Big\| \sum_{k\in\mathbf{Z}} r_k(t)(\zeta_{2^k} * f) \Big\|^2_{L^2(|x|^{-\alpha})} dt, \tag{5.5.15}$$

where r_k denotes a renumbering of the Rademacher functions (Appendix C.1 in [156]) indexed by the entire set of integers. A proof of (5.5.15) can be given by first restricting the L^2 norm to the ball $|x| \le M$ and using that $\sum_k \|\zeta_{2^k} * f\|_{L^\infty} < \infty$, a fact contained in Exercise 4.4.1(d), and then letting $M \uparrow \infty$.

For each $t \in [0,1]$ the operator

$$M_t(f) = \sum_{k\in\mathbf{Z}} r_k(t)(\zeta_{2^k} * f)$$

is associated with a multiplier m_t that satisfies $|\partial^\alpha m_t(\xi)| \le C_\alpha |\xi|^{-|\alpha|}$ uniformly in t. It follows that M_t is a singular integral operator bounded on all the L^p spaces for $1 < p < \infty$, and in view of Theorem 7.4.6 in [156], it is also bounded on $L^2(w)$ whenever $w \in A_2$. Since the weight $|x|^{-\alpha}$ is in A_2 whenever $-n < \alpha < n$, it follows that M_t is bounded on $L^2(|x|^{-\alpha})$ with a bound independent of $t > 0$. We deduce that

$$\big\|G^\delta(f)\big\|_{L^2(|x|^{-\alpha})} + \big\|\widetilde{G}^\delta(f)\big\|_{L^2(|x|^{-\alpha})} \le C'_{n,\alpha} \big(A_\alpha(\delta)\big)^{\frac{1}{2}} \big\|f\big\|_{L^2(|x|^{-\alpha})}.$$

We now recall estimate (5.5.12) to obtain

$$\big\|S_*^\delta(f)\big\|_{L^2(|x|^{-\alpha})} \le C'(n,\alpha)\big(\delta^{-1}A_\alpha(\delta)\big)^{1/2} \big\|f\big\|_{L^2(|x|^{-\alpha})}.$$

Taking $\delta = 2^{-k-3}$, recalling the value of $A_\alpha(\delta)$ from Lemma 5.5.4, and inserting this estimate in (5.5.10), we deduce Proposition 5.5.2. We note that the condition $\alpha < 1+2\lambda$ is needed to make the series in (5.5.10) converge when $1 < \alpha < n$.

5.5.3 Estimates for Radial Multipliers

It remains to prove Lemma 5.5.4. We restrict attention to $\alpha > 0$, since (5.5.13) holds for $\alpha = 0$ by Plancherel's identity and the observation that

$$\int_1^2 |m^\delta(t|\xi|)|^2 \frac{dt}{t} \le |\text{support } (m^\delta)| = 4\delta.$$

So we fix an $\alpha \in (0,n)$ and we reduce estimate (5.5.13) to an estimate for a single t with the bound $A_\alpha(\delta)/\delta$, which is worse than $A_\alpha(\delta)$. The reduction to a single t is achieved via duality. Estimate (5.5.13) says that the operator $f \mapsto \{S_t^\delta(f)\}_{1\le t\le 2}$ is bounded from $L^2(\mathbf{R}^n, |x|^{-\alpha}dx)$ to $L^2(L^2(\frac{dt}{t}), |x|^{-\alpha}dx)$. The dual statement of this fact is that the operator

$$\{g_t\}_{1\le t\le 2} \mapsto \int_1^2 S_t^\delta(g_t)\frac{dt}{t}$$

maps $L^2(L^2(\frac{dt}{t}), |x|^\alpha dx)$ to $L^2(\mathbf{R}^n, |x|^\alpha dx)$. Here we use the fact that the operators S_t are self-transpose and self-adjoint, since they have real and radial multipliers. Thus estimate (5.5.13) is equivalent to

$$\int_{\mathbf{R}^n}\left|\int_1^2 S_t^\delta(g_t)(x)\frac{dt}{t}\right|^2 |x|^\alpha\,dx \le C_{n,\alpha}A_\alpha(\delta)\int_{\mathbf{R}^n}\int_1^2 |g_t(x)|^2\frac{dt}{t}|x|^\alpha\,dx, \tag{5.5.16}$$

which by Plancherel's theorem is also equivalent to

$$\int_{\mathbf{R}^n}\left|\mathscr{D}^{\frac{\alpha}{2}}\left(\int_1^2 m^\delta(t|\cdot|)\widehat{g_t}(\cdot)\frac{dt}{t}\right)(\xi)\right|^2 d\xi \le C_{n,\alpha}A_\alpha(\delta)\int_{\mathbf{R}^n}\int_1^2 |\mathscr{D}^{\frac{\alpha}{2}}(\widehat{g_t})(\xi)|^2\frac{dt}{t}\,d\xi.$$

Here

$$\mathscr{D}^\beta(h)(x) = \left[\int_{\mathbf{R}^n}\frac{|D_y^{[\beta]+1}(h)(x)|^2}{|y|^{n+2\beta}}\,dy\right]^{\frac{1}{2}},$$

where $D_y(f)(x) = f(x+y) - f(x)$ is the difference operator introduced in Section 1.4 and $D_y^k = D_y \circ \cdots \circ D_y$ (k times). The operator $\mathscr{D}^\beta$ obeys the identity (see Exercise 1.4.9)

$$\|\mathscr{D}^\beta(\widehat{h})\|_{L^2}^2 = c_0(n,\beta)\int_{\mathbf{R}^n}|h(x)|^2\,|x|^{2\beta}\,dx.$$

Using the definition of $\mathscr{D}^{\alpha/2}$ we write

$$\left|\mathscr{D}^{\frac{\alpha}{2}}\left(\int_1^2 m^\delta(t|\cdot|)\widehat{g_t}(\cdot)\frac{dt}{t}\right)(\xi)\right|^2 = \int_{\mathbf{R}^n}\left|\int_1^2 D_\eta^{[\frac{\alpha}{2}]+1}(m^\delta(t|\cdot|)\widehat{g_t}(\cdot))(\xi)\frac{dt}{t}\right|^2\frac{d\eta}{|\eta|^{n+\alpha}}.$$

If the inner integrand on the right is nonzero, expressing D_y^{k+1} as in (1.4.2) and using the support properties of m^δ, we obtain that $1-5\delta \le t|\xi+s\eta| \le 1-\delta$ for some $s \in \{0,1,\dots,[\alpha/2]+1\}$; thus for each such s, t belongs to an interval of length $4\delta|\xi+s\eta|^{-1} \le 4\delta t(1-5\delta)^{-1}$. Since $t\le 2$ and $\delta \le 1/10$, it follows that t lies in a set of measure at most $16([\alpha/2]+2)\delta$. The Cauchy–Schwarz inequality then yields

$$\left|\mathscr{D}^{\frac{\alpha}{2}}\left(\int_1^2 m^\delta(t|\cdot|)\widehat{g_t}(\cdot)\frac{dt}{t}\right)(\xi)\right|^2 \le c_\alpha\,\delta\int_{\mathbf{R}^n}\int_1^2\left|D_\eta^{[\frac{\alpha}{2}]+1}\left(m^\delta(t|\cdot|)\widehat{g_t}(\cdot)\right)(\xi)\right|^2\frac{dt}{t}\frac{d\eta}{|\eta|^{n+\alpha}}.$$

In view of the preceding reduction, we deduce that (5.5.16) is a consequence of

$$\int_{\mathbf{R}^n}\int_{\mathbf{R}^n}\int_1^2\left|D_\eta^{[\frac{\alpha}{2}]+1}\left(m^\delta(t|\cdot|)\widehat{g_t}(\cdot)\right)(\xi)\right|^2\frac{dt}{t}\frac{d\eta}{|\eta|^{n+\alpha}}\,d\xi \le C_{n,\alpha}\frac{A_\alpha(\delta)}{c_\alpha\,\delta}\int_{\mathbf{R}^n}\int_1^2\left|\mathscr{D}^{\frac{\alpha}{2}}(\widehat{g_t})(\xi)\right|^2\frac{dt}{t}\,d\xi$$

which can also be written as

$$\int_{\mathbf{R}^n}\int_1^2\left|\mathscr{D}^{\frac{\alpha}{2}}\left(m^\delta(t|\cdot|)\widehat{g_t}(\cdot)\right)(\xi)\right|^2\frac{dt}{t}\,d\xi \le \frac{C_{n,\alpha}}{c_\alpha}\frac{A_\alpha(\delta)}{\delta}\int_{\mathbf{R}^n}\int_1^2\left|\mathscr{D}^{\frac{\alpha}{2}}(\widehat{g_t})(\xi)\right|^2\frac{dt}{t}\,d\xi.$$

This estimate is a consequence of

$$\int_{\mathbf{R}^n}\left|\mathscr{D}^{\frac{\alpha}{2}}\left(m^\delta(t|\cdot|)\widehat{g_t}(\cdot)\right)(\xi)\right|^2 d\xi \le \frac{C_{n,\alpha}}{c_\alpha}\frac{A_\alpha(\delta)}{\delta}\int_{\mathbf{R}^n}\left|\mathscr{D}^{\frac{\alpha}{2}}(\widehat{g_t})(\xi)\right|^2 d\xi \tag{5.5.17}$$

for all $t\in[1,2]$. A simple dilation argument reduces (5.5.17) to the single estimate

$$\int_{\mathbf{R}^n}\left|\mathscr{D}^{\frac{\alpha}{2}}\left(m^\delta(|\cdot|)\widehat{g}(\cdot)\right)(\xi)\right|^2 d\xi \le \frac{C_{n,\alpha}}{c_\alpha}\frac{A_\alpha(\delta)}{\delta}\int_{\mathbf{R}^n}\left|\mathscr{D}^{\frac{\alpha}{2}}(\widehat{g})(\xi)\right|^2 d\xi, \tag{5.5.18}$$

which is equivalent to

$$\int_{\mathbf{R}^n}|S_1^\delta(g)(x)|^2|x|^\alpha\,dx \le \frac{C_{n,\alpha}}{c_\alpha}\frac{A_\alpha(\delta)}{\delta}\int_{\mathbf{R}^n}|g(x)|^2|x|^\alpha\,dx$$

and also equivalent to

$$\int_{\mathbf{R}^n}|S_1^\delta(f)(x)|^2\frac{dx}{|x|^\alpha} \le \frac{C_{n,\alpha}}{c_\alpha}\frac{A_\alpha(\delta)}{\delta}\int_{\mathbf{R}^n}|f(x)|^2\frac{dx}{|x|^\alpha} \tag{5.5.19}$$

by duality. We have now reduced estimate (5.5.13) to (5.5.19).

We denote by $K^\delta(x)$ the kernel of the operator S_1^δ, i.e., the inverse Fourier transform of the multiplier $m^\delta(|\xi|)$. Certainly K^δ is a radial kernel on $\mathbf{R}^n$, and it is convenient to decompose it radially as

$$K^\delta = K_0^\delta + \sum_{j=1}^{\infty} K_j^\delta,$$

where $K_0^\delta(x) = K^\delta(x)\phi(\delta x)$ and $K_j^\delta(x) = K^\delta(x)\big(\phi(2^{-j}\delta x) - \phi(2^{1-j}\delta x)\big)$, for some radial smooth function ϕ supported in the ball $B(0,2)$ and equal to one on $B(0,1)$.

To prove estimate (5.5.19) we make use of the subsequent lemmas.

Lemma 5.5.5. *For all $M \geq 2n$ there is a constant $C_M = C_M(n,\phi)$ such that for all $j = 0,1,2,\dots$ we have*

$$\sup_{\xi\in\mathbf{R}^n} |\widehat{K_j^\delta}(\xi)| \leq C_M 2^{-jM} \tag{5.5.20}$$

and also

$$|\widehat{K_j^\delta}(\xi)| \leq C_M 2^{-(j+k)M} \tag{5.5.21}$$

whenever $\big||\xi|-1\big| \geq 2^k\delta$ and $k \geq 4$. Also

$$|\widehat{K_j^\delta}(\xi)| \leq C_M 2^{-jM}\,\delta^M (1+|\xi|)^{-M} \tag{5.5.22}$$

whenever $|\xi| \leq 1/8$ or $|\xi| \geq 15/8$.

Lemma 5.5.6. *Let $0 < \alpha < n$. Then there is a constant $C(n,\alpha)$ such that for all Schwartz functions f and all $\varepsilon > 0$ we have*

$$\int_{||\xi|-1|\leq\varepsilon} |\widehat{f}(\xi)|^2\, d\xi \leq C(n,\alpha)\,\varepsilon^{\alpha-1} A_\alpha(\varepsilon) \int_{\mathbf{R}^n} |f(x)|^2 |x|^\alpha dx \tag{5.5.23}$$

and also for $M \geq 2n$ there is a constant $C_M(n,\alpha)$ such that

$$\int_{\mathbf{R}^n} |\widehat{f}(\xi)|^2 \frac{1}{(1+|\xi|)^M}\, d\xi \leq C_M(n,\alpha) \int_{\mathbf{R}^n} |f(x)|^2 |x|^\alpha dx. \tag{5.5.24}$$

Assuming Lemmas 5.5.5 and 5.5.6 we prove estimate (5.5.19) as follows. Using Plancherel's theorem we write

$$\int_{\mathbf{R}^n} |(K_j^\delta * f)(x)|^2\, dx = \int_{\mathbf{R}^n} |\widehat{K_j^\delta}(\xi)|^2 |\widehat{f}(\xi)|^2\, d\xi \leq I + II + III,$$

where

$$I = \int_{|\xi|\leq\frac{1}{8},|\xi|\geq\frac{15}{8}} |\widehat{K_j^\delta}(\xi)|^2 |\widehat{f}(\xi)|^2\, d\xi,$$

$$II = \sum_{k=4}^{[\log_2 \frac{7}{16}\delta^{-1}]+1} \int_{2^k\delta \le ||\xi|-1| \le 2^{k+1}\delta} |\widehat{K_j^\delta}(\xi)|^2 |\widehat{f}(\xi)|^2 \, d\xi\,,$$

$$III = \int_{||\xi|-1|\le 16\delta} |\widehat{K_j^\delta}(\xi)|^2 |\widehat{f}(\xi)|^2 \, d\xi\,.$$

Using (5.5.22) and (5.5.24) we obtain that

$$I \le C'_M(n,\alpha) 2^{-jM} \delta^M \int_{\mathbf{R}^n} |f(x)|^2 \, |x|^\alpha dx\,.$$

In view of (5.5.21) and (5.5.23) we have

$$\begin{aligned} II &\le \sum_{k=4}^{[\log_2 \delta^{-1}]+1} C(n,\alpha)(2^{k+1}\delta)^{\alpha-1} A_\alpha(2^{k+1}\delta) 2^{-jM} 2^{-kM} \int_{\mathbf{R}^n} |f(x)|^2 \, |x|^\alpha dx \\ &\le C'_M(n,\alpha) 2^{-jM} \delta^{\alpha-1} A_\alpha(\delta) \int_{\mathbf{R}^n} |f(x)|^2 \, |x|^\alpha dx\,. \end{aligned}$$

Finally, (5.5.20) and (5.5.23) yield

$$III \le C'_M(n,\alpha) 2^{-jM} \delta^{\alpha-1} A_\alpha(\delta) \int_{\mathbf{R}^n} |f(x)|^2 \, |x|^\alpha dx\,.$$

Summing the estimates for I, II, and III we deduce

$$\int_{\mathbf{R}^n} |(K_j^\delta * f)(x)|^2 \, dx \le C_M(n,\alpha) 2^{-jM} \delta^{\alpha-1} A_\alpha(\delta) \int_{\mathbf{R}^n} |f(x)|^2 \, |x|^\alpha dx\,.$$

By duality, this estimate can be written as

$$\int_{\mathbf{R}^n} |(K_j^\delta * f)(x)|^2 \, \frac{dx}{|x|^\alpha} \le C_M(n,\alpha) 2^{-jM} \delta^{\alpha-1} A_\alpha(\delta) \int_{\mathbf{R}^n} |f(x)|^2 \, dx\,. \tag{5.5.25}$$

Given a Schwartz function f, we write $f_0 = f\chi_{Q_0}$, where Q_0 is a cube centered at the origin of side length $C 2^j/\delta$ for some C to be chosen. Then for $x \in Q_0$ we have $|x| \le C\sqrt{n}\, 2^j/\delta$; hence

$$\begin{aligned} \int_{\mathbf{R}^n} |(K_j^\delta * f_0)(x)|^2 \, \frac{dx}{|x|^\alpha} &\le \frac{C'_M(n,\alpha)\,\delta^{\alpha-1} A_\alpha(\delta)}{2^{jM}} \Big(C\sqrt{n}\, \frac{2^j}{\delta} \Big)^\alpha \int_{Q_0} |f_0(x)|^2 \, \frac{dx}{|x|^\alpha} \\ &= C''_M(n,\alpha) 2^{j(\alpha-M)} \frac{A_\alpha(\delta)}{\delta} \int_{Q_0} |f_0(x)|^2 \, \frac{dx}{|x|^\alpha}\,. \end{aligned} \tag{5.5.26}$$

Now write $\mathbf{R}^n \setminus Q_0$ as a mesh of cubes Q_i, indexed by $i \in \mathbf{Z} \setminus \{0\}$, of side lengths $2^{j+2}/\delta$ and centers c_{Q_i}. Since K_j^δ is supported in a ball of radius $2^{j+1}/\delta$, if f_i is supported in Q_i, then $f_i * K_j^\delta$ is supported in the cube $2\sqrt{n}\, Q_i$. If the constant C is large enough, say $C \ge 1000n$, then for $x \in Q_i$ and $x' \in 2\sqrt{n}\, Q_i$ we have

$$|x| \approx |c_{Q_i}| \approx |x'|\,,$$

which says that the moduli of x and x' are comparable in the following inequality:

$$\int_{2\sqrt{n}Q_i} |(K_j^\delta * f_i)(x')|^2 \frac{dx'}{|x'|^\alpha} \leq C_M' 2^{-jM} \int_{Q_i} |f_i(x)|^2 \frac{dx}{|x|^\alpha}. \tag{5.5.27}$$

Thus (5.5.27) is a consequence of

$$\int_{2\sqrt{n}Q_i} |(K_j^\delta * f_i)(x')|^2 \, dx' \leq C_M 2^{-jM} \int_{Q_i} |f_i(x)|^2 \, dx, \tag{5.5.28}$$

which is certainly satisfied, as seen by applying Plancherel's theorem and using (5.5.20). Since for $\delta < 1/10$ we have $A_\alpha(\delta)/\delta \geq 1$, it follows that

$$\int_{\mathbf{R}^n} |(K_j^\delta * f_i)(x)|^2 \frac{dx}{|x|^\alpha} \leq C_M 2^{-jM} \frac{A_\alpha(\delta)}{\delta} \int_{\mathbf{R}^n} |f_i(x)|^2 \frac{dx}{|x|^\alpha} \tag{5.5.29}$$

whenever f_i is supported in Q_i. We now pick $M = 2n$ and we recall that $\alpha < n$. We have now proved that

$$\int_{\mathbf{R}^n} |(K_j^\delta * f_i)(x)|^2 \frac{dx}{|x|^\alpha} \leq C''(n,\alpha) 2^{-jn} \frac{A_\alpha(\delta)}{\delta} \int_{Q_i} |f_i(x)|^2 \frac{dx}{|x|^\alpha}$$

for functions f_i supported in Q_i.

Given a general function f in the Schwartz class, write

$$f = \sum_{i \in \mathbf{Z}} f_i, \qquad \text{where} \qquad f_i = f \chi_{Q_i}.$$

Then

$$\begin{aligned}
\big\|K_j^\delta * f\big\|_{L^2(|x|^{-\alpha})}^2 &\leq 2\big\|K_j^\delta * f_0\big\|_{L^2(|x|^{-\alpha})}^2 + 2\Big\|\sum_{i \neq 0} K_j^\delta * f_i\Big\|_{L^2(|x|^{-\alpha})}^2 \\
&\leq 2\big\|K_j^\delta * f_0\big\|_{L^2(|x|^{-\alpha})}^2 + 2C_n \sum_{i \neq 0} \big\|K_j^\delta * f_i\big\|_{L^2(|x|^{-\alpha})}^2 \\
&\leq C'''(n,\alpha) 2^{-jn} \frac{A_\alpha(\delta)}{\delta} \Big[\big\|f_0\big\|_{L^2(|x|^{-\alpha})}^2 + \sum_{i \neq 0} \big\|f_i\big\|_{L^2(|x|^{-\alpha})}^2\Big] \\
&= C'''(n,\alpha) 2^{-jn} \frac{A_\alpha(\delta)}{\delta} \big\|f\big\|_{L^2(|x|^{-\alpha})}^2,
\end{aligned}$$

where we used the bounded overlap property of the family $\{K_j * f_i\}_{i \neq 0}$ in the second inequality of the preceding alignment (cf. Exercise 5.4.4). Taking square roots and summing over $j = 0, 1, 2, \dots$, we deduce (5.5.19).

We now address the proof of Lemma 5.5.5, which was left open.

Proof. For the purposes of this proof we set $\psi(x) = \phi(x) - \phi(2x)$. Then the inverse Fourier transform of the function $x \mapsto \psi(2^{-j}\delta x)$ is $\xi \mapsto 2^{jn}\delta^{-n}\widehat{\psi}(2^j\xi/\delta)$.

Convolving the latter with the function $\xi \mapsto m^\delta(|\xi|)$, we obtain $\widehat{K_j^\delta}(\xi)$. We may therefore write for $j \geq 1$,

$$\widehat{K_j^\delta}(\xi) = \int_{\mathbf{R}^n} m^\delta(|\xi - 2^{-j}\delta\eta|)\widehat{\psi}(\eta)\,d\eta\,, \tag{5.5.30}$$

while for $j = 0$ an analogous formula holds with ϕ in place of ψ. Since $|m^\delta| \leq 1$, (5.5.20) follows easily when $j = 0$. For $j \geq 1$ we expand the function

$$\xi \mapsto m^\delta(|\xi - 2^{-j}\delta\eta|)$$

in a Taylor series and we make use of the fact that $\widehat{\psi}$ has vanishing moments of all orders to obtain

$$\begin{aligned} |\widehat{K_j^\delta}(\xi)| &\leq \int_{\mathbf{R}^n} \sum_{|\gamma|=M} \frac{1}{\gamma!} \|\partial^\gamma m^\delta(|\cdot|)\|_{L^\infty} |2^{-j}\delta\eta|^M |\widehat{\psi}(\eta)|\,d\eta \\ &\leq C(M)\delta^{-M}\delta^M 2^{-jM} \int_{\mathbf{R}^n} |\eta|^M |\widehat{\psi}(\eta)|\,d\eta\,. \end{aligned}$$

This proves (5.5.20).

We turn now to the proof of (5.5.21). Suppose that $||\xi| - 1| \geq 2^k\delta$ and $k \geq 4$. Then for $|\xi| \leq 1$, recalling that m^δ is supported in $[1-5\delta, 1-\delta]$, we write

$$|2^{-j}\delta\eta| \geq |\xi - 2^{-j}\delta\eta| - |\xi| \geq (1-5\delta) - (1-2^k\delta) \geq 2^{k-1}\delta\,,$$

since $k \geq 4$. For $|\xi| \geq 1$ we have

$$|2^{-j}\delta\eta| \geq |\xi| - |\xi - 2^{-j}\delta\eta| \geq (1+2^k\delta) - (1-\delta) \geq 2^k\delta\,.$$

In either case we conclude that $|\eta| \geq 2^{k+j-1}$, and using (5.5.30) we deduce

$$|\widehat{K_j^\delta}(\xi)| \leq \int_{|\eta| \geq 2^{k+j-1}} |\widehat{\psi}(\eta)|\,d\eta \leq C_M 2^{-(j+k)M}\,.$$

The proof of (5.5.22) is similar. Since

$$|\xi - 2^{-j}\delta\eta| \geq 1 - 5\delta \geq 1/2,$$

if $|\xi| \leq 1/8$, it follows that $|2^{-j}\delta\eta| \geq 1/4$. Likewise, if $|\xi| \geq 15/8$, then

$$|2^{-j}\delta\eta| \geq |\xi| - 1 \geq |\xi|/4.$$

These estimates imply

$$|2^{-j}\delta\eta| \geq \frac{1}{8}(1+|\xi|) \implies |\eta| \geq 2^j \frac{1}{8\delta}(1+|\xi|)$$

in the support of the integral in (5.5.30). It follows that

$$|\widehat{K_j^\delta}(\xi)| \le \int_{|\eta|\ge 2^{j-3}(1+|\xi|)/\delta} |\widehat{\psi}(\eta)|\,d\eta \le C_M 2^{-jM}\delta^M(1+|\xi|)^{-M}$$

whenever $|\xi| \le 1/8$ or $|\xi| \ge 15/8$. □

We finish with the proof of Lemma 5.5.6, which had been left open.

Proof. We reduce estimate (5.5.23) by duality to

$$\int_{\mathbf{R}^n} |\widehat{g}(\xi)|^2 \frac{d\xi}{|\xi|^\alpha} \le C(n,\alpha)\varepsilon^{\alpha-1}A_\alpha(\varepsilon)\int_{||x|-1|\le\varepsilon} |g(x)|^2\,dx$$

for functions g supported in the annulus $||x|-1| \le \varepsilon$. Using that $(|\xi|^{-\alpha})^\wedge(x) = c_{n,\alpha}|x|^{\alpha-n}$ (cf. Theorem 2.4.6 in [156]), we write

$$\begin{aligned}
\int_{\mathbf{R}^n} |\widehat{g}(\xi)|^2 \frac{d\xi}{|\xi|^\alpha} &= \int_{\mathbf{R}^n} \widehat{g}(\xi)\overline{\widehat{g}(\xi)}\,\frac{1}{|\xi|^\alpha}\,d\xi \\
&= \int_{\mathbf{R}^n} \big(\widehat{g}\,\overline{\widehat{g}}\big)^\vee(x)\,\frac{c_{n,\alpha}}{|x|^{n-\alpha}}\,dx \\
&= \int_{\mathbf{R}^n} (g*\overline{\widetilde{g}})(x)\,\frac{dx}{|x|^{n-\alpha}} \\
&= \int_{||y|-1|\le\varepsilon}\int_{||x|-1|\le\varepsilon} g(x)\overline{\widetilde{g}}(y)\,\frac{c_{n,\alpha}}{|x-y|^{n-\alpha}}\,dx\,dy \\
&\le B(n,\alpha)\,\|g\|_{L^2}^2,
\end{aligned}$$

where $\widetilde{g}(x) = g(-x)$ and

$$B(n,\alpha) = \sup_{||x|-1|\le\varepsilon}\int_{||y|-1|\le\varepsilon} \frac{c_{n,\alpha}}{|y-x|^{n-\alpha}}\,dy.$$

The last inequality is proved by interpolating between the $L^1 \to L^1$ and $L^\infty \to L^\infty$ estimates with bound $B(n,\alpha)$ for the linear operator

$$L(g)(x) = \int_{\mathbf{R}^n} g(y)\,\frac{c_{n,\alpha}}{|x-y|^{n-\alpha}}\,dy.$$

It remains to establish that

$$B(n,\alpha) \le C(n,\alpha)\varepsilon^{\alpha-1}A_\alpha(\varepsilon).$$

Applying a rotation and a change of variables, matters reduce to proving that

$$\sup_{||x|-1|\le\varepsilon}\int_{||y-|x|e_1|-1|\le\varepsilon} \frac{c_{n,\alpha}}{|y|^{n-\alpha}}\,dy \le C(n,\alpha)\varepsilon^{\alpha-1}A_\alpha(\varepsilon),$$

where $e_1 = (1,0,\dots,0)$. This, in turn, is a consequence of

$$\int_{||y-e_1|-1|\le 2\varepsilon} \frac{c_{n,\alpha}}{|y|^{n-\alpha}}\,dy \le C(n,\alpha)\varepsilon^{\alpha-1}A_\alpha(\varepsilon), \tag{5.5.31}$$

since $||y-e_1|x||-1|\le\varepsilon$ and $||x|-1|\le\varepsilon$ imply $||y-e_1|-1|\le 2\varepsilon$. In proving (5.5.31), it suffices to assume that $\varepsilon < 1/100$; otherwise, the left-hand side of (5.5.31) is bounded from above by a constant, and the right-hand side of (5.5.31) is bounded from below by another constant. The region of integration in (5.5.31) is a ring centered at e_1 and width 4ε. We estimate the integral in (5.5.31) by the sum of the integrals of the function $c_{n,\alpha}|y|^{\alpha-n}$ over the sets

$$\begin{aligned} S_0 &= \{y\in\mathbf{R}^n:\ |y|\le\varepsilon,\quad ||y-e_1|-1|\le 2\varepsilon\},\\ S_\ell &= \{y\in\mathbf{R}^n:\ \ell\varepsilon\le|y|\le(\ell+1)\varepsilon,\quad ||y-e_1|-1|\le 2\varepsilon\},\\ S_\infty &= \{y\in\mathbf{R}^n:\ |y|\ge 1,\quad ||y-e_1|-1|\le 2\varepsilon\}, \end{aligned}$$

where $\ell = 1,\dots,[\frac{1}{\varepsilon}]+1$. The volume of each S_ℓ is comparable to

$$\varepsilon\left[((\ell+1)\varepsilon)^{n-1}-(\ell\varepsilon)^{n-1}\right]\approx\varepsilon^n\ell^{n-2}.$$

Consequently,

$$\int_{S_0}\frac{dy}{|y|^{n-\alpha}}\le\omega_{n-1}\int_0^\varepsilon\frac{r^{n-1}}{r^{n-\alpha}}\,dr=\frac{\omega_{n-1}}{\alpha}\varepsilon^\alpha,$$

whereas

$$\sum_{\ell=1}^{[\frac{1}{\varepsilon}]+1}\int_{S_\ell}\frac{dy}{|y|^{n-\alpha}}\le C'_{n,\alpha}\sum_{\ell=1}^{2/\varepsilon}\frac{\varepsilon^n\ell^{n-2}}{(\ell\varepsilon)^{n-\alpha}}\le C'_{n,\alpha}\,\varepsilon^\alpha\sum_{\ell=1}^{2/\varepsilon}\frac{1}{\ell^{2-\alpha}}.$$

Finally, the volume of S_∞ is about ε; hence

$$\int_{S_\infty}\frac{dy}{|y|^{n-\alpha}}\le|S_\infty|\le C''_{n,\alpha}\,\varepsilon.$$

Combining these estimates, we obtain

$$\int_{||y-e_1|-1|\le 2\varepsilon}\frac{c_{n,\alpha}}{|y|^{n-\alpha}}\,dy\le C_{n,\alpha}\left[\varepsilon^\alpha+\varepsilon^\alpha\sum_{\ell=1}^{2/\varepsilon}\frac{1}{\ell^{2-\alpha}}+\varepsilon\right],$$

and it is an easy matter to check that the expression inside the square brackets is at most a constant multiple of $\varepsilon^{\alpha-1}A_\alpha(\varepsilon)$.

We now turn attention to (5.5.24). Switching the roles of f and $\widehat{f}$, we rewrite (5.5.24) as

$$\begin{aligned}\int_{\mathbf{R}^n}\frac{|f(x)|^2}{(1+|x|)^M}\,dx &\le C'_M(n,\alpha)\int_{\mathbf{R}^n}|\widehat{(-\Delta)^{\frac{\alpha}{4}}(f)}(\xi)|^2\,d\xi\\ &= C'_M(n,\alpha)\int_{\mathbf{R}^n}|(-\Delta)^{\frac{\alpha}{4}}(f)(x)|^2\,dx,\end{aligned}$$

recalling the Laplacian introduced in (1.2.1). This estimate can also be restated in terms of the Riesz potential operator $I_{\alpha/2} = (-\Delta)^{-\alpha/4}$ as follows:

$$\int_{\mathbf{R}^n} \frac{|I_{\alpha/2}(g)(x)|^2}{(1+|x|)^M}\,dx \le C'_M(n,\alpha)\int_{\mathbf{R}^n} |g(x)|^2\,dx. \tag{5.5.32}$$

To show this, we use Hölder's inequality with exponents $q/2$ and n/α, where $q > 2$ satisfies

$$\frac{1}{2} - \frac{1}{q} = \frac{\alpha}{2n}.$$

Then we have

$$\begin{aligned}\int_{\mathbf{R}^n} \frac{|I_{\alpha/2}(g)(x)|^2}{(1+|x|)^M}\,dx &\le \left(\int_{\mathbf{R}^n} \frac{dx}{(1+|x|)^{Mn/\alpha}}\right)^{\frac{n}{\alpha}} \big\|I_{\alpha/2}(g)\big\|^2_{L^q(\mathbf{R}^n)} \\ &\le C'_M(n,\alpha)\big\|g\big\|^2_{L^2(\mathbf{R}^n)}\end{aligned}$$

in view of Theorem 1.2.3 and since $M > n$ and $\alpha < n$. This finishes the proof of the lemma. □

Exercises

5.5.1. Let $0 < r < p < \infty$ and $n(1-\frac{r}{p}) < \beta < n$. Show that $L^p(\mathbf{R}^n)$ is contained in $L^r(\mathbf{R}^n) + L^r(\mathbf{R}^n, |x|^{-\beta})$.
[*Hint:* Write $f = f_1 + f_2$, where $f_1 = f\chi_{|f|>1}$ and $f_2 = f\chi_{|f|\le 1}$.]

5.5.2. (a) With the notation of Lemma 5.5.4, use dilations to show that the estimate

$$\int_{\mathbf{R}^n}\int_1^2 |S_t^\delta(f)(x)|^2 \frac{dt}{t}\frac{dx}{|x|^\alpha} \le C_0 \int_{\mathbf{R}^n} |f(x)|^2 \frac{dx}{|x|^\alpha}$$

implies

$$\int_{\mathbf{R}^n}\int_a^{2a} |S_t^\delta(f)(x)|^2 \frac{dt}{t}\frac{dx}{|x|^\alpha} \le C_0 \int_{\mathbf{R}^n} |f(x)|^2 \frac{dx}{|x|^\alpha}$$

for any $a > 0$ and f in the Schwartz class.
(b) Use dilations to show that (5.5.18) implies (5.5.17).

5.5.3. Let h be a Schwartz function on $\mathbf{R}^n$. Prove that

$$\frac{1}{\varepsilon}\int_{||x|-1|\le\varepsilon} h(x)\,dx \to 2\int_{\mathbf{S}^{n-1}} h(\theta)\,d\theta$$

as $\varepsilon \to 0$. Use Lemma 5.5.6 to show that for $1 < \alpha < n$ we have

$$\int_{\mathbf{S}^{n-1}} |\widehat{f}(\theta)|^2\,d\theta \le C(n,\alpha)\int_{\mathbf{R}^n} |f(x)|^2 |x|^\alpha\,dx.$$

5.5.4. Let $w \in A_2$. Assume that the ball multiplier operator $B^0(f) = (\widehat{f}\chi_{B(0,1)})^\vee$ satisfies

$$\int_{\mathbf{R}^n} |B^0(f)(x)|^2 w(x)\,dx \le C_{n,\alpha} \int_{\mathbf{R}^n} |f(x)|^2 w(x)\,dx$$

for all $f \in L^2(w)$. Prove the same estimate for $\mathscr{B}(f) = \sup_{k\in\mathbf{Z}} |B^0_{2^k}(f)|$.
[*Hint:* Pick a smooth function $\widehat{\Phi}$ equal to 1 on $B(0,1)$ and zero outside $B(0,2)$ and define $\widehat{\Psi}(\xi) = \widehat{\Phi}(\xi) - \widehat{\Phi}(2\xi)$. Then $\chi_{B(0,1)}(\widehat{\Phi}(\xi) - \widehat{\Phi}(2\xi)) = \chi_{B(0,1)} - \widehat{\Phi}(2\xi)$; hence

$$\begin{aligned}\mathscr{B}(f) &\le \sup_k |\Phi_{2^{-k}} * f| + \Big(\sum_{k\in\mathbf{Z}} |B^0_{2^k}(f) - \Phi_{2^{-(k-1)}} * f|^2\Big)^{\frac{1}{2}} \\ &\le C_\Phi M(f) + \Big(\sum_{k\in\mathbf{Z}} |B^0_{2^k}(f * \Psi_{2^{-k}})|^2\Big)^{\frac{1}{2}}\end{aligned}$$

and show that each term is bounded on $L^2(w)$.]

5.5.5. Show that the Bochner–Riesz operator B^λ does not map $L^p(\mathbf{R}^n)$ to $L^{p,\infty}(\mathbf{R}^n)$ when $\lambda = \frac{n-1}{2} - \frac{n}{p}$ and $2 < p < \infty$. Derive the same conclusion for B^λ_*.
[*Hint:* Suppose the contrary. Then by duality it would follow that B^λ maps $L^{p,1}(\mathbf{R}^n)$ to $L^p(\mathbf{R}^n)$ when $1 < p < 2$ and $\lambda = \frac{n}{p} - \frac{n+1}{2}$. To contradict this statement test the operator on a Schwartz function whose Fourier transform is equal to 1 on the unit ball and argue as in Proposition 5.2.3.]

HISTORICAL NOTES

The geometric construction in Section 5.1 is based on ideas of Besicovitch, who used a similar construction to answer the following question posed in 1917 by the Japanese mathematician S. Kakeya: What is the smallest possible area of the trace of ink left on a piece of paper by an ink-covered needle of unit length when the positions of its two ends are reversed? This problem puzzled mathematicians for several decades until Besicovitch [33] showed that for any $\varepsilon > 0$ there is a way to move the needle so that the total area of the blot of ink left on the paper is smaller than ε. Fefferman [134] borrowed ideas from the construction of Besicovitch to provide the negative answer to the multiplier problem to the ball for $p \neq 2$ (Theorem 5.1.5). Prior to Fefferman's work, the fact that the characteristic function of the unit ball is not a multiplier on $L^p(\mathbf{R}^n)$ for $|\frac{1}{p} - \frac{1}{2}| \ge \frac{1}{2n}$ was pointed out by Herz [189], who also showed that this limitation is not necessary when this operator is restricted to radial L^p functions. The crucial Lemma 5.1.4 in Fefferman's proof is due to Y. Meyer.

The study of Bochner–Riesz means originated in the article of Bochner [40], who obtained their L^p boundedness for $\lambda > \frac{n-1}{2}$. Stein [322] improved this result to $\lambda > \frac{n-1}{2}|\frac{1}{p} - \frac{1}{2}|$ using interpolation for analytic families of operators. Theorem 5.2.4 was first proved by Carleson and Sjölin [72]. A second proof of this theorem was given by Fefferman [136]. A third proof was given by Hörmander [195]. The proof of Theorem 5.2.4 given in the text is due to Córdoba [102]. This proof elaborated the use of the Kakeya maximal function in the study of spherical summation multipliers, which was implicitly pioneered in Fefferman [136]. The boundedness of the Kakeya maximal function $\mathscr{K}_N$ on $L^2(\mathbf{R}^2)$ with norm $C(\log N)^2$ was first obtained by Córdoba [101]. The sharp estimate $C\log N$ was later obtained by Strömberg [331]. The proof of Theorem 5.3.5 is taken from this article of Strömberg. Another proof of the boundedness of the Kakeya maximal function without dilations on $L^2(\mathbf{R}^2)$ was obtained by Müller [273]. Barrionuevo [18] showed that for any subset Σ of $\mathbf{S}^1$ with N elements the maximal operator $\mathfrak{M}_\Sigma$ maps $L^2(\mathbf{R}^2)$ to itself with norm $CN^{2(\log N)^{-1/2}}$ for some absolute constant C. Note that this bound is $O(N^\varepsilon)$ for any $\varepsilon > 0$. Katz [212] improved this bound to $C\log N$ for some absolute constant C; see also Katz [213]. The latter is a sharp bound, as indicated in Proposition 5.3.4. Katz [211] also showed that the maximal operator $\mathfrak{M}_K$ associated with a set of unit vectors pointing along a Cantor set K of directions is unbounded on $L^2(\mathbf{R}^2)$. If Σ is an infinite set of vectors in $\mathbf{S}^1$ pointing in lacunary directions, then $\mathfrak{M}_\Sigma$ was studied by Strömberg [330], Córdoba and Fefferman [104], and Nagel, Stein, and Wainger [282]. The last authors obtained its L^p boundedness for all $1 < p < \infty$. Theorem 5.2.7 was first proved by Carleson [70]. For a short account on extensions of this theorem, the reader may consult the historical notes at the end of Chapter 5.

The idea of restriction theorems for the Fourier transform originated in the work of E. M. Stein around 1967. Stein's original restriction result was published in the article of Fefferman [132], which was the first to point out connections between restriction theorems and boundedness of the Bochner–Riesz means. The full restriction theorem for the circle (Theorem 5.4.7 for $p < \frac{4}{3}$) is due to Fefferman and Stein and was published in the aforementioned article of Fefferman [132]. See also the related article of Zygmund [378]. The present proof of Theorem 5.4.7 is based in that of Córdoba [103]. This proof was further elaborated by Tomas [350], who pointed out the logarithmic blowup when $p = \frac{4}{3}$ for the corresponding restriction problem for annuli. The result in Example 5.4.4 is also due to Fefferman and Stein and was initially proved using arguments from spherical harmonics. The simple proof presented here was observed by A. W. Knapp. The restriction property in Theorem 5.4.5 for $p < \frac{2(n+1)}{n+3}$ is due to Tomas [349], while the case $p = \frac{2(n+1)}{n+3}$ is due to Stein [325]. Theorem 5.4.6 was first proved by Fefferman [132] for the smaller range of $\lambda > \frac{n-1}{4}$ using the restriction property $R_{p\to 2}(\mathbf{S}^{n-1})$ for $p < \frac{4n}{3n+1}$. The fact that the $R_{p\to 2}(\mathbf{S}^{n-1})$ restriction property (for $p < 2$) implies the boundedness of the Bochner–Riesz operator B^λ on $L^p(\mathbf{R}^n)$ is contained in the work of Fefferman [132]. A simpler proof of this fact, obtained later by E. M. Stein, appeared in the subsequent article of Fefferman [136]. This proof is given in Theorem 5.4.6, incorporating the Tomas–Stein restriction property $R_{p\to 2}(\mathbf{S}^{n-1})$ for $p \le \frac{2(n+1)}{n+3}$. It should be noted that the case $n = 3$ of this theorem was first obtained in unpublished work of Sjölin. For a short exposition and history of this material consult the book of Davis and Chang [114]. Much of the material in Sections 5.2, 5.3, and 5.4 is based on the notes of Vargas [359].

There is an extensive literature on restriction theorems for submanifolds of $\mathbf{R}^n$. It is noteworthy to mention (in chronological order) the results of Strichartz [329], Prestini [300], Greenleaf [178], Christ [76], Drury [122], Barceló [16], [17], Drury and Marshall [124], [125], Beckner, Carbery, Semmes, and Soria [19], Drury and Guo [123], De Carli and Iosevich [115], [116], Sjölin and Soria [318], Oberlin [286], Wolff [373], and Tao [341].

The boundedness of the Bochner–Riesz operators on the range excluded by Proposition 5.2.3 implies that the restriction property $R_{p\to q}(\mathbf{S}^{n-1})$ is valid when $\frac{1}{q} = \frac{n+1}{n-1}\frac{1}{p'}$ and $1 \le p < \frac{2n}{n+1}$, as shown by Tao [340]; in this article a hierarchy of conjectures in harmonic analysis and interrelations among them is discussed. In particular, the aforementioned restriction property would imply estimate (5.3.32) for the Kakeya maximal operator $\mathscr{K}_N$ on $\mathbf{R}^n$, which would in turn imply that Besicovitch sets have Minkowski dimension n. (A Besicovitch set is defined as a subset of $\mathbf{R}^n$ that contains a unit line segment in every direction.) Katz, Laba, and Tao [214] have obtained good estimates on the Minkowski dimension of such sets in $\mathbf{R}^3$.

A general sieve argument obtained by Córdoba [101] reduces the boundedness of the Kakeya maximal operator $\mathscr{K}_N$ to the one without dilations $\mathscr{K}_N^a$. For applications to the Bochner–Riesz multiplier problem, only the latter is needed. Carbery, Hernández, and Soria [65] have proved estimate (5.3.30) for radial functions in all dimensions. Igari [202] proved estimate (5.3.31) for products of one-variable functions of each coordinate. The norm estimates in Corollary 5.3.7 can be reversed, as shown by Keich [216] for $p > 2$. The corresponding estimate for $1 < p < 2$ in the same corollary can be improved to $N^{\frac{2}{p}-1}$. Córdoba [102] proved the partial case $p \le 2$ of Theorem 5.3.10 on $\mathbf{R}^n$. This range was extended by Drury [121] to $p \le \frac{n+1}{n-1}$ using estimates for the x-ray transform. Theorem 5.3.10 (i.e., the further extension to $p \le \frac{n+1}{2}$) is due to Christ, Duoandikoetxea, and Rubio de Francia [83], and its original proof also used estimates for the x-ray transform; the proof of Theorem 5.3.10 given in the text is derived from that in Bourgain [42]. This article brought a breakthrough in many of the previous topics. In particular, Bourgain [42] showed that the Kakeya maximal operator $\mathscr{K}_N$ maps $L^p(\mathbf{R}^n)$ to itself with bound $C_\varepsilon N^{\frac{n}{p}-1+\varepsilon}$ for all $\varepsilon > 0$ and some $p_n > \frac{n+1}{2}$. He also showed that the range of p's in Theorem 5.4.5 is not sharp, since there exist indices $p = p(n) > \frac{2(n+1)}{n+3}$ for which property $R_{p\to q}(\mathbf{S}^{n-1})$ holds, and that Theorem 5.4.6 is not sharp, since there exist indices $\lambda_n < \frac{n-1}{2(n+1)}$ for which the Bochner–Riesz operators are bounded on $L^p(\mathbf{R}^n)$ in the optimal range of p's when $\lambda \ge \lambda_n$. Improvements on these indices were subsequently obtained by Bourgain [43], [44]. Some of Bourgain's results in $\mathbf{R}^3$ were re-proved by Schlag [311] using different geometric methods. Wolff [371] showed that the Kakeya maximal operator $\mathscr{K}_N$ maps $L^p(\mathbf{R}^n)$ to itself with bound $C_\varepsilon N^{\frac{n}{p}-1+\varepsilon}$ for any $\varepsilon > 0$ whenever $p \le \frac{n+2}{2}$. In higher dimensions, this range of p's was later extended by Bourgain [45] to $p \le (1+\varepsilon)\frac{n}{2}$ for some dimension-free positive constant ε. When $n = 3$, further improvements on the restriction and the Kakeya conjectures were obtained by Tao, Vargas, and Vega [343]. For further historical advances in the subject the reader is referred to the survey articles of Wolff [372] and Katz and Tao [215].

Regarding the almost everywhere convergence of the Bochner–Riesz means, Carbery [64] has shown that the maximal operator $B_*^\lambda(f) = \sup_{R>0} |B_R^\lambda(f)|$ is bounded on $L^p(\mathbf{R}^2)$ when $\lambda > 0$ and $2 \le p < \frac{4}{1-2\lambda}$, obtaining the convergence $B_R^\lambda(f) \to f$ almost everywhere for $f \in L^p(\mathbf{R}^2)$. For $n \ge 3$, $2 \le p < \frac{2n}{n-1-2\lambda}$, and $\lambda \ge \frac{n-1}{2(n+1)}$ the same result was obtained by Christ [77]. Theorem 5.5.3 is due to Carbery, Rubio de Francia, and Vega [66]. Theorem 5.5.1 is contained in Tao [339]. Tao [342] also obtained boundedness for the maximal Bochner–Riesz operators B_*^λ on $L^p(\mathbf{R}^2)$ whenever $1 < p < 2$ for an open range of pairs $(\frac{1}{p}, \lambda)$ that lie below the line $\lambda = \frac{1}{p} - \frac{1}{2}$.

On the critical line $\lambda = \frac{n}{p} - \frac{n+1}{2}$, boundedness into weak L^p for the Bochner–Riesz operators is possible in the range $1 \le p \le \frac{2n}{n+1}$. Christ [79], [78] first obtained such results for $1 \le p < \frac{2(n+1)}{n+3}$ in all dimensions. The point $p = \frac{2(n+1)}{n+3}$ was later included by Tao [338]. In two dimensions, weak boundedness for the full range of indices was shown by Seeger [313]; in all dimensions the same conclusion was obtained by Colzani, Travaglini, and Vignati [99] for radial functions. Tao [339] has obtained a general argument that yields weak endpoint bounds for B^λ whenever strong-type bounds are known above the critical line.

Chapter 6
Time–Frequency Analysis and the Carleson–Hunt Theorem

In this chapter we discuss in detail the proof of the almost everywhere convergence of the partial Fourier integrals of L^p functions on the line. The proof of this theorem is based on techniques involving both spatial and frequency decompositions. These techniques are referred to as time–frequency analysis. The underlying goal is to decompose a given function at any scale as a sum of pieces perfectly localized in frequency and well localized in space. The action of an operator on each piece is carefully studied and the interaction between different parts of this action is analyzed. Ideas from combinatorics are employed to organize the different pieces of the decomposition.

6.1 Almost Everywhere Convergence of Fourier Integrals

In this section we study the proof of one of the most celebrated theorems in Fourier analysis, Carleson's theorem on the almost everywhere convergence of Fourier series of square integrable functions on the circle. The same result is also valid for functions f on the line if the partial sums of the Fourier series are replaced by the (partial) Fourier integrals

$$\int_{|\xi|\le N} \widehat{f}(\xi)e^{2\pi i x\xi}\,d\xi\,.$$

The equivalence of these assertions can be obtained via transference; about this see Theorems 4.3.14 and 4.3.15 in [156].

For square-integrable functions f on the line, define the *Carleson operator*

$$\mathscr{C}(f)(x) = \sup_{N>0}\left|\left(\widehat{f}\chi_{[-N,N]}\right)^{\vee}(x)\right| = \sup_{N>0}\left|\int_{|\xi|\le N}\widehat{f}(\xi)e^{2\pi i x\xi}\,d\xi\right|. \tag{6.1.1}$$

We note that for f in $L^2(\mathbf{R})$ the functions $(\widehat{f}\chi_{[a,b]})^{\vee}$ are well defined and continuous when $-\infty < a < b < \infty$, and thus so is $\mathscr{C}(f)$. We have the following result about $\mathscr{C}$.

L. Grafakos, *Modern Fourier Analysis*, Graduate Texts in Mathematics 250,
DOI 10.1007/978-1-4939-1230-8_6, © Springer Science+Business Media New York 2014

Theorem 6.1.1. *There is a constant $C>0$ such that for all square-integrable functions f on the line the following estimate is valid:*

$$\big\|\mathscr{C}(f)\big\|_{L^{2,\infty}} \le C\big\|f\big\|_{L^2}. \tag{6.1.2}$$

It follows that for all f in $L^2(\mathbf{R})$ we have

$$\lim_{N\to\infty}\int_{|\xi|\le N}\widehat{f}(\xi)e^{2\pi i x\xi}\,d\xi = f(x) \tag{6.1.3}$$

for almost all $x\in\mathbf{R}$.

Suppose we know the validity of (6.1.2). Since by the Lebesgue dominated convergence theorem, (6.1.3) holds for Schwartz functions, applying Theorem 2.1.14 in [156] and using (6.1.2) we obtain that (6.1.3) holds for all square-integrable functions f on the line. It suffices therefore to prove (6.1.2).

Observe that it suffices to prove (6.1.2) for Schwartz functions f on the line. Indeed, suppose we know (6.1.2) for Schwartz functions h and we would like to prove it for all square-integrable functions f. Given f in $L^2(\mathbf{R})$ we pick a sequence of Schwartz functions h_j such that $h_j\to f$ in L^2 as $j\to\infty$. It follows from the Cauchy–Schwarz inequality that the sequence of continuous functions $\big(\widehat{h_j}\chi_{[-N,N]}\big)^\vee$ converges to the continuous function $\big(\widehat{f}\chi_{[-N,N]}\big)^\vee$ pointwise everywhere as $j\to\infty$ for all $N>0$. Then we have

$$\begin{aligned}
\big\|\mathscr{C}(f)\big\|_{L^{2,\infty}} &= \Big\|\sup_{N>0}\big|\big(\widehat{f}\chi_{[-N,N]}\big)^\vee\big|\Big\|_{L^{2,\infty}}\\
&= \Big\|\sup_{N>0}\big|\lim_{j\to\infty}\big(\widehat{h_j}\chi_{[-N,N]}\big)^\vee\big|\Big\|_{L^{2,\infty}}\\
&\le \Big\|\sup_{N>0}\liminf_{j\to\infty}\big|\big(\widehat{h_j}\chi_{[-N,N]}\big)^\vee\big|\Big\|_{L^{2,\infty}}\\
&\le \Big\|\liminf_{j\to\infty}\sup_{N>0}\big|\big(\widehat{h_j}\chi_{[-N,N]}\big)^\vee\big|\Big\|_{L^{2,\infty}}\\
&\le \liminf_{j\to\infty}\Big\|\sup_{N>0}\big|\big(\widehat{h_j}\chi_{[-N,N]}\big)^\vee\big|\Big\|_{L^{2,\infty}}\\
&\le \liminf_{j\to\infty} C\big\|h_j\big\|_{L^2}\\
&= C\big\|f\big\|_{L^2},
\end{aligned}$$

where we used Fatou's lemma in the third inequality. This proves (6.1.2) for all functions $f\in L^2(\mathbf{R})$.

We may therefore focus on the proof of (6.1.2) where $f\in\mathscr{S}(\mathbf{R})$. We devote the rest of this section to this task. Because of the simple identity

$$\int_{|\xi|\le N}\widehat{f}(\xi)e^{2\pi i x\xi}\,d\xi = \int_{-\infty}^{N}\widehat{f}(\xi)e^{2\pi i x\xi}\,d\xi - \int_{-\infty}^{-N}\widehat{f}(\xi)e^{2\pi i x\xi}\,d\xi,$$

it suffices to obtain $L^2 \to L^{2,\infty}$ bounds for the *one-sided maximal operators*

$$\mathscr{C}_1(f)(x) = \sup_{N>0} \left| \int_{-\infty}^{N} \widehat{f}(\xi) e^{2\pi i x \xi}\, d\xi \right|,$$
$$\mathscr{C}_2(f)(x) = \sup_{N>0} \left| \int_{-\infty}^{-N} \widehat{f}(\xi) e^{2\pi i x \xi}\, d\xi \right|,$$

acting on a Schwartz function f (with bounds independent of f). Note that

$$\mathscr{C}_2(f)(x) \le |f(x)| + \mathscr{C}_1(\widetilde{f})(-x),$$

where $\widetilde{f}(x) = f(-x)$ is the usual reflection operator. Therefore, it suffices to obtain bounds only for $\mathscr{C}_1$.

For $a > 0$ and $y \in \mathbf{R}$ we define the translation operator τ^y, the modulation operator M^a, and the dilation operator D^a as follows:

$$\tau^y(f)(x) = f(x-y),$$
$$D^a(f)(x) = a^{-\frac{1}{2}} f(a^{-1}x),$$
$$M^y(f)(x) = f(x) e^{2\pi i y x}.$$

These operators are isometries on $L^2(\mathbf{R})$.

We break down the proof of Theorem 6.1.1 into several steps.

6.1.1 Preliminaries

We denote rectangles of area 1 in the (x,ξ) plane by s, t, u, etc. All rectangles considered in the sequel have sides parallel to the axes. We think of x as the time coordinate and of ξ as the frequency coordinate. For this reason we refer to the (x,ξ) coordinate plane as the time–frequency plane. The projection of a rectangle s on the time axis is denoted by I_s, while its projection on the frequency axis is denoted by ω_s. Thus a rectangle s is just $s = I_s \times \omega_s$. Rectangles with sides parallel to the axes and area equal to one are called *tiles*.

The center of an interval I is denoted by $c(I)$. Also for $a > 0$, aI denotes an interval with the same center as I whose length is $a|I|$. Given a tile s, we denote by $s(1)$ its bottom half and by $s(2)$ its upper half defined by

$$s(1) = I_s \times \big(\omega_s \cap (-\infty, c(\omega_s))\big), \qquad s(2) = I_s \times \big(\omega_s \cap [c(\omega_s), +\infty)\big).$$

These sets are called *semitiles*. The projections of these sets on the frequency axes are denoted by $\omega_{s(1)}$ and $\omega_{s(2)}$, respectively. See Figure 6.1.

A dyadic interval is an interval of the form $[m2^k, (m+1)2^k)$, where k and m are integers. We denote by $\mathbf{D}$ the set of all rectangles $I \times \omega$ with I, ω dyadic intervals and $|I|\,|\omega| = 1$. Such rectangles are called *dyadic tiles*. We denote by $\mathbf{D}$ the set of all dyadic tiles. For every integer m, we denote by $\mathbf{D}_m$ the set of all tiles $s \in \mathbf{D}$ such that $|I_s| = 2^m$. We call these dyadic tiles *of scale m*.

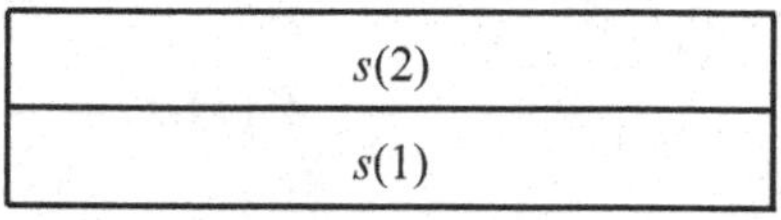

Fig. 6.1 The lower and the upper parts of a tile s.

We fix a Schwartz function φ such that $\widehat{\varphi}$ takes values in $[0,1]$ and supported in the interval $[-1/10, 1/10]$, and equal to 1 on the interval $[-9/100, 9/100]$. For each tile s, we introduce a function φ_s as follows:

$$\varphi_s(x) = |I_s|^{-\frac{1}{2}} \varphi\left(\frac{x - c(I_s)}{|I_s|}\right) e^{2\pi i c(\omega_{s(1)})x} . \tag{6.1.4}$$

This function is localized in frequency near $c(\omega_{s(1)})$. Using the previous notation, we have

$$\varphi_s = M^{c(\omega_{s(1)})} \tau^{c(I_s)} D^{|I_s|}(\varphi) .$$

Observe that

$$\widehat{\varphi_s}(\xi) = |\omega_s|^{-\frac{1}{2}} \widehat{\varphi}\left(\frac{\xi - c(\omega_{s(1)})}{|\omega_s|}\right) e^{2\pi i (c(\omega_{s(1)}) - \xi) c(I_s)} , \tag{6.1.5}$$

from which it follows that $\widehat{\varphi_s}$ is supported in $\frac{2}{5}\omega_{s(1)}$. Also observe that the functions φ_s have the same $L^2(\mathbf{R})$ norm.

Recall the complex inner product notation for $f, g \in L^2(\mathbf{R})$:

$$\langle f \,|\, g \rangle = \int_{\mathbf{R}} f(x) \overline{g(x)}\, dx . \tag{6.1.6}$$

Given a real number ξ and $m \in \mathbf{Z}$, we introduce an operator

$$A_\xi^m(f) = \sum_{s \in \mathbf{D}_m} \chi_{\omega_{s(2)}}(\xi) \langle f \,|\, \varphi_s \rangle \varphi_s , \tag{6.1.7}$$

for functions $f \in \mathscr{S}(\mathbf{R})$. The series in (6.1.7) converges absolutely and in L^2 for f in the Schwartz class (see Exercise 6.1.9) and thus A_ξ^m is well defined on $\mathscr{S}(\mathbf{R})$. Note that for a fixed m, the sum in (6.1.7) is taken over the row of dyadic rectangles of size $2^m \times 2^{-m}$ whose tops contain the horizontal line at height ξ. The Fourier transforms of the operators A_ξ^m are supported in a horizontal strip contained in $(-\infty, \xi]$ of width $\frac{2}{5}2^{-m}$. Notice that if the characteristic function were missing in (6.1.7), then for a suitable function φ, the sum would be equal to a multiple of $f(x)$; cf. Exercise 6.1.9. Thus for each $m \in \mathbf{Z}$ the operator $A_\xi^m(f)$ may be viewed as a "piece" of the multiplier operator $f \mapsto \big(\widehat{f}\chi_{(-\infty,\xi]}\big)^\vee$. Summing over m yields a better approximation to this half-line multiplier operator.

This discussion motivates the introduction of the operator

$$A_\xi(f) = \sum_{m\in\mathbf{Z}} A_\xi^m(f) = \sum_{s\in\mathbf{D}} \chi_{\omega_{s(2)}}(\xi)\langle f\,|\,\varphi_s\rangle\,\varphi_s \tag{6.1.8}$$

for $f \in \mathscr{S}(\mathbf{R})$ and $\xi \in \mathbf{R}$. We show in Lemma 6.1.2 that A_ξ is well defined on $\mathscr{S}(\mathbf{R})$ and that it admits a bounded extension on $L^2(\mathbf{R})$.

Lemma 6.1.2. *For any $\xi \in \mathbf{R}$, the operators A_ξ^m, initially defined on $\mathscr{S}(\mathbf{R})$, admit bounded extensions on $L^2(\mathbf{R})$, with norms uniformly bounded in m and ξ. Moreover, for any $g \in L^2(\mathbf{R})$, the series $\sum_{m\in\mathbf{Z}} A_\xi^m(g)$ converges in $L^2(\mathbf{R})$ to a function $A_\xi(g)$ in $L^2(\mathbf{R})$. The linear operator A_ξ defined in this way is bounded from $L^2(\mathbf{R})$ to itself with norm uniformly bounded in ξ. Moreover, when $\xi > 0$, for any $f \in L^1(\mathbf{R})$ the series in (6.1.8) converges absolutely pointwise and is bounded by a constant multiple of $\xi\|f\|_{L^1}$.*

Proof. Fix $\xi \in \mathbf{R}$. We first prove the results concerning A_ξ^m. We begin by observing that for $m \neq m'$ and $f,g \in \mathscr{S}(\mathbf{R})$ we have that $\langle A_\xi^m(f)\,|\,A_\xi^{m'}(g)\rangle = 0$. Indeed, given f and g in $\mathscr{S}(\mathbf{R})$ we have

$$\langle A_\xi^m(f)\,|\,A_\xi^{m'}(g)\rangle = \sum_{s\in\mathbf{D}_m}\sum_{s'\in\mathbf{D}_{m'}} \langle f\,|\,\varphi_s\rangle\overline{\langle g\,|\,\varphi_{s'}\rangle}\langle\varphi_s\,|\,\varphi_{s'}\rangle\chi_{\omega_{s(2)}}(\xi)\chi_{\omega_{s'(2)}}(\xi), \tag{6.1.9}$$

and this representation is possible, since the double sum converges absolutely (see Exercise 6.1.9). Suppose that $\langle\varphi_s\,|\,\varphi_{s'}\rangle\chi_{\omega_{s(2)}}(\xi)\chi_{\omega_{s'(2)}}(\xi)$ is nonzero. Then $\langle\varphi_s\,|\,\varphi_{s'}\rangle$ is also nonzero, which implies that $\omega_{s(1)}$ and $\omega_{s'(1)}$ intersect. Also, the function $\chi_{\omega_{s(2)}}(\xi)\chi_{\omega_{s'(2)}}(\xi)$ is nonzero; hence $\omega_{s(2)}$ and $\omega_{s'(2)}$ must intersect. Thus the dyadic intervals ω_s and $\omega_{s'}$ are not disjoint, and one must contain the other. If ω_s were properly contained in $\omega_{s'}$, then it would follow that ω_s is contained in $\omega_{s'(1)}$ or in $\omega_{s'(2)}$. But then either $\omega_{s(1)} \cap \omega_{s'(1)}$ or $\omega_{s(2)} \cap \omega_{s'(2)}$ would have to be empty, which does not happen, as observed. It follows that if $\langle\varphi_s\,|\,\varphi_{s'}\rangle\chi_{\omega_{s(2)}}(\xi)\chi_{\omega_{s'(2)}}(\xi)$ is nonzero, then $\omega_s = \omega_{s'}$, which is impossible if $m \neq m'$. Thus the expression in (6.1.9) has to be zero.

We first discuss the boundedness of each operator A_ξ^m. For $f \in \mathscr{S}(\mathbf{R})$ have

$$\begin{aligned}
\|A_\xi^m(f)\|_{L^2}^2 &= \sum_{s\in\mathbf{D}_m}\sum_{s'\in\mathbf{D}_m} \langle f\,|\,\varphi_s\rangle\overline{\langle f\,|\,\varphi_{s'}\rangle}\langle\varphi_s\,|\,\varphi_{s'}\rangle\chi_{\omega_{s(2)}}(\xi)\chi_{\omega_{s'(2)}}(\xi)\\
&= \sum_{s\in\mathbf{D}_m}\sum_{\substack{s'\in\mathbf{D}_m\\ \omega_{s'}=\omega_s}} \langle f\,|\,\varphi_s\rangle\overline{\langle f\,|\,\varphi_{s'}\rangle}\langle\varphi_s\,|\,\varphi_{s'}\rangle\chi_{\omega_{s(2)}}(\xi)\chi_{\omega_{s'(2)}}(\xi)\\
&\le \sum_{s\in\mathbf{D}_m}\sum_{\substack{s'\in\mathbf{D}_m\\ \omega_{s'}=\omega_s}} \big|\langle f\,|\,\varphi_s\rangle\big|^2\chi_{\omega_{s(2)}}(\xi)\big|\langle\varphi_s\,|\,\varphi_{s'}\rangle\big|\\
&\le C_1\sum_{s\in\mathbf{D}_m} \big|\langle f\,|\,\varphi_s\rangle\big|^2\chi_{\omega_{s(2)}}(\xi),
\end{aligned} \tag{6.1.10}$$

where we used an earlier observation about s and s', the Cauchy–Schwarz inequality, and the fact that

$$\sum_{\substack{s'\in\mathbf{D}_m\\ \omega_{s'}=\omega_s}} |\langle \varphi_s \mid \varphi_{s'}\rangle| \le C \sum_{\substack{s'\in\mathbf{D}_m\\ \omega_{s'}=\omega_s}} \left(1+\frac{\operatorname{dist}(I_s,I_{s'})}{2^m}\right)^{-10} \le C_1,$$

which follows from the result in Appendix B.1. To estimate (6.1.10), we use that

$$\begin{aligned}
|\langle f \mid \varphi_s\rangle| &\le C_2 \int_{\mathbf{R}} |f(y)|\,|I_s|^{-\frac{1}{2}}\left(1+\frac{|y-c(I_s)|}{|I_s|}\right)^{-10} dy\\
&\le C_3 |I_s|^{\frac{1}{2}} \int_{\mathbf{R}} |f(y)|\left(1+\frac{|y-z|}{|I_s|}\right)^{-10}\frac{dy}{|I_s|}\\
&\le C_4 |I_s|^{\frac{1}{2}} M(f)(z),
\end{aligned}$$

for all $z\in I_s$, in view of Theorem 2.1.10 in [156]. Since the preceding estimate holds for all $z\in I_s$, it follows that

$$\big|\langle f \mid \varphi_s\rangle\big|^2 \le (C_4)^2|I_s| \inf_{z\in I_s} M(f)(z)^2 \le (C_4)^2 \int_{I_s} M(f)(x)^2\,dx. \tag{6.1.11}$$

Next we observe that the rectangles $s\in\mathbf{D}_m$ with the property that $\xi\in\omega_{s(2)}$ are all disjoint. This implies that the corresponding time intervals I_s are also disjoint. Thus, summing (6.1.11) over all $s\in\mathbf{D}_m$ with $\xi\in\omega_{s(2)}$, we obtain that

$$\begin{aligned}
\sum_{s\in\mathbf{D}_m} \big|\langle f \mid \varphi_s\rangle\big|^2 \chi_{\omega_{s(2)}}(\xi) &\le (C_4)^2 \sum_{s\in\mathbf{D}_m} \chi_{\omega_{s(2)}}(\xi) \int_{I_s} M(f)(x)^2\,dx\\
&\le (C_4)^2 \int_{\mathbf{R}} M(f)(x)^2\,dx,
\end{aligned}$$

which establishes the required claim using the boundedness of the Hardy–Littlewood maximal operator M on $L^2(\mathbf{R})$. We conclude that each A_ξ^m, initially defined on $\mathscr{S}(\mathbf{R})$, admits an L^2-bounded extension and all these extensions have norms uniformly bounded in m and ξ. We denote these extensions also by A_ξ^m.

We now explain why $A_\xi = \sum_{m\in\mathbf{Z}} A_\xi^m$ is well defined on $L^2(\mathbf{R})$ and we examine its L^2 boundedness. For every fixed $m\in\mathbf{Z}$, the dyadic tiles that appear in the sum defining A_ξ^m have the form

$$s = [k2^m,(k+1)2^m)\times[\ell 2^{-m},(\ell+1)2^{-m}),$$

where $(\ell+\frac{1}{2})2^{-m}\le\xi<(\ell+1)2^{-m}$. Let $g\in L^2(\mathbf{R})$. Thus $\ell=[2^m\xi]$, and since $\widehat{\varphi_s}$ is supported in the lower part of the dyadic tile s, if g_m is defined via

$$\widehat{g_m} = \widehat{g}\chi_{[2^{-m}[2^m\xi],\,2^{-m}([2^m\xi]+\frac{1}{2}))},$$

then we have $A_\xi^m(g_m) = A_\xi^m(g)$. We use this observation to obtain

$$\begin{aligned}\sum_{m\in\mathbf{Z}} \big\|A_\xi^m(g)\big\|_{L^2}^2 &= \sum_{m\in\mathbf{Z}} \big\|A_\xi^m(g_m)\big\|_{L^2}^2 \\ &\le C_5 \sum_{m\in\mathbf{Z}} \big\|g_m\big\|_{L^2}^2 \\ &= C_5 \sum_{m\in\mathbf{Z}} \big\|\widehat{g_m}\big\|_{L^2}^2 \\ &\le C_5 \|g\|_{L^2}^2 < \infty .\end{aligned} \tag{6.1.12}$$

As already observed, the supports of the Fourier transforms of $A_\xi^m(g)$ are pairwise disjoint when $m \in \mathbf{Z}$. This implies that $\langle A_\xi^m(g) \,|\, A_\xi^{m'}(g)\rangle = 0$ whenever $m \neq m'$. Consequently, given $\varepsilon > 0$ there is an N_0 such that for $M > N \ge N_0$ we have

$$\Big\| \sum_{N\le |m|\le M} A_\xi^m(g)\Big\|_{L^2}^2 = \sum_{N\le |m|\le M} \big\|A_\xi^m(g)\big\|_{L^2}^2 < \varepsilon^2 . \tag{6.1.13}$$

Thus the series $\sum_{m\in\mathbf{Z}} A_\xi^m(g)$ is Cauchy and it converges to an element of $L^2(\mathbf{R})$ which we denote by $A_\xi(g)$. Combining (6.1.12) and (6.1.13) we obtain that A_ξ is bounded from $L^2(\mathbf{R})$ to itself with norm at most C_5.

We now address the last assertion about the absolute pointwise convergence of the series in (6.1.8) for all $x \in \mathbf{R}$ when $f \in L^1(\mathbf{R})$ and $\xi > 0$. For fixed $x \in \mathbf{R}$, $\xi > 0$, we pick $m_0 \in \mathbf{Z}$ such that $2^{-m_0-1} \le \xi < 2^{-m_0}$. We notice that for each $m \in \mathbf{Z}$ there is only one horizontal row of tiles of size $2^m \times 2^{-m}$ whose upper parts contain ξ and thus appearing in the sum in (6.1.8). Moreover, for all the tiles s that appear in the sum in (6.1.8), the size of ω_s cannot be bigger than 2^{-m_0} since the top part of ω_s contains ξ. Thus if $I_s = [2^m k, 2^m(k+1))$, we must have $m \ge m_0$. Combining these observations with the fact that $|\langle f \,|\, \varphi_s\rangle| \le \|f\|_{L^1} \|\varphi_s\|_{L^\infty}$, we estimate the sum of the absolute value of each term of the series in (6.1.8) by

$$C\,\|f\|_{L^1} \sum_{m\ge m_0} \sum_{k\in\mathbf{Z}} 2^{-\frac{m}{2}} \frac{2^{-\frac{m}{2}}}{(1+2^{-m}|x-2^m(k+\frac{1}{2})|)^2} \tag{6.1.14}$$

for some constant $C > 0$. Summing first over $k \in \mathbf{Z}$ and then over $m \ge m_0$, we obtain that the series in (6.1.8) converges absolutely for all $x \in \mathbf{R}$ and is bounded above by a constant multiple of $\xi \|f\|_{L^1}$. □

6.1.2 Discretization of the Carleson Operator

We let $h \in \mathscr{S}(\mathbf{R})$, $\xi \in \mathbf{R}$, and for each $m \in \mathbf{Z}$, $y, \eta \in \mathbf{R}$, and $\lambda \in [0,1]$ we introduce the operators

$$B^m_{\xi,y,\eta,\lambda}(h) = \sum_{s\in\mathbf{D}_m} \chi_{\omega_{s(2)}}(2^{-\lambda}(\xi+\eta)) \langle D^{2^\lambda} \tau^y M^\eta(h) \,|\, \varphi_s\rangle M^{-\eta}\tau^{-y} D^{2^{-\lambda}}(\varphi_s) .$$

It is not hard to see that for all $x \in \mathbf{R}$ and $\lambda \in [0,1]$, we have

$$B^m_{\xi,y,\eta,\lambda}(h)(x) = B^m_{\xi,y+2^{m-\lambda},\eta,\lambda}(h)(x) = B^m_{\xi,y,\eta+2^{-m+\lambda},\lambda}(h)(x);$$

in other words, the function $(y,\eta) \mapsto B^m_{\xi,y,\eta,\lambda}(h)(x)$ is periodic in $\mathbf{R}^2$ with period $(2^{m-\lambda}, 2^{-m+\lambda})$. See Exercise 6.1.1.

Using Exercise 6.1.2, we obtain that for $|m|$ large (with respect to ξ) we have

$$\begin{aligned}
&\Big|\sum_{s\in\mathbf{D}_m} \chi_{\omega_{s(2)}}(2^{-\lambda}(\xi+\eta)) \langle D^{2^\lambda}\tau^y M^\eta(h) \,|\, \varphi_s\rangle M^{-\eta}\tau^{-y}D^{2^{-\lambda}}(\varphi_s)(x)\Big| \\
&\qquad \le C_h \min(2^m,1) \sum_{s\in\mathbf{D}_m} \chi_{\omega_{s(2)}}(2^{-\lambda}(\xi+\eta))2^{-m/2}\Big|\varphi\Big(\frac{x+y-c(I_s)2^{-\lambda}}{2^{m-\lambda}}\Big)\Big| \\
&\qquad \le C_h \min(2^{m/2},2^{-m/2}) \sum_{k\in\mathbf{Z}} \Big|\varphi\Big(\frac{x+y-(k+\frac{1}{2})2^{m-\lambda}}{2^{m-\lambda}}\Big)\Big| \\
&\qquad \le C_h \min(2^{m/2},2^{-m/2}),
\end{aligned}$$

since the last sum is seen easily to converge to some quantity that remains bounded in x, y, η, and λ. It follows that for $h \in \mathscr{S}(\mathbf{R})$ we have

$$\sup_{x\in\mathbf{R}} \sup_{y\in\mathbf{R}} \sup_{\eta\in\mathbf{R}} \sup_{0\le\lambda\le 1} \big|B^m_{\xi,y,\eta,\lambda}(h)(x)\big| \le C_h \min(2^{m/2},2^{-m/2}). \tag{6.1.15}$$

In view of Exercise 6.1.3 and the periodicity of the functions $B^m_{\xi,y,\eta,\lambda}(h)$, we conclude that the averages

$$\frac{1}{2KL}\int_0^L\int_{-K}^K\int_0^1 B^m_{\xi,y,\eta,\lambda}(h)\,d\lambda\,dy\,d\eta$$

converge pointwise to some $\Pi^m_\xi(h)$ as $K,L\to\infty$. Estimate (6.1.15) implies the uniform convergence for the series $\sum_{m\in\mathbf{Z}} B^m_{\xi,y,\eta,\lambda}(h)$ and therefore

$$\begin{aligned}
&\lim_{\substack{K\to\infty\\L\to\infty}} \frac{1}{2KL}\int_0^L\int_{-K}^K\int_0^1 M^{-\eta}\tau^{-y}D^{2^{-\lambda}}A_{\frac{\xi+\eta}{2^\lambda}}D^{2^\lambda}\tau^y M^\eta(h)\,d\lambda\,dy\,d\eta \qquad (6.1.16)\\
&\quad = \lim_{\substack{K\to\infty\\L\to\infty}} \frac{1}{2KL}\int_0^L\int_{-K}^K\int_0^1 \sum_{m\in\mathbf{Z}} B^m_{\xi,y,\eta,\lambda}(h)\,d\lambda\,dy\,d\eta \\
&\quad = \sum_{m\in\mathbf{Z}} \lim_{\substack{K\to\infty\\L\to\infty}} \frac{1}{2KL}\int_0^L\int_{-K}^K\int_0^1 B^m_{\xi,y,\eta,\lambda}(h)\,d\lambda\,dy\,d\eta \\
&\quad = \sum_{m\in\mathbf{Z}} \Pi^m_\xi(h).
\end{aligned}$$

For $h \in \mathscr{S}(\mathbf{R})$ we now define

$$\Pi_\xi(h) = \sum_{m\in\mathbf{Z}} \Pi_\xi^m(h)$$

and we make some observations about this operator. First we observe that in view of Lemma 6.1.2 and Fatou's lemma, we have that Π_ξ is bounded on L^2 uniformly in ξ. Next we observe that Π_ξ commutes with all translations τ^z for $z \in \mathbf{R}$. To see this, we use the fact that $\tau^{-z}M^{-\eta} = e^{-2\pi i\eta z}M^{-\eta}\tau^{-z}$ to obtain

$$\begin{aligned}
&\sum_{s\in\mathbf{D}_m} \chi_{\omega_{s(2)}}(2^{-\lambda}(\xi+\eta))\langle D^{2^\lambda}\tau^y M^\eta \tau^z(h)\,|\,\varphi_s\rangle\, \tau^{-z}M^{-\eta}\tau^{-y}D^{2^{-\lambda}}(\varphi_s)\\
&\quad= \sum_{s\in\mathbf{D}_m} \chi_{\omega_{s(2)}}(2^{-\lambda}(\xi+\eta))\langle h\,|\,\tau^{-z}M^{-\eta}\tau^{-y}D^{2^{-\lambda}}(\varphi_s)\rangle\, \tau^{-z}M^{-\eta}\tau^{-y}D^{2^{-\lambda}}(\varphi_s)\\
&\quad= \sum_{s\in\mathbf{D}_m} \chi_{\omega_{s(2)}}(2^{-\lambda}(\xi+\eta))\langle h\,|\,M^{-\eta}\tau^{-y-z}D^{2^{-\lambda}}(\varphi_s)\rangle\, M^{-\eta}\tau^{-y-z}D^{2^{-\lambda}}(\varphi_s)\,.
\end{aligned}$$

Recall that $\tau^{-z}\Pi_\xi^m\tau^z(h)$ is equal to the limit of the averages of the preceding expressions over all $(y,\eta,\lambda) \in [-K,K]\times[0,L]\times[0,1]$. But in view of the previous identity, this is equal to the limit of the averages of the expressions

$$\sum_{s\in\mathbf{D}_m} \chi_{\omega_{s(2)}}(2^{-\lambda}(\xi+\eta))\langle D^{2^\lambda}\tau^{y'}M^\eta(h)\,|\,\varphi_s\rangle M^{-\eta}\tau^{-y'}D^{2^{-\lambda}}(\varphi_s) \tag{6.1.17}$$

over all $(y',\eta,\lambda) \in [-K+z,K+z]\times[0,L]\times[0,1]$. Since (6.1.17) is periodic in (y',η), it follows that its average over the set $[-K+z,K+z]\times[0,L]\times[0,1]$ is equal to its average over the set $[-K,K]\times[0,L]\times[0,1]$. Taking limits as $K,L\to\infty$, we obtain the identity $\tau^{-z}\Pi_\xi^m\tau^z(h) = \Pi_\xi^m(h)$. Summing over all $m\in\mathbf{Z}$, it follows that

$$\tau^{-z}\Pi_\xi\tau^z(h) = \Pi_\xi(h)\,.$$

Using averages over the shifted rectangles $[-K,K]\times[\theta,L+\theta]$, via a similar argument, we obtain the identity

$$M^{-\theta}\Pi_{\xi+\theta}M^\theta = \Pi_\xi \tag{6.1.18}$$

for all $\xi,\theta\in\mathbf{R}$. The details are left to the reader. Next, we claim that the operator $M^{-\xi}\Pi_\xi M^\xi$ commutes with dilations D^{2^a}, $a\in\mathbf{R}$. First we observe that for all integers k we have

$$A_\xi(h) = D^{2^{-k}}A_{2^{-k}\xi}D^{2^k}(h)\,, \tag{6.1.19}$$

which is simply saying that A_ξ is well behaved under change of scale. This identity is left as an exercise to the reader. Identity (6.1.19) may not hold for noninteger k, and this is exactly why we have averaged over all dilations 2^λ, $0\le\lambda\le1$, in (6.1.16).

Let us denote by $[a]$ the integer part of a real number a. Using the identities $D^bM^\eta = M^{\eta/b}D^b$ and $D^b\tau^z = \tau^{bz}D^b$, we obtain

$$
\begin{aligned}
&D^{2^{-a}}M^{-(\xi+\eta)}\tau^{-y}D^{2^{-\lambda}}A_{\frac{\xi+\eta}{2^\lambda}}D^{2^\lambda}\tau^{y}M^{\xi+\eta}D^{2^a} \qquad (6.1.20)\\
&=M^{-2^a(\xi+\eta)}\tau^{-2^{-a}y}D^{2^{-(a+\lambda)}}A_{\frac{\xi+\eta}{2^\lambda}}D^{2^{a+\lambda}}\tau^{2^{-a}y}M^{2^a(\xi+\eta)}\\
&=M^{-2^a(\xi+\eta)}\tau^{-y'}D^{2^{-\lambda'}}D^{2^{-[a+\lambda]}}A_{\frac{2^a(\xi+\eta)}{2^{\lambda'}2^{[a+\lambda]}}}D^{2^{[a+\lambda]}}D^{2^{\lambda'}}\tau^{y'}M^{2^a(\xi+\eta)}\\
&=M^{-2^a\xi}M^{-\eta'}\tau^{-y'}D^{2^{-\lambda'}}A_{\frac{2^a\xi+2^a\eta}{2^{\lambda'}}}D^{2^{\lambda'}}\tau^{y'}M^{\eta'}M^{2^a\xi}\\
&=M^{-\xi}M^{-\theta}\big(M^{-\eta'}\tau^{-y'}D^{2^{-\lambda'}}A_{\frac{\xi+\theta+\eta'}{2^{\lambda'}}}D^{2^{\lambda'}}\tau^{y'}M^{\eta'}\big)M^{\theta}M^{\xi}, \qquad (6.1.21)
\end{aligned}
$$

where we set $y'=2^{-a}y$, $\eta'=2^a\eta$, $\lambda'=a+\lambda-[a+\lambda]$, and $\theta=(2^a-1)\xi$. The average of (6.1.20) over all (y,η,λ) in $[-K,K]\times[0,L]\times[0,1]$ converges to the operator $D^{2^{-a}}M^{-\xi}\Pi_\xi M^\xi D^{2^a}$ as $K,L\to\infty$. But this limit is equal to the limit of the averages of the expression in (6.1.21) over all (y',η',λ') in $[-2^{-a}K,2^{-a}K]\times[0,2^aL]\times[0,1]$, which is

$$M^{-\xi}M^{-\theta}\Pi_{\xi+\theta}M^{\theta}M^{\xi}.$$

Using the identity (6.1.18), we obtain that

$$D^{2^{-a}}M^{-\xi}\Pi_\xi M^\xi D^{2^a}=M^{-\xi}\Pi_\xi M^\xi\,,$$

which says that the operator $M^{-\xi}\Pi_\xi M^\xi$ commutes with dilations.

Next we observe that if $\widehat{h}$ is supported in $[0,\infty)$, then $M^{-\xi}\Pi_\xi M^\xi(h)=0$. This is a consequence of the fact that the inner products

$$\big\langle D^{2^\lambda}\tau^yM^\eta M^\xi(h)\,|\,\varphi_s\big\rangle=\big\langle M^\xi(h)\,|\,M^{-\eta}\tau^{-y}D^{2^{-\lambda}}(\varphi_s)\big\rangle$$

vanish, since the Fourier transform of $\tau^{-z}M^{-\eta}\tau^{-y}D^{2^{-\lambda}}\varphi_s$ is supported in the set $(-\infty,2^\lambda c(\omega_{s(1)})-\eta+\frac{2^\lambda}{10}|\omega_s|)$, which is disjoint from the interval $(\xi,+\infty)$ whenever $2^{-\lambda}(\xi+\eta)\in\omega_{s(2)}$. Finally, we observe that Π_ξ is a positive semidefinite operator, that is, it satisfies

$$\big\langle \Pi_\xi(h)\,|\,h\big\rangle\geq 0. \qquad (6.1.22)$$

This follows easily from the fact that the inner product in (6.1.22) is equal to

$$\lim_{\substack{K\to\infty\\L\to\infty}}\frac{1}{2KL}\int_0^L\int_{-K}^K\int_0^1\sum_{s\in\mathbf{D}}\chi_{\omega_{s(2)}}\big(\tfrac{\xi+\eta}{2^\lambda}\big)\big|\big\langle D^{2^\lambda}\tau^yM^\eta(h)\,|\,\varphi_s\big\rangle\big|^2\,d\lambda\,dy\,d\eta. \qquad (6.1.23)$$

This identity also implies that Π_ξ is not the zero operator; indeed, notice that

$$\sum_{s\in\mathbf{D}_0}\chi_{\omega_{s(2)}}\big(\tfrac{\xi+\eta}{2^\lambda}\big)\big|\big\langle D^{2^\lambda}\tau^yM^\eta(h)\,|\,\varphi_s\big\rangle\big|^2=\big\langle h\,|\,B^0_{\xi,y,\eta,\lambda}(h)\big\rangle$$

is periodic with period $(2^{-\lambda}, 2^{\lambda})$ in (y,η), and consequently the limit in (6.1.23) is at least as big as

$$\int_0^{2^\lambda}\int_0^{2^{-\lambda}}\int_0^1 \sum_{s\in\mathbf{D}_0} \chi_{\omega_{s(2)}}\left(\tfrac{\xi+\eta}{2^\lambda}\right)\left|\left\langle D^{2^\lambda}\tau^y M^\eta(h)\,|\,\varphi_s\right\rangle\right|^2 d\lambda\,dy\,d\eta$$

(cf. Exercise 6.1.3). Since we can always find a Schwartz function h and a dyadic tile s such that $\left\langle D^{2^\lambda}\tau^y M^\eta(h)\,|\,\varphi_s\right\rangle$ is not zero for (y,η,λ) near $(0,0,0)$, it follows that the expression in (6.1.23) is strictly positive for some function h. The same is valid for the inner product in (6.1.22); hence the operators $M^{-\xi}\Pi_\xi M^\xi$ are nonzero for every ξ.

Let us summarize what we have already proved: The operator $M^{-\xi}\Pi_\xi M^\xi$ is nonzero, is bounded on $L^2(\mathbf{R})$, commutes with translations and dilations, and vanishes when applied to functions whose Fourier transform is supported in the positive semiaxis $[0,\infty)$. Using the result of Exercise 5.1.11(b) in [156], it follows that for some constant $c_\xi \neq 0$ we have

$$M^{-\xi}\Pi_\xi M^\xi(h)(x) = c_\xi \int_{-\infty}^0 \widehat{h}(\eta)e^{2\pi i x\eta}\,d\eta\,,$$

which identifies Π_ξ with the convolution operator whose multiplier is the function $c_\xi\chi_{(-\infty,\xi]}$. Using the identity (6.1.18), we obtain

$$c_{\xi+\theta} = c_\xi$$

for all ξ and θ, saying that c_ξ does not depend on ξ. We have therefore proved that for all Schwartz functions h the following identity is valid:

$$\Pi_\xi(h) = c\left(\widehat{h}\chi_{(-\infty,\xi]}\right)^\vee \tag{6.1.24}$$

for some fixed nonzero constant c.

6.1.3 Linearization of a Maximal Dyadic Sum

To prove (6.1.2) with $\mathscr{C}_1$ in place of $\mathscr{C}$ we first make some reductions. We notice that for a fixed $f\in\mathscr{S}(\mathbf{R})$, the function

$$(x,\xi)\mapsto \int_{-\infty}^{\xi} \widehat{f}(y)e^{2\pi i x y}\,dy$$

defined on $\mathbf{R}\times\mathbf{R}^+$ is continuous in both variables. This allows us to restrict the range of N in the supremum in (6.1.1) to $N\in\mathbf{Q}^+$. Using the Lebesgue monotone convergence theorem, we may also restrict N to a finite subset Q_0 of $\mathbf{Q}^+$ and obtain bounds independent of the size of this finite subset. Then for each $\xi_0 \in Q_0$ we have that

$$\left\{x \in \mathbf{R} : \max_{\xi \in Q_0} \left| \int_{-\infty}^{\xi} \widehat{f}(y) e^{2\pi i x y}\, dy \right| = \left| \int_{-\infty}^{\xi_0} \widehat{f}(y) e^{2\pi i x y}\, dy \right| \right\}$$

is closed and hence measurable. We may therefore select a measurable real-valued function $N_f : \mathbf{R} \to Q_0$ such that for all $x \in \mathbf{R}$ we have

$$\sup_{\xi \in Q_0} \left| \int_{-\infty}^{\xi} \widehat{f}(y) e^{2\pi i x y}\, dy \right| = \left| \int_{-\infty}^{N_f(x)} \widehat{f}(y) e^{2\pi i x y}\, dy \right|.$$

The appearance of this measurable function motivates the introduction of the operator

$$f \mapsto \int_{-\infty}^{N(x)} \widehat{f}(y) e^{2\pi i x y}\, dy \tag{6.1.25}$$

for a general measurable function $N : \mathbf{R} \to Q_0$. If we can prove an $L^2 \to L^{2,\infty}$ estimate for this operator applied to Schwartz functions with bounds independent of the measurable function N, then for a given $f \in \mathscr{S}(\mathbf{R})$ we pick $N = N_f$ and obtain the boundedness of

$$f \mapsto \sup_{\xi \in Q_0} \left| \int_{-\infty}^{\xi} \widehat{f}(y) e^{2\pi i x y}\, dy \right|$$

from $L^2 \to L^{2,\infty}$; then the boundedness of $\mathscr{C}_1$ follows as previously observed, replacing Q_0 by $\mathbf{Q}^+$ via the Lebesgue monotone convergence theorem, and then replacing $\mathbf{Q}^+$ by $\mathbf{R}^+$ by continuity.

For the rest of this section, we fix a measurable function N defined on the real line with finitely many positive rational values. We define a linear operator $\mathfrak{D}_N$ by setting for $f \in \mathscr{S}(\mathbf{R})$,

$$\mathfrak{D}_N(f)(x) = A_{N(x)}(f)(x) = \sum_{s \in \mathbf{D}} (\chi_{\omega_{s(2)}} \circ N)(x) \left\langle f \,|\, \varphi_s \right\rangle \varphi_s(x). \tag{6.1.26}$$

In view of Lemma 6.1.2, the series in (6.1.26) converges absolutely for all $x \in \mathbf{R}$.

It suffices to show that there exists $C > 0$ such that for all $f \in \mathscr{S}(\mathbf{R})$ and all measurable functions $N : \mathbf{R} \to \mathbf{Q}^+$ (with finitely many values) we have

$$\big\|\mathfrak{D}_N(f)\big\|_{L^{2,\infty}} \le C \big\|f\big\|_{L^2}. \tag{6.1.27}$$

Suppose we know the validity of (6.1.27). Then identity (6.1.16) gives

$$\Pi_\xi(f)(x) = \lim_{\substack{K \to \infty \\ L \to \infty}} \frac{1}{2KL} \int_0^L \int_{-K}^K \int_0^1 G_{\xi, y, \eta, \lambda}(f)(x)\, d\lambda\, dy\, d\eta,$$

for all $x \in \mathbf{R}$ and $\xi > 0$, where

$$G_{\xi, y, \eta, \lambda}(f)(x) = M^{-\eta} \tau^{-y} D^{2^{-\lambda}} A_{\frac{\xi+\eta}{2^\lambda}} D^{2^\lambda} \tau^y M^\eta (f)(x).$$

Taking $\xi = N(x)$, this gives for any $x \in \mathbf{R}$

$$\Pi_{N(x)}(f)(x) = \lim_{\substack{K\to\infty \\ L\to\infty}} \frac{1}{2KL} \int_0^L \int_{-K}^K \int_0^1 G_{N(x),y,\eta,\lambda}(f)(x)\, d\lambda\, dy\, d\eta$$

and hence

$$|\Pi_{N(x)}(f)(x)| \le \liminf_{\substack{K\to\infty \\ L\to\infty}} \frac{1}{2KL} \int_0^L \int_{-K}^K \int_0^1 |G_{N(x),y,\eta,\lambda}(f)(x)|\, d\lambda\, dy\, d\eta\,.$$

We now apply the $L^{2,\infty}$ quasi-norm on both sides and we use Fatou's lemma for weak L^2; see Exercise 1.1.12(d) in [156]. Since modulations, translations, and L^2-dilations are isometries on L^2, we reduce the sought estimate for the operator in (6.1.25) to the corresponding estimate for $f \mapsto A_{N(x)}(f)(x) = \mathfrak{D}_N(f)(x)$.

To justify certain algebraic manipulations we fix a finite subset $\mathbf{P}$ of $\mathbf{D}$ and we define

$$\mathfrak{D}_{N,\mathbf{P}}(f)(x) = \sum_{s\in\mathbf{P}} (\chi_{\omega_{s(2)}} \circ N)(x)\, \langle f \,|\, \varphi_s \rangle\, \varphi_s(x)\,. \tag{6.1.28}$$

To prove (6.1.27) it suffices to show that there exists a $C > 0$ such that for all f in $\mathscr{S}(\mathbf{R})$, all finite subsets $\mathbf{P}$ of $\mathbf{D}$, and all real-valued measurable functions N on the line we have

$$\big\|\mathfrak{D}_{N,\mathbf{P}}(f)\big\|_{L^{2,\infty}} \le C \big\|f\big\|_{L^2}\,. \tag{6.1.29}$$

The important point is that the constant C in (6.1.29) is independent of f, $\mathbf{P}$, and the measurable function N. Once (6.1.29) is known, then taking a sequence of sets $\mathbf{P}_L \to \mathbf{D}$, as $L \to \infty$ and using the absolute convergence of the series, we obtain (6.1.27).

To prove (6.1.29) we use duality. In view of the result of Exercises 1.4.12(c), it suffices to prove that for all $f \in \mathscr{S}(\mathbf{R})$ we have

$$\left| \int_{\mathbf{R}} \mathfrak{D}_{N,\mathbf{P}}(f)\, g\, dx \right| = \left| \sum_{s\in\mathbf{P}} \langle (\chi_{\omega_{s(2)}} \circ N)\varphi_s, g \rangle \langle \varphi_s \,|\, f \rangle \right| \le C \big\|g\big\|_{L^{2,1}} \big\|f\big\|_{L^2}\,. \tag{6.1.30}$$

Using the result of Exercise 1.4.7 in [156], (6.1.30) will follow from the fact that for all measurable subsets E of the real line with finite measure we have

$$\left| \int_E \mathfrak{D}_{N,\mathbf{P}}(f)\, dx \right| = \left| \sum_{s\in\mathbf{P}} \langle (\chi_{\omega_{s(2)}} \circ N)\varphi_s, \chi_E \rangle \langle \varphi_s \,|\, f \rangle \right| \le C |E|^{\frac{1}{2}} \big\|f\big\|_{L^2}\,. \tag{6.1.31}$$

We obtain estimate (6.1.31) as a consequence of

$$\sum_{s\in\mathbf{P}} \left| \langle (\chi_{\omega_{s(2)}} \circ N)\varphi_s, \chi_E \rangle \langle f \,|\, \varphi_s \rangle \right| \le C |E|^{\frac{1}{2}} \big\|f\big\|_{L^2} \tag{6.1.32}$$

for all f in $\mathscr{S}(\mathbf{R})$, all measurable functions N, all measurable sets E of finite measure, and all finite subsets $\mathbf{P}$ of $\mathbf{D}$. We therefore concentrate on estimate (6.1.32).

6.1.4 Iterative Selection of Sets of Tiles with Large Mass and Energy

We introduce a partial order in the set of dyadic tiles that provides a way to organize them. In this section, dyadic tiles are simply called tiles.

Definition 6.1.3. We define a *partial order* $<$ in the set of dyadic tiles $\mathbf{D}$ by setting

$$s < s' \iff I_s \subsetneqq I_{s'} \quad \text{and} \quad \omega_{s'} \subsetneqq \omega_s .$$

If two tiles $s, s' \in \mathbf{D}$ intersect, then we must have either $s < s'$ or $s' < s$. Indeed, both the time and frequency components of the tiles must intersect; then either $I_s \subsetneqq I_{s'}$ or $I_{s'} \subsetneqq I_s$. In the first case we must have $|\omega_s| \geq |\omega_{s'}|$, thus $\omega_{s'} \subsetneqq \omega_s$, which gives $s < s'$, while in the second case a similar argument gives $s' < s$. As a consequence of this observation, if $\mathbf{R}_0$ is a finite set of tiles, then all maximal elements of $\mathbf{R}_0$ under $<$ must be disjoint sets.

Definition 6.1.4. A finite set of tiles $\mathbf{P}$ is called a *tree* if there exists a tile $t \in \mathbf{P}$ such that all $s \in \mathbf{P}$ satisfy $s < t$. We call t the top of $\mathbf{P}$ and we denote it by $t = \text{top}(\mathbf{P})$. Observe that the top of a tree is unique.

We denote trees by $\mathbf{T}$, $\mathbf{T}'$, $\mathbf{T}_1$, $\mathbf{T}_2$, and so on.

We observe that every finite set of tiles $\mathbf{P}$ can be written as a union of trees whose tops are maximal elements. Indeed, consider all maximal elements of $\mathbf{P}$ under the partial order $<$. Then every nonmaximal element s of $\mathbf{P}$ satisfies $s < t$ for some maximal element $t \in \mathbf{P}$, and thus it belongs to a tree with top t.

Tiles can be written as a union of two *semitiles* $I_s \times \omega_{s(1)}$ and $I_s \times \omega_{s(2)}$. Since tiles have area 1, semitiles have area $1/2$.

Definition 6.1.5. A tree $\mathbf{T}$ is called a 1-tree if

$$\omega_{\text{top}(\mathbf{T})(1)} \subsetneqq \omega_{s(1)}$$

for all $s \in \mathbf{T}$. A tree $\mathbf{T}'$ is called a 2-tree if for all $s \in \mathbf{T}'$ we have

$$\omega_{\text{top}(\mathbf{T}')(2)} \subsetneqq \omega_{s(2)} .$$

We make a few observations about 1-trees and 2-trees. First note that every tree can be written as the union of a 1-tree and a 2-tree, and the intersection of these is exactly the top of the tree. Also, if $\mathbf{T}$ is a 1-tree, then the intervals $\omega_{\text{top}(\mathbf{T})(2)}$ and $\omega_{s(2)}$ are disjoint for all $s \in \mathbf{T}$ and similarly for 2-trees. See Figure 6.2.

Definition 6.1.6. Let $N : \mathbf{R} \to \mathbf{R}^+$ be a measurable function, let $s \in \mathbf{D}$, and let E be a set of finite measure. Then we introduce the quantity

$$\mathscr{M}(E;\{s\}) = \frac{1}{|E|} \sup_{\substack{u \in \mathbf{D} \\ s < u}} \int_{E \cap N^{-1}[\omega_u]} \frac{|I_u|^{-1}\, dx}{(1 + \frac{|x - c(I_u)|}{|I_u|})^{10}} .$$

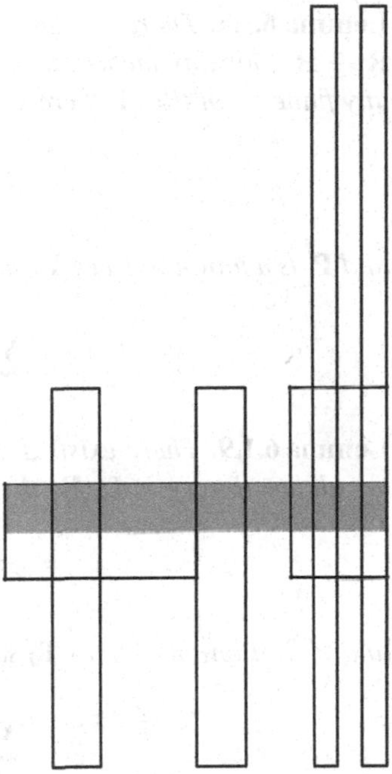

Fig. 6.2 A tree of seven tiles including the darkened top. The top together with the three tiles on the right forms a 1-tree, while the top together with the three tiles on the left forms a 2-tree.

We call $\mathscr{M}(E;\{s\})$ the *mass* of E with respect to $\{s\}$. Given a subset $\mathbf{P}$ of $\mathbf{D}$, we define the mass of E with respect to $\mathbf{P}$ as

$$\mathscr{M}(E;\mathbf{P}) = \sup_{s\in\mathbf{P}} \mathscr{M}(E;\{s\}).$$

We observe that the mass of E with respect to any set of tiles is at most

$$\frac{1}{|E|}\int_{-\infty}^{+\infty} \frac{dx}{(1+|x|)^{10}} \le \frac{1}{|E|}.$$

Definition 6.1.7. Given a finite subset $\mathbf{P}$ of $\mathbf{D}$ and a function g in $L^2(\mathbf{R})$, we introduce the quantity

$$\mathscr{E}(g;\mathbf{P}) = \frac{1}{\|g\|_{L^2}} \sup_{\mathbf{T}} \left(\frac{1}{|I_{\mathrm{top}(\mathbf{T})}|} \sum_{s\in\mathbf{T}} |\langle g\,|\,\varphi_s\rangle|^2 \right)^{\frac{1}{2}},$$

where the supremum is taken over all 2-trees $\mathbf{T}$ contained in $\mathbf{P}$. We call $\mathscr{E}(g;\mathbf{P})$ the *energy* of the function g with respect to the set of tiles $\mathbf{P}$.

We now state three important lemmas which we prove in the remaining three subsections, respectively.

Lemma 6.1.8. *There exists a constant C_1 such that for any measurable function $N: \mathbf{R} \to \mathbf{R}^+$, for any measurable subset E of the real line with finite measure, and for any finite set of tiles $\mathbf{P}$ there is a subset $\mathbf{P}'$ of $\mathbf{P}$ such that*

$$\mathscr{M}(E;\mathbf{P}\setminus\mathbf{P}') \le \frac{1}{4}\mathscr{M}(E;\mathbf{P})$$

and $\mathbf{P}'$ is a union of trees $\mathbf{T}_j$ satisfying

$$\sum_j |I_{\mathrm{top}(\mathbf{T}_j)}| \le \frac{C_1}{\mathscr{M}(E;\mathbf{P})}. \tag{6.1.33}$$

Lemma 6.1.9. *There exists a constant C_2 such that for any finite set of tiles $\mathbf{P}$ and for all functions g in $L^2(\mathbf{R})$ there is a subset $\mathbf{P}''$ of $\mathbf{P}$ such that*

$$\mathscr{E}(g;\mathbf{P}\setminus\mathbf{P}'') \le \frac{1}{2}\mathscr{E}(g;\mathbf{P})$$

and $\mathbf{P}''$ is a union of trees $\mathbf{T}_j$ satisfying

$$\sum_j |I_{\mathrm{top}(\mathbf{T}_j)}| \le \frac{C_2}{\mathscr{E}(g;\mathbf{P})^2}. \tag{6.1.34}$$

Lemma 6.1.10. *(The basic estimate) There is a finite constant C_3 such that for all trees $\mathbf{T}$, all functions g in $L^2(\mathbf{R})$, for any measurable function $N: \mathbf{R} \to \mathbf{R}^+$, and for all measurable sets E we have*

$$\begin{aligned}&\sum_{s\in\mathbf{T}} \big|\langle g\,|\,\varphi_s\rangle\langle \chi_{E\cap N^{-1}[\omega_{s(2)}]}\,|\,\varphi_s\rangle\big| \\ &\qquad\le C_3\,|I_{\mathrm{top}(\mathbf{T})}|\,\mathscr{E}(g;\mathbf{T})\mathscr{M}(E;\mathbf{T})\big\|g\big\|_{L^2}\,|E|.\end{aligned} \tag{6.1.35}$$

In the rest of this subsection, we conclude the proof of Theorem 6.1.1 assuming Lemmas 6.1.8, 6.1.9, and 6.1.10.

Given a finite set of tiles $\mathbf{P}$, a measurable set E of finite measure, a measurable function $N: \mathbf{R} \to \mathbf{R}^+$, and a function f in $\mathscr{S}(\mathbf{R})$, we find a very large integer n_0 such that

$$\begin{aligned}\mathscr{E}(f;\mathbf{P}) &\le 2^{n_0}, \\ \mathscr{M}(E;\mathbf{P}) &\le 2^{2n_0}.\end{aligned}$$

We shall construct by decreasing induction a sequence of pairwise disjoint sets

$$\mathbf{P}_{n_0}, \mathbf{P}_{n_0-1}, \mathbf{P}_{n_0-2}, \mathbf{P}_{n_0-3}, \dots$$

such that

$$\bigcup_{j=-\infty}^{n_0} \mathbf{P}_j = \mathbf{P} \tag{6.1.36}$$

and such that the following properties are satisfied:

(1) $\mathscr{E}(f;\mathbf{P}_j) \le 2^{j+1}$ for all $j \le n_0$;

(2) $\mathscr{M}(E;\mathbf{P}_j) \le 2^{2j+2}$ for all $j \le n_0$;

(3) $\mathscr{E}\big(f;\mathbf{P}\setminus(\mathbf{P}_{n_0}\cup\cdots\cup\mathbf{P}_j)\big) \le 2^j$ for all $j \le n_0$;

(4) $\mathscr{M}\big(E;\mathbf{P}\setminus(\mathbf{P}_{n_0}\cup\cdots\cup\mathbf{P}_j)\big) \le 2^{2j}$ for all $j \le n_0$;

(5) $\mathbf{P}_j$ is a union of trees $\mathbf{T}_{jk}$ such that for all $j \le n_0$ we have

$$\sum_k |I_{\mathrm{top}(\mathbf{T}_{jk})}| \le C_0\, 2^{-2j},$$

where $C_0 = C_1 + C_2$ and C_1 and C_2 are the constants that appear in Lemmas 6.1.8 and 6.1.9, respectively.

Assume momentarily that we have constructed a sequence $\{\mathbf{P}_j\}_{j\le n_0}$ with the described properties. Then to obtain estimate (6.1.32) we use (1), (2), (5), the observation that the mass of any set of tiles is always bounded by $|E|^{-1}$, and Lemma 6.1.10 to obtain

$$\begin{aligned}
&\sum_{s\in\mathbf{P}} \big|\langle f\,|\,\varphi_s\rangle\langle \chi_{E\cap N^{-1}[\omega_{s(2)}]}\,|\,\varphi_s\rangle\big| \\
&\quad= \sum_j\sum_{s\in\mathbf{P}_j} \big|\langle f\,|\,\varphi_s\rangle\langle \chi_{E\cap N^{-1}[\omega_{s(2)}]}\,|\,\varphi_s\rangle\big| \\
&\quad\le \sum_j\sum_k\sum_{s\in\mathbf{T}_{jk}} \big|\langle f\,|\,\varphi_s\rangle\langle \chi_{E\cap N^{-1}[\omega_{s(2)}]}\,|\,\varphi_s\rangle\big| \\
&\quad\le C_3\sum_j\sum_k |I_{\mathrm{top}(\mathbf{T}_{jk})}|\,\mathscr{E}(f;\mathbf{T}_{jk})\,\mathscr{M}(E;\mathbf{T}_{jk})\big\|f\big\|_{L^2}|E| \\
&\quad\le C_3\sum_j\sum_k |I_{\mathrm{top}(\mathbf{T}_{jk})}|\,2^{j+1}\min(|E|^{-1},2^{2j+2})\big\|f\big\|_{L^2}|E| \\
&\quad\le C_3\sum_j C_0 2^{-2j}2^{j+1}\min(|E|^{-1},2^{2j+2})\big\|f\big\|_{L^2}|E| \\
&\quad\le 8C_0C_3\sum_j \min(2^{-j}|E|^{-\frac12},2^j|E|^{\frac12})\big\|f\big\|_{L^2}|E|^{\frac12} \\
&\quad\le C|E|^{\frac12}\,\big\|f\big\|_{L^2}.
\end{aligned}$$

This proves estimate (6.1.32).

It remains to construct a sequence of disjoint sets $\mathbf{P}_j$ satisfying properties (1)–(5). The selection of these sets is based on decreasing induction. We start the induction at $j = n_0$ by setting $\mathbf{P}_{n_0} = \emptyset$. Then (1), (2), and (5) are clearly satisfied, while

$$\begin{aligned}
\mathscr{E}(f;\mathbf{P}\setminus\mathbf{P}_{n_0}) &= \mathscr{E}(f;\mathbf{P}) \le 2^{n_0}, \\
\mathscr{M}(E;\mathbf{P}\setminus\mathbf{P}_{n_0}) &= \mathscr{M}(E;\mathbf{P}) \le 2^{2n_0};
\end{aligned}$$

hence (3) and (4) are also satisfied for $\mathbf{P}_{n_0}$.

Suppose that we have selected pairwise disjoint sets $\mathbf{P}_{n_0}, \mathbf{P}_{n_0-1}, \dots, \mathbf{P}_n$ for some $n \le n_0$ such that (1)–(5) are satisfied for all $j \in \{n_0, n_0-1, \dots, n\}$. We construct a set of tiles $\mathbf{P}_{n-1}$ disjoint from all $\mathbf{P}_j$ with $j \ge n$ such that (1)–(5) are satisfied for $j = n-1$.

We define first an auxiliary set $\mathbf{P}'_{n-1}$. If $\mathscr{M}(E;\mathbf{P}\setminus(\mathbf{P}_{n_0}\cup\cdots\cup\mathbf{P}_n)) \le 2^{2(n-1)}$ set $\mathbf{P}'_{n-1} = \emptyset$. If $\mathscr{M}(E;\mathbf{P}\setminus(\mathbf{P}_{n_0}\cup\cdots\cup\mathbf{P}_n)) > 2^{2(n-1)}$ apply Lemma 6.1.8 to find a subset $\mathbf{P}'_{n-1}$ of $\mathbf{P}\setminus(\mathbf{P}_{n_0}\cup\cdots\cup\mathbf{P}_n)$ such that

$$\mathscr{M}(E;\mathbf{P}\setminus(\mathbf{P}_{n_0}\cup\cdots\cup\mathbf{P}_n\cup\mathbf{P}'_{n-1})) \le \frac{1}{4}\mathscr{M}(E;\mathbf{P}\setminus(\mathbf{P}_{n_0}\cup\cdots\cup\mathbf{P}_n)) \le \frac{2^{2n}}{4} = 2^{2(n-1)}$$

[by the induction hypothesis (4) with $j = n$] and $\mathbf{P}'_{n-1}$ is a union of trees $\mathbf{T}'_k$ satisfying

$$\sum_k |I_{\mathrm{top}(\mathbf{T}'_k)}| \le C_1\,\mathscr{M}(f;\mathbf{P}\setminus(\mathbf{P}_{n_0}\cup\cdots\cup\mathbf{P}_n))^{-1} \le C_1 2^{-2(n-1)}. \tag{6.1.37}$$

Likewise, if $\mathscr{E}(f;\mathbf{P}\setminus(\mathbf{P}_{n_0}\cup\cdots\cup\mathbf{P}_n)) \le 2^{n-1}$ set $\mathbf{P}''_{n-1} = \emptyset$; otherwise, apply Lemma 6.1.9 to find a subset $\mathbf{P}''_{n-1}$ of $\mathbf{P}\setminus(\mathbf{P}_{n_0}\cup\cdots\cup\mathbf{P}_n)$ such that

$$\mathscr{E}(f;\mathbf{P}\setminus(\mathbf{P}_{n_0}\cup\cdots\cup\mathbf{P}_n\cup\mathbf{P}''_{n-1})) \le \frac{1}{2}\mathscr{E}(f;\mathbf{P}\setminus(\mathbf{P}_{n_0}\cup\cdots\cup\mathbf{P}_n)) \le \frac{1}{2}2^n = 2^{n-1}$$

[by the induction hypothesis (3) with $j = n$] and $\mathbf{P}''_{n-1}$ is a union of trees $\mathbf{T}''_k$ satisfying

$$\sum_k |I_{\mathrm{top}(\mathbf{T}''_k)}| \le C_2\,\mathscr{E}(f;\mathbf{P}\setminus(\mathbf{P}_{n_0}\cup\cdots\cup\mathbf{P}_n))^{-2} \le C_2 2^{-2(n-1)}. \tag{6.1.38}$$

Whether the sets $\mathbf{P}'_{n-1}$ and $\mathbf{P}''_{n-1}$ are empty or not, we note that

$$\mathscr{M}(E;\mathbf{P}\setminus(\mathbf{P}_{n_0}\cup\cdots\cup\mathbf{P}_n\cup\mathbf{P}'_{n-1})) \le 2^{2(n-1)}, \tag{6.1.39}$$

$$\mathscr{E}(f;\mathbf{P}\setminus(\mathbf{P}_{n_0}\cup\cdots\cup\mathbf{P}_n\cup\mathbf{P}''_{n-1})) \le 2^{n-1}. \tag{6.1.40}$$

We set $\mathbf{P}_{n-1} = \mathbf{P}'_{n-1}\bigcup\mathbf{P}''_{n-1}$, and we verify properties (1)–(5) for $j = n-1$. Since $\mathbf{P}_{n-1}$ is contained in $\mathbf{P}\setminus(\mathbf{P}_{n_0}\cup\cdots\cup\mathbf{P}_n)$, we have

$$\mathscr{E}(f;\mathbf{P}_{n-1}) \le \mathscr{E}(f;\mathbf{P}\setminus(\mathbf{P}_{n_0}\cup\cdots\cup\mathbf{P}_n)) \le 2^n = 2^{(n-1)+1},$$

where the last inequality is a consequence of the induction hypothesis (3) for $j = n$; thus (1) holds with $j = n-1$. Likewise,

$$\mathscr{M}(E;\mathbf{P}_{n-1}) \le \mathscr{M}(E;\mathbf{P}\setminus(\mathbf{P}_{n_0}\cup\cdots\cup\mathbf{P}_n)) \le 2^{2n} = 2^{2(n-1)+2}$$

in view of the induction hypothesis (4) for $j = n$; thus (2) holds with $j = n-1$.

To prove (3) with $j = n-1$ notice that $\mathbf{P}\setminus(\mathbf{P}_{n_0}\cup\cdots\cup\mathbf{P}_n\cup\mathbf{P}_{n-1})$ is contained in $\mathbf{P}\setminus(\mathbf{P}_{n_0}\cup\cdots\cup\mathbf{P}_n\cup\mathbf{P}''_{n-1})$, and the latter has energy at most 2^{n-1} by (6.1.40). To prove (4) with $j = n-1$ note that $\mathbf{P}\setminus(\mathbf{P}_{n_0}\cup\cdots\cup\mathbf{P}_n\cup\mathbf{P}_{n-1})$ is contained in

$\mathbf{P}\setminus(\mathbf{P}_{n_0}\cup\cdots\cup\mathbf{P}_n\cup\mathbf{P}'_{n-1})$ and the latter has mass at most $2^{2(n-1)}$ by (6.1.39). Finally, adding (6.1.37) and (6.1.38) yields (5) for $j=n-1$ with $C_0=C_1+C_2$.

Pick $j\in\mathbf{Z}$ with $0<2^{2j}<\min_{s\in\mathbf{P}}\mathscr{M}(E;\{s\})$. Then $\mathscr{M}\big(E;\mathbf{P}\setminus(\mathbf{P}_{n_0}\cup\cdots\cup\mathbf{P}_j)\big)=0$, and since the only set of tiles with zero mass is the empty set, we conclude that (6.1.36) holds. It also follows that there exists an n_1 such that for all $n\le n_1$, $\mathbf{P}_n=\emptyset$. The construction of the $\mathbf{P}_j$'s is now complete.

6.1.5 Proof of the Mass Lemma 6.1.8

Proof. Given a finite set of tiles $\mathbf{P}$, we set $\mu=\mathscr{M}(E;\mathbf{P})$ to be the mass of $\mathbf{P}$. We define

$$\mathbf{P}'=\{s\in\mathbf{P}:\ \mathscr{M}(E;\{s\})>\tfrac{1}{4}\mu\}$$

and we observe that $\mathscr{M}(E;\mathbf{P}\setminus\mathbf{P}')\le\frac{1}{4}\mu$. We now show that $\mathbf{P}'$ is a union of trees whose tops satisfy (6.1.33).

It follows from the definition of mass that for each $s\in\mathbf{P}'$, there is a tile $u(s)\in\mathbf{D}$ such that $u(s)>s$ and

$$\frac{1}{|E|}\int_{E\cap N^{-1}[\omega_{u(s)}]}\frac{|I_{u(s)}|^{-1}\,dx}{(1+\frac{|x-c(I_{u(s)})|}{|I_{u(s)}|})^{10}}>\frac{\mu}{4}. \tag{6.1.41}$$

Let $\mathbf{U}=\{u(s):\ s\in\mathbf{P}'\}$. Also, let $\mathbf{U}_{\max}$ be the subset of $\mathbf{U}$ containing all maximal elements of $\mathbf{U}$ under the partial order of tiles $<$. Likewise define $\mathbf{P}'_{\max}$ as the set of all maximal elements in $\mathbf{P}'$. Tiles in $\mathbf{P}'$ can be grouped in trees

$$\mathbf{T}_j=\{s\in\mathbf{P}':\ s<t_j\}$$

with tops $t_j\in\mathbf{P}'_{\max}$. Observe that if $t_j<u$ and $t_{j'}<u$ for some $u\in\mathbf{U}_{\max}$, then ω_{t_j} and $\omega_{t_{j'}}$ intersect, and since t_j and $t_{j'}$ are disjoint sets, it follows that I_{t_j} and $I_{t_{j'}}$ are disjoint subsets of I_u. Consequently, we have

$$\sum_j|I_{t_j}|=\sum_{u\in\mathbf{U}_{\max}}\sum_{j:\,t_j<u}|I_{t_j}|\le\sum_{u\in\mathbf{U}_{\max}}|I_u|.$$

Therefore, estimate (6.1.33) will be a consequence of

$$\sum_{u\in\mathbf{U}_{\max}}|I_u|\le C\mu^{-1} \tag{6.1.42}$$

for some constant C. For $u\in\mathbf{U}_{\max}$ we rewrite (6.1.41) as

$$\frac{1}{|E|}\sum_{k=0}^{\infty}\int_{E\cap N^{-1}[\omega_u]\cap\left(2^kI_u\setminus 2^{k-1}I_u\right)}\frac{|I_u|^{-1}\,dx}{(1+\frac{|x-c(I_u)|}{|I_u|})^{10}}>\frac{\mu}{8}\sum_{k=0}^{\infty}2^{-k}$$

with the interpretation that $2^{-1}I_u = \emptyset$. It follows that for all u in $\mathbf{U}_{\max}$ there exists an integer $k \ge 0$ such that

$$|E|\frac{\mu}{8}|I_u|2^{-k} < \int_{E\cap N^{-1}[\omega_u]\cap\left(2^k I_u\setminus 2^{k-1}I_u\right)} \frac{dx}{(1+\frac{|x-c(I_u)|}{|I_u|})^{10}} \le \frac{|E\cap N^{-1}[\omega_u]\cap 2^k I_u|}{(\frac{4}{5})^{10}(1+2^{k-2})^{10}}.$$

We therefore conclude that

$$\mathbf{U}_{\max} = \bigcup_{k=0}^{\infty} \mathbf{U}_k,$$

where

$$\mathbf{U}_k = \{u \in \mathbf{U}_{\max} : |I_u| \le 8\cdot 5^{10} 2^{-9k}\mu^{-1}|E|^{-1}|E\cap N^{-1}[\omega_u]\cap 2^k I_u|\}.$$

The required estimate (6.1.42) will be a consequence of the sequence of estimates

$$\sum_{u\in\mathbf{U}_k} |I_u| \le C2^{-8k}\mu^{-1}, \qquad k \ge 0. \tag{6.1.43}$$

We now fix a $k \ge 0$ and we concentrate on (6.1.43). Select an element $v_0 \in \mathbf{U}_k$ such that $|I_{v_0}|$ is the largest possible among elements of $\mathbf{U}_k$. Then select an element $v_1 \in \mathbf{U}_k \setminus \{v_0\}$ such that the enlarged rectangle $(2^k I_{v_1}) \times \omega_{v_1}$ is disjoint from the enlarged rectangle $(2^k I_{v_0}) \times \omega_{v_0}$ and $|I_{v_1}|$ is the largest possible. Continue this process by induction. At the jth step select an element of

$$\mathbf{U}_k \setminus \{v_0, \dots, v_{j-1}\}$$

such that the enlarged rectangle $(2^k I_{v_j}) \times \omega_{v_j}$ is disjoint from all the enlarged rectangles of the previously selected tiles and the length $|I_{v_j}|$ is the largest possible. This process will terminate after a finite number of steps. We denote by $\mathbf{V}_k$ the set of all selected tiles in $\mathbf{U}_k$.

We make a few observations. Recall that all elements of $\mathbf{U}_k$ are maximal rectangles in $\mathbf{U}$ and therefore disjoint. For any $u \in \mathbf{U}_k$ there exists a selected $v \in \mathbf{V}_k$ with $|I_u| \le |I_v|$ such that the enlarged rectangles corresponding to u and v intersect. Let us associate this u to the selected v. Observe that if u and u' are associated with the same selected v, they are disjoint, and since both ω_u and $\omega_{u'}$ contain ω_v, the intervals I_u and $I_{u'}$ must be disjoint. Thus, tiles $u \in \mathbf{U}_k$ associated with a fixed $v \in \mathbf{V}_k$ have disjoint I_u's and satisfy

$$I_u \subseteq 2^{k+2} I_v.$$

Consequently,

$$\sum_{\substack{u\in\mathbf{U}_k \\ u \text{ associated with } v}} |I_u| \le |2^{k+2}I_v| = 2^{k+2}|I_v|.$$

Putting these observations together, we obtain

$$\begin{aligned}\sum_{u\in\mathbf{U}_k}|I_u| &\le \sum_{v\in\mathbf{V}_k}\sum_{\substack{u\in\mathbf{U}_k\\ u \text{ associated with } v}}|I_u|\\ &\le 2^{k+2}\sum_{v\in\mathbf{V}_k}|I_v|\\ &\le 2^{k+5}5^{10}\mu^{-1}|E|^{-1}2^{-9k}\sum_{v\in\mathbf{V}_k}|E\cap N^{-1}[\omega_v]\cap 2^kI_v|\\ &\le 32\cdot 5^{10}\mu^{-1}2^{-8k},\end{aligned}$$

since the enlarged rectangles $2^kI_v\times\omega_v$ of the selected tiles v are disjoint and therefore so are the subsets $E\cap N^{-1}[\omega_v]\cap 2^kI_v$ of E. This concludes the proof of estimate (6.1.43) and therefore of Lemma 6.1.8. □

6.1.6 Proof of Energy Lemma 6.1.9

Proof. Let $g\in L^2(\mathbf{R})$. We work with a finite set of tiles $\mathbf{P}$. For a 2-tree $\mathbf{T}'$, let us denote by

$$\Delta(g;\mathbf{T}') = \frac{1}{\|g\|_{L^2}}\left\{\frac{1}{|I_{\mathrm{top}(\mathbf{T}')}|}\sum_{s\in\mathbf{T}'}|\langle g\,|\,\varphi_s\rangle|^2\right\}^{\frac12}$$

the quantity associated with $\mathbf{T}'$ appearing in the definition of the energy. Consider the set of all 2-trees $\mathbf{T}'$ contained in $\mathbf{P}$ that satisfy

$$\Delta(g;\mathbf{T}') \ge \frac{1}{2}\mathscr{E}(g;\mathbf{P}) \tag{6.1.44}$$

and among them select a 2-tree $\mathbf{T}_1'$ with $c(\omega_{\mathrm{top}(\mathbf{T}_1')})$ as small as possible. We let $\mathbf{T}_1$ be the set of $s\in\mathbf{P}$ satisfying $s<\mathrm{top}(\mathbf{T}_1')$. Then $\mathbf{T}_1$ is the largest tree in $\mathbf{P}$ whose top is $\mathrm{top}(\mathbf{T}_1')$. We now repeat this procedure with the set $\mathbf{P}\setminus\mathbf{T}_1$. Among all 2-trees contained in $\mathbf{P}\setminus\mathbf{T}_1$ that satisfy (6.1.44) we pick a 2-tree $\mathbf{T}_2'$ with $c(\omega_{\mathrm{top}(\mathbf{T}_2')})$ as small as possible. Then we let $\mathbf{T}_2$ be the $s\in\mathbf{P}\setminus\mathbf{T}_1$ satisfying $s<\mathrm{top}(\mathbf{T}_2')$. Then $\mathbf{T}_2$ is the largest tree in $\mathbf{P}\setminus\mathbf{T}_1$ whose top is $\mathrm{top}(\mathbf{T}_2')$. We continue this procedure by induction until there is no 2-tree left in $\mathbf{P}$ that satisfies (6.1.44). We have therefore constructed a finite sequence of pairwise disjoint 2-trees $\mathbf{T}_1',\mathbf{T}_2',\mathbf{T}_3',\dots,\mathbf{T}_q'$, and a finite sequence of pairwise disjoint trees $\mathbf{T}_1,\mathbf{T}_2,\mathbf{T}_3,\dots,\mathbf{T}_q$, such that $\mathbf{T}_j'\subsetneqq\mathbf{T}_j$, $\mathrm{top}(\mathbf{T}_j)=\mathrm{top}(\mathbf{T}_j')$, and the $\mathbf{T}_j'$ satisfy (6.1.44). We now let

$$\mathbf{P}'' = \bigcup_j \mathbf{T}_j,$$

and observe that this selection of trees ensures that

$$\mathscr{E}(g;\mathbf{P}\setminus\mathbf{P}'') \le \frac{1}{2}\mathscr{E}(g;\mathbf{P}).$$

It remains to prove (6.1.34). Using (6.1.44), we obtain that

$$\begin{aligned}\frac{1}{4}\mathscr{E}(g;\mathbf{P})^2\sum_j |I_{\mathrm{top}(\mathbf{T}_j)}| &\le \frac{1}{\|g\|_{L^2}^2}\sum_j\sum_{s\in\mathbf{T}'_j}|\langle g\,|\,\varphi_s\rangle|^2\\ &= \frac{1}{\|g\|_{L^2}^2}\sum_j\sum_{s\in\mathbf{T}'_j}\langle g\,|\,\varphi_s\rangle\overline{\langle g\,|\,\varphi_s\rangle}\\ &= \frac{1}{\|g\|_{L^2}^2}\Big\langle g\,\Big|\sum_j\sum_{s\in\mathbf{T}'_j}\langle g\,|\,\varphi_s\rangle\varphi_s\Big\rangle\\ &\le \frac{1}{\|g\|_{L^2}}\Big\|\sum_j\sum_{s\in\mathbf{T}'_j}\langle\varphi_s\,|\,g\rangle\varphi_s\Big\|_{L^2},\end{aligned} \tag{6.1.45}$$

and we use this estimate to obtain (6.1.34). We set $\mathbf{U}=\bigcup_j\mathbf{T}'_j$. We shall prove that

$$\frac{1}{\|g\|_{L^2}}\Big\|\sum_{s\in\mathbf{U}}\langle\varphi_s\,|\,g\rangle\varphi_s\Big\|_{L^2} \le C\Big(\mathscr{E}(g;\mathbf{P})^2\sum_j|I_{\mathrm{top}(\mathbf{T}_j)}|\Big)^{\frac{1}{2}}. \tag{6.1.46}$$

Once this estimate is established, then (6.1.45) combined with (6.1.46) yields (6.1.34). (All involved quantities are finite, since $\mathbf{P}$ is a finite set of tiles.)

We estimate the square of the left-hand side in (6.1.46) by

$$\sum_{\substack{s,u\in\mathbf{U}\\ \omega_s=\omega_u}}\big|\langle\varphi_s\,|\,g\rangle\langle\varphi_u\,|\,g\rangle\langle\varphi_s\,|\,\varphi_u\rangle\big| + 2\sum_{\substack{s,u\in\mathbf{U}\\ \omega_s\subsetneqq\omega_u}}\big|\langle\varphi_s\,|\,g\rangle\langle\varphi_u\,|\,g\rangle\langle\varphi_s\,|\,\varphi_u\rangle\big|, \tag{6.1.47}$$

since $\langle\varphi_s\,|\,\varphi_u\rangle=0$ unless ω_s contains ω_u or vice versa. We now estimate the first term in (6.1.47) by the expression

$$\begin{aligned}&\sum_{\substack{s,u\in\mathbf{U}\\ \omega_s=\omega_u}}\big|\langle\varphi_s\,|\,g\rangle\big|\big|\langle\varphi_s\,|\,\varphi_u\rangle\big|^{1/2}\big|\langle\varphi_u\,|\,g\rangle\big|\big|\langle\varphi_s\,|\,\varphi_u\rangle\big|^{1/2}\\ &\le\Big(\sum_{\substack{s,u\in\mathbf{U}\\ \omega_s=\omega_u}}\big|\langle\varphi_s\,|\,g\rangle\big|^2\big|\langle\varphi_s\,|\,\varphi_u\rangle\big|\Big)^{1/2}\Big(\sum_{\substack{s,u\in\mathbf{U}\\ \omega_s=\omega_u}}\big|\langle\varphi_u\,|\,g\rangle\big|^2\big|\langle\varphi_s\,|\,\varphi_u\rangle\big|\Big)^{1/2}\\ &=\sum_{s\in\mathbf{U}}\big|\langle g\,|\,\varphi_s\rangle\big|^2\sum_{\substack{u\in\mathbf{U}\\ \omega_u=\omega_s}}\big|\langle\varphi_s\,|\,\varphi_u\rangle\big|\\ &\le\sum_{s\in\mathbf{U}}\big|\langle g\,|\,\varphi_s\rangle\big|^2\sum_{\substack{u\in\mathbf{U}\\ \omega_u=\omega_s}}C'\int_{I_u}\frac{1}{|I_s|}\Big(1+\frac{|x-c(I_s)|}{|I_s|}\Big)^{-100}dx\end{aligned}$$

$$\begin{aligned}
&\le C'' \sum_{s\in\mathbf{U}} |\langle g \,|\, \varphi_s\rangle|^2 \\
&= C'' \sum_j \sum_{s\in\mathbf{T}'_j} |\langle g \,|\, \varphi_s\rangle|^2 \\
&\le C'' \sum_j |I_{\mathrm{top}(\mathbf{T}_j)}| \, |I_{\mathrm{top}(\mathbf{T}_j)}|^{-1} \sum_{s\in\mathbf{T}'_j} |\langle g \,|\, \varphi_s\rangle|^2 \\
&\le C'' \sum_j |I_{\mathrm{top}(\mathbf{T}_j)}| \, \mathscr{E}(g;\mathbf{P})^2 \|g\|_{L^2}^2 , \qquad (6.1.48)
\end{aligned}$$

where in the derivation of the third inequality we used the fact that for fixed $s \in \mathbf{U}$, the intervals I_u with $\omega_u = \omega_s$ are pairwise disjoint.

Our next goal is to obtain a similar estimate for the second term in (6.1.47). That is, we need to prove that

$$\sum_{\substack{s,u\in\mathbf{U}\\ \omega_s \subsetneqq \omega_u}} \big|\langle g \,|\, \varphi_s\rangle \langle g \,|\, \varphi_u\rangle \langle \varphi_s \,|\, \varphi_u\rangle\big| \le C\mathscr{E}(g;\mathbf{P})^2 \|g\|_{L^2}^2 \sum_j |I_{\mathrm{top}(\mathbf{T}_j)}| . \qquad (6.1.49)$$

Then the required estimate (6.1.46) would follow by combining (6.1.48) and (6.1.49). To prove (6.1.49), we argue as follows:

$$\begin{aligned}
&\sum_{\substack{s,u\in\mathbf{U}\\ \omega_s \subsetneqq \omega_u}} \big|\langle g \,|\, \varphi_s\rangle \langle g \,|\, \varphi_u\rangle \langle \varphi_s \,|\, \varphi_u\rangle\big| \\
&\quad = \sum_j \sum_{s\in\mathbf{T}'_j} |\langle g \,|\, \varphi_s\rangle| \sum_{\substack{u\in\mathbf{U}\\ \omega_s \subsetneqq \omega_u}} |\langle g \,|\, \varphi_u\rangle \langle \varphi_s \,|\, \varphi_u\rangle| \\
&\quad \le \sum_j |I_{\mathrm{top}(\mathbf{T}_j)}|^{\frac12} \Delta(g;\mathbf{T}'_j) \|g\|_{L^2} \Bigg\{ \sum_{s\in\mathbf{T}'_j} \Bigg(\sum_{\substack{u\in\mathbf{U}\\ \omega_s \subsetneqq \omega_u}} |\langle g \,|\, \varphi_u\rangle \langle \varphi_s \,|\, \varphi_u\rangle| \Bigg)^2 \Bigg\}^{\frac12} \\
&\quad \le \mathscr{E}(g;\mathbf{P}) \|g\|_{L^2} \sum_j |I_{\mathrm{top}(\mathbf{T}_j)}|^{\frac12} \Bigg\{ \sum_{s\in\mathbf{T}'_j} \Bigg(\sum_{\substack{u\in\mathbf{U}\\ \omega_s \subseteqq \omega_{u(1)}}} |\langle g \,|\, \varphi_u\rangle \langle \varphi_s \,|\, \varphi_u\rangle| \Bigg)^2 \Bigg\}^{\frac12} ,
\end{aligned}$$

where we used the Cauchy–Schwarz inequality and the fact that if $\omega_s \subsetneqq \omega_u$ and $\langle \varphi_s \,|\, \varphi_u\rangle \neq 0$, then $\omega_s \subseteqq \omega_{u(1)}$. The proof of (6.1.49) will be complete if we can show that the expression inside the curly brackets is at most a multiple of $\mathscr{E}(g;\mathbf{P})^2$ $\|g\|_{L^2}^2 |I_{\mathrm{top}(\mathbf{T}_j)}|$. Since any singleton $\{u\} \subseteqq \mathbf{P}$ is a 2-tree, we have

$$\mathscr{E}(g;\{u\}) = \frac{1}{\|g\|_{L^2}} \left(\frac{|\langle g \,|\, \varphi_u\rangle|^2}{|I_u|} \right)^{\frac12} = \frac{1}{\|g\|_{L^2}} \frac{|\langle g \,|\, \varphi_u\rangle|}{|I_u|^{\frac12}} \le \mathscr{E}(g;\mathbf{P}) ;$$

hence

$$|\langle g \,|\, \varphi_u\rangle| \le \|g\|_{L^2} |I_u|^{\frac12} \mathscr{E}(g;\mathbf{P})$$

and it follows that

$$\sum_{s\in\mathbf{T}'_j}\Bigg[\sum_{\substack{u\in\mathbf{U}\\ \omega_s\subsetneqq\omega_{u(1)}}}\big|\langle g\,|\,\varphi_u\rangle\langle\varphi_s\,|\,\varphi_u\rangle\big|\Bigg]^2\le\mathscr{E}(g;\mathbf{P})^2\|g\|_{L^2}^2\sum_{s\in\mathbf{T}'_j}\Bigg[\sum_{\substack{u\in\mathbf{U}\\ \omega_s\subsetneqq\omega_{u(1)}}}|I_u|^{\frac12}\big|\langle\varphi_s\,|\,\varphi_u\rangle\big|\Bigg]^2.$$

Thus (6.1.49) will be proved if we can establish that

$$\sum_{s\in\mathbf{T}'_j}\Bigg(\sum_{\substack{u\in\mathbf{U}\\ \omega_s\subsetneqq\omega_{u(1)}}}|I_u|^{\frac12}\big|\langle\varphi_s\,|\,\varphi_u\rangle\big|\Bigg)^2\le C|I_{\mathrm{top}(\mathbf{T}_j)}|\,. \tag{6.1.50}$$

We need the following crucial lemma.

Lemma 6.1.11. *Let $\mathbf{T}_j$, $\mathbf{T}'_j$ be as previously. Let $s\in\mathbf{T}'_j$ and $u\in\mathbf{T}'_k$. Then if $\omega_s\subsetneqq\omega_{u(1)}$, we have $I_u\cap I_{\mathrm{top}(\mathbf{T}_j)}=\emptyset$. Moreover, if $u\in\mathbf{T}'_k$ and $v\in\mathbf{T}'_l$ are different tiles and satisfy $\omega_s\subsetneqq\omega_{u(1)}$ and $\omega_s\subsetneqq\omega_{v(1)}$ for some fixed $s\in\mathbf{T}'_j$, then $I_u\cap I_v=\emptyset$.*

Proof. We observe that if $s\in\mathbf{T}'_j$, $u\in\mathbf{T}'_k$, and $\omega_s\subsetneqq\omega_{u(1)}$, then the 2-trees $\mathbf{T}'_j$ and $\mathbf{T}'_k$ have different tops and therefore they cannot be the same tree; thus $j\neq k$.

Next we observe that the center of $\omega_{\mathrm{top}(\mathbf{T}'_j)}$ is contained in ω_s, which is contained in $\omega_{u(1)}$. Therefore, the center of $\omega_{\mathrm{top}(\mathbf{T}'_j)}$ is contained in $\omega_{u(1)}$, and therefore it must be smaller than the center of $\omega_{\mathrm{top}(\mathbf{T}'_k)}$, since $\mathbf{T}'_k$ is a 2-tree. This means that the 2-tree $\mathbf{T}'_j$ was selected before $\mathbf{T}'_k$, that is, we must have $j<k$. If I_u had a nonempty intersection with $I_{\mathrm{top}(\mathbf{T}_j)}=I_{\mathrm{top}(\mathbf{T}'_j)}$, then since

$$|I_{\mathrm{top}(\mathbf{T}'_j)}|=\frac{1}{|\omega_{\mathrm{top}(\mathbf{T}'_j)}|}\ge\frac{1}{|\omega_s|}\ge\frac{1}{|\omega_{u(1)}|}=\frac{2}{|\omega_u|}=2|I_u|\,,$$

I_u would have to be contained in $I_{\mathrm{top}(\mathbf{T}'_j)}$. Since also $\omega_{\mathrm{top}(\mathbf{T}'_j)}\subsetneqq\omega_s\subsetneqq\omega_u$, it follows that $u<\mathrm{top}(\mathbf{T}'_j)$; thus u would belong to the tree $\mathbf{T}_j$ [which is the largest tree with top $\mathrm{top}(\mathbf{T}'_j)$], since this tree was selected first. But if u belonged to $\mathbf{T}_j$, then it could not belong to $\mathbf{T}'_k$, which is disjoint from $\mathbf{T}_j$; hence we get a contradiction. We conclude that $I_u\cap I_{\mathrm{top}(\mathbf{T}_j)}=\emptyset$.

Next assume that $u\in\mathbf{T}'_k$, $v\in\mathbf{T}'_l$, $u\neq v$, and that $\omega_s\subsetneqq\omega_{u(1)}\cap\omega_{v(1)}$ for some fixed $s\in\mathbf{T}'_j$. Since the left halves of two dyadic intervals ω_u and ω_v intersect, three things can happen: (a) $\omega_u\subsetneqq\omega_{v(1)}$, in which case I_v is disjoint from $I_{\mathrm{top}(\mathbf{T}'_k)}$ and thus from I_u; (b) $\omega_v\subsetneqq\omega_{u(1)}$, in which case I_u is disjoint from $I_{\mathrm{top}(\mathbf{T}'_l)}$ and thus from I_v; and (c) $\omega_u=\omega_v$, in which case $|I_u|=|I_v|$, and thus I_u and I_v are either disjoint or they coincide. Since $u\neq v$, it follows that I_u and I_v cannot coincide; thus $I_u\cap I_v=\emptyset$. This finishes the proof of the lemma. □

We now return to (6.1.50). In view of Lemma 6.1.11, different $u\in\mathbf{U}$ that appear in the interior sum in (6.1.50) have disjoint intervals I_u, and all of these are contained in $(I_{\mathrm{top}(\mathbf{T}_j)})^c$. Set $t_j=\mathrm{top}(\mathbf{T}_j)$. Using Exercise 6.1.4, we obtain

$$\begin{aligned}
&\sum_{s\in\mathbf{T}'_j}\bigg(\sum_{\substack{u\in\mathbf{U}\\ \omega_s\subseteqq\omega_{u(1)}}}|I_u|^{\frac12}\big|\langle\varphi_s\,|\,\varphi_u\rangle\big|\bigg)^2\\
&\le C\sum_{s\in\mathbf{T}'_j}\bigg(\sum_{\substack{u\in\mathbf{U}\\ \omega_s\subseteqq\omega_{u(1)}}}|I_u|^{\frac12}\Big(\frac{|I_s|}{|I_u|}\Big)^{\frac12}\int_{I_u}\frac{|I_s|^{-1}\,dx}{(1+\frac{|x-c(I_s)|}{|I_s|})^{20}}\bigg)^2\\
&\le C\sum_{s\in\mathbf{T}'_j}|I_s|\bigg(\sum_{\substack{u\in\mathbf{U}\\ \omega_s\subseteqq\omega_{u(1)}}}\int_{I_u}\frac{|I_s|^{-1}\,dx}{(1+\frac{|x-c(I_s)|}{|I_s|})^{20}}\bigg)^2\\
&\le C\sum_{s\in\mathbf{T}'_j}|I_s|\bigg(\int_{(I_{t_j})^c}\frac{|I_s|^{-1}\,dx}{(1+\frac{|x-c(I_s)|}{|I_s|})^{20}}\bigg)^2\\
&\le C\sum_{s\in\mathbf{T}'_j}|I_s|\int_{(I_{t_j})^c}\frac{|I_s|^{-1}\,dx}{(1+\frac{|x-c(I_s)|}{|I_s|})^{20}},
\end{aligned}$$

since $\int_{\mathbf{R}}(1+|x|)^{-20}\,dx\le 1$. For each scale $k\ge 0$ the sets I_s, $s\in\mathbf{T}'_j$, with $|I_s|=2^{-k}|I_{t_j}|$ are pairwise disjoint and contained in I_{t_j}; therefore, we have

$$\begin{aligned}
\sum_{s\in\mathbf{T}'_j}|I_s|\int_{(I_{t_j})^c}\frac{|I_s|^{-1}\,dx}{(1+\frac{|x-c(I_s)|}{|I_s|})^{20}}
&\le\sum_{k=0}^{\infty}\frac{2^k}{|I_{t_j}|}\sum_{\substack{s\in\mathbf{T}'_j\\ |I_s|=2^{-k}|I_{t_j}|}}|I_s|\int_{(I_{t_j})^c}\frac{dx}{(1+\frac{|x-c(I_s)|}{|I_s|})^{20}}\\
&\le C\sum_{k=0}^{\infty}\frac{2^k}{|I_{t_j}|}\sum_{\substack{s\in\mathbf{T}'_j\\ |I_s|=2^{-k}|I_{t_j}|}}\int_{I_s}\int_{(I_{t_j})^c}\frac{dx}{(1+\frac{|x-y|}{|I_s|})^{20}}\,dy\\
&\le C\sum_{k=0}^{\infty}2^k|I_{t_j}|^{-1}\int_{I_{t_j}}\int_{(I_{t_j})^c}\frac{1}{(1+\frac{|x-y|}{2^{-k}|I_{t_j}|})^{20}}\,dx\,dy\\
&\le C'\sum_{k=0}^{\infty}2^k|I_{t_j}|^{-1}(2^{-k}|I_{t_j}|)^2\\
&=C''|I_{t_j}|,
\end{aligned}$$

in view of Exercise 6.1.5. This completes the proof of (6.1.50) and thus of Lemma 6.1.9. □

6.1.7 Proof of the Basic Estimate Lemma 6.1.10

Proof. In the proof of the required estimate we may assume that $\|g\|_{L^2}=1$, for we can always replace g by $g/\|g\|_{L^2}$. Throughout this subsection we fix a square-integrable function g with L^2 norm 1, a tree $\mathbf{T}$, a measurable function $N:\mathbf{R}\to\mathbf{R}^+$, and a measurable set E with finite measure.

Let $\mathscr{J}'$ be the set of all dyadic intervals J such that $3J$ does not contain any I_s with $s\in\mathbf{T}$. It is not hard to see that any point in $\mathbf{R}$ belongs to a set in $\mathscr{J}'$. Let $\mathscr{J}$ be the set of all maximal (under inclusion) elements of $\mathscr{J}'$. Then $\mathscr{J}$ consists of disjoint sets that cover $\mathbf{R}$; thus it forms a partition of $\mathbf{R}$. This partition of $\mathbf{R}$ is shown in Figure 6.3 when the tree consists of two tiles.

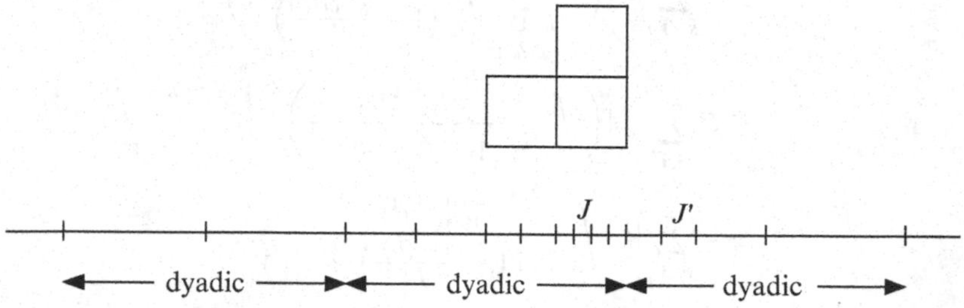

Fig. 6.3 A tree of two tiles and the partition $\mathscr{J}$ of $\mathbf{R}$ corresponding to it. The intervals J and J' are members of the partition $\mathscr{J}$.

For each $s\in\mathbf{T}$ pick an $\varepsilon_s\in\mathbf{C}$ with $|\varepsilon_s|=1$ such that

$$\big|\langle g\,|\,\varphi_s\rangle\langle\chi_{E\cap N^{-1}[\omega_{s(2)}]}\,|\,\varphi_s\rangle\big| = \varepsilon_s\langle g\,|\,\varphi_s\rangle\langle\varphi_s\,|\,\chi_{E\cap N^{-1}[\omega_{s(2)}]}\rangle\,.$$

We can now write the left-hand side of (6.1.35) as

$$\begin{aligned}\sum_{s\in\mathbf{T}}\varepsilon_s\langle g\,|\,\varphi_s\rangle\langle\varphi_s\,|\,\chi_{E\cap N^{-1}[\omega_{s(2)}]}\rangle &\le \Big\|\sum_{s\in\mathbf{T}}\varepsilon_s\langle g\,|\,\varphi_s\rangle\chi_{E\cap N^{-1}[\omega_{s(2)}]}\varphi_s\Big\|_{L^1(\mathbf{R})}\\ &= \sum_{J\in\mathscr{J}}\Big\|\sum_{s\in\mathbf{T}}\varepsilon_s\langle g\,|\,\varphi_s\rangle\chi_{E\cap N^{-1}[\omega_{s(2)}]}\varphi_s\Big\|_{L^1(J)}\\ &\le \Sigma_1+\Sigma_2\,,\end{aligned}$$

where

$$\Sigma_1 = \sum_{J\in\mathscr{J}}\Big\|\sum_{\substack{s\in\mathbf{T}\\ |I_s|\le 2|J|}}\varepsilon_s\langle g\,|\,\varphi_s\rangle\chi_{E\cap N^{-1}[\omega_{s(2)}]}\varphi_s\Big\|_{L^1(J)}, \tag{6.1.51}$$

$$\Sigma_2 = \sum_{J\in\mathscr{J}}\Big\|\sum_{\substack{s\in\mathbf{T}\\ |I_s|> 2|J|}}\varepsilon_s\langle g\,|\,\varphi_s\rangle\chi_{E\cap N^{-1}[\omega_{s(2)}]}\varphi_s\Big\|_{L^1(J)}. \tag{6.1.52}$$

We start with Σ_1. Observe that for every $s\in\mathbf{T}$, the singleton $\{s\}$ is a 2-tree contained in $\mathbf{T}$ and we therefore have the estimate

$$\big|\langle g\,|\,\varphi_s\rangle\big| \le |I_s|^{\frac{1}{2}}\mathscr{E}(g;\mathbf{T})\,. \tag{6.1.53}$$

Using this, we obtain

$$\begin{aligned}
\Sigma_1 &\le \sum_{J\in\mathscr{J}} \sum_{\substack{s\in\mathbf{T}\\ |I_s|\le 2|J|}} \mathscr{E}(g;\mathbf{T}) \int_{J\cap E\cap N^{-1}[\omega_{s(2)}]} |I_s|^{\frac{1}{2}} |\varphi_s(x)|\,dx \\
&\le C \sum_{J\in\mathscr{J}} \sum_{\substack{s\in\mathbf{T}\\ |I_s|\le 2|J|}} \mathscr{E}(g;\mathbf{T})|I_s| \int_{J\cap E\cap N^{-1}[\omega_{s(2)}]} \frac{|I_s|^{-1}}{\big(1+\frac{|x-c(I_s)|}{|I_s|}\big)^{20}}\,dx \\
&\le C \sum_{J\in\mathscr{J}} \sum_{\substack{s\in\mathbf{T}\\ |I_s|\le 2|J|}} \mathscr{E}(g;\mathbf{T})\,|E|\mathscr{M}(E;\mathbf{T})|I_s| \sup_{x\in J} \frac{1}{\big(1+\frac{|x-c(I_s)|}{|I_s|}\big)^{10}} \\
&\le C\mathscr{E}(g;\mathbf{T})\,|E|\mathscr{M}(E;\mathbf{T}) \sum_{J\in\mathscr{J}} \sum_{k=-\infty}^{\log_2 2|J|} 2^k \sum_{\substack{s\in\mathbf{T}\\ |I_s|=2^k}} \frac{1}{\big(1+\frac{\operatorname{dist}(J,I_s)}{2^k}\big)^5} \frac{1}{\big(1+\frac{\operatorname{dist}(J,I_s)}{2^k}\big)^5}\,.
\end{aligned}$$

But note that all I_s with $s\in\mathbf{T}$ and $|I_s|=2^k$ are pairwise disjoint and contained in $I_{\mathrm{top}(\mathbf{T})}$. Therefore, $2^{-k}\operatorname{dist}(J,I_s)\ge |I_{\mathrm{top}(\mathbf{T})}|^{-1}\operatorname{dist}(J,I_{\mathrm{top}(\mathbf{T})})$, and we have the estimate

$$\left(1+\frac{\operatorname{dist}(J,I_s)}{2^k}\right)^{-5} \le \left(1+\frac{\operatorname{dist}(J,I_{\mathrm{top}(\mathbf{T})})}{|I_{\mathrm{top}(\mathbf{T})}|}\right)^{-5}.$$

Moreover, the sum

$$\sum_{\substack{s\in\mathbf{T}\\ |I_s|=2^k}} \frac{1}{\big(1+\frac{\operatorname{dist}(J,I_s)}{2^k}\big)^5} \tag{6.1.54}$$

is controlled by a finite constant, since for every nonnegative integer m there exist at most two tiles $s\in\mathbf{T}$ with $|I_s|=2^k$ such that I_s are not contained in $3J$ and $m2^k\le \operatorname{dist}(J,I_s)<(m+1)2^k$. Therefore, we obtain

$$\begin{aligned}
\Sigma_1 &\le C\mathscr{E}(g;\mathbf{T})\,|E|\mathscr{M}(E;\mathbf{T}) \sum_{J\in\mathscr{J}} \sum_{k=-\infty}^{\log_2 2|J|} \frac{2^k}{\big(1+\frac{\operatorname{dist}(J,I_{\mathrm{top}(\mathbf{T})})}{|I_{\mathrm{top}(\mathbf{T})}|}\big)^5} \\
&\le C\mathscr{E}(g;\mathbf{T})\,|E|\mathscr{M}(E;\mathbf{T}) \sum_{J\in\mathscr{J}} \frac{|J|}{\big(1+\frac{\operatorname{dist}(J,I_{\mathrm{top}(\mathbf{T})})}{|I_{\mathrm{top}(\mathbf{T})}|}\big)^5} \qquad (6.1.55)\\
&\le C\mathscr{E}(g;\mathbf{T})\,|E|\mathscr{M}(E;\mathbf{T}) \sum_{J\in\mathscr{J}} \int_J \frac{1}{\big(1+\frac{|x-c(I_{\mathrm{top}(\mathbf{T})})|}{|I_{\mathrm{top}(\mathbf{T})}|}\big)^5}\,dx \\
&\le C|I_{\mathrm{top}(\mathbf{T})}|\,\mathscr{E}(g;\mathbf{T})\,|E|\mathscr{M}(E;\mathbf{T}),
\end{aligned}$$

since $\mathscr{J}$ forms a partition of $\mathbf{R}$. We need to justify, however, the penultimate inequality in (6.1.55). Since J and $I_{\mathrm{top}(\mathbf{T})}$ are dyadic intervals, there are only two possibilities: (a) $J\cap I_{\mathrm{top}(\mathbf{T})}=\emptyset$ and (b) $J\subsetneqq I_{\mathrm{top}(\mathbf{T})}$. [The third possibility $I_{\mathrm{top}(\mathbf{T})}\subseteqq J$ is excluded, since $3J$ does not contain $I_{\mathrm{top}(\mathbf{T})}$.] In case (a) we have $|J|\le \operatorname{dist}(J,I_{\mathrm{top}(\mathbf{T})})$, since $3J$ does not contain $I_{\mathrm{top}(\mathbf{T})}$. In case (b) we have $|J| \le |I_{\mathrm{top}(\mathbf{T})}|$. Thus in both cases we have $|J| \le \mathrm{dist}\,(J, I_{\mathrm{top}(\mathbf{T})}) + |I_{\mathrm{top}(\mathbf{T})}|$. Consequently, for any $x \in J$ one has

$$\begin{aligned} |x - c(I_{\mathrm{top}(\mathbf{T})})| &\le |J| + \mathrm{dist}\,(J, I_{\mathrm{top}(\mathbf{T})}) + \frac{1}{2}|I_{\mathrm{top}(\mathbf{T})}| \\ &\le 2\,\mathrm{dist}\,(J, I_{\mathrm{top}(\mathbf{T})}) + \frac{3}{2}|I_{\mathrm{top}(\mathbf{T})}|\,. \end{aligned}$$

Therefore, it follows that

$$\int_J \frac{dx}{\left(1+\frac{|x-c(I_{\mathrm{top}(\mathbf{T})})|}{|I_{\mathrm{top}(\mathbf{T})}|}\right)^5} \ge \frac{|J|}{\left(\frac{5}{2}+\frac{2\mathrm{dist}\,(J,I_{\mathrm{top}(\mathbf{T})})}{|I_{\mathrm{top}(\mathbf{T})}|}\right)^5} \ge \frac{\left(\frac{2}{5}\right)^5|J|}{\left(1+\frac{\mathrm{dist}\,(J,I_{\mathrm{top}(\mathbf{T})})}{|I_{\mathrm{top}(\mathbf{T})}|}\right)^5}\,.$$

In case (b) we have $J \subseteqq I_{\mathrm{top}(\mathbf{T})}$, and therefore, any point x in J lies in $I_{\mathrm{top}(\mathbf{T})}$; thus $|x - c(I_{\mathrm{top}(\mathbf{T})})| \le \frac{1}{2}|I_{\mathrm{top}(\mathbf{T})}|$. We conclude that

$$\int_J \frac{dx}{\left(1+\frac{|x-c(I_{\mathrm{top}(\mathbf{T})})|}{|I_{\mathrm{top}(\mathbf{T})}|}\right)^5} \ge \frac{|J|}{(3/2)^5} = \left(\frac{2}{3}\right)^5 \frac{|J|}{\left(1+\frac{\mathrm{dist}\,(J,I_{\mathrm{top}(\mathbf{T})})}{|I_{\mathrm{top}(\mathbf{T})}|}\right)^5}\,.$$

These observations justify the second-to-last inequality in (6.1.55) and complete the proof of the required estimate for Σ_1.

We now turn attention to Σ_2. We may assume that for all J appearing in the sum in (6.1.52), the set of s in $\mathbf{T}$ with $2|J| < |I_s|$ is nonempty. Thus, if J appears in the sum in (6.1.52), we have $2|J| < |I_{\mathrm{top}(\mathbf{T})}|$, and it is easy to see that J is contained in $3I_{\mathrm{top}(\mathbf{T})}$. [The intervals J in $\mathscr{J}$ that are not contained in $3I_{\mathrm{top}(\mathbf{T})}$ have size larger than $\frac{1}{2}|I_{\mathrm{top}(\mathbf{T})}|$.]

We let $\mathbf{T}_2$ be the 2-tree of all s in $\mathbf{T}$ such that $\omega_{\mathrm{top}(\mathbf{T})(2)} \subseteqq \omega_{s(2)}$, and we also let $\mathbf{T}_1 = \mathbf{T} \setminus \mathbf{T}_2$. Then $\mathbf{T}_1$ is a 1-tree minus its top. We set

$$\begin{aligned} F_{1J} &= \sum_{\substack{s\in\mathbf{T}_1 \\ |I_s|>2|J|}} \varepsilon_s \langle g \,|\, \varphi_s\rangle\, \varphi_s\, \chi_{E\cap N^{-1}[\omega_{s(2)}]}\,, \\ F_{2J} &= \sum_{\substack{s\in\mathbf{T}_2 \\ |I_s|>2|J|}} \varepsilon_s \langle g \,|\, \varphi_s\rangle\, \varphi_s\, \chi_{E\cap N^{-1}[\omega_{s(2)}]}\,. \end{aligned}$$

Clearly

$$\Sigma_2 \le \sum_{J\in\mathscr{J}} \|F_{1J}\|_{L^1(J)} + \sum_{J\in\mathscr{J}} \|F_{2J}\|_{L^1(J)} = \Sigma_{21} + \Sigma_{22}\,,$$

and we need to estimate both sums. We start by estimating F_{1J}. If the tiles s and s' that appear in the definition of F_{1J} have different scales, then the sets $\omega_{s(2)}$ and $\omega_{s'(2)}$ are disjoint and thus so are the sets $E \cap N^{-1}[\omega_{s(2)}]$ and $E \cap N^{-1}[\omega_{s'(2)}]$. Let us set

$$G_J = J \cap \bigcup_{\substack{s\in\mathbf{T} \\ |I_s|>2|J|}} E \cap N^{-1}[\omega_{s(2)}]\,.$$

Then F_{1J} is supported in the set G_J and we have

$$\begin{aligned}
\big\|F_{1J}\big\|_{L^1(J)} &\le \big\|F_{1J}\big\|_{L^\infty(J)}|G_J| \\
&= \Big\|\sum_{k>\log_2 2|J|}\sum_{\substack{s\in\mathbf{T}_1\\|I_s|=2^k}} \varepsilon_s\left\langle g\,|\,\varphi_s\right\rangle\varphi_s\,\chi_{E\cap N^{-1}[\omega_{s(2)}]}\Big\|_{L^\infty(J)}|G_J| \\
&\le \sup_{k>\log_2 2|J|}\Big\|\sum_{\substack{s\in\mathbf{T}_1\\|I_s|=2^k}} \varepsilon_s\left\langle g\,|\,\varphi_s\right\rangle\varphi_s\,\chi_{E\cap N^{-1}[\omega_{s(2)}]}\Big\|_{L^\infty(J)}|G_J| \\
&\le \sup_{k>\log_2 2|J|}\sup_{x\in J}\sum_{\substack{s\in\mathbf{T}_1\\|I_s|=2^k}} \mathscr{E}(g;\mathbf{T})2^{k/2}\frac{2^{-k/2}}{\big(1+\frac{|x-c(I_s)|}{2^k}\big)^{10}}|G_J| \\
&\le C\mathscr{E}(g;\mathbf{T})|G_J|,
\end{aligned}$$

using (6.1.53) and the fact that all the I_s that appear in the sum are disjoint. We now claim that for all $J\in\mathscr{J}$ we have

$$|G_J|\le C|E|\mathscr{M}(E;\mathbf{T})|J|\,. \tag{6.1.56}$$

Once (6.1.56) is established, summing over all the intervals J that appear in the definition of F_{1J} and keeping in mind that all of these intervals are pairwise disjoint and contained in $3I_{\text{top}(\mathbf{T})}$, we obtain the desired estimate for Σ_{21}.

To prove (6.1.56), we consider the unique dyadic interval $\widetilde{J}$ of length $2|J|$ that contains J. Then, by the maximality of $\mathscr{J}$, $3\widetilde{J}$ contains the time interval I_{s_J} of a tile s_J in $\mathbf{T}$. We consider the following two cases: (a) If I_{s_J} is either $\big(\widetilde{J}-|\widetilde{J}|\big)\cup\widetilde{J}$ or $\widetilde{J}\cup\big(\widetilde{J}+|\widetilde{J}|\big)$, we let $u_J=s_J$; in this case $|I_{u_J}|=2|\widetilde{J}|$. (This is the case for the interval J in Figure 6.3.) Otherwise, we have case (b), in which I_{s_J} is contained in one of the two dyadic intervals $\widetilde{J}-|\widetilde{J}|$, $\widetilde{J}+|\widetilde{J}|$. (This is the case for the interval J' in Figure 6.3.) Whichever of these two dyadic intervals contains I_{s_J} is also contained in $I_{\text{top}(\mathbf{T})}$, since it intersects it and has smaller length than it. In case (b) there exists a tile $u_J\in\mathbf{D}$ with $|I_{u_J}|=|\widetilde{J}|$ such that $I_{s_J}\subsetneqq I_{u_J}\subsetneqq I_{\text{top}(\mathbf{T})}$ and $\omega_{\text{top}(\mathbf{T})}\subsetneqq\omega_{u_J}\subsetneqq\omega_{s_J}$. In both cases we have a tile u_J satisfying $s_J<u_J<\text{top}(\mathbf{T})$ with $|\omega_{u_J}|$ being either $\frac{1}{4}|J|^{-1}$ or $\frac{1}{2}|J|^{-1}$.

Then for any $s\in\mathbf{T}$ with $|I_s|>2|J|$ we have $|\omega_s|\le|\omega_{u_J}|$. But since both ω_s and ω_{u_J} contain $\omega_{\text{top}(\mathbf{T})}$, they must intersect, and thus $\omega_s\subsetneqq\omega_{u_J}$. We conclude that any $s\in\mathbf{T}$ with $|I_s|>2|J|$ must satisfy $N^{-1}[\omega_s]\subsetneqq N^{-1}[\omega_{u_J}]$. It follows that

$$G_J\subsetneqq J\cap E\cap N^{-1}[\omega_{u_J}] \tag{6.1.57}$$

and therefore we have

$$|E|\mathscr{M}(E;\mathbf{T}) = \sup_{s\in\mathbf{T}}\sup_{\substack{u\in\mathbf{D}\\s<u}}\int_{E\cap N^{-1}[\omega_u]}\frac{|I_u|^{-1}}{\big(1+\frac{|x-c(I_u)|}{|I_u|}\big)^{10}}\,dx$$

$$\begin{aligned}
&\geq \int_{J\cap E\cap N^{-1}[\omega_{u_J}]} \frac{|I_{u_J}|^{-1}}{\left(1+\frac{|x-c(I_{u_J})|}{|I_{u_J}|}\right)^{10}}\,dx \\
&\geq c\,|I_{u_J}|^{-1}\,|J\cap E\cap N^{-1}[\omega_{u_J}]| \\
&\geq c\,|I_{u_J}|^{-1}\,|G_J|,
\end{aligned}$$

using (6.1.57) and the fact that for $x\in J$ we have $|x-c(I_{u_J})|\leq 4|J|\leq 2|I_{u_J}|$. It follows that

$$|G_J|\leq \frac{1}{c}|E|\mathcal{M}(E;\mathbf{T})|I_{u_J}|=\frac{2}{c}|E|\mathcal{M}(E;\mathbf{T})|J|,$$

and this is exactly (6.1.56), which we wanted to prove.

We now turn to the estimate for $\Sigma_{22}=\sum_{J\in\mathscr{J}}\big\|F_{2J}\big\|_{L^1(J)}$. All the intervals $\omega_{s(2)}$ with $s\in\mathbf{T}_2$ are nested, since $\mathbf{T}_2$ is a 2-tree. Therefore, for each $x\in J$ for which $F_{2J}(x)$ is nonzero, there exists a largest dyadic interval ω_{u_x} and a smallest dyadic interval ω_{v_x} (for some $u_x,v_x\in\mathbf{T}_2\cap\{s:\ |I_s|\geq 4|J|\}$) such that for $s\in\mathbf{T}_2\cap\{s:\ |I_s|\geq 4|J|\}$ we have $N(x)\in\omega_{s(2)}$ if and only if $\omega_{v_x}\subseteqq\omega_s\subseteqq\omega_{u_x}$. Then we have

$$\begin{aligned}
F_{2J}(x) &= \sum_{\substack{s\in\mathbf{T}_2\\ |I_s|\geq 4|J|}} \varepsilon_s\langle g\,|\,\varphi_s\rangle(\varphi_s\chi_{E\cap N^{-1}[\omega_{s(2)}]})(x)\\
&= \chi_E(x)\sum_{\substack{s\in\mathbf{T}_2\\ |\omega_{v_x}|\leq|\omega_s|\leq|\omega_{u_x}|}} \varepsilon_s\langle g\,|\,\varphi_s\rangle\,\varphi_s(x)\,.
\end{aligned}$$

Pick a Schwartz function ψ whose Fourier transform $\widehat{\psi}(t)$ is supported in $|t|\leq \frac{1}{2}+\frac{1}{100}$ and that is equal to 1 on $|t|\leq\frac{1}{2}$. We can easily check that for all $z\in\mathbf{R}$, if $|\omega_{v_x}|\leq|\omega_s|\leq|\omega_{u_x}|$, then

$$\left(\varphi_s*\left\{\frac{M^{c(\omega_{u_x})}D^{|\omega_{u_x}|^{-1}}(\psi)}{|\omega_{u_x}|^{-\frac{1}{2}}}-\frac{M^{c(\omega_{v_x(2)})}D^{|\omega_{v_x(2)}|^{-1}}(\psi)}{|\omega_{v_x(2)}|^{-\frac{1}{2}}}\right\}\right)(z)=\varphi_s(z) \tag{6.1.58}$$

by a simple examination of the Fourier transforms. Basically, the Fourier transform (in z) of the function inside the curly brackets is equal to

$$\widehat{\psi}\Big(\frac{\xi-c(\omega_{u_x})}{|\omega_{u_x}|}\Big)-\widehat{\psi}\Big(\frac{\xi-c(\omega_{v_x(2)})}{|\omega_{v_x(2)}|}\Big),$$

which is equal to 1 on the support of $\widehat{\varphi_s}$ for all s in $\mathbf{T}_2$ that satisfy $|\omega_{v_x}|\leq|\omega_s|\leq|\omega_{u_x}|$ but vanishes on $\omega_{v_x(2)}$. Taking $z=x$ in (6.1.58) yields

$$\begin{aligned}
F_{2J}(x) &= \sum_{\substack{s\in\mathbf{T}_2\\ |\omega_{v_x}|\leq|\omega_s|\leq|\omega_{u_x}|}} \varepsilon_s\langle g\,|\,\varphi_s\rangle\,\varphi_s(x)\chi_E(x)\\
&= \left[\sum_{s\in\mathbf{T}_2}\varepsilon_s\langle g\,|\,\varphi_s\rangle\,\varphi_s\right]*\left\{\frac{M^{c(\omega_{u_x})}D^{|\omega_{u_x}|^{-1}}(\psi)}{|\omega_{u_x}|^{-\frac{1}{2}}}-\frac{M^{c(\omega_{v_x(2)})}D^{|\omega_{v_x(2)}|^{-1}}(\psi)}{|\omega_{v_x(2)}|^{-\frac{1}{2}}}\right\}(x)\chi_E(x)\,.
\end{aligned}$$

Since all s that appear in the definition of F_{2J} satisfy $|\omega_s| \le (4|J|)^{-1}$, it follows that we have the estimate

$$\begin{aligned} |F_{2J}(x)| &\le 2\chi_E(x) \sup_{\delta > |\omega_{u_x}|^{-1}} \int_{\mathbf{R}} \Big| \sum_{s\in \mathbf{T}_2} \varepsilon_s \langle g \,|\, \varphi_s\rangle \varphi_s(z)\Big| \tfrac{1}{\delta} \big|\psi\big(\tfrac{x-z}{\delta}\big)\big|\, dz \\ &\le C \sup_{\delta > 4|J|} \frac{1}{2\delta}\int_{x-\delta}^{x+\delta} \Big| \sum_{s\in \mathbf{T}_2} \varepsilon_s \langle g \,|\, \varphi_s\rangle \varphi_s(z)\Big|\, dz\,. \end{aligned} \tag{6.1.59}$$

(The last inequality follows from Exercise 2.1.14 in [156].) Observe that the maximal function in (6.1.59) satisfies the property

$$\sup_{x\in J}\sup_{\delta>4|J|} \frac{1}{2\delta}\int_{x-\delta}^{x+\delta} |h(t)|\,dt \le 2\inf_{x\in J}\sup_{\delta>4|J|} \frac{1}{2\delta}\int_{x-\delta}^{x+\delta} |h(t)|\,dt\,.$$

Using this property, we obtain

$$\begin{aligned} \Sigma_{22} &\le \sum_{J\in \mathscr{J}} \big\|F_{2J}\big\|_{L^1(J)} \le \sum_{J\in \mathscr{J}} \big\|F_{2J}\big\|_{L^\infty(J)} |G_J| \\ &\le C \sum_{\substack{J\in \mathscr{J}\\ J\subseteqq 3I_{\mathrm{top}(\mathbf{T})}}} |E| \mathscr{M}(E;\mathbf{T})|J| \sup_{x\in J}\sup_{\delta>4|J|} \frac{1}{2\delta}\int_{x-\delta}^{x+\delta} \Big| \sum_{s\in \mathbf{T}_2} \varepsilon_s \langle g \,|\, \varphi_s\rangle \varphi_s(z)\Big|\, dz \\ &\le 2C|E|\mathscr{M}(E;\mathbf{T}) \sum_{\substack{J\in \mathscr{J}\\ J\subseteqq 3I_{\mathrm{top}(\mathbf{T})}}} \int_J \sup_{\delta>4|J|} \frac{1}{2\delta}\int_{x-\delta}^{x+\delta} \Big| \sum_{s\in \mathbf{T}_2} \varepsilon_s \langle g \,|\, \varphi_s\rangle \varphi_s(z)\Big|\, dz\,dx \\ &\le C|E|\mathscr{M}(E;\mathbf{T}) \Big\| M\Big(\sum_{s\in \mathbf{T}_2} \varepsilon_s \langle g \,|\, \varphi_s\rangle \varphi_s\Big)\Big\|_{L^1(3I_{\mathrm{top}(\mathbf{T})})}, \end{aligned}$$

where M is the Hardy–Littlewood maximal operator. Using the Cauchy–Schwarz inequality and the boundedness of M on $L^2(\mathbf{R})$, we obtain the following estimate:

$$\Sigma_{22} \le C|E|\mathscr{M}(E;\mathbf{T})\,|I_{\mathrm{top}(\mathbf{T})}|^{\frac12} \Big\| \sum_{s\in \mathbf{T}_2} \varepsilon_s \langle g \,|\, \varphi_s\rangle \varphi_s\Big\|_{L^2}.$$

Appealing to the result of Exercise 6.1.6(a), we deduce

$$\Big\| \sum_{s\in \mathbf{T}_2} \varepsilon_s \langle g \,|\, \varphi_s\rangle \varphi_s\Big\|_{L^2} \le C\Big(\sum_{s\in \mathbf{T}_2} |\varepsilon_s \langle g \,|\, \varphi_s\rangle|^2\Big)^{\frac12} \le C' |I_{\mathrm{top}(\mathbf{T})}|^{\frac12} \mathscr{E}(g;\mathbf{T})\,.$$

The first estimate was also shown in (6.1.46); the same argument applies here, and the presence of the ε_s's does not introduce any change. We conclude that

$$\Sigma_{22} \le C|E|\mathscr{M}(E;\mathbf{T})|I_{\mathrm{top}(\mathbf{T})}|\mathscr{E}(g;\mathbf{T})\,,$$

which is what we needed to prove. This completes the proof of Lemma 6.1.10. □

The proof of the theorem is now complete.

Exercises

6.1.1. Show that for every f in the Schwartz class, $x, \xi \in \mathbf{R}$, and $\lambda \in [0,1]$, the function $(y,\eta) \mapsto B^m_{\xi,y,\eta,\lambda}(f)(x)$ is periodic in y with period $2^{m-\lambda}$ and periodic in η with period $2^{-m+\lambda}$.

6.1.2. Fix a function h in the Schwartz class, $\xi, y, \eta \in \mathbf{R}$, $s \in \mathbf{D}_m$, and $\lambda \in [0,1]$. Suppose that $2^{-\lambda}(\xi+\eta) \in \omega_{s(2)}$.
(a) Assume that $m \le 0$ and that $2^{-m} \ge 40|\xi|$. Show that

$$\big|\langle D^{2^\lambda} \tau^y M^\eta(h) \,|\, \varphi_s\rangle\big| \le 2^{\frac{m}{2}} \big\|\widehat{h}\big\|_{L^1((-\infty,-\frac{1}{40\cdot 2^m})\cup(\frac{1}{40\cdot 2^m},\infty))}.$$

(b) Using the trivial fact that $\big|\langle D^{2^\lambda} \tau^y M^\eta(h) \,|\, \varphi_s\rangle\big| \le C\|h\|_{L^2}$, conclude that whenever $2^{|m|} \ge 40|\xi|$, we have

$$\chi_{\omega_{s(2)}}(2^{-\lambda}(\xi+\eta))|\langle D^{2^\lambda} \tau^y M^\eta(h) \,|\, \varphi_s\rangle| \le C_h \min(1, 2^m),$$

where C_h depends on h but is independent of y, ξ, η, and λ.

6.1.3. (a) Let g be a periodic function on $\mathbf{R}$ with period κ which is integrable on $[0,\kappa]$. Show that

$$\lim_{K\to\infty} \frac{1}{2K}\int_{-K}^{K} g(t)\,dt \to \frac{1}{\kappa}\int_0^{\kappa} g(t)\,dt.$$

(b) Let g be a periodic function on $\mathbf{R}^2$ with period (κ_1,κ_2) which is integrable over $[0,\kappa_1]\times[0,\kappa_2]$. Show that

$$\lim_{K_1,K_2\to\infty} \frac{1}{2K_1K_2}\int_0^{K_1}\int_{-K_2}^{K_2} g(t_1,t_2)\,dt_2\,dt_1 = \frac{1}{\kappa_1\kappa_2}\int_0^{\kappa_1}\int_0^{\kappa_2} g(t_1,t_2)\,dt_2\,dt_1$$

6.1.4. Use the result in Appendix B.1 to obtain the size estimate

$$\big|\langle \varphi_s \,|\, \varphi_u\rangle\big| \le C_M \frac{\min\left(\dfrac{|I_s|}{|I_u|}, \dfrac{|I_u|}{|I_s|}\right)^{\frac12}}{\left(1+\dfrac{|c(I_s)-c(I_u)|}{\max(|I_s|,|I_u|)}\right)^M}$$

for every $M > 5$. Conclude that if $|I_u| \le |I_s|$, then

$$\big|\langle \varphi_s \,|\, \varphi_u\rangle\big| \le C'_M \left(\frac{|I_s|}{|I_u|}\right)^{\frac12} \int_{I_u} \frac{|I_s|^{-1}\,dx}{(1+\frac{|x-c(I_s)|}{|I_s|})^M}.$$

[*Hint:* Use that

$$\left|\frac{|x-c(I_s)|}{|I_s|} - \frac{|c(I_u)-c(I_s)|}{|I_s|}\right| \le \frac12$$

for all $x \in I_u$.]

6.1.5. Prove that there is a constant $C > 0$ such that for any interval J and any $b > 0$,

$$\int_J \int_{J^c} \frac{1}{\left(1+\frac{|x-y|}{b|J|}\right)^{20}}\,dx\,dy = Cb^2|J|^2 .$$

[*Hint:* Translate J to the interval $\left[-\frac{1}{2}|J|, \frac{1}{2}|J|\right]$ and change variables.]

6.1.6. Let φ_s be as in (6.1.4). Let $\mathbf{T}_2$ be a 2-tree and $f \in L^2(\mathbf{R})$.
(a) Show that there is a constant C such that for all sequences of complex scalars $\{\lambda_s\}_{s\in\mathbf{T}_2}$ we have

$$\Big\|\sum_{s\in\mathbf{T}_2} \lambda_s\, \varphi_s\Big\|_{L^2(\mathbf{R})} \le C\Big(\sum_{s\in\mathbf{T}_2} |\lambda_s|^2\Big)^{\frac{1}{2}}.$$

(b) Use duality to conclude that

$$\sum_{s\in\mathbf{T}_2} \big|\langle f\,|\,\varphi_s\rangle\big|^2 \le C^2 \|f\|_{L^2}^2 .$$

[*Hint:* To prove part (a) define $\mathscr{G}_m = \{s \in \mathbf{T}_2 : |I_s| = 2^m\}$. Then for $s \in \mathscr{G}_m$ and $s' \in \mathscr{G}_{m'}$, the functions φ_s and $\varphi_{s'}$ are orthogonal to each other, and it suffices to obtain the corresponding estimate when the summation is restricted to a given $\mathscr{G}_m$. But for s in $\mathscr{G}_m$, the intervals I_s are disjoint, and we may use the idea of the proof of Lemma 6.1.2. Use that $\sum_{u:\, \omega_u=\omega_s} \big|\big\langle \varphi_s\,|\,\varphi_u\big\rangle\big| \le C$ for every fixed s.]

6.1.7. Fix $A \ge 1$. Let $\mathbf{S}$ be a finite collection of dyadic tiles such that for all s_1, s_2 in $\mathbf{S}$ we have either $\omega_{s_1} \cap \omega_{s_2} = \emptyset$ or $AI_{s_1} \cap AI_{s_2} = \emptyset$. Let $N_{\mathbf{S}}$ be the *counting function* of $\mathbf{S}$, defined by

$$N_{\mathbf{S}} = \sup_{x\in\mathbf{R}} \#\{I_s : s \in \mathbf{S} \text{ and } x \in I_s\}.$$

(a) Show that for any $M > 0$ there exists a $C_M > 0$ such that for all $f \in L^2(\mathbf{R})$ we have

$$\sum_{s\in\mathbf{S}} \left|\left\langle f, |I_s|^{-\frac{1}{2}}\left(1+\frac{\operatorname{dist}(\cdot, I_s)}{|I_s|}\right)^{-\frac{M}{2}}\right\rangle\right|^2 \le C_M N_{\mathbf{S}} \|f\|_{L^2}^2 .$$

(b) Let φ_s be as in (6.1.4). Show that for any $M > 0$ there exists a $C_M > 0$ such that for all finite sequences of scalars $\{a_s\}_{s\in\mathbf{S}}$ we have

$$\Big\|\sum_{s\in\mathbf{S}} a_s \varphi_s\Big\|_{L^2}^2 \le C_M(1+A^{-M}N_{\mathbf{S}}) \sum_{s\in\mathbf{S}} |a_s|^2 .$$

(c) Conclude that for any $M > 0$ there exists a $C_M > 0$ such that for all $f \in L^2(\mathbf{R})$ we have

$$\sum_{s\in\mathbf{S}} \big|\langle f, \varphi_s\rangle\big|^2 \le C_M(1+A^{-M}N_{\mathbf{S}}) \|f\|_{L^2}^2 .$$

[*Hint:* Use the idea of Lemma 6.1.2 to prove part (a) when $N_{\mathbf{S}} = 1$. Suppose now that $N_{\mathbf{S}} > 1$. Call an element $s \in \mathbf{S}$ *h*-maximal if the region in $\mathbf{R}^2$ that is directly horizontally above the tile s does not intersect any other tile $s' \in \mathbf{S}$. Let $\mathbf{S}_1$ be the set of all h-maximal tiles in $\mathbf{S}$. Then $N_{\mathbf{S}_1} = 1$; otherwise, some $x \in \mathbf{R}$ would belong to both I_s and $I_{s'}$ for $s \neq s' \in \mathbf{S}_1$, and thus the horizontal regions directly above s and s' would have to intersect, contradicting the h-maximality of $\mathbf{S}_1$. Now define $\mathbf{S}_2$ to be the set of all h-maximal tiles in $\mathbf{S} \setminus \mathbf{S}_1$. As before, we have $N_{\mathbf{S}_2} = 1$. Continue in this way and write $\mathbf{S}$ as a union of at most $N_{\mathbf{S}}$ families of tiles $\mathbf{S}_j$, each of which has the property $N_{\mathbf{S}_j} = 1$. Apply the result to each $\mathbf{S}_j$ and then sum over j. Part (b): observe that whenever $s_1, s_2 \in \mathbf{S}$ and $s_1 \neq s_2$ we must have either $\langle \varphi_{s_1}, \varphi_{s_2} \rangle = 0$ or $\operatorname{dist}(I_{s_1}, I_{s_2}) \geq (A-1)\max(|I_{s_1}|, |I_{s_2}|)$, which implies

$$\left(1+\frac{\operatorname{dist}(I_{s_1},I_{s_2})}{\max(|I_{s_1}|,|I_{s_2}|)}\right)^{-M} \leq A^{-\frac{M}{2}}\left(1+\frac{\operatorname{dist}(I_{s_1},I_{s_2})}{\max(|I_{s_1}|,|I_{s_2}|)}\right)^{-\frac{M}{2}}.$$

Use this estimate to obtain

$$\Big\|\sum_{s\in\mathbf{S}} a_s\varphi_s\Big\|_{L^2}^2 \leq \sum_{s\in\mathbf{S}} |a_s|^2 + \frac{C_M}{A^{\frac{M}{2}}}\Big\|\sum_{s\in\mathbf{S}} \frac{|a_s|}{|I_s|^{\frac{1}{2}}}\left(1+\frac{\operatorname{dist}(x,I_s)}{|I_s|}\right)^{-\frac{M}{2}}\Big\|_{L^2}^2$$

by expanding the square on the left. The required estimate follows from the dual statement to part (a). Part (c) follows from part (b) by duality.]

6.1.8. Let φ_s be as in (6.1.4) and let $\mathbf{D}_m$ be the set of all dyadic tiles s with $|I_s| = 2^m$. Show that there is a constant C (independent of m) such that for square-integrable sequences of scalars $\{a_s\}_{s\in\mathbf{D}_m}$ we have

$$\Big\|\sum_{s\in\mathbf{D}_m} a_s\varphi_s\Big\|_{L^2}^2 \leq C\sum_{s\in\mathbf{D}_m} |a_s|^2.$$

Conclude from this that

$$\sum_{s\in\mathbf{D}_m} \big|\langle f,\varphi_s\rangle\big|^2 \leq C\|f\|_{L^2}^2.$$

6.1.9. Fix $c_0 > 0$ and a Schwartz function φ whose Fourier transform is supported in the interval $[-\frac{3}{8},\frac{3}{8}]$ and that satisfies

$$\sum_{l\in\mathbf{Z}} |\widehat{\varphi}(t+\tfrac{l}{2})|^2 = c_0$$

for all real numbers t. Define functions φ_s as follows. Fix an integer m and set

$$\varphi_s(x) = 2^{-\frac{m}{2}}\varphi(2^{-m}x-k)e^{2\pi i 2^{-m}x\frac{l}{2}}$$

whenever $s = [k2^m,(k+1)2^m)\times[l2^{-m},(l+1)2^{-m})$ is a tile in $\mathbf{D}_m$.
(a) Prove that for all Schwartz functions f we have

$$\sum_{s\in\mathbf{D}_m} \big|\langle f\,|\,\varphi_s\rangle\big| < \infty.$$

(b) Show that for all $f \in \mathscr{S}(\mathbf{R})$ we have

$$\sum_{s\in\mathbf{D}_m} \langle f \mid \varphi_s\rangle \varphi_s = c_0 f .$$

(c) Prove that for every $f \in \mathscr{S}(\mathbf{R})$ the following identity holds

$$\|f\|_{L^2}^2 = \frac{\|\varphi\|_{L^2}^2}{c_0^2} \sum_{s\in\mathbf{D}_m} |\langle f \mid \varphi_s\rangle|^2 .$$

[*Hint:* Part (a): Use Appendix B. Part (b): First prove that

$$\sum_{s\in\mathbf{D}_m} \varphi_s(x)\overline{\widehat{\varphi_s}(y)} = c_0\, e^{2\pi i x y}$$

using the Poisson summation formula.]

6.1.10. This is a continuous version of Exercise 6.1.9. Fix a Schwartz function φ on $\mathbf{R}^n$ and define a *continuous wave packet*

$$\varphi_{y,\xi}(x) = \varphi(x-y)e^{2\pi i \xi\cdot x} .$$

Prove that for all f Schwartz functions on $\mathbf{R}^n$, the following identity is valid:

$$\|\varphi\|_{L^2}^2 f(x) = \int_{\mathbf{R}^n}\int_{\mathbf{R}^n} \varphi_{y,\xi}(x)\langle f \mid \varphi_{y,\xi}\rangle\, dy\, d\xi .$$

[*Hint:* Prove first that $\int_{\mathbf{R}^n}\int_{\mathbf{R}^n} \varphi_{y,\xi}(x)\overline{\widehat{\varphi_{y,\xi}}(z)}\, dy\, d\xi = \|\varphi\|_{L^2}^2 e^{2\pi i x\cdot z}$.]

6.2 Distributional Estimates for the Carleson Operator

In this section we derive estimates for the distribution function of the Carleson operator acting on characteristic functions of measurable sets. These estimates imply, in particular, that the Carleson operator is bounded on $L^p(\mathbf{R})$ for $1 < p < \infty$. To achieve this we build on the time–frequency analysis approach developed in the previous section. Working with characteristic functions of measurable sets of finite measure is crucial in obtaining an improved energy estimate, which is the key to the proof. Later in this section we obtain weighted estimates for the Carleson operator $\mathscr{C}$. These estimates are reminiscent of the corresponding estimates for the maximal singular integrals we encountered in the previous chapter.

6.2.1 The Main Theorem and Preliminary Reductions

In the sequel we use the notation introduced in Section 6.1. We begin by extending the definition of the operator $\mathscr{C}$ on L^p for $1 < p < \infty$. First we note that the linear operator

$$h \mapsto \left(\widehat{h}\chi_{[-\xi,\xi]}\right)^{\vee},$$

initially defined on Schwartz functions, admits a unique bounded extension on $L^p(\mathbf{R})$ for $1 < p < \infty$, which we denote by H_ξ. This extension H_ξ is given on $L^p(\mathbf{R})$ by convolution with the kernel

$$\left(\chi_{[-\xi,\xi]}\right)^{\vee}(y) = \frac{\sin(2\pi\xi y)}{\pi y}.$$

Notice that this kernel lies in $L^{p'}(\mathbf{R})$; hence $H_\xi(f)$ is well defined for $f \in L^p(\mathbf{R})$ as an absolutely convergent convolution of an L^p and an $L^{p'}$ function and is a continuous function. Then $f \in L^p(\mathbf{R})$, the action of the Carleson operator $\mathscr{C}$ on f,

$$\mathscr{C}(f) = \sup_{\xi>0} |H_\xi(f)|,$$

is well defined. The following is the main result of this section concerning $\mathscr{C}$.

Theorem 6.2.1. *(a) There exist finite constants $C, \kappa > 0$ such that for any measurable subset F of the reals with finite measure we have*

$$\big|\{x \in \mathbf{R} : \mathscr{C}(\chi_F)(x) > \alpha\}\big| \le C|F| \begin{cases} \frac{1}{\alpha}\left(1+\log\left(\frac{1}{\alpha}\right)\right) & \textit{when } 0 < \alpha < 1, \\ e^{-\kappa\alpha} & \textit{when } \alpha \ge 1. \end{cases} \tag{6.2.1}$$

(b) For any $1 < p < \infty$ there is a constant $C_p > 0$ such that for all f in $L^p(\mathbf{R})$ we have the estimate

$$\|\mathscr{C}(f)\|_{L^p(\mathbf{R})} \le C_p \|f\|_{L^p(\mathbf{R})}. \tag{6.2.2}$$

Proof. Assuming statement (a), we obtain

$$\|\mathscr{C}(\chi_F)\|_{L^p}^p = p\int_0^\infty \big|\{\mathscr{C}(\chi_F) > \alpha\}\big|\alpha^{p-1}\,d\alpha \le pC^p|F|\int_0^\infty \varphi(\alpha)\alpha^{p-1}\,d\alpha,$$

where $\varphi(\alpha) = \alpha^{-1}(1+\log(\alpha)^{-1})$ for $\alpha < 1$ and $\varphi(\alpha) = e^{-\kappa\alpha}$ for $\alpha \ge 1$. The last integral is convergent, and consequently one obtains a restricted strong type (p,p) estimate

$$\|\mathscr{C}(\chi_F)\|_{L^p(\mathbf{R})} \le C_p'|F|^{\frac{1}{p}}$$

for the Carleson operator. Fix $p \in (1,\infty)$ and select p_0, p_1 such that $1 < p_0 < p < p_1 < \infty$. Applying Theorem 1.4.19 in [156] we obtain that there is a constant $C_p < \infty$ such that $\mathscr{C}$ satisfies

$$\big\|\mathscr{C}(g)\big\|_{L^p} \le C_p \|g\|_{L^p} \tag{6.2.3}$$

for all functions g in $S_0(\mathbf{R})$, which is dense in $L^p(\mathbf{R})$. (Functions in $S_0(\mathbf{R})$ have the form $h_1 - h_2 + ih_3 - ih_4$, where each h_j is equal to $\sum_{k=m_1}^{m_2} 2^{-k}\chi_{A_k}$, where A_k are subsets of $\mathbf{R}$ of finite measure.) To extend (6.2.3) to all functions f in $L^p(\mathbf{R})$ we pick a sequence of functions g_j in $S_0(\mathbf{R})$ such that $g_j \to f$ in L^p as $j \to \infty$. We observe that H_ξ is given on a dense subset of L^p by multiplication on the Fourier transform by $\chi_{[-\xi,\xi]}$ or by convolution with the kernel $(\chi_{[-\xi,\xi]})^\vee(y) = \sin(2\pi\xi y)/\pi y$. Since this kernel lies in $L^{p'}(\mathbf{R})$, it follows that for all $f \in L^p(\mathbf{R})$, $H_\xi(f)$ is given as an absolutely convergent convolution with the same kernel. Then for any $\xi > 0$ and $x \in \mathbf{R}$ we have

$$\begin{aligned}
\big|H_\xi(f)(x) - H_\xi(g_j)(x)\big| &= \left| \int_{\mathbf{R}} (f(x-y) - g_j(x-y)) \frac{\sin(2\pi\xi y)}{\pi y}\, dy \right| \\
&\le \|f - g_j\|_{L^p} \left(\int_{-\infty}^{\infty} \left| \frac{\sin(2\pi\xi y)}{\pi y} \right|^{p'} dy \right)^{\frac{1}{p'}} \\
&= \|f - g_j\|_{L^p} c\, \xi^{\frac{1}{p}}
\end{aligned}$$

and this tends to zero as $j \to \infty$ for all $x \in \mathbf{R}$. This shows that the sequence of continuous functions $H_\xi(g_j)$ converges to $H_\xi(f)$ pointwise everywhere. Using this observation we write

$$\begin{aligned}
\big\|\mathscr{C}(f)\big\|_{L^p} &= \big\| \sup_{\xi>0} |H_\xi(f)| \big\|_{L^p} \\
&= \big\| \sup_{\xi>0} | \lim_{j\to\infty} H_\xi(g_j)| \big\|_{L^p} \\
&\le \big\| \sup_{\xi>0} \liminf_{j\to\infty} |H_\xi(g_j)| \big\|_{L^p} \\
&\le \big\| \liminf_{j\to\infty} \sup_{\xi>0} |H_\xi(g_j)| \big\|_{L^p} \\
&\le \liminf_{j\to\infty} \big\| \sup_{\xi>0} |H_\xi(g_j)| \big\|_{L^p} \\
&\le \liminf_{j\to\infty} C_p \big\|g_j\big\|_{L^p} \\
&= C_p \big\|f\big\|_{L^p},
\end{aligned}$$

where we used Fatou's lemma in the third inequality. Thus (a) implies (b).

It remains to prove the assertion in part (a) of the theorem. This is the goal of the rest of this section. Several reductions are made until the end of this subsection.

Since $(\widehat{\chi_F}\chi_{[-\xi,\xi]})^\vee(x)$ is continuous in ξ, we restrict the supremum in

$$\sup_{\xi>0}\left|\left(\widehat{\chi_F}\chi_{[-\xi,\xi]}\right)^\vee(x)\right|$$

to $\xi \in \mathbf{Q}^+$. In view of the Exercise 1.1.1(b) in [156] we may restrict the countable set $\mathbf{Q}^+$ to a finite subset Q_0 of $\mathbf{Q}^+$, obtain estimates independent of Q_0, and let $Q_0 \to \mathbf{Q}^+$. We pick a measurable function N_F from $\mathbf{R}$ to Q_0 such that

$$\sup_{\xi\in Q_0}\left|\left(\widehat{\chi_F}\chi_{[-\xi,\xi]}\right)^\vee(x)\right| = \left|\left(\widehat{\chi_F}\chi_{[-N_F(x),N_F(x)]}\right)^\vee(x)\right| \tag{6.2.4}$$

for all $x \in \mathbf{R}$. This motivates the introduction of the sublinear operator

$$\mathscr{C}_N(\chi_F)(x) = \left|\left(\widehat{\chi_F}\chi_{[-N(x),N(x)]}\right)^\vee(x)\right|$$

for a general and fixed measurable function $N : \mathbf{R} \to Q_0$. Once the assertion in part (a) is proved for $\mathscr{C}_N$ in place of $\mathscr{C}$, then for a given measurable set F, we replace $N(x)$ by $N_F(x)$ and we use (6.2.4) and a limiting argument to obtain (6.2.1).

For a Schwartz function h on the real line and $x \in \mathbf{R}$ we define an operator

$$\begin{aligned} G_{N,y,\eta,\lambda}(h)(x) &= \sum_{\substack{u\in\mathbf{D}\\ 2^{-\lambda}(N(x)+\eta)\in\omega_{u(2)}}} \left\langle h \,\middle|\, M^{-\eta}\tau^{-y}D^{2^{-\lambda}}\varphi_u\right\rangle M^{-\eta}\tau^{-y}D^{2^{-\lambda}}\varphi_u(x) \\ &= \sum_{\substack{s\in\mathbf{D}_{y,\eta,\lambda}\\ N(x)\in\omega_{s(2)}}} \left\langle h \,|\, \varphi_s\right\rangle \varphi_s(x), \end{aligned}$$

where $\mathbf{D}_{y,\eta,\lambda}$ is the set of all rectangles of the form

$$(2^{-\lambda}\otimes I_u - y)\times(2^{\lambda}\otimes\omega_u - \eta),$$

where u ranges over $\mathbf{D}$. Here $a\otimes I$ denotes the set $\{ax : x\in I\}$. For such s, φ_s is defined in (6.1.4). The rectangles in $\mathbf{D}_{y,\eta,\lambda}$ are formed by dilating the dyadic tiles in $\mathbf{D}$ by the amount $2^{-\lambda}$ in the time coordinate axis and by 2^{λ} in the frequency coordinate axis and then translating them by the amounts y and η, respectively.

Then we define $G^{\mathbf{P}}_{N,y,\eta,\lambda}(h)$ in a similar way except that the sum over $s\in\mathbf{D}_{y,\eta,\lambda}$ is replaced by $s\in\mathbf{P}$ for any subset $\mathbf{P}$ of $\mathbf{D}_{y,\eta,\lambda}$. Next for $M\in\mathbf{Z}^+$ we define

$$\mathbf{V}_M = \left\{s\in\mathbf{D}_{y,\eta,\lambda} : s = (2^{-\lambda}\otimes I_u - y)\times(2^{\lambda}\otimes\omega_u-\eta), u\in\mathbf{D}_m, m\in\mathbf{Z} \text{ and } |m|\le M\right\}.$$

Notice that for all $h\in\mathscr{S}(\mathbf{R})$ we have

$$G_{N,y,\eta,\lambda}(h) = \lim_{M\to\infty} G^{\mathbf{V}_M}_{N,y,\eta,\lambda}(h),$$

where the limit is taken in the pointwise sense.

Since $N(x)$ takes finitely many values, it follows from Lemma 6.1.2 (adapted to shifts and dilations of dyadic tiles by fixed amounts) that $G_{N,y,\eta,\lambda}$ maps L^2 to itself when restricted to $\mathscr{S}$ and also $G^{\mathbf{P}}_{N,y,\eta,\lambda}$ maps L^2 to itself uniformly in all subsets $\mathbf{P}$ of $\mathbf{D}_{y,\eta,\lambda}$. Then the operators $G_{N,y,\eta,\lambda}$ and $G^{\mathbf{P}}_{N,y,\eta,\lambda}$ have bounded extensions on $L^2(\mathbf{R})$ uniformly in $\mathbf{P}$; we denote these extensions in the same way.

We now observe that

$$G^{\mathbf{P}}_{N,y,\eta,\lambda}(g) \to G^{\mathbf{V}_M}_{N,y,\eta,\lambda}(g)$$

in L^2 whenever the finite subset $\mathbf{P}$ of $\mathbf{V}_M$ tends to $\mathbf{V}_M$. Indeed, this assertion is valid for Schwartz functions and an $\varepsilon/3$ argument can be used to establish it for all functions $g \in L^2(\mathbf{R})$.

Given $K, L > 0$ we define operators

$$\Pi^{\mathbf{P}}_{N,K,L}(h) = \frac{1}{2KL}\int_{-K}^{K}\int_{0}^{L}\int_{0}^{1} G^{\mathbf{P}}_{N,y,\eta,\lambda}(h)\,d\lambda\,dy\,d\eta$$

for any subset $\mathbf{P}$ of $\mathbf{D}_{y,\eta,\lambda}$ and $h \in \mathscr{S}(\mathbf{R})$. We claim that for any $M \in \mathbf{Z}^+$ and any $h \in \mathscr{S}(\mathbf{R})$ the sequence of functions $\Pi^{\mathbf{V}_M}_{N,K,L}(h)$ converges in L^2 as $K, L \to \infty$ to an operator we call $\Pi^{\mathbf{V}_M}_{N}(h)$. (The convergence is also pointwise since $h \in \mathscr{S}$).

Indeed, $G^{\mathbf{V}_M}_{N,y,\eta,\lambda}(h)$ is periodic with period $q = 2^{M+\lambda}$ in η and thus we have

$$\begin{aligned}
&\frac{1}{L'}\int_0^{L'} G^{\mathbf{V}_M}_{N,y,\eta,\lambda}(h)(x)\,d\eta - \frac{1}{L}\int_0^{L} G^{\mathbf{V}_M}_{N,y,\eta,\lambda}(h)(x)\,d\eta \\
&\quad = -\frac{\delta'}{q(k'q+\delta')}\int_0^{q} G^{\mathbf{V}_M}_{N,y,\eta,\lambda}(h)(x)\,d\eta + \frac{\delta}{q(kq+\delta)}\int_0^{q} G^{\mathbf{V}_M}_{N,y,\eta,\lambda}(h)(x)\,d\eta \\
&\qquad + \frac{1}{k'q+\delta'}\int_0^{\delta'} G^{\mathbf{V}_M}_{N,y,\eta,\lambda}(h)(x)\,d\eta - \frac{1}{kq+\delta}\int_0^{\delta} G^{\mathbf{V}_M}_{N,y,\eta,\lambda}(h)(x)\,d\eta\,,
\end{aligned}$$

where $L = kq + \delta$ and $L' = k'q + \delta'$ with $\delta, \delta' < q$ and $k, k' \in \mathbf{Z}^+$. It follows from this that the L^2 norm in x of the preceding expression is bounded by

$$\left(\frac{2}{k'} + \frac{2}{k}\right)\big\|G^{\mathbf{V}_M}_{N,y,\eta,\lambda}(h)\big\|_{L^2},$$

which becomes arbitrarily small as $k, k' \to \infty$. A similar argument with respect to the variable y, in which we have periodicity with period $2^{M-\lambda}$, shows that the averages in y are also Cauchy. Combining these observations we deduce that the sequence of functions $\Pi^{\mathbf{V}_M}_{N,K,L}(h)$ is Cauchy; hence it converges in L^2 as $K, L \to \infty$ to an operator which we call $\Pi^{\mathbf{V}_M}_{N}(h)$.

Therefore both $\Pi^{\mathbf{V}_M}_{N,K,L}$ and $\Pi^{\mathbf{V}_M}_{N}$ have L^2-bounded extensions that we denote in the same way. Moreover, an $\varepsilon/3$ argument yields that for all $g \in L^2$

$$\Pi^{\mathbf{V}_M}_{N,K,L}(g) \to \Pi^{\mathbf{V}_M}_{N}(g) \tag{6.2.5}$$

in L^2 as $K, L \to \infty$. Next we define

$$\Pi_{N,K,L}(h) = \frac{1}{2KL} \int_0^L \int_{-K}^K \int_0^1 \sum_{s \in \mathbf{D}_{y,\eta,\lambda}} \langle h \,|\, \varphi_s \rangle (\chi_{\omega_{s(2)}} \circ N) \varphi_s \, d\lambda \, dy \, d\eta \,.$$

and

$$\Pi_N(h) = \lim_{\substack{K \to \infty \\ L \to \infty}} \Pi_{N,K,L}(h)$$

for $h \in \mathscr{S}(\mathbf{R})$, where the limits are taken in the pointwise sense. The L^2 boundedness of $\Pi_{N,K,L}$ (uniformly in K, L) is a consequence of that of $G_{N,y,\eta,\lambda}$.

It follows from the L^2 boundedness of $\Pi_{N,K,L}$ and Fatou's lemma that Π_N is an L^2-bounded operator when restricted to Schwartz functions, and therefore it admits an L^2-bounded extension, which we denote in the same way. Another $\varepsilon/3$ argument yields that for all $g \in L^2$ we have

$$\Pi_N^{\mathbf{V}_M}(g) \to \Pi_N(g)$$

in L^2 as $M \to \infty$.

Next we show that for all $g \in L^2(\mathbf{R})$ and almost all $x \in \mathbf{R}$ we have

$$\mathscr{C}_N(g)(x) \le |g(x)| + \left|\Pi_N(g)(x)\right| + \left|\Pi_N(\widetilde{g})(-x)\right|, \tag{6.2.6}$$

where $\widetilde{g}(t) = g(-t)$ for all t. Indeed, (6.2.6) trivially holds for Schwartz functions. Given $g \in L^2(\mathbf{R})$, pick a sequence of Schwartz functions h_n such that $h_n \to g$ in L^2 as $n \to \infty$. Clearly we have

$$\left| \int_{|y| \le N(x)} (\widehat{g}(y) - \widehat{h_n}(y)) e^{2\pi i x y} dy \right| \le \left(2 \max\{q : q \in Q_0\}\right)^{1/2} \|g - h_n\|_{L^2(\mathbf{R})}$$

which implies that $\mathscr{C}_N(h_n)(x) \to \mathscr{C}_N(g)(x)$ everywhere as $n \to \infty$. Moreover, $\Pi_N(h_n)$ tends to $\Pi_N(g)$ in L^2, so there is a subsequence n_k such that $h_{n_k} \to g$ a.e., $\Pi_N(h_{n_k}) \to \Pi_N(g)$ a.e., and $\Pi_N(\widetilde{h_{n_k}}) \to \Pi_N(\widetilde{g})$ a.e. as $k \to \infty$. This establishes (6.2.6). In view of (6.2.6), it suffices to prove the claim in part (a) for Π_N with constants independent of the measurable function $N : \mathbf{R} \to Q_0$.

We will prove that there is a constant C' such that for any pair of measurable subsets (E, F) of the real line with nonzero finite measure there is a subset E' of E with $|E'| \ge \frac{1}{2}|E|$ such that

$$\left| \int_{E'} \Pi_N(\chi_F)(x) \, dx \right| \le 2C' \min(|E|, |F|) \left(1 + \left|\log \frac{|E|}{|F|}\right|\right). \tag{6.2.7}$$

We explain why (6.2.7) implies statement (a) of Theorem 6.2.1. Given $\alpha > 0$ we define sets

$$E_\alpha^1 = \{\operatorname{Re} \Pi_N(\chi_F) > \alpha\}, \qquad E_\alpha^2 = \{\operatorname{Re} \Pi_N(\chi_F) < -\alpha\},$$
$$E_\alpha^3 = \{\operatorname{Im} \Pi_N(\chi_F) > \alpha\}, \qquad E_\alpha^4 = \{\operatorname{Im} \Pi_N(\chi_F) < -\alpha\}.$$

We apply (6.2.7) to the pair (E_α^j, F) for any $j = 1,2,3,4$. We find a subset $(E_\alpha^j)'$ of E_α^j of at least half its measure so that (6.2.7) holds for this pair. Then we have

$$\frac{\alpha}{2}|E_\alpha^j| \le \alpha|(E_\alpha^j)'| \le \left| \int_{(E_\alpha^j)'} \Pi_N(\chi_F)(x)\,dx \right| \le 2C' \min(|E_\alpha^j|, |F|) \left(1 + \left| \log \frac{|E_\alpha^j|}{|F|} \right| \right). \tag{6.2.8}$$

If $|E_\alpha^j| \le |F|$, this estimate implies that

$$|E_\alpha^j| \le |F| e\, e^{-\frac{1}{4C'}\alpha}, \tag{6.2.9}$$

while if $|E_\alpha^j| > |F|$, it implies that

$$\alpha \le 4C' \frac{|F|}{|E_\alpha^j|} \left(1 + \log \frac{|E_\alpha^j|}{|F|} \right). \tag{6.2.10}$$

Case 1: $\alpha > 4C'$. If $|E_\alpha^j| > |F|$, setting $t = |E_\alpha^j|/|F| > 1$ and using the fact that $\sup_{1<t<\infty} \frac{1}{t}(1+\log t) = 1$, we obtain that (6.2.10) fails. In this case we must therefore have that $|E_\alpha^j| \le |F|$. Applying (6.2.9) four times, we deduce

$$|\{\Pi_N(\chi_F) > 4\alpha\}| \le 4e|F| e^{-\frac{1}{4C'}\alpha}. \tag{6.2.11}$$

Case 2: $\alpha \le 4C'$. If $|E_\alpha^j| > |F|$, we use the elementary fact that if $t > 1$ satisfies $t(1+\log t)^{-1} < \frac{B}{\alpha}$, then $t < \frac{2B}{\alpha}(1+\log \frac{2B}{\alpha})$; to prove this fact one may use the inequalities $t < \frac{2B}{\alpha}(1+\log\sqrt{t})$ and $\log\sqrt{t} \le \log t - \log(1+\log\sqrt{t}) \le \log\frac{2B}{\alpha}$ for $t > 1$. Taking t = $|E_\alpha^j|/|F|$ and $B = 4C'$ in (6.2.10) yields

$$\frac{|E_\alpha^j|}{|F|} \le \frac{8C'}{\alpha}\left(1 + \log \frac{8C'}{\alpha} \right). \tag{6.2.12}$$

If $|E_\alpha^j| \le |F|$, then we use (6.2.9), but we note that for some constant $c' > 1$ we have

$$e\, e^{-\frac{1}{4C'}\alpha} \le c' \frac{8C'}{\alpha}\left(1 + \log \frac{8C'}{\alpha} \right)$$

whenever $\alpha \le 4C'$. Thus, when $\alpha \le 4C'$, we always have

$$|\{\Pi_N(\chi_F) > 4\alpha\}| \le c' \frac{32C'}{\alpha}|F| \left(1 + \log \frac{8C'}{\alpha} \right). \tag{6.2.13}$$

Combining (6.2.11) and (6.2.13), we obtain (6.2.1) with Π_N in place of $\mathscr{C}$. Then (6.2.6) yields (6.2.1) with $\mathscr{C}_N$ in place of $\mathscr{C}$ and this suffices for the proof of the statement in part (a) of the theorem by a limiting argument, as observed before.

We now return to the proof of (6.2.7) which we obtain by making certain reductions. First we have that $\Pi_N^{\mathbf{V}_M}(\chi_F) \to \Pi_N(\chi_F)$ in $L^2(\mathbf{R})$ and thus by the Cauchy–Schwarz inequality we reduce the proof of (6.2.7) to

$$\left| \int_{E'} \Pi_N^{\mathbf{V}_M}(\chi_F)(x)\,dx \right| \le 2C' \min(|E|,|F|) \left(1 + \left|\log \frac{|E|}{|F|}\right|\right) \tag{6.2.14}$$

with constants independent of M. Then we reduce (6.2.14) to

$$\left| \int_{E'} \Pi_{N,K,L}^{\mathbf{V}_M}(\chi_F)(x)\,dx \right| \le 2C' \min(|E|,|F|) \left(1 + \left|\log \frac{|E|}{|F|}\right|\right) \tag{6.2.15}$$

by a similar argument using (6.2.5) (with constant C' independent of M,K,L). But since

$$\int_{E'} \frac{1}{2KL} \int_{-K}^{K} \int_0^L \int_0^1 |G_{N,y,\eta,\lambda}^{\mathbf{V}_M}(\chi_F)(x)|\,d\lambda\, d\eta\, dy\, dx < \infty$$

it suffices to show that

$$\left| \int_{E'} G_{N,y,\eta,\lambda}^{\mathbf{V}_M}(\chi_F)(x)\,dx \right| \le 2C' \min(|E|,|F|) \left(1 + \left|\log \frac{|E|}{|F|}\right|\right) \tag{6.2.16}$$

with constants independent of N, y, η, λ and M. Finally, $\mathbf{V}_M$ can be approximated by a finite set of tiles in $\mathbf{D}_{y,\eta,\lambda}$ and since

$$G_{N,y,\eta,\lambda}^{\mathbf{P}}(\chi_F) \to G_{N,y,\eta,\lambda}^{\mathbf{V}_M}(\chi_F)$$

in $L^2(\mathbf{R})$ as $\mathbf{P} \to \mathbf{V}_M$, we reduce (6.2.16) to

$$\left| \int_{E'} G_{N,y,\eta,\lambda}^{\mathbf{P}}(\chi_F)(x)\,dx \right| \le 2C' \min(|E|,|F|) \left(1 + \left|\log \frac{|E|}{|F|}\right|\right) \tag{6.2.17}$$

with constants independent of N, y, η, λ and $\mathbf{P}$. So we focus on the proof of (6.2.17).

To prove (6.2.17) we introduce the set

$$\Omega_{E,F} = \left\{ M(\chi_F) > 8 \min\left(1, \tfrac{|F|}{|E|}\right) \right\}.$$

It follows that $|\Omega_{E,F}| \le \frac{1}{2}|E|$, since the Hardy–Littlewood maximal operator is of weak type $(1,1)$ with norm 2. We conclude that the set

$$E' = E \setminus \Omega_{E,F}$$

satisfies $|E'| \ge \frac{1}{2}|E|$. (Notice that in the case $|F| \ge |E|$ the set $\Omega_{E,F}$ is empty.)

The required inequality (6.2.17) will be a consequence of the following two estimates:

$$\left|\int_{E'}\sum_{\substack{s\in\mathbf{P}\\ I_s\subseteqq\Omega_{E,F}}}\langle\chi_F\,|\,\varphi_s\rangle(\chi_{\omega_{s(2)}}\circ N)\,\varphi_s\,dx\right|\le C'\min(|E|,|F|)\tag{6.2.18}$$

and

$$\left|\int_{E'}\sum_{\substack{s\in\mathbf{P}\\ I_s\nsubseteqq\Omega_{E,F}}}\langle\chi_F\,|\,\varphi_s\rangle(\chi_{\omega_{s(2)}}\circ N)\varphi_s\,dx\right|\le C'\min(|E|,|F|)\left(1+\left|\log\frac{|E|}{|F|}\right|\right),\tag{6.2.19}$$

where the constant C' is independent of the sets E,F, of the measurable function N, of the finite subset $\mathbf{P}$ of $\mathbf{D}_{y,\eta,\lambda}$, and of y,η,λ. Estimates (6.2.18) and (6.2.19) are proved in the next three subsections. □

6.2.2 The Proof of Estimate (6.2.18)

In proving (6.2.18), we may assume that $|F|\le|E|$; otherwise, the set $\Omega_{E,F}$ is empty and there is nothing to prove.

Let $\mathbf{P}$ be a finite subset of $\mathbf{D}_{y,\eta,\lambda}$. We denote by $\mathscr{I}(\mathbf{P})$ the grid that consists of all the time projections I_s of tiles s in $\mathbf{P}$. For a fixed interval J in $\mathscr{I}(\mathbf{P})$ we define

$$\mathbf{P}(J)=\{s\in\mathbf{P}:\ I_s=J\}$$

and a function

$$\psi_J(x)=|J|^{-\frac12}\left(1+\frac{|x-c(J)|}{|J|}\right)^{-M},$$

where M is a large integer to be chosen momentarily. We note that for each $s\in\mathbf{P}(J)$ we have $|\varphi_s(x)|\le C_M\,\psi_J(x)$.

For each $k=0,1,2,\dots$, we introduce families

$$\mathscr{F}_k=\{J\in\mathscr{I}(\mathbf{P}):\ 2^kJ\subseteqq\Omega_{E,F},\ 2^{k+1}J\nsubseteqq\Omega_{E,F}\}.$$

We begin by writing the left-hand side of (6.2.18) as

$$\begin{aligned}&\sum_{\substack{J\in\mathscr{I}(\mathbf{P})\\ J\subseteqq\Omega_{E,F}}}\left|\sum_{s\in\mathbf{P}(J)}\int_{E'}\langle\chi_F\,|\,\varphi_s\rangle\,\chi_{\omega_{s(2)}}(N(x))\,\varphi_s(x)\,dx\right|\\ &=\sum_{k=0}^{\infty}\sum_{\substack{J\in\mathscr{I}(\mathbf{P})\\ J\in\mathscr{F}_k}}\left|\int_{E'}\sum_{s\in\mathbf{P}(J)}\langle\chi_F\,|\,\varphi_s\rangle\,\chi_{\omega_{s(2)}}(N(x))\,\varphi_s(x)\,dx\right|.\end{aligned}\tag{6.2.20}$$

Using the result of Exercise 7.2.8(b) in [156] we obtain the existence of a constant $C_0 < \infty$ such that for each $k = 0, 1, \ldots$ and $J \in \mathscr{F}_k$ we have

$$\begin{aligned}\langle \chi_F, \psi_J \rangle &\le |J|^{\frac{1}{2}} \inf_J M(\chi_F) \\ &\le |J|^{\frac{1}{2}} C_0^k \inf_{2^{k+1}J} M(\chi_F) \\ &\le 8 C_0^k |J|^{\frac{1}{2}} \frac{|F|}{|E|},\end{aligned} \tag{6.2.21}$$

since $2^{k+1}J$ meets the complement of $\Omega_{E,F}$.

For $J \in \mathscr{F}_k$ we also have that $E' \cap 2^k J = \emptyset$ and hence

$$\int_{E'} \psi_J(y)\, dy \le \int_{(2^k J)^c} \psi_J(y)\, dy \le |J|^{\frac{1}{2}} C_M 2^{-kM}. \tag{6.2.22}$$

Next we note that for each $J \in \mathscr{I}(\mathbf{P})$ and $x \in \mathbf{R}$ there is at most one $s = s_x \in \mathbf{P}(J)$ such that $N(x) \in \omega_{s_x(2)}$. Using this observation along with (6.2.21) and (6.2.22), we can therefore estimate the expression on the right in (6.2.20) as follows:

$$\begin{aligned}&\sum_{k=0}^{\infty} \sum_{\substack{J \in \mathscr{I}(\mathbf{P}) \\ J \in \mathscr{F}_k}} \left| \int_{E'} \langle \chi_F \,|\, \varphi_{s_x} \rangle \chi_{\omega_{s_x(2)}}(N(x)) \varphi_{s_x}(x)\, dx \right| \\ &\le C_M^2 \sum_{k=0}^{\infty} \sum_{\substack{J \in \mathscr{I}(\mathbf{P}) \\ J \in \mathscr{F}_k}} \int_{E'} \langle \chi_F, \psi_J \rangle \psi_J(x)\, dx \\ &\le C_M^2 4 \frac{|F|}{|E|} \sum_{k=0}^{\infty} C_0^k \sum_{J \in \mathscr{F}_k} |J|^{\frac{1}{2}} \int_{E'} \psi_J(x)\, dx \\ &\le 4 C_M^3 \frac{|F|}{|E|} \sum_{k=0}^{\infty} (C_0 2^{-M})^k \sum_{J \in \mathscr{F}_k} |J|,\end{aligned} \tag{6.2.23}$$

and we pick $M > \log C_0 / \log 2$. It remains to control

$$\sum_{J \in \mathscr{F}_k} |J|$$

for each nonnegative integer k. In doing this we let $\mathscr{F}_k^*$ be all elements of $\mathscr{F}_k$ that are maximal under inclusion. Then we observe that if $J \in \mathscr{F}_k^*$ and $J' \in \mathscr{F}_k$ satisfy $J' \subsetneqq J$ then

$$\operatorname{dist}(J', J^c) = 0,$$

otherwise $2J'$ would be contained in J and thus

$$2^{k+1} J' \subsetneqq 2^k J \subsetneqq \Omega_{E,F}.$$

Therefore, for any J in $\mathscr{F}_k^*$ and any scale m, there are at most two intervals J' from $\mathscr{F}_k$ contained in J with $|J'| = 2^m$. Summing over all possible scales, we obtain a bound of at most four times the length of J. We conclude that

$$\sum_{J\in\mathscr{F}_k} |J| = \sum_{J\in\mathscr{F}_k^*} \sum_{\substack{J'\in\mathscr{F}_k \\ J'\subsetneqq J}} |J'| \le \sum_{J\in\mathscr{F}_k^*} 4|J| \le 4|\Omega_{E,F}|,$$

since elements of $\mathscr{F}_k^*$ are disjoint and contained in $\Omega_{E,F}$. Inserting this estimate in (6.2.23), we obtain the required bound

$$C'_M \frac{|F|}{|E|} |\Omega_{E,F}| \le C''_M |F| = C''_M \min(|E|,|F|)$$

for the expression on the right in (6.2.20). This concludes the proof of (6.2.18).

6.2.3 The Proof of Estimate (6.2.19)

For fixed y, η, λ we define a partial order in the set of tiles in $\mathbf{D}_{y,\eta,\lambda}$ just as in Definition 6.1.3. All properties of dyadic tiles obtained in the previous section also hold for the tiles in $\mathbf{D}_{y,\eta,\lambda}$. Throughout this section, $\mathbf{P}$ is a finite subset of $\mathbf{D}_{y,\eta,\lambda}$.

To simplify notation, in the sequel we set

$$\mathbf{P}_{E,F} = \{s \in \mathbf{P} : I_s \nsubseteq \Omega_{E,F}\}.$$

Setting $N^{-1}[A] = \{x : N(x) \in A\}$ for a set $A \subseteq \mathbf{R}$, we note that (6.2.19) is a consequence of

$$\sum_{s\in\mathbf{P}_{E,F}} \left|\langle \chi_F, \varphi_s\rangle\langle \chi_{E'\cap N^{-1}[\omega_{s(2)}]}, \varphi_s\rangle\right| \le C \min(|E|,|F|)\left(1 + \left|\log\frac{|E|}{|F|}\right|\right). \tag{6.2.24}$$

The following lemma is the main ingredient of the proof and is proved in the next section.

Lemma 6.2.2. *There is a constant C such that for all measurable sets E and F of finite measure we have*

$$\mathscr{E}(\chi_F; \mathbf{P}_{E,F}) \le C|F|^{-\frac{1}{2}} \min\left(\frac{|F|}{|E|}, 1\right). \tag{6.2.25}$$

Assuming Lemma 6.2.2, we argue as follows to prove (6.2.19). Given the finite set of tiles $\mathbf{P}_{E,F}$, we write it as the union

$$\mathbf{P}_{E,F} = \bigcup_{j=-\infty}^{n_0} \mathbf{P}_j,$$

where the sets $\mathbf{P}_j$ satisfy properties (1)–(5) of page 431.

Given the sequence of sets $\mathbf{P}_j$, we use properties (1), (2), (5) on page 431, the observation that the mass is always bounded by $|E'|^{-1} \le 2|E|^{-1}$, and Lemmas 6.2.2 and 6.1.10 to obtain the following bound for the expression on the left in (6.2.19):

$$
\begin{aligned}
&\sum_{s\in\mathbf{P}_{E,F}} \left|\langle \chi_F \,|\, \varphi_s\rangle\right| \left|\langle \chi_{E'\cap N^{-1}[\omega_{s(2)}]}, \varphi_s\rangle\right| \\
&\quad = \sum_{j\in\mathbf{Z}} \sum_{s\in\mathbf{P}_j} \left|\langle \chi_F \,|\, \varphi_s\rangle\right| \left|\langle \chi_{E'\cap N^{-1}[\omega_{s(2)}]}, \varphi_s\rangle\right| \\
&\quad \le \sum_{j\in\mathbf{Z}} \sum_{k} \sum_{s\in\mathbf{T}_{jk}} \left|\langle \chi_F \,|\, \varphi_s\rangle\right| \left|\langle \chi_{E'\cap N^{-1}[\omega_{s(2)}]}, \varphi_s\rangle\right| \\
&\quad \le C_3 \sum_j \sum_k |I_{\mathrm{top}(\mathbf{T}_{jk})}| \, \mathscr{E}(\chi_F;\mathbf{T}_{jk}) \, \mathscr{M}(E',\mathbf{T}_{jk}) \, |E'| \, |F|^{\frac12} \\
&\quad \le C_3 \sum_{j\in\mathbf{Z}} \sum_k |I_{\mathrm{top}(\mathbf{T}_{jk})}| \min\Big(2^{j+1}, C\frac{|F|^{\frac12}}{|E|}, C|F|^{-\frac12}\Big) \min(|E'|^{-1}, 2^{2j+2}) \, |E| \, |F|^{\frac12} \\
&\quad \le C_4 \sum_{j\in\mathbf{Z}} 2^{-2j} \min\big(2^j, |F|^{\frac12}|E|^{-1}, |F|^{-\frac12}\big) \min(|E|^{-1}, 2^{2j}) \, |E| \, |F|^{\frac12} \\
&\quad \le C_5 \sum_{j\in\mathbf{Z}} \min\left(2^j|E|^{\frac12}, \min\left(\frac{|F|}{|E|}, \frac{|E|}{|F|}\right)^{\frac12}\right) \min\big((2^j|E|^{\frac12})^{-2}, 1\big) \, |E|^{\frac12} \, |F|^{\frac12} \\
&\quad \le C_6 \sum_{j\in\mathbf{Z}} \min\left(2^j, \min\left(\frac{|F|}{|E|}, \frac{|E|}{|F|}\right)^{\frac12}\right) \min(2^{-2j}, 1) \, |E|^{\frac12} \, |F|^{\frac12} \\
&\quad \le C_7 \min(|E|, |F|) \left(1 + \left|\log \frac{|E|}{|F|}\right|\right).
\end{aligned}
$$

The last estimate follows by a simple calculation considering the three cases $1 < 2^j$, $\min\left(\frac{|F|}{|E|}, \frac{|E|}{|F|}\right)^{\frac12} \le 2^j \le 1$, and $2^j < \min\left(\frac{|F|}{|E|}, \frac{|E|}{|F|}\right)^{\frac12}$.

6.2.4 The Proof of Lemma 6.2.2

It remains to prove Lemma 6.2.2.

Fix a 2-tree $\mathbf{T}$ contained in $\mathbf{P}_{E,F}$ and let $t = \mathrm{top}(\mathbf{T})$ denote its top. We show that

$$
\frac{1}{|I_t|} \sum_{s\in\mathbf{T}} \left|\langle \chi_F \,|\, \varphi_s\rangle\right|^2 \le C \min\left(\frac{|F|}{|E|}, 1\right)^2 \tag{6.2.26}
$$

for some constant C independent of F, E, and $\mathbf{T}$. Then (6.2.25) follows from (6.2.26) by taking the supremum over all 2-trees $\mathbf{T}$ contained in $\mathbf{P}_{E,F}$.

We decompose the function χ_F as follows:

$$
\chi_F = \chi_{F\cap 3I_t} + \chi_{F\cap(3I_t)^c}.
$$

We begin by observing that for s in $\mathbf{P}_{E,F}$ we have

$$\begin{aligned}\left|\langle\chi_{F\cap(3I_t)^c}\,|\,\varphi_s\rangle\right| &\le \frac{C_M|I_s|^{\frac{1}{2}}\inf\limits_{I_s} M(\chi_F)}{\left(1+\dfrac{\operatorname{dist}((3I_t)^c,c(I_s))}{|I_s|}\right)^M}\\ &\le 8C_M|I_s|^{\frac{1}{2}}\min\left(\frac{|F|}{|E|},1\right)\left(\frac{|I_s|}{|I_t|}\right)^M,\end{aligned}$$

since I_s meets the complement of $\Omega_{E,F}$ for every $s\in\mathbf{P}_{E,F}$. Square this inequality and sum over all s in $\mathbf{T}$ to obtain

$$\sum_{s\in\mathbf{T}}|\langle\chi_{F\cap(3I_t)^c}\,|\,\varphi_s\rangle|^2\le C\,|I_t|\,\min\left(\frac{|F|}{|E|},1\right)^2,$$

using Exercise 6.2.1.

We now turn to the corresponding estimate for the function $\chi_{F\cap 3I_t}$. At this point it is convenient to distinguish the simple case $|F|>|E|$ from the difficult case $|F|\le|E|$. In the first case the set $\Omega_{E,F}$ is empty and Exercise 6.1.6(b) yields

$$\begin{aligned}\sum_{s\in\mathbf{T}}\left|\langle\chi_{F\cap 3I_t}\,|\,\varphi_s\rangle\right|^2 &\le C\left\|\chi_{F\cap 3I_t}\right\|_{L^2}^2\\ &\le C|I_t|\\ &= C\,|I_t|\,\min\left(\frac{|F|}{|E|},1\right)^2,\end{aligned}$$

since $|F|>|E|$.

We may therefore concentrate on the case $|F|\le|E|$. In proving (6.2.25) we may assume that there exists a point $x_0\in I_t$ such that

$$M(\chi_F)(x_0)\le 8\,\frac{|F|}{|E|};$$

otherwise there is nothing to prove.

We write the set $\Omega_{E,F}=\left\{M(\chi_F)>8\frac{|F|}{|E|}\right\}$ as a disjoint union of dyadic intervals J'_ℓ such that the dyadic parent $\widetilde{J'_\ell}$ of J'_ℓ is not contained in $\Omega_{E,F}$ and therefore

$$|F\cap J'_\ell|\le|F\cap\widetilde{J'_\ell}|\le 16\,\frac{|F|}{|E|}\,|J'_\ell|.$$

Now some of these dyadic intervals may have size larger than or equal to $|I_t|$. Let J'_ℓ be such an interval. Then we split J'_ℓ into $\frac{|J'_\ell|}{|I_t|}$ intervals $J'_{\ell,m}$ each of size exactly $|I_t|$. Since there is an $x_0\in I_t$ with

$$M(\chi_F)(x_0)\le 8\,\frac{|F|}{|E|},$$

if K is the smallest interval that contains x_0 and $J'_{\ell,m}$, then

$$\frac{1}{|K|}\int_K \chi_F\,dx \le 8\frac{|F|}{|E|} \implies |F\cap J'_{\ell,m}| \le 8\frac{|F|}{|E|}|I_t|\frac{|K|}{|I_t|}.$$

We conclude that

$$|F\cap J'_{\ell,m}| \le c\frac{|F|}{|E|}|I_t|\left(1+\frac{\operatorname{dist}(I_t,J'_{\ell,m})}{|I_t|}\right). \tag{6.2.27}$$

We now have a new collection of dyadic intervals $\{J_k\}_k$ contained in $\Omega_{E,F}$ consisting of all the previous J'_ℓ when $|J'_\ell|<|I_t|$ and the $J'_{\ell,m}$'s when $|J'_\ell|\ge|I_t|$. In view of the construction we have

$$|F\cap J_k| \le \begin{cases} 2c\dfrac{|F|}{|E|}|J_k| & \text{when } |J_k|<|I_t|,\\[2ex] 2c\dfrac{|F|}{|E|}|J_k|\left(1+\dfrac{\operatorname{dist}(I_t,J_k)}{|I_t|}\right) & \text{when } |J_k|=|I_t|, \end{cases} \tag{6.2.28}$$

for all k. We now define the "bad functions"

$$b_k(x) = \left(e^{-2\pi i c(\omega_t)x}\chi_{F\cap 3I_t}(x) - \frac{1}{|J_k|}\int_{J_k} e^{-2\pi i c(\omega_t)y}\chi_{F\cap 3I_t}(y)\,dy\right)\chi_{J_k}(x),$$

which are supported in J_k, have mean value zero, and satisfy

$$\|b_k\|_{L^1} \le 2c\frac{|F|}{|E|}|J_k|\left(1+\frac{\operatorname{dist}(I_t,J_k)}{|I_t|}\right).$$

We also set

$$g(x) = e^{-2\pi i c(\omega_t)x}\chi_{F\cap 3I_t}(x) - \sum_k b_k(x),$$

the "good function" of this Calderón–Zygmund-type decomposition. We have therefore decomposed the function $\chi_{F\cap 3I_t}$ as follows:

$$\chi_{F\cap 3I_t}(x) = g(x)e^{2\pi i c(\omega_t)x} + \sum_k b_k(x)e^{2\pi i c(\omega_t)x}. \tag{6.2.29}$$

We show that $\|g\|_{L^\infty} \le C\frac{|F|}{|E|}$. Indeed, for x in J_k, we have

$$g(x) = \frac{1}{|J_k|}\int_{J_k} e^{-2\pi i c(\omega_t)y}\chi_{F\cap 3I_t}(y)\,dy,$$

which implies

$$|g(x)| \le \frac{|F \cap 3I_t \cap J_k|}{|J_k|} \le \begin{cases} \dfrac{|F \cap J_k|}{|J_k|} & \text{when } |J_k| < |I_t|, \\[2ex] \dfrac{|F \cap 3I_t|}{|I_t|} & \text{when } |J_k| = |I_t|, \end{cases}$$

and both of the preceding are at most a multiple of $\frac{|F|}{|E|}$; the latter is because there is an $x_0 \in I_t$ with $M(\chi_F)(x_0) \le 8\frac{|F|}{|E|}$. Also, for $x \in (\bigcup_k J_k)^c = (\Omega_{E,F})^c$ we have

$$|g(x)| = \chi_{F \cap 3I_t}(x) \le M(\chi_F)(x) \le 8\frac{|F|}{|E|}.$$

We conclude that $\|g\|_{L^\infty} \le C\frac{|F|}{|E|}$. Moreover,

$$\|g\|_{L^1} \le \sum_k \int_{J_k} \frac{|F \cap 3I_t \cap J_k|}{|J_k|}\, dx + \|\chi_{F \cap 3I_t}\|_{L^1} \le C|F \cap 3I_t| \le C\frac{|F|}{|E|}|I_t|,$$

since the J_k are disjoint. It follows that

$$\|g\|_{L^2} \le C\Big(\frac{|F|}{|E|}\Big)^{\frac{1}{2}}\Big(\frac{|F|}{|E|}\Big)^{\frac{1}{2}}|I_t|^{\frac{1}{2}} = C\frac{|F|}{|E|}|I_t|^{\frac{1}{2}}.$$

Using Exercise 6.1.6, we have

$$\sum_{s \in \mathbf{T}} \big|\big\langle g e^{2\pi i c(\omega_t)(\cdot)} \,\big|\, \varphi_s \big\rangle\big|^2 \le C\|g\|_{L^2}^2,$$

from which we obtain the required conclusion for the first function in the decomposition (6.2.29).

Next we turn to the corresponding estimate for the second function,

$$\sum_k b_k e^{2\pi i c(\omega_t)(\cdot)},$$

in the decomposition (6.2.29), which requires some further analysis. We have the following two estimates for all s and k:

$$\big|\big\langle b_k e^{2\pi i c(\omega_t)(\cdot)} \,\big|\, \varphi_s \big\rangle\big| \le \frac{C_M |F|\,|E|^{-1}\,|J_k|^2\,|I_s|^{-\frac{3}{2}}}{(1 + \frac{\operatorname{dist}(J_k, I_s)}{|I_s|})^M}, \tag{6.2.30}$$

$$\big|\big\langle b_k e^{2\pi i c(\omega_t)(\cdot)} \,\big|\, \varphi_s \big\rangle\big| \le \frac{C_M |F|\,|E|^{-1}\,|I_s|^{\frac{1}{2}}}{(1 + \frac{\operatorname{dist}(J_k, I_s)}{|I_s|})^M}, \tag{6.2.31}$$

for all $M > 0$, where C_M depends only on M.

To prove (6.2.30) we use the mean value theorem together with the fact that b_k has vanishing integral to write for some ξ_y,

$$
\begin{aligned}
&\big|\big\langle b_k\, e^{2\pi i c(\omega_t)(\cdot)} \,|\, \varphi_s\big\rangle\big| \\
&= \Big|\int_{J_k} b_k(y) e^{2\pi i c(\omega_t) y}\overline{\varphi_s(y)}\, dy\Big| \\
&= \Big|\int_{J_k} b_k(y)\big(e^{2\pi i c(\omega_t) y}\overline{\varphi_s(y)} - e^{2\pi i c(\omega_t) c(J_k)}\overline{\varphi_s(c(J_k))}\, dy\Big| \\
&\le |J_k| \int_{J_k} |b_k(y)| \Big[2\pi \tfrac{|c(\omega_s)-c(\omega_t)|}{|I_s|^{\frac12}} \big|\varphi\big(\tfrac{\xi_y - c(I_s)}{|I_s|}\big)\big| + |I_s|^{-\frac32}\big|\varphi'\big(\tfrac{\xi_y - c(I_s)}{|I_s|}\big)\big|\Big]\, dy \\
&\le \big\|b_k\big\|_{L^1} |J_k| \sup_{\xi\in J_k} \frac{C_M |I_s|^{-\frac32}}{(1+\frac{|\xi - c(I_s)|}{|I_s|})^{M+1}} \\
&\le C_M \frac{|F|}{|E|} |J_k| \Big(1+\frac{\operatorname{dist}(J_k, I_t)}{|I_t|}\Big) \frac{|J_k|\,|I_s|^{-\frac32}}{(1+\frac{\operatorname{dist}(J_k,I_s)}{|I_s|})^{M+1}} \\
&\le \frac{C_M |F|\,|E|^{-1}\,|J_k|^2\,|I_s|^{-\frac32}}{(1+\frac{\operatorname{dist}(J_k,I_s)}{|I_s|})^{M}},
\end{aligned}
$$

where we used the fact that

$$1+\frac{\operatorname{dist}(J_k, I_t)}{|I_t|} \le 1+\frac{\operatorname{dist}(J_k, I_s)}{|I_s|}.$$

To prove estimate (6.2.31) we note that

$$\big|\big\langle b_k e^{2\pi i c(\omega_t)(\cdot)} \,|\, \varphi_s\big\rangle\big| \le \frac{C_M |I_s|^{\frac12} \inf_{I_s} M(b_k)}{(1+\frac{\operatorname{dist}(J_k,I_s)}{|I_s|})^{M}}$$

and that

$$M(b_k) \le M(\chi_F) + \frac{|F\cap 3I_t \cap J_k|}{|J_k|} M(\chi_{J_k}),$$

and since $I_s \not\subseteq \Omega_{E,F}$, we have $\inf_{I_s} M(\chi_F) \le 8\frac{|F|}{|E|}$, while the second term in the sum was observed earlier to be at most $C\frac{|F|}{|E|}$.

Finally, we have the estimate

$$\big|\big\langle b_k e^{2\pi i c(\omega_t)(\cdot)} \,|\, \varphi_s\big\rangle\big| \le \frac{C_M |F|\,|E|^{-1}\,|J_k|\,|I_s|^{-\frac12}}{(1+\frac{\operatorname{dist}(J_k,I_s)}{|I_s|})^{M}}, \qquad (6.2.32)$$

which follows by taking the geometric mean of (6.2.30) and (6.2.31).

Now for a fixed $s \in \mathbf{P}_{E,F}$ we may have either $J_k \subsetneqq I_s$ or $J_k \cap I_s = \emptyset$ (since I_s is not contained in $\Omega_{E,F}$). Therefore, for fixed $s \in \mathbf{P}_{E,F}$, there are only three possibilities for J_k:

(a) $J_k \subsetneqq 3I_s$;

(b) $J_k \cap 3I_s = \emptyset$;

(c) $J_k \cap I_s = \emptyset$, $J_k \cap 3I_s \neq \emptyset$, and $J_k \nsubseteqq 3I_s$.

Observe that case (c) is equivalent to the following statement:

(c) $J_k \cap I_s = \emptyset$, $\operatorname{dist}(J_k, I_s) = 0$, and $|J_k| \geq 2|I_s|$.

Note that in case (c), for each I_s there exists exactly one $J_k = J_{k(s)}$ with the previous properties; but for a given J_k there may be a sequence of I_s's that lie on the left of J_k such that $|J_k| \geq 2|I_s|$ and $\operatorname{dist}(J_k, I_s) = 0$ and another sequence with similar properties on the right of J_k. The I_s's that lie on either side of J_k must be nested, and their lengths must add up to $|I_{s_k}^L| + |I_{s_k}^R|$, where $I_{s_k}^L$ is the largest one among them on the left of J_k and $I_{s_k}^R$ is the largest one among them on the right of J_k. Using (6.2.31), we obtain

$$\begin{aligned}
\sum_{s\in\mathbf{T}} \Big| \sum_{\substack{k:\, J_k\cap I_s=\emptyset \\ \operatorname{dist}(J_k,I_s)=0 \\ |J_k|\geq 2|I_s|}} \big\langle b_k e^{2\pi i c(\omega_t)(\cdot)} \,\big|\, \varphi_s \big\rangle \Big|^2 &= \sum_{s\in\mathbf{T}} \Big| \big\langle b_{k(s)} e^{2\pi i c(\omega_t)(\cdot)} \,\big|\, \varphi_s \big\rangle \Big|^2 \\
&\leq C \Big(\frac{|F|}{|E|}\Big)^2 \sum_{\substack{s\in\mathbf{T}:\, J_k\cap I_s=\emptyset \\ \operatorname{dist}(J_k,I_s)=0 \\ |J_k|\geq 2|I_s|}} |I_s| \\
&\leq C \Big(\frac{|F|}{|E|}\Big)^2 \sum_k \big(|I_{s_k}^L| + |I_{s_k}^R|\big).
\end{aligned}$$

But note that $I_{s_k}^L \subsetneqq 2J_k$, and since $I_{s_k}^L \cap J_k = \emptyset$, we must have $I_{s_k}^L \subsetneqq 2J_k \setminus J_k$ (and likewise for $I_{s_k}^R$). We define sets

$$\begin{aligned}
I_{s_k}^{L+} &= I_{s_k}^L + \frac{1}{2}|J_k|, \\
I_{s_k}^{R-} &= I_{s_k}^R - \frac{1}{2}|J_k|.
\end{aligned}$$

We have $I_{s_k}^{L+} \cup I_{s_k}^{R-} \subsetneqq J_k$, and hence the sets $I_{s_k}^{L+}$ are pairwise disjoint for different k, and the same is true for the $I_{s_k}^{R-}$. Moreover, since $\frac{1}{2}|J_k| \leq \frac{1}{2}|I_t|$ for all k, all the shifted sets $I_{s_k}^{L+}$, $I_{s_k}^{R-}$ are contained in $3I_t$. We conclude that

$$\begin{aligned}\sum_k |I_{s_k}^L| + \sum_k |I_{s_k}^R| &= \sum_k \left(|I_{s_k}^{L+}| + |I_{s_k}^{R-}|\right) \\ &\le \Big|\bigcup_k I_{s_k}^{L+}\Big| + \Big|\bigcup_k I_{s_k}^{R-}\Big| \\ &\le 2|3I_t|,\end{aligned}$$

which combined with the previously obtained estimate yields the required result in case (c).

We now consider case (a). Using (6.2.30), we can write

$$\bigg(\sum_{s\in\mathbf{T}}\Big|\sum_{k: J_k \subseteq 3I_s} \langle b_k e^{2\pi i c(\omega_t)(\cdot)} \,|\, \varphi_s\rangle\Big|^2\bigg)^{\frac{1}{2}} \le C_M \frac{|F|}{|E|} \bigg(\sum_{s\in T}\Big|\sum_{k: J_k \subseteq 3I_s} |J_k|^{\frac{1}{2}} \frac{|J_k|^{\frac{3}{2}}}{|I_s|^{\frac{3}{2}}}\Big|^2\bigg)^{\frac{1}{2}},$$

and we control the second expression by

$$\begin{aligned}&C_M \frac{|F|}{|E|}\bigg\{\sum_{s\in\mathbf{T}}\Big(\sum_{k: J_k \subseteq 3I_s} |J_k|\Big)\Big(\sum_{k: J_k \subseteq 3I_s} \frac{|J_k|^3}{|I_s|^3}\Big)\bigg\}^{\frac{1}{2}} \\ &\qquad\qquad \le C_M \frac{|F|}{|E|}\bigg\{\sum_{k: J_k \subseteq 3I_t} |J_k|^3 \sum_{\substack{s\in\mathbf{T} \\ J_k \subseteq 3I_s}} \frac{1}{|I_s|^2}\bigg\}^{\frac{1}{2}},\end{aligned}$$

having used the Cauchy–Schwarz inequality and the fact that the dyadic intervals J_k are disjoint. We note that the last sum is equal to at most $C|J_k|^{-2}$, since for every dyadic interval J_k there exist at most three dyadic intervals of a given length whose triples contain it. The required estimate $C|F|\,|E|^{-1}\,|I_t|^{\frac{1}{2}}$ now follows in case (a).

Finally, we deal with case (b), which is the most difficult case. We split the set of k into two subsets, those for which $J_k \subseteq 3I_t$ and those for which $J_k \nsubseteq 3I_t$ (recall that $|J_k| \le |I_t|$). Whenever $J_k \nsubseteq 3I_t$, we have

$$\operatorname{dist}(J_k, I_s) \approx \operatorname{dist}(J_k, I_t).$$

In this case we use Minkowski's inequality and estimate (6.2.32) to deduce

$$\begin{aligned}&\bigg(\sum_{s\in\mathbf{T}}\Big|\sum_{k: J_k \nsubseteq 3I_t} \langle b_k e^{2\pi i c(\omega_t)(\cdot)} \,|\, \varphi_s\rangle\Big|^2\bigg)^{\frac{1}{2}} \\ &\qquad \le \sum_{k: J_k \nsubseteq 3I_t}\bigg(\sum_{s\in\mathbf{T}}\big|\langle b_k e^{2\pi i c(\omega_t)(\cdot)} \,|\, \varphi_s\rangle\big|^2\bigg)^{\frac{1}{2}} \\ &\qquad \le C_M \frac{|F|}{|E|} \sum_{k: J_k \nsubseteq 3I_t} |J_k| \bigg(\sum_{s\in\mathbf{T}} \frac{|I_s|^{2M-1}}{\operatorname{dist}(J_k, I_s)^{2M}}\bigg)^{\frac{1}{2}} \\ &\qquad \le C_M \frac{|F|}{|E|} \sum_{k: J_k \nsubseteq 3I_t} \frac{|J_k|}{\operatorname{dist}(J_k, I_t)^M} \bigg(\sum_{s\in\mathbf{T}} |I_s|^{2M-1}\bigg)^{\frac{1}{2}}\end{aligned}$$

$$\le C_M \frac{|F|}{|E|} |I_t|^{M-\frac{1}{2}} \sum_{k:\, J_k \subsetneqq 3I_t} \frac{|J_k|}{\operatorname{dist}(J_k, I_t)^M}$$

$$\le C_M \frac{|F|}{|E|} |I_t|^{M-\frac{1}{2}} \sum_{l=1}^{\infty} \sum_{\substack{k: \\ \operatorname{dist}(J_k,I_t)\approx 2^l|I_t|}} \frac{|J_k|}{(2^l|I_t|)^M},$$

where $\operatorname{dist}(J_k, I_t) \approx 2^l |I_t|$ means that

$$\operatorname{dist}(J_k, I_t) \in [2^l |I_t|, 2^{l+1}|I_t|].$$

But note that all the J_k with $\operatorname{dist}(J_k, I_t) \approx 2^l |I_t|$ are contained in $2^{l+2}I_t$, and since they are disjoint, we estimate the last sum by $C2^l|I_t|(2^l|I_t|)^{-M}$. The required estimate $C_M|F|\,|E|^{-1}|I_t|^{\frac{1}{2}}$ follows.

Next we consider the case $J_k \subseteq 3I_t$, $J_k \cap 3I_s = \emptyset$, and $|J_k| \le |I_s|$, in which we use estimate (6.2.30). We have

$$\Bigg(\sum_{s\in\mathbf{T}} \Bigg| \sum_{\substack{k:\, J_k \subseteq 3I_t \\ J_k \cap 3I_s = \emptyset \\ |J_k| \le |I_s|}} \langle b_k e^{2\pi i c(\omega_t)(\cdot)} \,|\, \varphi_s \rangle \Bigg|^2 \Bigg)^{\frac{1}{2}}$$

$$\le C_M \frac{|F|}{|E|} \Bigg(\sum_{s\in\mathbf{T}} \Bigg| \sum_{\substack{k:\, J_k \subseteq 3I_t \\ J_k \cap 3I_s = \emptyset \\ |J_k| \le |I_s|}} |J_k|^2 |I_s|^{-\frac{3}{2}} \frac{|I_s|^M}{\operatorname{dist}(J_k, I_s)^M} \Bigg|^2 \Bigg)^{\frac{1}{2}}$$

$$\le C_M \frac{|F|}{|E|} \Bigg\{ \sum_{s\in\mathbf{T}} \Bigg[\sum_{\substack{k:\, J_k \subseteq 3I_t \\ J_k \cap 3I_s = \emptyset \\ |J_k| \le |I_s|}} \frac{|J_k|^3}{|I_s|^2} \Big(\frac{|I_s|}{\operatorname{dist}(J_k, I_s)} \Big)^M \Bigg]$$

$$\times \Bigg[\sum_{\substack{k:\, J_k \subseteq 3I_t \\ J_k \cap 3I_s = \emptyset \\ |J_k| \le |I_s|}} \frac{|J_k|}{|I_s|} \Big(\frac{\operatorname{dist}(J_k, I_s)}{|I_s|} \Big)^{-M} \Bigg] \Bigg\}^{\frac{1}{2}}$$

$$\le C_M \frac{|F|}{|E|} \Bigg\{ \sum_{s\in\mathbf{T}} \Bigg[\sum_{\substack{k:\, J_k \subseteq 3I_t \\ J_k \cap 3I_s = \emptyset \\ |J_k| \le |I_s|}} \frac{|J_k|^3}{|I_s|^2} \Big(\frac{|I_s|}{\operatorname{dist}(J_k, I_s)} \Big)^M \Bigg]$$

$$\times \Bigg[\sum_{\substack{k:\, J_k \subseteq 3I_t \\ J_k \cap 3I_s = \emptyset \\ |J_k| \le |I_s|}} \int_{J_k} \Big(\frac{|x - c(I_s)|}{|I_s|} \Big)^{-M} \frac{dx}{|I_s|} \Bigg] \Bigg\}^{\frac{1}{2}}$$

$$\le C_M \frac{|F|}{|E|}\Bigg\{\sum_{s\in\mathbf{T}}\Bigg[\sum_{\substack{k:\,J_k\subseteq 3I_t\\ J_k\cap 3I_s=\emptyset\\ |J_k|\le|I_s|}}\frac{|J_k|^3}{|I_s|^2}\Big(\frac{|I_s|}{\operatorname{dist}(J_k,I_s)}\Big)^M\Bigg] \times\Bigg[\int_{(3I_s)^c}\Big(\frac{|x-c(I_s)|}{|I_s|}\Big)^{-M}\frac{dx}{|I_s|}\Bigg]\Bigg\}^{\frac{1}{2}}$$

$$\le C_M \frac{|F|}{|E|}\Bigg\{\sum_{s\in\mathbf{T}}\sum_{\substack{k:\,J_k\subseteq 3I_t\\ J_k\cap 3I_s=\emptyset\\ |J_k|\le|I_s|}}|J_k|^3|I_s|^{-2}\Big(\frac{|I_s|}{\operatorname{dist}(J_k,I_s)}\Big)^M\Bigg\}^{\frac{1}{2}}.$$

But since the last integral contributes at most a constant factor, we can estimate the last displayed expression by

$$C_M \frac{|F|}{|E|}\Bigg\{\sum_{\substack{k:\,J_k\subseteq 3I_t\\ J_k\cap 3I_s=\emptyset\\ |J_k|\le|I_s|}}|J_k|^3\sum_{m\ge\log|J_k|}2^{-2m}\sum_{\substack{s\in\mathbf{T}\\ |I_s|=2^m}}\Big(\frac{\operatorname{dist}(J_k,I_s)}{2^m}\Big)^{-M}\Bigg\}^{\frac{1}{2}}$$

$$\le C_M \frac{|F|}{|E|}\Bigg\{\sum_{\substack{k:\,J_k\subseteq 3I_t\\ J_k\cap 3I_s=\emptyset\\ |J_k|\le|I_s|}}|J_k|^3\sum_{m\ge\log|J_k|}2^{-2m}\Bigg\}^{\frac{1}{2}}$$

$$\le C_M \frac{|F|}{|E|}\Bigg\{\sum_{\substack{k:\,J_k\subseteq 3I_t\\ J_k\cap 3I_s=\emptyset\\ |J_k|\le|I_s|}}|J_k|^3|J_k|^{-2}\Bigg\}^{\frac{1}{2}}$$

$$\le C_M \frac{|F|}{|E|}|I_t|^{\frac{1}{2}}.$$

There is also the subcase of case (b) in which $|J_k|>|I_s|$. Here we have the two special subcases $I_s\cap 3J_k=\emptyset$ and $I_s\subseteq 3J_k$. We begin with the first of these special subcases, in which we use estimate (6.2.31). We have

$$\Bigg(\sum_{s\in\mathbf{T}}\Bigg|\sum_{\substack{k:\,J_k\subseteq 3I_t\\ J_k\cap 3I_s=\emptyset\\ |J_k|>|I_s|\\ I_s\cap 3J_k=\emptyset}}\langle b_k e^{2\pi i c(\omega_t)(\cdot)}\,|\,\varphi_s\rangle\Bigg|^2\Bigg)^{\frac{1}{2}}$$

$$\le C_M \frac{|F|}{|E|}\Bigg(\sum_{s\in T}\Bigg|\sum_{\substack{k:\,J_k\subseteq 3I_t\\ J_k\cap 3I_s=\emptyset\\ |J_k|>|I_s|\\ I_s\cap 3J_k=\emptyset}}|I_s|^{\frac{1}{2}}\frac{|I_s|^M}{\operatorname{dist}(J_k,I_s)^M}\Bigg|^2\Bigg)^{\frac{1}{2}}$$

$$\le C_M \frac{|F|}{|E|}\Bigg\{\sum_{s\in\mathbf{T}}\Bigg[\sum_{\substack{k:\,J_k\subseteq 3I_t\\ J_k\cap 3I_s=\emptyset\\ |J_k|>|I_s|\\ I_s\cap 3J_k=\emptyset}}\frac{|I_s|^2}{|J_k|}\frac{|I_s|^M}{\operatorname{dist}(J_k,I_s)^M}\Bigg]\Bigg[\sum_{\substack{k:\,J_k\subseteq 3I_t\\ J_k\cap 3I_s=\emptyset\\ |J_k|>|I_s|\\ I_s\cap 3J_k=\emptyset}}\frac{|J_k|}{|I_s|}\frac{|I_s|^M}{\operatorname{dist}(J_k,I_s)^M}\Bigg]\Bigg\}^{\frac{1}{2}}.$$

Since $I_s \cap 3J_k = \emptyset$, we have that

$$\operatorname{dist}(J_k, I_s) \approx |x - c(I_s)|$$

for every $x \in J_k$, and therefore the second term inside square brackets satisfies

$$\sum_{\substack{k:\, J_k \subseteq 3I_t \\ J_k \cap 3I_s = \emptyset \\ |J_k| > |I_s| \\ I_s \cap 3J_k = \emptyset}} \frac{|J_k|}{|I_s|} \frac{|I_s|^M}{\operatorname{dist}(J_k, I_s)^M} \le \sum_k \int_{J_k} \left(\frac{|x - c(I_s)|}{|I_s|} \right)^{-M} \frac{dx}{|I_s|} \le C_M .$$

Using this estimate, we obtain

$$\begin{aligned}
& C_M \frac{|F|}{|E|} \left\{ \sum_{s \in \mathbf{T}} \left[\sum_{\substack{k:\, J_k \subseteq 3I_t \\ J_k \cap 3I_s = \emptyset \\ |J_k| > |I_s| \\ I_s \cap 3J_k = \emptyset}} \frac{|I_s|^2}{|J_k|} \frac{|I_s|^M}{\operatorname{dist}(J_k, I_s)^M} \right] \left[\sum_{\substack{k:\, J_k \subseteq 3I_t \\ J_k \cap 3I_s = \emptyset \\ |J_k| > |I_s| \\ I_s \cap 3J_k = \emptyset}} \frac{|J_k|}{|I_s|} \frac{|I_s|^M}{\operatorname{dist}(J_k, I_s)^M} \right] \right\}^{\frac{1}{2}} \\
& \le C_M \frac{|F|}{|E|} \left\{ \sum_{s \in \mathbf{T}} \left[\sum_{\substack{k:\, J_k \subseteq 3I_t \\ J_k \cap 3I_s = \emptyset \\ |J_k| > |I_s| \\ I_s \cap 3J_k = \emptyset}} \frac{|I_s|^2}{|J_k|} \frac{|I_s|^M}{\operatorname{dist}(J_k, I_s)^M} \right] \right\}^{\frac{1}{2}} \\
& = C_M \frac{|F|}{|E|} \left\{ \sum_{k:\, J_k \subseteq 3I_t} \frac{1}{|J_k|} \sum_{\substack{s \in \mathbf{T} \\ J_k \cap 3I_s = \emptyset \\ |J_k| > |I_s| \\ I_s \cap 3J_k = \emptyset}} |I_s|^2 \frac{|I_s|^M}{\operatorname{dist}(J_k, I_s)^M} \right\}^{\frac{1}{2}} \\
& \le C_M \frac{|F|}{|E|} \left\{ \sum_{k:\, J_k \subseteq 3I_t} \frac{1}{|J_k|} \sum_{m=-\infty}^{\log_2 |J_k|} 2^{2m} \sum_{\substack{s \in \mathbf{T}:\, |I_s| = 2^m \\ J_k \cap 3I_s = \emptyset \\ |J_k| > |I_s| \\ I_s \cap 3J_k = \emptyset}} \frac{|I_s|^M}{\operatorname{dist}(J_k, I_s)^M} \right\}^{\frac{1}{2}} \\
& \le C_M \frac{|F|}{|E|} \left\{ \sum_{k:\, J_k \subseteq 3I_t} \frac{1}{|J_k|} \sum_{m=-\infty}^{\log_2 |J_k|} 2^{2m} \right\}^{\frac{1}{2}} \\
& \le C_M \frac{|F|}{|E|} \left\{ \sum_{k:\, J_k \subseteq 3I_t} \frac{1}{|J_k|} |J_k|^2 \right\}^{\frac{1}{2}} \\
& \le C_M \frac{|F|}{|E|} |I_t|^{\frac{1}{2}} .
\end{aligned}$$

Finally, there is the subcase of case (b) in which $|J_k| \ge |I_s|$ and $I_s \subseteq 3J_k$. Here again we use estimate (6.2.31). We have

$$\left\{\sum_{s\in\mathbf{T}}\left|\sum_{\substack{k:\,J_k\subseteq 3I_t\\ J_k\cap 3I_s=\emptyset\\ |J_k|>|I_s|\\ I_s\subseteq 3J_k}}\langle b_k e^{2\pi i c(\omega_t)(\cdot)}\,|\,\varphi_s\rangle\right|^2\right\}^{\frac{1}{2}}$$

$$\leq C_M\frac{|F|}{|E|}\left\{\sum_{s\in\mathbf{T}}|I_s|\left|\sum_{\substack{k:\,J_k\subseteq 3I_t\\ J_k\cap 3I_s=\emptyset\\ |J_k|>|I_s|\\ I_s\subseteq 3J_k}}\frac{|I_s|^M}{\operatorname{dist}(J_k,I_s)^M}\right|^2\right\}^{\frac{1}{2}}. \tag{6.2.33}$$

Let us make some observations. For a fixed s there exist at most finitely many J_k's contained in $3I_t$ with size at least $|I_s|$. Let $J_L^1(s)$ be the interval that lies to the left of I_s and is closest to I_s among all J_k that satisfy the conditions in the preceding sum. Then $|J_L^1(s)|>|I_s|$ and

$$\operatorname{dist}(J_L^1(s),I_s)\geq|I_s|.$$

Let $J_L^2(s)$ be the interval to the left of $J_L^1(s)$ that is closest to $J_L^1(s)$ and that satisfies the conditions of the sum. Since $3J_L^2(s)$ contains I_s, it follows that $|J_L^2(s)|>2|I_s|$ and

$$\operatorname{dist}(J_L^2(s),I_s)\geq 2|I_s|.$$

Continuing in this way, we can find a finite number of intervals $J_L^r(s)$ that lie to the left of I_s and inside $3I_t$, satisfy $|J_L^r(s)|>2^{r-1}|I_s|$ and $\operatorname{dist}(J_L^r(s),I_s)\geq 2^{r-1}|I_s|$, and whose triples contain I_s. Likewise we find a finite collection of intervals $J_R^1(s),J_R^2(s),\dots$ that lie to the right of I_s and satisfy similar conditions. Then, using the Cauchy–Schwarz inequality, we obtain

$$\left|\sum_{\substack{k:\,J_k\subseteq 3I_t\\ J_k\cap 3I_s=\emptyset\\ |J_k|>|I_s|\\ I_s\subseteq 3J_k}}\frac{|I_s|^M}{\operatorname{dist}(J_k,I_s)^M}\right|^2$$

$$\leq 2\left|\sum_{r=1}^{\infty}\frac{|I_s|^{\frac{M}{2}}}{\operatorname{dist}(J_L^r(s),I_s)^{\frac{M}{2}}}\frac{1}{2^{\frac{(r-1)M}{2}}}\right|^2+2\left|\sum_{r=1}^{\infty}\frac{|I_s|^{\frac{M}{2}}}{\operatorname{dist}(J_R^r(s),I_s)^{\frac{M}{2}}}\frac{1}{2^{\frac{(r-1)M}{2}}}\right|^2$$

$$\leq C_M\sum_{r=1}^{\infty}\frac{|I_s|^M}{\operatorname{dist}(J_L^r(s),I_s)^M}+C_M\sum_{r=1}^{\infty}\frac{|I_s|^M}{\operatorname{dist}(J_R^r(s),I_s)^M}$$

$$\leq C_M\sum_{\substack{k:\,J_k\subseteq 3I_t\\ J_k\cap 3I_s=\emptyset\\ |J_k|>|I_s|\\ I_s\subseteq 3J_k}}\frac{|I_s|^M}{\operatorname{dist}(J_k,I_s)^M}.$$

We use this estimate to control the expression on the left in (6.2.33) by

$$C_M \frac{|F|}{|E|}\bigg\{\sum_{s\in\mathbf{T}} |I_s| \sum_{\substack{k:\, J_k\subseteq 3I_t\\ J_k\cap 3I_s=\emptyset\\ |J_k|>|I_s|\\ I_s\subseteq 3J_k}} \frac{|I_s|^M}{\operatorname{dist}(J_k,I_s)^M}\bigg\}^{\frac{1}{2}}$$

$$\leq C_M \frac{|F|}{|E|}\bigg\{\sum_{k:J_k\subseteq 3I_t} |J_k| \sum_{m=0}^{\infty} 2^{-m} \sum_{\substack{s:\, I_s\subseteq 3J_k\\ J_k\cap 3I_s=\emptyset\\ |I_s|=2^{-m}|J_k|}} \frac{|I_s|^M}{\operatorname{dist}(J_k,I_s)^M}\bigg\}^{\frac{1}{2}}.$$

Since the last sum is at most a constant, it follows that the term on the left in (6.2.33) also satisfies the estimate $C_M \frac{|F|}{|E|} |I_t|^{\frac{1}{2}}$. This concludes the proof of Lemma 6.2.2.

Exercises

6.2.1. Let $\mathbf{T}$ be a 2-tree with top I_t and let $M > 1$ and L be such that $2^L < |I_t|$. Show that there exists a constant $C_M > 0$ such that

$$\sum_{s\in\mathbf{T}} |I_s|^M \leq C_M |I_t|^M ,$$

$$\sum_{\substack{s\in\mathbf{T}\\ |I_s|\geq 2^L}} |I_s|^{-M} \leq C_M \frac{|I_t|}{(2^L)^{M+1}},$$

$$\sum_{\substack{s\in\mathbf{T}\\ |I_s|\leq 2^L}} |I_s|^M \leq C_M |I_t| (2^L)^{M-1} .$$

[*Hint:* Group the s that appear in each sum in families $\mathcal{G}_m$ such that $|I_s| = 2^{-m}|I_t|$ for each $s \in \mathcal{G}_m$.]

6.2.2. Show that the operator

$$g \mapsto \sup_{-\infty<a<b<\infty} \big|(\widehat{g}\chi_{[a,b]})^{\vee}\big|$$

defined on the line is bounded from $L^p(\mathbf{R})$ to itself for all $1 < p < \infty$.

6.2.3. On $\mathbf{R}^n$ fix a unit vector b and consider the maximal operator

$$T_b(g)(x) = \sup_{N>0} \bigg| \int_{|b\cdot\xi|\leq N} \widehat{g}(\xi) e^{2\pi i x\cdot\xi}\, d\xi \bigg| .$$

Show that T_b maps $L^p(\mathbf{R}^n)$ to $L^p(\mathbf{R}^n)$ for all $1 < p < \infty$.
[*Hint:* Apply a rotation.]

6.2.4. Define the *directional Carleson operators* by

$$\mathscr{C}^{\theta}(f)(x) = \sup_{a\in\mathbf{R}} \left| \lim_{\varepsilon\to 0+} \int_{\varepsilon<|t|<\varepsilon^{-1}} e^{2\pi i a t} f(x-t\theta)\, \frac{dt}{t} \right| ,$$

for functions f on $\mathbf{R}^n$. Here θ is a vector in $\mathbf{S}^{n-1}$.
(a) Show that $\mathscr{C}^{\theta}$ is bounded on $L^p(\mathbf{R}^n)$ for all $1 < p < \infty$.
(b) Let Ω be an odd integrable function on $\mathbf{S}^{n-1}$. Define an operator

$$\mathscr{C}^{\Omega}(f)(x) = \sup_{\xi\in\mathbf{R}^n} \left| \lim_{\varepsilon\to 0+} \int_{\varepsilon<|y|<\varepsilon^{-1}} e^{2\pi i \xi\cdot y} f(x-y)\, \frac{\Omega(\frac{y}{|y|})}{|y|^n}\, dy \right| .$$

Show that $\mathscr{C}^{\Omega}$ is bounded on $L^p(\mathbf{R}^n)$ for $1 < p < \infty$.
[*Hint:* Part (a): Reduce to the case $\theta = e_1 = (1,0,\dots,0)$ via a rotation and use Theorem 6.2.1(b). Part (b): Use the method of rotations and part (a).]

6.3 The Maximal Carleson Operator and Weighted Estimates

Recall the one-sided Carleson operator $\mathscr{C}_1$ defined in the previous section:

$$\mathscr{C}_1(f)(x) = \sup_{N>0} \left| \int_{-\infty}^{N} \widehat{f}(\xi) e^{2\pi i x\xi}\, d\xi \right| .$$

Recall also the modulation operator $M^a(g)(x) = g(x)e^{2\pi i a x}$. We begin by observing that the following identity is valid:

$$\big(\widehat{f}\chi_{(-\infty,b]}\big)^{\vee} = M^b \frac{I-iH}{2} M^{-b}(f) = \frac{1}{2} f - \frac{i}{2} M^b H M^{-b}(f) , \tag{6.3.1}$$

where H is the Hilbert transform. It follows from (6.3.1) that

$$\mathscr{C}_1(f) \le \frac{1}{2}|f| + \frac{1}{2}\sup_{\xi\in\mathbf{R}} |H(M^{\xi}(f))|$$

and that

$$\sup_{\xi\in\mathbf{R}} |H(M^{\xi}(f))| \le |f| + 2\mathscr{C}_1(f) .$$

We conclude that the L^p boundedness of the sublinear operator $f \mapsto \mathscr{C}_1(f)$ is equivalent to that of the sublinear operator

$$f \mapsto \sup_{\xi\in\mathbf{R}} |H(M^{\xi}(f))| .$$

Definition 6.3.1. The *maximal Carleson operator* is defined by

$$\begin{aligned}\mathscr{C}_*(f)(x) &= \sup_{\varepsilon>0}\sup_{\xi\in\mathbf{R}}\left|\int_{|x-y|>\varepsilon} f(y)e^{2\pi i\xi y}\frac{dy}{x-y}\right| \\ &= \sup_{\xi\in\mathbf{R}}\left|H^{(*)}(M^{\xi}(f))(x)\right|,\end{aligned} \tag{6.3.2}$$

where $H^{(*)}$ is the maximal Hilbert transform. Since $H^{(*)}$ are well defined on $\bigcup_{1\le p<\infty} L^p(\mathbf{R})$, then so are $\mathscr{C}(f)$ and $\mathscr{C}_*(f)$. Notice that the maximal Carleson operator controls the Carleson operator pointwise.

We begin with the following pointwise estimate, which reduces the boundedness of $\mathscr{C}_*$ to that of $\mathscr{C}$.

Lemma 6.3.2. *There is a positive constant $c>0$ such that for all functions f in $\bigcup_{1\le p<\infty} L^p(\mathbf{R})$ we have*

$$\mathscr{C}_*(f) \le cM(f) + M(\mathscr{C}(f)), \tag{6.3.3}$$

where M is the Hardy–Littlewood maximal function.

Proof. The proof of (6.3.3) is based on the classical inequality

$$H^{(*)}(g) \le cM(g) + M(H(g))$$

given in Theorem 5.3.4 in [156]. Applying this to the functions $M^{\xi}(f)$ and taking the supremum over $\xi\in\mathbf{R}$, we obtain

$$\mathscr{C}_*(f) \le cM(f) + \sup_{\xi\in\mathbf{R}} M\big(H(M^{\xi}(f))\big),$$

from which (6.3.3) easily follows by passing the supremum inside the maximal function. □

For every p in $(1,\infty)$ and for every $w\in A_p$, the operator $h\mapsto \big(\widehat{h}\chi_{[-N,N]}\big)^{\vee}$ has a unique bounded extension on $L^p(w)$ and thus $\mathscr{C}(f)$ is well defined on $L^p(w)$. Our next goal is to obtain the boundedness of the Carleson operator on weighted L^p spaces.

Theorem 6.3.3. *For every $p\in(1,\infty)$ and $w\in A_p$ there is a constant $C(p,[w]_{A_p})$ such that for all $f\in L^p(\mathbf{R})$ we have*

$$\big\|\mathscr{C}(f)\big\|_{L^p(w)} \le C(p,[w]_{A_p})\big\|f\big\|_{L^p(w)}, \tag{6.3.4}$$

$$\big\|\mathscr{C}_*(f)\big\|_{L^p(w)} \le C(p,[w]_{A_p})\big\|f\big\|_{L^p(w)}. \tag{6.3.5}$$

Proof. Fix a $1 < p < \infty$ and pick an $r \in (1,p)$ such that $w \in A_r$. It is convenient to work with a variant of the Hardy–Littlewood maximal operator. For $0 < r < \infty$ define

$$M_r(f) = M(|f|^r)^{\frac{1}{r}}$$

for f such that $|f|^r$ is locally integrable over the real line. Note that $M(f) \le M_r(f)$ for any $r \in (1,\infty)$. We show that for all $f \in L^p(w)$ we have the estimate

$$\int_{\mathbf{R}} \mathscr{C}(f)(x)^p w(x)\,dx \le C_p([w]_{A_p}) \int_{\mathbf{R}} M_r(f)(x)^p w(x)\,dx. \tag{6.3.6}$$

Then the boundedness of $\mathscr{C}$ on $L^p(w)$ is a consequence of the boundedness of the Hardy–Littlewood maximal operator on $L^{\frac{p}{r}}(w)$.

If we show that for any $w \in A_p$ there is a constant $C_p([w]_{A_p})$ such that

$$\int_{\mathbf{R}} M(\mathscr{C}(f))^p\, w\,dx \le C_p([w]_{A_p}) \int_{\mathbf{R}} M_r(f)^p\, w\,dx, \tag{6.3.7}$$

then the trivial fact $\mathscr{C}(f) \le M(\mathscr{C}(f))$, inserted in (6.3.7), yields (6.3.6).

Estimate (6.3.7) will be a consequence of the following two important observations:

$$M^\#(\mathscr{C}(f)) \le C_r M_r(f) \qquad \text{a.e.} \tag{6.3.8}$$

and

$$\big\|M(\mathscr{C}(f))\big\|_{L^p(w)} \le c_p([w]_{A_p})\big\|M^\#(\mathscr{C}(f))\big\|_{L^p(w)}, \tag{6.3.9}$$

where $c_p([w]_{A_p})$ depends on $[w]_{A_p}$ and C_r depends only on r.

We begin with estimate (6.3.8), which was obtained in Theorem 3.4.9 for singular integral operators. Here this estimate is extended to maximally modulated singular integrals. To prove (6.3.8) we use the result in Proposition 3.4.2 (2). We fix $x \in \mathbf{R}$ and we pick an interval I that contains x. We write $f = f_0 + f_\infty$, where $f_0 = f\chi_{3I}$ and $f_\infty = f\chi_{(3I)^c}$. We set $a_I = \mathscr{C}(f_\infty)(c_I)$, where c_I is the center of I. Then we have

$$\begin{aligned}\frac{1}{|I|}\int_I |\mathscr{C}(f)(y) - a_I|\,dx &\le \frac{1}{|I|}\int_I \sup_{\xi \in \mathbf{R}} \big|H(M^\xi(f))(y) - H(M^\xi(f_\infty))(c_I)\big|\,dy \\ &\le B_1 + B_2,\end{aligned}$$

where

$$\begin{aligned} B_1 &= \frac{1}{|I|}\int_I \sup_{\xi\in\mathbf{R}} \big|H(M^\xi(f_0))(y)\big|\,dy, \\ B_2 &= \frac{1}{|I|}\int_I \sup_{\xi\in\mathbf{R}} \big|H(M^\xi(f_\infty))(y) - H(M^\xi(f_\infty))(c_I)\big|\,dy. \end{aligned}$$

But

$$B_1 \le \frac{1}{|I|}\int_I \mathscr{C}(f_0)(y)\,dy$$

$$
\begin{aligned}
&\leq \frac{1}{|I|}\big\|\mathscr{C}(f_0)\big\|_{L^r}\big\|\chi_I\big\|_{L^{r'}} \\
&\leq \frac{\|\mathscr{C}\|_{L^r\to L^r}}{|I|}\big\|f_0\big\|_{L^r}|I|^{\frac{1}{r'}} \\
&\leq C_r M_r(f)(x)\,,
\end{aligned}
$$

where we used the boundedness of the Carleson operator $\mathscr{C}$ from L^r to L^r.

We turn to the corresponding estimate for B_2. We have

$$
\begin{aligned}
B_2 &\leq \frac{1}{|I|}\int_I\int_{\mathbf{R}^n}|f_\infty(z)|\left|\frac{1}{y-z}-\frac{1}{c_I-z}\right|dz\,dy \\
&= \frac{1}{|I|}\int_I\int_{(3I)^c}|f(z)|\left|\frac{y-c_I}{(y-z)(c_I-z)}\right|dz\,dy \\
&\leq \int_I\left(\int_{(3I)^c}|f(z)|\frac{C}{(|c_I-z|+|I|)^2}\,dz\right)dy \\
&\leq \int_I\frac{C}{|I|}M(f)(x)\,dy \\
&\leq CM(f)(x) \\
&\leq CM_r(f)(x)\,.
\end{aligned}
$$

This completes the proof of estimate (6.3.8). We now focus attention to the proof of (6.3.9). We derive estimate (6.3.9) as a consequence of Theorem 3.4.5, provided we have that

$$
\big\|M_d(\mathscr{C}(f))\big\|_{L^r(w)} \leq \big\|M(\mathscr{C}(f))\big\|_{L^r(w)} < \infty\,. \tag{6.3.10}
$$

Unfortunately, the finiteness estimate (6.3.10) for general functions f in $L^p(w)$ cannot be easily deduced without a priori knowledge of the sought estimate (6.3.4) for $p=r$. However, we can show the validity of (6.3.10) for functions f with compact support and weights $w\in A_p$ that are bounded. This argument requires a few technicalities, which we now present. For a fixed constant B we introduce a truncated Carleson operator

$$
\mathscr{C}^B(f) = \sup_{|\xi|\leq B}|H(M^\xi(f))|\,.
$$

Next we work with a weight w in A_p that is bounded. In fact, we work with $w_k = \min(w,k)$, which satisfies

$$
[w_k]_{A_p} \leq c_p[w]_{A_p}
$$

for all $k\geq 1$ (see Exercise 7.1.8 in [156]). Finally, we take $f=h$ to be a smooth function with support contained in an interval $[-R,R]$. Then for $|\xi|\leq B$ we have

$$
|H(M^\xi(h))(x)| \leq 2R\big\|(M^\xi(h))'\big\|_{L^\infty}\chi_{|x|\leq 2R} + \frac{\|h\|_{L^1}}{|x|+R}\chi_{|x|>2R} \leq \frac{BC_hR}{|x|+R}\,,
$$

where C_h is a constant that depends on h. This implies that the last estimate also holds for $\mathscr{C}^B(h)$. Using the result of the calculation in Example 2.1.8 in [156], we now obtain

$$M(\mathscr{C}^B(h))(x) \le BC_h \frac{\log\left(1+\frac{|x|}{R}\right)}{1+\frac{|x|}{R}}.$$

It follows that $M(\mathscr{C}^B(h))$ lies in $L^r(w_k)$, since $r>1$ and $w_k \le k$. Therefore,

$$\big\|M(\mathscr{C}^B(f))\big\|_{L^r(w_k)} < \infty,$$

and thus (6.3.10) holds in this setting. Applying the previous argument to $\mathscr{C}^B(h)$ and the weight w_k [in lieu of $\mathscr{C}(f)$ and w], we obtain (6.3.7) and thus (6.3.4) for $M(\mathscr{C}^B(h))$ and the weight w_k. This establishes the estimate

$$\big\|\mathscr{C}^B(h)\big\|_{L^p(w_k)} \le C(p,[w]_{A_p})\big\|h\big\|_{L^p(w_k)} \tag{6.3.11}$$

for functions h that are smooth and compactly supported, where the constant $C(p,[w]_{A_p})$ that is independent of B and k. Letting $k\to\infty$ in (6.3.11) and applying Fatou's lemma, we obtain (6.3.4) for smooth functions h with compact support. From this we deduce the validity of (6.3.4) for general functions f in $L^p(w)$ by density.

Finally, to obtain (6.3.5) for general $f\in L^p(w)$, we raise (6.3.3) to the power p, use the inequality $(a+b)^p \le 2^p(a^p+b^p)$, and integrate over $\mathbf{R}$ with respect to the measure $w\,dx$ to obtain

$$\int_{\mathbf{R}} \mathscr{C}_*(f)^p w\,dx \le 2^p c\int_{\mathbf{R}} M(f)^p w\,dx + 2^p\int_{\mathbf{R}} M(\mathscr{C}(f))^p w\,dx. \tag{6.3.12}$$

Then we use estimate (6.3.4) and the boundedness of the Hardy–Littlewood maximal operator on $L^p(w)$ to obtain the required conclusion. □

Exercises

6.3.1. (a) Let $\theta\in\mathbf{S}^{n-1}$. Define the *maximal directional Carleson operator*

$$\mathscr{C}_*^\theta(f)(x) = \sup_{a\in\mathbf{R}}\sup_{\varepsilon>0}\left|\int_{\varepsilon<|t|<\varepsilon^{-1}} e^{2\pi i a t} f(x-t\theta)\,\frac{dt}{t}\right|$$

for functions f on $\mathbf{R}^n$. Prove that $\mathscr{C}_*^\theta$ is bounded on $L^p(\mathbf{R}^n,w)$ for any weight $w\in A_p$ and $1<p<\infty$.
(b) Let Ω be an odd integrable function on $\mathbf{S}^{n-1}$. Obtain the same conclusion for the maximal operator

$$\mathscr{C}_*^\Omega(f)(x) = \sup_{\xi\in\mathbf{R}^n}\sup_{\varepsilon>0}\left|\int_{\varepsilon<|y|<\varepsilon^{-1}} e^{2\pi i\xi\cdot y} f(x-y)\,\frac{\Omega\left(\frac{y}{|y|}\right)}{|y|^n}\,dy\right|.$$

[*Hint:* Part (a): Reduce to the case $\theta = e_1 = (1,0,\dots,0)$ via a rotation and use Theorem 6.3.3 with $w = 1$. Part (b): Use the method of rotations and part (a).]

6.3.2. For a fixed $\lambda > 0$ and a function $f \in \cup_{1 \le p < \infty} L^p(\mathbf{R})$ write

$$\{x \in \mathbf{R} : \mathscr{C}_*(f)(x) > \lambda\} = \bigcup_j I_j ,$$

where $I_j = (\alpha_j, \alpha_j + \delta_j)$ are open disjoint intervals. Let $1 < r < \infty$. Show that there exists a $\gamma_0 > 0$ such that for every $0 < \gamma < \gamma_0$ there exists a constant $C_\gamma > 0$ such that $\lim_{\gamma \to 0} C_\gamma = 0$ and

$$\big|\{x \in I_j : \mathscr{C}_*(f)(x) > 3\lambda,\ M_r(f)(x) \le \gamma\lambda\}\big| \le C_\gamma |I_j| .$$

[*Hint:* Note that we must have $\mathscr{C}_*(f)(\alpha_j) \le \lambda$ and $\mathscr{C}_*(f)(\alpha_j + \delta_j) \le \lambda$ for all j. Set $I_j^* = (\alpha_j - 5\delta_j, \alpha_j + 6\delta_j)$, $f_1(x) = f(x)$ for $x \in I_j^*$, $f_1(x) = 0$ for $x \notin I_j^*$, and $f_2(x) = f(x) - f_1(x)$. We may assume that for all j there exists a z_j in I_j such that $M_r(f)(z_j) \le \gamma\lambda$. For a given $\varepsilon > 0$ we let $H^{(\varepsilon)}$ be the truncated Hilbert transform. For fixed $x \in I_j$ estimate $|H^{(\varepsilon)}(f_2)(x) - H^{(\varepsilon)}(f_2)(\alpha_j)|$ by the threefold sum

$$\begin{aligned} &\left| \int_{|\alpha_j - t| > \varepsilon} f_2(t) e^{2\pi i \xi t} \left(\frac{2}{\alpha_j - t} - \frac{2}{x - t} \right) dt \right| \\ &\qquad + \left| \int_{|x-t| > \varepsilon \ge |\alpha_j - t|} f_2(t) e^{2\pi i \xi t} \frac{1}{x-t}\, dt \right| \\ &\qquad + \left| \int_{|\alpha_j - t| > \varepsilon \ge |x-t|} f_2(t) e^{2\pi i \xi t} \frac{1}{\alpha_j - t}\, dt \right| , \end{aligned}$$

which is easily shown to be controlled by $c_0 M(f)(z_j)$ for some constant c_0. Thus $\mathscr{C}_*(f_2)(x) \le \mathscr{C}_*(f_2)(\alpha_j) + c_0 M(f)(z_j) \le \lambda + c_0 \gamma\lambda$. Select γ_0 such that $c_0 \gamma_0 < \frac{1}{2}$. Then $\lambda + c_0 \gamma\lambda < \frac{3}{2}\lambda$ for $\gamma < \gamma_0$; hence we have $\mathscr{C}_*(f)(x) \le \mathscr{C}_*(f_1)(x) + \frac{3}{2}\lambda$ for $x \in I_j$ and thus $I_j \cap \{\mathscr{C}_*(f) > 3\lambda\} \subsetneqq \{\mathscr{C}_*(f_1) > \lambda\}$. Using the boundedness of $\mathscr{C}_*$ on L^r and the fact that $M_r(f)(z_j) \le \gamma\lambda$, we obtain that the last set has measure at most a constant multiple of $\gamma^r |I_j|$.]

6.3.3. ([200]) Show that for every w in A_∞ there is a finite constant $\gamma_0 > 0$ such that for all $0 < \gamma < \gamma_0$ and all $1 < r < \infty$ there is a constant B_γ such that $\lim_{\gamma \to 0} B_\gamma = 0$ and

$$w\big(\{\mathscr{C}_*(f) > 3\lambda\} \cap \{M_r(f) \le \gamma\lambda\}\big) \le B_\gamma\, w\big(\{\mathscr{C}_*(f) > \lambda\}\big)$$

for all functions f in $\cup_{1 \le p < \infty} L^p(\mathbf{R})$.
[*Hint:* Start with positive constants C_0 and δ such that for all intervals I and any measurable set E we have $|E \cap I| \le \varepsilon |I| \implies w(E \cap I) \le C_0 \varepsilon^\delta w(I)$. Use the estimate of Exercise 6.3.2 with $I = I_j$ and sum over j to obtain the required estimate with $B_\gamma = C_0 (C_\gamma)^\delta$.]

6.3.4. Prove the following vector-valued version of Theorem 6.2.1:

$$\Big\|\Big(\sum_j |\mathscr{C}(f_j)|^r\Big)^{\frac{1}{r}}\Big\|_{L^p(w)} \le C_{p,r}(w)\Big\|\Big(\sum_j |f_j|^r\Big)^{\frac{1}{r}}\Big\|_{L^p(w)}$$

for all $1 < p, r < \infty$, all weights $w \in A_p$, and all sequences of functions f_j in $L^p(w)$. $\big[$*Hint:* You may want to use Corollary 7.5.7 in [156].$\big]$

HISTORICAL NOTES

A version of Theorem 6.1.1 concerning the maximal partial sum operator of Fourier series of square-integrable functions on the circle was first proved by Carleson [69]. An alternative proof of Carleson's theorem was provided by Fefferman [135], pioneering a set of ideas called time–frequency analysis. Lacey and Thiele [238] provided the first independent proof on the line of the boundedness of the maximal Fourier integral operator (6.1.1). The proof of Theorem 6.1.1 given in this text follows closely the one given in Lacey and Thiele [238], which improves in some ways that of Fefferman's [135], by which it was inspired. One may also consult the expository article of Thiele [347].

A version of Theorem 6.2.1 concerning the L^p boundedness, $1 < p < \infty$, of the maximal partial sum operator on the circle was obtained by Hunt [199]. Sjölin [317] extended this result to $L(\log^+ L)(\log^+\log^+ L)$ and Antonov [4] to $L(\log^+ L)(\log^+\log^+\log^+ L)$. Counterexamples of Kolmogorov [222], [223], Körner [227], and Konyagin [224] indicate that the everywhere convergence of partial Fourier sums (or integrals) may fail for functions in L^1 and in spaces near L^1. The exponential decay estimate for $\alpha \ge 1$ in (6.2.1) and the restricted weak type (p,p) estimate with constant $Cp^2(p-1)^{-1}$ for the maximal partial sum operator on the circle are contained in Hunt's article [199]. The estimate for $\alpha < 1$ in (6.2.1) appears in the article of Grafakos, Tao, and Terwilleger [173]; the proof of Theorem 6.2.1 is based on this article. This article also investigates higher-dimensional analogues of the theory which were initiated in Pramanik and Terwilleger [299]. The related article of Sawano [310] adapts the method in [173] to maximal operators associated with pseudodifferential operators with homogeneous symbols.

Theorem 6.3.3 was first obtained by Hunt and Young [200] using a good lambda inequality for the Carleson operator. A deep multilinear generalization of the Carleson-Hunt theorem was obtained by Li and Muscalu [245] An improved good lambda inequality for the Carleson operator is contained in of Grafakos, Martell, and Soria [166]. The particular proof of Theorem 6.3.3 given in the text is based on the approach of Rubio de Francia, Ruiz, and Torrea [308]. The books of Jørsboe and Mejlbro [206], Mozzochi [272], and Arias de Reyna [5] contain detailed presentations of the Carleson–Hunt theorem on the circle.

Chapter 7
Multilinear Harmonic Analysis

Multivariable calculus provides a robust approach into the study of functions of several variables that goes beyond the narrow perspective of studying a single variable by freezing the other ones. Analogously, multilinear analysis focuses on the study of operators that depend linearly on several functions by treating all inputs as variables and not just some as parameters. This study is based on multiple simultaneous decompositions and is naturally more complicated than its linear counterpart, but is also more far-reaching and yields more flexible results.

Examples of linear operators with fixed parameters that can be viewed as multilinear are plentiful in harmonic analysis: multiplier operators, homogeneous singular integrals associated with functions on the sphere, Littlewood–Paley operators, the Calderón commutators, and the Cauchy integral along Lipschitz curves.

Multilinear Fourier analysis provides a framework to study operations that depend linearly on several functions f_i whose frequencies $\widehat{f_i}$ are jointly altered by a common multiplier or symbol. These are called multilinear multiplier operators, and a big part of our study focuses on them. A very powerful tool, called multilinear interpolation, is also developed in this chapter. This tool makes it possible to obtain intermediate bounds for multilinear operators from a finite set of initial estimates.

7.1 Multilinear Operators

A multilinear operator $T(f_1,\dots,f_m)$ is an operator of several variables that is linear in each entry. In the special case $m=2$, T is called *bilinear*, when $m=3$ *trilinear*, and for $m\geq 4$, T is also called *m-linear*. In this chapter we study multilinear operators acting on m-tuples of functions defined on $\mathbf{R}^n$. Many such important operators arise from a linear operator L acting on functions $(\mathbf{R}^n)^m$ in the following way:

$$T(f_1,\dots,f_m)(x)=L(f_1\otimes\cdots\otimes f_m)(x,\dots,x),\quad x\in\mathbf{R}^n,$$

L. Grafakos, *Modern Fourier Analysis*, Graduate Texts in Mathematics 250,
DOI 10.1007/978-1-4939-1230-8_7, © Springer Science+Business Media New York 2014

where f_j are functions defined on $\mathbf{R}^n$ and $f_1 \otimes \cdots \otimes f_m$ is their *tensor product* defined on $(\mathbf{R}^n)^m$ by $(f_1 \otimes \cdots \otimes f_m)(x_1, \ldots, x_m) = f_1(x_1) \cdots f_m(x_m)$, $x_j \in \mathbf{R}^n$.

Multilinear operators arise in the linearization of certain nonlinear problems. Suppose that S is a linear operator acting on functions on $\mathbf{R}^n$. The nonlinear quantity $S(f^2)$ motivates the introduction of the bilinear operator

$$T(f_1, f_2) = S(f_1 f_2).$$

Bilinear estimates for $T(f_1, f_2)$ from $X \times X \to Y$ (for some function spaces X, Y) can be used to deduce the nonlinear estimate $\|S(f^2)\|_Y \le C\|f\|_X^2$ by taking $f = f_1 = f_2$.

7.1.1 Examples and initial results

We list a few examples that arise in the study of the theory of multilinear operators acting on m-tuples of functions on $\mathbf{R}^n$.

Examples 7.1.1. 1. The m-fold product

$$I(f_1, \ldots, f_m) = f_1 \cdots f_m$$

is the *identity* in the realm of m-linear operators. It indicates that natural inequalities between Lebesgue spaces are of the form $L^{p_1}(\mathbf{R}^n) \times \cdots \times L^{p_m}(\mathbf{R}^n) \to L^p(\mathbf{R}^n)$, where $1/p_1 + \cdots + 1/p_m = 1/p$, $0 < p_j, p \le \infty$.

2. A kernel of $m+1$ variables $K(x, y_1, \ldots, y_m)$ gives rise to an m-linear operator of the form

$$T(f_1, \ldots, f_m)(x) = \int_{\mathbf{R}^{mn}} K(x, y_1, \ldots, y_m) f_1(y_1) \cdots f_m(y_m)\, dy_1 \cdots dy_m,$$

where the integral may converge absolutely, in the principal value sense or even in the sense of distributions.

3. The special case in which the kernel $K(x, y_1, \ldots, y_m)$ in the previous case has the form $K_0(x - y_1, \ldots, x - y_m)$ corresponds to the so-called *m-linear convolution operator*

$$T_0(f_1, \ldots, f_m)(x) = \int_{\mathbf{R}^{mn}} K_0(x - y_1, \ldots, x - y_m) f_1(y_1) \cdots f_m(y_m)\, dy_1 \cdots dy_m,$$

in which the integral is taken in the principal value sense. This operator can also be expressed as an *m-linear multiplier* as follows:

$$\int_{\mathbf{R}^{mn}} m_0(\xi_1, \ldots, \xi_m) \widehat{f_1}(\xi_1) \cdots \widehat{f_m}(\xi_m)\, e^{2\pi i x \cdot (\xi_1 + \cdots + \xi_m)}\, d\xi_1 \cdots d\xi_m,$$

where m_0 is the distributional Fourier transform of K_0 on $\mathbf{R}^{mn}$.

4. The bilinear operator on $\mathbf{R}^n$ given by

$$B(f,g)(x) = \int_{|t|\le 1} f(x+t)\,g(x-t)\,dt\,. \tag{7.1.1}$$

This operator is *positive* and *local* in the sense that if f and g are supported in cubes of length 1 with sides parallel to the axes, then so is $B(f,g)$.

5. The bilinear fractional integral is another *positive* operator defined by

$$I_\alpha(f,g)(x) = \int_{\mathbf{R}^n} f(x+t)\,g(x-t)|t|^{\alpha-n}\,dt\,,$$

where $0<\alpha<n$ and f,g are bounded functions whose supports have finite measure, or they have sufficient decay at infinity, such as functions in $\mathscr{S}(\mathbf{R}^n)$.

Example 7.1.2. We show that the operator given in (7.1.1) is well defined on $L^1(\mathbf{R}^n)\times L^1(\mathbf{R}^n)$ and is bounded from $L^1(\mathbf{R}^n)\times L^1(\mathbf{R}^n)$ to $L^{1/2}(\mathbf{R}^n)$. Indeed, for $f,g\ge 0$ in $L^1(\mathbf{R}^n)$ we have

$$\begin{aligned}\int_{\mathbf{R}^n}\int_{\mathbf{R}^n} f(x+t)g(x-t)\,dt\,dx &= \int_{\mathbf{R}^n}\int_{\mathbf{R}^n} f(t')g(2x-t')\,dt'\,dx\\ &= \int_{\mathbf{R}^n} f(t')\int_{\mathbf{R}^n} g(2x-t')\,dx\,dt'\\ &= \frac{1}{2^n}\,\|f\|_{L^1}\|g\|_{L^1}\,. \end{aligned} \tag{7.1.2}$$

This implies that $B(f,g)(x)$ is well defined for almost all $x\in\mathbf{R}^n$ whenever f,g are nonnegative functions in $L^1(\mathbf{R}^n)$.

Next, we observe that for nonnegative integrable functions well defined f,g supported in cubes of length one with sides parallel to the axes, $B(f,g)$ is also supported in a cube of length one with sides parallel to the coordinate axes. For two such fixed functions f,g we have

$$\big\|B(f,g)\big\|_{L^{1/2}} \le \big\|B(f,g)\big\|_{L^1} \le \frac{1}{2^n}\,\|f\|_{L^1}\|g\|_{L^1}\,, \tag{7.1.3}$$

where the first inequality is a consequence of Hölder's inequality on a cube of measure 1 and the second inequality was proved in (7.1.2).

For general nonnegative integrable functions f and g, write $f_k = f\chi_{Q_k}$ and $g_m = g\chi_{Q_m}$, where Q_j is the cube $[j_1-\frac12, j_1+\frac12)\times\cdots\times[j_n-\frac12, j_n+\frac12)$, if $j=(j_1,\dots,j_n)$, i.e., the unit-length cube with sides parallel to the axes and center $j\in\mathbf{Z}^n$.

Let us fix $k\in\mathbf{Z}^n$. Then there exist at most 5^n points $m\in\mathbf{Z}^n$ such that $B(f_k,g_m)$ is nonzero. This is because the integral defining $B(f_k,g_m)$ is taken over the intersection of the sets $\{t:\ |t|<1\}$ and $\frac12(Q_k-Q_m)=\prod_{i=1}^n\big[\frac12(k_i-m_i-1),\frac12(k_i-m_i+1)\big)$, and this set is nonempty for at most 5^n points $m=(m_1,\dots,m_n)$. All these m satisfy $|m_i-k_i|\le 2$ for each coordinate $i\in\{1,\dots,m\}$, and thus $k-m\in 4\overline{Q_0}$.

Now write

$$B(f,g)=\sum_{k\in\mathbf{Z}^n}\sum_{m\in\mathbf{Z}^n}B(f_k,g_m)=\sum_{d\in\mathbf{Z}^n\cap\overline{4Q_0}}\sum_{k\in\mathbf{Z}^n}B(f_k,g_{k+d}).$$

Since the cardinality of the set $(\mathbf{Z}^n\cap\overline{4Q_0})$ is 5^n, we have

$$\begin{aligned}\|B(f,g)\|_{L^{1/2}} &\le 5^n\sum_{d\in\mathbf{Z}^n\cap\overline{4Q_0}}\left(\sum_{k\in\mathbf{Z}^n}\int_{\mathbf{R}^n}|B(f_k,g_{k+d})|^{1/2}dx\right)^2\\ &\le\frac{5^n}{2^n}\sum_{d\in\mathbf{Z}^n\cap\overline{4Q_0}}\left(\sum_{k\in\mathbf{Z}^n}\|f_k\|_{L^1}^{1/2}\|g_{k+d}\|_{L^1}^{1/2}\right)^2\\ &\le\frac{5^n}{2^n}5^n\|f\|_{L^1}\|g\|_{L^1},\end{aligned}$$

where the penultimate inequality above follows by applying (7.1.3) for the functions f_k and g_{k+d}, while the last inequality is an application of the Cauchy-Schwarz inequality and the fact that $\|f\|_{L^1}=\sum_{k\in\mathbf{Z}^n}\|f_k\|_{L^1}$.

Definition 7.1.3. Let $0<p_1,\dots,p_m,p\le\infty$. We say that an m-linear operator T is of *restricted weak type* $(p_1,\dots,p_m,p)$ if there is a constant $C=C(p_1,\dots,p_m,p)$ such that for all measurable subsets $A_1,\dots,A_m$ of finite measure we have

$$\sup_{\lambda>0}\lambda|\{x\in\mathbf{R}^n:\ |T(\chi_{A_1},\dots,\chi_{A_m})(x)|>\lambda\}|^{\frac{1}{p}}\le C|A_1|^{\frac{1}{p_1}}\cdots|A_m|^{\frac{1}{p_m}}$$

when $p<\infty$ and

$$\big\|T(\chi_{A_1},\dots,\chi_{A_m})\big\|_{L^\infty}\le C|A_1|^{\frac{1}{p_1}}\cdots|A_m|^{\frac{1}{p_m}}$$

when $p=\infty$.

Theorem 7.1.4. *Let* $0<\alpha<n$. *The* bilinear fractional integral

$$I_\alpha(f,g)(x)=\int_{\mathbf{R}^n}f(x+t)g(x-t)|t|^{\alpha-n}dt$$

is of restricted weak types $(\frac{n}{\alpha},\infty,\infty)$, $(\infty,\frac{n}{\alpha},\infty)$, $(1,\infty,\frac{n}{n-\alpha})$, $(\infty,1,\frac{n}{n-\alpha})$, *and* $(1,1,\frac{n}{2n-\alpha})$.

Proof. We first consider the point $(\frac{n}{\alpha},\infty,\infty)$. Let A and B be measurable subsets of $\mathbf{R}^n$ of finite measure. Let v_n be the volume of the unit ball in $\mathbf{R}^n$. We have

$$\begin{aligned}\|I_\alpha(\chi_A,\chi_B)\|_{L^\infty} &\le\sup_{x\in\mathbf{R}^n}\int_{-x+A}|t|^{\alpha-n}dt\\ &\le\sup_{x\in\mathbf{R}^n}\int_0^\infty(|t|^{\alpha-n})^*(s)(\chi_{-x+A})^*(s)\,ds\end{aligned}$$

$$= \int_0^\infty \left(\frac{s}{v_n}\right)^{\frac{\alpha}{n}-1} \chi_{[0,|A|)}(s)\,ds$$

$$= \frac{n v_n^{1-\frac{\alpha}{n}}}{\alpha} |A|^{\frac{\alpha}{n}}, \tag{7.1.4}$$

where g^* denotes the decreasing rearrangement of a function g; see Exercise 1.4.1 in [156] for the preceding inequality. Likewise, one obtains the claimed restricted weak type $(\infty, \frac{n}{\alpha}, \infty)$ estimate.

Next we show that I_α is of restricted weak type $(1, \infty, \frac{n}{n-\alpha})$. We observe that $I_\alpha(\chi_A, \chi_B) \le c_{\alpha,n} \mathcal{I}_\alpha(\chi_A)$, where $\mathcal{I}_\alpha$ is the classical fractional integral

$$\mathcal{I}_\alpha(f)(x) = 2^{-\alpha} \pi^{-\frac{n}{2}} \frac{\Gamma(\frac{n-\alpha}{2})}{\Gamma(\frac{\alpha}{2})} \int_{\mathbf{R}^n} f(x-y)|y|^{\alpha-n} dy$$

and

$$c_{\alpha,n} = 2^\alpha \pi^{\frac{n}{2}} \frac{\Gamma(\frac{\alpha}{2})}{\Gamma(\frac{n-\alpha}{2})}.$$

Given any measurable subset F of $\mathbf{R}^n$ with $0 < |F| < \infty$, we have

$$\begin{aligned}
\int_F |\mathcal{I}_\alpha(\chi_A)(x)|\,dx &= \frac{1}{c_{\alpha,n}} \int_F \int_{\mathbf{R}^n} \chi_A(x-y)|y|^{-n+\alpha}\,dy\,dx \\
&= \frac{1}{c_{\alpha,n}} \int_F \int_A |x-u|^{-n+\alpha}\,du\,dx \\
&= \frac{1}{c_{\alpha,n}} \int_A \int_F |x-u|^{-n+\alpha}\,dx\,du \\
&\le \frac{1}{c_{\alpha,n}} \frac{n v_n^{1-\frac{\alpha}{n}}}{\alpha} |F|^{\frac{\alpha}{n}} |A| \\
&= C_{\alpha,n} |F|^{\frac{\alpha}{n}} |A|,
\end{aligned}$$

where the last inequality was deduced in (7.1.4).

Since $\frac{n}{n-\alpha} > 1$, the result in Exercise 1.1.12(b) in [156] gives

$$\begin{aligned}
\|\mathcal{I}_\alpha(\chi_A)\|_{L^{\frac{n}{n-\alpha},\infty}} &\le \sup_{0<|F|<\infty} |F|^{-1+\frac{n-\alpha}{n}} \int_F |\mathcal{I}_\alpha(\chi_A)(x)|\,dx \\
&\le \sup_{0<|F|<\infty} |F|^{-\frac{\alpha}{n}} C_{n,\alpha} |A| |F|^{\frac{\alpha}{n}} \\
&= C_{n,\alpha} |A|.
\end{aligned}$$

It follows that

$$\|I_\alpha(\chi_A, \chi_B)\|_{L^{n/(n-\alpha),\infty}} \le C_{n,\alpha} |A|.$$

The fact that I_α is of restricted weak type $(\infty, 1, \frac{n}{n-\alpha})$ is obtained by symmetry.

Finally, we are left with the restricted weak type $(1,1,\frac{n}{2n-\alpha})$ estimate. For $j\in\mathbf{Z}$ and $f,g\geq 0$ we introduce operators

$$B_j(f,g)(x)=\int_{|t|\leq 2^j} f(x+t)g(x-t)dt$$

and we note that for $f,g\geq 0$ we have

$$I_\alpha(f,g)\leq \sum_{j\in\mathbf{Z}} 2^{j(\alpha-n)}B_{j+1}(f,g).$$

We observe that by an easy dilation argument, the result in Example 7.1.2 implies that B_j maps $L^1\times L^1\to L^{1/2}$ with norm bounded by $2^{-n}2^{jn}$ for each $j\in\mathbf{Z}$. This fact together with the observation that for all $f,g\geq 0$ we have

$$\left(\int_E (B_j(f,g)(x))^{1/2}\,dx\right)^2\leq |E|\int_{\mathbf{R}^n} B_j(f,g)(x)\,dx\leq 2^{-n}\|f\|_{L^1}\|g\|_{L^1}|E|,$$

implies that for any measurable set E with finite measure we have

$$\int_E (B_j(f,g)(x))^{1/2}\,dx\leq 2^{-n/2}\|f\|_{L^1}^{1/2}\|g\|_{L^1}^{1/2}\min(2^{jn},|E|)^{1/2}. \tag{7.1.5}$$

Now define

$$E_\lambda=\{x\in\mathbf{R}^n:\ |I_\alpha(f,g)(x)|>\lambda\}$$

for some fixed nonnegative functions $f,g\in L^\infty(\mathbf{R}^n)\cap L^1(\mathbf{R}^n)$. Then

$$|E_\lambda|\leq |\{x\in\mathbf{R}^n:\ \|g\|_{L^\infty}\tfrac{1}{c_{n,\alpha}}\mathcal{I}_\alpha(f)(x)>\lambda\}|<\infty$$

since, in view of Theorem 1.2.3, $\mathcal{I}_\alpha$ maps $L^1\to L^{n/(n-\alpha),\infty}$. Then Chebyshev's inequality and (7.1.5) give

$$\begin{aligned}
\lambda^{1/2}|E_\lambda| &\leq \int_{E_\lambda}|I_\alpha(f,g)(x)|^{\frac12}\,dx\\
&\leq \int_{E_\lambda}\Big|\sum_{j\in\mathbf{Z}} 2^{j(\alpha-n)}B_{j+1}(f,g)(x)\Big|^{\frac12}dx\\
&\leq \sum_{j\in\mathbf{Z}} 2^{\frac12(\alpha-n)j}\int_{E_\lambda}|B_{j+1}(f,g)(x)|^{\frac12}dx\\
&\leq 2^{-\frac{n}{2}}\sum_{j\in\mathbf{Z}} 2^{\frac12(\alpha-n)j}(\|f\|_{L^1}\|g\|_{L^1})^{\frac12}\min(2^{(j+1)n},|E_\lambda|)^{\frac12}\\
&\leq C(n,\alpha)\big(\|f\|_{L^1}\|g\|_{L^1}\big)^{\frac12}|E_\lambda|^{\frac{\alpha}{2n}}.
\end{aligned}$$

Using that $|E_\lambda|<\infty$, this implies that

$$\lambda|E_\lambda|^{\frac{2n-\alpha}{n}}\leq C(n,\alpha)^2\|f\|_{L^1}\|g\|_{L^1}$$

which proves the claimed weak type estimate at the point $(1, 1, \frac{n}{2n-\alpha})$. Here we have in fact shown that I_α maps $L^1 \times L^1 \to L^{n/(2n-\alpha),\infty}$ for all nonnegative bounded and integrable functions f, g, not only characteristic functions of sets of finite measure. By density, I_α has a unique bounded extension from $L^1 \times L^1$ to $L^{n/(2n-\alpha),\infty}$. $\square$

7.1.2 Kernels and Duality of m-linear Operators

We study multilinear operators defined in terms of some kernel. The precise relationship of the operator with the kernel is as follows. We assume that for a given m-linear operator T defined on $\mathscr{S}(\mathbf{R}^n) \times \cdots \times \mathscr{S}(\mathbf{R}^n)$ there is a tempered distribution W on $(\mathbf{R}^n)^{m+1}$ such that for all $\phi_1, \dots, \phi_m, \phi_0$ in $\mathscr{S}(\mathbf{R}^n)$

$$\langle T(\phi_1, \dots, \phi_m), \phi_0 \rangle = \langle W, \phi_0 \otimes \phi_1 \otimes \cdots \otimes \phi_m \rangle. \tag{7.1.6}$$

Here $\langle \cdot, \cdot \rangle$ denotes the action of a tempered distribution on a Schwartz function and $\phi_0 \otimes \phi_1 \otimes \cdots \otimes \phi_m$ denotes the function in $\mathscr{S}((\mathbf{R}^n)^{m+1})$ given by

$$(x, y_1, \dots, y_m) \to \phi_0(x)\phi_1(y_1) \cdots \phi_m(y_m).$$

If the tempered distribution W on $(\mathbf{R}^n)^{m+1}$ coincides with a function K on $(\mathbf{R}^n)^{m+1} \setminus \{(x, \dots, x) : x \in \mathbf{R}^n\}$, we will occasionally refer to W by K, assuming there is no confusion. We may also denote W by $W(x, y_1, \dots, y_m)$ to indicate the variables on which it acts.

We study m-linear operators defined on products of test functions and we seek conditions to extend them as bounded operators on certain products of Banach spaces. We use the notation

$$\|T\|_{X_1 \times \cdots \times X_m \to X} = \sup_{\substack{\|f_j\|_{X_j}=1 \\ 1 \le j \le m}} \|T(f_1, \dots, f_m)\|_X$$

to denote the *norm* of an m-linear operator T from a product of Banach spaces of functions $X_1 \times \cdots \times X_m$ into a quasi-Banach space X. We say that T is *bounded* from $X_1 \times \cdots \times X_m$ into X when the norm above is finite.

An m-linear operator $T : \mathscr{S}(\mathbf{R}^n) \times \cdots \times \mathscr{S}(\mathbf{R}^n) \to \mathscr{S}'(\mathbf{R}^n)$ is linear in every entry, and consequently it has m formal transposes. The jth *transpose* T^{*j} of T is defined as the unique operator that satisfies the identity

$$\langle T^{*j}(f_1, \dots, f_m), h \rangle = \langle T(f_1, \dots, f_{j-1}, h, f_{j+1}, \dots, f_m), f_j \rangle,$$

for all $f_1, \dots, f_m, h$ in $\mathscr{S}(\mathbf{R}^n)$.

It is easy to check that if the kernel of T is a locally integrable function K, then T^{*j} has a kernel K^{*j} that is related to the kernel K of T via the identity

$$K^{*j}(x,y_1,\dots,y_{j-1},y_j,y_{j+1},\dots,y_m) = K(y_j,y_1,\dots,y_{j-1},x,y_{j+1},\dots,y_m), \quad (7.1.7)$$

i.e., the first and jth entries are interchanged. If the kernel of T is a tempered distribution W, then the kernel of T^{*j} is another tempered distribution W^{*j} that is related to W via the identity

$$\langle W^{*j}, \phi_0 \otimes \phi_1 \otimes \cdots \otimes \phi_m \rangle = \langle W, \phi_j \otimes \phi_1 \otimes \cdots \otimes \phi_{j-1} \otimes \phi_0 \otimes \phi_{j+1} \otimes \cdots \otimes \phi_m \rangle,$$

with the obvious modification when $j=1$ or $j=m$. In the sequel, we will use the notation $W^{*0} = W$ and $T^{*0} = T$.

If a multilinear operator T maps a product of Banach spaces $X_1 \times \cdots \times X_m$ into another Banach space X, then the *transpose* T^{*j} maps the product of Banach spaces $X_1 \times \dots X_{j-1} \times X^* \times X_{j+1} \times \cdots \times X_m$ into X_j^*. Moreover, the norms of T and T^{*j} are equal.

It is sometimes customary to work with the adjoints of an m-linear operator T whose kernels are the complex conjugates of the kernels K^{*j} defined previously. Here we choose to work with the transposes, as defined earlier, to simplify the notation. This choice entails no differences in the study of these operators.

7.1.3 Multilinear Convolution Operators with Nonnegative Kernels

Given a nonnegative regular Borel measure μ on $\mathbf{R}^n \times \cdots \times \mathbf{R}^n$, we define the *m-linear convolution operator* by setting

$$T^\mu(f_1,\dots,f_m)(x) = \int_{\mathbf{R}^n \times \cdots \times \mathbf{R}^n} f_1(x-y_1)\cdots f_m(x-y_m)\,d\mu(y_1,\dots,y_m), \quad (7.1.8)$$

where $x \in \mathbf{R}^n$ and f_j are nonnegative measurable functions on $\mathbf{R}^n$. If

$$d\mu(y_1,\dots,y_m) = K(y_1,\dots,y_m)\,dy_1\cdots dy_m,$$

for some nonnegative measurable function K, then we use the notation

$$T^\mu(f_1,\dots,f_m) = T^K(f_1,\dots,f_m) \quad (7.1.9)$$

and we call K the *kernel* of T^K. We take a quick look at conditions that yield boundedness for operators of the form T^μ and T^K on products of Lebesgue spaces, when μ and K are nonnegative.

Proposition 7.1.5. *Let μ be a nonnegative regular Borel measure on $\mathbf{R}^n \times \cdots \times \mathbf{R}^n$, which is not identically equal to zero. Suppose that the m-linear operator T^μ maps $L^{p_1}(\mathbf{R}^n) \times \cdots \times L^{p_m}(\mathbf{R}^n)$ to $L^r(\mathbf{R}^n)$ for some $0 < p_j, r \le \infty$. Then we must have $1/p_1 + \cdots + 1/p_m \ge 1/r$.*

Proof. Fix $0 < p_1, \dots, p_m, r \le \infty$. By translating μ if necessary, we may assume that there exists a compact set $E \subset [1,M]^n \times \cdots \times [1,M]^n$ for some $M > 1$ such that $0 < \mu(E) < \infty$. Let $x = (x_1, \dots, x_n)$ in $\mathbf{R}^n$. For $j = 1, \dots, m$ we define

$$f_j(x) = \prod_{i=1}^n |x_i|^{-\alpha_j} \chi_{(1,\infty)^n}(x_1, \dots, x_n),$$

with $\alpha_j > 1/p_j$. Then, for $x \in (M+1, \infty)^n$, we have

$$\begin{aligned} T^\mu(f_1, \dots, f_m)(x) &\ge \int_E f_1(x-y_1) \cdots f_m(x-y_m)\, d\mu(y_1, \dots, y_m) \\ &\ge \mu(E) \prod_{i=1}^n (x_i - 1)^{-(\alpha_1 + \cdots + \alpha_m)}. \end{aligned}$$

Assuming that $T^\mu(f_1, \dots, f_m) \in L^r(\mathbf{R}^n)$, we obtain that $\alpha_1 + \cdots + \alpha_m > 1/r$ for all $\alpha_j > 1/p_j$; hence $1/p_1 + \cdots + 1/p_m \ge 1/r$. □

Next we show that in the diagonal case $1/p_1 + \cdots + 1/p_m = 1/r$, integrable functions are the only nonnegative kernels for which the associated operators are bounded.

Proposition 7.1.6. *Let μ be a nonnegative regular Borel measure on $(\mathbf{R}^n)^m$, where $m \ge 1$. Suppose that the operator T^μ maps $L^{p_1}(\mathbf{R}^n) \times \cdots \times L^{p_m}(\mathbf{R}^n)$ to $L^{r,\infty}(\mathbf{R}^n)$ for some $0 < p_1, \dots, p_m \le \infty$ satisfying $1/p_1 + \cdots + 1/p_m = 1/r$. Then μ is a finite measure; in particular, if $d\mu(y) = K(y)\,dy$ for some nonnegative measurable function K, then $K \in L^1((\mathbf{R}^n)^m)$.*

Proof. Let C be the norm of T^μ as a bounded operator from $L^{p_1}(\mathbf{R}^n) \times \cdots \times L^{p_m}(\mathbf{R}^n)$ to $L^{r,\infty}(\mathbf{R}^n)$. First we consider the case $0 < r < \infty$. For a given $R > 0$ let $B_R = B(0,R)$ be the ball of radius R centered at zero. Fix $R_0 > 0$ such that $\mu(B_R \times \cdots \times B_R) > 0$ for all $R \ge R_0$. Such an R_0 exists if μ is nonzero; otherwise, there is nothing to prove. Then, for every $R \ge R_0$ and $x \in B_R$, we have

$$T^\mu(\chi_{B_{2R}}, \dots, \chi_{B_{2R}})(x) = \mu\big(B(x,2R) \times \cdots \times B(x,2R)\big) \ge \mu(B_R \times \cdots \times B_R) = \lambda > 0.$$

Therefore, $B_R \subseteq \{T^\mu(\chi_{B_{2R}}, \dots, \chi_{B_{2R}}) > \lambda/2\}$, and

$$|B_R| \le \big|\{T^\mu(\chi_{B_{2R}}, \dots, \chi_{B_{2R}}) > \lambda/2\}\big| \le \frac{2^r C^r}{\lambda^r} |B_{2R}|^{\frac{r}{p_1}} \cdots |B_{2R}|^{\frac{r}{p_m}} = \frac{C^r 2^{r+n}}{\lambda^r} |B_R|.$$

Hence, for every $R > 0$, we have that $\lambda = \mu(B_R \times \cdots \times B_R) \le 2^{1+\frac{n}{r}} C$, which proves the result when $r < \infty$ by letting $R \to \infty$. When $r = \infty$, we have

$$\mu(B_R \times \cdots \times B_R) \le \big\|T^\mu(\chi_{B_{2R}}, \dots, \chi_{B_{2R}})\big\|_{L^\infty} \le C \big\|\chi_{B_{2R}}\big\|_{L^\infty}^m = C,$$

and the conclusion follows by letting $R \to \infty$ as well. □

Next we show that not every integrable function gives rise to a bounded operator in the endpoint case $L^1 \times L^1 \to L^{1/2,\infty}$.

Theorem 7.1.7. *There exists a nonnegative integrable function K on $\mathbf{R}\times\mathbf{R}$ such that the operator T^K defined in (7.1.9) via (7.1.8) does not map $L^1 \times L^1$ to $L^{1/2,\infty}$.*

Proof. For each $k = 1,2,3,\dots$ let R_k be a closed rectangle with vertices

$$\Big(\sum_{j=1}^{k-1}\frac{1}{j^3}, -\sum_{j=1}^{k-1}\frac{1}{j^3}\Big), \Big(\sum_{j=1}^{k}\frac{1}{j^3}, -\sum_{j=1}^{k}\frac{1}{j^3}\Big),$$

$$\Big(\sum_{j=1}^{k-1}\frac{1}{j^3}+\frac{2k}{\sqrt{2}}, -\sum_{j=1}^{k-1}\frac{1}{j^3}+\frac{2k}{\sqrt{2}}\Big), \Big(\sum_{j=1}^{k}\frac{1}{j^3}+\frac{2k}{\sqrt{2}}, -\sum_{j=1}^{k}\frac{1}{j^3}+\frac{2k}{\sqrt{2}}\Big),$$

where we use the convention that a sum over an empty set of indices is zero. Then the R_k have dimensions $\sqrt{2}k^{-3}$ and $2k$ and are contained in $\{(x,y) : -|x| \le y \le |x|\}$, and their boundaries touch. See Figure 7.1. Define a function K by setting

$$K(x) = \sum_{k=1}^{\infty} \chi_{R_k}(x)$$

for $x \in \mathbf{R}^2$. Then K lies in $L^1(\mathbf{R}^2)$ since

$$\|K\|_{L^1} = \Big|\bigcup_{k=1}^{\infty} R_k\Big| = \sum_{k=1}^{\infty}\frac{\sqrt{2}}{k^3}2k < \infty.$$

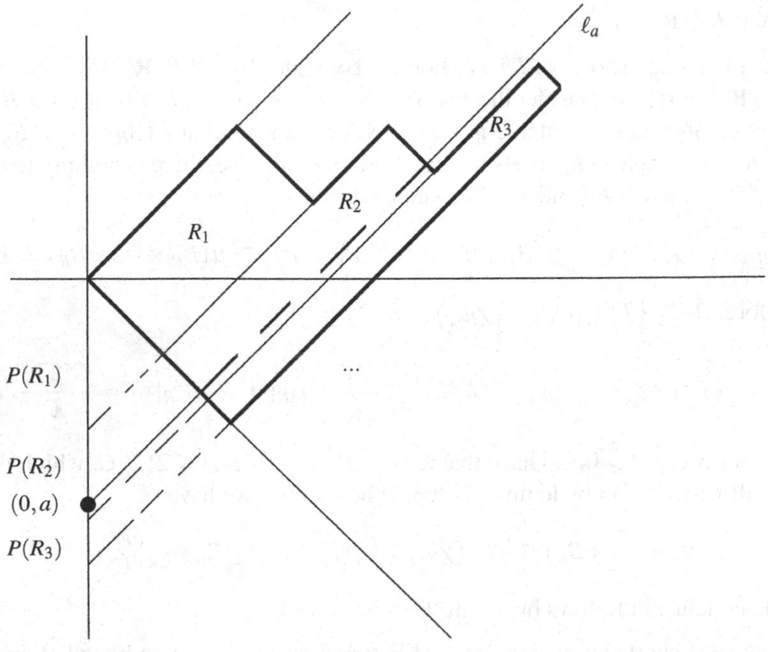

Fig. 7.1 The rectangles R_k,their projections $P(R_k)$, and the line ℓ_a.

For $a \le 0$ and $r > 0$ set

$$f_{a,r}(x) = \frac{1}{2r}\chi_{(a-r,a+r)}(x).$$

Then, for all $(x-a,x) \in \mathbf{R}^2$, we have

$$T^K(f_{a,r}, f_{0,r})(x) = \frac{1}{4r^2}\int_{(x-a-r,x-a+r)}\int_{(x-r,x+r)} K(y,z)\,dz\,dy,$$

and from this via the Lebesgue differentiation theorem we deduce that

$$T^K(f_{a,r}, f_{0,r})(x) \to K(x-a,x) \tag{7.1.10}$$

as $r \to 0$, for almost all $(x-a,x)$ in $\mathbf{R}^2$. Thus, (7.1.10) holds for almost every $a < 0$ and for almost every point on the line $\ell_a = \{(x-a,x):\ x \in \mathbf{R}\}$.

Suppose now that there is a constant C such that

$$\|T^K(f,g)\|_{L^{1/2,\infty}} \le C\|f\|_{L^1}\|g\|_{L^1}$$

for all $f,g \ge 0$ in $L^1(\mathbf{R})$. We let

$$P(R_k) = \Big[-2\sum_{j=1}^{k} j^{-3}, -2\sum_{j=1}^{k-1} j^{-3}\Big]$$

be the intersection of the y-axis with the smallest strip that contains R_k and is parallel to the line $y = x$. Then for any $k \ge 1$, there are an $a_k \in P(R_k)$ and a set $E_k \subsetneq \mathbf{R}$ of Lebesgue measure zero such that (7.1.10) holds with $a = a_k$ for all $x \in \mathbf{R} \setminus E_k$.

Now, given $x \in (0,k)$ and $a_k \in P(R_k)$, we have that

$$\sum_{j=1}^{k} j^{-3} \le 2\sum_{j=1}^{k-1} j^{-3} \le x - a_k \le k + 2\sum_{j=1}^{k} j^{-3} \le \frac{2k}{\sqrt{2}} + \sum_{j=1}^{k-1} j^{-3}$$

when $k \ge 5$. Thus, for all $x \in \mathbf{R}$ and $k \ge 5$ we have

$$\chi_{(0,k)}(x) \le \chi_{R_k}(x-a_k,x) \le K(x-a_k,x).$$

Using Fatou's lemma and (7.1.10), which holds for all $x \in \mathbf{R} \setminus E_k$, we deduce that

$$\|\chi_{(0,k)}\|_{L^{1/2,\infty}(\mathbf{R})} \le \liminf_{r\to 0}\|T^K(f_{a_k,r}, f_{0,r})\|_{L^{1/2,\infty}(\mathbf{R})} \le C\liminf_{r\to 0}\|f_{a_k,r}\|_{L^1(\mathbf{R})}\|f_{0,r}\|_{L^1(\mathbf{R})},$$

and since the preceding expression is equal to C for every integer $k \ge 5$, letting $k \to \infty$ we obtain a contradiction. □

Although boundedness from $L^1 \times L^1$ to $L^{1/2}$ may not hold for certain bilinear operators, boundedness from $L^1 \times L^1$ to $L^{1/2,\infty}$ is valid for a large class of bilinear operators (Section 7.4). Next, we show that there exist positive measures that provide examples of kernels with the same property.

Proposition 7.1.8. *There exists a nonnegative regular finite Borel measure μ on $\mathbf{R}\times\mathbf{R}$ with the property that T^μ maps $L^1\times L^1$ to $L^{1/2,\infty}$ but not $L^1\times L^1$ to $L^{1/2}$.*

Proof. If an operator T^μ maps $L^1(\mathbf{R})\times L^1(\mathbf{R})\to L^{1/2,\infty}(\mathbf{R})$, then necessarily μ must be a finite measure in view of Proposition 7.1.6.

We define a measure

$$\mu=\sum_{j=1}^{\infty}\frac{1}{j^2}\delta_{(j,j)},$$

where $\delta_{(j,j)}$ is the Dirac mass at the point (j,j). Clearly, $\mu(\mathbf{R}\times\mathbf{R})=\sum_{j=1}^{\infty}\frac{1}{j^2}<\infty$. Then we define a linear operator L by setting

$$L(h)(x)=\sum_{j=1}^{\infty}\frac{1}{j^2}h(x-j)$$

and a bilinear operator by

$$T^\mu(f,g)(x)=L(fg)(x)=\sum_{j=1}^{\infty}\frac{1}{j^2}f(x-j)g(x-j).$$

Using that $\{j^{-2}\}_{j\in\mathbf{Z}^+}\in\ell^{1/2,\infty}$ and Exercise 7.1.7, we have that L maps $L^{1/2}$ to $L^{1/2,\infty}$, and hence,

$$\|T^\mu(f,g)\|_{L^{1/2,\infty}}=\|L(fg)\|_{L^{1/2,\infty}}\le C\|fg\|_{L^{1/2}}\le C\|f\|_{L^1}\|g\|_{L^1}.$$

Now, since$\{j^{-2}\}_{j\in\mathbf{Z}^+}\notin\ell^{1/2}(\mathbf{Z})$, by Exercise 7.1.7, we have that L does not map $L^{1/2}$ to itself and

$$\|T^\mu\|_{L^1\times L^1\to L^{1/2}}\ge\sup_{f\neq0}\frac{\|T^\mu(f,f)\|_{L^{1/2}}}{\|f\|_{L^1}^2}=\sup_{h\neq0}\frac{\|L(h)\|_{L^{1/2}}}{\|h\|_{L^{1/2}}}=\infty.$$

□

The example of Proposition 7.1.8 is purely bilinear, i.e., it does not have a linear analog. Indeed, it follows from Proposition 7.1.6 (with $m=1$) that if a convolution operator with a positive Borel measure on $\mathbf{R}^n$ maps $L^1(\mathbf{R}^n)$ to $L^{1,\infty}(\mathbf{R}^n)$, then the measure is finite, and therefore it maps $L^1(\mathbf{R}^n)$ to itself.

Exercises

7.1.1. Prove that if I_α maps $L^p(\mathbf{R}^n)\times L^q(\mathbf{R}^n)$ to $L^r(\mathbf{R}^n)$, then we must necessarily have $1/p+1/q=\alpha/n+1/r$.

7.1.2. Let Z be a Banach space of functions on $\mathbf{R}^n$ with the property $\||f|\|_Z = \|f\|_Z$ for all $f \in Z$. Suppose that an m-linear operator satisfies

$$\|T(\chi_{A_1},\dots,\chi_{A_m})\|_Z \le C|A_1|^{\frac{1}{p_1}}\cdots|A_m|^{\frac{1}{p_m}}$$

for all characteristic functions of sets A_j of finite measure. Show that T has an extension that maps $L^{p_1,1}\times\cdots\times L^{p_m,1}$ to Z.

7.1.3. On the real line consider the bilinear operator given by

$$S(f,g)(x) = \int_{-\infty}^{x} f'(t)g(t)\,dt$$

defined for f,g smooth functions with compact support $\mathscr{C}_0^\infty$.
(a) Show that for any $0<p,q,r<\infty$ there is no constant C such that the estimate

$$\big\|S(f,g)\big\|_{L^r(\mathbf{R})} \le C\big\|f\big\|_{L^p(\mathbf{R})}\big\|g\big\|_{L^q(\mathbf{R})}$$

holds for all $\mathscr{C}_0^\infty$ functions f,g on the real line.
(b) Show that the estimate on $\mathscr{C}_0^\infty$ functions in part (a) also fails for any p, q, r, with $0<p,q,r<\infty$, if $\mathbf{R}$ is replaced by a compact interval $[a,b]$.
[*Hint:* Part (a): Take $f(x)=1$ when $x\in[-1,1]$ and $f(x)=0$ for $|x|\ge 1.1$, and let $g(x)=f(x-1/2)$. Part (b): Take $[a,b]=[-2,2]$. Given $\varepsilon>0$, let f_ε be the $\mathscr{C}_0^\infty$ function supported in $[-1,1]$ with $f_\varepsilon(t)=|t|^{-1/(2p)}$ for $\varepsilon\le|t|\le 1/2$ and $f_\varepsilon(t)=(\varepsilon/2)^{-1/(2p)}$ for $|t|\le\varepsilon/2$, and let $g_\varepsilon(t)$ be a smooth bump function supported in $[-\varepsilon/2,2]$, with $0\le g_\varepsilon\le 1$ and having $g_\varepsilon(t)=1$ on $\varepsilon/4\le t\le 1$.]

7.1.4. Let $K(y_0,y_1\dots,y_m)$ be a function on $\mathbf{R}^{(m+1)n}$ such that for all $0\le i\le m$ we have

$$\sup_{y_i\in\mathbf{R}^n}\int_{\mathbf{R}^{mn}}|K(y_0,y_1,\dots,y_m)|\,dy_0\cdots dy_{i-1}dy_{i+1}\cdots dy_m = A_i<\infty.$$

Then the m-linear operator

$$\mathscr{T}(f_1,\dots,f_m)(x) = \int_{\mathbf{R}^{mn}} K(x,y_1\dots,y_m)f_1(y_1)\cdots f_m(y_m)\,dy_1\cdots dy_m$$

maps $L^{p_1}(\mathbf{R}^n)\times\cdots\times L^{p_m}(\mathbf{R}^n)\to L^p(\mathbf{R}^n)$ with bound

$$A_0^{\frac{1}{p'}}A_1^{\frac{1}{p_1}}\cdots A_m^{\frac{1}{p_m}}$$

whenever $1/p_1+\cdots+1/p_m=1/p$ where $1\le p_1,\dots,p_m,p\le\infty$.

7.1.5. Let $X_1,\dots,X_m$ be σ-finite measure spaces equipped with nonnegative measures μ_j, $j=1,\dots,m$. Let (X,μ) be another σ-finite measure space. Suppose that $K(x,x_1,\dots,x_m)$ is a nonnegative measurable function on the product space $X\times X_1\times\cdots\times X_m$. Consider the m-linear operator T with kernel K, that is,

$$T(f_1,\dots,f_m)(x) = \int_{X_1}\cdots\int_{X_m} K(x,x_1,\dots,x_m)f_1(x_1)\cdots f_m(x_m)\,d\mu_1(x_1)\cdots d\mu_m(x_m),$$

defined for suitable measurable functions f_j on X_j. Fix indices $1 < p_1, \dots, p_m, p < \infty$ satisfying $\frac{1}{p_1} + \cdots + \frac{1}{p_m} = \frac{1}{p}$. Suppose that there exist measurable functions u_j on X_j, $j = 1, \dots, m$ and u on X, with $0 < u_1, \dots, u_m, u < \infty$ a.e., such that for some $B > 0$ we have

$$T(u_1^{p_1'}, u_2^{p_2'}, \dots, u_m^{p_m'}) \le B u^{p'} \qquad \mu\text{-a.e.}$$

$$T^{*1}(u^p, u_2^{p_2'}, \dots, u_m^{p_m'}) \le B u_1^{p_1} \qquad \mu\text{-a.e.}$$

$$\cdots$$

$$T^{*m}(u_1^{p_1'}, u_2^{p_2'}, \dots, u^p) \le B u_m^{p_m} \qquad \mu\text{-a.e.}$$

Show that T extends to a bounded operator from $L^{p_1}(X_1) \times \cdots \times L^{p_m}(X_m)$ to $L^p(X)$ with norm at most B.

[*Hint:* When $m = 2$ fix $f_1 \in L^{p_1}$, $f_2 \in L^{p_2}$ and $f \in L^{p'}$ nonnegative functions. Write $K(x,x_1,x_2)f(x)f_1(x_1)f_2(x_2) = L(x,x_1,x_2)M(x,x_1,x_2)N(x,x_1,x_2)$ where

$$L(x,x_1,x_2) = f(x)\frac{u_1(x_1)^{p_1'/p'}u_2(x_2)^{p_2'/p'}}{u(x)}K(x,x_1,x_2)^{1/p'},$$

$$M(x,x_1,x_2) = f_1(x_1)\frac{u(x)^{p/p_1}u_2(x_2)^{p_2'/p_1}}{u_1(x_1)}K(x,x_1,x_2)^{1/p_1}, \quad \text{and}$$

$$N(x,x_1,x_2) = f_2(x_2)\frac{u_1(x_1)^{p_1'/p_2}u(x)^{p/p_2}}{u_2(x_2)}K(x,x_1,x_2)^{1/p_2}$$

and apply Hölder's inequality.]

7.1.6. Let $1 \le p_j \le \infty$, $1/p_1 + \cdots + 1/p_m = 1/r \le 1$ and let μ be a nonnegative regular Borel measure. Then the following statements are equivalent:
(a) $T^\mu : L^{p_1}(\mathbf{R}^n) \times \cdots \times L^{p_m}(\mathbf{R}^n) \to L^r(\mathbf{R}^n)$.
(b) $T^\mu : L^{p_1}(\mathbf{R}^n) \times \cdots \times L^{p_m}(\mathbf{R}^n) \to L^{r,\infty}(\mathbf{R}^n)$.
(c) μ is a finite measure.

7.1.7. ([98]) Define a linear operator L acting on functions on the real line by setting

$$L(h)(x) = \sum_{j=1}^{\infty} \lambda_j h(x-j)$$

for some sequence of positive numbers λ_j. Let $0 < p < 1$.
(a) Show that if $\{\lambda_j\}_{j=1}^\infty$ lies in $\ell^{p,\infty}(\mathbf{Z}^+)$, then L maps $L^p(\mathbf{R})$ to $L^{p,\infty}(\mathbf{R})$. Here $\ell^{p,\infty}(\mathbf{Z}^+)$ denotes the space $L^{p,\infty}(\mathbf{Z}^+)$ equipped with counting measure.
(b) Show that if L maps $L^p(\mathbf{R})$ to $L^p(\mathbf{R})$, then $\{\lambda_j\}_{j=1}^\infty$ lies in $\ell^p(\mathbf{Z}^+)$.

7.1.8. ([171]) Suppose that $1 \le p, q < \infty$, $1/p + 1/q = 1/r \ge 1$, and that K is a nonnegative function on $\mathbf{R}^n \times \mathbf{R}^n$ that satisfies

$$\int_{\mathbf{R}^n}\int_{\mathbf{R}^n} \frac{|K(y_1,y_2)|^r}{(|y_1|^n|y_2|^n)^{1-r}}\,dy_1 dy_2 < \infty.$$

Moreover, assume that $|y_1| \le |y_1'|$ implies $|K(y_1,y_2)| \ge |K(y_1',y_2)|$, and $|y_2| \le |y_2'|$ implies $|K(y_1,y_2)| \ge |K(y_1,y_2')|$ for all $y_1,y_1',y_2,y_2' \in \mathbf{R}^n$. Show that T^K maps $L^p(\mathbf{R}^n) \times L^q(\mathbf{R}^n)$ to $L^r(\mathbf{R}^n)$.

[*Hint:* Define $\varphi(t,s) = K(y,z)$ whenever $t = |y|$ and $s = |z|$ and for each j_1, j_2 integers set $I_{j_l} = \{2^{j_l} < |y_l| \le 2^{j_l+1}\}$ and

$$K_{j_1,j_2}(y_1,y_2) = K(y_1,y_2)\chi_{I_{j_1}}(y_1)\chi_{I_{j_2}}(y_2).$$

Use that

$$T^K(f_1,f_2)(x) \le \sum_{j_1\in\mathbf{Z}}\sum_{j_2\in\mathbf{Z}} \varphi(2^{j_1},2^{j_2}) \int_{I_{j_1}} |f_1(x-y_1)|\,dy_1 \int_{I_{j_2}} |f_2(x-y_2)|\,dy_2$$

and pass the L^r quasi-norm through the double sum.]

7.1.9. (D. Bilyk) Show that the *bilinear Hilbert transform*

$$\mathscr{H}(f,g)(x) = \frac{1}{\pi}\lim_{\varepsilon\to 0}\int_{|t|\ge\varepsilon} f(x-t)g(x+t)\,\frac{dt}{t}$$

does not map $L^1(\mathbf{R}) \times L^p(\mathbf{R}) \to L^{\frac{p}{p+1}}(\mathbf{R})$ when $1 \le p \le \infty$.

[*Hint:* Apply this operator to the functions $f(x) = \chi_{|x|\le 1}$, $g_N(x) = x^{-\frac{1}{p}}\chi_{[1,N]}$ and estimate the $L^{\frac{p}{p+1}}$ quasi-norm of $\mathscr{H}$ on the interval $[2, \frac{N-1}{2}]$ for N large.]

7.2 Multilinear Interpolation

In this section we study topics concerning the real interpolation of multilinear operators. We focus on two results, m-linear interpolation for a single operator between estimates at $m+1$ points and multilinear interpolation between adjoint operators.

7.2.1 Real Interpolation for Multilinear Operators

Given a measure space (X,μ), we denote by $S(X)$ the space of functions of the form $\sum_{i=1}^N c_i\chi_{E_i}$, where each measurable subset E_i of X has finite measure. We call such functions *finitely simple*. We denote by $S_0^+(X)$ the space of all simple functions on X that have the form $f = \sum_{i=n_1}^{n_2} 2^{-i}\chi_{E_i}$, where E_i are subsets of X of finite measure, with $\mu(E_{n_1}) \ne 0$ and $\mu(E_{n_2}) \ne 0$, and n_1 and n_2 are integers such that $n_1 < n_2$. We also denote by $S_0^+(X) - S_0^+(X)$ the set of functions of the form $f-g$, where $f,g \in S_0^+(X)$, and we denote by $S_0(X) = (S_0^+(X) - S_0^+(X)) + i(S_0^+(X) - S_0^+(X))$ the space of all functions of the form $f_1 + if_2$, where $f_1, f_2 \in S_0^+(X) - S_0^+(X)$. It is shown in Proposition 1.4.21 in [156] that $S_0(X)$ is dense in the Lorentz space $L^{p,s}(X,\mu)$ whenever $0 < p,s < \infty$.

Let $X_1,\dots,X_m$ be measure spaces. An operator T defined on $S(X_1)\times\cdots\times S(X_m)$ and taking values in the set of complex-valued measurable functions on a measure space (Y,ν) is called *multisublinear* if for all $1\le j\le m$, all f_j,g_j in $S(X_j)$, and all $\lambda\in\mathbf{C}$ the statements

$$|T(f_1,\dots,\lambda f_j,\dots,f_m)| = |\lambda|\,|T(f_1,\dots,f_j,\dots,f_m)|, \tag{7.2.1}$$

$$|T(\dots,f_j+g_j,\dots)| \le |T(\dots,f_j,\dots)| + |T(\dots,g_j,\dots)| \tag{7.2.2}$$

hold ν-a.e.

We now state a multilinear extension of Lemma 1.4.20 in [156] concerning multisublinear operators.

Proposition 7.2.1. *Let (X_j,μ_j), $j=1,\dots,m$, (Y,ν) be σ-finite measure spaces. Let T be an operator defined on $S(X_1)\times\cdots\times S(X_m)$ and taking values into the set of measurable functions on Y that satisfies (7.2.1) and (7.2.2). For $j=1,\dots,m$, let $0<p_j<\infty$ and $0<q\le\infty$. Suppose that for some constant $M>0$ and for all measurable subsets E_j of X_j of finite measure we have*

$$\big\|T(\chi_{E_1},\dots,\chi_{E_m})\big\|_{L^{q,\infty}} \le M\prod_{j=1}^{m}\mu_j(E_j)^{\frac{1}{p_j}}.$$

Then for all δ satisfying $0<\delta<\min(1,q)$ and all f_j in $S_0(X_j)$ we have

$$\|T(f_1,\dots,f_m)\|_{L^{q,\infty}} \le C_0(m,\delta,p_1,\dots,p_m,q)\,M\prod_{j=1}^{m}\|f_j\|_{L^{p_j,\delta}}, \tag{7.2.3}$$

where

$$C_0(m,\delta,p_1,\dots,p_m,q) = 2^{4m+\frac{2m}{q}}\left(\tfrac{q}{q-\delta}\right)^{\frac{2}{\delta}}(1-2^{-\delta})^{-\frac{1}{\delta}}\,2^{\frac{2}{p_1}+\cdots+\frac{2}{p_m}}(\log 2)^{-\frac{m}{\delta}}.$$

Proof. The proof of (7.2.3) is based on a straightforward multilinear extension of Lemma 1.4.20 in [156]. Exercise 7.2.4 outlines the steps of the solution. □

We now introduce some notation that will be used in the main interpolation result of this section. First, $1/q$ is defined to be zero when $q=\infty$. Let m be a positive integer. For $1\le k\le m+1$ and $1\le j\le m$, we are given $p_{k,j}$, with $0<p_{k,j}\le\infty$ and $0<q_k\le\infty$. We introduce the determinants

$$\gamma_0 = \det\begin{pmatrix} 1/p_{1,1} & 1/p_{1,2} & \dots & \dots & 1/p_{1,m} & 1\\ 1/p_{2,1} & 1/p_{2,2} & \dots & \dots & 1/p_{2,m} & 1\\ \vdots & \vdots & \vdots & \vdots & \vdots & \vdots\\ 1/p_{m,1} & 1/p_{m,2} & \dots & \dots & 1/p_{m,m} & 1\\ 1/p_{m+1,1} & 1/p_{m+1,2} & \dots & \dots & 1/p_{m+1,m} & 1 \end{pmatrix}$$

and for each $j = 1, 2, \dots, m$

$$\gamma_j = \det \begin{pmatrix} 1/p_{1,1} & 1/p_{1,2} & \dots & -1/q_1 & \dots & 1/p_{1,m} & 1 \\ 1/p_{2,1} & 1/p_{2,2} & \dots & -1/q_2 & \dots & 1/p_{2,m} & 1 \\ \vdots & \vdots & \vdots & \vdots & \vdots & \vdots & \vdots \\ 1/p_{m,1} & 1/p_{m,2} & \dots & -1/q_m & \dots & 1/p_{m,m} & 1 \\ 1/p_{m+1,1} & 1/p_{m+1,2} & \dots & -1/q_{m+1} & \dots & 1/p_{m+1,m} & 1 \end{pmatrix}, \tag{7.2.4}$$

where the jth column of the determinant defining γ_j is obtained by replacing the jth column of the determinant defining γ_0 with the vector $(-1/q_1, \dots, -1/q_m, -1/q_{m+1})$.

We explain the geometric meaning of the determinant γ_0. For $k = 1, 2, \dots, m+1$ let

$$\vec{P}_k = \Big(\frac{1}{p_{k,1}}, \frac{1}{p_{k,2}}, \dots, \frac{1}{p_{k,m}}\Big)$$

be points in $\mathbf{R}^m$. Let $\mathbf{H}$ be the open convex hull of the points $\vec{P}_1, \dots, \vec{P}_{m+1}$. Then $\mathbf{H}$ is an open subset of $\mathbf{R}^m$ whose m-dimensional volume is $|\gamma_0|/m!$. Hence $\mathbf{H}$ is a nonempty set if and only if $\gamma_0 \neq 0$. Thus, the condition $\gamma_0 \neq 0$ is equivalent to the fact that the open convex hull of $\vec{P}_1, \dots, \vec{P}_{m+1}$ is a nontrivial open simplex in $\mathbf{R}^m$. The boundary of $\mathbf{H}$ is denoted by $\partial\mathbf{H}$.

We now state the multilinear version of the Marcinkiewicz interpolation theorem with initial restricted weak type conditions.

Theorem 7.2.2. *Let m be a positive integer, and let $X_1, \dots, X_m$ be σ-finite measure spaces. Suppose that T is a multisublinear operator defined on $S(X_1) \times \cdots \times S(X_m)$ that takes values in the set of measurable functions of a σ-finite measure space (Y, ν). For $1 \le k \le m+1$ and $1 \le j \le m$ we are given $p_{k,j}$, with $0 < p_{k,j} \le \infty$ and $0 < q_k \le \infty$. Suppose that the open convex hull of the points*

$$\vec{P}_k = \Big(\frac{1}{p_{k,1}}, \frac{1}{p_{k,2}}, \dots, \frac{1}{p_{k,m}}\Big)$$

is an open set in $\mathbf{R}^m$, in other words, $\gamma_0 \neq 0$. Assume that T satisfies

$$\|T(\chi_{E_1}, \dots, \chi_{E_m})\|_{L^{q_k,\infty}} \le B_k \prod_{j=1}^{m} \mu_j(E_j)^{\frac{1}{p_{k,j}}} \tag{7.2.5}$$

for all $1 \le k \le m+1$ and for all subsets E_j of X_j with $\mu_j(E_j) < \infty$. Let

$$\vec{P} = \Big(\frac{1}{p_1}, \dots, \frac{1}{p_m}\Big) = \sum_{k=1}^{m+1} \eta_k \vec{P}_k \tag{7.2.6}$$

for some $\eta_k \in (0,1)$ such that $\sum_{k=1}^{m+1} \eta_k = 1$, and define

$$\frac{1}{q} = \sum_{k=1}^{m+1} \frac{\eta_k}{q_k}. \tag{7.2.7}$$

For each $j \in \{1,2,\dots,m\}$ let s_j satisfy $0 < s_j \le \infty$, and let

$$\frac{1}{s} = \sum_{\substack{1\le j\le m\\ \gamma_j\neq 0}} \frac{1}{s_j}, \tag{7.2.8}$$

with the understanding that if there is no j, with $\gamma_j \neq 0$, then the sum in (7.2.8) is zero and $s = \infty$. Let

$$0 < \delta < \min\Big(\frac{q_1}{2},\frac{q_2}{2},\dots,\frac{q_{m+1}}{2},s_1,s_2,\dots,s_m,1\Big). \tag{7.2.9}$$

Under these assumptions, there is a finite constant $\mathbf{C}(m,\delta,p_{k,i},q_k,p_i,s_i)$ such that

$$\|T(f_1,\dots,f_m)\|_{L^{q,s}} \le \frac{\mathbf{C}(m,\delta,p_{k,i},q_k,p_i,s_i)}{\min(1,\operatorname{dist}(\vec{P},\partial\mathbf{H}))^{\frac{m}{\delta}}}\Big(\prod_{k=1}^{m+1} B_k^{\eta_k}\Big)\Big(\prod_{j=1}^{m}\|f_j\|_{L^{p_j,s_j}}\Big) \tag{7.2.10}$$

for all $f_j \in S_0(X_j)$, where for some other constant $\mathbf{C}_*(m,\delta,p_{k,i},q_k)$ we have

$$\mathbf{C}(m,\delta,p_{k,i},q_k,p_i,s_i) = \mathbf{C}_*(m,\delta,p_{k,i},q_k)\max(1,2^{\frac{m(1-s)}{s}})\prod_{\substack{1\le j\le m\\ \gamma_j\neq 0}}\Big|\frac{\gamma_0}{\gamma_j}\Big|^{\frac{1}{s_j}}\prod_{\substack{1\le j\le m\\ \gamma_j=0}}\Big(\frac{s_j}{p_j}\Big)^{\frac{1}{s_j}}.$$

This theorem is proved in Subsection 7.2.2. In what follows, we discuss some related remarks and corollaries. First, we notice that since $\gamma_0 \neq 0$ and $\eta_k > 0$ for all k, (7.2.6) implies that $p_j < \infty$ for all $j = 1,\dots,m$. Theorem 7.2.2 can be strengthened when combined with the result of the following proposition when $s_j < \infty$.

Proposition 7.2.3. *Let (X_j,μ_j), (Y,ν) be σ-finite measure spaces, and let T be a multilinear operator defined on $S(X_1)\times\cdots\times S(X_m)$ and taking values in the set of measurable functions on Y. Let $0 < q,s \le \infty$ and $0 < p_j,t_j < \infty$ for all $1 \le j \le m$. Suppose that*

$$\|T(f_1,\dots,f_m)\|_{L^{q,s}} \le M\prod_{j=1}^{m}\|f_j\|_{L^{p_j,t_j}} \tag{7.2.11}$$

holds for some fixed positive constant M and all f_j in $S_0(X_j)$. Then T has a unique bounded extension from $L^{p_1,t_1}(X_1)\times\cdots\times L^{p_m,t_m}(X_m)$ to $L^{q,s}(Y,\nu)$ that satisfies (7.2.11) for all functions $f_j \in L^{p_j,t_j}(X_j)$. Moreover, the hypothesis that T is multilinear can be replaced by the hypothesis that T is multisublinear with the property $T(f_1,\dots,f_m) \ge 0$ for all functions in its domain.

Proof. We show that (7.2.11) is valid for general functions in $L^{p_1,t_1}\times\cdots\times L^{p_m,t_m}$. For any $j = 1,2,\dots,m$ and $f_j \in L^{p_j,t_j}$, in view of the density of $S_0(X_j)$ in L^{p_j,t_j}, which is valid since $0 < t_j < \infty$, there exists a sequence $\{f_j^{(n)}\}_{n=1}^{\infty}$ contained in $S_0(X_j)$ such that

$$\lim_{n\to\infty}\|f_j^{(n)} - f_j\|_{L^{p_j,t_j}} = 0$$

and such that for all $j \in \{1,\dots,m\}$ we have

$$\|f_j^{(n)}\|_{L^{p_j,t_j}} \le 2\|f_j\|_{L^{p_j,t_j}}.$$

Note that the multilinearity (or multisublinearity and nonnegativity) of T implies that for all functions $g_1, f_1, \dots, f_m$ in its domain, we have

$$|T(f_1,f_2,\dots,f_m) - T(g_1,f_2,\dots,f_m)| \le |T(f_1-g_1,f_2,\dots,f_m)|$$

and an analogous inequality for the second through mth entries.

For all nonnegative integers n,i we use the multisublinearity of T to write

$$\begin{aligned}
&|T(f_1^{(n)},\dots,f_m^{(n)}) - T(f_1^{(i)},\dots,f_m^{(i)})| \\
&\quad \le \sum_{j=1}^{m} |T(f_1^{(i)},\dots,f_{j-1}^{(i)},f_j^{(n)},\dots,f_m^{(n)}) - T(f_1^{(i)},\dots,f_j^{(i)},f_{j+1}^{(n)},\dots,f_m^{(n)})| \\
&\quad \le \sum_{j=1}^{m} |T(f_1^{(i)},\dots,f_{j-1}^{(i)},f_j^{(n)}-f_j^{(i)},f_{j+1}^{(n)},\dots,f_m^{(n)})|.
\end{aligned}$$

This implies that

$$\begin{aligned}
&\|T(f_1^{(n)},\dots,f_m^{(n)}) - T(f_1^{(i)},\dots,f_m^{(i)})\|_{L^{q,s}} \\
&\quad \le 2^{\frac{m}{q}} \max(1,2^{\frac{m(1-s)}{s}}) \sum_{j=1}^{m} \|T(f_1^{(i)},\dots,f_{j-1}^{(i)},f_j^{(n)}-f_j^{(i)},f_{j+1}^{(n)},\dots,f_m^{(n)})\|_{L^{q,s}} \\
&\quad \le 2^{\frac{m}{q}} \max(1,2^{\frac{m(1-s)}{s}}) M \sum_{j=1}^{m} \left(\|f_j^{(n)} - f_j^{(i)}\|_{L^{p_j,t_j}} \prod_{\substack{1\le k\le m\\ k\ne j}} 2\|f_k\|_{L^{p_k,t_k}} \right),
\end{aligned}$$

which tends to 0 as $n,i\to\infty$. Thus, $\{T(f_1^{(n)},\dots,f_m^{(n)})\}_{n=1}^{\infty}$ is a Cauchy sequence in $L^{q,s}$, and it converges to some element in $L^{q,s}$, so it makes sense to define

$$\overline{T}(f_1,\dots,f_m) = \lim_{n\to\infty} T(f_1^{(n)},\dots,f_m^{(n)}) \quad \text{in } L^{q,s}. \tag{7.2.12}$$

A similar argument shows that if, for $j=1,2,\dots,m$, $\{g_j^{(n)}\}_{n=1}^{\infty}$ is another sequence contained in $S_0(X_j)$ that converges to f_j in L^{q_j,t_j}, then

$$\overline{T}(f_1,\dots,f_m) = \lim_{n\to\infty} T(g_1^{(n)},\dots,g_m^{(n)}) \quad \text{in } L^{q,s}.$$

Therefore, $\overline{T}$ is a well-defined operator. It follows from (7.2.12) and Exercise 1.4.11(b) in [156] that $\|T(g_1^{(n)},\dots,g_m^{(n)})\|_{L^{q,s}} \to \|\overline{T}(f_1,\dots,f_m)\|_{L^{q,s}}$ as $n\to\infty$ and thus for all functions $(f_1,\dots,f_m) \in L^{p_1,t_1}\times\cdots\times L^{p_m,t_m}$ we have

$$\begin{aligned}\big\|\overline{T}(f_1,\dots,f_m)\big\|_{L^{q,s}} &= \lim_{n\to\infty}\big\|T(f_1^{(n)},\dots,f_m^{(n)})\big\|_{L^{q,s}}\\ &\le M\limsup_{n\to\infty}\prod_{j=1}^m \|f_j^{(n)}\|_{L^{p_j,t_j}}\\ &= M\prod_{j=1}^m \|f_j\|_{L^{p_j,t_j}},\end{aligned}$$

where the last equality is also a consequence of Exercise 1.4.11(a) in [156]. To show that $\overline{T}$ is multilinear (or nonnegative and multisublinear) we note that each $\overline{T}(f_1,\dots,f_m)$ is the ν-a.e. limit of a subsequence of $T(f_1^{(n)},\dots,f_m^{(n)})$, and linearity, positivity, and sublinearity are preserved by limits. This shows that each T has a bounded extension $\overline{T}$ from $L^{p_1,t_1}(X_1)\times\cdots\times L^{p_m,t_m}(X_m)$ to $L^{q,s}(Y,\nu)$ with the same norm and concludes the proof. □

Corollary 7.2.4. *Under the hypotheses of Theorem 7.2.2, assume additionally that $\gamma_j\neq 0$ for all $j=1,\dots,m$ and that, instead of* (7.2.8), *the inequality holds:*

$$\frac{1}{q}\le\frac{1}{p_1}+\cdots+\frac{1}{p_m}. \tag{7.2.13}$$

Then, for any $0<\delta<\min\big(\frac{q_1}{2},\frac{q_2}{2},\dots,\frac{q_{m+1}}{2},1,p_1,\dots,p_m\big)$, there is a positive constant $\mathbf{C}_{**}(m,\delta,p_{k,i},q_k)$ such that T satisfies

$$\|T(f_1,\dots,f_m)\|_{L^q}\le\frac{\mathbf{C}_{**}(m,\delta,p_{k,i},q_k)}{\min(1,\operatorname{dist}(\vec{P},\partial\mathbf{H}))^{\frac{m}{\delta_0}}}\Big(\prod_{k=1}^{m+1}B_k^{\eta_k}\Big)\Big(\prod_{j=1}^m\|f_j\|_{L^{p_j}}\Big) \tag{7.2.14}$$

for all $f_j\in S_0(X_j)$. Moreover, T has a unique bounded extension that satisfies (7.2.14) *for all f_j in $L^{p_j}(X_j)$.*

Proof. Using (7.2.6), we see that if $p_i=\infty$ for some i, then $\gamma_0=0$. Thus $p_j<\infty$ for all j, and we pick $s_j=p_j<\infty$ in (7.2.10) and define s by $\frac{1}{s}=\frac{1}{p_1}+\cdots+\frac{1}{p_m}$, so that (7.2.8) is valid. Then, in view of (7.2.13), we have $q\ge s$, and thus

$$\|T(f_1,\dots,f_m)\|_{L^q}\le\Big(\frac{s}{q}\Big)^{\frac{1}{s}-\frac{1}{q}}\|T(f_1,\dots,f_m)\|_{L^{q,s}}\le\|T(f_1,\dots,f_m)\|_{L^{q,s}}.$$

Theorem 7.2.2 implies the assertion, but to derive the claimed form of the constant in (7.2.14), we observe the following. For $1\le j\le m$ we have

$$\frac{1}{p_j}\le\sum_{k=1}^{m+1}\frac{1}{p_{k,j}}, \tag{7.2.15}$$

which implies that

$$\max(1,2^{\frac{m(1-s)}{s}}) \prod_{\substack{1\le j\le m\\ \gamma_j\neq 0}} \left|\frac{\gamma_0}{\gamma_j}\right|^{\frac{1}{s_j}} \prod_{\substack{1\le j\le m\\ \gamma_j=0}} \left(\frac{s_j}{p_j}\right)^{\frac{1}{s_j}} \le \left[1+2^{m\sum_{j=1}^{m}\sum_{k=1}^{m+1}\frac{1}{p_{k,j}}}\right] \prod_{\substack{1\le j\le m\\ \gamma_j\neq 0}} \left|\frac{\gamma_0}{\gamma_j}\right|^{\sum_{k=1}^{m+1}\frac{1}{p_{k,j}}}.$$

Then we define the constant $\mathbf{C}_{**}(m,\delta,p_{k,i},q_k)$ in (7.2.14) to be the product of the constant on the right-hand side of the preceding inequality times $\mathbf{C}_{*}(m,\delta,p_{k,i},q_k)$, which appears in the statement of Theorem 7.2.2. Passing from $S_0(X_j)$ to $L^{p_j}(X_j)$ follows from Proposition 7.2.3 via the multisublinearity of T since $p_j<\infty$. □

Remark 7.2.5. Suppose that $\gamma_j=0$ for all $j\in\{1,2,\dots,m\}$ in Theorem 7.2.2. Then we have $q_1=q_2=\cdots=q_{m+1}=q$. Moreover, there is a positive constant $\mathbf{C}_{***}(m,\delta,p_{k,i},q)$ such that T satisfies

$$\|T(f_1,\dots,f_m)\|_{L^{q,\infty}} \le \frac{\mathbf{C}_{***}(m,\delta,p_{k,i},q)}{\min(1,\mathrm{dist}\,(\vec{P},\partial\mathbf{H}))^{\frac{m}{\delta}}} \Big(\prod_{k=1}^{m+1} B_k^{\eta_k}\Big)\Big(\prod_{j=1}^{m}\|f_j\|_{L^{p_j,\infty}}\Big) \tag{7.2.16}$$

for all $f_j\in S_0(X_j)$, where δ satisfies $0<\delta<\min\left(\frac{q}{2},1\right)$. Consequently, if $s_j<\infty$ for all $j\in\{1,2,\dots,m\}$, then the operator T has a unique bounded extension from $L^{p_1,s_1}(X_1)\times\cdots\times L^{p_m,s_m}(X_m)$ to $L^{q,\infty}(Y,\nu)$.

We first show that if $\gamma_j=0$ for all j, then $q_1=\cdots=q_{m+1}$. We define vectors

$$\vec{1}=(1,1,\dots,1),\qquad \vec{Q}=(1/q_1,\dots,1/q_{m+1}),$$

and for each $j\in\{1,2,\dots,m\}$ we also define

$$\vec{A}_j=(1/p_{1,j},1/p_{2,j},\dots,1/p_{m+1,j}).$$

Then $(\vec{A}_1,\vec{A}_2,\dots,\vec{A}_m,\vec{1})$ is linearly independent since $\gamma_0\neq 0$. If all $\gamma_j=0$, this means that for each $j\in\{1,2,\dots,m\}$, the set

$$\{\vec{A}_1,\vec{A}_2,\dots,\vec{A}_{j-1},\vec{Q},\vec{A}_{j+1},\dots,\vec{A}_m,\vec{1}\}$$

is linearly dependent. Therefore, for any $j\in\{1,2,\dots,m\}$ we can write

$$\vec{Q}=\sum_{\substack{1\le i\le m\\ i\neq j}} a_i^{(j)}\vec{A}_i+c^{(j)}\vec{1},$$

where $a_i^{(j)}$ and $c^{(j)}$ are constants. Equivalently,

$$\begin{cases} \vec{Q} = 0 + a_2^{(1)}\vec{A}_2 + a_3^{(1)}\vec{A}_3 + \cdots + a_{m-1}^{(1)}\vec{A}_{m-1} + a_m^{(1)}\vec{A}_m + c^{(1)}\vec{1} \\ \vec{Q} = a_1^{(2)}\vec{A}_1 + 0 + a_3^{(2)}\vec{A}_3 + \cdots + a_{m-1}^{(2)}\vec{A}_{m-1} + a_m^{(2)}\vec{A}_m + c^{(2)}\vec{1} \\ \vdots \\ \vec{Q} = a_1^{(m)}\vec{A}_1 + a_2^{(m)}\vec{A}_2 + a_3^{(m)}\vec{A}_3 + \cdots + a_{m-1}^{(m)}\vec{A}_{m-1} + 0 + c^{(m)}\vec{1}. \end{cases}$$

Consider $j = 1$ and $j = 2$. Then

$$\vec{0} = \vec{Q} - \vec{Q} = -a_1^{(2)}\vec{A}_1 + a_2^{(1)}\vec{A}_2 + \sum_{i=3}^{m}(a_i^{(1)} - a_i^{(2)})\vec{A}_i + (c^{(1)} - c^{(2)})\vec{1},$$

which combined with the fact that $(\vec{A}_1, \vec{A}_2, \dots, \vec{A}_m, \vec{1})$ is linearly independent implies that

$$a_2^{(1)} = 0.$$

Likewise, by considering $j = 1$ and $j = 3$ we obtain

$$\vec{0} = \vec{Q} - \vec{Q} = -a_1^{(3)}\vec{A}_1 + a_3^{(1)}\vec{A}_3 + \sum_{\substack{1 \le i \le m \\ i \ne 1, i \ne 3}} (a_i^{(1)} - a_i^{(3)})\vec{A}_i + (c^{(1)} - c^{(3)})\vec{1},$$

and consequently

$$a_3^{(1)} = 0.$$

Repeating the foregoing process implies that

$$a_4^{(1)} = \cdots = a_m^{(1)} = 0.$$

Therefore, $\vec{Q}$ is a constant multiple of the vector $\vec{1}$, that is, $q_1 = \cdots = q_{m+1}$. Then q is equal to these numbers as well. Thus, (7.2.16) holds.

The last assertion in Remark 7.2.5 is deduced from the embedding $\|f_j\|_{L^{p_j,\infty}} \le (s_j/p_j)^{1/s_j}\|f_j\|_{L^{p_j,s_j}}$ (see [156, Proposition 1.4.10]) and from the fact that $S_0(X_j)$ is dense in $L^{p_j,s_j}(X_j)$. Note that the distinction between $s_j = \infty$ and $s_j < \infty$ is due to the fact that $S_0(X_j)$ may not be dense in $L^{p_j,\infty}(X_j)$.

7.2.2 Proof of Theorem 7.2.2

Proof. If some $p_{j_0} = \infty$, then (7.2.6) implies that $p_{k,j_0} = \infty$ for all $k = 1, 2, \dots, m+1$; thus, $\gamma_0 = 0$, which is not assumed. Hence, we have $0 < p_j < \infty$ for all $j = 1, 2, \dots, m$.

Suppose that $0 < \rho_k < 1$ for all $1 \le k \le m+1$, and $\sum_{k=1}^{m+1} \rho_k = 1$. Let

$$\vec{R} = \left(\frac{1}{r_1}, \frac{1}{r_2}, \dots, \frac{1}{r_m}\right) = \sum_{k=1}^{m+1} \rho_k \vec{P}_k$$

be a point in $\mathbf{H}$, and define

$$\frac{1}{r} = \sum_{k=1}^{m+1} \frac{\rho_k}{q_k}.$$

It is a simple consequence of (7.2.5) that for all $E_j \subseteq X_j$, $1 \le j \le m$ of finite measure we have

$$\prod_{k=1}^{m+1} \|T(\chi_{E_1},\dots,\chi_{E_m})\|_{L^{q_k,\infty}}^{\rho_k} \le \left(\prod_{k=1}^{m+1} B_k^{\rho_k}\right) \prod_{j=1}^{m} \mu_j(E_j)^{\frac{1}{r_j}}.$$

But for any measurable function G, using $\sum_{k=1}^{m+1} \rho_k = 1$, we have

$$\|G\|_{L^{r,\infty}} \le \prod_{k=1}^{m+1} \|G\|_{L^{q_k,\infty}}^{\rho_k},$$

and this implies that

$$\|T(\chi_{E_1},\dots,\chi_{E_m})\|_{L^{r,\infty}} \le \left(\prod_{k=1}^{m+1} B_k^{\rho_k}\right) \prod_{j=1}^{m} \mu_j(E_j)^{\frac{1}{r_j}}. \tag{7.2.17}$$

Thus, T is of restricted weak type $(r_1,\dots,r_m,r)$, with constant proportional to $\prod_{k=1}^{m+1} B_k^{\rho_k}$.

In the sequel, we will make use of the set

$$S_m = \{(\sigma_{\ell,1},\sigma_{\ell,2},\dots,\sigma_{\ell,m}) : \ell = 1,2,\dots,2^m\}$$

of all possible m-tuples of the form $(\pm 1,\pm 1,\dots,\pm 1)$. Notice that elements of S_m lie in different 2^m-orthants of $\mathbf{R}^m$. Since all $p_j < \infty$, and since $\vec{P}$ lies in the open convex hull $\mathbf{H}$, we choose $\varepsilon > 0$ small enough such that $2\sqrt{m}\,\varepsilon$ is smaller than the distance from $\vec{P}$ to the boundary of the convex hull $\mathbf{H}$, i.e.,

$$\varepsilon < \min\left(1, \frac{\operatorname{dist}(\vec{P},\partial\mathbf{H})}{2\sqrt{m}}\right),$$

where $\partial\mathbf{H}$ is the set of all $(m-1)$-dimensional faces of $\mathbf{H}$.

Consider the system of equations

$$\begin{cases} \frac{1}{p_{1,1}}\theta_{\ell,1} + \frac{1}{p_{2,1}}\theta_{\ell,2} + \cdots + \frac{1}{p_{m+1,1}}\theta_{\ell,m+1} = \frac{1}{r_{\ell,1}} \\ \frac{1}{p_{1,2}}\theta_{\ell,1} + \frac{1}{p_{2,2}}\theta_{\ell,2} + \cdots + \frac{1}{p_{m+1,2}}\theta_{\ell,m+1} = \frac{1}{r_{\ell,2}} \\ \quad\vdots \qquad\qquad \vdots \qquad\quad \vdots \qquad\qquad \vdots \qquad\qquad \vdots \\ \frac{1}{p_{1,m}}\theta_{\ell,1} + \frac{1}{p_{2,m}}\theta_{\ell,2} + \cdots + \frac{1}{p_{m+1,m}}\theta_{\ell,m+1} = \frac{1}{r_{\ell,m}} \\ \theta_{\ell,1} + \theta_{\ell,2} + \cdots + \theta_{\ell,m+1} = 1, \end{cases}$$

which has a unique solution $(\theta_{\ell,1},\theta_{\ell,2},\dots,\theta_{\ell,m+1})$ since $\gamma_0 \neq 0$.

For all $\ell \in \{1,2,\dots,2^m\}$ and $j \in \{1,2,\dots,m\}$ define $r_{\ell,j}$ via

$$\frac{1}{r_{\ell,j}} - \frac{1}{p_j} = \varepsilon\sigma_{\ell,j} \tag{7.2.18}$$

and introduce vectors

$$\vec{R}_\ell = \left(\frac{1}{r_{\ell,1}}, \dots, \frac{1}{r_{\ell,m}}\right) = \sum_{k=1}^{m+1} \theta_{\ell,k} \vec{P}_k .$$

The choice of ε implies that the cube of side length 2ε centered at $\vec{P}$ belongs to the open convex hull $\mathbf{H}$. Moreover, since $\mathbf{H}$ lies in the orthant $[0,\infty)^m$, it follows that for all $j \in \{1,2,\dots,m\}$

$$2\sqrt{m}\,\varepsilon < \operatorname{dist}(\vec{P}, \partial\mathbf{H}) \le \frac{1}{p_j}. \tag{7.2.19}$$

From these and (7.2.18) we see that each $\vec{R}_\ell$ belongs to the open convex hull $\mathbf{H}$ and every $r_{\ell,j}$ is finite. Consequently, each $\theta_{\ell,k} \in (0,1)$.

Define the following matrix:

$$A = \begin{pmatrix} 1/p_{1,1} & 1/p_{2,1} & \dots & 1/p_{m+1,1} \\ 1/p_{1,2} & 1/p_{2,2} & \dots & 1/p_{m+1,2} \\ \vdots & \vdots & \vdots & \vdots \\ 1/p_{1,m} & 1/p_{2,m} & \dots & 1/p_{m+1,m} \\ 1 & 1 & \dots & 1 \end{pmatrix}.$$

For all $i,k \in \{1,2,\dots,m+1\}$ we denote by $D_{i,k}$ the determinant of the matrix obtained by deleting the ith row and kth column of matrix A. Since $\gamma_0 \neq 0$, it follows that not all these minor determinants are zero. Expanding the determinant (7.2.4) that defines γ_j along its jth column we obtain

$$\gamma_j = \sum_{k=1}^{m+1} (-1)^{j+k} \frac{1}{-q_k} D_{j,k} = -\sum_{k=1}^{m+1} (-1)^{j+k} \frac{1}{q_k} D_{j,k} . \tag{7.2.20}$$

For all $\ell = 1,2,\dots,2^m$, in view of (7.2.6) and $\sum_{k=1}^{m+1} \eta_k = 1$, we have that the $(m+1)$-tuple

$$(\theta_{\ell,1} - \eta_1, \theta_{\ell,2} - \eta_2, \dots, \theta_{\ell,m+1} - \eta_{m+1})$$

is a solution of the system

$$\begin{cases} \frac{1}{p_{1,1}}(\theta_{\ell,1} - \eta_1) + \dots + \frac{1}{p_{m+1,1}}(\theta_{\ell,m+1} - \eta_{m+1}) &= \frac{1}{r_{\ell,1}} - \frac{1}{p_1} \\ \frac{1}{p_{1,2}}(\theta_{\ell,1} - \eta_1) + \dots + \frac{1}{p_{m+1,2}}(\theta_{\ell,m+1} - \eta_{m+1}) &= \frac{1}{r_{\ell,2}} - \frac{1}{p_2} \\ \vdots \qquad \vdots \qquad \vdots & \vdots \\ \frac{1}{p_{1,m}}(\theta_{\ell,1} - \eta_1) + \dots + \frac{1}{p_{m+1,m}}(\theta_{\ell,m+1} - \eta_{m+1}) &= \frac{1}{r_{\ell,m}} - \frac{1}{p_m} \\ (\theta_{\ell,1} - \eta_1) + \dots + (\theta_{\ell,m+1} - \eta_{m+1}) &= 0. \end{cases}$$

This unique solution can be expressed as the ratio

$$\theta_{\ell,k}-\eta_k=\frac{\det\begin{pmatrix}1/p_{1,1} & 1/p_{2,1} & \dots & 1/r_{\ell,1}-1/p_1 & \dots & 1/p_{m+1,1}\\ 1/p_{1,2} & 1/p_{2,2} & \dots & 1/r_{\ell,2}-1/p_2 & \dots & 1/p_{m+1,2}\\ \vdots & \vdots & \vdots & \vdots & \vdots & \vdots\\ 1/p_{1,m} & 1/p_{2,m} & \dots & 1/r_{\ell,m}-1/p_m & \dots & 1/p_{m+1,m}\\ 1 & 1 & \dots & 0 & \dots & 1\end{pmatrix}}{\det\begin{pmatrix}1/p_{1,1} & 1/p_{2,1} & \dots & 1/p_{m+1,1}\\ 1/p_{1,2} & 1/p_{2,2} & \dots & 1/p_{m+1,2}\\ \vdots & \vdots & \vdots & \vdots\\ 1/p_{1,m} & 1/p_{2,m} & \dots & 1/p_{m+1,m}\\ 1 & 1 & \dots & 1\end{pmatrix}},$$

where these determinants are different only in the kth column. Expanding the determinant in the numerator, for all $k\in\{1,2,\dots,m+1\}$ and all $\ell\in\{1,2,\dots,2^m\}$, we deduce that

$$\theta_{\ell,k}-\eta_k=\sum_{j=1}^{m}\Big(\frac{1}{r_{\ell,j}}-\frac{1}{p_j}\Big)(-1)^{j+k}\frac{D_{j,k}}{\gamma_0}. \tag{7.2.21}$$

For any $\ell\in\{1,2,\dots,2^m\}$ we also define

$$\frac{1}{r_\ell}=\sum_{k=1}^{m+1}\frac{\theta_{\ell,k}}{q_k}. \tag{7.2.22}$$

Using these expressions and (7.2.7) we write

$$\begin{aligned}\frac{1}{q}-\frac{1}{r_\ell}=\sum_{k=1}^{m+1}\frac{\eta_k-\theta_{\ell,k}}{q_k}&=-\sum_{k=1}^{m+1}\frac{1}{q_k}\sum_{j=1}^{m}\Big(\frac{1}{r_{\ell,j}}-\frac{1}{p_j}\Big)(-1)^{j+k}\frac{D_{j,k}}{\gamma_0}\\ &=-\sum_{j=1}^{m}\Big(\frac{1}{r_{\ell,j}}-\frac{1}{p_j}\Big)\sum_{k=1}^{m+1}\frac{1}{q_k}(-1)^{j+k}\frac{D_{j,k}}{\gamma_0}\\ &=\sum_{j=1}^{m}\Big(\frac{1}{r_{\ell,j}}-\frac{1}{p_j}\Big)\frac{\gamma_j}{\gamma_0},\end{aligned} \tag{7.2.23}$$

where the last identity follows from (7.2.20).

We introduce some more notation. For any $j\in\{1,2,\dots,m\}$ and any k in $\{1,2,\dots,m+1\}$, set

$$t_{j,k}=(-1)^{j+k}\frac{D_{j,k}}{\gamma_0}$$

and then (7.2.21) can be written as

$$\eta_k = \theta_{\ell,k} - \sum_{j=1}^{m} \Big(\frac{1}{r_{\ell,j}} - \frac{1}{p_j}\Big) t_{j,k}\,. \tag{7.2.24}$$

Since the points $\vec{R}_\ell$ lie in the open convex hull $\mathbf{H}$ of the points $\vec{P}_j$, estimate (7.2.17) is valid for each $\vec{R}_\ell$, with $\theta_{\ell,k}$ in the place of ρ_k. Set

$$\widetilde{B}_\ell = \prod_{k=1}^{m+1} B_k^{\theta_{\ell,k}}\,.$$

In view of (7.2.17) we have

$$\|T(\chi_{E_1},\dots,\chi_{E_m})\|_{L^{r_\ell,\infty}} \le \widetilde{B}_\ell \prod_{j=1}^{m} \mu_j(E_j)^{\frac{1}{r_{\ell,j}}}$$

for all subsets E_j of X_j of finite measure. Let δ be a positive number satisfying (7.2.9). Observe that (7.2.9) and (7.2.22) imply

$$\delta < \min\Big(\frac{r_1}{2},\frac{r_2}{2},\dots,\frac{r_{2^m}}{2},1\Big)\,. \tag{7.2.25}$$

Then, it follows from Proposition 7.2.1 that

$$\|T(f_1,\dots,f_m)\|_{L^{r_\ell,\infty}} \le C_0(m,\delta,r_{\ell,i},r_\ell)\,\widetilde{B}_\ell \prod_{j=1}^{m} \|f_j\|_{L^{r_{\ell,j},\delta}} \tag{7.2.26}$$

for all functions $f_j \in S_0(X_j)$, where

$$C_0(m,\delta,p_1,\dots,r_{\ell_i},r_\ell) = 2^{4m+\frac{2m}{r_\ell}} \Big(\tfrac{r_\ell}{r_\ell-\delta}\Big)^{\frac{2}{\delta}} (1-2^{-\delta})^{-\frac{1}{\delta}} 2^{\frac{2}{r_{\ell,1}}+\dots+\frac{2}{r_{\ell,m}}} (\log 2)^{-\frac{m}{\delta}}\,.$$

Notice that (7.2.18) and (7.2.23), together with the fact $\varepsilon < 1$, imply that

$$\begin{aligned}
&\frac{1}{r_{\ell,1}}+\dots+\frac{1}{r_{\ell,m}}+\frac{m}{r_\ell}\\
&\le \frac{1}{p_1}+\dots+\frac{1}{p_m}+\frac{m}{q}+\Big|\frac{1}{r_{\ell,1}}-\frac{1}{p_1}\Big|+\dots+\Big|\frac{1}{r_{\ell,m}}-\frac{1}{p_m}\Big|+m\Big|\frac{1}{r_\ell}-\frac{1}{q}\Big|\\
&\le \frac{1}{p_1}+\dots+\frac{1}{p_m}+\frac{m}{q}+m+m\sum_{j=1}^{m}\frac{|\gamma_j|}{|\gamma_0|}\\
&\le \sum_{j=1}^{m}\sum_{k=1}^{m+1}\frac{1}{p_{k,j}}+m\sum_{k=1}^{m+1}\frac{1}{q_k}+m+m\sum_{j=1}^{m}\frac{|\gamma_j|}{|\gamma_0|},
\end{aligned}$$

where the last inequality is a consequence of the observation (7.2.15) and (7.2.7). Also, it follows from (7.2.25) that $\frac{r_\ell}{r_\ell-\delta} < 2$ for all $1 \le \ell \le 2^m$. Therefore, we can bound $C_0(m,\delta,r_{\ell,i},r_\ell)$ by

$$2^{\frac{2}{\delta}}(1-2^{-\delta})^{-\frac{1}{\delta}}\left(2^4(\log 2)^{-\frac{1}{\delta}}\right)^m 2^{\sum_{j=1}^m\sum_{k=1}^{m+1}\frac{2}{p_{k,j}}+\sum_{k=1}^{m+1}\frac{2m}{q_k}}\,2^{2m+2m\sum_{j=1}^m\frac{|\gamma_j|}{|\gamma_0|}} \tag{7.2.27}$$

for every ℓ. We denote the constant in (7.2.27) by $C_0'(m,\delta,p_{k,i},q_k)$. From this and (7.2.26) we obtain that for all functions $f_j \in S_0(X_j)$,

$$\|T(f_1,\dots,f_m)\|_{L^{r_\ell,\infty}} \le C_0'(m,\delta,p_{k,i},q_k)\,\widetilde{B}_\ell\prod_{j=1}^m \|f_j\|_{L^{r_{\ell,j},\delta}}. \tag{7.2.28}$$

For all $j=1,2,\dots,m$ fix functions f_j in $S_0(X_j)$, and for any $t>0$ write $f_j = f_{j,1,t}+f_{j,-1,t}$ by setting

$$f_{j,-1,t} = f_j\chi_{\{|f_j|>f_j^*(\lambda_j t^{-\frac{\gamma_j}{\gamma_0}})\}} \qquad\text{and}\qquad f_{j,1,t} = f_j\chi_{\{|f_j|\le f_j^*(\lambda_j t^{-\frac{\gamma_j}{\gamma_0}})\}} \tag{7.2.29}$$

for some $\lambda_j>0$ to be determined later. Here g^* is the decreasing rearrangement of g; see Definition 1.4.1 in [156].

Proposition 1.4.5 (6) in [156] and Exercise 1.1.5(c) in [156], together with the sublinearity of T and the quasilinearity of Lorentz norms, imply

$$\begin{aligned}
&\|T(f_1,\dots,f_m)\|_{L^{q,s}}\\
&= \|t^{\frac{1}{q}}T(f_1,\dots,f_m)^*(t)\|_{L^s(dt/t)}\\
&\le \Big\|t^{\frac{1}{q}}\Big(\sum_{i_1,\dots,i_m\in\{1,-1\}}|T(f_{1,i_1,t},\dots,f_{m,i_m,t})|\Big)^*(t)\Big\|_{L^s(dt/t)}\\
&\le \Big\|t^{\frac{1}{q}}\sum_{i_1,\dots,i_m\in\{1,-1\}}(|T(f_{1,i_1,t},\dots,f_{m,i_m,t})|)^*(t/2^m)\Big\|_{L^s(dt/t)}\\
&\le 2^{\frac{m}{q}}\max(1,2^{\frac{m(1-s)}{s}})\sum_{i_1,\dots,i_m\in\{1,-1\}}\|t^{\frac{1}{q}}(|T(f_{1,i_1,t},\dots,f_{m,i_m,t})|)^*(t)\|_{L^s(dt/t)}\\
&= 2^{\frac{m}{q}}\max(1,2^{\frac{m(1-s)}{s}})\sum_{\ell=1}^{2^m}\|t^{\frac{1}{q}}(|T(f_{1,\sigma_{\ell,1},t},\dots,f_{m,\sigma_{\ell,m},t})|)^*(t)\|_{L^s(dt/t)}
\end{aligned}$$

since each m-tuple $(i_1,\dots,i_m)$, with $i_j\in\{1,-1\}$, corresponds to a unique ℓ in $\{1,2,\dots,2^m\}$ such that $(i_1,\dots,i_m)=\sigma_\ell\in S_m$. It follows from (7.2.23) and (7.2.28) that for all $\ell\in\{1,2,\dots,2^m\}$ and $t>0$,

$$\begin{aligned}
&t^{\frac{1}{q}}(|T(f_{1,\sigma_{\ell,1},t},\dots,f_{m,\sigma_{\ell,m},t})|)^*(t)\\
&\le t^{\frac{1}{q}-\frac{1}{r_\ell}}\sup_{s>0}s^{\frac{1}{r_\ell}}(|T(f_{1,\sigma_{\ell,1},t},\dots,f_{m,\sigma_{\ell,m},t})|)^*(s)\\
&= t^{\frac{1}{q}-\frac{1}{r_\ell}}\big\|T(f_{1,\sigma_{\ell,1},t},\dots,f_{m,\sigma_{\ell,m},t})\big\|_{L^{r_\ell,\infty}}
\end{aligned}$$

$$\leq t^{\frac{1}{q}-\frac{1}{r_\ell}} C_0'(m,\delta,p_{k,i},q_k)\,\widetilde{B}_\ell \prod_{j=1}^{m} \|f_{j,\sigma_{\ell,j},t}\|_{L^{r_{\ell,j},\delta}}$$

$$= C_0'(m,\delta,p_{k,i},q_k)\,\widetilde{B}_\ell \prod_{j=1}^{m} t^{\frac{\gamma_j}{\gamma_0}(\frac{1}{r_{\ell,j}}-\frac{1}{p_j})} \|f_{j,\sigma_{\ell,j},t}\|_{L^{r_{\ell,j},\delta}}. \tag{7.2.30}$$

We now introduce the sets

$$\Lambda = \{1 \leq j \leq m : \gamma_j \neq 0\}$$
$$\Lambda' = \{1 \leq j \leq m : \gamma_j = 0\}$$

and we rewrite (7.2.30) as

$$t^{\frac{1}{q}}(|T(f_{1,\sigma_{\ell,1},t},\dots,f_{m,\sigma_{\ell,m},t})|)^*(t) \tag{7.2.31}$$
$$\leq C_0'(m,\delta,p_{k,i},q_k)\widetilde{B}_\ell\Big(\prod_{j\in\Lambda} t^{\frac{\gamma_j}{\gamma_0}(\frac{1}{r_{\ell,j}}-\frac{1}{p_j})} \|f_{j,\sigma_{\ell,j},t}\|_{L^{r_{\ell,j},\delta}}\Big)\Big(\prod_{j\in\Lambda'} \|f_{j,\sigma_{\ell,j},1}\|_{L^{r_{\ell,j},\delta}}\Big),$$

where we used the observation that for $j \in \Lambda'$ we have $\gamma_j = 0$, and hence for all $t > 0$

$$f_{j,\sigma_{\ell,j},t} = f_{j,\sigma_{\ell,j},1}\,.$$

To estimate the $L^s(dt/t)$ quasi-norm of (7.2.31), we need the following lemmas, whose proofs are presented in the next section.

Lemma 7.2.6. *For all $j \in \Lambda$ let s_j satisfy $0 < s_j \leq \infty$, and let $0 < \lambda_j < \infty$. Then, for all ℓ in $\{1,2,\dots,2^m\}$, the following inequalities are valid: when $p_j > r_{\ell,j}$, we have*

$$\Big\| t^{\frac{\gamma_j}{\gamma_0}(\frac{1}{r_{\ell,j}}-\frac{1}{p_j})} \|f_{j,-1,t}\|_{L^{r_{\ell,j},\delta}} \Big\|_{L^{s_j}(\frac{dt}{t})} \leq \frac{C_1(r_{\ell,j},p_j,\delta)}{|\frac{\gamma_j}{\gamma_0}|^{\frac{1}{s_j}}} \lambda_j^{\frac{1}{r_{\ell,j}}-\frac{1}{p_j}} \|f_j\|_{L^{p_j,s_j}} \tag{7.2.32}$$

and when $p_j < r_{\ell,j}$, we have

$$\Big\| t^{\frac{\gamma_j}{\gamma_0}(\frac{1}{r_{\ell,j}}-\frac{1}{p_j})} \|f_{j,1,t}\|_{L^{r_{\ell,j},\delta}} \Big\|_{L^{s_j}(\frac{dt}{t})} \leq \frac{C_1(r_{\ell,j},p_j,\delta)}{|\frac{\gamma_j}{\gamma_0}|^{\frac{1}{s_j}}} \lambda_j^{\frac{1}{r_{\ell,j}}-\frac{1}{p_j}} \|f_j\|_{L^{p_j,s_j}}, \tag{7.2.33}$$

where

$$C_1(r_{\ell,j},p_j,\delta) = \left[\frac{\max(1,\frac{r_{\ell,j}}{p_j})}{\delta|\frac{1}{p_j}-\frac{1}{r_{\ell,j}}|}\right]^{\frac{1}{\delta}} = \left[\frac{\max(1,\frac{r_{\ell,j}}{p_j})}{\delta\varepsilon}\right]^{\frac{1}{\delta}}.$$

We note that $C_1(r_{\ell,j},p_j,\delta)$ in Lemma 7.2.6 satisfies the following estimate:

$$C_1(r_{\ell,j},p_j,\delta) \leq \Big(\frac{2}{\delta\varepsilon}\Big)^{\frac{1}{\delta}}; \tag{7.2.34}$$

indeed, using (7.2.18) and the fact $\varepsilon p_j < \frac{1}{2\sqrt{m}}$ [see (7.2.19)] we have

$$\max\Big(1, \frac{r_{\ell,j}}{p_j}\Big) = \max\Big(1, \frac{1}{1+\varepsilon p_j \sigma_{\ell,j}}\Big) < \max\Big(1, \frac{1}{1-\frac{1}{2\sqrt{m}}}\Big) \le 2.$$

Lemma 7.2.7. *For all $j \in \Lambda'$ and all $\ell \in \{1,2,\dots,2^m\}$, when $p_j > r_{\ell,j}$, we have*

$$\|f_{j,-1,1}\|_{L^{r_{\ell,j},\delta}} \le C_1(r_{\ell,j}, p_j, \delta)\,\lambda_j^{\frac{1}{r_{\ell,j}}-\frac{1}{p_j}} \|f_j\|_{L^{p_j,\infty}}, \tag{7.2.35}$$

and when $p_j < r_{\ell,j}$, we have

$$\|f_{j,1,1}\|_{L^{r_{\ell,j},\delta}} \le C_1(r_{\ell,j}, p_j, \delta)\,\lambda_j^{\frac{1}{r_{\ell,j}}-\frac{1}{p_j}} \|f_j\|_{L^{p_j,\infty}}, \tag{7.2.36}$$

where $C_1(r_{\ell,j}, p_j, \delta)$ is as in Lemma 7.2.6.

Now we bound the $L^s(dt/t)$ quasi-norm of (7.2.31). First apply Lemma 7.2.7 when $j \in \Lambda'$, then apply Hölder's inequality with exponents s_j, $j \in \Lambda$, noting that $\frac{1}{s} = \sum_{j\in\Lambda} \frac{1}{s_j}$, and finally apply Lemma 7.2.6 to the product over $j \in \Lambda$. Summing over ℓ and invoking (7.2.34), we obtain that for all functions f_j in $S_0(X_j)$ the expression $\|T(f_1,\dots,f_m)\|_{L^{q,s}}$ is bounded by

$$2^{\frac{m}{q}} \max(1, 2^{\frac{m(1-s)}{s}}) \sum_{\ell=1}^{2^m} C_0'(m,\delta,p_{k,i},q_k)\widetilde{B}_\ell$$
$$\Bigg\{\Big(\prod_{j\in\Lambda}\Big(\frac{2}{\delta\varepsilon}\Big)^{\frac{1}{\delta}}\Big|\frac{\gamma_0}{\gamma_j}\Big|^{\frac{1}{s_j}} \lambda_j^{\frac{1}{r_{\ell,j}}-\frac{1}{p_j}} \|f_j\|_{L^{p_j,s_j}}\Big)\Big(\prod_{j\in\Lambda'}\Big(\frac{2}{\delta\varepsilon}\Big)^{\frac{1}{\delta}} \lambda_j^{\frac{1}{r_{\ell,j}}-\frac{1}{p_j}} \|f_j\|_{L^{p_j,\infty}}\Big)\Bigg\}.$$

To obtain (7.2.10), for each $j \in \{1,2,\dots,m\}$ we choose

$$\lambda_j = \Big(B_1^{t_{j,1}} B_2^{t_{j,2}} \cdots B_{m+1}^{t_{j,m+1}}\Big)^{-1}.$$

Then, for each $1 \le k \le m+1$, the dependence of the preceding expression on the B_k is

$$\prod_{k=1}^{m+1} B_k^{\theta_{\ell,k} - \sum_{j\in\Lambda}\big(\frac{1}{r_{\ell,j}}-\frac{1}{p_j}\big)t_{j,k} - \sum_{j\in\Lambda'}\big(\frac{1}{r_{\ell,j}}-\frac{1}{p_j}\big)t_{j,k}} = \prod_{k=1}^{m+1} B_k^{\eta_k}$$

in view of (7.2.24).

From this we conclude that for all $f_j \in S_0(X_j)$ the expression $\|T(f_1,\dots,f_m)\|_{L^{q,s}}$ is at most

$$\mathbf{C}_*'(m,\delta,p_{k,i},q_k,s_j,s)\varepsilon^{-\frac{m}{\delta}}\Big(\prod_{k=1}^{m+1} B_k^{\eta_k}\Big)\Big(\prod_{j\in\Lambda}\|f_j\|_{L^{p_j,s_j}}\Big)\Big(\prod_{j\in\Lambda'}\|f_j\|_{L^{p_j,\infty}}\Big),$$

where $\mathbf{C}'_*(m, p_{k,j}, q_k, s_j, s)$ is equal to

$$2^{\frac{m}{q}} \max(1, 2^{\frac{m(1-s)}{s}}) 2^m C'_0(m, \delta, p_{k,i}, q_k) \Big(\frac{2}{\delta}\Big)^{\frac{m}{\delta}} \prod_{j\in\Lambda} \Big|\frac{\gamma_0}{\gamma_j}\Big|^{\frac{1}{s_j}}.$$

If $j \in \Lambda'$, then it is a simple fact (see [156, Proposition 1.4.10]) that for any $s_j \in (0, \infty]$ we have

$$\|f_j\|_{L^{p_j,\infty}} \leq \Big(\frac{s_j}{p_j}\Big)^{\frac{1}{s_j}} \|f_j\|_{L^{p_j,s_j}},$$

with the obvious modification when $s_j = \infty$. Thus, for all functions $f_j \in S_0(X_j)$ we conclude

$$\|T(f_1, \dots, f_m)\|_{L^{q,s}} \leq \mathbf{C}''_*(m, \delta, p_{k,i}, q_k, s_i, s)\varepsilon^{-\frac{m}{\delta}} \Big(\prod_{k=1}^{m+1} B_k^{\eta_k}\Big) \prod_{j=1}^{m} \|f_j\|_{L^{p_j,s_j}}, \tag{7.2.37}$$

where $\mathbf{C}''_*(m, \delta, p_{k,i}, q_k, s_i, s)$ is equal to

$$2^{\frac{m}{q}} \max(1, 2^{\frac{m(1-s)}{s}}) 2^m C'_0(m, \delta, p_{k,i}, q_k) \Big(\frac{2}{\delta}\Big)^{\frac{m}{\delta}} \prod_{j\in\Lambda} \Big|\frac{\gamma_0}{\gamma_j}\Big|^{\frac{1}{s_j}} \prod_{j\in\Lambda'} \Big(\frac{s_j}{p_j}\Big)^{\frac{1}{s_j}}.$$

Since (7.2.37) is valid for all $\varepsilon < \min(1, \frac{\operatorname{dist}(\vec{P}, \partial\mathbf{H})}{2\sqrt{m}})$, letting $\varepsilon \to \min(1, \frac{\operatorname{dist}(\vec{P}, \partial\mathbf{H})}{2\sqrt{m}})$, and noticing that $\frac{1}{q} \leq \sum_{k=1}^{m+1} \frac{1}{q_k}$, we then obtain estimate (7.2.10) for all functions f_j in $S_0(X_j)$, $1 \leq j \leq m$, where

$$\mathbf{C}_*(m, \delta, p_{k,i}, q_k) = 2^{m\sum_{k=1}^{m+1} \frac{1}{q_k}} 2^m C'_0(m, \delta, p_{k,i}, q_k) \Big(\frac{2}{\delta}\Big)^{\frac{m}{\delta}} (2\sqrt{m})^{\frac{m}{\delta}}.$$

This concludes the proof of Theorem 7.2.2. □

7.2.3 Proofs of Lemmas 7.2.6 and 7.2.7

For each $j \in \{1, 2, \dots, m\}$ and $f_j \in S_0(X_j)$, with $f_{j,-1,t}$ and $f_{j,1,t}$ defined as in (7.2.29), using Definition 1.4.1 and Exercise 1.1.10 in [156], one can easily show that the following inequalities are valid:

$$f^*_{j,-1,t}(v) \leq \begin{cases} f^*_j(v) & \text{if } 0 < v < \lambda_j t^{-\frac{\gamma_j}{\gamma_0}}, \\ 0 & \text{if } v \geq \lambda_j t^{-\frac{\gamma_j}{\gamma_0}}, \end{cases} \tag{7.2.38}$$

and

$$f_{j,1,t}^*(v) \le \begin{cases} f_j^*(\lambda_j t^{-\frac{\gamma_j}{\gamma_0}}) & \text{if } 0 < v < \lambda_j t^{-\frac{\gamma_j}{\gamma_0}}, \\ f_j^*(v) & \text{if } v \ge \lambda_j t^{-\frac{\gamma_j}{\gamma_0}}. \end{cases} \tag{7.2.39}$$

First we prove Lemma 7.2.6.

Proof (Proof of Lemma 7.2.6). We first prove (7.2.32). In view of (7.2.38), we have

$$\begin{aligned} &\left\| t^{\frac{\gamma_j}{\gamma_0}(\frac{1}{r_{\ell,j}}-\frac{1}{p_j})} \|f_{j,-1,t}\|_{L^{r_{\ell,j},\delta}} \right\|_{L^{s_j}(\frac{dt}{t})} \\ &= \left[\int_0^\infty t^{s_j \frac{\gamma_j}{\gamma_0}(\frac{1}{r_{\ell,j}}-\frac{1}{p_j})} \left\{ \int_0^{\lambda_j t^{-\frac{\gamma_j}{\gamma_0}}} (f_{j,-1,t}^*(v))^\delta v^{\frac{\delta}{r_{\ell,j}}} \frac{dv}{v} \right\}^{\frac{s_j}{\delta}} \frac{dt}{t} \right]^{\frac{1}{s_j}} . \end{aligned}$$

We change variables $u = \lambda_j t^{-\frac{\gamma_j}{\gamma_0}}$ and use (7.2.38) to estimate the preceding expression by

$$\left|\frac{\gamma_0}{\gamma_j}\right|^{\frac{1}{s_j}} \lambda_j^{\frac{1}{r_{\ell,j}}-\frac{1}{p_j}} \left[\left\{ \int_0^\infty u^{-s_j(\frac{1}{r_{\ell,j}}-\frac{1}{p_j})} \left(\int_0^u (f_j^*(v))^\delta v^{\frac{\delta}{r_{\ell,j}}} \frac{dv}{v} \right)^{\frac{s_j}{\delta}} \frac{du}{u} \right\}^{\frac{\delta}{s_j}} \right]^{\frac{1}{\delta}} . \tag{7.2.40}$$

We now use the following inequality of Hardy (valid for $0 < \beta < \infty$, $1 \le p < \infty$):

$$\left(\int_0^\infty \left(\int_0^x |f(t)|\, dt \right)^p x^{-\beta} \frac{dx}{x} \right)^{\frac{1}{p}} \le \frac{p}{\beta} \left(\int_0^\infty |f(t)|^p t^{p-\beta} \frac{dt}{t} \right)^{\frac{1}{p}}$$

with

$$\beta = s_j\left(\frac{1}{r_{\ell,j}} - \frac{1}{p_j}\right) > 0$$

and $p = \frac{s_j}{\delta}$ since $p_j > r_{\ell,j}$ and $\delta \le s_j$. We obtain that (7.2.40) is at most

$$\begin{aligned} &\left(\frac{1}{\delta|\frac{1}{r_{\ell,j}} - \frac{1}{p_j}|} \right)^{\frac{1}{\delta}} \left|\frac{\gamma_0}{\gamma_j}\right|^{\frac{1}{s_j}} \lambda_j^{\frac{1}{r_{\ell,j}}-\frac{1}{p_j}} \left(\int_0^\infty ((f_j^*(v))^\delta v^{\frac{\delta}{r_{\ell,j}}} v^{-1})^{\frac{s_j}{\delta}} v^{\frac{s_j}{\delta} - s_j(\frac{1}{r_{\ell,j}}-\frac{1}{p_j})} \frac{dv}{v} \right)^{\frac{1}{s_j}} \\ &= \left(\frac{1}{\delta|\frac{1}{r_{\ell,j}} - \frac{1}{p_j}|} \right)^{\frac{1}{\delta}} \left|\frac{\gamma_0}{\gamma_j}\right|^{\frac{1}{s_j}} \lambda_j^{\frac{1}{r_{\ell,j}}-\frac{1}{p_j}} \|f_j\|_{L^{p_j,s_j}} . \end{aligned}$$

We now prove (7.2.33). We begin with

$$\left\| t^{\frac{\gamma_j}{\gamma_0}(\frac{1}{r_{\ell,j}}-\frac{1}{p_j})} \|f_{j,1,t}\|_{L^{r_{\ell,j},\delta}} \right\|_{L^{s_j}(\frac{dt}{t})} =$$

$$\left[\int_0^\infty t^{s_j\frac{\gamma_j}{\gamma_0}(\frac{1}{r_{\ell,j}}-\frac{1}{p_j})} \left(\int_0^{\lambda_j t^{-\frac{\gamma_j}{\gamma_0}}} (f_{j,1,t}^*(v))^\delta v^{\frac{\delta}{r_{\ell,j}}} \frac{dv}{v} + \int_{\lambda_j t^{-\frac{\gamma_j}{\gamma_0}}}^\infty (f_{j,1,t}^*(v))^\delta v^{\frac{\delta}{r_{\ell,j}}} \frac{dv}{v} \right)^{\frac{s_j}{\delta}} \frac{dt}{t} \right]^{\frac{1}{s_j}} .$$

In both integrals we first use (7.2.39) and then perform a change of variables $u = \lambda_j t^{-\frac{\gamma_j}{\gamma_0}}$ to estimate the preceding expression by

$$\left|\frac{\gamma_0}{\gamma_j}\right|^{\frac{1}{s_j}} \lambda_j^{\frac{1}{r_{\ell,j}}-\frac{1}{p_j}}$$

$$\times \left[\int_0^\infty u^{-s_j(\frac{1}{r_{\ell,j}}-\frac{1}{p_j})} \left\{ (f_j^*(u))^\delta \int_0^u v^{\frac{\delta}{r_{\ell,j}}} \frac{dv}{v} + \int_u^\infty (f_j^*(v))^\delta v^{\frac{\delta}{r_{\ell,j}}} \frac{dv}{v} \right\}^{\frac{s_j}{\delta}} \frac{du}{u} \right]^{\frac{\delta}{s_j}\frac{1}{\delta}} ,$$

which by Minkowski's inequality is at most

$$\left|\frac{\gamma_0}{\gamma_j}\right|^{\frac{1}{s_j}} \lambda_j^{\frac{1}{r_{\ell,j}}-\frac{1}{p_j}} \left[\left\{ \int_0^\infty u^{-s_j(\frac{1}{r_{\ell,j}}-\frac{1}{p_j})} \left((f_j^*(u))^\delta \int_0^u v^{\frac{\delta}{r_{\ell,j}}} \frac{dv}{v} \right)^{\frac{s_j}{\delta}} \frac{du}{u} \right\}^{\frac{\delta}{s_j}} \right.$$

$$\left. + \left\{ \int_0^\infty u^{-s_j(\frac{1}{r_{\ell,j}}-\frac{1}{p_j})} \left(\int_u^\infty (f_j^*(v))^\delta v^{\frac{\delta}{r_{\ell,j}}} \frac{dv}{v} \right)^{\frac{s_j}{\delta}} \frac{du}{u} \right\}^{\frac{\delta}{s_j}} \right]^{\frac{1}{\delta}} . \tag{7.2.41}$$

The first term of the sum is easily evaluated. For the second term of the sum we use the following inequality of Hardy (valid for $0 < \beta < \infty$, $1 \le p < \infty$):

$$\left(\int_0^\infty \left(\int_x^\infty |f(t)|\,dt \right)^p x^\beta \frac{dx}{x} \right)^{\frac{1}{p}} \le \frac{p}{\beta} \left(\int_0^\infty |f(t)|^p t^{p+\beta} \frac{dt}{t} \right)^{\frac{1}{p}} ,$$

with $\beta = -(\frac{1}{r_{\ell,j}} - \frac{1}{p_j})s_j > 0$ and $p = \frac{s_j}{\delta}$ since $p_j < r_{\ell,j}$ and $\delta \le s_j$. Then (7.2.41) can be estimated by

$$\left|\frac{\gamma_0}{\gamma_j}\right|^{\frac{1}{s_j}} \lambda_j^{\frac{1}{r_{\ell,j}}-\frac{1}{p_j}} \left[\frac{1}{\frac{\delta}{r_{\ell,j}}} \left\{ \int_0^\infty u^{\frac{s_j}{p_j}} (f_j^*(u))^{s_j} \frac{du}{u} \right\}^{\frac{\delta}{s_j}} \right.$$

$$\left. + \frac{1}{\delta(\frac{1}{p_j} - \frac{1}{r_{\ell,j}})} \left\{ \int_0^\infty \left((f_j^*(v))^\delta v^{\frac{\delta}{r_{\ell,j}}} v^{-1} \right)^{\frac{s_j}{\delta}} v^{\frac{s_j}{\delta}+(\frac{1}{p_j}-\frac{1}{r_{\ell,j}})s_j} \frac{dv}{v} \right\}^{\frac{\delta}{s_j}} \right]^{\frac{1}{\delta}}$$

$$= \left|\frac{\gamma_0}{\gamma_j}\right|^{\frac{1}{s_j}} \lambda_j^{\frac{1}{r_{\ell,j}}-\frac{1}{p_j}} \left[\frac{1}{\frac{\delta}{r_{\ell,j}}} \|f\|^{\delta}_{L^{p_j,s_j}} + \frac{1}{\delta(\frac{1}{p_j}-\frac{1}{r_{\ell,j}})} \|f\|^{\delta}_{L^{p_j,s_j}} \right]^{\frac{1}{\delta}}$$

$$= \left|\frac{\gamma_0}{\gamma_j}\right|^{\frac{1}{s_j}} \lambda_j^{\frac{1}{r_{\ell,j}}-\frac{1}{p_j}} \left[\frac{\frac{r_{\ell,j}}{p_j}}{\delta|\frac{1}{p_j}-\frac{1}{r_{\ell,j}}|} \right]^{\frac{1}{\delta}} \|f\|_{L^{p_j,s_j}},$$

which proves (7.2.33).

We now consider the case $s_j = \infty$. If $p_j > r_{\ell,j}$, then we change variables $u = \lambda_j t^{-\frac{\gamma_j}{\gamma_0}}$ and use (7.2.38) to obtain that for all $t > 0$,

$$t^{\frac{\gamma_j}{\gamma_0}(\frac{1}{r_{\ell,j}}-\frac{1}{p_j})} \|f_{j,-1,t}\|_{L^{r_{\ell,j},\delta}} \le \lambda_j^{\frac{1}{r_{\ell,j}}-\frac{1}{p_j}} u^{-(\frac{1}{r_{\ell,j}}-\frac{1}{p_j})} \left(\int_0^u (f_j^*(v))^{\delta} v^{\frac{\delta}{r_{\ell,j}}} \frac{dv}{v} \right)^{\frac{1}{\delta}}$$

$$\le \lambda_j^{\frac{1}{r_{\ell,j}}-\frac{1}{p_j}} u^{-(\frac{1}{r_{\ell,j}}-\frac{1}{p_j})} \left(\int_0^u v^{\frac{\delta}{r_{\ell,j}}-\frac{\delta}{p_j}} \frac{dv}{v} \right)^{\frac{1}{\delta}} \|f_j\|_{L^{p_j,\infty}}$$

$$= \left(\frac{1}{\delta|\frac{1}{r_{\ell,j}}-\frac{1}{p_j}|} \right)^{\frac{1}{\delta}} \lambda_j^{\frac{1}{r_{\ell,j}}-\frac{1}{p_j}} \|f_j\|_{L^{p_j,\infty}},$$

which implies (7.2.35).

If $p_j < r_{\ell,j}$, then again by the same change of variables $u = \lambda_j t^{-\frac{\gamma_j}{\gamma_0}}$ and via (7.2.39) we obtain for all $t > 0$

$$t^{\frac{\gamma_j}{\gamma_0}(\frac{1}{r_{\ell,j}}-\frac{1}{p_j})} \|f_{j,1,t}\|_{L^{r_{\ell,j},\delta}}$$

$$\le \lambda_j^{\frac{1}{r_{\ell,j}}-\frac{1}{p_j}} u^{-(\frac{1}{r_{\ell,j}}-\frac{1}{p_j})} \left(\int_0^u (f_j^*(u))^{\delta} v^{\frac{\delta}{r_{\ell,j}}} \frac{dv}{v} + \int_u^{\infty} (f_j^*(v))^{\delta} v^{\frac{\delta}{r_{\ell,j}}} \frac{dv}{v} \right)^{\frac{1}{\delta}}$$

$$\le \lambda_j^{\frac{1}{r_{\ell,j}}-\frac{1}{p_j}} u^{-(\frac{1}{r_{\ell,j}}-\frac{1}{p_j})} \left(\int_0^u u^{-\frac{\delta}{p_j}} v^{\frac{\delta}{r_{\ell,j}}} \frac{dv}{v} + \int_u^{\infty} v^{\frac{\delta}{r_{\ell,j}}-\frac{\delta}{p_j}} \frac{dv}{v} \right)^{\frac{1}{\delta}} \|f_j\|_{L^{p_j,\infty}}$$

$$\le \lambda_j^{\frac{1}{r_{\ell,j}}-\frac{1}{p_j}} \left(\frac{1}{\frac{\delta}{r_{\ell,j}}} + \frac{1}{\delta(\frac{1}{p_j}-\frac{1}{r_{\ell,j}})} \right)^{\frac{1}{\delta}} \|f_j\|_{L^{p_j,\infty}}$$

$$= \lambda_j^{\frac{1}{r_{\ell,j}}-\frac{1}{p_j}} \left(\frac{\frac{r_{\ell,j}}{p_j}}{\delta|\frac{1}{p_j}-\frac{1}{r_{\ell,j}}|} \right)^{\frac{1}{\delta}} \|f_j\|_{L^{p_j,\infty}}.$$

This concludes the proof of Lemma 7.2.6. □

Proof (Lemma 7.2.7). When $j \in \Lambda'$, we have $\gamma_j = 0$ and

$$f_{j,-1,1} = f_j \chi_{\{|f_j| > f_j^*(\lambda_j)\}}, \qquad f_{j,1,1} = f_j \chi_{\{|f_j| \le f_j^*(\lambda_j)\}}$$

and

$$f_{j,-1,1}^*(v) \le \begin{cases} f_j^*(v) & \text{if } 0 < v < \lambda_j, \\ 0 & \text{if } v \ge \lambda_j, \end{cases} \tag{7.2.42}$$

and

$$f_{j,1,1}^*(v) \le \begin{cases} f_j^*(\lambda_j) & \text{if } 0 < v < \lambda_j, \\ f_j^*(v) & \text{if } v \ge \lambda_j. \end{cases} \tag{7.2.43}$$

If $p_j > r_{\ell,j}$, then by (7.2.42) we obtain

$$\begin{aligned} \|f_{j,-1,1}\|_{L^{r_{\ell,j},\delta}} &\le \left[\int_0^{\lambda_j} v^{\frac{\delta}{r_{\ell,j}}} f_j^*(v)^\delta \frac{dv}{v}\right]^{1/\delta} \\ &\le \left[\int_0^{\lambda_j} v^{\frac{\delta}{r_{\ell,j}}-\frac{\delta}{p_j}} \frac{dv}{v}\right]^{1/\delta} \|f_j\|_{L^{p_j,\infty}} \\ &= \frac{\lambda_j^{\frac{1}{r_{\ell,j}}-\frac{1}{p_j}}}{\left|\frac{\delta}{r_{\ell,j}}-\frac{\delta}{p_j}\right|^{1/\delta}} \|f_j\|_{L^{p_j,\infty}}, \end{aligned}$$

which proves (7.2.35). Now we suppose $p_j < r_{\ell,j}$ and show (7.2.36). To this end, applying (7.2.43) yields that

$$\begin{aligned} \|f_{j,1,1}\|_{L^{r_{\ell,j},\delta}} &\le \left[\int_0^{\lambda_j} v^{\frac{\delta}{r_{\ell,j}}} \lambda_j^{-\frac{\delta}{p_j}} \lambda_j^{\frac{\delta}{p_j}} f_j^*(\lambda_j)^\delta \frac{dv}{v} + \int_{\lambda_j}^\infty v^{\frac{\delta}{r_{\ell,j}}-\frac{\delta}{p_j}} v^{\frac{\delta}{p_j}} f_j^*(v)^\delta \frac{dv}{v}\right]^{1/\delta} \\ &\le \left[\frac{\lambda_j^{\frac{\delta}{r_{\ell,j}}-\frac{\delta}{p_j}}}{\frac{\delta}{r_{\ell,j}}} + \frac{\lambda_j^{\frac{\delta}{r_{\ell,j}}-\frac{\delta}{p_j}}}{\frac{\delta}{p_j}-\frac{\delta}{r_{\ell,j}}}\right]^{1/\delta} \|f_j\|_{L^{p_j,\infty}} \\ &= \left[\frac{\frac{r_{\ell,j}}{p_j}}{\delta\left|\frac{1}{p_j}-\frac{1}{r_{\ell,j}}\right|}\right]^{\frac{1}{\delta}} \lambda_j^{\frac{1}{r_{\ell,j}}-\frac{1}{p_j}} \|f_j\|_{L^{p_j,\infty}}, \end{aligned}$$

and hence (7.2.36) holds. This concludes the proof of Lemma 7.2.7. □

Example 7.2.8. We recall the operator I_α of Theorem 7.1.4. Let $0 < \alpha < n$. It was shown that the bilinear fractional integral

$$I_\alpha(f,g)(x) = \int_{\mathbf{R}^n} f(x+t)g(x-t)|t|^{\alpha-n}dt,$$

is of restricted weak types $(\frac{n}{\alpha},\infty,\infty)$, $(\infty,\frac{n}{\alpha},\infty)$, $(1,\infty,\frac{n}{n-\alpha})$, $(\infty,1,\frac{n}{n-\alpha})$, $(1,1,\frac{n}{2n-\alpha})$. Consequently, I_α is bounded from $L^{p_1}\times L^{p_2}\to L^p$ whenever $(1/p_1,1/p_2,1/p)$ lies in the open convex hull of the points $(\frac{\alpha}{n},0,0)$, $(0,\frac{\alpha}{n},0)$, $(1,0,\frac{n-\alpha}{n})$, $(0,1,\frac{n-\alpha}{n})$, $(1,1,\frac{2n-\alpha}{n})$. Notice that each point (r,s,t) of the preceding five satisfies the equation $r+s=t+\alpha/n$, and thus any point $(1/p_1,1/p_2,1/p)$ in the open convex hull of these points also satisfies the equation $1/p_1+1/p_2=1/p+\alpha/n$. Moreover, the fact that $0 < \alpha < n$ implies that the determinant γ_0 associated with any three initial points among the five $(\frac{\alpha}{n},0)$, $(0,\frac{\alpha}{n})$, $(1,0)$, $(0,1)$, $(1,1)$ is nonzero.

Finally, since $1/p \le 1/p_1 + 1/p_2$, Corollary 7.2.4 applies and yields that I_α maps $L^{p_1}(\mathbf{R}^n) \times L^{p_1}(\mathbf{R}^n)$ to $L^p(\mathbf{R}^n)$ whenever $1 < p_1, p_2 < \infty$, $1/p_1 + 1/p_2 = 1/p + \alpha/n$, and $0 < \alpha < n$.

7.2.4 Multilinear Complex Interpolation

Interpolation for analytic families of linear operators can be easily extended to the case of multilinear operators.

We describe the setup for this theorem. A simple function is called *finitely simple* if it is supported in a set of finite measure. Finitely simple functions are dense in $L^p(X,\mu)$ for $0 < p < \infty$ whenever (X,μ) is a σ-finite measure space. Let (X_k,μ_k), $k = 1,\dots,m$, and (Y,ν) be σ-finite measure spaces. Suppose that for every z in the closed strip $\overline{S} = \{z \in \mathbf{C} : 0 \le \operatorname{Re} z \le 1\}$ there is an associated multilinear operator T_z defined on the space of finitely simple functions on $X_1 \times \cdots \times X_m$ and taking values in the space of measurable functions on Y such that

$$\int_Y |T_z(\chi_{A_1},\dots,\chi_{A_m})\chi_B| \, d\nu < \infty \tag{7.2.44}$$

whenever A_k are subsets of finite measure of X_k and B of Y. The family $\{T_z\}_z$ is said to be *analytic* if for all finitely simple functions f_k on X_k and g on Y we have that the mapping

$$z \mapsto \int_Y T_z(f_1,\dots,f_m)\, g \, d\nu \tag{7.2.45}$$

is analytic in the open strip $S = \{z \in \mathbf{C} : 0 < \operatorname{Re} z < 1\}$ and continuous on its closure. The analytic family $\{T_z\}_z$ is of *admissible growth* if there is a constant τ_0 with $0 \le \tau_0 < \pi$ such that for finitely simple functions f_k on X_k and g on Y there is a constant $C(f,g)$ such that

$$\log\left|\int_Y T_z(f_1,\dots,f_m)\, g \, d\nu\right| \le C(f_1,\dots,f_m,g)\, e^{\tau_0|\operatorname{Im} z|} \tag{7.2.46}$$

for all z satisfying $0 \le \operatorname{Re} z \le 1$. Note that if there is $\tau_0 \in (0,\pi)$ such that for all measurable subsets A_k of X_k and B of Y of finite measure there is a constant $c(A_1,\dots,A_m,B)$ such that

$$\log\left|\int_B T_z(\chi_{A_1},\dots,\chi_{A_m}) \, d\nu\right| \le c(A_1,\dots,A_m,B)\, e^{\tau_0|\operatorname{Im} z|}, \tag{7.2.47}$$

then (7.2.46) holds.

Theorem 7.2.9. *Let T_z be an analytic family of linear operators of admissible growth defined on the m-fold product of spaces of finitely simple functions of σ-finite measure spaces (X_i,μ_i) and taking values in the set of measurable functions of another σ-finite measure space (Y,ν). Let $1\le p_{0,k},p_{1,k},q_0,q_1\le\infty$ for all $1\le k\le m$ and suppose that M_0 and M_1 are positive functions on the real line such that for some τ_1 with $0\le\tau_1<\pi$ we have*

$$\sup_{-\infty<y<+\infty} e^{-\tau_1|y|}\log M_j(y)<\infty \tag{7.2.48}$$

for $j=0,1$. Let $0<\theta<1$ and for $k\in\{1,\dots,m\}$ define p_k and q by

$$\frac{1}{p_k}=\frac{1-\theta}{p_{0,k}}+\frac{\theta}{p_{1,k}} \qquad \text{and} \qquad \frac{1}{q}=\frac{1-\theta}{q_0}+\frac{\theta}{q_1}. \tag{7.2.49}$$

Suppose that for all finitely simple functions f_k on X_k, $k=1,2,\dots,m$, we have

$$\big\|T_{iy}(f_1,\dots,f_m)\big\|_{L^{q_0}}\le M_0(y)\big\|f_1\big\|_{L^{p_{0,1}}}\cdots\big\|f_m\big\|_{L^{p_{0,m}}}, \tag{7.2.50}$$
$$\big\|T_{1+iy}(f_1,\dots,f_m)\big\|_{L^{q_1}}\le M_1(y)\big\|f_1\big\|_{L^{p_{1,1}}}\cdots\big\|f_m\big\|_{L^{p_{1,m}}}. \tag{7.2.51}$$

Then for all finitely simple functions f_k on X_k we have

$$\big\|T_\theta(f_1,\dots,f_m)\big\|_{L^q}\le M(\theta)\big\|f_1\big\|_{L^{p_1}}\cdots\big\|f_m\big\|_{L^{p_m}} \tag{7.2.52}$$

where for $0<t<1$

$$M(t)=\exp\left\{\frac{\sin(\pi t)}{2}\int_{-\infty}^{\infty}\left[\frac{\log M_0(y)}{\cosh(\pi y)-\cos(\pi t)}+\frac{\log M_1(y)}{\cosh(\pi y)+\cos(\pi t)}\right]dy\right\}.$$

Moreover, when $p_1,\dots,p_m<\infty$, the operator T_θ admits a unique bounded extension from $L^{p_1}(X_1,\mu_1)\times\cdots\times L^{p_m}(X_m,\mu_m)$ to $L^q(Y,\nu)$.

Note that in view of (7.2.48) and of the fact that $0\le\tau_1<\pi$, the integral defining $M(t)$ converges absolutely.

To prove Theorem 7.2.9, we need the following lemma, whose proof can be found in [156] (Lemma 1.3.8).

Lemma 7.2.10. *Let F be analytic on the open strip $S=\{z\in\mathbf{C}:0<\operatorname{Re} z<1\}$ and continuous on its closure such that for some fixed $A<\infty$ and $0\le\tau_0<\pi$ we have*

$$\log|F(z)|\le A\,e^{\tau_0|\operatorname{Im} z|} \tag{7.2.53}$$

for all $z\in\overline{S}$. Then

$$|F(t)|\le\exp\left\{\frac{\sin(\pi t)}{2}\int_{-\infty}^{\infty}\left[\frac{\log|F(iy)|}{\cosh(\pi y)-\cos(\pi t)}+\frac{\log|F(1+iy)|}{\cosh(\pi y)+\cos(\pi t)}\right]dy\right\}$$

whenever $0<t<1$.

Assuming Lemma 7.2.10, we prove Theorem 7.2.9.

Proof. Fix $0<\theta<1$ and finitely simple functions f_k, g such that $\|f_k\|_{L^{p_k}} = \|g\|_{L^{q'}} = 1$ for all $1\le k\le m$. For $1\le k\le m$ let

$$f_k = \sum_{\ell=1}^{n_k} a_{k,\ell} e^{i\alpha_{k,\ell}} \chi_{A_{k,\ell}} \quad \text{and} \quad g = \sum_{j=1}^{n} b_j e^{i\beta_j} \chi_{B_j},$$

where for each k, $a_{k,\ell}>0$, $\alpha_{k,\ell}$ are real, $A_{k,\ell}$ are pairwise disjoint subsets of X_k with finite measure, $b_j>0$, β_j are real, and B_j are pairwise disjoint subsets of Y with finite measure. Let

$$P_k(z) = \frac{p_k}{p_{0,k}}(1-z) + \frac{p_k}{p_{1,k}} z \qquad \text{and} \qquad Q(z) = \frac{q'}{q_0'}(1-z) + \frac{q'}{q_1'} z \tag{7.2.54}$$

and

$$f_{k,z} = \sum_{\ell=1}^{n_k} a_{k,\ell}^{P_k(z)} e^{i\alpha_{k,\ell}} \chi_{A_{k,\ell}}, \quad g_z = \sum_{j=1}^{n} b_j^{Q(z)} e^{i\beta_j} \chi_{B_j}. \tag{7.2.55}$$

Define

$$F(z) = \int_Y T_z(f_{1,z},\dots,f_{m,z})\, g_z\, d\nu. \tag{7.2.56}$$

Multilinearity gives that $F(z)$ is equal to

$$\sum_{\ell_1=1}^{n_1} \cdots \sum_{\ell_m=1}^{n_m} \sum_{j=1}^{n} a_{1,\ell_1}^{P_1(z)} \cdots a_{m,\ell_m}^{P_m(z)} b_j^{Q(z)} e^{i\alpha_{1,\ell_1}} \cdots e^{i\alpha_{m,\ell_m}} e^{i\beta_j} \int_{B_j} T_z(\chi_{A_{1,\ell_1}},\dots,\chi_{A_{m,\ell_m}})\, d\nu,$$

and conditions (7.2.44), together with the analyticity of $\{T_z\}_z$, imply that $F(z)$ is a well-defined analytic function on the unit strip that extends continuously to its boundary.

Since $\{T_z\}_z$ is a family of admissible growth, (7.2.47) implies that for some constant $c(A_{1,\ell_1},\dots,A_{m,\ell_m},B_j)$ and $\tau_0\in(0,\pi)$ we have

$$\log\left| \int_{B_j} T_z(\chi_{A_{1,\ell_1}},\dots,\chi_{m,\ell_m})\, d\nu \right| \le c(A_{1,\ell_1},\dots,A_{m,\ell_m},B_j)\, e^{\tau_0 |\mathrm{Im}\, z|}.$$

Set $c_{\ell_1,\dots,\ell_m,j} = c(A_{1,\ell_1},\dots,A_{m,\ell_m},B_j)$. The preceding inequalities, combined with

$$|a_{k,\ell}^{P_k(z)}| \le a_{k,\ell}^{\frac{p_k}{p_{0,k}}+\frac{p_k}{p_{1,k}}} \qquad \text{and} \qquad |b_j^{Q(z)}| \le b_j^{\frac{q'}{q_0'}+\frac{q'}{q_1'}}$$

for all z with $0<\mathrm{Re}\, z<1$, imply (7.2.53), with τ_0 as in (7.2.47), and

$$A = A\Big(m,n_1,\dots,n_k,n,q',q_0',q_1',p_k,p_{0,k},p_{1,k},a_{k,\ell},b_j,A_{k,\ell},B_j\Big),$$

which is a function of all the preceding parameters. Thus, F satisfies the hypotheses of Lemma 7.2.10. Moreover, simple calculations show that (even when $p_{0,k}=\infty$, $q_0=1$, $p_{1,k}=\infty$, $q_1=1$)

$$\big\|f_{k,iy}\big\|_{L^{p_{0,k}}}=\big\|f_k\big\|_{L^{p_k}}^{\frac{p_k}{p_{0,k}}}=1=\big\|g\big\|_{L^{q'}}^{\frac{q'}{q'_0}}=\big\|g_{iy}\big\|_{L^{q'_0}} \tag{7.2.57}$$

$$\big\|f_{k,1+iy}\big\|_{L^{p_{1,k}}}=\big\|f_k\big\|_{L^{p_k}}^{\frac{p_k}{p_{1,k}}}=1=\big\|g\big\|_{L^{q'}}^{\frac{q'}{q'_1}}=\big\|g_{1+iy}\big\|_{L^{q'_1}} \tag{7.2.58}$$

when $y\in\mathbf{R}$ and $1\le k\le m$. Hölder's inequality, (7.2.57), and the hypothesis (7.2.50) now give

$$\begin{aligned}|F(iy)| &\le \big\|T_{iy}(f_{1,iy},\dots,f_{m,iy})\big\|_{L^{q_0}}\big\|g_{iy}\big\|_{L^{q'_0}}\\ &\le M_0(y)\big\|f_{1,iy}\big\|_{L^{p_{0,1}}}\cdots\big\|f_{m,iy}\big\|_{L^{p_{0,m}}}\big\|g_{iy}\big\|_{L^{q'_0}}\\ &= M_0(y)\end{aligned}$$

for all y real. Similarly, (7.2.58) and (7.2.51) imply

$$\begin{aligned}|F(1+iy)| &\le \big\|T_{1+iy}(f_{1,1+iy},\dots,f_{m,1+iy})\big\|_{L^{q_1}}\big\|g_{iy}\big\|_{L^{q'_1}}\\ &\le M_1(y)\big\|f_{1,1+iy}\big\|_{L^{p_{1,1}}}\cdots\big\|f_{m,1+iy}\big\|_{L^{p_{1,m}}}\big\|g_{iy}\big\|_{L^{q'_1}}\\ &= M_1(y)\end{aligned}$$

for all $y\in\mathbf{R}$. These inequalities and the conclusion of Lemma 7.2.10 yield

$$|F(t)|\le \exp\left\{\frac{\sin(\pi t)}{2}\int_{-\infty}^{\infty}\left[\frac{\log M_0(y)}{\cosh(\pi y)-\cos(\pi t)}+\frac{\log M_1(y)}{\cosh(\pi y)+\cos(\pi t)}\right]dy\right\}$$

for all $0<t<1$. But notice that

$$F(\theta)=\int_Y T_\theta(f_1,\dots,f_m)\,g\,d\nu. \tag{7.2.59}$$

Taking absolute values and the supremum over all finitely simple functions g on Y with $L^{q'}$ norm equal to one, we conclude the proof of (7.2.52) for finitely simple functions f_k with $L^{p_k}(X_k)$ norm one. □

Corollary 7.2.11. *Let T be an m-linear operator defined on the m-fold product of spaces of finitely simple functions of σ-finite measure spaces (X_i,μ_i) and taking values in the set of measurable functions of another σ-finite measure space (Y,ν). Let $1\le p_{0,k},p_{1,k},q_0,q_1\le\infty$ for all $1\le k\le m$, $0<\theta<1$, and for $k\in\{1,\dots,m\}$ define p_k and q by* (7.2.49). *Suppose that for all finitely simple functions f_k on X_k we have*

$$\big\|T(f_1,\dots,f_m)\big\|_{L^{q_0}}\le M_0\big\|f_1\big\|_{L^{p_{0,1}}}\cdots\big\|f_m\big\|_{L^{p_{0,m}}}, \tag{7.2.60}$$

$$\big\|T(f_1,\dots,f_m)\big\|_{L^{q_1}}\le M_1\big\|f_1\big\|_{L^{p_{1,1}}}\cdots\big\|f_m\big\|_{L^{p_{1,m}}}. \tag{7.2.61}$$

Then for all finitely simple functions f_k on X_k we have

$$\left\|T(f_1,\dots,f_m)\right\|_{L^q} \le M_0^{1-\theta} M_1^{\theta} \left\|f_1\right\|_{L^{p_1}} \cdots \left\|f_m\right\|_{L^{p_m}}. \tag{7.2.62}$$

Moreover, when $p_1,\dots,p_m < \infty$, the operator T_θ admits a unique bounded extension from $L^{p_1}(X_1,\mu_1)\times\cdots\times L^{p_m}(X_m,\mu_m)$ to $L^q(Y,\nu)$.

Proof. Take $T_z = T$ in Theorem 7.2.9, and use Exercise 1.3.8 in [156]. □

7.2.5 Multilinear Interpolation between Adjoint Operators

In this subsection we discuss a result that allows one to interpolate from a single estimate known for an operator and its adjoints. This theorem is useful in the setting where there is no duality, such as when an operator maps into L^q for $q<1$. For a number $q\in(0,\infty)$ set $q' = q/(q-1)$ when $q\neq 1$ and $\infty' = 1$.

Theorem 7.2.12. *Let $0<p<\infty$, $A,B>0$, and let f be a measurable function on a σ-finite measure space (X,μ).*
(i) Suppose that $\|f\|_{L^{p,\infty}} \le A$. Then for every measurable set E of finite measure there exists a measurable subset E' of E with $\mu(E') \ge \mu(E)/2$ such that f is bounded on E' and

$$\left|\int_{E'} f\,d\mu\right| \le 2^{\frac{1}{p}} A\,\mu(E)^{1-\frac{1}{p}}. \tag{7.2.63}$$

(ii) Suppose that a measurable function f on X has the property that for any measurable subset E of X, with $\mu(E)<\infty$, there is a measurable subset E' of E, with $\mu(E')\ge\mu(E)/2$, such that f is integrable on E' and

$$\left|\int_{E'} f\,d\mu\right| \le B\,\mu(E)^{1-\frac{1}{p}}.$$

Then we have that

$$\|f\|_{L^{p,\infty}} \le B\,2^{\frac{2}{p}+\frac{3}{2}}. \tag{7.2.64}$$

Proof. Define $E' = E\setminus\{|f| > A\,2^{\frac{1}{p}}\mu(E)^{-\frac{1}{p}}\}$. Since the set $\{|f|>A\,2^{\frac{1}{p}}\mu(E)^{-\frac{1}{p}}\}$ has measure at most $\mu(E)/2$, it follows that $\mu(E')\ge\mu(E)/2$. Obviously, (7.2.63) holds for this choice of E'. This proves (i).

To prove (ii), write $X=\bigcup_{n=1}^{\infty} X_n$, with $\mu(X_n)<\infty$. Given $\alpha>0$, note that the measurable set $\big\{|f|>\alpha\big\}$ is contained in

$$\big\{\mathrm{Re}\, f > \tfrac{\alpha}{\sqrt{2}}\big\} \cup \big\{\mathrm{Im}\, f > \tfrac{\alpha}{\sqrt{2}}\big\} \cup \big\{\mathrm{Re}\, f < -\tfrac{\alpha}{\sqrt{2}}\big\} \cup \big\{\mathrm{Im}\, f < -\tfrac{\alpha}{\sqrt{2}}\big\}. \tag{7.2.65}$$

Let E_n be any of the preceding four sets intersected with X_n. By hypothesis, there is a measurable subset E_n' of E_n with measure at least $\mu(E_n)/2$. Then

$$\frac{\alpha}{2\sqrt{2}}\mu(E_n) \le \left|\int_{E_n'} f\,d\mu\right| \le B\,\mu(E_n)^{1-\frac{1}{p}}$$

from which it follows that $\alpha\mu(E_n)^{1/p} \le B2\sqrt{2}$ since $\mu(E_n) < \infty$. Letting $n \to \infty$ we obtain that any of the four sets in (7.2.65) has measure bounded by $(B2\sqrt{2}/\alpha)^p$. Summing, we deduce that $\mu(\{|f| > \alpha\})$ is at most $4(B2\sqrt{2}/\alpha)^p$ and thus (7.2.64) follows. □

Theorem 7.2.13. *Let (X,μ), $(X_1,\mu_1),\dots,(X_m,\mu_m)$ be σ-finite measure spaces. Suppose that an m-linear operator T is defined on the space of simple functions on $X_1\times\cdots\times X_m$ and takes values in the space of measurable functions defined on X. Let $1 < p_1,\dots,p_m,p < \infty$ related by $1/p = 1/p_1+\cdots+1/p_m$. Assume that*

$$\sup_{A_0,A_1,\dots,A_m} \frac{1}{\mu(A_0)^{\frac{1}{p'}}\mu_1(A_1)^{\frac{1}{p_1}}\cdots\mu_m(A_m)^{\frac{1}{p_m}}}\left|\int_{A_0} T(\chi_{A_1},\dots,\chi_{A_m})\,d\mu\right| < \infty, \tag{7.2.66}$$

where the supremum is taken over all measurable subsets A_i of X_i with nonzero finite measure. Suppose that for each $j\in\{0,1,\dots,m\}$, T^{*j} is of restricted weak type $(1,1,\dots,1/m)$ with constant B_j. Then for every point $(1/q_1,\dots,1/q_m,1/q)$ on the line segment that joins $(1,\dots,1,m)$ to $(1/p_1,\dots,1/p_m,1/p)$, there is a constant $C_{q_1,\dots,q_m}$ such that T is of restricted weak type $(q_1,\dots,q_m,q)$ with norm at most

$$C_{q_1,\dots,q_m} B_0^{\frac{1}{mq}} B_1^{\frac{1}{mq_1'}}\cdots B_m^{\frac{1}{mq_m'}}\,. \tag{7.2.67}$$

Proof. First we prove the claim for $(1/q_1,\dots,1/q_m,1/q) = (1/p_1,\dots,1/p_m,1/p)$. Let M be the supremum in (7.2.66). For notational uniformity we set $\mu_0 = \mu$.
Case 1: Suppose $\frac{\mu(A_0)}{\sqrt[m]{B_0}} = \max\left\{\frac{\mu_i(A_i)}{\sqrt[m]{B_i}}, i = 0,1,\dots,m\right\}$. Since T maps $L^1\times\cdots\times L^1$ to weak $L^{1/m}$ when restricted to characteristic functions, by Theorem 7.2.12 (i) there exists a subset A_0' of A_0 of measure $\mu(A_0') \ge \frac{1}{2}\mu(A_0)$ such that

$$\left|\int_{A_0'} T(\chi_{A_1},\dots,\chi_{A_m})\,d\mu\right| \le CB_0\,\mu_1(A_1)\cdots\mu_m(A_m)\,\mu(A_0)^{1-\frac{1}{1/m}}$$

for some constant C. Then

$$\begin{aligned}
&\left|\int_{A_0} T(\chi_{A_1},\dots,\chi_{A_m})\,d\mu\right| \\
&\le \left|\int_{A_0'} T(\chi_{A_1},\dots,\chi_{A_m})\,d\mu\right| + \left|\int_{A_0\setminus A_0'} T(\chi_{A_1},\dots,\chi_{A_m})\,d\mu\right| \\
&\le CB_0\,\mu(A_0)^{1-m}\mu_1(A_1)\cdots\mu_m(A_m) + M\Big(\frac{1}{2}\mu(A_0)\Big)^{\frac{1}{p'}}\mu_1(A_1)^{\frac{1}{p_1}}\cdots\mu_m(A_m)^{\frac{1}{p_m}} \\
&\le CB_0\,\mu(A_0)^{1-m}\mu_1(A_1)^{\frac{1}{p_1}}\left(\frac{\sqrt[m]{B_1}}{\sqrt[m]{B_0}}\right)^{\frac{1}{p_1'}}\mu(A_0)^{\frac{1}{p_1'}}\cdots\mu_m(A_m)^{\frac{1}{p_m}}\left(\frac{\sqrt[m]{B_m}}{\sqrt[m]{B_0}}\right)^{\frac{1}{p_m'}}\mu(A_0)^{\frac{1}{p_m'}} \\
&\qquad + M2^{-\frac{1}{p'}}\mu_1(A_1)^{\frac{1}{p_1}}\cdots\mu_m(A_m)^{\frac{1}{p_m}}\mu(A_0)^{\frac{1}{p'}} \\
&= \left[CB_0\left(\frac{\sqrt[m]{B_1}}{\sqrt[m]{B_0}}\right)^{\frac{1}{p_1'}}\cdots\left(\frac{\sqrt[m]{B_m}}{\sqrt[m]{B_0}}\right)^{\frac{1}{p_m'}} + M2^{-\frac{1}{p'}}\right]\mu_1(A_1)^{\frac{1}{p_1}}\cdots\mu_m(A_m)^{\frac{1}{p_m}}\mu(A_0)^{\frac{1}{p'}},
\end{aligned}$$

since $\frac{1}{p_1'}+\cdots+\frac{1}{p_m'}+1-m=\frac{1}{p'}$. It follows that M must be less than or equal to

$$CB_0\left(\frac{\sqrt[m]{B_1}}{\sqrt[m]{B_0}}\right)^{\frac{1}{p_1'}}\cdots\left(\frac{\sqrt[m]{B_m}}{\sqrt[m]{B_0}}\right)^{\frac{1}{p_m'}}+M2^{-\frac{1}{p'}}$$

and consequently

$$M\le\frac{C}{1-2^{-1/p'}}B_0^{\frac{1}{mp}}B_1^{\frac{1}{mp_1'}}\cdots B_m^{\frac{1}{mp_m'}}.$$

Case 2: Suppose that $\frac{\mu_j(A_j)}{\sqrt[m]{B_j}}=\max\left\{\frac{\mu_i(A_i)}{\sqrt[m]{B_i}},i=0,1,\dots,m\right\}$ for some $j\in\{1,\dots,m\}$. Here we use that T^{*j} maps $L^1\times\cdots\times L^1$ to weak $L^{1/m}$ when restricted to characteristic functions.

For notational simplicity, in the following argument we take $j=1$. Then there exists a subset A_1' of A_1 of measure $\mu(A_1')\ge\frac{1}{2}\mu(A_1)$ such that

$$\left|\int_{A_1'}T^{*1}(\chi_{A_0},\chi_{A_2},\dots,\chi_{A_m})\,d\mu_1\right|\le CB_1\,\mu_1(A_1)^{1-m}\mu(A_0)\,\mu_2(A_2)\cdots\mu_m(A_m)$$

for some constant C. Equivalently, we write this statement as

$$\left|\int_{A_0}T(\chi_{A_1'},\chi_{A_2},\dots,\chi_{A_m})\,d\mu\right|\le CB_1\,\mu_1(A_1)^{1-m}\mu(A_0)\,\mu_2(A_2)\cdots\mu_m(A_m)$$

by the definition of the first dual operator T^{*1}. Therefore, we obtain

$$\begin{aligned}
&\left|\int_{A_0}T(\chi_{A_1},\chi_{A_2},\dots,\chi_{A_m})\,d\mu\right|\\
&\le\left|\int_{A_0}T(\chi_{A_1'},\chi_{A_2},\dots,\chi_{A_m})\,d\mu\right|+\left|\int_{A_0}T(\chi_{A_1\setminus A_1'},\chi_{A_2},\dots,\chi_{A_m})\,d\mu\right|\\
&\le CB_1\,\mu_1(A_1)^{1-m}\mu(A_0)\prod_{j=2}^m\mu_j(A_j)+M\,\mu(A_0)^{\frac{1}{p'}}\left(\frac{1}{2}\mu_1(A_1)\right)^{\frac{1}{p_1}}\prod_{j=2}^m\mu_j(A_j)^{\frac{1}{p_j}}\\
&\le CB_1\mu_1(A_1)^{1-m}\mu(A_0)^{\frac{1}{p'}}\mu_1(A_1)^{\frac{1}{p}}\left(\frac{\sqrt[m]{B_0}}{\sqrt[m]{B_1}}\right)^{\frac{1}{p}}\prod_{j=2}^m\mu_j(A_j)^{\frac{1}{p_j}}\mu_1(A_1)^{\frac{1}{p_j'}}\left(\frac{\sqrt[m]{B_j}}{\sqrt[m]{B_1}}\right)^{\frac{1}{p_j'}}\\
&\qquad+M\,\mu(A_0)^{\frac{1}{p'}}2^{-\frac{1}{p_1}}\mu_1(A_1)^{\frac{1}{p_1}}\prod_{j=2}^m\mu_j(A_j)^{\frac{1}{p_j}}\\
&=\left[CB_1\left(\frac{\sqrt[m]{B_0}}{\sqrt[m]{B_1}}\right)^{\frac{1}{p}}\prod_{j=2}^m\left(\frac{\sqrt[m]{B_j}}{\sqrt[m]{B_1}}\right)^{\frac{1}{p_j'}}+M2^{-\frac{1}{p_1}}\right]\mu(A_0)^{\frac{1}{p'}}\prod_{j=1}^m\mu_j(A_j)^{\frac{1}{p_j}},
\end{aligned}$$

since $\frac{1}{p}+\frac{1}{p_2'}+\cdots+\frac{1}{p_m'}+1-m=\frac{1}{p_1}$. By the definition of M, it follows that

$$M\le\frac{C}{1-2^{-1/p_1}}B_0^{\frac{1}{mp}}B_1^{\frac{1}{mp_1'}}\cdots B_2^{\frac{1}{mp_2'}}.$$

The statement of the theorem follows with

$$C_{p_1,\dots,p_m} = C\max\left(\frac{1}{1-2^{-1/p_1}},\dots,\frac{1}{1-2^{-1/p_m}},\frac{1}{1-2^{-1/p'}}\right).$$

Now fix a point $(1/q_1,\dots,1/q_m,1/q)$ (for which we may have $q \le 1$) on the line segment that joins $(1/p_1,\dots,1/p_m,1/p)$ to $(1,\dots,1,m)$. We showed that T is of restricted weak type $(p_1,\dots,p_m,p)$ with constant $C_{p_1,\dots,p_m} B_0^{1/mp} B_1^{1/mp_1'} \cdots B_m^{1/mp_m'}$. Using Exercise 7.2.1 we obtain that T is also of restricted weak type $(q_1,\dots,q_m,q)$ with constant

$$\left(C_{p_1,\dots,p_m} B_0^{\frac{1}{mp}} B_1^{\frac{1}{mp_1'}} \cdots B_m^{\frac{1}{mp_m'}}\right)^{\theta} B_0^{1-\theta} \tag{7.2.68}$$

for some $0<\theta<1$, where $1/q_j = 1-\theta+\theta/p_j$ for all $j=1,\dots,m$. Since we have $1/p = 1/p_1+\cdots+1/p_m$ and $1/q = 1/q_1+\cdots+1/q_m$, it follows that the constant in (7.2.68) has the form claimed in (7.2.67). □

Remark 7.2.14. We note that condition (7.2.66) in Theorem 7.2.13 holds if μ and μ_j are equal to Lebesgue measure and T is an m-linear operator acting on functions defined on $\mathbf{R}^n$ whose kernel $K(x,y_1,\dots,y_m)$ is bounded and supported in $\{(x,y_1,\dots,y_m):\ |x-y_1|,\dots,|x-y_m| \le M\}$ for some $M>0$. Indeed, in this case we control the absolute value of the integral in (7.2.66) via Hölder's inequality by

$$|A_0|^{\frac{1}{p'}} \|K\|_{L^\infty} \left(\int_{\mathbf{R}^n} \left[\prod_{j=1}^m \int_{|x-y_j|\le M} \chi_{A_j}(y_j)\,dy_j\right]^p dx\right)^{\frac{1}{p}}$$

and applying Hölder's inequality again and the fact that convolution with an L^1 function preserves L^{p_j}, $p_j>1$, the preceding expression is bounded by

$$|A_0|^{\frac{1}{p'}} \|K\|_{L^\infty} (v_n M)^m |A_1|^{\frac{1}{p_1}} \cdots |A_m|^{\frac{1}{p_m}} < \infty .$$

Nevertheless, the bound in (7.2.67) is independent of $\|K\|_{L^\infty}$ and of the constant M.

Exercises

7.2.1. Suppose that an m-linear operator is of restricted weak type $(p_1,\dots,p_m,p)$ with constant B_0 and of restricted weak type $(q_1,\dots,q_m,q)$ with constant B_1 for some $0<p_i,q_i,p,q\le\infty$. Show that for any $\theta\in[0,1]$ the operator is of restricted weak type $(r_1,\dots,r_m,r)$ with constant at most $B_0^{1-\theta}B_1^{\theta}$, where

$$\left(\frac{1}{r_1},\dots,\frac{1}{r_m},\frac{1}{r}\right) = (1-\theta)\left(\frac{1}{p_1},\dots,\frac{1}{p_m},\frac{1}{p}\right) + \theta\left(\frac{1}{q_1},\dots,\frac{1}{q_m},\frac{1}{q}\right).$$

7.2.2. For $\alpha,\beta \in \mathbf{R}\setminus\{0\}$ with $\alpha \neq \beta$, consider the family of operators

$$B_{\alpha,\beta}(f,g)(x) = \int_{|t|\le 1} f(x-\alpha t)\,g(x-\beta t)\,dt$$

defined for Schwartz functions f,g on $\mathbf{R}^n$.
(a) Show that $B_{\alpha,\beta}$ maps $L^p \times L^q \to L^r$ when $1 \le p,q,r \le \infty$ and $1/p+1/q=1/r$.
(b) Use Theorem 7.2.13 and the result in Example 7.1.2 to show that $B_{\alpha,\beta}$ is of restricted weak type (p,q,r) when $1<p,q<\infty$, $1/2<r<1$, and $1/p+1/q=1/r$.
(c) Apply Corollary 7.2.4 to conclude that $B_{\alpha,\beta}$ maps $L^p \times L^q \to L^r$ when $1/2<r<1$, $1<p,q<\infty$, and $1/p+1/q=1/r$.

7.2.3. Suppose $0<s<\infty$, $1<p_1,p_2<\infty$ and $2-1/s=1/p_1+1/p_2-1/p$. Assume that a bilinear operator T defined on finitely simple functions of $\mathbf{R}\times\mathbf{R}$ has the property that

$$\sup_{A_0,A_1,A_2} |A_0|^{-1/p'}|A_1|^{-1/p_1}|A_2|^{-1/p_2}\left|\int_{A_0} T(\chi_{A_1},\chi_{A_2})(x)\,dx\right| < \infty$$

where the supremum is taken over all subsets subsets A_0,A_1,A_2 of $\mathbf{R}$ of finite measure. Suppose that T, T^{*1}, and T^{*2} are of restricted weak type $(1,1,s)$ with constants B_0,B_1,B_2, respectively. Then there is a constant $C(p_1,p_2,p,s)$ such that T is of weak type (p_1,p_2,p) with norm at most $C(p_1,p_2,p,s)B_0^{s/p}B_1^{s/p_1'}B_2^{s/p_2'}$.

7.2.4. Follow the steps below to prove Proposition 7.2.1.
(a) Fix $0<\delta<\min(1,q)$. Then for f_j^i in $S_0^+(X_j)$ we have

$$|T(f_1^1+\cdots+f_1^{N_1},\dots,f_m^1+\cdots+f_m^{N_m})|^\delta \le \sum_{j_1=1}^{N_1}\cdots\sum_{j_m=1}^{N_m} |T(f_1^{j_1},\dots,f_m^{j_m})|^\delta .$$

(b) Show that for functions f_j in $S_0^+(X_j)$ and A_j measurable subsets of X_j with finite measure we have

$$\big\|T(f_1\chi_{A_1},\dots,f_m\chi_{A_m})\big\|_{L^{q,\infty}} \le \big(\tfrac{q}{q-\delta}\big)^{\frac{1}{\delta}}(1-2^{-\delta})^{-\frac{1}{\delta}} M \prod_{j=1}^m \mu(A_j)^{\frac{1}{p_j}} \big\|f_j\chi_{A_j}\big\|_{L^\infty} .$$

(c) For each $f_j \in S_0^+(X_j)$ find measurable sets A_j^k, $k=-N_j,\dots,N_j$, such that $\mu(A_j^{k_j}) \le d_f(f_j^*(2^{k_1+1})) \le 2^{k_j+1}$ and that $\|f_j\chi_{A_j^{k_j}}\| \le f_j^*(2^{k_j})$.
(d) Prove that for functions $f_j \in S_0^+(X_j)$ we have

$$\big\|T(f_1,\dots,f_m)\big\|_{L^{q,\infty}(Y)} \le C'(p_1,\dots,p_m,q,\delta)\,M\,\|f_1\|_{L^{p_1,\delta}(X_1)}\cdots\|f_m\|_{L^{p_m,\delta}(X_m)}$$

where

$$C'(p_1,\dots,p_m,q,\delta) = \big(\tfrac{q}{q-\delta}\big)^{\frac{2}{\delta}}(1-2^{-\delta})^{-\frac{1}{\delta}}(\log 2)^{-\frac{m}{\delta}}\,2^{\frac{2}{p_1}+\cdots+\frac{2}{p_m}} .$$

(e) Express each f_j in $S_0(X_j)$ as $f_j = f_j^1 - f_j^2 + i(f_j^3 - f_j^4)$, where f_j^k lies in $S_0^+(X_j)$ and $\|f_j^i\|_{L^{p,\delta}(X)} \le \|f_j\|_{L^{p,\delta}(X)}$.
(f) Extend the result in part (d) to all $f_j \in S_0(X_j)$.
[*Hint:* Parts (b) and (d): Use the normability of $L^{q/\delta,\infty}$ when $q > \delta$. Parts (c), (e): Use the idea of Lemma 1.4.20 in [156]. Part (f): Use sublinearity.]

7.2.5. Suppose that μ is a nonnegative measure on $\mathbf{R}^{2n}$ such that

$$|\widehat{\mu}(\xi,-\xi)| + |\widehat{\mu}(\xi,0)| + |\widehat{\mu}(0,\xi)| \le C_0|\xi|^{-\gamma}$$

for some $C_0 < \infty$ and some γ satisfying $0 < \gamma < n/2$. Then the bilinear operator

$$T^{\mu}(f_1,f_2)(x) = \int_{\mathbf{R}^n}\int_{\mathbf{R}^n} f_1(x-y_1)f_2(x-y_2)\,d\mu(y_1,y_2)$$

maps $L^{p_1}(\mathbf{R}^n) \times L^{p_2}(\mathbf{R}^n)$ to $L^p(\mathbf{R}^n)$ when

$$\frac{1}{p} = \frac{1}{p_1} + \frac{1}{p_2} - \frac{\gamma}{n}$$

and $(1/p_1, 1/p_2)$ lies in the closed hexagon with vertices

$$\Big(\frac{n+2\gamma}{2n},\frac{1}{2}\Big), \Big(\frac{n+2\gamma}{2n},0\Big), \Big(\frac{1}{2},\frac{n+2\gamma}{2n}\Big), \Big(0,\frac{n+2\gamma}{2n}\Big), \Big(0,\frac{1}{2}\Big), \Big(\frac{1}{2},0\Big).$$

[*Hint:* Consider first the point $p_1 = 2$, $p_2 = 2n/(n+2\gamma)$, $p = 1$ and obtain the remaining points by duality, symmetry, and Corollary 7.2.11.]

7.2.6. Suppose that m is a bounded function on $\mathbf{R}^n$ and let $T_m(f) = (\widehat{f}m)^\vee$ for $f \in \mathscr{S}(\mathbf{R}^n)$.
(a) Show that if $m = m_1 * m_2$, then

$$\|T_m\|_{L^p\to L^p} \le \|m_1\|_{L^{(\frac{3}{2}-\frac{1}{p})^{-1}}} \|m_2\|_{L^{(\frac{1}{p}-\frac{1}{2})^{-1}}}.$$

(b) Prove that for $1 < p < 2$ we have

$$\|T_m\|_{L^p\to L^p} \le C(n,p,\gamma)\|(I-\Delta)^{\gamma/2}m\|_{L^{(\frac{1}{p}-\frac{1}{2})^{-1}}}$$

whenever $\gamma > n(\frac{1}{p}-\frac{1}{2})$.
[*Hint:* Part (a): Use Corollary 7.2.11 starting from $\|T_m\|_{L^2\to L^2} \le \|m_1\|_{L^1}\|m_2\|_{L^\infty}$ and $\|T_m\|_{L^1\to L^1} \le \|m_1\|_{L^2}\|m_2\|_{L^2}$. Part (b): Write $m = J_\gamma * ((I-\Delta)^{\gamma/2}m)$ and use part (a).]

7.3 Vector-valued Estimates and Multilinear Convolution Operators

In this section we study basic properties of multilinear convolution operators or translation invariant operators. A fundamental result in this section says that regularizations of multilinear symbols are also multilinear symbols, a fact that requires vector-valued extensions of multilinear operators.

7.3.1 Multilinear Vector-valued Inequalities

A classical result (Theorem 5.5.1 [156]) says that every linear operator that maps L^p to L^q for some $0 < p, q < \infty$ admits an ℓ^2-bounded extension. Here we discuss a version of this theorem in the multilinear setting defined on Lebesgue spaces of general measure spaces.

Theorem 7.3.1. *Let (X_j, μ_j), (Y, ν) be σ-finite measure spaces.*
(a) Let T be an m-linear operator that maps

$$L^{p_1}(X_1, \mu_1) \times \cdots \times L^{p_m}(X_m, \mu_m) \to L^q(Y, \nu)$$

for some $0 < p_1, p_2, \dots, p_m, q < \infty$ with norm $\|T\|$. Then there is a constant $C = C(m, p_1, \dots, p_m, q)$ such that for all sequences $\{f_j^k\}_{k\in\mathbf{Z}}$ in $L^{p_j}(X_j)$, $1 \le j \le m$, we have

$$\Big\|\Big(\sum_{k_1\in\mathbf{Z}} \cdots \sum_{k_m\in\mathbf{Z}} |T(f_1^{k_1}, \dots, f_m^{k_m})|^2\Big)^{\frac{1}{2}}\Big\|_{L^q} \le C\|T\| \prod_{j=1}^{m} \Big\|\Big(\sum_{k\in\mathbf{Z}} |f_j^k|^2\Big)^{\frac{1}{2}}\Big\|_{L^{p_j}}. \tag{7.3.1}$$

(b) Suppose that an m-linear operator T maps $L^{p_1}(X_1, \mu_1) \times \cdots \times L^{p_m}(X_m, \mu_m)$ to $L^{q,\infty}(Y, \nu)$ for some $0 < p_1, p_2, \dots, p_m, q < \infty$ with norm $\|T\|_w$. Then T has an ℓ^2-valued extension, i.e., for all sequences $\{f_j^k\}_{k\in\mathbf{Z}}$ it satisfies

$$\Big\|\Big(\sum_{k_1\in\mathbf{Z}} \cdots \sum_{k_m\in\mathbf{Z}} |T(f_1^{k_1}, \dots, f_m^{k_m})|^2\Big)^{\frac{1}{2}}\Big\|_{L^{q,\infty}} \le C' \|T\|_w \prod_{j=1}^{m} \Big\|\Big(\sum_{k\in\mathbf{Z}} |f_j^k|^2\Big)^{\frac{1}{2}}\Big\|_{L^{p_j}} \tag{7.3.2}$$

for some constant $C' = C'(m, p_1, \dots, p_m, q)$.

Proof. To prove the inequality in part (a), we recall the Rademacher functions r_j that satisfy the following inequality:

$$B_q^m \Big(\sum_{k_1} \cdots \sum_{k_m} |c_{k_1,\dots,k_m}|^2\Big)^{\frac{1}{2}} \le \|F_m\|_{L^q([0,1]^m)} \le D_q^m \Big(\sum_{k_1} \cdots \sum_{k_m} |c_{k_1,\dots,k_m}|^2\Big)^{\frac{1}{2}}, \tag{7.3.3}$$

where $0 < q < \infty$, $0 < D_q, B_q < \infty$, $c_{k_1,\dots,k_n}$ is a sequence of complex numbers, and

$$F_m(t_1,\dots,t_m) = \sum_{k_1\in\mathbf{Z}} \cdots \sum_{k_m\in\mathbf{Z}} c_{k_1,\dots,k_m} r_{k_1}(t_1)\cdots r_{k_m}(t_m),$$

for $t_j \in [0,1]$; see Appendix C.5 in [156].

We define p by setting $1/p = 1/p_1 + \cdots + 1/p_m$, and we consider two cases as follows.

Case 1: $q \le p$. In this case we fix a positive integer N. Using both estimates in (7.3.3) and the multilinearity of T we obtain

$$\begin{aligned}
&\Big\|\Big(\sum_{|k_1|\le N} \cdots \sum_{|k_m|\le N} |T(f_1^{k_1},\dots,f_m^{k_m})|^2\Big)^{\frac12}\Big\|_{L^q}^q \\
&\le B_q^{-qm} \int_Y \int_{[0,1]^m} \Big| \sum_{|k_1|\le N} \cdots \sum_{|k_m|\le N} T(f_1^{k_1},\dots,f_m^{k_m}) r_{k_1}(t_1)\dots r_{k_m}(t_m)\Big|^q dt_1\cdots dt_m\, d\nu \\
&\le B_q^{-qm} \int_{[0,1]^m} \int_Y \Big| T\Big(\sum_{|k_1|\le N} r_{k_1}(t_1) f_1^{k_1},\dots, \sum_{|k_m|\le N} r_{k_m}(t_m) f_m^{k_m}\Big)\Big|^q d\nu\, dt_1\cdots dt_m \\
&\le B_q^{-qm} \|T\|^q \int_{[0,1]^m} \prod_{j=1}^m \Big\| \sum_{|k_j|\le N} r_{k_j}(t_j) f_j^{k_j} \Big\|_{L^{p_j}(X_j)}^q dt_1\cdots dt_m \\
&\le B_q^{-qm} \|T\|^q \prod_{j=1}^m \Big(\int_0^1 \Big\| \sum_{|k_j|\le N} r_{k_j}(t_j) f_j^{k_j} \Big\|_{L^{p_j}(X_j)}^{p_j} dt_j \Big)^{\frac{q}{p_j}} \\
&\le B_q^{-qm} \|T\|^q \prod_{j=1}^m \Big(D_{p_j}^{p_j} \Big\| \Big(\sum_{|k_j|\le N} |f_j^{k_j}|^2 \Big)^{\frac12} \Big\|_{L^{p_j}(X_j)}^{p_j} \Big)^{\frac{q}{p_j}} \\
&\le B_q^{-qm} D_{p_1}^q \cdots D_{p_m}^q \|T\|^q \prod_{j=1}^m \Big\| \Big(\sum_{k_j\in\mathbf{Z}} |f_j^{k_j}|^2 \Big)^{\frac12} \Big\|_{L^{p_j}(X_j)}^q
\end{aligned}$$

where we used the fact that $p_j \ge p \ge q$ in Hölder's inequality in the fourth inequality above. Letting $N \to \infty$ yields the required conclusion in Case 1, with constant $C = B_q^{-m} D_{p_1} \cdots D_{p_m}$.

Case 2: $p < q$. Using duality we can write

$$\begin{aligned}
&\Big\|\Big(\sum_{k_1\in\mathbf{Z}} \cdots \sum_{k_m\in\mathbf{Z}} |T(f_1^{k_1},\dots,f_m^{k_m})|^2\Big)^{\frac12}\Big\|_{L^q} \\
&\qquad = \sup_{\|g\|_{L^{(q/p)'}}\le 1} \Big(\int_Y \Big(\sum_{k_1\in\mathbf{Z}} \cdots \sum_{k_m\in\mathbf{Z}} |T(f_1^{k_1},\dots,f_m^{k_m})|^2 \Big)^{\frac{p}{2}} |g|\, d\nu \Big)^{\frac1p}
\end{aligned} \tag{7.3.4}$$

and we define an m-linear operator T_g by setting

$$T_g(f_1,\dots,f_m) = |g|^{\frac{1}{p}} T(f_1,\dots,f_m)$$

for some fixed function g in $L^{(q/p)'}$ with norm at most 1. We can easily verify that T_g is bounded from $L^{p_1} \times \cdots \times L^{p_m}$ into L^p with norm at most $\|T\|$. Indeed, for all $\|f_j\|_{L^{p_j}} \le 1$, we have

$$\begin{aligned}
\big\|T_g(f_1,\dots,f_m)\big\|_{L^p} &= \left\{ \int_Y |g|\, |T(f_1,\dots,f_m)|^p d\nu \right\}^{\frac{1}{p}} \\
&\le \|g\|_{L^{(\frac{q}{p})'}}^{\frac{1}{p}} \big\| |T(f_1,\dots,f_m)|^p \big\|_{L^{\frac{q}{p}}}^{\frac{1}{p}} \\
&\le \|T\|
\end{aligned}$$

since $\|g\|_{L^{(q/p)'}} \le 1$. In view of the result of Case 1 applied to T_g, we have

$$\left\{ \int_Y \Big(\sum_{k_1\in\mathbf{Z}} \cdots \sum_{k_m\in\mathbf{Z}} |T(f_1^{k_1},\dots,f_m^{k_m})|^2 \Big)^{\frac{p}{2}} |g| d\nu \right\}^{\frac{1}{p}} \le C_{p_j,q,m} \|T\| \prod_{j=1}^m \Big\| \Big(\sum_{k\in\mathbf{Z}} |f_j^k|^2 \Big)^{\frac{1}{2}} \Big\|_{L^{p_j}},$$

and this estimate, combined with (7.3.4), gives (7.3.1) in Case 2.

Part (b). We recall the following well-known characterization of weak L^q (Exercise 1.1.12 in [156]), which holds for σ-finite measure spaces:

$$\|f\|_{L^{q,\infty}} \le \sup_{0<\nu(E)<\infty} \nu(E)^{\frac{1}{q}-\frac{1}{r}} \left(\int_E |f|^r d\nu \right)^{\frac{1}{r}} \le \left(\frac{q}{q-r} \right)^{\frac{1}{r}} \|f\|_{L^{q,\infty}}, \tag{7.3.5}$$

where $0 < r < q$. For a measurable set E with finite measure we define an operator $T_E(f_1,\dots,f_m) = \chi_E\, T(f_1,\dots,f_m)$. Using (7.3.5) we obtain

$$\begin{aligned}
&\Big\| \Big(\sum_{k_1\in\mathbf{Z}} \cdots \sum_{k_m\in\mathbf{Z}} |T(f_1^{k_1},\dots,f_m^{k_m})|^2 \Big)^{\frac{1}{2}} \Big\|_{L^{q,\infty}(\nu)} \\
&\le \sup_{0<\nu(E)<\infty} \nu(E)^{\frac{1}{q}-\frac{1}{r}} \left(\int_E \Big(\sum_{k_1\in\mathbf{Z}} \cdots \sum_{k_m\in\mathbf{Z}} |T(f_1^{k_1},\dots,f_m^{k_m})|^2 \Big)^{\frac{r}{2}} d\nu \right)^{\frac{1}{r}} \\
&= \sup_{0<\nu(E)<\infty} \nu(E)^{\frac{1}{q}-\frac{1}{r}} \left(\int_Y \Big(\sum_{k_1\in\mathbf{Z}} \cdots \sum_{k_m\in\mathbf{Z}} |\chi_E\, T(f_1^{k_1},\dots,f_m^{k_m})|^2 \Big)^{\frac{r}{2}} d\nu \right)^{\frac{1}{r}} \\
&\le \sup_{0<\nu(E)<\infty} \nu(E)^{\frac{1}{q}-\frac{1}{r}} \|T_E\|_{L^{p_1}\times\cdots\times L^{p_m}\to L^r} \prod_{j=1}^m \left(\int_{X_j} \Big(\sum_{k\in\mathbf{Z}} |f_j^k|^2 \Big)^{\frac{p_j}{2}} d\mu_j \right)^{\frac{1}{p_j}},
\end{aligned} \tag{7.3.6}$$

where we used the result in part (a). But since for all m-tuples of functions $(f_1,\dots,f_m) \in L^{p_1} \times \cdots \times L^{p_m}$ we have

$$\begin{aligned} v(E)^{\frac{1}{q}-\frac{1}{r}} \|T_E(f_1,\dots,f_m)\|_{L^r} &\le \Big(\frac{q}{q-r}\Big)^{\frac{1}{r}} \|T(f_1,\dots,f_m)\|_{L^{q,\infty}} \\ &\le \Big(\frac{q}{q-r}\Big)^{\frac{1}{r}} \|T\|_w \prod_{j=1}^m \|f_j\|_{L^{p_j}}, \end{aligned}$$

it follows that for any measurable set E of finite measure the estimate

$$v(E)^{\frac{1}{q}-\frac{1}{r}} \|T_E\|_{L^{p_1}\times\cdots\times L^{p_m}\to L^r} \le \Big(\frac{q}{q-r}\Big)^{\frac{1}{r}} \|T\|_w \tag{7.3.7}$$

is valid. Now returning to (7.3.6) and using (7.3.7) we obtain (7.3.2). □

7.3.2 Multilinear Convolution and Multiplier Operators

For $h \in \mathbf{R}^n$, let $\tau^h(f)(x) = f(x-h)$ be the translation by h. Recall that $\mathscr{S}(\mathbf{R}^n)$ denotes the space of Schwartz functions on $\mathbf{R}^n$ and $\mathscr{S}'(\mathbf{R}^n)$ the space of tempered distributions. We say that a multilinear operator T from $\mathscr{S}(\mathbf{R}^n) \times \cdots \times \mathscr{S}(\mathbf{R}^n) \to \mathscr{S}'(\mathbf{R}^n)$ *commutes with simultaneous translations*, or that it is *translation invariant*, if for all $f_1,\dots,f_m \in \mathscr{S}(\mathbf{R}^n)$ and all $h \in \mathbf{R}^n$ we have

$$\tau^h(T(f_1,\dots,f_m)) = T(\tau^h f_1,\dots,\tau^h f_m). \tag{7.3.8}$$

When $m=1$, bounded operators from L^p to L^q that commute with translations are exactly the convolution operators (Theorem 2.5.2 in [156]), i.e., they have the form

$$Tf(x) = (f*K)(x)$$

for some tempered distribution K on $\mathbf{R}^n$. These operators play a very important role in linear analysis and it is quite natural to introduce their multilinear extensions.

Definition 7.3.2. For a given K_0 in $\mathscr{S}'((\mathbf{R}^n)^m)$, let T be a multilinear operator from $\mathscr{S}(\mathbf{R}^n) \times \cdots \times \mathscr{S}(\mathbf{R}^n)$ to $\mathscr{S}'(\mathbf{R}^n)$ that satisfies, for all $\psi_1,\dots,\psi_m$ in $\mathscr{S}(\mathbf{R}^n)$,

$$T(\psi_1,\dots,\psi_m)(x) = \big((\psi_1\otimes\cdots\otimes\psi_m)\star K_0\big)(x,\dots,x), \tag{7.3.9}$$

where $\star$ denotes convolution on $(\mathbf{R}^n)^m$, and recall that

$$(\psi_1\otimes\cdots\otimes\psi_m)(y_1,\dots,y_m) = \psi_1(y_1)\dots\psi_m(y_m)$$

for all $y_1,\dots,y_m \in \mathbf{R}^n$. Then we say that T is an *m-linear convolution operator*.

Notice that m-linear convolution operators commute with simultaneous translations. In this chapter we investigate when such operators admit bounded extensions from $L^{p_1}(\mathbf{R}^n)\times\cdots\times L^{p_m}(\mathbf{R}^n)$ into $L^p(\mathbf{R}^n)$ for some indices $p_1,\dots,p_m,p$.

For $\xi_k\in\mathbf{R}^n$ we introduce the notation $\vec{\xi}=(\xi_1,\dots,\xi_m)$ for a vector in $(\mathbf{R}^n)^m$ and we denote by $d\vec{\xi}=d\xi_1\cdots d\xi_m$ the combined differential of all variables.

A multilinear convolution or translation-invariant operator can be written in the form

$$T(f_1,\dots,f_m)(x)=\int_{(\mathbf{R}^n)^m}K_0(x-y_1,\dots,x-y_m)f_1(y_1)\cdots f_m(y_m)\,d\vec{y}, \tag{7.3.10}$$

where K_0 is a function on $(\mathbf{R}^n)^m$ or a tempered distribution, in which case the integral in (7.3.10) is interpreted in the sense of distributions. Let σ be the distributional Fourier transform of K_0. If σ is a function, i.e., a locally integrable function such that there are $R,N,C>0$ so that

$$|\sigma(\vec{\xi}\,)|\le C(1+|\vec{\xi}\,|)^N \tag{7.3.11}$$

for all $|\vec{\xi}\,|>R$, then the operator in (7.3.10) can also be expressed in the form

$$T_\sigma(f_1,\dots,f_m)(x)=\int_{(\mathbf{R}^n)^m}\sigma(\vec{\xi}\,)\widehat{f_1}(\xi_1)\cdots\widehat{f_m}(\xi_m)e^{2\pi i x\cdot(\xi_1+\cdots+\xi_m)}d\vec{\xi}. \tag{7.3.12}$$

Definition 7.3.3. Fix $0<p_1,\dots,p_m\le\infty$ and $0<p<\infty$ satisfying

$$\frac{1}{p_1}+\cdots+\frac{1}{p_m}=\frac{1}{p}. \tag{7.3.13}$$

A locally integrable function σ defined on $(\mathbf{R}^n)^m$ that satisfies (7.3.11) is called a $(p_1,\dots,p_m,p)$ *multilinear multiplier* if the associated operator T_σ given by (7.3.12) extends to a bounded operator from $L^{p_1}(\mathbf{R}^n)\times\cdots\times L^{p_m}(\mathbf{R}^n)$ to $L^p(\mathbf{R}^n)$. The function σ is also called the *multilinear symbol* of T_σ. We denote by $\mathscr{M}_{p_1,\dots,p_m}(\mathbf{R}^n)$, or simply by $\mathscr{M}_{p_1,\dots,p_m}$, the space of all $(p_1,\dots,p_m,p)$ multilinear multipliers on $\mathbf{R}^n$, and we define the quasi-norm of σ in $\mathscr{M}_{p_1,\dots,p_m}(\mathbf{R}^n)$ as the quasi-norm of T_σ from $L^{p_1}\times\cdots\times L^{p_m}$ into L^p, i.e.,

$$\|\sigma\|_{\mathscr{M}_{p_1,\dots,p_m}}=\|T_\sigma\|_{L^{p_1}\times\cdots\times L^{p_m}\to L^p}.$$

Thus if a multilinear convolution operator is bounded from $L^{p_1}\times\cdots\times L^{p_m}$ to L^p for some indices related as in Hölder's inequality, then we call it a multilinear multiplier operator and we call its symbol a multilinear multiplier. We have the following list of basic properties of multilinear multipliers.

Proposition 7.3.4. *Let $0<p_1,\dots,p_m\le\infty$. Then the following statements are valid:*

(i) If $\lambda\in\mathbf{C}$, σ,σ_1 and σ_2 are in $\mathscr{M}_{p_1,\dots,p_m}$, then so are $\lambda\sigma$ and $\sigma_1+\sigma_2$, and

$$\|\lambda\sigma\|_{\mathscr{M}_{p_1,\dots,p_m}}=|\lambda|\|\sigma\|_{\mathscr{M}_{p_1,\dots,p_m}},$$

$$\|\sigma_1+\sigma_2\|_{\mathcal{M}_{p_1,\dots,p_m}} \le C_p\big(\|\sigma_1\|_{\mathcal{M}_{p_1,\dots,p_m}} + \|\sigma_2\|_{\mathcal{M}_{p_1,\dots,p_m}}\big).$$

(ii) If $\sigma(\vec{\xi}\,) \in \mathcal{M}_{p_1,\dots,p_m}$ and $\vec{a} \in \mathbf{R}^n$, then $\sigma(\vec{\xi}+\vec{a})$ is in $\mathcal{M}_{p_1,\dots,p_m}$ and

$$\|\sigma\|_{\mathcal{M}_{p_1,\dots,p_m}} = \|\sigma(\cdot+\vec{a})\|_{\mathcal{M}_{p_1,\dots,p_m}}.$$

(iii) If $\sigma(\vec{\xi}\,) \in \mathcal{M}_{p_1,\dots,p_m}$ and $\delta > 0$, then $\sigma(\delta\vec{\xi}\,)$ is in $\mathcal{M}_{p_1,\dots,p_m}$ and

$$\|\sigma\|_{\mathcal{M}_{p_1,\dots,p_m}} = \|\sigma(\delta(\cdot))\|_{\mathcal{M}_{p_1,\dots,p_m}}.$$

(iv) If $\sigma(\xi_1,\dots,\xi_m) \in \mathcal{M}_{p_1,\dots,p_m}$ and A is an orthogonal matrix in $\mathbf{R}^n$, then $\sigma(A\xi_1,\dots,A\xi_m)$ is in $\mathcal{M}_{p_1,\dots,p_m}$ with the same quasi-norm.
(v) Let σ_j be a sequence of functions in $\mathcal{M}_{p_1,\dots,p_m}$ such that $\|\sigma_j\|_{\mathcal{M}_{p_1,\dots,p_m}} \le C$ for all $j = 1,2,\dots$. If σ_j are uniformly bounded and they converge pointwise to σ a.e. as $j\to\infty$, then σ is in $\mathcal{M}_{p_1,\dots,p_m}$ with quasi-norm bounded by C.

Proof. Item *(i)* is trivial while *(ii)-(iv)* are proved by a straightforward change of variables: translation, dilation, and rotation. Item *(v)* easily follows by applying Fatou's lemma on L^p (where p is related to $p_1,\dots,p_m$ via (7.3.13)) since

$$T_\sigma(f_1,\dots,f_m)(x) = \lim_{j\to\infty} T_{\sigma_j}(f_1,\dots,f_m)(x)$$

for all $x \in \mathbf{R}^n$ and all $f_k \in \mathscr{S}(\mathbf{R}^n)$. □

7.3.3 Regularizations of Multilinear Symbols and Consequences

Next we show that certain regularizations of the symbols σ of operators T_σ preserve boundedness. Recall that $\star$ denotes convolution on $(\mathbf{R}^n)^m$.

Theorem 7.3.5. *Let $0 < p_1,\dots,p_m < \infty$ and $0 < p < \infty$, and let σ be a locally integrable function defined on $(\mathbf{R}^n)^m$ that satisfies (7.3.11) for some $N \ge 0$. If $N = 0$, suppose that φ lies in $L^1(\mathbf{R}^n)$, and if $N > 0$, suppose that $|\varphi(\xi)| \le C'(1+|\xi|)^{-N-n-1}$ for all $\xi \in \mathbf{R}^n$, so that $(\varphi\otimes\cdots\otimes\varphi)\star\sigma$ is well defined. Assume that the multilinear convolution operator T_σ associated with σ maps $L^{p_1}\times\cdots\times L^{p_m} \to L^p$. Then $T_{(\varphi\otimes\cdots\otimes\varphi)\star\sigma}$ is also bounded and satisfies*

$$\big\|T_{(\varphi\otimes\cdots\otimes\varphi)\star\sigma}\big\|_{L^{p_1}\times\cdots\times L^{p_m}\to L^p} \le C_{m,p_1,\dots,p_m,p}\big\|\varphi\big\|_{L^1}^m \big\|T_\sigma\big\|_{L^{p_1}\times\cdots\times L^{p_m}\to L^p}$$

for some constant $C_{m,p_1,\dots,p_m,p}$.

Proof. Let us denote by $M^b(f)(x) = e^{2\pi i b\cdot x}f(x)$ the modulation operator acting on a function f. For functions $f_1,\dots,f_m \in \mathscr{S}(\mathbf{R}^n)$, an easy calculation based on a change of variables gives that for all x in $\mathbf{R}^n$ we have

$$T_{(\varphi\otimes\cdots\otimes\varphi)\star\sigma}(f_1,\dots,f_m)(x)$$
$$=\int_{\mathbf{R}^n}\cdots\int_{\mathbf{R}^n}\varphi(y_1)\cdots\varphi(y_m)T_\sigma\big(M^{-y_1}(f_1),\dots,M^{-y_m}(f_m)\big)(x)e^{2\pi i x\cdot(y_1+\cdots+y_m)}\,d\vec{y},$$

from which it follows that

$$|T_{(\varphi\otimes\cdots\otimes\varphi)\star\sigma}(f_1,\dots,f_m)(x)|\le S(f_1,\dots,f_m)(x),$$

where

$$S(f_1,\dots,f_m)(x) \tag{7.3.14}$$
$$=\int_{\mathbf{R}^n}\cdots\int_{\mathbf{R}^n}|\varphi(y_1)|\cdots|\varphi(y_m)|\,\big|T_\sigma\big(M^{-y_1}(f_1),\dots,M^{-y_m}(f_m)\big)(x)\big|\,d\vec{y}.$$

In view of this fact, the conclusion follows when $p\ge 1$ by applying Minkowski's integral inequality since L^p is a normed space in this case. We may therefore focus our attention on the case $p<1$.

We first assume that φ is supported in a compact cube $[-K,K]^n$. For every $x\in\mathbf{R}^n$ we define a function

$$F_x(y_1,\dots,y_m)=T_\sigma\big(M^{-y_1}(f_1),\dots,M^{-y_m}(f_m)\big)(x).$$

Then for any $\vec{y}=(y_1,y_2,\dots,y_m)$ and $\vec{z}=(z_1,y_2,\dots,y_m)$, with $z_1,y_1y_2,\dots,y_m$ in $[-K,K]^n$, there is a Ξ_{y_1,z_1} on the line segment joining y_1 to z_1 such that

$$\begin{aligned}
\big|F_x(\vec{y})-F_x(\vec{z})\big| &\le \int_{(\mathbf{R}^n)^m}|\sigma(\vec{\xi}\,)|\,\big|\widehat{f_1}(\xi_1+y_1)-\widehat{f_1}(\xi_1+z_1)\big|\prod_{l\neq 1}|\widehat{f_l}(\xi_l+y_l)|\,d\vec{\xi}\\
&\le |y_1-z_1|\int_{(\mathbf{R}^n)^m}|\sigma(\vec{\xi}\,)|\,\big|\nabla\widehat{f_1}(\xi_1+\Xi_{y_1,z_1})\big|\prod_{l\neq 1}|\widehat{f_l}(\xi_l+y_l)|\,d\vec{\xi}\\
&\le |y_1-z_1|\int_{(\mathbf{R}^n)^m}\frac{C_{n,M,K}\,|\sigma(\vec{\xi}\,)|}{(2\sqrt{n}K+|\xi_1+\Xi_{y_1,z_1}|)^M}\prod_{l\neq 1}|\widehat{f_l}(\xi_l+y_l)|\,d\vec{\xi}\\
&\le |y_1-z_1|\int_{(\mathbf{R}^n)^m}|\sigma(\vec{\xi}\,)|\frac{C_{n,M,K}}{(\sqrt{n}K+|\xi_1|)^M}\prod_{l\neq 1}|\widehat{f_l}(\xi_l+y_l)|\,d\vec{\xi},
\end{aligned}$$

and this expression can be made arbitrarily small uniformly in $y_1,\dots,y_m,z_1$ in $[-K,K]^n$, as long as y_1 and z_1 are close to each other. Analogous estimates hold when $\vec{y}$ and $\vec{z}$ are different only in the jth coordinate for some j in $\{2,\dots,m\}$.

Fix $k\in\mathbf{Z}^+$. By a compactness argument, there are pairwise disjoint subcubes $V_1,\dots,V_{L_k}$ of $[-K,K]^n$ whose union is $[-K,K]^n$ such that if c_ℓ is the center of V_ℓ, then for all $\ell_1,\dots,\ell_m$ in $\{1,\dots,L_k\}$ we have

$$\sup_{x\in\mathbf{R}^n}\sup_{y_1\in V_{\ell_1}}\sup_{\substack{y_i\in\mathbf{R}^n\\ i\geq 2}}\big|F_x(y_1,y_2,\dots,y_m)-F_x(c_{\ell_1},y_2,\dots,y_m)\big|<\frac{1}{k} \tag{7.3.15}$$

$$\sup_{x\in\mathbf{R}^n}\sup_{y_2\in V_{\ell_2}}\sup_{\substack{y_i\in\mathbf{R}^n\\ i\geq 3}}\big|F_x(c_{\ell_1},y_2,\dots,y_m)-F_x(c_{\ell_1},c_{\ell_2},y_3,\dots,y_m)\big|<\frac{1}{k} \tag{7.3.16}$$

$$\vdots$$

$$\sup_{x\in\mathbf{R}^n}\sup_{y_{m-1}\in V_{\ell_{m-1}}}\sup_{y_m\in\mathbf{R}^n}\big|F_x(c_{\ell_1},\dots,c_{\ell_{m-2}},y_{m-1},y_m)-F_x(c_{\ell_1},c_{\ell_2},\dots,c_{\ell_{m-1}},y_m)\big|<\frac{1}{k}$$

$$\sup_{x\in\mathbf{R}^n}\sup_{y_m\in V_{\ell_m}}\big|F_x(c_{\ell_1},\dots,c_{\ell_{m-1}},y_m)-F_x(c_{\ell_1},c_{\ell_2},\dots,c_{\ell_m})\big|<\frac{1}{k}.$$

We find the V_ℓ in the following order. First we find cubes so that (7.3.15) holds, then each of these cubes is partitioned into subcubes so that (7.3.16) holds, and so on.

Then for every $y_1\in[-K,K]^n$ there is a unique cube V_ℓ such that y_1 lies in V_ℓ. It follows from (7.3.14) that

$$S(f_1,\dots,f_m)(x)$$
$$\leq\frac{\|\varphi\|_{L^1}^m}{k}+\sum_{\ell_1=1}^{L_k}\lambda_{\ell_1}\underbrace{\int_{\mathbf{R}^n}\cdots\int_{\mathbf{R}^n}}_{m-1\text{ times}}\prod_{i=2}^m|\varphi(y_i)|\,\big|F_x(c_{\ell_1},y_2,\dots,y_m)\big|\,dy_2\cdots dy_m\,,$$

where $\lambda_\ell=\int_{V_\ell}|\varphi(t)|\,dt$. Applying the same argument to the remaining variables $y_2,\dots,y_m$ we deduce that

$$S(f_1,\dots,f_m)(x)\leq\frac{m\|\varphi\|_{L^1}^m}{k}+\sum_{\ell_1=1}^{L_k}\cdots\sum_{\ell_m=1}^{L_k}\lambda_{\ell_1}\cdots\lambda_{\ell_m}\big|F_x(c_{\ell_1},\dots,c_{\ell_m})\big|.$$

Setting

$$B=\Big(\sum_{\ell_1=1}^{L_k}\cdots\sum_{\ell_m=1}^{L_k}|\lambda_{\ell_1}\cdots\lambda_{\ell_m}|\Big)^{\frac12}=\big(\|\varphi\|_{L^1}^m\big)^{\frac12},$$

we obtain that

$$S(f_1,\dots,f_m)(x)$$
$$\leq\liminf_{k\to\infty}\sum_{\ell_1=1}^{L_k}\cdots\sum_{\ell_m=1}^{L_k}\lambda_{\ell_1}\cdots\lambda_{\ell_m}\big|T_\sigma\big(M^{-c_{\ell_1}}(f_1),\dots,M^{-c_{\ell_m}}(f_m)\big)(x)\big|$$
$$\leq\liminf_{k\to\infty}B\Big\{\sum_{\ell_1=1}^{L_k}\cdots\sum_{\ell_m=1}^{L_k}\big|[\lambda_{\ell_1}\cdots\lambda_{\ell_m}]^{\frac12}T_\sigma\big(M^{-c_{\ell_1}}(f_1),\dots,M^{-c_{\ell_m}}(f_m)\big)(x)\big|^2\Big\}^{\frac12},$$

where in the last step we used the Cauchy-Schwarz inequality.

It follows from Fatou's lemma that

$$\begin{aligned}&\big\|S(f_1,\dots,f_m)\big\|_{L^p}\\&\le \big\|\varphi\big\|_{L^1}^{\frac{m}{2}}\liminf_{k\to\infty}\bigg\|\bigg\{\sum_{\ell_1=1}^{L_k}\cdots\sum_{\ell_m=1}^{L_k}\big|T_\sigma\big(\lambda_{\ell_1}^{\frac12}M^{-c_{\ell_1}}(f_1),\dots,\lambda_{\ell_m}^{\frac12}M^{-c_{\ell_m}}(f_m)\big)\big|^2\bigg\}^{\frac12}\bigg\|_{L^p}\end{aligned}$$

and applying Theorem 7.3.1 we obtain that the preceding expression is bounded by

$$\big\|\varphi\big\|_{L^1}^{\frac{m}{2}}C_{m,p_1,\dots,p_m,p}\big\|T_\sigma\big\|_{L^{p_1}\times\cdots\times L^{p_m}\to L^p}\liminf_{k\to\infty}\prod_{j=1}^m\bigg\|\bigg\{\sum_{\ell_j=1}^{L_k}\big|\lambda_{\ell_j}^{\frac12}M^{-c_{\ell_j}}(f_j)\big|^2\bigg\}^{\frac12}\bigg\|_{L^{p_j}}.$$

This is in turn bounded by

$$\begin{aligned}&\big\|\varphi\big\|_{L^1}^{\frac{m}{2}}C_{m,p_1,\dots,p_m,p}\big\|T_\sigma\big\|_{L^{p_1}\times\cdots\times L^{p_m}\to L^p}\liminf_{k\to\infty}\prod_{j=1}^m\Big\||f_j|\big(\sum_{\ell_j=1}^{L_k}\lambda_{\ell_j}\big)^{\frac12}\Big\|_{L^{p_j}}\\&\le\big\|\varphi\big\|_{L^1}^{m}C_{m,p_1,\dots,p_m,p}\big\|T_\sigma\big\|_{L^{p_1}\times\cdots\times L^{p_m}\to L^p}\prod_{j=1}^m\big\|f_j\big\|_{L^{p_j}}.\end{aligned}$$

This was the required conclusion. To dispose of the assumption that φ is compactly supported, we set $\varphi^K=\varphi\chi_{[-K,K]^n}$ and

$$\begin{aligned}&S^K(f_1,\dots,f_m)(x)\\&=\int_{\mathbf{R}^n}\cdots\int_{\mathbf{R}^n}|\varphi^K(y_1)|\cdots|\varphi^K(y_m)|\,\big|T_\sigma\big(M^{-y_1}(f_1),\dots,M^{-y_m}(f_m)\big)(x)\big|\,d\vec{y}.\end{aligned}$$

The preceding argument shows that

$$\big\|S^K(f_1,\dots,f_m)\big\|_{L^p}\le C_{m,p_1,\dots,p_m,p}\big\|\varphi^K\big\|_{L^1}^m\big\|T_\sigma\big\|_{L^{p_1}\times\cdots\times L^{p_m}\to L^p}\prod_{j=1}^m\big\|f_j\big\|_{L^{p_j}},$$

so letting $K\to\infty$ and using the Lebesgue monotone theorem, we obtain the conclusion for a general function φ in $L^1(\mathbf{R}^n)$ that satisfies $|\varphi(\xi)|\le C'(1+|\xi|)^{-N-n-1}$ when $N>0$ [if $N=0$ in (7.3.11), then no extra assumption on φ is required.] This assumption is needed so that $(\varphi\otimes\cdots\otimes\varphi)\star\sigma$ is well defined. □

When $m=1$, elements of $\mathscr{M}_p$ $(1<p<\infty)$ are necessarily elements of $\mathscr{M}_2$, and thus they are automatically bounded. We obtain a similar result when $m\ge 2$.

Proposition 7.3.6. *Given $0<p_1,\dots,p_m<\infty$, we have that $\mathscr{M}_{p_1,\dots,p_m}$ embeds in L^∞ and there is a constant $C'_{p_1,\dots,p_m}$ such that for all $\sigma\in\mathscr{M}_{p_1,\dots,p_m}$ we have*

$$\big\|\sigma\big\|_{L^\infty}\le C'_{p_1,\dots,p_m}\big\|\sigma\big\|_{\mathscr{M}_{p_1,\dots,p_m}}.\tag{7.3.17}$$

Proof. Suppose first that σ is continuous. For fixed $\vec{a} \in (\mathbf{R}^n)^m$ and $f_1, \dots, f_m$ in $\mathscr{S}(\mathbf{R}^n)$ we have that

$$\int_{(\mathbf{R}^n)^m} \widehat{f_1}(\xi_1) \cdots \widehat{f_m}(\xi_m) \sigma(\vec{a} + \tfrac{1}{k}\vec{\xi}) e^{2\pi i x \cdot (\xi_1 + \cdots + \xi_m)} d\vec{\xi}$$

converges to $\sigma(\vec{a}) f_1(x) \cdots f_m(x)$ as $k \to \infty$. Moreover, the functions $\sigma(\vec{a} + \frac{1}{k}\vec{\xi})$ have the same $\mathscr{M}_{p_1,\dots,p_m}$ quasi-norm for all $\vec{a} \in (\mathbf{R}^n)^m$ and $k \in \mathbf{Z}^+$. Fatou's lemma and the fact that σ is in $\mathscr{M}_{p_1,\dots,p_m}$ give that

$$|\sigma(\vec{a})| \, \|f_1 \cdots f_m\|_{L^p} \le \big\|\sigma\big\|_{\mathscr{M}_{p_1,\dots,p_m}} \|f_1\|_{L^{p_1}} \cdots \|f_m\|_{L^{p_m}} ,$$

where p is defined via (7.3.13). We now choose $f_j(x) = e^{-\pi \frac{1}{p_j} |x|^2}$, $j = 1, \dots, m$ and use (7.3.13) to obtain that

$$|\sigma(\vec{a})| \le \big\|\sigma\big\|_{\mathscr{M}_{p_1,\dots,p_m}}$$

for all $\vec{a}$ in $(\mathbf{R}^n)^m$. This proves (7.3.17) when σ is continuous.

If σ is not continuous, we let $\varphi = \frac{1}{v_n}\chi_{B(0,1)}$ and define $\sigma_\varepsilon = \Phi_\varepsilon \star \sigma$, where $\Phi_\varepsilon(\vec{y}) = \varepsilon^{-mn}\Phi(\vec{y}/\varepsilon)$ and $\Phi = \varphi \otimes \cdots \otimes \varphi$. Note that the convolution is well defined since σ is assumed to satisfy (7.3.11) and Φ_ε has compact support. Moreover, if $\vec{\xi}$ lies in the ball of radius M centered at $\vec{0}$, then in the convolution

$$(\Phi_\varepsilon \star \sigma)(\vec{y}) = \int_{\mathbf{R}^n} \cdots \int_{\mathbf{R}^n} \Phi_\varepsilon(\xi_1 - y_1) \cdots \Phi_\varepsilon(\xi_m - y_m) \sigma(y_1, \dots, y_m) \, d\vec{y}$$

the vector $\vec{y}$ is restricted to the compact set $B(0, M+\varepsilon)^m$ on which $\sigma(\vec{y})$ is bounded by $(1 + m\varepsilon + mM)^N$. Then we have the convolution of an integrable function with a bounded one. This produces a continuous function when $\vec{\xi}$ lies in the ball of radius M centered at $\vec{0}$, and since M was arbitrary, $\Phi_\varepsilon \star \sigma$ is continuous everywhere. It follows from Theorem 7.3.5 that

$$\big\|\sigma_\varepsilon\big\|_{\mathscr{M}_{p_1,\dots,p_m}} \le C'_{m,p_1,\dots,p_m} \big\|\varphi_\varepsilon\big\|_{L^1}^m \big\|\sigma\big\|_{\mathscr{M}_{p_1,\dots,p_m}} = C'_{m,p_1,\dots,p_m} \big\|\sigma\big\|_{\mathscr{M}_{p_1,\dots,p_m}} .$$

The continuous case applied to σ_ε implies that

$$|\sigma_\varepsilon(\vec{a})| \le C'_{m,p_1,\dots,p_m} \big\|\sigma\big\|_{\mathscr{M}_{p_1,\dots,p_m}}$$

uniformly in $\varepsilon > 0$. Letting $\varepsilon \to 0$ and using the fact that $\sigma_\varepsilon \to \sigma$ a.e. (Corollary 2.1.16 in [156]) we deduce (7.3.17). □

Next we have the following relationship between indices for bounded operators of the form T_σ.

Proposition 7.3.7. *Let σ be a locally integrable function defined on $(\mathbf{R}^n)^m$ that satisfies* (7.3.11). *Suppose that the multilinear convolution operator T_σ is nonzero and* *maps the m-fold product $L^{p_1} \times \cdots \times L^{p_m}$ to L^p, where $0 < p_1, \dots, p_m, p < \infty$. Then the following relationship is valid:*

$$\frac{1}{p} \le \frac{1}{p_1} + \cdots + \frac{1}{p_m}. \tag{7.3.18}$$

Proof. Assume first that the kernel K_0 of T_σ is supported in $\overline{B(0,M)}^m$. Fix $f_1, \dots, f_m$ in $\mathscr{C}_0^\infty(\mathbf{R}^n)$, and suppose that all f_j are supported in $\overline{B(0,M')}$. Then

$$\begin{aligned} T_\sigma(f_1 + \tau^h f_1, \dots, f_m + \tau^h f_m) &= T_\sigma(f_1, \dots, f_m) + T_\sigma(\tau^h f_1, \dots, \tau^h f_m) \\ &= T_\sigma(f_1, \dots, f_m) + \tau^h(T_\sigma(f_1, \dots, f_m)) \end{aligned}$$

for $|h| > 2M + 2M'$, since all the remaining terms vanish by an easy calculation. Taking L^p quasi-norms, letting h tend to infinity, and using the fact that for g in $L^p(\mathbf{R}^n)$ we have $\|\tau^h g + g\|_{L^p} \to 2^{1/p}\|g\|_{L^p}$ (Exercise 2.5.1 in [156]), we obtain

$$2^{\frac{1}{p}} \|T_\sigma(f_1, \dots, f_m)\|_{L^p} \le 2^{\frac{1}{p_1} + \cdots + \frac{1}{p_m}} \|T_\sigma\| \, \|f_1\|_{L^{p_1}} \cdots \|f_m\|_{L^{p_m}}.$$

This implies (7.3.18) since T_σ is a nonzero operator.

If the kernel K_0 of T_σ is not necessarily compactly supported, then we pick a Schwartz function φ on $\mathbf{R}^n$ with integral 1 whose inverse Fourier transform has compact support and define $\sigma_0 = (\varphi \otimes \cdots \otimes \varphi) * \sigma$. Then σ_0 has a compactly supported inverse Fourier transform, and it follows from Theorem 7.3.5 that T_{σ_0} is also bounded from $L^{p_1} \times \cdots \times L^{p_m}$ to L^p. But T_{σ_0} has a compactly supported kernel, and the preceding case yields the validity of (7.3.18). □

We conclude that the relationship (7.3.18) is the natural one for bounded m-linear convolution operators from $L^{p_1} \times \cdots \times L^{p_m}$ to L^p.

Proposition 7.3.8. *Let $0 < p_1, \dots, p_m, p < \infty$ be indices that satisfy (7.3.13). Then the spaces $\mathscr{M}_{p_1,\dots,p_m}(\mathbf{R}^n)$ are complete, and thus they are Banach spaces when $p \ge 1$ and quasi-Banach spaces when $p < 1$.*

Proof. Let $\{\sigma_j\}_{j \in \mathbf{Z}^+}$ be a Cauchy sequence in $\mathscr{M}_{p_1,\dots,p_m}$. Then there is a constant C_0 such that $\sup_{j \in \mathbf{Z}^+} \|\sigma_j\|_{\mathscr{M}_{p_1,\dots,p_m}} \le C_0$. Since $\mathscr{M}_{p_1,\dots,p_m}$ embeds in L^∞, the sequence σ_j is Cauchy in L^∞, and thus it converges to a σ in the L^∞ sense. Clearly, for $f_1, \dots, f_m$ in $\mathscr{S}(\mathbf{R}^n)$ we have that $T_{\sigma_j}(f_1, \dots, f_m) \to T_\sigma(f_1, \dots, f_m)$ pointwise as $j \to \infty$; then using Fatou's lemma used below we obtain

$$\begin{aligned} \int_{\mathbf{R}^n} |T_\sigma(f_1, \dots, f_m)|^p \, dx &= \int_{\mathbf{R}^n} \liminf_{j \to \infty} |T_{\sigma_j}(f_1, \dots, f_m)|^p \, dx \\ &\le \liminf_{j \to \infty} \int_{\mathbf{R}^n} |T_{\sigma_j}(f_1, \dots, f_m)|^p \, dx \\ &\le \liminf_{j \to \infty} \|\sigma_j\|^p_{\mathscr{M}_{p_1,\dots,p_m}} \|f_1\|^p_{L^{p_1}} \cdots \|f_m\|^p_{L^{p_m}} \\ &\le C_0^p \|f_1\|^p_{L^{p_1}} \cdots \|f_m\|^p_{L^{p_m}}, \end{aligned}$$

which implies that $\sigma \in \mathscr{M}_{p_1,\dots,p_m}$. This argument shows that if $\sigma_j \in \mathscr{M}_{p_1,\dots,p_m}$ and $\sigma_j \to \sigma$ uniformly, then σ is in $\mathscr{M}_{p_1,\dots,p_m}$ and satisfies

$$\left\|\sigma\right\|_{\mathscr{M}_{p_1,\dots,p_m}} \le \liminf_{j\to\infty} \left\|\sigma_j\right\|_{\mathscr{M}_{p_1,\dots,p_m}}.$$

Apply this inequality to $\sigma_k - \sigma_j$ in place of σ_j and $\sigma_k - \sigma$ in place of σ for some fixed k. We obtain

$$\left\|\sigma_k - \sigma\right\|_{\mathscr{M}_{p_1,\dots,p_m}} \le \liminf_{j\to\infty} \left\|\sigma_k - \sigma_j\right\|_{\mathscr{M}_{p_1,\dots,p_m}} \tag{7.3.19}$$

for each k. Given $\varepsilon > 0$, by the Cauchy criterion, there is an N such that for $j,k > N$ we have $\|\sigma_k - \sigma_j\|_{\mathscr{M}_{p_1,\dots,p_m}} < \varepsilon$. Using (7.3.19) we conclude that

$$\|\sigma_k - \sigma\|_{\mathscr{M}_{p_1,\dots,p_m}} \le \varepsilon$$

when $k > N$; thus, σ_k converges to σ in $\mathscr{M}_{p_1,\dots,p_m}$ as $k \to \infty$. Hence, $\mathscr{M}_{p_1,\dots,p_m}$ is complete, and thus it is a Banach space when $p \ge 1$ and a quasi-Banach space when $p < 1$. □

7.3.4 Duality of Multilinear Multiplier Operators

We know that m-linear convolution operators (in particular multilinear multiplier operators) have m transposes. These are also m-linear convolution operators whose symbols are related to the symbol of the original operator in simple ways.

Let σ be a locally integrable function on $\mathbf{R}^n$ that satisfies (7.3.11). For each j in $\{1,\dots,m\}$ we introduce another locally integrable function σ^{*j} on $\mathbf{R}^n$ that satisfies (7.3.11) by setting

$$\sigma^{*j}(\xi_1,\dots,\xi_m) = \sigma(\xi_1,\dots,\xi_{j-1},-(\xi_1+\cdots+\xi_m),\xi_{j+1},\dots,\xi_m) \tag{7.3.20}$$

that is, all variables of σ remain the same except for the jth variable, which is replaced by $-(\xi_1+\cdots+\xi_m)$. We call σ^{*j} the *jth transpose symbol of* σ.

Then the *jth dual* of the multilinear convolution operator T_σ is $T_{\sigma^{*j}}$, which is another multilinear convolution operator. To verify this assertion, we fix Schwartz functions $f_0, f_1, \dots, f_m$. Then we have

$$\begin{aligned}
&\big\langle T_\sigma(f_1,\dots,f_m), f_0\big\rangle \\
&= \int_{\mathbf{R}^n}\int_{(\mathbf{R}^n)^m} \sigma(\vec{\xi}\,)\widehat{f_1}(\xi_1)\cdots\widehat{f_m}(\xi_m)e^{2\pi i x\cdot(\xi_1+\cdots+\xi_m)}\,d\vec{\xi}\, f_0(x)\,dx \\
&= \int_{(\mathbf{R}^n)^m} \sigma(\vec{\xi}\,)\widehat{f_1}(\xi_1)\cdots\widehat{f_m}(\xi_m)\widehat{f_0}(-(\xi_1+\cdots+\xi_m))\,d\vec{\xi}
\end{aligned}$$

$$= \int_{(\mathbf{R}^n)^m} \sigma^{*j}(\vec{\xi}\,)\widehat{f_0}(\xi_j) \prod_{k\neq j} \widehat{f_k}(\xi_k)\widehat{f_j}(-(\xi_1+\cdots+\xi_m))\,d\vec{\xi}$$

$$= \int_{\mathbf{R}^n}\int_{(\mathbf{R}^n)^m} \sigma^{*j}(\vec{\xi}\,)\widehat{f_0}(\xi_j) \prod_{k\neq j} \widehat{f_k}(\xi_k)e^{2\pi i x\cdot(\xi_1+\cdots+\xi_m)}\,d\vec{\xi}\, f_j(x)\,dx$$

$$= \big\langle T_{\sigma^{*j}}(f_1,\dots,f_{j-1},f_0,f_{j+1},f_m), f_j\big\rangle\,.$$

This calculation shows that the jth transpose of T_σ is $T_{\sigma^{*j}}$. Often in the literature the terms *adjoint* and *transpose* have the same meaning. In this text, the former refers to the dual operator with respect to the complex inner product $\langle\cdot\,|\,\cdot\rangle$ and the latter to that with respect to the real inner product $\langle\cdot,\cdot\rangle$. In the case of multilinear operators, there does not seem to exist a nice formula, such as that in (7.3.20), for the jth adjoint operator.

Exercises

7.3.1. Suppose that $\sigma(\xi_1,\dots,\xi_m)$ is the symbol of an m-linear convolution operator T_σ that maps $L^{p_1}(\mathbf{R}^n)\times\cdots\times L^{p_m}(\mathbf{R}^n)$ to $L^p(\mathbf{R}^n)$ with norm B. Show that for any $\delta>0$, the m-linear operator T_{σ_δ} associated with $\sigma_\delta(\vec{\xi}\,)=\delta^{n(\frac{1}{p_1}+\cdots+\frac{1}{p_m}-\frac{1}{p})}\sigma(\delta\vec{\xi}\,)$ also maps $L^{p_1}(\mathbf{R}^n)\times\cdots\times L^{p_m}(\mathbf{R}^n)$ to $L^p(\mathbf{R}^n)$ with the same norm.

7.3.2. Show that m-linear operators L satisfy the identity

$$L(f_1,\dots,f_m)-L(g_1,\dots,g_m)=\sum_{j=1}^m L(g_1,\dots,g_{j-1},f_j-g_j,f_{j+1},\dots,f_m),$$

for all functions f_k,g_k in their domain.

7.3.3. (G. Diestel) (a) Let $r_j=e^{2\pi i j/m}\in\mathbf{C}$, $j=0,1,\dots,m-1$, be the mth roots of unity. Show that for any $y\in\mathbf{R}^n$, an m-linear convolution operator T satisfies

$$mT(f_1,\dots,f_m)+mT(\tau^y f_1,\dots,\tau^y f_m)=\sum_{j=1}^m T(f_1+r_j\tau^y f_1,\dots,f_m+r_j\tau^y f_m)$$

for all functions f_j in its domain.
(b) Use this property to obtain another proof of Proposition 7.3.7 when $p\geq 1$.
$\big[$*Hint:* Part (a): Use that $\sum_{i=1}^m r_i^s=\sum_{i=1}^m r_i=0$, $\sum_{i=1}^m r_i^m=m$. Part (b): Take L^p norms of both sides, use that T commutes with simultaneous translations and the fact that for $g\in L^p(\mathbf{R}^n)$ we have $\|\tau^h g+g\|_{L^p}\to 2^{1/p}\|g\|_{L^p}$.$\big]$

7.3.4. ([126]) Suppose that K is a kernel in $\mathbf{R}^{2n}$ (which may be a distribution), and let T^K be the bilinear convolution operator associated with K:

$$T^K(f,g)(x) = \int_{\mathbf{R}^n}\int_{\mathbf{R}^n} K(x-y,x-z)f(y)g(z)\,dy\,dz.$$

Assume that T^K is bounded from $L^p(\mathbf{R}^n)\times L^q(\mathbf{R}^n)\to L^r(\mathbf{R}^n)$, with norm $\|T\|$, when $1p+1/q=1/r$. Let M be an $n\times n$ invertible matrix. Define a $2n\times 2n$ invertible matrix

$$\widetilde{M} = \begin{pmatrix} M & O \\ O & M \end{pmatrix},$$

where O is the $n\times n$ zero matrix. Show that the operator $T^{K\circ\widetilde{M}}$ is also bounded from $L^p(\mathbf{R}^n)\times L^q(\mathbf{R}^n)\to L^r(\mathbf{R}^n)$ with norm exactly $\|T^K\|$.

7.3.5. Suppose that a bounded function σ on $\mathbf{R}^{2n}$ is written as $\sigma=\sigma_0*(\sigma_1\otimes\sigma_2)$, where σ_0 is a function on $\mathbf{R}^{2n}$ and σ_1, σ_2 are functions on $\mathbf{R}^n$.
(a) Show that for $2<p<\infty$ we have

$$\|T_\sigma\|_{L^p\times L^{p'}\to L^1} \le \|\sigma_0\|_{L^{(\frac{1}{2}+\frac{1}{p})^{-1}}} \|\sigma_1\|_{L^{(\frac{1}{2}-\frac{1}{p})^{-1}}} \|\sigma_2\|_{L^{(\frac{1}{2}-\frac{1}{p})^{-1}}}.$$

(b) Prove that when $\gamma>n(\frac{1}{2}+\frac{1}{p})$, we have

$$\big\|T_\sigma\big\|_{L^p\times L^{p'}\to L^1} \le C(n,\gamma,p)\big\|(I-\Delta_{\xi_1})^{\gamma/2}(I-\Delta_{\xi_2})^{\gamma/2}\sigma\big\|_{L^{(\frac{1}{2}+\frac{1}{p})^{-1}}}.$$

(c) Show that for $2<p<\infty$, $1<q<\infty$, $1/p+1/q=1/r<1$, and $\gamma>n(\frac{1}{2}+\frac{1}{p})$ we have

$$\big\|T_\sigma\big\|_{L^p\times L^q\to L^r} \le C(n,\gamma,p,q)\big\|(I-\Delta)^{\gamma}\sigma\big\|_{L^{(\frac{1}{2}+\frac{1}{p})^{-1}}}.$$

[*Hint:* Part (a): Apply Corollary 7.2.11 using $\|T_\sigma\|_{L^2\times L^2\to L^1}\le\|\sigma_0\|_{L^1}\|\sigma_1\|_{L^\infty}\|\sigma_2\|_{L^\infty}$ and $\|T_\sigma\|_{L^\infty\times L^1\to L^1}\le\|\sigma_0\|_{L^2}\|\sigma_1\|_{L^2}\|\sigma_2\|_{L^2}$. Part (c): Use duality and interpolation.]

7.3.6. ([120]) Given a unit vector v in $\mathbf{R}^2$, define a half-space of $\mathbf{R}^4$ by setting $\mathscr{H}_v=\{(\xi,\eta)\in\mathbf{R}^2\times\mathbf{R}^2:\ (\xi+\eta)\cdot v>0\}$. Let v_j be a unit vector in $\mathbf{R}^2$ for each $j\in\mathbf{Z}$. For $\rho>0$ define the sets

$$\begin{aligned}
B_\rho &= \{(\xi,\eta)\in\mathbf{R}^2\times\mathbf{R}^2:\ |\xi|^2+|\eta|^2<2\rho^2\}\\
B_{j,\rho} &= \{(\xi,\eta)\in\mathbf{R}^2\times\mathbf{R}^2:\ |\xi-\rho v_j|^2+|\eta-\rho v_j|^2<2\rho^2\}\\
B_\rho^{*1} &= \{(\xi,\eta)\in\mathbf{R}^2\times\mathbf{R}^2:\ |\xi+\eta|^2+|\eta|^2<2\rho^2\}\\
B_{j,\rho}^{*1} &= \{(\xi,\eta)\in\mathbf{R}^2\times\mathbf{R}^2:\ |\xi-\sqrt{2}\rho v_j+\eta|^2+|\eta|^2<2\rho^2\}\\
B_\rho^{*2} &= \{(\xi,\eta)\in\mathbf{R}^2\times\mathbf{R}^2:\ |\xi|^2+|\xi+\eta|^2<2\rho^2\}\\
B_{j,\rho}^{*2} &= \{(\xi,\eta)\in\mathbf{R}^2\times\mathbf{R}^2:\ |\xi|^2+|\xi+\eta-\sqrt{2}\rho v_j|^2<2\rho^2\}.
\end{aligned}$$

Assume that one of the three functions χ_{B_ρ}, $\chi_{B_\rho^{*1}}$, $\chi_{B_\rho^{*2}}$ lies in $\mathscr{M}_{p,q}(\mathbf{R}^2)$ with norm $C(p,q)$, when $\rho = 1$. Let $r = (1/p + 1/q)^{-1}$. Prove that

$$\Big\|\Big(\sum_{j\in\mathbf{Z}}|T_{\chi_{\mathscr{H}_{v_j}}}(f_j,g_j)|^2\Big)^{1/2}\Big\|_{L^r} \le C(p,q)\Big\|\Big(\sum_{j\in\mathbf{Z}}|f_j|^2\Big)^{1/2}\Big\|_{L^p}\Big\|\Big(\sum_{j\in\mathbf{Z}}|g_j|^2\Big)^{1/2}\Big\|_{L^q}$$

for all Schwartz functions f_j and g_j.
[*Hint:* Use that $B_{j,\rho}$, $B_{j,\rho}^{*1}$, $B_{j,\rho}^{*2}$ tend to $\mathscr{H}_{v_j}$ as $\rho \to \infty$ and apply Theorem 7.3.1.]

7.3.7. Given a rectangle R in $\mathbf{R}^2$, let R' be the union of the two copies of R adjacent to R in the direction of its longest side. Let R be a rectangle in $\mathbf{R}^2$ and let v be a unit vector in $\mathbf{R}^2$ parallel to the longest side of R. Fix a unit vector v in $\mathbf{R}^2$ and define $\mathscr{H}_v = \{(\xi,\eta)\in\mathbf{R}^2\times\mathbf{R}^2 :\ (\xi+\eta)\cdot v > 0\}$. Prove that for all $x\in\mathbf{R}^2$ we have

$$\left|\int_{\mathbf{R}^2}\int_{\mathbf{R}^2}\chi_{\mathscr{H}_v}(\xi,\eta)\widehat{\chi_R}(\xi)\widehat{\chi_R}(\eta)e^{2\pi i x\cdot(\xi+\eta)}\,d\xi\,d\eta\right| \ge \frac{1}{10}\chi_{R'}(x).$$

[*Hint:* Introduce a rotation $\mathscr{O}$ of $\mathbf{R}^2$ such that $\mathscr{O}(v) = (1,0)$. Then $\mathscr{O}[R]$ has sides parallel to the axes, and the conclusion follows. See also Proposition 5.1.2.]

7.3.8. ([120]) Let $1 < p,q,r < \infty$ satisfy $1/p + 1/q = 1/r$ and suppose that one of p,q, or r' is less than 2. Let B be the unit ball in $\mathbf{R}^4$. Show that $\chi_B \notin \mathscr{M}_{p,q}(\mathbf{R}^2)$.
[*Hint:* It suffices to show that χ_B, $\chi_{B^{*1}}$, $\chi_{B^{*2}}$ is not in $\mathscr{M}_{p,q}(\mathbf{R}^2)$ for $p,q,r > 2$, where we denote $B^{*1} = \{(\xi,\eta)\in\mathbf{R}^2\times\mathbf{R}^2 :\ |\xi+\eta|^2+|\eta|^2 < 1\}$ and analogously $B^{*2} = \{(\xi,\eta)\in\mathbf{R}^2\times\mathbf{R}^2 :\ |\xi|^2+|\xi+\eta|^2 < 1\}$. Assuming the opposite prove that

$$\sum_j\int_E\big|T_{\mathscr{H}_{v_j}}(\chi_{R_j},\chi_{R_j})(x)\big|^2dx \le \|T_{\chi_B}\|^2_{L^p\times L^q\to L^r}\,\delta^{\frac{r-2}{r}}\sum_j|R_j|,$$

where E and R_j are as in Lemma 5.1.1 and v_j is a unit vector parallel to the longest side of R_j. Reach a contradiction by proving that

$$\sum_j\int_E\big|T_{\mathscr{H}_{v_j}}(\chi_{R_j},\chi_{R_j})(x)\big|^2dx \ge \frac{1}{1200}\sum_j|R_j|,$$

using the result of the preceding exercise and the idea of Theorem 5.1.5.]

7.3.9. ([120]) Suppose that $m(\xi_1,\eta_1,\xi_2,\eta_2)$ lies in $\mathscr{M}_{p_1,p_2}(\mathbf{R}^{n+k})$, where $\xi_1,\xi_2\in\mathbf{R}^n$ and $\eta_1,\eta_2\in\mathbf{R}^k$ and $1 < p = (p_1^{-1}+p_2^{-1})^{-1} < \infty$. Show that for almost every (ξ_1,ξ_2) in $(\mathbf{R}^n)^2$ the function $(\eta_1,\eta_2)\mapsto m(\xi_1,\eta_1,\xi_2,\eta_2)$ lies in $\mathscr{M}_{p_1,p_2}(\mathbf{R}^k)$, with norm

$$\|m(\xi_1,\cdot,\xi_2,\cdot)\|_{\mathscr{M}_{p_1,p_2}(\mathbf{R}^k)} \le \|m\|_{\mathscr{M}_{p_1,p_2}(\mathbf{R}^{n+k})}.$$

[*Hint:* Use duality.]

7.3.10. ([120]) Combine the results of the preceding two exercises to prove that if B is the characteristic function of the unit ball in $\mathbf{R}^{2n}$ with $n \geq 2$, then $\chi_B \notin \mathscr{M}_{p,q}(\mathbf{R}^{2n})$ when one of p, q, or $(1-1/p-1/q)^{-1}$ is less than 2.

7.4 Calderón-Zygmund Operators of Several Functions

Let $K(y_0, y_1, \dots, y_m)$ be a function defined away from the diagonal $y_0 = y_1 = \cdots = y_m$ on $(\mathbf{R}^n)^{m+1}$ that satisfies the *size estimate*

$$|K(y_0, y_1, \dots, y_m)| \leq \frac{A}{(\sum_{k,l=0}^{m} |y_k - y_l|)^{nm}} \tag{7.4.1}$$

for some $A > 0$ and all $(y_0, y_1, \dots, y_m) \in (\mathbf{R}^n)^{m+1}$ not in the diagonal of $(\mathbf{R}^n)^{m+1}$. Furthermore, assume that for some $\varepsilon > 0$ we have the *smoothness estimates*

$$|K(y_0, \dots, y_j, \dots, y_m) - K(y_0, \dots, y_j', \dots, y_m)| \leq \frac{A|y_j - y_j'|^{\varepsilon}}{(\sum_{k,l=0}^{m} |y_k - y_l|)^{nm+\varepsilon}} \tag{7.4.2}$$

for each $j \in \{0, 1, \dots, m\}$ and all $(y_0, y_1, \dots, y_m)$ not in the diagonal of $(\mathbf{R}^n)^{m+1}$ whenever

$$|y_j - y_j'| \leq \frac{1}{m+1}\big(|y_0 - y_j| + |y_1 - y_j| + \cdots + |y_m - y_j|\big). \tag{7.4.3}$$

Interchanging the roles of y_j and y_j', assuming that $(y_0, y_1, \dots, y_{j-1}, y_j', y_{j+1}, \dots, y_m)$ is not in the diagonal of $(\mathbf{R}^n)^{m+1}$, we can replace the denominator in (7.4.2) with $(\sum_{k,l=0}^{m} |y_k - y_l'|)^{nm+\varepsilon}$, where $y_l' = y_l$ if $l \neq j$ when we replace (7.4.3) with

$$|y_j - y_j'| \leq \frac{1}{m+1}\big(|y_0 - y_j'| + |y_1 - y_j'| + \cdots + |y_m - y_j'|\big).$$

More importantly, we should observe that conditions (7.4.1) and (7.4.2) are invariant under the operation of interchanging y_0 with some y_l, and thus they hold for every transpose K^{*l} of K.

We notice that in applications, it is easier to check estimates (7.4.1) and (7.4.2) with the expression $(|y_0 - y_1| + \cdots + |y_0 - y_m|)^{mn+\varepsilon}$ in the denominator since

$$|y_0 - y_1| + \cdots + |y_0 - y_m| \geq \frac{1}{2m} \sum_{k,l=0}^{m} |y_k - y_l|.$$

Definition 7.4.1. Let $A, \varepsilon > 0$ and $m \in \mathbf{Z}^+$. Functions defined on $(\mathbf{R}^n)^{m+1} \setminus \{(x, \dots, x) : x \in \mathbf{R}^n\}$ that satisfy conditions (7.4.1) and (7.4.2) are called *m-linear standard Calderón–Zygmund kernels*. The class of all such kernels is denoted by m-$CZK(A, \varepsilon)$.

Example 7.4.2. Let $K_0(\vec{u})$ be a function on $(\mathbf{R}^n)^m \setminus \{\vec{0}\}$ that satisfies the size condition $|K_0(\vec{u})| \le A' |\vec{u}|^{-mn}$ and the regularity condition $|\nabla K_0(\vec{u})| \le A' |\vec{u}|^{-mn-1}$. Then the function

$$K(x,y_1,\dots,y_m) = K_0(x-y_1,\dots,x-y_m) \tag{7.4.4}$$

satisfies (7.4.1) and (7.4.2). We verify the second assertion. For each j in $\{1,\dots,m\}$, fix y_j' and take points y_k, $k=1,\dots,m$ and x not all equal to each other, satisfying (7.4.3). Then by the mean value theorem, we bound the left-hand side of (7.4.2) by

$$\frac{CA'|y_j-y_j'|}{\big(|x-\theta y_j'-(1-\theta)y_j|+\sum_{k\in\{1,\dots,m\}\setminus\{j\}}|x-y_k|\big)^{mn+1}} \tag{7.4.5}$$

for some $\theta\in[0,1]$. Since

$$\begin{aligned}
|x-\theta y_j'-(1-\theta)y_j| &\ge |x-y_j|-|y_j-y_j'| \\
&\ge |x-y_j|-\frac{1}{m+1}\big(|x-y_j|+|y_1-y_j|+\dots+|y_m-y_j|\big) \\
&\ge |x-y_j|-\frac{1}{m+1}\sum_{k=1}^m |x-y_k|-\frac{m-1}{m+1}|x-y_j| \\
&= \frac{1}{m+1}|x-y_j|-\frac{1}{m+1}\sum_{\substack{k=1\\k\neq j}}^m |x-y_k|,
\end{aligned}$$

the denominator in (7.4.5) is at least

$$\left(\frac{1}{m+1}\sum_{k=1}^m |x-y_k|\right)^{mn+1}.$$

Thus, we deduce (7.4.2) with $\varepsilon=1$ and $A\le C'A'$. When $j=0$, assuming (7.4.3), the corresponding estimate for the left-hand side of (7.4.2) is

$$\frac{CmA'|x-x'|}{\big(|y_1-\theta x'-(1-\theta)x|+\dots+|y_m-\theta x'-(1-\theta)x|\big)^{mn+1}}$$

which, combined with $|y_k-\theta x'-(1-\theta)x| \ge |y_k-x|-|x-x'|$, yields a similar estimate. Analogous estimates are valid for every transpose K^{*l} of K defined by $K^{*l}(y_0,y_1,\dots,y_m) = K_0(y_l-y_1,\dots,y_l-y_{l-1},y_l-y_0,y_l-y_{l+1},\dots,y_l-y_m)$.

Finally, we observe that the same proof gives that if

$$|K_0(u_1,\dots,u_j,\dots,u_m)-K_0(u_1,\dots,u_j',\dots,u_m)| \le \frac{A'|u_j-u_j'|^{\varepsilon}}{|(u_1,\dots,u_m)|^{nm+\varepsilon}},$$

whenever $|u_j-u_j'| < \frac{1}{m+1}\sum_{k=1}^m |u_j-u_k|$, then (7.4.1) and (7.4.2) hold for K defined by (7.4.4), with $A\le C'A'$.

Definition 7.4.3. Let $A, \varepsilon > 0$ and $m \in \mathbf{Z}^+$ and K be in m-$CZK(A,\varepsilon)$. Let T be an m-linear operator defined on $\mathscr{S}(\mathbf{R}^n) \times \cdots \times \mathscr{S}(\mathbf{R}^n)$ and taking values in $\mathscr{S}'(\mathbf{R}^n)$. We say that T *is associated with* K if the following relationship holds

$$\langle T(f_1,\dots,f_m), f_0\rangle = \int_{(\mathbf{R}^n)^{m+1}} K(x,y_1,\dots,y_m) f_0(x) f_1(y_1)\cdots f_m(y_m)\, d\vec{y}\, dx \quad (7.4.6)$$

for all functions $f_0, f_1, \dots, f_m$ in $\mathscr{S}(\mathbf{R}^n)$, with $\cap_{j=1}^m \operatorname{supp} f_j = \emptyset$.

The fact that $\cap_{j=1}^m \operatorname{supp} f_j = \emptyset$ implies that there is no $x \in \mathbf{R}^n$ such that $|x-y_j| = 0$ for all $j = 1,\dots,m$ whenever y_j lies in the support of f_j, and thus the kernel $K(x,\vec{y})$ is bounded when $(x,y_1,\dots,y_m) \in \mathbf{R}^n \times \operatorname{supp} f_1 \times \cdots \times \operatorname{supp} f_m$; hence, the integral in (7.4.6) converges absolutely.

We are interested in bounded extensions of m-linear operators associated with kernels of class m-$CZK(A,\varepsilon)$. For integrable functions $g_1,\dots,g_m$ with compact support in $\mathbf{R}^n$ we note that the integral

$$\int_{(\mathbf{R}^n)^m} K(x,y_1,\dots,y_m) g_1(y_1)\cdots g_m(y_m)\, d\vec{y} \quad (7.4.7)$$

converges absolutely whenever $x \notin \cap_{j=1}^m \operatorname{supp} g_j$ and in this situation one can define $T(g_1,\dots,g_m)(x)$ as the expression in (7.4.7).

Example 7.4.4. We define an operator $\mathscr{R}_1$ acting on m-tuples of Schwartz functions on the line as follows:

$$\mathscr{R}_1(f_1,\dots,f_m)(x) = \text{p.v.} \int_{\mathbf{R}^m} \frac{\frac{\Gamma(\frac{n+1}{2})}{\pi^{\frac{n+1}{2}}}(x-y_1)}{|(x-y_1,\dots,x-y_m)|^{m+1}} f_1(y_1)\cdots f_m(y_m)\, dy_1\cdots dy_m.$$

We call $\mathscr{R}_1$ the *m-linear Riesz transform in the first variable*. If all f_j are smooth and supported in the interval $[-1,1]$, then for $|x| \ge 2$ we have that $\mathscr{R}_1(f_1,\dots,f_m)(x)$ behaves at infinity like $|x|^{-m}$. This function is not in $L^{1/m}$, but it lies in the weak space $L^{1/m,\infty}$.

This and other examples like it suggest that operators associated with kernels of class m-$CZK(A,\varepsilon)$ lie in weak $L^{1/m}$ when acting on certain functions; this provides motivation for the main theorem of the next subsection.

Example 7.4.4 provides a special case of a kernel that is antisymmetric in the first variable. A more general situation is discussed in what follows.

Example 7.4.5. Suppose that $K(x,\vec{y})$ satisfies (7.4.1) and (7.4.2) and is *antisymmetric*, in the sense that for some $j \in \{1,\dots,m\}$ we have

$$K(y_0,y_1,\dots,y_m) = -K(y_j,y_1,\dots,y_{j-1},y_0,y_{j+1},\dots,y_m)$$

for all $(y_0, y_1, \dots, y_m)$ in $(\mathbf{R}^n)^{m+1}$ away from its diagonal. Then there is a distribution W on $\mathbf{R}^{n(m+1)}$ that extends K. Indeed, we show that the limit

$$\langle W, F\rangle = \lim_{\delta\to 0} \int\cdots\int_{\sum_{j,k=0}^m |y_j - y_k|>\delta} K(y_0,\vec{y})F(y_0,\vec{y})\,d\vec{y}\,dy_0$$

exists for all F in $\mathscr{S}(\mathbf{R}^{n(m+1)})$ and defines a tempered distribution on $\mathbf{R}^{n(m+1)}$. Let $\vec{y}^j = (y_1, \dots, y_{j-1}, y_0, y_{j+1}, \dots, y_m)$ be the vector in $\mathbf{R}^{nm}$ obtained from $(y_1, \dots, y_m)$ by replacing y_j by y_0. In view of antisymmetry, we may write

$$\begin{aligned}&\int\cdots\int_{\sum_{j,k=0}^m |y_j - y_k|>\delta} K(y_0,\vec{y})F(y_0,\vec{y})\,d\vec{y}\,dy_0\\ &\qquad = \frac{1}{2}\int\cdots\int_{\sum_{j,k=0}^m |y_j - y_k|>\delta} K(y_0,\vec{y})\big(F(y_0,\vec{y}) - F(y_j,\vec{y}^j)\big)\,d\vec{y}\,dy_0 .\end{aligned} \tag{7.4.8}$$

Let $N > (m+1)n$. Combining

$$|F(y_0,\vec{y}) - F(y_j,\vec{y}^j)| \le \frac{2\,|y_0 - y_j|}{(1+|y_0|^2+|\vec{y}|^2)^N} \sup_{y_0,\vec{y}} \Big|\nabla_{y_0,\vec{y}}\Big[(1+|y_0|^2+|\vec{y}|^2)^N F(y_0,\vec{y})\Big]\Big|$$

with condition (7.4.1), we obtain that the integrand in (7.4.8) is bounded by

$$C'\,|y_0 - y_j|^{-n+\frac{2m+1}{3m}} \prod_{l=0}^m (1+|y_l|^2)^{-\frac{N}{m+1}} \prod_{\substack{l=1\\ l\neq j}}^m |y_0 - y_l|^{-n+\frac{1}{3m}} \Big(\sum_{|\alpha|,|\beta|\le 2N} \rho_{\alpha,\beta}(F)\Big)$$

and thus the integral in (7.4.8) has a limit as $\delta\to 0$. Here $\rho_{\alpha,\beta}$ are the Schwartz seminorms of F on $\mathbf{R}^{n(m+1)}$. We can therefore define a bounded m-linear operator $T : \mathscr{S}(\mathbf{R}^n)\times\cdots\times\mathscr{S}(\mathbf{R}^n) \to \mathscr{S}'(\mathbf{R}^n)$ by setting

$$\langle T(f_1,\dots,f_m),\varphi\rangle = \lim_{\delta\to 0} \int\cdots\int_{\sum_{j,k=0}^m |y_j - y_k|>\delta} K(y_0,\vec{y}) f_1(y_1)\cdots f_m(y_m)\varphi(y_0)\,d\vec{y}\,dy_0 .$$

Then T defined in this way is associated with the antisymmetric (in the jth variable) function K in the sense of Definition 7.4.3.

7.4.1 Multilinear Calderón–Zygmund Theorem

The first fundamental result of this section is the multilinear extension of the Calderón–Zygmund theorem.

Theorem 7.4.6. *Let T be an m-linear operator associated with a kernel K in m-$CZK(A,\varepsilon)$, where $m \geq 2$. Assume that for some $1 \leq q_1, q_2, \dots, q_m \leq \infty$ and some $0 < q < \infty$ satisfying*

$$\frac{1}{q_1} + \frac{1}{q_2} + \cdots + \frac{1}{q_m} = \frac{1}{q}$$

T extends to a bounded operator from $L^{q_1} \times \cdots \times L^{q_m}$ to $L^{q,\infty}$. Then T can be extended to a bounded operator from $L^1 \times \cdots \times L^1$ to $L^{1/m,\infty}$ such that for some constant $C_{n,m,\varepsilon}$ we have

$$\|T\|_{L^1 \times \cdots \times L^1 \to L^{1/m,\infty}} \leq C_{n,m,\varepsilon}\big(A + \|T\|_{L^{q_1} \times \cdots \times L^{q_m} \to L^{q,\infty}}\big). \tag{7.4.9}$$

Proof. Set $B = \|T\|_{L^{q_1} \times \cdots \times L^{q_m} \to L^{q,\infty}}$. For $1 \leq j \leq m$ fix step functions f_j. Assume that each f_j is a step function given by a finite linear combination of characteristic functions of disjoint dyadic cubes. In proving (7.4.9), by a simple scaling argument, we may assume that

$$\|f_1\|_{L^1} = \cdots = \|f_m\|_{L^1} = 1.$$

Fix $\lambda > 0$ and set $E_\lambda = \{x : |T(f_1, \dots, f_m)(x)| > \lambda\}$. We need to show that for some constant $C = C_{m,n}$ we have

$$|E_\lambda| \leq C(A+B)^{1/m} \lambda^{-1/m}. \tag{7.4.10}$$

Once (7.4.10) is established for f_j with L^1 norm one, the general case follows immediately by scaling. Let γ be a positive real number to be determined later. For each $j = 1, \dots, m$ apply the Calderón–Zygmund decomposition to the function f_j at height $(\lambda\gamma)^{1/m}$ to obtain "good" and "bad" functions g_j and b_j, and finite families of dyadic cubes $\{Q_{j,k}\}_{k \in \mathbf{Z}}$ with disjoint interiors such that

$$f_j = g_j + b_j \qquad \text{and} \qquad b_j = \sum_k b_{j,k}.$$

Then for each $j = 1, \dots, m$ we have

$$\begin{gathered}
\operatorname{support}(b_{j,k}) \subseteq Q_{j,k} \\
\int_{Q_{j,k}} b_{j,k}(x)\,dx = 0 \\
\int_{Q_{j,k}} |b_{j,k}(x)|\,dx \leq 2^{n+1}(\lambda\gamma)^{1/m}|Q_{j,k}| \\
|\cup_k Q_{j,k}| \leq (\lambda\gamma)^{-1/m} \\
\|g_j\|_{L^1} \leq \|f_j\|_{L^1} = 1 \\
\|g_j\|_{L^\infty} \leq 2^n(\lambda\gamma)^{1/m}.
\end{gathered}$$

Now let

$$\begin{aligned}
E_1 &= \{x \in \mathbf{R}^n : |T(g_1,g_2,\dots,g_m)(x)| > \lambda/2^m\} \\
E_2 &= \{x \in \mathbf{R}^n : |T(b_1,g_2,\dots,g_m)(x)| > \lambda/2^m\} \\
E_3 &= \{x \in \mathbf{R}^n : |T(g_1,b_2,\dots,g_m)(x)| > \lambda/2^m\} \\
&\dots \\
E_{2^m} &= \{x \in \mathbf{R}^n : |T(b_1,b_2,\dots,b_m)(x)| > \lambda/2^m\},
\end{aligned}$$

where each E_s has the form $\{x \in \mathbf{R}^n : |T(h_1,h_2,\dots,h_m)(x)| > \lambda/2^m\}$, with $h_j = b_j$ for $j \in S$ and $h_j = g_j$ for $j \in \{1,\dots,m\} \setminus S$ for a subset S of $\{1,\dots,m\}$. Since

$$|\{x \in \mathbf{R}^n : |T(f_1,\dots,f_m)(x)| > \lambda\}| \le \sum_{s=1}^{2^m} |E_s|,$$

it will suffice to prove estimate (7.4.10) for each of the 2^m sets E_s.

Let us start with set E_1 which is the easiest. The $L^{q_1} \times \cdots \times L^{q_m} \to L^{q,\infty}$ boundedness of T gives

$$\begin{aligned}
|E_1| &\le \frac{(2^m B)^q}{\lambda^q} \|g_1\|_{L^{q_1}}^q \cdots \|g_m\|_{L^{q_m}}^q \\
&\le \frac{CB^q}{\lambda^q} \prod_{j=1}^m (\lambda\gamma)^{\frac{q}{mq_j}} \\
&= \frac{CB^q}{\lambda^q} (\lambda\gamma)^{(m-\frac{1}{q})\frac{q}{m}} \\
&= CB^q \lambda^{-\frac{1}{m}} \gamma^{q-\frac{1}{m}}.
\end{aligned} \tag{7.4.11}$$

Consider a set E_s as above with $2 \le s \le 2^m$. Suppose that for some $1 \le l \le m$ we have l bad functions and $m-l$ good functions in the set $\{h_1,\dots,h_m\}$, where $h_j \in \{g_j, b_j\}$, and by permuting the variables, without loss of generality, we assume that the bad functions appear in the entries $1,\dots,l$ and the good functions in the entries $l+1,\dots,m$. We will show that for any $s \in \{2,3,\dots,2^m\}$ we have

$$|E_s| \le C\lambda^{-\frac{1}{m}} \big(\gamma^{-\frac{1}{m}} + \gamma^{-\frac{1}{m}} (A\gamma)^{\frac{1}{l}}\big). \tag{7.4.12}$$

Let $\ell(Q)$ denote the side length of a cube Q, and let Q^* have the same center and orientation as Q and

$$\ell(Q^*) = ((m+1)\sqrt{n}+1)\ell(Q).$$

Fix an $x \notin \cup_{j=1}^m \cup_k (Q_{j,k})^*$. Also, fix for the moment the cubes $Q_{1,k_1},\dots,Q_{l,k_l}$, and without loss of generality, suppose that Q_{1,k_1} has the smallest size among them. Let c_{i,k_i} be the center of Q_{i,k_i}. For fixed $y_2,\dots,y_l$ in $\mathbf{R}^n$, the mean value property of b_{1,k_1} gives

$$
\begin{aligned}
&\left|\int_{Q_{1,k_1}} K(x,y_1,\dots,y_m)\,b_{1,k_1}(y_1)\,dy_1\right| \\
&=\left|\int_{Q_{1,k_1}} \big(K(x,y_1,y_2,\dots,y_m)-K(x,c_{1,k_1},y_2,\dots,y_m)\big)\,b_{1,k_1}(y_1)\,dy_1\right| \\
&\le \int_{Q_{1,k_1}} |b_{1,k_1}(y_1)|\,\frac{A\,|y_1-c_{1,k_1}|^{\varepsilon}}{(|x-y_1|+\dots+|x-y_m|)^{mn+\varepsilon}}\,dy_1 \\
&\le \int_{Q_{1,k_1}} |b_{1,k_1}(y_1)|\,\frac{CA\,\ell(Q_{1,k_1})^{\varepsilon}}{(|x-y_1|+\dots+|x-y_m|)^{mn+\varepsilon}}\,dy_1,
\end{aligned}
$$

where the preceding penultimate inequality is due to the facts that $x\notin Q^*_{1,k_1}$ and

$$
|y_1-c_{1,k_1}|\le \frac{\sqrt{n}}{2}\ell(Q_{1,k_1})\le \frac{1}{m+1}|x-y_1|\le \frac{1}{m+1}\sum_{r=1}^{m}|x-y_r|.
$$

Fix $k_2,\dots,k_l$ in $\mathbf{Z}$. We integrate the inequality

$$
\left|\int_{Q_{1,k_1}} K(x,\vec{y})\,b_{1,k_1}(y_1)\,dy_1\right| \le \int_{Q_{1,k_1}} \frac{CA\,|b_{1,k_1}(y_1)|\,\ell(Q_{1,k_1})^{\varepsilon}}{(|x-y_1|+\dots+|x-y_m|)^{mn+\varepsilon}}\,dy_1
$$

with respect to the measure $dy_{l+1}\cdots dy_m$ over $(\mathbf{R}^n)^{m-l}$ and we obtain the estimate

$$
\int_{\mathbf{R}^{n(m-l)}}\left|\int_{Q_{1,k_1}} K(x,\vec{y})\,b_{1,k_1}(y_1)\,dy_1\right| dy_{l+1}\cdots dy_m \tag{7.4.13}
$$

$$
\begin{aligned}
&\le \int_{Q_{1,k_1}}\int_{(\mathbf{R}^n)^{m-l}} \frac{CA\,|b_{1,k_1}(y_1)|\,\ell(Q_{1,k_1})^{\varepsilon}}{(|x-y_1|+\dots+|x-y_m|)^{mn+\varepsilon}}\,dy_{l+1}\cdots dy_m\,dy_1 \\
&= \int_{Q_{1,k_1}} |b_{1,k_1}(y_1)|\,\frac{AC'\,\ell(Q_{1,k_1})^{\varepsilon}}{(\sum_{j=1}^{l}|x-y_j|)^{mn+\varepsilon-(m-l)n}}\,dy_1 \\
&\le C'A\int_{Q_{1,k_1}} |b_{1,k_1}(y_1)|\,\frac{\ell(Q_{1,k_1})^{\varepsilon}}{\big(\sum_{j=1}^{l}(\ell(Q_{j,k_j})+|x-c_{j,k_j}|)\big)^{nl+\varepsilon}}\,dy_1 \\
&\le C'A\,\|b_{1,k_1}\|_{L^1}\prod_{j=1}^{l}\frac{\ell(Q_{j,k_j})^{\frac{\varepsilon}{l}}}{(\ell(Q_{j,k_j})+|x-c_{j,k_j}|)^{n+\frac{\varepsilon}{l}}}\,.
\end{aligned} \tag{7.4.14}
$$

The preceding penultimate inequality is due to the fact that for $x\notin \cup_{j=1}^{m}\cup_k (Q_{j,k})^*$ and $y_j\in Q_{j,k}$ we have that $|x-y_j|\approx \ell(Q_{j,k_j})+|x-c_{j,k_j}|$, while the last inequality is due to our assumption that the cube Q_{1,k_1} has the smallest side length.

Multiplying (7.4.13) and (7.4.14) by

$$
\prod_{i=l+1}^{m}|g_i(y_i)|\prod_{j=2}^{l}|b_{j,k_j}(y_j)|
$$

and integrating with respect to $dy_2 \cdots dy_l$ over $(\mathbf{R}^n)^{l-1}$, for any fixed point $x \notin \cup_{j=1}^m \cup_k (Q_{j,k})^*$ we obtain

$$\begin{aligned}
&\int_{\mathbf{R}^{n(m-1)}} \prod_{i=l+1}^{m} |g_i(y_i)| \prod_{j=2}^{l} |b_{j,k_j}(y_j)| \left| \int_{Q_{1,k_1}} K(x,\vec{y}) b_{1,k_1}(y_1)\, dy_1 \right| dy_2 \cdots dy_m \\
&\le C'A \prod_{i=l+1}^{m} \|g_i\|_{L^\infty} \prod_{j=1}^{l} \frac{\ell(Q_{j,k_j})^{\frac{\varepsilon}{l}}}{(\ell(Q_{j,k_j}) + |x - c_{j,k_j}|)^{n+\frac{\varepsilon}{l}}} \|b_{1,k_1}\|_{L^1} \int_{\mathbf{R}^{n(l-1)}} \prod_{j=2}^{l} |b_{j,k_j}(y_j)| dy_2 \cdots dy_l \\
&\le C'A \prod_{i=l+1}^{m} \|g_i\|_{L^\infty} \prod_{j=1}^{l} \left(\frac{\|b_{j,k_j}\|_{L^1} \ell(Q_{j,k_j})^{\frac{\varepsilon}{l}}}{(\ell(Q_{j,k_j}) + |x - c_{j,k_j}|)^{n+\frac{\varepsilon}{l}}} \right) \\
&\le C''A(\lambda\gamma)^{\frac{m-l}{m}} \prod_{j=1}^{l} \left(\frac{(\lambda\gamma)^{\frac{1}{m}} \ell(Q_{j,k_j})^{n+\frac{\varepsilon}{l}}}{(\ell(Q_{j,k_j}) + |x - c_{j,k_j}|)^{n+\frac{\varepsilon}{l}}} \right).
\end{aligned}$$

The preceding estimate, combined with the fact that for $x \notin \cup_{j=1}^m \cup_k (Q_{j,k})^*$ we have

$$\begin{aligned}
&\sum_{k_1 \in \mathbf{Z}} \cdots \sum_{k_l \in \mathbf{Z}} T(b_{1,k_1}, \dots, b_{l,k_l}, g_{l+1}, \dots, g_m)(x) \\
&\qquad = T\Big(\sum_{k_1 \in \mathbf{Z}} b_{1,k_1}, \dots, \sum_{k_l \in \mathbf{Z}} b_{l,k_l}, g_{l+1}, \dots, g_m \Big)(x),
\end{aligned}$$

yields that

$$|T(b_1, \dots, b_l, g_{l+1}, \dots, g_m)(x)| \le C''A\lambda\gamma \prod_{j-1}^{l} M_{j,\varepsilon/l}(x), \tag{7.4.15}$$

where

$$M_{j,\varepsilon/l}(x) = \sum_{k_j \in \mathbf{Z}} \frac{\ell(Q_{j,k_j})^{n+\frac{\varepsilon}{l}}}{(\ell(Q_{j,k_j}) + |x - c_{j,k_j}|)^{n+\frac{\varepsilon}{l}}}$$

is the Marcinkiewicz function associated with the union of the cubes $\{Q_{j,k_j}\}_{k_j \in \mathbf{Z}}$. Naturally, (7.4.15) is also valid for any other permutation of the functions b_j and g_i. It is a known fact (Exercise 5.6.6 in [156]) that

$$\int_{\mathbf{R}^n} M_{j,\varepsilon/l}(x)\, dx \le C \Big| \bigcup_{k_j \in \mathbf{Z}} Q_{j,k_j} \Big| \le C(\lambda\gamma)^{-\frac{1}{m}}.$$

Now, since

$$\Big| \bigcup_{j=1}^{m} \bigcup_{k} (Q_{j,k})^* \Big| \le C(\lambda\gamma)^{-\frac{1}{m}},$$

inequality (7.4.12) will be a consequence of the estimate

$$\Big| \Big\{ x \notin \bigcup_{j=1}^{m} \bigcup_{k} (Q_{j,k})^* :\ |T(h_1, \dots, h_m)(x)| > \lambda/2^m \Big\} \Big| \le C(\lambda\gamma)^{-\frac{1}{m}} (A\gamma)^{\frac{1}{l}}, \tag{7.4.16}$$

where h_j is either b_j or g_j and at least one of them is a b_j. We prove claim (7.4.16) using an $L^{1/l}$ estimate outside $\cup_{j=1}^m \cup_k (Q_{j,k})^*$; recall that we are considering the situation where l is not zero. Using the size estimate previously derived for $|T(h_1,\dots,h_m)(x)|$ outside the exceptional set, we obtain

$$\begin{aligned}
&\Big|\Big\{x \notin \bigcup_{j=1}^m \bigcup_k (Q_{j,k})^* : |T(h_1,\dots,h_m)(x)| > \lambda/2^m\Big\}\Big| \\
&\le C\lambda^{-1/l} \int_{\mathbf{R}^n \setminus \cup_{j=1}^m \cup_k (Q_{j,k})^*} (\lambda\gamma A M_{1,\varepsilon/l}(x)\cdots M_{l,\varepsilon/l}(x))^{\frac{1}{l}}\, dx \\
&\le C(A\gamma)^{1/l} \Big(\int_{\mathbf{R}^n} M_{1,\varepsilon/l}(x)dx\Big)^{\frac{1}{l}} \cdots \Big(\int_{\mathbf{R}^n} M_{l,\varepsilon/l}(x)dx\Big)^{\frac{1}{l}} \\
&\le C'(A\gamma)^{\frac{1}{l}} \big(\underbrace{(\lambda\gamma)^{-\frac{1}{m}}\cdots(\lambda\gamma)^{-\frac{1}{m}}}_{l \text{ times}}\big)^{\frac{1}{l}} \\
&= C'\lambda^{-\frac{1}{m}} (A\gamma)^{\frac{1}{l}} \gamma^{-\frac{1}{m}},
\end{aligned}$$

which proves (7.4.16), and thus (7.4.12).

We have now proved (7.4.12) for any $\gamma > 0$. Selecting $\gamma = (A+B)^{-1}$ in both (7.4.11) and (7.4.12) we obtain that all the sets E_s satisfy (7.4.10). Summing over all $1 \le s \le 2^m$ we obtain the conclusion of the theorem.

We have now shown that T satisfies the estimate

$$\big\|T(f_1,\dots,f_m)\big\|_{L^{1/m,\infty}} \le C_{n,m,\varepsilon}(A+B) \prod_{j=1}^m \|f_j\|_{L^1}$$

for all functions f_j given by finite linear combinations of characterstic functions of dyadic cubes. It follows that T has bounded extension from $L^1 \times \cdots \times L^1$ to $L^{1/m,\infty}$, which satisfies (7.4.9). □

Proposition 7.4.7. *Given K in m-CZK(A,ε) and $0 < \delta < 1/4$, define*

$$K_\delta(y_0,y_1,\dots,y_m) = K(y_0,y_1,\dots,y_m)\big[\chi_{\{\sum_{j,k=0}^m |y_j-y_k|>\delta\}} - \chi_{\{\sum_{j,k=0}^m |y_j-y_k|>1/\delta\}}\big]$$

and let T^{K_δ} be m-linear operators associated with K_δ. Suppose that T^{K_δ} admit bounded extensions from $L^{q_1} \times \cdots \times L^{q_m}$ to $L^{q,\infty}$ for some $1 \le q_1,q_2,\dots,q_m,q \le \infty$ with $q < \infty$ satisfying $\frac{1}{q_1} + \cdots + \frac{1}{q_m} = \frac{1}{q}$ such that

$$\sup_{\delta>0} \|T^{K_\delta}\|_{L^{q_1}\times\cdots\times L^{q_m} \to L^{q,\infty}} \le B < \infty. \tag{7.4.17}$$

Then T^{K_δ} extend to bounded operators from $L^{p_1} \times \cdots \times L^{p_m}$ to L^p for all $p_1,\dots,p_m,p$ satisfying $1 < p_1,\dots,p_m < \infty$, $0 < p < \infty$, and

$$\frac{1}{p_1} + \cdots + \frac{1}{p_m} = \frac{1}{p}. \tag{7.4.18}$$

Moreover, there is a constant $C_{n,m,\varepsilon,p_1,\dots,p_m}$ independent of δ such that

$$\sup_{\delta>0} \|T^{K_\delta}\|_{L^{p_1}\times\cdots\times L^{p_m}\to L^p} \le C_{n,m,p_1,\dots,p_m}(A+B) < \infty .$$

Proof. We pick a smooth function $\Phi(t)$ on the real line with values in $[0,1]$, which is equal to 1 when $t \ge 2$ and which vanishes when $t \le 1$. We define the function

$$s(y_0,y_1,\dots,y_m) = s(y_0,\vec{y}) = \sum_{j,k=0}^{m} |y_j - y_k|$$

and we consider the kernels

$$L_\delta(y_0,\vec{y}) = K(y_0,\vec{y})[\Phi(s(y_0,\vec{y})/\delta) - \Phi(s(y_0,\vec{y})\delta)] .$$

We observe that

$$\begin{aligned}
\big|L_\delta(y_0,\vec{y}) - K_\delta(y_0,\vec{y})\big| &\le |K(y_0,\vec{y})|\big[\chi_{\delta\le s(y_0,\vec{y})\le 2\delta} + \chi_{1/\delta\le s(y_0,\vec{y})\le 2/\delta}\big] \\
&\le A\, s(y_0,\vec{y})^{-mn}\big[\chi_{\delta\le s(y_0,\vec{y})\le 2\delta} + \chi_{1/\delta\le s(y_0,\vec{y})\le 2/\delta}\big] \\
&\le c_{n,m}A\,\big[\delta^{-mn}\chi_{\delta\le s(y_0,\vec{y})\le 2\delta} + \delta^{mn}\chi_{1/\delta\le s(y_0,\vec{y})\le 2/\delta}\big] \\
&\le c_{n,m}A\,\Big[\delta^{-mn}\prod_{j=1}^{m}\chi_{|y_0-y_j|\le 2\delta} + \delta^{mn}\prod_{j=1}^{m}\chi_{|y_0-y_j|\le 2/\delta}\Big] .
\end{aligned}$$

Consequently, the operators T^{K_δ} and T^{L_δ} associated with K_δ and L_δ satisfy

$$\big|T^{K_\delta}(f_1,\dots,f_m) - T^{L_\delta}(f_1,\dots,f_m)\big| \le C'_{n,m}A\prod_{j=1}^{m} M(f_j)(x) ,$$

where M is the Hardy–Littlewood maximal operator. Thus, by Hölder's inequality and the boundedness of M on L^{p_j}, the claimed boundedness for T^{K_δ} is equivalent to that for T^{L_δ}.

Next we observe that the kernels L_δ are of class m-$CZK(A,\varepsilon)$ uniformly in $\delta > 0$. Indeed, condition (7.4.1) is trivial while (7.4.2) is a consequence of the fact that for any $j \in \{0,1,\dots,m\}$ when $|y_j - y_j'| \le \frac{1}{m+1}\sum_{k=0}^{m}|y_j - y_k|$ we have that

$$\big|\Phi(s(y_0,y_1,\dots,y_m)/\delta) - \Phi(s(y_0,\dots,y_{j-1},y_j',y_{j+1},\dots,y_m)/\delta)\big|$$

is bounded by

$$m\,\|\Phi'\|_{L^\infty}\min\big(2,\tfrac{|y_j-y_j'|}{\delta}\big)\chi_{\frac{1}{2}\delta\le s(y_0,y_1,\dots,y_m)\le\frac{5}{2}\delta} .$$

Certainly an analogous estimate holds with $1/\delta$ in place of δ.

The next observation is that the jth dual of L_δ is $(K^{*j})_\delta$ since the function $s(y_0,\vec{y})$ remains unchanged if y_0 and y_j are permuted. Thus, if L_δ satisfies (7.4.1) and (7.4.2), then so do all its duals and with the same constants. Moreover, by duality, it follows that (7.4.17) is valid for all the transposes $(T^{L_\delta})^{*j}$ of T^{L_δ}, $j=1,\dots,m$. Applying Theorem 7.4.6 we obtain that $(T^{L_\delta})^{*j}$ are bounded from $L^1\times\cdots\times L^1$ to $L^{1/m,\infty}$ for all $j=0,1,\dots,m$.

Note that L_δ is supported in the set $\{(y_0,y_1,\dots,y_m): |y_0-y_j|\le 2/\delta\}$ and is bounded. It follows that condition (7.2.66) holds whenever $1<p_1,\dots,p_m,p<\infty$ satisfy (7.4.18) (Remark 7.2.14). Thus, Theorem 7.2.13 is applicable and yields that T^{L_δ} is of restricted weak type $(p_1,\dots,p_m,p)$ for all indices that satisfy (7.4.18) and $1<p_1,\dots,p_m,p<\infty$. Interpolating between all such points and the point $(1,\dots,1,1/m)$ via Corollary 7.2.4 yields the required conclusion. □

7.4.2 A Necessary and Sufficient Condition for the Boundedness of Multilinear Calderón–Zygmund Operators

We start with an endpoint estimate for operators with kernels in m-$CZK(A,\varepsilon)$.

Proposition 7.4.8. *Suppose that an m-linear operator T with kernel in m-$CZK(A,\varepsilon)$ is bounded from $L^{q_1}\times\cdots\times L^{q_m}$ to L^q with norm B for some $1<q_1,\dots,q_m,q<\infty$ satisfying*

$$\frac{1}{q_1}+\cdots+\frac{1}{q_m}=\frac{1}{q}. \tag{7.4.19}$$

Then there is a constant $C=C(n,m,\varepsilon)$ such that

$$\|T(f_1,\dots,f_m)\|_{BMO}\le C\,(A+B)\|f_1\|_{L^\infty}\cdots\|f_m\|_{L^\infty} \tag{7.4.20}$$

for all compactly supported and bounded functions f_j.

We note that compactly supported and bounded functions lie in L^r for all r with $1<r<\infty$, and thus the action of T on them is well defined.

Proof. Let us fix a cube Q. We write each function $f_j=f_j^0+f_j^1$, where $f_j^0=f_j\chi_{Q^*}$, where Q^* is a cube with the same orientation and center as Q and side length

$$\ell(Q^*)=((m+1)\sqrt{n}+1)\ell(Q).$$

Let x_Q be the center of Q. Let F be the set of all sequences of length m consisting of zeros and ones. The cardinality of F is $|F|=2^m$. For each sequence $\vec{k}=(k_1,\dots,k_m)$ in F we will find a constant $C_{\vec{k}}$ such that

$$\frac{1}{|Q|}\int_Q |T(f_1^{k_1},\dots,f_m^{k_m})(x)-C_{\vec{k}}|\,dx\le C\,(A+B)\|f_1\|_{L^\infty}\cdots\|f_m\|_{L^\infty}.$$

Then, using Proposition 3.1.2(3), with $C_Q=\sum_{\vec{k}\in F}C_{\vec{k}}$, we obtain the required claim.

If $\vec{k} = (0,\dots,0)$, then we pick $C_{\vec{0}} = 0$. Then we have

$$\begin{aligned}
\frac{1}{|Q|}\int_Q |T(f_1^0,\dots,f_m^0)(x)|\,dx &\le \left(\frac{1}{|Q|}\int_Q |T(f_1^0,\dots,f_m^0)(x)|^q\,dx\right)^{\frac{1}{q}}\\
&\le \left(\frac{1}{|Q|}\int_{\mathbf{R}^n} |T(f_1^0,\dots,f_m^0)(x)|^q\,dx\right)^{\frac{1}{q}}\\
&\le B|Q|^{-\frac{1}{q}}\|f_1^0\|_{L^{q_1}}\cdots\|f_m^0\|_{L^{q_m}}\\
&\le B|Q|^{-\frac{1}{q}}|Q^*|^{\frac{1}{q_1}+\cdots+\frac{1}{q_m}}\|f_1\|_{L^\infty}\cdots\|f_m\|_{L^\infty}\\
&= CB\|f_1\|_{L^\infty}\cdots\|f_m\|_{L^\infty}.
\end{aligned}$$

Suppose that $\vec{k} = (\overbrace{1,\dots,1}^{l\text{ times}},\overbrace{0,\dots,0}^{m-l\text{ times}})$ for some $l \in \mathbf{Z}^+$, with $1 \le l \le m$. Then we set $C_{\vec{k}} = T(f_1^{k_1},\dots,f_m^{k_m})(x_Q)$. For $x \in Q$ and $y_1 \notin Q^*$ we have that

$$|x - x_Q| \le \frac{\sqrt{n}}{2}\ell(Q) \le \frac{1}{m+1}|x-y_1| \le \frac{1}{m+1}\sum_{j=1}^m |x-y_j|$$

and thus

$$\begin{aligned}
&\frac{1}{|Q|}\int_Q |T(f_1^{k_1},\dots,f_m^{k_m})(x) - C_{\vec{k}}|\,dx\\
&\le \frac{1}{|Q|}\int_Q\int_{(\mathbf{R}^n)^m} |K(x,\vec{y}) - K(x_Q,\vec{y})|\,|f_1^{k_1}(y_1)|\dots|f_m^{k_m}(y_m)|\,d\vec{y}\,dx\\
&\le \frac{1}{|Q|}\int_Q\int_{(\mathbf{R}^n)^m} \frac{A\,|x-x_Q|^\varepsilon}{(|x-y_1|+\cdots+|x-y_m|)^{mn+\varepsilon}}|f_1^{k_1}(y_1)|\dots|f_m^{k_m}(y_m)|\,d\vec{y}\,dx\\
&\le \prod_{i=1}^m \|f_i\|_{L^\infty}\frac{1}{|Q|}\int_Q\int_{(\mathbf{R}^n)^l}\frac{C'A\,\ell(Q)^{\varepsilon+(m-l)n}}{(\ell(Q)+|x-y_1|+\cdots+|x-y_l|)^{mn+\varepsilon}}\,dy_1\cdots dy_l\,dx\\
&\le C''A\|f_1\|_{L^\infty}\cdots\|f_m\|_{L^\infty}.
\end{aligned}$$

The second inequality is due to (7.4.2) and the third to the fact that $y_1,\dots,y_l \notin Q^*$. By permuting the places of ones and zeros, the same estimate holds for any other $\vec{k} \in F$ that has at least one nonzero entry. This concludes the proof of (7.4.20). □

Theorem 7.4.9. *Fix a $\mathscr{C}^\infty$ function η on $\mathbf{R}^n$ supported in $B(0,2)$ that satisfies $0 \le \eta(x) \le 1$ and $\eta(x) = 1$ when $|x| \le 1$. Let $\eta_k(x) = \eta(x/k)$ for $k > 0$. Let T be an m-linear operator that maps $\mathscr{S}(\mathbf{R}^n)\times\cdots\times\mathscr{S}(\mathbf{R}^n)$ continuously into $\mathscr{S}'(\mathbf{R}^n)$, which is associated with a kernel K in m-CZK(A,ε). Then T has a bounded extension from $L^{q_1}\times\cdots\times L^{q_m}$ to L^q for all $1 < q_1,\dots,q_m,q < \infty$ satisfying*

$$\frac{1}{q_1}+\cdots+\frac{1}{q_m} = \frac{1}{q} \tag{7.4.21}$$

if and only if the following condition is satisfied:

$$\sum_{j=0}^{m} \sup_{k_i>0} \sup_{\xi_i \in \mathbf{R}^n} \|T^{*j}(\eta_{k_1} e^{2\pi i \xi_1 \cdot (\cdot)}, \dots, \eta_{k_m} e^{2\pi i \xi_m \cdot (\cdot)})\|_{BMO} \le B < \infty. \tag{7.4.22}$$

Moreover, if (7.4.22) holds, then we have that

$$\|T\|_{L^{q_1} \times \cdots \times L^{q_m} \to L^q} \le C_{n,m,\varepsilon,q_j}(A+B),$$

for some constant C_{n,m,ε,q_j}, depending only on the parameters indicated.

Proof. We begin by observing that the necessity of the conditions in (7.4.22) is a consequence of Proposition 7.4.8. Thus, the main implication in the proof is contained in their sufficiency, i.e., the fact that if (7.4.22) holds, then T extends to a bounded operator from $L^{q_1} \times \cdots \times L^{q_m}$ to L^q.

We show the required sufficiency by induction on m. We start with $m = 1$.

Recall from Definition 4.3.1 that a normalized bump is a smooth function φ supported in the ball $B(0,10)$ that satisfies $|\partial_x^\alpha \varphi(x)| \le 1$ for all multi-indices $|\alpha| \le 2[\frac{n}{2}]+2$. Let us fix a normalized bump φ. We use the notation $\varphi^{x_0,R}(x) = R^{-n}\varphi(R^{-1}(x-x_0))$ for $R>0$ and $x_0 \in \mathbf{R}^n$. Then it is easy to see that $\eta_k \varphi^{x_0,R}$ converges to $\varphi^{x_0,R}$ in the topology of $\mathscr{S}(\mathbf{R}^n)$, and thus, given a function ψ in $\mathscr{S}(\mathbf{R}^n)$, with integral zero, since $T^t(\psi)$ is an element of $\mathscr{S}'(\mathbf{R}^n)$ we have

$$\langle T^t(\psi), \eta_k \varphi^{x_0,R} \rangle \to \langle T^t(\psi), \varphi^{x_0,R} \rangle. \tag{7.4.23}$$

We show that for any fixed $k>0$, $x_0 \in \mathbf{R}^n$, and $R>0$ we have

$$\langle T^t(\psi), \eta_k \varphi^{x_0,R} \rangle = \int_{\mathbf{R}^n} \widehat{\varphi^{x_0,R}}(\xi) \langle T^t(\psi), \eta_k(\cdot) e^{2\pi i (\cdot)\cdot \xi} \rangle \, d\xi. \tag{7.4.24}$$

This will be a consequence of the facts that as the positive integers M,N tend to infinity, we have that

$$\sum_{\substack{m \in \mathbf{Z}^n \\ |m_j| \le MN}} \widehat{\varphi^{x_0,R}}(\tfrac{m}{N}) \eta_k(x) e^{2\pi i x \cdot \frac{m}{N}} \frac{1}{N^n} \tag{7.4.25}$$

converges to

$$\int_{\mathbf{R}^n} \widehat{\varphi^{x_0,R}}(\xi) \eta_k(x) e^{2\pi i x \cdot \xi} \, d\xi \tag{7.4.26}$$

in the topology of $\mathscr{S}$ and that

$$\Big\langle T^t(\psi), \sum_{\substack{m \in \mathbf{Z}^n \\ |m_j| \le MN}} \widehat{\varphi^{x_0,R}}(\tfrac{m}{N}) \eta_k e^{2\pi i (\cdot)\cdot \frac{m}{N}} \frac{1}{N^n} \Big\rangle = \sum_{\substack{m \in \mathbf{Z}^n \\ |m_j| \le MN}} \widehat{\varphi^{x_0,R}}(\tfrac{m}{N}) \langle T^t(\psi), \eta_k e^{2\pi i (\cdot)\cdot \frac{m}{N}} \rangle \frac{1}{N^n}$$

converges pointwise to

$$\int_{\mathbf{R}^n} \widehat{\varphi^{x_0,R}}(\xi)\langle T^t(\psi), \eta_k(\cdot)e^{2\pi i(\cdot)\cdot\xi}\rangle \, d\xi \,. \tag{7.4.27}$$

Of these assertions, the second one is straightforward in view of the fact that the integrand in (7.4.27) is a continuous[1] and integrable function of ξ and the associated Riemann sums converge to the integral in (7.4.27) as $M,N \to \infty$. To prove the first assertion, for given α, β multi-indices we set

$$F(x,\xi) = \partial_x^\alpha \big(x^\beta \eta_k(x) e^{2\pi i x\cdot\xi}\big)\,.$$

Note that

$$|F(x,\xi)| \le C_{k,\alpha,\beta}\,(1+|\xi|)^{|\alpha|}$$

and

$$|\nabla_\xi F(x,\xi)| \le C_{k,\alpha,\beta}\,(1+|\xi|)^{|\alpha|}\,.$$

Given $\delta > 0$ find $M \in \mathbf{Z}^+$ (depending on R,k,α,β) such that for all $N \ge 1$ we have

$$\int_{([-MN,MN]^n)^c} \big|\widehat{\varphi^{x_0,R}}(\xi)F(x,\xi)\big|\, d\xi < \delta\,.$$

Breaking up the complementary integral and applying the mean value theorem yields

$$\begin{aligned}
&\sup_{x\in\mathbf{R}^n}\bigg|\int_{[-MN,MN]^n} \widehat{\varphi^{x_0,R}}(\xi)F(x,\xi)\,d\xi - \sum_{\substack{m\in\mathbf{Z}^n\\ |m_j|\le MN}} \widehat{\varphi^{x_0,R}}(\tfrac{m}{N})F(x,\tfrac{m}{N})\tfrac{1}{N^n}\bigg| \\
&\le \tfrac{\sqrt{n}}{N} \sum_{\substack{m\in\mathbf{Z}^n\\ |m_j|\le MN}} \sup_{x\in\mathbf{R}^n} \sup_{\xi\in\frac{m}{N}+[0,\frac{1}{N}]^n} \big|\nabla_\xi\big[F(x,\xi)\widehat{\varphi^{x_0,R}}(\xi)\big]\big|\tfrac{1}{N^n} \\
&\le \tfrac{C}{N} \sum_{m\in\mathbf{Z}^n} (1+|\tfrac{m}{N}|)^{|\alpha|}\Big[\big|\widehat{\varphi^{x_0,R}}(\tfrac{m}{N})\big| + \big|\nabla\widehat{\varphi^{x_0,R}}(\tfrac{m}{N})\big|\Big]\tfrac{1}{N^n} \\
&\le \tfrac{C'}{N}\,\big\|(1+|\cdot|)^{|\alpha|}\widehat{\varphi^{x_0,R}}\big\|_{L^1} + \tfrac{C'}{N}\,\big\|(1+|\cdot|)^{|\alpha|}\big|\nabla\widehat{\varphi^{x_0,R}}\big|\big\|_{L^1}
\end{aligned}$$

which can also be made less than δ if N is large enough. This shows that every Schwartz seminorm $\rho_{\alpha,\beta}$ of the difference of (7.4.25) and (7.4.26) tends to zero as $N,M \to \infty$. This proves (7.4.24).

Taking absolute values in (7.4.24), applying the H^1-BMO duality, and using (7.4.22), we obtain that

$$\sup_{k>0}\big|\langle T^t(\psi), \eta_k \varphi^{x_0,R}\rangle\big| \le cR^{-n}B\,\|\psi\|_{H^1}\,, \tag{7.4.28}$$

[1] The Schwartz seminorms of $\eta_k(\cdot)(e^{2\pi i(\cdot)\cdot\xi} - e^{2\pi i(\cdot)\cdot\xi'})$ tend to zero as $\xi' \to \xi$.

where $c = R^n \|\widehat{\varphi^{x_0,R}}\|_{L^1}$, which is a constant independent of x_0 and R, in view of (4.3.1). Combining (7.4.28) and (7.4.23) yields

$$\big|\langle T(\varphi^{x_0,R}), \psi\rangle\big| \le cR^{-n}B\|\psi\|_{H^1}.$$

Since this is valid for all ψ Schwartz functions with mean value zero, we conclude that $T(\varphi^{x_0,R})$ is a linear functional on H^1, and thus it can be identified with an element of BMO that satisfies

$$R^n\big\|T(\varphi^{x_0,R})\big\|_{BMO} \le c_n' B,$$

where the constant c_n' is uniform in $x_0 \in \mathbf{R}^n$, $R > 0$, and the normalized bump φ. Also, the same conclusion is valid for T^t in place of T. It follows from Theorem 4.3.3 (vi) that T admits a bounded extension on L^2 and, thus, on any L^q, $1 < q < \infty$, with constant at most a multiple of $A+B$.

We now consider the case $m \ge 2$. Suppose by induction that the claimed assertion holds for all $k \in \mathbf{Z}^+$, with $1 \le k \le m-1$. Given an operator T associated with a kernel in m-$CZK(A,\varepsilon)$, for $j = 0,1,\dots,m$, we consider the linear operators

$$L_j(g) = T^{*j}(\eta_{k_1}e^{2\pi i\xi_1\cdot(\cdot)},\dots,\eta_{k_{m-1}}e^{2\pi i\xi_{m-1}\cdot(\cdot)},g)$$

which are associated with kernels of class 1-$CZK(A,\varepsilon)$; see Exercise 7.4.2. Applying the case $m=1$, we obtain that each L_j has a bounded extension on L^q for all $1 < q < \infty$ with norm bounded by a constant a multiple of $A+B$. Thus, by Proposition 7.4.8, each L_j satisfies (uniformly on $\xi_i \in \mathbf{R}^n$ and $k_i > 0$)

$$\big\|T^{*j}(\eta_{k_1}e^{2\pi i\xi_1\cdot(\cdot)},\dots,\eta_{k_{m-1}}e^{2\pi i\xi_{m-1}\cdot(\cdot)},g)\big\|_{BMO} \le C_n(A+B)\,\|g\|_{L^\infty} \qquad (7.4.29)$$

for every compactly supported and bounded function g on $\mathbf{R}^n$. Now fix such a function g and consider the $(m-1)$-linear operator

$$T'(f_1,\dots,f_{m-1}) = T(f_1,\dots,f_{m-1},g)$$

which is associated with a kernel in $(m-1)$-$CZK(C_{n,m}A\|g\|_{L^\infty},\varepsilon)$; see Exercise 7.4.2. Estimate (7.4.29) provides condition (7.4.22) for T' for all $j = 0,1,\dots,m-1$. Indeed, for $j=0$ it is straightforward. Also, Exercise 7.4.1 gives that

$$\begin{aligned}(T')^{*k}&(\eta_{k_1}e^{2\pi i\xi_1\cdot(\cdot)},\dots,\eta_{k_{m-1}}e^{2\pi i\xi_{m-1}\cdot(\cdot)})\\ &= T^{*(k+1)}(\eta_{k_1}e^{2\pi i\xi_1\cdot(\cdot)},\dots,\eta_{k_{m-1}}e^{2\pi i\xi_{m-1}\cdot(\cdot)},g)\end{aligned}$$

for all $k = 1,\dots,m-1$. Thus, the induction hypothesis holds. Applying the result in the case $m-1$ yields that T' admits a bounded extension from $L^{q_1}\times\cdots\times L^{q_{m-1}}$ to L^q (when $1/q_1+\cdots+1/q_{m-1} = 1/q$, $1 < q_1,\dots,q_{m-1},q < \infty$) with norm at most a multiple of $(A+B)\,\|g\|_{L^\infty} + A\|g\|_{L^\infty}$. In other words, T has an extension on products of bounded and compactly supported functions that satisfies

$$\big\|T(g_1,\dots,g_m)\big\|_{L^q} \le C(n,m,\varepsilon,q_j)(A+B)\bigg(\prod_{j=1}^{m-1}\|g_j\|_{L^{q_j}}\bigg)\|g_m\|_{L^\infty}.$$

By symmetry, we can replace the role of the last variable by any other variable, and thus we obtain

$$\big\|T(g_1,\dots,g_m)\big\|_{L^r} \le C(n,m,\varepsilon,r_j)(A+B)\prod_{j=1}^{m}\|g_j\|_{L^{r_j}}$$

where $1/r_1+\cdots+1/r_m = 1/r$, exactly one $r_j=\infty$, and the functions g_j are bounded and compactly supported.

Every point in the open convex set $H=\{(\frac{1}{q_1},\dots,\frac{1}{q_m}):\ 1<q_1,\dots,q_m,q<\infty,\ 1/q_1+\cdots+1/q_m=1/q\}$ can be written as a convex combination of points of the form $(\frac{1}{r_1},\dots,\frac{1}{r_m})$, where exactly one $r_i=\infty$; note that these points lie on the ∂H.

Next, consider the measure spaces $X_1=\cdots=X_m=\overline{B(0,M)}$ for some fixed $M>0$. Applying Corollary 7.2.4 we obtain that T admits an extension that satisfies

$$\big\|T(f_1,\dots,f_m)\big\|_{L^q} \le C(n,m,\varepsilon,q_j)(A+B)\prod_{j=1}^{m-1}\|f_j\|_{L^{q_j}}$$

whenever $1<q_1,\dots,q_m,q<\infty$ and (7.4.21) holds for all functions f_j in L^{q_j} that are supported in $\overline{B(0,M)}$. Since the bound does not depend on M, by density it also holds for all functions f_j in L^{q_j}, as claimed. □

Corollary 7.4.10. *Let $K_0(u_1,\dots,u_m)$ be a function on $(\mathbf{R}^n)^m\setminus\{0\}$ that satisfies the size estimate*

$$|K_0(u_1,\dots,u_m)| \le A|(u_1,\dots,u_m)|^{-nm}, \tag{7.4.30}$$

the cancellation condition

$$\left|\int_{R_1<|(u_1,\dots,u_m)|<R_2} K_0(u_1,\dots,u_m)\,d\vec{u}\right| \le A<\infty \tag{7.4.31}$$

for all $0<R_1<R_2<\infty$, and the smoothness condition

$$|K_0(u_1,\dots,u_j,\dots,u_m)-K_0(u_1,\dots,u_j',\dots,u_m)| \le A\frac{|u_j-u_j'|^{\varepsilon}}{|(u_1,\dots,u_m)|^{nm+\varepsilon}} \tag{7.4.32}$$

whenever $|u_j-u_j'|<\frac{1}{m+1}\sum_{k=1}^m|u_j-u_k|$. Suppose that for some sequence $\varepsilon_j\downarrow 0$ the limit

$$\lim_{j\to\infty}\int_{\varepsilon_j<|\vec{u}|\le 1} K_0(u_1,\dots,u_m)\,d\vec{u}$$

exists, and therefore K_0 extends to a tempered distribution W on $(\mathbf{R}^n)^m$. Then the multilinear operator

$$S(f_1,\dots,f_m)(x)=\lim_{\varepsilon_j\to 0}\int_{|x-y_1|+\cdots+|x-y_m|>\varepsilon_j} f_1(y_1)\cdots f_m(y_m)K_0(x-y_1,\dots,x-y_m)\,d\vec{y},$$

initially defined on $\mathscr{S}(\mathbf{R}^n)\times\cdots\times\mathscr{S}(\mathbf{R}^n)$, admits a bounded extension from $L^{p_1}(\mathbf{R}^n)\times\cdots\times L^{p_m}(\mathbf{R}^n)$ to $L^p(\mathbf{R}^n)$ when $1<p_j<\infty$ and $1/p_1+\cdots+1/p_m=1/p$.

Proof. Let W be a tempered distribution that coincides with K_0 on $\mathbf{R}^n \setminus \{0\}$. In view of Theorem 5.4.1 in [156], we have that the Fourier transform of W is a bounded function whose L^∞ norm is controlled by a multiple of A. We note that condition (7.4.30) implies (7.4.1) and (7.4.32) implies (7.4.2); see Example 7.4.2.

We now obtain the claimed boundedness by applying Theorem 7.4.9. An easy calculation shows that

$$S(\eta_{k_1} e^{2\pi i \xi_1 \cdot (\cdot)}, \dots, \eta_{k_m} e^{2\pi i \xi_m \cdot (\cdot)})(x) = \int_{(\mathbf{R}^n)^m} \widehat{W}(\vec{u}) e^{2\pi i x \cdot (u_1 + \dots + u_m)} \prod_{j=1}^{m} k_j^n \widehat{\eta}(k_j(u_j - \xi_j))\, d\vec{u}.$$

This function has L^∞ norm bounded by $\|\widehat{W}\|_{L^\infty} \|\widehat{\eta}\|_{L^1}^m < \infty$ and thus it belongs to BMO; hence, condition (7.4.22) holds uniformly on ξ_i and k_i.

The calculation with the jth transpose is similar; the only difference is that $\widehat{W}(u)$ is replaced by $\widehat{W}(v)$, where $v_l = u_l$ if $l \neq j$ and $v_j = -(u_1 + \dots + u_m)$. Thus the functions $S^{*j}(\eta_{k_1} e^{2\pi i \xi_1 \cdot (\cdot)}, \dots, \eta_{k_m} e^{2\pi i \xi_m \cdot (\cdot)})$, $j = 1, \dots, m$, are in L^∞ with the same bounds as in the case $j = 0$ and thus they are in BMO.

In view of Theorem 7.4.9, we obtain the boundedness when $p > 1$. To extend this corollary to the case where $p \le 1$, we first use Theorem 7.4.6 to obtain a weak type $(1, \dots, 1, 1/m)$ estimate, and then we apply multilinear interpolation (Theorem 7.2.2) to obtain the claimed boundedness in the case $p \le 1$. □

We now apply Corollary 7.4.10 in a specific case.

Example 7.4.11. Suppose that K_0 has the form

$$K_0(u_1, \dots, u_m) = \frac{\Omega\left(\frac{(u_1, \dots, u_m)}{|(u_1, \dots, u_m)|}\right)}{|(u_1, \dots, u_m)|^{mn}},$$

where Ω is a continuous function with mean value zero on the sphere $\mathbf{S}^{nm-1}$ which is Lipschitz of order $\varepsilon > 0$. This means that there is a constant C such that for all $\vec{w}, \vec{w}' \in \mathbf{S}^{nm-1}$ we have

$$|\Omega(\vec{w}) - \Omega(\vec{w}')| \le C |\vec{w} - \vec{w}'|^\varepsilon.$$

It is not hard to show that K_0 satisfies (7.4.32). Then the m-linear *homogeneous singular integral* operator

$$T_\Omega(f_1, \dots, f_m)(x) = \text{p.v.} \int_{(\mathbf{R}^n)^m} f_1(x_1 - y_1) \cdots f_m(x - y_m) \frac{\Omega\left(\frac{(y_1, \dots, y_m)}{|(y_1, \dots, y_m)|}\right)}{|(y_1, \dots, y_m)|^{mn}}\, d\vec{y},$$

initially defined for $f_j \in \mathscr{S}(\mathbf{R}^n)$, admits a bounded extension from the product $L^{p_1}(\mathbf{R}^n) \times \dots \times L^{p_m}(\mathbf{R}^n)$ to $L^p(\mathbf{R}^n)$ when $1 < p_1, \dots, p_m < \infty$ and $1/m < p < \infty$.

In particular the m-linear Riesz transform of Example 7.4.4 is a special case of a multilinear homogeneous singular integral operator.

Exercises

7.4.1. Let $m \geq 2$. Given an m-linear operator T and a fixed function f_j for some $1 \leq j \leq m$, we define $(m-1)$-linear operators

$$T_{f_j}(f_1,\dots,f_{j-1},f_{j+1},\dots,f_m) = T(f_1,\dots,f_{j-1},f_j,f_{j+1},\dots,f_m).$$

Show that the transposes of T_{f_j} are

$$\begin{aligned}(T_{f_j})^{*k} &= (T^{*k})_{f_j} &&\text{when}\quad k=1,\dots,j-1,\\ (T_{f_j})^{*k} &= (T^{*(k+1)})_{f_j} &&\text{when}\quad k=j,\dots,m-1.\end{aligned}$$

7.4.2. Let K be in m-$CZK(A,\varepsilon)$, and let $2 \leq l \leq m$, $f_l,\dots,f_m \in L^\infty$. For $(x,y_1,\dots,y_{l-1})$ not in the diagonal of $(\mathbf{R}^n)^l$ define

$$K_{f_l,\dots,f_m}(x,y_1,\dots,y_{l-1}) = \int_{(\mathbf{R}^n)^{m-l+1}} K(x,y_1,\dots,y_m) f_l(y_l)\cdots f_m(y_m)\,dy_l\cdots dy_m.$$

Then $K_{f_l,\dots,f_m}$ lies in $(l-1)$-$CZK(c_{n,m,l}\|f_l\|_{L^\infty}\cdots\|f_m\|_{L^\infty}A,\varepsilon)$ for some constant $c_{n,m,l} > 0$.

7.4.3. Let T be an m-linear operator that maps $\mathscr{S}(\mathbf{R}^n)\times\cdots\times\mathscr{S}(\mathbf{R}^n)$ continuously into $\mathscr{S}'(\mathbf{R}^n)$ which is associated with a kernel K in m-$CZK(A,\varepsilon)$. Then T has a bounded extension from $L^{q_1}\times\cdots\times L^{q_m}$ to L^q for all $1 < q_1,\dots,q_m,q < \infty$ satisfying $\frac{1}{q_1}+\cdots+\frac{1}{q_m} = \frac{1}{q}$ if and only if the condition

$$\sup_{\varphi_1,\dots,\varphi_m}\sum_{j=0}^{m}\sup_{R_1,\dots,R_m>0}\sup_{x_i\in\mathbf{R}^n}(R_1\cdots R_m)^n\|T^{*j}(\varphi_1^{x_1,R_1},\dots,\varphi_m^{x_m,R_m})\|_{BMO} \leq B' < \infty,$$

is satisfied, where the first supremum is taken over all m-tuples of normalized bumps. Moreover, if this condition holds, then

$$\|T\|_{L^{q_1}\times\cdots\times L^{q_m}\to L^q} \leq C_{n,m,\varepsilon,q_j}(A+B')$$

for some constant C_{n,m,ε,q_j}.

7.4.4. Let σ be a smooth function on $(\mathbf{R}^n)^{m+1}$ satisfying

$$|\partial_x^\alpha\partial_{\xi_1}^{\beta_1}\dots\partial_{\xi_m}^{\beta_m}\sigma(x,\xi_1,\dots,\xi_m)| \leq C_{\alpha,\beta}(1+|\xi_1|+\cdots+|\xi_m|)^{|\alpha|-(|\beta_1|+\cdots+|\beta_m|)}$$

for all $\alpha, \beta_1, \dots, \beta_m$ n-tuples of nonnegative integers. Let $\sigma^\vee(x, \vec{z})$ be the inverse Fourier transform of $\sigma(x, \vec{\xi}\,)$ in the ξ variable.
(a) Show that the function

$$K(y_0, y_1, \dots, y_m) = \sigma^\vee(y_0, y_0 - y_1, \dots, y_0 - y_m)$$

satisfies

$$\left|\partial_{y_0}^{\alpha_0} \cdots \partial_{y_m}^{\alpha_m} K(y_0, y_1, \dots, y_m)\right| \le \frac{C_{\alpha_0, \dots, \alpha_m}}{(|y_0 - y_1| + \cdots + |y_0 - y_m|)^{mn + |\alpha_0| + \cdots + |\alpha_m|}},$$

in particular, that it is of class m-$CZK(A, 1)$.
(b) Consider a *multilinear pseudodifferential operator*

$$T(f_1, \dots, f_m)(x) = \int_{\mathbf{R}^n} \cdots \int_{\mathbf{R}^n} \sigma(x, \vec{\xi}) \widehat{f_1}(\xi_1) \dots \widehat{f_m}(\xi_m) e^{2\pi i x \cdot (\xi_1 + \cdots + \xi_m)} d\xi_1 \dots d\xi_m .$$

Suppose that all of the transposes T^{*j} also have symbols that satisfy the same estimates as σ. Then T extends as a bounded operator from $L^{p_1} \times \cdots \times L^{p_m}$ to L^p for $1 < p_j < \infty$ and $\frac{1}{p_1} + \cdots + \frac{1}{p_m} = \frac{1}{p}$.

7.4.5. Let Ψ be a Schwartz function on $\mathbf{R}^n$ whose Fourier transform is supported in the annulus $\frac{6}{7} \le |\xi \le 2$ and is equal to 1 on $1 \le |\xi \le \frac{12}{7}$, and let $\Delta_j^\Psi(f) = f * \Psi_{2^{-j}}$ and $S_j(f) = \sum_{k \le j} \Delta_j^\Psi(f)$. Show that the *paraproduct*

$$\Pi_2(f, g) = \sum_{j \in \mathbf{Z}} \Delta_j^\Psi(f) S_j(g),$$

defined for $f, g \in \mathscr{S}(\mathbf{R}^n)$, has a bounded extension that maps $L^1 \times L^1$ to $L^{1/2, \infty}$.
[*Hint:* Show that Π_2 maps $L^4 \times L^4 \to L^2$ by writing $\Pi_2(f, g) = \sum_{j \in \mathbf{Z}} \Delta_j^\Psi(f) S_{j-2}(g) + \Pi_2'(f, g)$, where $\Pi_2'(f, g) = \sum_{j \in \mathbf{Z}} \Delta_j^\Psi(f) \Delta_{j-1}^\Psi(g) + \sum_{j \in \mathbf{Z}} \Delta_j^\Psi(f) \Delta_j^\Psi(g)$. Then prove that Π_2 has a kernel in 2-$CZK(A, 1)$ for some $A > 0$ and use Theorem 7.4.6.]

7.4.6. With the notation of the preceding exercise, show that the trilinear *paraproduct*

$$\Pi_3(f, g, h) = \sum_{j \in \mathbf{Z}} \Delta_j^\Psi(f) S_j(g) S_j(h)$$

maps $L^1(\mathbf{R}^n) \times L^1(\mathbf{R}^n) \times L^1(\mathbf{R}^n) \to L^{1/3, \infty}(\mathbf{R}^n)$.

7.5 Multilinear Multiplier Theorems

We begin this section with a fact concerning the Littlewood–Paley theorem. Given a Schwartz function Θ, we denote by Δ_k^Θ the operator given by convolution with the function $\Theta_{2^{-k}}(x) = 2^{kn}\Theta(2^k x)$.

7.5.1 Some Preliminary Facts

Proposition 7.5.1. *Let $m \in \mathbf{Z}^n$ and $\Theta(x) = \theta(x+m)$ for some Schwartz function θ whose Fourier transform is supported in an annulus of the form $2^{b_1} \le |\xi| \le 2^{b_2}$, where $-\infty < b_1 < b_2 < \infty$. Let $1 < p < \infty$. Then there is a constant C_{n,p,b_1,b_2} such that for every Schwartz function f on $\mathbf{R}^n$ we have*

$$\Big\| \Big(\sum_{j\in\mathbf{Z}} |\Delta_j^\Theta(f)|^2 \Big)^{\frac{1}{2}} \Big\|_{L^p(\mathbf{R}^n)} \le C_{n,p,b_1,b_2} \log(2+|m|) \big\| f \big\|_{L^p(\mathbf{R}^n)} \tag{7.5.1}$$

and

$$\Big\| \sup_{k\in\mathbf{Z}} \Big| \sum_{j\le k} \Delta_j^\Theta(f) \Big| \Big\|_{L^p(\mathbf{R}^n)} \le C_{n,p,b_1,b_2} \log(2+|m|) \big\| f \big\|_{L^p(\mathbf{R}^n)}. \tag{7.5.2}$$

Proof. We recall the following version of the Littlewood–Paley theorem (Theorem 6.1.2 in [156]). Suppose that Θ is an integrable function on $\mathbf{R}^n$ that satisfies

$$\sum_{j\in\mathbf{Z}} |\widehat{\Theta}(2^{-j}\xi)|^2 \le B^2 \tag{7.5.3}$$

and

$$\sup_{y\in\mathbf{R}^n\setminus\{0\}} \sum_{j\in\mathbf{Z}} \int_{|x|\ge 2|y|} \big| \Theta_{2^{-j}}(x-y) - \Theta_{2^{-j}}(x) \big| dx \le B. \tag{7.5.4}$$

Then there exists a constant $C_n < \infty$ such that for all $1 < p < \infty$ and all f in $L^p(\mathbf{R}^n)$,

$$\Big\| \Big(\sum_{j\in\mathbf{Z}} |\Delta_j^\Theta(f)|^2 \Big)^{\frac{1}{2}} \Big\|_{L^p(\mathbf{R}^n)} \le C_n B \max\big(p, (p-1)^{-1}\big) \big\| f \big\|_{L^p(\mathbf{R}^n)}. \tag{7.5.5}$$

In proving the claimed estimates (7.5.1) and (7.5.2), it suffices to assume $m \neq 0$. Note that

$$\widehat{\Theta}(\xi) = \widehat{\theta}(\xi) e^{2\pi i \xi\cdot m}.$$

The fact that $\widehat{\theta}$ is supported in an annulus implies (7.5.3) with some constant B that depends on b_1, b_2, and we now focus on (7.5.4). We fix a nonzero y in $\mathbf{R}^n$ and $j \in \mathbf{Z}$ and we examine

$$\int_{|x|\ge 2|y|} \big| \Theta_{2^{-j}}(x-y) - \Theta_{2^{-j}}(x) \big| dx = \int_{|x|\ge 2|y|} 2^{jn} \big| \theta(2^j x - 2^j y + m) - \theta(2^j x + m) \big| dx.$$

Changing variables, we write the preceding expression as

$$I_j = \int_{|x|\ge 2|y|} \big| \Theta_{2^{-j}}(x-y) - \Theta_{2^{-j}}(x) \big| dx = \int_{|x-m|\ge 2^{j+1}|y|} \big| \theta(x - 2^j y) - \theta(x) \big| dx.$$

Case 1: $2^j \geq 2\,|m|\,|y|^{-1}$. In this case, we estimate I_j by

$$\int_{|x-m|\geq 2^{j+1}|y|} \frac{c}{(1+|x-2^jy|)^{n+2}}dx + \int_{|x-m|\geq 2^{j+1}|y|} \frac{c}{(1+|x|)^{n+2}}dx$$
$$= \int_{|x+2^jy-m|\geq 2^{j+1}|y|} \frac{c}{(1+|x|)^{n+2}}dx + \int_{|x-m|\geq 2^{j+1}|y|} \frac{c}{(1+|x|)^{n+2}}dx. \tag{7.5.6}$$

Suppose that x lies in the domain of integration of the first integral in (7.5.6). Then

$$|x| \geq |x+2^jy-m| - 2^j|y| - |m| \geq 2^{j+1}|y| - 2^j|y| - \frac{1}{2}2^j|y| = \frac{1}{2}2^j|y|.$$

If x lies in the domain of integration of the second integral in (7.5.6), then

$$|x| \geq |x-m| - |m| \geq 2^{j+1}|y| - |m| \geq 2^{j+1}|y| - \frac{1}{2}2^j|y| = \frac{3}{2}2^j|y|.$$

In both cases we have

$$I_j \leq 2\int_{|x|\geq\frac{1}{2}2^j|y|} \frac{c}{(1+|x|)^{n+2}}dx \leq \frac{C}{2^j|y|}\int_{\mathbf{R}^n} \frac{1}{(1+|x|)^{n+1}}dx \leq \frac{C_n}{2^j|y|},$$

and clearly

$$\sum_{j:\, 2^j|y|\geq 2|m|} I_j \leq \sum_{j:\, 2^j|y|\geq 2} I_j \leq C_n.$$

Case 2: $|y|^{-1} < 2^j < 2\,|m|\,|y|^{-1}$. The number of j in this case is $O(1+\log|m|)$. Thus, uniformly bounding I_j by a constant, we obtain

$$\sum_{j:\, 1<2^j|y|<2|m|} I_j \leq C_n(1+\log|m|).$$

Case 3: $2^j \leq |y|^{-1}$. In this case we have

$$\big|\theta(x-2^jy)-\theta(x)\big| = \left|\int_0^1 2^j\nabla\theta(x-2^jty)\cdot y\,dt\right| \leq 2^j|y|\int_0^1 \frac{c}{(1+|x-2^jty|)^{n+1}}dt.$$

Integrating over $x\in\mathbf{R}^n$ gives the bound $I_j \leq C_n 2^j|y|$. Thus, we deduce

$$\sum_{j:\, 2^j|y|\leq 1} I_j \leq C_n.$$

Overall, we obtain the desired bound $C_n\log(2+|m|)$ in (7.5.4). This proves (7.5.1).

We now turn to the proof of (7.5.2). Define a Schwartz function Φ whose Fourier transform is equal to 1 on the ball $B(0,2^{b_2})$ and vanishes off the ball $B(0,2^{b_2+1})$ and define an operator S_k given by multiplication on the Fourier transform side by $\widehat{\Phi}(2^{-k}\xi)$. Then we have

$$\sum_{j\leq k}\Delta_j^\Theta = S_k\Big(\sum_{j\leq k}\Delta_j^\Theta\Big) = S_k\Big(\sum_{j\in\mathbf{Z}}\Delta_j^\Theta\Big) - S_k\Big(\sum_{j>k}\Delta_j^\Theta\Big).$$

We also have

$$S_k\Big(\sum_{j>k}\Delta_j^{\Theta}\Big)=S_k\Big(\sum_{j=k+1}^{b_2-b_1+k+1}\Delta_j^{\Theta}\Big)=(S_k-S_{k+b_1+1})\Big(\sum_{r=1}^{b_2-b_1-b_2}\Delta_{k+r}^{\Theta}\Big)$$

in view of the support properties of these functions. Set $\widetilde{\Delta}_k=S_k-S_{k+b_1-b_2}$. Combining all these facts, for every Schwartz function f on $\mathbf{R}^n$ we deduce that

$$\sup_{k\in\mathbf{Z}}\Big|\sum_{j\le k}\Delta_j^{\Theta}(f)\Big|\le M\Big(\sum_{j\in\mathbf{Z}}\Delta_j^{\Theta}(f)\Big)+\sum_{r=1}^{b_2-b_1+1}\Big(\sum_{k\in\mathbf{Z}}|\widetilde{\Delta}_k\Delta_{k+r}^{\Theta}(f)|^2\Big)^{\frac12}, \tag{7.5.7}$$

where M is the Hardy–Littlewood maximal operator. Using duality and Theorem 6.1.2 in [156] we obtain that $\big\|\sum_{j\in\mathbf{Z}}\Delta_j^{\Theta}(f)\big\|_{L^p}\le c\big\|\big(\sum_{j\in\mathbf{Z}}|\Delta_j^{\Theta}(f)|^2\big)^{1/2}\big\|_{L^p}$ and hence the L^p norm of the first term on the right-hand side in (7.5.7) is bounded by a constant multiple of $\log(2+|m|)\|f\|_{L^p}$ in view of (7.5.1). For the second term in (7.5.7) we use Proposition 6.1.4 (with $r=2$) in [156] and then again (7.5.1) to derive the desired conclusion. □

Next we have an orthogonality lemma for L^p, which is especially useful when $p<1$ due to the lack of duality in this case.

Lemma 7.5.2. *Let Ψ be a Schwartz function whose Fourier transform is supported in the set $\frac67\le|\xi|\le 2$, equals 1 on the annulus $1\le|\xi|\le\frac{12}{7}$, and satisfies $\sum_{j\in\mathbf{Z}}\widehat{\Psi}(2^{-j}\xi)=1$ for $\xi\neq 0$. Fix b_1,b_2, with $b_1<b_2$, and define a Schwartz function Ω via*

$$\widehat{\Omega}(\xi)=\sum_{j=b_1}^{b_2}\widehat{\Psi}(2^{-j}\xi). \tag{7.5.8}$$

Define $\Delta_k^{\Omega}(g)\widehat{\ }(\xi)=\widehat{g}(\xi)\widehat{\Omega}(2^{-k}\xi)$, $k\in\mathbf{Z}$. Let $q=b_2-b_1+1$ and fix $0<p<\infty$.

(a) For any $r\in\{0,1,\dots,q-1\}$ there is a constant $c=c(n,p,b_1,b_2,\Psi)$ such that for all L^2 functions F we have

$$\|F\|_{L^p}\le c\Big\|\Big(\sum_{k=r \bmod q}|\Delta_k^{\Omega}(F)|^2\Big)^{1/2}\Big\|_{L^p}. \tag{7.5.9}$$

(b) Let F_k be L^2 functions that satisfy $\sum_{k\in\mathbf{Z}}\|F_k\|_{L^2}^2<\infty$. Suppose that the Fourier transforms of F_k are supported in the annulus $2^{k+b_1}\le|\xi|\le 2^{k+b_2}$. Then for any $0<p<\infty$ there is a constant $C=C(n,p,b_1,b_2,\Psi)$ such that

$$\Big\|\sum_{k\in\mathbf{Z}}F_k\Big\|_{L^p}\le C\Big\|\Big(\sum_{k\in\mathbf{Z}}|F_k|^2\Big)^{1/2}\Big\|_{L^p}. \tag{7.5.10}$$

Proof. (*a*) Assume that the expression on the right-hand side in (7.5.9) is finite; otherwise there is nothing to prove. It follows from Corollary 2.2.10 that there is a polynomial Q such that $F-Q$ lies in $H^p(\mathbf{R}^n)$ and that

$$\|F-Q\|_{H^p} \le c'(n,p,b_1,b_2,\Psi)\Big\|\Big(\sum_{k=r \bmod q} |\Delta_k^{\Omega}(F)|^2\Big)^{1/2}\Big\|_{L^p} < \infty .$$

By the characterization of H^p quasi-norms, this implies

$$\Big\|\sup_{t>0}|(F-Q)*\Phi_t|\Big\|_{L^p} \le c''(n,p,b_1,b_2,\Psi)\Big\|\Big(\sum_{k=r \bmod q} |\Delta_k^{\Omega}(F)|^2\Big)^{1/2}\Big\|_{L^p}, \tag{7.5.11}$$

where Φ is a fixed Schwartz function with integral equal to 1. We clearly have

$$|F-Q| \le \sup_{t>0}|(F-Q)*\Phi_t|$$

since the family $\{\Phi_t\}_{t>0}$ is an approximate identity and F is locally integrable. Taking L^p quasi-norms and using (7.5.11) we obtain

$$\|F-Q\|_{L^p} \le c(n,p,b_1,b_2,\Psi)\Big\|\Big(\sum_{k=r \bmod q} |\Delta_k^{\Omega}(F)|^2\Big)^{1/2}\Big\|_{L^p} < \infty . \tag{7.5.12}$$

Next, we show that Q is the zero polynomial. If Q is nonconstant, then

$$\infty = |\{|Q|>2\lambda\}| \le \{|F|>\lambda\}|+|\{|F-Q|>\lambda\}| \le \frac{1}{\lambda^2}\|F\|_{L^2}^2 + \frac{1}{\lambda^p}\|F-Q\|_{L^p}^p \to 0$$

as $\lambda \to \infty$, which is a contradiction. Thus, Q is a constant, and since $F \in L^2$ and $F-Q \in L^p$, we obtain that $Q=0$. Now (7.5.9) follows from (7.5.12).

(*b*) Notice that the function $\xi \mapsto \widehat{\Omega}(2^{-k}\xi)$ is equal to 1 on the support of $\widehat{F_k}$, and thus we have $F_k = \Delta_k^{\Omega}(F_k)$ for all k. Set $q = b_2 - b_1 + 1$. Moreover, for any $r \in \{0,1,\dots,q-1\}$ we have $\sum_{k=r \bmod q}\widehat{\Omega}(2^{-k}\xi) = 1$ as long as $\xi \neq 0$. For r in $\{0,1,2,\dots,q-1\}$ we define the functions

$$G_r = \sum_{l=r \bmod q} F_l$$

and observe that

$$\sum_{j\in\mathbf{Z}} F_j = \sum_{r=0}^{q-1} G_r .$$

Then, if $k = r \bmod q$, we have

$$\Delta_k^{\Omega}(G_r) = \Delta_k^{\Omega}\Big(\sum_{l=r \bmod q} F_l\Big) = \Delta_k^{\Omega}(F_k) = F_k \tag{7.5.13}$$

since the intersection of the annuli $\frac{6}{7}2^{k+b_1} \le |\xi| \le 2^{k+b_2+1}$ and $2^{b_1+l} \le |\xi| \le 2^{b_2+l}$ has measure zero if $l = r \bmod q$, $k = r \bmod q$, and l is not equal to k modulo q. The function G_r lies in L^2 by the assumption $\sum_k \|F_k\|_{L^2}^2 < \infty$, and thus part (a) yields that

$$\big\|G_r\big\|_{L^p} \le c(n,p,b_1,b_2,\Psi)\Big\|\Big(\sum_{k=r \bmod q} |\Delta_k^{\Omega}(G_r)|^2\Big)^{1/2}\Big\|_{L^p}.$$

This inequality, combined with (7.5.13), implies (7.5.10), with G_r in place of $\sum_{k\in\mathbf{Z}} F_k$. Summing over $r \in \{0,1,\dots,q-1\}$ yields (7.5.10) with a bigger constant. □

7.5.2 Coifman-Meyer Method

In this subsection we describe a method to obtain boundedness for a bilinear multiplier operator using Fourier series expansions.

Theorem 7.5.3. *Suppose that a bounded function σ on $(\mathbf{R}^n)^2 \setminus \{(0,0)\}$ satisfies*

$$\big|\partial^{\alpha_1}\partial^{\alpha_2}\sigma(\xi_1,\xi_2)\big| \le C_{\alpha_1,\alpha_2}(|\xi_1|+|\xi_2|)^{-|\alpha_1|+|\alpha_2|} \tag{7.5.14}$$

for all $(\xi_1,\xi_2) \neq (0,0)$ and all multi-indices α_1,α_2, with $|\alpha_1|+|\alpha_2| \le 2n$. Given p_1,p_2,p such that $1 < p_1,p_2 \le \infty$ and $1/2 < p < \infty$ satisfying $1/p = 1/p_1 + 1/p_2$, the bilinear operator T_σ is bounded from $L^{p_1}(\mathbf{R}^n) \times L^{p_2}(\mathbf{R}^n)$ to $L^p(\mathbf{R}^n)$.

Proof. We first assume that $p_1,p_2 < \infty$. We fix a Schwartz function Ψ whose Fourier transform is nonnegative, supported in the set $\{\xi \in \mathbf{R}^n : \frac{6}{7} \le |\xi| \le 2\}$, is equal to 1 on the set $\{\xi \in \mathbf{R}^n : 1 \le |\xi| \le \frac{12}{7}\}$, and satisfies

$$\sum_{j\in\mathbf{Z}} \widehat{\Psi}(\xi/2^j) = 1 \tag{7.5.15}$$

for all $\xi \neq 0$. We set $\widehat{\Phi}(\xi) = \sum_{j\le 0} \widehat{\Psi}(2^{-j}\xi)$ and define $\widehat{\Phi}(0) = 1$. Then $\widehat{\Phi}(\xi)$ is a smooth bump with compact support that is equal to 1 when $|\xi| \le \frac{12}{7}$ and is equal to zero when $|\xi| \ge 2$.

We introduce the Littlewood–Paley operators Δ_j^{Ψ} associated with Ψ via $\Delta_j^{\Psi}(f) = f * \Psi_{2^{-j}}$, and we fix Schwartz functions f_1, f_2 on $\mathbf{R}^n$. We express each f_j as

$$f_j = \sum_{k\in\mathbf{Z}} \Delta_k^{\Psi}(f_j)$$

where the sum is rapidly converging. We write

$$T_\sigma(f_1,f_2) = \sum_{k\in\mathbf{Z}} \Big[T_{\sigma_k^1}(f_1,f_2) + T_{\sigma_k^2}(f_1,f_2) + T_{\sigma_k^3}(f_1,f_2)\Big], \tag{7.5.16}$$

where

$$\sigma_k^1(\xi_1,\xi_2) = \sigma(\xi_1,\xi_2)\widehat{\Psi}(2^{-k}\xi_1)\widehat{\Phi}(2^{-k+6}\xi_2)$$
$$\sigma_k^2(\xi_1,\xi_2) = \sigma(\xi_1,\xi_2)\widehat{\Psi}(2^{-k}\xi_1)\sum_{s=-5}^{5}\widehat{\Psi}(2^{-k+s}\xi_2)$$
$$\sigma_k^3(\xi_1,\xi_2) = \sigma(\xi_1,\xi_2)\widehat{\Phi}(2^{-k+6}\xi_1)\widehat{\Psi}(2^{-k}\xi_2).$$

We start with σ_k^1 which is supported in $\overline{B(0,2^{k+1})}\times\overline{B(0,2^{k-5})}$ and is $\mathscr{C}^\infty$ since its support does not contain the origin. Thus the $\mathscr{C}^\infty$ function

$$(\xi_1,\xi_2)\mapsto\sigma_k^1(2^{k+3}\xi_1,2^{k+3}\xi_2) \tag{7.5.17}$$

is supported in the cube $[-\frac14,\frac14]^{2n}$. Set $\widehat{\psi}(\xi)=\widehat{\Psi}(\frac12\xi)+\widehat{\Psi}(\xi)+\widehat{\Psi}(2\xi)$. Expanding the function in (7.5.17) in Fourier series over the cube $[-\frac12,\frac12]^{2n}$ we write

$$\sigma_k^1(2^{k+3}\xi_1,2^{k+3}\xi_2)=\sum_{l_1\in\mathbf{Z}^n}\sum_{l_2\in\mathbf{Z}^n}C_k(l_1,l_2)e^{2\pi i(\xi_1\cdot l_1+\xi_2\cdot l_2)}\widehat{\psi}(2^3\xi_1)\widehat{\Phi}(2^8\xi_2), \tag{7.5.18}$$

where the factor $\widehat{\psi}(2^3\xi_1)\widehat{\Phi}(2^8\xi_2)$ is equal to 1 on the support of the function $(\xi_1,\xi_2)\mapsto\sigma_k^1(2^{k+3}\xi_1,2^{k+3}\xi_2)$ and is itself supported in $[-\frac12,\frac12]^{2n}$. (Notice that without this factor the series produces a periodic function on $\mathbf{R}^{2n}$.) Here $C_k(l_1,l_2)$ is the Fourier coefficient given by

$$C_k(l_1,l_2)=\iint_{[-\frac12,\frac12]^{2n}} e^{-2\pi i(y_1\cdot l_1+y_2\cdot l_2)}\sigma_k^1(2^{k+3}y_1,2^{k+3}y_2)\,dy_1dy_2. \tag{7.5.19}$$

Fix $N\in\mathbf{Z}^+$. We estimate the expression in (7.5.19) using integration by parts with respect to the differential operator $(1-\Delta_{y_1})^N(1-\Delta_{y_2})^N$, which can also be expressed as a sum of derivatives of the form $\partial_{y_1}^{\alpha_1}\partial_{y_2}^{\alpha_2}$ with $|\alpha_j|\le 2N$. Since

$$\big|\partial_{y_1}^{\alpha_1}\partial_{y_2}^{\alpha_2}\sigma_k^1(2^{k+3}y_1,2^{k+3}y_2)\big|\le\frac{C_N'C_{\alpha_1,\alpha_2}}{(|y_1|+|y_2|)^{|\alpha_1|+|\alpha_2|}}\chi_{\frac67\frac18\le|y_1|\le2\frac18}\le C_NC_{\alpha_1,\alpha_2},$$

which is a consequence of Leibniz's rule and (7.5.14), it follows that

$$\sup_{k\in\mathbf{Z}}|C_k(l_1,l_2)|\le\frac{C_N\sup_{|\alpha_1|,|\alpha_2|\le2N}C_{\alpha_1,\alpha_2}}{(1+|l_1|^2)^N(1+|l_2|^2)^N}. \tag{7.5.20}$$

Dilating back in (7.5.18) we write

$$\sigma_k^1(\xi_1,\xi_2)=\sum_{l_1\in\mathbf{Z}^n}\sum_{l_2\in\mathbf{Z}^n}C_k(l_1,l_2)e^{2\pi i2^{-k-3}\xi_1\cdot l_1}\widehat{\psi}(2^{-k}\xi_1)e^{2\pi i2^{-k-3}\xi_2\cdot l_2}\widehat{\Phi}(2^{-k+5}\xi_2).$$

Notice that $\left(e^{2\pi i 2^{-k-3}(\cdot)\cdot l_1}\widehat{\psi}(2^{-k}(\cdot))\right)^\vee = (\tau^{-l_1/8}\psi)_{2^{-k}}$, where $\tau^h f(x) = f(x-h)$, and thus

$$\sum_{k\in\mathbf{Z}} T_{\sigma_k^1}(f_1,f_2) = \sum_{l_1\in\mathbf{Z}^n}\sum_{l_2\in\mathbf{Z}^n}\sum_{k\in\mathbf{Z}} C_k(l_1,l_2)\Delta_k^{\tau^{-l_1/8}\psi}(f_1)S_k^{l_2}(f_2)$$

where $\Delta_k^{\tau^{-l_1/8}\psi}(f_1)$ is the associated Littlewood–Paley operator and

$$S_k^{l_2}(f_2) = 2^{-5n}\left[\tau^{-l_2/8}\Phi(2^{-5}(\cdot))\right]_{2^{-k}} * f_2 = 2^{-5n}\sum_{j\le k}\Delta_j^{\tau^{-l_2/8}\psi(2^{-5}(\cdot))}(f_2).$$

Note that $\sup_{k\in\mathbf{Z}}|S_k^{l_2}(\cdot)|$ is bounded on L^{p_2} in view of (7.5.2).

Next we fix l_1,l_2 and obtain $L^{p_1}\times L^{p_2}\to L^p$ bounds for the bilinear operator

$$\sum_{k\in\mathbf{Z}} C_k(l_1,l_2)\Delta_k^{\tau^{-l_1/8}\psi}(f_1)\, S_k^{l_2}(f_2) \tag{7.5.21}$$

with constant $C'(1+|l_1|^2)^{-N}(1+|l_2|^2)^{-N}\log(2+|l_1|)\log(2+|l_2|)$. These bounds can be trivially summed in l_1,l_2 when $p\ge 1$ by Minkowski's inequality as long as $2N>n$. When $p\le 1$, we use the subadditivity of the quantity $\|\cdot\|_{L^p}^p$ to derive the same conclusion, provided $2Np>n$.

We notice that the Fourier transform of $S_k^{l_2}(f_2)$ is supported in the ball $B(0,2^{k-4})$, whereas that of $\Delta_k^{\tau^{-l_1/8}\psi}(f_1)$ is contained in the annulus $\frac{1}{2}\frac{6}{7}2^k\le|\xi|\le 2^2 2^k$. Thus, the support of the Fourier transform of

$$F_k = \Delta_k^{\tau^{-l_1/8}\psi}(f_1)S_k^{l_2}(f_2)$$

is contained in the algebraic sum of these sets, which is contained in the annulus $2^{-2}2^k\le|\xi|\le 2^3 2^k$. Moreover, we have that

$$\sum_k \|F_k\|_{L^2}^2 \le c\,\|M(f_2)\|_{L^\infty}^2\|f_1\|_{L^2}^2<\infty.$$

Applying Lemma 7.5.2(b) and (7.5.20) we obtain that

$$\begin{aligned}&\Big\|\sum_{k\in\mathbf{Z}} C_k(l_1,l_2)\Delta_k^{\tau^{-l_1/8}\psi}(f_1)S_k^{l_2}(f_2)\Big\|_{L^p}\\ &\le\frac{C_{p,n}C_N\sup_{|\alpha_1|,|\alpha_2|\le 2N}}{(1+|l_1|^2)^N(1+|l_2|^2)^N}\Big\|\Big(\sum_{k\in\mathbf{Z}}|\Delta_k^{\tau^{-l_1/8}\psi}(f_1)S_k^{l_2}(f_2)|^2\Big)^{\frac12}\Big\|_{L^p}.\end{aligned} \tag{7.5.22}$$

The L^p quasi-norm in (7.5.22) is bounded by

$$\Big\|\Big(\sum_{k\in\mathbf{Z}}|\Delta_k^{\tau^{-l_1/8}\psi}(f_1)|^2\Big)^{\frac12}\sup_{k\in\mathbf{Z}}|S_k^{l_2}(f_2)|\Big\|_{L^p}\le C_{n,p}\prod_{i=1}^{2}\log(2+|l_i|)\,\|f_i\|_{L^{p_i}} \tag{7.5.23}$$

in view of Hölder's inequality and Proposition 7.5.1. Combining (7.5.23) and (7.5.22) yields the claimed bound for the bilinear operator in (7.5.21).

Since in σ_k^1 and σ_k^3 the roles of ξ_1 and ξ_2 are symmetric, a similar decomposition is valid for $T_{\sigma_k^3}$.

Finally, for σ_k^2 we apply a similar Fourier series expansion technique to write

$$\sum_{k\in\mathbf{Z}} T_{\sigma_k^2}(f_1,f_2) = \sum_{l_1\in\mathbf{Z}^n}\sum_{l_2\in\mathbf{Z}^n}\left[\sum_{k\in\mathbf{Z}} C_k'(l_1,l_2)\Delta_k^{\tau^{-l_1/2^8}\theta}(f_1)\Delta_k^{\tau^{-l_2/2^8}\theta}(f_2)\right], \tag{7.5.24}$$

where θ is a Schwartz function whose Fourier transform is equal to 1 on the smallest annulus that contains $[-\frac{1}{4},\frac{1}{4}]^n$ and vanishes outside another annulus contained in $[-\frac{1}{2},\frac{1}{2}]^n$; here $C_k'(l_1,l_2)$ is a constant that satisfies estimates similar to the estimates of $C_k(l_1,l_2)$. Then the expression inside the square brackets in (7.5.24) is bounded by the product of two square functions such as those in (7.5.1) times a constant multiple of $(1+|l_1|^2)^{-N}(1+|l_2|^2)^{-N}$. Applying Hölder's inequality and Proposition 7.5.1 we obtain that the L^p quasi-norm of the product of these square functions is bounded by $C_{p,n}\log(2+|l_1|)\log(2+|l_2|)\|f_1\|_{L^{p_1}}\|f_2\|_{L^{p_2}}$. The series in l_1,l_2 in (7.5.24) are summable, and thus the claimed conclusion follows.

This argument requires $2Np>n$ and $2N>n$. Since $p>1/2$, we may take $N=n$. But $2N$ was the maximum number of derivatives required of σ. This is compatible with (7.5.14) which holds for all multi-indices α_1,α_2, with $|\alpha_1|+|\alpha_2|\le 2n=2N$.

We now dispose of the assumption that $p_1,p_2<\infty$. Assume, for instance, that $p_1=\infty$; in this case we should have $p_2<\infty$ since $p_2=p<\infty$ by assumption. The fact that T_σ maps $L^\infty\times L^p$ to L^p is equivalent to the fact that T_σ^{*1} maps $L^{p'}\times L^p$ to L^1. But $T_\sigma^{*1}=T_{\sigma^{*1}}$, and $\sigma^{*1}(\xi_1,\xi_2)=\sigma(-\xi_1-\xi_2,\xi_2)$ is another multiplier that satisfies (7.5.14); see Exercise 7.5.1. The conclusion of the theorem proved when $p_1,p_2,p<\infty$ implies the claim by duality in the case $p_1=\infty$. □

7.5.3 Hörmander-Mihlin Multiplier Condition

In this section we extend Theorem 7.5.3 in two ways. First, we reduce the number of derivatives required of the symbol. Secondly, we extend it to multilinear operators. Multi-indices in $(\mathbf{R}^n)^m$ are denoted by $\vec{\alpha}=(\alpha_1,\dots,\alpha_m)$, where each α_j is a multi-index in $\mathbf{R}^n$. We denote by $|\vec{\alpha}|=|\alpha_1|+\cdots+|\alpha_m|$ the total length of $\vec{\alpha}$.

Definition 7.5.4. We denote by $\mathscr{S}_*(\mathbf{R}^d)$ the space of all Schwartz functions Ψ on $\mathbf{R}^d$ whose Fourier transforms are supported in an annulus of the form $c_1<|\xi|<c_2$, are nonvanishing in a smaller annulus $c_1'\le|\xi|\le c_2'$, for some choice of constants $0<c_1<c_1'<c_2'<c_2<\infty$ and satisfy for some nonzero constant b

$$\sum_{j\in\mathbf{Z}}\widehat{\Psi}(2^{-j}\xi)=b \tag{7.5.25}$$

for all $\xi\in\mathbf{R}^d\setminus\{0\}$.

Recall that the Sobolev L^r_γ norm of a function g is defined as the L^r norm of the function $(I-\Delta)^{\gamma/2}(g)$. The main result of this section is as follows.

Theorem 7.5.5. *Let $1<r\le 2$. Suppose that σ is a bounded function on $(\mathbf{R}^n)^m\setminus\{0\}$. Let Ψ be in $\mathscr{S}_*((\mathbf{R}^n)^m)$. Suppose that for some γ satisfying $\frac{mn}{r}<\gamma\le mn$ we have*

$$\sup_{k\in\mathbf{Z}}\big\|\sigma^k\widehat{\Psi}\big\|_{L^r_\gamma((\mathbf{R}^n)^m)}=K<\infty, \tag{7.5.26}$$

where

$$\sigma^k(\xi_1,\dots,\xi_m)=\sigma(2^k\xi_1,\dots,2^k\xi_m). \tag{7.5.27}$$

Then the m-linear operator T_σ is bounded from $L^{p_1}(\mathbf{R}^n)\times\cdots\times L^{p_m}(\mathbf{R}^n)$ to $L^p(\mathbf{R}^n)$, whenever $\frac{mn}{\gamma}<p_j<\infty$ for all $j=1,\dots,m$, and p satisfies

$$\frac{1}{p}=\frac{1}{p_1}+\cdots+\frac{1}{p_m}.$$

Before we prove this theorem we discuss some preliminary facts.

Definition 7.5.6. For $s\in\mathbf{R}$, we introduce the weight

$$w_s(x)=(1+4\pi^2|x|^2)^{s/2}.$$

For $1\le p<\infty$ the weighted Lebesgue space $L^p(w_s)$ is defined as the set of all measurable functions f on $\mathbf{R}^n$ such that

$$\|f\|_{L^p(w_s)}=\left(\int_{\mathbf{R}^n}|f(x)|^p\,w_s(x)\,dx\right)^{1/p}<\infty.$$

We note that for $1<r\le 2$ one has

$$\begin{aligned}\|\widehat{g}\|_{L^{r'}(w_s)}&=\left(\int_{\mathbf{R}^n}|\widehat{g}|^{r'}w_s\,d\xi\right)^{\frac{1}{r'}}\\&=\left(\int_{\mathbf{R}^n}|\widehat{g}\,w_{s/r'}|^{r'}\,d\xi\right)^{\frac{1}{r'}}\\&=\left(\int_{\mathbf{R}^n}\big|[(I-\Delta)^{\frac{s}{2r'}}g]\widehat{\ }\big|^{r'}\,d\xi\right)^{\frac{1}{r'}}\\&\le\left(\int_{\mathbf{R}^n}\big|(I-\Delta)^{\frac{s}{2r'}}g\big|^{r}\,dx\right)^{\frac{1}{r}}\\&=\|g\|_{L^r_{s/r'}}\end{aligned} \tag{7.5.28}$$

via the Hausdorff-Young inequality (Proposition 2.2.16 in [156]).

Lemma 7.5.7. *Let $1 < p < q < \infty$. Let $R_0 > 0$. Then for every $s \geq 0$ there exists a constant $C = C(p,q,s,n,R_0) > 0$ such that for all functions g in L^q_s that are supported in a ball of radius R_0 in $\mathbf{R}^n$ we have*

$$\|g\|_{L^p_s(\mathbf{R}^n)} \leq C\|g\|_{L^q_s(\mathbf{R}^n)}\,. \tag{7.5.29}$$

Proof. We fix a smooth and compactly supported function φ that is equal to one on the ball of radius R_0. It will suffice to prove that

$$\|\varphi g\|_{L^p_s(\mathbf{R}^n)} \leq C\|g\|_{L^q_s(\mathbf{R}^n)}\,. \tag{7.5.30}$$

Since Schwartz functions are dense in L^q_s (Exercise 2.2.4), it will suffice to prove (7.5.30) for a Schwartz function g. If g is a Schwartz function, then so are $(I-\Delta)^{s/2}(g)$ and $(I-\Delta)^{-s/2}(g)$; thus, we may write (7.5.30) equivalently as

$$\big\|(I-\Delta)^{s/2}\big[\varphi(I-\Delta)^{-s/2}(\psi)\big]\big\|_{L^p(\mathbf{R}^n)} \leq C\|\psi\|_{L^q(\mathbf{R}^n)}\,, \tag{7.5.31}$$

where ψ is a Schwartz function.

We fix an index r such that $1/p = 1/q + 1/r$. We fix a Schwartz function h with $L^{p'}$ norm equal to one. For $z \in \mathbf{C}$ we define an entire function

$$\begin{aligned} G(z) &= \int_{\mathbf{R}^n} (I-\Delta)^{z/2}\big[\varphi(I-\Delta)^{-z/2}(\psi)\big](x)\overline{h(x)}\,dx \\ &= \int_{\mathbf{R}^n} (1+4\pi^2|\xi|^2)^{z/2} \int_{\mathbf{R}^n} \widehat{\varphi}(\xi-\eta)(1+4\pi^2|\eta|^2)^{-z/2}\widehat{\psi}(\eta)\,d\eta\,\overline{\widehat{h}(\xi)}\,d\xi\,. \end{aligned}$$

We show that

$$\big|G(z)\big| \leq C(1+|\mathrm{Im}\,z|)^c\|\psi\|_{L^q(\mathbf{R}^n)} \tag{7.5.32}$$

where C,c are positive constants and z is a complex number that satisfies either $\mathrm{Re}\,z = 0$ or $\mathrm{Re}\,z = 2[s]+2$. Note that in view of the Mihlin multiplier theorem (Theorem 6.2.7 in [156]), we have that

$$\big\|(I-\Delta)^{-it/2}(g)\big\|_{L^q} \leq C_{n,q}(1+|t|)^{[n/2]+1}\|g\|_{L^q}\,.$$

When $z = 0+it$, using Hölder's inequality we obtain

$$\begin{aligned} |G(it)| &\leq \big\|(I-\Delta)^{it/2}\big[\varphi(I-\Delta)^{-it/2}(\psi)\big]\big\|_{L^p}\|h\|_{L^{p'}} \\ &\leq c\,(1+|t|)^{[\frac{n}{2}]+1}\big\|\varphi(I-\Delta)^{-it/2}(\psi)\big\|_{L^p}\|h\|_{L^{p'}} \\ &\leq c\,\|\varphi\|_{L^r}(1+|t|)^{[\frac{n}{2}]+1}\big\|(I-\Delta)^{-it/2}(\psi)\big\|_{L^q} \\ &\leq c'\,(1+|t|)^{2[\frac{n}{2}]+2}\|\psi\|_{L^q}\,. \end{aligned}$$

Set $N = [s]+1$. When $z = it+2N$, we have

$$\begin{aligned} &|G(it+2N)| \\ &\leq \big\|(I-\Delta)^{it/2+N}\big[\varphi(I-\Delta)^{-it/2-N}(\psi)\big]\big\|_{L^p}\|h\|_{L^{p'}} \end{aligned}$$

$$
\begin{aligned}
&\leq c(1+|t|)^{[\frac{n}{2}]+1}\big\|(I-\Delta)^N\big[\varphi(I-\Delta)^{-it/2-N}(\psi)\big]\big\|_{L^p}\\
&\leq c(1+|t|)^{[\frac{n}{2}]+1}\sum_{|\alpha|\leq 2N} C_\alpha\big\|\partial^\alpha\big[\varphi(I-\Delta)^{-it/2-N}(\psi)\big]\big\|_{L^p}\\
&\leq c(1+|t|)^{[\frac{n}{2}]+1}\sum_{|\beta|+|\gamma|\leq 2N} C_{\beta,\gamma}\big\|\partial^\beta\varphi\partial^\gamma(I-\Delta)^{-it/2-N}(\psi)\big]\big\|_{L^p}\\
&\leq c(1+|t|)^{[\frac{n}{2}]+1}\sum_{|\beta|+|\gamma|\leq 2N} C_{\beta,\gamma}\big\|\partial^\beta\varphi\big\|_{L^r}\big\|(I-\Delta)^{-it/2}\partial^\gamma(I-\Delta)^{-N}(\psi)\big]\big\|_{L^q}\\
&\leq c'(1+|t|)^{2[\frac{n}{2}]+2}\sum_{|\beta|+|\gamma|\leq 2N} C_{\beta,\gamma}\big\|\partial^\beta\varphi\big\|_{L^r}\big\|\partial^\gamma(I-\Delta)^{-N}(\psi)\big]\big\|_{L^q}\\
&\leq c''(1+|t|)^{2[\frac{n}{2}]+2}\|\psi\|_{L^q}
\end{aligned}
$$

since $\partial^\gamma(I-\Delta)^{-N}$ is an L^q multiplier operator as long as $|\gamma|\leq 2N$ by another application of the Mihlin multiplier theorem (Theorem 6.2.7 in [156]).

Now consider the function $F(z)=G(2Nz)$ defined on the strip $[0,1]\times\mathbf{R}$. We observe that F is analytic on the open strip and continuous on its closure. Moreover, it satisfies $|F(z)|\leq C(\operatorname{Re} z)$ as it easily follows by applying Parserval's identity; thus, (7.2.53) holds. Lemma 7.2.10 applies and, combined with the result of Exercise 1.3.8 in [156], yields that

$$|F(t)|\leq C\,\|\psi\|_{L^q}$$

for any $t\in(0,1)$. Taking $t=s/2N$ we obtain the proof of (7.5.31). □

Lemma 7.5.8. *Suppose that $s\geq 0$ and $1<r<\infty$. Assume that φ lies in $\mathscr{S}(\mathbf{R}^n)$. Then there is a constant c_φ such that for all $g\in L^r_s(\mathbf{R}^n)$ we have*

$$\|g\,\varphi\|_{L^r_s}\leq c_\varphi\,\|g\|_{L^r_s}\,. \tag{7.5.33}$$

Proof. Since Schwartz functions are dense in L^q_s (Exercise 2.2.4), it will suffice to prove (7.5.33) for a Schwartz function g. We have

$$(I-\Delta)^{s/2}(g\,\varphi)=\int_{\mathbf{R}^n}\widehat{\varphi}(\tau)(I-\Delta)^{s/2}(g\,e^{2\pi i\tau\cdot(\cdot)})\,d\tau\,.$$

We will show that the L^r norm of $(I-\Delta)^{s/2}(g\,e^{2\pi i\tau\cdot(\cdot)})$ is controlled by $C_{s,n}\,(1+|\tau|)^s$ times the L^r norm of $(I-\Delta)^{s/2}(g)$. This statement will be a consequence of the fact that the function

$$\left(\frac{1+4\pi^2|\xi+\tau|^2}{1+4\pi^2|\xi|^2}\right)^{\frac{s}{2}} \tag{7.5.34}$$

is an L^r Fourier multiplier with norm that is bounded by a multiple of $(1+|\tau|)^B$, where $B=s+n+4$. But this is an easy consequence of the Mihlin multiplier theorem (Theorem 6.2.7 in [156]) in view of the fact that for all multi-indices α, with $|\alpha|\leq[n/2]+1$, we have

$$\left|\partial_\xi^\alpha\left(\frac{1+|\xi+\tau|^2}{1+|\xi|^2}\right)^{\frac{s}{2}}\right|\leq C_{\alpha,s}(1+|\tau|)^B(1+|\xi|)^{-|\alpha|} \tag{7.5.35}$$

and so the $\mathscr{M}_r$ norm of the function in (7.5.34) is bounded by $C_{s,n}(1+|\tau|)^B$. To verify (7.5.35), we argue as follows. Set

$$F(\xi,\tau) = \frac{1+|\xi+\tau|^2}{1+|\xi|^2} = \frac{|\xi|^2}{1+|\xi|^2} + \sum_{j=1}^{n} \frac{\xi_j}{1+|\xi|^2} 2\tau_j + \frac{1+|\tau|^2}{1+|\xi|^2}$$

and note that

$$\frac{1}{2}(1+|\tau|^2)^{-1} \le F(\xi,\tau) \le 2(1+|\tau|^2)$$

and that

$$\big|\partial_\xi^\alpha F(\xi,\tau)\big| \le C_\alpha (1+|\tau|)^2 (1+|\xi|)^{-|\alpha|}.$$

Then for $1 \le |\alpha| \le [n/2]+1$ we have

$$\big|\partial_\xi^\alpha (F(\xi,\tau)^{\frac{s}{2}})\big| \le \sum_{k=1}^{|\alpha|} F(\xi,\tau)^{\frac{s}{2}-k} C_{k,\alpha} \frac{(1+|\tau|)^2}{(1+|\xi|)^{|\alpha|}} \le C_{\alpha,s} \frac{(1+|\tau|)^{s+2[\frac{n}{2}]+2+2}}{(1+|\xi|)^{|\alpha|}}.$$

This proves (7.5.35) and concludes the proof of the lemma. □

Corollary 7.5.9. *Assume that $r = 2$ in Theorem 7.5.5. Then T_σ is bounded from $L^{p_1}(\mathbf{R}^n) \times \cdots \times L^{p_m}(\mathbf{R}^n)$ to $L^p(\mathbf{R}^n)$ whenever $1 < p_1, \dots, p_m, p < \infty$, and the relationship $1/p_1 + \cdots + 1/p_m = 1/p$ holds.*

We prove Corollary 7.5.9 assuming Theorem 7.5.5.

Proof. We first prove that condition (7.5.26) is invariant under the transposes, that is, it is also valid for the symbols of the dual operators. Indeed, the symbol of the kth transpose operator is

$$\sigma^{*k}(\xi_1,\dots,\xi_m) = \sigma(\xi_1,\dots,\xi_{k-1},-(\xi_1+\cdots+\xi_m),\xi_{k+1},\dots,\xi_m),$$

with the obvious modification if $k=1$ or $k=m$. This is equal to $\sigma(A_k\vec{\xi}\,)$, where $\vec{\xi}$ is the column vector $(\xi_1,\dots,\xi_m)$ and A_k is a modified $m \times m$ identity matrix whose kth row has been replaced by the row $(-1,\dots,-1)$. Notice that $A_k^{-1} = A_k$. Condition (7.5.26) for σ^{*k} is

$$\sup_{j\in\mathbf{Z}} \int_{(\mathbf{R}^n)^m} \big|[\sigma(2^j A_k\vec{\xi}\,)\widehat{\Psi}(\vec{\xi}\,)]^\frown(\vec{y})\big|^2 w_\gamma(\vec{y})\,d\vec{y} < \infty, \tag{7.5.36}$$

where the hat denotes Fourier transform in the $\vec{\xi}$ variable. We note that the function Ψ_k whose Fourier transform is the function $\vec{\xi} \mapsto \widehat{\Psi}(A_k\vec{\xi}\,)$, lies in $\mathscr{S}_*((\mathbf{R}^n)^m)$ since it satisfies (7.5.25).

By a change of variables inside the Fourier transform, (7.5.36) transforms into

$$\sup_{j\in\mathbf{Z}} \int_{(\mathbf{R}^n)^m} \big|[\sigma(2^j\vec{\xi}\,)\widehat{\Psi_k}(\vec{\xi}\,)]^\frown(A_k^t\vec{y})\big|^2 w_\gamma(\vec{y})\,d\vec{y} < \infty, \tag{7.5.37}$$

where A_k^t is the transpose of A_k. But $(A_k^t)^{-1} = A_k^t$ and $|A_k^t\vec{y}| \approx |\vec{y}|$; thus we have $w_\gamma(A_k^t\vec{y}) \approx w_\gamma(\vec{y})$. Therefore, by another change of variables, condition (7.5.37) is equivalent to

$$\sup_{j\in\mathbf{Z}} \int_{(\mathbf{R}^n)^m} \left|[\sigma(2^j\vec{\xi}\,)\widehat{\Psi_k}(\vec{\xi}\,)]^\frown(\vec{y})\right|^2 w_\gamma(\vec{y})\,d\vec{y} < \infty. \tag{7.5.38}$$

Thus, condition (7.5.26) for σ^{*k} holds.

We now have that (7.5.26) holds for σ^{*k} for all Ψ in $\mathscr{S}_*((\mathbf{R}^n)^m)$. Theorem 7.5.5 implies that $(T_\sigma)^{*k}$, the kth adjoint of T_σ, is bounded from $L^{p_1}(\mathbf{R}^n)\times\cdots\times L^{p_m}(\mathbf{R}^n)$ to $L^p(\mathbf{R}^n)$ whenever $2 < p_j < \infty$, in particular when $2 \le m < p_j < \infty$. In this case, each $(T_\sigma)^{*k}$ is bounded from $L^{p_1}(\mathbf{R}^n)\times\cdots\times L^{p_m}(\mathbf{R}^n)$ to $L^p(\mathbf{R}^n)$, with $1 < p < \infty$. By duality we obtain that T_σ is bounded from $L^{p_1}(\mathbf{R}^n)\times\cdots\times L^{p_m}(\mathbf{R}^n)$ to $L^p(\mathbf{R}^n)$, where $m < p_j < \infty$ when $j \neq k$ and $1 < p_k < m/(m-1)$. This is also valid when $m = 1$.

We now have boundedness for T_σ from $L^{q_1}(\mathbf{R}^n)\times\cdots\times L^{q_m}(\mathbf{R}^n)$ to $L^q(\mathbf{R}^n)$ in the following $m+1$ cases: (a) when all indices q_j are near infinity and (b) when the $m-1$ indices q_j, $j\neq k$ are near infinity and q_k is near 1 for all $k \in \{1,\dots,m\}$. Applying Corollary 7.2.4 we obtain that T_σ is bounded from $L^{p_1}(\mathbf{R}^n)\times\cdots\times L^{p_m}(\mathbf{R}^n)$ to $L^p(\mathbf{R}^n)$ for indices p_j satisfying $1 < p_1,\dots,p_m,p < \infty$. □

7.5.4 Proof of Main Result

In this section we discuss the proof of Theorem 7.5.5.

Proof. For each $j = 1,\dots,m$ we let R_j be the set of points $(\xi_1,\dots,\xi_m)$ in $(\mathbf{R}^n)^m$ such that $|\xi_j| = \max\{|\xi_1|,\dots,|\xi_m|\}$. For $j = 1,\dots,m$ we introduce nonnegative smooth functions ϕ_j on $[0,\infty)^{m-1}$ that are supported in $[0,\frac{11}{10}]^{m-1}$ such that

$$1 = \sum_{j=1}^m \phi_j\Big(\frac{|\xi_1|}{|\xi_j|},\dots,\widehat{\frac{|\xi_j|}{|\xi_j|}},\dots,\frac{|\xi_m|}{|\xi_j|}\Big)$$

for all $(\xi_1,\dots,\xi_m) \neq 0$, with the understanding that the variable with the hat is missing. These functions introduce a partition of unity of $(\mathbf{R}^n)^m \setminus \{0\}$ subordinate to a conical neighborhood of the region R_j. See Exercise 7.5.4.

Each region R_j can be written as the union of sets

$$R_{j,k} = \{(\xi_1,\dots,\xi_m) \in R_j : \ |\xi_k| \ge |\xi_s| \quad \text{for all } s \neq j\}$$

over $k = 1,\dots,m$. We need to work with a finer partition of unity, subordinate to each $R_{j,k}$. To achieve this, for each j we introduce smooth functions $\phi_{j,k}$ on $[0,\infty)^{m-2}$ supported in $[0,\frac{11}{10}]^{m-2}$ such that

$$1 = \sum_{\substack{k=1\\k\neq j}}^{m} \phi_{j,k}\Big(\frac{|\xi_1|}{|\xi_k|},\dots,\widehat{\frac{|\xi_k|}{|\xi_k|}},\dots,\widehat{\frac{|\xi_j|}{|\xi_k|}},\dots,\frac{|\xi_m|}{|\xi_k|}\Big)$$

for all $(\xi_1,\dots,\xi_m)$ in the support of ϕ_j with $\xi_k \neq 0$.

We now have obtained the following partition of unity of $(\mathbf{R}^n)^m$ minus a set of measure zero

$$1 = \sum_{j=1}^{m}\sum_{\substack{k=1\\k\neq j}}^{m} \phi_j\Big(\frac{|\xi_1|}{|\xi_j|},\dots,\widehat{\frac{|\xi_j|}{|\xi_j|}},\dots,\frac{|\xi_m|}{|\xi_j|}\Big)\phi_{j,k}\Big(\frac{|\xi_1|}{|\xi_k|},\dots,\widehat{\frac{|\xi_k|}{|\xi_k|}},\dots,\widehat{\frac{|\xi_j|}{|\xi_k|}},\dots,\frac{|\xi_m|}{|\xi_k|}\Big),$$

where the dots indicate the variables of each function.

We now introduce a nonnegative smooth bump ψ supported in the interval $[(10m)^{-1},2]$ and equal to 1 on the interval $[(5m)^{-1},\frac{12}{10}]$, and we decompose the identity on $(\mathbf{R}^n)^m \setminus E$, where $|E| = 0$, as follows:

$$1 = \sum_{j=1}^{m}\sum_{\substack{k=1\\k\neq j}}^{m} \big[\Phi_{j,k} + \Psi_{j,k}\big],$$

where $\Phi_{j,k}(\xi_1,\dots,\xi_m)$ is equal to

$$\phi_j\Big(\frac{|\xi_1|}{|\xi_j|},\dots,\widehat{\frac{|\xi_j|}{|\xi_j|}},\dots,\frac{|\xi_m|}{|\xi_j|}\Big)\phi_{j,k}\Big(\frac{|\xi_1|}{|\xi_k|},\dots,\widehat{\frac{|\xi_k|}{|\xi_k|}},\dots,\widehat{\frac{|\xi_j|}{|\xi_k|}},\dots,\frac{|\xi_m|}{|\xi_k|}\Big)\Big(1-\psi\Big(\frac{|\xi_k|}{|\xi_j|}\Big)\Big)$$

and $\Psi_{j,k}(\xi_1,\dots,\xi_m)$ is equal to

$$\phi_j\Big(\frac{|\xi_1|}{|\xi_j|},\dots,\widehat{\frac{|\xi_j|}{|\xi_j|}},\dots,\frac{|\xi_m|}{|\xi_j|}\Big)\phi_{j,k}\Big(\frac{|\xi_1|}{|\xi_k|},\dots,\widehat{\frac{|\xi_k|}{|\xi_k|}},\dots,\widehat{\frac{|\xi_j|}{|\xi_k|}},\dots,\frac{|\xi_m|}{|\xi_k|}\Big)\psi\Big(\frac{|\xi_k|}{|\xi_j|}\Big).$$

This partition of unity induces the following decomposition of σ:

$$\sigma = \sum_{j=1}^{m}\sum_{\substack{k=1\\k\neq j}}^{m} \big(\sigma\,\Phi_{j,k} + \sigma\,\Psi_{j,k}\big). \tag{7.5.39}$$

We will prove the required assertion for each piece of this decomposition, i.e., for the multipliers $\sigma\,\Phi_{j,k}$ and $\sigma\,\Psi_{j,k}$ for each pair (j,k) in the previous sum. In view of the symmetry of the decomposition, it suffices to consider the case of a fixed pair (j,k) in the sum in (7.5.39). To simplify the notation, we fix the pair $(m,m-1)$; thus, for the rest of the proof we fix $j=m$ and $k=m-1$, and we prove boundedness for the m-linear operators whose symbols are $\sigma_1 = \sigma\,\Phi_{m,m-1}$ and $\sigma_2 = \sigma\,\Psi_{m,m-1}$. These correspond to the m-linear operators T_{σ_1} and T_{σ_2}, respectively. An important note is that σ_1 is supported in the set where

$$\max(|\xi_1|,\dots,|\xi_{m-2}|) \le \tfrac{11}{10}\,|\xi_{m-1}| \le \tfrac{11}{10}\cdot\tfrac{1}{5m}\,|\xi_m|.$$

Also σ_2 is supported in the set where

$$\max(|\xi_1|,\dots,|\xi_{m-2}|) \le \tfrac{11}{10}\,|\xi_{m-1}|$$

and

$$\frac{1}{10m} \le \frac{|\xi_{m-1}|}{|\xi_m|} \le 2\,.$$

We first consider $T_{\sigma_1}(f_1,\dots,f_m)$, where f_j are fixed Schwartz functions. We fix a Schwartz radial function η whose Fourier transform is supported in the annulus $1-\frac{1}{25} \le |\xi| \le 2$ and satisfies

$$\sum_{j\in\mathbf{Z}} \widehat{\eta}(2^{-j}\xi) = 1, \qquad \xi \in \mathbf{R}^n \setminus \{0\}.$$

Associated with η we define the Littlewood–Paley operator $\Delta_j^\eta(f) = f * \eta_{2^{-j}}$, where $\eta_t(x) = t^{-n}\eta(t^{-1}x)$ for $t>0$. We decompose

$$f_m = \sum_{j\in\mathbf{Z}} \Delta_j^\eta(f_m)$$

and we note that the support of the Fourier transform of $T_{\sigma_1}(f_1,\dots,f_{m-1},\Delta_j^\eta(f_m))$ is contained in the set

$$\big\{\xi_1 : |\xi_1| \le \tfrac{3\cdot 2^j}{5m}\big\} + \dots + \big\{\xi_{m-1} : |\xi_{m-1}| \le \tfrac{3\cdot 2^j}{5m}\big\} + \big\{\xi_m : \tfrac{24}{25}\cdot 2^j \le |\xi_m| \le 2\cdot 2^j\big\}.$$

The algebraic sum of these sets is contained in the annulus

$$\{z \in \mathbf{R}^n :\, 2^{j-2} \le |z| \le 2^{j+3}\}.$$

Next, we define an operator S_j by setting

$$S_j(g) = g * \zeta_{2^{-j}}\,,$$

where ζ is a smooth function whose Fourier transform is equal to 1 on the ball $|z| < 3/5m$ and vanishes outside the double of this ball.

The Fourier transforms of $T_{\sigma_1}(S_j(f_1),\dots,S_j(f_{m-1}),\Delta_j^\eta(f_m))$ are supported in the annuli $\{z \in \mathbf{R}^n :\, 2^{j-2} \le |z| \le 2^{j+3}\}$, and we claim that

$$\sum_{j\in\mathbf{Z}} \Big\| T_{\sigma_1}(S_j(f_1),\dots,S_j(f_{m-1}),\Delta_j^\eta(f_m))\Big\|_{L^2}^2 < \infty. \tag{7.5.40}$$

Indeed, using Exercise 7.5.3 we have that $T_{\sigma_1}(S_j(f_1),\dots,S_j(f_{m-1}),\Delta_j^\eta(f_m))\widehat{\ }(\xi)$ is bounded by

$$\|\sigma_1\|_{L^\infty}\int_{(\mathbf{R}^n)^{m-1}} \Big|\widehat{S_j(f_1)}(\xi_1)\cdots\widehat{S_j(f_{m-1})}(\xi_{m-1})\widehat{\Delta_j^\eta(f_m)}\Big(\xi-\sum_{k=1}^{m-1}\xi_k\Big)\Big|\,d\xi_1\cdots d\xi_{m-1}\,.$$

In view of the Cauchy-Schwarz inequality, the square of the L^2 norm of the preceding expression is at most

$$\|\sigma_1\|_{L^\infty}^2 \prod_{i=1}^{m-1} \int_{\mathbf{R}^n} |\widehat{S_j(f_i)}(\xi_i)|\,d\xi_i \left[\int_{(\mathbf{R}^n)^m} \prod_{i=1}^{m-1} |\widehat{S_j(f_i)}(\xi_i)|\,|\widehat{\Delta_j^\eta(f_m)}(\xi - \sum_{k=1}^{m-1}\xi_k)|^2 d\vec{\xi}'d\xi\right],$$

where $d\vec{\xi}' = d\xi_1 \cdots d\xi_{m-1}$, and this expression is bounded by

$$\|\sigma_1\|_{L^\infty}^2 \prod_{i=1}^{m-1} \|\widehat{S_j(f_i)}\|_{L^1}^2 \|\widehat{\Delta_j^\eta(f_m)}\|_{L^2}^2 < \infty,$$

thus (7.5.40) holds. Lemma 7.5.2(b) now yields that for some constant C we have

$$\|T_{\sigma_1}(f_1,\dots,f_{m-1},f_m)\|_{L^p} \le C\Big\|\Big[\sum_j |T_{\sigma_1}(S_j(f_1),\dots,S_j(f_{m-1}),\Delta_j^\eta(f_m))|^2\Big]^{\frac12}\Big\|_{L^p}.$$

By definition, we have

$$\begin{aligned}&T_{\sigma_1}(S_j(f_1),\dots,S_j(f_{m-1}),\Delta_j^\eta(f_m))(x)\\&= \int_{(\mathbf{R}^n)^m} e^{2\pi i x\cdot(\xi_1+\cdots+\xi_m)}\sigma_1(\xi_1,\dots,\xi_m)\prod_{k=1}^{m-1}\widehat{S_j(f_k)}(\xi_k)\,\widehat{\Delta_j^\eta(f_m)}(\xi_m)\,d\xi_1\cdots d\xi_m.\end{aligned}$$

A simple calculation yields that the support of the integrand in the previous integral is contained in the annulus

$$\Big\{(\xi_1,\dots,\xi_m)\in(\mathbf{R}^n)^m:\ \tfrac{7}{10}\cdot 2^j < |(\xi_1,\dots,\xi_m)| < \tfrac{21}{10}\cdot 2^j\Big\},$$

so one may introduce in the previous integral the factor $\widehat{\Psi}(2^{-j}\xi_1,\dots,2^{-j}\xi_m)$, where Ψ is a radial function in $\mathscr{S}_*((\mathbf{R}^n)^m)$ whose Fourier transform is supported in some annulus and is equal to 1 on the annulus

$$\Big\{(z_1,\dots,z_m)\in(\mathbf{R}^n)^m:\ \tfrac{7}{10}\le |(z_1,\dots,z_m)|\le\tfrac{21}{10}\Big\}.$$

Inserting this factor and taking the inverse Fourier transform, we obtain that

$$T_{\sigma_1}(S_j(f_1),\dots,S_j(f_{m-1}),\Delta_j^\eta(f_m))(x)$$

is equal to

$$\int_{(\mathbf{R}^n)^m} 2^{mnj}(\sigma_1^j\widehat{\Psi})^\vee(2^j(x-y_1),\dots,2^j(x-y_m))\prod_{i=1}^{m-1}S_j(f_i)(y_i)\,\Delta_j^\eta(f_m)(y_m)\,d\vec{y},$$

where $d\vec{y} = dy_1\cdots dy_m$, the check indicates the inverse Fourier transform in all variables, and

$$\sigma_1^j(\xi_1,\xi_2,\dots,\xi_m) = \sigma_1(2^j\xi_1,2^j\xi_2,\dots,2^j\xi_m).$$

Recall our assumptions that $r > 1$, $\frac{mn}{\gamma} < r$, and $1 \le \frac{mn}{\gamma} < p_j$ for $j = 1, \dots, m$. We pick a ρ such that $\max(\frac{mn}{\gamma}, 1) = \frac{mn}{\gamma} < \rho < \min(p_1, \dots, p_m, r)$. We now have

$$
\begin{aligned}
&|T_{\sigma_1}(S_j(f_1), \dots, S_j(f_{m-1}), \Delta_j^\eta(f_m))(x)| \\
&\le \int_{(\mathbf{R}^n)^m} w_\gamma(2^j(x-y_1), \dots, 2^j(x-y_m))\, |(\sigma_1^j \widehat{\Psi})^\vee(2^j(x-y_1), \dots, 2^j(x-y_m))| \\
&\qquad\qquad \times \frac{2^{mnj}|S_j(f_1)(y_1)\cdots S_j(f_{m-1})(y_{m-1})\Delta_j^\eta(f_m)(y_m)|}{w_\gamma(2^j(x-y_1), \dots, 2^j(x-y_m))}\, d\vec{y} \\
&\le \left[\int_{(\mathbf{R}^n)^m} |(w_\gamma(\sigma_1^j \widehat{\Psi})^\vee)(2^j(x-y_1), \dots, 2^j(x-y_m))|^{\rho'} d\vec{y}\right]^{\frac{1}{\rho'}} \\
&\qquad \times 2^{mnj}\left(\int_{(\mathbf{R}^n)^m} \frac{|S_j(f_1)(y_1)\cdots S_j(f_{m-1})(y_{m-1})\Delta_j^\eta(f_m)(y_m)|^\rho}{w_{\gamma\rho}(2^j(x-y_1), \dots, 2^j(x-y_m))}\, d\vec{y}\right)^{\frac{1}{\rho}} \\
&\le C\left(\int_{(\mathbf{R}^n)^m} w_{\gamma\rho'}(y_1, \dots, y_m)|(\sigma_1^j \widehat{\Psi})^\vee(y_1, \dots, y_m)|^{\rho'} d\vec{y}\right)^{\frac{1}{\rho'}} \\
&\qquad \times \left(\int_{(\mathbf{R}^n)^m} \frac{2^{mnj}|S_j(f_1)(y_1)\cdots S_j(f_{m-1})(y_{m-1})\Delta_j^\eta(f_m)(y_m)|^\rho}{(1+2^j|x-y_1|)^{\gamma\rho/m}\cdots(1+2^j|x-y_m|)^{\gamma\rho/m}}\, d\vec{y}\right)^{\frac{1}{\rho}} \\
&\le C\|(\sigma_1^j \widehat{\Psi})^\vee\|_{L^{\rho'}(w_{\gamma\rho'})} \prod_{i=1}^{m-1}\left(\int_{\mathbf{R}^n} \frac{2^{jn}|S_j(f_i)(y_i)|^\rho}{(1+2^j|x-y_i|)^{\gamma\rho/m}}\, dy_i\right)^{\frac{1}{\rho}} \\
&\qquad\qquad \times \left(\int_{\mathbf{R}^n} \frac{2^{jn}|\Delta_j^\eta(f_m)(y_m)|^\rho}{(1+2^j|x-y_m|)^{\gamma\rho/m}}\, dy_m\right)^{\frac{1}{\rho}} \\
&\le C'\|(\sigma_1^j \widehat{\Psi})^\vee\|_{L^{\rho'}(w_{\gamma\rho'})} c^{m/\rho} \prod_{i=1}^{m-1} (M(M(f_i)^\rho)(x))^{\frac{1}{\rho}} \left(M(|\Delta_j^\eta(f_m)|^\rho)(x)\right)^{\frac{1}{\rho}},
\end{aligned}
$$

where we used that

$$\int_{\mathbf{R}^n} \frac{2^{jn}|h(y)|}{(1+2^j|x-y|)^{\gamma\rho/m}}\, dy \le cM(h)(x),$$

a consequence of the fact that $\gamma\rho/m > n$.

We now have the following sequence of inequalities:

$$\|(\sigma_1^j \widehat{\Psi})^\vee\|_{L^{\rho'}(w_{\gamma\rho'})} \le \|\sigma_1^j \widehat{\Psi}\|_{L^\rho_\gamma} \le C'' \|\sigma_1^j \widehat{\Psi}\|_{L^r_\gamma} \le C' \|\sigma^j \widehat{\Psi}\|_{L^r_\gamma} < CK,$$

justified by the result in calculation (7.5.28) for the first inequality, Lemma 7.5.7 together with the facts that $1 < \rho < r$ and σ_1^j is supported in a ball of fixed radius for the second inequality, Lemma 7.5.8 for the third inequality, and the hypothesis of Theorem 7.5.5 for the last inequality.

Thus, we have obtained the estimate

$$|T_{\sigma_1}(S_j(f_1),\dots,S_j(f_{m-1}),\Delta_j^\eta(f_m))| \le CK\prod_{i=1}^{m-1}\left(M(M(f_i)^\rho)\right)^{\frac{1}{\rho}}\left(M(|\Delta_j^\eta(f_m)|^\rho)\right)^{\frac{1}{\rho}}.$$

We now square the previous expression, sum over $j\in\mathbf{Z}$, and take square roots. Since by the choice of ρ we have $p_j>\rho$, each function $(M(M(f_i)^\rho))^{\frac{1}{\rho}}$ lies in $L^{p_i}(\mathbf{R}^n)$. We obtain

$$\begin{aligned}
&\left\|T_{\sigma_1}(f_1,\dots,f_{m-1},f_m)\right\|_{L^p(\mathbf{R}^n)}\\
&\le CK\left\|\Big\{\sum_j|T_{\sigma_1}(S_j(f_1),\dots,S_j(f_{m-1}),\Delta_j^\eta(f_m))|^2\Big\}^{\frac12}\right\|_{L^p(\mathbf{R}^n)}\\
&\le C'K\left\|\Big\{\sum_j M(|\Delta_j^\eta(f_m)|^\rho)^{\frac{2}{\rho}}\Big\}^{\frac12}\right\|_{L^{p_m}(\mathbf{R}^n)}\prod_{i=1}^{m-1}\left\|\left(M(M(f_i)^\rho)\right)^{\frac{1}{\rho}}\right\|_{L^{p_i}(\mathbf{R}^n)}\\
&\le C''K\left\|\Big\{\sum_j M(|\Delta_j^\eta(f_m)|^\rho)^{\frac{2}{\rho}}\Big\}^{\frac{\rho}{2}}\right\|_{L^{p_m/\rho}(\mathbf{R}^n)}^{\frac{1}{\rho}}\prod_{i=1}^{m-1}\|f_i\|_{L^{p_i}(\mathbf{R}^n)}
\end{aligned}$$

and this is bounded by

$$C'''K\prod_{i=1}^{m}\|f_i\|_{L^{p_i}(\mathbf{R}^n)}$$

in view of Theorem 5.6.6 in [156] with $q=2/\rho$ (note $q>1$ since $\rho<r\le 2$) and the Littlewood–Paley theorem (c.f. Theorem 6.1.2 in [156]).

Next we deal with σ_2. Using the notation introduced earlier, we write

$$T_{\sigma_2}(f_1,\dots,f_{m-1},f_m)=\sum_{j\in\mathbf{Z}}T_{\sigma_2}(f_1,\dots,f_{m-1},\Delta_j^\eta(f_m)).$$

We introduce another Littlewood–Paley operator Δ_j^θ, which is given on the Fourier transform by multiplication with a bump $\widehat{\theta}(2^{-j}\xi)$, where $\widehat{\theta}$ is equal to one on the annulus $\{\xi\in\mathbf{R}^n:\frac{24}{25}\cdot\frac{1}{10m}\le|\xi|\le 4\}$ and vanishes on a larger annulus. Also, we define S_j' as the operator given by convolution with $\zeta_{2^{-j}}$, where ζ is a smooth function whose Fourier transform is equal to 1 on the ball $|z|<\frac{22}{10}$ and vanishes outside the double of this ball.

The key observation in this case is that for each $j\in\mathbf{Z}$ we have

$$T_{\sigma_2}(f_1,\dots,f_{m-1},\Delta_j^\eta(f_m))=T_{\sigma_2}\left(S_j'(f_1),\dots,S_j'(f_{m-2}),\Delta_j^\theta(f_{m-1}),\Delta_j^\eta(f_m)\right).$$

As in the previous case, one has that in the support of the integral

$$\begin{aligned}
&T_{\sigma_2}\left(S_j'(f_1),\dots,S_j'(f_{m-2}),\Delta_j^\theta(f_{m-1}),\Delta_j^\eta(f_m)\right)(x)\\
&=\int_{(\mathbf{R}^n)^m}e^{2\pi i x\cdot(\xi_1+\cdots+\xi_m)}\sigma_2(\vec{\xi})\prod_{t=1}^{m-2}\widehat{S_j'(f_t)}(\xi_t)\,\widehat{\Delta_j^\theta(f_{m-1})}(\xi_{m-1})\widehat{\Delta_j^\eta(f_m)}(\xi_m)\,d\vec{\xi}
\end{aligned}$$

we have that

$$|\xi_1|+\cdots+|\xi_m| \approx 2^j;$$

thus one may insert into the integrand the factor $\widehat{\Psi}(2^{-j}\xi_1,\dots,2^{-j}\xi_m)$ for some Ψ in $\mathscr{S}_*((\mathbf{R}^n)^m)$ such that $\widehat{\Psi}$ is equal to one on a sufficiently wide annulus.

A calculation similar to that in the case for σ_1 yields the estimate

$$\begin{aligned}&|T_{\sigma_2}(S_j'(f_1),\dots,S_j'(f_{m-2}),\Delta_j^\theta(f_{m-1}),\Delta_j^\eta(f_m))|\\ \le & CK\prod_{i=1}^{m-2}(M(M(f_i)^\rho))^{\frac{1}{\rho}}\Big(M(|\Delta_j^\theta(f_{m-1})|^\rho)\Big)^{\frac{1}{\rho}}\Big(M(|\Delta_j^\eta(f_m)|^\rho)\Big)^{\frac{1}{\rho}}.\end{aligned}$$

Summing over j and taking L^p norms yields

$$\begin{aligned}&\big\|T_{\sigma_2}(f_1,\dots,f_{m-1},f_m)\big\|_{L^p(\mathbf{R}^n)}\\ \le & CK\Big\|\prod_{i=1}^{m-2}(M(M(f_i)^\rho))^{\frac{1}{\rho}}\sum_{j\in\mathbf{Z}}\Big(M\Big(|\Delta_j^\theta(f_{m-1})|^\rho\Big)\Big)^{\frac{1}{\rho}}\Big(M\Big(|\Delta_j^\eta(f_m)|^\rho\Big)\Big)^{\frac{1}{\rho}}\Big\|_{L^p}\\ \le & CK\Big\|\prod_{i=1}^{m-2}(M(M(f_i)^\rho))^{\frac{1}{\rho}}\Big\{\prod_{i=m-1}^{m}\sum_{j\in\mathbf{Z}}|M(|\Delta_j(f_i)|^\rho)|^{\frac{2}{\rho}}\Big\}^{\frac{1}{2}}\Big\|_{L^p(\mathbf{R}^n)}\end{aligned}$$

where the last step follows by the Cauchy-Schwarz inequality and Δ_j is Δ_j^θ if $j=m-1$ and Δ_j^η if $j=m$. Applying Hölder's inequality and using that $\rho<2$ and Theorem 5.6.6 in [156] we obtain the conclusion that the preceding expression is bounded by

$$C'K\|f_1\|_{L^{p_1}(\mathbf{R}^n)}\cdots\|f_m\|_{L^{p_m}(\mathbf{R}^n)}.$$

This concludes the proof of the theorem. □

Exercises

7.5.1. Suppose that a function σ defined on $(\mathbf{R}^n)^m\setminus\{\vec{0}\}$ satisfies

$$|\partial_{\xi_1}^{\alpha_1}\cdots\partial_{\xi_m}^{\alpha_m}\sigma(\xi_1,\dots,\xi_m)|\le C_{\alpha_1,\dots,\alpha_m}(|\xi_1|+\cdots+|\xi_m|)^{-(|\alpha_1|+\cdots+|\alpha_m|)}$$

for all multi-indices α_j satisfying $|\alpha_1|+\cdots+|\alpha_m|\le N$. Show that the same is valid for each function σ^{*j} defined in (7.3.20).

7.5.2. Let $1<r<\infty$. Fix a function σ on $\mathbf{R}^n$. Prove that if the condition

$$\sup_{k\in\mathbf{Z}}\big\|\sigma(2^k(\cdot))\widehat{\Psi}\big\|_{L^r_\gamma(\mathbf{R}^n)}=K<\infty$$

holds for some function Ψ in $\mathscr{S}_*(\mathbf{R}^n)$, then it holds for all functions Θ in $\mathscr{S}_*(\mathbf{R}^n)$.

7.5.3. (a) Let f_j be in $\mathscr{S}(\mathbf{R}^n)$. Show that for an m-linear multiplier operator T_σ we have that $T_\sigma(f_1,\dots,f_m)\widehat{\ }(\xi)$ is equal to

$$\int_{(\mathbf{R}^n)^{m-1}} \sigma\Big(\xi_1,\dots,\xi_{m-1},\xi-\sum_{k=1}^{m-1}\xi_k\Big)\prod_{l=1}^{m-1}\widehat{f_l}(\xi_l)\widehat{f_m}\Big(\xi-\sum_{k=1}^{m-1}\xi_k\Big)\,d\xi_1\cdots d\xi_{m-1}.$$

(b) Show that for functions $f_j \in \mathscr{S}(\mathbf{R}^n)$ we have

$$\int_{\mathbf{R}^n} T_\sigma(f_1,\dots,f_m)f_0\,dx = \int_{(\mathbf{R}^n)^m}\sigma(\vec{\xi}\,)\widehat{f_0}(-(\xi_1+\cdots+\xi_m))\widehat{f_1}(\xi_1)\cdots\widehat{f_m}(\xi_m)\,d\vec{\xi}\,.$$

7.5.4. Let $a>1$. Construct nonnegative smooth functions ϕ_j^m on $[0,\infty)^{m-1}$ that are supported in $[0,a^{m-1}]^{m-1}$ such that

$$1=\sum_{j=1}^{m}\phi_j^m\Big(\frac{|\xi_1|}{|\xi_j|},\dots,\widehat{\frac{|\xi_j|}{|\xi_j|}},\dots,\frac{|\xi_m|}{|\xi_j|}\Big)$$

for all $(\xi_1,\dots,\xi_m)\neq 0$, with the understanding that the variable with the hat is missing, and that $|\xi_i|/|\xi_j|=\infty$, when $\xi_j=0$ regardless of the value of ξ_i.
[*Hint:* Use induction. When $m=2$, pick a smooth function ϕ_1^1 that is equal to 1 on the set $[0,1/a]$ and supported in $[0,a]$, and define $\phi_2^1(t)=1-\phi_1^1(1/t)$. Assume that

$$1=\sum_{j=1}^{m-1}\phi_j^{m-1}\Big(\frac{|\xi_1|}{|\xi_j|},\dots,\widehat{\frac{|\xi_j|}{|\xi_j|}},\dots,\frac{|\xi_{m-1}|}{|\xi_j|}\Big)$$

for some smooth functions ϕ_j^{m-1} supported in $[0,a^{m-2}]^{m-2}$, $j=1,\dots,m-1$. Define

$$\phi_j^m\Big(\frac{|\xi_1|}{|\xi_j|},\dots,\widehat{\frac{|\xi_j|}{|\xi_j|}},\dots,\frac{|\xi_m|}{|\xi_j|}\Big)=\phi_j^{m-1}\Big(\frac{|\xi_1|}{|\xi_j|},\dots,\widehat{\frac{|\xi_j|}{|\xi_j|}},\dots,\frac{|\xi_{m-1}|}{|\xi_j|}\Big)\phi_1^1\Big(\frac{|\xi_m|}{|\xi_j|}\Big)$$

when $j=1,\dots,m-1$ and

$$\phi_m^m\Big(\frac{|\xi_1|}{|\xi_m|},\dots,\frac{|\xi_j|}{|\xi_m|},\dots,\frac{|\xi_{m-1}|}{|\xi_m|}\Big)=\sum_{j=1}^{m-1}\phi_j^{m-1}\Big(\frac{|\xi_1|}{|\xi_j|},\dots,\widehat{\frac{|\xi_j|}{|\xi_j|}},\dots,\frac{|\xi_{m-1}|}{|\xi_j|}\Big)\phi_2^1\Big(\frac{|\xi_j|}{|\xi_m|}\Big).$$

Then the function

$$\phi_m^m(u_1,\dots,u_j,\dots,u_{m-1})=\sum_{j=1}^{m-1}\phi_j^{m-1}\Big(\frac{u_1}{u_j},\dots,\widehat{\frac{u_j}{u_j}},\dots,\frac{u_{m-1}}{u_j}\Big)\phi_2^1(u_j)$$

is supported in $[0,a^{m-1}]^{m-1}$ since each ϕ_j^{m-1} is supported in $[0,a^{m-2}]^{m-2}$.]

7.5.5. Consider the following function on $\mathbf{R}^{2n}$:

$$m(\xi_1,\xi_2)=\int_{\mathbf{R}^n}\int_{\mathbf{R}^n} m_0(y_1,y_2)m_1(\xi_1-y_1,y_2)m_2(y_1,\xi_2-y_2)\,dy_1dy_2,$$

where m_0,m_1,m_2 are functions on $\mathbf{R}^{2n}$ such that the preceding integral is absolutely convergent.
(a) Prove that

$$\big\|T_m\big\|_{L^2\times L^2\to L^1}\le \|m_0\|_{L^1}\|m_1\|_{L^\infty}\|m_2\|_{L^\infty}.$$

(b) Suppose, moreover, that the Fourier transform of m is compactly supported. Then show that for any indices $1\le p_1,p_2,p\le\infty$, with $1/=1/p_1+1/p_2$, we have

$$\big\|T_m\big\|_{L^{p_1}\times L^{p_2}\to L^p}\le\sqrt{K}\,\|m_0\|_{L^2}\|m_1\|_{L^2}\|m_2\|_{L^2},$$

where K is the measure of the support of $m^\vee$.

7.5.6. ([161]) Let $2\le p_1,p_2<\infty$, $1<p\le 2$, and $1/p_1+1/p_2=1/p$ and suppose that $\{L_m\}_{m\in\mathbf{Z}}$ is a family of bilinear operators that satisfy

$$\sup_{m\in\mathbf{Z}}\|L_m\|_{L^{p_1}(\mathbf{R})\times L^{p_2}(\mathbf{R})\to L^p(\mathbf{R})}=M<\infty.$$

Let $\{A_m\}_{m\in\mathbf{Z}}$ be a sequence of disjoint intervals of equal length, and let $\{B_m\}_{m\in\mathbf{Z}}$ be another sequence of disjoint intervals of equal length. Suppose that for all m in $\mathbf{Z}$ and all Schwartz functions f,g on the line we have that $L_m(f,g)=L_m(\Delta_m^1(f),\Delta_m^2(g))$, where $\Delta_m^1(f)=(\widehat{f}\chi_{A_m})^\vee$, $\Delta_m^2(g)=(\widehat{f}\chi_{B_m})^\vee$, and suppose, moreover, that the Fourier transforms of $L_m(f,g)$ are supported in disjoint intervals of equal length. Prove that there is a constant $C=C(p_1,p_2,p)$ such that for all Schwartz functions f,g we have

$$\Big\|\sum_{m\in\mathbf{Z}}L_m(f,g)\Big\|_{L^p(\mathbf{R})}\le CM\,\big\|f\big\|_{L^{p_1}(\mathbf{R})}\big\|g\big\|_{L^{p_2}(\mathbf{R})}.$$

[*Hint:* First show that $\|\sum_{m\in\mathbf{Z}}L_m(f,g)\|_{L^p}\le C_{p'}\|(\sum_{m\in\mathbf{Z}}|L_m(f,g)|^2)^{1/2}\|_{L^p}$, using duality and Theorem 5.2.7. Then use the embeddings $\ell^p\subseteq\ell^2$, $\ell^2\subseteq\ell^{p_1}\cap\ell^{p_2}$, and the uniform boundedness of L_m to derive the conclusion.]

7.6 An Application Concerning the Leibniz Rule of Fractional Differentiation

In this section we use the theory of m-linear multiplier operators to obtain a version of Hölder's inequality for the Leibniz rule of fractional differentiation. We consider the following differentiation operators for $s\ge 0$:

$$D^s f=\big(\widehat{f}(\xi)|\xi|^s\big)^\vee$$

defined for Schwartz functions f on $\mathbf{R}^n$. This operator is singular at the origin on the Fourier transform side unless s is a nonnegative even integer. We have the following result concerning it.

Theorem 7.6.1. *Let* $\frac{1}{2} < r < \infty$, $1 < p_1, p_2, q_1, q_2 \le \infty$ *satisfy* $\frac{1}{r} = \frac{1}{p_1} + \frac{1}{p_2} = \frac{1}{q_1} + \frac{1}{q_2}$. *Given* $s > \max\left(0, \frac{n}{r} - n\right)$ *or* $s \in 2\mathbf{Z}^+$, *there exists a* $C = C(n,s,r,p_1,q_1,p_2,q_2) < \infty$ *such that for all* $f, g \in \mathscr{S}(\mathbf{R}^n)$ *we have*

$$\|D^s(fg)\|_{L^r(\mathbf{R}^n)} \le C\left[\|D^s f\|_{L^{p_1}(\mathbf{R}^n)}\|g\|_{L^{p_2}(\mathbf{R}^n)} + \|f\|_{L^{q_1}(\mathbf{R}^n)}\|D^s g\|_{L^{q_2}(\mathbf{R}^n)}\right]. \quad (7.6.1)$$

This result is sharp in the sense that for $0 \le s \le \max(\frac{n}{r} - n, 0)$ *and* $s \notin 2\mathbf{Z}^+ \cup \{0\}$, *inequality* (7.6.1) *fails for any* $1 < p_1, q_1, p_2, q_2 \le \infty$.

The inequality also fails for $s < 0$; see Exercise 7.6.3.

7.6.1 Preliminary Lemma

We begin with the following useful lemma.

Lemma 7.6.2. *Fix a function* $f \in \mathscr{S}(\mathbf{R}^n)$, *and for a given* $s > 0$ *define* $f_s = D^s f$. *Then* f_s *lies in* $L^\infty(\mathbf{R}^n)$, *and there exists a constant* $C(n,s,f)$ *such that*

$$|f_s(x)| \le C(n,s,f)|x|^{-n-s} \qquad \text{for all } |x| \ge 2. \quad (7.6.2)$$

Moreover, for $s \notin 2\mathbf{Z}^+$, *if* $f(x) \ge 0$ *for all* $x \in \mathbf{R}^n$ *and* $f \not\equiv 0$, *then there exist a constant* $R > 0$ *and a constant* $C(n,s,f,R)$ *such that*

$$|f_s(x)| \ge C(n,s,f,R)|x|^{-n-s} \qquad \text{for all } |x| > R. \quad (7.6.3)$$

Proof. For any $z \in \mathbf{C}$ with $\operatorname{Re} z > -n$ and $g \in \mathscr{S}(\mathbf{R}^n)$, define the distribution u_z by

$$\langle u_z, g\rangle = \int_{\mathbf{R}^n} \frac{\pi^{\frac{z+n}{2}}}{\Gamma\left(\frac{z+n}{2}\right)}|x|^z g(x)\,dx, \quad (7.6.4)$$

where $\Gamma(\cdot)$ denotes the gamma function. In view of the discussion in Subsection 2.4.3 in [156], we know that u_z admits an extension to an entire function with values in the space of tempered distributions. This means that for each $g \in \mathscr{S}(\mathbf{R}^n)$, the map $z \mapsto \langle u_z, g\rangle$ is an entire function. Moreover, using Theorem 2.4.6 in [156], we have $\widehat{u_z} = u_{-n-z}$, i.e., $\langle u_z, \widehat{g}\rangle = \langle u_{-n-z}, g\rangle$, for all $g \in \mathscr{S}(\mathbf{R}^n)$. Notice that both u_z and u_{-n-z} lie in L^1_{loc} if and only if $-n < \operatorname{Re} z < 0$.

Now, fix $f \in \mathscr{S}(\mathbf{R}^n)$, write for $s > 0$

$$f_s(x) = \int_{\mathbf{R}^n} |\xi|^s \widehat{f}(\xi)\, e^{2\pi i \xi \cdot x}\, d\xi = \frac{\Gamma\left(\frac{s+n}{2}\right)}{\pi^{\frac{s+n}{2}}} \left\langle u_s, \widehat{f}(\cdot) e^{2\pi i (\cdot)\cdot x} \right\rangle ,$$

and note that $\Gamma\left(\frac{s+n}{2}\right)/\pi^{\frac{s+n}{2}} \neq 0$ when $s > 0$. Now it suffices to prove estimates (7.6.2) and (7.6.3) for $\langle u_s, \widehat{f}(\cdot) e^{2\pi i (\cdot)\cdot x} \rangle$.

Applying Theorem 2.4.6 in [156] we write

$$\left\langle u_z, \widehat{f}(\cdot) e^{2\pi i (\cdot)\cdot x} \right\rangle = \left\langle u_{-n-z}, f(\cdot + x) \right\rangle .$$

Using the notation in Subsection 2.4.3 in [156] we write

$$\left\langle u_{-n-z}, f(\cdot + x) \right\rangle = I_0(x,z) + I_1(x,z) + I_2(x,z),$$

where

$$I_0(x,z) = \sum_{|\alpha| \le N} b(n,\alpha,z)(-1)^{|\alpha|} \langle \partial^\alpha \delta_0, f(\cdot + x) \rangle ,$$

$$I_1(x,z) = \int_{|y|\le 1} \frac{\pi^{-\frac{z}{2}}}{\Gamma\left(-\frac{z}{2}\right)} \left\{ f(x+y) - \sum_{|\alpha|\le N} \frac{(\partial^\alpha f)(x)}{\alpha!} y^\alpha \right\} |y|^{-n-z}\, dy,$$

$$I_2(x,z) = \frac{\pi^{-\frac{z}{2}}}{\Gamma\left(-\frac{z}{2}\right)} \int_{|y|>1} |y|^{-n-z} f(x+y)\, dy,$$

for some $N \in \mathbf{Z}^+$ and suitable entire functions $b(n,\alpha,z)$. Notice that $I_1(x,z)$ is holomorphic in the region $\operatorname{Re} z < N+1$ and that there is a constant $C(z,n,N)$ such that

$$|I_0(x,z)| + |I_1(x,z)| \le C(z,n,N) \left(\sum_{|\alpha|\le N} |\partial^\alpha f(x)| + \sup_{|y|\le 1} \sum_{|\beta|=N+1} \sup |\partial^\beta f(x+y)| \right).$$

It follows that $I_0(x,z) + I_1(x,z)$ decays like a Schwartz function for any fixed $z \in \mathbf{C}$.

Now we consider $I_2(x,z)$, which is an entire function in z and can be written as

$$I_2(x,z) = \int_{|x-y|>1} \frac{\pi^{-\frac{z}{2}}}{\Gamma\left(-\frac{z}{2}\right)} |x-y|^{-n-z} f(y)\, dy. \tag{7.6.5}$$

We notice that the constant $C_z = \pi^{-\frac{z}{2}}/\Gamma\left(-\frac{z}{2}\right)$ vanishes when z is a nonnegative even integer since the Gamma function has poles at the points $0, -1, -2, \dots$. However, if $z \notin \{0\} \cup 2\mathbf{Z}^+$, then $C_z \neq 0$. Thus, assertion (7.6.2) follows when s is an even integer.

Now fix $s \in \mathbf{R}^+ \setminus 2\mathbf{Z}^+$. It is easily seen from (7.6.5) that $I_2(x,s)$ is bounded for all $x \in \mathbf{R}^n$. Next we examine the decay rate of $I_2(x,s)$ for $|x| > 2$. Split the integral in (7.6.5) as a sum $I_2^1(x,s) + I_2^2(x,s)$, where

$$I_2^1(x,s) = \int_{\substack{|x|\le 2|y| \\ |x-y|>1}} \frac{\pi^{-\frac{s}{2}}}{\Gamma\left(-\frac{s}{2}\right)} |x-y|^{-n-s} f(y)\,dy$$

$$I_2^2(x,s) = \int_{\substack{|x|> 2|y| \\ |x-y|>1}} \frac{\pi^{-\frac{s}{2}}}{\Gamma\left(-\frac{s}{2}\right)} |x-y|^{-n-s} f(y)\,dy.$$

Then for every $M>0$ there is a constant $C_f(M)>0$ satisfying

$$|I_2^1(x,s)| \le |C_s| \int_{|y|\ge\frac{|x|}{2}} |f(y)|\,dy \le \frac{|C_s|}{(1+\frac{1}{2}|x|)^M}\int_{\mathbf{R}^n}(1+|y|)^M|f(y)|\,dy \le \frac{|C_s|\,C_f(M)}{(1+|x|)^M}.$$

For the integrand of $I_2^2(x,s)$ we notice that $\frac{1}{2}|x| \le |x-y| \le \frac{3}{2}|x|$ when $|x|>2|y|$. Thus, $|I_2^2(x,s)| \le C_s 2^{n+s}\|f\|_{L^1}|x|^{-n-s}$ for $|x|>2$. This fact, combined with the preceding estimates for $I_2^1(x,s)$ and $I_0(x,s)+I_1(x,s)$, yields (7.6.2).

Moreover, if $f(y)\ge 0$ for all $y\in\mathbf{R}^n$ but $f\not\equiv 0$, then for $|x|\ge 2$ we have

$$|I_2^2(x,s)| \ge \left(\frac{3}{2}\right)^{n+s}\frac{|C_s|}{|x|^{n+s}}\int_{|x|>2|y|} f(y)\,dy.$$

Taking $|x|$ large enough so that $f\not\equiv 0$ on the ball $B(0,|x|/2)=\{y\in\mathbf{R}^n: |y|<|x|/2\}$, the preceding integral is bounded from below. This proves (7.6.3). □

7.6.2 Proof of Theorem 7.6.1

Proof. Note that (7.6.3) states that for all s satisfying $s\ne 2k$ for some $k\in\mathbf{Z}^+$ and $0<s\le\frac{n}{r}-n$, $D^s f\notin L^r(\mathbf{R}^n)$. In particular, we can choose any nonzero function $f\in\mathscr{S}(\mathbf{R}^n)$ so that $D^s(|f|^2)\notin L^r(\mathbf{R}^n)$. On the other hand, (7.6.2) says that $D^s f$ lies in $L^{p_1}(\mathbf{R}^n)$ for any $s>0$ and $1<p_1\le\infty$. This disproves inequality (7.6.1) when $0<s\le\frac{n}{r}-n$. We remark also that if $s>\frac{n}{r}-n$, then $D^s f\in L^r(\mathbf{R}^n)$ for any $f\in\mathscr{S}(\mathbf{R}^n)$.

We now prove (7.6.1). Fix $\varphi\in\mathscr{S}(\mathbf{R}^n)$ whose Fourier transform is supported in $B(0,3/2)$ and is equal to 1 on $B(0,1)$. Also, let $\widehat{\psi}(\xi)=\widehat{\varphi}(\xi)-\widehat{\varphi}(2\xi)$, and note that $\widehat{\psi}$ is supported on an annulus $1/2\le|\xi|\le 3/2$ and $\sum_{k\in\mathbf{Z}}\widehat{\psi}(2^{-k}\xi)=1$ for all $\xi\ne 0$.

Given $f,g\in\mathscr{S}(\mathbf{R}^n)$, we decompose $D^s(fg)(x)$ as follows:

$$\begin{aligned}
&\int_{\mathbf{R}^{2n}}|\xi+\eta|^s\widehat{f}(\xi)\widehat{g}(\eta)e^{2\pi i(\xi+\eta)\cdot x}\,d\xi\,d\eta\\
=&\int_{\mathbf{R}^{2n}}|\xi+\eta|^s\left(\sum_{j\in\mathbf{Z}}\widehat{\psi}(2^{-j}\xi)\widehat{f}(\xi)\right)\left(\sum_{k\in\mathbf{Z}}\widehat{\psi}(2^{-k}\eta)\widehat{g}(\eta)\right)e^{2\pi i(\xi+\eta)\cdot x}\,d\xi\,d\eta\\
=&\,\Pi_1[f,g](x)+\Pi_2[f,g](x)+\Pi_3[f,g](x),
\end{aligned}$$

where

$$\Pi_1[f,g](x) = \sum_{j\in\mathbf{Z}}\sum_{k<j-1}\int_{\mathbf{R}^{2n}} |\xi+\eta|^s \widehat{\psi}(2^{-j}\xi)\widehat{f}(\xi)\widehat{\psi}(2^{-k}\eta)\widehat{g}(\eta)e^{2\pi i(\xi+\eta)\cdot x}\,d\xi\,d\eta$$

$$\Pi_2[f,g](x) = \sum_{k\in\mathbf{Z}}\sum_{j<k-1}\int_{\mathbf{R}^{2n}} |\xi+\eta|^s \widehat{\psi}(2^{-j}\xi)\widehat{f}(\xi)\widehat{\psi}(2^{-k}\eta)\widehat{g}(\eta)e^{2\pi i(\xi+\eta)\cdot x}\,d\xi\,d\eta$$

$$\Pi_3[f,g](x) = \sum_{k\in\mathbf{Z}}\sum_{j=k-1}^{k+1}\int_{\mathbf{R}^{2n}} |\xi+\eta|^s \widehat{\psi}(2^{-j}\xi)\widehat{f}(\xi)\widehat{\psi}(2^{-k}\eta)\widehat{g}(\eta)e^{2\pi i(\xi+\eta)\cdot x}\,d\xi\,d\eta.$$

The treatments of the terms Π_1 and Π_2 are identical in view of the symmetry between ξ and η, so it suffices to consider Π_1 and Π_3. We write $\Pi_1[f,g](x)$ as

$$\int_{\mathbf{R}^{2n}} \left\{ \sum_{j\in\mathbf{Z}} \widehat{\psi}(2^{-j}\xi)\widehat{\varphi}(2^{-j+2}\eta)\frac{|\xi+\eta|^s}{|\xi|^s} \right\} \widehat{D^s f}(\xi)\widehat{g}(\eta)e^{2\pi i(\xi+\eta)\cdot x}\,d\xi\,d\eta$$

and we note that the symbol inside the curly brackets in the preceding expression is supported in the set $\{(\xi,\eta):\ |\eta|\le\frac{3}{4}|\xi|\}$, is $\mathscr{C}^\infty$ on $\mathbf{R}^{2n}\setminus\{(0,0)\}$, and satisfies (7.5.14); hence, $\Pi_1[f,g]$ satisfies

$$\big\|\Pi_1[f,g]\big\|_{L^r} \le C\,\|D^s f\|_{L^{p_1}(\mathbf{R}^n)}\|g\|_{L^{p_2}(\mathbf{R}^n)}\,.$$

Likewise, we have

$$\big\|\Pi_2[f,g]\big\|_{L^r} \le C\,\|f\|_{L^{q_1}(\mathbf{R}^n)}\|D^s g\|_{L^{q_2}(\mathbf{R}^n)}$$

and thus (7.6.1) holds for $\Pi_1[f,g]+\Pi_2[f,g]$.

For $\Pi_3[f,g]$, note that the summation in j is finite; thus, it suffices to show estimate (7.6.1) for one of the three terms, say for

$$\sum_{k\in\mathbf{Z}}\int_{\mathbf{R}^{2n}} |\xi+\eta|^s \widehat{\psi}(2^{-k}\xi)\widehat{f}(\xi)\widehat{\psi}(2^{-k}\eta)\widehat{g}(\eta)e^{2\pi i(\xi+\eta)\cdot x}\,d\xi\,d\eta$$

which can be written as

$$\int_{\mathbf{R}^{2n}} \left\{ \sum_{k\in\mathbf{Z}} \frac{|\xi+\eta|^s}{|\eta|^s}\widehat{\psi}(2^{-k}\xi)\widehat{\psi}(2^{-k}\eta) \right\} \widehat{f}(\xi)\widehat{D^s g}(\eta)e^{2\pi i(\xi+\eta)\cdot x}\,d\xi\,d\eta. \tag{7.6.6}$$

The symbol in the curly brackets is supported in $\{(\xi,\eta):\ |\eta|/3\le|\xi|\le 3|\eta|\}$, and it may not be smooth along the line $\xi+\eta=0$ if s is not in $2\mathbf{Z}^+$. However, if $s\in 2\mathbf{Z}^+$ then the symbol in (7.6.6) is $\mathscr{C}^\infty$ on $\mathbf{R}^{2n}\setminus\{(0,0)\}$ and satisfies (7.5.14), and thus the corresponding operator is bounded in the claimed range of exponents, in view of Theorem 7.5.3.

We therefore fix $s\in\mathbf{R}^+\setminus 2\mathbf{Z}^+$ and focus on the estimate for Π_3, which requires a more careful study. We consider the following cases.

Case 1: $\frac{1}{2} < r < \infty$ and $1 < p_1, p_2, q_1, q_2 < \infty$.

Notice that when $|\xi|, |\eta| \le \frac{3}{2} 2^k$, then we have $|\xi + \eta| \le 2^{k+2}$, and thus $\widehat{\varphi}(2^{-k-2}(\xi+\eta)) = 1$. Using this fact we write $\Pi_3[f,g](x)$ as

$$\iint_{\mathbf{R}^{2n}} \sum_{k\in\mathbf{Z}} |\xi+\eta|^s \widehat{\psi}(2^{-k}\xi)\widehat{f}(\xi)\widehat{\psi}(2^{-k}\eta)\widehat{g}(\eta)e^{2\pi i(\xi+\eta)\cdot x}\, d\xi\, d\eta$$

$$= \iint_{\mathbf{R}^{2n}} \sum_{k\in\mathbf{Z}} |\xi+\eta|^s \widehat{\varphi}(2^{-k-2}(\xi+\eta))\widehat{\psi}(2^{-k}\xi)\widehat{f}(\xi)\widehat{\psi}(2^{-k}\eta)\widehat{g}(\eta)e^{2\pi i(\xi+\eta)\cdot x} d\xi d\eta$$

$$=2^{2s} \sum_{k\in\mathbf{Z}} \iint_{\mathbf{R}^{2n}} \widehat{\varphi_s}(2^{-k-2}(\xi+\eta))\widehat{\psi}(2^{-k}\xi)\widehat{f}(\xi)\widehat{\psi_s}(2^{-k}\eta)\widehat{D^s(g)}(\eta)e^{2\pi i(\xi+\eta)\cdot x} d\xi d\eta$$

where $\widehat{\psi_s}(\cdot) = |\cdot|^{-s}\widehat{\psi}(\cdot)$ and $\widehat{\varphi_s}(\cdot) = |\cdot|^s\widehat{\varphi}(\cdot)$.

The function $\xi \mapsto \widehat{\varphi_s}(2^{-2}\xi)$ is supported in $[-8,8]^n$ and can be expressed in terms of its Fourier series multiplied by the characteristic function of the set $[-8,8]^n$. Specifically, we have

$$\widehat{\varphi_s}(2^{-2}(\xi+\eta)) = \sum_{m\in\mathbf{Z}^n} c_m^s e^{\frac{2\pi i}{16}(\xi+\eta)\cdot m} \chi_{[-8,8]^n}(\xi+\eta), \tag{7.6.7}$$

where

$$c_m^s = \frac{1}{16^n} \int_{[-8,8]^n} |y|^s\, \widehat{\varphi}(2^{-2}y)e^{-\frac{2\pi i}{16}y\cdot m}\, dy.$$

Due to the support properties we also have

$$\chi_{[-8,8]^n}(2^{-k}(\xi+\eta))\,\widehat{\psi}(2^{-k}\xi)\widehat{\psi_s}(2^{-k}\eta) = \widehat{\psi}(2^{-k}\xi)\widehat{\psi_s}(2^{-k}\eta). \tag{7.6.8}$$

Inserting a dilation into (7.6.7) we have

$$\widehat{\varphi_s}(2^{-k-2}(\xi+\eta)) = \sum_{m\in\mathbf{Z}^n} c_m^s e^{\frac{2\pi i}{16}2^{-k}(\xi+\eta)\cdot m} \chi_{[-8,8]^n}(2^{-k}(\xi+\eta)).$$

Using this identity and (7.6.8) we express $\Pi_3[f,g](x)$ as

$$2^{2s} \sum_{m\in\mathbf{Z}^n} \sum_{k\in\mathbf{Z}} \iint_{\mathbf{R}^{2n}} c_m^s e^{\frac{2\pi i}{16}2^{-k}(\xi+\eta)\cdot m}\widehat{\psi}(2^{-k}\xi)\widehat{f}(\xi)\widehat{\psi_s}(2^{-k}\eta)\widehat{D^s g}(\eta)e^{2\pi i(\xi+\eta)\cdot x} d\xi d\eta$$

which can also be written as

$$2^{2s} \sum_{m\in\mathbf{Z}^n} c_m^s \sum_{k\in\mathbf{Z}} \Delta_k^m(f)(x)\Delta_k^{m,s}(D^s g)(x), \tag{7.6.9}$$

where Δ_k^m is the Littlewood–Paley operator given by multiplication on the Fourier transform side by $e^{2\pi i 2^{-k}\xi\cdot\frac{m}{16}}\widehat{\psi}(2^{-k}\xi)$, whereas $\Delta_k^{m,s}$ is the Littlewood–Paley operator given by multiplication on the Fourier transform side by $e^{2\pi i 2^{-k}\xi\cdot\frac{m}{16}}\widehat{\psi_s}(2^{-k}\xi)$. Both Littlewood–Paley operators have the form

$$\int_{\mathbf{R}^n} 2^{nk}\Theta(2^k(x-y)+\tfrac{1}{16}m)f(y)\,dy$$

for some Schwartz function Θ whose Fourier transform is supported in some annulus centered at zero.

Let $r_* = \min(r,1)$. Taking the L^r quasi-norm of (7.6.9) we obtain

$$\begin{aligned}
&\big\|\Pi_3[f,g]\big\|_{L^r}^{r_*} \\
&\le 2^{2sr_*}\sum_{m\in\mathbf{Z}^n}|c_m^s|^{r_*}\Big\|\sum_{k\in\mathbf{Z}}\Delta_k^m(f)\,\Delta_k^{m,s}(D^s g)\Big\|_{L^r(\mathbf{R}^n)}^{r_*} \\
&\le 2^{2sr_*}\sum_{m\in\mathbf{Z}^n}|c_m^s|^{r_*}\Big\|\sqrt{\sum_{k\in\mathbf{Z}}|\Delta_k^m(f)|^2}\Big\|_{L^{q_1}(\mathbf{R}^n)}^{r_*}\Big\|\sqrt{\sum_{k\in\mathbf{Z}}|\Delta_k^{m,s}(D^s g)|^2}\Big\|_{L^{q_2}(\mathbf{R}^n)}^{r_*}
\end{aligned}$$

since $1/q_1 + 1/q_2 = 1/r$. By Proposition 7.5.1, the preceding expression is bounded by a constant multiple of

$$\sum_{m\in\mathbf{Z}^n}|c_m^s|^{r_*}[\log(2+|m|)]^{2r_*}\|f\|_{L^{q_1}}^{r_*}\|D^s g\|_{L^{q_2}}^{r_*}$$

since $1 < q_1, q_2 < \infty$, and this term yields the required inequality, provided we can show that the preceding series converges.

Applying Lemma 7.6.2 we obtain

$$\begin{aligned}
c_m^s &= \frac{1}{16^n}\int_{[-8,8]^n}|\xi|^s\widehat{\varphi}(2^{-2}\xi)e^{-\frac{2\pi i}{16}m\cdot\xi}\,d\xi \\
&= cD^s(\varphi(4(\cdot)))(-\tfrac{m}{16}) \\
&= O((1+|m|)^{-n-s})
\end{aligned}$$

as $|m|\to\infty$ and c_m^s is uniformly bounded for all $m\in\mathbf{Z}$. Thus, since $r_*(n+s) > n$, the series

$$\sum_{m\in\mathbf{Z}^n}|c_m^s|^{r_*}[\log(2+|m|)]^{2r_*}$$

converges. This concludes the proof in Case 1.

Case 2: $1 < r < \infty$, p_1 or p_2 equals infinity, q_1 or q_2 equals infinity.

The treatment of the terms $\Pi_1[f,g]$ and $\Pi_2[f,g]$ was based on Theorem 7.5.3, which allows one of p_1 or p_2 to be equal to infinity, and likewise with q_1 or q_2. To treat the term $\Pi_3[f,g]$, we use the Littlewood–Paley theorem (Theorem 6.1.2 in [156]) to write

$$\big\|\Pi_3[f,g]\big\|_{L^r(\mathbf{R}^n)} \le C(r,n)\Big\|\Big(\sum_{j\in\mathbf{Z}}|\Delta_j^{\Psi}(\Pi_3[f,g])|^2\Big)^{\frac{1}{2}}\Big\|_{L^r(\mathbf{R}^n)}$$

for the given function ψ. We have the following estimate:

$$
\begin{aligned}
&|\Delta_j^{\psi}(\Pi_3[f,g])(x)| \\
&= \left| \int_{\mathbf{R}^{2n}} |\xi+\eta|^s \psi\big(\tfrac{\xi+\eta}{2^j}\big) \sum_{k\geq j-2} \widehat{\psi}(2^{-k}\xi)\widehat{f}(\xi)\widehat{\psi}(2^{-k}\eta)\widehat{g}(\eta)\, e^{2\pi i(\xi+\eta)\cdot x}\, d\xi\, d\eta \right| \\
&= \left| \int_{\mathbf{R}^{2n}} 2^{js}\widehat{\psi_s}\big(\tfrac{\xi+\eta}{2^j}\big)\widehat{\psi}\big(\tfrac{\xi}{2^k}\big)\widehat{f}(\xi) \sum_{k\geq j-2} 2^{-ks}\widehat{\psi_{-s}}(2^{-k}\eta)\widehat{D^s g}(\eta)\, e^{2\pi i(\xi+\eta)\cdot x}\, d\xi\, d\eta \right| \\
&= 2^{js}\left| \sum_{k\geq j-2} 2^{-ks}\Delta_j^{\psi_s}\left[\Delta_k^{\psi}(f)\Delta_k^{\psi_{-s}}(D^s g)\right](x)\right| \\
&\leq 2^{js}\Big(\sum_{k\geq j-2} 2^{-2ks}\Big)^{\frac12}\Big(\sum_{k\geq j-2} |\Delta_j^{\psi_s}\left[\Delta_k^{\psi}(f)\Delta_k^{\psi_{-s}}(D^s g)\right]|^2\Big)^{\frac12} \\
&\leq C(s)\Big(\sum_{k\geq j-2} \Big|\Delta_j^{\psi_s}\left[\Delta_k^{\psi}(f)\Delta_k^{\psi_{-s}}(D^s g)\right]\Big|^2\Big)^{\frac12},
\end{aligned}
$$

where $\widehat{\psi_s}(\cdot) = |\cdot|^s\widehat{\psi}(\cdot)$ and $\widehat{\psi_{-s}}(\cdot) = |\cdot|^{-s}\widehat{\psi}(\cdot)$. Thus, we have

$$
\big\|\Pi_3[f,g]\big\|_{L^r} \leq C(r,n,s)\Big\|\Big(\sum_{j\in\mathbf{Z}}\sum_{k\in\mathbf{Z}} |\Delta_j^{\psi_s}\left[\Delta_k^{\psi}(f)\Delta_k^{\psi_{-s}}(D^s g)\right]|^2\Big)^{\frac12}\Big\|_{L^r}.
$$

We now apply Proposition 6.1.4 in [156] (with $r=2$), which yields

$$
\begin{aligned}
\big\|\Pi_3[f,g]\big\|_{L^r} &\leq C_1(r,n,s)\Big\|\Big(\sum_{k\in\mathbf{Z}} |\Delta_k^{\psi}(f)\,\Delta_k^{\psi_{-s}}(D^s g)|^2\Big)^{\frac12}\Big\|_{L^r} \\
&\leq C_1(r,n,s)\Big\|\sup_{k\in\mathbf{Z}}\Delta_k^{\psi_{-s}}(D^s g)\Big\|_{L^\infty}\Big\|\Big(\sum_{k\in\mathbf{Z}} |\Delta_k^{\psi}(f)|^2\Big)^{\frac12}\Big\|_{L^r} \\
&\leq C_2(r,n,s)\|M(D^s g)\|_{L^\infty}\|f\|_{L^r} \\
&\leq C_3(r,n,s)\|D^s g\|_{L^\infty}\|f\|_{L^r},
\end{aligned}
$$

where M is the Hardy–Littlewood maximal operator. This proves the case where $q_2=\infty$, whereas the case where $q_1=\infty$ follows by symmetry. □

Exercises

7.6.1. Show that for all functions $\varphi \in \mathscr{S}(\mathbf{R}^n)$ we have

$$
\sum_{j=1}^{n} \|\partial_j\varphi\|_{L^\infty} \leq 2n\,\big[\|\varphi\|_{L^\infty}\big]^{\frac12}\Big[\sum_{j,k=1}^{n} \|\partial_j\partial_k\varphi\|_{L^\infty}\Big]^{\frac12}.
$$

[*Hint:* Start with the identity $\varphi(x+h)-\varphi(x) = \nabla\varphi(x)\cdot h + \frac12\sum_{j,k=1}^{n}\partial_j\partial_k\varphi(\xi)h_ih_j$, take absolute values and optimize in h.]

7.6.2. Given positive integers Q,N, with $1 \le Q \le N$, show that there is a constant $C = C(n,Q,N)$ such that for all $\varphi \in \mathscr{S}(\mathbf{R}^n)$ we have

$$\sum_{j_1=1}^{n} \cdots \sum_{j_Q=1}^{n} \|\partial_{j_1} \cdots \partial_{j_Q} \varphi\|_{L^\infty} \le C\Big[\|\varphi\|_{L^\infty}\Big]^{\frac{N-Q}{N}} \Bigg[\sum_{k_1=1}^{n} \cdots \sum_{k_N=1}^{n} \|\partial_{k_1} \cdots \partial_{k_N} \varphi\|_{L^\infty}\Bigg]^{\frac{Q}{N}}.$$

[*Hint:* Prove by induction on N that the preceding statement is valid for all $Q \le N$; use the previous exercise.]

7.6.3. (S. Oh) Let $\frac{1}{2} < r < \infty$, $1 < p_1, p_2, q_1, q_2 \le \infty$ satisfy $\frac{1}{r} = \frac{1}{p_1} + \frac{1}{q_1} = \frac{1}{p_2} + \frac{1}{q_2}$. Show that when $s < 0$, the inequality

$$\|D^s(fg)\|_{L^r(\mathbf{R}^n)} \le C\left[\|D^s f\|_{L^{p_1}(\mathbf{R}^n)}\|g\|_{L^{q_1}(\mathbf{R}^n)} + \|f\|_{L^{p_2}(\mathbf{R}^n)}\|D^s g\|_{L^{q_2}(\mathbf{R}^n)}\right],$$

fails even for $f, g \in \mathscr{S}(\mathbf{R}^n)$.
[*Hint:* Try $f(x) = e^{2\pi i 2^k e_1 \cdot x}\Phi(x)$ and $g(x) = e^{-2\pi i 2^k e_1 \cdot x}\Phi(x)$, where $\Phi \in \mathscr{S}(\mathbf{R}^n)$ has Fourier transform supported in the unit ball. Here $e_1 = (1,0,\dots,0)$.]

7.6.4. ([164]) Let $0 \le r < s < t$. Prove that there exist constants C_1, C_2 such that for all $f, g \in \mathscr{S}(\mathbf{R}^n)$ we have

$$\|D^s(fg)\|_{L^\infty} \le C_1\Big[\big(\|D^r f\|_{L^\infty} + \|D^t f\|_{L^\infty}\big)\|g\|_{L^\infty} + \big(\|D^r g\|_{L^\infty} + \|D^t g\|_{L^\infty}\big)\|f\|_{L^\infty}\Big],$$

$$\|D^s(fg)\|_{L^\infty} \le C_2\Big[\|D^r f\|_{L^\infty}^{\frac{t-s}{t-r}}\|D^t f\|_{L^\infty}^{\frac{s-r}{t-r}}\|g\|_{L^\infty} + \|D^r f\|_{L^\infty}^{\frac{t-s}{t-r}}\|D^t f\|_{L^\infty}^{\frac{s-r}{t-r}}\|f\|_{L^\infty}\Big].$$

[*Hint:* Obtain the second inequality by applying the first inequality to functions of the form $f_\lambda(x) = f(\lambda x)$ and $g_\lambda(x) = g(\lambda x)$ and optimizing over $\lambda > 0$. To obtain the first inequality, split the function $|\xi + \eta|^s$ as $\sum_{j\in\mathbf{Z}}\sum_{k<j}|\xi+\eta|^s\widehat{\psi}(2^{-j}\xi)\widehat{\psi}(2^{-k}\eta)$ plus $\sum_{j\in\mathbf{Z}}\sum_{k\ge j}|\xi+\eta|^s\widehat{\psi}(2^{-j}\xi)\widehat{\psi}(2^{-k}\eta)$, where ψ is as in the proof of Theorem 7.6.1 and show that the operator corresponding to the first symbol has L^∞ norm bounded by a multiple of $(\|D^r f\|_{L^\infty} + \|D^t f\|_{L^\infty})\|g\|_{L^\infty}$ by considering the terms $j \le 0$ and $j > 0$ in the sum.]

7.6.5. (a) Let Θ and Ω be Schwartz functions whose Fourier transforms are supported in an annulus that does not contain the origin. Let $s > 0$. Show that the function

$$\sigma(\xi,\eta) = \sum_{j\ge -2} 2^{-sj}\sum_{k\in Z}\widehat{\Theta}(2^{-k}(\xi+\eta))\widehat{\Omega}(2^{-(j+k)}\xi)\widehat{\Omega}(2^{-(j+k)}\eta)$$

satisfies $|\partial_\xi^\alpha\partial_\eta^\beta\sigma(\xi,\eta)| \le C_{\alpha,\beta}(|\xi|+|\eta|)^{-|\alpha|-|\beta|}$ for all multi-indices α,β satisfying $|\alpha|+|\beta| < s$.

(b) Let Ψ be a Schwartz function on $\mathbf{R}^{2n}$ whose Fourier transform is supported in an annulus that does not contain the origin and let $s>0$. Show that for $1<r<\infty$ and $0<\gamma<s$ we have

$$\sup_{l\in\mathbf{Z}}\left\|\sigma^l\,\widehat{\Psi}\right\|_{L^r_\gamma(\mathbf{R}^{2n})}<\infty,$$

where $\sigma^l(\xi,\eta)=\sigma(2^l\xi,2^l\eta)$.
[*Hint:* Part (b): Use the 3-lines lemma to show that

$$\sup_{l\in\mathbf{Z}}\left\|\sum_{k\in Z}\widehat{\Theta}(2^{l-k}(\xi+\eta))\widehat{\Omega}(2^{l-(j+k)}\xi)\widehat{\Omega}(2^{l-(j+k)}\eta)\widehat{\Psi}(\xi,\eta)\right\|_{L^r_\gamma(\mathbf{R}^{2n})}\leq C2^{j\gamma},$$

where the norm is taken in the variables (ξ,η).]

HISTORICAL NOTES

The systematic study of multilinear operators originated in the work of Coifman and Meyer [91], [92] in connection with the study of commutators of singular integrals, initiated by Calderón [54]. It was not until about a quarter century later, when Lacey and Thiele [236], [237] obtained the boundedness of the bilinear Hilbert transform $H_{\alpha,\beta}(f_1,f_2)(x)=\frac{1}{\pi}\text{p.v.}\int_{\mathbf{R}}f_1(x-\alpha t)f_2(x-\beta t)\,\frac{dt}{t}$, that this area attracted significant research attention. In their fundamental work Lacey and Thiele showed that $H_{\alpha,\beta}$ is bounded from $L^{p_1}(\mathbf{R})\times L^{p_2}(\mathbf{R})$ to $L^p(\mathbf{R})$ when $1<p_1,p_2\leq\infty$, $2/3<p<\infty$, and $1/p=1/p_1+1/p_2$, whenever $\alpha\neq\beta$. The family $H_{\alpha,\beta}$ arose in early attempts of A. Calderón to show that the first commutator (Example 4.3.8, $m=1$) is bounded on L^2 when A' is in L^∞, via the relationship $\mathscr{C}_1(f;A)=\int_0^1 H_{1,\alpha}(f,A')\,d\alpha$; this approach was completed only using the uniform (in α,β) boundedness of $H_{\alpha,\beta}$, obtained by Thiele [346], Grafakos and Li [160], and Li [244]. The less singular bilinear fractional integrals were studied by Kenig and Stein [219] and Grafakos and Kalton [157], who proved Theorem 7.1.4 independently. On multilinear fractional integrals, see also the work of Bak [15], Moen [269], Kuk and Lee [231], and of Kokilashvili, Mastyło, and Meskhi [221]. The material in Subsection 7.1.3 is based on Grafakos and Soria [171]. The version of Schur's lemma in Exercise 7.1.5 appeared in Bekollé, Bonami, Peloso, and Ricci [20] and in Grafakos and Torres [174]. The interplay between distributional estimates and the boundedness of multilinear operators was studied by Bilyk and Grafakos [36]; these ideas have been used to obtain distributional estimates for the bilinear Hilbert transform by the same authors [37].

Multilinear complex interpolation, even for analytic families of operators as in Theorem 7.2.9, is a straightforward adaptation of linear interpolation; see Zygmund [377, 21, Chapter XII, (3.3)] and Berg and Löfström [27, Theorem 4.4.2]. The multilinear real interpolation method is more involved. References on the subject include the articles of Strichartz [328], Sharpley [315] and [316], Zafran [375], Christ [76], Janson [204], Grafakos and Kalton [157]. The present exposition of Theorem 7.2.2 is based on Grafakos, Liu, Lu, and Zhao [162]. Theorem 7.2.9 is the main result in Grafakos and Tao [172]. Multilinear complex interpolation with respect to analytic families of operators has been developed by Grafakos and Mastyło [167].

Theorem 7.3.1 appears in Grafakos and Martell [165]. Basic properties of m-linear multipliers are listed in Grafakos and Torres [177]. The regularization process of Theorem 7.3.5 is due to Rodríguez-López [306]. This result makes it possible to extend the basic properties of $\mathscr{M}_{p_1,\dots,p_m}(\mathbf{R}^n)$ multipliers to situations where the target space is L^p, with $p=(1/p_1+\cdots+1/p_m)^{-1}<1$. The characteristic function of the two-dimensional unit ball lies in $\mathscr{M}_{p_1,p_2}(\mathbf{R})$ when

$1 < p_1, p_2 < 2$ and $2 < p = (1/p_1 + 1/p_2)^{-1} < \infty$; this was shown by Grafakos and Li [161]. The analogous result fails in dimensions $n \geq 2$ when p_1, p_2, or p' is greater than 2, as shown by Diestel and Grafakos [120]. The last reference contains a version of de Leeuw's theorem in the multilinear setting (Exercise 7.3.10).

The m-linear version of the Calderón–Zygmund theorem (Theorem 7.4.6) is due to Grafakos and Torres [177], although the bilinear case was independently obtained by Kenig and Stein [219]. Both sets of authors employed the Calderón–Zygmund decomposition, which first appeared in the trilinear setting as a lemma in the work of Coifman and Meyer [91]. Proposition 7.4.8 (the m-linear Peetre–Spanne–Stein theorem) and a version of Theorem 7.4.9 is also from [177]. The latter provides a $T(1)$ type theorem for m-linear operators associated with Calderón–Zygmund kernels. Other results of this type are due to Christ and Journé [84], Bényi, Demeter, Nahmod, Thiele, Torres, and Villaroya [23], Bényi [22], and Hart [186]. The action of multilinear Calderón–Zygmund singular integrals on Hardy spaces was studied by Grafakos and Kalton [159].

The case $m = 1$ of Theorem 7.5.5 is essentially contained in Hörmander's article [193]. The m-linear version was obtained by Tomita [351] when the target index is L^p for $p > 1$ and $r = 2$ (Corollary 7.5.9). The extension to indices $p \leq 1$ is due to Grafakos and Si [170], while the extension to the endpoint cases where L^∞ is allowed in the domain (but not in all spaces) is due to Grafakos, Miyachi, and Tomita [168]. Miyachi and Tomita [267] extended Theorem 7.5.5 to situations where Lebesgue spaces are replaced by Hardy spaces in the domain and obtained minimal smoothness conditions for the multipliers; see also the related work [268]. The formulation of Theorem 7.5.5 in the text was suggested by Tomita and is natural according to the viewpoint of a corresponding theorem in Kurtz and Wheeden [232]. Theorem 7.5.3 was first proved by Coifman and Meyer [92] via Fourier series techniques; see Coifman and Meyer [93] for extensions. Bernicot and Germain [30] obtained boundedness for bilinear multipliers whose symbols have narrow support.

Gilbert and Nahmod [154] obtained boundedness for bilinear multipliers on $\mathbf{R} \times \mathbf{R}$ whose derivatives of order α blow up like the distance to a line (with slope not taking three values) raised to the power $-|\alpha|$. Muscalu, Tao, and Thiele [280] provided an analogous m-linear version of this result. A maximal function related to the bilinear Hilbert transform was shown by Lacey [234] to be bounded on the same products of spaces as the bilinear Hilbert transform. More general singular multilinear maximal operators were studied by Demeter, Tao, and Thiele [118]. Deep counterexamples for trilinear operators were devised by Christ [82] and Demeter [117]. On the topic of multilinear Littlewood–Paley theory, one may consult the articles of Lacey [233], Diestel [119], Bernicot [29], Bernicot and Shrivastava [32], and Mohanty and Shrivastava [270], [271].

Paraproducts provide important examples of multilinear operators with specific properties. They first emerged in Bony's theory of paradifferential operators [41], which took the pseudodifferential operator theory of Coifman and Meyer [93] a step further. The boundedness of paraproducts on L^p spaces for $p > 1$ is easily achieved via duality, but the extension to indices $p \leq 1$ is more delicate and was proved independently by Grafakos and Kalton [158] and by Auscher, Hofmann, Muscalu, Thiele, and Tao [10]; subsequently this result was reproved by Lacey and Metcalfe [235], while a different proof was given by Bényi, Maldonado, Nahmod, and Torres [24]. The articles of Bernicot [28], Bilyk, Lacey, Li, Wick [38] Muscalu, Pipher, Thiele, and Tao [278], [279] study certain forms of paraproducts in depth. The expository article of Bényi, Maldonado, Naibo [25] makes a strong case for the use of paraproducts in analysis and partial differential equations.

A large body of literature on the topic of multilinear weighted norm inequalities appeared after the initial work of Grafakos and Torres [175]. A natural class of multiple weights that satisfies a vector A_p condition suitable for the multilinear Calderón–Zygmund theory was developed by Lerner, Ombrosi, Pérez, Torres and Trujillo-González [241]. Other weighted estimates were obtained by Bui and Duong [51], Hu [198], Li, Xue, and Yabuta [243]. Fujita and Tomita [147] and Li and Sun [242] obtained weighted estimates for multilinear Fourier multipliers.

The commutator estimate $\|\mathcal{J}_s(fg) - f\mathcal{J}_s(g)\|_{L^p} \leq C\|\nabla f\|_{L^\infty}\|\mathcal{J}_{s-1}(g)\|_{L^p} + C\|\mathcal{J}_s(f)\|_{L^p}\|g\|_{L^\infty}$, where $1 < p < \infty$ and $s > 0$, was proved by Kato and Ponce [210], where $\mathcal{J}_s = (1-\Delta)^{s/2}$ is the Bessel potential on $\mathbf{R}^n$. Kenig, Ponce, Vega [218] obtained the homogeneous commutator estimate $\|D^s(fg) - fD^s f - gD^s f\|_{L^r} \leq C\|D^{s_1} f\|_{L^p}\|D^{s_2} g\|_{L^q}$, with $D^s = (-\Delta)^{s/2}$ in place of $\mathcal{J}_s$, with $s = s_1 + s_2$ for $s, s_1, s_2 \in (0,1)$, and $1 < p, q, r < \infty$ satisfying $1/r = 1/p + 1/r$. The inequality in

Theorem 7.6.1 is referred to as the Kato–Ponce inequality. Proofs of this inequality when $r > 1$ were given by Christ and Weinstein [85], Gulisashvili and Kon [179], Muscalu and Schlag [276], and Bae and Biswas [13]. Bernicot, Maldonado, Moen, and Naibo [31] proved the Kato–Ponce inequality in weighted Lebesgue spaces under certain restrictions on the weights and they considered indices $r < 1$ under the assumption $s > n$. Muscalu and Schlag [277] and, independently, Grafakos and Oh [169] extended the Kato–Ponce inequality to indices $r < 1$ under the sharp restriction $s > n/r - n$. Bourgain and Li [46] have obtained the endpoint case $r = \infty$ in Theorem 7.6.1. Muscalu, Pipher, Tao, and Thiele, [278] extended the Kato–Ponce inequality to allow for partial fractional derivatives in $\mathbf{R}^2$. A more general multiparameter situation was considered in [169]. Cordero and Zucco [100] showed that the homogeneous Kato–Ponce inequality can be derived by its inhomogeneous counterpart, where $(I-\Delta)^{s/2}$ is in place of D^s.

The following expositions contain aspects of the theory of multilinear operators: Coifman and Meyer [93] focusing on multilinear pseudodifferential operators, Thiele [348] centering around wave packet analysis, a chapter in Meyer and Coifman [264] with emphasis on multilinear Calderón–Zygmund operators, Grafakos, Liu Maldonado and Yang [163] extending the theory to the setting of metric spaces, and Muscalu and Schlag [276], [277] encompassing a rather general study of singular multilinear theory.

The subject of Fourier analysis is currently enjoying a surge of activity. Emerging connections with number theory, combinatorics, geometric measure theory, and partial differential equations, have introduced new dynamics into the field and present promising developments. These connections are also creating new research directions that extend beyond the scope and level of this book.

Appendix A
The Schur Lemma

Schur's lemma provides sufficient conditions for linear operators to be bounded on L^p. Moreover, for positive operators it provides necessary and sufficient such conditions. We discuss these situations.

A.1 The Classical Schur Lemma

We begin with an easy situation. Suppose that $K(x,y)$ is a locally integrable function on a product of two σ-finite measure spaces $(X,\mu) \times (Y,\nu)$, and let T be a linear operator given by

$$T(f)(x) = \int_Y K(x,y) f(y)\, d\nu(y)$$

when f is bounded and compactly supported. It is a simple consequence of Fubini's theorem that for almost all $x \in X$ the integral defining T converges absolutely. The following lemma provides a sufficient criterion for the L^p boundedness of T.

Lemma. *Suppose that a locally integrable function $K(x,y)$ satisfies*

$$\sup_{x \in X} \int_Y |K(x,y)|\, d\nu(y) = A < \infty,$$

$$\sup_{y \in Y} \int_X |K(x,y)|\, d\mu(x) = B < \infty.$$

Then the operator T extends to a bounded operator from $L^p(Y)$ to $L^p(X)$ with norm $A^{1-\frac{1}{p}} B^{\frac{1}{p}}$ for $1 \le p \le \infty$.

Proof. The second condition gives that T maps L^1 to L^1 with bound B, while the first condition gives that T maps L^∞ to L^∞ with bound A. It follows by the Riesz–Thorin interpolation theorem that T maps L^p to L^p with bound $A^{1-\frac{1}{p}} B^{\frac{1}{p}}$. □

This lemma can be improved significantly when the operators are assumed to be positive.

A.2 Schur's Lemma for Positive Operators

We have the following necessary and sufficient condition for the L^p boundedness of positive operators.

L. Grafakos, *Modern Fourier Analysis*, Graduate Texts in Mathematics 250,
DOI 10.1007/978-1-4939-1230-8, © Springer Science+Business Media New York 2014

Lemma. *Let (X,μ) and (Y,ν) be two σ-finite measure spaces, where μ and ν are positive measures, and suppose that $K(x,y)$ is a nonnegative measurable function on $X\times Y$. Let $1<p<\infty$ and $0<A<\infty$. Let T be the linear operator*

$$T(f)(x)=\int_Y K(x,y)f(y)\,d\nu(y)$$

and T^t its transpose operator

$$T^t(g)(y)=\int_X K(x,y)g(x)\,d\mu(x).$$

To avoid trivialities, we assume that there is a compactly supported, bounded, and positive ν-a.e. function h_1 on Y such that $T(h_1)>0$ μ-a.e. Then the following are equivalent:

(i) T maps $L^p(Y)$ to $L^p(X)$ with norm at most A.
(ii) For all $B>A$ there is a measurable function h on Y that satisfies $0<h<\infty$ ν-a.e., $0<T(h)<\infty$ μ-a.e., and such that

$$T^t\big(T(h)^{\frac{p}{p'}}\big)\le B^p h^{\frac{p}{p'}}.$$

(iii) For all $B>A$ there are measurable functions u on X and v on Y such that $0<u<\infty$ μ-a.e., $0<v<\infty$ ν-a.e., and such that

$$\begin{aligned} T(u^{p'}) &\le B v^{p'},\\ T^t(v^p) &\le B u^p. \end{aligned}$$

Proof. First we assume (ii) and we prove (iii). Define u,v by the equations $v^{p'}=T(h)$ and $u^{p'}=Bh$ and observe that (iii) holds for this choice of u and v. Moreover, observe that $0<u,v<\infty$ a.e. with respect to the measures μ and ν, respectively.

Next we assume (iii) and we prove (i). For g in $L^{p'}(X)$ we have

$$\int_X T(f)(x)\,g(x)\,d\mu(x)=\int_X\int_Y K(x,y)f(y)g(x)\frac{v(x)}{u(y)}\frac{u(y)}{v(x)}\,d\nu(y)\,d\mu(x).$$

We now apply Hölder's inequality with exponents p and p' to the functions

$$f(y)\frac{v(x)}{u(y)}\qquad\text{and}\qquad g(x)\frac{u(y)}{v(x)}$$

with respect to the measure $K(x,y)\,d\nu(y)\,d\mu(x)$ on $X\times Y$. Since

$$\left(\int_Y\int_X f(y)^p\frac{v(x)^p}{u(y)^p}K(x,y)\,d\mu(x)\,d\nu(y)\right)^{\frac{1}{p}}\le B^{\frac{1}{p}}\|f\|_{L^p(Y)}$$

and

$$\left(\int_X\int_Y g(x)^{p'}\frac{u(y)^{p'}}{v(x)^{p'}}K(x,y)\,d\nu(y)\,d\mu(x)\right)^{\frac{1}{p'}} \le B^{\frac{1}{p'}}\|g\|_{L^{p'}(X)},$$

we conclude that

$$\left|\int_X T(f)(x)g(x)\,d\mu(x)\right| \le B^{\frac{1}{p}+\frac{1}{p'}}\|f\|_{L^p(Y)}\|g\|_{L^{p'}(X)}.$$

Taking the supremum over all g with $L^{p'}(X)$ norm 1, we obtain

$$\|T(f)\|_{L^p(X)} \le B\|f\|_{L^p(Y)}.$$

Since B was any number greater than A, we conclude that

$$\|T\|_{L^p(Y)\to L^p(X)} \le A,$$

which proves (i).

We finally assume (i) and we prove (ii). Without loss of generality, take here $A=1$ and $B>1$. Define a map $S:\ L^p(Y)\to L^p(Y)$ by setting

$$S(f)(y) = \left(T^t\left(T(f)^{\frac{p}{p'}}\right)\right)^{\frac{p'}{p}}(y).$$

We observe two things. First, $f_1 \le f_2$ implies $S(f_1)\le S(f_2)$, which is an easy consequence of the fact that the same monotonicity is valid for T. Next, we observe that $\|f\|_{L^p}\le 1$ implies that $\|S(f)\|_{L^p}\le 1$ as a consequence of the boundedness of T on L^p (with norm at most 1).

Construct a sequence h_n, $n=1,2,\dots$, by induction as follows. Pick $h_1>0$ on Y as in the hypothesis of the theorem such that $T(h_1)>0$ μ-a.e. and such that $\|h_1\|_{L^p}\le B^{-p'}(B^{p'}-1)$. (The last condition can be obtained by multiplying h_1 by a small constant.) Assuming that h_n has been defined, we define

$$h_{n+1} = h_1 + \frac{1}{B^{p'}}S(h_n).$$

We check easily by induction that we have the monotonicity property $h_n\le h_{n+1}$ and the fact that $\|h_n\|_{L^p}\le 1$. We now define

$$h(x) = \sup_n h_n(x) = \lim_{n\to\infty} h_n(x).$$

Fatou's lemma gives that $\|h\|_{L^p}\le 1$, from which it follows that $h<\infty$ ν-a.e. Since $h\ge h_1>0$ ν-a.e., we also obtain that $h>0$ ν-a.e.

Next we use the Lebesgue dominated convergence theorem to obtain that $h_n \to h$ in $L^p(Y)$. Since T is bounded on L^p, it follows that $T(h_n) \to T(h)$ in $L^p(X)$. It follows that $T(h_n)^{\frac{p}{p'}} \to T(h)^{\frac{p}{p'}}$ in $L^{p'}(X)$. Our hypothesis gives that T^t maps $L^{p'}(X)$ to $L^{p'}(Y)$ with norm at most 1. It follows $T^t\big(T(h_n)^{\frac{p}{p'}}\big) \to T^t\big(T(h)^{\frac{p}{p'}}\big)$ in $L^{p'}(Y)$. Raising to the power $\frac{p'}{p}$, we obtain that $S(h_n) \to S(h)$ in $L^p(Y)$.

It follows that for some subsequence n_k of the integers we have $S(h_{n_k}) \to S(h)$ a.e. in Y. Since the sequence $S(h_n)$ is increasing, we conclude that the entire sequence $S(h_n)$ converges almost everywhere to $S(h)$. We use this information in conjunction with $h_{n+1} = h_1 + \frac{1}{B^{p'}} S(h_n)$. Indeed, letting $n \to \infty$ in this identity, we obtain

$$h = h_1 + \frac{1}{B^{p'}} S(h)\,.$$

Since $h_1 > 0$ ν-a.e. it follows that $S(h) \le B^{p'} h$ ν-a.e., which proves the required estimate in (ii).

It remains to prove that $0 < T(h) < \infty$ μ-a.e. Since $\|h\|_{L^p} \le 1$ and T is L^p bounded, it follows that $\|T(h)\|_{L^p} \le 1$, which implies that $T(h) < \infty$ μ-a.e. We also have $T(h) \ge T(h_1) > 0$ μ-a.e. □

A.3 An Example

Consider the Hilbert operator

$$T(f)(x) = \int_0^\infty \frac{f(y)}{x+y}\,dy,$$

where $x \in (0,\infty)$. The operator T takes measurable functions on $(0,\infty)$ to measurable functions on $(0,\infty)$. We claim that T maps $L^p(0,\infty)$ to itself for $1 < p < \infty$; precisely, we have the estimate

$$\int_0^\infty T(f)(x)\,g(x)\,dx \le \frac{\pi}{\sin(\pi/p)} \|f\|_{L^p(0,\infty)} \|g\|_{L^{p'}(0,\infty)}\,.$$

To see this we use Schur's lemma. We take

$$u(x) = v(x) = x^{-\frac{1}{pp'}}\,.$$

We have that

$$T(u^{p'})(x) = \int_0^\infty \frac{y^{-\frac{1}{p}}}{x+y}\,dy = x^{-\frac{1}{p}} \int_0^\infty \frac{t^{-\frac{1}{p}}}{1+t}\,dt = v(x)^{p'} \int_0^1 (1-s)^{\frac{1}{p'}-1} s^{\frac{1}{p}-1}\,ds\,,$$

where last identity follows from the change of variables $s=(1+t)^{-1}$. Now an easy calculation yields

$$\int_0^1 (1-s)^{\frac{1}{p'}-1} s^{\frac{1}{p}-1}\,ds = B\big(\tfrac{1}{p'},\tfrac{1}{p}\big) = \frac{\pi}{\sin(\pi/p)},$$

so the lemma in Appendix I.2 gives that $\|T\|_{L^p\to L^p} \le \frac{\pi}{\sin(\pi/p)}$. The sharpness of this constant follows by considering the sequence of functions

$$h_\varepsilon(t) = t^{-\frac{1+\varepsilon}{p}} \chi_{(1,\infty)}(t)$$

for $\varepsilon>0$. To verify the last assertion notice that for $x>1$ and $0<\varepsilon<p-1$ we have

$$\begin{aligned} T(h_\varepsilon)(x) &= \int_0^\infty \frac{t^{-\frac{1+\varepsilon}{p}}}{x+t}\,dt - \int_0^1 \frac{t^{-\frac{1+\varepsilon}{p}}}{x+t}\,dt \\ &= x^{-\frac{1+\varepsilon}{p}} \int_0^\infty \frac{t^{-\frac{1+\varepsilon}{p}}}{1+t}\,dt - x^{-\frac{1+\varepsilon}{p}} \int_0^{\frac{1}{x}} \frac{t^{-\frac{1+\varepsilon}{p}}}{1+t}\,dt \\ &\ge x^{-\frac{1+\varepsilon}{p}} \int_0^\infty \frac{t^{-\frac{1+\varepsilon}{p}}}{1+t}\,dt - x^{-\frac{1+\varepsilon}{p}} \frac{x}{x+1} \int_0^{\frac{1}{x}} t^{-\frac{1+\varepsilon}{p}}\,dt \\ &= x^{-\frac{1+\varepsilon}{p}} \int_0^\infty \frac{t^{-\frac{1+\varepsilon}{p}}}{1+t}\,dt - \frac{1}{x+1}\,\frac{p}{p-1-\varepsilon}. \end{aligned}$$

Notice that the expression directly after the $\ge$ sign is nonnegative, and so is the last expression. It follows that

$$\big\|T(h_\varepsilon)\big\|_{L^p(1,\infty)} \ge \int_0^\infty \frac{t^{-\frac{1+\varepsilon}{p}}}{1+t}\,dt\,\|h_\varepsilon\|_{L^p(1,\infty)} - \frac{p}{p-1-\varepsilon}\,\Big\|\frac{1}{(\cdot)+1}\Big\|_{L^p(1,\infty)}.$$

Dividing both sides of this inequality by $\|h_\varepsilon\|_{L^p(0,\infty)} = \|h_\varepsilon\|_{L^p(1,\infty)} = \varepsilon^{-1/p}$, and letting $\varepsilon\to 0$ we obtain

$$\liminf_{\varepsilon\to 0} \frac{\|T(h_\varepsilon)\|_{L^p(0,\infty)}}{\|h_\varepsilon\|_{L^p(0,\infty)}} \ge \int_0^\infty \frac{t^{-\frac{1}{p}}}{1+t}\,dt = B\big(\tfrac{1}{p'},\tfrac{1}{p}\big) = \frac{\pi}{\sin(\pi/p)}.$$

Since

$$\limsup_{\varepsilon\to 0} \frac{\|T(h_\varepsilon)\|_{L^p(0,\infty)}}{\|h_\varepsilon\|_{L^p(0,\infty)}} \le \frac{\pi}{\sin(\pi/p)}$$

as already shown, it follows that

$$\lim_{\varepsilon\to 0} \frac{\|T(h_\varepsilon)\|_{L^p(0,\infty)}}{\|h_\varepsilon\|_{L^p(0,\infty)}} = \frac{\pi}{\sin(\pi/p)}.$$

A.4 Historical Remarks

We make some comments related to the history of Schur's lemma. Schur [312] first proved a matrix version of the lemma in Appendix I.1 when $p=2$. Precisely, Schur's original version was the following: If $K(x,y)$ is a positive decreasing function in both variables and satisfies

$$\sup_m \sum_n K(m,n) + \sup_n \sum_m K(m,n) < \infty,$$

then

$$\sum_m \sum_n a_m K(m,n) b_n \le C \|\{a_m\}_m\|_{\ell^2} \|\{b_n\}_n\|_{\ell^2} .$$

Hardy–Littlewood and Pólya [185] extended this result to L^p for $1 < p < \infty$ and disposed of the condition that K be a decreasing function. Aronszajn, Mulla, and Szeptycki [6] proved that (iii) implies (i) in the lemma of Appendix I.2. Gagliardo in [149] proved the converse direction that (i) implies (iii) in the same lemma. The case $p=2$ was previously obtained by Karlin [208]. Condition (ii) was introduced by Howard and Schep [197], who showed that it is equivalent to (i) and (iii). A multilinear analogue of the lemma in Appendix I.2 was obtained by Grafakos and Torres [174]; the easy direction (iii) implies (i) was independently observed by Bekollé, Bonami, Peloso, and Ricci [20]. See also Cwikel and Kerman [107] for an alternative multilinear formulation of the Schur lemma.

The case $p=p'=2$ of the application in Appendix I.3 is a continuous version of Hilbert's double series theorem. The discrete version was first proved by Hilbert in his lectures on integral equations (published by Weyl [367]) without a determination of the exact constant. This exact constant turns out to be π, as discovered by Schur [312]. The extension to other p's (with sharp constants) is due to Hardy and M. Riesz and published by Hardy [181].

Appendix B
Smoothness and Vanishing Moments

B.1 The Case of No Cancellation

Let $a,b \in \mathbf{R}^n$, $\mu,\nu \in \mathbf{R}$, and $M,N > n$. Set

$$I(a,\mu,M;b,\nu,N) = \int_{\mathbf{R}^n} \frac{2^{\mu n}}{(1+2^{\mu}|x-a|)^M} \frac{2^{\nu n}}{(1+2^{\nu}|x-b|)^N}\,dx.$$

Then we have

$$I(a,\mu,M;b,\nu,N) \le C_0 \frac{2^{\min(\mu,\nu)n}}{\big(1+2^{\min(\mu,\nu)}|a-b|\big)^{\min(M,N)}},$$

where

$$C_0 = v_n \left(\frac{M4^N}{M-n} + \frac{N4^M}{N-n}\right)$$

and v_n is the volume of the unit ball in $\mathbf{R}^n$.

To prove this estimate, first observe that

$$\int_{\mathbf{R}^n} \frac{dx}{(1+|x|)^M} \le \frac{v_n M}{M-n}.$$

Without loss of generality, assume that $\nu \le \mu$. Consider the cases $2^\nu|a-b| \le 1$ and $2^\nu|a-b| \ge 1$. In the case $2^\nu|a-b| \le 1$ we use the estimate

$$\frac{2^{\nu n}}{(1+2^\nu|x-b|)^N} \le 2^{\nu n} \le \frac{2^{\nu n}2^{\min(M,N)}}{(1+2^\nu|a-b|)^{\min(M,N)}},$$

and the claimed inequality is a consequence of the estimate

$$I(a,\mu,M;b,\nu,N) \le \frac{2^{\nu n}2^{\min(M,N)}}{(1+2^\nu|a-b|)^{\min(M,N)}} \int_{\mathbf{R}^n} \frac{2^{\mu n}}{(1+2^\mu|x-a|)^M}\,dx.$$

In the case $2^\nu|a-b| \ge 1$ let H_a and H_b be the two half-spaces, containing the points a and b, respectively, formed by the hyperplane perpendicular to the line segment $[a,b]$ at its midpoint. Split the integral over $\mathbf{R}^n$ as the integral over H_a and the integral over H_b. For $x \in H_a$ use that $|x-b| \ge \frac{1}{2}|a-b|$. For $x \in H_b$ use a similar inequality and the fact that $2^\nu|a-b| \ge 1$ to obtain

L. Grafakos, *Modern Fourier Analysis*, Graduate Texts in Mathematics 250,
DOI 10.1007/978-1-4939-1230-8, © Springer Science+Business Media New York 2014

$$\frac{2^{\mu n}}{(1+2^{\mu}|x-a|)^{M}} \leq \frac{2^{\mu n}}{(2^{\mu}\frac{1}{2}|a-b|)^{M}} \leq \frac{4^{M}2^{(\nu-\mu)(M-n)}2^{\nu n}}{(1+2^{\nu}|a-b|)^{M}}.$$

The claimed estimate follows.

B.2 One Function has Cancellation

Fix $a,b \in \mathbf{R}^n$, $M > 0$, $\mu,\nu \in \mathbf{R}$, and $L \in \mathbf{Z}^+$. Assume that $\nu \geq \mu$ and that $N > L+M+n$.

Given a function Ψ on $\mathbf{R}^n$ and another function $\Phi \in \mathscr{C}^L(\mathbf{R}^n)$ consider the quantities

$$K_{\mu,a}^{M,L}(\Phi) = \sup_{|\beta|=L} \sup_{x\in\mathbf{R}^n} (1+2^{\mu}|x-a|)^{M} |\partial^{\beta}\Phi(x)|,$$
$$K_{\nu,b}^{N}(\Psi) = \sup_{x\in\mathbf{R}^n} (1+2^{\nu}|x-b|)^{N} |\Psi(x)|$$

and assume they are both finite. Suppose, moreover, that

$$\int_{\mathbf{R}^n} \Psi(x) x^{\beta}\, dx = 0 \qquad \text{for all } |\beta| \leq L-1.$$

Then there is a constant $C_{M,N,L,n}$ such that

$$\left| \int_{\mathbf{R}^n} \Phi(x)\Psi(x)\, dx \right| \leq C_{M,N,L,n}\, K_{\mu,a}^{M,L}(\Phi)\, K_{\nu,b}^{N}(\Psi) \frac{2^{-\nu L-\nu n}}{(1+2^{\mu}|a-b|)^{M}}.$$

To prove this claim, we subtract the Taylor polynomial of order $L-1$ of Φ at the point a from the function Φ using the cancellation of Ψ. Then we write

$$\begin{aligned}
&\left| \int_{\mathbf{R}^n} \Phi(x)\Psi(x)\, dx \right| \\
&= \left| \int_{\mathbf{R}^n} \left[\Phi(x) - \sum_{|\gamma|\leq L-1} \frac{\partial^{\gamma}\Phi(b)}{\gamma!}(x-b)^{\gamma} \right] \Psi(x)\, dx \right| \\
&= \left| \int_{\mathbf{R}^n} \sum_{|\beta|=L} \frac{\partial^{\beta}\Phi(\xi_{b,x})}{\beta!}(x-b)^{\beta}\, \Psi(x)\, dx \right| \\
&\leq K_{\mu,a}^{M,L}(\Phi) K_{\nu,b}^{N}(\Psi) \sum_{|\beta|=L} \frac{1}{\beta!} \int_{\mathbf{R}^n} \frac{|x-b|^{L}}{(1+2^{\mu}|\xi_{b,x}-a|)^{M}} \frac{1}{(1+2^{\nu}|x-b|)^{N}}\, dx \\
&\leq K_{\mu,a}^{M,L}(\Phi) K_{\nu,b}^{N}(\Psi) \sum_{|\beta|=L} \frac{1}{\beta!} \int_{\mathbf{R}^n} \frac{2^{-\nu L}}{(1+2^{\mu}|\xi_{b,x}-a|)^{M}} \frac{1}{(1+2^{\nu}|x-b|)^{N-L}}\, dx
\end{aligned}$$

where $\xi_{b,x}$ lies on the open segment joining b to x.

Now since $\nu \geq \mu$, the triangle inequality and the fact $\mu \leq \nu$ give

$$\begin{aligned}1+2^{\mu}|a-b| &\leq 1+2^{\mu}|a-\xi_{b,x}|+2^{\mu}|\xi_{b,x}-b| \\ &\leq 1+2^{\mu}|a-\xi_{b,x}|+2^{\nu}|x-b| \\ &\leq (1+2^{\mu}|\xi_{b,x}-a|)(1+2^{\nu}|x-b|),\end{aligned}$$

hence

$$\frac{1}{1+2^{\mu}|\xi_{b,x}-a|} \leq \frac{1+2^{\nu}|x-b|}{1+2^{\mu}|a-b|}.$$

Thus we obtain the estimate

$$\begin{aligned}&\left|\int_{\mathbf{R}^n} \Phi(x)\Psi(x)\,dx\right| \\ &\leq K_{\mu,a}^{M,L}(\Phi)\,K_{\nu,b}^{N}(\Psi)\frac{2^{-\nu L}}{(1+2^{\mu}|a-b|)^M}\Big(\sum_{|\beta|=L}\frac{1}{\beta!}\Big)\int_{\mathbf{R}^n}\frac{1}{(1+2^{\nu}|x-b|)^{N-L-M}}\,dx \\ &= K_{\mu,a}^{M,L}(\Phi)\,K_{\nu,b}^{N}(\Psi)\frac{2^{-\nu n}2^{-\nu L}}{(1+2^{\mu}|a-b|)^M}C_{M,N,L,n},\end{aligned}$$

since the last integral produces a constant in view of the assumption $N > L+M+n$.

B.3 One Function has Cancellation: An Example

Fix $L \in \mathbf{Z}^+$, $A,B,M,N > 0$, and $a,b \in \mathbf{R}^n$ satisfy $N > M+L+n$ and $\nu \geq \mu$. Let $\Phi \in \mathscr{C}^L(\mathbf{R}^n)$ and Ψ be another function on $\mathbf{R}^n$. Let

$$A = \sup_{|\alpha|=L}\sup_{x\in\mathbf{R}^n} |\partial^{\alpha}\Phi(x)|(1+|x|)^M$$

and

$$B = \sup_{x\in\mathbf{R}^n} |\partial^{\alpha}\Psi(x)|(1+|x|)^N$$

and suppose that $A+B<\infty$. Suppose moreover that

$$\int_{\mathbf{R}^n}\Psi(x)\,x^{\beta}\,dx = 0 \qquad \text{for all } |\beta| \leq L-1.$$

Then there is a constant $C'_{M,N,L,n}$ such that

$$\left|\int_{\mathbf{R}^n}\Phi_{2^{-\mu}}(x-a)\Psi_{2^{-\nu}}(x-b)\,dx\right| \leq C'_{M,N,L,n}\,A\,B\,\frac{2^{\mu n}2^{-(\nu-\mu)L}}{(1+2^{\mu}|a-b|)^M}.$$

In particular, we have

$$\big|(\Phi_{2^{-\mu}} * \Psi_{2^{-\nu}})(x)\big| \le C'_{M,N,L,n} A B \frac{2^{\mu n} 2^{-(\nu-\mu)L}}{(1+2^{\mu}|x|)^M}$$

Let $\Phi_t(x) = t^{-n}\Phi(t^{-1}x)$ and $\Psi_s(x) = s^{-n}\Psi(s^{-1}x)$ for $t,s>0$. Set $2^{-\mu}=t$ and $2^{-\nu}=s$. The assumption $\nu \ge \mu$ can be equivalently stated as $s \ge t$.

The preceding inequalities can also be written equivalently as

$$\left|\int_{\mathbf{R}^n} \Phi_t(x-a)\Psi_s(x-b)\,dx\right| \le C'_{M,N,L,n} A B \frac{t^{-n}\left(\frac{s}{t}\right)^L}{(1+t^{-1}|a-b|)^M}.$$

and

$$\big|(\Phi_t * \Psi_s)(x)\big| \le C'_{M,N,L,n} A B \frac{t^{-n}\left(\frac{s}{t}\right)^L}{(1+2^{\mu}|x|)^M}$$

for all $x \in \mathbf{R}^n$.

These results are easy consequences of the inequality in Appendix B.2. If Ψ has no cancellation (i.e., $L=0$), then the estimate reduces to that in Appendix B.1.

B.4 Both Functions have Cancellation: An Example

Let $L \in \mathbf{Z}^+$, $A,B,N>0$ and $\mu,\nu \in \mathbf{R}$. Suppose that $N > L+n$. Let Ω,Ψ be $\mathscr{C}^L$ functions on $\mathbf{R}^n$ such that

$$A = \sup_{|\gamma|\le L} \sup_{x\in\mathbf{R}^n} |\partial^\gamma \Omega(x)|(1+|x|)^N < \infty$$

$$B = \sup_{|\gamma|\le L} \sup_{x\in\mathbf{R}^n} |\partial^\gamma \Psi(x)|(1+|x|)^N < \infty$$

and moreover, for all multi-indices β with $|\beta| \le L-1$ we have

$$\int_{\mathbf{R}^n} \Omega(x)\,x^\beta\,dx = \int_{\mathbf{R}^n} \Psi(x)\,x^\beta\,dx = 0.$$

Then given $M>0$ satisfying $M < N-L-n$ there is a constant $C''_{N,M,L,n}$ such that for all $x,a,b \in \mathbf{R}^n$ we have

$$\left|\int_{\mathbf{R}^n} \Omega_{2^{-\mu}}(x-a)\Psi_{2^{-\nu}}(x-b)\,dx\right| \le C''_{N,M,L,n} A B \frac{\min(2^{\mu n},2^{\nu n})2^{-|\nu-\mu|L}}{(1+\min(2^{\mu},2^{\nu})|a-b|)^M}.$$

In particular, we have

$$\left|(\Omega_{2^{-\mu}} * \Psi_{2^{-\nu}})(x)\right| \leq C''_{N,M,L,n}\, A\, B \frac{\min(2^{\mu n}, 2^{\nu n}) 2^{-|\nu-\mu|L}}{(1+\min(2^{\mu}, 2^{\nu})|x|)^M}$$

for all $x \in \mathbf{R}^n$ and for all $\mu, \nu \in \mathbf{R}$.

Let $\Omega_t(x) = t^{-n}\Omega(t^{-1}x)$ and $\Psi_s(x) = s^{-n}\Psi(s^{-1}x)$ for $t, s > 0$. Then if $2^{-\mu} = t$ and $2^{-\nu} = s$, the preceding statements can also be written as

$$\left| \int_{\mathbf{R}^n} \Omega_t(x-a)\Psi_s(x-b)\, dx \right| \leq C''_{N,M,L,n}\, A\, B \frac{\max(t,s)^{-n} \min\left(\frac{s}{t}, \frac{t}{s}\right)^L}{(1+\max(t,s)^{-1}|a-b|)^M}.$$

and

$$\left|(\Omega_t * \Psi_s)(x)\right| \leq C''_{N,M,L,n}\, A\, B \frac{\max(t,s)^{-n} \min\left(\frac{s}{t}, \frac{t}{s}\right)^L}{(1+\max(t,s)^{-1}|x|)^M}$$

for all $x \in \mathbf{R}^n$ and for all $t, s > 0$.

These assertions follow from the results in Appendix B.3 by interchanging the roles of Ω and Ψ, noting that

$$A \geq \sup_{|\gamma| \leq L} \sup_{x \in \mathbf{R}^n} |\partial^{\gamma}\Omega(x)|(1+|x|)^M$$

$$B \geq \sup_{|\gamma| \leq L} \sup_{x \in \mathbf{R}^n} |\partial^{\gamma}\Psi(x)|(1+|x|)^M$$

since $M < N$.

B.5 The Case of Three Factors with No Cancellation

Given three numbers a, b, c we denote by $\operatorname{med}(a,b,c)$ the number with the property $\min(a,b,c) \leq \operatorname{med}(a,b,c) \leq \max(a,b,c)$.

Let $x_\nu, x_\mu, x_\lambda \in \mathbf{R}^n$. Suppose that ψ_ν, ψ_μ, ψ_λ are functions defined on $\mathbf{R}^n$ such that for some $N > n$ and some $A_\nu, A_\mu, A_\lambda < \infty$ we have

$$|\psi_\nu(x)| \leq A_\nu \frac{2^{\nu n/2}}{(1+2^{\nu}|x-x_\nu|)^N},$$

$$|\psi_\mu(x)| \leq A_\mu \frac{2^{\mu n/2}}{(1+2^{\mu}|x-x_\mu|)^N},$$

$$|\psi_\lambda(x)| \leq A_\lambda \frac{2^{\lambda n/2}}{(1+2^{\lambda}|x-x_\lambda|)^N},$$

for all $x \in \mathbf{R}^n$. Then the following estimate is valid:

$$\int_{\mathbf{R}^n} |\psi_\nu(x)|\,|\psi_\mu(x)|\,|\psi_\lambda(x)|\,dx$$
$$\leq \frac{C_{N,n} A_\nu A_\mu A_\lambda\, 2^{-\max(\mu,\nu,\lambda)n/2}\, 2^{\mathrm{med}(\mu,\nu,\lambda)n/2}\, 2^{\min(\mu,\nu,\lambda)n/2}}{\big((1+2^{\min(\nu,\mu)}|x_\nu - x_\mu|)(1+2^{\min(\mu,\lambda)}|x_\mu - x_\lambda|)(1+2^{\min(\lambda,\nu)}|x_\lambda - x_\nu|)\big)^N}$$

for some constant $C_{N,n} > 0$ independent of the remaining parameters.

Analogous estimates hold if some of these three factors are assumed to have cancellation and the others vanishing moments; see Grafakos and Torres [176] for precise statements and applications. Similar estimates with m factors, $m \in \mathbf{Z}^+$, are studied in Bényi and Tzirakis [26].

Glossary

$A \subseteq B$	A is a subset of B (also denoted by $A \subseteq B$)
$A \subsetneqq B$	A is a proper subset of B
$A \supset B$	B is a proper subset of A
A^c	the complement of a set A
χ_E	the characteristic function of the set E
d_f	the distribution function of a function f
f^*	the decreasing rearrangement of a function f
$f_n \uparrow f$	f_n increases monotonically to a function f
$\mathbf{Z}$	the set of all integers
$\mathbf{Z}^+$	the set of all positive integers $\{1,2,3,\dots\}$
$\mathbf{Z}^n$	the n-fold product of the integers
$\mathbf{R}$	the set of real numbers
$\mathbf{R}^+$	the set of positive real numbers
$\mathbf{R}^n$	the Euclidean n-space
$\mathbf{Q}$	the set of rationals
$\mathbf{Q}^n$	the set of n-tuples with rational coordinates
$\mathbf{C}$	the set of complex numbers
$\mathbf{C}^n$	the n-fold product of complex numbers
$\mathbf{T}$	the unit circle identified with the interval $[0,1]$
$\mathbf{T}^n$	the n-dimensional torus $[0,1]^n$,
$\lvert x\rvert$	$\sqrt{\lvert x_1\rvert^2+\cdots+\lvert x_n\rvert^2}$ when $x=(x_1,\dots,x_n)\in\mathbf{R}^n$

L. Grafakos, *Modern Fourier Analysis*, Graduate Texts in Mathematics 250,
DOI 10.1007/978-1-4939-1230-8, © Springer Science+Business Media New York 2014

$\mathbf{S}^{n-1}$	the unit sphere $\{x \in \mathbf{R}^n : \lvert x \rvert = 1\}$
e_j	the vector $(0,\dots,0,1,0,\dots,0)$ with 1 in the jth entry and 0 elsewhere
$\log t$	the logarithm to base e of $t > 0$
$\log_a t$	the logarithm to base a of $t > 0$ $(1 \neq a > 0)$
$\log^+ t$	$\max(0, \log t)$ for $t > 0$
$[t]$	the integer part of the real number t
$x \cdot y$	the quantity $\sum_{j=1}^n x_j y_j$ when $x = (x_1,\dots,x_n)$ and $y = (y_1,\dots,y_n)$
$B(x,R)$	the ball of radius R centered at x in $\mathbf{R}^n$
ω_{n-1}	the surface area of the unit sphere $\mathbf{S}^{n-1}$
v_n	the volume of the unit ball $\{x \in \mathbf{R}^n : \lvert x \rvert < 1\}$
$\lvert A \rvert$	the Lebesgue measure of the set $A \subseteq \mathbf{R}^n$
dx	Lebesgue measure
$\mathrm{Avg}_B f$	the average $\frac{1}{\lvert B \rvert}\int_B f(x)\,dx$ of f over the set B
$\langle f, g \rangle$	the real inner product $\int_{\mathbf{R}^n} f(x)g(x)\,dx$
$\langle f \mid g \rangle$	the complex inner product $\int_{\mathbf{R}^n} f(x)\overline{g(x)}\,dx$
$\langle u, f \rangle$	the action of a distribution u on a function f
p'	the number $p/(p-1)$, whenever $0 < p \neq 1 < \infty$
$1'$	the number ∞
∞'	the number 1
$f = O(g)$	means $\lvert f(x) \rvert \leq M \lvert g(x) \rvert$ for some M for x near x_0
$f = o(g)$	means $\lvert f(x) \rvert \, \lvert g(x) \rvert^{-1} \to 0$ as $x \to x_0$
A^t	the transpose of the matrix A
A^*	the conjugate transpose of a complex matrix A
A^{-1}	the inverse of the matrix A
$O(n)$	the space of real matrices satisfying $A^{-1} = A^t$
$\lVert T \rVert_{X \to Y}$	the norm of the (bounded) operator $T : X \to Y$
$A \approx B$	means that there exists a $c > 0$ such that $c^{-1} \leq \frac{B}{A} \leq c$
$\lvert \alpha \rvert$	indicates the size $\lvert \alpha_1 \rvert + \cdots + \lvert \alpha_n \rvert$ of a multi-index $\alpha = (\alpha_1,\dots,\alpha_n)$
$\partial_j^m f$	the mth partial derivative of $f(x_1,\dots,x_n)$ with respect to x_j
$\partial^\alpha f$	$\partial_1^{\alpha_1} \cdots \partial_n^{\alpha_n} f$

$\mathscr{C}^k$	the space of functions f with $\partial^\alpha f$ continuous for all $\lvert\alpha\rvert \le k$
$\mathscr{C}_0$	space of continuous functions with compact support
$\mathscr{C}_{00}$	the space of continuous functions that vanish at infinity
$\mathscr{C}_0^\infty$	the space of smooth functions with compact support
$\mathscr{D}$	the space of smooth functions with compact support
$\mathscr{S}$	the space of Schwartz functions
$\mathscr{C}^\infty$	the space of smooth functions $\bigcap_{k=1}^\infty \mathscr{C}^k$
$\mathscr{D}'(\mathbf{R}^n)$	the space of distributions on $\mathbf{R}^n$
$\mathscr{S}'(\mathbf{R}^n)$	the space of tempered distributions on $\mathbf{R}^n$
$\mathscr{E}'(\mathbf{R}^n)$	the space of distributions with compact support on $\mathbf{R}^n$
$\mathscr{P}$	the set of all complex-valued polynomials of n real variables
$\mathscr{S}'(\mathbf{R}^n)/\mathscr{P}$	the space of tempered distributions on $\mathbf{R}^n$ modulo polynomials
$\ell(Q)$	the side length of a cube Q in $\mathbf{R}^n$
∂Q	the boundary of a cube Q in $\mathbf{R}^n$
$L^p(X,\mu)$	the Lebesgue space over the measure space (X,μ)
$L^p(\mathbf{R}^n)$	the space $L^p(\mathbf{R}^n, \lvert\cdot\rvert)$
$L^{p,q}(X,\mu)$	the Lorentz space over the measure space (X,μ)
$L^p_{\text{loc}}(\mathbf{R}^n)$	the space of functions that lie in $L^p(K)$ for any compact set K in $\mathbf{R}^n$
$\lvert d\mu\rvert$	the total variation of a finite Borel measure μ on $\mathbf{R}^n$
$\mathscr{M}(\mathbf{R}^n)$	the space of all finite Borel measures on $\mathbf{R}^n$
$\mathscr{M}_p(\mathbf{R}^n)$	the space of L^p Fourier multipliers, $1 \le p \le \infty$
$\mathscr{M}^{p,q}(\mathbf{R}^n)$	the space of translation-invariant operators that map $L^p(\mathbf{R}^n)$ to $L^q(\mathbf{R}^n)$
$\lVert\mu\rVert_{\mathscr{M}}$	$\int_{\mathbf{R}^n} \lvert d\mu\rvert$ the norm of a finite Borel measure μ on $\mathbf{R}^n$
$\mathcal{M}$	the centered Hardy–Littlewood maximal operator with respect to balls
M	the uncentered Hardy–Littlewood maximal operator with respect to balls
$\mathcal{M}_c$	the centered Hardy–Littlewood maximal operator with respect to cubes
M_c	the uncentered Hardy–Littlewood maximal operator with respect to cubes
$\mathcal{M}_\mu$	the centered maximal operator with respect to a measure μ
M_μ	the uncentered maximal operator with respect to a measure μ

M_s	the strong maximal operator
M_d	the dyadic maximal operator
$M^{\#}$	the sharp maximal operator
$\mathscr{M}$	the grand maximal operator
$L^p_s(\mathbf{R}^n)$	the inhomogeneous L^p Sobolev space
$\dot{L}^p_s(\mathbf{R}^n)$	the homogeneous L^p Sobolev space
$\Lambda_\alpha(\mathbf{R}^n)$	the inhomogeneous Lipschitz space
$\dot{\Lambda}_\alpha(\mathbf{R}^n)$	the homogeneous Lipschitz space
$H^p(\mathbf{R}^n)$	the real Hardy space on $\mathbf{R}^n$
$B^p_{s,q}(\mathbf{R}^n)$	the inhomogeneous Besov space on $\mathbf{R}^n$
$\dot{B}^p_{s,q}(\mathbf{R}^n)$	the homogeneous Besov space on $\mathbf{R}^n$
$\dot{B}^p_{s,q}(\mathbf{R}^n)$	the homogeneous Besov space on $\mathbf{R}^n$
$F^p_{s,q}(\mathbf{R}^n)$	the inhomogeneous Triebel–Lizorkin space on $\mathbf{R}^n$
$\dot{F}^p_{s,q}(\mathbf{R}^n)$	the homogeneous Triebel–Lizorkin space on $\mathbf{R}^n$
$BMO(\mathbf{R}^n)$	the space of functions of bounded mean oscillation on $\mathbf{R}^n$

References

1. Adams, D. R., *A note on Riesz potentials*, Duke Math. J. **42** (1975), no. 4, 765–778.
2. Adams, R. A., *Sobolev Spaces*, Pure and Applied Mathematics, Vol. 65, Academic Press, New York–London, 1975.
3. Alexopoulos, G., *La conjecture de Kato pour les opérateurs différentiels elliptiques à coefficients périodiques*, C. R. Acad. Sci. Paris Sér. I Math. **312** (1991), no. 2, 263–266.
4. Antonov, N. Yu., *Convergence of Fourier series*, Proceedings of the XX Workshop on Function Theory (Moscow, 1995), East J. Approx. **2** (1996) no. 2, 187–196.
5. Arias de Reyna, J., *Pointwise Convergence of Fourier Series*, Lecture Notes in Mathematics, 1785, Springer-Verlag, Berlin, 2002.
6. Aronszajn, N., Mulla, F., Szeptycki, P., *On spaces of potentials connected with L^p-classes*, Ann. Inst. Fourier (Grenoble) **12** (1963), 211–306.
7. Aronszajn, N., Smith, K. T., *Theory of Bessel potentials, I*, Ann. Inst. Fourier (Grenoble) **11** (1961), 385–475.
8. Auscher, P., Hofmann, S., Lacey, M., McIntosh, A., Tchamitchian, P., *The solution of the Kato square root problem for second order elliptic operators on $\mathbb{R}^n$*, Ann. of Math. (2nd Ser.) **156** (2002), no. 2, 633–654.
9. Auscher, P., Hofmann, S., Lewis, J. L., Tchamitchian, P., *Extrapolation of Carleson measures and the analyticity of Kato's square-root operators*, Acta Math. **187** (2001), no. 2, 161–190.
10. Auscher, P., Hofmann, S., Muscalu, C., Tao, T., Thiele, C., *Carleson measures, trees, extrapolation and Tb theorems*, Publ. Mat. **46** (2002), no. 2, 257–325.
11. Auscher, P., McIntosh, A., Nahmod, A., *Holomorphic functional calculi of operators, quadratic estimates, and interpolation*, Indiana Univ. Math. J. **46** (1997), no. 2, 375–403.
12. Auscher, P., Tchamitchian, P., *Square root problem for divergence operators and related topics*, Astérisque No. 249, Société Mathématique de France, 1998.
13. Bae, H., Biswas, A., *Gevrey regularity for a class of dissipative equations with analytic nonlinearity*, to appear.
14. Baernstein, A., II, Sawyer, E. T., *Embedding and multiplier theorems for $H^p(\mathbf{R}^n)$*, Mem. Amer. Math. Soc. 53 (1985), no. 318.
15. Bak, J.-G., *An interpolation theorem and a sharp form of a multilinear fractional integration theorem*, Proc. Amer. Math. Soc. **120** (1994), no. 2, 435–441.
16. Barceló, B., *On the restriction of the Fourier transform to a conical surface*, Trans. Amer. Math. Soc. **292** (1985), no. 1, 321–333.
17. Barceló, B., *The restriction of the Fourier transform to some curves and surfaces*, Studia Math. **84** (1986), no. 1, 39–69.
18. Barrionuevo, J., *A note on the Kakeya maximal operator*, Math. Res. Lett. **3** (1996), no. 1, 61–65.
19. Beckner, W., Carbery, A., Semmes, S., Soria, F., *A note on restriction of the Fourier transform to spheres*, Bull. London Math. Soc. **21** (1989), no. 4, 394–398.

L. Grafakos, *Modern Fourier Analysis*, Graduate Texts in Mathematics 250,
DOI 10.1007/978-1-4939-1230-8,

20. Bekollé, D., Bonami, A., Peloso, M., Ricci, F., *Boundedness of Bergman projections on tube domains over light cones*, Math. Z. **237** (2001), no. 1, 31–59.
21. Bennett, C., DeVore, R. A., Sharpley, R., *Weak L^∞ and BMO*, Ann. of Math. (2nd Ser.) **113** (1981), no. 3, 601–611.
22. Bényi, Á., *Bilinear singular integral operators, smooth atoms and molecules*, J. Fourier Anal. Appl. **9** (2003), no. 3, 301–319.
23. Bényi, Á., Demeter, C., Nahmod, A., Thiele, C., Torres, R. H., Villaroya, P., *Modulation invariant bilinear $T(1)$ theorem*, J. Anal. Math. **109** (2009), 279–352.
24. Bényi, Á., Maldonado, D., Nahmod, A., Torres, R. H., *Bilinear paraproducts revisited*, Math. Nachr. **283** (2010), no. 9, 1257–1276.
25. Bényi, Á., Maldonado, D., Naibo, V., *What is a paraproduct?* Notices Amer. Math. Soc. **57** (2010), no. 7, 858–860.
26. Bényi, Á., Tzirakis, N., *Multilinear almost diagonal estimates and applications*, Studia Math. **164** (2004), no. 1, 75–89.
27. Bergh, J., Löfström, J., *Interpolation Spaces, An Introduction*, Grundlehren der Mathematischen Wissenschaften, 223, Springer-Verlag, Berlin–New York, 1976.
28. Bernicot, F., *Uniform estimates for paraproducts and related multilinear operators*, Revista Mat. Iberoamer. **25** (2009), no. 3, 1055–1088.
29. Bernicot, F., *L^p estimates for non-smooth bilinear Littlewood-Paley square functions on $\mathbb{R}$*, Math. Ann. **351** (2011), no. 1, 1–49.
30. Bernicot, F., Germain, P., *Boundedness of bilinear multipliers whose symbols have a narrow support*, J. Anal. Math. **119** (2013), 165–212.
31. Bernicot, F., Maldonado, D., Moen, K., Naibo, V., *Bilinear Sobolev-Poincaré inequalities and Leibniz-type rules,* J. Geom. Anal. **24** (2014), no. 2, 1144–1180.
32. Bernicot, F., Shrivastava, S., *Boundedness of smooth bilinear square functions and applications to some bilinear pseudo-differential operators*, Indiana Univ. Math. J. **60** (2011), no. 1, 233–268.
33. Besicovitch, A., *On Kakeya's problem and a similar one*, Math. Z. **27** (1928), no. 1, 312–320.
34. Besov, O. V., *On some families of function spaces. Imbedding and extension theorems* (Russian), Dokl. Akad. Nauk SSSR **126** (1959), 1163–1165.
35. Besov, O. V., *Investigation of a class of function spaces in connection with imbeddings and extension theorems* (Russian), Trudy Mat. Inst. Steklov. **60** (1961), 42–81.
36. Bilyk, D., Grafakos, L., *Interplay between distributional estimates and boundedness in harmonic analysis*, Bull. London Math. Soc. **37** (2005), no. 3, 427–434.
37. Bilyk, D., Grafakos, L., *Distributional estimates for the bilinear Hilbert transforms*, J. Geom. Anal. **16** (2006), no. 4, 563–584.
38. Bilyk, D., Lacey, M., Li, X., Wick, B., *Composition of Haar paraproducts: the random case*, Anal. Math. **35** (2009), 1–13.
39. Birnbaum, Z. W., Orlicz, W., *Über die Verallgemeinerung des Begriffes der Zueinander konjugierten Potenzen*, Studia Math. **3** (1931), 1–67; reprinted in W. Orlicz, "Collected Papers," pp. 133–199, PWN, Warsaw, 1988.
40. Bochner, S., *Summation of multiple Fourier series by spherical means*, Trans. Amer. Math. Soc. **40** (1936), no. 2, 175–207.
41. Bony, J. M., *Calcul symbolique et propagation des singularités pour les équations aux dérivées partielles non linéaires*, Ann. Sci. École Norm. Sup. (4) **14** (1981), no. 2, 209–246.
42. Bourgain, J., *Besicovitch type maximal operators and applications to Fourier analysis*, Geom. Funct. Anal. **1** (1991), no. 2, 147–187.
43. Bourgain, J., *On the restriction and multiplier problems in $\mathbb{R}^3$*, Geometric Aspects of Functional Analysis (1989–90), pp. 179–191, Lecture Notes in Math. 1469, Springer, Berlin, 1991.
44. Bourgain, J., *Some new estimates on oscillatory integrals*, Essays on Fourier Analysis in Honor of Elias M. Stein (Princeton, NJ, 1991) pp. 83-112, Princeton Math. Ser. 42, Princeton Univ. Press, Princeton, NJ, 1995.
45. Bourgain, J., *On the dimension of Kakeya sets and related maximal inequalities*, Geom. Funct. Anal. **9** (1999), no. 2, 256–282.

46. Bourgain, J., Li D., *On an endpoint Kato-Ponce inequality*, to appear.
47. Bownik, M., *Anisotropic Hardy spaces and wavelets*, Mem. Amer. Math. Soc. 164 (2003), no. 781.
48. Bownik, M., *Boundedness of operators on Hardy spaces via atomic decompositions*, Proc. Amer. Math. Soc. **133** (2005), no. 12, 3535–3542.
49. Bownik, M., Li, B., Yang, D., Zhou, Y., *Weighted anisotropic Hardy spaces and their applications in boundedness of sublinear operators*, Indiana Univ. Math. J. **57** (2008), no. 7, 3065–3100.
50. Bui, H. Q., *Some aspects of weighted and non-weighted Hardy spaces*, Kôkyûroku Res. Inst. Math. Sci. **383** (1980), 38–56.
51. Bui, T. A., Duong, X. T., *Weighted norm inequalities for multilinear operators and applications to multilinear Fourier multipliers*, Bull. Sci. Math. **137** (2013), no. 1, 63–75.
52. Burkholder, D. L., Gundy, R. F., Silverstein, M. L., *A maximal characterization of the class* H^p, Trans. Amer. Math. Soc. **157** (1971), 137–153.
53. Calderón, A. P., *Lebesgue spaces of differentiable functions and distributions*, Proc. Sympos. Pure Math. **4** (1961), 33–49, Amer. Math. Soc., Providence, RI.
54. Calderón, A. P., *Commutators of singular integral operators*, Proc. Nat. Acad. Sci. U.S.A. **53** (1965), 1092–1099.
55. Calderón, A. P., *Cauchy integrals on Lipschitz curves and related operators*, Proc. Nat. Acad. Sci. U.S.A. **74** (1977), no. 4, 1324–1327.
56. Calderón, A. P., *An atomic decomposition of distributions in parabolic* H^p *spaces*, Advances in Math. **25** (1977), no. 3, 216–225.
57. Calderón, A. P., Zygmund, A., *A note on the interpolation of linear operations*, Studia Math. **12** (1951), 194–204.
58. Calderón, A. P., Torchinsky, A., *Parabolic maximal functions associated with a distribution*, Advances in Math. **16** (1975), 1–64.
59. Calderón, A. P., Torchinsky, A., *Parabolic maximal functions associated with a distribution, II*, Advances in Math. **24** (1977), no. 2, 101–171.
60. Calderón, A. P., Vaillancourt, R., *A class of bounded pseudo-differential operators*, Proc. Nat. Acad. Sci. U.S.A. **69** (1972), 1185–1187.
61. Calderón, A. P., Zygmund, A., *Singular integrals and periodic functions*, Studia Math. **14** (1954), 249–271.
62. Campanato, S., *Proprietà di hölderianità di alcune classi di funzioni*, Ann. Scuola Norm. Sup. Pisa (3) **17** (1963), 175–188.
63. Campanato, S., *Proprietà di una famiglia di spazi funzionali*, Ann. Scuola Norm. Sup. Pisa (3) **18** (1964), 137–160.
64. Carbery, A., *The boundedness of the maximal Bochner–Riesz operator on* $L^4(\mathbb{R}^2)$, Duke Math. J. **50** (1983), no. 2, 409–416.
65. Carbery, A., Hernández, E., Soria, F., *Estimates for the Kakeya maximal operator on radial functions in* $\mathbf{R}^n$, Harmonic Analysis (Sendai, 1990), pp. 41–50, Springer, Tokyo, 1991.
66. Carbery, A., Rubio de Francia, J.-L., Vega, L., *Almost everywhere summability of Fourier integrals*, J. London Math. Soc. (2) **38** (1988), no. 3, 513–524.
67. Carleson, L., *An interpolation problem for bounded analytic functions*, Amer. J. Math. **80** (1958), 921–930.
68. Carleson, L., *Interpolation by bounded analytic functions and the corona problem*, Ann. of Math. (2nd Ser.) **76** (1962), no. 3, 547–559.
69. Carleson, L., *On convergence and growth of partial sums of Fourier series*, Acta Math. **116** (1966), no. 1, 135–157.
70. Carleson, L., *On the Littlewood–Paley Theorem*, Mittag-Leffler Institute Report, Djurs- holm, Sweden 1967.
71. Carleson, L., *Two remarks on* H^1 *and B.M.O*, Advances in Math. **22** (1976), no. 3, 269–277.
72. Carleson, L., Sjölin, P., *Oscillatory integrals and a multiplier problem for the disc*, Studia Math. **44** (1972), 287–299.
73. Chang, D.-C., Krantz, S. G., Stein, E. M., H^p *theory on a smooth domain in* R^N *and elliptic boundary value problems*, J. Funct. Anal. **114** (1993), no. 2, 286–347.

74. Chiarenza, F., Frasca, M., *Morrey spaces and Hardy–Littlewood maximal function*, Rend. Mat. Appl. Series 7, **7** (1987), no. 3–4, 273–279.
75. Christ, M., *Estimates for the k-plane transform*, Indiana Univ. Math. J. **33** (1984), no. 6, 891–910.
76. Christ, M., *On the restriction of the Fourier transform to curves: endpoint results and the degenerate case*, Trans. Amer. Math. Soc. **287** (1985), no. 1, 223–238.
77. Christ, M., *On almost everywhere convergence of Bochner–Riesz means in higher dimensions*, Proc. Amer. Math. Soc. **95** (1985), no. 1, 16–20.
78. Christ, M., *Weak type endpoint bounds for Bochner–Riesz multipliers*, Rev. Mat. Iberoamericana **3** (1987), no. 1, 25–31.
79. Christ, M., *Weak type* $(1,1)$ *bounds for rough operators*, Ann. of Math. (2nd Ser.) **128** (1988), no. 1, 19–42.
80. Christ, M., *A* $T(b)$ *theorem with remarks on analytic capacity and the Cauchy integral*, Colloq. Math. **60/61** (1990), no. 2, 601–628.
81. Christ, M., *Lectures on singular integral operators*, CBMS Regional Conference Series in Mathematics, 77, American Mathematical Society, Providence, RI, 1990.
82. Christ, M., *On certain elementary trilinear operators*, Math. Res. Lett. **8** (2001), no. 1–2, 43–56.
83. Christ, M., Duoandikoetxea, J., Rubio de Francia, J.-L., *Maximal operators related to the Radon transform and the Calderón–Zygmund method of rotations*, Duke Math. J. **53** (1986), no. 1, 189–209.
84. Christ, M., Journé, J.-L., *Polynomial growth estimates for multilinear singular integral operators*, Acta Math. **159** (1987), no. 1–2, 51–80.
85. Christ, M., Weinstein, M., *Dispersion of small-amplitude solutions of the generalized Korteweg-de Vries equation*, J. Funct. Anal. (1991), **100**, no. 1, 87–109.
86. Coifman, R. R., *A real variable characterization of* H^p, Studia Math. **51** (1974), 269–274.
87. Coifman, R. R., Deng, D. G., Meyer, Y., *Domaine de la racine carée de certaines opérateurs différentiels acrétifs*, Ann. Inst. Fourier (Grenoble) **33** (1983), no. 2, 123–134.
88. Coifman, R. R., Jones, P., Semmes, S., *Two elementary proofs of the* L^2 *boundedness of Cauchy integrals on Lipschitz curves*, J. Amer. Math. Soc. **2** (1989), no. 3, 553–564.
89. Coifman, R. R., Lions, P. L., Meyer, Y., Semmes, S., *Compensated compactness and Hardy spaces*, J. Math. Pures Appl. (9) **72** (1993), no. 3, 247–286.
90. Coifman, R. R., McIntosh, A., Meyer, Y., *L' intégrale de Cauchy définit un opérateur borné sur* L^2 *pour les courbes lipschitziennes*, Ann. of Math. (2nd Ser.) **116** (1982), no. 2, 361–387.
91. Coifman, R. R., Meyer, Y., *On commutators of singular integral and bilinear singular integrals*, Trans. Amer. Math. Soc. **212** (1975), 315–331.
92. Coifman, R. R., Meyer, Y., *Commutateurs d' intégrales singulières et opérateurs multilinéaires*, Ann. Inst. Fourier (Grenoble) **28** (1978), no. 3, 177–202.
93. Coifman, R. R., Meyer, Y., *Au délà des opérateurs pseudo-différentiels*, Astérisque No. 57, Societé Mathematique de France, 1979.
94. Coifman, R. R., Meyer, Y., *A simple proof of a theorem by G. David and J.-L. Journé on singular integral operators*, Probability Theory and Harmonic Analysis, (Cleveland, Ohio, 1983) pp. 61–65, Monogr. Textbooks Pure Appl. Math., 98 Dekker, New York, 1986.
95. Coifman, R. R., Meyer, Y., Stein, E. M., *Some new function spaces and their applications to harmonic analysis*, J. Funct. Anal. **62** (1985), no. 2, 304–335.
96. Coifman, R. R., Rochberg, R., Weiss, G., *Factorization theorems for Hardy spaces in several variables*, Ann. of Math. (2nd Ser.) **103** (1976), no. 3, 611–635.
97. Coifman, R. R., Weiss, G., *Extensions of Hardy spaces and their use in analysis*, Bull. Amer. Math. Soc. **83** (1977), no. 4, 569–645.
98. Colzani, L., *Translation invariant operators on Lorentz spaces*, Ann. Scuola Norm. Sup. Pisa Cl. Sci. (4) **14** (1987), no. 2, 257–276.
99. Colzani, L., Travaglini, G., Vignati, M., *Bochner–Riesz means of functions in weak-*L^p, Monatsh. Math. **115** (1993), no. 1–2, 35–45.
100. Cordero, E., Zucco, D., *Strichartz estimates for the vibrating plate equation*, J. Evol. Equ. **11** (2011), 827–845.

101. Córdoba, A., *The Kakeya maximal function and the spherical summation multipliers*, Amer. J. Math. **99** (1977), no. 1, 1–22.
102. Córdoba, A., *A note on Bochner–Rise operators*, Duke Math. J. **46** (1979), no. 3, 505–511.
103. Córdoba, A., *Multipliers of $F(L^p)$*, Euclidean harmonic analysis (Proc. Sem., Univ. Maryland, College Park, MD, 1979), pp. 162–177, Lecture Notes in Math., 779, Springer, Berlin, 1980.
104. Córdoba, A., Fefferman, R., *On differentiation of integrals*, Proc. Nat. Acad. Sci. U.S.A. **74** (1977), no. 6, 2211–2213.
105. Cotlar, M., *A combinatorial inequality and its applications to L^2 spaces*, Rev. Mat. Cuyana, **1** (1955), 41–55.
106. Cunningham, F., *The Kakeya problem for simply connected and for star-shaped sets*, Amer. Math. Monthly **78** (1971), 114–129.
107. Cwikel, M., Kerman, R., *Positive multilinear operators acting on weighted L^p spaces*, J. Funct. Anal. **106** (1992), no. 1, 130–144.
108. Dafni, G., *Hardy spaces on some pseudoconvex domains*, J. Geom. Anal. **4** (1994), no. 3, 273–316.
109. Dafni, G., *Local VMO and weak convergence in h^1*, Canad. Math. Bull. **45** (2002), no. 1, 46–59.
110. David, G., *Opérateurs intégraux singuliers sur certains courbes du plan complexe*, Ann. Sci. École Norm. Sup. (4) **17** (1984), no. 1, 157–189.
111. David, G., Journé, J.-L., *A boundedness criterion for generalized Calderón–Zygmund operators*, Ann. of Math. (2nd Ser.) **120** (1984), no. 2, 371–397.
112. David, G., Journé, J.-L., Semmes, S., *Opérateurs de Calderón–Zygmund, fonctions para-accrétives et interpolation*, Rev. Math. Iberoamericana **1** (1985), no. 4, 1–56.
113. David, G., Semmes, S., *Singular integrals and rectifiable sets in $\mathbf{R}^n$: Beyond Lipschitz graphs*, Astérisque No. 193, Societé Mathematique de France, 1991.
114. Davis, K. M., Chang, Y. C., *Lectures on Bochner–Riesz Means*, London Mathematical Society Lecture Notes Series, 114, Cambridge University Press, Cambridge, 1987.
115. De Carli, L., Iosevich, A., *A restriction theorem for flat manifolds of codimension two*, Illinois J. Math. **39** (1995), no. 4, 576–585.
116. De Carli, L., Iosevich, A., *Some sharp restriction theorems for homogeneous manifolds*, J. Fourier Anal. Appl. **4** (1998), no. 1, 105–128.
117. Demeter, C., *Divergence of combinatorial averages and the unboundedness of the trilinear Hilbert transform*, Ergodic Theory Dynam. Systems **28** (2008), no. 5, 1453–1464.
118. Demeter, C., Tao, T., Thiele, C., *Maximal multilinear operators*, Trans. Amer. Math. Soc. **360** (2008), no. 9, 4989–5042.
119. Diestel, G., *Some remarks on bilinear Littlewood-Paley theory*, J. Math. Anal. Appl. **307** (2005), no. 1, 102–119.
120. Diestel, G., Grafakos, L., *Unboundedness of the ball bilinear multiplier operator*, Nagoya Math. J. **185** (2007), 151–159.
121. Drury, S. W., *L^p estimates for the X-ray transform*, Illinois J. Math. **27** (1983), no. 1, 125–129.
122. Drury, S. W., *Restrictions of Fourier transforms to curves*, Ann. Inst. Fourier (Grenoble) **35** (1985), no. 1, 117–123.
123. Drury, S. W., Guo, K., *Some remarks on the restriction of the Fourier transform to surfaces*, Math. Proc. Cambridge Philos. Soc. **113** (1993), no. 1, 153–159.
124. Drury, S. W., Marshall, B. P., *Fourier restriction theorems for curves with affine and Euclidean arclengths*, Math. Proc. Cambridge Philos. Soc. **97** (1985), no. 1, 111–125.
125. Drury, S. W., Marshall, B. P., *Fourier restriction theorems for degenerate curves*, Math. Proc. Cambridge Philos. Soc. **101** (1987), no. 3, 541–553.
126. Duong, X. T., Grafakos, L., Yan, L., *Multilinear operators with non-smooth kernels and commutators of singular integrals*, Trans. Amer. Math. Soc. **362** (2010), no. 4, 2089–2113.
127. Duren, P. L., *Theory of H^p Spaces*, Dover Publications Inc., New York, 2000.
128. Duren, P. L., Romberg, B. W., Shields, A. L., *Linear functionals on H^p spaces with $0 < p < 1$*, J. Reine Angew. Math. **238** (1969), 32–60.

129. Fabes, E., Jerison, D., Kenig, C., *Multilinear Littlewood–Paley estimates with applications to partial differential equations*, Proc. Nat. Acad. Sci. U.S.A. **79** (1982), no. 18, 5746–5750.
130. Fabes, E., Jerison, D., Kenig, C., *Multilinear square functions and partial differential equations*, Amer. J. Math. **107** (1985), no. 6, 1325–1367.
131. Fabes, E., Mitrea, I., Mitrea, M., *On the boundedness of singular integrals*, Pacific J. Math. **189** (1999), no. 1, 21–29.
132. Fefferman, C., *Inequalities for strongly singular convolution operators*, Acta Math. **124** (1970), no. 1, 9–36.
133. Fefferman, C., *Characterizations of bounded mean oscillation*, Bull. Amer. Math. Soc. **77** (1971), 587–588.
134. Fefferman, C., *The multiplier problem for the ball*, Ann. of Math. (2nd Ser.) **94** (1971), no. 2, 330–336.
135. Fefferman, C., *Pointwise convergence of Fourier series*, Ann. of Math. (2nd Ser.) **98** (1973), no. 3, 551–571.
136. Fefferman, C., *A note on spherical summation multipliers*, Israel J. Math. **15** (1973), 44–52.
137. Fefferman, C., Riviere, N., Sagher, Y., *Interpolation between H^p spaces: The real method*, Trans. Amer. Math. Soc. **191** (1974), 75–81.
138. Fefferman, C., Stein, E. M., *Some maximal inequalities*, Amer. J. Math. **93** (1971), 107–115.
139. Fefferman, C., Stein, E. M., *H^p spaces of several variables*, Acta Math. **129** (1972), no. 3–4, 137–193.
140. Flett, T. M., *Lipschitz spaces of functions on the circle and the disc*, J. Math. Anal. Appl. **39** (1972), 125–158.
141. Folland, G. B., Stein, E. M., *Estimates for the $\overline{\partial}_b$ complex and analysis on the Heisenberg group*, Comm. Pure Appl. Math. **27** (1974), 429–522.
142. Folland, G. B., Stein, E. M., *Hardy Spaces on Homogeneous Groups*, Mathematical Notes 28, Princeton University Press, Princeton, NJ, 1982.
143. Frazier, M., Jawerth, B., *Decomposition of Besov spaces*, Indiana Univ. Math. J. **34** (1985), no. 4, 777–799.
144. Frazier, M., Jawerth, B., *A discrete transform and decompositions of distribution spaces*, J. Funct. Anal. **93** (1990), no. 1, 34–170.
145. Frazier, M., Jawerth, B., *Applications of the φ and wavelet transforms to the theory of function spaces*, Wavelets and Their Applications, pp. 377–417, Jones and Bartlett, Boston, MA, 1992.
146. Frazier, M., Jawerth, B., Weiss, G., *Littlewood–Paley Theory and the Study of Function Spaces*, CBMS Regional Conference Series in Mathematics, 79, American Mathematical Society, Providence, RI, 1991.
147. Fujita, M., Tomita, N., *Weighted norm inequalities for multilinear Fourier multipliers* Trans. Amer. Math. Soc. **364** (2012), no. 12, 6335–6353.
148. Gagliardo, E., *Proprietà di alcune classi di funzioni in più variabili*, Ricerche Mat. **7** (1958), 102–137.
149. Gagliardo, E., *On integral transformations with positive kernel*, Proc. Amer. Math. Soc. **16** (1965), 429–434.
150. García-Cuerva, J., Rubio de Francia, J.-L., *Weighted Norm Inequalities and Related Topics*, North-Holland Mathematics Studies, 116, North-Holland Publishing Co., Amsterdam, 1985.
151. Garnett, J., *Bounded Analytic Functions*, Pure and Applied Mathematics, 96, Academic Press, Inc., New York–London, 1981.
152. Garnett, J., Jones, P., *The distance in BMO to L^∞*, Ann. of Math. (2nd Ser.) **108** (1978), no. 2, 373–393.
153. Garnett, J., Jones, P., *BMO from dyadic BMO*, Pacific J. Math. **99** (1982), no. 2, 351–371.
154. Gilbert, J. E., Nahmod, A. R., *Bilinear operators with non-smooth symbol, I*, J. Fourier Anal. Appl. **7** (2001), no. 5, 435–467.
155. Goldberg, D., *A local version of real Hardy spaces*, Duke Math. J. **46** (1979), no. 1, 27–42.
156. Grafakos, L., *Classical Fourier Analysis*, Third edition, Graduate Texts in Math. 249, Springer, New York, 2014.

157. Grafakos, L., Kalton, N., *Some remarks on multilinear maps and interpolation*, Math. Ann. **319** (2001), no. 1, 151–180.
158. Grafakos, L., Kalton, N. J., *The Marcinkiewicz multiplier condition for bilinear operators*, Studia Math. **146** (2001), no. 2, 115–156.
159. Grafakos, L., Kalton, N. J., *Multilinear Calderón–Zygmund operators on Hardy spaces*, Collect. Math. **52** (2001), no. 2, 169–179.
160. Grafakos, L., Li, X., *Uniform bounds for the bilinear Hilbert transforms I*, Ann. of Math. (2nd Ser.) **159** (2004), no. 3, 889–933.
161. Grafakos, L., Li, X., *The disc as a bilinear multiplier*, Amer. J. Math. **128** (2006), no. 1, 91–119.
162. Grafakos, L., Liu, L., Lu, S., Zhao, F., *The multilinear Marcinkiewicz interpolation theorem revisited: the behavior of the constant* J. Funct. Anal. **262** (2012), no. 5, 2289–2313.
163. Grafakos, L., Liu L., Maldonado, D., Yang, D., *Multilinear analysis on metric spaces*, Dissertationes Math. (Rozprawy Mat.) 497 (2014), 121 pp.
164. Grafakos, L., Maldonado, D., Naibo, V., *A remark on an endpoint Kato-Ponce inequality*, Differential and Integral Equations, **27**, no. 5–6 (2014), 415–424.
165. Grafakos, L., Martell, J. M., *Extrapolation of weighted norm inequalities for multivariable operators and applications* J. Geom. Anal. **14** (2004), no. 1, 19–46.
166. Grafakos, L., Martell, J. M., Soria, F., *Weighted norm inequalities for maximally modulated singular integral operators*, Math. Ann. **331** (2005), no. 2, 359–394.
167. Grafakos, L., Mastyło, M., *Analytic families of multilinear operators*, Nonlinear Anal. **107** (2014), 47–62.
168. Grafakos, L., Miyachi, A., Tomita, N., *On Multilinear Fourier Multipliers of Limited Smoothness*, Canad. J. Math. **65** (2013), no. 2, 299–330.
169. Grafakos, L., Oh, S., *The Kato Ponce inequality*, Comm. in PDE, Comm. Partial Differential Equations **39** (2014), no. 6, 1128–1157.
170. Grafakos, L., Si, Z., *The Hörmander multiplier theorem for multilinear operators*, J. Reine Angew. Math. **668** (2012), 133–147.
171. Grafakos, L., Soria, J., *Translation-invariant bilinear operators with positive kernels*, Integral Equations Operator Theory **66** (2010), no. 2, 253–264.
172. Grafakos, L., Tao, T., *Multilinear interpolation between adjoint operators*, J. Funct. Anal. **199** (2003), no. 2, 379–385.
173. Grafakos, L., Tao, T., Terwilleger, E., *L^p bounds for a maximal dyadic sum operator*, Math. Z. **246** (2004), no. 1–2, 321–337.
174. Grafakos, L., Torres, R., *A multilinear Schur test and multiplier operators*, J. Funct. Anal. **187** (2001), no. 1, 1–24.
175. Grafakos, L., Torres, R., *Maximal operator and weighted norm inequalities for multilinear singular integrals*, Indiana Univ. Math. J. **51** (2002), no. 5, 1261–1276.
176. Grafakos, L., Torres, R., *Discrete decompositions for bilinear operators and almost diagonal conditions*, Trans. Amer. Math. Soc. **354** (2002), no. 3, 1153–1176.
177. Grafakos, L., Torres, R. H., *Multilinear Calderón–Zygmund theory*, Adv. Math. **165** (2002), no. 1, 124–164.
178. Greenleaf, A., *Principal curvature and harmonic analysis*, Indiana Univ. Math. J. **30** (1981), no. 4, 519–537.
179. Gulisashvili, A., Kon, M., *Exact smoothing properties of Schrdinger semigroups* Amer. J. Math. **118** (1996), no. 6, 1215–1248.
180. Hardy, G. H., *The mean value of the modulus of an analytic function*, Proc. London Math. Soc. **14** (1915), 269–277.
181. Hardy, G. H., *Note on a theorem of Hilbert concerning series of positive terms*, Proc. London Math. Soc. **23** (1925), Records of Proc. XLV–XLVI.
182. Hardy, G. H., Littlewood, J. E., *Some properties of fractional integrals I*, Math. Z. **27** (1927), no. 1, 565–606.
183. Hardy, G. H., Littlewood, J. E., *Some properties of fractional integrals II*, Math. Z. **34** (1932), no. 1, 403–439.

184. Hardy, G. H., Littlewood, J. E., *Generalizations of a theorem of Paley*, Quarterly Jour. **8** (1937), 161–171.
185. Hardy, G. H., Littlewood, J. E. , Pólya, G., *The maximum of a certain bilinear form*, Proc. London Math. Soc. **S2-25** (1926), no. 1, 265–282.
186. Hart, J., *A new proof of the bilinear $T(1)$ theorem*, Proc. Amer. Math. Soc. **142** (2014), no. 9, 3169–3181.
187. He, D., *Square function characterization of weak Hardy spaces*, J. Fourier Anal. Appl. **20** (2014), no. 5, 1083–1110.
188. Hedberg, L., *On certain convolution inequalities*, Proc. Amer. Math. Soc. **36** (1972), 505–510.
189. Herz, C., *On the mean inversion of Fourier and Hankel transforms*, Proc. Nat. Acad. Sci. U.S.A. **40** (1954), 996–999.
190. Hewitt, E., Stromberg, K., *Real and Abstract Analysis* Springer-Verlag, New York, 1965.
191. Hofmann, S., Lacey, M., M^c^Intosh, A., *The solution of the Kato problem for divergence form elliptic operators with Gaussian heat kernel bounds*, Ann. of Math. (2nd Ser.) **156** (2002), no. 2, 623–631.
192. Hofmann, S., M^c^Intosh, A., *The solution of the Kato problem in two dimensions*, Proceedings of the 6th International Conference on Harmonic Analysis and Partial Differential Equations (El Escorial, 2000), pp. 143–160, Publ. Mat. 2002, Vol. Extra.
193. Hörmander, L., *Estimates for translation invariant operators in L^p spaces*, Acta Math. **104** (1960), no. 1–2, 93–140.
194. Hörmander, L., *Linear Partial Differential Operators*, Academic Press, Inc., Publishers, New York; Springer-Verlag, Berlin–Göttingen–Heidelberg, 1963.
195. Hörmander, L., *Oscillatory integrals and multipliers on FL^p*, Ark. Mat. **11** (1973), 1–11.
196. Hörmander, L., *The Analysis of Linear Partial Differential Operators I, Distribution theory and Fourier Analysis*, Second edition, Springer-Verlag, Berlin, 1990.
197. Howard, R., Schep, A. R., *Norms of positive operators on L^p spaces*, Proc. Amer Math. Soc. **109** (1990), no. 1, 135–146.
198. Hu, G., *Weighted norm inequalities for the multilinear Calderón-Zygmund operators*, Sci. China Math. **53** (2010), no. 7, 1863–1876.
199. Hunt, R., *On the convergence of Fourier series*, Orthogonal Expansions and their Continuous Analogues (Proc. Conf. Edwardsville, IL, 1967), pp. 235–255, Southern Illinois Univ. Press, Carbondale, IL, 1968.
200. Hunt, R., Young, W.-S., *A weighted norm inequality for Fourier series*, Bull. Amer. Math. Soc. **80** (1974), 274–277.
201. Igari, S., *An extension of the interpolation theorem of Marcinkiewicz, II*, Tôhoku Math. J. (2) **15** (1963), 343–358.
202. Igari, S., *On Kakeya's maximal function*, Proc. Japan Acad. Ser. A Math. Sci. **62** (1986), no. 8, 292–293.
203. Janson, S., *Mean oscillation and commutators of singular integral operators*, Ark. Math. **16** (1978), no. 2, 263–270.
204. Janson, S., *On interpolation of multilinear operators*, Function Spaces and Applications (Lund, 1986), pp. 290–302, Lect. Notes in Math., 1302, Springer, Berlin–New York, 1988.
205. John, F., Nirenberg, L., *On functions of bounded mean oscillation*, Comm. Pure Appl. Math. **14** (1961), 415–426.
206. Jørsboe, O. G., Mejlbro, L., *The Carleson–Hunt Theorem on Fourier Series*, Lecture Notes in Mathematics, 911, Springer-Verlag, Berlin–New York, 1982.
207. Journé, J.-L., *Calderón–Zygmund Operators, Pseudo-Differential Operators and the Cauchy Integral of Calderón*, Lecture Notes in Mathematics, 994, Springer-Verlag, Berlin, 1983.
208. Karlin, S., *Positive operators*, J. Math. Mech. **8** (1959), 907–937.
209. Kato, T., *Fractional powers of dissipative operators*, J. Math. Soc. Japan **13** (1961), 246–274.
210. Kato, T., Ponce, G., *Commutator estimates and the Euler and Navier-Stokes equations*, Comm. Pure Appl. Math. **41** (1988), no. 7, 891–907.
211. Katz, N., *A counterexample for maximal operators over a Cantor set of directions*, Math. Res. Lett. **3** (1996), no. 4, 527–536.

212. Katz, N., *Maximal operators over arbitrary sets of directions*, Duke Math. J. **97** (1999), no. 1, 67–79.
213. N. Katz, *Remarks on maximal operators over arbitrary sets of directions*, Bull. London Math. Soc. **31** (1999), no. 6, 700–710.
214. Katz, N., Łaba, I., Tao, T., *An improved bound on the Minkowski dimension of Besicovitch sets in $\mathbb{R}^3$*, Ann. of Math. (2nd Ser.) **152** (2000), no. 2, 383–446.
215. Katz, N., Tao, T., *Recent progress on the Kakeya conjecture*, Proceedings of the 6th International Conference on Harmonic Analysis and Partial Differential Equations (El Escorial, 2000), pp. 161–179, Publ. Mat. 2002, Vol. Extra.
216. Keich, U., *On L^p bounds for Kakeya maximal functions and the Minkowski dimension in $\mathbb{R}^2$*, Bull. London Math. Soc. **31** (1999), no. 2, 213–221.
217. Kenig, C., Meyer, Y., *Kato's square roots of accretive operators and Cauchy kernels on Lipschitz curves are the same*, Recent progress in Fourier analysis (El Escorial, 1983), pp. 123–143, North-Holland Math. Stud. 111, North-Holland, Amsterdam, 1985.
218. Kenig, C., Ponce, G., Vega, L., *Well-posedness and scattering results for the generalized Korteweg-de-Vries equation via the contraction principle*, Comm. Pure Appl. Math. (1993), **46**, no. 4, 527–620.
219. Kenig, C., Stein, E. M., *Multilinear estimates and fractional integration*, Math. Res. Lett. **6** (1999), no. 1, 1–15.
220. Knapp, A., Stein, E. M., *Intertwining operators for semisimple groups*, Ann. of Math. (2nd Ser.) **93** (1971), no. 3, 489–578.
221. Kokilashvili, V., Mastyło, M., Meskhi, A., *On the boundedness of the multilinear fractional integral operators*, Nonlinear Anal. **94** (2014), 142–147.
222. Kolmogorov, A. N., *Une série de Fourier–Lebesgue divergente presque partout*, Fund. Math. **4** (1923), 324–328.
223. Kolmogorov, A. N., *Une série de Fourier–Lebesgue divergente partout*, C. R. Acad. Sci. Paris **183** (1926), 1327–1328.
224. Konyagin, S. V., *On the divergence everywhere of trigonometric Fourier series*, Sb. Math. **191** (2000), no. 1–2, 97–120.
225. Koosis, P., *Sommabilité de la fonction maximale et appartenance à H_1*, C. R. Acad. Sci. Paris Sér A–B **286** (1978), no. 22, A1041–A1043.
226. Koosis, P., *Introduction to H_p Spaces*, Second edition, Cambridge Tracts in Mathematics, 115, Cambridge University Press, Cambridge, 1998.
227. Körner, T., *Everywhere divergent Fourier series*, Colloq. Math. **45** (1981), no. 1, 103–118.
228. Krantz, S. G., *Fractional integration on Hardy spaces*, Studia Math. **73** (1982), no. 2, 87–94.
229. Krantz, S. G., *Lipschitz spaces, smoothness of functions, and approximation theory*, Exposition. Math. **1** (1983), no. 3, 193–260.
230. Krein, M. G., *On linear continuous operators in functional spaces with two norms*, Trudy Inst. Mat. Akad. Nauk Ukrain. SSRS **9** (1947), 104–129.
231. Kuk, S., Lee, S., *Endpoint bounds for multilinear fractional integrals*, Math. Res. Lett. **19** (2012), no. 5, 1145–1154.
232. Kurtz, D., Wheeden, R., *Results on weighted norm inequalities for multipliers*, Trans. Amer. Math. Soc. **255** (1979), 343–362.
233. Lacey, M., *On bilinear Littlewood–Paley square functions*, Publ. Mat. **40** (1996), no. 2, 387–396.
234. Lacey, M. T., *The bilinear maximal functions map into L^p for $2/3 < p \leq 1$*, Ann. of Math. (2nd Ser.) **151** (2000), no. 1, 35–57.
235. Lacey, M., Metcalfe, J., *Paraproducts in one and several parameters*, Forum Math. **19** (2007), no. 2, 325–351.
236. Lacey, M. T., Thiele, C. M., *L^p bounds for the bilinear Hilbert transform for $2 < p < \infty$*, Ann. of Math. (2nd Ser.) **146** (1997), no. 3, 693–724.
237. Lacey, M., Thiele, C., *On Calderón's conjecture*, Ann. of Math. (2nd Ser.) **149** (1999), no. 2, 475–496.
238. Lacey, M., Thiele, C., *A proof of boundedness of the Carleson operator*, Math. Res. Lett. **7** (2000), no. 4, 361–370.

239. Latter, R. H., *A characterization of $H^p(\mathbf{R}^n)$ in terms of atoms*, Studia Math. **62** (1978), no. 1, 92–101.
240. Latter, R. H., Uchiyama, A., *The atomic decomposition for parabolic H^p spaces*, Trans. Amer. Math. Soc. **253** (1979), 391–398.
241. Lerner, A., Ombrosi, S., Pérez, C., Torres, R. H., Trujillo-González, R., *New maximal functions and multiple weights for the multilinear Calderón–Zygmund theory*, Adv. Math. **220** (2009), no. 4, 1222–1264.
242. Li, K., Sun, W., *Weighted estimates for multilinear Fourier multipliers*, Forum Math., to appear.
243. Li, W., Xue, Q., Yabuta, K., *Multilinear Calderón-Zygmund operators on weighted Hardy spaces*, Studia Math. **199** (2010), no. 1, 1–16.
244. Li, X., *Uniform bounds for the bilinear Hilbert transforms II*, Rev. Mat. Iberoamericana **22** (2006), no. 3, 1069–1126.
245. Li, X., Muscalu, C., *Generalizations of the Carleson-Hunt theorem. I. The classical singularity case,* Amer. J. Math. **129** (2007), no. 4, 983–1018.
246. Lieb, E. H., *Sharp constants in the Hardy–Littlewood–Sobolev and related inequalities*, Ann. of Math. (2nd Ser.) **118** (1983), no. 2, 349–374.
247. Lieb, E. H., Loss, M., *Analysis*, Graduate Studies in Mathematics 14, American Mathematical Society, Providence, RI, 1997.
248. Lions, J.-L., *Espaces d' interpolation and domaines de puissances fractionnaires d'opératerus*, J. Math. Soc. Japan **14** (1962), 233–241.
249. Lions, J.-L., Lizorkin, P. I., Nikol'skij, S. M., *Integral representation and isomorphic properties of some classes of functions*, Ann. Scuola Norm. Sup. Pisa (3) **19** (1965), 127–178.
250. Liu, L., Yang, D., *Boundedness of sublinear operators in Triebel–Lizorkin spaces via atoms*, Studia Math. **190** (2009), no. 2, 163–183.
251. Lizorkin, P. I., *Properties of functions of the spaces $\Lambda^r_{p\Theta}$* (Russian), Studies in the theory of differentiable functions of several variables and its applications, V, Trudy Mat. Inst. Steklov. **131** (1974), 158–181, 247.
252. Lu, S.-Z., *Four Lectures on Real H^p Spaces*, World Scientific Publishing Co., Inc., River Edge, NJ, 1995.
253. M^c^Intosh, A., *On the comparability of $A^{1/2}$ and $A^{*1/2}$*, Proc. Amer. Math. Soc. **32** (1972), 430–434.
254. McIntosh, A., *On representing closed accretive sesquilinear forms as $(A^{1/2}u, A^{*1/2}v)$*, Nonlinear partial differential equation and their applications, Collège de France Seminar, Vol. III (Paris, 1980/1981), pp. 252–267, Res. Notes in Math., 70, Pitman, Boston, Mass.–London, 1982.
255. McIntosh, A., *Square roots of operators and applications to hyperbolic PDE*, Miniconference on Operator Theory and Partial Differential Equations (Canberra, 1983), *Proc. Centre Math. Anal. Austral. Nat. Univ.*, **5**, Austral. Nat. Univ., Canberra, 1984.
256. McIntosh, A., *Square roots of elliptic operators*, J. Funct. Anal. **61** (1985), no. 3, 307–327.
257. McIntosh, A., Meyer, Y., *Algèbres d' opérateurs définis par des intégrales singulières* C. R. Acad. Sci. Paris Sér I Math. **301** (1985), no. 8, 395–397.
258. Maldonado D., Naibo, V., *Weighted norm inequalities for paraproducts and bilinear pseudodifferential operators with mild regularity*, J. Fourier Anal. Appl. **15** (2009), no. 2, 218–261.
259. Mauceri, G., Picardello, M., Ricci, F., *A Hardy space associated with twisted convolution*, Adv. in Math. **39** (1981), no. 3, 270–288.
260. Maz'ya, V. G., *Sobolev Spaces*, Springer Series in Soviet Mathematics, Springer-Verlag, Berlin, 1985.
261. Meda, S., Sjögren, P., Vallarino, M., *On the $H^1 - L^1$ boundedness of operators*, Proc. Amer. Math. Soc. **136** (2008), no. 8, 2921–2931.
262. Melnikov, M., Verdera, J., *A geometric proof of the L^2 boundedness of the Cauchy integral on Lipschitz graphs*, Internat. Math. Res. Notices **7** (1995), 325–331.

263. Meyer, Y., *Régularité des solutions des équations aux dérivées partielles non linéaires (d'aprés J.-M. Bony)*, Bourbaki Seminar, Vol. 1979/80, pp. 293–302, Lecture Notes in Math., 842, Springer, Berlin-New York, 1981.
264. Meyer, Y., Coifman, R. R., Wavelets. Calderón-Zygmund and multilinear operators, Cambridge Studies in Advanced Mathematics, 48, Cambridge University Press, Cambridge, 1997.
265. Meyer, Y., Taibleson, M., Weiss, G., *Some functional analytic properties of the spaces B_q generated by blocks*, Indiana Univ. Math. J. **34** (1985), no. 3, 493–515.
266. Meyers, N. G., *Mean oscillation over cubes and Hölder continuity*, Proc. Amer. Math. Soc. **15** (1964), 717–721.
267. Miyachi, A., Tomita, N., *Minimal smoothness conditions for bilinear Fourier multipliers*, Rev. Mat. Iberioamerican **29** (2013), no. 2, 495–530.
268. Miyachi, A., Tomita, N., *Boundedness criterion for bilinear Fourier multiplier operators*, Tohoku Math. J. (2) **66** (2014), no. 1, 55–76.
269. Moen, K., *Weighted inequalities for multilinear fractional integral operators*, Collect. Math. **60** (2009), no. 2, 213–238.
270. Mohanty, P., Shrivastava, S., *A note on the bilinear Littlewood–Paley square function*, Proc. Amer. Math. Soc. **138** (2010), no. 6, 2095–2098.
271. Mohanty, P., Shrivastava, S., *Bilinear Littlewood–Paley for circle and transference*, Publ. Mat. **55** (2011), no. 2, 501–519.
272. Mozzochi, C. J., *On the pointwise convergence of Fourier Series*, Lecture Notes in Mathematics, 199, Springer-Verlag, Berlin–New York 1971.
273. Müller, D., *A note on the Kakeya maximal function*, Arch. Math. (Basel) **49** (1987), no. 1, 66–71.
274. Murai, T., *Boundedness of singular integral integral operators of Calderón type*, Proc. Japan Acad. Ser. A Math. Sci. **59** (1983), no. 8, 364–367.
275. Murai, T., *A real variable method for the Cauchy transform and analytic capacity*, Lecture Notes in Mathematics, 1307, Springer-Verlag, Berlin, 1988.
276. Muscalu, C., Schlag, W., *Classical and Multilinear Harmonic Analysis, Vol. I,* Cambridge Studies in Advanced Mathematics, 137, Cambridge University Press, Cambridge, 2013.
277. Muscalu, C., Schlag, W., *Classical and Multilinear Harmonic Analysis, Vol. II,* Cambridge Studies in Advanced Mathematics, 138, Cambridge University Press, Cambridge, 2013.
278. Muscalu, C., Pipher, J., Tao, T., Thiele, C., *Bi-parameter paraproducts*, Acta Math. **193** (2004), no. 2, 269–296.
279. Muscalu, C., Pipher, J., Tao, T., Thiele, C., *Multi-parameter paraproducts*, Rev. Mat. Iberoam. **22** (2006), no. 3, 963–976.
280. Muscalu, C., Tao, T., Thiele, C., *Multi-linear operators given by singular multipliers*, J. Amer. Math. Soc. **15** (2002), no. 2, 469–496.
281. Muskhelishvili, N. I., *Singular Integral Equations*, Boundary Problems of Functions Theory and their Applications to Mathematical Physics [Revised translation from the Russian, edited by J. R. M. Radok], Wolters-Noordhoff Publishing, Groningen, 1972.
282. Nagel, A., Stein, E. M., Wainger, S., *Differentiation in lacunary directions*, Proc. Nat. Acad. Sci. U.S.A. **75** (1978), no. 3, 1060–1062.
283. Nazarov, F., Treil, S., Volberg, A., *Cauchy integral and Calderón–Zygmund operators on nonhomogeneous spaces*, Internat. Math. Res. Notices **15** (1997), 703–726.
284. Nazarov, F., Treil, S., Volberg, A., *Weak type estimates and Cotlar inequalities for Calderón–Zygmund operators on nonhomogeneous spaces*, Internat. Math. Res. Notices **9** (1998), 463–487.
285. Nirenberg, L., *On elliptic partial differential equations*, Ann. di Pisa **13** (1959), 116–162.
286. Oberlin, D., *Fourier restriction for affine arclength measures in the plane*, Proc. Amer. Math. Soc. **129** (2001), no. 11, 3303–3305.
287. Orlicz, W., *Über eine gewisse Klasse von Räumen vom Typus B*, Bull. Int. Acad. Pol. de Science, Ser A (1932), 207–220; reprinted in W. Orlicz "Collected Papers," pp. 217–230, PWN, Warsaw, 1988.
288. Orlicz, W., *Über Räume* (L^M), Bull. Int. Acad. Pol. de Science, Ser A (1936), 93–107; reprinted in W. Orlicz "Collected Papers," pp. 345–359, PWN, Warsaw, 1988.

289. Peetre, J., *On convolution operators leaving $L^{p,\lambda}$ spaces invariant*, Ann. Mat. Pura Appl. (4) **72** (1966), 295–304.
290. Peetre, J., *Sur les espaces de Besov*, C. R. Acad. Sci. Paris **264** (1967), 281–283.
291. Peetre, J., *Remarques sur les espaces de Besov. Le cas* $0 < p < 1$, C. R. Acad. Sci. Paris **277** (1973), 947–950.
292. Peetre, J., *H_p Spaces*, Lecture Notes, University of Lund and Lund Institute of Technology, Lund, Sweden, 1974.
293. Peetre, J., *On spaces of Triebel–Lizorkin type*, Ark. Math. **13** (1975), 123–130.
294. Peetre, J., *New Thoughts on Besov Spaces*, Duke University Mathematical Series, No. 1, Mathematics Department, Duke University, Durham, NC, 1976.
295. Pérez, C., *Endpoint estimates for commutators of singular integral operators*, J. Funct. Anal. **128** (1995), no. 1, 163–185.
296. Pérez, C., *Sharp estimates for commutators of singular integrals via iterations of the Hardy–Littlewood maximal function*, J. Fourier Anal. Appl. **3** (1997), no. 6, 743–756.
297. Pérez, C., Wheeden, R., *Uncertainty principle estimates for vector fields*, J. Funct. Anal. **181** (2001), no. 1, 146–188.
298. Plemelj, J., *Ein Ergänzungssatz zur Cauchyschen Integraldarstellung analytischer Functionen, Randwerte betreffend*, Monatsh. Math. Phys. **19** (1908), no. 1, 205–210.
299. Pramanik, M., Terwilleger, E., *A weak L^2 estimate for a maximal dyadic sum operator on* $\mathbf{R}^n$, Illinois J. Math. **47** (2003), 775–813.
300. Prestini, E., *A restriction theorem for space curves*, Proc. Amer. Math. Soc. **70** (1978), no. 1, 8–10.
301. Privalov, J., *Sur les fonctions conjuguées*, Bull. Soc. Math. France **44** (1916), 100–103.
302. Rao, M. M., Ren, Z. D., *Theory of Orlicz spaces*, Monographs and Textbooks in Pure and Applied Mathematics, 146, Marcel Dekker, Inc., New York, 1991.
303. Riesz, F., *Über die Randwerte einer analytischen Funktion*, Math. Z. **18** (1923), no. 1, 87–95.
304. Riesz, M., *L' intégrale de Riemann-Liouville et le problème de Cauchy*, Acta Math. **81** (1949), no. 1, 1–223.
305. Riviere, N., Sagher, Y., *Interpolation between L^∞ and H^1, the real method*, J. Funct. Anal. **14** (1973), 401–409.
306. Rodríguez-López, S., *A homomorphism theorem for bilinear multipliers*, J. London Math. Soc., (2) 88 (2013), no. 2, 619–636.
307. Rubio de Francia, J.-L., *Estimates for some square functions of Littlewood–Paley type*, Publ. Sec. Mat. Univ. Autónoma Barcelona **27** (1983), no. 2, 81–108.
308. Rubio de Francia, J.-L., Ruiz, F. J., Torrea, J. L., *Calderón–Zygmund theory for operator-valued kernels*, Adv. in Math. **62** (1986), no. 1, 7–48.
309. Sarason, D., *Functions of vanishing mean oscillation*, Trans. Amer. Math. Soc. **207** (1975), 391–405.
310. Sawano, Y., *Maximal operator for pseudodifferential operators with homogeneous symbols*, Michigan Math. J. **59** (2010), no. 1, 119–142.
311. Schlag, W., *A geometric inequality with applications to the Kakeya problem in three dimensions*, Geom. Funct. Anal. **8** (1998), no. 3, 606–625.
312. Schur, I., *Bemerkungen zur Theorie der beschränkten Bilinearformen mit unendlich vielen Veränderlichen*, J. Reine Angew. Math. **140** (1911), 1–28.
313. Seeger, A., *Endpoint inequalities for Bochner–Riesz multipliers in the plane*, Pacific J. Math. **174** (1996), no. 2, 543–553.
314. Semmes, S., *Square function estimates and the $T(b)$ theorem*, Proc. Amer. Math. Soc. **110** (1990), no. 3, 721–726.
315. Sharpley, R., *Interpolation of n pairs and counterexamples employing indices*, J. Approximation Theory **13** (1975), 117–127.
316. Sharpley, R., *Multilinear weak type interpolation of mn-tuples with applications*, Studia Math. **60** (1977), no. 2, 179–194.
317. Sjölin, P., *On the convergence almost everywhere of certain singular integrals and multiple Fourier series*, Ark Math. **9** (1971), 65–90.

318. Sjölin, P., Soria, F., *Some remarks on restriction of the Fourier transform for general measures*, Publ. Mat. **43** (1999), no. 2, 655–664.
319. Sobolev, S. L., *On a theorem in functional analysis* [in Russian], Mat. Sob. **46** (1938), 471–497.
320. Spanne, S., *Sur l' interpolation entre les espaces* $\mathscr{L}_k^{p,\Phi}$, Ann. Scuola Norm. Sup. Pisa (3) **20** (1966), 625–648.
321. Stefanov, A., *Characterizations of* H^1 *and applications to singular integrals*, Illinois J. Math. **44** (2000), 574–592.
322. Stein, E. M., *Interpolation of linear operators*, Trans. Amer. Math. Soc. **83** (1956), 482–492.
323. Stein, E. M., *On limits of sequences of operators*, Ann. of Math. (Ser. 2) **74** (1961), no. 1, 140–170.
324. Stein, E. M., *Singular integrals, harmonic functions, and differentiability properties of functions of several variables*, Singular Integrals (Proc. Sympos. Pure Math., Chicago, IL, 1966), pp. 316–335, Amer. Math. Soc., Providence, RI, 1967.
325. Stein, E. M., *Oscillatory integrals in Fourier analysis*, Beijing Lectures in Harmonic Analysis (Beijing, 1984), pp. 307–355, Ann. of Math. Stud. 112, Princeton Univ. Press, Princeton, NJ, 1986.
326. Stein, E. M., *Harmonic Analysis, Real Variable Methods, Orthogonality, and Oscillatory Integrals*, With the assistance of Timothy S. Murphy, Princeton Mathematical Series, 43, Monographs in Harmonic Analysis, III, Princeton University Press, Princeton, NJ, 1993.
327. Stein, E. M., Weiss, G., *On the theory of harmonic functions of several variables, I: The theory of* H^p *spaces*, Acta Math. **103** (1960), no. 1–2, 25–62.
328. Strichartz, R., *A multilinear version of the Marcinkiewicz interpolation theorem*, Proc. Amer. Math. Soc. **21** (1969), 441–444.
329. Strichartz, R., *Restrictions of Fourier transforms to quadratic surfaces and decay of solutions of wave equations*, Duke Math. J. **44** (1977), no. 3, 705–714.
330. Strömberg, J.-O., *Maximal functions for rectangles with given directions*, Doctoral dissertation, Mittag-Leffler Institute, Djursholm, Sweden, 1976.
331. Strömberg, J.-O., *Maximal functions associated to rectangles with uniformly distributed directions*, Ann. of Math. (Ser. 2) **107** (1978), no. 2, 399–402.
332. Strömberg, J.-O., *Bounded mean oscillation with Orlicz norms and duality of Hardy spaces*, Indiana Univ. Math. J. **28** (1979), no. 3, 511–544.
333. Strömberg, J.-O., Torchinsky, A., *Weighted Hardy spaces*, Lecture Notes in Mathematics, 1381, Springer-Verlag, Berlin, 1989.
334. Taibleson, M. H., *The preservation of Lipschitz spaces under singular integral operators*, Studia Math. **24** (1964), 107–111.
335. Taibleson, M. H., *On the theory of Lipschitz spaces of distributions on Euclidean n-space, I*, J. Math. Mech. **13** (1964), 407–479.
336. Taibleson, M. H., *On the theory of Lipschitz spaces of distributions on Euclidean n-space, II*, J. Math. Mech. **14** (1965), 821–839.
337. Taibleson, M. H., *On the theory of Lipschitz spaces of distributions on Euclidean n-space, III*, J. Math. Mech. **15** (1966), 973–981.
338. Tao, T., *Weak-type endpoint bounds for Riesz means*, Proc. Amer. Math. Soc. **124** (1996), no. 9, 2797–2805.
339. Tao, T., *The weak-type endpoint Bochner–Riesz conjecture and related topics*, Indiana Univ. Math. J. **47** (1998), no. 3, 1097–1124.
340. Tao, T., *The Bochner–Riesz conjecture implies the restriction conjecture*, Duke Math. J. **96** (1999), no. 2, 363–375.
341. Tao, T., *Endpoint bilinear restriction theorems for the cone, and some sharp null form estimates*, Math. Z. **238** (2001), no. 2, 215–268.
342. Tao, T., *On the Maximal Bochner–Riesz conjecture in the plane, for* $p < 2$, Trans. Amer. Math. Soc. **354** (2002), no. 5, 1947–1959.
343. Tao, T., Vargas, A., Vega, L., *A bilinear approach to the restriction and Kakeya conjectures*, J. Amer. Math. Soc. **11** (1998), no. 4, 967–1000.

344. Taylor, M., *Pseudodifferential Operators and Nonlinear PDE*, Progress in mathematics, 100, Birkhäuser Boston, Inc., Boston, MA, 1991.
345. Tchamitchian, P., *Ondelettes et intégrale de Cauchy sur les courbes lipschitziennes*, Ann. of Math. (Ser. 2) **129** (1989), no. 3, 641–649.
346. Thiele, C., *A uniform estimate*, Ann. of Math. (Ser. 2) **156** (2002), no. 2, 519–563.
347. Thiele, C., *Multilinear singular integrals*, Proceedings of the 6th International Conference on Harmonic Analysis and Partial Differential Equations (El Escorial, Spain, 2000), Publ. Mat., 2002, Vol. Extra, pp. 229–274.
348. Thiele, C., *Wave Packet Analysis*, CBMS Regional Conference Series in Mathematics, 105, American Mathematical Society, Providence, RI, 2006.
349. Tomas, P. A., *A restriction theorem for the Fourier transform*, Bull. Amer. Math. Soc. **81** (1975), 477–478.
350. Tomas, P. A., *A note on restriction*, Indiana Univ. Math. J. **29** (1980), no. 2, 287–292.
351. Tomita, N., *A Hörmander type multiplier theorem for multilinear operators*, J. Funct. Anal. **259** (2010), no. 8, 2028–2044.
352. Torres, R. H., *Boundedness results for operators with singular kernels on distribution spaces*, Mem. Amer. Math. Soc. 90 (1991), no. 442.
353. Triebel, H., *Spaces of distributions of Besov type on Euclidean n-space. Duality, interpolation*, Ark. Math. **11** (1973), 13–64.
354. Triebel, H., *Theory of function spaces*, Monographs in Mathematics, 78, Birkhäuser Verlag, Basel, 1983.
355. Uchiyama, A., *A constructive proof of the Fefferman-Stein decomposition of* $BMO(\mathbf{R}^n)$, Acta Math. **148** (1982), no. 1, 215–241.
356. Uchiyama, A., *Characterization of* $H^p(\mathbf{R}^n)$ *in terms of generalized Littlewood–Paley g-function*, Studia Math. **81** (1985), no. 2, 135–158.
357. Uchiyama, A., *On the characterization of* $H^p(\mathbf{R}^n)$ *in terms of Fourier multipliers*, Proc. Amer. Math. Soc. **109** (1990), no. 1, 117–123.
358. Uchiyama, A., *Hardy Spaces on the Euclidean Space*, Springer Monographs in Mathematics, Springer-Verlag, Tokyo, 2001.
359. Vargas, A., *Bochner–Riesz multipliers, Maximal operators, Restriction theorems in* $\mathbf{R}^n$, Notes from lectures given at MSRI, August 1997.
360. Varopoulos, N., *BMO functions and the* $\overline{\partial}$*-equation*, Pacific J. Math. **71** (1977), no. 1, 221–273.
361. Verbitsky, I. E., *A dimension-free Carleson measure inequality*, Complex Analysis, Operators, and Related Topics, pp. 393–398, Oper. Theory: Adv. Appl. 113, Birkhäuser, Basel, 2000.
362. Verdera, J., L^2 *boundedness of the Cauchy Integral and Menger curvature*, Harmonic Analysis and Boundary Value Problems (Fayetteville, AR, 2000), 139–158, Contemp. Math. 277 Amer. Math. Soc., Providence, RI, 2001.
363. Walsh, T., *The dual of* $H^p(\mathbb{R}^{n+1}_+)$ *for* $p < 1$, Canad. J. Math. **25** (1973), 567–577.
364. Wang, S., *A note on characterization of Hardy space* H^1, (English summary) Sci. China Ser. A **48** (2005), no. 4, 448–455.
365. Weiss, G., *An interpolation theorem for sublinear operators on* H_p *spaces*, Proc. Amer. Math. Soc. **8** (1957), 92–99.
366. Welland, G. V., *Weighted norm inequalities for fractional integrals*, Proc. Amer. Math. Soc. **51** (1975), 143–148.
367. Weyl, H., *Singuläre Integralgleichungen mit besonderer Berücksichtigung des Fourierschen Integraltheorems*, Inaugural-Dissertation, Göttingen, 1908.
368. Weyl, H., *Bemerkungen zum Begriff der Differentialquotienten gebrochener Ordnung*, Viertel Natur. Gesellschaft Zürich **62** (1917), 296–302.
369. Whitney, H., *Analytic extensions of differentiable functions defined in closed sets*, Trans. Amer. Math. Soc. **36** (1934), no. 1, 63–89.
370. Wilson, J. M., *On the atomic decomposition for Hardy spaces*, Pacific J. Math. **116** (1985), no. 1, 201–207.

371. Wolff, T. H., *An improved bound for Kakeya type maximal functions*, Rev. Mat. Iberoamericana **11** (1995), no. 3, 651–674.
372. Wolff, T. H., *Recent work connected with the Kakeya problem*, Prospects in Mathematics (Princeton, NJ, 1996), pp. 129–162, Amer. Math. Soc., Providence, RI, 1999.
373. Wolff, T., *A sharp bilinear cone restriction estimate*, Ann. of Math. (2nd Ser.) **153** (2001), no. 3, 661–698.
374. Yang, D, Zhou, Y., *A boundedness criterion via atoms for linear operators in Hardy spaces*, Constr. Approx. **29** (2009), no. 2, 207–218.
375. Zafran, M., *A multilinear interpolation theorem*, Studia Math. **62** (1978), no. 2, 107–124.
376. Zygmund, A., *On a theorem of Marcinkiewicz concerning interpolation of operators*, J. Math. Pures Appl. (9) **35** (1956), 223–248.
377. Zygmund, A., *Trigonometric Series*, Vol. II, Second edition, Cambridge University Press, New York, 1959.
378. Zygmund, A., *On Fourier coefficients and transforms of functions of two variables*, Studia Math. **50** (1974), 198–201.

Index

L. Grafakos, *Modern Fourier Analysis*, Graduate Texts in Mathematics 250,
DOI 10.1007/978-1-4939-1230-8, © Springer Science+Business Media New York 2014